Engineering Sciences and Fundamentals 2016

Core Programming Area at the 2016 AIChE Annual Meeting

San Francisco, California, USA
13 - 18 November 2016

Volume 1 of 2

ISBN: 978-1-5108-3434-7

Printed from e-media with permission by:

Curran Associates, Inc.
57 Morehouse Lane
Red Hook, NY 12571

Some format issues inherent in the e-media version may also appear in this print version.

Printed by Curran Associates, Inc. (2017)

For permission requests, please contact AIChE
at the address below.

AIChE
120 Wall Street, FL 23
New York, NY 10005-4020

Phone: (800) 242-4363
Fax: (203) 775-5177

www.aiche.org

Additional copies of this publication are available from:

Curran Associates, Inc.
57 Morehouse Lane
Red Hook, NY 12571 USA
Phone: 845-758-0400
Fax: 845-758-2633
Email: curran@proceedings.com
Web: www.proceedings.com

TABLE OF CONTENTS

VOLUME 1

VOLUME 2

472770 A Microfluidic Study of Liquid-Liquid Extraction Mediated By Carbon Dioxide

Abstract:

Liquid-liquid extraction is an important separation and purification method, however it faces a challenge in reducing the energy consumption and the environmental impact of solvent (extractant) recovery. The reversible chemical reactions of switchable solvents (nitrogenous bases) with carbon dioxide (CO_2) can be implemented in reactive liquid-liquid extraction to reduce significantly the cost and energy requirements of solvent recovery. The development of new effective switchable solvents reacting with CO_2 and the optimization of extraction conditions rely of the ability to evaluate and screen the performance of switchable solvents in extraction processes. We report a microfluidic strategy for time- and labour-efficient studies of CO_2-mediated solvent extraction. The platform utilizes a liquid segment containing an aqueous extractant droplet and a droplet of a solution of a switchable solvent in a non-polar liquid, with gaseous CO_2 supplied to the segment from both sides. Following the reaction of the switchable solvent with CO_2, the solvent becomes hydrophilic and transfers from the non-polar solvent to the aqueous droplet. By monitoring the time-dependent variation in droplet volumes, we determined the efficiency and extraction time for the CO_2-mediated extraction of different nitrogenous bases in a broad experimental parameter space. The platform enables a significant reduction in the amount of switchable solvents used in these studies; provides accurate temporal characterization of the liquid-liquid extraction process, and offers the capability of high-throughput screening of switchable solvents.

465291 Modeling of Water Quality Downgradient of Mulch Biowall with Nearby Surface Water Receptor

Abstract:

Groundwater treatment biowalls are commonly placed at a property boundary to prevent contaminant transport off-site or adjacent to a surface water body to protect from contaminant discharges to the surface water. Groundwater contaminants passing through the biowall are treated within the biowall or immediately downgradient of the biowall. A biowall typically contains bark mulch or other agricultural waste material that serves as an electron donor to the degradation process. One of the results of biowall treatment activity is the generation of anaerobic groundwater, which may result in increased sulfide levels if influent groundwater contains moderate to high sulfate concentrations. Sulfide represents a degraded water quality of the groundwater exiting the biowall, which is of significant concern if the biowall is located in close proximity to a groundwater discharge to surface water. A time-dependent one-dimensional fate and transport model was developed to evaluate concentrations of sulfide that may be generated in such a situation. The model was developed by applying mass governing equations over a control volume of the aquifer between a biowall and a surface water body receptor. COMSOL Multiphysics was used to solve the family of equations for a timescale extending to approximately 4 years, using appropriate parameters and initial / boundary condition values. The mass species glucose (surrogate for carbon source upon degradation of solid carbonaceous material), lactic acid (surrogate for fatty acids), and electron donor sulfate were modeled with the partial differential equations. The model results provide insight into the advance of sulfide levels in groundwater downgradient of a biowall, and in close proximity of a surface water receptor. The carbon source, sulfate and sulfide data is compared to actual site data obtained from literature or other sources.

464439 Influence of Silica Nanoparticles on Transport and Interfacial Phenomena in Liquid/Liquid Systems

Abstract:

The hydroformylation of organic reactants through the Ruhrchemie/RhÃ´ne-Poulenc process is one of the most important homogenous catalytic reactions in chemical industry. The catalyst is designed to be soluble in water to provide a good separation from the likewise organic products, leading to a liquid/liquid system. In these systems, the reaction is controlled by the interface through interfacial and transport phenomena. The process fails for long chained olefins â€“ vegetable oils and fats from renewable resources â€“ due to their strong insolubility in water. Nanoparticles can be used as emulsifiers to create Pickering emulsions and increase the reaction speed. The particles adsorb at the interface and affect the transport processes and thus the reaction yield and selectivity. In this study the influence of nanoparticles on mass transfer and drop size distributions is investigated and the occurring interfacial phenomena are characterized. As material system water and the reactant 1-Dodecen with added silica nanoparticles is used. For simplicity reasons no catalyst is considered. The mass transfer is measured in a droplet swarm by using a model reaction in a stirred reactor. The occurring drop size distribution is detected with an endoscope camera. The properties of the interface are characterized by pendant drop and spinning drop tensiometry. With increasing particle mass fraction, the system shows rising interfacial tensions until a constant value is reached. Nonetheless, decreasing drop sizes were found, suggesting a better mass transfer due to larger interfacial area. Contrary to these observations, the mass transfer is reduced. The results lead to the assumption of an increasing particle coverage at the interface that hinders coalescence and forms an additional transport resistance. To get a better understanding of the occurring effects, the influence of the surface rheology and coverage has to be quantified. Therefore, the dilatational viscosity was measured using the oscillation drop method. A hardening of the drop interface could be determined.

470253 Molecular Dynamic Study of Hydrodyanmic Drag and Diffusion for Nanoparticles at Liquid Vapour Interface

Abstract:

Particle tracking experiments on the mean square displacement of micron-sized, spherical, Brownian colloids diffusing at an air/water interface indicate that the colloidâ€™s translational drag coefficient is greater, by a factor of two or more, than the value obtained by a continuum (Stokes flow) calculation on a flat surface. To rationalize this difference, additional dissipative forces arising from the pinning of the contact line at the colloid surface as it moves have been suggested. A further complication is the fact that measurements in which a non-Brownian deterministic force (e.g. a magnetic or capillary attraction force) acts on the particle, and the drag is calculated from the resulting velocity, are in agreement with the Stokes simulations.

To provide insight, we undertake molecular dynamic simulations of a colloid diffusing along a solvent interface based on generic Lennard-Jones interactions. The colloid is a section of a rigid atomic lattice to create a rough surface and allow contact line pinning. The solvent is a tetramer chain molecule which has a relatively sharp vapor/liquid interface. By varying the interaction strength between the solvent and colloid atoms, different equilibrium immersion depths are achieved corresponding to conditions between complete wetting (strong interaction) and non-wetting (weak interaction). We calculate the drag coefficient of a colloid calculated from its mean square displacement due to Brownian diffusion, and for a colloid dragged along the interface at constant velocity. We find that the MD calculated drag coefficients for each method are within 10 percent of each other. The drag coefficients monotonically increase with increasing depth of immersion. While in qualitative agreement with Stokes flow simulations of the drag, the MDcalculated coefficients are, for each immersion depth, smaller than the continuum calculations. We attribute the discrepancy to the fact that the interfacial thickness is not negligibly smaller than the size of the colloid, as would be the case in the continuum calculation. Measurements of the orientation of the colloid as it moves along the surface under Brownian or applied forces indicate that it rotates and sometimes becomes pinned. However, the fact that the drag coefficients by each method of MD simulation agree with each other establishes that despite pinning, no additional forces are exerted on the colloid during Brownian motion.

453493 Human Tear-Film Evaporation Rate from Simultaneous Ocular-Surface Temperature and Tear-Breakup Area

Abstract:

Tear-film evaporation rate is critical to eye health. If the 3-μm human tear film evaporates at close to the rate of pure water, the tear film completely disappears by the end of a blink. Fortunately, human tear film is covered by a 100-nm lipid film that is thought to retard tear-evaporation rates by approximately a factor of 10 [1,2]. Transient infrared ocular surface temperature (OST) provides a convenient, non-invasive, and simple *in-situ* method to gage tear evaporation rates. Current analyses of OST results, however, suggest little reduction of the tear evaporation rate compared to that of pure water.

For the first time, we devise a straightforward methodology to assess quantitative *in-vivo* tear-evaporation rates (TERs) from simultaneous measurement of transient infrared ocular surface temperature (OST) and tear-breakup area (TBA) on 9 human subjects. Specifically, we analyze simultaneously measured transient OST and transient TBA based on an extended heat-transfer model. Key to the model is explicit recognition of expanding tear breakup spots [3] that are proposed to be lipid free and, accordingly, exhibit higher evaporation rate than that for lipid-covered tear. Transient tear-breakup areas follow from image analysis. Remaining parameters include an overall heat transfer coefficient to capture natural-convective and radiative heat loss to the environment, and a temperature-dependent evaporation rate of pure water. These two parameters are obtained by independent *in-vitro* measurement of simultaneous temperature decline and mass loss for small volumes of heated water or agar gels undergoing cooling via natural convection. In all cases (9 subjects), model-predicted *in-vivo* OSTs agree well with experiment. Percent reductions in lipid-covered tear evaporation rate range from 62 to 95% of that for pure water, in agreement with accepted literature values [1-3]. Our new methodology for obtaining *in-vivo* human tear evaporation rate is simple and obtains realistic TERs.

1. Liat Rosenfeld, Colin Cerretani, Danielle L. Leiske, Michael F. Toney, Clayton J. Radke, and Gerald G. Fuller,

"Structural and Rheological Properties of Meibomian Lipid," Investigative Ophthalmology and Vision Science, 54(4), 2720-2732 (2013).

2. C. Cerretani, N. Ho, and C. J. Radke, "Water-Evaporation Reduction by Duplex Films: Application to the Human Tear Film," Advances in Colloid and Interface Science, 197-198, 33-57 (2013).

3. C.C. Peng, C. Cerretani, R.J. Braun, and C.J. Radke, "Evaporation-Driven Instability of the Precorneal Tear Film," Advances in Colloid and Interface Science, Velarde Honorary Issue, 206, 250-264 (2014).

457374 Enhancing the Spreading Behavior of Drug Formulations on Pulmonary Airway: Effect of Catanionic Surfactant Mixtures

Abstract:

Lungs carry a great potential to be used in drug delivery since they have a large surface area and possess low catabolic enzymatic activity. Pulmonary airways are covered with mucus layer, which acts as a biological barrier against the foreign particles, dusts and pathogens and while this natural barrier protects the lungs, it also obstructs the drug delivery through the lungs[1]. Efficient spreading of the therapeutic agents on the airways is the one of the key parameters for successful drug delivery through the lungs. Particularly in case of lung diseases such as cystic fibrosis, the viscosity and the thickness of the mucus layer increases and it becomes harder for drug formulations to spread on the airways. Therefore, importance of effective spreading of the therapeutics on the airway increases. Commonly, aerosolized formulations are used for the treatment of such diseases. Although spreading experiments have been conducted with different types of formulations on mucus-mimicking subphases, an effective formulation could not have been developed, yet[2]. Basically, spreading of the formulations on a viscous surface such as mucus, depend on both the surface tension difference between the phases and surface tension gradient between the center and the edge of the spreading film in the radial direction[3]. This gradient triggers the Marangoni forces and as a result, spreading occurs in the direction from lower surface tension to higher. As lower surface tension provides higher surface tension gradient, spreading behavior can be enhanced by adding surface-active molecules to the formulations.

From this point of view, in this study catanionic (cationic+anionic) surfactant mixtures are used owing to their advantageous interfacial and bulk properties such as possessing lower CMC and providing lower surface tension value than its individual components. Catanionic surfactant mixtures composed of Dodecyltrimethyl ammonium bromide (DTAB) and Dioctyl sulfosuccinate sodium salt (AOT) mixed at various mole ratios are prepared and their spreading performances on both mucin and cystic fibrosis mucus model are investigated for the first time in the literature. Characterizations of the catanionic mixtures are implemented with the surface tension measurements and light

scattering analysis. Spreading experiments are utilized via tracking the movements of tracer particles by a tracking tool and analyzing their spreading characteristics by fitting the spreading radius of the tracking particles to a power-law. Also, contact angle measurements of the mixtures are performed.

Surface tension measurements of the mixtures with different ratios of DTAB/AOT have showed that a synergistic interaction is obtained between the components of the mixtures, as expected. Synergistic interaction provided lower CMC values for the mixtures, and accordingly the equilibrium surface tension of the solutions had reached to lower values than the surface tension of the molecules contributing to the mixture. Decrement of the surface tension value had directly affected the spreading behavior of the mixtures. This interaction had also seen as an increment in the micelle sizes and the increasing asymmetry at the structures.

Spreading behaviors of the mixtures are analyzed by comparing the spreading radii of the chosen tracer particle when pure DTAB, pure AOT or the mixtures are applied to the surface of mucin. Due to its cationic structure, DTAB solution has not spread at all, whereas the spreading radius of the tracer particle, when pure AOT solution had been introduced to the surface, is determined as 11.67±0.23 mm. It is obtained that catanionic surfactant mixtures at all mixing ratios spread more than 10 mm. Especially, mixtures with the mole ratios of X_{DTAB}=0.8 and X_{DTAB}=0.7 provided superior spreading performances than both of the individual components of the mixtures even though the mole fraction of DTAB in the mixture is at least 7 times higher than AOTs. Spreading radii for these mixtures are obtained as 13.21±0.35 and 13.96±0.43 mm, respectively. Moreover, for every mixing ratio, both DTAB-rich and AOT-rich mixtures have spread faster than the pure AOT and pure DTAB. Above all, the spreading regime of the DTAB-rich mixtures is determined as Marangoni-dominant, which means that superspreading of the droplets on the mucin is observed. Even on the cystic fibrosis sputum model, the spreading of the droplet has been enhanced compared to pure surfactants and the spreading of the mixtures are again in the Marangoni-dominant region.

In the light of the results, we propose that this type of mixtures can be used successfully for enhancing the spreading behavior of the formulations that will be used in the treatment of lung diseases.

REFERENCES

[1] B.S. Schuster, J.S. Suka, G.F. Woodworth, J. Hanes, Biomaterials, 34 (2013) 3439-3446

[2] K. Koch, B. Dew, T.E. Corcoran, T.M. Przybycien, R.D. Tilton, S. Garoff, Molecular Pharmaceutics, 8 (2011) 387â€“394.

[3] R. Sharma, T.E. Corcoran, S. Garoff, T.M. Przybycien, E.R. Swanson, R. D. Tilton, ACS Applied Materials & Interfaces, 5 (2013) 5542-5549.

471949 Dynamic Surface Tension Measurements with Maximum Bubble Pressure Tensiometry

Abstract:

Dynamic surface tension refers to the time dependent variation in surface tension, and is intimately linked with the rate of mass transfer of a surfactant from liquid sub-phase to the interface. The diffusion- or adsorption-limited kinetics of mass transfer to the liquid/gas interface is said to impact the so-called foamability and the Gibbs-Marangoni elasticity of surfaces. Dynamic surface tension measurements carried out with conventional methods like pendant drop analysis, Wilhelmy plate, etc are limited in their temporal resolution (> 50 ms). In this study, we describe the design and application of maximum bubble pressure tensiometry for the measurement of dynamic surface tension effects at extremely short (1-50 ms) timescales. Using experiments and theory, we discuss the challenges and experimental constraints related with the maximum bubble pressure tensiometry measurement.

451196 Dynamic Contact Angles in Liquid-Liquid Systems

Abstract:

When a plate is withdrawn or inserted through a liquid-liquid interface, the additional forced required due to surface effects, is measured and reported in form of a dynamic contact angle. In the first set of experiments the case of water advancing into very viscous oils is shown to have dynamic contact angles somewhat larger than the equilibrium values. In the receding case, the meniscus is seen to be unstable. In the second set, water is replaced with an aqueous polymer (PEO) solution. Significant changes in contact angles are observed but with no obvious signs of shear thinning effects of polymer solutions. The receding contact lines are unstable but at higher velocities, with probably an increased effect due to elastic recoil.

470090 Experimental Study of the Contact Line Region for a Pure Fluid and Binary Fluid Mixture on the ISS

Abstract:

The Constrained Vapor Bubble (CVB) is a wickless heat pipe that is designed to produce a simple, light, and reliable heat transfer system that can be used for cooling critical components of spacecraft. A full-scale fluid experiment flown on the International Space Station (ISS) studied the thermophysical principles underlying change-of-phase heat transfer systems. The CVB system consists of a relatively simple setup - a quartz cuvette with sharp corners partially filled with either pentane (CVB1) or an ideal mixture of pentane and isohexane (CVB2) as the working fluids. The Light Microscopy Module (LMM) was used to determine the two-dimensional thickness profile of the menisci formed at the corners of the cuvette, along with temperature and pressure measurements. Interfacial forces dominate in these extremely small Bond number systems. The transport processes were found to be complex despite being conceptually simple. At the heated end of the CVB, due to a high temperature gradient, we observed "Marangoni flow" in the direction of increasing surface tension toward the cooler end when pure pentane was used as the working fluid. This temperature induced Marangoni flow prevents liquid from recirculating to the heater end, and therefore reduces the effectiveness of the heat pipe. Recently, several research groups used a water and alcohol mixture, with a low concentration of alcohol, resulting in better performance of the heat pipe [1, 2]. The alcohol/water combinations were peculiar in that for a certain composition range, the surface tension increases with increasing temperature thereby driving liquid toward the hotter end. It was believed that changing the direction of the Marangoni stress or reducing its magnitude by differential evaporation of an ideal binary mixture would also improve the performance of the heat pipe. In our experiment, in order to mitigate Marangoni flow, we use a mixture of 94 vol%-pentane and 6 vol%-isohexane. This mixture forms an ideal liquid containing a low concentration of lower vapor-pressure and higher-interfacial-tension, isohexane. With isohexane added, the composition of the liquid will change with position in the heat pipe. The normal change in interfacial tension due to the temperature gradient will be opposed by an opposite change in interfacial tension due to the change in mixture composition.

On the large scale, to study the heat transport phenomena in the CVB heat pipe, we developed a one-dimensional heat transfer model which was validated using data from the ISS experiments. Based on this model, the internal heat transfer coefficient in the mixture case was improved to almost twice that of the pure pentane case. We also found that the Marangoni stress reduced by about five times [3].

On the small scale, the junction of the liquid with the vapor and the solid substrate is of great interest. This junction has been called the contact line region. The concept of phase change heat and mass transfer in the three-phase contact line region and the dynamics of how that contact line behaves are essential to understanding and controlling the wide variety of processes including boiling, evaporation, condensation, and coating processes. The CVB module is an ideal experimental setup for studying complex fluid flow, evaporative heat transfer, and interfacial phenomena at multiple scales. Using interferometry technique and high resolution (50X) images from the LLM, we were able to measure the two-dimensional thickness profile of the menisci formed at the corners of the cuvette. This data allowed us to obtain, with high accuracy, the curvature gradient at the liquid-vapor interface which is one of the main parameters that controls fluid flow toward the contact line region. Our results showed that the curvature gradient is a strong function of local heat flux and concentration. For both the pure pentane and the mixture cases, the curvature gradient increases with increasing local heat flux in the evaporation region, and approaches constant value moving toward the condensation region. For the mixture case, the curvature gradient increases with decreasing pentane concentration. Comparing the pure pentane and the mixture cases at the same local heat flux, we found that the curvature gradient is larger for the mixture case (up to five times larger); and the difference between the two cases increases as pentane concentration decreases. Our data also showed that the curvature gradient profile along the evaporation region of the heat pipe is divided into two regions: the region from the heater end to "the central drop" and the region from the central drop to the end of the evaporation section. The division location of the two regions corresponds to the location of maximum internal heat transfer rate. We also saw the change in liquid film distribution in the region from the heater end to the central drop before and after the maximum point is observed in the internal heat transfer rate profile. Before the maximum point is observed, the thin liquid film moves away from the wall as we move toward the heater; after the maximum point is observed, it moves toward the wall.

Supported by NASA under Grant number NNX13AQ78G.

References

1. N. di Francescantonio, R. Savino, Y. Abe, New alcohol solutions for heat pipes: Marangoni effect and heat transfer enhancement, Int. J. Heat Mass Transfer 51 (2008) 6199–6207.

2. K.M. Armijo, V.P. Carey, An experimental study of heat pipe performance using binary mixture fluids that exhibit strong concentration Marangoni effects, J. Therm. Sci. Eng. Appl. 3 (3) (2011) 031003.

3. T.T.T. Nguyen, A. Kundan, P.C. Wayner Jr., J.L. Plawsky, D.F. Chao, R.J. Sicker, The effect of an ideal fluid mixture on the evaporator performance of a heat pipe in microgravity, Int. J. Heat Mass Transf. 95 (2016) 765–772.

452125 Transport Properties in Ternary Liquid-Liquid Systems

Abstract:

In chemical engineering extraction plays a crucial role for the purification of products ranging from bulk products to biopharmaceuticals. For the extraction, the mass transfer of the target product across the interface is essential as it determines the residence time and so the dimensions of the extraction unit. The nature of the interface between two fluids has been the subject of science since Young, Laplace and Gauss in the 1800 [1]. All three have the assumption of an interface between two fluids as a surface of zero thickness. On this surface there would be unsteadiness in all physical properties. The idea of an interface with a non-zero thickness was developed in detail by Lord Rayleigh and van der Waals proposing gradient theories for the interface based on thermodynamic principles [1]. Based on the ideas of the gradient theory by Van der Waals, Cahn and Hilliard developed the density gradient theory (DGT), which can be used to calculate interfacial properties in equilibrium and the mass transfer across the interface in non-equilibrium [2]. In contrast to other mass transfer models, the thermodynamics can be directly considered by the inclusion of the Helmholtz free energy in the DGT. Up to now the DGT was used in binary system for the calculations of the mass transfer across the interface [3].

In this work this approach was expanded to ternary mixtures. First, the mass transfer across the interface was experimentally determined using a Nitsch-Cell [4] for the systems toluene-acetone-water (System 1) and heptane-hexane-methanol (System 2). Afterwards the phase equilibra were successfully described by the Koningsveld-Kleintjens model and the interfacial tensions of these systems were modeled by adjusting the influence parameters of the DGT to experimental data of interfacial tension for each system. In addition to the interfacial tension, also the concentration profiles of the components across the interface could be modelled. As a result a strong accumulation of acetone in the interface of System 1 could be observed, but there was only a light accumulation of hexane in System 2. The problem is that this accumulation cannot be verified experimentally. Moreover, the experimental results show, that System 1 shows three times longer equilibration times than System 2. By the modelling of the mass transfer across the interface, it could be shown, that the DGT in combination with Koningsveld-Kleintjens approach can be used to model the mass transfer in good agreement with experimental data. Hereby, the so called mobility coefficients have to be adjusted to experimental mass transfer data. Comparing the resulting mobility coefficients, it can be seen that in both systems the mobility coefficients are in the same range even though there is a large difference in the equilibration times. This can be a hint for the accumulation at the interface and that it forms a resistance for the mass transfer.

Here, the DGT in combination with a thermodynamic model was firstly compared to experimental data.

References

[1] D. M. Anderson, G. B. McFadden, A. A. Wheeler "Diffuse-interface methods in fluid mechanics", Annual review of fluid mechanics 1998, 30, 139-165.

[2] J. W. Cahn, J. E. Hilliard "Free energy of a nonuniform system. I. Interfacial free energy", The Journal of chemical physics 1958, 28, 258-267.

[3] A. Kulaguin Chicaroux, A. Gorak, T. Zeiner " Demixing of Behaviour of binary Polymer Mixtures", Journal of Molecular Liquids, 2015, 209, 42-49.

[4] Plucinski, P., W. Nitsch, „Mechanism of mass transfer between aqueous phase and water-in-oil microemulsion" Langmuir, 1994, 10, 371-376.

463883 Hydrodynamics and Mechanism Study of Foam Column Trays with Various Contact Angles

Abstract:

Foam materials have been widely used as equipment internals, a typical application being the column tray for distillation[1-3]. Hydrodynamic phenomena are very important fundamentals for further design, optimization and operation of process equipments, which are extremely affected by surface wettability. In this work, cylindrical foam silicon carbide (SiC) devices, with the diameter of 44-mm, the height of 6-mm and the mean pore size of 3-mm, having various contact angle () were fabricated and investigated as fixed valves. A detailed fabrication process was put forward for preparing uniform hydrophobic and superhydrophobic surfaces with tunable contact angles inside and outside of foam SiC materials considering the specialty of foam structure. Subsequently, systematic study of gas distribution, pressure drop and weeping on foam SiC fixed valves with various contact angles was conducted to evaluate the effects of surface wettability. Hydrophilicity, hydrophobicity and superhydrophobicity present remarkably different influences on hydrodynamic performance of foam SiC fixed valves. Research results prove that with the increase of surface contact angles\, nonuniformity of gas distribution above the valves owing to relatively wide distribution of pore size is weaken greatly, especially in the condition of lower gas velocity, which is extremely important for stable, long-periodic and high-efficient operation in industry.

Except for gas distribution, pressure drop is also an extremely parameter for process equipment, which determines energy consumption to some extent. Pressure drop of foam SiC fixed valves is divided into four parts: net dry pressure drop, pressure drop caused by the clear liquid height above the valve, pressure drop by surface tension in the process of bubble formation and growth and additional pressure drop generated by surface wettability of foam SiC fixed valves. Additional pressure drop are generated by different types for hydrophilic, hydrophobic and superhydrophobic surfaces, that is to say, pore blocking and shrinkage because of the spreading of liquids for hydrophilic surfaces, pore blocking and shrinkage because of liquid adhesion for hydrophobic surfaces and pore blocking because of liquid dropsâ€™ entrance for superhydrophbobic surfaces. The experimental results indicate that pressure drop is vastly decreased along with the increase of surface contact angles, about 30% decrease at most for hydrophobic surfaces and about 43% decrease at most for superhydrophobic surfaces compared with hydrophilic surfaces under the operation condition of 20-mm weir height. In addition, weeping measurement indicates that quantity of liquid leakage decreases as surface contact angles increase and follows a regular pattern.

In order to gain deep insights into the underlying physical process, an analytical model and mechanism were proposed, which consists of three sections based on the process of gas flowing through foam materials, solid-liquid upper interface and liquid layer. Critical F-factors and critical contact angle values of were used as characteristic parameters to distinguish flow mechanism and process features that mainly were described by phase slip and non-slip, flow patterns and interfacial forces.

References

1. Zhang LH, Liu XK, Li H, et al. Hydrodynamic and mass transfer performances of a new SiC foam column tray. *Chem Eng Technol.* 2012;35(12):2075-2083.

2. Gao X, Li X, Liu X, Li H, Yang Z, Zhang J. A novel potential application of SiC ceramic foam material to distillation: foam monolithic tray. *Chem Eng Sci.* 2015;135:489-500.

3. Li H, Fu L, Li X, Gao X. Mechanism and analytical models for the gas distribution on the SiC foam monolithic tray. *AIChE J.* 2015;61(12):4509-4516.

473009 Density Gradient Theory for Modeling Interfacial Properties of Surfactant Systems

Abstract:

Density gradient theory (DGT) allows a fast and accurate prediction of surface tension and stress profile through an interface. When combined with the diffusion equation, as in Cahn-Hilliard theory, the dynamics of the system can be determined. This approach has been widely used but limited to model systems where molecules have only hard sphere structures. Of considerable interest is to include a surface active component in the system which normally features a chain-like structure. However, even for equilibrium systems, the inclusion of a surface active component is considered as a challenge for density gradient theory.

In our work, the interfacial properties of an oil-water-surfactant system were calculated by DGT for the first time. While the conventional DGT algorithm requires a reference substance and can only handle hard sphere molecules, a modified â€œspace stabilized algorithmâ€ is introduced and extended to be applied in a multiphase multicomponent system. This algorithm makes it possible to calculate interfacial properties at any domain size without choosing a reference substance or assuming the functional form of the density profile.

As part of DGT inputs, perturbed chain statistical associating fluid theory (PC-SAFT) equation of state was employed in the calculation, which has an excellent performance in predicting liquid phase properties and phase behavior. The space stabilized algorithm is applied and compared with laboratory data for several mixture systems. The numerical stability analysis is also included in each calculation to verify the reliability of this algorithm for future applications.

462440 A Simple-to-Apply Predictive Wetting Model for Textured (Rough/Patterned) Surfaces and the Role of Re-Entrant Cavities

Abstract:

Rough/patterned/textured surfaces with nano/micro- cavities that broaden below the surface – known as re-entrant cavities – can be omniphobic (macroscopic contact angle greater than 90° for both water and oils). The underlying physical principles/models that explain texture-driven omniphobicity have been studied extensively. However, existing models do not provide a simple procedure for predicting the thermodynamically stable and, in particular, the kinetically-trapped metastable states and contact angles (for example, wetting states that involve partially-filled cavities). Here, we develop a model that allows for deriving general conclusions and demonstrate the applicability of the model to analyze/determine/predict the macroscopic contact angle(s) on any textured surface.

In general, when liquid is placed on any textured surface, the liquid can either (1) partially- or (2) fully-fill the cavities. Both states (partially- or fully-filled) can occur inside and/or outside (by condensation) of the droplet (bulk liquid). Importantly, the macroscopic contact angles manifested for each state can be significantly different, and either of these states can be a transient (unstable) state, a kinetically-trapped (short or long-lived) metastable state, or the thermodynamic equilibrium state. Utilizing an energy minimization approach, we derive a 'general wetting model' that (1) predicts *apriori* the state (partially- or fully-filled) of the cavities both under (in contact with) and outside of the liquid droplet, and the corresponding macroscopic contact angles on any type of textured surface; (2) allows for determination of the conditions under which metastable states exist; and (3) allows for engineering of specific nano/micro-textures that yield any desired macroscopic contact angle, θ_t, for a given intrinsic contact angle θ_0 (contact angle on smooth and flat surface).

During this talk, our 'general wetting model' will be discussed. In addition, we will demonstrate how this model can be used to design specific nano- and micro- textures to yield any desired macroscopic contact angle. Controlling the macroscopic contact angle, whether above or below the intrinsic contact angle, is desirable for many applications including both non-wetting, self-cleaning and anti-fouling surfaces, and completely-wetting/spreading applications, such as cosmetics and lubricant fluids.

459840 Obtaining Solid-Liquid Interfacial Free Energy for Realistic Systems

Abstract:

The solid-liquid interfacial free energy γ_{sl} is a critical quantity in understanding the thermodynamic aspects of shape-selective nanocrystal formation. Unlike the fluid-fluid interfacial free energy, the solid-liquid interfacial free energy is facet-specific and not identical to surface tension, which makes direct measurements with high enough resolution to resolve the anisotropy extremely challenging in experiments. Several simulation methods have been developed to obtain facet-specific γ_{sl} for systems with simple structures and potentials, but in the current stage, these methods are not applicable to systems with complicated configurations and complex force fields, and are also not compatible with commonly used open source simulation packages. Here we present a novel multi-scheme thermodynamic integration (TI) method that can be used to obtain γ_{sl} for real and complex systems using open source simulation packages. This is a six-step simulation method that obtains γ_{sl} by calculating the negative free energy change when transforming a solid-liquid coexistence to the individual bulk phases. This method is developed based on the previous "cleaving wall" method [1, 2] and is advanced in two major aspects: two different TI schemes to deal with periodic boundary condition without locally modifying the simulation package, and a numerical approximation is introduced when the analytical form of the thermodynamic integrand is difficult to differentiate. These features successfully resolve simulation difficulties when using the previous methods, thus our method can be applied to a broad range of systems and force fields. To demonstrate its capability, we use our method to obtain γ_{sl} for Ag(100) and Ag(111) surfaces in contact with a solution used in shape-selective Ag nanocrystal synthesis – polyvinylpyrrolidone (PVP)-dissolved ethylene glycol (EG) solution. By having a higher binding affinity to Ag(100), PVP is an effective structure-directing agent (SDA) that helps to form {100}-faceted Ag nanocrystals, such as nanocubes and nanowires. Since bare Ag(111) surfaces are more thermodynamically stable than Ag(100) by possessing lower surface free energy, a hypothesis for observing {100}-faceted structures is that PVP can make $\gamma_{Ag(100)} < \gamma_{Ag(111)}$ by tightly bound to Ag(100)

surfaces and the produced {100}-faceted structures are thermodynamically favored. We obtain anisotropic γ_{sl} for Ag(100) and Ag(111) surfaces in contact with the PVP-EG solution using the multi-scheme TI method and show that although PVP can lower $\gamma_{Ag(100)}$ to a larger extent than $\gamma_{Ag(111)}$, it does not make $\gamma_{Ag(100)} < \gamma_{Ag(111)}$. Combining with our findings in a previous "relative deposition flux" study [3], in which we quantified the relative deposition fluxes to these surfaces, we demonstrate that these {100}-faceted structure are not thermodynamic shapes and confirm the kinetic role of PVP in shape-selective Ag nanocrystal synthesis.

[1] Broughton, J. Q., and Gilmer, G. H., J. Chem. Phys. 1986, 84, 5759.
[2] Benjamin, R. and Horbach, J., J. Chem. Phys. 2014, 141, 044715.
[3] Qi, X., Balankura, T. Zhou, Y., Fichthorn, K. A., Nano Lett. 2015, 15, 7711.

472689 Protein Interfacial Behavior Studied with Multiscale Simulations

Abstract:

Protein Interfacial Behavior Studied with Multiscale Simulations

Mohammadreza Samieegohar, Tao Wei,

Dan F. Smith Department of Chemical Engineering, Lamar University, Texas, 77710.

Understanding of the protein adsorption dynamic process plays a prominent role in chemical, materials, biotechnologies and medical industries. In this work, protein interfacial dynamic behavior (i.e., adsorption, aggregation and surface mobility) on Au (111) surface in a micro-channel has been investigated with a multi-scale simulation approach, which consists of atomistic Molecular Dynamic (MD) simulation, free energy computation, Langevin Dynamics (LD) simulation with coarse-gained model and computational fluid dynamics (CFD). Atomistic simulation demonstrates that the gold surface has strong attraction to lysozyme proteins mainly due to gold surface's dehydration. The protein-surface desorption free energy is up to 150 kJ/mol. LD simulation reveals that protein-protein interaction and Brownian dynamic motion are the main driving forces for protein desorption from the attractive gold surface, and desorption time scale can be up to microseconds. Protein molecules tightly adsorbed onto the surface in the first layer display large surface mobility affected by shear stress. CFD demonstrates protein adsorption different kinetics regions correlated with protein-surface interactions, bulk fluid velocity, bulk concentration and microchannel dimensions.

470811 Modeling Surface-Anchored Nematic Liquid Crystals for Chemoresponsive Application

Abstract:

Surface-anchored nematic liquid crystals can undergo an orientational response to changes in their chemical environment, providing new opportunities for designing selective and sensitive chemoresponsive systems. We have previously shown that the liquid crystal 4-cyano-4â€™-pentylbiphenyl (5CB) can be an interesting starting candidate for future sensing applications because its orientational change can be triggered by several important small molecules such as dimethyl methylphosphonate (DMMP), a common component of nerve gas production. The anchoring properties of 5CB also vary greatly with surface-functionality, making it possible to tune the orientational response.[1-3] We developed a series of theoretical models, based on quantum mechanics, to capture the effect of important experimentally adjustable parameters affecting orientational transitions and formulated an integrated high throughput-like materials discovery approach exploiting the synergistic effects between fast theoretical prediction, synthesis of new liquid crystal materials, and their experimental characterization.

In this presentation, we will show how our new computational models can predict orientational transitions induced by the displacement of liquid crystals by a variety of small molecules. We will summarize how our integrated approach could lead to new, improved candidate materials in several different examples such as detection of NO_2, acetone, or ClO_2.

1. Shah, R. R.; Abbott, N. L., *Science* **2001**, *293*, 1296.

2. Hunter, J.T.; Abbott, N.L., *Applied Materials and Interfaces*, **2013**, *6*, 2362

3. Yang, K. L.; Cadwell, K.; Abbott, N. L., *Journal of Physical Chemistry B*, **2004**, *108*, 20180.

466266 Molecular Dynamics Simulations of HIV-1 Regulatory Protein on Graphene

Abstract:

The fragment of viral protein R (Vpr), Vpr13-33, plays an important role in regulating nuclear importing of HIV genes through channel formation in which it adopts a leucine-zipper-like alpha-helical conformation. A recent experimental study reported that helical Vpr13-33 would transform to Î²-sheet or random coil structures and aggregate on the surface of graphene or graphene oxide through hydrophobic interactions. Due to experimental limitations, however, there is still a considerable lack of understanding on the adsorption dynamics at the early stage of the conformational transition at water-graphene interface and the underlying driving force at molecular level. In this study, atomistic molecular dynamics simulations were used to explore the conformation transition phenomena. Vpr13-33 kept Î±-helical structure in solution, but changed to Î²-sheet structure when strongly adsorbed onto graphene. Preferential adsorption of Vpr13-33 on graphene is dominated by hydrophobic interactions. The cluster analysis identified the most significant populated conformation and the early stage of structure conversion from Î±-helical to Î²-sheet was found, but the full Î²-sheet propagation was not observed. Free energy landscape analysis further complemented the transformation analysis of peptide conformations. These findings are consistent with experimental results, and give a molecular level interpretation for the reduced cytotoxicity of Vpr13-33 to some extent upon graphene exposure. Meanwhile, this study provides some significant insights into the detailed mechanism of graphene-induced protein conformation transition.

467845 A Thermodynamic Approach to Interfacial Transport Phenomen

Abstract:

Formulating balance equations for extensive quantities at interfaces in terms of excess densities is a subtle problem because these densities change with the precise location of the dividing surface within the interfacial region. We propose to handle such ambiguities for dynamic interfaces by introducing a gauge degree of freedom associated with the location of the dividing surface and by studying all the normal velocities associated with different excess densities. Unambiguous interfacial balance equations can then be obtained directly from the differences between normal velocities. By considering the interfacial entropy balance along with the local equilibrium assumption, and identifying the entropy production rate at the interface, we are naturally led to thermodynamically consistent constitutive equations for the fluxes characterizing transport both within and at the interface. The usefulness and generality of the proposed approach is illustrated in the context of several examples.

470819 Adhesion and Translocation of Nanoparticles through Lipid Bilayers Studied By Mesoscale Simulations

Abstract:

Understanding of the mechanisms of nanoparticle (NP) adhesion to lipid bilayer (LB) membranes, which constitute the foundation of cell membranes, is of paramount importance for the design of biomedical nanotechnologies, as well as for evaluating health threats related to nanoparticle manufacturing. NPs are employed as intracellular delivery vehicles for controlled release of genes and membrane impermeable chemicals. This process involves internalization, or intake of a drug containing NP, where the NP is first engulfed by the cell membrane, then transferred through the membrane into the cell. Since the NP adhesion is governed by multiple factors (NP size and shape, surface modification, hydrophobicity and charge) that cannot be easily varied in the experiments, the use of in silico modeling helps guide the experiments and optimize NP structure and surface properties. Despite the tremendous progress made recently in molecular simulations of biological systems, modeling of NP-lipid interactions is still difficult due to a complex interplay of distinct characteristic scales: LB thickness of ~5 nm, to NP size of 10-100 nm, to cell size of 5- 50 mm. Also, the dynamic processes of membrane rupture and NP trans-membrane transport involve metastable states and irreversible transformations, modeling of which requires non-trivial calculations of free energies and nucleation barriers.

The current study investigates the mechanisms of nanoparticle adhesion to and penetration through LB membranes. For this purpose, we construct a soft-core coarse-grained models of hydrophobic nanoparticle and 1,2-dimyristoyl-sn-glycero-3-phosphocholine (DMPC) bilayer membrane maintained under constant surface tension conditions using a plank, to which a constant force is applied. In a series of dissipative particle dynamics simulations, we consider NP transport across the LB membrane. The free energy landscape of the NP in the bilayer vicinity is explored using the ghost field method that emulates a lab experiment performed with optical tweezers. Hydrophobic particles adsorb a self-assembled monolayer of lipid. As the NP approaches the LB, the latter deforms. The deformation is followed by a spontaneous fusion of the freestanding bilayer and the adsorbed monolayer and particle incorporation inside the hydrophobic inner space of LB. The encapsulation stage where NP is captured by LB corresponds to a free energy minimum. The transition between a free and encapsulated NP is associated with a free energy barrier. The barrier is insignificant when NP is smaller than LB width but increases rapidly with the NP diameter. NP escape from the LB membrane is also associated with deformation of the latter and a free energy penalty. The free energy barrier associated with NP escape decreases with the NP size, that is, shows a tendency opposite to the entry barrier.

This work was supported by NSF grant 1264702

470243 Calculating the Interfacial Properties of CO_2/Water/Silica Systems Using Monte Carlo Molecular Simulations

Abstract:

Geological sequestration has the potential to occupy a strategic role in the reduction of carbon dioxide emissions. Before injecting CO_2 into a potential storage site, one must first assess the capability of the site to retain the injected CO_2. Understanding the interactions between the fluid CO_2-rich and water-rich phases and the caprock mineral of a site is particularly important. These interactionsâ€"in the form of the CO_2/water/caprock contact angle and CO_2/water interfacial tensionâ€"dictate the breakthrough pressure, that is, the pressure at which the CO_2-rich phase may leak due to capillary failure. We present here the use of Monte Carlo methods to measure the interfacial tension and contact angles of CO_2/water mixtures near a silica surface. We apply a combination of interface potential free energy techniques and expanded ensemble simulations to measure differences in these interfacial properties over a range of temperature and pressure conditions for several combinations of available water and CO_2 molecular models.

466525 Compositional Effect of Varying Feedstocks on Heavy Oil Upgrading By Supercritical Water

Abstract:

Crude oil prices have shown big fluctuations around 150 and 35 $/barrel during 2008-2016. Such a volatile and unpredictable oil price tendency, which is being drastically affected by political issues, forces energy policy makers to focus on alternative energy conversion technologies for energy security. On the other hand, one third of Worldâ€™s recoverable oil reserves are in a form of heavy oil. Therefore developing alternative oil upgrading technologies with improved conversion and tunable selectivity providing low coke and improved light liquid yield are of desire. However, the variations in feedstock chemistry due to the source of the heavy oil affect economics of investments to a great extent. Supercritical fluids can be used for the development of alternative and innovative heavy oil upgrading processes and supercritical water (SCH_2O) having decreased dielectric constant and increased ionic product values is a potential candidate for the reactive extraction and thereby upgrading of heavy oils in an intensified process scheme. These altered physicochemical properties of water enable it to mimic hydrocarbons for reactive extraction purposes and also to increase $[H_3O^+]$ concentrations for triggering acid-catalyzed cracking reactions. Domestic tar sand samples together with feedstocks of state-of-the art heavy oil upgrading units (Delayed Cokers - DCU), namely 50/70 bitumen from 2 different refineries in Turkey, 70/100 pen bitumen, and by product of Base Oil Unit, PDA bitumen, were used for the experiments. The reactions were carried out at 440 °C and 30MPa for 1 hour with water/oil ratio of 2 in a home-made 10 cm^3, SS316, bomb reactor. Characterization was done by GPC, H-NMR, DSC, TGA, FTIR, XRD, GC-Simdist, API, SEM images and SARA analysis. The coke-liquid-gas yield, coke type and boiling point distribution of liquids were shown to be influenced from the feedstock chemistry. Varying conversions with scattered selectivity values have been obtained. Modeling studies will be carried out to derive a tool for the prediction of conversion and selectivity of the reactions for a given feed under certain conditions as a part of a complementary study.

467087 *Product Formation and Kinetics of the Non-Isothermal Hydrothermal Liquefaction of Soy Protein Isolate*

Abstract:

James Sheehan - Graduate Student

Dr. Phil Savage – Adviser

The Pennsylvania State University, Department Chemical Engineering

Product Formation and Kinetics of the Non-isothermal Hydrothermal Liquefaction of Soy Protein Isolate

Hydrothermal liquefaction (HTL) is a thermochemical process capable of producing biocrude oil from biomass. HTL of various biomass feedstocks including microalgae, bacteria, lignocellulose and model compounds has been studied extensively under subcritical temperatures and long holding times (>60 min).[1–7] More recently, HTL of microalgae under rapid heating conditions and short reaction times (<1 min) has achieved higher biocrude yields achieved than HTL of microalgae under subcritical conditions and longer holding times.[8,9] Protein is a major biochemical constituent of microalgae and we performed HTL of soy protein isolate to understand better the incorporation of protein derived products into biocrude.

We conducted non-isothermal, fast HTL of soy protein isolate (SPI) at temperatures ranging 140 °C to 500 °C and residence times of 10 s to 300 s. SPI solids rapidly transform into aqueous phase products and biocrude. The highest biocrude yields (38-40wt%) were obtained within 45 s to 120 s and at temperatures ranging from 375 °C to 435 °C. The largest recovery of atomic N in the aqueous phase products was observed prior to the formation of substantial yields of biocrude. Ammonia formation was significant when the hydrothermal reaction medium reached supercritical conditions. Over 50% of the atomic N was observed as ammonia under such conditions. We developed a kinetic model and determined Arrhenius parameters that describe the fast HTL of SPI. The results of the modeling are shown in Figure 1. As can be seen, the proposed reaction network and optimized Arrhenius parameters can accurately model the major product yields from fast HTL of SPI.

Figure 1: Experimental (discrete points) and model predicted yields (continuous curves) for fast HTL of SPI at set point temperatures 400°C.

References

(1) Valdez, P. J.; Nelson, M. C.; Wang, H. Y.; Lin, X. N.; Savage, P. E. *Biomass and Bioenergy* **2012**, *46*, 317–331.

(2) Valdez, P. J.; Nelson, M. C.; Faeth, J. L.; Wang, H. Y.; Lin, X. N.; Savage, P. E. *Energy & Fuels* **2014**, *28* (2013), 67–75.

(3) Biller, P.; Ross, A. B. *Bioresour. Technol.* **2011**, *102*, 215–225.

(4) Teri, G.; Luo, L.; Savage, P. E. *Energy & Fuels* **2014**, *28*, 7501–7509.

(5) Changi, S.; Zhu, M.; Savage, P. E. *ChemSusChem* **2012**, *5*, 1743–1757.

(6) Luo, L.; Sheehan, J. D.; Dai, L.; Savage, P. E. *ACS Sustain. Chem. Eng.* **2016**.

(7) Toor, S. S.; Rosendahl, L.; Rudolf, A. *Energy* **2011**, *36* (5), 2328–2342.

(8) Faeth, J. L.; Valdez, P. J.; Savage, P. E. *Energy and Fuels* **2013**, *27*, 1391–1398.

(9) Hietala, D. C.; Faeth, J. L.; Savage, P. E. *Bioresour. Technol.* **2016**, *214*, 102–111.

James Sheehan - Graduate Student

Dr. Phil Savage – Adviser The Pennsylvania State University, Department Chemical Engineering

Product Formation and Kinetics of the Non-isothermal Hydrothermal Liquefaction of Soy Protein Isolate

-

Hydrothermal liquefaction (HTL) is a thermochemical process capable of producing biocrude oil from biomass. HTL of various biomass feedstocks including microalgae, bacteria, lignocellulose and model compounds has been studied extensively under subcritical temperatures and long holding times (>60 min).[1–7] More recently, HTL of microalgae under rapid heating conditions and short reaction times (<1 min) has achieved higher biocrude yields achieved than HTL of microalgae under subcritical conditions and longer holding times.[8,9] Protein is a major biochemical constituent of microalgae and we performed HTL of soy protein isolate to understand better the incorporation of protein derived products into biocrude.

We conducted non-isothermal, fast HTL of soy protein isolate (SPI) at temperatures ranging 140 °C to 500 °C and residence times of 10 s to 300 s. SPI solids rapidly transform into aqueous phase products and biocrude. The highest biocrude yields (38-40wt%) were obtained within 45 s to 120 s and at temperatures ranging from 375 °C to 435 °C. The largest recovery of atomic N in the aqueous phase products was observed prior to the formation of substantial yields of biocrude. Ammonia formation was significant when the hydrothermal reaction medium reached supercritical conditions. Over 50% of the atomic N was observed as ammonia under such conditions. We developed a kinetic model and determined Arrhenius parameters that describe the fast HTL of SPI. The results of the modeling are shown in Figure 1. As can be seen, the proposed reaction network and optimized Arrhenius parameters can accurately model the major product yields from fast HTL of SPI.

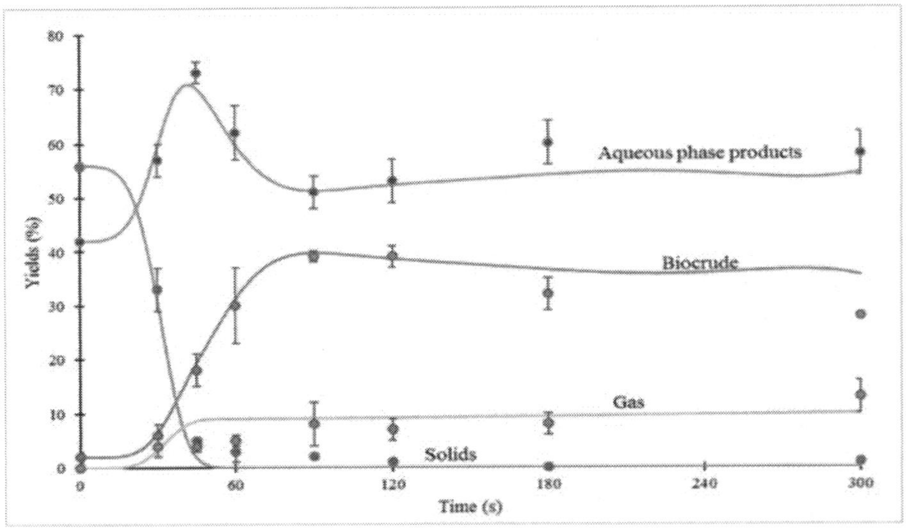

Figure 1: Experimental (discrete points) and model predicted yields (continuous curves) for fast HTL of SPI at set point temperatures 400°C.

References

(1) Valdez, P. J.; Nelson, M. C.; Wang, H. Y.; Lin, X. N.; Savage, P. E. *Biomass and Bioenergy* **2012**, *46*, 317–331.

(2) Valdez, P. J.; Nelson, M. C.; Faeth, J. L.; Wang, H. Y.; Lin, X. N.; Savage, P. E. *Energy & Fuels* **2014**, *28* (2013), 67–75.

(3) Biller, P.; Ross, A. B. *Bioresour. Technol.* **2011**, *102*, 215–225.

(4) Teri, G.; Luo, L.; Savage, P. E. *Energy & Fuels* **2014**, *28*, 7501–7509.

(5) Changi, S.; Zhu, M.; Savage, P. E. *ChemSusChem* **2012**, *5*, 1743–1757.

(6) Luo, L.; Sheehan, J. D.; Dai, L.; Savage, P. E. *ACS Sustain. Chem. Eng.* **2016**.

(7) Toor, S. S.; Rosendahl, L.; Rudolf, A. *Energy* **2011**, *36* (5), 2328–2342.

(8) Faeth, J. L.; Valdez, P. J.; Savage, P. E. *Energy and Fuels* **2013**, *27*, 1391–1398.

(9) Hietala, D. C.; Faeth, J. L.; Savage, P. E. *Bioresour. Technol.* **2016**, *214*, 102–111.

467990 An Investigation on Short Residence Time Macro Algae Hydro Thermal Liquefaction in a Batch Reactor

Abstract:

AIChE annual meeting, November 2016 San Francisco

An investigation on short residence time macro algae hydro thermal liquefaction in a batch reactor

Arash Izadpanah [a], Niccol¨ Le Brun [a], Klaus Hellgardt[a]

[a] *Department of Chemical Engineering, Imperial College London, South Kensington Campus, London, SW7 2AZ, United Kingdom*

Bio crude oil production from macro algae may become a key technology to tackle tomorrowÕs renewable fuels challenge. In order to seed this technology it is important to identify and evaluate the right process pathways to convert macro algae to biofuels. In conventional algal biofuels processing, drying of the algae is a costly procedure, which could be bypassed by using Hydrothermal Liquefaction (HTL) technology with substantial advantages. For an efficient implementation of a continuous HTL process, the effect of process variables on the conversion and selectivity towards liquid biofuels needs to be understood. This is often a challenge because the timescale of the HTL process is of the order of minutes and it is difficult to be captured properly by ordinary methods. In this study we employed a molten salt bath to facilitate rapid heat transfer to the batch reactor and thus study HTL for precisely controlled residence times between 1 and 25 min. Under experimental conditions the conversion of macro algae to bio crude oil was found to be complete within 15 min reaching a maximum and then decreasing slightly (23% maximum yield after 11 min at 390 ¡C). Using different analytical techniques including CHNS-O, GC-MS, Simulated Distillation and Size Exclusion chromatography the bio crude from macro algae is characterized and compared with North Sea crude oil.

Keywords: Macro algae, Hydrothermal liquefaction, Conversion rate, Filtration, Centrifugation

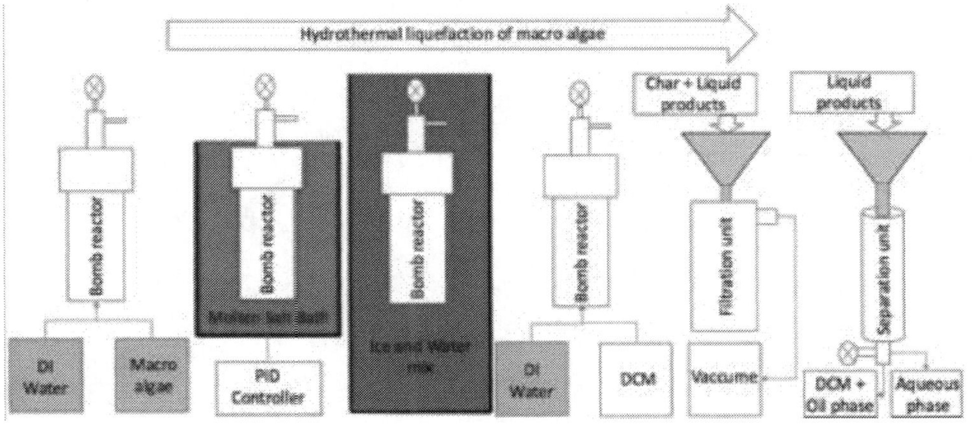

Figure 1:Graphical summery of the experiment

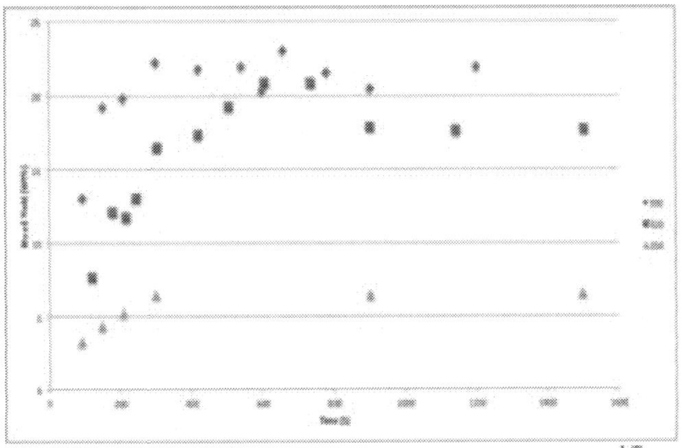

Figure 2: Bio crude oil yield as a function of residence time at 250 ℃, 320 ℃, 390 ℃.

459274 Hydrothermal Liquefaction (HTL) of High-Ash Algal Biomass: The Effect of Ash Contents in HTL Reactions

Abstract:

Previous studies demonstrate that the high ash contents appeared to inhibit the formation of biocrude oil in the hydrothermal liquefaction (HTL) processes. Mixed-culture algal biomass from wastewater systems (AW) were screened into three sizes (<106 μm, 106-180 μm and >180 μm) and their ash contents were measured. AW biomass with different ash contents was converted into biocrude oil via HTL at 300 °C for a 60 minute reaction time, which is the previously determined optimum condition for producing biocrude oil from AW. Compared to AW biomass before screen pretreatments, the ash contents of AW of 106-180 μm and <106 μm were respectively decreased from 53.3 % to 38.5% and 41.8 %, while the biocrude oil yields remained almost constant. When the ash contents were reduced from 53.3 % to 38.5 % in the AW biomass, the HHV of biocrude oil was increased from 27.5 MJ/kg to 32.3 MJ/kg and the fraction of light oil (boiling point of 100-300 °C) was increased from 31% to 49%. On the other hand, GC-MS analyses of algal biocrude oil and aqueous products show that the ash contents could promote denitrogenation, catalyze the formation of hydrocarbons, as well as mitigate the recalcitrant compounds in aqueous products under the HTL processes. The above results indicate that algal biomass with certain amounts of ash contents can be converted into biocrude oil with reasonable quality and quantity. In order to quantify the limit of the ash contents under the HTL processes, further HTL conversions with model algal feedstock and representative ash contents were conducted in this study. Elemental and thermogravimetric analyses of the resulting biocrude oil both demonstrate that when the ash content in the algal feedstock was below 40%, the HHV and boiling point distribution of the algal biocrude oil could be hardly affected. This fact substantiates the feasibility of using high-ash algal biomass as HTL feedstock and diminishes the necessity of multi-step pretreatments and modifications of high-ash algal biomass for biofuel application.

456792 Simultaneous Solid, Phosphorus Recovery and Biocrude Production through Hydrothermal Liquefaction of Algae

Abstract:

Subcritical water has become a highly regarded method for converting biomass, primarily algae, to an energy dense biocrude product. This conversion technique, more commonly referred to as hydrothermal liquefaction (HTL), still faces challenges of nutrient recovery, producing a practical biocrude substitute, and generating value-added coproducts before becoming a sustainable and commercially viable method. Recovering the phosphorus (P) in the solid biochar could be beneficial for secondary applications.

Previous work has shown that the inorganic content of the growth media, in particular Ca, P, and Si, and pH during harvesting dictate the fate of P in the HTL products. This paper expands on the idea of solid P-recovery via *in situ* synthesis of calcium phosphates present in the biochar, as seen previously in mixed culture, wastewater-cultivated algae. HTL was performed on diatoms, a unique group of algae that harness Si naturally within their cell walls, and mixed cultures of diatoms and *Chlorella kessleri*. Powder x-ray diffraction (XRD) characterized both the starting algal solids and resulting HTL biochar while acid digestions and inductively coupled plasma optical emission spectrometry (ICP) closed the P balance of HTL reaction.

In addition, it is hypothesized the algal carbohydrates, proteins, and lipids also play a role in the morphology and structure of the biochar. Model compounds were individually reacted via HTL, in the presence and absence of calcium phosphates, and calcium and phosphate ions. The resulting biochar, aqueous coproduct, and biocrude were characterized. Gas chromatography mass spectrometry, CHN/O elemental analysis, and simulated distillation analysis of the biocrude product were analyzed to identify potential catalytic upgrading properties of the calcium phosphate biochar.

453154 Supercritical Water Treatment of Fractions of Crude Oil: Quantification of the Products

Abstract:

The Arabian Heavy (AH) crude oil was fractionated into distillate ($\leq C_{20}$) fraction and bottoms fraction. The distillate and the bottoms fraction of crude oil were treated with supercritical water (SCW) in a batch reactor setup. The gaseous and liquid phase products were analyzed using gas chromatography (GC-MS, GCxGC, FID, SCD). The change in structural composition of maltenes and asphaltenes in the bottoms fraction before and after supercritical water treatment was investigated using proton NMR, FTIR and Raman spectroscopy.

The distillate fraction primarily consisted of n-alkanes ($\leq C_{20}$) plus a minor fraction of alkyl aromatics and sulfur compounds. SCW treatment of the distillate fraction resulted in cracking of higher carbon number alkanes to lower carbon number alkanes, and production of gaseous species such as methane, ethane and propane.

SCW treatment of the bottoms fraction caused much more dramatic changes. Much of the heavy components in the crude oil cracked to form compounds in the gasoline and diesel range which are observed by GCxGC. The higher carbon number aliphatics ($\geq C_{20}$) and long chain ($\geq C_9$) alkyl aromatic compounds in the bottoms fractions underwent cracking and produced C_6-C_{18} alkanes and C_3-C_8 alkylbenzenes. Supercritical water treatment of the bottoms fraction cracked the heavy sulfur compounds, and resulted in increased recovery of lighter sulfur compounds such as alkylated benzothiophene and dibenzothiophenes. The SCW-treated bottoms also showed increased recovery of polycyclic aromatic hydrocarbons such as naphthalene. Despite being carried out at temperatures similar to those used in conventional thermal coking processes, the SCW process did not form significant amounts of coke.

The maltenes and asphaltenes were separated from the bottoms fraction and the SCW-treated bottoms fraction. The

proton NMR of the SCW-treated maltenes fraction showed a decrease in the protons corresponding to aliphatic CH_2 and CH_3 groups. There was also an increase in protons in single ring and polycyclic aromatic hydrocarbons. The proton NMR results support the observations made by the GCxGC measurements.

The proton NMR, FTIR and Raman spectroscopy on untreated and SCW-treated asphaltenes showed changes in the structural composition of asphaltenes post-SCW treatment. The proton NMR experiments showed a decrease in the protons which correspond to aliphatic CH_2 and CH_3 groups. The decrease in aliphatic CH_2 and CH_3 functional groups could be due to scission of the long alkyl chains attached to the aromatic core structure of the asphaltenes. The proton NMR results also showed 49 % increase in the mono-aromatic protons and 1 % decrease in di-aromatic and poly-aromatic protons, which suggests formation of new aromatic rings and/or perhaps conversion of maltenes containing single aromatic rings into asphaltenes (e.g. by the cleavage of alkyl tails, affecting the solubility).

FT-IR spectroscopy showed similar results to proton NMR, wherein an increase in aromatic and heteroatom content (O-H/N-H) was observed in the SCW-treated asphaltenes. Raman spectroscopy also showed evidence of chemical rearrangement and higher aromatic content in the SCW-treated asphaltenes when compared to asphaltenes from untreated bottoms fraction.

Overall, the SCW treatment of the crude oil fractions resulted in cracking of heavier hydrocarbon components into more valuable lighter molecules, which could be used in transportation fuel or as chemicals.

Acknowledgement: This research was supported through a Research Agreement with Saudi Aramco, a Founding Member of the MIT Energy Initiative.

458338 Supercritical Water Process - Finding a Way to Accomodate into Petroleum Industry

Abstract:

Supercritical fluid is not new to petroleum industry. Solvent extraction process utilizes supercritical fluid for enhancing selectivity and yield. However, supercritical water has not been applied in industrial scale due to its harsh condition. With progress of material and process technologies, supercritical water has been studied and evaluated in various areas to resolve technical challenges of conventional technologies. Recent research activities on heavy oil upgrading by supercritical water resulted in serious consideration of accommodating such a nonconventional way in petroleum industry. This presentation will introduce potential application areas of supercritical water technology as well as its implication in petroleum industry.

457095 AÂ High Pressure Viscometer for the Assessment of Viscosity and Density of Lubricants and Polymer Solutions Under Pressure

Abstract:

A unique rotational viscometer has been developed to simultaneously determine viscosity, density, and phase behavior of fluids across a wide range of pressures, temperatures, and shear rates. This high pressure instrument utilizes a magnetically coupled inner rotating shaft for the measurement of viscosity across multiple shear rates. It uses two pistons with accompanying LVDT (Linear Variable Differential Transformer) sensors to apply pressure to the fluid being evaluated by altering its volume and thus generating density data across a range of pressures at a given temperature. The instrument uses two pairs of sapphire windows for direct visual observation of the phase behavior of a fluid being examined. The system is capable of reaching pressures up to 70 MPa at temperatures up to 100°C and rotational speeds of the inner shaft up to 700 rpm. Using this versatile system, we have explored the effects of pressure, temperature, shear rate, and composition on the viscosity and density of mixtures of PDMS (polydimethylsiloxane) and CO_2. We have conducted similar measures also for poly (alpha olefin) base oils (PAOs) at high pressures. Volumetric data for both the polymer solutions and lubricants have been fitted to the Sanchez-Lacombe equation of state and thermophysical properties, such as isothermal compressibility, isobaric expansivity, and internal pressure, have been calculated. The viscosities of the examined fluids have been related to density by using free volume approaches, such as the Doolittle Equation, and scaling factors.

454421 Exploring the Temperature and Pressure Dependence of the Viscosities of Two Potential Deepwater Viscosity References

Abstract:

Deeper drilling for hydrocarbon fuels has elevated the need for viscosity reference fluids for the purpose of calibrating and validating of viscometers and rheometers at high temperatures and high pressures. The International Association for Transport Properties (IATP) established the nominal value of the high-pressure viscosity standard at about 20 cP at a temperature of 200°C and pressure of 200 MPa. The desirable attributes of the high-pressure viscosity standard candidate include thermal stability, inertness, insensitivity to UV radiation, hydrophobicity, monodispersity, lot-to-lot consistency, occurrence as a liquid at ambient conditions and commercial availability. Further, the candidate should be safe to use in the laboratory and it should be environmentally benign. DuPontâ€™s perï¬‚uoropolyether oil Krytox GPL 102 was proposed by our team and several members of the high pressure transport property community as a viable deepwater viscosity standard (DVS) that meets most of the aforementioned requirements. However, lot-to-lot variation of this polydisperse perfluoropolyether could be a serious hurdle preventing its long term acceptance as a viscosity standard. Tris(2-ethylhexyl) trimellitate (TOTM) has also been proposed by Diogo and coworkers as a promising alternative to Krytox GPL 102. This study explores the viability of Krytox GPL 102 and tris(2-ethylhexyl) trimellitate (TOTM) as deepwater viscosity standards. High-pressure viscosity measurements for both fluids are reported at pressures between 7 and 242 MPa and temperatures between 314 and 527 K with an expanded uncertainty of 3% at a 95% confidence level. In addition, the viscosity results are correlated with an empirical temperature/pressure-dependent function and a modified Vogel-Fulcher-Tammann (VFT) Equation.

459308 Widom Line, Dynamical and Structural Crossovers in Supercritical Oxygen Via Molecular Dynamics Simulations

Abstract:

Supercritical oxygen, a cryogenic fluid, is widely used as an oxidizer in jet propulsion systems. Understanding and characterizing the supercritical state is therefore of paramount importance in providing physical insights into processes such as cryogenic jet break-up, seen in combustion chambers of rockets, and jets. Recently, a 'pseudo-boiling' phenomenon was described for trancritical injection processes, prompting the discussion on a through characterization of the supercritical state of oxygen.

It is well known from existing literature that the locus of the maxima of thermodynamic response functions converge to a single line, referred to as 'The Widom line', on approaching the critical point, for a number of substances such as argon, water, nitrogen and others. In this study, we identify the Widom line for oxygen from atomistic molecular dynamics simulations of supercritical oxygen and compare them with experimental data of thermodynamic response functions. Additionally, we also analyze the crossover in transport properties from liquid-like to gas-like behavior with increasing temperature, which appears to coincide with the Widom line. In addition to the transition in shear viscosity and translational diffusion coefficient, we also determine the crossover in the rotational diffusion constant by analyzing the behavior of the rotational auto-correlation function of supercritical oxygen, which shows a marked change from diffusion dominated hindered rotation to a free rotation dominated state with increasing temperature. Finally, we identify the structural crossover for supercritical oxygen across the Widom line, by analyzing the pair correlation function and the structure factor. Our results indicate that in addition to the thermodynamic response functions, transport properties and structural properties show a marked change in the vicinity of the critical point, and that supercritical oxygen is far more complex than originally perceived.

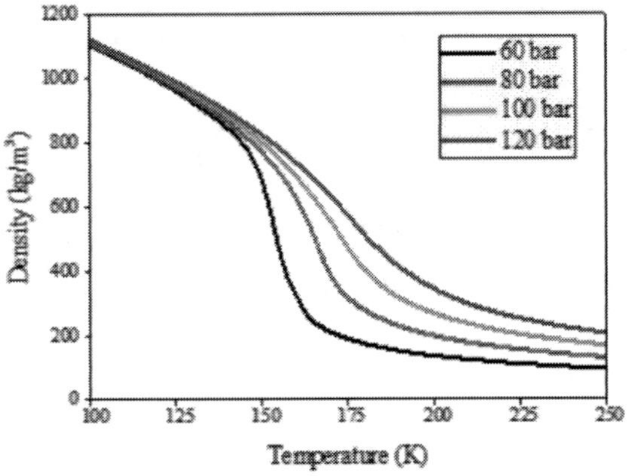

Fig 1. Supercritical isobars for oxygen showing transition in density with increasing temperature, obtained via atomistic MD simulations.

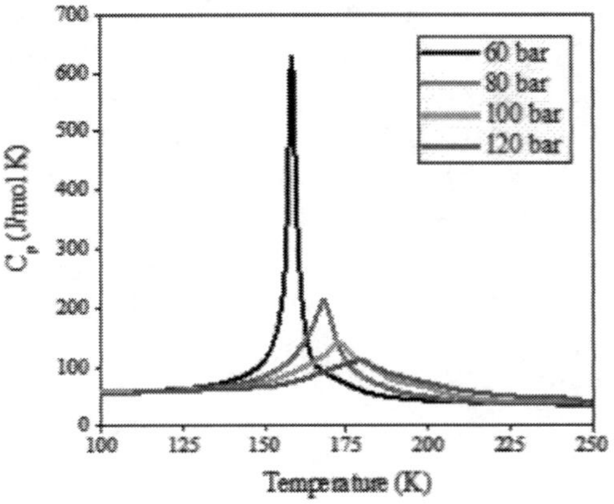

Fig 2. Supercritical isobars for oxygen showing maxima in the isobaric heat capacity with increasing temperature, obtained via atomistic MD simulations.

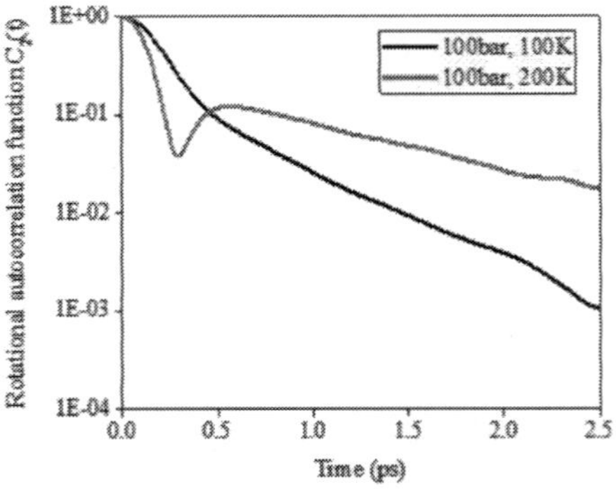

Fig 3. Rotational correlation function $C_2(t)$ of supercritical oxygen at different thermodynamic states showing different rotational behavior, obtained via atomistic MD simulations.

461221 Development and Evaluation of Molecular Models for the Calculation of Thermodynamic and Transport Properties of H2O+CO2+NaCl System

Abstract:

Understanding the thermo-physical properties of the H_2O+CO_2+NaCl mixture is of great importance to geochemistry and design of CO_2 geologic sequestration. Thermodynamic and transport properties of binary H_2O+NaCl and H_2O+CO_2 mixtures were studied over a wide range of temperature and pressure conditions using a series of fixed point charge non-polarizable and Drude oscillator based polarizable molecular models. For the H_2O+NaCl mixture, the mean ionic activity coefficient, salt solubility, vapor pressure, interfacial tension, and viscosity were obtained as functions of temperature, pressure and salt concentration. Among the studied models, the BK3 [1,2] set of polarizable water and ions force fields give reasonable prediction for all properties of interest, and the explicit inclusion of polarization in molecular models was found to be crucial for accurate description of the binary H_2O+NaCl mixture [3]. For the H_2O+CO_2 mixture, the phase equilibrium compositions and diffusion coefficients were obtained from ambient conditions to elevated temperatures and pressures. Since prior simulation studies [4,5] showed that non-polarizable molecular models have significant inaccuracy for the calculation of H_2O compositions in CO_2 rich phase, we focused on the performance of several polarizable H_2O and CO_2 models. In particular, the BK3, GCPM [6] and SWM4-NDP [7] H_2O models were studied. TraPPE [8] and a recently developed Gaussian charge polarizable models [9] were used to represent CO_2. The cross interactions between H_2O and CO_2 models were optimized to equilibrium phase compositions of the binary H_2O+CO_2 mixture. The polarizable models yield more accurate calculation of phase compositions compared to non-polarizable ones, however, the representation of H_2O compositions in CO_2 rich phase

was still unsatisfactory, especially at high pressures. H_2O molecules were found to form small clusters in CO_2 rich environment, which may indicate the H_2O composition in CO_2 rich phase may be strongly affected by the H_2O-H_2O interactions instead of the CO_2-H_2O cross interactions. A new polarizable model was developed for H_2O, and its performance with respect to the calculation of H_2O-CO_2 phase equilibria and diffusion coefficients was investigated. The proposed new polarizable H_2O model is extended to study the H_2O-CO_2-NaCl mixture.

[1] Kiss, P.; Baranyai, A. *J. Chem. Phys*. **2013**, 138, 204507.

[2] Kiss, P.; Baranyai, A. *J. Chem. Phys*. **2014**, 141, 114501.

[3] Jiang, H.; Mester, Z.; Moultos, O. A.; Economou, I. G.; Panagiotopoulos, A. Z. *J. Chem. Theory. Comput.* **2015**, 11, 3802.

[4] Vlcek, L.; Chialvo, A. A.; Cole, D. R. *J. Phys. Chem. B* **2011**, 115, 8775.

[5] Orozco, G. A.; Economou, I. G.; Panagiotopoulos, A. Z. *J. Phys. Chem. B* **2014**, 118, 11504.

[6] Paricaud, P.; Predota, M.; Chialvo, A. A.; Cummings, P. T. J. Chem. Phys. **2005**, 122, 244511.

[7] Lamoureus, G.; Harder, E.; Vorobyov, I. V.; Roux, B.; MacKerell, A. D. *Chem. Phys. Lett.* **2006**, 418, 245.

[8] Potoff, J. J.; Siepmann, J. I. *AIChE J.* **2001**, 47, 1676

[9] Jiang, H.; Moultos, O. A.; Economou, I. G.; Panagiotopoulos, A. Z. *J. Phys. Chem. B* **2016**, 120, 984.

468336 Isobaric Heat Capacity of Supercritical Fluid Mixtures: Experimental Measurements and Molecular Simulations

Abstract:

Isobaric heat capacity of supercritical fluid mixtures: experimental measurements and molecular simulations

Mitchell Ishmael, Lauren Stutzman, Maciej Lukawski, and Jefferson Tester

Cornell Energy Institute and CBE and MSE

Supercritical fluids offer advantageous, tunable properties, making them attractive candidates for a diverse set of applications. Our study was motivated by the potential of using supercritical pure fluids and fluid mixtures as media for thermal energy storage and as working fluids in thermodynamic power cycles for low enthalpy heat sources. We hope to take advantage of the thermophysical property changes that occur near the critical point, in particular the large isobaric heat capacities in the near critical and supercritical region. For a small set of pure fluids and very few fluid mixtures, sufficiently accurate equations of state (EOS) exist for predicting heat capacities in the supercritical region. In order to obtain accurate isobaric heat capacity values for supercritical pure fluids and mixtures, our group has built an accurate (±1%) high temperature (20-150 °C), high pressure (1-300 bar) calorimeter. We have made measurements on a number of binary mixtures (X + carbon dioxide), including methanol, decane, and various refrigerants. As experimental measurements can be costly and time intensive, we also used molecular simulations to estimate heat capacities of supercritical fluid mixtures. A thorough comparison was conducted between predicted and experimental values over a wide range of temperatures, pressures, and compositions. For the majority of these comparisons, quantitative agreement was achieved. As a result, molecular simulation has been demonstrated as a powerful complement to direct experimental measurements and as a distinct alternative to using equations of state for predictions.

472863 High-Pressure, Global Phase Behavior for the Guaiacol–CO_2 System

Abstract:

The global phase behavior of guaiacol, a model lignin monomer, with CO_2, has been determined. Lignin, an abundant, naturally occurring source of aromaticity in biomass, can potentially be used as an inexpensive replacement for petroleum-derived polymers; however, phase-behavior measurements of lignin or lignin model compounds with supercritical fluids are practically nonexistent. Guaiacol was chosen for this study, as it contains both methoxyl and hydroxyl groups on the aromatic ring, as do most lignins. Furthermore, guaiacol is a liquid at ambient conditions, and our research program with lignin carries out the fractionation of lignin in a "liquid-lignin" phase. In this work, a variable-volume, high-pressure view cell was used to observe both liquid–liquid cloud points as well as bubble points along the liquid–liquid–vapor (LLV) three-phase line for this system. CO_2 + guaiacol was found to exhibit Type III phase behavior according to the classification system of Scott and Van Konynenburg, indicating that even at high pressures the mutual solubility of the two components is relatively low. However, the system does exhibit a pressure minimum in the liquid–liquid critical curve at 15 °C and pressures just under 500 bar.

457561 Volumetric Properties of Cyclohexane + CO_2 Mixtures at High Pressures

Abstract:

Processing polymers at high pressures in mixtures of organic solvents with CO_2 is of high interest for the production of particles, fibers, and foams.Â Of particular interest is the ability to tune the solvent power of these mixtures with pressure.Â In this study the thermophysical properties of mixtures of cyclohexane and CO_2 were explored at pressures up to 350 bar. Volumetric properties were determined for mixtures containing 0, 10, 20, 30, 40, and 50 % by mass CO_2 and at 40, 60, 80, 100, and 120 °C at pressures ranging from 50 to 350 bar.

Measurements were conducted in a special high pressure view-cell which utilizes dual movable pistons, the positions of which are monitored by computer-interfaced Linear Variable Differential Transformers (LVDT) which allow real-time recording of the cellâ€™s internal volume. Since the initial mass loading of the cell is known, the real-time volume data generated during a pressure scan allows the generation of continuous density profiles of the fluid mixture inside the view-cell. The density data are then modeled with lattice-fluid models such as the Sanchez-Lacombe equation of state and the derived thermodynamic properties such as the isothermal compressibility, isobaric expansivity, internal pressure, and the solubility parameters of the mixtures are evaluated. The density data that have been generated were found to be within 1% of the NIST values for pure cyclohexane. For mixtures of cyclohexane and carbon dioxide, the present measurements extend the data base beyond what is available in the open literature which is limited to 70 bar. We are exploring these mixtures as pressure-tunable processing solvents for polyolefin-based copolymers.

465873 Prediction of Solid-Liquid-Gas Equilibrium for Binary Mixtures of Carbon Dioxide + Organic Compounds

Abstract:

The supercritical carbon dioxide is considered as a solvent for extraction of high-value chemicals and recrystallization and micronization of drugs. The solid-supercritical fluid equilibrium (SFE) and solidâ€“liquidâ€“gas equilibrium (SLGE) for binary mixtures of carbon dioxide + organic compounds are important information to understand and design processes with supercritical carbon dioxide. Both phase behaviors are investigated with experiments and thermodynamic models in literatures. Since measurements of the SLGE and SFE, especially at high pressure conditions, are challenging and costly, it is useful to estimate such information from a reliable thermodynamic models. In this study, the combination of Peng-Robinson-Stryjek-Vera equation of state (PRSV EOS) with the COSMO-SAC model through Gex-based mixing rules is applied to predict the SLGE for binary mixtures of carbon dioxide + organic compounds. Two different versions of COSMO-SAC models, the COSMO-SAC(2002) model originally proposed by Lin and Sandler and the COSMO-SAC(2010) model proposed by Hsieh et al., and two Gex-based mixing rules, Modified Huronâ€“Vidal 1st-order (MHV1) and Wong-Sandler (WS) mixing rules, are studied. A total of eight combinations of methods and 16 binary mixtures are considered in this work to understand the effects of activity coefficient model and mixing rule on the prediction of SLGE.

466857 Simultaneous Description of Thermodynamic and Transport Properties Using the Extended Soft-SAFT Equation of State

Abstract:

The optimization of current industrial processes requires detailed knowledge of the thermophysical properties of fluids at highly non-ideal conditions. In fact, high-pressure process applications are common at an industry level. At these conditions, most of the theoretical approaches fail into the description of basic thermodynamic and transport properties, such as density, solubility, compressibility, heat capacity, viscosity and conductivity, among others. An additional problem comes from the use of independent equations/correlations to evaluate each thermophysical property. The use of integrated equations able to provide accurate values of a group of thermophysical properties for practical applications is of high interest to find a link between the different methodologies/correlations available for each particular property (heat capacity, speed of sound, viscosity, thermal conductivity) in order to have a common framework, avoiding data inconsistencies.

The goal of this contribution is to highlight the advantages of an extended version of the molecular-based soft-SAFT equation of state [1] for high-pressure processes. Particular attention is paid to the study of transport properties. The Free-Volume Theory (FVT) formulism [2] is coupled into soft-SAFT for the description of the viscosity of fluids. The approach includes 3 additional parameters related to the viscosity, normally fitted to viscosity data of the pure fluid at several isotherms or isobars [3]. The effect of these parameters is investigated in order to identify trends with the molecular weight and the number of carbons of the compound. The extension to mixtures is evaluated by testing several mixing rules [4]. In addition, the Bridgmann equation [5] is also included in the same framework to estimate the thermal conductivity of a series of compounds of interest. The accuracy of this equation is dependent on the ability of soft-SAFT to predict accurate values for the derivative properties. Hence, apart from the classical parameters optimization method, where only density and vapor pressure data are used, a new method for determining the soft-SAFT compound molecular parameters, including derivative properties data, is investigated and developed. With this integrated approach, all the data is consistent and there is no need to use additional independent correlations. Several examples including *n*-alkanes, hydrofluorocarbons, fatty acid esters (biodiesel), 1-alkanols and ionic liquids in a wide range of temperature and pressure are highlighted [4,6-8].

Acknowledgments

This work has been partially financed by the Catalan government (2014SGR-1582) and the Air Products Group.

References

[1]. F. J. Blas, L. F. Vega, Mol. Phys. 92 (1997) 135-150.

[2]. A. Allal, C. Boned, A. Baylaucq, Phys. Rev. E, 64 (2001) 011203.

[3]. F. Llovell, R.M. Marcos, L.F. Vega, J. Phys. Chem. B, 117 (2013) 8159–8171.

[4]. F. Llovell, R.M. Marcos, L.F. Vega, J. Phys. Chem. B, 117 (2013) 5195–5205.

[5]. J. D. DeSain, B.B. Brady, J. Thermophys. Heat Transfer, 23 (2009) 828-835.

[6]. F. Llovell, O. Vilaseca, N. Jung, L.F. Vega, Fluid Phase Equilib., 360 (2013) 367–378.

[7]. M.B. Oliveira, S.V.D. Freitas, F. Llovell, L. F. Vega, J.A.P. Coutinho, Chem. Eng. Res. Des., 92 (2014) 1898–1911.

[8]. F. Llovell, L.F. Vega, J. Chem. Eng. Data, 59 (2014) 3220–3231.

471565 Explanation of Asphaltene Phase Behavior

Abstract:

Asphaltenes represents one of the major flow assurance challenges in upstream oil production. Asphaltene precipitation and deposition can occur at different stages of production causing reservoir formation damage and plugging of pipeline and production equipment. Asphaltenes are also a challenge in refining operations both in plugging of heat exchangers and contributing to coke formation in catalytic crackers. Part of the challenge is that asphaltenes represent a polydisperse distribution of molecules that are defined primarily by their solubility in certain solvents since their molecular structures are difficult to determine.

Predicting asphaltene flow assurance issues requires the ability to model phase behavior of asphaltenes as a function of temperature, pressure, and composition. In this presentation, we review some recent approaches to model asphaltene phase behavior. We also present a method to characterize crude oil including asphaltenes using the Perturbed Chain form of the Statistical Associating Fluid Theory (PC-SAFT). The theory accurately predicts crude oil bubble point and density as well as asphaltene precipitation conditions. The theory is used to examine the effects of gas injection, oil based mud contamination, and asphaltene polydispersity on the phase behavior of asphaltenes. The analysis produces some interesting insights into field and laboratory observations of asphaltene phase behavior.

460737 Thermodynamic Modeling of Hydrogen-Bonded Mixtures

Abstract:

Separation process design requires high quality phase equilibria predictions. Interest in bio-derived chemicals and fuels has inspired new pathways for chemical manufacture using polar raw materials, and thus almost all streams involve hydrogen-bonding components. Today's engineers are faced with the challenge of quantitatively modeling hydrogen bonding (or association) effects when predicting the phase equilibria of polar solutions. Traditional gamma models such as NRTL[1] do not account explicitly for association but instead attribute deviations to strong dispersion forces, misrepresenting the physical characteristics of the hydrogen bond. As a result, predictions for dilute streams and multicomponent mixtures are often unsatisfactory.

Several thermodynamic models have been developed that incorporate association either through chemical theory or Wertheim's perturbation theory. While the two approaches differ in the underlying assumptions, the apparent fugacities modeled by both approaches depend on the fraction of molecules that remain nonbonded — the 'monomers'. The majority of modern association models are equations of state using the Wertheim method, such as SAFT[2], CPA[3] and ESD[4]. Equations of state association parameters are fitted to macroscopic pure component properties and subsequently used to predict mixture behavior. The model parameters depend on the quality/quantity of pure component data available. An equation of state requires accurate vapor pressure and excess Gibbs energy from the same model. A gamma model requires only the excess Gibbs energy from the mixture model. Activity coefficient (gamma) models have been the favored by industry for many applications (typically low or moderate pressures) because the pure-component vapor pressures are usually known to good accuracy and attention can be focused on the mixture nonideality. However, to date, the Wertheim method has not been widely applied in activity coefficient models.

This project is a collaboration between academic and industrial partners to merge the Wertheim association approach with an activity coefficient model. Microscopic tools including spectroscopy and quantum simulations are used to gain insight into association at the molecular level. The fundamental physical measurements and simulations provide additional insights into the hydrogen bonding.

References

[1] H. Renon, J.M. Prausnitz, On the thermodynamics of alcohol-hydrocarbon solutions, Chem. Eng. Sci. 22 (1967) 299–307. doi:10.1016/0009-2509(67)80116-7.

[2] W.G. Chapman, K.E. Gubbins, G. Jackson, M. Radosz, New reference equation of state for associating liquids, Ind. Eng. Chem. Res. 29 (1990) 1709–1721. doi:10.1021/ie00104a021.

[3] G.M. Kontogeorgis, E.C. Voutsas, I.V. Yakoumis, D.P. Tassios, An Equation of State for Associating Fluids, Ind. Eng. Chem. Res. 35 (1996) 4310–4318. doi:10.1021/ie9600203.

[4] J.R. Elliott, S.J. Suresh, M.D. Donohue, A simple equation of state for nonspherical and associating molecules, Ind. Eng. Chem. Res. 29 (1990) 1476–1485. doi:10.1021/ie00103a057.

458773 Modelling of Phase Behavior of Polyethylene Solutions

Abstract:

Modelling of Phase Behavior of Polyethylene Solutions

Christoph Walowski, Sabine Enders

Karlsruhe Institute of Technology, Institute of Technical Thermodynamics and Refrigeration Engineering, Engler-Bunte-Ring 21, D-76131 Karlsruhe, Germany.

Email: christoph.walowski@kit.edu; sabine.enders@kit.edu

Key words: Liquid-Liquid Equilibria, Impact of branching, Impact of asymmetric molecular weight distribution functions, Impact of Solvents

Abstract:

Polyolefins, i.e. polyethylene (PE), polypropylene (PP), poly-1-butene, ethylene/1-alkene copolymers (also known as linear-low density PE) and propene/alkene copolymers, are the world's most widely used synthetic polymers and their production still continues to grow exponentially. They are known to be cost-effective and good performing materials used in a broad range of commodity applications that influence our everyday lives. Folie and Radosz [1] explained in detail the significance of the involved phase behavior for the industrial production. These authors [1] emphasize the complexity of the phase behavior of this system, which is caused by a) the materials behaving highly non-ideal at high pressure, b) the polymer and the solvent greatly differing in size, and c) commercial polymers being composed of many molecules differing in molar mass and chemical composition. This contribution is closely related to an earlier work by us [2], in which we proposed an alternative method for phase equilibria calculations for the system polyethylene + ethylene.

The basic idea of the new method [2] is the combination of a cubic equations of state, namely the Sako-Wu-Prausnitz-equation of state, with the Lattice Cluster Theory [3], where the last is used for the development of the combining rules for the pure-component parameters. This method permits to consider the degree of branching by the use of the involved architecture coefficient's given by the experimental degree of branching. Applying continuous thermodynamics [4] allows to take into account the molecular weight distribution function directly in the thermodynamics framework.

However, until now only the Schulz-Flory distribution function were incooperated in the model. Usually, synthetic polyethylene can have a very broad and asymmetric distribution function with respect to the molecular weight. One important feature of this contribution is the investigation of the impact of an asymmetric distribution function on the phase behavior. High asymmetric distribution functions can be coursed a shoulder in the cloud point curve and three phase equilibria. This situation is related to the occurrence of heterogeneous double plait points that separate the branch of stable critical points and the branch of unstable critical points.

The second feature of this contribution consists in the application of the developed framework to other systems. We focus our attention to the system polyethylene + hexane, because several experimental data are available in the literature [5,6]. These experimental data could be modelled with the Sanchez-Lacombe EOS, where the polydispersity was approximated by the use of pseudo-components [6]. Unfortunately, the binary interaction parameter depends on temperature and molecular weight. In this contribution, we will compare both methods, where we pay special attention to the required model parameter.

[1] B. Folie, M. Radosz, Ind. Eng. Chem. Res. 34 (1995) 1501-1516.

[2] C. Walowski, K. Langenbach, D. Browarzik, S. Enders, Fluid Phase Equilibria, submitted.

[3] K.F. Freed, J. Dudowicz, Adv. Polym. Sci. 183 (2005) 63-126.

[4] M.T. Rätzsch, H. Kehlen, Prog. Polym. Sci. 14 (1989) 1-46.

[5] M. Haruki, S. Mano, Y. Koga, S.I. Kihara, S. Takishima, Fluid Phase Equilibria 295 (2010) 137–147.

[6] M. Haruki, K. Nakanishi, S. Mano, S.I. Kihara, S.Takishima, Fluid Phase Equilibria 305 (2011) 152–160.

458626 Multi-Criteria Optimization of Thermodynamic Models: Describing Water with PCP-SAFT As an Example

Abstract:

Equations of state are important in process engineering for describing thermo-physical properties of the pure components and mixtures. In spite of the effort that has been done in the past years to develop sophisticated physically-sound approaches such as the statistical associating fluid theory (SAFT), the accuracy of the equation of state will ultimately depend on the choice of the model parameters used for describing the pertinent compounds. It is challenging to find appropriate parameter sets which are able to represent various properties at various conditions with the same quality. The parameter estimation is an optimization problem with conflicting objectives representing the deviations of the different properties from experimental data. Typically, a single objective is formulated by weighting of the conflicting objectives resulting in a single set of parameters. In this work, the problem is considered using multi-objective optimization and the calculation of Pareto solutions; i.e., those parameter sets with best compromises between the objectives. Thus, not only one model but a whole set of optimal models is obtained, where optimality means that no improvement in any single objective is possible without deteriorating at least one other objective. By considering and comparing all members of the Pareto set, the decision maker is greatly supported in understanding single solutions in an appropriate context and finding models which are particularly suited for individual applications. To the authorâ€™s knowledge this is the first application of such technique for the parameterization of equations of state.

The importance of computing Pareto sets in this field as a powerful tool to support decisions in model development is highlighted. As an example the perturbed-chain polar PCP-SAFT equation of state is used and applied to model the vapor-liquid equilibria of water. The PCP-SAFT equation of state explicitly takes into account the non-sphericity and directional interactions (such as hydrogen-bonding) of molecules; as well as dipole-dipole interactions. The objectives are set to minimize deviations in both the vapor pressure and the liquid density. The equation of state is coupled with

an efficient algorithm[1] that yields a finite set of Pareto solutions, interpolation of which approximates the whole Pareto set within a given accuracy. PCP-SAFT water models with different levels of molecular detail, and hence different parameters, are studied and discussed against previously published models in the literature. Comparing the Pareto sets obtained for the different models proves to be valuable to assess which molecular features of the models (e.g. number of association sites, polarity, ...) are effectively improving the model quality. For the first time, an overview is given of what can be achieved with PCP-SAFT in its different variants regarding the modeling of different thermodynamic properties of water. Pareto-optimal models are proposed that represent the studied properties of water more accurately than the any existing model from the literature.

(1) Bortz, M.; Burger, J.; Asprion, N.; Blagov, S.; BÃ¶ttcher, R.; Nowak, U.; Scheithauer, A.; Welke, R.; KÃ¼fer, K.-H.; Hasse, H. Multi-Criteria Optimization in Chemical Process Design and Decision Support by Navigation on Pareto Sets. *Comput. Chem. Eng.* **2014**, *60*, 354â€“363.

469258 Thermodynamic Property Needs for the Oleochemical Industry

Abstract:

The oleochemical industry cover mainly the food and pharmaceutical reactions but production of fuels (biodiesel) and other speciality chemical production processes also handle oleochemicals (in other words, lipids). The core of process synthesis and design depend on availability of properties data and/or reliable thermodynamic models for the chemicals involved. Limited availability of consistent physical and thermodynamic properties of lipids compounds and their mixtures lead to difficulties with the use of process simulators for process synthesis and design, since all the models to be used require apriori estimated parameters from experimental data.

Lipids applications such as bio-diesel, edible applications, and production of other important chemical compounds from the fatty acids include a series of processes such as: fat splitting, esterification, epoxidation, hydrogenation, amidation, sulfonation, ethoxilation, etc [1]. Each of these processes involves several types of unit operations. The lipids hydrolysis process (high pressure, high temperature conditions – Colgate Emery) [2], which is an old process, still applied worldwide nowadays, involves several phase equilibria: first a liquid-liquid equilibrium is involved for the fat splitting column where the main components are: unreacted triglycerides, fatty acids, glycerol and water. Further, the process consist in separating the two liquid phases (fatty acids-water, and glycerol-water) by distillation. This involves the vapour-liquid equilibria of the mixture species (fatty acids-water, glycerol-water). Further, if separation of fatty acids mixture into high purity individual components is considered, several methods can be applied: fractional distillation/steam stripping within vacuum condition [3] or fractional crystallization with or without using a solvent [1]. For this process/separations several phase equilibria are involved: VLE equilibria of fatty acids and water, if we are talking about steam stripping, and several SLE equilibria (fatty acid-fatty acid, or fatty acid-solved), in the case of fractional crystallization. In order to be able to design these type of processes, involving different types of phase equilibria, one need to have knowledge about and to be able to predict the behaviour of these type of systems.

A database and a properties library have been developed for chemical product and process design. The database includes 45500 compounds, and according to the compound type (e.g.: normal fluids, polar non-associating compounds, multifunctional group compounds, polymers, electrolytes, etc) different properties are given according to group contribution approach for primary properties (e.g.: acentric factor, critical properties, triple point temperature and pressure, normal boiling point, liquid volume, ideal gas enthalpy, ideal gas Gibbs energy, ideal gas entropy, solubility parameters, van der Waals surface area and volume, dipole moment, etc.), temperature dependent properties (solid density, liquid density, vapour pressure, heat of vaporization, solid heat capacity, liquid and vapour viscosity, liquid and vapour thermal conductivity, etc.), and mixture properties (e.g.: binary and ternary data for VLE, LLE and SLE, infinite dilution activity coefficients, enthalpies of mixing, etc.).

For this, available property and thermodynamic models needs to be fine-tuned for this type of compounds. Regarding the use of models for mixtures properties, such as phase equilibria, available binary interactions parameters for predictive models are not sufficiently good to describe them. New parameters for UNIFAC original model have been calculated for lipids compounds. In order to be able to use all this data and models for process design and simulation, development of a lipids database is required. Our developed lipids database, SPEED LIPIDS Database, together with our chemicals database, includes all this options: one part of the database is dedicated to the pure compounds and it includes: 330 pure compounds and 27 models to describe the pure compound properties (e.g.: single values properties – critical properties, heat of formation; temperature dependent properties: vapour pressure, liquid heat capacity, liquid density, liquid viscosity, and surface tension, etc.). Safe extrapolation of pure compound properties is enabled. The other part of the database is dedicated to phase equilibria for the mixtures and it is composed from 4500 measure data points for 332 different phase equilibrium data-sets (92 VLE, 91 LLE, 70 SLE and 79 solubility data). For all the available VLE and SLE data sets, consistency tests are performed. It is seen that many mixture data published related to lipids do not pass these tests: only 3% of the data sets have quality factors over 0.5 (where the quality factor varies between 0 – minimum, and 1 – maximum) [4].

The database has the following features: property data consistency test; optimization based data regression to estimate the model parameters; knowledge representation and search engine for database use knowledge representation and search engine for database; computer aided modelling tool to quickly and efficiently develop the necessary property prediction models; and ability to integrated with process simulation tools. The application of the developed properties models will be illustrated through case studies involving different lipid compound processing steps. It allows a better and easier utilization of available thermodynamic models.

References

[1] Shahidi F., Ed., *Bailey ' S Industrial Oil and Fat*, 6th ed. New Jersey, USA: John Wiley & Sons, Inc., 2005.

[2] Barnebey H. L., Brown A. C., "Continuous Fat Splitting Plants Using the Colgate-Emery Process," Journal of the American Oil Chemists' Society, 1948, 25(3), 95–99.

[3] Berger R., McPherson W., "Fractional distillation," Journal of the American Oil Chemists' Society, 1979, 56(11), 743A–744A.

[4] Cunico L. P., Ceriani R., Sarup B., O'Connell J. P., Gani R., "Data, analysis and modeling of physical properties for process design of systems involving lipids," Fluid Phase Equilibria, 2014, 362, 318-327.

450378 Multi-Property Optimization of Vapor Pressure, Heat of Vaporization, and Liquid and Ideal Gas Heat Capacities

Abstract:

Vapor pressure, heat of vaporization, liquid heat capacity, and ideal gas heat capacity can be measured for pure organic compounds between the triple point and critical point. Additionally, these properties can be connected using temperature derivatives of thermodynamic relationships which provides an avenue to check consistency among data sets. These relationships and experimental data were compared for several compounds to fill gaps in available data and increase consistency amongst these properties. Liquid heat capacity data have been used to improve low temperature vapor pressure [1], but a methodology for assessing the thermodynamic consistency amongst data sets and optimizing the accepted property values has been further developed [2] so that this procedure can be applied to compounds for which there are fewer experimental data available. The process involves critically evaluating available experimental data and critical point predictions, manipulating the correlating vapor pressure equations, and improving low temperature ideal gas heat capacity correlations to improve accuracy of triple point pressure, low temperature vapor pressure, low temperature heat of vaporization, and high temperature liquid heat capacity in particular. The various data are weighted based on their accuracy and on the perceived relative importance of the properties in process design. This process achieves better thermodynamic consistency than fitting experimental data for each property independent of its relationship to the others.

[1] Poling, B.E., J.M. Prausnitz, and J.P. O'Connell, The Properties of Gases and Liquids. 5thed. 2001: McGraw-Hill.

[2] J.W. Hogge, R. Messerly, N. Giles, T. Knotts, R. Rowley, W.V. Wilding, â€œImproving Thermodynamic Consistency among Vapor Pressure, Heat of Vaporization, and Liquid and Ideal Gas Isobaric Heat Capacities Using Multi-Property Optimizationâ€ Fluid Phase Equilibria, 2015.

463461 Modeling Minimum Free Energy Path for Crystallization of Drug Molecules in Different Solvents

Abstract:

Crystallization from solution is fundamental to drug production in pharmaceutical industries where about 90% of the products contain one or more crystalline materials. As it is a combined result of nucleation and crystal growth, crystallization is affected by different precipitation conditions, such as solvents, presence of additives, etc.. Studying the mechanisms that how drug molecules aggregate and organize themselves to crystallize in different solvents is important to many aspects in drug design and production, including polymorph selection and nucleation rate. While such molecular motions are possible to be observed via computational simulations, it remains a challenge for standard molecular modeling with nucleation being a rare event. In addition, to characterize the transition from liquid to solid state requires an effective measurement of local molecule order which is not an easy task. In this work, we implemented the string method in collective variables in combination with the construction of order parameters to sketch a minimum free energy path for crystallization of drug molecules in different solvents. Two types of drug molecules, sulfadiazine and sulfamerazine (Form II), and three different solvents, water, methanol and acetonitrile, were selected.

Firstly, force field parameters for sulfadiazine and sulfamerazine were developed using the CHARMM general force field (CGenFF). The partial charges were modified to reproduce interactions with explicit TIP3P water molecules at quantum mechanical level of theory. The bond, angle and dihedral parameters were optimized by fitting them to yield a minimum energy geometry while capturing the potential energy surface calculated by quantum mechanics. The structural error of the unit cell parameters for sulfadiazine decreased from 3-8% before optimization to 0.3-1.5% after refitting the parameters. For sulfamerazine, the error decreased from 6.7-11.6% to 0.1-1.8%. Secondly, we constructed order parameters (OPs) from a generalized pair distribution function which gives the probability that a molecule with a particular internal configuration has a neighboring molecule with a given internal configuration, a given center of mass distance away from the first molecule and a given orientation with respect to the first molecule. The OPs of bond orientation and relative orientation have well separated distributions and both are able to quantitatively distinguish between liquid and solid states of the drug molecules. Finally, the minimum free energy paths were determined by using string method in collective variables while the constructed OPs parameterized the â€œreaction coordinateâ€ during the crystallization of sulfadiazine and sulfamerazine from their liquid states to solid states.

463137 John Quinn: Master Weaver of a Remarkable Professional Tapestry

Abstract:

Professor John A. Quinn first came to prominence in chemical engineering circles for his fundamental research on mass transfer and interfacial phenomena. His work featured experiments thoughtfully conceived to permit accurate determination and rigorous analysis of interphase mass transfer rates at freshly created interfaces – an achievement made possible by his design of apparatus that brought immiscible fluids together in well-defined geometries at unprecedentedly short phase contact times. These experiments enabled the intrinsic mass transfer resistance of the interface itself to be explored, thus providing John with the opportunity to test the prevailing assumption that solutes diffusing from one immiscible phase into another would always experience equilibrium across the infinitely thin gas/liquid and liquid/liquid interfaces separating them.

While in the process of elucidating the role of the interface in mass transport, John Quinn and his students would initially be plagued by -- and then later intrigued with -- the thin films of surface-active materials that invariably concentrated at phase boundaries. Exploration of these insoluble monolayers revealed that their influence frequently extended well beyond the immediate vicinity of the interface; under some circumstances these interfacial films promoted convective instabilities while at other times they exerted a stabilizing effect on the bulk phases on either side of them.

Given his preoccupation with these thin films or interfacial "membranes" that spontaneously accumulated at phase boundaries, it was inevitable that John Quinn's attention would turn to synthetic membranes, inasmuch as these afforded controllable properties (e.g., porosities and thicknesses) that made them compelling platforms for fundamental transport studies. Track-etched mica membranes provided particularly fertile ground for the investigation of various steric, hydrodynamic, and electrodynamic effects key to the transport of solutes and colloids through submicron pores. Out of this work would come a much-improved theoretical description of sterically hindered diffusion, as well as critical insights on filtration and adsorption in finely porous media. Synthetic membrane composites also informed studies in his group of gas permeation across the skin.

By virtue of his fundamental mass transfer work and his keen observations of transport across thin films, John Quinn was well-positioned to witness and participate in the birth and subsequent growth of a membrane separation industry initially focused on simple polymer films. But John's vision extended far beyond the capabilities of those conventional synthetic membranes to what might be possible if engineers were to mimic elements of the structure and function of vastly more powerful biological membranes. For instance, John noted that cell membranes achieved unparalleled selectivity by exploiting reversible reactions between permeants and membrane-localized carriers, and he was fascinated by the ability of membrane-bound biomolecules to deftly orchestrate key biochemical reactions. He responded by materially advancing the theory of carrier-mediated or facilitated transport and by inspiring the development of enzyme membrane bioreactors capable of chiral separations important to the pharmaceutical industry.

More significantly, John Quinn's enthusiasm for the intellectual interface between biology and engineering – and his influence on the direction that his department at Penn and his chosen field of chemical engineering would take -- amply justify his recognition as a pioneer in the then-emerging field of biomolecular engineering. John and his students would go on to explore several phenomena mediated by biomolecule-rich cell membranes, including, for instance, the attachment of cells to surfaces, the motility of cells on those surfaces, and the directed transport of motile bacteria.

This presentation traces the logic and serendipity that underlie the evolution of John Quinn's varied research interests, identifying the threads that tie together the body of his work into a coherent whole – a truly remarkable "professional tapestry". It documents the considerable impact that his scientific and technical contributions have had on our profession – and it provides an occasion to honor the thoughtful guidance and warm friendship that John so generously offered to so many, his students and colleagues alike.

457447 John Quinn: Advisor, Mentor, and Friend

Abstract:

Much of John Quinn's interest and curiosity centered on surfaces and interfacial phenomena. My thesis on interfacial resistance to mass transfer at liquid/liquid interfaces was based on a most unlikely experiment that somehow John knew had a chance of working. This work inspired much of my career. For example, early in my career I made membranes as thin as 10nm, not much more than two surfaces. Later I helped find a catalytic needle in a haystack that saved a business based on a reaction at a solid surface.

456402 Facilitated Transport Membranes for CO_2 Separation and Capture

Abstract:

This presentation is in honor of Dr. John Quinn for his pioneering work on facilitated transport membranes including those for CO_2 separation, and it covers CO_2 facilitated transport membranes for: (1) hydrogen purification for fuel cells and (2) CO_2 capture from flue gas in power plants and from <1% CO_2 concentration sources. We have synthesized CO_2-selective membranes by incorporating amino groups into polymer networks. The membranes have shown high CO_2 permeability and selectivity vs. hydrogen, carbon monoxide and nitrogen. We have elucidated the effect of amine steric hindrance in the solid membrane, showing significant enhancement for CO_2 transport. The membranes were scaled up to 14 inches in width by using a continuous roll-to-roll machine equipped with a thin-film-coating assembly. For the CO_2 separation from a synthesis gas stream containing CO_2 and H_2S, i.e., hydrogen purification for fuel cells or pre-combustion CO_2 capture in IGCC (integrated gasification combined cycle) technology, thicker amine-containing membranes with around 15 ± 1 microns of selective layer thicknesses were fabricated. For the post-combustion CO_2 capture from flue gas, thinner amine-containing membranes with < 200 ± 20 nm of selective layer thicknesses were scaled up. From gas transport measurements, the flat-sheet samples of the scale-up membranes for both applications exhibited similar performances compared to the membranes synthesized in lab scale. CO_2 permeance of approximately 200 GPU and CO_2/H_2 selectivity > 200 were for the thicker membranes for the purification of hydrogen and synthesis gas along with CO_2 capture (e.g., pre-combustion capture), and CO_2 permeance of about 800 GPU and CO_2/N_2 selectivity > 140 were for the thinner membranes for CO_2 capture from flue gas. The scale-up membrane has been fabricated into spiral-wound membrane modules. Techno-economic analysis has shown that the post-combustion capture process using the thin membrane has approached DOEâ€™s

capture cost target for 2025.

455619 The Force on a Body in Active Matter

Abstract:

The statistical mechanics and microhydrodynamics of active matter systems have seen a dramatic increase in interest in the past several years. Much attention has focused on the fascinating nonequilibrium behaviors of active matter not observed in equilibrium thermodynamic systems, such as spontaneous collective motion and swarming. Even minimal kinetic models of active Brownian particles exhibit self-assembly that resembles a gas-liquid phase separation. Self propulsion allows active systems to generate internal stresses that enable them to control and direct their own behavior and that of their surroundings. Recent work involving the unique 'swim pressure' exerted by active systems offers a perspective on the basic underlying physical mechanism responsible for self-assembly and pattern formation in active matter [1-3].

In this work we take a more microscopic view and present a general theory for determining the force (and torque) exerted on a body (or boundary) in active matter [4]. The theory extends the description of passive Brownian colloids to self-propelled active particles and applies for all ratios of the thermal energy k_BT to the swimmer's activity $k_sT_s = \zeta U^2{}_0 \tau R/6$, where ζ is the Stokes drag coefficient, U_0 is the swim speed and τR is the reorientation time of the active particles. The theory, which is valid on all length and time scales, has a natural microscopic length scale over which concentration and orientation distributions are confined near boundaries, but the microscopic length does not appear in the force. The swim pressure emerges naturally and dominates the behavior when the body size is large compared to the swimmer's run length $l = U_0 \tau R$. The theory is used to predict the motion of bodies of all sizes immersed in active matter.

References
[1] S.C. Takatori, W. Yan and J.F. Brady, Phys. Rev. Lett. 113, 028103 (2014)
[2] S.C. Takatori and J.F. Brady, Phys. Rev. E 91, 032117 (2015)
[3] S.C. Takatori and J.F. Brady, Current Opinion in Colloid & Interface Science, 21, 24 (2016)
[4] W. Yan and J.F. Brady, J. Fluid Mech. 785, R1 (2015)

459226 A Talk on Bijels, in Honor of John A. Quinn

Abstract:

John A. Quinn was a pioneer who advanced our fundamental understanding of processes that occur at fluid interfaces in the field of chemical engineering with practical implications in separations and reaction processes. John was also an outstanding colleague who formed and nurtured at the University of Pennsylvania our extraordinary department culture of intellectual exchange and interaction. In this talk, I briefly highlight a few examples of his work in the field. Thereafter, I discuss an emerging vehicle for study by chemical engineers in these contexts: Bijel membranes. Bijels, first predicted in simulation (1) and realized by a delicate experimental process, (2) comprise bicontinuous streams of immiscible fluids stabilized by jammed layers of nanoparticles at their interface. We have recently developed a robust and versatile means of fabricating these membranes that exploits a suspension of nanoparticles in a ternary aqueous/oil/co-solvent system. (3) This platform allows a broad range of particles and fluid pairs to be formed into fibers or membranes with tremendous promise in the fields of separations and interphase catalysis. Further, the potential importance of these vehicles in practical application will spur fundamental questions of trapped states and interfacial transport worthy of detailed analysis in the style of Quinn. Finally, this work, performed with colleague Daeyeon Lee at the University of Pennsylvania, embodies Johnâ€™s spirit of collaborative endeavor.

(1) **Colloidal Jamming at Interfaces: A Route to Fluid-Bicontinuous Gels, K. Stratford et al** *ScienceÂ* 2005,**309**, 2198-2201.

Â (2) **Bijels: a new class of soft materials, Michael E. Cates and Paul S. Clegg** *Soft Matter*, 2008, **4**, 2132-2138.

Â (3) **Continuous Fabrication of Hierarchical and Asymmetric Bijel Microparticles, Fibers, and Membranes by Solvent Transfer-Induced Phase Separation (STRIPS)**Â M.F. **Continuous Fabrication of Hierarchical and Asymmetric Bijel Microparticles, Fibers, and Membranes by Solvent Transfer-Induced Phase Separation**

(STRIPS) Haase, D. Lee, K.J. Stebe, Advanced Materials 2015, **27**, 7065-7071.

466714 Reverse Osmosis As a Tool in Water Treatment Strategies

Abstract:

Reverse osmosis is a workhorse tool commonly leveraged in water treatment flowsheets to remove highly soluble salts. This paper will briefly review state of the art reverse osmosis membrane chemistry and the manufacturing process, and then discuss how RO spiral wound modules are employed in conjunction with suitable pre-treatment technologies to create very high quality water for a wide variety of municipal and industrial applications.

463011 Recombinant Oleosin As a Functional Surfactant

Abstract:

Our laboratory is interested in novel surfactants that can be self-assembled into technologically useful suprastructures and can be equipped with novel biofunctionality. Oleosin is a naturally occurring surfactant protein found in plants. The molecule has three domains – two terminal hydrophilic domains, and a central hydrophobic domain with an embedded proline knot, which forces the molecule into a turn. We have obtained the gene for oleosin and expressed it in E.coli, and, using the tools of molecular biology, made hundreds of variants of the molecule, including numerous truncations and additions of functional or targeting domains. Depending on the truncation, oleosin can assemble into vesicles, worm-like micelles, sheets, or spherical micelles – assemblies which have utility in drug delivery and tissue engineering. In one application, we show how spherical micelles equipped with receptor targeting domains can be activated by a protease by protecting receptor-targeting domains with protease cleavable domains. The corresponding spherical micelles bind to target cells, but only when both the protease and the receptor are present, effectively creating an "and" switch for cellular recognition. We also show how oleosin's properties as a surfactant can be exploited to stabilize mondisperse air bubbles for ultrasound imaging. We will also provide a further update on ongoing efforts to incorporate other functional domains into oleosin for a variety of applications.

473326 Driving Change in Pharmaceutical Development & Manufacturing through Innovation and Standards

Abstract:

We are all familiar with the old adage â€œbetter-faster-cheaper, Pick Twoâ€. However in todayâ€™s world of product development and manufacturing, all three are needed and needed now. Organizations are tasked with moving diverse and complex pipelines forward in record time with undivided attention required for product quality, safety and reliability of manufacturing. These challenges do not require incremental change in the operating model but revolutionary change in our model. In this talk I will discuss activities, specifically process and product innovation geared to customer value and standards-based thinking for product lifecycle management and partnership models. I will discuss the implications for the processes, plants and people to ensure successful commercial products.

459070 Single-File Diffusion of Gas Mixtures in Dipeptide Nanochannels: High Field Diffusion NMR Study

Abstract:

Molecular diffusion in one dimension inside non-intersecting channels so narrow that they forbid mutual passage of molecules is referred to as single-file diffusion (SFD). An important consequence of SFD is a significant slowdown in the growth of MSD with time in comparison to that for normal Fickian diffusion. The potential use of this property in highly-selective separations and controlled catalysis has motivated studies of SFD in mixtures of different types of molecules using molecular-level simulations.[1-4] Our recent communication reported, for the first time, experimental observation of SFD of a gas mixture.[5]

Here, we report the observation of single-file diffusion of CO/CH_4 and CO/CO_2 mixtures as well as the corresponding pure gases in L-Ala-L-Val channels by C-13 pulsed field gradient (PFG) NMR. High magnetic field gradients up to 23 T/m were used to study diffusion in a broad range of diffusion times. Under the single file conditions, both studied mixtures were observed to diffuse faster than the slowest pure component forming the mixture. This is in stark contrast to the trend often reported in the case of normal diffusion in microporous materials where the addition of a faster-diffusing component to a slower-diffusing component does not change significantly the diffusivity of the slower-diffusing component. Molecular clustering is proposed to explain the observed relationship between the mixture and one-component mobilities and to reconcile the experimental SFD data with the predictions of a random walk model reported earlier.

References:

1. Sholl, D. S.; Fichthorn, K. A., Normal, single-file, and dual-mode diffusion of binary adsorbate mixtures in AlPO4-

5. J. Chem. Phys. 1997, 107, 4384-4389.

2. Adhangale, P.; Keffer, D., Exploiting single - file motion in one-dimensional nanoporous materials for hydrocarbon separation. Sep. Sci. Technol. 2003, 38, 977-998.

3. Rao, S. M.; Coppens, M.-O., Predicting Mixture Diffusion in Zeolites: Capturing Strong Correlations with a Simplified Theory and Examining the Role of Adsorption Thermodynamics. J. Phys. Chem. C 2012, 116 (51), 26816-26821.

4. Chen, Q.; Moore, J. D.; Liu, Y.-C.; Roussel, T. J.; Wang, Q.; Wu, T.; Gubbins, K. E., Transition from single-file to Fickian diffusion for binary mixtures in single-walled carbon nanotubes. J. Chem. Phys. 2010, 133 (9), 094501.

5. Dutta, A. R.; Sekar, P.; Dvoyashkin, M.; Bowers, C. R.; Ziegler, K. J.; Vasenkov, S., Relationship between single-file diffusion of mixed and pure gases in dipeptide nanochannels by high field diffusion NMR. Chem. Commun. 2015, 51 (69), 13346-13349.

467587 Exploring Diffusion in Hierarchical Zeolites with Molecular Dynamics Simulations

Abstract:

Introducing mesoporosity to conventional microporous sorbents or catalysts is often proposed as a solution to enhance their mass transport rates. Here we show that diffusion in these hierarchical materials is more complex and exhibits non-monotonic dependence on sorbate loading. Our atomistic simulations of n-hexane, ethanol, and water in a model system containing microporous nanosheets and mesopore channels indicate that diffusivity can be smaller than in a conventional zeolite with the same micropore structure, and this observation holds true even if we confine the analysis to molecules completely inside the microporous nanosheets. Our model system is free of structural defects, such as pore blocking or surface disorder, that are typically invoked to explain slower-than-expected diffusion phenomena in experimental measurements. Examination of free energy profiles and visualization of molecular diffusion pathways allows us to rationalize the unique transport behavior in hierarchical nanoporous materials.

460375 Compressibility of Confined Fluid and Its Dependence on Pressure

Abstract:

Ultrasonic experiments allow one to measure the compressibility (or elastic modulus) of bulk solid or fluid samples. Recent ultrasonic experiments carried out on fluid-saturated nanoporous glass shed light on compressibility of a confined fluid [1]. In particular those experiments showed that the elastic modulus of confined argon is a linear function of Laplace pressure in the pores. Our calculations based on classical density functional theory and Monte Carlo simulations confirmed this observation qualitatively [2,3]. Moreover, we found that the compressibility of a confined fluid is a function of the pore size.

Here we focus on the pressure dependence of the modulus. For bulk materials, this relationship is linear over a wide range of pressures and is known as the Tait-Munaghan equation. Using transition matrix Monte Carlo simulations [4] we calculated the isothermal modulus of bulk argon and confined in silica mesopores of various sizes. Our calculations show that although the elastic modulus itself is strongly affected by confinement, the slope of the Tait-Murnaghan equation is not, and is practically a universal thermodynamic property of a fluid. In addition to revealing a new fundamental regularity for confined fluids, our results provide additional information which can be applied for the analysis of experimental ultrasonic data, and allows one to estimate the elastic parameters of an unknown porous medium.

1. Schappert, K. & Pelster, R. Influence of the Laplace pressure on the elasticity of argon in nanopores, Europhys. Lett., 2014, 105, 56001
2. Gor, G. Y. Adsorption Stress Changes the Elasticity of Liquid Argon Confined in a Nanopore, Langmuir, 2014, 30, 13564-13569
3. Gor, G. Y.; Siderius, D. W.; Rasmussen, C. J.; Krekelberg, W. P.; Shen, V. K. & Bernstein, N. Relation Between Pore Size and the Compressibility of a Confined Fluid, J. Chem. Phys., 2015, 143, 194506

4. Siderius, D. W. & Shen, V. K. Use of the grand canonical transition-matrix Monte Carlo method to model gas adsorption in porous materials, J. Phys. Chem. C, 2013, 117, 5861-5872

463307 Supercooled Water in Small Nanotubes As a Shear-Thinning Fluid

Abstract:

Water has many unusual properties under bulk conditions. Thus, the question is whether such unusual properties also hold in confined media, and in particular in nanotubes with small diameters. In particular, an important problem is the behavior of water in confined media below the bulk freezing point of 273 K. The problem is relevant to a variety of biological phenomena. We have shown [1,2] that water in small nanotubes does not freeze, even at temperatures well below 273 K. Using extensive molecular dynamics (MD) simulations, we show in this study that under the same conditions water is not a Newtonian fluid, but behaves as a shear-thinning one. We compute temperature- and shear-rate dependence of the shear stress and apparent viscosity of water over the temperature range 230 K - 273 K.

464196 Rich Phase Behavior of Confined Water and Ice at the Nanoscale

Abstract:

Understanding phase behavior of highly confined water, ice, amorphous ice, and clathrate hydrates (or gas hydrates) not only enriches our view of phase transitions and structures of low-dimensional solids not seen in the bulk phases but also has important implications for diverse phenomena at the intersection between physical chemistry, chemical engineering and nanoscience. Classical molecular simulation (MD) is an indispensable tool to explore states and properties of highly confined water and ice. It also has the advantage in precisely monitoring the time and spatial domain in the sub-picosecond and sub-nanometer scales, which are difficult to control in laboratory experiments, and yet allows relatively long simulation at 10^2 nanosecond timescale that is impractical to first-principles molecular dynamics simulations. In this talk, I shall present an overview of MD simulation studies of the structures and phase behaviors of highly confined water, ice, amorphous ice, and clathrate in slit and cylinder nanopores. I will survey a variety of crystalline phases of low-dimensional ices and 2D ice clathrates revealed from MD simulations, as well as possible simulation evidence of 2D very-low-density, low-density, and high-density amorphous (2D VLDA, LDA HDA) ices.

462633 Deep Eutectic Solvents in Confinement: Structure, Dynamics and CO2 Solubility

Abstract:

Molecular dynamics (MD) simulations were performed to study systems of deep eutectic solvents (DESs) inside slit-like nanopores. DESs share many of the physical and chemical properties of room temperature ionic liquids (ILs) while being considerably cheaper and (mostly) nontoxic. Here we first studied the structure and dynamics of common DESs inside slit-like titania and graphitic nanopores. Our results show that the properties of the DES can be significantly affected by the pore width and the chemical composition of the pore walls. In particular, when the species forming the DES can form hydrogen bonds with atoms in the pore walls, we observe important inhomogeneities in local composition, significant changes in liquid structure and very slow dynamics in the confined DES when compared to its bulk counterpart. In the second part of our study, we considered systems of CO_2 and common DESs inside slit-like nanopores. This part of our work was motivated by previous studies exploring the use of bulk DESs in CO_2 capture, as well as by recent reports of the â€œoversolubilityâ€ phenomenon, where it has been observed that gas molecules exhibit increased solubility in solvents that are confined within a nanoporous material. Here we performed molecular simulations of systems of CO_2 and DESs in the bulk and inside slit graphitic nanopores at different conditions; detailed comparisons of the properties of these systems will be presented and discussed.

471615 Molecular Simulation Study on Interfacial Humid Ionic Liquids in Supercapacitors

Abstract:

Electrical double layer capacitors (EDLCs), also called supercapacitors, have attracted a great deal of attention in present electrical energy community, owing to their advantageous properties, such as high-power density and exceptional cycle life. Room-temperature ionic liquids are potentially ideal materials for electrochemical applications such as supercapacitors and batteries, and impressive progress has recently been demonstrated in well-controlled laboratory systems. However, complete removal of water from ionic liquids is nearly impossible, and such water content may bring significant effects on electrical double layers in IL-based supercapacitors since the electrolysis of water possibly arises due to the potential applied on IL supercapacitors larger than water decomposition voltage.

In this talk, using molecular dynamics (MD) simulation, we show the first work on the adsorption of water on carbon electrode surfaces in contact with humid ILs. We found that water is generally enriched within sub-nanometer distance from charged electrodes. The level of enrichment depends on the type of ions and typically increases as the surface charge density of electrodes increases. We clarified the mechanisms of these electrosorption behaviors. We highlighted the key role of ion-water association and interactions between water dipole and inhomogeneous electrical fields in double layers in determining water electrosorption, and rationalized the ion specificity and the asymmetrical dependence of electrosorption on the polarity of electrode charge. We also investigate influence of electrode materials (carbon vs. gold) on the interfacial humid ionic liquids. The comparison of MD between carbon and gold electrodes with the humid ILs and observed different water and ion accumulation, which may be attributed to the interaction between gold and ILs as well as water content.

470636 Competitive Adsorption of Linear Alkanes on Nanoporous Carbon: Density Functional Theory and Molecular Dynamics Simulation

Abstract:

Competitive adsorption of linear alkanes in microporous materials has been observed and studied in experiments and molecular simulation. While it is well known that larger molecules have higher affinity to attractive pore surfaces, the entropy penalty of confining flexible chain molecules will also critically influence the partitioning of molecules in a mixture. As understanding experiments and even molecular simulations of multicomponent systems can be challenging at the molecular scale, a theoretical model that accounts for both effects accurately is thus invaluable to predict the pore selectivity for different components. Our earlier work in graphite pores has qualitatively shown the transition of partitioning coefficients for longer alkanes as pressure increases by employing inhomogeneous statistical associating fluid theory (iSAFT). The entropic effect due to confinement is mainly accounted by the chain connectivity in iSAFT, however, the external potential is not physical enough given that most external potential models (e.g. Steele 10-4-3 potential) are defined for spherical molecules. Although Lorentz-Berthelot rule is often used in many density functional theory applications, the real interaction energy still needs to be well understood.

In this work, we take the molecular dynamics (MD) simulation of alkanes on graphite pore system as a reference to quantify the attraction between graphite surface and alkanes. Using inputs from MD simulation within iSAFT, we then study the competitive adsorption among alkanes in graphite pores. The significance for predicting the composition of shale gas in nanoscale pores is further discussed.

449922 Shifting Isotropic-Nematic Transition in Strongly Confined Semi-Flexible Polymer Solutions

Abstract:

We investigated the phase behavior of semi-flexible polymer solutions under sub-persistence length slit confinement of height *H*. Strong confinement leads to increased intra-polymer segmental correlation length ($l_{//}$) in the unconfined dimensions that could reach up to twice the polymer persistence length P. The increased $l_{//}$ results in better alignment of segments and shifts the isotropic-nematic transition density f_{IN}. We found that in unconfined systems ($H >> P$), the dependence of f_{IN} on *P* follows Onsager theory predictions. Enhanced segmental alignment is also observed near walls. For strongly confined polymer solutions, f_{IN} decreases as *H* decreases due to the increased segmental correlation length, and enhanced segmental alignment is found throughout the solution. Qualitative agreements with mean-field theory are found. The shift in f_{IN} also affects the confined polymer solution dynamics.

460959 Investigation of the Interactions of Drug Delivery Vehicles with Pulmonary Mucus Layer

Abstract:

Pulmonary route is considered to be advantageous in drug delivery owing to very large surface area and absorption characteristics of lungs. However, the airways are covered with the mucus layer, which serves as a barrier to exogenous matters and provides a natural defense to immune system with its self-cleaning mechanism. Therefore, it is difficult to achieve a successfully working drug delivery system through the lungs. Moreover, during the course of several lung diseases such as Chronic Obstructive Pulmonary Disease (COPD) or cystic fibrosis (CF), the thickness and the viscosity of the mucus layer increases and this change at the physicochemical parameters of the barrier turns out to be a real challenge towards the diffusion and the delivery of the therapeutics[1]. In this case, traditional aerosolized drugs can be ineffective and a more efficient way to provide a better treatment should be considered. It is known that the diffusion of drugs, particles or nanocarriers is considerably hindered by the mucus via several interactions between the particles and mucus layer. Most of them are removed from the airways via the mucus clearance mechanism before they reach to underlying epithelium. In the literature, different approaches have been developed in order to protect the drug carriers from both the macrophages and clearance mechanism and to enhance the penetration of drugs. These include the encapsulation of the drugs in synthetic or natural degradable polymeric particles/capsules, stealth design of the particles, usage of liposomal delivery systems, etc. Mostly, the surface of the particles within the size range of 100-500 nm modified with non-ionic surfactants, carboxyl or amine groups, which provided the opportunity to study with different surface charges[2]. However, the penetration behavior of the drug delivery systems through the mucus layer is still needs to be enhanced. In order to achieve better diffusion of drug carriers, primarily interactions between the drug carriers and mucus should be understood well.

In this study, solid lipid nanoparticles (SLN) with different size and surface properties are used as the model drug delivery vehicle and their interactions with mucin, which is the key glycoprotein of mucus, are investigated in detail.

SLNs are biocompatible structures and their lipid contents makes them advantageous in drug delivery. In addition, their stability is considerably high and they can be used as nanocarriers by encapsulating drugs. These and many other properties makes SLNs as preferred carriers to be used in a lot of areas especially in biomedical[3]. Interactions of SLNs with mucin are firstly determined by using Dynamic Light Scattering and Zeta Potential before and after incubating the SLNs to mucin solution. This method provided us to detect a possible evidence for interactions between particles and mucin by investigating the changes at the surface charge and size of the particles. Within the study, in order to determine the effect of zeta potential and size of the particles, three different SLNs are used. Investigations are studied further with Langmuir Through at the air/PBS interface by introducing the particles beneath the mucin layer that is previously formed and compressed to a chosen surface pressure. This lead us to study with mucin layer at different thicknesses and helped us to understand if the particles can adsorb, integrate and penetrate through mucin membrane. Moreover, the interfaces obtained at different surface pressures are transferred to substrates and analyzed by using Atomic Force Microscopy (AFM).

In the literature, it is reported that the penetration behavior of the particles enhance when their charge is close to zero due to their less interactions with mucus[4]. In this study, zeta potential values of the SLNs in 1 mM NaCl solution are obtained as -9.49 ±1.41 mV, -22.14 ±2.08 mV and -25.17 ±3.18 mV, respectively. It is obtained that, after incubation in mucin solution, the potential values of the particles have not changed significantly even when the mucin concentration was increased. This proves the fact that mucin does not stick on the particles and particles have no interaction with adhesive mucin. This result has been also supported by the DLS and AFM measurements: size of the particles has not increased significantly. In order to mimic the mucus membrane, mucin solution is spread on air/NaCl solution interface and to understand if the particles can adsorb or create a change at the membrane, interfacial behaviors of both the mucin membrane and SLNs have also been obtained by measuring the surface pressure with time. Cyclic compression-expansion behavior of the interfaces formed by mucin and SLNs are also obtained. From the surface pressure-time isotherms, for all surface pressures, it is determined that SLNs have adsorbed to the mucin layer, diffused through the membrane and reached to the interface even at high surface pressures and increased mucin thickness. The results of this study are promising in terms of obtaining a mucus penetrating drug delivery vehicle and understanding its interactions with mucin.

REFERENCES

[1] S.K. Lai, YY. Wang, J. Hanes, Advanced Drug Delivery Reviews, 61 (2009) 158â€“171.

[2] J.S. Suk, S.K. Lai, YY. Wang, L.M. Ensign, P.L. Zeitlin, M.P. Boyle, J. Hanes, Biomaterials, 30 (2009) 2591â€“2597.

[3] L. Battaglia, M. Gallarate, E. Peira, D. Chirio, E. Muntoni, E. Biasibetti, M.T. Capucchio, A. Valazza, P.P. Panciani, M. Lanotte, D. Schiffer, L. Annovazzi, V. Caldera, M. Mellai, C. Riganti, Journal of Pharmaceutical Sciences, 103 (2014) 2157â€“2165.

[4] Y. Cu, W.M. Saltzman , Molecular Pharmaceutics, 6 (2008) 173-181.

464944 Interactions and Mechanical Characterization of Spiny Particles

Abstract:

In comparison to smooth particles with simple shapes, the interaction between particles decorated by intricate surface features, such as spines, is not understood well. However, these types of structural features can be used to tailor particle adhesion by adjusting contact area and standoff distance. Plant pollen grains serve as a convenient model and template for such finely structured particles. They are also showing promise as reinforcing fillers for polymers, because sporopollenin, the major component of the pollen exine, is one of the most chemically and thermally stable natural polymers. Meanwhile, the modulus of these particleâ€™s spiny ornamentations has a significant influence on particle-particle and particle-surface interactions and the behavior as a filler. However, there are no literature reports of mechanical characterization of individual pollen grains. Here, the interaction between sunflower pollen particles (SFP) was characterized by colloidal probe atomic force microscopy (AFM). Moreover, the mechanical properties of three pollen species were also characterized by AFM, including ragweed, pecan and Kentucky blue grass.

For the interaction between SFPs, the results indicated that the presence of spiny protrusions on a particle surface could either enhance or reduce the interaction force by a factor of 6 times, compared to SFP interaction with a flat substrate. The spiny particle interactions were classified into five distinctive types of force-distance (f-d) curves. These results suggest that the interaction force is highly dependent on the orientation of spines on the surface, which can control whether frictional forces contribute to the observed effective adhesion. A model was developed to compare the effects of pure spine-spine adhesion with spine-spine friction at different orientations.

For the mechanical characterization, the elastic moduli of pollen shells were mapped quantitatively on 5Ã—5 Î¼m^2 areas. The values of modulus were determined to be between 7~9Â GPa for ragweed, 3~7 GPa for pecan, and 8~18 GPa for Kentucky blue grass pollen particles. Saturation at a relative humidity of 90% reduces the Kentucky blue grass pollen modulus to 3~6 GPa after absorbing water. These values are used to place in context recent reports of pollen

behavior as a filler. To the best of our knowledge, this is the first time that the interactions between complex spiny microstructured particles have been directly measured by AFM. Moreover, pollen grains have excellent mechanical strength as an environmentally-friendly filler for high-performance composites.

466694 Generating Charged Liquid Droplets By Contact Electrification

Abstract:

Contact electrification is the phenomenon in which static charge is generated when two surfaces come into contact and are then separated. This phenomenon has been widely observed and investigated for contacts that involve two solid surfaces. Here, we showed that charge can also be generated when a liquid comes into contact with a solid surface via contact electrification. In particular, we found that when water droplets flowed down a solid (e.g., polymeric) surface under gravity, the droplets became charged. When a continuous flow of charged liquid was collected in a centralized location (e.g., a vessel), we found that a constant electric current was generated. Therefore, electricity can be generated directly by harvesting the gravitational energy of water (e.g., from sources such as rain or streams of water) through contact electrification. After optimizing the conditions of the system, we found that the efficiency of energy conversion (i.e., the amount of electrical energy generated divided by the amount of gravitational energy lost by the droplets) can reach up to around 3-4%. In addition, we found that the droplets generated have high electric potentials. Hence, contact electrification may be a simple, clean, and inexpensive way to generate charged liquid droplets.

466115 Adhesion Forces Between Individual Oxide Nanoparticles in Gas-Phase Processes

Abstract:

Nanoparticles serve as the building block in an increasing amount of advanced materials and devices: they are often coated on a surface or dispersed in a matrix. Practical examples are fluidization, coating, or mixing of nanoparticles. In order to do this properly, it is crucial to have good control over the contact forces between the nanoparticles. However, in the scale down to the nano-range, contact forces are relatively strong and determine the interplay of individual nanoparticles. Hence, a fundamental understanding of the contact forces between single nanoparticles is necessary to achieve highly efficient processes and products.

Here we present a comprehensive force spectroscopy study of various metal oxide nanopowders under different gas-phase conditions to preserve an understanding of adhesion forces between individual nanoparticles. Therefore, highly porous agglomerates in the size of hundreds of microns were fixed onto a flat substrate with double sided tape. These spread agglomerates were penetrated with an AFM-tip. During retracting the tip a chain of nanoparticles unfolds between the tip and the agglomerate which finally detaches at a contact between two single nanoparticles [1]. By investigating hundreds of these measurements we collected force informationâ€™s for most relevant oxides such as TiO_2, Al_2O_3, and SiO_2. These quantitative analysis was verified with a special AFM-TEM set-up recording the unfolding process between the tip and the agglomerate *in situ*[2]. For a comprehensive analysis we measured the adhesion forces in a humidity range from 0-90 %, in vacuum, and in isopropanol enriched nitrogen. The results of the experimental studies suggest that the adhesion forces between individual nanoparticles (sizes of 10s of nanometers) cannot only be interpreted by state-of-the-art continuum theories such as van der Waals or capillary forces. Rather the molecular nature of the physisorbed water molecules strongly influences the particle-particle interaction [3]. An extensive surface characterization of the investigated nanoparticles by FTIR, TGA, and TEM reveals that a noticeable

amount of physisorbed water molecules is always present on the surface which only can be removed by using high temperatures. The molecules lead to the formation of a water meniscus between contacting nanoparticles in which the molecules take discrete positions: it cannot be treated as a continuous meniscus. These findings were confirmed by all-atom MD simulations of two interacting nanoparticles. Furthermore, these structuring is strongly influenced by the surface chemistry of the nanoparticles as shown experimentally by using surface modificated nanoparticles. Hydrophobic nanoparticles show decreased adhesion forces which we related to the weaker interaction of the organic surface groups with the always present water molecules.

Summarizing, we will show that adhesion forces between individual oxide nanoparticles can be described by a combination of continuum (capillary) and non-continuum (structuring) forces which has a large impact on the handling and processing of nanoparticles.

[1] S. Salameh, J. Schneider, J. Laube, A. Alessandrini, P. Facci, J.W. Seo, L. Colmbi Ciacchi, L. Mädler, Adhesion Mechanisms of the Contact Interface of TiO_2Nanoparticles in Films and Aggregates, Langmuir, 28 (2012) 11457-11464.

[2] S. Salameh, R. Scholz, J.W. Seo, L. Mädler, Contact behavior of size fractionated TiO_2nanoparticle agglomerates and aggregates, Powder Technol., 256 (2014) 345-351.

[3] J. Laube, S. Salameh, M. Kappl, L. Mädler, L. Colombi Ciacchi, Contact Forces between TiO_2 Nanoparticles Governed by an Interplay of Adsorbed Water Layers and Roughness, Langmuir, 31 (2015) 11288-11295.

469238 Interfacial Activity of Isotropic Particles in Fluid-Fluid Interfaces

Abstract:

Surfactants can adsorb in fluid-fluid interfaces and lower the interfacial tension. Like surfactants, particles with appropriate wettability can also adsorb in fluid-fluid interfaces. Despite many studies of particle adsorption at fluid interfaces, there has been a long ongoing debate whether isotropic particles can reduce the interfacial tension. In the present work, the interfacial activity of isotropic silica nanoparticles for air-water and hexadecane-water interfaces was analyzed through pendant drop tensiometry. Our measurements show that isotropic particles can change interfacial tensions provided that they have a strong affinity to an interface by virtue of their wettability and that no energy barriers to adsorption prevent them from reaching the interface. In addition, a simple thermodynamic model was used to estimate the effective interfacial tension accounting for particle adsorption; its predictions are consistent with experimental data presented. Finally, we suggest that dynamic interfacial tension measurements, when combined with information about the particlesâ€™ wetting properties, provide a convenient and accurate way to assess the packing density of particles in fluid-fluid interfaces. These results will be useful for understanding and controlling the assembly of isotropic nanoparticles at fluid-fluid interfaces relevant to applications ranging from surfactant-free formulations and food technology to oil recovery.Â

469528 Adhesion Characterization of Energetic Microparticles to Functionalized Surfaces

Abstract:

The improved elucidation of energetics adhesion behavior will improve munitions formulation and the performance of collection materials (*e.g.*, swabs or traps) used to detect trace explosive residue. Resolving the role that surface chemistry plays in the adhesion of energetic materials is a critical step towards the rational design of these materials. In particular, the use of specific chemical functional groups at the interface allows for the modulation of the adhesion force. This will directly impact the performance and reliability of munitions and collection materials. In order to evaluate the effects of specific functional groups on adhesion, atomic force microscopy pull-off force measurements were performed using energetic (and non-energetic) particles against self-assembled monolayers (SAMs). SAMs on gold were selected to evaluate surface chemistries due to their reproducibility and facile production. These SAMs were characterized using contact angle measurements, cyclic voltammetry, and ellipsometry. A suite of surface chemistries are analyzed in order to develop a database of the influence the functional groups have on adhesion of energetic (and non-energetic) materials. Furthermore, these results will facilitate the rational design of energetic particle collection materials and munitions through chemical-tailoring in order to enhance performance.

471217 Pattern Formation of Amphiphilic Particles and Polymers Under 2D Confinement

Abstract:

In nature, quasi-ordered nanostructures yield structural color in many species such as Male Eastern Bluebirds(1), the Hawaiian Bobtail Squid(2), and Blue Spikemoss(3). Assemblies with this kind of precise meso-structure are difficult to replicate in the laboratory, and our work attempts to engineer quasi-ordered nanostructures in a scalable manner using two components: (1) an amphiphilic polymer and (2) amphiphilic particles under 2D confinement.

We have examined the behavior of the first component, amphiphilic polymers, at the air-water interface and found our rationally-designed amphiphilic polypeptides self-assemble under confinement in a manner that could be useful for serving as an organic template capable of rearrangement (4). Additionally, we have examined the behavior of the second component, dichlorodimethyl silane surface-modified silica particles and Janus particles, revealing packing behavior at the air-water interface that includes buckling, folding and subduction.

This work investigates the combination of the amphiphilic polypeptide and Janus particle assembly and the possibility for it to lead to structural color. Our studies highlights the non-equilibrium behavior of anisotropic particles coupled to amphiphilic co-polymers such as pluronic, which have also demonstrated unique conformational changes under compression and are well documented in the literature. Combination of the particle systems with pluronic provides control over the spacing and patterning of the particle layers. Using a Langmuir trough, surface pressure isotherms of mixed particle-pluronic systems are obtained, revealing composite-like behavior at the interface. Additionally, Brewster angle microscopy is used to observe the phase behavior of the two-component systems under compression and expansion. Current experiments investigate these mixed systems under varying compression rates.

Works Cited:

1. Dufresne, E.R.; Noh, H.; Saranathan, V.; Mochrie, S. G. J.; Cao, H.; Prum R. O. Self-Assembly of amorphous biophotonic nanostructures by phase separation, *Soft Matter* **2009**, *5*, 1792-1795.
2. Crookes, W. J.; Ding, L. L.; Huang, Q. L.; Kimbell, J. R.; Horwitz, J.; McFall-Ngai, M. J. Refelctins: The unusual proteins of squid reflective tissues, *Science* **2004**, *303*, 235-238.
3. Vukusic, P.; Sambles, J. R. Photonic structures in biology, *Nature* **2004**, *429*, 680-680.
4. Leon, L.; LoGrippo, P.; Tu, R. Self-Assembly of Rationally Designed Peptides Under 2-D confinement, *Biophys. J.* **2010**, *99*.

469955 Interactions Between Peptide-Mimetic Nanoparticles and Synthetic Cells

Abstract:

Via the Dissipative Particle Dynamics simulation technique, we examine the interaction between peptide mimetic nanoparticles, or nanopins, and multicomponent vesicles. We study the role of nanopin architecture and cholesterol concentration on the capture of the nanopins by the bilayer, their insertion and post-insertion self-organization. We find the insertion to be triggered by unfavorable enthalpic interactions between the hydrophobic components of the nanopins and the hydrophilic solvent. The nanopins are observed to form aggregates in solution, insert into the bilayer and disassemble into the individual nanopins following the insertion process. We examine factors that influence the orientation of the nanopins in the host vesicle. We report the length of the hydrophilic segment of the nanopins to regulate their orientation in the clusters prior to the insertion process and in the bilayer, after the post-insertion disassembly of the aggregates. The orientation angle distribution for a given nanopin architecture is found to be driven by energy minimization. In addition, higher concentration of cholesterol is observed to constrain the orientation of the nanopins. We also report thermal fluctuations to induce transverse diffusion of one of the nanopin architectures. The incidence of transverse diffusion is observed to reduce with the concentration of cholesterol. Our results can provide guidelines for designing peptide-mimetic nanoparticles or macromolecules which can interface with living cells to serve as sensors for applications in medicine, sustainability and energy.

472570 Interfacial Thermal Active Modes and Dispersion Forces

Abstract:

Thermal active modes, i.e., molecular and submolecular mobilities, inherent to the relaxation behavior of solid condensed organic matter, such as polymers and molecular self-assemblies, and, interfacial and surface free energies, are of great importance for many practical applications. For instance, in organic second-order nonlinear optical materials (NLO), pursued for photonic device applications, the molecular mobility plays a pivotal role in producing stable NLO materials of high poling efficiency [1]. Other applications such as frictional sliding systems or nanocomposites, both involving nanoconstrained materials, depend on the surface/interfacial energies. Common to thermal modes and interfacial forces are that they are difficult to quantify, in particular, in systems that either possess an amorphous complex entropic structure, or are of small size.

From scanning force microscopy (SFM) an approach has been devised over the years, providing direct and local insight into thermal modes and interfacial forces via an energetic analysis that is based on the time-temperature superposition principle [2]. Dubbed â€œintrinsic friction analysisâ€ (IFA), it utilizes the mechanical scattering process between a sliding SFM tip in contact with the thermal active modes of the scanned sample. In the past, in particular, rotational and translational modes have been investigated involving complex organic systems, such as polymers and organic molecular glasses [2-5]. Recently, also molecular binding interactions [1] and surface dispersion interactions [6-7] could be energetically analyzed with IFA. In this paper, we will focus on both aspects, namely the molecular mobility in self-assembled systems depending on the molecular interaction strength, and the quantum electro-dynamic binding fluctuations between Van der Waals interacting surfaces.

[1] D. B. Knorr, S. J. Benight, B. Krajina, C. Zhang, L. R. Dalton, R.M. Overney, J. Phys. Chem. B, **116**, (2012) 13793.

[2] D.B. Knorr, R.M. Overney, J. Chem. Phys., **129**, (2008) 074504.

[3] S.E. Sills, T. Gray, R.M. Overney, J. Chem. Phys., **123**, (2005) 134902.

[4] T. Gray et al., NanoLetters, **8**, (2008) 754.

[5] D.B. Knorr, Jr., L. S. Kocherlakota, J. P. Killgore, and R. M. Overney, J. Membrane Sci., **346**, (2010) 302.

[6] B.A. Krajina, L.S. Kocherlakota, R.M. Overney, J. Chem. Phys., **141**, (2014) 164707

[7] L. S. Kocherlakota, Brad A. Krajina, R. M. Overney, Local Energetic Analysis of the Surface Energies of Graphene from the Single Layer to Graphite, J. Chem. Phys., 143, 241105 (2015)

468339 Electrostatic Origin of Enhanced Rate of Silica Surface Dissolution

Abstract:

Using an electrochemical cell in a Surface Forces Apparatus (EC-SFA) we found an enhanced dissolution rate of silica surfaces in close proximity to either muscovite mica or a gold surface as functions of time and electrochemical potential. The surfaces was pressed together at relatively low pressures (a few atm) while submerged in weak electrolyte solutions at 25°C and over a wide range of pH. Typical dissolution rates varies from 0.1 – 4 Å/ min, depending on the "effective" potential difference $\Delta\psi$. For the silica-mica interface the origin of the $\Delta\psi$ is interpreted as arising from the overlapping of the double layers of the surfaces. The $\Delta\psi$ of the gold surface was controlled by a potentiostat. We find that changes in the dissolution rates of silica correlate quantitatively with the $\Delta\psi$ via a Butler-Volmer type equation, suggesting the dissolution is driven, at least in part, by electrochemical reactions. Our finding may have implications for a wide variety of applications, from geological processes to chemical mechanical polishing.

474046 Shaping the Behavior of Colloidal Fluids

Abstract:

The behavior of colloidal suspensions depends upon many factors, including the shape of the colloidal particles. In this talk, we discuss the influence of shape on the thermodynamics and kinetics of hard particle fluids, including how shape is manifested in equations of state and in phase transitions from fluid to solid form. We present new results on the nucleation and growth of colloidal crystals, where subtle changes in particle shape can have profound influence on thermodynamics and kinetic processes.
*with J. Anderson, J. Antonaglia, A. Cadotte, E. Chen, P. Damasceno, J. Dshemuchadse, M. Engel, J. Glaser, E. Harper, E. Irrgang, A. Karas, D. Klotsa, D. Kofke, R. Newman, S. Nola, A. Schultz, and W. Shen.

474912 Modeling Biosynthetic Active Matter

Abstract:

New active materials are being synthesized that use cellular constituents such as biopolymers and motor-proteins, as well as synthetic components such as depletants or drugs that modify polymerization. These self-driven fluids will stretch, roil, or contract, depending upon the polymers and motor-proteins involved, and could be used to drive microscale transport processes. These systems also teach us about complex biological processes such as cell division. I'll discuss how we model these materials as multi-scale complex fluids driven by internal stresses generated by motor-protein activity, how they behave dynamically, and a bit on what they may teach us about biology.

474050 Gas-Solid Clustering: Not Your Ordinary Instability

Abstract:

In high velocity, gas-solid flows, the presence of the clustering instability – collections of highly-concentrated particles that continuously form, rearrange and breakup – are the norm rather than the exception. For decades, such instabilities have been predicted qualitatively by dozens of researchers using a kinetic-theory analogy for the continuum description. The ability to quantitatively predict clusters, however, has received relatively scant attention. Achieving such accuracy is non-trivial given the complexity of the problem – transient, three-dimensional flows with a flow instability that evolves over several length scales – as well as the seemingly inappropriateness of a "small-gradient" starting assumption. In this talk, the mechanisms for clustering will be discussed along with the competing influences that impact stability/instability, as well as recent progress on assessing the quantitative accuracy of a recent theory.

474656 Dynamic Solidification of Dense Suspensions Under Shear

Abstract:

Dense suspensions of small solid particles in a simple liquid exhibit a range of intriguing non-Newtonian flow behaviors [1]. One of the most striking and counter-intuitive of these is the transformation from overall liquid-like to solid-like behavior under rapid impact, shear or extension. This talk discusses some of the mechanisms driving such behavior, emphasizing the differences between steady-state responses to applied forcing, such as shear thickening, and dynamic responses, such as impact-activated solidification. We recently showed that impact-activated solidification involves rapidly moving fronts that emanate from the impact point and convert fluid-like, unjammed suspension into rigidly jammed material in their wake [2]. Measuring the speed of sound, we find that such fronts do not involve any significant increase in local packing fraction, while imaging of the flow field with high-speed ultrasound directly links the evolving front position to regions of high shear intensity. Taken together, this provides evidence for a solidification process we call dynamic shear jamming. We discuss the relationship to shear jamming in frictional granular systems, and we introduce a state diagram that delineates dynamic shear jamming from discontinuous shear thickening (DST) on the one hand and traditional jamming by densification on the other [3].

[1] E. Brown and H. M. Jaeger, Shear thickening in concentrated suspensions: phenomenology, mechanisms, and relations to jamming, Reports On Progress In Physics 77, 046602 (2014).

[2] S. R. Waitukaitis and H. M. Jaeger, Impact-activated solidification of dense suspensions via dynamic jamming fronts, Nature 487, 205-209 (2012).

[3] I. R. Peters, S. Majumdar, and H. M. Jaeger, Direct observation of dynamic shear jamming in dense suspensions, Nature 532, 214â€"217 (2016).

474049 Rheological Modeling of Granular Flows

Abstract:

Granular matter is very common --- sands, soils, raw materials, food stuffs, pills, powders --- but the challenge of predicting the motion of a collection of flowing grains has proven to be a difficult one, from both computational and theoretical perspectives. Grain-by-grain discrete element methods can be used, but these become computationally unrealistic for large systems and long times. A broadly accurate continuum model would be ideal as it could provide a much faster means of calculating flows and stresses in real-world problems, such as those encountered in industry and geotechnical applications. Motivated by this challenge, in this talk we present a new constitutive relation for granular matter, which is shown to produce quantitatively accurate predictions. The model emphasizes the key importance of cooperativity in the granular flow process through the inclusion of a grain-size-dependent length-scale within the rheology. Upon calibration of the model's three material parameters, solutions of the model are shown to be in quantitative agreement with over 160 experiments in several different families of geometries, including the notorious split-bottom cell. The same model is then shown to reconcile a number of other unusual features of granular media, such as the motion-induced "quicksand" effect, wherein flow at one location effectively removes the yield stress everywhere, and the observation that thinner granular layers behave as if they are stronger.

465215 Polymer-Surfactant Layer Structure Investigated with Soft Matter Imaging and Single Molecule Force Spectroscopy

Abstract:

A wide range of solution behavior for polymer-surfactant (PS) complexes can be achieved in response to changes in bulk surfactant and electrolyte concentration, which has driven their use in applications in personal care products, food formulations and pharmaceuticals. The association of the surfactant to the polymer provides a means to mediate a range of polymer properties including microstructure phase behavior and surface adsorption. PS complexes have along history of study from surface tension measurements and phase diagrams to more sophisticated scattering methods to probe solution and interfacial structure. Yet, there are still challenges in directly measuring the microscopic structure of adsorbed PS complex in a wide range of solution conditions as well as the molecular configurations of these complexes. In this work, a combination of colloidal probe force microscopy, force spectroscopy and a novel soft matter imaging method were used to determine the microstructure of oppositely charged PS complexes at a solid-liquid interface. Both colloidal probe measurements and a novel soft matter imaging approach were used to directly measure the PS complex layer thickness immobilized on surfaces with carefully controlled surface chemistry as a function of surfactant and electrolyte concentration. In addition, through both physical adsorption and covalent bonding of the PS complex to surfaces, force spectroscopy was used to probe the configuration of individual PS complexes. These data provide insight into the microscopic and molecular response of PS complexes to changes in surfactant and ionic strength when adsorbed on surfaces.

469648 Insights into Surface Hydration through Parallel Measurements of Water Diffusivity and Surface Forces

Abstract:

The fundamental molecular origins of solvent-structural forces, such as hydration and hydrophobic interactions, remain a mystery. Surface forces measurements, while accurate, are complicated by the need to distinguish between simultaneous electrostatic, steric, and solvent-structural forces, a task that is particularly challenging for silica and lipid bilayer surfaces. Here, I will discuss our efforts to use two complementary nanoscale measurements of surface water diffusivity, using Overhauser Dynamic Nuclear Polarization (ODNP), and surface forces measurements, using the surface forces apparatus (SFA), to elucidate the mechanisms of surface hydration at silica and lipid bilayer surfaces. Bilayer surfaces were studied in the presence of varying bulk concentrations of dimethyl sulfoxide (DMSO), a commonly used cellular cryoprotectant. The surface water diffusivity increases with DMSO concentration, as does the adhesion force between bilayers. Both trends appear to be caused by a shrinking of the hydrated volume of lipid head groups by DMSO, which allows for extraction of quantitative surface hydration length scales. Similar insight was gained for silica surfaces by adjusting surface, rather than solution, properties. Silica (SiO_2) surfaces are composed of silanol (Si-OH) groups, which impart hydrophilic character, and siloxane (Si-O-Si) groups, which give rise to modest hydrophobicity. Over a range of silica surface compositions, water diffusivity near the surface increases with decreasing silanol density (increasing hydrophobicity), and the surfaces exhibit a corresponding decrease in the range of the hydration repulsion. None of the silica surfaces adhere in water due to a short-range (1-4 nm) repulsion, likely arising from the hydration of uncharged (deprotonated) silanols. Overall, our work suggests two contrasting paradigms for surface hydration: hydrated excluded volume of individual surface groups, and a collective hydrogen bond network.

467590 The Structure of Polyelectrolyte Brushes in the Presence of Multivalent Counterions

Abstract:

Polyelectrolyte brushes are of great importance to a wide range of fields, ranging from colloidal stabilization to responsive and tunable materials to lubrication. We synthesized high-density polystyrene sulfonate (PSS) brushes using surface initiated atom-transfer radical polymerization, and performed neutron reflectivity (NR) and surface force measurements using a surface forces apparatus (SFA) to investigate the effect of monovalent Na^{+}, divalent Ca^{2+}, Mg^{2+}, and Ba^{2+} ions, to the structure of the PSS brushes. NR and SFA results demonstrate that in monovalent salt solution, the behavior PSS brush agrees with scaling theory well, exhibiting two distinct regimes, the *osmotic* and *salted* brush regimes. Divalent cations, Mg^{2+}, Ca^{2+}, and Ba^{2+}, while all significantly affecting the structure of PSS brushes, show strong ion specific effects that are related to the specific interactions between the divalent cations and the sulfonate groups. Our results demonstrate that the presence of multivalent counterions, even at relatively low concentrations, can strongly affect the structure of polyelectrolyte brushes. The results also highlight the importance of ion specificity to the structure of polyelectrolyte brushes in solution.

458654 Understanding the Conformation of a Nanometer-Thick Environmentally Friendly Hydrocarbon with Short Fluorinated Side-Chains on the Solid Substrate

Abstract:

Understanding the conformation of a nanometer-thick environmentally friendly hydrocarbon with short fluorinated side-chains on the solid substrate

Andrew Kozbial and Lei Li

Department of Chemical and Petroleum Engineering, University of Pittsburgh, 3700 Oâ€™Hara Street, Pittsburgh, PA 15261

Long-chain fluorocarbons are the state-of-the-art materials as nanometer-thick coatings in the growing nanotechnology industries. The low surface energy, resulting from the molecular nature of C-F bonds, is the key feature of the perfluoro-materials that no other materials can provide. However, the research in the past decades showed that long-chain fluorocarbons pose serious toxicological and environmental concerns because their degradation products, which has at least six fluorocarbons, are bioaccumulative, toxic and have high global warming potential. One possible solution for the dilemma is to develop hydrocarbons with short fluorocarbon side chains (HC-SFSC), which has been demonstrated to be environmentally much more friendly. Unfortunately, to date there has been few experimental study of the nanometer-thick HC-SFSC on the solid substrate. In the current study, nanometer-thick HC-SFSC has been deposited on the silica substrate by dip-coating. Contact angle testing results indicated that the as-coated HC-SFSC/silica has high surface energy and, interestingly, a simple thermal annealing reduces the surface energy to the value close to that of Teflon. X-ray photoelectron spectroscopy (XPS) results suggested that the thermodynamic equilibrium conformation of HC-SFSCs on a solid substrate is actually in favor of low surface energy and the low mobility of HC-SFSCs induced by the solid confinement traps the molecule in a non-equilibrium conformation with higher surface energy. This finding here potentially provides a viable approach to make the surface energy of HC-SFSCs as low as the conventional perfluoro-materials

463251 Combustion Ash As a Potential Electrode Material: Modifying Surface Structures Via Electrical Treatment

Abstract:

Ash, a byproduct of waste to energy plants, typically has a relatively high heavy metal content and as such safe means for its disposal or reutilization are under investigation. It is intrinsically hydrophobic and highly oxidized, which leads to high melting points and, in general, very low conductivities. The research presented here seeks to explore the potential of combustion ash to be used as an electrode material for a microbial fuel cell (MFC). This application requires an increase in conductivity, hydrophilicity, and, ideally, low toxicity.

Three bottom ash samples were investigated, namely Essex-2, Essex-6, and HP3-BA. By applying an electrical potential across the ash in the presence of deionized water, several key property changes can be achieved. Using a variety of analytical techniques including SEM, XRD, TGA/DSC, and tensiometry, we find that not only is the melting point lowered for all three ash samples, a color change indicative of chemical reactions occurs within the ash upon application of the potential. SEM analysis provides evidence of changes in crystal structure and water-ash contact angles are altered from hydrophobic to completely wetting. Unlike in electrowetting, these changes are found to last even after the electric potential is removed. The color change of the ash could be indicative of a change in oxidation state, which may assist in metal recovery processes and in other disposal methods. The ability to alter these key characteristics of the ash opens up an array of possibilities for its safe reuse, including applicability as an electrode in a MFC.

469956 Characterizing the Electrochemically Enhanced Dissolution of Silica and Alumina in Alkaline Environments

Abstract:

Alumina (Al_2O_3) and silica (SiO_2) materials are prevalent in geological environments and complex materials used for device fabrication, structural materials, and catalysts. In each of these systems, the solid-liquid-solid interface plays a crucial role in determining material properties and interactions in aqueous environments, especially with regards to dissolution. In particular, both pressure solution and chemical-mechanical polishing demonstrate that the dissolution of alumina and silica materials can be drastically enhanced in aqueous environments however the cause of enhancement is highly debated. Dissolution enhancement in both of these systems share the same setup: asymmetric, or distinct, materials in close proximity in a saline, aqueous environment. In this work, we study the enhanced dissolution of alumina and silica in alkaline environments due to the presence of asymmetric surfaces. Using the surface forces apparatus (SFA) to characterize the enhanced dissolution of alumina and silica in proximity to muscovite mica surfaces, we found that the dissolution is enhanced to varying degrees depending on the relative surface potentials of the asymmetric surfaces. The impact of key parameters of the electrostatic double layer, such as the decay length and surface potential, were explored to provide insight into the electrochemical enhancement. Our findings highlight the importance of the asymmetric solid-liquid-solid interface and have implications in a variety of technological applications, such as structural materials development and chemical mechanical polishing.

469297 Load Induced Hydrodynamic Lubrication of Porous Polymers: An Explanatory Study of Underlying Mechanisms

Abstract:

We present an exploratory study of the tribological properties and mechanisms of porous polymer surfaces under applied loads in aqueous media. We demonstrate that it is possible to change the lubrication regime from boundary lubrication to hydrodynamic lubrication even at relatively low shearing velocities by the addition of vertical pores to a compliant polymer. It is hypothesized that the compressed, pressurized liquid in the pores produces a repulsive hydrodynamic force as it extrudes from the pores. The presence of the fluid between two shearing surfaces results in low coe cients of friction ($\mu \approx 0.31$). The coe cient of friction is reduced further by using a boundary lubricant. Probe material was varied to show that the effect is independent of the shearing surface. The curvature of the probe, the compliance of the polymer and the depth of the pores, were changed to enhance the understanding of underlying mechanisms of the proposed lubrication mechanism. Finally a surface with non- uniform distribution of pores was tested to show that the effect can be observed over a range of surface designs and patterns. By using simulations to optimize the design, these surfaces can be used as a coating for total joint replacement implants and other application where ultra-low and tunable friction is desired.

We will also be studying the effect of surface patterning on the effective modulus of PDMS. We can modify the effective modulus of polymer (PDMS) sample by varying the degree of porosity. By increasing the degree of porosity for a hard polymer and by reducing the degree of porosity for soft polymer, we can reach the same value of effective modulus for both the samples. We aim to establish a relation between the effective modulus of PDMS and the Coefficient of Friction values for different samples.

465355 Role of Surface Microstructure and Elasticity in Promoting Lubricated Contact and Adhesion

Abstract:

Despite its importance in many applications and processes, a complete and unified view on how deformable nano- and micro-scale asperities influence hydrodynamic interactions has yet to be reached. In particular, the competing effects of surface structure and elastic compliance can be expected to become more dominant with soft materials when the length scale of the asperities or textures become comparable to that of the fluid flow. By combining experiments with the surface forces apparatus (SFA) with experiments performed with a custom-built peeling instrument[1], we characterized and modeled the hydrodynamic forces between a smooth plane and lyophilic surfaces decorated with hexagonal arrays of isolated (holes) and interconnected (pillars) microstructures in a viscous silicone oil. We have also investigated how elastic compliance competes with drainage through asperities in the viscous forces during both approach and retraction, especially in the peeling configuration[1]. For all the microstructures studied, the periodicity of the surface features was much larger than the separation range of our measurements, hence formed a thin channel geometry.[2]

In this talk, the approach and separation forces measured for both normal and peeling configurations with different microstructures and elasticity will be compared to predictions for existing theoretical boundary condition models for smooth surfaces (Reynolds theory), interfacial slip and a shifted no-slip plane. We also highlight the contribution of surface deformation[3] and compare our peeling results to a model based on hydrodynamics and bending beam. Finally we determine the surface, mechanical, and fluid properties that are necessary for surface structures to enhance adhesion under flooded conditions. In particular we highlight the importance of van der Waals forces and viscoelasticity. We show that in absence of these two effects the presence of surface structure does not lead to enhancement in adhesion (and in many cases reduces the adhesion), even with highly compliant surfaces.

1. Dhong, C. and J. Fréchette, *Coupled effects of applied load and surface structure on the viscous forces during peeling*. Soft matter, **2015**. 11(10): p. 1901-1910.

2. Pilkington, G.A., R. Gupta, and J. Fréchette, *Scaling Hydrodynamic Boundary Conditions of Microstructured Surfaces in the Thin Channel Limit.* Langmuir, **2016**. 32(10): p. 2360-2368.

3. Wang, Y., C. Dhong, and J. Frechette, *Out-of-Contact Elastohydrodynamic Deformation due to Lubrication Forces.* Physical review letters, **2015**. 115(24): p. 248302.

468960 Bioinspired Multiphase Humidity- and Rate-Dependent Capillary Adhesives

Abstract:

It is well-known that viscous dissipation plays an important role in many natural and synthetic adhesives. Many liquid biological adhesive systems utilize capillary forces, which can be either static or hydrodynamic, including fluids on spider silk and on insect legs. In such cases, it is important to maintain viscosity within a defined range to preserve the adhesive characteristics, such as the liquid spreading rate and the magnitude and rate-dependence of the adhesive forces. Thus, natural liquid adhesives must employ methods to counteract changes in physical properties due to water loss or uptake associated with variations in humidity. Here, we report on the ability of an important natural capillary adhesive, pollenkitt (the liquid coating found on pollen particles), to stabilize viscosity relative to humidity changes. Pollenkitt consists of an aqueous phase and an oily phase, but the contributions of these phases to adhesion was previously unknown. In this study, we measured the wet adhesive force required to separate a pollen particle from a pollenkitt drop as a function of separation rate and relative humidity (%RH). To investigate the function of the oily phase coat on wet adhesion, pollenkitt drops without the oily phase coat are prepared as the control group. A remarkable rate-dependent adhesive forces is observed, and shown to be attributed to strong capillary hydrodynamics above a critical separation rate. At lower rates, the adhesion is due to static capillary forces. We find that the aqueous phase is the main contributor to this rate-dependent adhesion. The oil phase forms a thin coating around the aqueous phase droplet. Interestingly, the oily phase contributes very little to the observed capillary adhesion, but rather it serves to prevent excessive water absorption under elevated humidity (or drying under reduced humidity). This action stabilizes the viscosity of the aqueous phase during humidity changes, thereby stabilizing the capillary static and hydrodynamic adhesion contributed by the aqueous phase. This result provides inspiration for future development of novel humidity-stabilized and rate-dependent adhesive materials.

469727 Influence of Humidity on Gecko-Inspired Adhesives

Abstract:

Geckos have developed foot pads that allow them to maintain their supreme climbing ability despite vast differences in environment, from dry desert to humid rainforest. Successful gecko-inspired mimics should exhibit necessary adhesive and frictional performance across a similarly diverse range of climates. In this work we focus on the effect of relative humidity (RH) on the frictional adhesion behavior of gecko-inspired adhesive pads. A surface forces apparatus (SFA) was used to quantitatively measure adhesive and frictional forces of an anisotropic (tilted half-cylinder) microfibrillar polydimethylsiloxane (PDMS) surface against a smooth hemispherical glass disk at varying levels of relative humidity from 0-100 %, including fully submerged under water. The results are also compared with a flat PDMS control. Changes in the relative contributions of van der Waals and capillary forces with shearing direction and relative humidity have significant implications for frictional adhesion and â€˜reversibilityâ€™. These results can be extended to formulate design principles for reversible adhesive platforms in humid environments.

453538 Development of Resorcinol Formaldehyde Aerogels with Enhanced Mechanical Properties Via Improved Particle Necking

Abstract:

A new highly porous resorcinol-formaldehyde (RF) aerogel with improved particle necking is presented in this work. This RF aerogel is developed under CO_2 supercritical drying conditions without any structural shrinkage. The water content and the catalyst percentage are varied to modify the solution's initial pH value and enhance particles connections. Firstly, the water content is fixed while the catalyst percentage increases in the solution to modify particles number and size. The number of particles increases with the catalyst content, and this generates a fine pearl necklace configuration with point-to-point connections that have brittle characteristics and a very low total thermal conductivity of 24.8 mW/m.K. The solid conductivity in this configuration is reduced at a high catalyst concentration due to the increased interface resistance and the reduced solid path. As the water content increases in the solution, the solution shifts towards the acidic region, and the chance to form a network decreases due to the increased void fraction. But as the catalyst percentage increases, the solution' pH value shifts towards the basic, and the particle number increases, resulting in strong particle-to-particle connections. This transforms the particle-to-particle point connections to neck connections (termed to be a semi-fibril-like structure in this paper) with a similar thermal conductivity (24.6 mW/m.K) as the pearl necklace configuration. The semi-fibril-like structural elasticity is improved substantially due to the neck formation between the particles. The study concluded that at high catalyst concentration the structure thermal conductivity is fixed while its elasticity is improved with the semi-fibril-like structural.

463063 Adsorptive Precipitation from Supercritical Solutions in Aerogels: A Way Towards Stable Amorphous Drugs

Abstract:

Properties of solid materials are fundamentally determined by their crystallinity. In pharmaceutical technology the transition from amorphous to crystalline state plays an essential role, since the bioavailability of a drug is influenced by its state. Amorphous drugs often have higher bioavailability compare to the corresponding crystalline form. Tendency towards recrystallization is a major factor precluding more wide use of the amorphous drugs in solid dosage systems.

To overcome this problem, we study adsorptive precipitation of drugs and drug-like organic solids in aerogels. Adsorptive precipitation consists in immersion of a porous carrier in a drug solution in supercritical carbon dioxide ($scCO_2$) followed by rapid expansion of the solution. The latter causes precipitation of the drug in the carrier.

To reveal the influence of the drug nature, adsorptive precipitation experiments with various drugs were performed. Biopolymer and silica aerogels were used as porous carriers. Crystallinity and loading were assessed by XRD and UV spectrophotometry, respectively. The results along with literature data are discussed in order to reveal a key parameter that governs the stabilization of the amorphous state. The conclusion is that the drug–carrier interactions are crucial for the stabilization of amorphous state.

To quantify the interaction in $scCO_2$/drug/aerogel system, supercritical fluid chromatography (SFC) experiments were employed. Retention time was interpreted in the frame of van't Hoff analysis. Due to intense interactions, elution by pure $scCO_2$ was often not possible. In this case, a modifier (alcohol) was used. Foe such quaternary systems ($scCO_2$/modifier/drug/aerogel) a novel thermodynamic model (mixed retention model) was developed and validated. SFC experiments were carried out in the CO_2 density range of 0.40–0.85 g/cm^3. It was shown that CO_2 density plays an important role in drug–carrier interactions and changes the adsorption enthalpy by a factor of 2. Thus, by changing the density of the drug solution in sc-CO_2, it is possible to change the adsorption enthalpy of the drug on the carrier. Furthermore, it was found that the crystallinity of a model compound, naphthalene, measured by XRD is in an inverse relation to the adsorption enthalpy: the higher the enthalpy the lower the crystallinity.

To reveal the influence of the drug nature, SFC experiments were performed with various drugs. Biopolymer aerogels were used as carriers. The mixed retention model was applied to acquire enthalpy of adsorption from elution curves. Correlation between enthalpy of adsorption and crystallinity of the drugs is found and discussed along with literature data that considered to support our experimental findings.

463315 Protein Based Aerogels: Preparation, Applications and Potential for Food Engineering

Abstract:

In the last decades the interest toward porous materials based on natural polymers is continuously growing. A promising way to achieve mesoporous materials, having porosity up to 99% and correspondingly a huge internal surface area is the supercritical drying of gels resulting in aerogels. Natural polymer-based aerogels consisting of polysaccharides or proteins are promising for life science and food applications because of their biocompatibility and biodegradability. However, also the area of bio-based insulation could be targeted as well, since very low thermal conductivity is given.

In this work we focus on aerogels based on natural proteins, such as milk and egg white proteins, which are well-known as staple foods. Producing aerogels from natural food sources can open a new market for the aerogel in food, for instance as food additives. In this work mainly an application of protein-based aerogels as carrier material for sensitive and sensorial unpleasant compounds in food products is aimed. The carrier material is produced in different shapes, depending on the targeted application. Primarily, spherical particles are preferred due to simple handling and a later possible coating. The main requirements for such a carrier material are high specific surface areas coupled with high mechanical stability as well as a homogeneous particle size. Protein aerogels exhibit BET surface area up to 390 m^2/g depending on pH during the gelation. Aerogels formed at very low and very high pH values are found to have highest surface areas. Since the proteins have positive and negative charges the orientation and arrangement of the proteins vary with pH. Ordered proteins generate more ordered gel structures yielding aerogels with higher BET surface area compare to those gelated at the isoelectric point. The mechanical stability of the protein aerogels depends on the reactivity of S-H-groups. At higher pH more $-S^-$ anions are formed and hence more S–S-bonds lead to stronger gels and aerogels.

The gel particle size distribution is depending on the droplet size distribution in the emulsion before gelation. It was optimized to obtain aerogel particle sizes between 50 to 100 µm.

Further, aerogels with the desired properties were loaded by two compounds, fish oil and ascorbic acid (vitamin C), *via* adsorption from supercritical CO_2. Remarkably high fish oil loadings were achieved without losing the free-flowing properties of the aerogel particles.

Further, storage and release behavior in simulated gastrointestinal fluids were investigated showing the potential of protein-based aerogels as carrier materials for food applications.

466138 Alginic Acid Foams with Hierarchical Porosity: Promising Materials for Dyes Adsorption

Abstract:

Alginic acid foams with hierarchical porosity: Promising materials for dyes adsorption

Nathalie Tanchoux[1*], Asja Pettignano[1,2], Luca Bernardi[2], Thierry Vincent[3], Eric Guibal[3], Françoise Quignard[1]

[1] Institut Charles Gerhardt, CNRS-UM2-ENSC-UM1, 8 Rue Ecole Normale, 34090, Montpellier, France. nathalie.tanchoux@enscm.fr

[2] Department of Industrial Chemistry, UNIBO, V. Risorgimento, 4, 40136, Bologna, Italy.

[3] Ecole des Mines d'Alès, C2MA/MPA, 6 Avenue de Clavières, 30319, Alès, France.

Keywords: foams, hierarchical porosity, biopolymers, alginic acid, biocompatible adsorbent

Water pollution due to organic dyes is a serious environmental problem. Their release, even in low quantities, is highly undesirable because of their high visibility, low biodegradability and toxic effects on humans and marine organisms. Among the treatments proposed for water decontamination, adsorption is a popular method, because of its high efficiency and low operating cost. Therefore, the development of low-cost adsorbents is becoming extremely attractive and natural biopolymers are often presented as abundant, renewable and biodegradable resources for this application.[1]

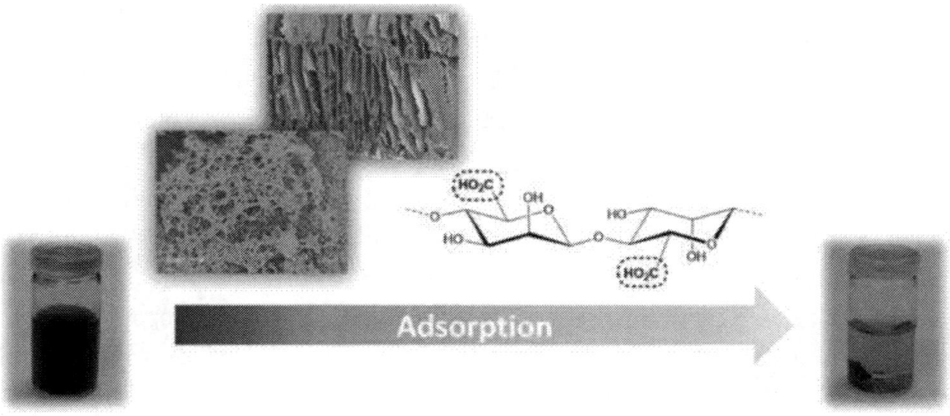

In our study, alginic acid, a biopolymer derived from brown algae, has been studied as a cost-effective replacement for conventional adsorbents. The presence of hydroxyl and carboxylic groups in its polymeric backbone confers: (a) the ability to associate, by physical and chemical interactions, with a wide variety of molecules [2] and (b) an excellent selectivity towards cationic dyes. Employing the tremendous manipulability of this polysaccharide, alginate foams with macro-meso pores have been developed with a simple procedure. The influence of preparation parameters (alginate composition, temperature, concentration, acidification procedure) on the final texture have been studied by SEM microscopy and N_2 adsorption, which confirmed the presence of a hierarchical porosity. The applicability of the newly developed material have been tested by examining its sorption behavior towards cationic dyes (such as methylene blue), by equilibrium and kinetic experiments, proving that alginic acid foams can be employed as biocompatible, green and low-cost adsorbents for wastewater remediation.

[1]. G. Crini, *Prog. Polym. Sci.* **2005**, 30, 38-70.

[2]. F. Quignard, F. Di Renzo, E. Guibal, *Top. Curr. Chem.* **2010**, 294, 165-197.

Thermoplastic Foaming Assisted by Microwave

Elham Rezvanpanah[1, 2], S.Reza Ghaffarian Anbaran[1], Ernesto Di Maio[2]*
[1]Department of Polymer Engineering & Color Technology, AmirKabir University of Technology, Tehran, Iran.
[2]Department of Chemical, Materials and Production Engineering, University of Napoli Federico II, Napoli, Italy.

Abstract

Cell collapse can reduce the cell density and change the cell size of a foam structure. In particular, in conventional heating an optimal foaming temperature is selected. This temperature is a compromise between increasing the temperature to increase the polymer compliance and gas supersaturation; and decreasing the temperature to avoid weakening the polymer and limiting cell collapse. All of these processes must occur concurrently in the short time available for foaming. When attempting to foam large pieces, the described technique manifests its limits and, actually, the impossibility to achieve acceptable foam morphologies.

In this work, microwave heating is used, in comparison conventional heating, since it offers several advantages over conventional heating methods such as the use of a remote source, the relatively high speed of the process, and volume- and material-selectivity.

We used carbon nanotubes as the dispersed source of heating as well as their well-known ability as bubble nucleators. In this way, foaming will occur near the hot particles, where the polymer is sufficiently compliant, while remaining strong elsewhere. As a model system, we tested polystyrene/carbon nanotube foamed with carbon dioxide and achieved thick foams characterized by a uniform morphology and low density.

Keywords: Thermoplastic foaming, Microwave heating, Carbon nano tube

Introduction

Polymeric foams are important materials in a variety of applications due to their inherent low density, giving high specific strength, lightweight, materials savings, low thermal conductivity and good sound insulation and absorption [1]. Among the processing methods to produce foams, Physical Blowing Agent (PBA) foaming is one of the most utilized for its versatility and high throughput. In this method, the polymer is first saturated by a gas at a high pressure and then, after a pressure drop, it experiences a supersaturation state, which may result in the PBA escape to form of a new phase (gas bubble) [2]. To allow bubble formation, the polymer is softened above the melting point or the glass transition. In conventional heating systems (typically, by submerging the polymer/gas solution in an oil bath), heat is transferred from the hot oil to the material, through its surface, until the required temperature has been achieved throughout the sample (in particular, also at the very center of the sample). As plastics have low thermal conductivity, this process takes time and there often is a temperature gradient in the sample thickness. Inevitably, this induces strong differences in the foam thickness and restricts the maximum thickness, for foams processed with said method, to 1-2 mm. Above this limit, there is often a risk of having a collapsed structure at the hot, external sample parts or an unfoamed, bulk structure at the cold, internal part.

Preservation of cell density and prevention of the cell collapse or cell coalescence are key to maintain good foam properties and to keep the foam density low. To avoid this phenomenon, in the literature, different methods such as crystallinity, branching or nanoparticles are utilized to increase the melt index and elasticity [3-5]. Among these methods, polymer nanocomposite have received increasing attention in both scientific and industrial communities to control foam formation in terms of reduction of cell size and preventing cell collapse, and to include the functional properties brought about by the nanometric additive [6].

Microwave heating offers several advantages over conventional heating methods such as the use of a remote source, the relatively high speed of the process, and volumetric heating and material-selectivity [7]. Carbon Nano Tubes (CNTs) have already been investigated for microwave absorption applications due to their great thermal stability, reluctance toward chemical agents and excellent mechanical and electrical properties [7, 8].

Herein we report microwave heating as a new source of heating in thermoplastic foaming. As a model system, we tested polystyrene (PS) with CNT, foamed with carbon dioxide. We also proved effective MW foaming with thick samples (1cm), in comparison with oil bath heating.

Experimental

Materials:

PS (N2380) was supplied by Versalis S.p.A, Mantova, Italy. The average molecular weight, density and melt flow index, are 300 kDa, 1.05 g/cm3 and 2.0 g/10 min at 200°C and 10 kg, respectively, as reported by the producer.

Multi walled CNT (MWCNT) (average outer diameter: 10 nm, average length: 1.5 mm, surface area: 250–300m^2/g, carbon purity 90%, density = 1.44 g/cm^3) were supplied by Nanocyl^{TM}, Belgium, and were produced by Catalytic Carbon Vapor Deposition (CCVD).

CO_2 (99.95% pure) supplied by Sol Group S.p.A., Monza, Italy, was used as the physical blowing agent.

Preparation of PS/MWCNT nanocomposites:

PS/MWCNT 10 wt% master batch was supplied by Nanocyl^{TM}, Belgium. The nanocomposites were then prepared by mixing the master batch with pristine PS using a DSM twin-screw compounder (DSM Xplore MC15) at 200 °C and 100 rpm for 10 min. The final samples were prepared by compression molding at 200°C in 1mm and 10 mm thickness.

Foaming:

The samples were saturated with CO_2 at 10 MPa and 80 °C for 3h and 90h, for the 1mm and 10mm thick samples, respectively, in a high-pressure autoclave [9]. After the saturation, the pressure vessel was cooled to room temperature and pressure released to allow vessel opening. Saturated samples were then exposed to microwave radiation in a microwave oven for a determined time. All samples were placed exactly in middle of microwave cavity which base on literatures could receive the most uniform field [10]. To compare the microwave source system with conventional system, the water bath was considered. In this case, samples were kept in boiling water for a determined time and then air cooled.

Results and Discussion

As the MW radiation power was turned on, CNTs absorb microwave radiation and its temperature increases. Actually by using CNT as the sole microwave absorber particles in a clear polymer media, a localized heating system is introduced. In this case, the polymer close to the CNT gets warm and the supersaturated CO_2 can escape the polymer solution to nucleate the bubbles. The polymer/gas solution far from the particles is colder, conversely and no cell nucleation is possible. With time, the induced heat can transfer to the neighboring regions and cell growth can invade most of the polymer. In this way, foaming will occur only near the hot particles, with reduced risk of collapse. The Figure 1 is a schematic of cell nucleation and growth during MW assisted foaming.

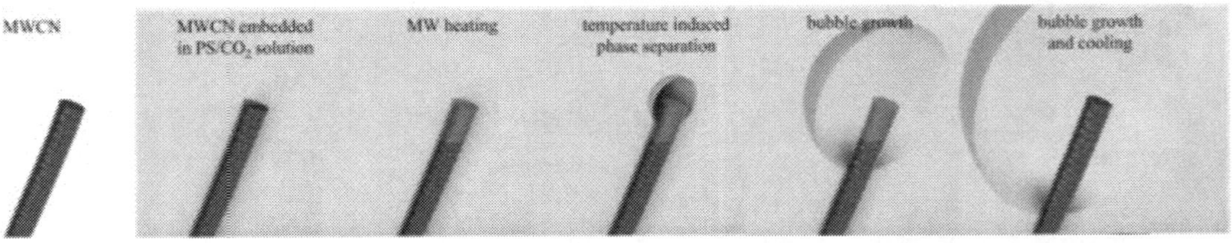

Figure 1. Schematic of cell nucleation and growth with MW radiation

Each unit (CNT and neighbor polymer matrix) has a very small size, is independent from others and is exposed to microwave simultaneously in all of the sample volume. This, of course, means more uniform cell structure, faster foaming and, most importantly, no dependency on sample thickness. As mentioned before the thickness is a limiting factor in other methods and usually thickness is around 1mm. In these experiment we successfully foamed, with uniform morphology, foams samples with 1cm thickness.

In the conventional heating system the heat must transfer from surface to the center of samples, which, by considering the low conductivity of polymers, takes time and induces a gradient in both local foam density and morphology. This phenomenon is obvious in water bath method (Figure 2- a1) foams. These samples experienced a notable gradient in the thickness, in contrast with foaming by microwave.

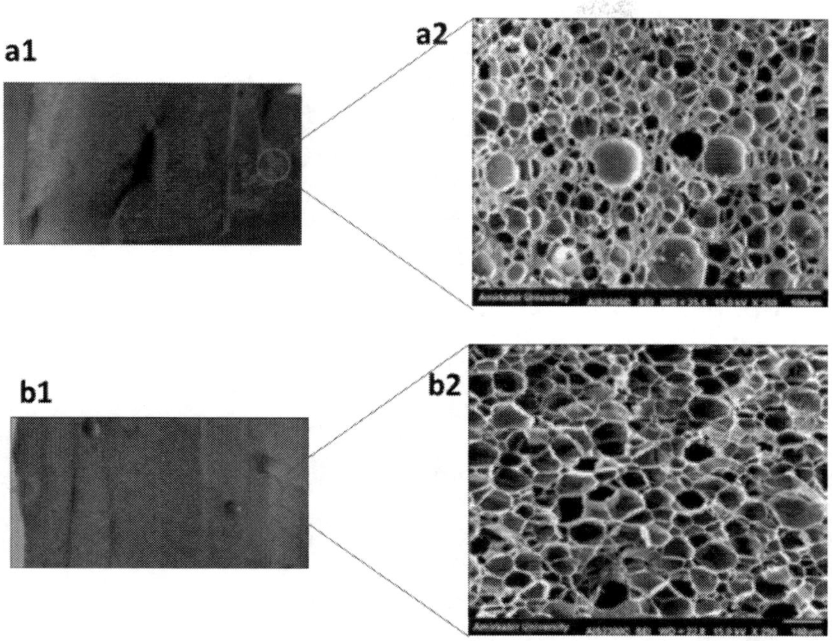

Figure 2 a: results from the water bath method, **1** optical image and **2** SEM image in the center of the foam. **b:** results from the MW method, **1** optical image and **2** SEM image in the center of the foam.

Table 1. Presents the density of foams of samples with different thickness. If we consider the density of foams as to verify foam quality and expansion, a nice improvement in expansion in MW method is evident. In both sample thicknesses, the foams with MW have lower density. We can conclude that with MW method the efficiency of the foaming process is almost independent of thickness.

Table 1.Foam densities for the two analyzed methods, sample thickness of 1 and 10 mm

	Density (g/cm^3)	
Thickness	1mm	10mm
Boling Water Bath@ 100 °C	0.14	0.34
Microwave Radiation @ 40 s	0.11	0.05

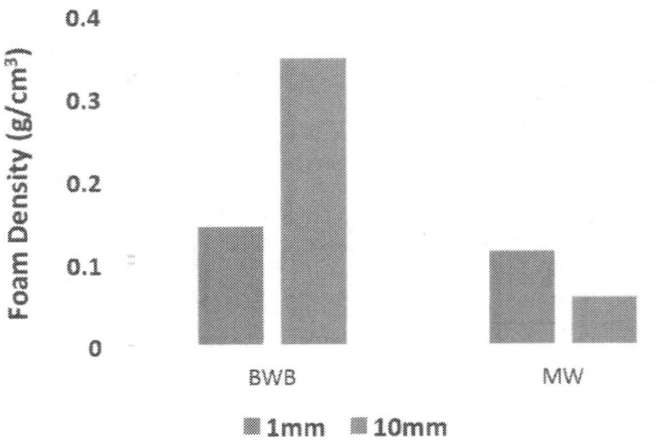

Figure 3. Density of samples at 1 and 10 mm thickness in water bath (BWB) and MW methods.

Conclusions

We used for the first time microwave heating as source of energy in PS/CNT nanocomposite foaming. CNTs react to microwave and transduce the radiation energy to heat. As this method used tiny dispersed heating source, more uniform morphologies are achieved with respect to neat PS and lighter foam are achieved. The MW foams have more uniform cell structure and can reach higher expansion ratio than the water bath foams, particularly at increased foam thickness.

References

1. Zeng, C., N. Hossieny, C. Zhang, and B. Wang (2010), "Synthesis and processing of PMMA carbon nanotube nanocomposite foams," *Polymer*, 51(3), pp. 655-664.
2. Tammaro, D., A. Astarita, E. Di Maio, and S. Iannace (2016), "Polystyrene foaming at high pressure drop rates," *Industrial & Engineering Chemistry Research*, 55(19), pp. 5696-5701.
3. Sharudin, R.W.B. and M. Ohshima (2011), "CO2-Induced Mechanical Reinforcement of Polyolefin-Based Nanocellular Foams" *Macromolecular Materials and Engineering*, 296 (11): p. 1046-1054.
4. Garancher, J.-P. and A. Fernyhough (2012), "Crystallinity effects in polylactic acid-based foams". *Journal of Cellular Plastics*,48 (5), pp. 387-397.
5. Moon, Y. and S.W. Cha (2007), "Study on viscosity changes with talc in microcellular foaming process" *Fibers and Polymers*, 8 (4), pp. 393-398.
6. Lee, L.J., C. Zeng, X. Cao, X. Han, J. Shen, and G. Xu (2005), "Polymer nanocomposite foams" *Composites science and technology*, 65 (15), pp. 2344-2363.
7. Wu, T., Y. Pan, E. Liu, and L. Li (2012), "Carbon nanotube/polypropylene composite particles for microwave welding" *Journal of Applied Polymer Science*, 126 (S2).
8. Savi, P., M. Miscuglio, M. Giorcelli, and A. Tagliaferro (2014), "Analysis of microwave absorbing properties of epoxy MWCNT composites" *Progress In Electromagnetics Research Letters*, 44, pp. 63-69.

9. Tammaro, D., Contaldi, V., Pastore Carbone, M.G., Di Maio, E., Iannace, S. (2016), "A novel lab-scale batch foaming equipment: the mini-batch, Journal of Cellular Plastics", *Journal of Cellular Plastics,* 52 pp. 533-543

10. Cheng, J., R. Roy, and D. Agrawal (2001), "Experimental proof of major role of magnetic field losses in microwave heating of metal and metallic composites" *Journal of materials science letters*, 20 (17), pp. 1561-1563.

471406 Polyurethane Foam Expansion, Polymerization and Bubble Pressurization

Abstract:

In this presentation, we discuss our new, experimentally driven, approach to foam expansion and polymerization modeling. We are working to develop high-fidelity computational models to be used to design and troubleshoot mold filling for polyurethane structural foam parts. We are studying PMDI polyurethane with a fast catalyst, such that filling and polymerization occur simultaneously. The foam is over-packed to twice or more of its free rise density, to reach the density of interest. Developing a relevant model to represent the expansion, filling, curing, and final foam properties is quite challenging. PMDI is chemically blown foam, where carbon dioxide is produced via the reaction of water and isocyanate. The isocyanate also reacts with polyol in a competing reaction, which produces the polymer. A new kinetic model is implemented, which follows a simplified mathematical formalism that decouples these two reactions. The model predicts the polymerization reaction via condensation chemistry, where vitrification and glass transition temperature evolution must be included to correctly predict this quantity. The foam expansion kinetics are determined by tracking the molar concentration of water and carbon dioxide in both the gas and polymer phases. A bubble-scale mathematical model has been developed that looks at the super saturation of carbon dioxide in the polymer and the concomitant bubble pressurization. The model is calibrated with bubble-expansion data from bubble growth experiments. It is used to predict the density of a foam bar and compared to to Xâ€"Ray CT data.

473436 The Role of Viscoelasticity in Bubble Breaking

Abstract:

The rupture dynamics of a bubble affects many phenomena in nature, e.g., magma fragmentation in volcanology, the opening of vesicles in medicine, and many processing technologies from the production of polymeric foams to the cooking of pancakes[1-2]. If the bubble is made of a viscoelastic fluid, it can store elastic energy during inflation that can be released during rupture, in turn affecting the breaking dynamics. In the present contribution, we investigate experimentally the way elastic energy (stored as a consequence of fast bubble inflation) drives rupture in a model viscoelastic fluid. A viscoelastic bubble is created via a pump and a homemade inflating system, a hot needle induces rupture and a fast camera is used to acquire images of the hole opening over time. The roles of the extensional rate and of the stored elastic energy on bubble rupture are investigated, and a new mathematical model to describe the rupture phenomenon is developed.

[1] de Gennes PG, Brochard-Wyart F and Quéré D, Capillarity and Wetting Phenomena, Springer (2004).

[2] Eggers J and Villermaux E, Physics of liquid jets, Rep. Prog. Phys., 71, 036601 (2008).

473667 The Use of the Viscoelasticity in Polymer Foaming to Obtain a Fully Opened Cell Structure

Abstract:

In this work a design tool to control cell-opening in gas foaming of thermoplastic semicrystalline polymers is developed. The sequence of events following bubble nucleation, namely, bubble growth and impingement, are modeled to gain a comprehensive, perspective view on the mechanisms of bubble wall rupture and on the conditions for achieving a fully opened cell morphology by flow induced crystallization. In particular, unlike the previously published literature, the polymer elastic recovery is recognized as an important factor for wall retraction, often dominant with respect to surface tension, which is typically considered as the only driving force determining the retraction of ruptured bubble walls. The new approach is experimentally validated on poly(e-caprolactone) (PCL), foamed with CO_2, as a model polymer/gas system.

467298 Sprial Vortex Instability in Microfluidic Cross-Slot Flow

Abstract:

Improved understanding and characterization of stability conditions for flows through intersecting geometries is vital for the optimization of many laboratory microfluidic experiments and also practical lab-on-a-chip designs, including for the specific goal of enhancing the mixing of fluids in channels with small dimensions operating at low Re.

In this work, we report the results of detailed experimental studies of the spiral vortex flow instability of Newtonian fluids and dilute polymer solutions in cross-slots with a range of aspect ratios and over a wide range of Re. In contrast to previous studies, we identify appropriate order parameters that characterize the instability as a function of Re in each case. At small Reynolds numbers, Re, the flow is two-dimensional and a sharp symmetric boundary exists between fluid streams entering the cross-slot from opposite directions. Above an ÒaÓ (aspect ratio) dependent critical value $Re_c \sim$ 20-100, the flow bifurcates to an asymmetric state (though remains steady and laminar), and a single three-dimensional spiral vortex structure develops around the central axis of the outflow channel. Image analysis allows an assessment of the mixing quality between the two incoming fluid streams (one stream fluorescently-dyed with rhodamine b), which undergoes a significant increase following the onset of the instability. For $Re > Re_c$, the mixing parameter grows according to a sixth-order Landau potential. Fitting parameters indicate the transition is second order at aspect ratio a = 0.5, and passes through a tricritical point, becoming first order for a > 1. A simple scaling of the fitting parameters with allows full collapse of the experimental data. This instability can be used to drive enhanced mixing at the moderate Re that can be achieved in microfluidic devices and we show that further mixing enhancement can be achieved by patterning the surfaces of the channel walls. The effect of adding a small concentration (~0.01 wt%) of high molecular weight polymer is to reduce the value of Re_c in comparison to the Newtonian solvent.

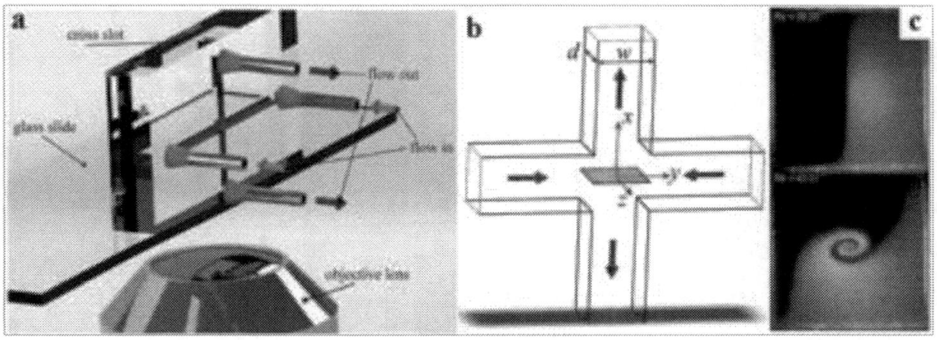

Fig.1 (a) Schematic drawing of the experimental set-up showing the cross-slot device vertically mounted on an inverted microscope. (b) Detail of the coordinate system showing the fluid flow direction and the measurement plane (x = 0 plane, green). (c) Confocal microscope images obtained in the x = 0 plane for the flow of water at $Re < Re_c$ (top) and at $Re > Re_c$ (bottom). Water with fluorescent dye enters from the left, undyed water from right; outflow is normal to the page.

472964 Multicomponent Polymer Gels with Enhanced Elasticity

Abstract:

We report rheology and small-angle neutron scattering (SANS) studies of a multicomponent gel consisting of the biopolymer alginate together with a temperature-sensitive copolymer, Pluronic® F127, and silicate nanoparticles, laponite®. Although the F127 systems on their own form relatively weak gels in water, the multicomponent gels exhibit significantly higher storage and loss modulus (*G′*, *G″*) values than alginate-laponite gels. The *G′* of the nanocomposite gels containing F-127 Pluronic® are two orders of magnitude higher than the control containing just alginate and laponite. The interaction between the components appears to be synergistic; that is, the resulting multicomponent hydrogels are much more elastic than the individual components or binary combinations. The F127 also imparts some thermoresponsive character to the gels, although the effects are not as dramatic as seen in many other thermoresponsive polymers. SANS studies show that the alginate chains impact formation of the cubic micellar structure that is typically seen in Pluronic® copolymers.

461896 Understanding Structural Changes in Complex Fluids at High Shear Rates: Developing μrheosans

Abstract:

Industrial applications, such as lubrication, mixing, spraying and injection, involve the flow complex fluids at extreme deformation rates. Clogging, fluid degradation, and other processing challenges can arise in these extreme contexts and are often driven by structural changes in the fluid. As an example, there are concerns that high shear rate delivery of protein therapeutics via injection with a very small bore syringe needle could alter the protein structure and drive aggregation, leading to toxic and potentially fatal health effects. For shear rates $<10^4$ s^{-1}, the rheology and structure of a flowing fluid can be assessed using an established technique such as RheoSANS, RheoSAXS, confocal rheometry. High shear rate rheological characterization can be carried out using capillary and slit rheometers, but measuring the structure of complex fluids under these conditions is more challenging. It could potentially be achieved using light, x-ray, or neutron scattering; with neutron scattering providing particular advantages: neutrons probe the length scales of interest for many complex fluids, penetrate the materials used to construct capillary/slit rheometers, and provide the ability to adjust contrast to selectively probe the structure of particular components of the fluid. We have recently begun developing a series of microfluidic devices that will enable simultaneous rheological and structural characterization of complex fluids at high shear rates using small-angle neutron scattering (SANS). Prototype devices have achieved effective shear rates exceeding 10^5 s^{-1}, and an apparatus can achieve shear rates $>10^6$ s^{-1} is under develpment. Additionally, a data analysis method has been developed to isolate scattering from the high-shear region located near the wall of the microfluidic channel, helping to better resolve shear rate dependent scattering. Here, we will discuss both the development of our new μRheoSANS apparatus and the data analysis approach in the context of initial measurements of a cetylpyridinium chloride-sodium salicylate wormlike micelle system.

469047 The Effect of Branching on Shear Banding in Wormlike Micelles (WLMs) Under Large Amplitude Oscillatory Shear (LAOS)

Abstract:

Self-assembled wormlike micelles (WLMs) are of particular scientific and technological interest due to their ability to branch, break, and reform under shear, which can lead to nonlinear flow phenomena and instabilities such as shear banding [1,2]. As measuring chain branching in polymer physics is a long-standing scientific challenge, branched WLMs are also used as a model system for studying branched polymers. Nonlinear rheology and large amplitude oscillatory shear (LAOS) have shown to be useful in the study of chain branching [3], but quantitative relationships are still lacking. Thus, understanding the coupling between flow behavior and molecular topology requires experimental methods that combine time- and spatially-resolved small angle neutron scattering (SANS) measurements of structure with rheometry for nonlinear deformations [4]. Here, we explore the relationship between branching, microstructure and nonlinear responses using a series of well-characterized, branched WLM solutions [1, 5-7]. While micellar branching is physically distinct from chemical branching in polymers [8], we observe similar responses under shear startup and LAOS [3, 9].

By altering WLM solution composition or temperature, a maximum in the zero-shear viscosity, Î·$_0$, is often observed, which corresponds to changes in micellar topology. The increase in viscosity with added surfactant or salt results from micellar growth and entanglement, whereas branching has been shown to lower Î·$_0$ [3] by providing another mechanism for stress relaxation. Micellar branching leads to deviations from Maxwellian behavior in the LVE rheology [2], and can alter or eliminate steady shear banding [2,3]. Although there are no theoretical predictions of the effects of branching on shear banding, work by Vasquez, Cook and McKinley (VCM model) has made predictions incorporating reversible chain scission that can help interpret nonlinear rheological experiments on shear banding WLMs [10].

The rheology and shear-induced microstructures of a model series of WLMs is studied to determine the effect of branching on shear flow phenomena. The degree of branching in the mixed cationic/anionic surfactant (CTAT/SDBS) solutions is controlled via the addition of the hydrotropic salt sodium tosylate [2, 6]. The phase behavior has been mapped extensively and we have explored the degree of branching and network formation using SANS, cryo-TEM and rheo-optical methods [2, 6-8]. The shear-induced micellar alignment is spatially and temporally characterized under steady shear, shear startup, and large amplitude oscillatory shear (LAOS) by flow-SANS in the flow-gradient (1-2) and flow-vorticity (1-3) planes. Advanced methods of time-resolved data analysis improve the resolution of the experiments by several-fold [7].

By employing non-linear rheological techniques with concurrent SANS, we determine rheological and scattering signatures of branching. Segmental orientation and alignment in the 1-3 and 1-2 planes is a complex function of the branching level and deformation type. Shear banding is verified in low and mildly branched solutions (â‰¤0.05% NaTos), but is mitigated in highly branched solutions (â‰¥0.1% NaTos). In the mildly branched solution, steady shear banding is observed between 2.6< Wi<100. Significant, discontinuous decreases in the scattering anisotropy from the inner to outer wall indicate shear banding (Wi=25, 45, 75). Similar responses at the inner and outer walls indicate shear thinning (Wi=1, 250). An alignment factor (A_f) is calculated at each gap position, and when A_f>0.2 at all gap positions [12], the material no longer shear bands. Startup measurements in both planes verify shear banding, where a long transient is observed in the A_f response in the low and mildly branched solutions during banding. Differences in this transient suggest different mechanisms of banding between the solutions. No transient is observed in shear thinning conditions and in branched solutions.

Branching also inhibits shear banding under LAOS. In order to experimentally validate the predictions of the VCM model [10], seven LAOS conditions were examined in the mildly branched solution, of which six exhibited shear banding under steady shear. Only four conditions displayed LAOS shear banding, however. Two distinct forms of dynamic shear banding were identified by focusing on two LAOS conditions: De=0.17, Wi=75 and De=0.58, Wi=75. When De=0.17, the $A_f(r/H)$ during LAOS at Wi=75 is similar to $A_f(r/H)$ in the steady shear case, indicating a similar form of shear banding. The long oscillation period gives the material several relaxation times near the stress overshoot before reaching Wi=75, resulting in a LAOS stress that is similar to the steady shear stress at Wi=75. At De=0.58, Wi=75 a different form of shear banding is seen, where the material is trapped in a metastable banded state. The material exhibits over-orientation, where the maximum A_f under LAOS is larger than in steady shear. The stress and time scales allow little time for relaxation after the stress overshoot, leading to over-orientation and a larger LAOS stress than steady shear stress. Over-orientation is also observed in the non-shear banding conditions at higher De, indicating that shear banding is not a requirement for over-orientation. Further, this over-orientation is magnified by branching.

Novel flow-SANS methods have enabled quantitative differences in branched and linear WLMs to be determined. Branching inhibits flow alignment and shear banding under both steady and dynamic (LAOS) deformations. SANS measurements during startup confirm shear banding, and suggest a different mechanism of shear banding between low and mildly branched solutions. Under LAOS, shear banding has been experimentally verified in Couette flow for the first time, in agreement with VCM model predictions [10]. Distinct shear banding mechanisms are identified, which can be explained by analyzing LAOS time scales. In the first mechanism, the oscillation period is long and the material can relax during the cycle, leading to an alignment similar to steady shear. In the second, the over-orientation is a consequence of the faster period, and thereby incomplete relaxation during the cycle. This multi-technique approach, combining nonlinear rheology and spatially-and temporally-resolved SANS has enabled us to link the micellar microstructure and topology to the macroscopic flow properties of WLM solutions, providing a more complete data set for the development and testing of microstructure-based constitutive equations that explicitly incorporate branching and breakage.

References:

[1] M. A. Calabrese, et al., *J. Rheol.,* **59**, 5 (2015).

[2] S. A. Rogers, M.A. Calabrese and N. J. Wagner, *Curr. Opin. Colloid Interface Sci.,* **19**, 6(2014).

[3] K. Hyun, et al., *Prog. Polym. Sci.,* **36**, 12 (2011).

[4] A. K. Gurnon, et al., *J. Vis. Exp.,* **84** (2014).

[5] B. Schubert, N.J. Wagner, and E.W. Kaler, *Langmuir* **19**, 10 (2003).

[6] M. A. Calabrese, et al., *J. Rheol.(under review),* (2016).

[7] M. A. Calabrese, N. J. Wagner, and S. A. Rogers, *Soft Matter,* (2016).

[8] S. J. Candau, and R. Oda, *Colloid Surface A,* **183** (2001).

[9] F. Snijkers, et al., *J. Rheol.,* **57**, 4 (2013).

[10] L. Zhou, L. P. Cook and G. H. McKinley, *J. Non-Newton Fluid*, **165**, 21 (2010).

[11] M.E. Helgeson, et al., *J. Rheol.*, **53**, 3 (2009).

469290 Structure and Rheology of Branched Micellar Fluids from Molecular Dynamics Simulations

Abstract:

We extend our previous coarse-grained molecular dynamics (CGMD) simulations of cationic surfactant micelles of cetyltrimethyl ammonium chloride (CTAC) and organic salt sodium salicylate (NaSal) in presence of explicit solvent (water) to understand the shear and extensional rheology of branched micellar fluids. Our simulations preserve the underlying physico-chemical, electrostatic and hydrodynamic interactions of the system; simulation methods are discussed in detail elsewhere [1-3]. CGMD simulations were performed for various surfactant (c_D) and salt (c_s) concentrations over micro second time scales to study equilibrium structure, and over deformation rates spanning several decades to understand the underlying structure-dynamics-rheology relationships.

It has been shown in previous studies [4-6] that the addition of highly binding organic salt, such as NaSal, results in a double peak in the variation of the zero-shear viscosity (Î·$_0$) of the solution with increasing salt to surfactant concentration ratio R. From the analysis of MD simulation trajectories, we offer a mechanism underlying this experimentally observed non-monotonic variation in the zero-shear viscosity. Specifically, branch formation at relatively large salt concentration (R»1) results in a significant reduction in the solution viscosity. The second maximum in the viscosity vs. R curve is attributed to the formation of a fully connected micellar network [3]. In this presentation, we will discuss how the hyperbolic interfaces at branch points are energetically stabilized by a relatively larger density of condensed Sal^- counter ions as well as various mechanisms of branch formation. Moreover, the simulations revealed pronounced *dynamic heterogeneity* associated with the motion of the surfactant molecules. Specifically, the ensemble averaged mean square displacement of all the surfactant molecules in the system suggests sub-diffusive dynamics. However, time averaged mean square displacements of individual surfactants exhibit a broad range of dynamical behavior including super-diffusion as proposed in Ref. [5], which has been attributed to the presence of *Lévy flight*-like trajectories [5]. This is the first time such super-diffusive motion has been observed directly in molecular simulations. This motion stems from the sliding of micelle branches along the main contour which has been proposed in the literature as an alternative mechanism for stress relaxation in branched micellar fluids. At higher surfactant concentrations, micelle entanglements inhibit the motion of surfactant molecules, resulting in sub-diffusive behavior.

In the second part of the talk, we will discuss the deformation of a micellar network in a steady uniform shear and uniaxial extensional flow. First, we will briefly discuss the dynamics of (flexible) wormlike and cylindrical micelles in an extensional flow [7]. Under uniaxial extension, transition from a folded to a stretched state is observed beyond a Weissenberg number-independent critical strain, after which micelle scission occurs through a mid-plane thinning mechanism. The effect of the binding counter ions on micelle stiffness will also be discussed. Extending our previous single micelle simulations [7], we have performed a large number of simulations to probe the dynamics of micelle networks under elongational flow. The effects of the rate of deformation and salt concentration on the extensional rheology and breakage of micelle networks will be discussed.

Acknowledgements: The authors gratefully acknowledge NSF grants CBET-1049454 and CBET-1049489 as well as the computational resources provided by XSEDE.

References:

[1] A. V. Sangwai and R. Sureshkumar, Langmuir, 27, 6628 (2011); 28, 1127 (2012)

[2] A. Sambasivam, A. V. Sangwai, and R. Sureshkumar, Phys. Rev. Lett., 114, 158302 (2015); Langmuir, 32, 1214 (2016)

[3] S. Dhakal and R. Sureshkumar, J. Chem. Phys., 143, 024905 (2015)

[4] D. Sachsenheimer, C. Oelschlaeger, S. Müller, J. Küstner, S. Bindgen and N. Willenbacher, J. Rheol., 58 (6), 2017 (2014)

[5] J. P. Bouchaud, A. Ott, D. Langevin and W. Urbach, Journal de Physique II, 1(12), 1465 (1991)

[6] S. A. Rogers, M.A. Calabrese and N.J. Wagner, Curr. Opin. Colloid Interface Sci., 19, 530 (2014)

[7] S. Dhakal and R. Sureshkumar, ACS Macro. Lett. , 5, 108 (2016)

469128 Distinguishing Shear Banding from Shear Thinning in Flows with a Shear Stress Gradient

Abstract:

Shear banding of complex fluids, including wormlike micelles (WLMs), involves spontaneous separation of a flow field into regions with distinct velocity gradient upon shear, and is typically attributed to mechanical instability due to a non-monotonic constitutive curve. The detailed signatures of steady state banding, including an apparent stress plateau between lower and upper critical shear rates as well as the simple â€œlever ruleâ€ followed by the location of the interface between bands, have been thoroughly studied for narrow gap Taylor-Couette flows that approximate simple shear. However, geometries with significant spatial gradients in the shear stress (e.g., large gap Taylor-Couette flows) can produce banding-like features, even when the fluid possesses a monotonic constitutive curve. This raises important fundamental questions. Most importantly, how does shear banding manifest in geometries with significant stress gradients, and how do we identify whether a fluid is truly shear banding, as opposed to strongly shear thinning?

Here, we report a model-free approach to experimentally distinguish shear banding from strong shear thinning, and apply it to rheological and flow velocimetry measurements on WLMs in a Taylor-Couette device with systematically varying curvature (i.e., the ratio of gap size to cylinder radius). The method, which involves numerical analysis of various derivatives of the velocity profile, is verified through one-dimensional flow calculations using fluids with non-monotonic and monotonic constitutive curves. We then apply the method to experimental data obtained through high-resolution velocimetry measurements. In both simulation and experiment, we find that shear banding WLMs exhibit several distinctive features in the velocity profile, whose presence and quantitative behavior are relatively insensitive both to the applied shear rate and the degree of curvature of the flow geometry. By contrast, the shear thinning WLMs exhibit features that are strongly dependent on the geometry and applied shear rate. Our results provide a clear, unambiguous criterion for identifying shear banding in complex fluids.

471653 Associating Colloids and Active Surfaces; A Density Functional Theory Approach with Multi-Body Correlations

Abstract:

Predicting the thermodynamic properties and self-assembly of patchy colloids and of hydrogen bonding molecules (e.g., hydrogen fluoride and water) continues to be a challenge of statistical mechanics based theories. The most commonly applied approach is Wertheimâ€™s perturbation theory and its extensions for associating particles. The first order limit of this theory treats associating sites independently and assumes each site can form only one bond. This approach allows for the development of a very general equation of state for associating molecules in homogeneous (SAFT) and inhomogeneous systems (iSAFT) with any number of association sites. However, in case of large associating sites, the assumption of only bond per site is not accurate anymore. Therefore, a higher order theory is needed to take such effect into account. Here we apply Wertheimâ€™s thermodynamic perturbation theory beyond first order for a patchy colloidal fluid near a hard wall with active sites on its surface using a density functional theory approach where the sites on colloids or wall can be large enough to form two bonds, simultaneously. The effect of steric hindrance between two associating species bonded to a third site is included, and the equation of state derived for this system is dependent on the angular size of the sites. To verify the theory predictions, new Monte Carlo simulations have been performed. The theory predictions for distribution of associated species near the hard wall as a function of the size of the sites are in a good agreement with MC simulation.

468021 Single Molecule Studies of DNA-PNIPAM Copolymers

Abstract:

Polymer chain architecture encompasses a wide range of shapes and chain types, including linear polymers, branched polymers, ring polymers, and polymer networks. In addition to chain topology, the chemical composition of polymers also plays a key role in determining emergent properties. To this end, copolymers and block polymers with simple linear chain architectures can give rise to intricate structural ordering and assembly across multiple length scales. Polymer architectures play a critical role in supramolecular assembly of advanced materials with applications in encapsulation for drug delivery and catalysis, surfactants for emulsions, and materials for chemical separation. For example, PEG-PLA is an important drug delivery carrier that can encapsulate hydrophobic drugs such as camptothecin when mixed with drug solutions. Recently, the development of hybrid polymers that contain both synthetic and natural components have enabled new materials with higher compatibility, enhanced biodegradation, wide functionalities, and unprecented levels of sequence-control. A recent area of interest has focused on the self-assembly of hybrid polymers, especially DNA-based hybrid polymers, into various architectures such as micelles, vesicles, and tubes. Despite recent work, however, we still lack a complete understanding of this process.

In this work, we study the real-time transition of trigger-responsive hybrid polymers using single molecule techniques. In particular, we synthesize DNA-PNIPAM copolymers and study the supramolecular assembly of these materials using fluorescence microscopy. First, we use polymerase chain reaction (PCR) to produce 10 kb DNA with 1% substitution of chemically modified nucleotides containing DBCO, which is a strained alkyne for Cu-free click chemistry. In a separate reaction, we use reversible addition-fragmentation chain-transfer (RAFT) polymerization to synthesize fluorescent PNIPAM copolymerized with rhodamine monomers. In addition, the PNIPAM polymers contain a terminal azide group, which facilitates click reaction with the DBCO-modified DNA. In this way, we generate DNA-PNIPAM comb and bottle-brush polymers, and after purification, these materials show a sizable shift via agarose gel electrophoresis denoting the formation of the desired product. In this way, we effectively synthesize

semi-sequence controlled DNA-PNIPAM polymers with different architectures, and we can directly study these materials using single molecule fluorescence microscopy, including the dynamic transition process for temperature responsive materials. In particular, upon increasing the temperature above the lower critical solution temperature (LCST) of PNIPAM around 32C, the material transitions from hydrophilic to hydrophobic character. In this way, we directly observe the spontaneous formation of higher order structures in solution by dynamically increasing the temperature above the LCST. As a control, we also used scanning electron microscopy (SEM) to characterize the self-assembled structures as a function of temperature, and we found micron-sized vesicle structures compared with the control group. Finally, we study the single chain dynamics of DNA-PNIPAM polymers using fluorescence microscopy, both in equilibrium conditions and under flow. Using this approach, we characterize single chain dynamics as a function of the temperature-sensitive dynamic transition of DNA-PNIPAM above and below the LCST of PNIPAM.

460967 System Size Dependent Rheology of Sheared Lamellar Phases

Abstract:

The rheology of a lamellar liquid crystalline medium under shear is studied using a two-dimensional mesoscale simulation technique which distinguishes between the hydrophobic and hydrophilic components of the lamellar phase. The dimensionless parameters in this model are the Reynolds number which is the ratio of inertia and viscosity, the Schmidt number which is the ratio of momentum and mass diffusion, an Ericksen number which provides the ratio of viscous stress which tend to deform the layers an elastic restoring stresses, the viscosity contrast between the hydrophilic and hydrophobic components and the ratio of the system size and the layer spacing. Two distinct types of structural evolution are observed depending on the Schmidt number. At high Schmidt number where momentum diffusion is faster than mass diffusion, shear tends to break and reform the layers; this results in a well-aligned background lamellar phase with embedded defects. The defects are annealed by the shear, resulting in a well ordered lamellar phase in the long time limit. At low Schmidt number where mass diffusion is faster than momentum diffusion, the layers are well formed locally in domains with random orientations. Shear tends to rotate and align the layers, but there is also the creation of defects due to shear, which results in a disordered steady state determined by a balance between the annealing and creation of defects. Complete ordering is not observed even at very long times. The average viscosity (ratio of stress and macroscopic strain rate) shows a strong dependence on system size, proportional to (L/ Î»)$^{3/2}$. The reason for this strong system size dependence is the divergence in the length scale of the region over which there is a distortion of layers due to a point defect, and the consequent divergence in the secondary flow caused due to the disturbance of the linear shear flow by the defect.

469969 From Microstructure to Macro-Rheology in Complex Fluids: Towards Targeted Rheology-Assisted Structural Assembly of Attractive Systems

Abstract:

The field of complex fluids encompasses a wide class of materials, which exhibit unusual mechanical responses to an applied stress or strain. In virtually all complex fluids, this rich and unusual mechanical response originates from a microstructure that responds to different applied stress or strain in specific and varied ways. Thus understanding the microstructure – macroscopic behavior relationship is a crucial step for systematically designing complex fluid materials for novel applications. The complex fluid landscape can be subdivided based on the particle-level interactions that govern their underlying microstructure and the resulting macro rheology. Attractive particulate-based systems are one such general class, with ever-growing applications in a wide range of consumer-product, food and other industries. These are, in their simplest form, liquid-liquid (emulsions) or solid-liquid (suspensions) binary mixtures in which the nature of interaction between the dispersed species is primarily attractive. As a result, nearest-neighbor pairs of particles are able to establish local stable energy minima. At the same time, the random nature of thermal fluctuations in a system, also referred to as Brownian motion, provides the particles with sufficient energy to experience other states and slowly build larger structures with other particles within their attraction range. However, any applied stress or strain effectively drives rearrangement and changes the energy landscape of every particle and thus the rheological behavior of these systems becomes complex under imposed flow conditions.

Another way to categorize complex fluids is based on their mechanical response. A sub-class of complex and structured fluids including (but not limited to) colloidal gels, nano emulsions, micellar solutions and crude oils can be identified as *Thixotropic Elasto-Visco-Plastic* (TEVP) materials. TEVPs, as suggested by their name, show a rich and complex set of rheological responses to imposed deformations in different regimes: below the yield stress, the microstructural network formed by individual particles remains intact and resists large deformations by external loading and the material acts as a viscoelastic solid. By increasing the applied stresses above the yielding point, the material starts to flow and undergoes plastic deformation and microstructural rearrangement over a wide range of length scales. Plastic flow results in a viscous-like response; however, due to constant formation and breakage of interparticle bonds that form the network, thixotropic behavior also begins to emerge. These time- and rate-dependent responses lead to other secondary effects including micro-phase separation, vorticity-aligned structure formation, shear banding, rigid plug development as well as shear-induced rejuvenation of the particle network. Continuum-level descriptions of TEVP fluids resort to describing such responses using *ad hoc* evolution equations for one or more scalar measures of this microstructural complexity. Although essential for defining the rheological response of any thixotropic fluid, these internal model parameters do not commonly correlate to specific physical processes in the test fluid. In this work we employ a mesoscale numerical simulation method that captures a canonical anisotropic and weakly attractive material microstructure at a sufficiently coarse-grained level that we can firstly reproduce characteristic rheological features of a TEVP fluid under flow, and secondly identify the sequence of microstructural changes that give rise to these macroscopic features.

In order to precisely understand the significance of microstructural rearrangements of an attractive system under flow on its rheology, accurate control over the initial configuration of particulate structure, as well as descriptive methods of visualizing the microstructure is required. This, in addition to difficulties in accurately controlling & quantifying the interaction between the particles, makes experimental study of the subject very challenging. Hence, computational methods have developed over the last few decades in order to mimic experimental conditions, with the advantage of being able to control the interaction types/strengths between the constituents as well as providing a detailed view of the microstructure under different conditions. As a result of the particle-level bond formation, and collective motion of attractive particles, multi-body and hydrodynamic interactions are essential in reproducing a realistic description of these systems. Thus, only during the past few years and after recent advances in computational capabilities, have these types of studies become possible on the dynamics of weakly-attractive systems.

In the first part of this work, we present a mesoscale simulation study of a model TEVP fluid. We perform Dissipative Particle Dynamics (DPD) simulations on model TEVP fluids consisting of 10 *vol*% 2D weakly attractive hexagonal solid particles reminiscent of the waxy crystalline particles observed in a waxy crude oil, which is known to exhibit many complex TEVP rheological signatures [1]. We substitute the traditional conservative force in DPD with a Morse potential in order to mimic the attractive nature of inter-particle interactions in these fluids.

$$F_{ij}^{C}=-(\partial U(r_{ij})/2\partial r_{ij})e_{ij} \text{ with } U(r_{ij})=-\Gamma(2e^{-\kappa r_{ij}}-e^{-2\kappa r_{ij}}) \quad (1)$$

Steady state flow curves for different values of the attraction strengths between the hexagonal disks are presented in Figure 1(a), confirming that the yielding behavior of TEVPs is captured in our DPD simulations. In order to correlate the microstructural changes of a TEVP fluid to its rheological response, we define a *fabric tensor* [2] formed by an ensemble average of the spatial configuration of particle-particle bonds:

$$Z=(\Sigma Z^{p})/N \text{ with } Z^{p}=\Sigma n_i \times n_i \quad (2)$$

where $\mathbf{n}_i$ represents the unit vector along the center-center line of two bonded neighboring particles. The trace of this tensor gives the average number of bonds/links or the *coordination number* of the microstructure. The off-diagonal components of **Z** correspond to the average preferred orientation of particle bonds with respect to different directions, and the deviatoric normal components give insight with regards to structural anisotropies that develop in the sheared fluid. Figure 1(b) shows the evolution in tr{**Z**} (*i.e.* the coordination number) and the deviatoric components of the fabric tensor as a function of the accumulated strain in the start-up of steady shear flow over a wide range of shear rates.

After validating the capability of the model in reproducing realistic and precise microstructural information and macroscopic properties of a model TEVP fluid, we present in the second part of this work a Time-Peclet-Transformation (TPT) procedure to construct dynamical phase diagram for attractive systems of different mechanical properties. Using spherical particles, and the dimensionless ratio of the shear forces to attractive forces between the particles (Peclet number) as the key dimensionless group, we identify the interdependent roles of different deformation parameters such as strain and strain rate, on the final state of the system.

In this work we will show that a mesoscale model that conserves mass and momentum both locally and globally, and thus preserves multi-body hydrodynamics intrinsically, successfully reproduces the key features of a model TEVP fluid including flow instabilities and coarsening dynamics at rest. In addition to reproducing the experimental results for these fluids, our model provides a detailed description of the microstructure that results in such rich rheological behavior. The fabric tensor provides a quantitative microstructural measure of the system that correlates to the macroscopic stress response in a TEVP fluid in both steady and transient states. The evolution in the components of **Z** with strain provides quantitative insight on the flow instabilities and secondary structures that develop in the sheared microstructure. The TPT procedure provides a mechanism to systematically develop phase map for attractive systems, based on their flow and mechanical properties. This will pave the way for exploiting the largely-unexplored rheology of such systems, for designing novel materials with engineered properties.

References:

1. C. J. Dimitriou and G. H. McKinley, *Soft Matter*, **10**(35):6619-6644 (2014).

2. T. Olsen and K. Kamrin, *Soft Matter*, **11**(19):3875-3883 (2015).

467784 Enhancement of Van Der Waals Mediated Adhesion of Mosquito Leg to Fibrous Surfaces

Abstract:

The aim of our project is to introduce a novel path that can increase the efficiency of Long Lasting Insecticidal Nets (LLINs). LLINs are one of the most effective ways to control and reduce malaria transmitted by mosquitoes. Mosquito leg consists of micro-nanostructures which can be exploited to enhance their adhesion to other surfaces. We have identified that increasing the roughness of a LLIN surface enhances the contact area and generates larger adhesion between the net surface and micro-nanostructures present in mosquito leg. This increased duration leads to a greater lethal dose of the insecticide delivered to the mosquito body. We focused our initial efforts in obtaining quantitative data for measuring force of adhesion between the mosquito leg and High Density Polyethylene (HDPE) sheets (same as LLIN material) using Atomic Force Microscope (AFM). Adhesion force between the mosquitoâ€™s leg and a flat polyethylene sheet is compared with the force generated between the mosquitoâ€™s leg and rough sheets. We create rough polymer surfaces using mechanical drilling. We repeat the experiment for different surface roughness and humidity in order to clarify relative contribution of van der Waals forces compared to capillary forces in providing adhesion of mosquito leg to the surface. We aim to establish a model understanding how mosquitoes adhere to a surface depending on their micro scale roughness.

469604 Filling of Initially Empty Cavities By Advancing Water Fronts

Abstract:

Phenomena involving the wetting of liquids, especially water, on rough or patterned surfaces arise in many different areas of surface science, personal care products, etc. Many fundamental issues remain unresolved, such as the energetics of complex surface structures, and the kinetics/dynamics of cavity-filling both under the wetting liquid and ahead of it, which manifest themselves in â€˜contact angle hysteresisâ€™ and non-uniform (e.g., stick-slip) motion of advancing or receding three-phase boundaries.

Â We have investigated the factors affecting the rate at which water fills micro-scale, cavities. Three-dimensional imaging with confocal microscopy coupled with fluorescence microscopy were used in these studies. We find that the concentration of dissolved air in the water significantly changes the rate at which cavities get filled: deaerated water filled the cavities within 15 minutes, while 2 weeks was required for aerated water. Therefore, the water available for air to dissolve into plays a very significant role in cavity filling. We also tested the effects of various surfactants in the water: surfactants (dissolved at their CMC) shortened the filling time, taking less than 15 minutes in *all* cases, due more to the surface tension lowering than the changing intrinsic contact angle. More generally, reducing Î³$_{LV}$ always speeded up the filling rates. We also observed capillary condensates forming and slowly growing on the bottom of the cavities while the cavities were still only in the partially-filled state, due to the diffusion of water through the vapor to the (hydrophilic) surfaces.

472766 Control of Microstructure during Roll-to-Roll Deposition of Self-Assembled Colloidal Monolayers

Abstract:

We describe a general method to fabricate well-ordered nanoparticle films by combining Langmuir-Blodgett deposition and continuous Landau-Levich thin film coating. Our approach is a scaled-up process for preparing nanoparticle films compared to regular batchwise methods. In a continuously flowing trough, silica nanoparticles are injected onto a floating water surface and the lateral pressure of the circulating water helps the particles assemble into high density film. At the edge of the trough, a rolling system handles with dispensing spare substrate, transferring assembled nanoparticles film onto the substrate and collecting the samples with nanoparticles film afterwards. An appropriate combination of particle mass flow rate and the web speed is a crucial factor in this approach so as to control the organization of particles in the film. Moreover, binary suspensions of micron-scale particles with nanoparticles can result in highly ordered domains spanning large areas of the film. The mechanism of this process was investigated both experimentally and theoretically.

469736 Scalable Nanomanufacturing Platforms Enabled By Bottom-up Colloidal Self-Assembly

Abstract:

Photonic crystals and plasmonics are two key techniques that could ultimately enable all-optical integrated circuits and quantum information processing. Unfortunately, the development and implementation of these techniques have been greatly impeded by expensive and painstaking top-down nanomanufacturing approaches (e.g., electron-beam lithography). By contrast, bottom-up colloidal self-assembly and subsequent templating nanofabrication provide a much simpler, faster, and inexpensive alternative to nanolithography in creating 3-D highly ordered photonic crystals and plasmonic nanostructures. However, traditional colloidal self-assembly and templating nanofabrication technologies suffer from low throughput, incompatibility with standard microfabrication, and limited crystal structures, which greatly hamper the mass-production and on-chip integration of practical optoelectronic devices. Here we present two scalable bottom-up nanomanufacturing platforms for large-scale production of high-quality photonic crystals and a large variety of periodic metal nanostructures with tunable plasmonic properties for next-generation integrated nanooptics. Shear forces generated by a batch spin-coating technique and a continuous doctor blade coating process are utilized to align concentrated colloidal suspensions to rapidly form large-area photonic crystals with remarkably large domain sizes and unusual non-close-packed structures. These platform technologies combine the simplicity and cost benefits of bottom-up colloidal self-assembly with the scalability and compatibility of standard top-down microfabrication. Additionally, they enable scalable templating nanofabrication of a myriad of nanostructured functional materials ranging from plasmonic nanostructures to bioinspired broadband antireflection coatings.

453526 Effect of the Surface Charge Distribution on the Fluid Phase Separation of Charged Colloids and Proteins

Abstract:

Within the last decade, there has been an increasing interest in studying the phase behavior and self-assembly of molecules and particles with directional interactions such as proteins and "smart" colloids [1]. Significant progress has been achieved in the theoretical characterization of anisotropic molecules via "patchy" colloidal models with short-range, adhesive interactions [2]. Although such models successfully capture the general behavior of many colloidal systems, they fail to represent the behavior of other systems like proteins and para- or ferro-magnetic particles where long-range electrostatic interactions play a predominant role [3,4]. Here, a generic but simple colloidal model is presented to evaluate the effect of the heterogeneous surface charge distribution of proteins and zwitterionic nanoparticles on their rich thermodynamic phase behavior via Wang-Landau Transition Matrix Monte Carlo [5]. By considering surface charges as continuous "patches", the rich set of surface patterns that is embedded in proteins and charged patchy particles can readily be described. This model is used to study the fluid phase separation of charged particles where the screening length is of the same order of magnitude than the molecular/particle size. In particular, two type of charged particles are studied: dipolar fluids and protein-like fluids. The former represents the simplest case of zwitterionic particles, whose charge distribution can be described by the dipole moment. The latter system corresponds to molecules/particles with complex surface charge arrangements such as those in biomolecules. The results for both systems suggest a relation between the critical temperature, the strength of interparticle interactions, and the dispersion of charged patches, where the critical temperature is strongly correlated to the magnitude of the dipole moment. Additionally, competition between attractive and repulsive charge–charge interactions seems to be related to the formation of transient clusters along the dilute phase in dipolar fluids, as well as to the broadening of the binodal curve in protein-like fluids. Finally, a variety of self-assembled architectures are detected for dipolar fluids upon small changes in the charge distribution, providing the groundwork for studying the self-assembly of charged patchy particles.

[1] S. Sacanna et al., Soft Matter, 2013, 9, 8096

[2] E. Zaccarelli, J. Phys.: Condens. Matter, 2007, 19, 323101

[3] F. Kogler et al., Soft Matter, 2015, 11, 7356

[4] D. Staneva and D. Frenkel, J. Chem. Phys., 2015, 143, 194511

[5] V. K. Shen and D. W. Siderius, J. Chem. Phys., 2014, 140, 244106

458849 Industrial Viability Study of the Synthesis of Nanoparticles for Use As Lubricant Additives

Abstract:

With an ever increasing demand for fossil fuels and an ever decreasing supply, global environmental regulations continue to push for higher fuel economy and lower emission in all forms of vehicular transportation. This push has created the need for new engine oils that can produce better fuel economy and wear properties, while operating under harsher conditions and at lower viscosities. Traditionally, organic friction modifiers like glycerol monooleate (GMO) or molybdenum dithiocarbamate (MoDTC) have been used to reduce engine oil friction. More recently, researchers have shown that inorganic nanoparticles of cerium (Ce), zinc (Zn), or alloys of Ce and Zn that have been modified to be oil-soluble and self-dispersing can be used to reduce friction in engine oils (SAE Technical Paper 2008-01-2460 (2008)) . In this talk, we will review a variety of techniques for synthesizing these nanoparticles and discuss the feasibility of scale up from the lab bench to the industrial scale.

469896 Dual-Responsive Block-Arm Star Copolymers for on-Demand Reduction or Enhancement of Friction and Adhesion

Abstract:

Achieving low friction between moving parts undergoing sliding contact is of considerable and obvious importance in the efficient operation of machinery. It is equally desirable for the comfortable, low wear functioning of natural or artificial joints in the body. The importance of achieving low friction has motivated many studies of novel lubricating systems, including boundary lubrication by swollen polymer brushes. The attainment of very low friction coefficients by planar brushes in a good solvent is well established.Â There are situations where higher friction is the desirable condition. For example, greater traction between surfaces requires larger frictional forces. Consumer perception of the â€œfeelâ€ of a surface is influenced by friction, and friction among suspended particles has been implicated as a possible source of shear thickening rheology.Â While it is possible to modulate the friction between planar brushes for a given load by modulating solvent quality, the dynamic range of accordingly modified frictional forces that have been reported is not large. We have developed a system based on adsorption of responsive block-arm star copolymers as friction control agents. These are designed to be triggered to reversibly decrease or increase friction in response to pH changes. As such, they include a weak polycation block among the arms. The magnitude of the frictional dynamic response to stimuli can be manipulated by controlling the non-equilibrium packing of the adsorbed layer, for example by controlling temperature during the initial adsorption of the star copolymers to the surfaces. An approximately two order of magnitude change in friction can be triggered for a given star copolymer system, with load-dependent friction forces that exceed those of the bare surfaces under one condition, transitioning to frictional forces much weaker than those of the bare surface under changed conditions. The response is based in part on the change in apparent friction coefficient and in part on the change in adhesion.

The manipulation of adhesion is particularly important. The full dynamic range is tuned by adjusting the ability of star copolymer arm segments to bridge between opposing surfaces. This takes advantage of the ability of adsorbed star copolymer arms to adapt their chain extensions both normal and parallel to the underlying surface. Star copolymers adsorbed at high coverage may fully block bridging access to arms emanating from star copolymers adsorbed to the opposing surface when the stars are swollen under lower pH conditions that yield strong polycation charging. Increasing pH to de-swell the star copolymers exposes underlying surfaces to bridging adhesion and large friction. A series of block-arm star copolymers was synthesized, consisting of an innermost temperature-responsive block of poly(di(ethylene glycol) methylether methacrylate), a second block of temperature- and pH-responsive poly(2-(dimethylamino)ethyl methacrylate), and in some cases an outermost zwitterionic poly(2-methacryloyloxyethyl phosphorylcholine). The first two blocks are surface active on silica, the model surface for this study, and provide stimulus response. The latter block is not pH- or temperature-triggerable under the range of conditons examined here, and provides a high degree of hydration. The bulk swelling characteristics of these materials are monitored by dynamic light scattering, and adsorption is monitored by quartz crystal microbalance with dissipation and/or ellipsometry. Normal and tangential forces between opposing surfaces decorated by adsorbed star copolymers are measured by colloidal probe atomic force microscopy. This presentation will focus on the exploitation of thermal responsiveness to control adsorbed layer packing, the subsequent reversible manipulation of star copolymer swelling and bridging induced by pH changes and the resulting quantitative and qualitative changes in dynamic friction forces.

467411 Self-Assembly of Thermoresponsive Colloidal Clusters of Patchy Particles Via Capillary Binding

Abstract:

Janus and patchy particles are emerging as models for studying complex directed assembly patterns and as precursors of new structured materials and composites. Here, we introduce lipid induced capillary bridging as a new and nonâ €‘conventional method of assembling patchy particles into ordered structures. We used the selective wetting of iron oxide surface patches on latex microspheres to drive the patchy microsphere assembly into 2D and 3D clusters via interparticle capillary bridge formation. The liquid nature of the bridges allows local reorganization of the particles within the clusters and assists in forming *true* equilibrium configurations. The temperature driven fluid-to-gel and gel-to-fluid phase transition of the fatty acids within the bridge acts as a thermal switch for cluster assembly and disassembly. By comparing our experimental data to Monte-Carlo simulations we show that the equilibrium cluster morphology is determined by the patch size, number, and shape. The results demonstrate the ability of capillary bridging as a unique tool to assemble thermoresponsive clusters and aggregates. This simple method of binding particles is very robust and generic, and it can be extended further to assemble particles with non-spherical shapes and complex surface chemistries leading to the realization of colloidal clusters of sophisticated structure.

468099 Synchrotron X-Ray Reflectivity and Computational Simulations of Structural Organization of the Binary Mixture 5CB/8CB Liquid Crystals at Liquid Crystal-Air Interface

Abstract:

Eutectic mixtures of liquid crystals are commercially important because of their extended operating temperature range in the nematic phase. However, little is known about the precise molecular structure of these materials at their free surfaces. In this work, we have performed synchrotron X-ray reflectivity measurements accompanied by an advanced theoretical and computational analysis to study the structural organization of liquid crystals at the air-liquid crystal interface. The X-ray reflectivity was measured from two nematic (5CB) and smectic (8CB) liquid crystals and their binary mixtures at several temperatures, in the nematic phase and above the nematic-isotropic transition. Our computational simulations and X-ray reflectivity results indicate that in the case of 8CB nematic phase, incipient bulk smectic fluctuations are induced at the interface to form temperature-dependent multilayers at the interface. Such layers can extend far from the interface. However, the interface of 5CB in the nematic phase exhibits a relatively small number of layers. For the binary mixture 5CB/8CB liquid crystals, our results show that the molecular organization depends strongly on the 5CB content. Above 20% 5CB, the smectic properties of 8CB are suppressed and the surface induced smectic layering and the correlation length of the liquid crystal order decrease dramatically.

472241 Interfacial Properties of Nanoconfined Ionic Liquids

Abstract:

Ionic liquids (ILs) are a relatively new class of liquids that consist only of cations and anions. Negligibly low vapor pressure, fire resistance, excellent chemical and thermal stability, wide liquid temperature ranges, and wide electrochemical windows are examples of the useful properties of typical ILs. Because of these excellent properties, ILs have been used or considered for use in organic synthesis, catalysis, chemical separation, fuel and solar cells, and their applications continue to expand. In spite of a plethora of emerging processes involving ionic liquids, the understanding of interfacial properties of ILs under confinement such as rheological properties and glass transition temperatures, Tg is still very limited compared to that of other confined material classes such as polymers, electrolyte solutions, and liquid crystals. In this work, we studied the rheological and thermal properties of ILs under nanoconfiment using the surface forces apparatus (SFA) and template-assisted difference scanning calorimetry (DSC). It was found that Newtonian behavior of IL breaks down when the separation between shearing surfaces fall below 15 nm. The ILs with larger molecular size of cation (more number of carbons in cation) showed a higher Tg in bulk. The Tg of the confined ILs was found to be strongly dependent on the pore sizes regardless of types of ILs studied and such a dependence appeared to be relatively more pronounced (e.g. sharper increase in Tg as decreasing a pore size) for the ILs with smaller cation size. These results can be interpreted based on the surface-enhanced molecular organization/ordering under nanoconfinement that requires more thermal energy to induce a liquid state, thereby increasing Tg.

459852 Thermodynamic Models for Solid – Liquid – Gas Equilibrium of CO_2 mixtures

Abstract:

Fossil fuels are currently the most widely used sources for power and heat generation. Their extensive consumption contributes significantly to the increased levels of greenhouse gases in the atmosphere, which subsequently leads to environmental problems such as global warming. The most important greenhouse gas, in terms of quantity and impact, is carbon dioxide (CO_2). As the global energy demand has increased, CO_2 levels have risen significantly, from the preindustrial levels of 280 ppm to 405 ppm in March 2016. Moreover, fossil fuels will continue to play an important role in power and heat production and also be used in large industrial operations in the foreseeable future. Unless major measures are taken for the reduction of CO_2 emissions, the CO_2 concentration is projected to rise even more over the next 25 years as global demands for energy are anticipated to increase.

Significant amount of research has been conducted for the development of new technologies that aim to reduce the levels of CO_2 in the atmosphere. The most mature technology today is Carbon Capture and Sequestration (CCS), which is the process of capturing CO_2 from the flue gas of a large point source (typically a power plant), transporting it to a sequestration site and then depositing it to a geological formation, which can be a saline aquifer or a depleted oil well. The CCS process can be divided into three main parts: CO_2 capture, transport and storage. There are several different methodologies for the capture process, including pre-combustion, post-combustion and oxy-combustion process. These processes result in streams that contain different gases such as N_2, CH_4, O_2, Ar, SO_2, H_2S and H_2 at various concentrations.

An important part of the CCS process is the transportation of the CO_2-rich stream from the capture plant to the sequestration site. In most cases, transporting CO_2 via pipelines is the most cost effective way of transport. Preliminary conceptual design, detailed design, simulation and optimization of the transport process require, among others, accurate knowledge of the physical properties of the chemical system involved as a function of temperature, pressure and composition. Quite often, the system exists in more than one phase (i.e., liquid, vapor and/or solid) and as a result process design calculations have to take into account the phase equilibrium conditions and also the composition of the relevant phases and the respective physical property values. The two challenges that arise are the accurate prediction or correlation of the physical properties of the system and the conditions of instability, where the system is going to split into two or more coexisting phases.

Vapor – liquid equilibrium (VLE) of CO_2 mixtures has attracted lots of attention both in terms of experimental measurements and modeling using equations of state (EoS). On the other hand, relatively little work has been performed to measure and predict solid-fluid equilibrium (SFE) of CO_2 mixtures, which is critical to the design and operation of CO_2 pipelines and storage facilities. Hazard assessment studies associated with CO_2 transport include scenarios of accidental releases with sharp expansion, where solid – vapor (SVE) as well as solid – liquid (SLE) equilibria may occur. CO_2 exhibits a relatively high Joule – Thomson expansion coefficient and a pipeline depressurization will lead in rapid cooling which can reach very low temperatures and as a result, solid formation can be expected. Taking this into account, it is easily understood that the formation of dry-ice resulting from SFE can largely affect the safety of CCS facilities during equipment depressurization, process shutdown or other process upsets.

In this work, solid phase thermodynamic models of different complexity are applied to model the SFE of pure CO_2 and also of CO_2 mixtures with other compounds. These models include an empirical correlation model, a model based on thermodynamic integration and a solid phase EoS for pure CO_2. The different models are coupled with three fluid phase EoS (PR, SRK and PC-SAFT) and the performance of each combined model is evaluated for various binary mixtures. In total, 7 different models are examined i.e. the empirical correlation and the thermodynamic integration models each one coupled with SRK, PR and PC-SAFT EoS and an EoS developed for solids coupled with an EoS developed for fluids. The solid phase EoS used for this purpose was the EoS proposed by Jäger and Span (*J. Chem. Eng. Data,* 57, 590, 2012). Finally, in order to examine the general validity of the proposed scheme, mixtures where the solid forming component is not CO_2 are examined.

Our calculations reveal that a model that successfully reproduces the pure solid former triple or normal melting point will predict more accurately the solid – liquid – gas (SLG) locus of the mixture. In this context, the thermodynamic integration model and the Jäger and Span EoS provide in general better predictions of the SLGE, when all the binary interaction parameters (BIPs) are zero for the mixtures of CO_2 with N_2 and H_2. For these two mixtures, the empirical correlation model for the solid phase is comparable to the other two models only when coupled with PC-SAFT which accurately reproduces the pure CO_2 triple point. The use of BIPs, regressed from binary VLE data at low temperature, significantly improves the prediction of the SLG behavior for most models. All models provide very similar results for the mixture of CO_2 with CH_4 and very low deviations have been achieved with the use of BIPs regressed from binary VLE data over a wide temperature range. The use of these BIPs also enabled a successful unified description of both the SLGE and the LG critical locus of the mixture with the PR EoS.

The thermodynamic integration model shows clear superiority in the mixtures where naphthalene and phenanthrene form the solid phase, where the successful representation of the normal melting point of the heavy compound is crucial for the overall fit of the SLG line. The empirical correlation model in these cases is unsuccessful in reproducing the normal melting point of the pure component except when coupled with SRK EoS for the two mixtures of naphthalene. For the mixture of phenanthrene, the overall fit of the SLGE is very poor with this model, but a satisfactory correlation of the behavior was achieved with the thermodynamic integration model coupled with PC-SAFT.

452745 High Pressure Multiphase Equilibria Modeling with GCA-EoS

Abstract:

The Group-Contribution with Association equation of state (GCA-EoS) has been successfully applied to represent the global phase behavior of carbon dioxide mixtures with normal and branched alkanes and alcohols under a wide range of conditions. Using a single set of parameters, the model is able to predict vapor-liquid, liquid-liquid and solid-liquid equilibria of binary mixtures not included in the parameterization procedure. In addition, liquid-vapor and liquid-liquid critical locus were included in the optimization. Special emphasis was set on describing the transformation between types of fluid phase behavior in each homologous series of organic compounds to reduce the jeopardy of predicting incorrect liquid split. This family of binary systems has been investigated in previous works; however, this is the first time that GCA-EoS was shown to be able to represent the transformation between types of phase behavior as the length of the alkyl chain increases. The wide range of conditions and the parameterization strategy are keys to develop a robust thermodynamic tool for predicting multiphase behavior.

The GCA-EoS is the first equation of state of the SAFT family that uses a group contribution approach of the Wertheim association thoery. The free volume and attractive contributions are based on Carnahan-Starling and Non Random Two Liquids (NRTL) models respectively. The density-dependent mixing rules of this model are laborious to fit. However, they give enough flexibility to reproduce the global phase diagrams without affecting the accuracy of bubble and dew curves calculations. It is worth to highlight that the same set of parameters accomplishes accurate results, both in subcritical and critical regions, for the evaluated homologous series. Prediction of multiphase behavior with a single set of parameters is a key point when developing a model for new processes and products design.

464266 Modeling of Near-Critical and Supercritical Properties Via the Virial Equation of State

Abstract:

The virial equation of state is a foundational method in the statistical mechanics of fluids. It is unique in its ability to provide a thermodynamic equation of state with parameters that are rigorously given in terms of a detailed molecular model. As such, it has capability as a modeling method of real predictive capability, in some cases rivaling experiment in its accuracy. It also provides a valuable tool for assessing strengths and weaknesses in molecular models.

We consider application of the virial equation of state to the description of supercritical and near-critical fluid systems, considering both pure substances and mixtures. We report coefficients up to seventh order, computed using Mayer sampling Monte Carlo and Wheatleyâ€™s recursive method for cluster summation, for a variety of realistic molecular model potentials. These results are used to generate expressions for volumetric (e.g., pressure) and thermal (e.g., heat capacity) properties, and others derived from them (e.g., speed of sound), as a function of density and temperature. We also examine prediction of the critical point and (for mixtures) critical lines. The ability to describe properties accurately from the virial coefficients is enhanced through the application of an approximant that is formulated to have correct scaling in the critical region. The property calculations are tested against available data from molecular simulation for the same model potentials, and from experiment.

470518 Molecular Simulation of Double Retrograde Vaporization

Abstract:

Liquid dropout due to retrograde condensation during production of rich natural gas leaves valuable condensate fluids in the oil reservoirs, leading to significant loss of well productivity and potentially unfeasible recoveries. Thus, accurate modeling of fluid properties across a wide range of conditions is vital for the design and optimization of enhanced oil and gas recovery processes. In this work, Gibbs-ensemble Monte Carlo simulations using transferable force fields are applied to investigate double retrograde vaporization in fluid mixtures near the critical point of the lower-boiling compound. Here, methane/n-butane, carbon dioxide/n-alkane, and carbon dioxide/(benzene or toluene) mixtures are discussed. The molecular simulations allow one to provide molecular-level understanding of the phase behavior of these mixtures at relevant reservoir operation conditions.

462872 Modelling of Phase and Interfacial Behavior of Ternary Mixtures

Abstract:

Modelling of Phase and Interfacial Behavior of Ternary Mixtures

Sabine Enders

Karlsruhe Institute of Technology, Institute of Technical Thermodynamics and Refrigeration Engineering, Engler-Bunte-Ring 21, D-76131 Karlsruhe, Germany.

Email: sabine.enders@kit.edu

Key words: vapor-liquid equilibria, oil recovery, density gradient theory, interfacial tension, density profiles, PCP-SAFT-EOS

Abstract:

My stay at the TU Delft in the group of Applied Thermodynamics in 1996 inspired me to study interfacial phenomena. The reason for this situation was the well-known paper by Cornelisse, Peters and de Swaan Arons [1] dealing with the application of the density gradient theory in combination with the Peng-Robinson Equation of state (PR-model). In this paper, is written â€œHowever, in the previous section it was concluded, that the scaling behavior of gradient theory with the PR model did not agree with the experiments. Although from the gradient theory interfacial tension calculations have been illustrated by examples for pure the carbon dioxide and carbon dioxide/butane system, the conclusions are believed to be generally applicable. This was assumed, because the error in the results is probably caused by the PR model. In any event, the results presented here demonstrate that accurate predictions will at least require an equation of state with proper scaling behavior in the critical region.â€œ

In the last twenty years, I studied phase- and interfacial properties related to different phase equilibria (liquid-liquid [2] or liquid-vapor [3]), different binary [4] and ternary mixtures [5] and the application of different equations of state (SAFT [5], PCP-SAFT [6,7], lattice cluster theory [8], Sanchez-Lacombe [2]).

The interfacial properties play an important role in the field of oil recovery and in separation technology, especially for extraction. In this contribution the phase equilibria and the involved interfacial properties for binary and ternary mixture representing the situation for oil recovery will discussed, where the focus is the dependence of interfacial properties, like interfacial tension and relative enrichment in the interface, on pressure. Additionally, the impact of sulphur containing components, like H_2O and SO_2, on the interfacial tension will be discussed [9]. For the investigation of this effect, thermodynamic modelling is very important, because no experimental data are available in the public literature. The following ternary mixtures: CO_2+H_2S+CH_4, CO_2+H_2S+C_7H_{16}, CO_2+SO_2+CH_4, CO_2+SO_2+C_7H_{16} were investigated over a large pressure and temperature range. It turns out that the presence of the sulfur containing molecules leads to a decrease of the interfacial tension in comparison to the corresponding binary system [9] at the same temperature and pressure. At low pressures, CO_2 as well as H_2S will enriched at the interface in the CO_2+H_2S+C_7H_{16} mixture, where the enrichment effect of CO_2 is more pronounced.

References

[1] P.M.W. Cornelisse, C.J. Peters, J. de Swaan Arons, Fluid Phase Equilibria 82 (1993) 119.

[2] S. Enders, K. Quitzsch, Langmuir 14 (1998) 4606.

[3] O.G. Nino-Amezquita, S. Enders, Ph. T. Jaeger, R. Eggers, Ind. Eng. Chem. Res. 49 (2010) 592.

[4] S. Enders, H. Kahl, Fluid Phase Equilibria 263 (2008) 160.

[5] H. Kahl, S. Enders, Fluid Phase Equilibria 172 (2000) 27.

[6] E. Schäfer, F. Horbach, S. Enders, Journal of Chemical & Engineering Data 59 (2014) 3003.

[7] E. Schäfer, S. Enders, G. Sadowski, Fluid Phase Equilibria 362 (2014) 151.

[8] T. Zeiner, P. Schrader, S. Enders, D. Browarzik, Fluid Phase Equilibria 302 (2011) 321.

[9] A. Danzer, C. Bühl, S. Enders, Fluid Phase Equilibria 416 (2016) 94.

470776 Interfacial Investigations of Gas Hydrates in Flow Assurance & Energy Applications

Abstract:

Interfacial Interactions of Gas Hydrates in Flow Assurance and Energy Applications

Ahmad A.A. Majid, Erika Brown, Vishal Srivastava, Jose Dapena, M. Naveed Khan, Thomas Charlton, Carolyn A. Koh

Colorado School of Mines, Center for Hydrate Research, Golden, CO 80401, USA

Abstract

Natural gas hydrates are crystalline guest-host structures that form in the presence of water and small gas molecules at high pressure and low temperatures. The formation of gas hydrates in deepwater flowlines can lead to blockages of the flowlines and hence significant potential safety and environmental risks. On the other hand hydrates have a number of interesting energy related technological applications, including energy transportation and storage. Understanding the interfacial interactions of gas hydrates in multiphase systems is critical to being able to control and tailor hydrate formation in flow assurance or energy applications. This paper presents the application of a micromechanical force apparatus to measure hydrate particle-particle interactions and rheological and flowloop tests to investigate the hydrate slurry properties. Reducing particle-particle interactions can lead to the production of a transportable hydrate slurry, which is an important factor when assessing hydrate blockage mitigation strategies for fully and partially dispersed systems, as well as the continuous production of hydrates for different energy applications.

474935 The Prediction of Environmentally Important Properties for Hazardous Chemicals

Abstract:

In this talk the successes in prediction of environmentally important properties of hazardous chemicals using modern computational methods will be viewed. Some of the properties considered are solubility in water, Henryâ€™s law constants, octanol-water partition coefficients, and soil-water partition coefficients. The species include industrial chemicals, explosives and chemical warfare agents.

454967 Corresponding States Theory for Property Prediction at the Nanoscale

Abstract:

The Principle of Corresponding States was first proposed by van der Waals in 1881, based on his equation of state. The Principle made it possible to predict the vapor pressure curve, including the critical and normal boiling points, of liquids for which no measurements exist, based on experimental data for other liquids. It quickly found use in this way for predicting liquid properties for hydrogen and helium, neither of which had been liquefied at the time. When these light gases were later liquefied the predictions were found to be accurate. A more powerful derivation of the Principle is based on statistical thermodynamics, which makes clear the assumptions involved and provides a path to improve the theory so that it is more generally applicable. Subsequent generalizations of the theory of corresponding states, in particular that due to Pitzer, provide the basis for the prediction of many physical properties of homogeneous gases and liquids, and are widely used in process simulation. However, these correlations are not applicable to highly inhomogeneous or nanoscale systems, such as small drops, nanoparticles, adsorbed thin films on solid substrates, polymer brushes, etc.

A statistical mechanical analysis of the properties of a highly inhomogeneous nanoscale system shows that the equilibrium properties (adsorption isotherms, isosteric heats, phase change conditions, etc.) are determined by a small number of dimensionless parameters, including the dimensionless width, H, of the nanophase and a *microscopic wetting parameter*, α_w, that is defined in terms of intermolecular forces and solid structure; α_w is a measure of wetting that applies at all scales and for any kind of adsorbed film (gas, liquid or solid) [1,2].

We illustrate the usefulness of this approach using examples drawn from both experimental and molecular simulation studies of gas-liquid and liquid-solid phase separations [3,4], and pressure enhancement in nanopores [5-7], with emphasis on simple pore geometries. These examples illustrate the central role

played by wetting, and also the breakdown of some concepts and macroscopic laws, such as the equations of Kelvin, Laplace and Gibbs-Thomson, for nano-phases confined within small pores. They also suggest consistency tests for experimental data that should be applied in the event of capillary condensation or other phase changes in mesoporous materials.

The approach can be extended to account for geometric or energetic wall roughness [8], and to adsorbate molecules that interact with site-site potentials having both Lennard-Jones and point charge sites.

1. R .Radhakrishnan, K.E. Gubbins and M. Śliwinska-Bartkowiak, "Global Phase Diagrams for Freezing in Porous Media", *Journal of Chemical Physics*, **116**, 1147-1155 (2002).
2. Keith E. Gubbins, Yun Long and Malgorzata Sliwinska-Bartkowiak, "Thermodynamics of Confined Nano-Phases", *Journal of Chemical Thermodynamics*, **74**, 169-183 (2014).
3. L.D. Gelb, K.E. Gubbins, R. Radhakrishnan and M. Sliwinska-Bartkowiak, "Phase Separation in Confined Systems", *Reports on Progress in Physics*, **62**, 1573-1659 (1999).
4. Ç. Alba-Simionesco, B. Coasne, G. Dosseh, G. Dudziak, K.E. Gubbins, R. Radhakrishnan and M. Śliwinska-Bartkowiak, "Effects of Confinement on Freezing and Melting", *Journal of Physics: Condensed Matter*, **18**, R15-R68 (2006).
5. Yun Long, Jeremy C. Palmer, Benoit Coasne, Małgorzata Śliwinska-Bartkowiak and Keith E. Gubbins, "Pressure enhancement in carbon nanopores: A major confinement effect", *Physical Chemistry Chemical Physics*, **13**, 17163-17170 (2011).
6. H. Drozdowski, M. Kempinski, M. Śliwinska-Bartkowiak, M. Jazdzewska, Y. Long, J.C. Palmer and K.E. Gubbins, „Structural Analysis of the Behavior of Water Adsorbed in Activated Carbon Fibers", *Physical Chemistry Chemical Physics*, **14**, 7145-7153 (2012).
7. Yun Long, Jeremy C. Palmer, Benoit Coasne, Małgorzata Śliwinska-Bartkowiak, George Jackson, Erich A. Müller and Keith E. Gubbins, "On the Molecular Origin of High Pressure Effects in Nanoconfinement: Effects of Surface Chemistry and Roughness", *Journal of Chemical Physics*, **139**, 144701 (2013).
8. M. Śliwinska-Bartkowiak, A. Sterczyńska, Y. Long and K.E. Gubbins, "Influence of Microroughness on the Wetting Properties of Nano-Porous Silica Matrices", *Molecular Physics*, **112**, 2365-2371 (2014).

464668 Heat Induced and UV Induced Grafting of Poly(glycidyl methacrylate) on PBT Nonwovens for Bioseparations

Abstract:

Polybutylene terephthalate (PBT) nonwovens were successfully grafted with poly(GMA) using a heat induced grafting (HIG) approach using the thermal initiator benzoyl peroxide. The grafting method resulted in uniform and conformal grafted layers around the PBT fibers that were functionalized as anion and cation exchangers for protein capture. Equilibrium protein binding capacities as high as 200 mg of protein per gram of fabric were observed. The equilibrium binding capacities for protein adsorption to the HIG grafted materials were reached within minutes, compared to UV grafted polyGMA ion exchange fabrics which reached equilibrium protein binding capacities in hours, even though they had the same weight % of grafted layers. However, UV grafted ion exchange nonwoven fabrics were capable of binding between 5 to 7 times more protein per mass of fabric at equilibrium than the HIG materials. It was found that the HIG materials showed a decrease in binding capacity with increased molecular weight of the solute that was much steeper than the drop observed with UV grafted materials. These observations indicate that even though the weight % of grafting is the same for HIG and UV grafting, the HIG grafted layers are significantly thinner and have a lower porosity in solution than the UV grafted brush layers.

466177 Thermodynamics and Statistical Mechanics of Drying at the Nano-Scale

Abstract:

Hydrophobically-induced drying transitions at the nano-scale are believed to play an important role in protein folding and transport in ion channels. In this talk I will present recent results on the thermodynamics and statistical mechanics of drying transitions in the presence of flexible hydrophobic substrates. The results suggest that substrate flexibility can play a significant role in both the thermodynamics and kinetics of this process, increasing both the driving force and the rate with respect to the corresponding rigid substrate reference case. The calculations to be discussed include analytical derivations for the underlying thermodynamics, as well as path-sampling simulations for rate calculations and free energy barrier determination. The path-sampling simulations also offer useful insights into the evolution of the evaporation mechanism as a function of substrate flexibility.

471679 Harmonically-Mapped Averaging Applied to Lennard-Jones Crystal Phase

Abstract:

Mapped averaging is a technique we developed to improve ensemble averages during a molecular simulation[1]. When applied to crystals, we map based on the behavior of a harmonic description of the solid and compute the anharmonic contribution to properties of interest with high precision. We can then add in the harmonic contribution and obtain the actual property value. The anharmonic contribution is insensitive to system size and truncation effects while the harmonic contribution can be computed for very large systems with long truncation. The resulting property values are both very precise and accurate with relatively small computational effort.

We have applied mapped averaging to collect potential energy and pressure data for the Lennard-Jones model in the FCC and HCP phases. We fit an equation of state to each set of data to obtain a description of both phases from low to high density (the soft-sphere limit) and up to melting temperatures. We find the HCP phase to be more stable at low temperatures and densities. Coupled with the virial equation of state for the vapor and an equation of state fit to more limited simulation data for the liquid phase, we predict the phase boundaries including the triple point with unprecedented accuracy and precision. We also discuss the effect of vacancies on the properties and phase diagram.

(1) Moustafa, S. G.; Schultz, A. J.; and Kofke, D. A. Very fast averaging of thermal properties of crystals by molecular simulation. Phys. Rev. E 2015, 92, 043303.

458803 Computational-Hydrodynamic Studies of the Noh Compressible Flow Problem Using Non-Ideal Equations of State

Abstract:

The â€œNoh problemâ€ is classic verification problem in the field of computational hydrodynamics of compressible flows. A strong outward facing shockwave is created by impinging a uniformly flowing ideal gas on a hard wall (or imploding it onto the axis of a cylinder or the center of a sphere, depending on the geometry of interest). A simple problem to conceptualize and one which admits an exact analytical solution, it is nonetheless devilishly difficult for numerical codes to predict correctly, making it an ideal code-verification test bed. In its original incarnation, and in nearly all applications since, the fluid is a simple ideal gas; once validated, however, these codes are often used to study highly non-ideal fluids and solids. In this work the classic Noh problem is extended beyond the commonly-studied polytropic ideal gas to more realistic equations of state (EOS) including the stiff gas, the Nobel-Abel gas, and the Carnahan-Starling hard-sphere fluid, thus enabling verification studies to be performed on more physically-realistic fluids. Self-similarity methods are used to solve the Euler compressible flow equations, which in combination with the Rankine-Hugoniot jump conditions, provide a tractable general solution. In the planar case, this solution can be applied to any equation of state and does not necessarily have to exhibit strong shocks; for cylindrical and spherical geometries it is necessary that the analysis be restricted to strong shocks. The exact solutions are compared with numerical results obtained from the Lagrangian hydrocode FLAG, developed at Los Alamos. For these more realistic EOSs, the simulation errors decreased in magnitude both at the origin and at the shock, but also spread more broadly about these points compared to the ideal EOS. The overall spacial convergence rate remained first order. (LA-UR-16-23103)

465378 Lattice Models, Polymers and DNA Topology

Abstract:

The standard Statistical Mechanics models for linear and ring polymers in dilute solution are respectively the self-avoiding walk (SAW) and self-avoiding polygon (SAP) lattice models. For a SAP model, a ring polymer configuration is represented by a random polygon on a lattice such as the simple cubic lattice, and under good solvent conditions, each equal-length polygon is considered to be an equally likely configuration. These models have the advantages that the excluded volume property is easily incorporated and that combinatorial and asymptotic analysis is possible; they have proved to be very useful for understanding critical phenomena, such as phase transitions, for polymer systems. At the same time, despite the simplicity of the models, there are many challenging questions that remain open about them. In this talk, I will review progress made regarding using SAP models to address questions about the entanglement complexity (knotting and linking) of polymer systems. The work has been motivated in part by experimental studies of: (i) enzyme action on DNA, where observed knot and link frequencies characterize the action of the enzyme, and (ii) DNA confined to viral capsids, where observed knot frequencies characterize how the DNA is packed in the capsid.

464879 Design of Macromolecular Biomaterials Using Atomistic and Coarse-Grained Molecular Simulations

Abstract:

In this talk, I will present our recent work using molecular simulations to design novel oligonucleic acids (ONAs) with varying backbone chemistries to tune ONA hybridization and melting. Since double stranded DNA (ds-DNA) is the basis of various bio- and nano- technologies, the need for cheaper alternatives and significant synthetic advances have led to the design of DNA mimics with new backbone chemistries (e.g. click nucleic acids, locked nucleic acids). A fundamental understanding of how these backbone modifications in the oligo-nucleic acids impact the hybridization and melting behavior of the double strands is still lacking. We use coarse-grained (CG) and atomistic molecular dynamics simulations along with well-tempered metadynamics calculations to capture the effects of varying backbone chemistries on the H-bonding between complementary nucleobases and the intra-strand base-base stacking, and ONA hybridization/melting. I will also present our recent results on how conjugation of these novel ONAs with biocompatible polymers (e.g. PEG) impacts the polymer-conjugated ONA hybridization. These polymer-conjugated ONAs will serve as building blocks for thermo-responsive gels and networks in biomaterials.

468481 Rational Design of Peptide-Based Stimuli-Responsive Biomaterials Via Multiscale Modeling

Abstract:

Dynamic, self-assembling peptide-based nanomaterials that can change shape and structure in response to environmental stimuli hold promise to revolutionize medicine and biotechnology. However, the current discovery process is slow and often serendipitous due to the enormously large design space and lack of systematic knowledge as well as predictive models. To tackle the challenge and harness the full potential of â€œsmartâ€ materials, we have built a platform of computer-aided design of nanomaterials by performing *in silico* discovery in collaboration with *in vitro* and *in vivo* experimentalists in facilitating rapid development of novel stimuli-responsive nanomaterials for biomedical applications. Towards this goal, we integrate multi-scale modeling techniques to design and discover novel peptide-polymer conjugates called peptide amphiphiles (PA, comprised of a bioactive peptide sequence and an alkyl tail) that can self-assemble into nanostructures upon a change in a physiological stimulus. We have developed a fundamental understanding of how environmental conditions impact the formation, functionality and performance of these biomaterials.

464886 Dynamic Self-Assembly of Human Cells Using Inertial Microflow

Abstract:

The inertial migration phenomenon of particles in a dilute suspension can be utilized to place the particles at precise positions inside a conduit. In microchannels, particles self-assemble into ordered trains at inertial focusing positions close to the channel walls. Therefore, engineering the inertial fluid flow at microscale using the inertial microfluidic platform has emerged as a powerful method for dynamic self-assembly of micron-sized particles. Although the number and location of inertial focusing positions can be controlled by changing the aspect ratio of the channel and fluid inertia, engineering the spacing between particles necessitates understanding the hydrodynamic interaction between particles. Forming ordered trains of human cells inside the microchannel poses new challenges due to polydispersity and deformability of cell populations. Regardless of the challenges in controlling the space between cells in microchannel trains of particles, dynamic self-assembly of cells comprises a wide range of important applications such as single cell droplet encapsulation, flow cytometry and epigenomics. In the present work, we employ microchannel experiments and lattice-Boltzmann simulations in order to study the dynamic self-assembly of a multi-modal suspension of hard spheres and human cancer cells. It will be discussed that particles equilibrate at preferred spacings, or attractors, from one another in their inertial flow inside the channel. We demonstrate the eect of ow inertia, the number density of particles in the suspension and the channel dimensions on the location and the number of attractors.

465579 Theoretical Model for the Aspect Ratio-Based Separation of Ellipsoidal Particles in Field-Flow Fractionation

Abstract:

We have developed a new theoretical model for describing the separation behaviors of ellipsoidal particle in flow field-flow fractionation (Fl-FFF), which is a particle separation technique which utilizes a parabolic channel flow and a cross-flow field. Previous model on the aspect ratio-based separation of rod-like particles [Alfi & Park, J. Sep. Sci. (2014) 37:876-883; Park & Mittal, Chromatography (2015) 2: 472-487], has been extended for prolate ellipsoids, which has a lower aspect ratio of less than 5. As in the previous model, the cross-sectional concentration distribution is derived considering the particle orientation and the configuration near the accumulation wall. The derived concentration distribution is then used for the evaluation of the retention ratio, which is equivalent to the normalized elution time. The effects of the aspect ratio, axial flow rate, and cross flow rate on the retention ratio will be presented to explain the elution order. Comparison with the experimental data will also be discussed.

468616 A Library of Soft Particles for the Study of Lateral Migration in Low Reynolds Number Flow

Abstract:

The cross stream distribution of suspension components in pressure-driven flow in a channel depends on a range of factors, including particle size, shape, deformability, volume fraction, and Reynolds number[1]. However, how each of these factors results in lateral migration is not well understood. In this work, we attempt to elucidate the roles of particle deformability and shape on lateral migration in low Reynolds number, pressure-driven channel flows through the use of model suspensions. Deformable, uniformly gelled alginate microspheres, particles with soft cores, and capsules, all with a broad range of deformabilities, were generated using a modified flow-focusing microfluidic device. Particle deformability was then quantified using capillary micromechanics in a tapered capillary[2]. Average particle shear moduli range from 6 ± 1 kPa to 26 ± 8 kPa for the solid microspheres and from 0.6 ± 0.1 kPa to 3.1 ± 0.5 kPa for the particles with soft cores. An assortment of non-spherical alginate particles was also generated by allowing alginate emulsion drops to fall into an external bath containing crosslinking ions[3]. By changing the properties, e.g. viscosity, of the bath, the shapes of the resulting particles were altered. This library of particles is currently being used for lateral migration studies in long channels, the results of which will be discussed.

References:

(1) Kumar, A. and Graham, M. D. *Soft Matter,* **2012**, *8*, 10536.

(2)Â Guo, M. and Wyss, H. M. *Macromol. Mater. Eng.* **2011**, *296*, 223â€"229.

(3) Hu, Y.; Wang, Q.; Wang, J.; Zhu, J.; Wang, H.; Yang, Y. *Biomicrofluidics*. **2012,** *6*, 026502.

469405 Microfluidic Particle Operations Based on One-Way Particle Transport in Asymmetric Traps

Abstract:

Utilizing micro-particles can significantly reduce the cost and time for life science and medical diagnosis with advantages including multiplexing capability, easy separation, and flexible surface modification. While research-based commercial platforms are successfully capitalizing on these benefits, portable point-of-care microdevices have showed limited processing ability of micro-particles. One possible difficulty of incorporating micro-particles into portable devices is the external devices necessary to accurately control particle positioning and transport. The development of simple, passive manipulation techniques could expand the use of particle-based devices.

In this presentation, we report on flexible, efficient particle operations. The one-way particle transport is based on hydrodynamic and steric interactions between micro-particles and an array of physical obstacles in an asymmetric trap. This technique uses asymmetric transport of particles in oscillatory flow, deterministically displacing hundreds of micro-particles across tens of array rows in a few fluid oscillations (oscillation frequency $> 2\ s^{-1}$) at low Reynolds number ($Re < 0.1$). We found comprehensive theoretical relationships between array dimension and the trap-particle interaction dynamics by using simple mass balance equations and Finite Element Methods. Experimental validation using polystyrene microbeads (diameter = 20.3 μm) confirmed that actual asymmetric trap-particle interaction dynamics compared favorably with our theoretical predictions. The one-way particle transport could successfully isolate target particles from non-target particles. Other flexible, efficient particle operations such as particle distribution and concentration using one-way particle transport will also be presented.

457997 Criteria for Drop Generation in Microfluidic Devices

Abstract:

Drop generation in microfluidic devices is an active area of research because of its relevance to a variety of applications, including high-throughput disease screening and combinatorial experimental investigations. Many applications rely on the controlled generation of drops; others use co-flowing streams of immiscible fluids. Controlling the transition between drop generation and co-flowing fluids is crucial to the performance of these devices. We present a quantitative criterion for this transition. The criterion is derived from an analysis of the critical state for unidirectional two-phase flow. Drop generation is not possible beyond a critical ratio of the volumetric flow rates of the dispersed-phase and continuous-phase fluids. Our theory shows that the critical state depends on the ratio of the fluid viscosities, the cross-section geometry of the microchannel, and the contact angle formed between the dispersed-phase fluid and the channel walls. The magnitude of surface tension does not affect the transition. Theoretical predictions are compared to experimental measurements. The theory helps explain the hysteretic behavior that is sometimes observed.

469787 Droplet Locomotion over Oil Immersed Superhydrophobic Surfaces

Abstract:

Microfluidic devices provide a technology for engineering and manipulating droplets at the micro- and nano-scales and open up new avenues for research. In particular, nano- and picoliter volumes of aqueous droplets of reagents and samples are generated in an oil continuous phase on a chip and then maneuvered, merged and divided to undertake chemical diagnostic analyses in dropet microfludics. The droplet microfluidic platform offers the abiity to encapsuate samples into droplet "packets" to prevent cross contamination. Additionally, this format has high surface area to volume ratios, shorter diffusion times and lower reaction volumes thereby enabling precise control over reactions as an alternative to conventional analytical techniques. Apart from these new applications, droplet microfluidic devices present new opportunities to conduct detailed studies of interfacial phenomena such as accurate measurement of slip lengths and ultralow values of interfacial tension which might otherwise be challenging. Devices in droplet microfluidic arrangements are typically fabricated from glass or polydimethyl- siloxane (PDMS), and water droplets immersed in oil drag over these surfaces. This surface movement has distinct disadvantages because such surfaces are not smooth and droplets can stick onto the surface. Sticking interferes with the trafficking of droplets and can cause droplets to break into pieces which can lead to cross contamination of samples.

In this presentation, we study a novel solution in which the water droplets are designed to move along superhydrophobic surfaces which are submerged in the continuous oil phase of the microfluidic channel. A superhydrophobic surface is a substrate which forms a high contact angle($>150^0$) when a water droplet is placed on the surface. Such high contact angles are achieved by tuning the chemistry and the surface roughness simultaneously. The low energy coating renders the surface hydrophobic and the surface roughness enhances the contact angle to achieve superhydrophobicity. High contact angles are obtained because a water droplet placed on such a surface lies on top of the asperities in a "Cassie-Baxter" wetting state leaving a plastron of air between the drop and the surface. Along with a high contact angle, an superhydrophobic surface also possesses low contact angle hysteresis (CAH). Low CAH eliminates pinning of water droplets to a substrate and ensures low adhesion between the substrate and the aqueous drop. When a superhydrophobic surface is submerged in oil, the oil wets into the surface topology. Since oils are known to have a lower surface energy than water, they are retained in the topology to create an oil infused surface. Hence a superhydrophobic surface can, in principle, function optimally and repel water even when immersed in oil. Water droplets easily move over the oil cushion of these surfaces, with reduced adhesion and friction, and importantly reduced contact angle hysteresis. A number of investigations have examined the movement of water droplets over superhydrophobic surfaces in air, in which case the droplets move over an air cushion. In addition, the movement of water droplets over superhydrophobic surfaces situated in air and infused with an oil layer have been studied. But there is very little understanding about the physics of the motion of water droplets over superhydrophobic surfaces in which the droplet and surface are completely immersed in a continuous oil phase. Our aim is to measure the mobility of the drops on the oil infused surface, and obtain a slip coefficient for water moving over an oil infused superhydrophobic surface by simulating the droplet hydrodynamics. Dielectrophoretic forces are used to actuate motion of droplet pairs over the surface.

Superhydrophobic surfaces are first fabricated by dip-coating glass slides into a nano particle polymer composite solution. The polymeric resin binds the nano particles to the substrate and to each other and creates a resilient surface topology. The superhydrophobic substrates are subsequently incorporated into the microuidic devices as the bottom wall and flow focusing is used to form a train of monodisperse droplets of water in a mineral oil. After attaining steady state, the flow is stopped allowing the initially suspended droplets to gravitationally settle to the bottom and equilibrate. An alternating current electric field is then applied by insertion of copper electrodes across the channel. The applied electric field induces surface charges of opposite signs on droplet pairs. This generates a dielectrophoretic force between the drops which eventually leads to their coalescence. The droplet motion and merging was controlled by adjusting the electric field. The trajectories of the merging droplet pairs were recorded with a high speed camera. With the use of image visualization and droplet tracking, the distance between the droplets was measured as a function of time. Comparison with control experiments (microfluidic device with a plain glass bottom) provides substantial evidence about the superior mobility of water drops on superhydrophobic surfaces. The slip coeffcient is obtained from the experimental trajectories by simulating the hydrodynamic motion of the droplets towards each other under inuence of the electrophoretic driving force and comparing the trajectories from this motion to the experimental trajectories. The measured values of the slip length of water moving over the oil infused surfaces are shown to be two orders of magnitude larger than over control surfaces of glass. Finite element simulations were undertaken in the context of computing the hydrodynamic drag and the di-electrophoretic forces. Since the droplets under consideration are micron sized, the shapes are approximated as spherical caps of dierent dimensions in the limit of a low Bond number. The computed values are hydrodynamic drag on a single drop are reported as a function of the contact angle and the slip length while the dielectrophoretic force is evaluated numerically for a droplet pair. An approximation for the viscous drag due to the approach of the identical droplets is formulated as the product of the isolated droplet drag multiplied by the resistance of two approaching spheres in an innite medium (Jerey-Stimson formula). The finite element simulations are validated by comparison with well asymptotic expressions (Davis force) in the limit of a completely non-wetting droplet. The measured values of drag and slip length are found to be in agreement with the corresponding calculated values.

472095 Controlled Oil Entrapment through Photo-Patterned Obstacles

Abstract:

There is a growing interest in trapping liquid inside a microchannel to create isolated nanolitre chambers, much like a multi-well plate system, for applications in diagnostics, drug discovery, reaction engineering and material synthesis. The compartmentalization of the liquid chambers at small length scale enables large number of parallel experiments in a high concentration environment. In this work, we develop a novel method to photopattern obstacles with controlled size, shape and wetting properties. We demonstrate oil entrapment around oleophilic obstacles by sequentially injecting oil and water in a microfluidic channel. We present a simple geometrical approach to explain the experimentally observed evolution of water/oil interface as it flows over the obstacle. Our model provides an analytic solution of the interface evolution, and successfully predicts the dependence of pinching off position and the amount of trapped oil with the obstacle contact angle. We also show that our model can be extended to two posts where we are able to successfully predict the critical distance for the formation of a liquid bridge. We further demonstrate effects such as interface pinning by using obstacles of different shapes.

472970 Measuring Partition Coefficients of Single Walled Carbon Nanotubes in Aqueous Two Phase Microliter Droplets

Abstract:

The distribution of a dispersed species between two immiscible liquid phases, expressed as the partition coefficient, is a critical parameter for designing and evaluating the performance of laboratory and industrial-scale liquid-liquid separation processes. The lack of data relating the partition coefficient of a compound to experimental parameters continues to present a major roadblock to the large-scale optimization of aqueous two phase separations. This is particularly apparent in the purification of single wall carbon nanotubes, and has restricted their wide-spread application. State of the art methods for screening partition coefficients as a function of experimental parameters such as temperature and composition require milliliters of sample per data point, in addition to labor intensive handling and detection procedures. In this talk, we demonstrate a high throughput, low volume semi-automated platform for mapping the partition coefficient of carbon nanotubes as a function of matrix composition. Our approach uses a microliter droplet-based fluid handling system coupled with inline absorption spectroscopy. Monodisperse droplet slugs are generated that contain aqueous two phase polymer mixtures along with specific surfactants and carbon nanotube samples of interest. Upon phase separation the carbon nanotubes preferentially partition into one of the polymer-rich phases, and inline absorption spectroscopy is used to identify and quantify the concentration in each phase. In this way, the critical parameters for aqueous two phase extraction of nanotubes can be quantified at high compositional resolution.

469703 Deterministic Lateral Displacement of Semiflexible Chain Dynamics in Microfluidic Pillar Arrays

Abstract:

Deterministic lateral displacement (DLD) is a high resolution size-based separation technique showing great potential for particle separation, especially for biological particle sorting. Over the past decade, rigid spherical particle separation using DLD method has been widely explored to facilitate the design of new DLD devices, while particles with different shapes and deformability have not been fully studied. Here, we investigate the migration dynamics of semi-flexible chains with high aspect ratio in gravity-driven DLD devices. Specifically, DNA linked semi-flexible colloidal chains serve as model high aspect ratio deformable particles and transport through a microfluidic pillar array device induced by a gravitational body force. Chain trajectories are presented using both experiments and simulation results. Similar to rigid spherical particles in the pillar arrays, zigzag and laterally displaced trajectories are observed for the longer and more flexible chains. In addition, the high aspect ratio and deformable nature of the chains lead to a richer local dynamical behavior, including bouncing, hairpin shape bending, sliding and hooking. We discuss how the deformability and size of chains affect their migration trajectories through microfluidic posts.

471903 3D Chaotic Printing: Using Simple Chaotic Flows to Fabricate Complex Microstructure

Abstract:

Chaos has the ability to create complex and predictable structures. We demonstrate the use of simple chaotic flows for the fabrication of complex 3D microstructures in cross-linkable or curable liquids, using a process that we refer to as 3D chaotic printing. We inject a drop of â€œinkâ€ (i.e., a drop of a miscible liquid, fluorescent beads, or cells) into a viscous Newtonian liquid and then apply a chaotic mixing recipe. This generates a complex structure in just a few flow applications. This structure is then preserved with high fidelity and reproducibility by rapid crosslinking or curing of the material. The 3D structure is the result of the rapid alignment of the injected material to the flow manifold. Therefore, its main features are reproducible and the overall process of fabrication is quite robust. Moreover, since the process is deterministic, it is amenable to computational modeling using Computational Fluid Dynamics (CFD).

Â In 3D chaotic printing, the interface between the injected material and the fluid matrix grows exponentially with time, and the volume of the injection is finite. Consequently, the average thickness of the lamellae of the material rapidly decreases from a scale of millimeters (the diameter of the injected drop) to one of nanometers. This exponentially fast increase in the interface, as well as the accompanying rapid decrease in the relevant length scales of the microstructure (which still preserves high resolution), is not currently achievable by any other 3D printing technique. We illustrate potential applications for this technology, including the rational reinforcement of constructs by the chaotic alignment of cells and nanoparticles, the fabrication of cell-laden fibers, the development of highly complex multi-lamellar and multi-cellular tissue-like structures for biomedical applications, and the fabrication of bioinspired catalytic surfaces.

472052 Kinetic Modeling of Propene Hydroformylation in Propane-Expanded Liquid with Rh-Based Complexes

Abstract:

Conventional propylene hydroformylation to produce butyraldehyde requires polymer grade propylene as feedstock.[1] High purity propylene is obtained by energy intensive separation from propane/propylene mixtures. If propane/propylene mixtures can be used effectively as feed, process economics would be significantly enhanced. Previously, we had reported preliminary results on the beneficial effects of propane expanded liquids (with propylene/propane mixtures instead of polymer grade propylene as feed) for propylene hydroformylation with Rhodium-based catalysts. At sufficiently high pressures (>10 bar), the pressurized propane/propylene mixture forms a gas-expanded liquid phase that promotes regio-selective hydroformylation of propylene with homogenous Rh complex.[2] In this talk, we report hydroformylation of propene in propane expanded solvents in the kinetic regime in a semi batch stainless steel reactor at 70–90 °C and 6–18 bar overpressure in 2,2,4-trimethyl-1,3-pentanediol monoisobutyrate (NX-795), toluene and *n*-butanol solvent with rhodium based complexes (Rh/TPP, Rh/ CHDPP, Rh/TPPA and Rh/BIPHEPHOS). The *n/i* ratio and TOF achieved during Rh/BIPHEPHOS catalyzed hydroformylation in propane-expanded solvents is comparable with conventional processes in neat solvent. By tuning the propane/propylene partial pressure, the *n/i* ratio in propane-expanded solvents could be enhanced by 30% in Rh-catalyzed reaction with mono-phosphine ligands while maintaining a reaction rate comparable to that without propane addition. Vapor–liquid equilibria of the reaction system were also measured in a variable-volume view cell at temperatures ranging from (343.15 to 363.15) K and pressures up to 20 Bar. The measured solubilities of CO and H_2 in the liquid phases were consistent with literature values.[3] The presence of propane was found to enhance the solubilities of both CO and H_2 in the liquid phase. The enhancement factor is up to 1.56 for carbon monoxide and 1.91 for hydrogen. Based on experimentally recorded kinetic data, a kinetic model was developed to explain the effects of propane-expanded solvent on the observed rate and selectivity.

1. Krokoszinski, R., Hammon, U., & Todd, K. (2005). U.S. Patent No. 6,864,391. Washington, DC: U.S. Patent and Trademark Office.

2. Liu, D., Xie Z., Chaudhari, R.V., Subramaniam, B. Enhanced Hydroformylation of Propylene in Propane-Expanded Liquids with Rh-Based Complexes, 2015 AIChE annual meeting. Salt Lake City, UT

3. Nair, V. S., Mathew, S. P., & Chaudhari, R. V. (1999). Kinetics of hydroformylation of styrene using homogeneous rhodium complex catalyst. Journal of Molecular Catalysis A: Chemical, 143(1), 99-110.

469755 Catalytic Cracking of Dodecane in Supercritical Water

Abstract:

Catalytic Cracking of Dodecane in Supercritical Water

Azadeh Zaker, Patricia Guerra, Geoffrey Tompsett, Michael T. Timko

Department of Chemical Engineering

Worcester Polytechnic Institute

Worcester MA 10609

An abstract to be submitted to the National Meeting of the AIChE (2016)

Technical Session: Reaction Engineering of Biomass and Hydrocarbons in Supercritical Water

Sponsored by: Catalysis and Reaction Engineering Division

In this work, we investigated dodecane cracking in supercritical water (SCW) using ZSM-5 as an acid catalyst. We studied the reaction in the presence and absence of water by tracking reaction progress in a batch reactor held at 400 ĪŠC and 24.1 MPa. For the same catalyst loading (5 wt. % oil-based), addition of water reduced formation rates of gas, liquid, and solid products. We identified and quantified liquid products using GC/MS and GC×GC chromatography. GC/MS was used primarily to track production of specific compounds, including toluene, xylenes, and ethylbenzene. GC×GC was used to monitor both individual compounds as well as classes of compounds: aliphatics, 1-ring aromatics, 2-ring aromatics, and 3-ring aromatics. SCW changes the observed product distribution of dodecane cracking, with increased production of aliphatics observed in the presence of water compared to increased production of aromatic compounds in its absence. Furthermore, in the absence of water, GC×GC analysis indicates increased yields of 2-ring and 3-ring products, including molecules such as naphthalene and alkyl-naphthalene. In addition to liquid phase analysis, we characterized the coke deposited on the catalyst using Temperature Programmed Oxidation (TPO) technique. Water reduces coke formation by an order of magnitude and alters its chemical characteristics. Coke deposited in the presence of water is mostly soft coke while coke formed in the absence of water is mostly graphitic hard coke. Finally, we developed a reaction network model to interpret our results. The reaction network consisted of aliphatic, aromatic, gas, and coke products. Future work will investigate catalysts other than ZSM-5 and more sophisticated molecular reaction networks.

469368 Demetallation of Biocrude from Hydrothermal Liquefaction of Microalgae

Abstract:

Hydrothermal liquefaction (HTL) converts wet biomass into a crude bio-oil that could be further treated to be acceptable for processing in petroleum refineries. HTL takes advantage of the properties of hot, compressed water near its critical point to break down biomacromolecules in the feedstock into smaller molecules. Several issues presently prevent the substitution of HTL biocrude for petroleum and one is the presence of metals in the crude bio-oil. Metals could cause problems such as corrosion, fouling, catalyst deactivation during downstream processing, etc. Metal impurities in crude bio-oil are unavoidable and metals will need to be removed if biocrude is used as transportation fuels. The most abundant metals in algal biocrude after HTL include Na and Fe. Na is present in typical growth media and Fe is an element essential for microalgae cultivation.

We hydrothermally treated microalgae using different processing conditions (isothermal, rapid heating and non-isothermal conditions, different temperatures and times) and determined the metal content in the biocrude oil as well as in the other product phases. We then studied the ability of different interventions to reduce this metal content. Process options explored include treatment in supercritical water in the absence of catalyst, use of different solvents for biocrude oil recovery after HTL, and use of heterogeneous catalysts during hydrothermal treatment. The Fe and Na content was reduced over 70% by use of specific organic solvents and reduced over 98% via application of heterogeneous catalysts. Additional catalytic studies with Fe model compounds permitted post-reaction catalyst characterization via techniques such as scanning electron microscopy and Mössbauer Spectroscopy.

463588 Pilot-Scale Continuous Flow Reactor for Hydrothermal Liquefaction of Algae

Abstract:

Algae-based biofuels have attracted much attention due to their advantages of not competing with land for food production, fewer terrestrial and weather limitations, higher growth rates, and strong CO_2-mitigation abilities. In recent years, research has focused on hydrothermal liquefaction (HTL) of algae since HTL does not require that the algae be dried before processing. Hydrothermal liquefaction uses high temperatures (270-350 °C) and high pressures (80-173 bar) to produce relatively high yields of bio-oil that can be upgraded to liquid transportation fuels. Most research to date has been conducted in batch reactors; shifting from batch reactors to continuous flow systems is an important next step towards commercial-scale processing.

This study presents the development and preliminary testing of a pilot-scale continuous flow reactor for hydrothermal liquefaction of fresh water microalgae at low solids loadings. A significant challenge to be overcome is the tendency of the algae feedstock and product slurries to create clogs inside filters and at bends and joints in the tubing even at very low solids loadings (<1 wt.%). To address such issues, a dual high-pressure cylinder filter system was designed to allow one filter to run while the other is being cleaned. Solid particles (char and ash), ranging from 6 to 25 μm, are captured by the cylinder filters and collected in blowdown pots. The filter elements can be cleaned in place by flushing back the product fluid. Tubing size and configuration was carefully considered to accommodate a narrow diameter desirable for the high pressure and to prevent unnecessary clogging opportunities. The goal of the reactor is to be able to perform continuous HTL on slurries with solid algae contents of 5-10 wt.%, and to produce char-free bio-oils.

457442 Production of Biodiesel Fuel Via Sub/Supercritical Transesterification Reactions with Trace Amount of Homogeneous Catalysts

Abstract:

2016 AIChE Annual Meeting

November 13th to 18th, San Francisco, CA

PRODUCTION OF BIODIESEL FUEL VIA SUB/SUPERCRITICAL TRANSESTERIFICATION REACTIONS WITH TRACE AMOUNT OF HOMOGENEOUS CATALYSTS

*Jiuxu Liu, Yue Nan and Lawrence L. Tavlarides**

Department of Biomedical and Chemical Engineering, Syracuse University,

329 Link Hall, Syracuse, NY 13244, USA

* Corresponding author. Tel.: 315-443-1883; Fax: 315-443-9175

Email: lltavlar@syr.edu (Tavlarides); jliu23@syr.edu(Liu)

Abstract

Refined oils and anhydrous alcohols are required in conventional biodiesel industry since impurities as free fatty acids (FFA) and water poison the catalysts. One of the most promising advantages of the non-catalytic sub/supercritical technology (SCTE) over the conventional method is that it does not require high quality feedstocks, so low-cost waste oils which contains FFA and water, and hydrous alcohols which contains water can be used. However, it is difficult to reach a high biodiesel yield since extending reaction time at those required temperatures will thermally decompose biodiesel product, which has been proved to lower the fuel qualities such as viscosity and cold flow properties. We have known that employing catalysts in the SCTE reactions with high quality feedstocks (refined oils and anhydrous alcohol) will complete the reactions in shorter residence times to avoid decomposition, but we have not determined if those catalysts can still function under the presence of impurities such as FFA and water. Another issue related to biodiesel studies is that much fewer investigations have been done on kinetics of synthesizing FAEEs, which is a potential replacement of FAMEs, than FAMEs. Presently, the kinetic analyses of SCTE ethanolysis reactions have been based on the overall reaction and the model of first-order reaction with respect to triglycerides, and the temperature range is very limited.

In this study, we are investigating FAEEs and FAMEs biodiesel synthesis under sub and supercritical alcohol conditions catalyzed by trace amounts of homogeneous catalysts including sulfuric acid, potassium hydroxide, and potassium m/ethoxide. The objectives include: a) a kinetic study of FAEEs biodiesel synthesis at SCTE conditions with 0.1 wt% sulfuric acid, b) comparison of catalytic ability to improve yields of FAEEs and FAMEs biodiesel between sulfuric acid and other catalysts including potassium hydroxide and potassium m/ethoxide under sub and supercritical conditions, c) effect of impurities such as FFA (up to 30 wt%) and water (up to 10 v%) on these catalysts, and d) effect of reaction pressure on product yield.

At this writing we have completed the first aim which is the kinetic study of synthesizing FAEEs biodiesel under sub and supercritical conditions. The reactions were conducted at temperatures from 175 to 350 °C, residence times from 1 to 120 min, pressure of 200 bar, and 0.1 wt% sulfuric acid based on oil mass. A three-step partially reversible second-order reaction model was employed to best fit the data, and the activation energy for each reaction step was calculated. It was found that the catalytic activity of sulfuric acid decreased when switching from subcritical to supercritical conditions. Comparison of the kinetic parameters for sub and supercritical reactions will be shown, and the mechanism will be discussed. High biodiesel yields (> 90%) were reached at 175, 200, 225, 300, and 325 °C with 0.1 wt% sulfuric acid concentration.

Experiments to complete the remaining objectives are in progress and will also be reported.

472237 Molecular and Microkinetic Modeling of Supercritical Fluid and Liquid-Phase Fischer-Tropsch Synthesis

Abstract:

Due to the specific and tunable properties of supercritical fluids (SCFs), they have been considered as unique media for different technical processes of extraction, separation, particle formation, and chemical reactions. In the particular case of Fischer-Tropsch synthesis (FTS), experiments have shown that utilizing supercritical solvent in this heterogeneous process would result in several distinguished improvements. Accelerating the conversion of reactants, providing more favorable product distributions, as well as preventing catalyst deactivation are some of the mentioned advantages.[1] While previous experimental studies have discussed that all these known enhancements (such as better heat distribution, improved mass transfer, selective solubility of coking products, etc.[2]) relate to solution thermodynamics and transport, not many investigations have considered the solvent effect(s) on heterogeneous kinetics of FTS reactions. In this study, we have analyzed the influence of hexane solvent on the reaction rates of FTS process over Co(0001) catalyst using different theoretical approaches of molecular dynamics and quantum mechanics. By adopting an existing microkinetic model[3] and modifying it, potential of mean forces (PMF) and density functional theory (DFT) calculation have been performed in order to estimate the equilibrium conditions and rate parameters of individual elementary reactions under different phase conditions. Ultimately, these kinetic and thermodynamic parameters determined from molecular modeling and simulation are used as input into a micro-kinetic model, to characterize the differences in catalysis in SCF and liquid hexane versus conventional gas-phase processing. Microkinetic modeling results include yield and selectivity versus reaction time and net rate and sensitivity analyses of the elementary reaction mechanisms in all the considered reaction phases. The simulation results show that the SCF and liquid phase solvent not only have significant effects on the thermodynamics of the adsorption/desorption reactions, but also they may alter the kinetics of surface reactions toward selecting certain reaction pathways.

Keywords: Fischer-Tropsch synthesis, supercritical fluid, microkinetic modeling, PMF, DFT

---

[1] D. B. Bukur, X. Lang, and L. Nowicki, Ind. Eng. Chem. Res. 44, 6038-6044 (2005).

[2] P. E. Savage, S. Gopalan, T. I. Mizan, C. J. Martino, and E. E. Brock, AIChE journal 7, 1723-1778 (1995).

[3] S, Storsæter, D. Chen, A. Holmen, Surf. Sci. 600, 2051-2063(2006).

466897 Intensified Hydrolysis By Employing the Synergy of Mixed Subcritical H_2o and Supercritical CO_2

Abstract:

Hydrolysis of naturally occuring compounds into more bioactive and bioavailable materials are gaining popularity in food and pharmaceutical industries. However, conventional method employs harmful catalysts such as sulfuric acid with low yield, while separation of the target compounds from the reaction products requires rather tedious approaches. To overcome these problems, this research aims to develop, optimize and evaluate by experimental approaches the combined reaction and separation process employing the synergy of subcritical H_2O and supercritical CO_2. Application to hydrolysis of citrus bioflavonoids (hesperidin) into its aglycones (hesperitin) was elucidated. Using a batch reactor, we were successful in hydrolyzing hesperidin into hesperetin under hydrothermal conditions with supercritical CO_2. The highest yield of hesperetin was obtained at T=160 °C, P=15 MPa and t=4 h. Based on the results, a semi-batch reactor system was developed to study reaction and separation process employing the mixture of supercritical CO_2 and subcritical H_2O. After the reaction, extraction of hesperetin was carried out by passing supercritical CO_2 through the reaction products. Results indicated the possibility of separating hesperetin from the mixture, and higher yield can be obtained by simply manipulating the pressure and temperature of CO_2.

474745 Session Keynote - Co-Extrusion: Advanced Manufacturing for Energy Devices

Abstract:

The development of mass markets for large-format batteries, including electric vehicles (EVs) and grid support, depends on both cost reductions and performance enhancements to improve their economic viability. Palo Alto Research Center (PARC) has developed a multi-material, advanced manufacturing process called co-extrusion (CoEx) to remove multiple steps in a conventional battery coating process with the potential to simultaneously increase battery energy and power density. CoEx can revolutionize battery manufacturing across most chemistries, significantly lowering end-product cost and shifting the underlying economics to make EVs and other battery applications a reality.

PARCâ€™s scale-up of CoEx for electric vehicle (EV) batteries builds on a solid base of experience in applying CoEx to solar cell manufacturing, deposition of viscous ceramic pastes, and Li-ion battery chemistries. In the solar application, CoEx has been deployed commercially at production scale where multi-channel CoEx printheads are used to print viscous silver gridline pastes at full production speeds (>40 ft/min). This operational scale-up provided invaluable experience with the nuances of speed, yield, and maintenance inherent in taking a new technology to the factory floor. PARC has leveraged this experience, adapting the CoEx process for Lithium-ion (Li-ion) battery manufacturing. To date, PARC has worked with Li-ion battery materials and structured cathodes with high-density Li-ion regions and low-density conduction regions, documenting both energy and power performance. Modeling results for a CoEx cathode show a path towards a 10-20% improvement in capacity for an EV pouch cell. Experimentally, we have realized a co-extruded battery structure with a Lithium Nickel Manganese Cobalt (NMC) cathode at print speeds equivalent to conventional roll coating processes. The heterogeneous CoEx cathode enables improved capacity in thick electrodes at higher C-rates. The proof-of-principle coin cells demonstrate the feasibility of the CoEx technology and a path towards higher energy and higher power EV pouch cells.

463732 Instability of Reversible Pump Turbines and Its Physical Origin

Abstract:

Currently, reversible pump turbines are being developed worldwide in the pumped hydro energy storage power plants. One of the paramount technical problems of the reversible pump turbines is the instabilities during the off-design working conditions. For example, in generating modes, significant pressure fluctuation has been observed in the reversible pump turbines with prominent systematic vibrations. In the present paper, several examples of the measured instabilities of reversible pump turbines are given for both the prototype and model type turbines. Based on those experimental data, the basic features of the instabilities are revealed together with the paramount influencing parameters (e.g. guide vane opening). Numerical simulations has also been performed to further investigate the physical origins of the instability. Finally, some recent progress are also briefly discussed for relieving the instability.

473631 Simplified Model of a Redox Flow Battery for Deriving Design and Dispatch Strategies in a Resource Planning Tool

Abstract:

Energy resource planning tools (ERPTs) that integrate an energy storage technology with renewable energy sources require representative models for the energy storage resources. ERPTs may involve use of optimization framework to select the optimal mix of resources and dispatch strategies. Such optimization problems can involve a yearlong time frame and several energy resources. Hence use of linear models is desirable for faster convergence of the optimization problem.

Flow battery technology has emerged as one of the potential storage candidates which can be used for stationary energy storage systems. The flexibility of scaling energy and power independently may give these systems an advantage over other energy storage technologies for applications where the energy is high and power demand is low. Flow battery systems exhibit nonlinear behavior during charging and discharging which is typical of electrochemical energy storage systems. Hence there is a need to develop a simplified (preferably linear) model without undue compromise of the underlying physics of the batteries.

Rather than modeling a flow battery as an ideal, finite source of energy with constant energy efficiency, a simple model incorporating the electrochemical behavior of a redox flow battery was developed. This model enables the optimizer to (i) choose energy and power independently (design decision) and (ii) access the performance of the chosen battery size (dispatch decision) in order to derive the optimal energy mix. The talk will also focus on quantification of errors associated with the simple model compared with the nonlinear representation.

Acknowledgements

The work presented herein was funded by the Office of Naval Research.

474162 Selection of Ion-Exchange Membrane for Vanadium Redox Flow Battery

Abstract:

The increasing share of renewable energy sources in the total energy production emphasizes the need for a reliable and cheap stationary energy storage in order to compensate for hardly predictable non-stabilities in output power of photovoltaics or wind turbines on the power transmission level. Vanadium redox flow battery (VRFB) appears as a promising solution for the stationary energy storage as it offers high efficiency (80% DC-DC), versatile arrangement of decoupled power (kW) to capacity (kWh), extended durability and fast demand response. However, the broader commercialization of the technology is still obstructed by relatively high investment costs.

Ion-exchange membrane is a key component of VRFB cell as it directly influences the power and the efficiency of the stack - the resistance of the typically used cation-exchange membrane counts for up to 50 % of the total internal resistance of the battery stack. In the same time, price of the membrane may exceed 25 % of the overall costs. Thus, the identification of highly conductive, low-cost membrane with minimal permeability for vanadium ions and sufficient durability in acidic electrolytes is vitally needed for VRFB developers.

In our contribution we present the results of a broad systematic study focused on the effect of membrane properties (charge, ion-exchange capacity, thickness, method of preparation and preconditioning) on VRFB operation. The various types of commercially available ion-exchange membranes were characterized with the respect to the properties relevant for VRFB operation such as: i) through-plane ionic conductivity in the environment of VRFB cell, ii) permeability for vanadium ions of different oxidation states, iii) performance in VRFB single-cell and iv) durability in the presence of oxidizing V^{V}electrolyte.

From the results obtained from permeation measurement it is evident that the effective diffusion coefficient of V^{3+} ions is significantly lower than of VO^{2+}ions for all the studied membranes, which corresponds with the differences in the solvation numbers of the ions. Generally higher conductivities were found for cation-exchange membranes when compared to anion-exchange ones, which is caused by the different mobility of the transported ions (hydronium vs. sulfate ions). According to our assumptions, the conductivity and selectivity of membrane significantly depend on the concentration of ion-exchange groups in the membrane. However, these properties are also influenced by the specific inner arrangement of hydrophobic and hydrophilic domains in the material. The observed trends were also confirmed by the characterization of membranes in VRFB single-cell. Based on these results we identified optimum membranes for VRFB with the respect to the energy efficiency of charging discharging cycle and maximal power output. Perfluorinated cation-exchange membranes are the solution when high power and long durability is required. Anion-exchange membranes show lower vanadium crossover and thus appears more suitable for the application when low maintenance is required.

463885 Energy Storage for the Enhancement of Penetration Level of Wind Energy: A Case Study

Abstract:

In the past five years, wind energy increases significantly in China. In terms of the equipped capacity of wind turbine, China serves as the top country in the world. However, serious problems occur due to the connection of the wind energy with the electrical grid. For example, the fluctuation nature of the wind energy causes significant instabilities in the grid, leading to the rejection of the wind energy. In some local area of China, the rejection rate of the wind energy could reach 40% or even more. To solve this dilemma, it is quite necessary to develop the energy storage facilities (in terms of both large-scale and small-scale capacities). In the present paper, effects of the energy storage on the penetration level of wind energy are investigated based on the wind farms in the western parts of the Inner Mongolia province of China. The detailed analysis are given based on the quantitative data obtained from the government and industries. The importance of the energy storage are categorized into several groups and several emerging trends are briefly discussed.

474663 An Investigation of a Polysulfide Polyiodide Aqueous Redox Flow Battery

Abstract:

Redox flow batteries (RFBs) have attracted considerable academic and industrial interest based on their favorable combination of performance and safety in the widespread integration of renewable, non-dispatchable energy sources and in the improvement of energy efficiency of the electric grid. However, at present, RFBs are broadly considered too expensive for widespread deployment. A cost-effective aqueous RFB based on alkaline metal polysulfide (PS) and polyiodide (PI) is shown here for the application grid-scale energy storage. The proposed RFB applies earth-abundant elements, which shows great potential to meet the stringent cost target for RFBs ($100/kWh). The OCV of the proposed RFB is ~ 0.88 V with a theoretical energy density as high as ~ 46 Wh/L that is ~84% higher than that of all vanadium redox flow batteries. Galvanostatic and potentiostatic studies were carried out on both the negative and positive sides, which sheds light on the electrochemistry of sulfide and iodide species, respectively. The cell performance of the proposed PS-PI chemistry was further studied using a flow battery architecture operated at 5 mL/min and 20 degree Celsius. The resultant RFB displayed stable performance at 15 mA/cm^2 for over 150 hours (25 cycles) with relatively small capacity decay. A system diagnosis after the cycling experiment revealed that membrane (sodiated Nafion membrane) degradation is the primary reason for the significantly decreased energy efficiency of the cell after 25 cycles.

461432 Liquid Polyamorphism: From Metallic Hydrogen to Supercooled Water

Abstract:

"Liquid polyamorphism" is the existence of two alternative amorphous structures in a single-component fluid. Liquid polyamorphism is found in a broad group of very different materials, such as triphenyl phosphite, silicon, silicon dioxide, cerium, and hydrogen, usually at extreme conditions. For example, at extremely high pressures, between 100 and 400 GPa and much above the liquid-gas critical temperature, there exists a fluid-fluid transition between metallic monoatomic hydrogen and non-metallic molecular hydrogen. This transition is projected to be terminated by a fluid-fluid critical point at about 100 GPa and 1500 K. Under another extreme condition, in deeply supercooled metastable water a liquid-liquid transition is hypothesized. Cold and supercooled water is assumed to exist in two forms, a low-density or high-density liquid. The existence of these two alternative structures could, at certain conditions, result in a metastable liquid-liquid separation in pure water. The hypothesized liquid-liquid metastable coexistence is not directly accessible in bulk-water experiments because it is presumably located a few degrees below the empirical limit of homogeneous ice formation. We consider thermodynamic constrains imposed on the existence of liquid polyamorphism in a single-component fluid. The first fundamental question to be addressed is the separation of time scales: a system with two inter-convertible fluid structures can be thermodynamically treated as a single-component fluid if the time of observation is longer than the time of conversion (fast conversion). In the opposite limit (slow conversion) the system thermodynamically behaves as a two-component mixture. The second fundamental question is the difference between the order parameters describing the liquid-gas transition and the liquid-liquid transition and the corresponding critical points. In particular, the order parameter for the liquid-gas transition is associated with density, while for the liquid-liquid transition, this is the "reaction coordinate", the fraction of conversion. The reaction coordinate is a function of temperature and pressure. The field, conjugate to the order parameter, is the logarithm of the reaction equilibrium constant. Correspondingly, while the relevant thermodynamic potential for the liquid-gas transition is the density of the Helmholtz energy, for the liquid-liquid transition in a single-component fluid, this is the Gibbs energy. The developed equation of state with a "chemical reaction" between two alternative strictures generates a global fluid phase behavior with liquid-gas and liquid-liquid transitions in the same single-component fluid.

464033 Modeling of Alternative Solvents for Sustainable Processes: Supercritical CO_2 and Ionic Liquids

Abstract:

A great effort has been devoted in recent years to develop â€œgreenâ€ processes, searching for lowering their impact in the environment with respect to existing ones. One of the key steps towards the development of sustainable chemical processes is the replacement of traditional solvents by others with lesser environmental impact. In this regard, supercritical fluid technology has allowed significant improvements in several industrial processes, and there are still many opportunities for further applications in new emergent areas. Ionic liquids (ILs) and deep eutectic solvents (DESs also appear as promising alternative solvents for some of the existing processes, as well as emerging ones. Additionally, supercritical CO_2 in combination with ILs can be used in continuous green biocatalytic processes. In this manner, CO_2 is converted from a residue into an added value compound or into a catalytic co-solvent.

This contribution provides some perspectives on opportunities for new or advanced applications of supercritical fluids and ILs for sustainable processes. The particular choice of the systems is due to the inspiring work of Prof. Cor J. Peters in these areas. After a general overview, we will present some specific applications that have been modeled using an advanced molecular-based equation of state (EoS), the soft-SAFT EoS [1,2]. In particular, several examples on CO_2-water mixtures [3], the solubility of supercritical CO_2 in fatty acid esters [3], the removal of acid gases by studying the ternary diagram of these mixtures with CO_2 and ILs [4], and the application of ILs for biomass deconstruction processes [5] will be characterized.

This work highlights how the molecular modeling can assess the choice of the right solvent in a specific application, a very useful tool when dealing with the design of new sustainable processes.

Partial financial support for this work has been provided by the Catalan government under project 2014-SGR1582.

References:

[1] F.J. Blas, L.F. Vega, Mol. Phys. 92 (1), 135-150 (1997).

[2] F.J. Blas, L.F. Vega, Ind & Eng. Chem. Res. 37 (2), 660-674 (1998).

[3] F Llovell, LF Vega, J. Supercrit. Fluids 96, 86-95 (2015)

[4] F. Llovell, M.B. Oliveira, J.A.P. Coutinho, L.F. Vega, Catalysis Today 255, 87-96 (2015).

[5] N. Mac Dowell, F. Llovell, N. Sun, J.P. Hallett, A. George, P.A. Hunt, T. Welton, B.A. Simmons, L.F. Vega, J. Phys. Chem. B 118 (23), 6206-6221 (2014).

457149 Volumetric Properties and Modeling of the Binary Mixtures of Ethanol with Ionic Liquids [BMIM]Ac and [BMIM]Cl

Abstract:

Ionic liquids and their mixtures are under consideration for their utility as processing fluids for bio-based polymers of lignocellulosic materials. Volumetric properties of mixtures of two different ionic liquids, namely 1-butyl-3-methylimidazolium acetate ([BMIM]Ac) and 1-butyl-3-methylimidazolium chloride ([BMIM]Cl), with ethanol were determined in a variable-volume view-cell equipped with a motorized pressure generator. Continuous density data were generated from 10 to 40 MPa at temperatures ranging from 298 to 398 K. Mixtures compositions covered a wide range from 5 to 75 wt % ionic liquid. The density data were modeled using the Sanchez-Lacombe and Simha-Somcynsky Lattice-Fluid models. Derived thermodynamic properties such as the isothermal compressibility, isobaric expansivity, internal pressure, and for the first time the solubility parameters of these mixtures and their pressure dependent variations were evaluated. The effectiveness of these mixtures in dissolution and separation of lignocellulosic materials is being explored with softwood and hardwood species as well as woody biomass such as bamboo.

454130 High-Pressure Hydroformylation Reaction Equilibria

Abstract:

Hydroformylation converts an olefin into an aldehyde via reaction with CO and H_2 (syngas). The reaction is performed in suitable solvents/solvent mixtures to ensure the intense contact of the hydrophobic olefin and the hydrophilic catalyst. On the one hand, the solvents also determine the syngas solubility in the reaction mixture and on the other hand have an influence on the reaction equilibria and reaction kinetics.

This work investigated the influence of solvent composition in DMF/decane solvent mixtures on gas solubilities as well as on reaction equilibria of the hydroformylation of dodecene to tridecanal.

The syngas solubilities in the reaction mixtures at varying solvent compositions were calculated using the thermodynamic model PC-SAFT. The same model was also used to determine the activity coefficients of reactants (dodecene, CO, and H_2) and product (tridecanal) in the different solvent systems. For a given ratio of the DMF/decane solvent mixture, equilibrium concentrations of dodecene and tridecanal were measured after sampling form a high-pressure reactor by means of gas chromatography and nuclear magnetic resonance spectroscopy. These values together with the syngas solubilities and reagent/product activity coefficients obtained from PC-SAFT were used to determine the thermodynamic equilibrium constant K_f and therewith the standard Gibbs energy of the dodecene hydroformylation reaction. The latter was found in very good agreement with the value obtained from quantum-chemical calculations.

Applying the so-obtained K_f and accounting for the solvent influence on the reagent/product activity coefficients, the equilibrium concentrations for varying solvent compositions could be predicted in almost quantitative agreement with the experimental data.

UNIVERSITI TEKNOLOGI PETRONAS

Optimization of Hydrothermal Liquefaction of Palm Kernel Shell to Bio-oil by Response Surface Methodology and Synergistic Effect of Supercritical Carbon Dioxide

Yi Herng Chan, Suzana Yusup, Armando T. Quitain, Yoshimitsu Uemura, Mitsuru Sasaki, Tetsuya Kida

UNIVERSITI TEKNOLOGI PETRONAS

Research Objectives

1. To optimize reaction parameters for sub- and supercritical hydrothermal liquefaction of PKS using Response Surface Methodology (RSM)

2. To investigate the effect of supercritical CO_2 in the hydrothermal liquefaction of PKS on the bio-oil yield

3. To characterize bio-oils produced by GC-MS

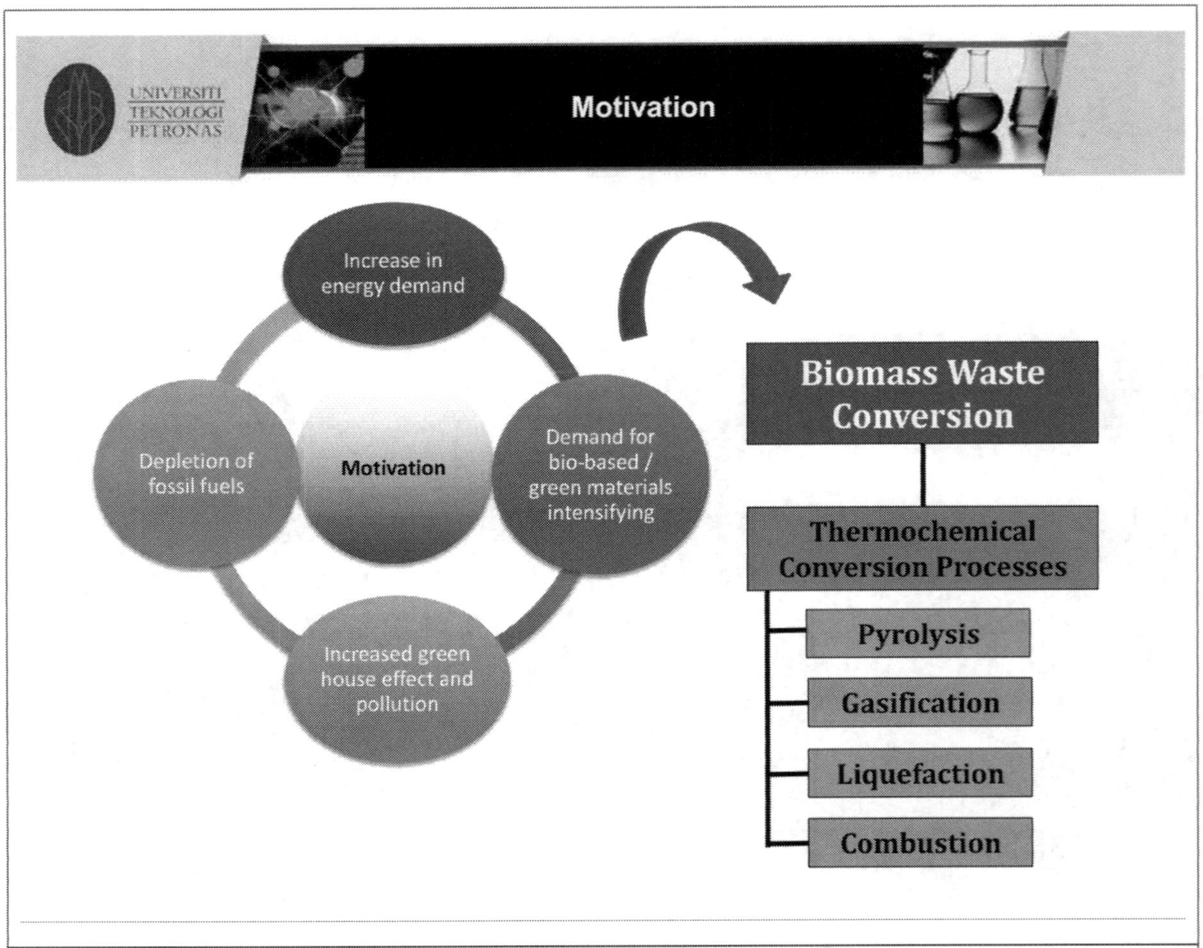
UNIVERSITI TEKNOLOGI PETRONAS
Motivation
Increase in energy demand
Depletion of fossil fuels
Motivation
Demand for bio-based / green materials intensifying
Increased green house effect and pollution
Biomass Waste Conversion
Thermochemical Conversion Processes
Pyrolysis
Gasification
Liquefaction
Combustion

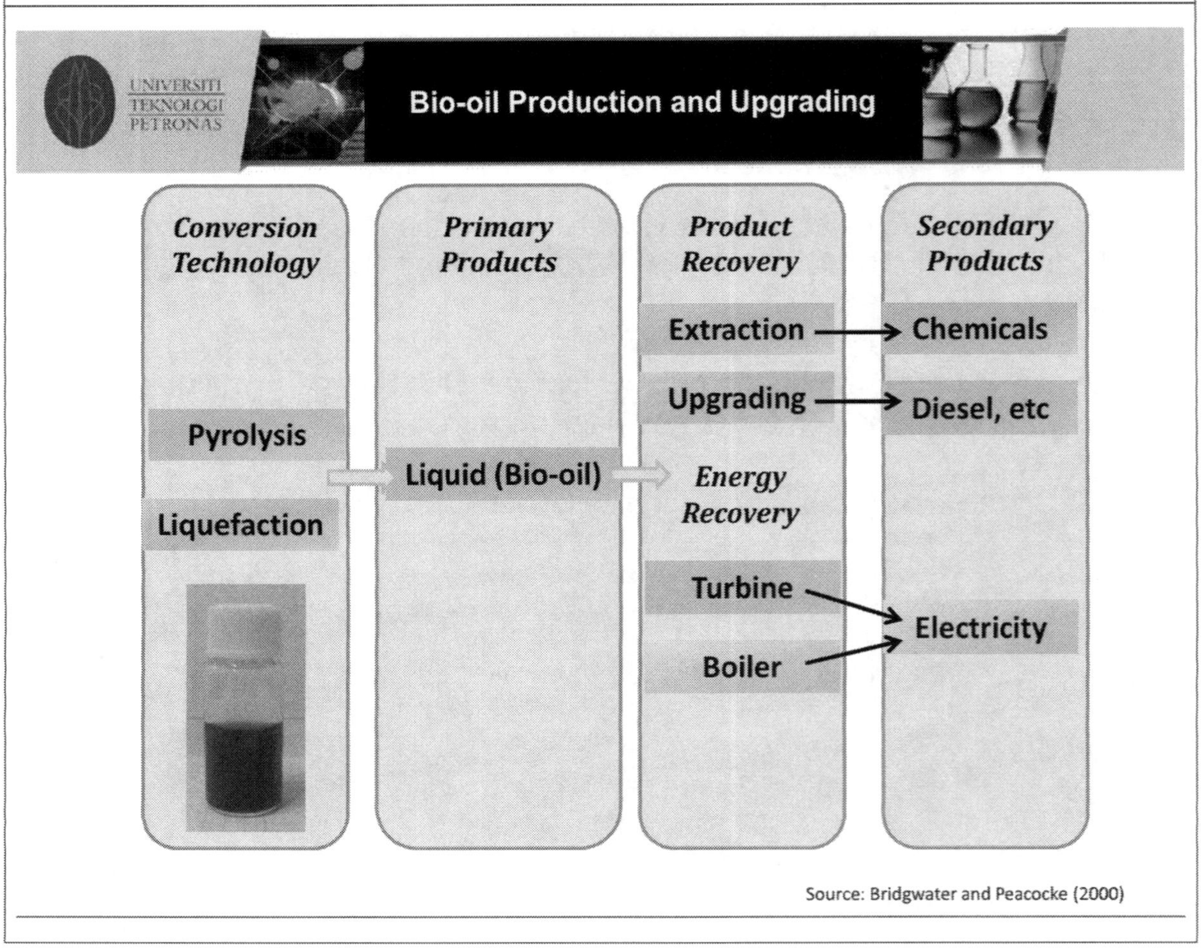
UNIVERSITI TEKNOLOGI PETRONAS
Bio-oil Production and Upgrading
Conversion Technology
Primary Products
Product Recovery
Secondary Products
Extraction
Chemicals
Upgrading
Diesel, etc
Pyrolysis
Liquid (Bio-oil)
Energy Recovery
Liquefaction
Turbine
Electricity
Boiler
Source: Bridgwater and Peacocke (2000)

Palm Kernel Shell as Feedstock

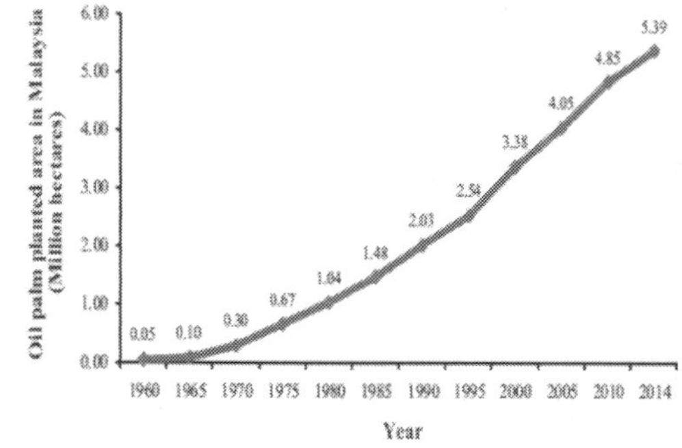

Oil Palm Residues	Quantity (Mt/y)	Net calorific value (MJ/t)
Empty fruit bunch	21.27	18795
Mesocarp fibre	10.80	19055
Palm kernel shell	4.98	20093
Palm oil mill effluent	49.85	16992

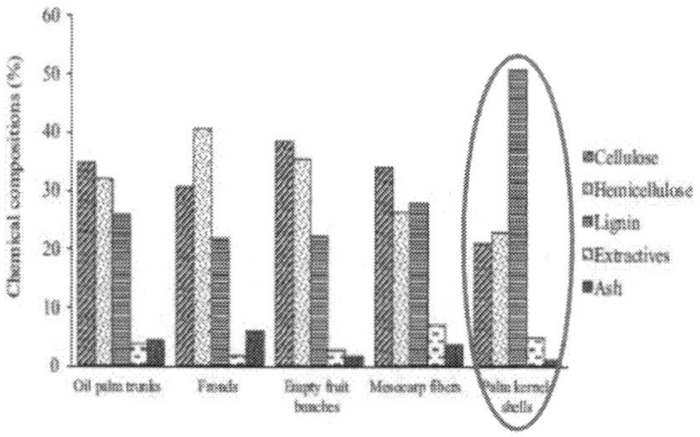

- Among various oil palm biomass materials, **palm kernel shell (PKS)** is selected as the feedstock in this research because:
  - ✓ Have the highest energy content.
  - ✓ Can be processed and undergo pre-treatment easily.
  - ✓ Rich in lignin (source of phenolic and aromatic compounds).

Source: Awalludin et al. (2015), Ng et al. (2012)

Introduction

Bio-oil

- Dark brown liquid.
- Comprises of oxygenated compounds, organic acids and organic compounds such as aldehydes, ketones, phenols, alcohols and polyaromatic hydrocarbons (PAHs).
- Can further be upgraded into usable liquid fuels (transportation, boilers, turbines) and value-added products, such as food flavourings, resins, agri-chemicals, fertilizers and fine chemicals.

Introduction

Motivation ➡ ***Liquefaction of biomass waste using water***

Liquefaction involves a set of complex thermochemical reactions that converts solid biomass to a mixture of organic products using solvent(s).

Why Liquefaction ?

Diversification **of prime energies for fuel production**

High moisture content **of raw biomass feedstocks**

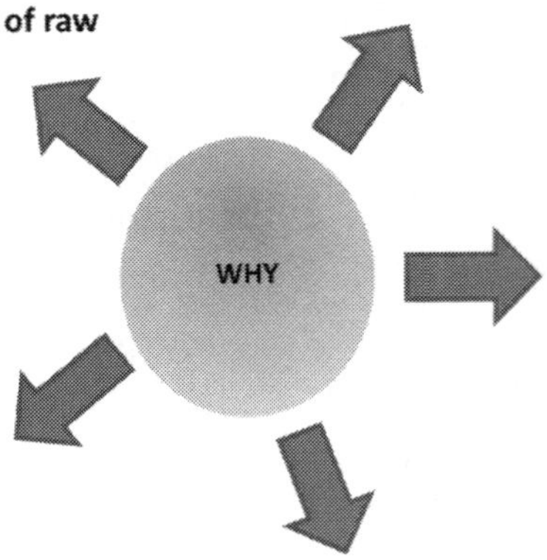

Presence of solvent dilutes concentrations of products, **prevents formation of char.**

Operation at lower temperatures **compared to pyrolysis**

Replacement **of conventional methods (landfill and incineration)**

Source: Akhtar et al. (2010).

Selection of solvent

- **Sub- and Supercritical H_2O**

➢ is able to dissolve and extract materials which are normally water insoluble

➢ promotes higher reactivity and better separation and extraction

➢ cheap, non-toxic, non-flammable and advantageous compared to other organic solvents

- **Supercritical CO_2**

➢ does not leave chemical residue in the products

➢ the reuse of carbon dioxide gas is possible

➢ an energy conserving, selective and waste reducing alternative

➢ reduced or eliminate employees' exposure to hazardous organic solvents

➢ Increase the acidity of water – may help in the degradation of biomass

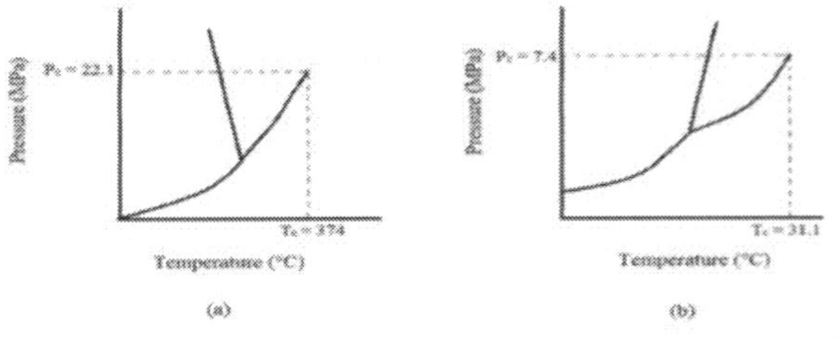

Source: Ruen-ngam et al. (2012), Sanghi and Srivastava (2003)

Literature Review (Liquefaction of biomass)

Researchers	Feedstock	Optimized Parameters (Range)	Optimum condition	Remarks (Solvent, catalyst, technology)
Mazaheri et al. (2010)	Oil palm fruit press fiber	RSM with CCRD: ➤Temperature (230 – 310 °C) ➤Residence time (15 – 75 min) ➤Particle size (<250, 710-1000 μm) ➤Specimen loading (5 – 15 g) ➤Additive loading (1-9 wt%)	278 °C 40 min 710-1000 μm Specimen loading 5 g Additive loading 9 wt% Liquid product yield: 76.16 %	Subcritical water liquefaction + NaOH
Gan and Yuan (2013)	Corncob	RSM with CCRD: ➤Temperature (280 – 340 °C) ➤Retention time (12 – 48 min) ➤Biomass content (9 – 21 %) ➤Catalyst loading (0.76 – 2.25 %)	280 °C 12 min 21 % biomass solid content 1.03 – 1.56 % catalyst loading Bio-oil yield: 41.3 %	Hydrothermal liquefaction + alkaline catalyst
Liu et al. (2013)	Macroalgae (Ulva prolifera)	RSM with CCRD: ➤Catalyst dosage (2 – 6 %) ➤Solvent-to-feedstock ratio (12:1 – 20:1) ➤Temperature (160 – 180 °C)	Microwave power: 600 W 30 min 165 °C Solvent:feedstock = 18.87:1 H_2SO_4 loading: 4.93 % Liquefaction yield: 93.17 %	Microwave-assisted direct liquefaction in ethylene glycol with H_2SO_4
Guo et al. (2012)	Macroalgae (Sargassum polycystum C.Agardh)	RSM with CCRD ➤Solvent-to-feedstock ratio (12:1 – 20:1 w/w) ➤Temperature (160 – 180 °C) ➤Catalyst content (6 – 10 w/w%)	Microwave power: 400 W Solvent:feedstock = 18.50:1 (w/w) 170 °C 15 min Catalyst/solvent: 9.6 % (w/w%) Liquefaction yield: 87.70 %	Microwave-assisted direct liquefaction in ethylene glycol with H_2SO_4

Optimization using RSM

Optimization of bio-oil production from PKS using Design of Experiment (Response Surface Methodology)

Variable	Coded Variable	Unit	Lower Axial Point (-α = -2)	Lower Factorial Point (-1)	Center Point (0)	Higher Factorial Point (+1)	Higher Axial Point (+α = +2)
Temperature	X_1	°C	300	330	360	390	420
Pressure	X_2	MPa	20	25	30	35	40
Reaction Time	X_3	min	30	60	90	120	150
Biomass to Water Ratio	X_4	wt/wt	0.05	0.2	0.35	0.5	0.65

Experimental Procedures

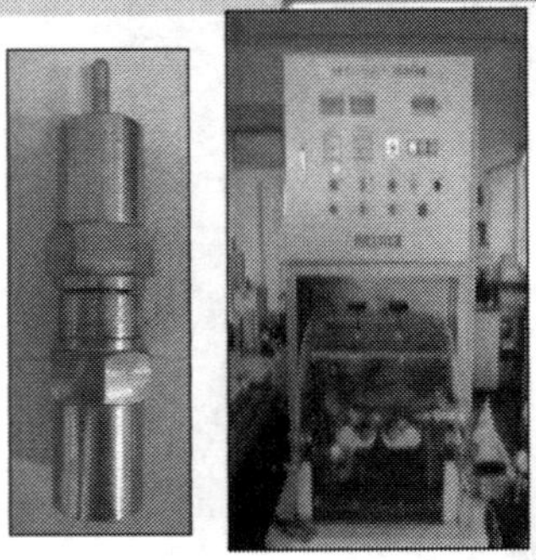

8.8 ml Inconel Batch Reactor

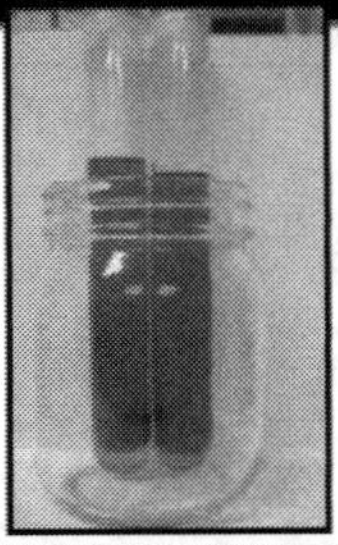

Separation to 3 phases
(Toluene-soluble, solid residue, water)

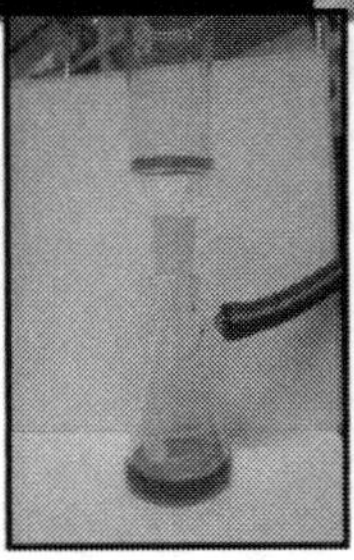

Filtration using fritted disc filters of pore size 40-100 µm to remove suspended particles.

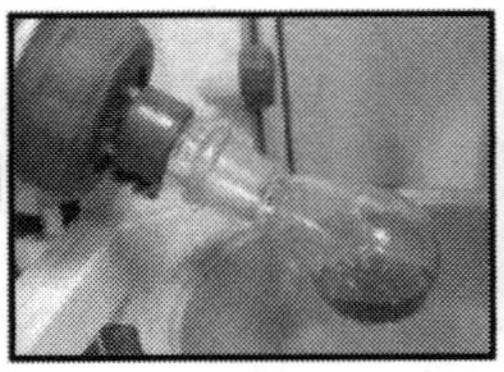

Evaporation of toluene
Pressure: 88 hPa
Water bath: 57 °C

Bio-oil

$$Bio-oil\ yield\ (wt\%) = \frac{Weight\ of\ bio\text{-}oil\ obtained\ (g)}{Weight\ of\ biomass\ (g)} \times 100$$

Results

Run No.	Temperature (°C)	Pressure (MPa)	Reaction Time (min)	Biomass:Water Ratio (wt/wt)	Bio-oil yield (wt%)
1	330	35	120	0.5	3.89
2	330	25	120	0.5	3.55
3	330	35	120	0.2	6.12
4	360	30	90	0.65	5.45
5	360	20	90	0.35	11.44
6	360	30	150	0.35	7.66
7	360	40	90	0.35	5.89
8	390	25	120	0.2	13.10
9	330	35	60	0.5	3.59
10	360	30	90	0.35	7.83
11	360	30	90	0.35	7.25
12	330	25	60	0.5	3.44
13	390	25	60	0.2	15.55
14	390	35	120	0.5	8.48
15	300	30	90	0.35	3.00

UNIVERSITI TEKNOLOGI PETRONAS

Results

Run No.	Temperature (°C)	Pressure (MPa)	Reaction time (min)	Biomass:Water Ratio (wt/wt)	Bio-oil yield (wt%)
16	420	30	90	0.35	12.76
17	360	30	90	0.35	9.88
18	330	35	60	0.2	6.33
19	390	35	60	0.5	10.05
20	390	35	120	0.2	14.73
21	360	30	30	0.35	7.00
22	360	30	90	0.35	7.97
23	360	30	90	0.35	7.80
24	330	25	60	0.2	6.48
25	360	30	90	0.05	13.74
26	390	35	60	0.2	15.85
27	390	25	60	0.5	14.06
28	360	30	90	0.35	7.05
29	390	25	120	0.5	13.45
30	330	25	120	0.2	6.59

UNIVERSITI TEKNOLOGI PETRONAS

Results

Optimization using Design Expert 8.0.4 Software

Parameters	Units	Prediction	Experimental validation			
			1	2	3	Average
Temperature	°C	390	390	390	390	390
Pressure	MPa	25	25	25	25	25
Reaction time	min	60	60	60	60	60
Biomass:water ratio	wt/wt	0.20	0.20	0.20	0.20	0.20
Bio-oil yield	wt%	**15.48 ± 1.62**	**14.01**	**15.13**	**14.18**	**14.44 ± 0.60**

Results

Analysis of Variance (ANOVA)

Source	Sum of squares	Degree of freedom	Mean square	F value	p-value	Comment
Model	405.01	14	28.93	10.96	< 0.0001	**Significant**
X_1 - Temperature	299.63	1	299.63	113.48	< 0.0001	**Significant**
X_2 - Pressure	13.92	1	13.92	5.27	0.0365	**Significant**
X_3 – Reaction time	0.71	1	0.71	0.27	0.6123	
X_4 – Biomass:water ratio	69.43	1	69.43	26.30	0.0001	**Significant**
X_1X_2	2.99	1	2.99	1.13	0.3039	
X_1X_3	2.30	1	2.30	0.87	0.3659	
X_1X_4	0.29	1	0.29	0.11	0.7465	
X_2X_3	3.60×10^{-3}	1	3.60×10^{-3}	1.36×10^{-3}	0.9710	
X_2X_4	6.00	1	6.00	2.27	0.1524	
X_3X_4	0.23	1	0.23	0.085	0.7740	
X_1^2	0.26	1	0.26	0.099	0.7572	
X_2^2	2.37	1	2.37	0.90	0.3584	
X_3^2	0.043	1	0.043	0.016	0.8997	
X_4^2	7.60	1	7.60	2.88	0.1104	
Residual	39.60	15	2.64			
Lack of fit	34.54	10	3.45	3.41	0.0939	**Not significant**
Pure error	5.06	5	1.01			
Corrected total	444.62	29				
R^2	**0.9109**					
Adj-R^2	0.8278					

Results

Response surface plots:

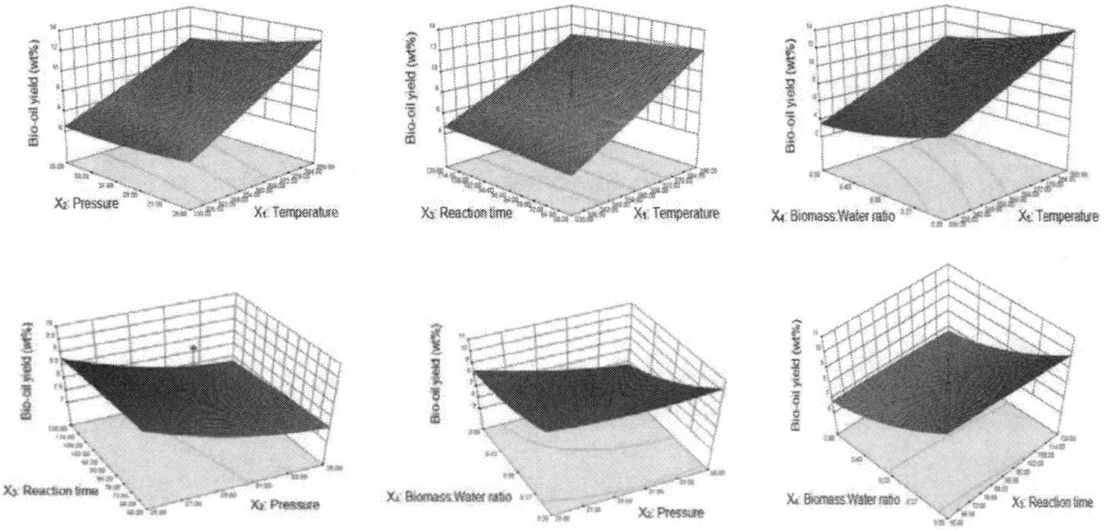

Results

Regression model in terms of coded variables:

$$Y = 7.96 + 3.53X_1 - 0.76X_2 - 0.17X_3 - 1.70X_4 - 0.43X_1X_2 - 0.38X_1X_3 - 0.13X_1X_4 + 0.015X_2X_3 - 0.61X_2X_4 + 0.12X_3X_4 + 0.098X_1^2 + 0.29X_2^2 - 0.040X_3^2 + 0.53X_4^2$$

where:
Y = bio-oil yield
X_1 = Temperature
X_2 = Pressure
X_3 = Reaction time
X_4 = Biomass:water ratio

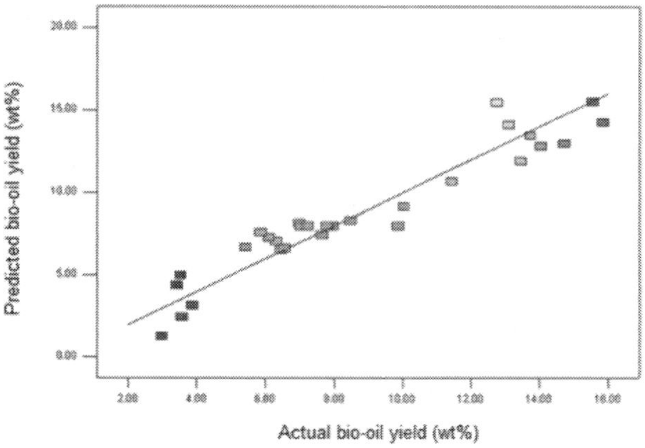

Discussion

Effect of temperature

- Bio-oil yield **increased with increasing temperature** from 330 °C to 390 °C as the **rate of cracking, hydrolysis and decomposition of lignocellulosic components in the biomass increased**, and hence **increased formation of low-polar compounds**, which were captured as bio-oil.

Effect of pressure

- Higher pressure caused a **slight decrease** in bio-oil yield which might be due to **increased caging effect for the C-C bonds** in biomass inhibiting the bonds cleavage and fragmentation, and **increased gasification** due to increased free radical reactions.

Effect of reaction time

- Negligible impact on bio-oil yield. **Prolonged reaction at higher temperature** (390 °C) **decreased** bio-oil yield due to **secondary decomposition** of bio-oil to gaseous products and **char formation** by condensation, cyclization and repolymerization reactions.

Effect of biomass:water ratio

- **Decreased** bio-oil yield with **increasing biomass-to-water ratio**. Increased amount of biomass per unit mass of water **decreased the efficiency** of liquefaction, hence yielding less bio-oil.

Literature Review
(Role of Supercritical CO_2 in hydrothermal processes)

Addition of supercritical carbon dioxide into subcritical and supercritical water **promotes in situ formation and dissociation of carbonic acid** from the reaction of water and carbon dioxide at elevated temperature and pressure, which served as a **natural catalyst** in hydrothermal processes.

Researchers	Summary of Research
Brunner (2009)	➢ Improved efficiency of hydrolysis of hesperidin aided with supercritical CO_2.
Ruen-ngam et al. (2012)	➢ Synergistic effect of supercritical CO_2 on rate of liquefaction of cellulose.
Gosselink (2012)	➢ Depolymerized of lignin in supercritical CO_2 / acetone / water for production of aromatic chemicals.
Singh et al. (2015)	➢ Studied the product distributions of bio-oils from hydrothermal liquefaction of rice straw in the presence of supercritical CO_2.

Synergistic effect of sc-CO_2

Dissociation of carbonic acid into ions acts as natural catalyst to enhance the decomposition of PKS

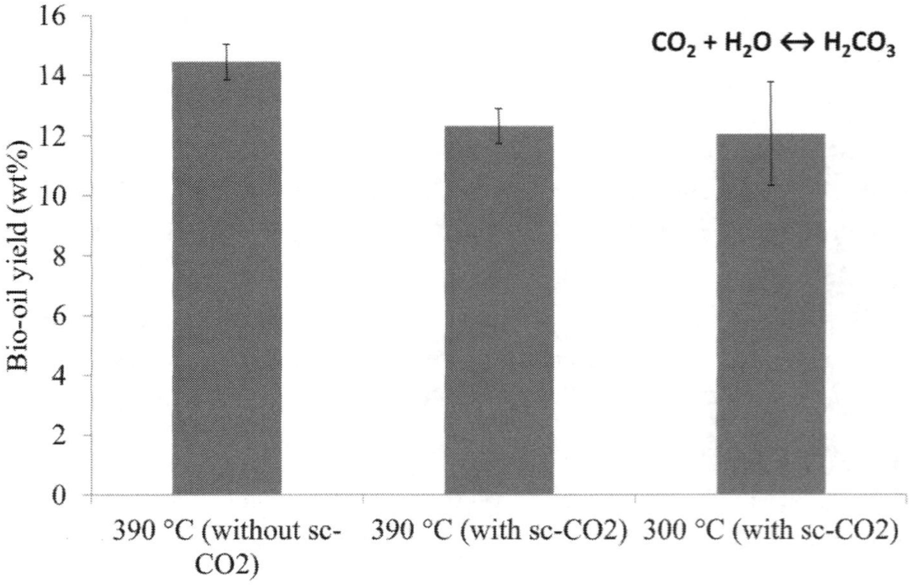

UNIVERSITI TEKNOLOGI PETRONAS

Chemical Composition of Bio-oils

Peak no.	RT (min)	Compound	Percentage peak area (%)	
			390 ° C (without sc-CO_2)	300 ° C (with sc-CO_2)
1	8.012	2-cyclopenten-1-one, 3-methyl-	3.59	-
2	8.440	phenol	49.60	67.39
3	10.106	2,3-dimethylcyclopent-2-en-1-one	3.27	-
4	10.536	phenol, 2-methyl-	5.14	-
5	11.117	phenol, 4-methyl-	5.40	-
6	11.577	phenol, 2-methoxy- / mequinol	6.78	8.78
7	12.948	phenol, 2-ethyl-	1.90	-
8	13.750	phenol, 4-ethyl-	1.42	-
9	14.614	1,2-benzenediol	4.08	-
10	16.365	1,2-benzenediol, 3-methyl-	0.92	-
11	16.413	1,4-benzenediol, 2-methoxy- / 1,2-benzenediol, 3-methoxy-	1.60	-
12	16.952	2-(2,2-dimethylcyclopropyl)thiophene / phenol, 4-ethyl-2-methoxy	2.15	-
13	17.143	1,2-benzenediol, 4-methyl-	4.29	-
14	18.869	phenol, 2,6-dimethoxy-	4.05	20.09
15	19.632	1,3-benzenediol, 4-ethyl- / benzene, 1,4-dimethoxy-	2.75	-
16	24.055	dodecanoic acid	3.05	3.74

UNIVERSITI TEKNOLOGI PETRONAS

Chemical Composition of Bio-oil

GC-MS chromatogram of bio-oil (a) 390 °C (without sc-CO_2); (b) 300 °C (with sc-CO_2)

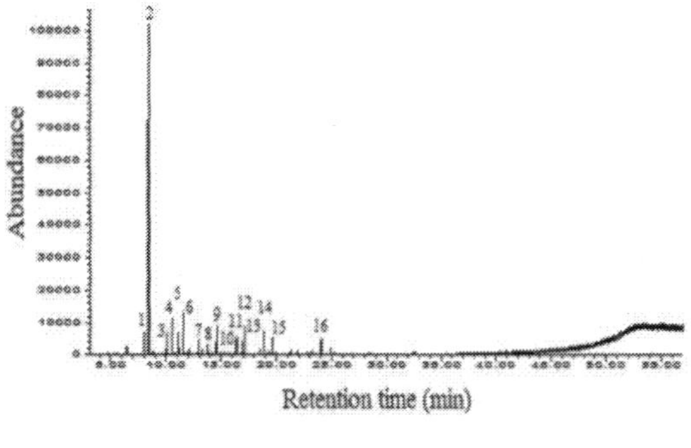

(a)

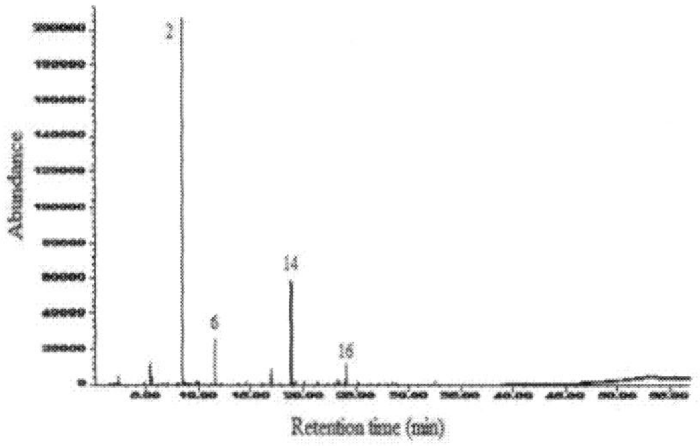

(b)

Discussion

- Bio-oil was comprised of various organic compounds, with phenolic compounds as the major components, followed by ketones and aromatic compounds (benzene-containing compounds), and acid.
- Phenols and its derivatives were the **major compounds** found in bio-oils, due to the high lignin content of PKS.
- **Hydrolysis of the ether bonds** in the lignin structure produced a variety of **phenol derivatives and methoxy phenols**.
- The formation of **ketones** was attributed to the **decomposition of hemicellulose, cellulose and lignin monomers**.
- **Aromatic compounds** originated from the **decomposition of lignin**.
- **Acid** was the degradation product of **cellulose, hemicellulose and lignin**.

Conclusion

- **Optimum condition for hydrothermal liquefaction of PKS was determined.**
  - 390 °C, 25 MPa, 1 hour reaction time, biomass:water ratio of 0.2
  - Optimum condition predicted by RSM was verified.
  - Optimum bio-oil yield: 14.44 wt%
- **Effect of supercritical CO_2 on bio-oil yield**
  - Synergistic effect of supercritical CO_2 was significant at 300 °C, and insignificant at 390 °C.
- **Characterization of bio-oils (GC-MS)**
  - PKS-derived bio-oils contained phenolic compounds as the major constituents, as a result of lignin degradation in the hydrothermal liquefaction process.

UNIVERSITI
TEKNOLOGI
PETRONAS
THANK YOU

DEVELOPMENT OF SOLVENT SELECTION GUIDE FOR CO_2 SPRAY COATING SYSTEM

Yoshiyuki Sato, Tomoki Shimada, Kohei Abe, Hiroshi Inomata, Tohoku University, Sendai, Japan
Shin-ichiro Kawasaki, National Institute of Advanced Industrial Science and Technology, Sendai, Japan

Abstract

CO_2 spray coating system is a promising technique for VOC reduction in paint industry. Precipitation of polymer component in the systems has been pointed out as a problem since CO_2 is a poor solvent for polymer solutions. In this work, a solvent selection guide for the CO_2 spray coating system is proposed. The phase behavior and phase equilibria of polymer + solvent + CO_2 systems are measured. Furthermore, relationship between solubility parameter of the solvent and the phase behavior was investigated.

Introduction

The paint industry is the largest source of volatile organic compound (VOC) emissions in Japan[1]. It is an urgent issue to reduce VOC emissions. UNICARB® CO_2 spray system[2] was developed in 1990 by Union Carbide with the aim of reducing the concentration of VOC in coating formulations. Lewis et al.[3] reported that the CO_2 spray systems had some technological advantages such as superior leveling compared with conventional ones. Recently, Kawasaki et al.[4] proposed a CO_2 spray system equipped with a micromixer, which enables CO_2 to dissolve into organic paint solutions rapidly and homogeneously.

Conventional paint consists of polymer, solvent, and diluents with a large portion being organic solvents. CO_2 spray coating system, which uses CO_2 as a diluent, has the potential to reduce the VOC amount as well as energy consumption in paint industry. However, precipitation of polymer is induced by CO_2 dissolution into some paints because CO_2 can act as anti-solvent for polymers. It is important to prevent precipitation of polymer in paint solution for spray coating. Since it is difficult to replace proven polymers, solvents should be changed to prevent precipitation in CO_2 spray coating systems.

In this work, a solvent selection guide (methodology) for the CO_2 spray coating system is proposed. Acrylic resin was chosen as a polymer component and eight solvents were used. Phase behavior and phase equilibria of polymer + solvent + CO_2 system were experimentally studied at 40°C. Furthermore, relationship between solubility parameter of the solvent and the phase behavior was investigated based on the phase behaviors.

Experiment

Sample

Acrylic resin (ACRYDIC 52-666-BA, DIC Corporation) was used as a polymer component in paints. This polymer resin is used for isocyanate curing and was supplied as a butyl acetate solution. Characteristics and composition of the acrylic resin obtained by the supplier are shown in Table 1. Chemical structure of the polymer was not available, so the chemical structure was estimated from the information given in Table 1 as shown in Figure 1. Solubility parameter of the acrylic resin was estimated to be 21.9 - 23.2 $MPa^{0.5}$ by using Fedors group contribution method[5].

Table 1. Characteristics and composition of acrylic resin used in this work

component	fraction [wt%]
acrylic resin	45-55
butyl acetate	45-55
methyl methacrylate	1.0
acid value	3.5 mgKOH/g
hydroxyl number	73.3 mgKOH/g

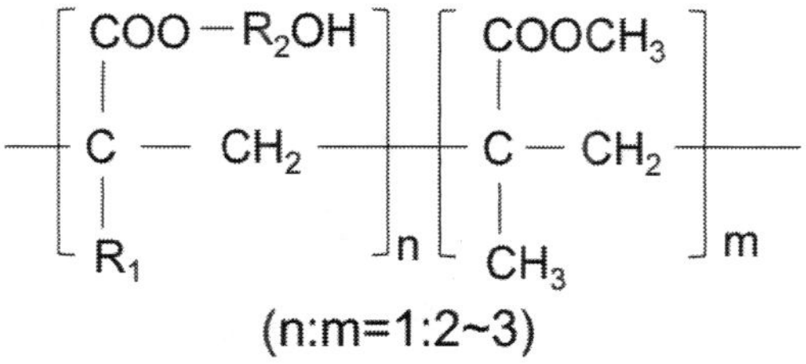

Figure 1. Estimated chemical structure of the acrylic resin

Eight solvents were used for compatibility tests described below. The data on source, purity, and solubility parameter of solvents used in this work are given in Table 2. Sample (acrylic resin dissolved in butyl acetate) was diluted with a solvent (sample / solvent =5/3 wt.) listed in Table 2. Polymer component in diluted solution was prepared so as to have a concentration of 10 – 20 wt% when CO_2 dissolved into the solution. The diluted solutions were used for compatibility test and also CO_2 solubility measurement. CO_2 (purity>99.5 vol%) was obtained from Showa Denko Gas Products Co., Ltd. All chemicals were used as received.

Table 2. Source, purity, and solubility parameter of solvents used

Chemical name	Source	Purity (wt%)[a]	Solubility Parameter ($MPa^{0.5}$)[6]
n-hexane	Wako Pure Chemicals	96.0	14.9
diisobutyl ketone (DIBK)	Wako Pure Chemicals	90.0	16.9
diethyl ketone (DEK)	Wako Pure Chemicals	98.0	18.2
pentanol	Wako Pure Chemicals	98.0	21.5
Butanol	Wako Pure Chemicals	99.0	23.2
Ethanol	Wako Pure Chemicals	99.85	26.5
Methanol	Wako Pure Chemicals	99.8	29.6
ethylene glycol	Wako Pure Chemicals	99.5	33.0

[a] as reported by the supplier

Compatibility Test

Compatibility of solvent was carried out by using a phase behavior observation apparatus shown in Figure 2. Observation cell was equipped with a cylindrical sapphire window (i.d ϕ10, o.d. ϕ14, length 18 mm) and a test tube (i.d. ϕ8.3, length 27 mm) and was set in a thermostated water bath. Diluted solution (0.3 g) was filled into the test tube. CO_2 was supplied with a regulator (≤5 MPa) or a syringe pump (>5 MPa) and CO_2 pressure was increased at a 0.5 MPa step. When polymer precipitation was observed, the pressure was recorded. Equilibration of gas dissolution was confirmed with height change of the liquid meniscus in the test tube. In experiments, equilibration was reached mostly in about 10 minute.

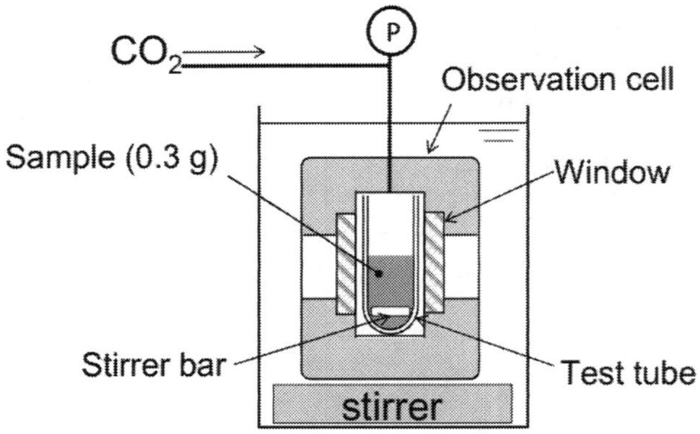

Figure 2. Phase behavior observation apparatus

CO_2 Solubility measurement

Solubility of CO_2 in polymer solutions was measured with a synthetic method apparatus (not shown). The apparatus and the details of measuring bubble point pressure (BP) are described in our previous paper[7]. BP was determined from the relationship between the pressure and the inner volume of the cell based on compressibility difference of the liquid phase and the vapor-liquid phase. Estimated uncertainties of temperature and BP are ±0.06 K and ±0.028 MPa, respectively. A sample of known composition was prepared gravimetrically. Estimated uncertainties of CO_2 weight fraction for the BP experiments are less than 0.0005.

Results and Discussion

Compatibility Test

Experimental results of compatibility test of the diluted solutions are shown in Figure 3. Precipitation pressure in the diluted solutions increased as the diluent solubility parameter was close to that of the acrylic resin. It would be noted that the precipitation pressures for DIBK and DEK cases were fairly low compared with those for ethanol and methanol cases of which solubility parameter differences from the resin were comparable or slightly larger than whose of DIBK and DEK. This would be attributable to the CO_2 dissolution into the solution. The solubility parameter of CO_2[8] is calculated as 3.4 $MPa^{0.5}$ at temperature of 40°C and pressure of 7 MPa so that the solubility parameter of the diluted solution decreases as increasing of CO_2 dissolution amount. Solubility parameter of the diluents, which have higher solubility parameter than that of acrylic resin, decreases and is close to that of acrylic resin. Therefore, methanol showed a wide soluble pressure range.

From these results, it can be noticed that it is important for solvent selection to choose a solvent having higher solubility parameter than that of polymer component. For further quantitative discussion on this issue, we decided to measure CO_2 solubility in the diluted solution.

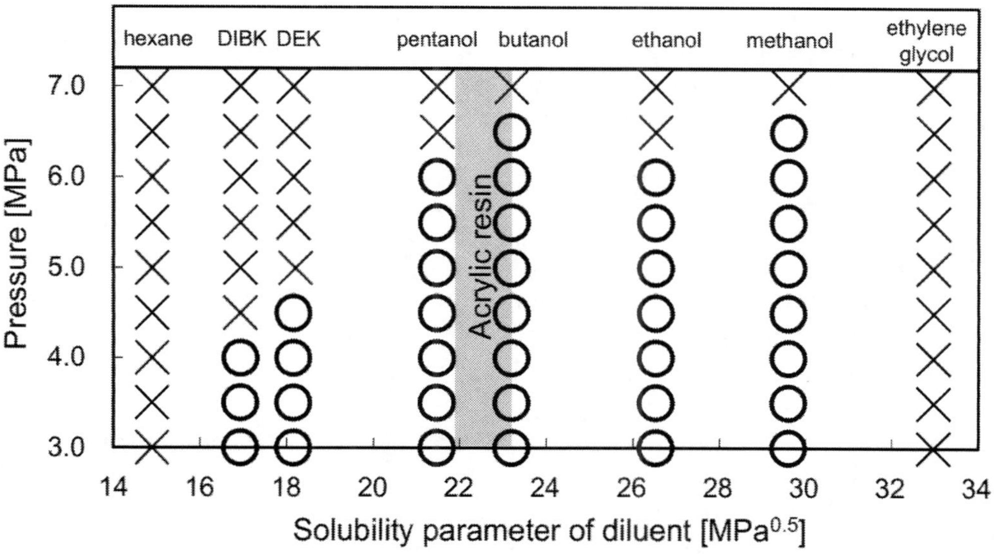

Figure 3. Experimental results of compatibility for polymer (acrylic resin) + solvent (butyl acetate) + diluent + CO_2 (40 ºC), ○: dissolution; ×: precipitation

CO_2 Solubility measurement

Solubility of CO_2 in the diluted solution was measured at a temperature of 40ºC. The diluted solutions of DEK, butanol, and methanol were used for the solubility measurements. Experimental results of the solubility are shown in Figure 4. The solubility of CO_2 in these diluted solutions was directly proportioned the pressure. The solubility of CO_2 in the butanol solution is slightly lower than that in the other diluted solutions.

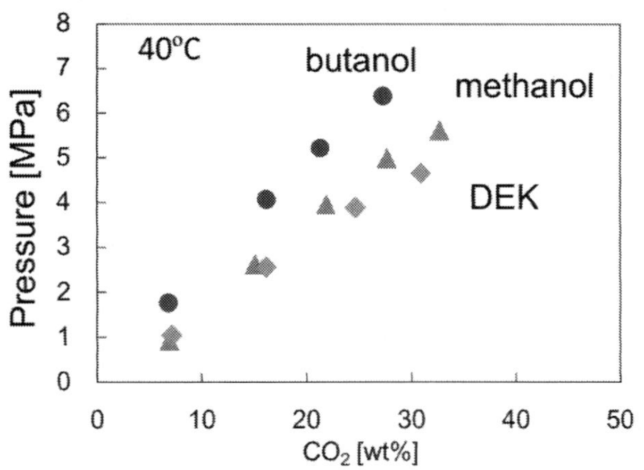

Figure 4. Solubility of CO_2 in diluted solution of acrylic resin butyl acetate solution with butanol, methanol, and diethyl ketone at temperature of 40ºC

Solubility Parameter

Solubility parameter was used to evaluate the compatibility results. Namely, the solubility parameter of the mixture was considered to account for the amounts of solvent, diluent, and also the measured CO_2 solubility. Thus, the solubility parameter of the mixture, $\delta_{mixture}$, was determined from eq. (1):

$$\delta_{mixture} = \phi_1\delta_1 + \phi_2\delta_2 + \phi_3\delta_3 \quad (1)$$

where ϕ is the volume fraction. Subscripts 1, 2, 3 denote butyl acetate (solvent), diluent, and CO_2, respectively. The volume fractions were evaluated by using (saturated) liquid density at 25°C and composition. The solubility parameter of pure components was obtained from literature data[6] at 25°C. In the case of CO_2, the solubility parameter (=4.32 $MPa^{0.5}$) was estimated with Williams' method[8] for the saturated liquid state at 25°C. This is because we assumed that dissolved CO_2 acted as a liquid in solution.

Figure 6 shows the plots of the solubility parameter versus CO_2 solubility. The solubility parameters clearly decreased with increasing CO_2 concentration as expected. It would be noted that the solubility parameter of the mixture at precipitation starting had similar values (11-14 $MPa^{0.5}$). This means that the precipitation occurs when the solubility parameter difference between the mixture and acrylic resin is about 8 $MPa^{0.5}$. If a generalized rule is found on the value for the solubility parameter difference between resin and CO_2 dissolved mixture, it can be a useful guide (methodology) for selecting a solvent in CO_2 spray painting for given paint/resin condition.

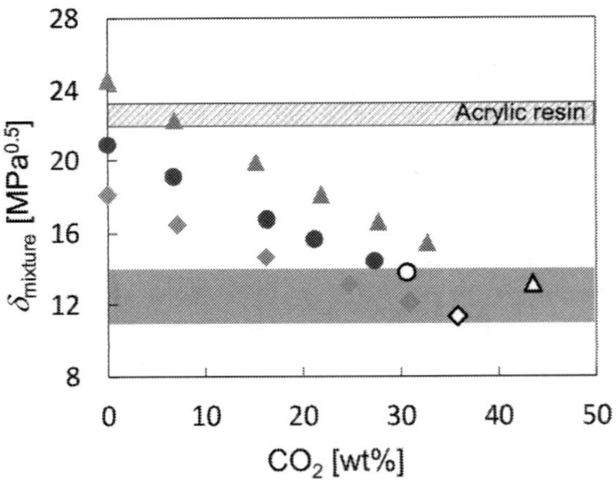

Figure 6. Solvent selection guide for CO_2 spry coating system.
Polymer: acrylic resin; ▲: methanol; ● : butanol; ♦: diethyl ketone; open symbols: extrapolated solubility at precipitation pressure

Conclusion

A solvent selection guide was presented for the CO_2 spray coating systems. It was suggested that solubility parameter of the mixture is helpful in solvent selection and conditions. It is important for solvent selection to choose of solvent having higher solubility parameter than that of polymer component. Precipitation of the polymer from the diluent solution will occur, when solubility parameter of solvent and CO_2 mixture differs by about 8 $MPa^{0.5}$. Similar experiments with different polymers will be needed to generalize the conclusions in future work.

Acknowledgment

This research was supported by JSPS KAKENHI JP15K06534.

References

1. Ministry of Environment Japan, VOC emission inventory report 2015 (2016)
2 H.F. Bok, K.L. Hoy, K.A. Nielsen, Patent US 5 057 342 (1989).
3 J. Lewis, J.N. Argyropoulos, K.A. Nielson, Supercritical Carbon Dioxide Spray Systems, Metal Finishing, 98, 254-262 (2000).
4 S.-I. Kawasaki, K. Ishida, Y. Sakurai, T. Fujii, A. Suzuki, CO_2 atomizing technology "CAT" for green coating process using high pressure micromixer", 14th European Meeting on Supercritical Fluids, MS21, Marseilles (2014).
5 R. F. Fedors, "A Method for Estimating Both the Solubility Parameters and Molar Volumes of Liquids", Jet Propul. Lab. Quart. Tech. Rev., 3, 45 (1973)
6 C.M. Hansen, Hansen solubility parameters: A user's handbook, 2nd edn., CRC Press, Boca Raton (2007)
7 Y. Sato, N. Hosaka, K. Yamamoto, H. Inomata, Compact apparatus for rapid measurement of high-pressure phase equilibria of carbon dioxide expanded liquids, Fluid Phase Equilib., 296 (2010) 25-29.
8 L.L. Williams, J.B. Rubin, H.W. Edwards, Calculation of Hansen Solubility Parameter Values for a Range of Pressure and Temperature Conditions, Including the Supercritical Fluid Region, Ind. Eng. Chem. Res., 43, 4967-4972 (2004)

452704 Dilute Solution Thermodynamic Study of Phosphatidylcholine - Organic Solute Systems Using Inverse Gas Chromatography (IGC)

Abstract:

A fundamental understanding of solute-solvent interactions is critical in the design of separation processes such as liquid solvent or supercritical fluid extraction. Toward this end, dilute solute-solvent interactions have been determined using inverse gas chromatography (IGC) for twenty-three solute – phosphatidylcholine (PC) systems over the temperature range of 55 – 105°C. Utilizing a purified soya-derived PC (Biolipon 95), several IGC columns were prepared using 45/60 Mesh, DMCS-treated Chromasorb G column packing, allowing a matrix of IGC-derivable data to be accumulated at infinite dilution, such as mole and weight fraction activity coefficients, Henry's Law and partition coefficients, interaction and solubility parameters, as well as heats of solution and mixing. Mole fraction activity coefficients of the organic solute probes in PC show both positive and negative deviations from Raoult's Law and can be used to understand and optimize the choice of extraction-separation conditions. Activity coefficients less than unity are exhibited by both aromatic and chlorinated classes of solutes and particularly ethanol – all solute activity coefficients showing the excepted trend with increasing temperature. The experimental determined solubility parameters (δ) for PC over the above-stated temperature range vary from 20.5 – 18.3 $MPa^{1/2}$, δ data which has not been available previously. Enthalpy data, i.e., heats of solution and mixing indicate that at this concentration, that the recorded heats of solution are dominated by the heats of solute vaporization. The utilization of this IGC-derived data base will be discussed with respect to using it in the supercritical fluid extraction, dissolution, and purification of phospholipids.

MEASUREMENT AND CORRELATION OF SOLUBILITY AND DIFFUSION COEFFICIENT OF ETHYLENE IN MOLTEN PROPYLENE-CO-POLYMERS

Suiri Takizawa, Yoshiyuki Sato, Hiroshi Inomata, Tohoku University, Sendai, Japan

Abstract

Polypropylene copolymer is commonly used in our life and solubility and diffusion coefficient of monomer in polypropylene copolymer are important parameters to optimize the gas phase polymerization process. The objective of this work was to measure solubilities and diffusion coefficients of ethylene in four molten polypropylene based (ethylene and 1-butene) copolymer (cPP) samples at 140 °C and 180 °C and up to 2.5 MPa and to correlate solubilities of ethylene in cPPs with the Sanchez-Lacombe equation of state and diffusion coefficients of ethylene in cPPs with the free-volume-theory reported by Kulkarni and Stern. For the solubility, the Henry's constants at each temperature were to within 1.84% of deviation for all cPPs. The experimental solubilities were reproduced by the Sanchez-Lacombe equation of state to within 1.72% of the average deviation. For the diffusivity, the diffusion coefficients of ethylene in cPPs increased with increasing 1-butene content in cPPs. The difference of the diffusion coefficients can be explained by the difference of free volume fraction estimated from Pressure-Volume-Temperature data and glass transition temperature of the cPPs. In addition, the Kulkarni's free-volume-theory reproduced the experimental diffusion coefficients to within 4.96% of the average deviation.

Introduction

Polypropylene copolymer is widely used in the society due to its favorable mechanical and molding characteristics. Large amount of polypropylene copolymer is manufactured by the gas phase polymerization process which is progressed via monomer molecules dissolving and diffusing into polymer and reaching catalysts. It has been well known that polypropylene copolymer is crystalline polymer having both crystalline and amorphous regions in it and it has reported that monomer dissolution and diffusion only occur in the amorphous region.[1] Therefore crystallinity of polypropylene copolymer strongly affects solubility and diffusivity of monomer in polypropylene copolymer. Furthermore, elucidation of the relationship between crystallinity and solubility and diffusivity of monomer is important to optimize the polymerization process. However, it is difficult to obtain phase equilibrium data under a constant crystallinity since crystallinity under the gas phase polymerization condition has temperature and pressure dependences and is changeable by thermal histories. Although there are a few researches which have been reported the solubility and diffusivity of α-olefine in polypropylene (copolymer)[2, 3, 4], the solubility and diffusion coefficient of ethylene in molten polypropylene copolymers which have different crystallinity at solid state has not been reported.

In this work, solubility and diffusion coefficient of ethylene in molten polypropylene based copolymers (cPPs) which have different crystallinities at solid state were measured at 140 °C and 180 °C. Since the experimental temperatures are above melting point of cPPs, crystalline region in cPPs does not exist under experimental conditions and cPPs can be treated as amorphous polymer. In addition, solubility and diffusion coefficient of ethylene in cPPs were correlated by the Sanchez-Lacombe equation of state and the Kulkarni's free-volume-theory, respectively. Moreover, the difference of diffusion coefficient was discussed with free volume fraction of cPPs which was derived from the PVT data.

Materials and Methods

Ethylene (purity > 99.5%, Takachiho Chemical Industrial Co., LTD) was used as monomer and four polypropylene based copolymers with different crystallinities at solid state were used as polymer samples. Table 1 shows the characteristics of polymer samples. As seen in Table 1, PP1 is crystalline poly(propylene-*rand*-ethylene) random copolymer and PP4 is non-crystalline poly(propylene-*rand*-1-butene) random copolymer. PP2 and PP3 are poly(PP1-*brend*-PP4) blend copolymers with different weight fraction. Crystallinity was determined by density at solid state. Melting point (T_m, peak top) and glass transition temperature (T_g, midpoint temperature) were determined by DSC according to JIS K7121. Measurement of solubility and diffusion coefficient was carried out with the gravimetric method using a Magnetic Suspension Balance at temperatures of 140 °C and 180 °C and pressures up to 2.5 MPa with 0.5 MPa steps, according to the procedure Sato *et al.*[5] The PVT was obtained with the bellows method[6, 7]. The measurement condition was up to 240 °C and 200 MPa. The specific volumes of cPPs at atmosphere were extrapolated from isothermal PVT data in pressure range from 10 MPa to 200 MPa with the same method as reported in Sato *et al.*[6]

Table 1. The Characteristics of cPPs

cPP Sample	C3C2-RCP[*1] [wt%]	C3C4-RCP[*2] [wt%]	Crystallinity[*3] [wt%]	T_m[*4] [°C]	T_g[*4] [°C]
PP1	100	0	46.6	133.8	0.0
PP2	60	40	39.3	133.3	-5.7
PP3	20	80	29.1	133.1	-9.3
PP4	0	100	27.3	76.2	-11.3

*1: poly(propylene-*rand*-ethylene) random copolymer (propylene: 97.8 wt%, ethylene: 2.2 wt%)
*2: poly(propylene-*rand*-1-butene) random copolymer (propylene: 70.0 wt%, 1-butene: 30.0 wt%)
*3: determined from density at solid state
*4: determined by DSC

Models

The Sanchez-Lacombe equation of state for correlation of solubility

Solubility of ethylene in cPPs was correlated by the Sanchez-Lacombe equation of state[8, 9] (SL-EoS) shown in Equations 1 and 2.

$$\tilde{P} = -\tilde{\rho}^2 - \tilde{T}\left[\ln(1-\rho) + \left(1 - \frac{1}{r}\right)\tilde{\rho}\right] \quad (1)$$

$$\tilde{P} = P/P^*, \quad \tilde{\rho} = \rho/\rho^*, \quad \tilde{T} = T/T^* \quad (2)$$

P^*, ρ^*, and T^* are characteristic parameters of the SL-EoS. The characteristic parameters of ethylene were reported by Sato *et al.*[2] and those of cPPs were determined by correlating the molten PVT data with the SL-EoS. The characteristic parameters used in this work are shown in Table 2.

In order to apply the SL-EoS to binary systems, the mixing rule reported by Sanchez and Lacombe[8] was used. The binary interaction parameter k_{ij} in Equation 3 was a fitting parameter for solubility correlation.

$$P^* = \sum_i \sum_j \phi_i \phi_j \left(1 - k_{ij}\right)\sqrt{P_i^* P_j^*} \tag{3}$$

Table 2. The Characteristic Parameter of the SL-EoS

Sample	P^* [MPa]	T^* [K]	ρ^* [kg/m^3]
PP1	318.6	660.0	896.9
PP2	320.5	660.8	897.4
PP3	327.5	657.5	898.6
PP4	343.5	650.2	905.1
Ethylene[2]	335.0	285.1	648.0

Free Volume Theory for correlation of the diffusion coefficient

Diffusion coefficient was correlated by the Kulkarni's free-volume-theory[10] (FVT) expressed in Equations 4, 5 and 6.

$$D_{mtl} = \frac{x_2 D_{self}}{RT}\left(\frac{\partial \mu_1^P}{\partial \ln x_1}\right)_{T,P} \tag{4}$$

$$D_{self} = RTA_d \exp\left\{\frac{-B_d}{\phi_a v_f(T,P,v_1)}\right\} \tag{5}$$

$$v_f(T,P,v_1) = v_{fs}(T_g,P_s,0) + \alpha(T - T_s) - \beta(P - P_s) + \gamma v_1 \tag{6}$$

$$v_{fs} = \frac{V_g - V_{g,occ}}{V_g} = 0.025 \tag{7}$$

$$V = V_{occ} + V_f \tag{8}$$

A_d, B_d and γ are fitting parameters of the FVT and determined for each ethylene/cPP system as temperature and penetrant concentration independent parameters. The chemical potential of monomer in cPPs, μ_1^P, expressed in Equation 4 was calculated by the SL-EoS. The thermal expansion coefficient, α, and the compressibility coefficient, β, were evaluated by the method which Kulkarni and Stern.[10] suggested with the PVT data. The standard free volume fraction of polymer, v_{fs}, in Equation 6 is defined in Equation 7. In the free volume theory, the volume of polymer, V, is the total of the volume actually occupied by the polymer molecules themselves (the occupied volume, V_{occ},) and the free volume, V_f expressed in Equation 8. In this work, the universal constant, $v_{fs} = 0.025$, which was suggested by Wiliams *et al.*[11] was used. Additionally, the occupied volume of the polymer was assumed to be temperature independent.

Results and Discussion

Solubility

Figure 1 shows solubility of ethylene in cPPs at 140 °C and 180 °C. Solubility of ethylene increased with decreasing temperature and was proportional to pressures. The deviation of solubility of ethylene in PP1 and PP4 defined by Equation 9 was $Dev\_Sol_{PP4\text{-}PP1}$=1.69% in 140 °C and $Dev\_Sol_{PP4\text{-}PP1}$=0.93% in 180 °C. As considering the experimental uncertainty of solubility of ethylene in cPPs was to within ±5%, it can be said that there is no significant difference in the solubility between cPPs. This means that solubility of ethylene in crystalline polymer above its melting point is not affected by crystallinity of polymer. In addition, the Henry's constant for all ethylene/cPP systems was to within 1.84% of deviation at 140 °C and 1.47% of deviation at 180 °C.

Solid lines in Figure 1 denote the correlation with the SL-EoS. Experimental solubility of ethylene in cPPs was correlated with the SL-EoS to within the AAD defined by Equation 10 of 1.72%. Table 3 shows k_{ij} value and AAD for all cPPs and temperatures.

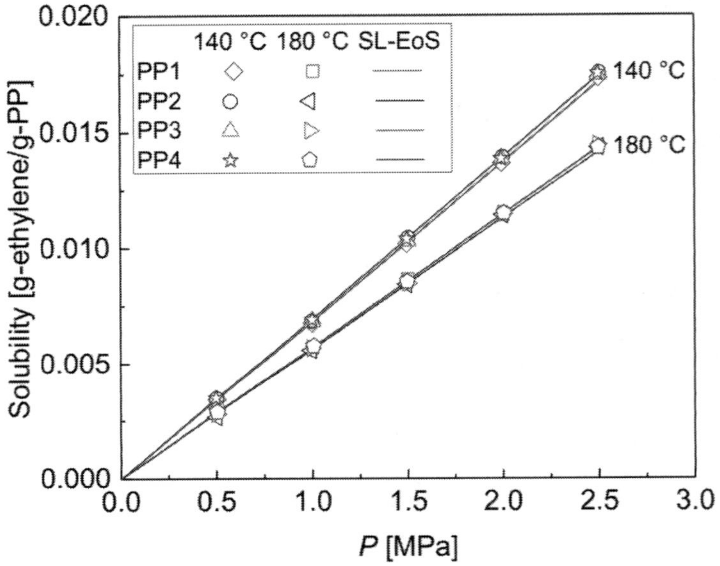

Figure 1. Solubility of Ethylene in Molten cPPs at 140 °C and 180 °C (Solid lines: Correlation with the SL-EoS)

$$Dev\_Sol_{PP4-PP1} = \frac{100}{N}\sum\frac{Sol_{PP4} - Sol_{PP1}}{Sol_{PP1}} \tag{9}$$

$$AAD = \frac{100}{N}\sum\frac{\left|Sol_{calc} - Sol_{\exp}\right|}{Sol_{\exp}} \tag{10}$$

Table 3. k_{ij} of the SL-EoS and AAD

cPP Sample	140 °C		180 °C	
	k_{ij} [-]	AAD [%]	k_{ij} [-]	AAD [%]
PP1	-0.0281	0.45	-0.0410	0.31
PP2	-0.0307	0.32	-0.0393	1.60
PP3	-0.0274	1.72	-0.0394	0.76
PP4	-0.0270	0.40	-0.0366	0.15

Diffusion Coefficient

Figure 2 shows the diffusion coefficients of ethylene in cPPs at 140 °C and 180 °C. Diffusion coefficients of ethylene in cPPs increased with increasing temperature and concentration of ethylene. The deviation of diffusion coefficients of ethylene in PP1 and PP4 derived by Equation 11 was about $Dev\_Diff_{PP4\text{-}PP1}$=24% at both temperatures. According to the sensitivity analysis, uncertainty of diffusion coefficient of ethylene in cPPs was 20%. Therefore it can be argued that the diffusion coefficient of ethylene in PP4 is larger than that in PP1, while there is quite different behavior of the solubility in the previous section.

Solid lines in Figure 2 denote correlation by the FVT. Experimental diffusion coefficients of ethylene in cPPs were correlated with the FVT to within the AAD defined in Equation 12 of 4.96% at both temperatures.

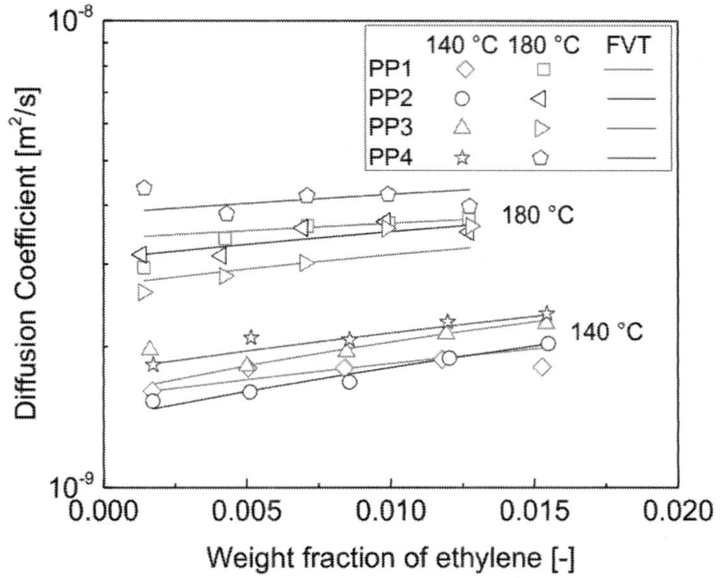

Figure 2. Diffusion Coefficient of Ethylene in cPPs at 140 °C and 180 °C (Solid lines: Correlation with the FVT)

$$Dev\_Diff_{PP4-PP1} = \frac{100}{N}\sum\frac{D_{PP4}-D_{PP1}}{D_{PP1}} \tag{11}$$

$$AAD = \frac{100}{N} \sum \frac{\left| D_{calc} - D_{\exp} \right|}{D_{\exp}} \tag{12}$$

Table 4. The Fitting Parameters of the FVT and AAD

cPP Sample	A_d [m^2 · mol/(s · J)]	B_d [-]	γ [-]	AAD [%]
PP1	2.13×10^{-11}	0.392	0.584	3.73
PP2	2.16×10^{-11}	0.417	0.763	2.69
PP3	6.11×10^{-12}	0.273	1.206	4.96
PP4	2.75×10^{-11}	0.434	0.627	3.52

Figure 3 shows the relationship of diffusion coefficients of ethylene in cPPs at infinite dilution and free volume fraction of cPPs. The diffusion coefficients at infinite dilution were discussed to neglect the interaction between penetrant molecules in polymer. The diffusion coefficients of ethylene in cPPs at infinite dilution were calculated by extrapolating the FVT to zero concentration. Free volume fractions of cPPs were derived by the PVT data. In Figure 3, free volume fraction of cPPs at each temperature increases with increasing 1-butene content in cPPs. In order to discuss this behavior, the free volume of cPPs was calculated with Equations 7 and 8. The specific volume of cPP at its T_g was extrapolated from the isobaric data of molten cPPs with the Tait equation. Table 5 shows the calculated occupied volume and the free volume of cPPs at 140 °C and 180 °C. From Table 5, the occupied volume of cPP decreases with decreasing its T_g. Besides, the free volume of PP4 at 140 °C and 180 °C is about 3% larger than that of PP1 as assuming the occupied volume of cPPs is temperature independent. This result is that free volume fraction of cPPs above T_m increased with increasing 1-butene content in cPPs. According to the FVT, diffusion coefficient of ethylene in polymer increases with increasing free volume fraction of polymer. In this study, free volume fraction of PP4 is larger than that of PP1. Therefore diffusion coefficient of ethylene in PP4 is larger than that in PP1.

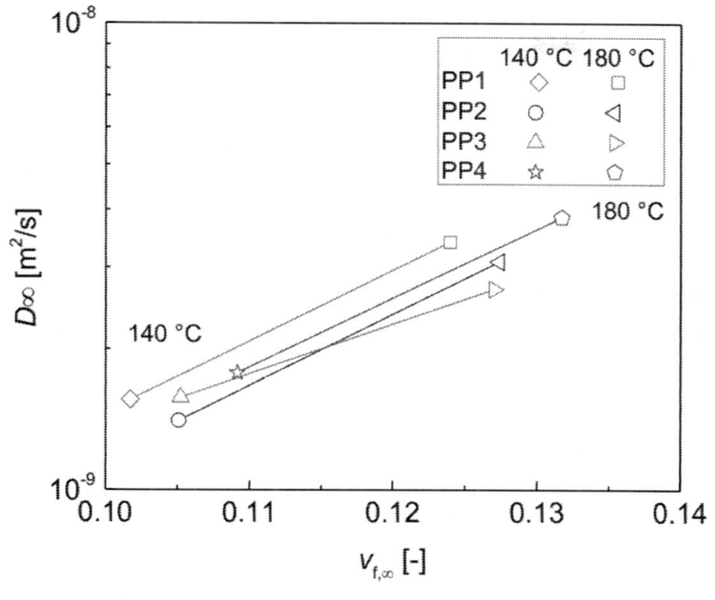

Figure 3. Diffusion Coefficient vs Free Volume Fraction of cPPs at Infinite Dilution

Table 5. The Occupied Volume and the Free Volume of cPPs

cPP Sample	T_g [°C]	V_{occ} [cm^3/g]	V_f [cm^3/g]	
			T=140 °C	T=180 °C
PP1	0.0	1.124	0.149	0.185
PP2	-5.7	1.119	0.152	0.188
PP3	-9.3	1.119	0.154	0.188
PP4	-11.3	1.115	0.155	0.190

Conclusions

Solubilities and diffusion coefficients of ethylene in molten cPPs were measured at 140 °C and 180 °C and up to 2.5 MPa. For the solubility, the Henry's constants at each temperature were to within 1.84% of deviation for all cPPs. For the diffusivity, the diffusion coefficients of ethylene in PP4 was 24% higher than that in PP1. In addition, diffusion coefficient of ethylene in cPPs at infinite dilution increased as 1-butene content in cPPs increased. The difference of diffusion coefficients of ethylene in cPPs was caused by the difference of the free volume of cPPs. Both experimental solubilities and diffusion coefficients of ethylene in cPPs were correlated by the SL-EoS and the FVT, respectively.

Symbols

P [MPa]: pressure, ρ [kg/m^3]: density, T [K]: temperature, P^* [MPa], ρ^* [kg/m^3], T^* [K]: characteristic parameters in the SL-EoS, r [-]: number of segment, ϕ [-]: volume fraction, k_{ij} [-]: interaction parameter, D_{mutual} [m^2/s]: mutual diffusion coefficient, D_{self} [m^2/s]: self diffusion coefficient, x [-]: mole fraction, R [J/mol/K]: gas constant, μ^P [J/mol]: chemical potential of polymer, A_d [m$^2 \cdot$ mol/(s $\cdot$ J)], B_d [-], γ [-]: characteristic parameters, T_g [K]: glass transition temperature of polymer, α [K^{-1}]: thermal expansion coefficient, β [MPa^{-1}]: compressibility coefficient, v_f [-]: free volume fraction of polymer, v_{fs} [-]: free volume fraction of polymer at standard state, V [m^3]: volume of polymer, V_{occ} [m^3]: occupied volume of polymer, V_f [m^3]: free volume of polymer, V_g [m^3]: volume of polymer at glass transition temperature, $v_{f,\infty}$ [-]: free volume fraction of polymer at infinite dilution, *Sol* [g_ethylene/g_polymer]: solubility, D [m^2/s]: diffusion coefficient

References

1. Micheals, A. S. and Bixler, H. J. (1961), "Flow of gases through polyethylene," *J. Polym. Sci.*, **50**, pp. 413-439.
2. Sato, Y., Tsuboi, A., Sorakubo, A., Takishima, S., Masuoka, H., Ishikawa, T. (2000), "Vapor-liquid equilibrium ratios for hexane at infinite dilution in ethylene+impact polypropylene copolymer and propylene+impact polypropylene copolymer," *Fluid Phase Equilibria*, **170**, pp. 49-67.
3. Tsuboi, A., Kolar, P., Ishikawa, T., Kamiya, Y., Masuoka, H. (2001), "Sorption and Partial Molar Volumes of C_2 and C_3 Hydrocarbons in Polypropylene Copolymers," *Journal of Polymer Science: Part B: Polymer Physics*, **39**, pp. 1255-1262.
4. Palamara, J. E., Mulcahy, K. A., Jones, A. T., Danner, R. P., Duda, J. L. (2005), "Solubility and Diffusivity of Propylene and Ethylene in Atactic Polypropylene by the Static Sorption Technique," *Ind. Eng. Chem. Res.*, **44**, pp. 9943-9950.
5. Sato, Y., Takikawa, T., Takishima, S., Masuoka, H. (2001), "Solubilities and diffusion coefficients of carbon dioxide in poly(vinyl acetate) and polystylene," *Journal of Supercritical Fluids*, **19**, pp. 187-198.
6. Sato, Y., Yamasaki, Y., Takishima, S., Masuoka, H. (1997), "Precise Measurement of the PVT of Polypropylene and Polycarbonate up to 330 °C and 200 MPa," *J. Appl. Polym. Sci.*, **66**, pp. 141-150.
7. Machida, H., Sato, Y., Smith Jr., R. L. (2008), "Pressure-volume-temperature (PVT) measurement of ionic liquids ([bmin$^+$][PF_6^-], [bmin$^+$][BF_4^-], [bmin$^+$][$OcSO_4^-$]) and analysis with the Sanchez-Lacombe equation of state," *Fluid Phase Equilibria*, **264**, pp. 147-155.
8. Sanchez, I. C. and Lacombe, R. H. (1976), "An elementary molecular theory of classical fluids. Pure fluids," *J. Phys. Chem.*, **80**, pp. 2352-2362.
9. Sanchez, I. C. and Lacombe, R. H. (1978), "Statistical thermodynamics of Polymer Solutions," *Macromolecules*, **11**, pp. 1145-1156.
10. Kulkarni, S. S. and Stern, S. A. (1983), "The diffusion of CO_2, CH_4, C_2H_4, and C_3H_8 in polyethylene at elevated pressures," *Journal of Polymer Science: Polymer Physics Edition*, **21**, pp. 441-465.
11. Williams, M. L., Landel, R. F., Ferry, J. D. (1955), "The Temperature Dependence of Relaxation Mechanisms in Amorphous Polymers and Other Glass-forming Liquids," *J. Am. Chem. Soc.*, **77**, pp. 3701-3707.

466050 Molecular Structure and Dynamics in Room Temperature Ionic Liquids Insights from Molecular Experiment and Simulation

Abstract:

One the focuses of Cor Peterâ€™s professional career has been on the thermodynamic properties of room temperature ionic liquids (RTILs) and phase equilibria in mixtures of RTILs and other components. In our group, we have a long-term interest in RTILs for their application in capacitative energy storage systems (specifically, so-called electrical double layer capacitors or supercapacitors). To this end, we have investigated, both computationally and experimentally, the RTIL/nanoporous carbon electrolyte/electrode combinations, in which the key issue for capacitative energy storage is the structure and dynamics in the electrical double layer formed at the electrolyte/electrode interface. Additionally, to contrast with the behavior at the interface, using a combination of molecular simulation and molecular experimental probes (including NMR, SAXS, and SANS), we have investigated the structure and dynamics of neat RTILs, RTIL mixtures and dicationic RTILs to understand their bulk behavior. We will provide an overview of this latter work in this presentation.

465142 Slip Slidin' Away: Three Decades of Adventures in Computational Rheology and Lubrication

Abstract:

Carol Hall has been a defining influence on the lives and careers of many of us. As an academic younger brother (Carol received her PhD under the direction of George Stell, with whom Cummings would later perform postdoctoral research), Cummings has enjoyed career-long benefit of Carol's advice and unwavering support. As a young assistant professor new to the discipline that Carol helped him to enter, Cummings sought Carol's advice about where to make a unique impact in chemical engineering. Transport properties was one of the areas that came to the forefront of these discussions, and so began – with the first paper published in 1986 – a three decades long focus on applying non-equilibrium molecular simulation to the prediction and understanding of transport properties in systems ranging from the simple to the complex. We will briefly review some of the highlights of this work, concluding with our recent work exploring the hypothesized hydration lubrication mechanism in knee joints that purports to explain extremely low friction coefficients in a biosystem subject to very high normal pressures.

471438 How to Make Particles Interact in Multiple Directions and Form Interconnected Networks: Colloidal Assembly Based on Field-Induced Multipoles

Abstract:

The assembly of new materials based on particle networks and ordered phases requires the application of "interaction toolbox" that allows independent control of the particle interactions in multiple directions by external field(s). It also necessitates a closely coordinated experimental and theoretical effort, as the enormous parameter space of particle types, interactions, densities, equilibrium/non-equilibrium assembled states and corresponding dynamics are nearly impossible to explore by experimental means only. We will present the experimental and theoretical basis of two strategies for such multidirectional assembly. The first strategy is based on using complex asymmetric particles. Metallo-dielectric particles, such as Janus and patchy spheres and cubes, acquire complex polarization patterns in AC electric and magnetic fields, which can lead to multidirectional interactions. They can be controlled through the field frequency and strength. We will demonstrate how quadrupolar, hexapolar and multipolar polarization interactions induced by external electric and magnetic fields can be used to assemble new types of staggered chains, networks of unusual symmetry and crystals. We will show how the particle assembly motifs and long-scale structures can be analyzed via numerical modeling. The second strategy for multidirectional assembly is to use simple particles that polarize and interact in complex external fields. We show how this can be achieved by simultaneous application of electric and magnetic fields to dispersions of superparamagentic microspheres. We investigate the dynamics of the assembly process, and identify and classify the intermediate states formed. The concurrent field application lead to the formation of bidirectional chains, colloidal networks, and discrete crystals. The morphologies of these structures are in good correlation with theoretical predictions based on Brownian Dynamics simulations. We distinguish two separate kinetic routes for bidirectional assembly, colloidal "polymerization" and "re-polymerization", leading to the formation of structures with specific symmetry, connectivity, directional response or unusual transport properties.

474934 Polymers in Ionic Liquids

Abstract:

Ionic liquids have generated considerable excitement for their varied potential applications and their interesting physical properties. The viability of ionic liquids (ILs) in materials applications is limited by their lack of mechanical integrity, which may be provided by mixing them with a polymeric material. Recent experiments on polymers in ILs have unearthed a wealth of interesting phenomena that raise fundamental questions. This talk focuses on computational studies of PEO in imidazolium ILs. We develop a physically motivated first principles force field; this force field is in quantitative agreement with experiment with no adjustable parameters. The calculations emphasize the importance of polarization but show little evidence for charge transfer. Based on the same quantum calculations we develop a hierarchy of united atom models with decreasing resolution and increasing computational efficiency. Atomistic simulations show a lower critical solution temperature in the PEO in [BMIM] [BF_4] system and provide physical insight into the mechanism based on entropic effects.

454964 *Collective Dynamics of Dipolar and Multipolar Colloids: From Passive to Active Systems*

Abstract:

In this talk I will present recent research on the collective dynamical behavior of colloids with dipolar or multipolar interactions. Starting from "passive" dipolar particles exhibiting network formation and various types of cluster aggregation, we will then move towards systems composed of dipolar "microswimmers" with unidirectional propulsion. These systems display a transition towards a laned state reminiscent of lane formation in driven systems with isotropic repulsive interactions. We interpret our findings by a phase separation process and force balance arguments. Finally, we discuss first results on the collective behavior of dipolar particles with random self-propulsion.

462510 Promising Vistas for Applied Chemical Thermodynamics: Statistical Mechanics As a First Step Toward Thermodynamic Properties

Abstract:

When statistical mechanics is coupled with concepts from physical chemistry, useful results for condensed systems can often be obtained; perhaps the best-known example is Floryâ€˜s derivation of a remarkably sample formula for the configurational entropy of mixing a long-chain polymer with a solvent. For obtaining quantitative thermodynamic properties, statistical mechanics alone is not sufficient; it is also essential to construct a carefully-considered physical model that is simple enough for easy calculations but also sufficiently realistic to represent physical phenomena.

While this procedure is not new, during the last generation, its application for chemical engineering has grown extensively in part, due to the leadership of Carol Hall. To illustrate, some examples from the literature are reviewed. While these examples may or may not refer directly to Carolâ€™s published work, they all exhibit the beneficial influence of her research philosophy.

472566 A Biased Review about Computer Simulations of Protein Folding and Aggregation

Abstract:

Amyloid fibrils are ordered aggregates of misfolded peptides or proteins associated with type 2 diabetes mellitus, Alzheimer's disease, Parkinson's disease, etc. Amyloid diseases are apparently unconnected; however, intermolecular secondary structure (mostly β-sheets) is present in all amyloid aggregates.

Assumptions about the dominance of a particular type of force are common in computer simulations of protein aggregation –most of the times these simplifications are based on experimental observations. However, sometimes the simplifications are rooted on the knowledge of other systems or simply on accepted rules of thumb. Two common simplifications are to overemphasize electrostatic interactions (at one end) or hydrophobic interactions (at the other end). Even though the promoters of each approach make good points in defending them, they (for the most part) ignore the role that solvent may have in stabilizing the structure. Most simulations (are performed with very few peptides, a single protein, a single monomer added to an existing fibril. The fact that aggregation happens at multiple time scales adds complexity to the theoretical work. Prof. Hall's group has used a simplified model for peptides consisting of a few united atoms bearing a square-well potential that represents either hydrogen bonding or hydrophobic forces to study the self-assembly of peptides of alanine and glutamic acid. It is fascinating that such a simple model predicts the correct assembly, even after incorporating the solvent into an oversimplified potential of mean force. Another mechanism that has been explored is molecular crowding; this approach allows for the use of a peptide at infinite dilution in a crowding media. These studies support the idea of incorporating interfaces --we have already shown that interfaces act as an enrichment site in the study of peptide aggregation.

The few reports devoted to study the effect of surfaces on amyloid deposits formation are mostly dedicated to finding a link between the interaction of the deposits with cell surfaces and their toxicity. Concentration

and misfolding of the peptides at the solid/liquid interface support the hypothesis that the surface serves as a catalyst. Recently, we have studied the fibrillation of a series of diblock peptides in the presence of poly(styrene) latex with various surface chemistries and of liposomes with different phospholipid compositions. It was concluded that fibrillation depends on the composition of the diblock and on the presence or absence of interfaces but no adsorption was observed. Interestingly, the replacing of an isoleucine block by a leucine one in diblock peptides of the same length inhibits fibrillation. The consequences of these findings on current modeling efforts as well as the development of a new model are discussed.

464055 "Grafting through" Polymerization

Abstract:

We use a combination of computer simulation and experiments to comprehend "grafting through" polymerization, i.e., copolymerization of free monomers in solution and polymerizable units bound chemically to a substrate. Free polymer chains are formed initially in bulk solution and if they incorporate at least one surface-bound monomer, the macromolecules get covalently bonded to the surface during the polymerization process. We concentrate specifically on addressing the effect of spatial density of the surface-bound monomers on the formation of the surface-attached polymers. We employ a lattice-based Monte Carlo model based on the bond fluctuation model scheme to provide molecular-level insight into the grafting process. For experimental validation, we create gradients of density of bound methacrylate units on flat silicon wafers using organosilane chemistry and carry out "grafting through" free radical polymerization initiated in bulk. The proximity of the surface-bound polymerizable units promotes the "grafting through" process but prevents more free growing chains to "graft through" the polymerizable units.

468105 Understanding Temperature-Responsive Behavior of Elastin-like Polypeptides

Abstract:

The elastin-like polypeptide (ELP) is temperature sensitive biopolymer which undergoes conformational transition at critical inverse transition temperature or LCST; the feature which has led to their use in a variety of applications. ELPs are defined by a pseudo-periodic, low complexity sequence with repeat motifs, as in a model system with repeat sequence of VPGXG. Substitution of X residue with various amino acids or modifications of chain length and solute quality allows for tuning the inverse transition temperature. However, the molecular origin of LCST and the observed differences as a function of various factors are not yet completely understood. In order to elucidate the factors driving the conformational transitions of ELPs we performed large scale all-atom molecular dynamics simulations of ELPs with different sequence and length and at a range of temperatures. Comprehensive analyses of the hydrophobicity, shape, size, hydration and dynamics of ELPs suggest that the transition behavior includes the thermal disruption of water network, loss of hydration, and increase in peptide hydrophobicity. In this talk I will discuss the temperature sensitive behavior of polypeptides as a function of sequence, sequence direction, length, and concentration. The results from our study provide an atomic-level description of the thermo-responsive conformational properties of elastin-like peptides and explain experimental observations.

477518 Utilizing Multivariate Analysis to Enhance Bioprocess Development and Operation

Abstract:

Advances in the development of cell culture bioprocesses have resulted in an extraordinary amount of data, which spreads over all stages of development and operation. This accumulation of data is commonly overlooked but often contains untapped resources and information. This talk aims to outline high level strategic approaches for multivariate analysis and visualization techniques which can extract this knowledge and be further applied to enhance all facets of biopharmaceutical production.

477948 Self-Assembly of Colloids in Ordered Carrier Fluids

Abstract:

The talk focuses on the complex interplay of molecular scale effects, nonlinearity of the orientational field, and long-range elastic forces which make the liquid-crystal physics very challenging. A consistent way to extract information from the microscopic, molecular scale up to the meso- and macroscopic scale is still missing. Here, we develop a hybrid procedure that bridges this gap by combining extensive Monte Carlo simulations, a local Landau-de Gennes theory, classical density functional theory, and finite-size scaling theory. As a test case to demonstrate the power and validity of our novel approach we study the effective interaction among colloids with Boojum defect topology immersed in a nematic liquid crystal. In particular, at sufficiently small separations colloids attract each other if the angle between their center-of-mass distance vector and the far-field nematic director is about 30ï,°. Using the effective potential in coarse-grained two-dimensional MC simulations we show that self-assembled structures formed by the colloids are in excellent agreement with experimental data.

464201 The Relationship Between Chitinase Binding Affinity, Processivity, and Work Required to Decrystallize β -Chitin

Abstract:

Structural polysaccharides, such as cellulose and chitin, are highly crystalline in nature owing to their complex networks of intra- and inter-molecular hydrogen bonds; this provides strong resistance to microbial and chemical attack. Nature evolves enzyme cocktails, including processive glycoside hydrolases (GHs), to hydrolyze these recalcitrant polysaccharides to soluble sugars. Between subsequent hydrolytic reactions, processive GHs remain attached to the polymeric substrate, and hence, are responsible for the majority of hydrolytic bond cleavage. However, our understanding of the relationship between processive enzyme function and the crystalline polymorph and morphology is limited. We hypothesize that the free energy of binding an oligosaccharide to the active site of a processive GH is directly related to the ability of an enzyme to processively hydrolyze a polymer chain from a crystal; moreover, the ligand binding free energy of a processive enzyme must also be more thermodynamically favorable than the work required to decrystallize the polymer. We have selected the *Serratia marcescens* Family 18 chitinase model system, including processive chitinases, ChiA and ChiB and a non-processive chitinase, ChiC, to test our hypothesis. Combining molecular dynamics simulations and free energy calculations, we find that processive ChiA and ChiB exhibit ligand binding free energies that are more thermodynamically favorable than the work to decrystallize a chito-oligosaccharide from the crystalline Î²-Chitin surface, which is essential for forward processive movement. The non-processive ChiC binds chito-oligosaccharides with a free energy that is significantly less favorable than the work of decrystallization. In general, our findings suggest that processive GH function necessitates tight binding within the enzyme active site. Comparisons to cellulases aid in understanding the transferability of these findings to other processive GHs.

468643 Studying Peptoid Side-Chain/Structure Relationships Using Metadynamics and the Generalized Charmm Force Field

Abstract:

Peptoids (poly-n-substituted glycines) are peptide-like macromolecules composed of modified glycine units which have side chains attached to the nitrogen atoms. Peptoids are biocompatible, resistant to protease degradation, and they can be synthesized using more than 300 commercially available amines, making peptoids versatile and appealing compounds for combinatorial materials design. Applications for peptoids include antimicrobial coatings, binding inhibitors, drug carriers, and lung surfactants.

We have previously fit peptoid-specific parameters to for the CHARMM General Force Field (CGENFF). We briefly describe the important considerations to observe when fitting peptoid parameters with CHARMM. Finally, we show the effectiveness of these CGENFF peptoid parameters in reproducing experimentally observed peptoid structures.

We present molecular simulation studies of the relationship between peptoid side chains and the peptoid backbone folding. We show the results of atomistic molecular dynamics simulations, Well-Tempered Metadynamics, and Bias Exchange Metadynamics that explore the free energy surfaces of peptoid dimers and short peptoid chains. Finally, we suggest new combinations of peptoid side chains and offer predictions on their backbone structure.

This work presents a framework for studying peptoid side-chain/structure relationships at the atomistic level using a CHARMM based force-field. In the near future, the methods here will be combined with additional NMR studies to study novel, non-protein-like peptoid side chains and their effect on peptoid backbone conformations.

472877 Reactive Molecular Dynamics Simulation of Liquid Water and Ice Crystal (Ice-1h) Â Using a New Generation of Reactive Force Fields Based on Polarized Charge Distributions and Valance Bond Concepts

Abstract:

Water is probably the most important molecule to biological systems. Although it has a simple molecular structure, it is considered to be one of the most complex substances. Particular properties of water such as hydrogen bond and polarization effects are responsible for its complex behaviors. Despite of extensive atomistic simulations that have been performed during last decades, there is a growing interest to improve water potentials to describe various phenomena.

Accurately describing the electrostatic interaction between water molecules has been a longstanding challenge due to polarization effects. To address this problem, we propose a new generation of reactive force field based on polarized charge distribution (PQEq) and valance bond concepts. Our potential model allows an accurate description of the local polarization needed to describe dynamic dielectric properties of water. In PQEq methodology each atom is represented by a variable charge plus two fixed charges (core and shell), modeled with a Gaussian charge distribution. This enables the charges to evolve in response to the dynamics while including damping. In addition, in our new approach we use valence bond concepts to understand and simulate proton solvation and transport in water. These together provide all essential ingredients to model various dissociation pathways and formation of chemical bonds in water simulations.

Here, we first validate our model by describing the experimental density and equation of state of the H_2O crystal (ice-1h). We also compare the H_2O molecule and H_2O dimer dissociations with the best available quantum mechanics computations.

We also report reactive molecular dynamics of bulk water and Ice-1h crystal for large time and size scales to study structural and dynamical properties of water such as melting point, phase diagram, dielectric constants, etc. We validate our results by comparing them with available experimental data.

472131 Effect of Functionalization on Surfactant Activity of Amphipathic Surfactants â€“ Physics of Bile Salts and Counter-Ions in Aqueous Suspensions of Single-Walled Carbon Nanotubes

Abstract:

Bile Salts constitute a diverse class of multifunctional amphipathic surfactants. Recently, bile salts have been employed in analytical centrifugation (AUC) to efficiently enrich and separate single-walled carbon nanotubes (SWCNT) based on chirality. Thus, bile salts seem capable of amplifying the relatively minor differences in hydrodynamic behavior of SWCNTs based on chirality. Further, it was found that subtle single-point mutation of functional groups on bile salts had a drastic impact on the hydrodynamic behavior of the aqueous suspensions of SWCNTs. These results provide strong motivation for understanding the relationship between molecular architecture and thermodynamics of the bile salts at the aqueous SWCNT interface. We use all-atom molecular dynamics simulations to study models of SWCNT systems containing different bile salts. Our findings indicate that the overall hydrodynamic behavior of the SWCNT-surfactant complex in water is influenced by an interplay of interactions between (a) SWCNTs and surfactants, (b) surfactants between themselves, (c) surfactant and the counter-ion, and (d) the counter-ion and its hydration layer.

472623 Investigating the Planarity of Aromatic Rings in Biomolecular Modeling

Abstract:

The accuracy of biomolecular modeling results is based in part on Force field (FF) parameters. Charmm FF is one of the best FFs in describing the intermolecular interactions of biological systems. In this work, Monte Carlo (MC) simulations were conducted of a novel antibacterial peptide on the basis of Charmm27 FF. Implementation of Charmm FF led to some structural deficiencies in aromatic rings from their expected physical structure. As the simulations progressed to equilibrium, nonplanarity was seen for the aromatic rings of Tryptophan (TRP) and Tyrosine (TYR). It was found that Charmm27 does not provide any improper torsion for carbon atoms in aromatic rings of these two units. These atoms have an SP2 hybridization and thus are typically planar. In addition, certain minimum angle parameters in Charmm27 induce these rings in both units to deviate from being in a plane. Considering improper torsion and amending the minimum angle parameters in MC simulation brings the aromatic rings in TYR and TRP closer to planar; however, the aromatic ring of TRP still has unacceptable deficiencies from planarity. Nevertheless, it cannot be stated that current Charmm FF (without improper torsion parameters) induces nonplanar structures of planar rings over simulation in TYR and TRP. Natural vibration of atoms in molecule might be the reason of this nonplanarity. All atom Normal Mode Analysis (NMA) provides the vibrational frequencies of atoms in those rings. Comparing the results of this calculation to vibrational spectra data can clarify to a certain extent the effect of Charmm FF in the deficiencies of aromatic ring from planarity. To that end, molecular dynamics (MD) simulations of TYR and TRP in the water are carried out on the basis of either the last version of Charmm(36) or modified Charmm by considering some improper torsion parameters and amending minimum angles parameters for atoms in the aromatic rings of TYR and TRP. Analytical and numerical NMA is applied on the configuration results of MD simulation to calculate the average vibrational frequency and normal mode of each atom in the system. The results of NMA are compared to infrared spectra of TRP and TYR. This comparison of simulation and experimental data can show how considering improper torsion or adding amending minimum angle parameters can change the calculated vibrational spectra of short peptides.

471348 Probing the Structure and Aggregation Features of 1-N-alkyl-3-Methyl Imidazolium Octylsulfate Ionic Liquid Homologous Series Using Molecular Dynamics Simulations

Abstract:

Ionic liquids (ILs) are novel chemical substances composed entirely of ions and are attractive candidates for replacement of volatile organic compounds used in chemical industry. One of the widely research topics in the field of IL is their ability to absorb polar substances such as water, CO_2 and SO_2 owing to their ionic nature. At the same time, conventional ILs exhibit rather poor solubility for nonpolar substances such as alkanes commonly obtained from crude oil. One strategy, to enhance the solubility of nonpolar compounds, is to increase the nonpolar content, for example, by increasing the alkyl chain length, in either the cation or the anion. It has been established that such changes induce structural heterogeneities resulting in the formation of nanodomains composed of polar and nonpolar regions in which alkanes can be dissolved. It is believed that the origin of such molecular level segregation is due to the competition between the electrostatic interactions of the ionic moieties in ILs and short-range attractive interactions between the nonpolar regions. An increasing amount of research studies are being carried out by carefully selecting alkyl chain lengths either on cation or anion to tune the size and shape of these domains. In this study, mesoscopic segregation of ILs containing long alkyl chains on both the ions is probed with molecular dynamic simulations. The ionic liquids studied here belong to the 1-*n*-alkyl-3-methyl-Imidazolium [$C_n$mim] (n=1,2,4,6,8,10,12) family with octylsulfate [C_8SO_4] as the anion. The results of total static structure factors and radial distribution functions will be presented in an effort to interpret the morphology of polar and non-polar domains. Further, the nonpolar region will be characterized in terms of aggregate size and shape using domain analysis.

470789 Bulk Property Changes in Glassy Materials Induced By Photoisomerization of Azobenzene

Abstract:

Changes induced by the photo-isomerization reaction for various azobenzene derivatives have been used in various applications, including light activated switches in polymer systems and surface modified materials. In recent years, studies have shown that the addition of azobenzene derivatives can allow for photo-induced changes in the properties of glassy materials. For instance, the photo-isomerization reaction within a glass matrix can cause the material to become malleable far below the glass transition (T_g) of the material. Glassy materials are of particular interest since they exhibit many mechanical properties similar to a crystalline solid, but the underlying structure is heavily dependent upon the process and conditions under which it was formed. Vapor deposited (VD) organic glasses have been of interest in recent years, exhibiting exceptionally enhanced properties depending on the temperature of the substrate it was deposited. Using a coarse-grained azobenzene model, we formed vapor deposited films at various substrate temperatures and investigated the dependence of photo-reactivity on substrate temperature. In addition to the other enhanced properties already observed in VD film, using a procedure that mimics the photo-excitation process of shining polarized light, we have shown that VD glasses are more photo-stable and exhibit a two-stage behavior on the way to the stationary state, which is not observed in the conventional glasses. Similar observations have been seen in vapor deposited films of an azobenzene derivative, disperse orange 37. During the initial stages of irradiation, the vapor deposited films change little in structure but a negative stress builds in the bulk of the film. The film then rapidly relaxes to the stationary state after enough stress has built up. There are some cases where this photo-induced malleability is of interest. For instance, we demonstrate that by using this process in a PEG-Azo-PLA polymersome system the glassy matrix of the PLA can be used as a gate for drug delivery. By plasticizing the PLA hydrophobic layer through the photo-isomerization reaction, the vesicles will release the drug and then reseal once the irradiation ceases and the glassy dynamics are restored.

462949 Multi-Scale Modeling of Periodic Mesoporous Silica Materials: Exploring the Role of Silica Oligomers

Abstract:

Periodic mesoporous silicas (PMSs) have shown great potential in many applications such as catalysis, membrane separation, drug vehicles, etc. This kind of materials is usually fabricated through the self-assembly of surfactants and silica precursors in solutions, followed by the rearrangement of those species and thus forming different mesostructures. Several studies have been carried out to investigate various aspects of MCM-41 materials, one of the most popular PMSs, including their pore structure, pore properties, and formation mechanisms.[1] However, the elucidation of the formation process is hampered by the high complexity of the self-assembly between organic-inorganic species in solutions. To unravel the MCM-41 formation mechanism, a multi-scale modeling simulation approach is proposed herein to simulate mesostructure formation at large length and time scales.

Our simulations show that the silica oligomers play an important role during the MCM-41 synthesis. A substantial degree of condensation is required to promote the formation of hexagonal array. Those multiply charged silica oligomers are able to bridge adjacent micelles, thus allowing them to overcome their mutual repulsion and form aggregates. They also induce a phase separation that contains a dilute solution and a silica/surfactant-rich mesophase, which leads to MCM-41 formation at a very low surfactant content.[2] In addition, the system with a larger surfactant content shows that liquid crystal templating mechanism is not viable for MCM-41 synthesis. The hexagonal phase found in pure surfactant solution may collapse while the silica monomers are added, and it forms again after a higher degree of condensation of silica is reached. This study provides new insight into the formation mechanism of PMS materials, enabling tailored design of nanoporous materials using computational models.

[1] Beck, J. S., Vartuli, J. C., Roth, W. J., Leonowicz, M. E., Kresge, C. T., Schmitt, K. D., Chu, C. T. W., Olson, D. H., Sheppard, E. W., McCullen, S. B., Higgins, J. B., and Schlenker, J. L., *J. Am. Chem. Soc.* **1992**, *114*, 10834.

[2] Peİrez-Saİnchez, G., Chien, S.-C., Gomes, J. R. B., Cordeiro, M. N. D. S., Auerbach, S.M., Monson, P.A., and Jorge, M., *Chem. Mater.* **2016**, *28*, 2715.

DSMC SIMULATIONS OF LEADING EDGE FLAT- PLATE BOUNDARY LAYER FLOWS AT HIGH MACH NUMBER

S. Pradhan
Department of Chemical Engineering
Indian Institute of Science
Bangalore- 560 012
India

ABSTRACT

The flow over a 2D leading-edge flat plate is studied at Mach number $Ma = (U_{inf} / \sqrt{\gamma\ k_B\ T_{inf} / m})$ in the range $3 < Ma < 10$, and at Reynolds number number $Re = (L_T\ U_{inf}\ \rho_{inf})/\mu_{inf}$ equal to 10^4 using two-dimensional (2D) direct simulation Monte Carlo (DSMC) simulations to understand the flow phenomena of the leading-edge flat plate boundary layer at high Mach number [1, 2, 3, 4, 5, 6, 7, 8, 9, 10]. Here, L_T is the characteristic dimension, U_{inf} and T_{inf} are the free stream velocity and temperature, ρ_{inf} is the free stream gas density, m is the molecular mass, μ_{inf} is the molecular viscosity based on the free stream temperature T_{inf}, and k_B is the Boltzmann constant.

The variation of streamwise velocity, temperature, number-density, and mean free path along the wall normal direction away from the plate surface is studied. The qualitative nature of the streamwise velocity at high Mach number is similar to those in the incompressible limit (parabolic profile). However, there are important differences. The amplitudes of the streamwise velocity increase as the Mach number increases and turned into a more flatter profile near the wall. There is significant velocity and temperature slip ((Pradhan and Kumaran, *J. Fluid Mech*-2011); (Kumaran and Pradhan, *J. Fluid Mech*-2014)) at the surface of the plate, and the slip increases as the Mach number is increased. It is interesting to note that for the highest Mach numbers considered here, the streamwise velocity at the wall exceeds the sound speed, and the flow is supersonic throughout the flow domain. The subsonic region near the wall, expected when a no-slip boundary condition is applied, is not present when there is wall slip at sufficiently high Mach number.

In a compressible leading-edge flat plate boundary layer flows we determine the mean free path at different streamwise direction (x = 0.2, and 0.8 m) away from the plate surface, and found significant dependence on the streamwise location. This is due to the variation in the local temperature along the streamwise direction across the boundary layer by the viscous heating. The leading edge shock wave is evidently captured in the present DSMC simulations. An important finding is that the wall heating (by increasing the wall temperature in the simulations) increases the local mean free path at the wall. However away from the wall the mean free path profiles remains unaffected by the wall heating.

COMPUTATIONAL DOMAIN

A rectangular domain of total length (L_C + L_T) and height H is considered, where the x, *and* y axes are in the streamwise, and wall-normal directions respectively. The velocity components in the streamwise & wall-normal directions, density, pressure and temperature are denoted by u, v, ρ, p, and T respectively. The free-stream gas density is denoted by ρ_{inf}, while the free-stream gas velocity and temperature are U_{inf} *and* T_{inf} respectively.

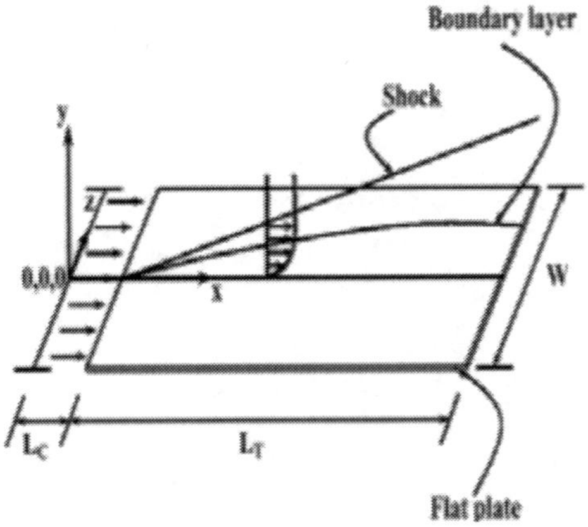

Computational domain and coordinate system

The fluid is a hard sphere gas with the molecular diameter $d = 4.17 \times 10^{-10}$ m, molecular mass $m = 4.6480 \times 10^{-26}$ kg/molecule, viscosity-temperature index 0.72. It is assumed that the gas molecules undergo diffuse reflection from all the solid surfaces with the thermal accommodation coefficient equal to 1.

DSMC SIMULATION

The Direct Simulation Monte Carlo method (DSMC) is a numerical method for modeling rarefied gas flows, where the Boltzmann equation for the probability distribution function for gas molecules is solved using a probabilistic (Monte Carlo) simulation technique. The DSMC method uses simulated molecules which represent a large number of real gas molecules. The translation of molecules between collisions is deterministically modeled, while the collisions between two molecules or a molecule and a bounding surface are treated stochastically. The domain is divided into a large number of cells, and properties are evaluated in these cells. The cells are further divided into 'sub-cells' for collision localization.

In the DSMC simulations, inter-molecular collisions are executed at random from all molecules within 'sub-cells' within the simulation cell, whose linear dimension is much smaller than the mean free path. The probability of collisions between a pair of molecules within a sub-cell is selected equal to that for a dilute gas with the same temperature as that in the cell. In this manner, the temperature (mean square velocity), mean free path and collision frequency for the simulated molecules are equal to those for the real molecules, provided the sub-cell dimension is much smaller than the mean free path. Due to the temperature in-homogeneities in a turbulent flow, an additional precaution required is that the collision probability should be chosen based on the local temperature in the cell, and not the average temperature.

The most common molecular model is the hard-sphere model, but other models such as the variable hard sphere and variable soft sphere models are also used in order to accurately reproduce the density and temperature dependence of the viscosity. At the boundaries, it is necessary to prescribe interaction laws between the molecules and the solid surfaces.

Here, we adopt the commonly used diffuse reflection condition with wall accommodation coefficient set equal to 1. In this, the velocity of a molecule after a wall collisions is chosen randomly from a Maxwell-Boltzmann distribution with the same mean velocity and temperature as that of the wall. The diffuse reflection boundary condition results in a non-zero velocity slip at the wall. This is because even though the mean velocity of the reflected molecules is equal to that of the wall, the mean velocity of the incident molecules could be different from the wall velocity. This results in a nonzero velocity slip. In our previous study on lid-driven cavity, it was shown that there could also be a temperature slip at the wall due to the diffuse reflection boundary condition. The velocity and temperature slip become important in our comparison with the earlier DNS results on compressible turbulent flows. The hydrodynamic fields (density, velocity, temperature) are obtained by averaging over the molecular distribution functions in the DSMC simulations.

The details of the DSMC simulation procedure are discussed in Bird (1994), and so we do not repeat the discussion here, but restrict attention to the parameters in the simulations. The DSMC simulations are advantageous because they provide access to molecular parameters, such as the mean free path, in addition to continuum parameters. Temperature and velocity slip, which need to be modelled in continuum simulations, follow naturally from the particle-wall interactions specified in the DSMC simulations.

RESULTS

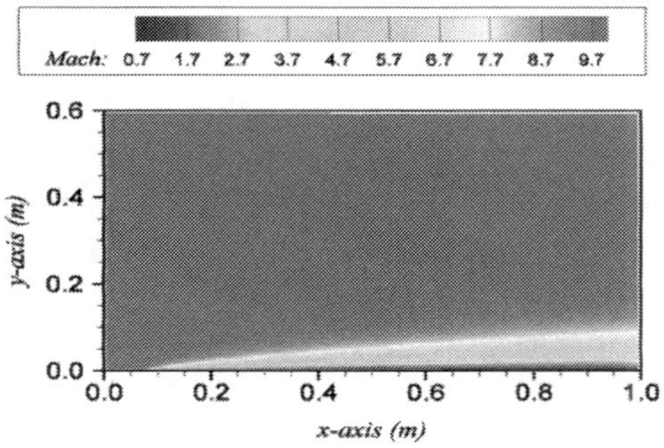

Mach Contour at $Ma = 10;\ Re = 10^4$

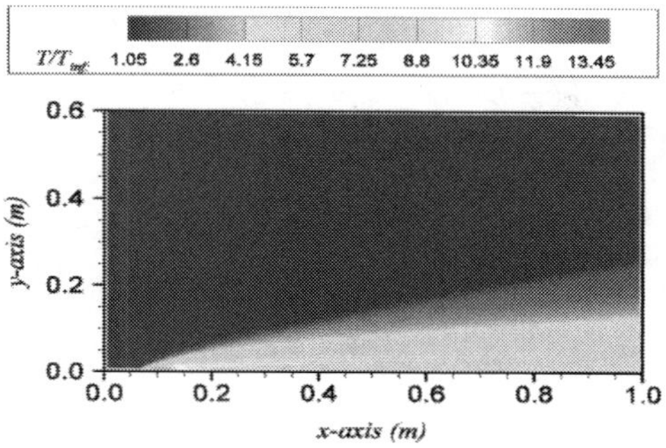

Temperature Contour at $Ma = 10;\ Re = 10^4$

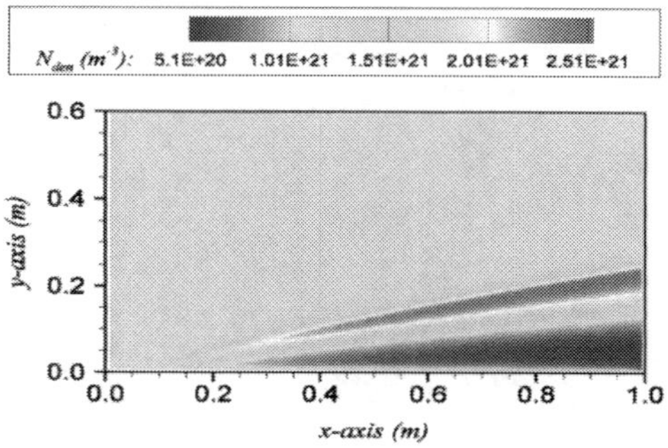

Number density Contour at $Ma = 10;\ Re = 10^4$

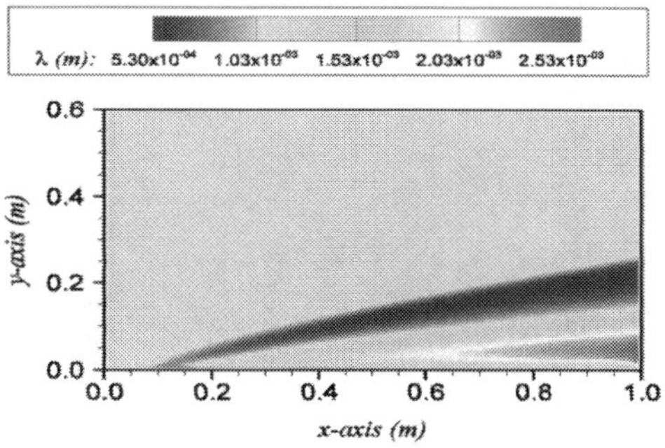

Mean Free Path Contour at $Ma = 10;\ Re = 10^4$

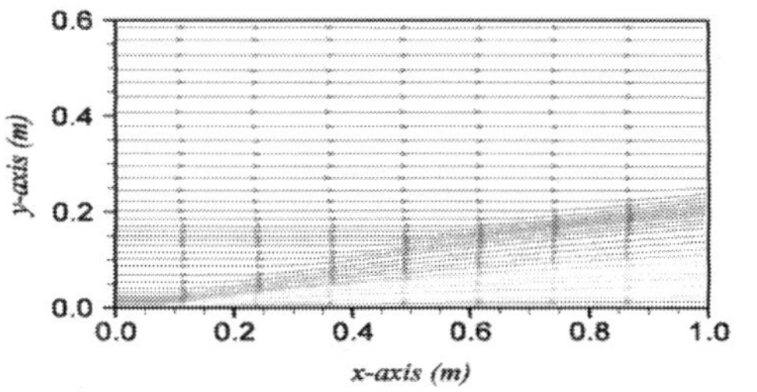

Streamlines at $Ma = 10;\ Re = 10^4$

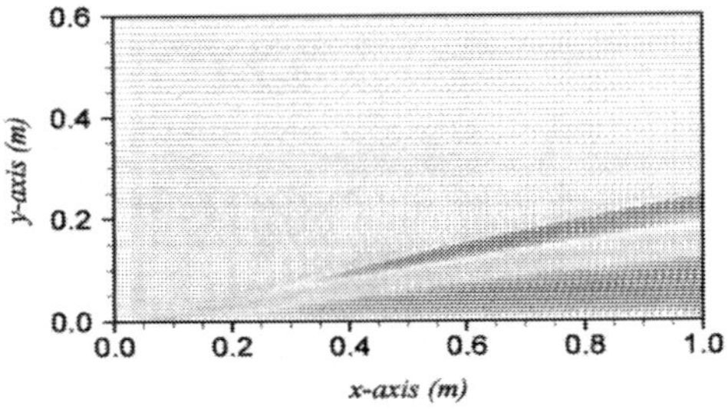

Velocity Vector at *Ma = 10; Re = 10^4*

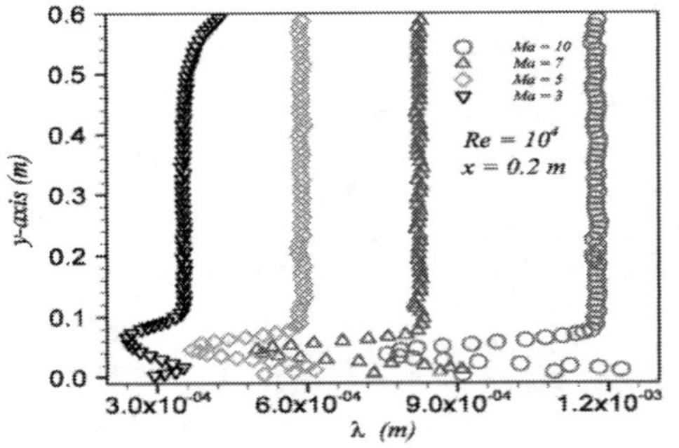

Mean Free Path profiles at *Ma = 10, 7, 5, 3; Re = 10^4*

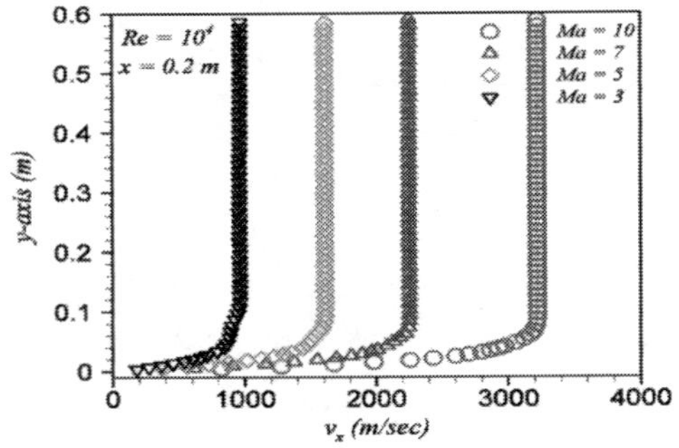

X- Component velocity profiles at *Ma = 10, 7, 5, 3; Re = 10^4*

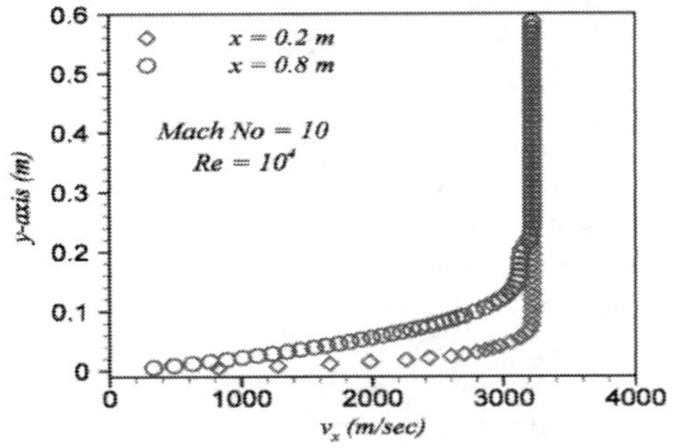

X- Component velocity profiles at *Ma = 10; Re = 10^4; x = 0.2 m , 0.8 m*

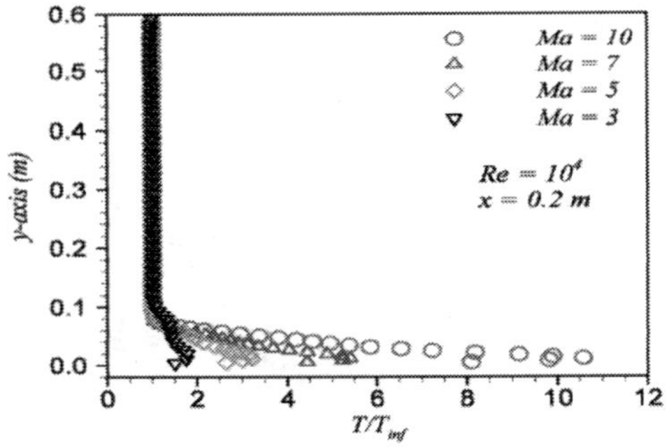

Temperature profiles at *Ma = 10, 7, 5, 3; Re = 10^4*

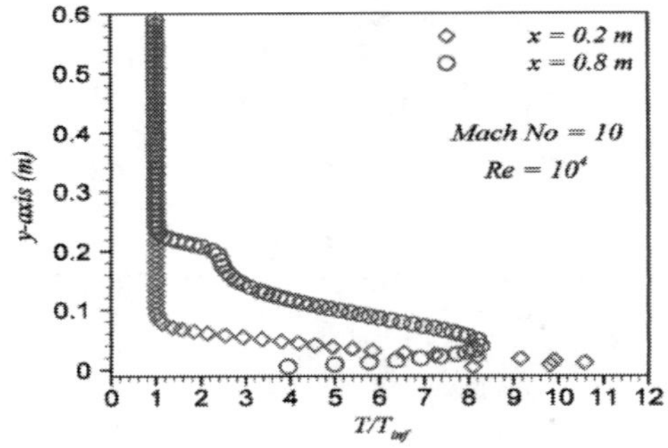

Temperature profiles at *Ma = 10; Re = 10^4; x = 0.2 m , 0.8 m*

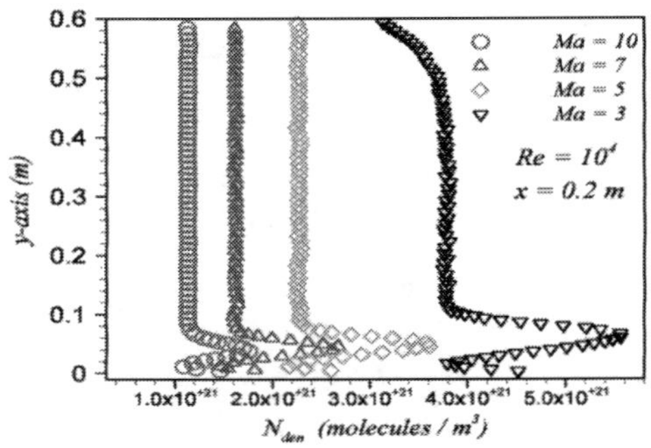

Number Density profiles at *Ma = 10, 7, 5, 3; Re = 10^4*

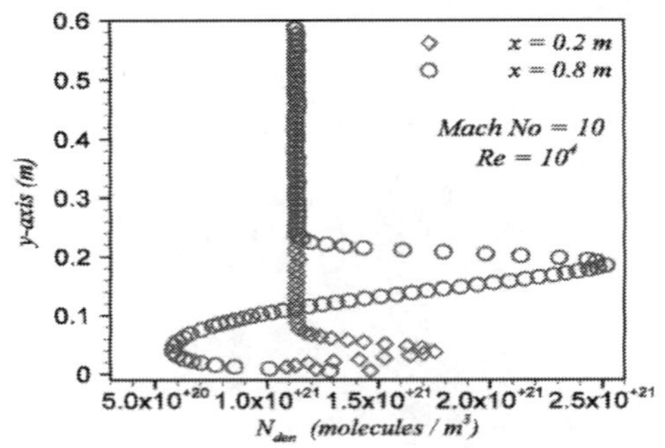

Number Density profiles at *Ma = 10; Re = 10^4; x = 0.2 m , 0.8 m*

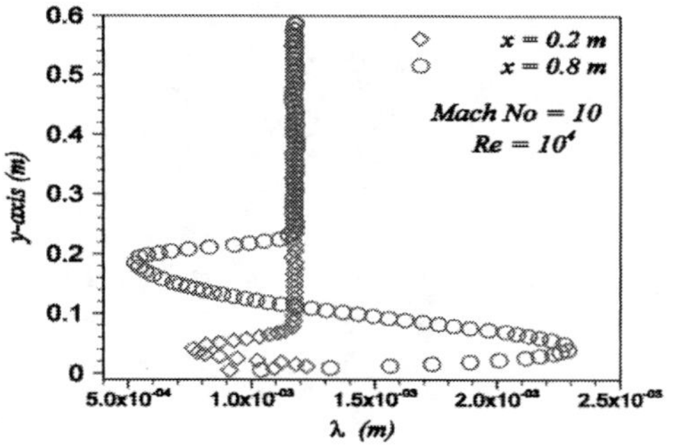

Mean Free Path profiles at $Ma = 10; Re = 10^4; x = 0.2$ m, 0.8 m

REFERENCES

[1] Pradhan, S. & Kumaran, V. 2011 The generalized Onsager model for the secondary flow in a high-speed rotating cylinder. J. Fluid Mech. 686, 109.

[2] Kumaran, V. & Pradhan, S. 2014 The generalized Onsager model for a binary gas mixture. J. Fluid Mech. 753, 307.

[3] Bird, G. A. 1994 Molecular gas dynamics and the direct simulation of gas flows. Clarendon Press, Oxford.

[4] Bird, G. A. 1963 Approach to translational equilibrium in a rigid sphere gas. Physics of fluids 6, 1518.

[5] Cercignani, C. 2000 Rarefied gas dynamics. From basic concepts to actual calculations. Cambridge University Press.

[6] Chapman, S. & Cowling, T. G. 1970 The Mathematical Theory of Non-Uniform Gases. Cambridge University Press.

[7] Gaviglio, J. 1987 Reynolds analogies and experimental study of heat transfer in the supersonic boundary layer. Intl J. Heat Mass Transfer 30, 911.

[8] Thompson, K. W. 1987 Time dependent boundary conditions for hyperbolic systems. J. Comput. Phys. 68, 1, 24.

[9] Garcia, A. L. & Wagner, W. 2000 Time step truncation error in direct simulation Monte Carlo. Phys. Fluids. 12 , 2621.

[10] Vincenti, W. G. and Kruger, C. H. 1967 Introduction to Physical Gas Dynumics. Wiley, New York.

AIChE Annual Meeting, San Francisco, November 13-18, 2016

Multiscale modelling of reacting polymer foams via computational fluid dynamics and molecular dynamics

Mohsen Karimi[1], Erik Laurini[2], Maurizio Fermeglia[2], Sabrina Pricl[2], Daniele Marchisio[1]

[1] Department of Applied Science and Technology, Politecnico di Torino, Torino, Italy
[2] Molecular Simulation Engineering (MOSE) Laboratory, University of Trieste, Trieste, Italy

Objectives

In this work we want to develop a fully predictive multi-scale model for the simulation of reacting and expanding polyurethane (PU) foams.
The adoption of a multi-scale modelling strategy allows to reduce the number of experiments to be performed for the development of the model.

Industrial context

PU foams are characterized by interesting mechanical and thermal (insulating) properties (see Fig. 1.)

Fig. 1. PU foam products.

PU foams are manufactured via mold filling processes. It is therefore very important that the model keeps tracks of the macroscopic fluid dynamics of the foam and in particular of the foam-air interface (see Fig. 2.)

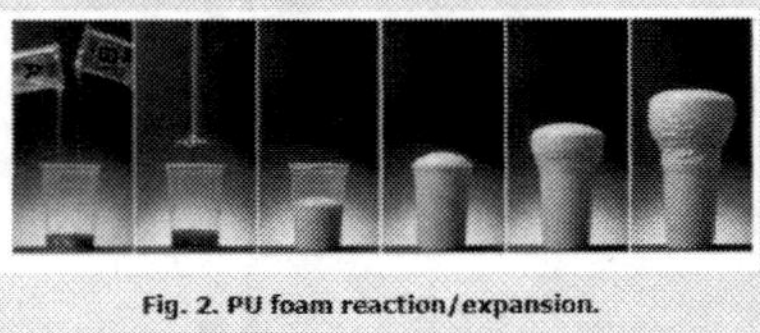

Fig. 2. PU foam reaction/expansion.

The model must account for the polymerization kinetics (i.e. reactions between polyols and isocyanates), the blowing kinetics (i.e. reaction between water and isocyanates producing gaseous carbon dioxide), the evaporation of the physical blowing agent (n-penthane) and the changes in the PU foam apparent density, viscosity and thermal conductivity.

Multiscale modelling

The PU foam expansion is simulated with a macro-scale fluid dynamics model based on the volume of fluid method (VOF) implemented in Computational Fluid Dynamics (CFD). The macroscale model needs as input the density of the liquid mixture undergoing polymerization, the solubility of physical and chemical (i.e. carbon dioxide) blowing agents as a function of temperature and polymerization conversion (or cross-linking). These are evaluated via Molecular Dynamics (MD).

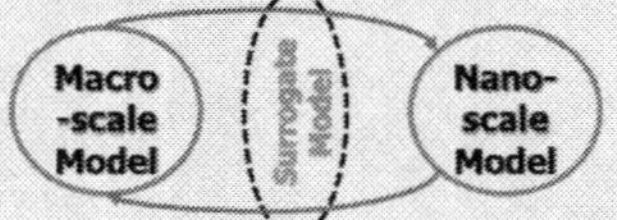

Fig 3. Schematic illustration of models communication.

Computational Fluid Dynamics

The macro-scale CFD model is implemented in ANSYS Fluent. In VOF two fluids are considered: PU foam and surrounding air. The properties of the PU foam are specified via user-defined subroutines: density, viscosity and thermal conductivity. These depend on the degree of polymerization (cross-linking), advancement of blowing reaction, amount of evaporated physical blowing agent, determined by solving user-defined scalars, resulting in the PU foam expansion as depicted in Fig. 4.

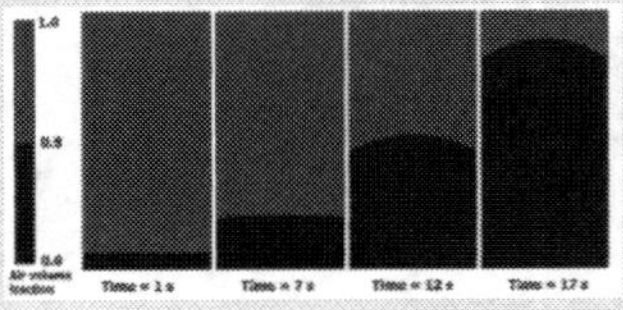

Fig. 4. PU foam expansion predicted by CFD.

Molecular Dynamics

MD simulations are run using the comercial software Materials Studio. The COMPASS force field was selected to perform the calculations. MD was used to estimate the density of PU (prior to foaming) constituted by polyether polyol (with an OH value of 365 mg KOH/g polyol) and polymethyline-poly(phenyl isocyanate) with an equivalent weight of 135. Calculations were performed at different temperatures and polyol conversions (cross-linking). MD calculations were also used to estimate the solubility of carbone dioxide and n-penthane.

Test Cases

Four different test cases are simulated using chemical (i.e. water) and physical (i.e. n-pentane) blowing agents (Table 1).

Table 1. Summary of the operating conditions

Batch	C_{poly} (mol m^{-3})	C_{iso} (mol m^{-3})	C_w (mol m^{-3})	$C_{n\text{-}pentane}$ (kg kg^{-1})
1	4400	4400	305	-
2	4400	4400	610	-
3	4400	4400	915	-
4	5140	4455	671	0.057

Simulations are performed on cylindrical geometries and the evolutions of foam density and temperature are compared with measurements, obtained as shown in Fig 5.

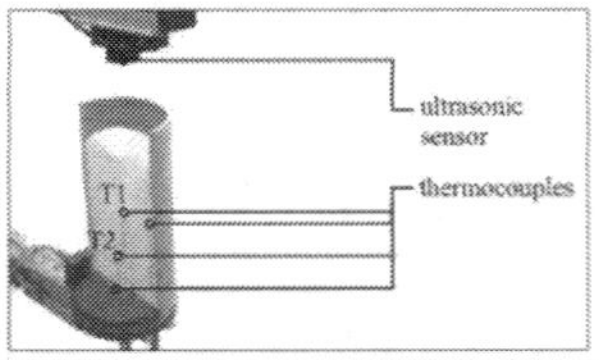

Fig 5. Foam rising heights (apparent density) and point-measurements of temperature.

Results and Discussion

Results from MD simulations, including density and carbon dioxide solubility under different temperatures and degrees of polymerization (cross-linking), are used to create surrogate models, which are then implemented in CFD. To ensure the fast and accurate approximation of polymer density and carbon dioxide solubility the mathematical forms of the surrogate models are restricted to explicit algebraic expressions. One example (density of the liquid mixture undergoing polymerization) is reported in Fig 6.

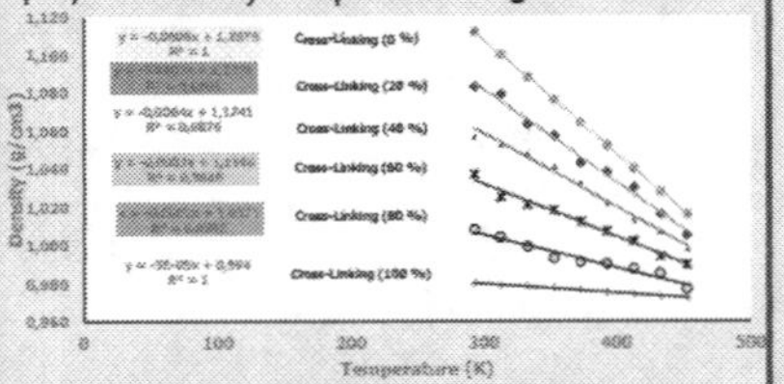

Fig 6. Surrogate model for polymer density.

The comparison of CFD predictions (line) and experimental measurements (symbols) for batch 4 are shown in Fig. 7. The comparison for the other batches is instead reported in Fig. 8.

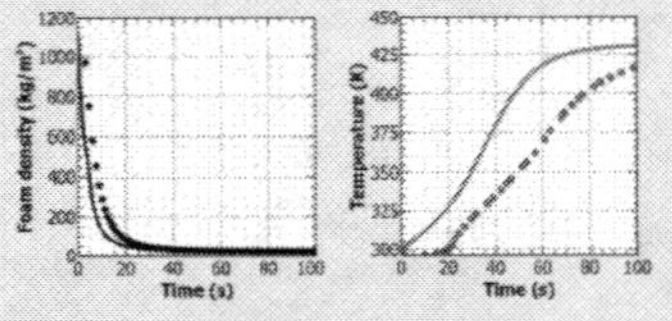
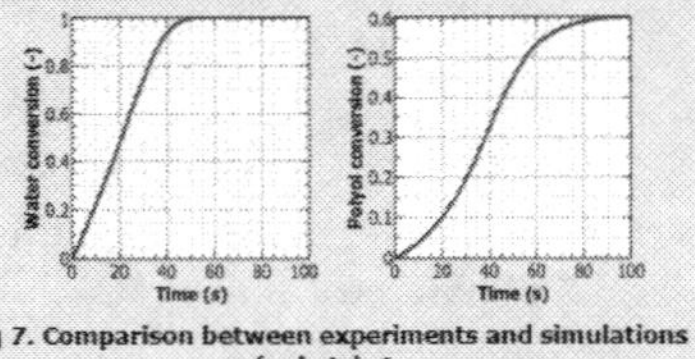

Fig 7. Comparison between experiments and simulations for batch 4.

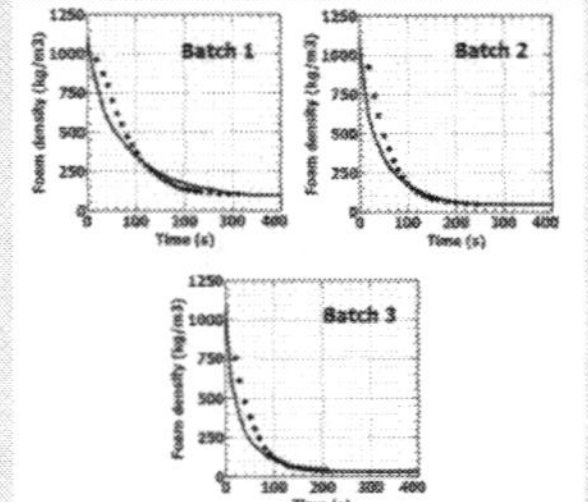

Fig 8. Comparison between experiments and simulations for batch 1, 2, 3.

Acknowledgements

This work was funded by the European Commission under the grant agreement number 604271 (project acronym: MoDeNa; call identifier: FP7-NMP-2013-SMALL-7)

Imperial College London

Qatar Petroleum

QATAR SCIENCE & TECHNOLOGY PARK

Member of Qatar Foundation

Qatar Carbonates and Carbon Storage Research Centre

Improving reservoir recovery and carbon storage potential

Capillary Disconnect during Evaporation in Porous Media

Ayorinde Rufai[1], John Crawshaw[1], and Geoffery Maitland[1]

[1] Department of Chemical Engineering, Imperial College London, South Kensington, London, UK. SW7 2AZ

Research Objective

During CO_2 injection into an underground reservoir, salt is precipitated which greatly reduces the permeability and porosity of the reservoir by blocking the pore bodies and throat in the reservoir formation. These salt depositions happen mainly near the point of injection (injector wellhead) threatening the entire sequestration process, a process known as wellbore dry-out. This decrease in net injectivity often lead to increased injection pressure, increased injection costs, reduced carbon efficiency, and reduced CO_2 storage potential. Understanding this blockage process on the pore-scale is imperative for us to be able to enhance and evaluate our carbon sequestration processes.

Micro-models

Why micro-models?

1. Real porous media are very complex
2. Allows direct visual observation of pore scale processes in a network of pores
3. Synthetic systems: can be designed to address specific research questions

Micro-model Characteristics

- Grains range in size from 6 – 1250 µm
- Silicon etch depth: 25 µm
- Porosity: 0.28
- Total pore area: 0.7cm²
- Two fracture channels, 500 µm in width
- Coordination number: 1~ 4
- Fracture and Matrix permeability ≈ 700mD and 300mD respectively

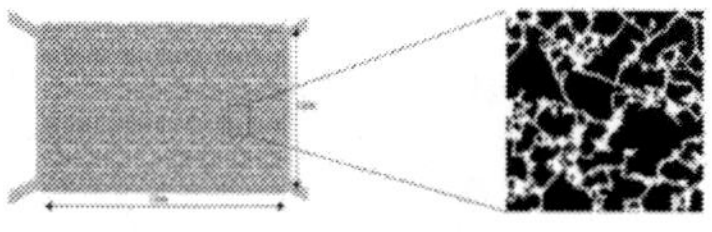

Figure 1. Micromodel dimension with the repeated pattern.

Methodology

Evaporative drying

- Inject micro-model network with NaCl brine solution and pass dry air along fracture
- Observe salt precipitation in the matrix during drying process
- Process micro-visual data for quantitative analyses

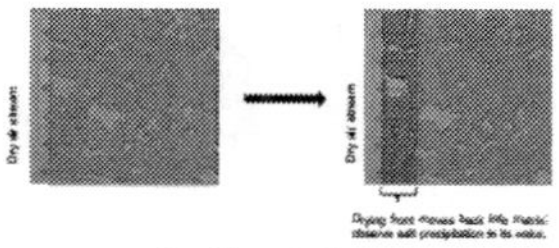

Figure 2: Evaporative drying procedure

Qualitative Analyses

- The images in Figures 3 and 4 provides visual evidence for salt precipitation mainly in the evaporating surface and in some pores just below it for saturated brine solution. The raw images captured with a DSLR camera are shown in Figure 3 while higher resolution images with a smaller FOV taken with an Inverted light Microscope are shown in Figure 4.

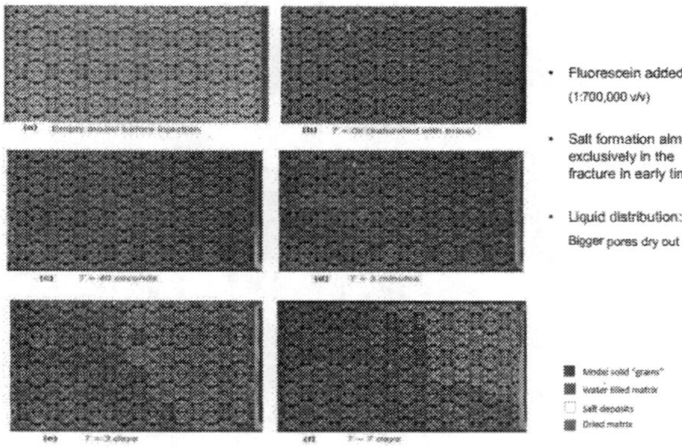

- Fluorescein added (1:700,000 v/v)
- Salt formation almost exclusively in the fracture in early time
- Liquid distribution: Bigger pores dry out first

Figure 3: Evolution of drying

- The "greyish-dark" particles in each of the images in Figure 4 are precipitated salt in the evaporating surface (vertical channel) and in the matrix near the evaporating surface. For Saturated brine, we expect evaporation rate to remain relatively high (although lower than that of deionized water because of the decrease in vapour pressure) since capillary continuity is maintained for a significant time into the experiment. However, the reverse is what we observe, evaporation rate falls sharply as soon as drying starts and we propose this is due to the appearance of dry patches above the deposited salt at the evaporating surface thereby reducing the surface area available for evaporation.

Figure 4: Images of deposited salt in the evaporating surface and in the matrix below at the end of complete drying

Quantitative Analyses

- Micro-visual data were processed with ImageJ, an image processing software, to extract some quantitative data.

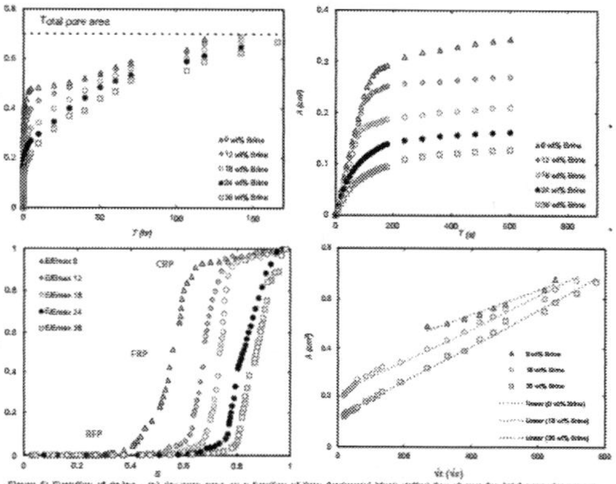

Liquid Mass Transport

- CRP: Hydraulic connectivity, Capillarity through corner wetting films
- RFP: Vapour Diffusion (for DI Water)

- **CRP**: Constant Rate Period
- **FRP**: Falling Rate Period
- **FRP**: Receding Front Period

Figure 5: Evolution of drying- (a) dry pore area as a function of time (horizontal black dotted line shows the total pore dry area ≈ 0.7cm²), (b) early time dry pore area as a function of time, (c) evaporation as a function of liquid saturation, (d) dry pore area as a function of square root of time.

- Pseudo-constant early time slopes for all brine concentration, transition period (rate falls sharply) and then a much lower-slope late evaporation region. For DI water, these correspond to the classical periods of evaporation: CRP, FRP, and RFP respectively.
- During CRP, liquid remains connected to the matrix surface and evaporation rate is high.
- The FRP is the transition period which is characterize by a sharp fall in evaporation rate.
- Capillary connection is broken and liquid transport becomes dominated by vapour diffusion during RFP.

Saturated Brine

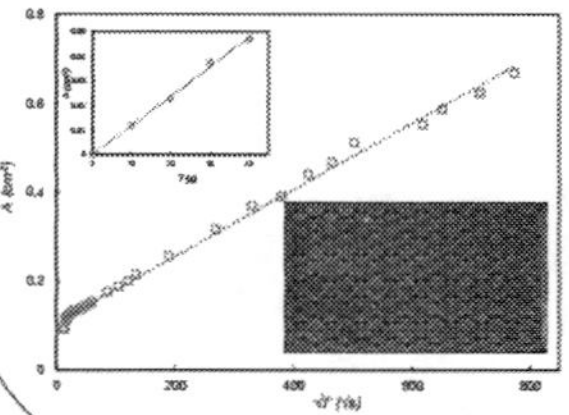

- Very Short CRP
- Much longer low rate period linear in $\sqrt{T}$
- $\sqrt{T} \neq$ capillary disconnection

- $\sqrt{T}$ likely due to deposited salt impeding hydraulic connectivity & High viscous resistance to flow
- Peclet number >> 1 at high evaporation rate.
- Image at t ≈ 200s

Figure 6. Plot of pore dry area as a function of $\sqrt{T}$ for saturated brine (Upper inset is a plot of pore dry area versus time for early time (CRP) evaporation. Lower inset is an image of salt deposition during the falling rate period)

Conclusion

- We have carried out for the first time, evaporation studies on a 2D micro-model based on a thin section of a carbonate (dolomite) rock.
- For deionised water, we observed the three classical periods of evaporation expected from earlier drying experiments in porous media; the CRP, FRP and RFP.
- For brines, we observed that the length of the CRP decreased with increasing brine concentration and became almost non-existent for saturated brine. Dry pore area (equivalent to mass lost by evaporation) became linear with the square root of time after a short CRP.
- This behavior is not due to capillary disconnection from the evaporating surface (because salt continued to deposit at the surface during this time) as usually the case for deionised water.
- We propose that this behavior is likely due to salt deposited in the fracture channel and matrix below it greatly impeding hydraulic connectivity, lower vapour pressure of brines, and the relatively high viscosity of brines increasing viscous resistance to flow.

457203 Visualization Phases Distributions in the Pilot-Scale Bubble Column with Internals Via γ-Ray Computed Tomography (CT) Technique

Abstract:

Visualization Phases Distributions in the Pilot-Scale Bubble Column with Internals via Î³-Ray Computed Tomography (CT) Technique

Abbas J. Sultan[1], Laith S. Sabri[1] and Muthanna H. Al-Dahhan[1]

[1] *Department of Chemical and Biochemical Engineering, Missouri University of Science and Technology, Rolla, MO 65409-1230. USA*

Gamma-ray computed tomography (CT) technique have been utilized to assess the influence of the vertical internals and superficial gas velocities on the gas -liquid behavior in the bubble column with internals. The time-averaged cross-sectional gas holdup distributions and their radial profiles for the first time have been investigated in an 18-inch bubble column with and without internals for the air-water system at a wide range of superficial gas velocity up to 45 cm/s. Bundle of seventy-five internals covers 25% of the total cross-sectional area representing the intensity of heat exchange tubes employed in Fischer-Tropsch (FT) synthesis have been arranged uniformly by using circular configuration and equipped with the bubble column.

The experimental results will provide, for the first time, benchmark data to validate reactor models and computational fluid dynamics (CFD) simulations and consequently will facilitate the design and scale-up of bubble column with internals. The experimental results and findings will be presented at the conference.

Keywords: computed tomography (CT), bubble column, internals, cross-sectional gas holdup distributions.

457206 Microalgae's Cells Mapping By Radioactive Particle Tracking (RPT)

Abstract:

Microalgae's Cells Mapping by Radioactive Particle Tracking (RPT)

Laith S. Sabri, Abbas J. Sultan and Muthanna H. Al-Dahhan

Multiphase Reactors and Applications Laboratory M-REAL. *Department of Chemical and Biochemical Engineering, Missouri University of Science and Technology, Rolla, MO 65409-1230. USA*

Movement of the microalgae cells inside the photobioreactors is the key to enhancing the biomass productivity. This movement has affected not only by the hydrodynamics (flow dynamics) inside the reactor, but also the light intensity which their delivered. In fact, when the algae's cell delivered, the light from the illuminated surface and moving to the dark zone, biomass productivity could frequently become high. Moreover, the availability of the light in all the zones inside the reactor plays a critical factor for photobioreactor performance due to the effect on the biomass enrichment. Unfortunately, the interactions between the hydrodynamics parameters and the light availability remain unclear yet. Therefore, the request for both the mechanics of cell motion and for the mathematical model which contact between the variable concentrations of microalgae cells with irradiance spread inside the reactor are need. This work focuses on the integrated of the cells' movements (cell positions) which is tracked by using advanced non-invasive technique called Radioactive Particle Tracking (RPT) with a mathematical model for growth rate. This model linked the irradiance distribution to the concentrations of cells through the period of growing. A Plexiglas split photobioreactor column carried out in this experimental work with 5.5-inch (0.14m) inner diameter. A green algae (Scenedesmus) used for the cultivation at different superficial gas velocities (1, 2, and 3 cm/sec). The advanced information which obtained from this investigation crystallize the understanding and knowledge of photobioreactor behavior. The experimental results and conclusions will present at the

conference.

Keywords: RPT, Photobioreactor, Microalgae

459652 Discrete Element Model of Concentrated Colloidal Dispersions: Linking Viscosity to Cluster Properties

Abstract:

Colloidal dispersions are common in a large variety of applications, such as food, pharmaceutical products, cosmetics or polymer latexes. The current understanding of colloidal latexes subjected to a shear is restricted to the limit of diluted or moderately concentrated dispersions. However, in real applications (e.g., emulsion polymerization) volume fractions of particles up to 50 % are commonly encountered. There is therefore a large need to broaden the theoretical understanding of these systems in terms of their viscosity and stability to the region of large concentrations. Mathematical modeling can be a useful tool to achieve this goal.
We use the Discrete Element Method (DEM) to model the dynamic behavior of waterborne polystyrene latex subjected to a high shear. The interaction of the particles with the fluid is accounted for using the two-way coupling technique. The interaction between the particles that are stabilized, adhesive and elastic is described by the combination of the Derjaguin-Landau-Verwey-Overbeek (DLVO) and Johnson-Kendall-Roberts (JKR) theories. The model is spatially three-dimensional and allows therefore capturing not only the characteristic coagulation time, but also the size and structure of the produced clusters. Moreover, we developed a method for the evaluation of the viscosity of concentrated dispersions.
The results of the model in terms of both the characteristic coagulation time and the size of the produced clusters are in a good agreement with literature experimental data and established theories. The coagulation is shown to be faster for more concentrated and less stabilized dispersions and the clusters tend to be larger for more adhesive particles subjected to a smaller shear. Based on the balance of forces acting on a single cluster, we generalized these results into a simple expression relating the cluster size to the key parameters, such as the shear rate, primary particle size and surface energy. The viscosity of the dispersion is shown to be affected primarily by the particle volume fraction, which is in an agreement with the semi-empirical description by the Krieger-Dougherty equation for a simplified hard-sphere system. For more complicated systems, there is no established description. However, our model predicts the rise of viscosity for adhesive coagulating particles and it is also able to capture the viscosity growth during the autocatalytic coagulation of the stabilized latex. Combining these results together, it is possible to link the viscosity of the coagulating dispersion to the properties of the produced clusters, such as their size and structure. This provides a powerful tool for the prediction of the rheological properties of colloidal dispersions based on their structure.
The results of this work bring a theoretical insight into the complicated behavior of concentrated dispersions. Moreover, they can serve as a basis for the operation of real processes that handle colloidal latexes.

References

[1] Kroupa, M.; Vonka, M.; Kosek, J. Modeling the Mechanism of Coagulum Formation in Dispersions. Langmuir 2014, 30, 2693â€"2702.
[2] Kroupa, M.; Vonka, M.; Soos, M.; Kosek, J. Size and Structure of Clusters Formed by Shear Induced Coagulation: Modeling by Discrete Element Method. Langmuir 2015, 31, 7727â€"7737.
[3] Kroupa, M.; Vonka, M.; Soos, M.; Kosek, J. Utilizing the Discrete Element Method for the Modeling of Viscosity in Concentrated Suspensions. Submitted 2016.

461418 Relationships Between Structure and Permeability in Colloidal Networks

Abstract:

The resistance to flow through colloidal networks is characterized by the permeability. The permeability influences many practical applications of colloidal gels, and in particular controls the rate of (gravitational) collapse of non-neutrally buoyant networks. The permeability is generally accepted to be a power-law function of volume fraction, where the power is dependent on the fractal dimension of the network. To probe microscopically the influence of gel structure on permeability, we investigate permeability in structures generated by diffusion-limited cluster aggregation (DLCA) and related simulation techniques. We apply a geometric analysis to determine the local pore size associated with every point in the pore network, from which we then extract the pore size distribution. The overall network permeability is then computed by assigning permeabilities based on the local pore size field and applying Darcy's Law. Remarkably, predictions based on the full-field calculations and just based on the mean pore size agree extremely well for volume fractions from 0.2% up to 20%, indicating that there is no effect of void-space connectivity on flow. This modeling approach is shown to agree well with available experimental data at low volume fraction [1]. We then consider the effects of modifying the gel structure on pore size distribution and permeability. Reaction-limited cluster aggregation (RLCA) generates networks with a different power-law dependence of pore size on volume fraction and larger pores at any volume fraction. Incorporating some relaxation into the network-formation simulations likewise leads to larger pores and higher permeabilities. These findings have significant consequences for network stability: networks which can relax through Brownian motion will collapse faster than networks which cannot, both because they have lower moduli and because buoyancy forces have a greater effect due to their higher permeability. Finally, we discuss the relationship of permeability to volume fraction in networks composed of non-spherical particles, and consider the application of our numerical tools to determining a priori the permeability of network structures measured by 3D microscopy.

[1] S. Manley, J. M. Skotheim, L. Mahadevan, and D. A. Weitz, Gravitational Collapse of Colloidal Gels, PRL 94, 218302 (2005).

457705 Scalable Tools for Generating Synthetic Isotropic Turbulence with Arbitrary Spectra

Abstract:

A scalable, parallel, isotropic turbulence generator is developed and deployed online at http://turbulence.utah.edu (password: utahturbogen) for immediate access by the fluids, aerosapce, and aeroacoustics communities. The generated isotropic turbulence can be used for a priori and a posteriori verification and validation of turbulence codes and models as well as aeroacoustics calculations. This synthetic generator takes an arbitrary energy spectrum as input and produces a velocity field that matches the given spectrum up to the Nyquist limit. The Nyquist limit is dictated by the domain size and grid resolution, which are specified by the user. The velocity field is stored on a staggered grid and satisfies the discrete divergence-free requirement on the vast majority of grid points without the need for a mass correction procedure. The turbulence generator was conveniently coded in Python, C, and C++ using parallel computing concepts and is freely available to execute online in a web browser. The errors produced by the generator are examined for canonical spectra, such as the Comte-Bellot-Corrsin and the von Karman-Pao spectra. The scalability of the codes are verified and compared for various grid resolutions. Finally, an example application is presented for decaying isotropic turbulence.

459742 Head Loss in Gate and Globe Valves on Pumping Thixotropic Polymer Solutions

Abstract:

Pumping is one of the most important unit operations process, being used in chemical as well as petrochemical procedures, such as, transportation of polymeric solutions, paints, resins, hydrocarbons, oils and pastes. The project of a pumping system is based on the determination of the pump motor power, which is expressed as a function of the potential and kinetical energies, taking into consideration head losses for straight sections and accessories (valves and connections). In particular, the determination of head losses in accessories is done through abacus obtained by empiric methods, such as equivalent lengths, K and 2K methods, in the way that the literature is rich in information on the Newtonian fluidâ€™s transportation (e.g. water, hydrocarbons and oils). Nevertheless, around 80% of the fluids that are industrially transported are featured as Non-Newtonian, and, in most cases, thixotropic. The polymeric solutions follow a thixotropic behavior according to their rheology, meaning that their apparent viscosity decrease as time passes and shear stress rises alongside temperature and pressure. The existing literature provides some information on the K method for gate valves and globe valves as they transport pseudoplastic fluids, however, lacks consistent data on the transport of thixotropic fluidâ€™s. This study aimed to determine empirical equations for K coefficient for a gate valve and for a globe valve, with two diameters, during the transport of sodium carboxymethylcellulose of high viscosity at two concentrations based on the modified Reynolds number of Metzner and Reed. The experimental unit consists, basically, on a centrifugal pump with closed impeller and a break horse power equal 0, 75 cv, 2 gate valves (0, 5 in. and 1 in.), 2 globe valves (0, 5 in. and 1 in.), both screwed down and transparent PVC piping of 0, 5 in. and 1 in. diameter. It were used two aqueous solutions of carboxymethylcellulose (CMC) in two different mass concentrations: 0,5% and 1%. The mentioned solutions were analyzed according to their rheology in a Stormer viscometer, varying the shear stress and temperature. The current time of experiments were registered. The rheological behavior was based on the power law, however, the consistency factor and behaviour index were expressed as a function of time and temperature, due to the thixotropic solutions behavior. The experiments were carried measuring

head losses through the differences between two piezometerâ€™s heights, being one of them placed downstream the accessory and the other upstream. The head loss varied according to the employed flow in each trial. Tests were performed for each valve in terms of its diameter, and polymer solution concentration. All tests conducted using accessories, were also produced using straight sections of pipe. Deducting the values obtained with the straight sections from the values obtained with the accessories, it was possible to obtain the head loss for each of the valves isolatedly. The modified Reynolds number of Metzner and Reed was calculated according to each volumetric flow rate measured and it was noted that any of the four valves reached the transition flow, so that the entire modeling was performed into laminar flow. The values of the coefficient K were calculated and plotted using two regressions between the values of K and its corresponding Reynolds numbers, one for the gate valve and other for the globe valve. In order to provide a practical model for projects, this article modeled equations for Reynolds, in different ranges, for the two valves studied as a function of CMC concentrations and the pipe diameter (0.5 in. and 1 in.). Based on the K equations depending on the Reynolds number for the pseudoplastic fluids, using gate valves; literature contains a functional relationship for K as follows: K = A / Re, similar for globe valves. The results of this article demonstrated for the gate valve, two expressions depending on the range of the Reynolds number: for values lower than 231, K = 294.36 / Re, with a deviation of 14.2% compared to the experimental data, and for Reynolds between 373 and 2804 (below the transition zone, Reynolds less than 3100), K = 717.73 / Re, with a deviation of 2.9%. Similarly, for the globe valve, with lower Reynolds 176, K = 1163.96 / Re, providing a deviation of 14.2% and for Reynolds between 324 and 872, the value of the K coefficient was almost constant at 18.49. It was concluded that the expressions obtained for the determination of the coefficient K are adjusted to the experimental data, due to the low experimental error, allowing engineers to design more reliable pumping systems for thixotropic polymer solutions.

461826 Bubble and Droplet Coalescences in Microchannels

Abstract:

Coalescence is a significant fundamental issue in chemical engineering process, which appeals to scientists for a long time. Random collision usually made it hard to investigate the basic rules of droplet/bubble coalescence in flows for the uncontrolled collision results. In recent years, microfluidics shows its potential in droplet and bubble manipulation. So that the investigations of droplet/bubble coalescence under laminar condition can be well implemented in microchannels [1]. Cross-shaped microchannel was effective for the research on head-on collision [2]; Y-junction microchannel can be used to study the effect of collision angle [3,4], and the expansion microchannel was useful for the study of coalescence in chasing flow [5]. Experimental results showed that the regular collision results can be divided to absolute coalescence, probabilistic coalescence and non-coalescence. Critical capillary number (Ca*) was important for distinguishing coalescence and non-coalescence. For the water/alcohol (with alkyl chains varying between 5 and 10 carbon atoms in molecule) systems, Ca* is about 0.005 in head-on collision processes. Investigations showed that the critical capillary number varied with droplet size, collision angle and physical properties [3]. Furthermore, the microfluidic platform was fit for the investigation on the liquid-liquid systems containing nanoparticles [5]. Model equations were proposed to predict the critical capillary numbers based on the droplet contact time and liquid film drainage time, which may provide more valuable information for understanding the basic laws of coalescence phenomenon.

References

[1] Bremond N, Bibette J. Exploring emulsion science with microfluidics. Soft Matter, 2012, 8(41): 10549-10559.

[2] Wang K, Lu YC, Tostado C P, et al. Coalescences of microdroplets at a cross-shaped microchannel junction without strictly synchronism control. Chemical Engineering Journal, 2013, 227: 90-96.

[3] Wang K, Lu YC, Yang L, et al. Microdroplet coalescences at microchannel junctions with different collision angles. AIChE Journal, 2013, 59(2): 643-649.

[4] Yang L, Wang K, Tan J, et al. Experimental study of microbubble coalescence in a T-junction microfluidic device. Microfluidics and Nanofluidics. 2012, 12(5): 715-722.

[5] WANG K, YI ST, ZHOU QQ, et al. Effect of nano-particles on droplet coalescence in microchannel device. CIESC Journal, 2016, 67(2): 469-475.

[6] Zhou QQ, Sun Y, Yi ST, et al. Investigation of droplet coalescence in nanoparticle suspensions by a microfluidic collision experiment. Soft Matter, 2016, 12, 1674-1682.

462836 Collective Dielectrophoretic Trapping Assisted By AC Electrokinetic Flow

Abstract:

Dielectrophoresis (DEP) is the migration of a polarized object under the action of a non-uniform electric field. As DEP is capable of trapping and distinguishing samples according to their properties, it has become a useful tool for achieving on-chip screening and diagnosis using microfluidics. However, because DEP trapping often occurs in close proximity of electrodes, it generally lacks a long-range trapping capability. In addition, because the DEP mobility is proportional to square of the sample size, it is also difficult to use DEP to trap nano-sized objects.

In this work we demonstrate that the above shortcomings of DEP can be overcome by using a proper electrode design. Our electrode design is a quadrupole one. But unlike a standard quadrupole design, the four electrodes here are made to be converging shaped with width being gradually decreased toward the central region. Using this design to trap microparticles, we find that not only the particles can be rapidly trapped by DEP around the central region but also the trapped particle assembly can keep growing in size due to the continuous particle supply from the AC electrokinetic flow (ACEK) set up by the converging electrodes. A successful trapping of nano-sized particles and biomolecules can also be realized, offering a promising means for rapid sensing and detection in chip scales.

462169 Label Free Size Separation Strategies for Polydisperse Vesicle Suspensions

Abstract:

Microfluidic devices can be used to separate suspensions of deformable particles with different intrinsic characteristics (e.g. size, deformability, and shape) with reasonable throughputs and without requiring external labeling to do so. Vesicles serve as a model for cells and using these suspensions to test microfluidic separation schemes provides insight into cell separation and isolation for diagnostic purposes as well as separation of suspended contaminants from drinking water[1]. Two schemes for separating deformable vesicle suspensions by size are discussed: size exclusion and inertial focusing. The size exclusion separation device makes use of a filter that prevents many of the larger vesicles from passing through. The filtrate is collected at one outlet and the larger vesicles are collected at a separate outlet. This device showed good size separation between the two collected suspensions. It also was able to reduce the polydispersity of the collected suspensions relative to the original vesicle suspension. The inertial separation device was based on a design studied by Di Carlo et al[2]. This design was modified for our suspension parameters and showed an ability to separate the suspension by size; however, the separated suspensionâ€™s polydispersity was either unchanged or only reduced slightly relative to the starting polydispersity. The advantage of the inertial separation device relative to the size exclusion filter device was its greatly increased throughput. Ultimately, one may select a separation strategy based on the relative importance of high throughput versus reduced polydispersity. Future work includes investigating cascading separation schemes to further optimize the separation.

(1) Di Carlo, D. *Lab Chip* **2009**, *9*, 3038-3046.

(2) Di Carlo, D.; Edd, J. F.; Irimia, D.; Tompkins, R. G.; Toner, M. *Anal. Chem.* **2008**, *80*, 2204-2211.

463210

A Graph-Based Method for Structure Classification in Amorphous Materials

Abstract:

Identifying structures of amorphous systems has received a great deal of interest because of its theoretic importance for understanding the nature of glass as well as its potential application in developing glassy materials, including polymers and metallic glasses. In this work, we propose a graph theory-based tool to identify structural motifs in glassy system and also compute the structural entropy. In short, we group together local structures by comparing the associated Voronoi diagrams using a graph theoretic technique -- graph isomorphism. In comparison with other cluster identification approaches that rely on predefined structures, our method is completely automated. We apply the method to dense colloidal suspensions via computer simulation and identify a plenty of distinct clusters, among which icosahedron-like structures are the most common. In contrast to crystals that dominated by fcc order, less than 10 % atoms of the system belong to the center of icosahedral clusters, as a result of geometric frustration. These common clusters grow in general as glass transition is approached through densification, with icosahedral order showing the most significant increase. Finally, we obtain the structural entropy for hard disks and hard spheres by counting the number of distinct graphs of subregions. Our results are consistent with the configurational entropy via thermodynamic integration.

463724 Two-Layer Spin Coating Flow of Newtonian Liquids: A Computational Study

Abstract:

Spin coating is the most commonly used method in industries to form coating films of desired thickness and functionality. In case of two-layer spin coating process, stratified layers of two immiscible liquids are deposited onto the substrate simultaneously. When the substrate is rotated, these layers spread radially to form thin two-layer coating film of finite thickness. Questions concerning the effect of surface/interfacial tension on uniformity of these films and contact line evolution need thorough investigation. Here, we have developed an axisymmetric model governing the flow of two-layer spin coating process. Liquids used for the study are assumed to be Newtonian and fully wetting. The contact line singularity is resolved using a precursor layer model and the governing equations are simplified using lubrication approximation. The resulting fourth order non-linear PDEs are solved using a numerical technique based on Galerkin finite-element method (G/FEM). It is observed that a uniform two-layer film surrounded by thin single layer film is formed when the viscosity ratio is small. On the contrary, when viscosity ratio is large, a thin two-layer film surrounded by bulky capillary ridges is formed. Similarly, the results also show that sharpness of capillary ridge increases with decrease in the surface tension ratio. Further, it is found that increase in the precursor layer thickness increases the spreading rate, thereby making the film more uniform. Finally, the uniformity of the final two-layer film does not get affected by the initial volume of fluid present in the upper layer. The results of our numerical study may serve as a basis for the design and optimization of multi-layer spin coating process for a given application. [1]

Reference:

[1] S. Sahoo, A. Arora, and P. Doshi, â€œTwo-layer spin coating flow of Newtonian liquids: A computational study,â€ Comput. Fluids 131, 180 (2016).

464073 High Throughput Microfluidic Platform to Characterize Flow Resistance of Trapped Deformable Particles

Abstract:

Measurement of pressure drop versus flow rate relations in microfluidic geometries is of broad interest in a variety of applications including rheology of complex fluids and mechanics of deformable particles. Experimental approaches to determine these relations in microscale geometries are limited. This is particularly evident for the case of trapped deformable particles where the non-circular cross-section of the microchannel, shape of the bounding walls and degree of occlusion, leads to a large control parameter space that can influence their flow resistance. Accessing this large parameter space requires methods that can determine pressure drop flow rate relations in a parallelized fashion. In this study, we investigate the use of co-flowing laminar streams in microfluidic parking networks to both trap deformable particles and measure their flow resistance. To validate this method, we fabricate microfluidic parking networks in which we vary the geometry, size and shape of the trap. Using Newtonian fluids, we quantify pressure drop flow rate relations for different trap geometries in a parallelized fashion and find that the data match well with the results from computational fluid dynamics simulations of single-phase flows. Next, we trap droplets in the microfluidic parking network and impose co-flowing laminar streams to measure flow resistance due to the occluding droplets. We find that the trap geometry and occlusion determine the strength of the gutter flows - which plays an important role in determining the flow resistance of the trapped droplets. Finally, we explore the capability of this approach in measuring the mechanical properties of giant unilamellar vesicles.

464754 Dynamics of Linear and Comb DNA Solutions Using Efficient Brownian Dynamics Techniques

Abstract:

Excluded volume (EV) and hydrodynamic interactions (HI) play a central role in macromolecular dynamics under equilibrium and non-equilibrium settings, specifically in determining the concentration dependence of static and dynamic properties of semidilute polymer solutions. The high computational cost of incorporating the influence of HI in mesoscale Brownian dynamics simulations (BDS) of polymeric solutions has motivated much research on development of high-fidelity and cost efficient techniques. In this work, a Krylov subspace based technique is implemented to enable fast calculations of single chain dynamics with computational expense scaling as $O(N_b^2)$ where N_b is the number of beads in the bead-spring micromechanical model [1]. For simulations of coupled multichain systems, a matrix-free approach for calculation of HI is implemented which leads to $O(N \log N)$ scaling of simulation time, where N is the total number of beads in the periodic box [2].

To investigate the properties of DNA solutions using coarse-grained bead-spring chains, a new technique is developed on the basis of properties of exact worm-like chain (WLC). In this approach, the correct correlation along the chain is modeled through the introduction of a bending potential and the spring force-law is obtained using asymptotic behavior of WLC at small and large forces. Using this model, the relaxation and stretching dynamics of linear and comb DNA molecules are evaluated and compared against experimental data [3].

References:

[1] A. Saadat and B. Khomami, *J. Chem. Phys.*, **140**, 184903 (2014).

[2] A. Saadat and B. Khomami, *Phys. Rev. E*, **92**, 033307 (2015).

[3] D. J. Mai, A. B Marciel, C. E. Sing, C. M. Schroeder, *ACS Macro Lett.*, **4**, 446 (2015).

465665 Mechanism Driving Fluid Circulation in a Microcavity Junction Due to Immiscible Displacement Flow

Abstract:

Understanding the fluid dynamics of displacement of a non-wetting phase by a wetting continuous phase is relevant for a wide range of applications including oil recovery, environmental remediation, and droplet-based microfluidics. In this study, we investigate fluid circulation patterns due to an immiscible displacement flow occurring at a micrometer-scale cavity located at a channel bifurcation. Using particle-image velocimetry, we map both the direction and strength of fluid circulation as a function of the capillary number. At low capillary numbers, we find that fluid circulation is in a direction that is opposite of what is expected for lid-driven single-phase cavity flow. However, at higher capillary numbers, the circulation direction is the same as that of single-phase cavity flow. We explain these counterintuitive observations by considering the competition between viscous stressesâ€”1) shear stress at the cavity entrance due to the displacement flow, and 2) shear stress along the perimeter of the non-wetting phase within the cavity due to continuous-phase gutter flowsâ€”as well as the dependence of gutter size on capillary number. We propose a scaling analysis that can qualitatively describe the cavity circulation direction and strength as a function of capillary number. Our study highlights the importance of gutter flows associated with immiscible displacement flows in non-circular microchannels. The boundary-driven advection of fluid by the gutter flows could potentially be harnessed as a new hydrodynamic mechanism for pumping out fluid trapped in microcavities and porous media.

465775 Gelation Process of *in Situ* Crosslinkable Hydrogels in the Static Mixer

Abstract:

Static mixers have been widely used in industry to mix various kinds of fluids and are recently utilized to fabricate hydrogels. Hydrogels have attracted much attention in the biomedical field as tissue adhesions, drug carriers or tissue engineering scaffolds. It was reported that the fast gelation rate made hydrogel inhomogeneous, which resulted in the decrease of mechanical strength. However, few studies have been done on the effect of mixing on the gelation process.

In the present study, we studied the gelation process of in situ crosslinkable hydrogels, which can be formed simply by mixing two reactive solutions, in a Kenics static mixer. In situ crosslinkable hydrogels are promising material for biomedical application, because they can be administered into any desired location in the body without large incision. The effects of process parameters, e.g., flow rate, the number of mixing elements, and the concentration of the pre-gel polymer solution, on the pressure drop during mixing and the homogeneity of formed hydrogels were investigated.

We used the static mixers with 4, 8, 12, 16 elements. Green dye was used as a tracer to visualize the flow inside the static mixer. To organize the results, we calculated Damköhler number (Da), a dimensionless number expressed as the ratio of the residence time in the static mixer to the batch gelation time. The pressure drop between the inlet and outlet of the static mixer was measured in fabricating in situ crosslinkable hydrogels. The flow rate was controlled using a syringe pump. As an indicator of homogeneity, we observed the formed hydrogels via confocal microscopy and calculated coefficient of variance (COV) by dividing the standard deviation by the average of the obtained pixel values of confocal images (N=3).

As a model material for ionically crosslinkable hydrogels, calcium alginate was utilized in this study. Alginate is a natural polysaccharide comprising repeating units of alpha-L-guluronate and beta-D mannuronate. When mixed with

calcium ions, alginates are crosslinked with calcium ions via the so-calledgegg box modelh, producing an insoluble hydrogel. Without calcium ions, the fluid was sol and the stream line got thinner along with elements in the static mixer. On the other hand, in the presence of enough amount of calcium ions, hydrogels formed in the mixer interrupted the flow, resulting in green stagnations. The pressure drop increased with the number of elements for both with and without calcium ions, but it showed different tendency against the axial velocity because the mixing efficiency differed. These results were organized by using Da. The normalized pressure drop, normalized by that in 0 mM calcium ions case in the same operation condition, increased monotonically with Da until Da was 10. COVs of the hydrogels showed similar tendency, first increased and then decreased with increasing Da. When Da was low, gelation occurs after the extrusion rather than inside the static mixer. With higher Da, the hydrogels were more likely to be formed inside the static mixer, resulting in the increased pressure drop and decreased homogeneity of formed hydrogel. When Da is further increased, hydrogel was not completely formed inside the static mixer due to insufficient mixing at very low axial velocity, leading to the increase in the homogeneity (T. Hozumi, S. Ohta and T. Ito, Ind. Eng. Chem. Res., 2015, 54, 2099-2107).

In addition to ionic crosslinking, various kinds of crosslinking reactions can be utilized for producing in situ crosslinkable hydrogels. It has been reported that the types of crosslinking reaction significantly affect the gelation rate and property of in situ crosslinkable hydrogels. In this study, the above experimental examination was also applied for other types of crosslinking reaction, such as Schiff base formation and Michael addition. The difference in the gelation process in the static mixer and consequent property of formed hydrogels will be discussed.

465815 Structure of Internal Circulations in Slug Flow through a Rectangular Micro-Channel

Abstract:

Mixing in the commonly observed slug flow regime in a liquid-liquid system in micro-channels is enhanced by internal circulations induced by shear. This helps improve mass transfer in this flow regime. In this work we exploit the low Re characteristic of the flow and seek a semi analytical solution to understand the structure of the vortex patterns formed in the two phases in the slug flow regime. The system is analysed in a moving reference frame under both creeping flow and the low Re flow-regimes which prevail in micro-channels. To get an analytical solution we assume the slugs to be rectangular. The governing equations of the model are solved using two approaches: 1) primitive variable formulation with the Chebyshev collocation method, 2) a stream function vorticity formulation based on finite differences. Results obtained from the two approaches are validated with each other. Specifically we focus on formation of vortex structures in the different phases and understanding the role of viscosity of the fluids. The system is analysed for two cases (i) when there is no thin film surrounding the slug and (ii) there is a thin film surrounding the slug. In the former secondary vortices are always induced in the fluid with the lower viscosity. In the latter the secondary vortices are always induced in the slug phase or discrete phase.

466744 Hierarchical Emulsion Networks from Endoskeletal Droplets

Abstract:

The shape of liquid-liquid interfaces at small length scales is dictated by surface tension. In the colloidal size regime, isotropic interfaces are highly favorable to anisotropic ones due to the large energy cost of increasing interfacial area. As a result, emulsions consist of spherical droplets that coarsen through coalescence into larger spheres and eventually a fully phase-separated system. The coarsening behavior of emulsions implies that they are intrinsically incapable of building larger structures, limiting them to forming isotropic, discontinuous dispersions.

Producing emulsion droplets that resist coarsening requires a structural scaffold to be introduced to counteract surface tension. In endoskeletal droplets, the scaffold consists of an internal network of intercalated wax crystallites with a yield stress strong enough to resist the driving force for droplet collapse. As a result, the droplets can be molded into anisotropic shapes. However, the internal elasticity is not insurmountable; multiple droplets can partially coalesce with others, forming larger suprastructures that retain the shapes of the component droplets. Many droplets can be coalesced into linear secondary structures akin to polymer chains, which can then assemble tertiary folded structures similar to electrospun fiber networks, resulting in hierarchical assemblies of bicontinuous fluids that preserve a continuous, anisotropic liquid interface. The size scales of these networks can be tuned during formation or once formed by applying external fields or tuning coalescence between droplets, allowing diverse networks to be formed. We present a microfluidic method for producing droplet suprastructures continuously and demonstrate control of three-dimensional structures during and after production.

467657 Microrheological Measurements of the Degradation of Covalently Adaptable Hydrogel Scaffolds

Abstract:

Synthetic hydrogel scaffold are designed for a broad array of applications, including wound healing, tissue regeneration and 3D stem cell culture platforms. The lack of knowledge of the material microstructure and its effect on the performance of the scaffold is a growing challenge especially in complex matrices. Covalently adaptable (CA) hydrogels mimic the native extracellular matrix that cells experience in vivo due to their ability to physically adapt to their environment. The CA hydrogels we are studying is made from multi-arm poly(ethylene glycol) (PEG) molecules that form reversible bis-aliphatic hydrazone bonds. This unique chemistry creates a material that yields when a stress is applied and reforms covalent bonds once the stress is released. In this work, we use multiple particle tracking microrheology (MPT) to measure dynamic material properties during scaffold degradation due to a change in pH or shift in reaction equilibrium. MPT measures the thermal motion of embedded probe particles to extract rheological properties using the Generalized Stokes-Einstein Relation. The CA hydrogel in a pH 4.3 buffer degrades over 2 days and transitions through time from a gel to a sol, with a single reformation of bonds in the middle of the degradation reaction. Reverse time-cure superposition is used to pinpoint the critical degradation time, t_c=0.2 days, and critical relaxation exponent, n=0.43. The pH 7.1 progresses through the gel-sol transition, but breaks and reforms bonds throughout the entire gelation reaction until complete dissolution of the matrix occurs. This occurs over a much longer time scale, on the order of tens of days. These measurements also show that the scaffold degrades homogeneously throughout the entire process, determined from ensemble-averaged van Hove correlation functions. In depth information about the material microstructure will enhance design of implantable scaffolds that accelerate tissue regeneration by enabling facile motility and differentiation of encapsulated cells.

467976 Modeling of Ultra-High Density Bio-Reactor

Abstract:

Ultra High Density well plates are one of the leading concepts using small foot print formats towards culture of large number of cells. In this presentation, we present a CFD model for modeling the fluid flow characteristics for a single pit and orifice of an ultra- high density well plate bioreactor. An analytical tool was developed based on the principle of conservation of energy to estimate the minimum velocity needed to dislodge the biological cell spheroids from the pits. The design changes from the CFD model were suggested based on the criteria of increasing the fluid velocities in the pit to increase convective mass transfer along with retaining the spheroids inside them. Some of the designs concepts tested appear to be feasible, thereby presenting the scope of improvement for this particular cell culture device. Another challenge in the design of Ultra High Density bioreactors is air-entrapment in the pit of the bioreactor. This phenomenon which is known as â€˜air lockâ€™ can be detrimental to the functionality of the bioreactor as the entrapped air prevents complete filling of the pits with culture media. We have developed a transient two-dimensional computational fluid dynamics model to analyze the effect of UHD pit geometry, pit wall surface energy, and filling rate on air-entrapment. Through modeling, we identified conditions that allow us to completely fill the pits with culture media and calculated the amount of time that is required to fill the pit. Our analysis showed that making the pit surface hydrophilic and therefore decreasing surface energy, is an efficient means to ensure complete filling of the pits and reduce the time of filling. This modification can also make steeper pit wall designs work which usually show severe air entrapment issues.

468166 Single Molecule Studies of Ring Polymer Dynamics in Dilute and Semi-Dilute Solutions

Abstract:

Chain topology has a profound impact on the flow behavior of single macromolecules. For circular polymers, the absence of free ends results in a unique chain architecture compared to linear or branched chains, thereby generating distinct dynamics. In this work, we use single molecule techniques to investigate the dynamics of ring polymers in dilute and semi-dilute solutions in planar extensional flow, including polymer relaxation from high stretch, transient stretching dynamics in step-strain experiments, and steady-state stretching in flow. In this work, we use fluorescently labeled circular DNA as ring polymers, which naturally provides monodisperse rings for single molecule experiments. In the first set of studies, we preformed a series of controlled strain-rate experiments on ring polymers in ultra-dilute solutions in planar extensional flow using precision microfluidics. [1,2] Our results show that the longest relaxation times of rings follow a power-law scaling relation with molecular weight that differs from linear chains. Also, relative to their linear counterparts, circular DNA molecules show a shifted coil-to-stretch transition and less diverse "molecular individualism" behavior as evidenced by their conformational stretching pathways. These results show the impact of chain topology on dynamics and reveal commonalities in the steady state behavior of circular and linear DNA that extends beyond chain architecture. We further compared our experimental results with Brownian dynamics (BD) simulations, and we confirmed the shifted coil-to-stretch transition is due to coupling of intra-molecular hydrodynamic interactions in the presence of architectural constraints. In a second set of studies, we extended single polymer experiments to study the dynamics of ring polymers in background solutions of semi-dilute linear polymers and semi-dilute ring polymers. This work represents the first single molecule measurement and direct observation of ring polymer dynamics in semi-dilute solutions under flow. Interestingly, we observe distinct "threading behavior" of single ring polymers in a sea of semi-dilute linear polymers, which results in qualitatively different dynamics compared to linear chains. We observe strong inhibition of polymer extension and *strikingly large fluctuations*in steady-state polymer extension for ring polymers in flow, demonstrating the strong interplay between polymer topology and polymer concentration. Taken together, our work aims to provide a molecular-level understanding of the role of polymer concentration and topology in non-dilute polymer solutions via direct observation of single chain dynamics in strong flow.

[1] Y. Li, **K. Hsiao**, C. A. Brockman, D. Y. Yates, R. M. Robertson-Anderson, J. A. Kornfield,â€¯M. J. San Francisco, C. M. Schroeder, and G. B. McKenna, "When ends meet: Circular DNA stretches differently in elongational flows,"Â *Macromolecules*, **48**, 5997-6001 (2015).

[2]Â **K. Hsiao,** C. M. Schroeder, and C. E. Sing, "Ring polymer dynamics are governed by a coupling between architecture and hydrodynamic interactions," *Macromolecules*, **49**, 1961-1971 (2016).

468201 *Lda Measurement of a Simplified Nasal Cavity*

Abstract:

LDA (Laser Doppler anemometry) [1] is a noninvasive flow measurement technique which allows evaluating the velocity components in a flow field. In contrast to other methods the temporal resolution is much higher, so it is possible to get information about the intensity of turbulence to validate time-varying structures in a LB (Lattice Boltzmann) [2] simulation of a simplified nasal cavity. The comparison of LDA and LB should determine whether the numerical calculation fits well with the flow phenomena by nature.In order to get a reliable and successful measurement it is necessary to define the boundaries of the flow. Since breathing process is complex and slightly different at every person, for simplification the maximum flow rate of an adult human from the rhinomanometry [3] (800 ml/s) is taken into account for the measurement. Therefore, a closed loop control system adjusts rotational speed of a fan by the use of a thermal anemometer [4]. Furthermore, the simplified nose model has a bounding box like an ordinary nose, though, the cross section is rectangular. In the middle a flow resistance with the dimensions of 12.5 x 10 mm is positioned. However, the same obstacle can be found within the frontal sinus near the nose entrance. Since the measurement technique uses a sensor to detect interference fringes of two laser lights, reflections would impede the execution heavily. Therefore, the model is made of glass and Plexiglas® elements. The principle of LDA needs moving particles which are in this case generated by an air humidifier. As the LDA setup has two moveable axes the domain will be analyzed in planes. Since the main flow is two dimensional the third velocity direction is neglected due to the simpler setup. Nevertheless, there is the need to do the measurement twice, since only one velocity component can be ascertained. Subsequently, the velocity magnitude is calculated for further investigations and post processing. In order to get out of the measurement points a continuous interpolated velocity function the Delaunay algorithm [5] is applied.This preparatory work helps to find boundary conditions and other numerical parameters for the LB simulation.

[1] Boutier, A., Laser velocimetry in fluid mechanics, 2012, ISBN: 978-1-84821-397-5.

[2] Mohamad, A., Lattice Boltzmann Method, Fundamentals and Engineering Applications with computer codes, 2011.

[3] Anniko M., Bernal-Sprekelsen M.,Bonkowsky V.,Bradley P., Iurato S., Otorhinolaryngology, Head and Neck Surgery, Springer Science & Business Media, 2010, ISBN: 978-3-54068-940-9.

[4] Goldstein R., Fluid Mechanics Measurements, Second Edition, 1996, ISBN: 978-1-56032-306-8.

[5] Edelsbrunner, H., Geometry and Topology for Mesh Generation. Cambridge University Press, Cabridge, 2001.

467803 Shear-Induced Structural Changes in Electrostatically Gelled Nanoparticle Dispersions

Abstract:

Mixtures of oppositely-charged fumed metal oxide nanoparticles will heteroaggregate into a fluid spanning network when dispersed in water. The gels produced by this process exhibit a yield stress at nanoparticle volume fractions greater than 0.015. Phase behavior studies and corresponding rheological characterization have demonstrated that the nanoparticle network only exists when the solution pH is in the range where the different nanoparticle species have opposite surface charges. Additionally, the strongest gel networks, i.e. the gels with the highest yield stress, were formed at particle ratios where the number of positive and negative surface charges are approximately equal. This work uses small angle neutron scattering to characterize the structure of the gel network in binary mixtures of fumed silica (SiO_2) and fumed alumina (Al_2O_3) and how it evolves under flow. Static gels were characterized at various particle concentrations, solution pHs, and mixture ratios. RheoSANS and RheoUSANS is used to investigate how the gel networks respond under flow and gain insight into network rearrangements and shear-induced phase separation at certain shear rates.

468448 Theory of Margination in Blood and Other Multicomponent Suspensions

Abstract:

A mechanistic theory is developed to describe distribution and segregation of particles in confined multicomponent suspensions such as blood. It incorporates the two key phenomena arising in these systems at low Reynolds number: hydrodynamic pair collisions and wall-induced migration. In simple shear flow, an analytical relationship is found for the depletion layer or â€œcell-free layerâ€ thickness of deformable-particle suspensions in terms of confinement ratio and volume fraction. Based on this result, a master curve collapsing a wide variety of experimental and direct simulation results can be found. In multicomponent suspensions several regimes of segregation arise, depending on the value of a ``margination parameter" M. Most importantly, there is a critical value of M below which a sharp ``drainage transition" occurs: one component is completely depleted from the bulk flow to the vicinity of the walls. Direct simulations also exhibit this transition as the size or flexibility ratio of the components changes. Results are presented for both Couette and plane Poiseuille flow. Experiments performed in the laboratory of Wilbur Lam indicate the physiological and clinical importance of these observations.

468804 Gravitational Collapse of Colloidal Gels: Structure, Dynamics, and Rheology

Abstract:

Reconfigurable soft solids such as viscoelastic gels have emerged in the past decade as a promising material in numerous applications ranging from engineered tissue to drug delivery to injectable sensors. These include colloidal gels, which microscopically comprise a scaffoldlike network of interconnected particles embedded in a solvent. Network bonds can be permanent or reversible, depending on the nature and strength of interparticle attractions. When attractions are on the order of just a few kT, bonds easily rupture and reform. On a macroscopic scale, bond reversibility allows a gel to transition from solidlike behavior during storage, to liquidlike behavior during flow (e.g., injection or shear), and back to solidlike behavior in situ. On a microscopic scale, thermal fluctuations of the solvent are occasionally strong enough to break colloidal bonds, temporarily allowing particles to migrate and exchange neighbors before rebonding to the network, leading to structural evolution over time. In experiments, gel aging continues and the gel may remain stable for several hours, days, or weeks, but then the gel may suddenly undergo a catastrophic collapse, sedimenting to the bottom of the container and eliminating any intended functionality of the network scaffold. Although this phenomenon has been studied extensively in the experimental literature, the microscopic mechanism underlying the collapse transition is not well understood. To study this behavior, we conduct large-scale dynamic simulation to model the structural and rheological evolution of colloidal gels subjected to various gravitational stresses, examining the detailed micromechanics in three temporal regimes: slow sedimentation prior to collapse; rapid sedimentation after the onset of rapid collapse; and the final slow compaction of the material as it approaches its final bed height. We also examine in detail the effects of gel age (structural lengthscales), interparticle attraction strength, and wall effects on the collapse behavior.

469144 Water-in-Diesel Fuel Emulsions Using Non-Ionic Surfactants: An Experimental Study

Abstract:

This study reports the water-in-diesel fuel emulsions prepared using non-ionic surfactants combined with different co-surfactants. Different concentrations of surfactants (S) and co-surfactants (C) with different Hydrophilic-Lipophilic Balance (HLB) were used individually and in mixture to prepare emulsions. To prepare these emulsions, different concentrations of water (5% - 15%) and SC (5 â€“ 10 wt%) were used using high speed Silverson Laboratory Mixer (nominal maximum speed 6000 rpm). Kinematic viscosities (cSt) of the diesel fuel and the prepared emulsions were measured using Anton Parr Viscometer (SVM300) and droplet size by laser diffraction analyzer (Microtrac S3500). The stability of the emulsion systems was studied by reading the amount of water separated from the emulsion. Using 5 wt% water and 10 wt% SC showed stable emulsion system with uniform particle size distribution. The particle size obtained using 10 wt% SC, HLB value of 9.6 at 4000 rpm (1 min) was 3.07 Âµm and decreased to 0.497 Âµm at 6000 rpm (2 min). Similarly, using 10 wt% SC, HLB value of 9.9 formed emulsion with a particle size of 3.97µm at 4000 rpm (1 min) and 0.536 µm at 6000 rpm (2 min). The results obtained on the effect of SC concentration, HLB of SC blend, water content, mixing speed as a function of time and temperature on the stability, droplet size, density and kinematic viscosity of water-in-diesel fuel stable emulsion system will be presented.

468881 Stretching Behavior of Red Blood Cells at High Strain Rates

Abstract:

Most work on the mechanical behavior of red blood cells (RBCs) has focused on simple shear flows. Relatively little work has examined RBC deformations in the physiologically important extensional flow that occurs at the entrance to a constriction. In particular, previous work suggests that RBCs rapidly stretch out and then retract upon entering the constriction, but to date no model predicts this behavior for the extremely high strain rates typically experienced there. In this work, we use high speed video to perform systematic measurements of the dynamic stretching behavior of RBCs as they enter a microfluidic constriction. We demonstrate that a simple viscoelastic model captures the observed stretching dynamics, up to strain rates as high as 1000 $s^{\text{â€“}1}$. The results indicate that the effective elastic modulus of the RBC membrane at these strain rates is an order of magnitude larger than moduli measured by micropipette aspiration or other low strain rate techniques.

469295 Capillary Flow Dynamics in 3D Printed, Open Microchannels

Abstract:

Three-dimensional (3D) printing is a potentially valuable tool for micro-scale manufacturing processes due to its relative speed and low cost as well as the high degree of part customization that it affords its users. Microchannels are an important component of many micro-manufactured parts, including microfluidic devices and micromolds. In this work, we investigate open microchannels printed with two different 3D printing technologies: fused deposition modeling (FDM) with a Stratasys Dimension printer and stereolithography (SLA) with a FormLabs Form 1+ printer. First, the ability of these two technologies to print open microchannels is assessed, both with respect to minimum printable feature size and dimensional accuracy. Microchannels as small as 250 µm x 250 µm (width x height) were printed using both FDM and SLA. Second, the flow dynamics of liquid traveling through these microchannels is investigated over timescales spanning five orders of magnitude, from milliseconds to hundreds of seconds, using high speed optical microscopy. The surface roughness characteristic of many 3D printing processes (including FDM and SLA) is found to cause local pinning of the contact line during capillary flow, leading to inhomogeneous wetting and a reduction in the capillary pressure driving the flow. Flow dynamics are evaluated with respect to print technology, print orientation, channel dimensions, and the properties of the liquid. Based on these results, specific recommendations are made for designing 3D printed microchannels for flow. 3D printing is shown to be appropriate and convenient in many scenarios for rapidly manufacturing viable microchannels.

469941 Time Evolution of Shear-Induced Particle Margination and Migration in a Cellular Suspension

Abstract:

The inhomogeneous center-of-mass distributions of red blood cells and platelets normal to the flow direction in small vessels play a significant role in hemostasis, drug delivery and microfluidics. Under pressure-driven flow in channels, the migration of deformable red blood cells at steady state is characterized by a concentration peak at the channel center and a cell-free layer or Fahraeus-Lindqvist layer near the vessel wall. Rigid particles such as platelets, however, "marginate" and thus develop a near-wall excess concentration. Though many efforts have been invested into determining these steady state behaviors of shear-induced migration and margination, little progress has been made on the time evolution of such processes. In the vascular system, branching occurs on the length scale of approximately 1mm for arterioles. In the design of microfluidic devices, short flow paths are preferred in order to reduce the volume of blood used for diagnostic purposes. Therefore, it is worth investigating the time evolution of particle margination and migration from a non-equilibrium state and determine the corresponding entrance lengths.

In this talk, we investigate the time-dependent concentration distribution of red blood cells and platelets in pressure-driven flow by solving a Boltzmann model, advection-diffusion equation for both species. From a fluid mechanics point of view, deformability-induced hydrodynamic lift and shear-induced diffusion are essential mechanisms for the cross-flow particle migration and margination. The governing Boltzmann equation for red blood cells includes both lift flux away from the wall and shear-induced diffusion due to cell-cell "collisions". On the other hand, the governing transport equation for platelets includes shear-induced diffusion from cell-platelet "collisions" and platelet-platelet "collisions". We verify our model by, first, solving for the steady concentration profile of red blood cells and platelets, and demonstrating that these predictions are in good agreement with full boundary element simulation and experimental results. We then explore the time evolution and report entrance lengths for red blood cell migration and platelet margination. Our theory serves as a fast and computationally-efficient alternative to large-scale simulation.

469154 A New Dielectric Rheosans Instrument for the Simultaneous Interrogation of Rheology, Microstructure, and Electronic Properties of Flow Battery Electrodes

Abstract:

Inexpensive grid-scale storage remains a major hurdle slowing the wide-scale adoption of renewable energy technologies. Flow Batteries have been proposed as one viable solution to this problem. Whereas, the capacity of a traditional battery is limited by its packaging, flow batteries pump redox active fluids from external storage tanks through a flow cell where electricity is extracted or stored via reversible redox reactions. One promising flow battery technology is the semi-solid flow battery (SSFB). Instead of using a soluble electrolyte, SSFBs incorporate dispersions of electrochemically active particles as the active material allowing for higher specific power densities. In addition to the electrochemically active species, carbon black is typically incorporated into the flow battery electrode as a conductive additive to provide the necessary electrical percolation throughout the electrode. While carbon black can allow for electrical percolation at relatively low volume fractions, the rheological properties of these dense slurries have a detrimental effect on the flow battery performance. In order to understand the intrinsic link between microstructure, rheology and conductivity in carbon black suspensions, we have developed a new dielectric RheoSANS instrument. We have fabricated a Couette geometry capable of performing RheoSANS measurements while the electrical properties of these carbon black slurries are monitored. By measuring the microstructure and electrical properties of carbon black network as function of the shear rate we have begun to understand the intimate link between the flow properties and electrical percolation in these materials. Our findings will open up new pathways for the optimization of these promising energy storage technologies.

Using Microfluidic Device to Study Rheological Properties of Heavy Oil

*Kiarash Keshmiri [a], Saeed Mozaffari [a, b], Plamen Tchoukov [a], Haibo Huang [c], Neda Nazemifard [a, *]*
[a] Department of Chemical and Materials Engineering, University of Alberta, Edmonton, AB, Canada
[b] Department of Chemical Engineering, Virginia Polytechnic Institute and State University, Blacksburg, VA, United States
[c] Reservoir and Geosciences, Alberta Innovates Technology Futures, Edmonton, AB, Canada

Abstract

In this study, capillary-driven flow of different pure liquids and diluted bitumen samples were studied using microfluidic channel (width of 30 μm and depth of 9 μm). Capillary filling kinetics of liquids as a function of time were evaluated and compared with theoretical predictions. For pure liquids including water, toluene, hexane, and methanol experimental results agreed well with theoretical predictions. However, for bitumen samples, as concentration of bitumen increased the deviation between theoretical and experimental results became larger. The higher deviation for high concentrations (i.e. above 30%) can be due to the difference between dynamic contact angle and bulk contact angle. Microchannels are suitable experimental devices to study the flow of heavy oil and bitumen in porous structure such as those of reservoirs.

1. Introduction

Microfluidics is an emerging technology that deals with the fluid flow in micro-scale channels [1]. Microfluidic devices has been used as a good representative of porous media to evaluate dynamic aspects of fluid flow at pore-scale [2]. Micro and nanofluidic devices have received attention during the last few decades due to high control of hydrodynamics, fast and cheap fabrication, possibility of working at high temperature and pressure, and visualization of the fluid flow [3]. Prediction of capillary flow in micro and nanochannels are essential to develop emerging nanofluidic methods.

Capillary action drives fluid motion in microchannel without external driving forces. Capillary driven flow refers to spontaneous movement of interface due to curved liquid-liquid or gas-liquid interface [4]. In this study, glass etched microchannels with the depth of 10 μm and width of 30 μm were used to monitor the capillary-driven flow of different liquids. The size of channels is close to the size of real porous media of oil sands in Alberta, Canada. Capillary filling speed of hexane, methanol, toluene, and water as well as solutions of bitumen in hexane (10% to 50%) were experimentally monitored using inverted microscope with a digital charged coupled device (CCD) camera. In each chip, six parallel microchannels were fabricated which enables us to conduct several experiments with the similar experimental condition. This is important for repeatability of the data and having average values for each run.

Theoretical viscosity of each sample was calculated with the assumption of constant contact angle which was then compared with measured bulk viscosities. In our previous work in our group [5], capillary filling kinetics of bitumen solutions in nanochannel (depth ~ 47 nm) was investigated, where theoretical results were significantly deviated from experimental values specially for higher concentrations of bitumen samples. It seems that sharp variation of advancing contact angle and interface shape is the reason of this deviation (Figure 1). In fact, advancing contact angle is completely different from bulk contact angle and considering a constant value is not a reliable approach for high concentration of bitumen. Therefore, classical model failed to explain the filling kinetics of concentrated bitumen in the nanochannel.

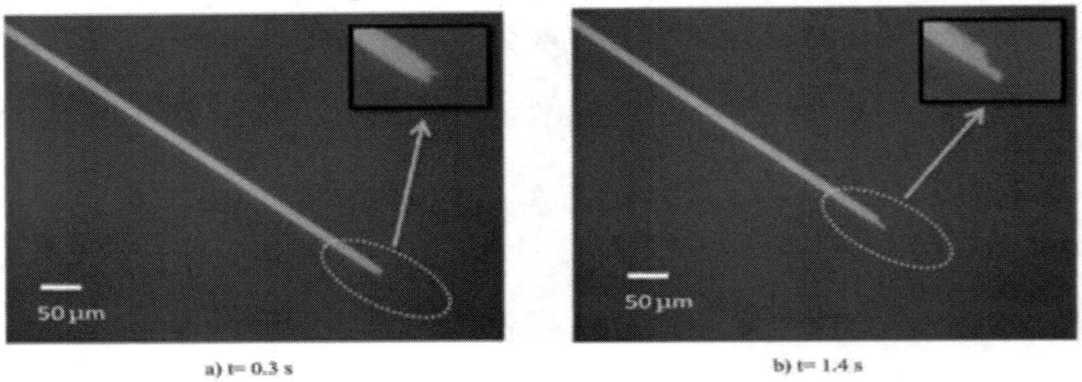

Figure 1. Capillary filling of 40% bitumen diluted in heptol (80:20) [5]

These findings further suggest that channel size can affect the wettability, where decreased wettability in nano-scale separations leads to large deviation between theoretical and experimental results. On the other hand, the experimental results of microchannels showed less deviation from classical model compared with nanoscale flows. It seems that microscale flow shows more predictable flow behavior compared with nanoscale using classical models while filling kinetics of viscous fluids in micro-scale require more study in order to have better prediction of real porous media structure.

2. Material and method

For capillary-driven flow study, methanol (HPLC grade, Fisher Scientific), hexane (HPLC grade, Fisher Scientific), toluene (HPLC grade Fisher Scientific) were supplied. Athabasca bitumen was also provided by Alberta Innovative Technology Feature (AITF) diluted with hexane (10% to 50%).

2.1. Microchannel fabrication

Photolithographic method was used for microchannel fabrication on 4 inch × 4 inch × 0.043 inch borofloat (81 % SiO_2, 13 % B_2O_2, 4 % Na_2O/K_2O, 2 % Al_2O_3) glass wafers. Wafers were cleaned at cleanroom using piranha solution (H2SO4 and H2O2 with 3:1 volume ratio) and rinsed to remove piranha solution using de-ionized water and then dried in spin rinse. A layer of Cr (75 nm) and Au (180 nm) was coated on the wafers as the masking layer. Photoresist (HPR 504) was spun onto the wafer and backed for 30 min. photoresist was exposed to light through a photomask that was previously designed using L-edit software in order to transfer the designed pattern to the photoresist. Then Au and Cr were wet etched with KI etchant based on photomask pattern. Borofloat glass was etched using hydrofluoric acid (HF) for the specific time based on desire depth of channel (1.4 µm/min). The remained Au and Cr and photoresist were removed from the wafer. Finally, the etched and unpatterned substrates were cleaned using piranha solution and pressed together to form temporary bonding. The advantage of temporary bonding is that it enables us to reuse the fabricated channels which significantly lowers the cost of fabrication. A general schematic of the fabrication process is illustrated in Figure 2.

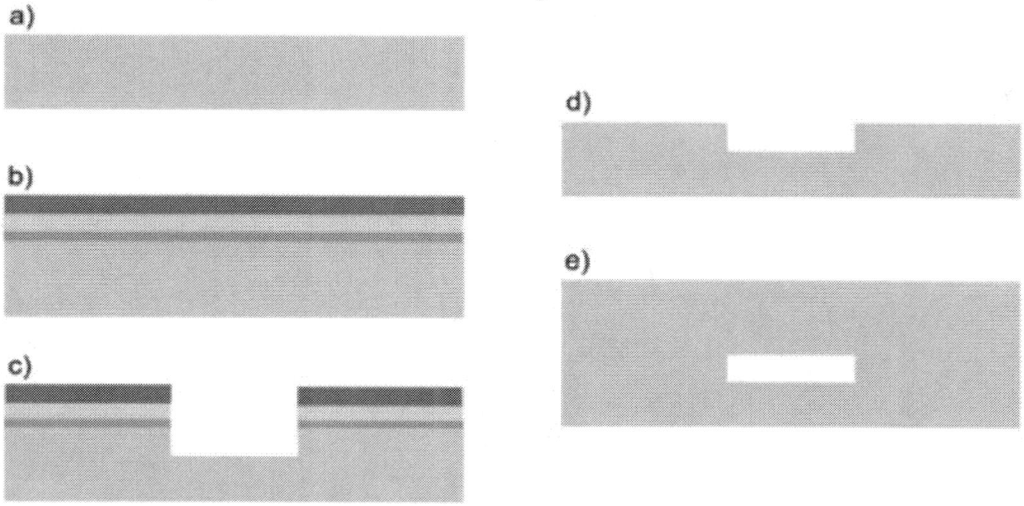

Figure 2. Microchannel fabrication process, a) Piranha cleaning, b) Metal layer and photoresist coating (Cr and Au), c) Developing of photoresist, etching the metals and borofloat, d) Removing the metal layers and photoresist, e) Glass-glass temporary bonding

2.2. Viscosity, contact angle, and surface tension ent

Viscosity of diluted bitumen samples and other pure liquids was measured using rheometer (Rheolab QC, Anton Paar, Austria) with a double gap measurement system. Shear rate was ranged from 1 to 1000 s^{-1}. Measurement for each sample was repeated at least three times and temperature was maintained at 22 $\pm$ 0.5 °C. Bitumen samples showed Newtonain fluid behavior in agreement with our previous work which was done for slightly different condition [6]. Bulk contact angle of diluted bitumen samples and pure liquids was measured on the etched glass surface using FTA 200 equipped with camera and image analytical program. The same system (FTA 200) was used for air-liquids surface tension measurement using Young-Laplace equation to symmetric pendant drop. Bulk contact angles and surface tensions were measured three times at temperature of 22 $\pm$ 0.5 °C.

3. Theory

Experimental analysis of capillary filling kinetic of heavy oil in microchannels and further comparison with the theoretical models improves our understanding of fluid transport and wettability phenomena in the microscale. Fluid-fluid and fluid-surface interaction, geometry, wettability (oil-wet to water-wet), and surface tension are among the effective parameters in filling analysis. Capillary filling of liquids can be evaluated and compared with theoretical models such as Lucas-Washburn classical model [7]. This equation applies physical properties of the liquid and microchannel surface to predict the meniscus location as a function of time. Considering moderately constant contact angle and absence of gravity forces on the basis of Navier- Stokes law for an incompressible, laminar, and viscose flow the governing equation is written as [8]:

$$-\frac{\partial P}{\partial x} = -\mu \frac{\partial^2 V_x}{\partial y^2} \quad (1)$$

Where P is fluid pressure, x is penetration distance, μ the viscosity of the fluid, and V_x fluid velocity. Pressure drop across the free surface of the meniscus is calculated based on Young-Laplace equation (Eq. 2) and combining with Eq. 1, leads to Lucas-Washburn equation (Eq. 3):

$$\Delta P = \frac{2\sigma \cos \theta}{r} \quad (2)$$

$$x = \left(\frac{r\sigma \cos \theta}{2\mu} t\right)^{1/2} \quad (3)$$

Where r is the hydraulic radius, σ the air-fluid surface tension, and θ the fluid advancing contact angle. Average filling speed of advancing meniscus as a function of position can be written as (Eq. 4):

$$V_x = \frac{h\sigma \cos \theta}{4\mu x} \quad (4)$$

4. Results and Discussion

Advancing liquid meniscus inside the microchannel was monitored using white light microscope coupled with high speed CCD camera. Recorded images were processed using ImageJ software. For all pure liquids and bitumen samples linear relation between square of position of advancing liquids and propagation time was found as expected in the case of Newtonian fluids. The experimental results are illustrated in Figure 3 and compared with the Washburn model. It can be seen that theoretical prediction agrees well with experimental values.

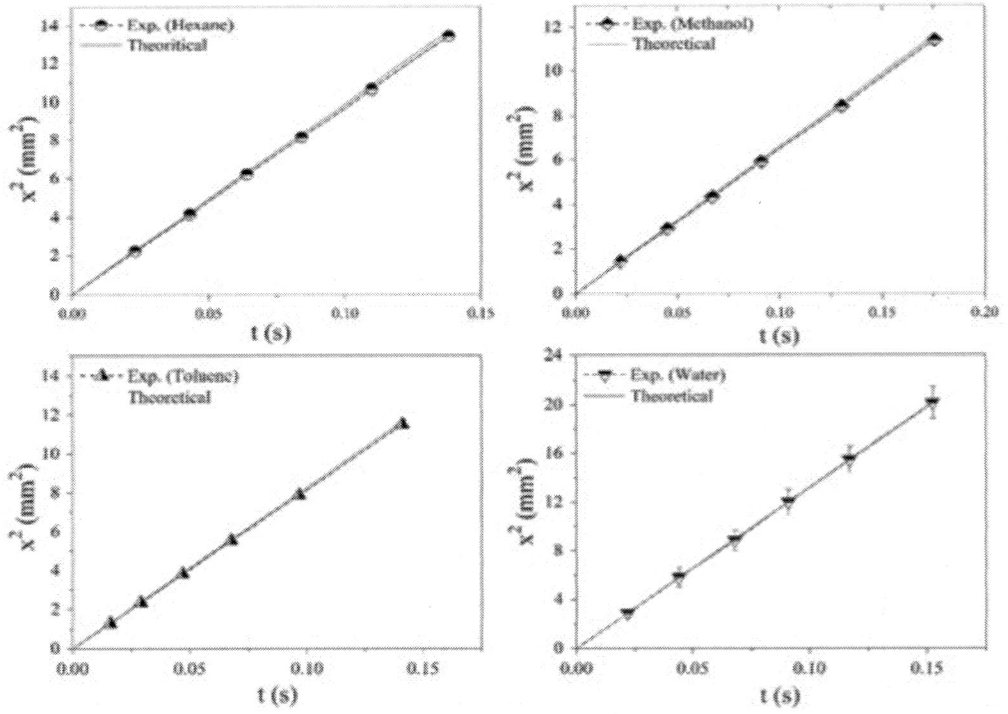

Figure 3. Experimental and theoretical comparison of pure fluid filling in microchannel

Averaged experimental data for the filling of water, methanol, toluene, and hexane in microscale channels were used to calculate fluid viscosity using the Washburn equation (Eq. 4). To determine the difference between theoretical and experimental values Average Absolute Relative Deviation (AARD%) is calculated using Eq. (6).

$$AARD\ (\%) = \frac{100}{N}\sum_{i=1}^{N}\left|\frac{x^{exp} - x^{cal}}{x^{exp}}\right| \tag{6}$$

In this equation, N is the number of data points, x^{exp} is the experimental position of the advancing interface and x^{calc} is the calculated value obtained from Eq. 4. The summary of the results for pure liquids is shown in Table 1.

Table 1. Theoretical and experimental viscosity and AARD pure fluids

	Line slope x^2/t (mm^2/s)	**Theoretical viscosity (mPa.s)**	**Experimental viscosity (mPa.s)**	**AARD (%)**
Water	132.37	0.887	0.89	0.27
Toluene	81.58	0.567	0.56	0.64
Hexane	96.74	0.316	0.31	0.99
Methanol	64.67	0.596	0.58	0.8

Figure 4 shows time-lapsed images of capillary-driven flow of 40% bitumen in microchannel. The interesting observation during capillary flow in microscale is that the shape of interface is similar at different penetration distances as opposed to the continuous changes in the shape and contact angle of bitumen observed during capillary flow of bitumen in nanochannel [9]. It seems that effective forces exerted on the liquid in nanoscale are different from the micro scale. For instance, Phan et al [10] showed the importance of electrical double layer (EDL) in nano-scale, where the thickness of EDL is comparable to the length of channel and because of that the observed viscosity in the nanoscale was larger than the bulk value.

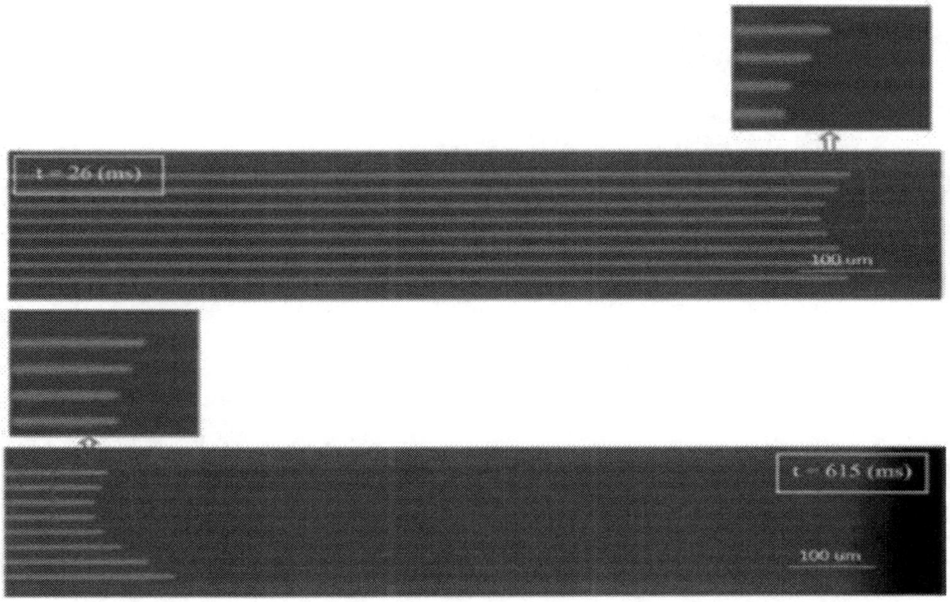

Figure 4. Time-lapsed images of capillary filling of diluted bitumen (40%)

In the case of diluted bitumen samples (10 wt.% to 50 wt.%), a linear relation between position squared of advancing liquids and propagation time was found (Figure 5). In order to compare experimental results with theoretical predictions, x^2/m as a function of time is plotted as shown in Figure 6 ($m = r\sigma\cos\theta/2\mu$). Despite the good agreement between experimental and theoretical predictions of pure liquids, there was an increasing deviation when bitumen concentration increased from 10% to 50%. According to table 2, theoretical viscosities were calculated and compared with the experimental viscosities. Moreover, experimental and theoretical values for square meniscus position were compared based on AARD. This deviation between theory and experiment for higher bitumen fractions can be due to the change in the wettability and dynamic contact angle being different from the bulk.

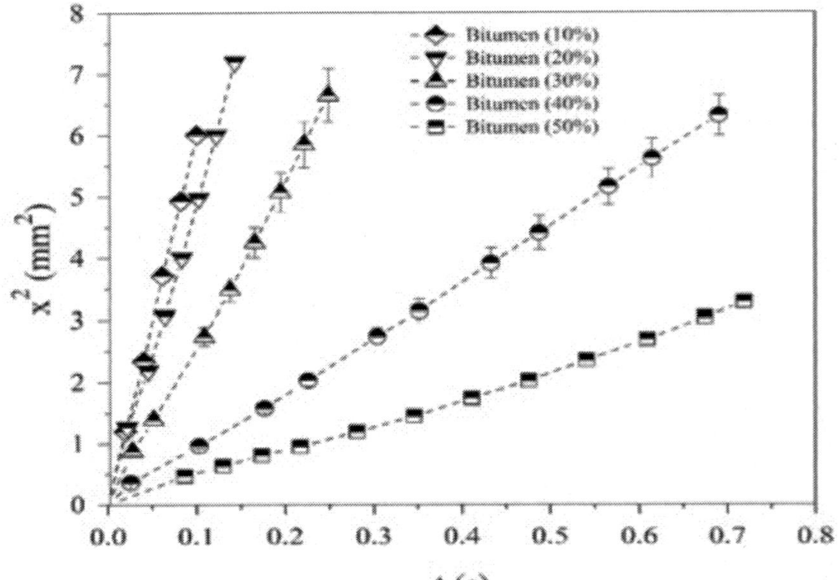

Figure 5. Square capillary flow of diluted bitumen vs time

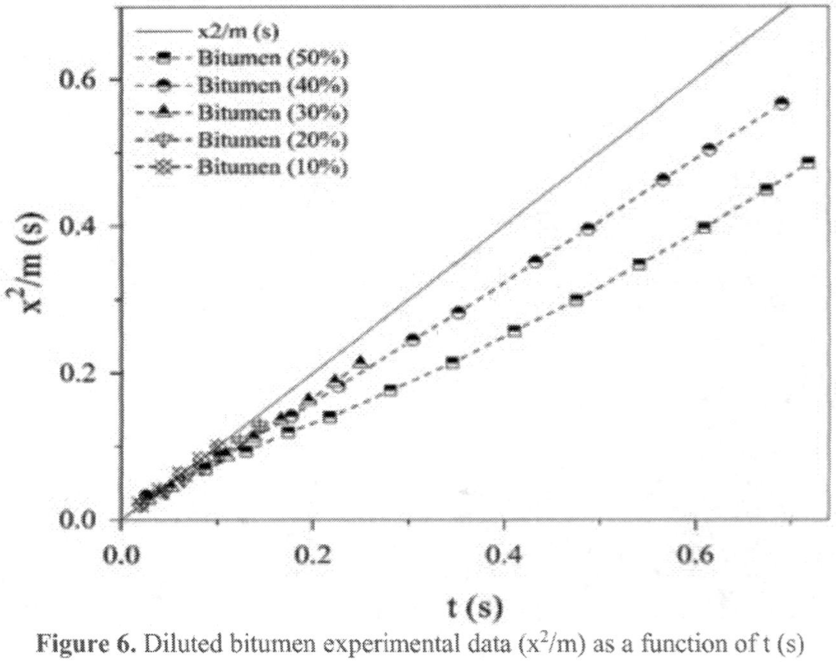

Figure 6. Diluted bitumen experimental data (x^2/m) as a function of t (s)

Table 2. Theoretical and experimental viscosity and AARD for diluted bitumen

	Line slope x^2/t (mm^2/s)	Theoretical viscosity (mPa.s)	Experimental viscosity (mPa.s)	AARD (%)
Bitumen (10%)	59.70	0.578	0.557	3.31
Bitumen (20%)	48.91	0.672	0.584	12.84
Bitumen (30%)	26.12	1.21	0.988	19.23
Bitumen (40%)	9.03	3.09	2.51	23.43
Bitumen (50%)	4.41	5.68	3.75	53.9

5. Conclusion

In this study, capillary filling of pure liquids and diluted bitumen (10%-50%) were studied and compared with theoretical predictions. For all the pure liquids and bitumen samples, a linear dependency between square position and time was observed which is an indication of Newtonian fluid type behavior. Capillary filling kinetics of pure liquids (i.e. methanol, toluene, hexane, and water) were in good agreement with the theoretical model (i.e. Washburn) with AARD less than 1%. For bitumen samples, theoretical predictions deviated from experimental results. This deviation was increased at higher concentrations of bitumen. It could be a result of surface wettability changes or difference between dynamic contact angle and bulk contact angle.

References

[1] Sinton, D. (2014), "Energy: the microfluidic frontier," *Lab Chip,* pp. 3127-3134.

[2] Rossi, M. P., Ye, H., Gogotsi, Y., Babu, S., Ndungo, P. and Bradeley, J. C. (2004), "Environmental Scanning Electron Microscopy Study of Water in Carbon Nanopipes," *Nano Lett,* pp. 989-993.

[3] Mcdonal, J. C., Duffy, D. C., Anderson, J. R., Chiu, D. T., Wu, H., Shueller, O. J. and Whitesides, G. M. (2000), "Fabrication of microfluidic systems in poly(dimethylsiloxane)," *Electrophoresis,* vol. 21 (1), pp. 27-40.

[4] kuo, J. N., and Lin, Y. K. (2012), "Capillary-Driven Dynamics of Water in Hydrophilic Microscope Coverslip Nanochannels," *Japanese Journal of Applied Physics,* p. 51.

[5] Mozaffari, S. (2015), "Rheology of Bitumen at the Onset of Asphaltene Aggregation and its Effects on the Stability of Water-in-Oil Emulsion," MSc thesis, University of Alberta.

[6] Mozaffari, S., Tchoukov, P., Atias, J., and Nazemifard, N. (2015), "Effect of Asphaltene Aggregation on Rheological Properties of Diluted Athabasca Bitumen," *Energy & Fuels,* vol. 29(9), pp. 5595–5599.

[7] Washburn, E. W., (1921), "The Dynamics of Capillary Flow," vol. 17, p. 273.

[8] Chebbi, R. (2011), "Electroosmotic-capillary penetration in circular microchannels," *Journal of Colloid and Interface Science,* vol. 361, pp. 639-642.

[9] Mozaffari, S., Tchoukov, P., Mozaffari, A., Atias, J., Czarnecki, J. and Nazemifard, N. (2016), "Capillary Driven Flow in Nanochannels - Application to Heavy Oil Rheology Studies," *Colloids and Surfaces A: Physicochemical and Engineering Aspects,* Under review.

[10] Phan, V. N., Yang, C. and Nguyen, N. T. (2009), "Analysis of capillary filling in nanochannels with electroviscous effects," *Microfluidics and Nanofluidics,* vol. 7, p. 519.

470678 Early Atherogenesis: How Blood-Pressure-Driven Flow and Membrane Proteins Influence Artery Wall Mechanics and Mass Transfer Across the Artery Wall

Abstract:

Atherosclerosis has long been the leading cause of death in the United States and all Western countries. It begins with the transmural (across the wall)-pressure (DP)-driven advection of macromolecular aggregates such as low-density lipoprotein (LDL) cholesterol from the blood across the monolayer of endothelial cells that lines the vessel and into the artery wall. Particles of the size of LDL (~22nm) cross the endothelium through the junctions around rare (~1/2000-5000) endothelial cells whose junctions are sufficiently wide and enter the wallâ€™s thin subendothelial intima (SI) layer where they can bind to SI extracellular matrix (ECM) and accumulate. The blood pressure also drives much smaller molecules such as water and small solutes across the entire endothelium, much of it likely across the normal inter-endothelial tight junctions, and through the wall. This flow may be critical determinant of prelesion events since it can dilute the local SI LDL concentration, which likely slows its kinetics of binding to ECM, and flushes unbound LDL from the wall.

We have earlier shown that bovine and rat aortic endothelial cells (ECs) express the membrane water channel protein, aquaporin-1 (AQP1) and that up/down-regulating the number of EC AQP1s increases/decreases the transmural water flow at fixed DP, i.e., increases/decreases the ratio, the hydraulic conductivity, Lp, of flux to DP. Thus a portion of the transmural water flow goes through the cells, in parallel with the portion that goes between them. Further we have shown earlier that at fixed EC AQP1 expression, Lp vs DP is flat at low DP, drops over a narrow DP window and then remains flat to high DP, but the position of this window depends on DEC AQP1 expression. We have shown that this Lp-dependence at fixed EC AQP1 is due to compression of the SI region of the vessel wall, which blocks certain fluid transport pathways into deeper regions of the wall. Here we present a hydrodynamic theory to explain the EC AQP!

dependence of the position of the transition window and how to shift this window. Time permitting, we shall show how using this flow field and the hgihg AQP1 selectivity for pure water transport in an advective-diffusion theory for small solute, e.g., albumin, transport yields a measurable oncotic effect. The theory shows how measured Lp values include mass transfer effects, how one can extract intrinsic Lp values from these measurements and a strategy to potentially slow pre-lesion lipid entrainment in the artery wall.

471275 Viscoelasticity of Mucus Layers Secreted By Intestinal Epithelial Cells

Abstract:

A small-gap rheometry technique is developed to evaluate the viscoelasticity of the adherent mucus secreted by living cell cultures of human intestinal epithelial cells as a function of position above the substrate surface. Non-mucus producing (HT29) and mucus producing (HT29-MTX) intestinal epithelial cells are grown to confluence over 2.5 cm glass cover slips mounted in polystyrene wells, which are then mounted directly to the lower platen of a parallel plate rheometer (Rheometrics ARES), using cyanoacrylate adhesive. A reference point for setting the gap of the rheometer is established by "zeroing" the instrument against an ultra-high precision 3mm stainless steel ball placed at the center of the cover slip. The linear viscoelastic responses G'(ω) and G"(ω) are then measured at frequencies ω ranging from 0.16-16 Hz and 0.5% strain, for gaps varying from 0.050-0.150mm above the cover slip surface. The results[1] show that the adherent mucus layer over the mucus-producing (H29-MTX) cells is approximately 0.10 mm thick, and that the viscoelastic response is dominated by the storage modulus G'(ω), which varies strongly with position within the mucus layer. At a fixed frequency of ω=1.6 Hz, G' varies from approximately 180 Pa to 37000Pa as the gap (height above the cover slip) decreases from 0.100mm to 0.050mm. As expected, the non-mucus producing (HT29) cell culture does not exhibit a significant viscoelastic response at any of the gap positions tested. An addition set of experiments performed with HT29-MTX mucus producing cells exposed to estrogen shows that estrogen-treated cells produce mucus layers with significantly higher viscoelasticity than those not treated with estrogen.

1. Diebel, M. E, Diebel, L. N., Liberati, D. M. and Manke, C. W., "Estrogen modulates intestinal mucus physiochemical properties and protects against oxidant injury", Journal of Trauma and Acute Care Surgery, 78, 94-99 (2015).

471384 Dynamics of Thrombus Formation in a Microfluidic Network Mimicking Vasculature

Abstract:

The reaction dynamics of a complex mixture of cells and proteins, such as blood, in branched networks within microvasculature, as for instance in brain, remains ill-defined. In this work, we utilize an interconnected microfluidic network with dimensions mimicking venules to study patterns of platelet aggregation and fibrin formation under complex shear. Devices were coated with fibrillar collagen and tissue factor to facilitate platelet aggregation and fibrin formation. Blood was perfused through the microfluidic device at a physiologically relevant shear rate. We observed distinct flow and thrombus formation patterns near channel intersections and stagnation points using fluorescent microscopy. A non-Newtonian fluid mechanics model blood was developed for blood flow in microfluidic networks. Red blood cells and platelets were assumed to be non-interacting particles transported by the bulk fluid flow through visous drag force. Based on the insights from these models, we were able to provide metrics that quantify the spatial and temporal locations of thombus formation in these networks. Findings from this proof-of-principle microfluidic network model suggests a specific correlation between microvascular geometry and thrombus formation dynamics under shear perturbed flow. We believe that this integrative approach has a huge potential in identifying thrombus-susceptible vascular regions within a complex network, such as human organ vasculature.

471662 Nonlinear Relaxation Modulus Via Dual-Frequency Medium Amplitude Oscillatory Shear (MAOS): General Framework and a Case Study for a Dilute Suspension of Brownian Spheroids

Abstract:

We present a framework for determining the viscoelastic relaxation moduli of a complex fluid from an oscillatory shear deformation. Knowledge of these moduli allows one to predict the stress response under an arbitrary transient deformation via a memory integral expansion. Our framework is demonstrated by determining the first nonlinear relaxation modulus, referred to here as the medium amplitude oscillatory shear (MAOS) relaxation modulus, for a dilute suspension of Brownian spheroids subject to a dual-frequency oscillatory simple shear flow. Specifically, we first determine the second normal stress difference for such a deformation from a co-rotational memory integral expansion. Second, the microstructural stress response of the model system of Brownian spheroids is determined via a regular perturbation expansion of the orientation distribution function at small dimensionless strain-rate amplitude, or Weissenberg number. An analytical expression for the MAOS relaxation modulus is resolved by comparing the second normal stress difference results of the memory integral expansion and microstructural stress response. Finally, using the MAOS relaxation modulus, we reconstruct the stress response of our model system for the start-up and cessation of steady shear and steady uniaxial extension. In summary, our work offers a route to utilizing medium (and large) amplitude oscillatory shear data to predict stress dynamics in other transient, nonlinear flows.

471882 Modeling the Viscosity of Polydisperse Suspensions

Abstract:

Polydisperse particulate suspensions are commonplace in many industrial processes and modeling their rheological behavior is important towards understanding their flow properties. The present study develops a more consistent extension of the approach pioneered by Farris (1968) to model the viscosity in polydisperse suspensions. A key element outcome of this work is the development of formulae that by construction preserve all know limiting rheological behaviors of suspensions and is therefore compatible with previous arguments on binary suspension behavior. This work outlines the development of the model in the limit of binary suspensions and presents a generalization to polydisperse suspensions. An important property of the model is that once parameterized for non-colloidal hard sphere binary suspension behavior, assumed to represent ideal binary suspensions, it may be used to predict more complex suspension behavior. The systematic extension of the model to ternary suspensions is also presented. Following Wagner and Woutersen (1994), we demonstrate that polydisperse suspensions may be modeled by representing them as equivalent ternary suspension. This is accomplished by matching the first 6 moments of the polydisperse distribution to those of a ternary suspension. Convergence studies are also presented to demonstrate the robustness of the approach. The model is applied toward predicting the viscosity of non-colloidal polydisperse suspensions and demonstrates satisfactory agreement with experimental data on the viscosity of suspensions with continuous size distributions. In addition, we also demonstrate the assumptions behind the resultant model through carefully selected and designed experiments involving polydiserse suspensions. Moving beyond ideal binary suspension behavior, we also present an extension to the viscosity model for complex suspensions in which the two or more particle sizes obey different monodisperse viscosity correlations. This demonstrates the model's robustness and applicability to an even wider variety of suspension such as a binary suspensions of spheres and cubes or colloidal and non-colloidal particles. Comparison against experimental data will also be presented.

References

1. Farris, R. J., "Prediction of the viscosity of multimodal suspensions from unimodal viscosity data," Transactions of the Society of Rheology **12**, 281-301 (1968).
2. Wagner, N. J., and A. T. J. M. Woutersen, "The viscosity of binary and polydisperse suspensions of hard spheres in the dilute limit," Journal of Fluid Mechanics **278**, 267-287 (1994).
3. Mwasame, P. M., N. J. Wagner, and A. N. Beris. "Modeling the effects of polydispersity on the viscosity of noncolloidal hard sphere suspensions," Journal of Rheology **60**, 225-240 (2016).

470398 Impact of Colloidal Stability on the Electric Field Driven Break-up of Non-Aqueous Suspension Drops

Abstract:

We examine the breakup of an oil drop containing a colloidal suspension of carbon black particles under a uniform DC electric field. A drop of squalane containing a fixed concentration 3.3 g/L of carbon black and varying amounts of polyisobutylene succinimide dispersant (OLOA) is exposed to field strengths that yield drop breakup. When a particle-loaded drop containing an appreciable concentration of OLOA (30 parts OLOA per 100 parts carbon black, pph OLOA) is subjected to an electric field of 2.5 kV/cm, the drop exhibits homogeneous breakup via the pinch-off of bulbous ends. Increasing the applied field to 5.3 kV/cm yields an inhomogeneous, nonaxisymmetric drop breakup mode. Since the added dispersant stabilizes the colloidal suspension of carbon black particles in this system, we observe no appreciable qualitative differences between the breakup of a drop containing both particles and dispersant, and a drop containing dispersant alone. When the concentration of dispersant is decreased to a value of 2 pph OLOA, the colloidal suspension becomes unstable within the timescale of the experiment, and the drop achieves inhomogeneous breakup via the formation of lobes that turn into fingers and eventually disintegrate. Here, the drop containing both particles and dispersant achieves a radically different field-induced breakup mode from a drop containing dispersant alone. The experimental systems in this study are modeled using the boundary integral method, which assumes that both the drop and medium phases are homogeneous. We implement our computations to assess the importance of surface charge transport and the heterogeneity of the particle-loaded drop. The method depicted in this study may be implemented to deduce the colloidal stability of particle-loaded drops exposed to electric fields.

471983 Microstructure-Rheology Relationships in Strongly Shear Thickening Colloidal Dispersions

Abstract:

While continuous shear thickening in colloidal dispersions can be quantitatively attributed to lubrication hydrodynamic interactions in strong shear flows(1), questions remain concerning the mechanism(s) underlying discontinuous shear thickening and dilation. To explore this, the shear rheology and colloidal microstructure are measured for a model, near-hard sphere, shear thickening fluid (STF) comprised of near hard sphere silica nanoparticles at high volume fraction (0.52) in PEG/glycol. This suspension exhibits discontinuous shear thickening at stresses above ~200 Pa. The shear and normal stress differences are compared against prior reports(1) and quantitatively compared with the colloidal microstructure using recently expanded stress-SANS rules (2). These measurements provide evidence for the mechanism of shear thickening by relating the microstructure to the flow conditions of the material, and test prevailing ideas concerning the role of specific surface interactions versus hydrodynamic interactions in colloidal dispersions at high concentrations and strong flow conditions.

1. Cwalina, C. D. & Wagner, N.J. â€œMaterial properties of the shear thickened state in concentrated near hard-sphere colloidal dispersion.â€ Journal of Rheology. 2014;58:949.
2. Gurnon, A.K. & Wagner N.J., Microstructure and rheology relationships for shear thickening colloidal dispersions. Journal of Fluid Mechanics. 2015;769:242-76.

472276 A Universal Interpretation of Transient Viscoelastic Extensional Dynamics: Application to Single Polymer Experiments

Abstract:

Advances in fluorescence imaging and particle manipulation have enabled the direct observation of single polymer dynamics in model flows such as shear flow and planar extensional flow. The majority of single polymer studies in the literature have focused on chain dynamics using simple step forcing functions. In order to study single polymer dynamics in non-idealized 'model' flows, there is a clear need to implement more complicated flow types and transient flow forcing functions. For example, large amplitude oscillatory shear flow (LAOS) has been widely used in recent years to elicit nonlinear viscoelastic responses of materials in bulk rheology, but complementing molecular techniques have not yet been applied.

In this work, we directly probe linear and nonlinear single polymer responses using large amplitude oscillatory extensional flow (LAOE) in precisely controlled microfluidic devices. We are able to generate both small and large amplitude extensional flows in a cross-slot microfluidic device while imaging the conformational dynamics of a single λ-DNA polymers trapped at the stagnation point. Using this technique, the projected extension of the polymer is measured. Our experimental results are in good agreement with Brownian dynamics (BD) simulations. We characterize the projected polymer extension as a function of both the dimensionless input strain rate $Wi_0 sin(\omega t)$ and its temporal derivative $Wi_0 \omega cos(\omega t)$, and have constructed a series of single polymer Lissajous plots. Responses have been mapped over a range of Weissenberg number amplitudes (*Wi*, flow strength) and Deborah numbers (*De*, probing frequency). By applying the sequence of physical processes (SPP) framework, we quantify the instantaneous polymer responses under LAOE through a set of newly-defined rheological parameters that are readily extendable from the single polymer scale up to bulk extensional measurements. The transition dynamics of a single polymer from its linear extensional response to a nonlinear response can also be rigorously quantified

through this framework. Additionally, the contribution to overall projected extension from both of the input flow parameters is rigorously determined. These techniques provide a stringent test of extensional theories of polymers on a single-chain basis.

472030 Free Surface Flows and Extensional Rheology of Polymer Solutions

Abstract:

Free-surface flows â€“ jetting, spraying, atomization during fuel injection, roller-coating, gravure printing, nanoimprint hot embossing, several microfluidic drop/particle formation techniques, and screen-printing â€“ all involve the formation of axisymmetric fluid elements that spontaneously break into droplets by a surface-tension-driven instability. The growth of the capillary-driven instability and pinch-off dynamics are dictated by a complex interplay of inertial, viscous and capillary stresses for simple fluids. Additional contributions by elasticity, extensibility and extensional viscosity play a role for complex fluids.Â We show that visualization and analysis of capillary-driven thinning and pinch-off dynamics of the columnar neck in an asymmetric liquid bridge created by dripping-onto-substrate can be used for characterizing the extensional rheology of complex fluids. Using a particular example of dilute, aqueous PEO solutions, we show the measurement of both the extensional relaxation time and extensional viscosity of weakly elastic, polymeric complex fluids with low shear viscosity (< 20 mPa.s) and relatively short relaxation time, (< 1 ms). Characterization of elastic effects and extensional relaxation times in these dilute solutions is beyond the range measurable in the standard geometries used in commercially available shear and extensional rheometers (including CaBER, capillary breakup extensional rheometer). As the radius of the neck that connects a sessile drop to a nozzle is detected optically, and the extensional response for viscoelastic fluids is characterized by analyzing their elastocapillary self-thinning, we refer to this technique as dripping-onto-substrate (DoS) extensional rheometry. Using DoS experiments and theory of polymer dynamics, we elucidate how polymer composition, flexibility, and molecular weight as well as coil-stretch transition determine the thinning and pinch-off dynamics, and thus influence printability, spinnability, sprayability and jettability of polymeric complex fluids.

472419 *Drosophila Melanogaster* Embryo As an Active Granular Fluid:Intercellular Coordination Via Mechanical Feedback during Morphogenesis

Abstract:

Embryogenesis hinges on seamlessly proceeding multiple morphogenetic processes that require both local and embryo-wide cellular coordination. Chemical signaling, whose importance to biological processes is incontrovertible, is not capable of controlling the compatibility of cellular shape changes and motions. We believe that the precise coordination occurring during the embryo development is accomplished through communication via mechanical stress fields and feedback. Therefore, we propose an active granular fluid (AGF) modeling platform that describes a tissue as a collection of mechanically active, stress-responsive objects. As a proof of concept, we provide two applications of the AGF approach to Ventral Furrow Formation (VFF) during the gastrulation of a *Drosophila* embryo. The first application, focused on the ventral face of the embryo during the initiation of VFF, has provided strong evidence that the observed patterns of constricted cells are the result of the cellular sensitivity to tensile stresses. The other application considers a cross-section of the embryo throughout the invagination of the ventral face, which is completed during VFF. Our modeling has revealed that the process is more robust when cells can actively deform to relieve excess stresses or to provide an additional push to help complete a perturbed invagination process.

472308 Two Spheres and a Spring Make a Good Swimmer

Abstract:

Rigid spherical particles in oscillating fluid flows form interesting patterns as a result of fluid mediated interactions, at intermediate Reynolds numbers. Here, through both experiments and simulations, we show that two spheres under horizontal vibration align themselves at right angles to the oscillation and sit with a gap between them, which scales in a non-classical way with the boundary layer thickness. A large number of spherical particles form strings perpendicular to the direction of oscillation. Investigating the details of the interactions we find that the driving force is the nonlinear hydrodynamic effect of steady streaming. We then design a simple swimmer (two-spheres-and-a-spring) that utilizes steady streaming in order to propel itself and discuss the nature of the transition at the onset of swimming as the Reynolds number gradually increases. We discuss our latest results on the collective behavior of simple swimmers at intermediate Reynolds numbers.

472552 Predicting Fractional Flow Reserve Using Computational Fluid Dynamics and Response Surface Method Statistics

Abstract:

Cardiovascular diseases remain the number one cause of death above cancer in the US. Invasive measurement of fractional flow reserve (FFR) is the gold standard of care to determine if a patient has significant blockage (stenosis) that limits blood flow to the heart muscle in the coronary arteries. FFR is determined invasively by inserting a pressure measurement wire across a stenotic coronary lesion, and then computing the ratio of pressure distal to the stenosis during maximum blood flow to pressure during normal flow in the same artery. FFR < 0.8 suggests that the stenosis limits blood flow to the down-stream heart muscle and should be treated by stent placement. In this study, we modeled blood flow through coronary arteries exhibiting varying degrees of stenosis using computational fluid dynamics (CFD). A second-order central-composite design with analysis by RSM identified the effects of blood pressure, flow rate, and degree of stenosis on computationally determined FFR values. Analysis of variance (ANOVA) showed a high variance coefficient (R^2) value of 0.935, thus ensuring a satisfactory adjustment of the model with the CFD results. According to the RSM analysis, flow rate had a significant effect on FFR ($p<0.05$). Increasing flow rate decreased FFR, which is consistent with clinical data. On the other hand, the effect of pressure over a physiologically relevant range was not significant. From the statistical analysis, FFR can be predicted for a given coronary anatomy as a function of blood pressure and flow rate.

472985 Numerical Study of the Effect of Wall Wettability on CO_2-Water Two-Phase Flow with Mass Transfer in Microchannels

Abstract:

The advancements of micro-scale fabrication and integration of microchannels into functional micro-process system have attracted the research interest on gas-liquid two-phase flow in microchannels featuring high surface to volume ratio. The typical low EÃ¶tvÃ¶sÂ number (Eo) in microchannels indicates that surface tension dominates fluid flow. It is critical to understand the effect of the surface force acting at channel boundaries on the multiphase flow along with fluid interfacial tension. This work developed an numerical model based on advanced Eulerianâ€"Volume of Fluid using computational fluid dynamics (CFD) of Ansys Fluent with fine grids at very small time step 10^{-6} to simulate carbon dioxide (CO_2) and water flow in 500 Î¼m ID microchannels with mass transfer. The volumetric mass transfer coefficients of CO_2 to water corresponding from different gas/liquid flow rates were calculated by the published correlations determined from experiments. The effects of wall wettability were investigated with the static contact angle (SCA) from 9° to 110°. The continuum surface force model (CSF) was used to represent surface tension into an equivalent body force by a source term in the momentum equation. The transitions of flow regime and the slug formation mechanism were studied. The critical hydrodynamic parameters including bubble length, bubble velocity, and pressure drop were analyzed. Pressure drops achieved from this researchare in good agreement with available experimental data reported. Further, the two-phase frictional multiplier was proposed and a correlation was developed to predict bubble length of gas-liquid flow in microchannels.

472999 Miscible, Unstable Viscous Fingering Patterns in a Radial Hele-Shaw Cell

Abstract:

Viscous fingering experiments were performed by injection of a shear-thinning non-Newtonian fluid into a radial Hele-Shaw cell filled with a less viscous fluid. Shear-thinning is a term used in rheology to describe a fluid that displays decreasing viscosity when subjected to shear (non-Newtonian properties). Many engineering applications related to enhanced recovery system, hydraulic fracturing and remediation take advantage of specific shear-thinning fluids in their process. The objective was to develop a method for testing fluids of differing dynamic viscosity to provide further insight into the size distribution of apertures carrying flow in aquifers. It is also of interest to understand unstable viscous displacement of a non-Newtonian fluid in radial injection/withdrawal configuration while investigating the dynamic interface subject to advective and diffusive processes (length of the finger, width of the channel and mixing zone).

The effect of flow rate, polymer concentration, and salt addition was also demonstrated. Dye tracer was mixed with the non-Newtonian fluid (Xanthan Gum (XG) T622, CP Kelco) before injection/withdrawal to capture images of resulting viscous fingering patterns when a less viscous fluid displaces a more viscous fluid (unstable case). The Hele-Shaw cell was constructed from two parallel smooth glass plates with an initial displacement of 5 mm. Shims were placed at the outer edge of the circular plate to create a variable aperture system for studying displacement and flow path selectivity. The effect of flow rate, polymer concentration, and salt addition was also demonstrated. Rheological characterization of guar gum (GG) and XG at various concentrations was also performed in order to constrain rheological behavior when it is used for future in-field measurements utilizing ground penetrating radar (GPR) to reveal aquifer system structure. These polymers were also characterized with the addition of salt, which is used as the conductive tracer for GPR, and low density glass bubbles for density adjustments. Rheological data for both GG and XG with and without salt additions will be presented, along with characterization and parameterization of unstable viscous fingering patterns. Guar gum was shown to have high error and poor experimental reproducibility while xanthan gum showed better reproducibility and optimal shear-thinning behavior.

473187 Probing Dynamics of a Magnetic Colloidal Chain Confined in a Spherical Cavity

Abstract:

Behaviors and conformational properties of semi-flexible macromolecular chains can provide insight of understanding some fundamental polymer dynamics such as chain entanglement or the cellular processes such as DNA replication or viral transfection. We conduct a study of such confined system via magnetically driven colloidal chains under confinement. We present a bead-spring model to describe the confinement dynamics inside a spherical cavity using coarse-grained Brownian Dynamics simulation. Radial contraction is employed as the confining procedure, and a possible rotating magnetic field will be introduced to actuate dynamics in the system. We aim to study the conformational properties of confined semiflexible colloidal chain with respect to the contraction rate in both weak and strong confinement regimes, and discuss the change of dynamics along the chain when a short-ranged wall interaction is present.

473686 Estimation of the Drying Length during Particle Assembly By Convective Deposition

Abstract:

Convective deposition is a widely used tool to create well-ordered particle arrays by drawing a suspension across a substrate utilizing evaporation-driven flow to localize and order particles in a thin film. We re-examine the assumption that the length scale associated with the drying front in convective deposition is constant. Over a wide range of parameters, we measure the drying length scale and note that it is inversely proportional to the deposition speed. Using a model of Darcy flow through the deposited particles, we develop a model that is more robust in describing convective deposition than those previously suggested by Nakayama et al.

474310 Shear-Induced Clustering of Brownian Colloids in Associative Polymer Networks at Moderate Peclet Number

Abstract:

Shear-induced clustering of Brownian particles has been widely observed in viscoelastic polymer solutions, and has been hypothesized to arise from the influence of polymer normal stresses on particle-particle interactions. In principle, such interactions depend on two characteristic time scales: the relaxation time of the viscoelastic polymer network, t_r, and the Brownian time scale for relaxation of the suspension microstructure to equilibrium, t_{Br}. We have developed thermoresponsive colloid-polymer mixtures that form associative polymer-colloid networks, in which the relative magnitudes of these time scales can be tuned over several orders of magnitude relative. This allows for systematic study of the behavior of shear-induced clustering in various regimes of the two corresponding shear rate scales: the Weissenberg number, Wi, and the Peclet number, Pe. The steady shear behavior of these fluids exhibit two regimes of shear thinning: one atr small Pe where the rheology scales with Wi, and another above Pe~0.1 where the steady shear rheology scales with Pe.

Rheo-SANS measurements reveal that anisotropic scattering develops in both the flow-vorticity (1-3) and flow-gradient (1-2) planes above Pi ~ 0.1, and becomes enhanced with increasing Pe. The orientation of the anisotropic scattering indicates concentration fluctuations oriented along both the vorticity and compressional axes of shear. The degree of anisotropy over a wide range of material parameters and shear rates collapse onto a master curve, which follows a scaling of $A_f \sim Pe^{1/3}$. From this, we propose and validate a simple empirical model to capture the steady shear rheology of the fluid, in which the clustering of particles imparts a perturbative contribution to the shear stress that modifies the underlying viscoelastic shear stress. This model accurately captures the shear rheology of the system over a wide range of relative values of Wi and Pe, and shows that clustering exerts a shear-thickening contribution to the total stress. Finally, we report time-resolved rheo-SANS measurements to elucidate the mechanism(s) of cluster

formation, which reveal that the clusters are dynamic, transient structures. These results provide clear evidence for the role of interparticle hydrodynamic interactions in the formation of anisotropic particle clusters in viscoelastic fluids, and elucidate means by which hydrodynamic interactions influence and control the shear-induced clustering of Brownian suspensions in viscoelastic liquids.

Formatting Issue - Best Available Paper/Graphic

On the Electrical Double Layers of Multi-Component Electrolytes

Kenji Kiyohara[1], Masaki Yamagata[2] and Masashi Ishikawa[2]
[1] Inorganic Functional Material Research Institute, National Institute of Advanced Industrial Science and Technology (AIST), Ikeda, Osaka, Japan
[2] Faculty of Chemistry, Materials and Bioengineering, Kansai University, Osaka, Japan

Introduction

Yamagata et al. has reported in their electrochemical measurement using carbon materials as the electrode and ionic liquids as the electrolyte that **the capacitance dramatically increases in the negative potential range when lithium salt is added to the electrolyte as the second component** [1]. They have also shown that, with the addition of a lithium salt, **the potential window is enhanced**. We clarify **the advantage of using multi-component** electrolytes for electrochemical capacitors at the molecular level by using the Monte Carlo simulation [2].

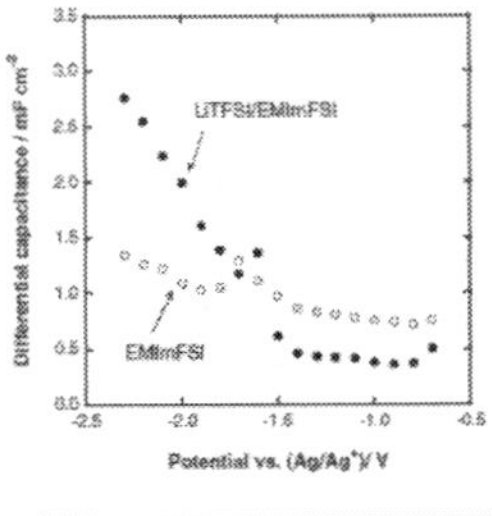

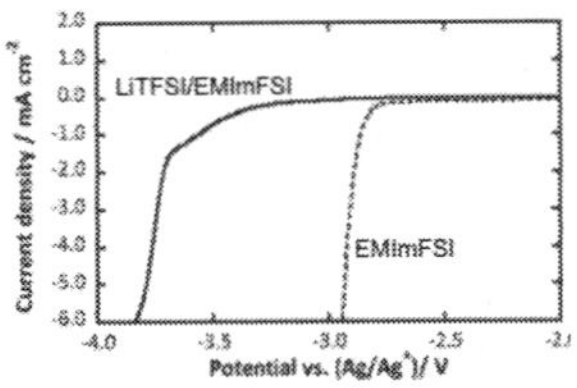

Monte Carlo Simulation

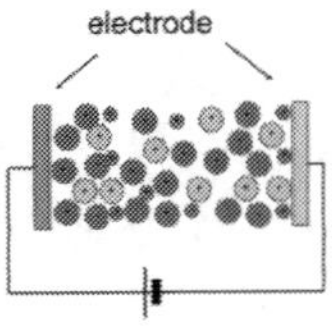

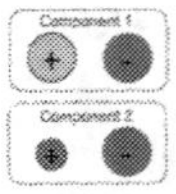

The smaller the ion, the stronger the electrostatic interaction.

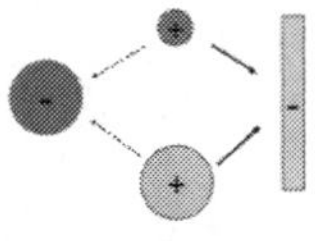

Competition of interaction between electrode and counterions.

One-component electrolyte

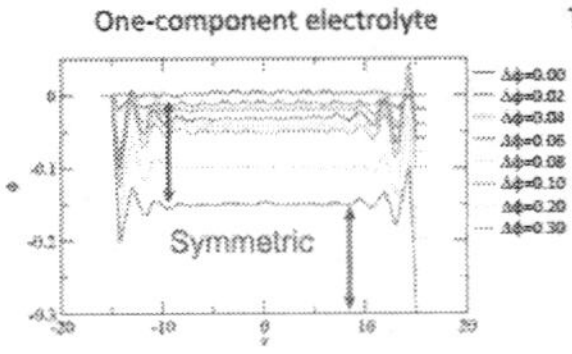

Two-component electrolyte (1.00 : 0.25)

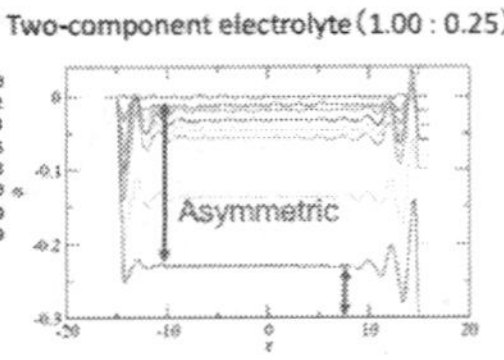

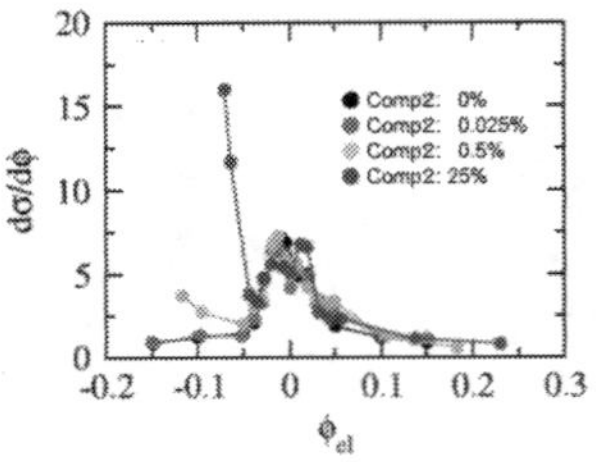

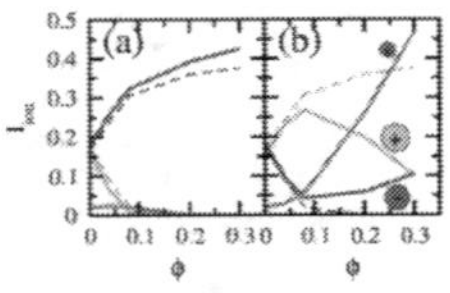

Number of ions in the first layer of EDL near the anode (a) and the cathode (b).

Conclusions

- The electrical double layer on the cathode is dominantly formed by the larger cations under small negative potentials, while they are replaced by the smaller cations under large negative potentials.
- The mechanism of the observed structural transition with potential can be explained in terms of **the competition between the electrostatic interaction of ions with the electrode and that among the ions**.

References

[1] M. Yamagata et al., *Electrochim. Acta*, **110**, 181 (2013).
[2] K. Kiyohara et al., *J. Chem. Phys.*, **138**, 234704 (2013); K. Kiyohara et al., *J. Chem. Phys.*, **144**, 134701 (2016).

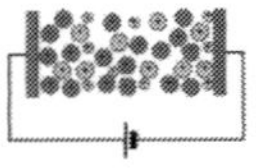

	[illegible]	[illegible]
[illegible]	[illegible]	[illegible]
Component 1 cation	1.5 d	0.1 ε
Component 1 anion	1.5 d	0.1 ε
Component 2 cation	1.0 d	0.1 ε
Component 2 anion	1.5 d	0.1 ε

Surface Charge Density
(two-electrode cell)

voltage surface charge

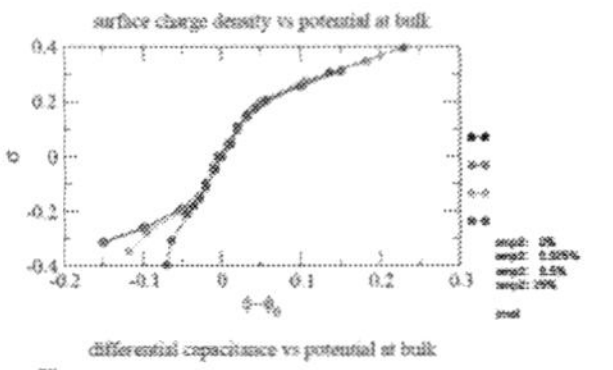

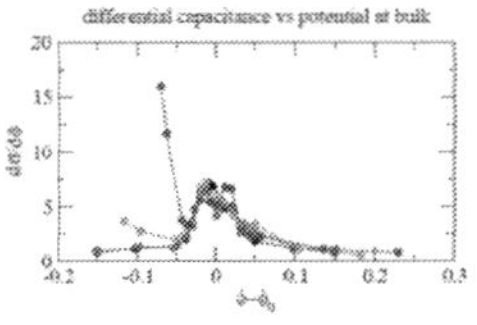

	[illegible]
[illegible]	[illegible]
1 comp	[illegible]
2 comp	[illegible]
[illegible]	[illegible]
[illegible]	[illegible]

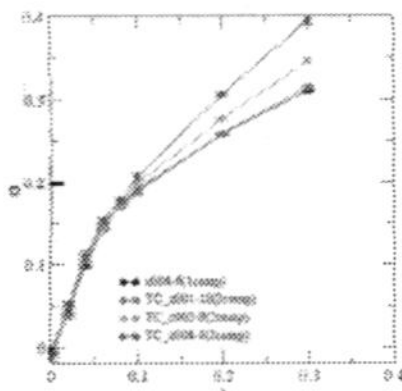

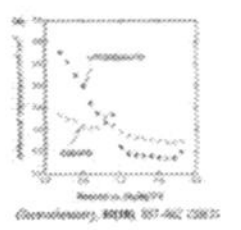

Electrochemistry, **81(10)**, 857–862 (2013)

Strong electrode potential

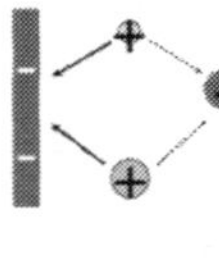

ion of ions in the bulk

$$W(\text{Boltzmann factor}) = \exp\left[-\frac{E - \sum_i \mu_i N_i - \phi q}{k_B T}\right]$$

Monte Carlo simulation

Constant-voltage Grand-canonical Ensemble

453831 Heat Transfer Enhancement from a Hot Steel Plate Using Triton X-100 Surfactant Additive in Water Jet

Abstract:

Abstract

Heat transfer is one of the most widely used transport processes in industries. It is the most important phenomena in steel industries for the production of high strength steel. The mechanical properties of steel depend upon the finished roll temperature, run-out table cooling and coiling temperature. Heat transfer rate at run-out table (ROT) during hot rolling process in steel industries determines the metallurgical microstructure for specified mechanical properties required for particular applications. The phase transformation of steel occurs between a rolling temperature of 900 °C and a coiling temperature of 600 °C. The phase microstructure and corresponding mechanical properties depend upon the cooling intensity between the aforementioned temperature ranges. Ultrafast cooling is a technique used to obtain high strength steel.

The current work attempts to achieve ultrafast cooling by introducing different cooling methods for steel. It focuses on the enhancement of heat transfer from a hot steel plate by introducing non-ionic surfactant additive, Triton X-100 in jet cooling. The characterization of surfactant solution has been done in order to understand the reason for increment in heat transfer. The concentrations of surfactant have been varied from 60 ppm to 140 ppm in intervals of 20 ppm and the results indicate that the addition of Triton in water jet leads to an appreciable increase in cooling rate compared to that of water. The maximum cooling rate is achieved at a surfactant concentration of 120 ppm which is attributed to the low values of surface tension (as shown in Figure 1 below) and contact angle (as shown in Figure 2 below) obtained at this concentration leading to higher spreadability and heat removal rate. The highest cooling rate achieved is 170 °C/sec (as shown in Figure 3 below) which is 56% more than that attained for pure water. The study reveals that ultrafast cooling can be attained by the addition of Triton in water. Thus heat transfer plays an essential role in the production of high strength steel for various industrial applications.

Keywords: Heat transfer, jet cooling, ultrafast cooling, Triton X-100, run-out table, cooling rate.

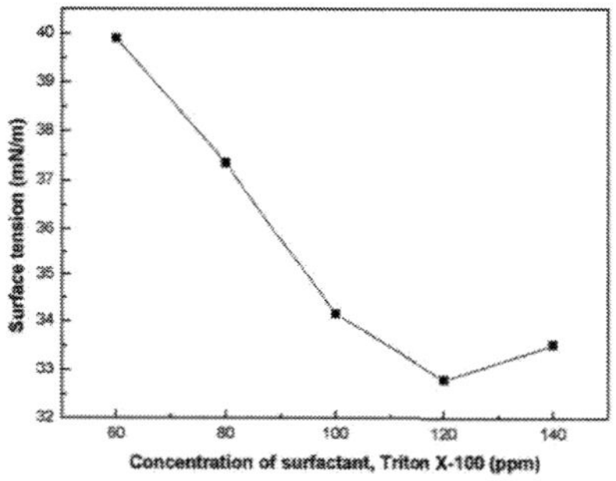

Figure 1: Variation of surface tension with varying concentration of Triton X-100

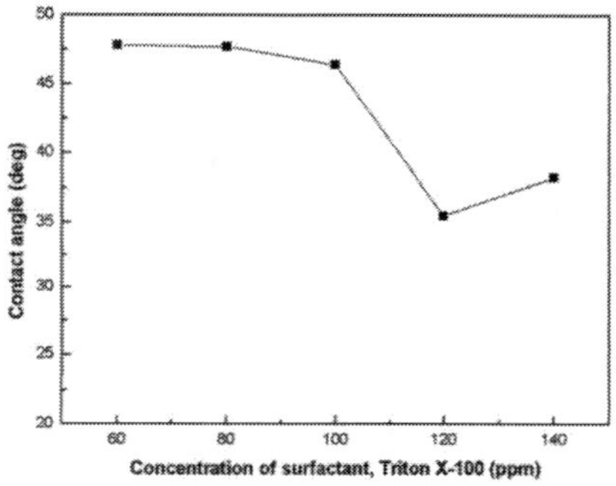

Figure 2: Variation of contact angle with varying concentration of Triton X-100

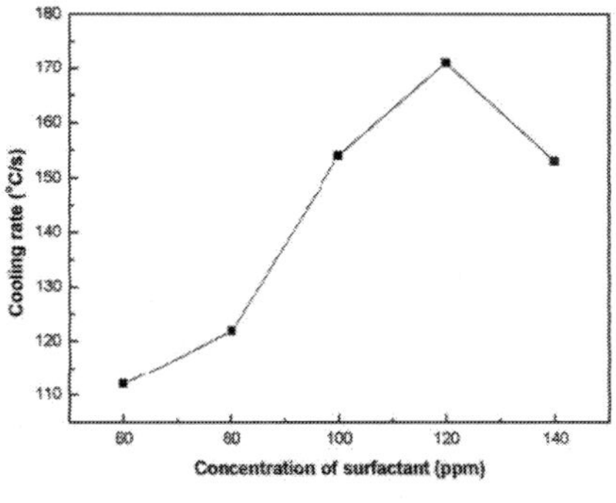

Figure 3: Variation of cooling rate with varying concentration of Triton X-100

References:

1. D.A. Zumbrunnen, R. Viskanta, F.P. Incropera, The Effect of Surface Motion on Forced Convection, Film Boiling Heat Transfer, Trans. ASME Journal of Heat Transfer,, 111 (1989) 760-766.
2. L. Cheng, D. Mewes, A. Luke, Boiling phenomena with surfactants and polymeric additives: A state-of-the-art review, International Journal of Heat and Mass Transfer, 50 (2007) 2744-2771.
3. T. Stoebe, Z. Lin, R.M. Hill, M.D. Ward, H.T. Davis, Surfactant-enhanced spreading, Langmuir, 12(2) (1996) 337-344.
4. G. Karapetsas, R.V. Craster, O.K. Matar, On surfactant-enhanced spreading and superspreading of liquid drops on solid surfaces, Journal of Fluid Mechanics, 670 (2011) 5-37.
5. Ravikumar S.V., Jha J.M., Mohapatra S.S., Pal S.K., Chakraborty S., Influence of Ultrafast Cooling on Microstructure and Mechanical Properties of Steel, steel research international, 84 (2013) 1157-1170.

456250 Phase Behaviors and Physical Properties of Microemulsion of Palm Oil and Ethanol Emulsified By Low HLB Type of Anionic and Cationic Surfactants

Abstract:

Microemulsion-based biodiesel could be a promising fuel product because of the limitation of petroleum sources in the future. The preparation technique requires less equipment and energy, no chemical reaction involved and catalyze-free. It is related to how to mix vegetable oil, ethanol, surfactant and co-surfactant with proper volumetric ratio to obtain the microemulsion (thermodynamical stability, transparent, and 10-100 nm droplet size) with good physical properties such as, lower kinematic viscosity and lower surface tension regarding to the standard or commercial biodiesel (FAME) prepared by esterification reaction. This work presents a feasibility study of biodiesel development by microemulsion technique using palm oil and ethanol emulsified by low HLB type of anionic surfactant (sodium dioctyl sulfosuccinate, AOT) and cationic surfactant (benzyldimethylhexadecyl ammonium chloride, BMHAC) together with co-surfactants (1-butanol and 1-octanol) at 40 °C. The ternary phase diagrams reveal that surfactants used in this work could produce w/o and o/w types of microemulsion. The higher surfactant/co-surfactant, the more solubilize between palm oil and ethanol. The AOT/1-octanol mixture shows ability to solubilize palm oil and ethanol at the relatively low composition compared with BMHAC/1-butanol. The experimental results also reveal physical properties regarding to the ASTM standard temperature. For AOT/1-octanol mixture, it can be observed that increasing amount of ethanol results in increasing droplet size and lowering surface tension of microemulsion system due to AOT is more water dispersible (w/o) and surface tension of ethanol is relative low. However, increasing of palm oil results in higher viscosity. For BMHAC/1-butanol mixture, the physical properties show that increasing amount of ethanol results in lowering surface tension and viscosity while increasing amount of palm oil results in larger droplet size due to BMHAC is more oil dispersible (o/w).

457699 Removal of Endotoxins Using Polymer Nanoparticles

Abstract:

The control of microbial infections is critical for the preparation of biological medium including water to prevent lethal septic shock. Sepsis is one of the leading causes of death in the United States. More than half a million patients suffer from sepsis every year. Both gram-positive and gram-negative bacteria are responsible for septic infection by the most common organisms *i.e.*, *Escherichia coli* and *Pseuodomonas aeruginosa*. The bacterial cell membrane releases negatively charged endotoxins upon death and enzymatic destruction, which stimulate antigenic response in humans to gram-negative infections. Several methods including distillation, ethylene oxide treatment, filtration and irradiation have been employed to remove endotoxins from contaminated samples, however, the reduction efficiency remains low, and presents a challenge. Polymer nanoparticles can be used to overcome the current inability to effectively sequester endotoxins from water. This process is termed *endotoxin hitchhiking*. The binding of endotoxin on polymer nanoparticles *via* electrostatic and hydrophobic interactions offers efficient removal from water. However, the effect of polymer nanoparticles and its surface areas has not been investigated for removal of endotoxins. Poly(É›-caprolactone) (PCL) polymer was tested for its ability to effectively bind and remove endotoxins from water. By employing a simple one-step phase separation technique, we were able to synthesize PCL nanoparticles of 398.3 Â± 95.13 nm size and a polydispersity index of 0.2. PCL nanoparticles showed ~78.8% endotoxin removal efficiency, the equivalent of 3.9Â ´10^5 endotoxin units (EU) per ml. This is 8.34-fold more effective than that reported for commercially available membranes. Transmission electron microscopic (TEM) images confirmed binding of multiple endotoxins to the nanoparticle surface. The concept of using nanoparticles may be applicable not only to eliminate gram-negative bacteria, but also for any gram-positive bacteria, fungi and parasites.

459878 Shape-Controlled Synthesis of Ag Nanocrystals Mediated By Polyvinylpyrrolidone: Thermodynamics Vs. Kinetics

Abstract:

Functional nanomaterials, especially metallic materials, are highly shape sensitive, which drives the development of shape-controlled nanomaterial synthesizing techniques. Structure-directing agent (SDA)-mediated solution-phase metal nanocrystal synthesis has shown its ability to produce a large variety of nanostructures, but to further increase the product homogeneity and quantity, fundamental understanding of the workings of SDAs is a priority. Computational means can serve as an excellent tool to investigate such phenomena in low length and time scales. We use molecular dynamic (MD) simulations to study the role of polyvinylpyrrolidone (PVP) in PVP-mediated formation of {100}-faceted Ag nanocrystals in ethylene glycol (EG) solution. Though not fully understood in previous studies, the consensus is that PVP promotes the expression of Ag(100) facets via stronger binding to Ag(100) than Ag(111). In investigating the role of PVP, we consider both thermodynamic and kinetic control. We compute the solid-liquid interfacial free energy γ_{sl} of Ag(100) and Ag(111) in contact with the PVP- EG solution using the multi-scheme thermodynamic integration (TI) method we have developed based on the previous "cleaving wall" method [1, 2]. We find that although PVP lowers $\gamma_{Ag(100)}$ more than $\gamma_{Ag(111)}$ comparing to the bare surfaces (i.e. in contact with vacuum), Ag(111) still has the lower γ_{sl}, thus more thermodynamically stable. For the kinetic control, we compare the facet growth rates of Ag(100) and Ag(111) by looking at the deposition flux to these surfaces and predict the kinetic Wulff shapes [3]. We calculate the mean first-passage time of a free Ag atom going from the solution phase to the Ag surface using the Smoluchowski equation, and find that the difference in growth rates is a combined effect of both the probability a PVP segment attracts a solution-phase Ag atom and the diffusion of the Ag atom inside the PVP layer adsorbed on the surface. Due to stronger binding affinity to Ag(100) facets, PVP chains pack more loosely on Ag(111) and extend longer into the solution phase; thus, PVP layer on Ag(111) can have larger probability to attract a solution-phase

Ag and the diffusion is faster for the Ag atom inside such PVP layer. Our findings in the kinetic study are consistent with experimental observations. Combining the interfacial free energy study and kinetic Wulff shape study, we resolve the role of PVP in Ag nanocube formation and demonstrate the dominance of kinetic control in such formation. Our findings can be indicative for other similar SDA/metal combinations and the methods used in this work can be applied to a broader range of systems.

[1] Broughton, J. Q., and Gilmer, G. H., J. Chem. Phys. 1986, 84, 5759.
[2] Benjamin, R. and Horbach, J., J. Chem. Phys. 2014, 141, 044715.
[3] Qi, X., Balankura, T. Zhou, Y., Fichthorn, K. A., Nano Lett. 2015, 15, 7711.

462133 Controlled Synthesis of Organic Nanowires on Gold Nanoparticle Seeds for Sensors Applications

Abstract:

This research focuses on the fabrication of inorganic-organic crystallization using gold nanoparticles (AuNPs) as seeds in order to control potassium tetracyanoplatinate (KCP) or tetrathiafulvalene bromide charge transfer salt (TTFBr) deposition. The AuNPs were first deposited on highly oriented pyrolytic graphite (HOPG) for 0.01 s to 1 s. Then, either 1 to 10 mM TTF or 0.07 mM KCP was deposited on the AuNP decorated HOPG following literature procedures. Both steps were accomplished by electrochemistry. All the depositions were monitored by cyclic voltammetry and atomic force microscope. Field emission scanning electron microscope was employed to image the inorganic-organic system. For depositing AuNPs, the results show that concentration is the key factor to control the size and shape of AuNPs while the applied overpotential and deposition time play a secondary role. The size of the organic crystals grown on AuNP decorated HOPG is smaller comparing with those grown on bare HOPG. Using the same method, we successfully deposited TTFBr and KCP crystals on a gold nanopatterns designed by our lab. The electrical impedance of the sensor was measured by exposing the sensor surface to different gases at different concentrations. The results show that both sensors can detect ppm level toxic or flammable gases, for example, ammonia, within one minute. We plan to develop our sensor further for specific applications. Our work contributes both fundamental understanding of seed-mediated nucleation and technology of low-cost manufacturing of chemical sensors.

465488 Customizing Polyelectrolyte Complex Shape through Photodirected Assembly

Abstract:

Polyelectrolyte complexes (PECs) form when oppositely-charged polymers are mixed in water and are routinely used in diverse applications such as drug delivery, sensors and separation technologies. Despite their broad use, however, current fabrication techniques typically limit their geometries to planar, cylindrical or spherical structures. To prepare PECs with a broader range of shapes, we have developed two methods of directing PEC formation through photopatterning. In the first method, poly(allylamine) (PAH) was mixed with poly(styrene sulfonate) (PSS) and a photoacid generator (PAG) at a pH that greatly exceeded the effective pK_a of PAH and (since the PAH amine groups were mostly neutralized) prevented insoluble PEC formation. These precursor solutions were then photopatterned through photomasks, which caused localized photochemical ionization of PAH (due to the PAG-triggered pH reduction) and, consequently, formation of insoluble PAH/PSS PECs, whose size and shape reflected the photoirradiation pattern. Since this technique relied on the development of pH gradients, however, the resolution of this photopatterned PEC formation was limited by the diffusion of OH^- ions into the irradiated region (which, when irradiation zones with sub-millimeter dimensions were used, impeded the PAG-mediated pH change). To overcome this limitation, we have developed a second photodirected assembly method, where PAH was mixed with acrylic acid (AA) monomers and a photoinitiator, and where photopatterning instead initiated site-specific polymerization of AA into poly(acrylic acid) (PAA). This localized polyanion formation led to the assembly of photopatterned PAH/PAA PECs with finer structural features than those formed through a localized pH reduction. The photopatterned PECs were characterized by microscopy, spectroscopy and gravimetry and demonstrated ability to either alter their dimensions or dissolve in response to changes in their external environment. This ability to form environmentally responsive structures with customized shapes suggests photodirected PEC assembly as a potentially attractive route to stimulus-responsive soft device fabrication.

455511 Self-Assembly of Anionic Surfactants in Aprotic Imidazolium Ionic Liquids

Abstract:

Self-assembly of anionic surfactants in aprotic imidazolium ionic liquids

B. Kusiak,[a] P. Bernatowicz,[a] P. Korczyk,[b] J. Gregorowicz[a]

[a]*Institute of Physical Chemistry, Polish Academy of Sciences, Kasprzaka 44/52, 01-224 Warsaw, Poland*

[b]*Institute of Fundamental Technological Research, Polish Academy of Science, Pawninskiego 5B, 02-106 Warsaw, Poland*

There is a very limited number of known solvents possessing capability to support self-assembly of surfactants. Fortunately, quite a few ionic liquids possess this ability. Moreover, apart from the protic imidazolium ionic liquids, some aprotic imidazolium ionic liquids were found to be capable of supporting amphiphile self-assembly. Investigation of amphiphile behavior in ILs is important for better understanding of the effect of solvents on self-assembly and is also necessary for expanding applications of these nonaqueous solvents.

Ionic liquids (ILs) provide an attractive medium for a preparation of nanocomposites in particular cellulose/carbon nanotubes (CNTs) composites (patent number P-392221 and GB2483158). The limiting step for the process of the composite preparation with an ionic liquid as a solvent is dispersion of CNTs (patent application P-395835). The most common approach to circumventing this problem is to disperse the CNTs in the solvent containing some amount of an anionic surfactant. The disadvantages of this method is formation of the surfactant micelles. That is why in this work

we concentrate our attention on ILs which dissolve cellulose, like AMIMCl, EMIMDEP or BMIMCl.

Aggregation behavior of anionic surfactant sodium dodecyl sulfate (SDS) and sodium n-octadecylsulfate (SOS) in 1-allyl-3-methylimidazolium chloride (AMIMCl), 1-butyl-3-methylimidazolium chloride (BMIMCl), 1-hexyl-3-methylimidazolium chloride (HMIMCl) and 1-ethyl-3-methylimidazolium diethylphosphate (EMIMDEP) was investigated by the surface tension measurement. Additionally NMR spectroscopy was used to analyze micelle formation in the SDS/AMIMCl and SDS/EMIMDEP systems.

It was established that SDS and SOS formed micelles in AMIMCl and BMIMCl, while no aggregation of the surfactant molecules occurred in EMIMDEP. The cmcs estimated from surface tension were 54 mM, 53 mM, 87 mM and 72.5 mM for SDS/AMIMCl, SDS/BMIMCl, SOS/AMIMCl and SOS/BMIMCl, respectively. The free energy of micellization is negative: -28.3 kJ/mol for SDS/AMIMCl, -25.8 kJ/mol for SDS/BMIMCl, -31.8 kJ/mol for SOS/AMIMCl and -30.0 kJ/mol for SOS/BMIMCl. Therefore the formation of micelles becomes spontaneous in this systems. The critical micelle concentration of SDS and SOS in AMIMCl and BMIMCl are higher than in aqueous solution mostly due to higher solubility of the surfactants in ILs, which leads to weaker solvophobic interactions, and also increased viscosity of the system, limiting diffusion of the molecules.

The impact of water present in ILs on self-assembly of surfactants was also investigated. The results of the ^{1}H NMR measurements revealed significant differences in interactions of water with AMIMCl and EMIMDEP. Water interactions with EMIMDEP is strong enough to cause even distribution of water molecules in the system. In AMIMCl mutual interactions of water molecules was preferred and as a result water domains within the IL were formed.

Micelle formation is an indication that also liquid crystalline phases are formed in a system. The differential scanning calorimetry (DSC), X-ray diffraction (XRD) and polarized optical microscopy (POM) were used to study formation of liquid crystalline phases in the SDS/AMIMCl, SDS/BMIMCl, SDS/HMIMCl, SOS/AMIMCl, SOS/BMIMCl and SOS/HMIMCl systems. In all investigated surfactant/ionic liquid systems transitions from a lamellar to isotropic phase were observed. In three systems (SOS/AMIMCl, SOS/BMIMCl and SOS/HMIMCl) transitions from a solid phase to liquid crystalline lamellar phase were also observed. The solid phase was formed by pure surfactants. The POM observations showed that in the systems the phase transitions from the lamellar phases to the isotropic phases are reversible. This is confirmed by DSC, where only small temperature hysteresis upon consecutive heating and cooling runs was observed.

458822 A Theoretical Approach to Predict Adsorption Isotherms of Isomers Using Density-Functional-Theory and Lattice-Cluster-Theory

Abstract:

A Theoretical Approach to Predict Adsorption Isotherms of Isomers Using Density-Functional-Theory and Lattice-Cluster-Theory

Patrick Zimmermann[1], Thomas Goetsch[2], Tim Zeiner[2], Sabine Enders[1]

[1]Karlsruhe Institute of Technology, Institute of Technical Thermodynamics and Refrigeration Engineering, Engler-Bunte-Ring 21, D-76131 Karlsruhe, Germany.

Email: patrick.zimmermann@kit.edu; sabine.enders@kit.edu

[2]Technical University Dortmund, Laboratory of Fluid Separations, Emil-Figge-Straße 70, 44227 Dortmund, Germany.

Email: Thomas.Goetsch@bci.tu-dortmund.de, Tim.Zeiner@bci.tu-dortmund.de

Key words: Adsorption Isothermes of Binary Mixtures, Density Functional Theory, Separation of Close Boiling Mixtures

Abstract:

The separation of mixtures having components with very similar vapor pressures, like isomers, using distillation is always a very hard task and requires a large amount of invest and energy cost. For some purposes, e.g. exact measurements of thermophysical properties, a high grade of purity of the compound is required. One possible way to achieve almost pure substances is given by adsorption. For the development of industrial adsorption processes, adsorption isotherms, describing the relation of the excess fluid density in the pores over wide pressure ranges are applied. McCabe-Thiele plots showing the composition of the adsorbed phase against the feed composition can be helpful, too. Experiments in this area require a high effort of money and time so that it would be great to have the possibility to reduce the number of experiments and predict adsorption isotherms in a wide range of conditions using parameters that are adjusted just to a few experimental data points. Therefore, this contribution deals with the prediction of partial density profiles in narrow pores applying density functional theory (DFT) [1,2,3,4] in combination with a suitable equation of state, which should be able to describe different isomers. The analysis of these partial densities profiles results in the knowledge of the adsorption isotherms, phase transitions in pores, hysteresis effects and wetting as well as prewetting phenomena.

The DFT applied in the grand canonical ensemble starts with an expression for the grand potential as a functional of the density profile of the fluid, where we assume the fluid is inhomogeneous in only the direction perpendicular to the wall of the narrow slit pore. When considering a fluid in an external field at a fixed temperature and at fixed chemical potentials of the components in the fluid mixture, the density of the fluid is not spatially constant. At thermodynamic equilibrium the grand potential function is in its global minimum with respect to variations of the density profiles. The key step in DFT is to specify an approximate form for grand potential that is both tractable and accurate.

The properties of the fluid mixture can be described with the inhomogeneous Helmholtz energy, which can be calculated using any equation of state (EOS). Recently, the different equations of state (EOS) (e.g. the Peng-Robinson EOS [5], SAFT-VR-EOS [6], PC-SAFT-EOS [7]) were applied in the DFT framework for mixtures, whereupon the local density approach was used. However, all used equations are to join a common feature, they are not able to model isomers in terms of branching. In order to model mixtures of several isomers differing in the degree of branching the Lattice-Cluster-Theory (LCT) introduced by Freed et al. [8,9] and reformulated as equation of state (LCT-EOS) by Langenbach et al. [10] can be applied. The advantage of the LCT-EOS is given by the use of three pure component parameters to describe different isomers with the same number of united atom groups. These parameters can be fitted to the physical bulk properties of the e.g. n-alkane (vapor-pressure or/and liquid densities) and then the properties of the branched alkanes can be predicted by involving architecture coefficients given by the chemical formula. The binary interaction parameters of mixtures can be estimated by fitting to vapor-liquid equilibria. The external potential occurring in the grand potential describes the interaction of the fluid with solid wall and can be expressed by the Steele 10-4-3 potential [11]. To obtain the equilibrium density profile, the grand potential functional is minimized with respect to the local density.

The DFT using the Steele potential in combination with the LCT requires three parameters for one class of isomers to describe the bulk fluid and another triplet of parameters for the external field. Given a few points of an adsorption isotherm of a pure substance we are able to calculate the adsorption behavior of mixtures. By integration of the achieved density profiles and comparing this values with corresponding bulk densities, we can establish adsorption isotherms as well as McCabe-Thiele plots. Additionally, the impact of the porosity on the adsorption isotherm is implemented by the use of experimental pore size distributions.

Literature:

[1] M.J. de Oliveira, R.B. Griffiths, Surf. Sci. 71 (1978) 687-94.

[2] C. Ebner, Physical Rev. A 22 (1980) 2776-81.

[3] B.K. Peterson, K.E. Gubbins, G.S. Heffelfinger, U.M.B. Marconi, F. van Swol, J. Chem. Phys. 88 (1988) 6487-500.

[4] C. Lastoskie, K.E. Gubbins, N. Quuirke, J. Phys. Chem. 97 (1993) 4786-96.

[5] P. Zimmermann, T. Goetsch, T. Zeiner, S. Enders , Fluid Phase Equilibria, in press, doi:10.1016/j.fluid.2016.04.006.

[6] M. Castro, A. Martinez, A. Gil-Villegas, Adsorption Sci. & Techn. 29 (2011) 59.

[7] G. Shen, X. Ji, X. Lu, J. Chem. Phys. 138 (2013) 224706.

[8] K.F. Freed, Journal of Physics A: Mathematical and General, 18 (1985) 871.

[9] J. Dudowicz, K.F. Freed, W.G. Madden 23(22) (1990) 4803-4819.

[10] K. Langenbach, S. Enders, Fluid Phase Equilibria 331 (2012) 58-79.

[11] W.A. Steele, Surf. Sci. 36 (1973) 317-324.

460457 Shape Driven Colloidal Crystal—Crystal Transitions

Abstract:

Recent experimental, theoretical, and computational work has shown that anisotropic colloids exhibit a rich phase behavior, including the self assembly of complex crystal phases. Here, we study the thermodynamics of crystal—crystal transitions that are driven by changes in colloid shape. Using rare event sampling techniques, we find that shape driven crystal—crystal transitions that have similar diffusionless pathways can have unexpectedly complex thermodynamic behaviors. We discuss implications of our results for reconfigurable colloidal materials.

Penetration of a non-swelling liquid into a tableted porous medium

Jesús Esteban[1], Serafim Bakalis[1], Claire Duckitt[2], Hossam Tantawy[2], Peter J. Fryer[1]

[1] School of Chemical Engineering.University of Birmingham. Edgbaston. Birmingham B15 2TT, United Kingdom.

[2] Procter & Gamble Ltd, Newcastle Innovation Centre. Whitley Road, Longbenton, Newcastle upon Tyne, NE12 9BZ, United Kingdom.

J.EstebanSerrano@bham.ac.uk. (Phone number: +44 (0) 121 414 5081)

Abstract

The imbibition of a liquid into porous substrates is a phenomenon of interest in a wide array of processes both in nature and industrial applications. Among the latter are several of relevance in the fast moving consumer goods industry, such as the application of coating surfactants to the formulation of powder detergents. The structure of the substrate and the properties of the liquid as well as the type of interaction will determine the kinetics of imbibition. This work presents a methodology to compact powders of relevant composition in the formulation of detergents into tablet preparations. Non-invasive techniques are used to characterise the tablet surface (White-light interferometry) and internal structure (micro-computed tomography) of the substrate. Imbibition of n-hexadecane, which does not swell the substrate, has been studied. The initial contact angle has been studied as a function of roughness, showing that this variable did not affect greatly this variable. Finally, the penetration kinetics of droplets of the liquid showed a strong dependence with respect to the porosity of the substrate.

Keywords: tableting, micro-CT, dynamic contact angle, droplet spreading, droplet penetration

1. Introduction and background

Imbibition of liquids into porous substrate plays an important role in many events in nature like the absorption of water by seeds or dry wood and irrigation. The phenomenon is important in many processes and applications, including multiphase catalytic reactions, flotation in the mining industry, ink-jet printing, cooling of electronic devices, fabric dying or coating of porous materials. In these and other activities, the effectiveness of the process or the final quality of the product is in many cases directly related to the rate and extent of the penetration of the liquid into the porous medium [1].

Several experimental studies have dealt with the imbibition of liquids over porous substrates through the evaluation of droplet spreading and penetration. Media such as loosely packed powder beds [2, 3], fiber mats [4], or other thin, thick and layered porous substrates [5-7] have been studied. These studies used high-speed cameras to monitor the fast occurring interactions between liquid and solid phases.

Tablets are widely used porous media in the pharmaceutical, food and consumer goods industries among others, in which assessment of liquid penetration and uptake is of importance to evaluate processes like coating. The properties of the porous substrate and the liquid as well as the type of interaction will determine the imbibition behaviour. It has been observed that roughness of materials can affect the sessile drop contact angle (CA) [8] and also it is well known that porosity is a decisive factor in imbibition phenomena [1].

Accordingly, an evaluation of the surface and the internal structure of the solid is relevant. Here, white-light interferometry (WLI) and micro-computed tomography (µ-CT) will be the techniques employed, for they have proven valid in the evaluation of the topography [9] [10] and internal structure of solids, respectively [11].

Depending on the chemical interaction between the liquid and the solid phases, phenomena like swelling may occur simultaneously to the imbibition [12] or even dissolution of the components within the solid matrix. For this reason, a non-polar non-volatile liquid (n-hexadecane) has been selected for the study.

2. Experimental section

2.1. Materials

The powder used for this experimental work was supplied by the Procter & Gamble Company and are porous spray-dried granules of a relevant composition for the manufacturing process of powder detergents, featuring different salts as well as linear alkylbenzene sulfonate as surfactant. The granules were dried in a thermal balance prior to manipulation. The liquid selected for this study was n-hexadecane (purity > 99%), supplied by Sigma-Aldrich Co Ltd and was used as received.

2.2. Preparation of tablets

To prepare the porous substrates, 0.25 g of powder were weighed. Compaction into circular tablets was performed with a punch and a die of 13 mm of diameter making use of an Instron Microtester model 5848 (Norwood, MA, USA) equipped with a load cell of up to 2000 N.

2.3. Characterisation of the surface roughness

To assess the surface topography of the compacted powder preparations, interferometry was the technique used. A three-dimensional profilometer was used consisting in a white light interferometric microscope, a KLA Tencor model MicroXAM2 (Wokingham, UK). For surface characterisation, e

Fivefold scanning of areas of 125 µm x 175 µm throughout depths of 50 µm was made at fixed positions within the surface of the tablet. The data and images were acquire with the complimentary software Mapvue AE v. 2.27.1 and further processed and analysed after transfer to the software Scanning Probe Image Processor, SPIP version 4.4.3.0 by Image Metrology (Hoersholm, Denmark).

2.3. Characterisation of the internal structure of the tablets

Tablets were scanned by means of an X-ray micro-computed tomography (µ-CT) Burker Skyscan 1172 (Brussels, Belgium) device. Acquisition of images was realized with a mid-camera (pixel size 10.96 µm) of round scanning trajectory (0.400 degrees per step) with no filter. The scanning parameters were set at 50 kV maximum X-ray energy, a current of 90 mA and an exposure time of 350 ms. Image reconstruction was made with the SkyScan NRecon package applying ring artifacts and beam hardening correction.

Subsequent analysis for characterisation of the internal structure of the compacted powder preparation from the reconstructed stack of images was completed with the software CTAn (v.1.15.4.0). First, a region of interest (ROI) was defined for a cross-sectional image and then interpolated across the entire image stack to create a volume of interest (VOI) for analysis. Operations with the images included the following steps: global thresholding for binarisation, removal of speckles constituting noise pixels, shrink-wrap of the ROI to fit to the physical borders of the sample and finally 3D analysis to determine the structural parameters of the solid. Open, close and overall porosity can be quantified together with the distribution of pore size. Images can be three-dimensionally visualized on CTVox v.3.0.

2.4. Droplet spreading and penetration studies

Liquid penetration experiments of n-hexadecane into tablets of different porosities were performed using a Krüss drop size analyser model DSA30S (Hamburg, Germany) goniometer. The setup consists of a XYZ manually controlled stage for substrate positioning, a high-speed camera and a syringe for droplet control and deposition. The software allows accurate control of droplet volume (15 μL selected for this study) and positioning onto the substrate. An automated program for droplet dosing and deposition, syringe withdrawal and video recording (150 fps) was created for reproducibility purposes. Subsequent analysis was made fitting the droplet boundaries to a circle function.

3. Results and discussion

Compaction of powders in the particle microtester device could be done by applying a range of forces between 100 and 2000 N, corresponding to 0.75 MPa and 15 MPa in accordance with the surface area of the tablet. Below 100 N force, preparations were too brittle to be manipulated. Hence, the following discussion of results will be made on the basis of structurally consistent preparations.

3.1. Roughness characterisation

Surfaces were evaluated according to a series of amplitude parameters whose definitions have been established as representative of their roughness [13], namely: maximum distance between peak and valley, S_z; arithmetic mean of the absolute height, S_a; root mean square value of the ordinate values within a sampling area, S_q; skewness of topography height distribution, S_{sk} and kurtosis of topography height distribution, S_{ku}. On the basis of these parameters, Table 1 compiles the roughness parameters for the samples subject of study.

It can be easily observed that increasing forming pressures lowers the values of S_z, S_a and S_q. In the case of the skewness, the tablets tested showed a slight negative value, indicative of the bulk of the material being above the mean plane, yet no significant differences are perceived as a function of the pressure applied. As for kurtosis, S_{ku} can detect whether the profile spikes are evenly distributed and also a measure of the spikiness of the area, with high values of this parameter corresponding to spiky surfaces and low to bumpy surfaces. In this case, no clear trend can be observed and most of the kurtosis correspond to about 20-35% of the maximum distances between peaks and valleys, which could qualify as relatively smooth [13].

Table 1. Roughness parameters for the tested tablets

Forming pressure (MPa)	S_z (μm)	S_a (μm)	S_q (μm)	S_{sk} (μm)	S_{ku} (μm)
0.75	36.38 ± 7.32	3.85 ±0.43	5.52 ± 1.13	-1.60± 0.21	7.39± 3.40
3.01	40.96 ± 10.21	3.38 ±0.12	4.83 ± 0.30	-2.33± 0.87	14.30± 3.99
4.52	20.66 ± 3.49	1.43 ±0.39	2.08 ± 1.01	-2.84± 0.77	17.82 ± 4.11
7.53	24.80 ± 2.11	1.92 ±0.32	2.74 ± 0.44	-2.13± 0.69	10.87 ± 3.15
15.07	13.54 ± 3.77	1.01 ± 0.11	1.36 ± 0.21	-1.89±0.17	13.01 ± 5.21

The aforementioned roughness comments are confirmed by topographic events like those featured in Figure 1, which presents characteristic three-dimensional maps of solid samples pressed under the extreme conditions of (a) 0.75 MPa and (b) 15 MPa. In the former case, roughness is higher due to a higher presence of consistently occurring gaps present among powder granules, which are even identifiable even with the naked eye. In Figure 1b, however, gaps between individual granules are less and less frequent, leading to a smoother surface.

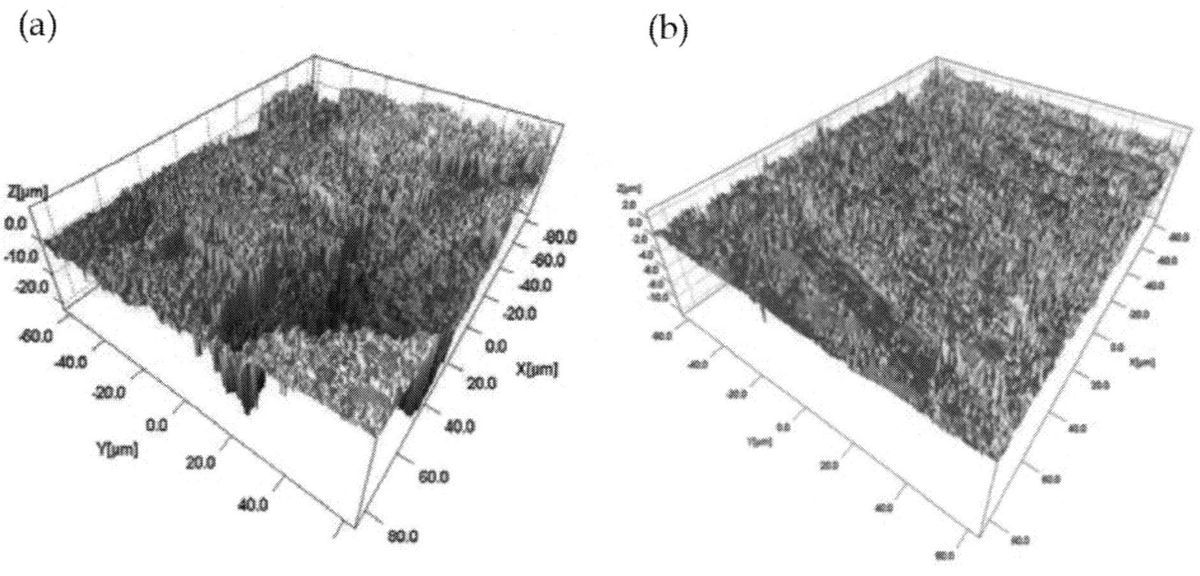

Figure 1. Topographic profile of the surface of tablets compressed at (a) 0.75 MPa and (b) 15 MPa.

3.2. Internal porous structure visualization

μ-CT analysis allows the observation of the internal structure of the substrates. Figure 2 depicts the three-dimensional reconstruction of the tablets, where assembling of the individual powder granules can be seen in different extents. Comparing both images, in the case of Figure 2a it can be seen that the granules are more separated within the structure, which also leads to a thicker tablet when compared to Figure 2b.

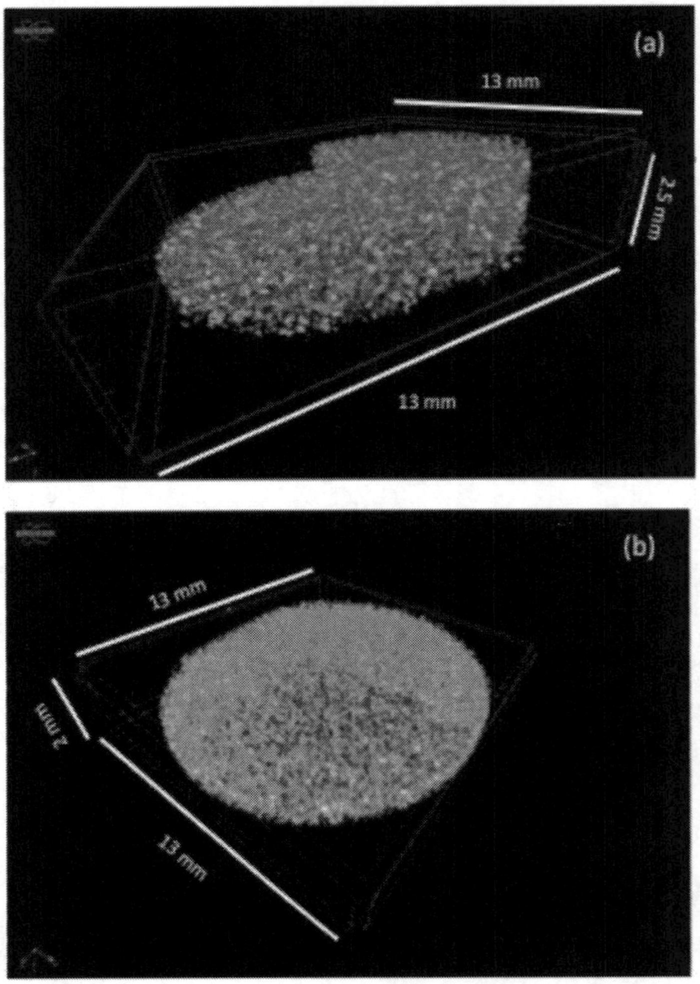

Figure 2. 3D reconstruction of μ-CT scans of tablets compressed at (a) 0.75 MPa and (b) 15 MPa.

Figure 3 shows the dependency of the overall porosity with the pressure applied to make the tablet. The evaluation of the porosity of the tablets was made for triplicate samples and shows the average value with the standard deviation of the measurements. The overall porosity of the tablet decreases from approximately 55% tablet made using the minimum pressure to have a consistent tablet (0.75 MPa) and 20% for the highest pressure (15 MPa). These overall porosity values were validated with helium pycnometry, with errors in the determination of this variable always below 4.3%.

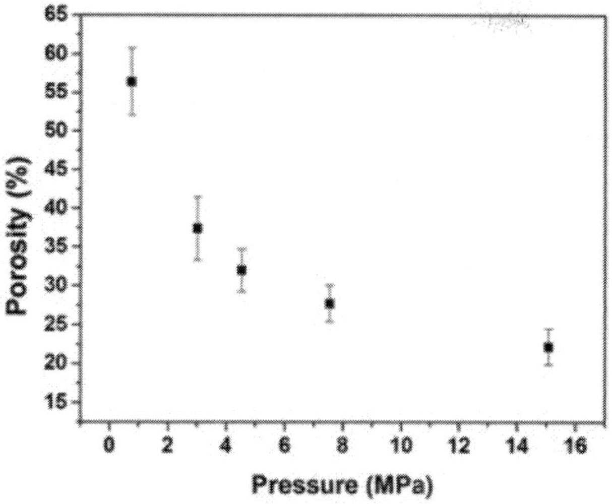

Figure 3. Overall porosity of the porous tablets applying different pressure with the Instron Microtester 5848 device (data points are the average of 3 replicates)

In all cases, the open porosity (defined as the void spaces connected to the outer surface) amounted to the practical the totality of the overall porosity observed for tablets pressed up to 4.5 MPa, whereas in the case of 7.5 MPa it was about 95% and in the case of 15 MPa, it decreased down to about 87%.

μ-CT can visualize the internal structure of the samples, so a pore size distribution can be obtained. Drawing an arbitrary border and defining pores below 55 μm as small and above 100 μm as large, Figure 5 shows the relative amount of (a) small and (b) large pores with respect to the porosity of each sample. From the plots, it can be easily inferred that the greater the pressure applied, the higher the relative amount of smaller pores and, accordingly, the smaller the amount of large pores.

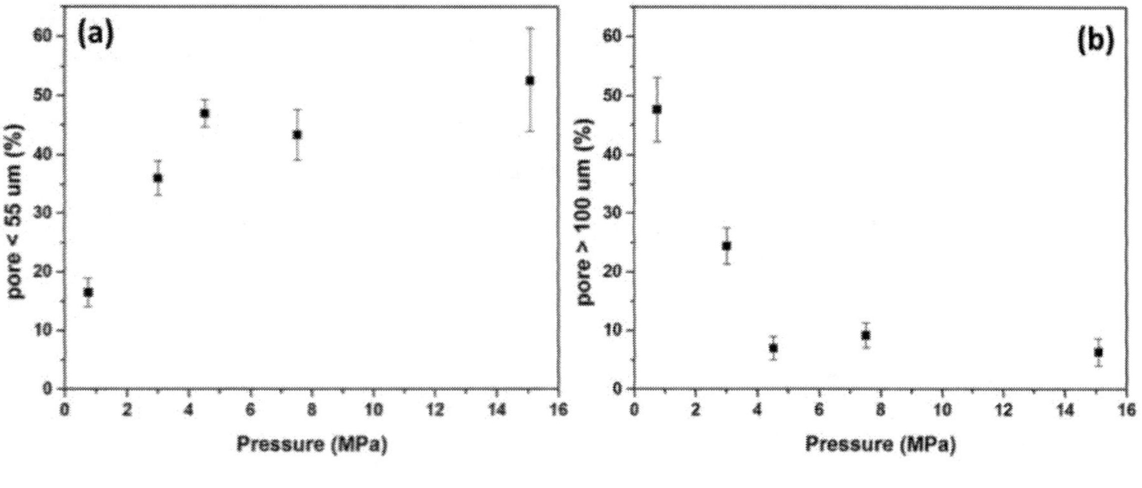

Figure 4. Relative amount of pores (a) below 55 μm and (b) above 100 μm as a function of the pressure used to make the tablets (data points are the average of 3 replicates)

3.3. Initial contact angles

Figure 5 displays the initial CA (i.e, the angle upon contact of the droplet with the solid surface) of n-hexadecane tested for tablets pressed under different pressures (Table 1 and Figure 3), and thus different degrees of roughness and porosities. Figure 5 shows the dependence of the initial CA with roughness. Nevertheless, the values of the CA overlap in some cases, so a solid conclusion cannot be reached.

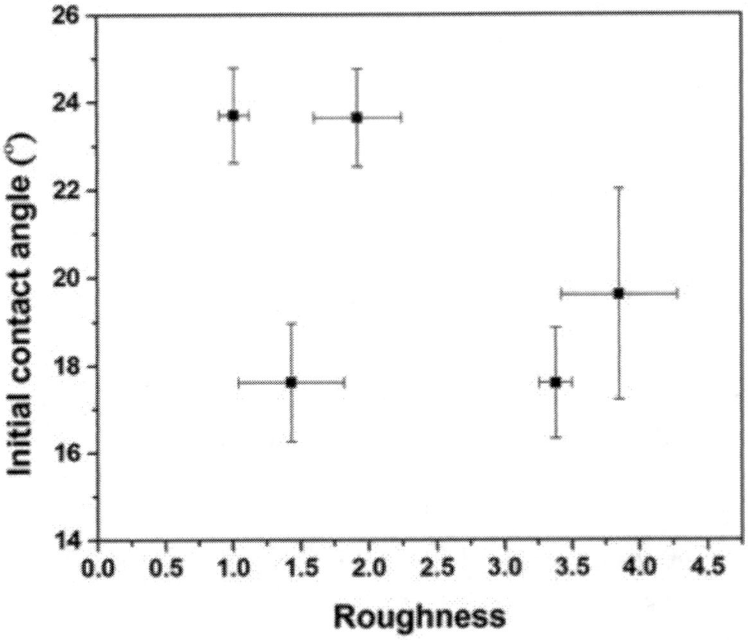

Figure 5. Initial contact angles as a function of the roughness of the porous substrates

The dependence of CA of nanofluids based on water solutions had previously been analysed for the case on different metallic surfaces showing some dependence with the surface material in most cases [8]. This suggests that it is very related to the material properties surface material and not only to the degree of roughness, as it is well known.

3.4. Kinetic evaluation of droplet imbibition

The sessile drop method was employed for the assessment of the kinetics of droplet imbibition into the porous tablet. Figure 6 captures the evolution of the droplet after different periods after the initial contact for a tablet of a porosity of (a) 54.89% and (b) 19.84 %.

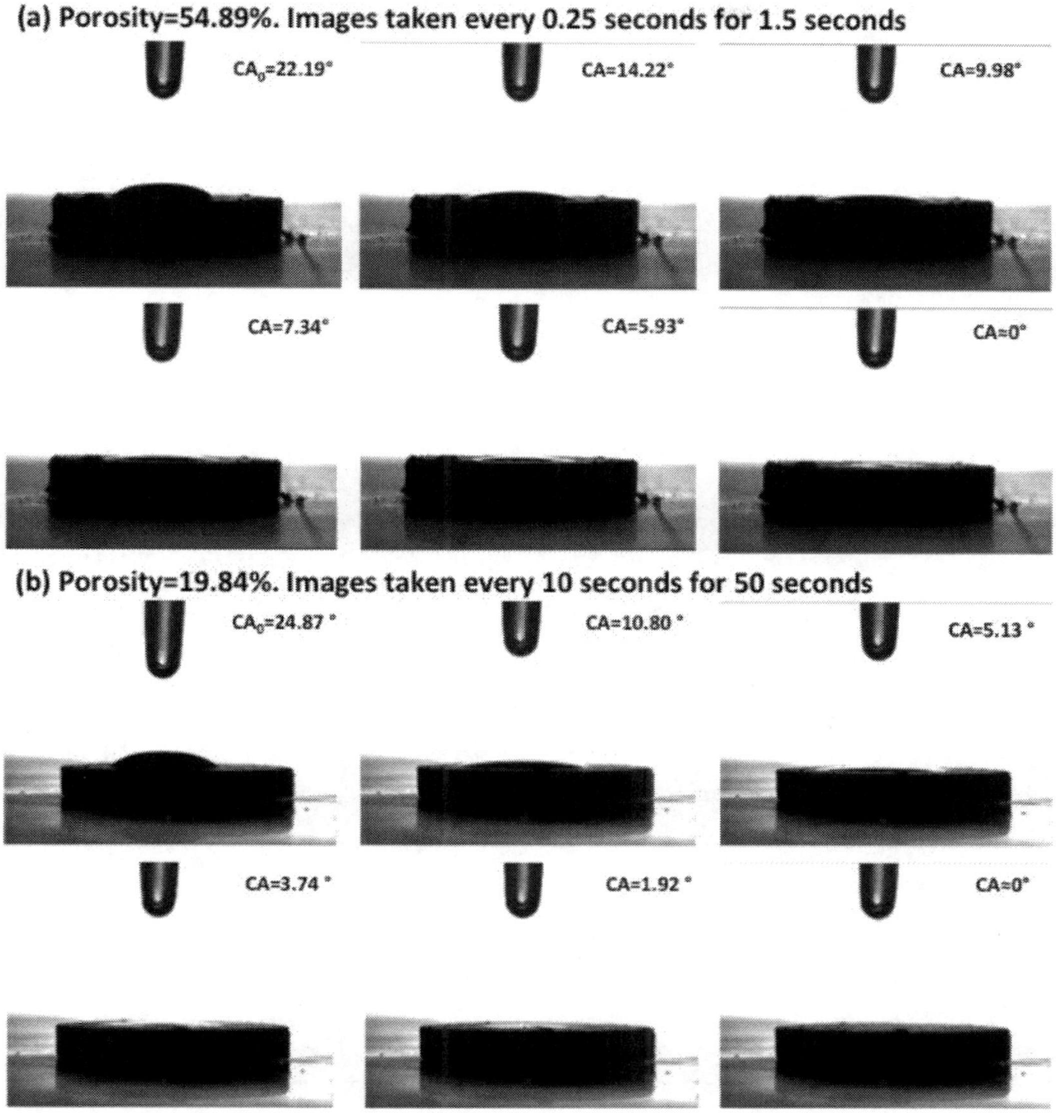

Figure 6. Evolution of a droplet of n-hexadecane on tablets of (a) 54.89% and (b) 19.84% of porosity

Virtually complete penetration of the droplet took place for all the tablets tested, i.e, the value of the contact angle eventually reached a value very close to zero due to the entire droplet being soaked into the substrate. Penetration and spreading of the droplet occurs rapidly in both cases, which is clearly affected by the viscosity (3.070 cP) and surface tension (27.327 mN m^{-1}) of n-hexadecane, although porosity is a much more decisive factor.

Figure 7 depicts the progress of the dynamic contact angle of droplets tested on tablets of the same examples above. The decrease takes place much more rapidly as porosity increases if one considers the respective time scales.

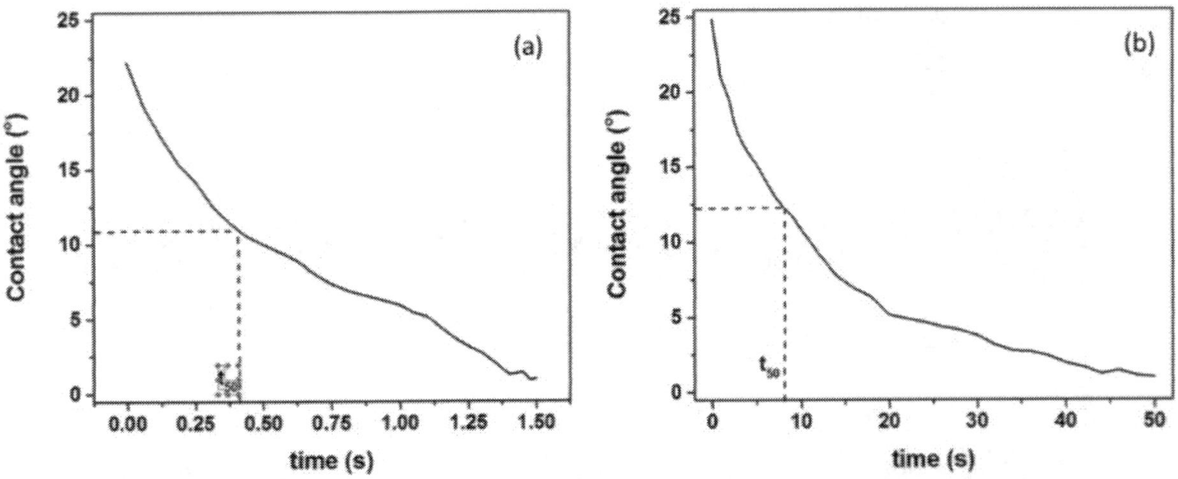

Figure 7. Plots of the evolution of the contact angles of droplets of n-hexadecane as they penetrate tablets of (a) 54.89% and (b) 19.84% of porosity.

To deepen on the relationship between the penetration rate and porosity, an exponential curve has been fitted to approximately the first 10% initial data points of each set of dynamic contact angle profile. Subsequent differentiation of the curves has been made, with their evaluation at time zero corresponding to the initial rates of decrease of the contact angle, CA_0', as the liquid imbibes into the porous substrate. Table 2 compiles the CA_0' values and the time necessary for a decrease of 50% of the initial contact angle, t_{50}, illustrating the great dependence of these values on porosity.

Table 2. Initial rates of contact angle decrease and t_{50} for tablets of different porosities.

Porosity of tablet (%)	CA'_0 (^{0} s^{-1})	t^{50} (s)
54.89	-27.31	8.02
37.93	-16.28	4.43
28.57	-8.93	1.89
26.89	-5.44	1.54
19.84	-1.82	0.41

4. Conclusions

A methodology has been presented to elaborate porous tablets from granules and analyse the surface roughness and internal porous structure with non-invasive techniques. Surface roughness showed a decreasing trend with increasing pressure applied ranging from 3.85 to 1.01 µm of arithmetic mean of the absolute height. . Overall porosity could be controlled within a range from 22 to 57 % using pressures from 15 to 0.75 MPa, respectively during preparation, most of which was found to correspond to open pores.Penetration of n-hexadecane droplets experiments were completed employing a goniometer to such effect and it was found that (i) roughness of the tablet barely showed any effect on the kinetics of the imbibition process and (ii) that both the penetration times and were greatly influenced by the internal porous structure of the tablet.

Acknowledgement

The authors would like to thank the Advanced Manufacturing Supply Chain Initiative (AMSCI) for funding the CHARIOT Consortium. AMSCI is a government supply chain fund which is helping to rebuild British manufacturing prowess.

References

1. Gambaryan-Roisman, T., *Liquids on porous layers: wetting, imbibition and transport processes*. Current Opinion in Colloid & Interface Science, 2014. **19**(4): p. 320-335.
2. Hapgood, K.P., et al., *Drop penetration into porous powder beds*. Journal of Colloid and Interface Science, 2002. **253**(2): p. 353-366.
3. Nguyen, T., W. Shen, and K. Hapgood, *Drop penetration time in heterogeneous powder beds*. Chemical Engineering Science, 2009. **64**(24): p. 5210-5221.
4. Wang, Z., et al., *Water droplet spreading and imbibition on superhydrophilic poly(butylene terephthalate) melt-blown fiber mats*. Chemical Engineering Science, 2016. **146**: p. 104-114.
5. Starov, V.M., et al., *Spreading of liquid drops over dry porous layers: Complete wetting case*. Journal of Colloid and Interface Science, 2002. **252**(2): p. 397-408.
6. Alleborn, N. and H. Raszillier, *Spreading and sorption of droplets on layered porous substrates*. Journal of Colloid and Interface Science, 2004. **280**(2): p. 449-464.
7. Desie, G., et al., *Printing of dye and pigment-based aqueous inks onto porous substrates*. Journal of Imaging Science and Technology, 2004. **48**(5): p. 389-397.
8. Cieslinski, J.T. and K.A. Krygier, *Sessile droplet contact angle of water-Al2O3, water-TiO2 and water-Cu nanofluids*. Experimental Thermal and Fluid Science, 2014. **59**: p. 258-263.
9. Green, N.C., et al., *Assessment of non-contacting optical methods to measure wear and surface roughness in ceramic total disc replacements*. Proceedings of the Institution of Mechanical Engineers Part H-Journal of Engineering in Medicine, 2015. **229**(3): p. 245-254.
10. Xu, C., C. Reece, and M. Kelley, *Characterization of Nb SRF cavity materials by white light interferometry and replica techniques*. Applied Surface Science, 2013. **274**: p. 15-21.
11. Cox, S.C., et al., *Adding functionality with additive manufacturing: Fabrication of titanium-based antibiotic eluting implants*. Materials Science & Engineering C-Materials for Biological Applications, 2016. **64**: p. 407-415.
12. Anderson, D.M., *Imbibition of a liquid droplet on a deformable porous substrate*. Physics of Fluids, 2005. **17**(8).
13. Leach, R., et al., *Measurement Good Practice Guide No. 108. Guide to the Measurement of Smooth Surface Topography using Coherence Scanning Interferometry*. 2008, National Physical Laboratory: Teddington, Middlesex.

465808 Cyclopentane Hydrate Formation Equilibria with Varying HLB Numbers in Sorbitan Monoester Surfactants

Abstract:

Sorbitan monoesters are nonionic surfactants having sorbitan functional group as a hydrophilic head. Their hydrophilic-lipophilic balance numbers (HLB numbers) can be determined by carbon numbers or structure of hydrophobic tails. The hydrophobicity of surfactants is one of the important factors on the emulsion formation, and thus it is significant that the variation of the HLB number of surfactants affects the hydrate formation equilibria from the water-in-oil emulsion. Span 20, 40, 60, and 80, which have the HLB number as 8.6, 6.7, 4.7, and 4.3, were utilized in this study, then we analyzed the effect of HLB numbers on the size of emulsion droplets and equilibrium temperature of cyclopentane hydrate formed from water-in-cyclopentane emulsion. The size of emulsion and equilibrium temperature of cyclopentane hydrate monotonically increased with increasing the HLB number of sorbitan monoesters. Especially, these values were suddenly decreased at the 4.3 of HLB number, because Span 80 uniquely have cis double bond on their hydrophobic tails. With 3.5 wt% sodium chloride, the trend of equilibrium temperature of cyclopentane hydrate was removed due to the significant effect of salt ions on the interaction between surfactant and water molecules.

467204 Ionic Conductance in Alumina Nanochannels: Temperature and pH Effects

Abstract:

Inspired by biological ion channels, nanochannels embedded in a thin alumina membrane have captured much attention of scientists due to their promising applications in energy conversion, ionic diodes, and biosensors. The working principle of these emerging applications is highly related to their ionic current (or conductance) behaviors, which depend on the surface charge property of nanochannels. It is known that the surface charge property of alumina nanochannels is exquisitely dependent on the surface chemistry reaction of dissociable functional groups (e.g., Al-OH) on the channel wall and, therefore, the variation in solution temperature might play a role. Although the relevant studies on the ion transport in alumina nanochannels are ample, the influence of solution temperature has not been discussed so far. Herein, we experimentally investigate the temperature effect on the ionic conductance in alumina nanochannels, which were fabricated by the electrochemical anodic oxidation process. SEM images reveal that the diameter of cylindrical alumina nanochannels we fabricated is about 20-30 nm. Results obtained show that temperature sensitive of nanochannel conductance depends apparently on the levels of pH and salt concentration, which provides useful information for the design of ionic devices used in temperature sensitive applications such as ionic diodes and gates.

467542 Current Rectification in a Nanopore with Biomimetic Functional Groups

Abstract:

Biological Ion channels, which generally have asymmetric, non-cylindrical structure, in cell membranes play an essential role, such as gate to regulate water, ionic species, and small molecules through the membrane. Inspired by actual physiologic environments, the nanoscale channels can be stimulated by pH and salt concentration differences through two sides of channels, and hence might have interesting and unique ion transport phenomena. Herein, we fabricate a conical nanopore with biomimetic lysine groups by two step functionalization processes. The polyacrylic acid polymer chains were first coated on the inner surface of conical nanopore using the surface plasma polymerization technique, followed by the grafted modification of lysine groups by wet solution chemical reaction. Results show that the ion current rectification in the prepared biomimetic nanopore is dependent on the solution pH and salt concentration. Moreover, the behaviors of current rectification is different for two reservoirs connected to the nanopore with different solution pHs.

467544 Fabrication of pH-Sensitive Ionic Diodes in Highly Ordered Alumina Nanochannel Arrays

Abstract:

Fabrication of current rectification ionic diodes to mimic biological ion channels in cell membranes is of circular significance in the biomimetic application technology. In this study, we fabricate a novel ionic diode device, which comprises highly ordered nanochannels of anodic alumina membrane with heterogeneous surface charge properties by patterning the carboxyl functional groups on one side of nanochannels through silanization process. Several factors key to the system under consideration, such as pH, salt concentration, and pH gradient, were experimentally examined for their influences on the ion current rectification. We find that the magnitude of current rectification depends not only on the solution pH and salt concentration but also on the pHs at different sides of reservoirs. In general, the present ionic diode device exhibits high rectification behavior when the nanochannels have bipolar charges. The findings are informative and provide necessary information for further applications such as nanoscale energy conversion.

465912 Effect of Surfactant Architecture on Micelle and Reverse Micelle from iSAFT Molecular Density Functional Theory

Abstract:

Surfactants reduce the interfacial tension making them an important additive in a number of industrial and commercial applications from enhanced oil recovery to personal care products (e.g., shampoo, detergents, etc). Understanding the phase behavior of surfactants is of vital importance to effectively and economically utilizing surfactants for various applications. For example, producing ultra-low interfacial tensions needed for enhanced oil recovery requires the formation of a middle phase microemulsion. Determining the surfactant architecture (e.g., type of surfactant and head/tail size) and fluid conditions (e.g., temperature and salinity) to produce the middle phase microemulsion involves trial and error experiments. A predictive molecular model would provide guidance to effectively screen surfactants. As the basic surfactant aggregates, micelle and reverse micelle structures as well as surface stress calculations should lead to improved understanding of the middle phase microemulsion.

Interfacial Statistical Associating Fluid Theory (iSAFT) classical density functional theory has been developed and utilized to study surfactants and other complex fluids. In current work, iSAFT is applied to study the effect of surfactant architecture on the formation of micelle and reverse micelle structures in solution. Further, the elements necessary to produce a middle phase microemulsion are investigated.

468469 Rectifying Ionic Current in Polyelectrolyte-Functionalized Bipolar Nanopores

Abstract:

Nanopore-based devices have emerged as promising tools for regulating the transport of nanoentities, sensing single molecules, and rectifying ion current. Inspired by biological ion channel, synthetic nanopores functionalized with biomimetic structures or polyelectrolyte (PE) brushes attracted lots of scientific interests in using them as novel platforms for both biosensors and nanofluidic systems. One of these phenomena is ionic current rectification (ICR), a diode-like current-voltage behavior, in a nanopore with bipolar charge. The specific distribution of the surface charge in such a nanopore brings about an asymmetric transport of ions through it, thereby exhibiting a strong ICR behavior. In this work, the ion transport in a nanopore functionalized with bipolar PE brushes is theoretically investigated by adopting a continuum-based model comprising a set of Poisson-Nernst-Planck, Stokes, and Brinkman equations. We show that the degree of ICR decreases with increasing bulk salt concentration, and increases with increasing repeat unit of PE chain. In addition, the higher the charge density the more significant the ion enrichment/depletion, resulting in a larger degree of ICR. However, if the positive and negative charges are asymmetrically distributed on either side of a nanopore, the degree of ICR has a local maximum as charge density varies. The results obtained suggest that the ICR behavior in a PE-functionalized nanopore with bipolar charges can be regulated by its surface modification.

467977 Development of Approaches to Couple Aerosol Charging with Aerosol Coagulation

Abstract:

Aerosol charging can influence the time-evolution of the size distribution of aerosols, which is an important parameter needed to understand their fate in the atmosphere. Electrical charge may be easily accumulated on aerosols through various charging mechanisms, such as ion diffusion. Charge on aerosols can affect their coagulation rates by creating electrostatic interactions. Because the number of elementary charges on aerosols varies, the coagulation rate of aerosols depends not only on their size but also the charge distribution. However, few investigations have been focused on examining the influence of the charge distribution of aerosols on their size distribution, as well as simultaneously tracking both distributions. This study is aimed at developing three modeling approaches to simultaneously investigate charging and coagulation kinetics of aerosols. The modeling approaches are developed on the basis of population balance (PB), but the level of their complexity varies. The first approach is based on bivariate PB equations that have charge and size as variables. For this approach, the bivariate PB model can be used to simultaneously predict the time-evolution of both aerosol charge and size distributions. To reduce computational cost, in the second approach, the bivariate PB equations are replaced by the monovariate PB equations that have particle size as the variable. A set of charge balance (CB) equations is separately derived and then coupled with the monovariate PB equations to include aerosol charging effects. Similarly to the second approach, the last approach includes the monovariate PB equations, but excludes the CB equations by assuming that the charge distribution of aerosols instantaneously reaches steady-state. Case studies using the modeling approaches will be provided and the limitations of each approach will be discussed.

469847 Molecular Dynamics Study of the Capacitive Performance of Oxidized Graphene

Abstract:

Electrochemical capacitors, which are commonly referred to as supercapacitors, offer relatively high energy and power densities and extended cycle lifetimes. Subsequently, they are attracting substantial interest as a new class of electrical energy storage devices. Graphene layers improve the specific surface areas and electrical conductivity of electrodes and have been implemented in novel material composites in these systems. Although recent efforts have broadened the understanding of electrosorption of charges on heterogeneous features of these materials, most modeling studies approximate the graphene electrodes as pristine surfaces without functional groups or defects. To reconcile these efforts with non-idealized graphene allotropes that are more common in supercapacitors, we present a comprehensive study of the effects of oxygen groups on the structure of the electric double layer (EDL) and the resulting capacitance.

We conducted molecular dynamics (MD) simulations of different electrolytes on pristine/functionalized graphene, and the electrolytes included both neat ionic liquids and ionic liquids in acetonitrile (ACN) solution. We complemented computational results with experimental cyclic voltammetry data to further clarify the effects of the functional groups. The findings show that ions have higher interaction energies with functionalized surfaces due to the increased contribution of electrostatic interactions. Higher electrode-electrolyte interaction energies attract co-ions and/or expel counter-ions, thus decreasing the capacitance of neat ionic liquids on oxidized surfaces. The functional groups have lower influence on the interaction energy of the electrode with organic ACN solvents. However, solvated electrolytes exhibit different electrochemical behaviors than their solvent-free counterparts. Since the molecular size of the cation in ionic liquids is much larger than ACN, inner-most layers at the cathode are primarily occupied by the solvent. Subsequently, capacitance of cathode was similar for both functionalized and pristine graphene. However, since the anions aggregate at the innermost layer of the anode, functional groups decreased the capacitance of the anode. These results corresponded well with the electrochemical tests, and improved our fundamental understanding of the effects of surface chemistry on the performance of different electrolytes in supercapacitors.

470169 Peg-Coated Silver Nanoparticle Surface Activity and Lipid Monolayer Interactions

Abstract:

The use of engineered nanoparticles (ENPs) has increased rapidly over the last few decades. As ENPs enter the environment, determining their interaction with biological molecules is a key aspect to understanding their impact on environmental health and safety. ENPs can interact with cell membranes by adsorbing onto their surface and compromising their integrity, permeability, and function. Fundamental information on these interaction mechanisms can be used to create ENPs with minimal or controlled membrane activity. The surface activity of coated ENPs is an important parameter in determining how they behave in the environment and controlling how a nanoparticle will interact with cells membranes.

This study investigates the interactions between polyethylene glycol (PEG)-coated silver NPs and lipid monolayers using dynamic surface-pressure measurements and hyperspectral microscopy. PEG-coated nanoparticles are generally considered to be inert, however, there have been a number of studies showing that PEG and PEG-coated NPs are surface active, which would promote their insertion into lipid monolayers or membranes. Langmuir monolayer techniques were employed to measure the surface pressure-area (π–A) isotherms of the lipid monolayer as a function of NP concentration. Monolayers were formed by depositing solutions of dioleoylphosphocholine/dioleoylphosphoglycerol (DOPC/DOPG; 1:1 mole ratio) at the air-water interface to mimic an anionic membrane. A suspension of PEG-coated silver NPs was injected below the lipid monolayer in the water subphase and the interaction with the monolayer was examined as a function of time.

Isotherms were conducted in the absence of the lipid monolayer to determine the NPs surface activity and to distinguish between the surface activity of the excess coating polymer coating from coated NPs themselves. Our results show that the PEG-coated NPs are surface-active and compete with the lipids for adsorption at the air/water interface. Hyperspectral microscopy and TEM imaging of the nanoparticles and lipid monolayer at the interface confirm these observations. In the absence of lipids, the NPs increase the surface pressure up to 45 mN/m upon compression before the NPs surface layer collapsed. In the presence of lipids, the NPs and lipids act cooperatively to increase the π. At high NPs concentrations, corresponding to the NPs completely covering the interface, the π–A isotherms more closely resembled that for NPs than for lipids. NP surface activity is predominantly (~80%) due to the excess coating material rather than the NPs themselves. Low NP concentrations lead to an increase in π due to increase lipid packing and the inherent surface activity of the NPs, and high NP concentrations lead to monolayer collapse due to lipid desorption from the interface.

470394 Continuous Flow Synthesis of Noble Metal Patchy Particles

Abstract:

The syntheses of noble metal nanoparticles of gold, silver or platinum have attracted high interest in research since several decades. Known potential applications can be found e.g. in the fields of biomedicine, plasmonics, sensors or catalysis. Especially, the synthesis and the control of asymmetrical particle morphologies, which often bring enhanced functionality, by simple and scalable methods is one of the major hurdles to their exploitation. By this poster a simple continuous one-step flow process will be shown to synthesize noble metal patchy particles. In the context of this work, patchy particles are defined as particles having at least one surface region – the so called patch – of a different material which provides an additional surface chemistry or functionality. A well-known example of these relatively new types of particles are Janus particles which show a surface contribution of different material and/-or functionality of 50%. Most of the fabrication methods for combining noble metals and non-metallic core-particles are based on the immobilization of spherical core particles onto a substrate to shadow a desired fraction of the particles' surface against further chemical or physical treatments like the functionalization of the surface by capping molecules or the deposition of different materials like gold by e.g. sputtering. Such methods include several steps which lead to a high complexity and a limited scalability of the whole process. In our group we have developed an approach which is based on the cooperative interaction of metal precursors and other reagents with a core particle's surface. We have shown that this leads to a strong preference for heterogeneous nucleation and surface diffusion, leading to the surface conformal growth of noble metal patches. By this poster the three patch/core particles systems of Ag@SiO_2, Au@PS (polystyrene) and Pt@PS will be presented. Our results show that in case of Au@PS and Pt@PS ascorbic acid plays an essential surface guiding function in addition to its' role as reducing agent. Independent of the combination of noble metal and core particle, the synthesis is processed at aqueous conditions by a simple T-mixer based flow process in the millimeter scale. This allows the production of larger quantities of patchy particles at lab scale for the exploration of potential applications and further shows an option of process design

which has the potential of classical scalability up to a certain limit or the possibility to increase the throughput of product particles by numbering up method.

471610 Gelation of Polymer-Grafted Silica Nanoparticles Studied with X-Ray Photon Correlation Spectroscopy (XPCS) and Rheology

Abstract:

We report a combined XPCS study of particle dynamics and rheological study of moderately concentrated suspensions of silica colloids and their binary mixtures that form a gel on cooling. The suspensions are comprised of silica colloids (32nm, 86nm and 185 nm) coated with octadecyl-hydrocarbon chains and suspended in decalin at colloidal volume fractions (f) ranging from 0.2 to 0.4. Different scaling relationships are explored to predict the elastic modulus and the limit of linearity. Gel temperatures of the different size particles and the elastic moduli, though different, collapse onto a single curve when scaled as Gâ€™D^3/fkT and plotted versus (1/T â€“ 1/T_{gel}), where Gâ€™ is the elastic modulus and T_{gel} is the gel temperature. This scaling is in line with predictions of mode coupling theory which emphasizes the importance of local structure (localization length) over longer-range correlations in determining the dynamical and mechanical properties of such gels. The dynamics of nanoparticles during the gelation process is measured using X-ray Photon Correlation Spectroscopy (XPCS) at Argonne National Labs. Using a newly developed ultrafast frame rate (11.8 kHz) pixel-array-detector, we have, for the first time, captured the complete transition of the dynamics of the suspension from tens of seconds in the gel state to hundreds of μs in the liquid state. The transition is triggered when the gel is slowly heated. Our nanoscale measurements of the colloid dynamics as a function of temperature and particle size can be compared to macroscopic viscoelastic properties probed by rheology under the same conditions and suggest that smaller colloids form stronger networks when they gel. This is in line with the measurements of the gel boundary with smaller particles gelling at higher temperatures. Efforts to extend the scaling relationships of elastic modulus to smaller particle sizes and their mixtures and limitations of the theories in predicting the mechanical properties will also be discussed.

472473 Understanding Optimal Oil-in-Water Emulsifer Formulation Using Interfacial Tensiometry and Microstructural Studies

Abstract:

Marine oil dispersants are concentrated, low-viscosity, low-toxicity surfactant-solvent blends which are applied to crude oil spills in order to disperse them into the ocean. Despite their widespread use, the fundamental principles underlying oil dispersantsâ€™ formulation have remained poorly understood for decades. In our work, the dispersion effectiveness of various blends of Tween 80, Span 80, and DOSS (dioctyl sodium sulfosuccinate), three of the most widely used surfactants in marine oil dispersants, has been correlated with the dynamic crude oil-seawater interfacial tension produced by each blend. DOSS-rich blends which produce very low interfacial tensions ($<10^{-4}$ mN/m) exhibit high dispersion effectiveness; DOSS-poor blends producing comparably low interfacial tensions exhibit much lower effectiveness, because DOSS-poor dispersantsâ€™ initial adsorption to the oil-water interface is much slower than that of DOSS-rich dispersants. Cryo-transmission electron microscopy (cryo-TEM) and dynamic light scattering (DLS) show that these solutions contain spherical W/O microstructures, the size and dispersity of which vary with surfactant blend composition and surfactant:seawater molar ratio. Span 80-rich microstructures exhibit high polydispersity index (PDI > 0.2) and large diameters (â‰¥100 nm), whereas DOSS-rich microstructures exhibit smaller diameters (20-40 nm) and low polydispersity index (PDI < 0.1), indicating a narrow microstructure size distribution. The smaller diameters of DOSS-rich microstructures, as well as the fact that DOSS molecules, being oil-soluble, can diffuse to a bulk oil-water interface as monomers much faster than any of these microstructures, may explain why DOSS-rich blends adsorb to the oil-water interface more quickly than Span 80-rich blends, a phenomenon which has been linked in prior work to the higher effectiveness of DOSS-rich Tween/Span/DOSS-based oil dispersants.

470322 Simulating Total Internal Reflection Microscopy of Anisotropic Particles

Abstract:

Total Internal Reflection Microscopy (TIRM) has been developed over the past three decades to be the most sensitive technique for measuring weak ~kT scale interactions between a single spherical colloidal particle and substrate. The measurement consists of a micron-scale particle nearby, but not adhered to, a substrate over which an evanescent wave is traveling. Classic TIRM has been used to measure very weak ~kT scale colloidal interactions, including electrostatic double layer repulsion, van der Waals attraction, steric interactions, depletion interactions, electrophoretic forces, hydrodynamic interactions, sedimentation deposition, and even the Casimir interaction have been measured with classic TIRM. Unfortunately, one major restriction currently imposed by TIRM is that the particle needs to be spherical. This poster will summarize our groupâ€™s work on expanding the capabilities of TIRM to anisotropic particles.

450334 Ionic Current Rectification in a Polyelectrolyte-Modified Conical Nanopore

Abstract:

Due to their specific geometry, conical nanopores are capable of yielding several interesting electrokinetic phenomena, such as ion concentration polarization (ICP) and ion current rectification (ICR). Extending previous analyses, we consider two types of nanopore: only the inner surface of a nanopore is functionalized by a polyelectrolyte (PE) layer in type I nanopore, and both its outer and inner surfaces are functionalized in type II nanopore. Numerical simulation is conducted to examine the influences of the double layer thickness and the thickness of the PE layer of a nanopore on its ICR behavior. We show that the ICP in type I nanopore is more significant than that in type II nanopore. In addition, as the bulk salt concentration varies, the rectification factors R_f for type I and II nanopore are different significantly. In particular, if the salt concentration is low, an inversion in R_f is observed in type I nanopore.

450347 Diffusiophoresis of a pH-Regulated Toroidal Polyelectrolyte in a Solution Containing Multiple Ionic Species

Abstract:

Considering recent application of diffusiophoresis in biosensing technology, we extend previous analysis for the diffusiophoresis of a toroidal PE at a fixed charge density to that at a pH-regulated condition in the presence of multiple ionic species so that the underlying assumptions are more realistic. The diffusiophoretic behaviors of the toroidal PE considered under various conditions are examined by varying its functional group density, size, and softness, and the solution pH and bulk salt concentration. The results of numerical simulation reveal that due to the competition between the hydrodynamic force and the electric force acting on a particle, the behavior of a toroidal PE can be different both quantitatively and qualitatively from the corresponding spherical PE. This implies that using an equivalent sphere to simulate the actual entity can yield significant deviation and, therefore, knowing the exact shape of an entity is necessary both in interpreting experimental data and in designing relevant devices.

449111 Study on the Transient Interfacial Tension in a Microfluidic Droplet Formation Coupling Interphase Mass Transfer Process

Abstract:

In two-phase dispersion coupling interphase mass transfer process, the variation of interfacial tension is an important factor affecting the dispersion. In this study, we described a microfluidic method for the determination of the transient interfacial tension (TIFT). The method has the advantage of determining TIFT during the whole droplet formation process, rather than only at the rupture moment as reported in previous studies. The TIFTs of several systems were determined. In certain systems, it has been found that the droplet size decreased with the increase of the dispersed phase flow rate, which is obviously different from the constant interfacial tension system. It has also been found that TIFT was mainly affected by two-phase flow rates, solute type and concentration, and droplet size. A semi-empirical equation was finally established to predict TIFT. It has the potential to be used in a variety of industrial equipment with dispersion – mass transfer coupling process.

455254 Equation of State Selection for Organic Rankine Cycle Modeling Under Uncertainty

Abstract:

In recent years there has been a great interest in the design and selection of working fluids for low-temperature Organic Rankine Cycles (ORC), to efficiently produce electrical power from waste heat from chemical engineering applications, as well as from renewable energy sources such as biomass combustion, geothermal and solar heat sources. The working fluid is essential to the performance of the cycle. In order to evaluate and test promising fluid candidates, an appropriate Equation of State (EoS) [1] is necessary.

For ORC applications, an EoS is commonly selected based on goodness-of-fits to data, width of range of availability of fluid data and complexity of formulation, which is closely related to numerical expenses. We have explored an additional criterion for the selection of a particular EoS, namely the influence on the input uncertainty of the fluid parameters on the ORC model output.

We have recently presented a methodology [2] to propagate and quantify the impact of input property uncertainty and fluid property parameter uncertainty on the ORC model output. It is applied using different EoS: Cubic EoS such as Soave-Redlich-Kwong (SRK), Peng-Robinson (PR) and Perturbed Chain Statistical Association Fluid Theory (PC-SAFT). The different EoS are assessed based on the uncertainty propagated in the model output.

The study demonstrates that the range of property parameter uncertainty, the number of parameters, the sensitivity of the property parameter w.r.t to the EoS and the overall cycle, all influence the model output uncertainty.

The procedure is highlighted for an ORC for with a low-temperature heat source from exhaust gas from a marine diesel engine.

[1] Saleh B, Koglbauer G, Wendland M, Fischer J. Working fluids for lowtemperature organic Rankine cycles. Energy 2007;32:1210–21.

[2] Frutiger J, Andreasen JG, Liu W, Spliethoff H, Haglind F, Abildskov J, Sin G. Working fluid selection for organic Rankine cycles - impact of uncertainty of fluid properties. Energy (accepted s.t. revision).

16AIChE
Annual Meeting, San Francisco, CA

Computational Thermodynamic Analysis of a Solar Thermochemical Manganese Oxide - Manganese Sulfate Water Splitting Cycle

(Principal Author)
Rahul Bhosale
Assistant Professor
Department of Chemical Engineering
College of Engineering
Qatar University, Doha, Qatar
rahul.bhosale@qu.edu.qa

(Other Authors)
Parag N. Sutar
Postdoctoral Fellow
Department of Chemical Engineering
College of Engineering
Qatar University, Doha, Qatar
parag.sutar@qu.edu.qa

Fares A. AlMomani
Associate Professor
Department of Chemical Engineering
College of Engineering
Qatar University, Doha, Qatar
falmomani@qu.edu.qa

Dareen Dardor
Undergraduate Student
Department of Chemical Engineering
College of Engineering
Qatar University, Doha, Qatar
dd1202223@student.qu.edu.qa

Marc Rosen
Professor
Department of Automotive, Mechanical and Manufacturing Engineering
Faculty of Engineering and Applied Science
University of Ontario Institute of Technology, Oshawa, Ontario L1H 7K4, Canada
Marc.Rosen@uoit.ca

Prepared for Presentation at
American Institute of Chemical Engineers
2016 Annual Meeting
San Francisco, California
November 13-18, 2016

Computational Thermodynamic Analysis of a Solar Thermochemical Manganese Oxide - Manganese Sulfate Water Splitting Cycle

Rahul Bhosale
Department of Chemical Engineering
College of Engineering
Qatar University, Doha, Qatar
rahul.bhosale@qu.edu.qa

Parag Sutar[1], Fares AlMomani[1], Dareen Dardor[1], Marc Rosen[2]

[1]Department of Chemical Engineering
College of Engineering
Qatar University, Doha, Qatar

[2]Department of Automotive, Mechanical and Manufacturing Engineering
Faculty of Engineering and Applied Science
University of Ontario Institute of Technology, Oshawa, Ontario L1H 7K4, Canada

Keywords: Mn Oxide – Mn Sulfate Cycle, Water Splitting, Solar Hydrogen, Computational Thermodynamic Modeling

Abstract

In this paper computational thermodynamics modeling of the solar H_2 production via two-step thermochemical manganese oxide – manganese sulfate (MnO-MnS) water splitting cycle is reported. In this cycle, the first step corresponds to the exothermic oxidation of manganese oxide via SO_2 and H_2O producing $MnSO_4$ and H_2 and second step is related to the endothermic reduction of $MnSO_4$ into MnO, SO_2, and O_2. The MnO and SO_2 produced in the second step can be recycled to the step 1 and hence, reutilized for the production of H_2 via water splitting reaction. The equilibrium compositions for the $MnSO_4$ thermal reduction and water splitting steps were determined by performing thermodynamic analysis and are reported here. Furthermore, exergy analysis was performed to determine the cycle efficiency and solar to fuel energy conversion efficiency attainable without heat recuperation.

1. Introduction

Several water splitting thermochemical cycles were investigated until now for H_2 production, for instance, such as iron oxide cycle, zinc/zinc oxide cycle, tin/tin oxide cycle, mixed ferrite cycle, ceria cycle, sulfur-iodine cycle, and hybrid sulfur cycle [1-59]. Due to the lower operating temperatures, the sulfur-iodine cycle (reaction set I) and its variation the hybrid sulfur cycle (reaction set II) are more appealing

Reaction set I: sulfur-iodine cycle

$H_2SO_4 \rightarrow SO_3 + H_2O$	(673K)	(1)
$SO_3 \leftrightarrow SO_2 + {}^1/_2\, O_2$	(1123 – 1273K)	(2)
$SO_2 + 2H_2O + I_2 \rightarrow H_2SO_4 + 2HI$	(393K)	(3)
$2HI \rightarrow H_2 + I_2$	(573 – 723K)	(4)

Reaction set II: hybrid sulfur cycle

$H_2SO_4 \rightarrow SO_3 + H_2O$	(673K)	(5)
$SO_3 \leftrightarrow SO_2 + {}^1/_2\, O_2$	(1123 – 1273K)	(6)
$SO_2 + 2H_2O \rightarrow H_2SO_4$	(353 – 393K)	(7)

In this paper, we have attempted to investigate the manganese oxide – manganese sulfate (MnO-MnS) water splitting cycle thermodynamically. The computational thermodynamic analysis was performed and the results obtained are presented in detail. The results section includes the equilibirum compositions associated with this cycle. In addition, the cycle and solar-to-fuel conversion efficiency values were determined by performing the second law thermodynamic analysis using HSC Chemistry software and its thermodynamic database.

2. Chemical Thermodynamic Equilibrium

According to the data reported in Fig. 1, the solar thermal reduction of $MnSO_4$ into MnO, SO_2 (g), and O_2 (g) is possible at 1335 K in presence of Ar = 30 mol/sec.

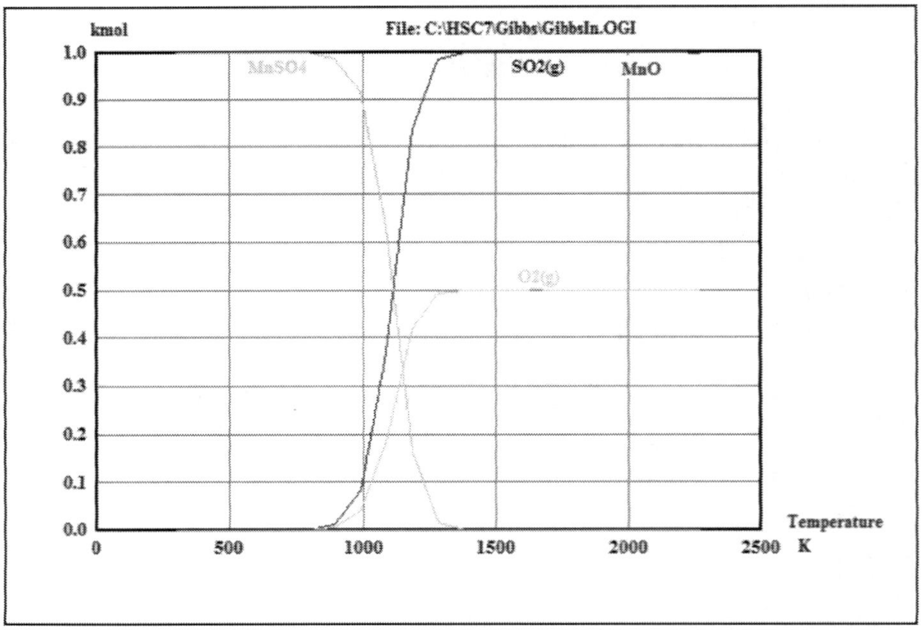

Fig 1. Solar thermal reduction of $MnSO_4$ (Ar = 30 mol/sec).

As per the Fig. 2 data, the H_2 production via MnO–MnS water splitting reaction (in absence of Ar) is possible below 550K.

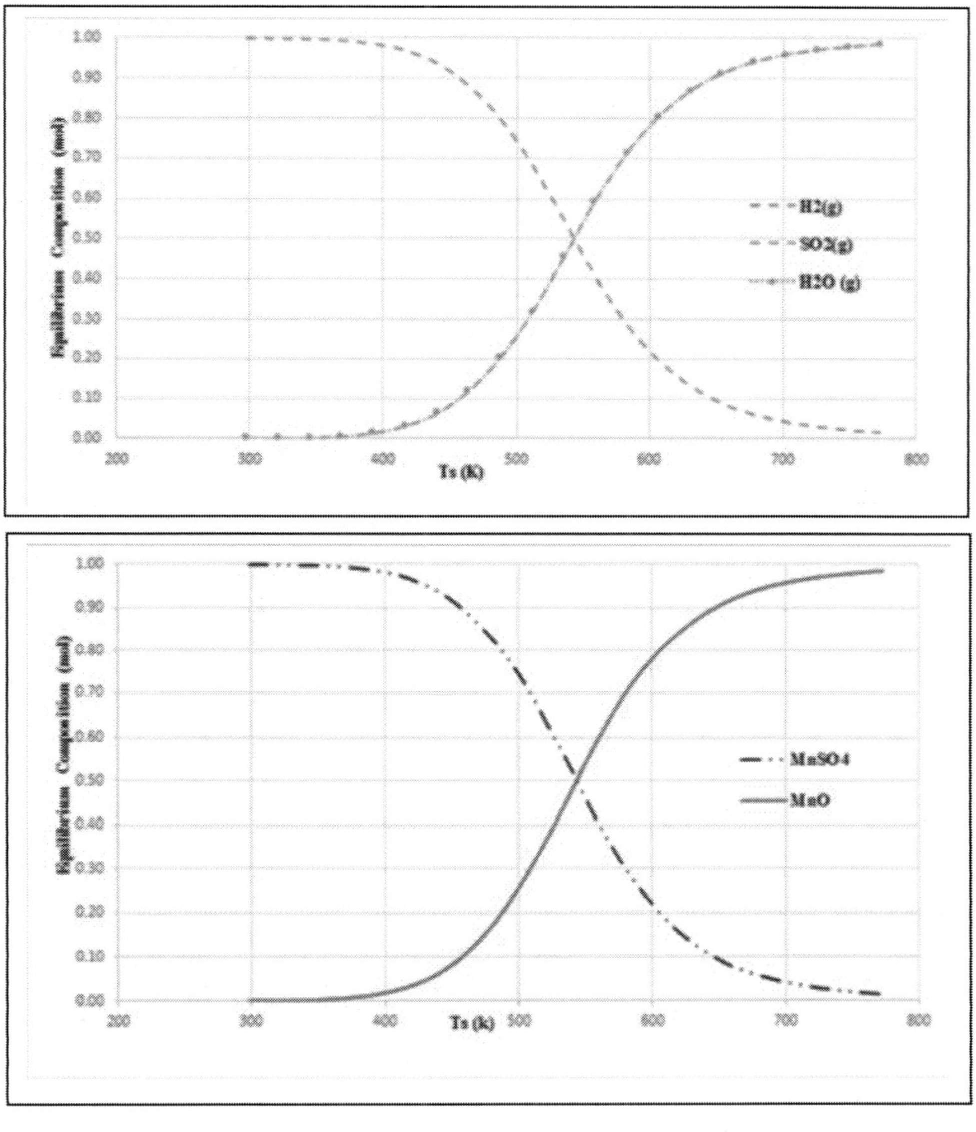

Fig. 2. Water splitting reaction using MnO-MnS cycle.

3. Exergy Analysis

Efficiencies associated with the MnO-MnS water splitting cycle were calculated via second law thermodynamic analysis. The process flow diagram for the MnO-MnS water splitting cycle is presented in Fig. 3 and it is consists of a) solar reactor, b) MnO oxidizer, c) Theoretical H_2/O_2 ideal fuel cell, d) two coolers, e) gas separator.

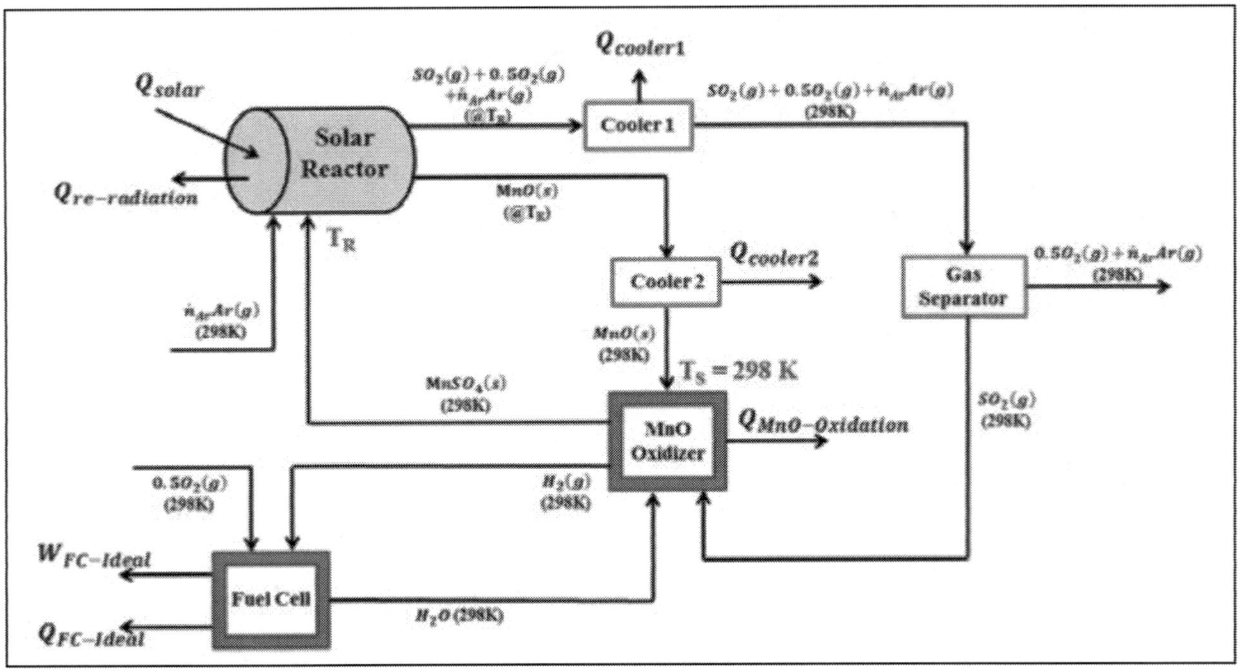

Fig. 3. Process configuration for H_2 production via two-step solar thermochemical MnO-MnS water splitting cycle.

The second law efficiency analysis was performed by following below mentioned assumptions:

- ✓ The solar reactor is assumed to be a perfectly insulated blackbody absorber
- ✓ No convective/conductive heat losses
- ✓ Effective emissivity and absorptivity equal to unity
- ✓ All the products separate naturally without expending any work
- ✓ All reactions reach 100% completion
- ✓ H_2 production is carried out at atmospheric pressure and at steady state conditions
- ✓ Viscous losses and variation in the kinetic and potential energies are neglected
- ✓ Energy required for the separation of gaseous products such as separation of SO_2 from O_2 and Ar are not accounted during the efficiency calculation
- ✓ Heat exchangers required for recovering the sensible latent heat are omitted from thermodynamic considerations

The exergy analysis was carried out by following the methodology and governing equations derived and used previously for other MO based solar thermochemical cycles [7]. Thermodynamic properties were extracted from HSC software and databases and the analysis is normalized to the $CuSO_4$ molar flow rate (1 mol/sec) entering the solar reactor.

The solar energy absorption efficiency is defined as the net rate at which energy is being absorbed by the solar reactor divided by the solar power input through the aperture of the solar reactor. The $\eta_{absorption}$ can be calculated as:

$$\eta_{absorption} = 1 - \left(\frac{\sigma T_R^4}{IC}\right) \qquad (3)$$

Where, σ = Stefan-Boltzmann constant, I = normal beam insolation $1(kW/m^2)$, C= solar flux concentration ratio of the solar concentrating system (5000 suns), $Q_{reactor-net}$= net energy absorbed in the solar reactor, Q_{solar} = solar energy input. For MnO-MnS water splitting cycle, at T_R= 1355 K, the $\eta_{absorption}$ is observed to be equal to 96.18%.

Energies required for the heating of Ar from ambient conditions upto T_R and thermal reduction of $MnSO_4$ must be determined as these are the solar energy inputs required for the operation of solar thermochemical MnO-MnS water splitting cycle.

$$Q_{reactor-net} = Q_{MnSO_4-reduction} + Q_{Ar-heating} \quad (4)$$

For complete thermal reduction of $MnSO_4$ at temperatures of 1355 K, 30 mol/sec of inert *Ar* flow rate is required as a carrier gas inside the solar reactor. Equation (5) indicates that 659.12 kW of heat energy is needed to heat the inert Ar from 298 K upto 1355 K.

$$Q_{Ar-heating} = \dot{n}\Delta H|_{Ar(g)@298K \rightarrow Ar(g)@T_R} \quad (5)$$

The Energy required for the complete reduction of $MnSO_4$ (at 1355 K) into CuO, SO_2 (g), and O_2 (g) is given by following equation and observed to be equal to 446.74 kW.

$$Q_{MnSO_4-reduction} = \dot{n}\Delta H|_{MnSO_4@298K \rightarrow MnO + SO_2(g) + {}^1/_2 O_2(g)@T_R} \quad (6)$$

At the end, the total net solar energy absorbed for the operation of MnO-MnS cycle is observed to be equal to 1182.78 kW (according to Eq. 4).

Based on the calculated $\eta_{absorption}$ and $Q_{reactor-net}$, the amount of solar energy input required to operate the solar reactor (Q_{solar}) dealing with the CO-SS water splitting cycle can be determined by equation (7) and observed to be equal to 1229.80 kW.

$$Q_{solar} = \frac{Q_{reactor-net}}{\eta_{absorption}} \quad (7)$$

Radiation heat losses and % re-radiation from the solar reactor conducting the thermal reduction of $MnSO_4$ can be calculated by equations (8) and (9):

$$Q_{re-radiation} = Q_{solar} - Q_{reactor-net} \quad (8)$$

$$\%\, re - radiation\ losses = \frac{Q_{re-radiation}}{Q_{solar}} \times 100 \quad (9)$$

Computational thermodynamic modeling indicate that radiation heat losses = 47.02 kW and % re-radiation = 3.82%.

The product components coming out of the solar reactor includes solid MnO and gaseous SO_2, O_2, and Ar. The solid MnO was cooled from 1355 K to water splitting temperature (298 K) and during this step 739.04 kW of heat energy was rejected to the surroundings. Similarly, the gaseous mixture of SO_2, O_2, and Ar was also cooled to 298 K and during this operation 60.52 kW of energy is released by the cooler – 2.

MnO generated via thermal reduction of $MnSO_4$ is transferred to the oxidizer and reacted with the combined stream of SO_2 and H_2O at 298 K for the production of H_2. By assuming 100% oxidation of MnO producing $MnSO_4$ and H_2 via water splitting reaction, the rate of heat rejected to the surrounding by MnO oxidizer is estimated as 97.39 kW according to equation (10).

$$Q_{MnO-oxidizer} = -\dot{n}\Delta H|_{MnO+H_2O(g)+SO_2(g)@298K \rightarrow MnSO_4+H_2(g)@298K} \quad (10)$$

Solar thermochemical MnO-MnS water splitting cycle is completed with the addition of an ideal H_2/O_2 fuel cell with 100% work efficiency (assumption). This ideal fuel cell is added to the cycle to extract the maximum work from the net H_2 produced by this process. The rate of theoretical work performed by the ideal fuel cell is calculate by equation (11) and observed to be equal to 237.05 kW. Likewise, the rate of heat energy released by the ideal fuel cell is determined as equal to 48.56 kW [according to equation (12)].

$$W_{FC-Ideal} = -\dot{n}\ \Delta G|_{H_2(g)+0.5O_2(g)@298K \rightarrow H_2O(l)@298K} \quad (11)$$

$$Q_{FC-Ideal} = -(298) \times \dot{n}\ \Delta S|_{H_2(g)+0.5O_2(g)@298K \rightarrow H_2O(l)@298K} \quad (12)$$

Cycle efficiency of the solar thermochemical MnO-MnS water splitting cycle is defined as the ratio of theoretical work performed by the ideal fuel cell to the solar energy input:

$$\eta_{cycle} = \frac{W_{FC-Ideal}}{Q_{solar}} \quad (13)$$

Furthermore, the solar-to-fuel energy conversion efficiency of the solar thermochemical MnO-MnS water splitting process is defined as the ratio of higher heating value (HHV) of the H_2 produced to the solar energy input:

$$\eta_{solar-to-fuel} = \frac{HHV_{H_2}}{Q_{solar}} \quad (14)$$

Where,

$$HHV_{H_2} = -\dot{n}\ \Delta H|_{H_2(g)+0.5O_2(g)@298K \rightarrow H_2O(l)@298K} \quad (15)$$

As per the HSC computation thermodynamic modeling, η_{cycle} and $\eta_{solar-to-fuel}$ equal to 19.28% and 23.24%, respectively.

4. Conclusion

Two-step solar thermochemical cycles for H_2 production via MnO-MnS based H_2O splitting reactions have been thermodynamically examined. Thermodynamic equilibrium calculations performed at 1 atm pressure and it indicates that the thermal reduction of $MnSO_4$ into MnO, SO_2 and O_2 can be achieved at 1355 K when 30 mol/sec of inert Ar flow rate is used.

The second law thermodynamic analysis indicate that indicate $\eta_{absorption}$ = 96.18%, Q_{solar}= 1229.80 kW, $Q_{reactor-net}$ = 1182.78 kW, $Q_{re-radiation}$ = 47.02 kW, $Q_{cooler-1}$= 739.04 kW, $Q_{MnO-oxd}$ = 97.38 kW, and η_{cycle}= 19.28%, and $\eta_{solar-to-fuel}$= 23.24%, respectively.

5. Acknowledgments

This publication was made possible by the NPRP grant (NPRP8-370-2-154) from the Qatar National Research Fund (a member of Qatar Foundation). The statements made herein are solely the responsibility of author(s).

References

[1] Agrafiotis C, Pagkoura C, Lorentzou S, Kostoglou M, Konstandopoulos AG. Hydrogen production in solar reactors. Catalysis Today 2007;127:265–77.

[2] Steinfeld A, Sanders S, Palumbo R. Design aspects of solar thermochemical engineering – A case study: two-step water-splitting cycle using the Fe3O4/FeO redox system. Solar Energy 1999;43:43–53.

[3] Kodama T, Nakamuro Y, Mizuno T. A two-step thermochemical water spitting by iron-oxide on stabilized zirconia. Journal of Solar Energy Engineering 2006;128:3–7.

[4] Gokon N, Murayama H, Umeda J, Hatamachi T, Kodama T. Monoclinic zirconia-supported Fe3O4 for the two-step water-splitting thermochemical cycle at high thermal reduction temperatures of 1400–1600°C. International Journal of Hydrogen Energy 2009;34:1208–17.

[5] Ishihara H, Kaneko H, Hasegawa N, Tamaura Y. Two-step water-splitting at 1273–1623K using yttria-stabilized zirconia-iron oxide solid solution via co-precipitation and solid-state reaction. Energy 2008;33:1788–93.

[6] Miller J, Allendorf M, Diver R, Evans L, Seigel N, Stuecker J. Metal oxide composites and structures for ultra-high temperature solar thermochemical cycles. Journal of Material Science 2008;43:4714–28.

[7] Steinfeld A. Solar hydrogen production via a two-step water-splitting thermochemical cycle based on Zn/ZnO redox reactions. International Journal of Hydrogen Energy 2002;37:611–19.

[8] Abanades S. Thermogravimetric analysis of CO2 and H2O reduction from solar nanosized Zn powder for thermochemical fuel production. Industrial & Engineering Chemistry Research 2012;51:741–50.

[9] Ernst F, Steinfeld A, Pratsinis S. Hydrolysis rate of submicron Zn particles for solar H2 synthesis. International Journal of Hydrogen Energy 2009;34:1166–75.

[10] Vishnevetsky I, Epstein M. Production of hydrogen from solar zinc in steam atmosphere. International Journal of Hydrogen Energy 2007;32:2791–2802.

[11] Abanades S, Charvin P, Lemont F, Flamant G. Novel two-step SnO2/SnO water-splitting cycle for solar thermochemical production of hydrogen. International Journal of Hydrogen Energy 2008;33:6021–30.

[12] Charvin P, Abanades S, Lemont F, Flamant G. Experimental study of SnO2/SnO/Sn thermochemical systems for solar production of hydrogen. AIChE Journal 2008;54:2759–67.

[13] Vishnevetsky I, Epstein M. Tin as a possible candidate for solar thermochemical redox process for hydrogen production. Journal of Solar Energy Engineering 2009;131:021007-1–021007-8.

[14] Scheffe JR, Li J, Weimer AW, A spinel ferrite/hercynite water-splitting redox cycle. International Journal of Hydrogen Energy 2010;35:3333–3340.

[15] Bhosale RR, Shende RV, Puszynski JA, H2 generation from thermochemical water-splitting using sol-gel derived Ni-ferrite. Journal of Energy and Power Engineering 2010;4:27–38.

[16] Neises M, Roeb M, Schmucker M, Sattler C, Pitz-Paal R. Kinetic investigations of the hydrogen production step of a thermochemical cycle using mixed iron oxides coated on ceramic substrates. International Journal of Energy Research 2010;34:651–61.

[17] Bhosale RR, Shende RV, Puszynski JA, H2 generation from thermochemical water-splitting using sol-gel synthesized Zn/Sn/Mn-doped Ni-ferrite. International Review of Chemical Engineering 2010;2:852–62.

[18] Lorentzou S, Zygogianni A, Tousimi K, Agrafiotis C, Konstandopoulos AG. Advanced synthesis of nanostructured material for environmental applications. Journal of Alloys and Compounds 2009;483:302–05.

[19] Bhosale RR, Khadka RP, Shende RV, Puszynski JA. H2 generation from two-step thermochemical water-splitting reaction using sol-gel derived SnxFeyOz. Journal of Renewable and Sustainable Energy 2011;3:063104-1–063104-16.

[20] Varsano F, Padella F, Alvani C, Bellusci M, Barbera A. Chemical aspects of the water-splitting thermochemical cycle based on sodium manganese ferrite. International Journal of Hydrogen Energy 2012;37:11595–601.

[21] Bhosale RR, Shende RV, Puszynski JA. Thermochemical water-splitting for H2 generation using sol-gel derived Mn-ferite in a packed bed reactor. International Journal of Hydrogen Energy 2012;37:2924–34.

[22] Scheffe JR, Steinfeld A. Thermodynamic analysis of cerium-based oxides for solar thermochemical fuel production. Energy & Fuels 2012;26:1928–36.

[23] Abanades S, Flamant G. Thermochemical hydrogen production from a two-step solar-driven water-splitting cycle based on cerium oxides. Solar Energy 2006;80:1611–23.

[24] Gal A, Abanades S. Dopant incorporation in ceria for enhanced water-splitting activity during solar thermochemical hydrogen generation. Journal of Physical Chemistry C 2012;116:13516–523.

[25] Lin X, Zhang Y, Wang Z, Wang R, Zhou J, Cen K. Hydrogen production by HI decomposition over nickel–ceria–zirconia catalysts via the sulfur–iodine thermochemical water-splitting cycle. Energy Conversion and Management 2014;84:664–70.

[26] Murphy IV J, O'Connell J. Process simulations of HI decomposition via reactive distillation in the sulphur-iodine cycle for hydrogen manufacture. International Journal of Hydrogen Energy 2012;37:4002–11.

[27] Dehghani S, Sayyaadi H. Energy and exergetic evaluations of Bunsen section of the sulfur–iodine thermochemical hydrogen production plant. International Journal of Hydrogen Energy 2013;38:9074–84.

[28] Lin S, Flaherty R. Design studies of the sulphur trioxide decomposition reactor for the sulphur cycle hydrogen production process. International Journal of Hydrogen Energy 1983;8:589–96.

[29] Hinkley J, O'Brein J, Fell C, Lindquist S. Prospects for solar only operation of the hybrid sulphur cycle for hydrogen production. International Journal of Hydrogen Energy 2011;8:589–96.

[30] Bhosale RR, Dardor D, Gharbia S, Folady J, Jilani M, Kumar A, van den Broke LJP, Lin F, Alexneit I, Thermochemical Conversion of CO_2 into Solar Fuels Using Ferrite Nanomaterials. In: Proceedings of 4th International Gas Processing Symposium 2015;141-185.

[31] Bhosale RR, Yousefi S, Folady J, Jilani M, Dardor D, Gharbia S, Kumar A, van den Broke LJP, Alexneit I, Shende R, Thermodynamic Analysis of Solar Fuel Production via Thermochemical H_2O and/or CO_2 Splitting Using Tin Oxide Based Redox Reactions. In: Proceedings of 4th International Gas Processing Symposium 2015;39 – 48.

[32] Bhosale RR, Jilani M, Dardor D, Gharbia S, Folady J, Kumar A, van den Broke LJP, Lin F, Alexneit I, Solar Fuel Production via Non-Stoichiometric $Ce_xZr_yHf_zO_{2-\delta}$ Based Two-Step Thermochemical Redox Cycle. In: Proceedings of 4th International Gas Processing Symposium 2015;117 – 124.

[33] Kumar A, Ashok A, Kuti YO, Bhosale RR, van den Broke LJP, Single Step Synthesis of Transition Metal Nanoparticles in Aqueous Phase for Catalytic Applications. In: Proceedings of 4th International Gas Processing Symposium 2015;69 – 80.

[35] Dowass NA, Alazewi NS, Shurair MS, Bhosale RR, van den Broke LJP, Kumar A, Modeling of reaction front movement in combustion synthesis for catalyst preparation. In: Proceedings of 4th International Gas Processing Symposium 2015;391 – 399.

[37] Dardor D, Bhosale RR, Gharbia S, Kumar A, AlMomani F, Solar Carbon Production via Thermochemical ZnO/Zn Carbon Dioxide Splitting Cycle. Journal of Emerging Trends in Engineering & Applied Sciences 2015;6:129 – 135.

[38] Bhosale RR, Kumar A, AlMomani F, Alexneit I, Propylene Oxide Assisted Sol-Gel Synthesis of Zinc Ferrite Nanoparticles for Solar Fuel Production. Ceramics International 2015 (in press).

[39] Dardor D, Bhosale RR, Gharbia S, Kumar A, AlMomani F, Solar Thermochemical Conversion of CO_2 into C via SnO_2/SnO Redox Cycle: Thermodynamic Study. International Journal of Engineering Research & Applications 2015;5:134 – 140.

[40] Bhosale RR, Kumar A, AlMomani F, Gharbia S, Dardor D, Ali M, Folady J, Yousefi S, Jilani M, Solar Thermochemical Reforming of Natural Gas via Metal Oxide Based Redox Reactions. In: Proceedings of 2015 AIChE Spring Meeting (ISBN: 978-0-8169-1089-2) 2015.

[41] Bhosale RR, Kumar A, AlMomani F, Gharbia S, Dardor D, Ali M, Folady J, Yousefi S, Jilani M, Al-Fakih M, A Novel Two-Step Metal Oxide – Metal Sulfate Water Splitting Cycle for the Production of Solar Hydrogen. In: Proceedings of 2015 AIChE Spring Meeting (ISBN: 978-0-8169-1089-2) 2015.

[42] Bhosale RR, Kumar A, AlMomani F, Dardor D, Gharbia S, Ali M, Folady J, Yousefi S, Jilani M, Abuarja D, Angre P, Solar Hydrogen Production via Thermochemical Metal Oxide – Metal Sulfate Water Splitting Cycle. In: Proceedings of 2015 AIChE Spring Meeting (ISBN: 978-0-8169-1089-2) 2015.

[43] Dardor D, Bhosale RR, Kumar A, AlMomani F, Gharbia S, Ali M, Folady J, Yousefi S, Jilani M, Computational Thermodynamic Analysis of Solar Fuel Production via Metal Oxide Based H_2O and CO_2 Splitting Thermochemical Cycles. In: Proceedings of 2015 AIChE Spring Meeting (ISBN: 978-0-8169-1089-2) 2015.

[44] Gharbia S, Bhosale RR, Kumar A, AlMomani F, Dardor D, Ali M, Folady J, Yousefi S, Jilani M, Al-Fakih M, Angre P, Thermochemical Conversion of H_2O and CO_2 into Solar Fuels via Metal Oxide Based Redox Reaction. In: Proceedings of 2015 AIChE Spring Meeting (ISBN: 978-0-8169-1089-2) 2015.

[45] Ashok A, Kumar A, Bhosale RR, Saleh MAH, van den Broeke LJP, Cellulose Assisted Combustion Synthesis of Porous Cu-Ni Nanopowders. RSC Advances 2015;5:28703 – 28712.

[46] Kumar A, Cross A, Manukyan K, Bhosale RR, van den Broeke LJP, Miller JT, Mukasyan AS, Wolf EE, Combustion Synthesis of Copper-Nickel Catalysts for Hydrogen Production from Ethanol. Chemical Engineering Journal 2015;278:46 – 54.

[47] Bhosale RR, Kumar A, van den Broeke LJP, Dardor D, Gharbia S, Jilani M, Folady J, Al-Fakih M, Tarsad M, Solar Hydrogen Production via Thermochemical Iron Oxide – Iron Sulfate Water Splitting Cycle. International Journal of Hydrogen Energy 2015;40:1639 – 1650.

[48] Bhosale RR, Alxneit I, Kumar A, van den Broeke LJP, Folady J, Dardor D, Gharbia S, Jilani M, Al-Fakih M, Elshrbajy NN, Tarsad M, Almkhyari FA, Solar Thermochemical Conversion of CO_2 into Synthetic Fuels via Ferrite Based Redox Reactions. In: Proceedings of 2014 AIChE Annual Meeting (ISBN: 978-0-8169-1086-1) 2014.

[49] Bhosale RR, Alxneit I, Kumar A, van den Broeke LJP, Jilani M, Folady J, Dardor D, Gharbia S, Dejan R, Al-Emadi N, Al-Massih F, Al-Yazedi S, Thermochemical Dissociation of CO_2 into Renewable Fuels via $Ce_xZr_yHf_zO_2$ Based Redox Reactions Using Concentrated Solar Energy. In: Proceedings of 2014 AIChE Annual Meeting (ISBN: 978-0-8169-1086-1) 2014.

[50] Kumar A, Ashok A, Kuty YO, Bhosale van den Broeke LJP, Combustion Synthesis of Copper-Nickel Catalysts for Hydrogen Production from Ethanol. In: Proceedings of 2014 AIChE Annual Meeting (ISBN: 978-0-8169-1086-1) 2014

[51] Bhosale RR, Alxneit I, van den Broeke LJP, Kumar A, Jilani M, Gharbia S, Folady J, Dardor D, Sol-Gel Synthesis of Nanocrystalline Ni-Ferrite and Co-Ferrite Redox Materials for Thermochemical Production of Solar Fuels. Materials Research Society Symposium Proceedings 2014;1675:mrss14-1675-rr06-10.

[52] Bhosale RR, Shende RV, Puszynski JA, Sol-Gel Derived $NiFe_2O_4$ Modified With ZrO_2 for Hydrogen Generation from Solar Thermochemical Water-Splitting Reaction. Materials Research Society Symposium Proceedings 2012;1387:mrsf11-1387-e09-07.

[53] Pasala X, Bhosale RR, Amar VS, Puszynski JA, Shende RV, Thermally Stabilized Ferrite Nanoparticles for Hydrogen Generation from Thermochemical Water-Splitting Reaction. In: Proceedings of 2012 AIChE Annual Meeting (ISBN 978-0-8169-1073-1) 2012.

[54] Bhosale RR, Pasala X, Yelakanti S, Puszynski JA, Shende RV, Solar Thermochemical Water-Splitting for H_2 Generation Using Sol-Gel Derived Ferrite Nanomaterials. In: Proceedings of NSTI – Nanotech Conference & Expo., (ISBN: 978-1-4665-6276-9) 2012;3:555.

[55] Bhosale RR, Shende RV, Puszynski JA, Synthesis of Nano-Sized Ferrite Materials for H2 Generation from High Temperature Water-Splitting Reaction. In: Proceedings of NSTI – Nanotech Conference & Expo., (ISBN: 978-1-4398-7142-3) 2011;1:237.

[56] Bhosale RR, Shende RV, Puszynski JA, H_2 Generation from Thermochemical Water-Splitting Using Sol-Gel Synthesized Ferrites ($M_xFe_yO_z$, M = Ni, Zn, Sn, Co, Mn, Ce). In: Proceedings of 2010 AIChE Annual Meeting (ISBN: 978-0-8169-1065-6) 2010.

[57] Bhosale RR, Shende RV, Puszynski JA, Sol-Gel Synthesis of Ferrite Foam Materials for H_2 Generation from Water-Splitting Reaction. In: Proceedings of NSTI – Nanotech Conference & Expo., (ISBN: 978-1-4398-3401-5) 2010;1:368.

[58] Bhosale RR, Shende RV, Opoku MK, Puszynski JA, H_2 Generation from Water-Splitting Using Novel Ferrite Foam Materials. In: Proceedings of 2009 AIChE Annual Meeting (ISBN: 978-0-8169-1058-6) 2009.

[59] Shende RV, Puszynski JA, Opoku MK, Bhosale RR, Synthesis of Novel Ferrite Foam Materials for Water-Splitting Application. In: Proceedings of NSTI – Nanotech Conference & Expo., (ISBN: 978-1-4398-1782-7) 2009;1:201.

455103 Phase Equilibria Prediction for Systems Containing Lipids

Abstract:

In recent years, utilization of fats and oils has started to shift towards the bio-industry making them one of the most important renewable materials for the future chemical industry, and leading to oils and fats industry expansion along with a scale reorientation from local to large-scale industry [1]. Such developments have led to new challenges regarding the design and development of better performing processes and products. Despite the advances in property modelling and process design techniques available via different computer-aided methods and tools for the chemical and petrochemical industries, the oleochemical industry is not able to exploit this knowledge due to a lack of experimental data and property models within commercial software applications and the ability to describe accurately the phase behaviour of systems with lipids. Over the past years, new methods and models for predicting single properties and temperature dependent properties (e.g. critical properties [2], viscosity [3], heat capacity [4], heat of vaporization [4], [5], vapour pressure [5]) have been reported. Likewise, another important modelling task is phase equilibria prediction which is directly related to process synthesis, modelling and simulation. An important aspect in phase equilibria prediction is represented by quality of the data used for regression of model parameters. In previous work, Cunico et al. [5] applied several consistency tests for VLE data sets involving lipids that are available in open literature and their results show that only 3% of the analysed data sets have quality factors over 0.5 (where the quality factor varies between 0 – minimum, and 1 – maximum) [5].

In this work, our available extended CAPEC Lipids Database and CAPEC Lipids Mixtures Database is used for revising the Original UNIFAC model group contribution parameters for lipids by proposing new values, aimed to offer a better prediction of phase equilibria calculation (vapour-liquid equilibrium VLE, solid-liquid equilibrium, SLE). The regression of the new parameters is done using carefully selected VLE data sets, screened out for possible erroneous data. VLE data selection is performed based on the quality factor given

by the different consistency tests available in ThermoData Engine (TDE) from NIST. More than 60 VLE data sets consisting of over 600 data points, available in CAPEC LIPIDS Mixture Database, are used for the regression of the 54 binary interaction parameters corresponding to 10 groups for Original UNIFAC model. Note that only 10 groups are needed to represent all the lipids data sets. However, to allow a better performance of the model for this type of systems, two new groups were introduced: one group is describing the behaviour of hydroxyl within mono and diglycerides (OHacyl) and another one is describing the glycerol molecule (GLY). The parameters are tested and evaluated on VLE and SLE data and by using a cross validation method. Compared to original UNIFAC, the performance of the new parameters for the lipids systems present a substantial improvement in phase equilibria predictions.

Keywords: Lipids, phase equilibria, original UNIFAC

Refrences:

1. De Greyt W., "Edible Oil Refining: Current and Future Technologies" in Edible Oil Processing, 2nd ed., Hamm W., Hamilton R. J., Calliauw G. editors. Oxford: John Wiley & Sons, 2013

2. Marrero J., Gani R., "Group-contribution based estimation of pure component properties", Fluid Phase Equilibria, 2001, 183–184, 183–208.

3. Ceriani R., Gonçalves C. B., Coutinho J. A. P., "Prediction of viscosities of fatty compounds and biodiesel by group contribution", Energy and Fuels, 2009, 25(8), 3712–3717.

4. Ceriani R., Gani R., Meirelles J. A., "Prediction of heat capacities and heats of vaporization of organic liquids by group contribution methods", Fluid Phase Equilibria, 2009, 283(1-2), 49-55.

5. Cunico L., Ceriani R., Sarup B., O'Connell J. P., Gani R., "Data, analysis and modelling of physical properties for process design of systems involving lipids", Fluid Phase Equilibria, 2014, 362, 318-327.

455666 Viscosity and Density of (CO_2 + Synthetic Crude Oil) Mixtures at Temperatures from (298 to 423) K and at Pressures up to 100 Mpa

Abstract:

Knowledge of the viscosity and density of CO_2-hydrocarbon mixtures at reservoir conditions is very important for designing and operating enhanced oil recovery processes using CO_2, and also for large-scale carbon storage in depleted hydrocarbon reservoirs. However, very few data have been reported in the literature for the viscosity and density of compositionally-characterized multi-component CO_2-hydrocarbon mixtures. To help address this deficit, we have measured the viscosity and density of a synthetic crude oil mixture with and without dissolved CO_2. The synthetic dead oil was identical in composition to that studied by Al Ghafri et al.[1], who matched the chemical and physical properties of the synthetic mixture to those of a Qatari bottom-hole crude-oil sample. The dead synthetic crude contained 17 components including linear and branched alkanes, naphthenes and aromatics. Live oils with different gas-to-oil ratios were obtained from this dead oil by adding (0.813 CH_4 + 0.126 C_2H_6 + 0.061 C_3H_8) as solution gas. The measurements of viscosity and density were made in the single-phase compressed liquid region at temperatures between (298 and 423) K at pressures up to 100 MPa. The CO_2 mole fractions of the mixtures studied were x_1 = 0.1, 0.2, 0.4, 0.6 and 0.8. The viscosity was measured with a vibrating-wire viscometer while the density was measured by means of a vibrating U-tube densimeter; the viscometer was calibrated with toluene and nitrogen, while the densimeter was calibrated with toluene, water and vacuum following a new calibration method presented by May et al.[2]. Measurements of the viscosity and density of the mixtures are associated with relative uncertainties of 0.2 % for density and 2 % for viscosity. The results for both properties have been correlated as functions of temperature, pressure and the mole fraction of dissolved CO_2.

Acknowledgment

We gratefully acknowledge the funding of QCCSRC provided jointly by Qatar Petroleum, Shell, and the Qatar Science and Technology Park, and their permission to publish this research.

References

1. S. Z. Al Ghafri, G. C. Maitland, and J. P. M. Trusler, Fluid Phase Equilibria 365, 2014, 20-40.
2. E. F. May, W. J. Tay, M. Nania, A. Aleji, S. Z. Al Ghafri, and J. P. M. Trusler, Review of Scientific Instruments 85, 2014, 095111.

456315 A Fundamental Understanding of Cation Transport in Mixed Metal Ferrite Spinels

Abstract:

Magnetite (Fe_3O_4) and its alloys are used in wide variety of materials and chemical engineering applications and reactions. In several of these applications the performance of the ferrite spinels are controlled by the rates of cationic diffusion through the lattice. In this work we develop a fundamental understanding of the diffusion process using *ab initio* calculations based on density functional theory. Specifically, we investigate the diffusion of Fe, Co and Ni cations through three difference ferrite spinels (Fe_3O_4, $CoFe_2O_4$, and $NiFe_2O_4$). The cations diffuse via a vacancy mediated octahedral-octahedral hopping mechanism where the cations hop through a trigonal planer transition state to a tetrahedral meta-stable state. The individual hopping rate depends on both the diffusing cation and the material. The cations have relative activation energies of Co â‰ˆ< Fe < Ni. This ordering is controlled by the change in cation coordination along the reaction path which leads to changes in electronic ordering of the diffusant cationsâ€™ t_{2g} and e_g orbitals due to crystal field effects. Diffusing cations with higher occupancy of the t_{2g} and e_g orbitals have higher activation energies because more electrons must be promoted to higher energy orbitals during hopping events. An inverse relationship exists between the lattice constant of a material and the hopping activation energy, leading to relative activation energies for the various spinels of $Fe_3O_4 < CoFe_2O_4 < NiFe_2O_4$. This behavior is attributed to increased steric interactions between the diffusing cation and the O anions in the materials with smaller lattices.

456866 Designing an Extraterrestrial Submarine for Titan. III. Estimating Titan Sea Mixture Properties

Abstract:

An extraterrestrial submarine is currently being designed to explore the depths of the hydrocarbon rich seas on the moon Titan under the NASA Innovative Advanced Concepts Phase 2 study. The submarine must be designed to cycle and operate for one year in the cold, cryogenic hydrocarbon seas of Titan. Critical to the design and operation of the submarine is knowledge of the amount of dissolved nitrogen gas within the cryogenic liquid ethane and methane seas. Solubility is defined as a property of a solute to dissolve in a particular solvent. Based on available binary and ternary solubility data available, gaseous nitrogen (GN_2) solubility in the Titan seas is relatively high. Dissolved GN_2 that comes out of solution will hinder operation of the submarine during quiescent, horizontal cruising, and vertical ascent and descent stages of the mission. Therefore, robust analytical solubility models for predicting the amount of dissolved nitrogen as a function of temperature, pressure (depth), and mole fraction of ethane and methane (location within the seas) are required to mature the design of the submarine. Aside from needing precise knowledge of the amount of dissolved nitrogen, accurate estimations of Titan sea mixture properties are required. Mixture properties can be readily computed with fluid property software such as REFPROP, but the software requires as initial conditions the mole fractions of all mixture components, which is determined through solubility modeling.

This presentation presents an analytical solubility model fit to over 2,000 methane/ethane/nitrogen vapor/liquid equilibrium (VLE) data points. Data is first fit to the following proposed functional form:

$$x = p^*\exp((a_0 + a_1T + a_2T^2) + (b_0 + b_1T + b_2T^2)p^*),\ p^* = (P - P_0)/(1\ \text{MPa}).$$

where x is the mole fraction solubility of nitrogen in either methane or ethane in units of moles/moles, is

the vapor pressure of the pure liquid solvent at the given temperature , is the pressure, and and are the coefficients to be fitted. For $b_i = 0$, the solubility equation reduces to Henry's Law. Next, parity plots are shown to illustrate the goodness of the fit of the solubility models. Then, solubility models are used to estimate dissolved GN_2 in two mixtures which bound the Titan sea liquid ethane/liquid methane concentrations. Since the solubility model only specifies a value at the sea/atmosphere interface, two models are used to estimate the amount of dissolved GN_2 as a function of depth. One method assumes the surface solubility value is constant throughout, while the second method assumes a new solubility value at each depth, corresponding to an increased pressure. Finally, the solubility values at each depth are fed into REFPROP, using a detailed equation of state, to estimate relevant thermodynamic properties, such as viscosity, speed of sound, and dielectric constant. While a wide variety of effects could occur in the Titan seas, complexities such as freezing are for the time being ignored, and properties are estimated by assuming constant enthalpy at each depth. Mixture properties can then be used as inputs into much more complex modeling later, such as the ballast tank system and effervescence models.

This work was funded through the NASA Innovative Advanced Concepts (NIAC) Phase 2 Titan Submarine Project.

457461 Solubility of Gases in Conventional Solvents and an Ionic Liquid for Synthesis Gas and Natural Gas Sweetening from Monte Carlo Simulations

Abstract:

The aim of the present work is to study the removal of acid gases from the synthesis and natural gases using physical solvents. Synthesis gas (syngas) is made from the gasification of a fuel to produce a mixture of carbon monoxide (CO) and hydrogen (H_2). However, the syngas also contains several impurities like carbon dioxide (CO_2), nitrogen (N_2), and sulfur compounds (mainly, hydrogen sulfide, H_2S). Similar impurities are also present in raw natural gas. These sour gases should be removed to meet customer and pipeline specifications. CO_2 should be removed to avoid dry ice formation during the liquefaction of the gas and due to the low caloric value of the gas, while H_2S is highly corrosive for the pipelines. The removal of the acid gases is typically performed in an absorber-stripper configuration using either a physical, a chemical solvent or a mixture of both (hybrid) solvents. The choice of the solvent is case specific and depends on many factors like the type and concentration of the impurities, the composition, the temperature, the pressure and ultimately the product specifications. However, chemical solvents (e.g., amines) are most widely used in the natural gas industry. Although amines are very effective in reducing the acid gas concentration, especially at low acid gas partial pressures, it has several serious drawbacks. These include their high energy requirement for solvent regeneration, volatility, corrosivity and the low CO_2/H_2S selectivity. Since the natural gas and syngas sweetening operates at high pressures, physical solvents can be applied for acid gas removal. In the past decade, ionic liquids (ILs) have emerged as a promising physical solvent for acid gas removal. ILs are defined as salts with melting point lower than 100 °C and characterized by a very low vapor pressure, high thermal and chemical stability.

The solubility data of acid gases are extremely important for the optimal design of absorption columns. However, experimental data of poorly soluble gases (e.g., CH_4, N_2 and H_2) and toxic gases (e.g., CO and H_2S) are scarcely reported in the literature for a wide range of temperatures and pressures. Moreover, experimental data for the solubility of gas mixtures in solvents, which requires an increased experimental effort, are even more scarce. Here we use force field based Monte Carlo simulations to compute the solubility of the pure gases CH_4, C_2H_6, H_2, CO, N_2, H_2S, and CO_2 in several conventional solvents (namely Selexol, Purisol, Rectisol and Fluor solvent) and an ionic liquid [bmim][Tf_2N]. The Henryâ€™s coefficients of the gases in the solvents are calculated from the solubility data. Both the calculated solubilities and Henryâ€™s coefficients of the gases from Monte Carlo simulations are in good agreement with available experimental data. The results show that molecular simulation can be a powerful tool, in the absence of experimental data, to obtain gas solubilities in complex solvents. Furthermore, the ideal selectivity of relevant gases is calculated from the Henryâ€™s coefficients. The computed selectivities show the potential of the investigated solvents in removing the acid gases, namely CO_2 and H_2S. In conclusion, Monte Carlo simulations can be employed to obtain the required solubility data for designing natural gas or synthesis gas sweetening processes when no experimental data is available.

We have submitted a paper to the special issue of Fluid Phase Equilibria honoring Theo de Loos.

458407 Prediction of Flash Points: A Review and Practice on Ethanol-Blends

Abstract:

Flash point is one of the most important characteristics of flammable liquids such as fuels. It is needed to determine the conditions for safe handling and storage of fuels. According to the definition of American Society for Testing and Materials (ASTM), flash point is the lowest temperature at which the vapour over the liquid under testing can be ignited by an ignition source. Flash points can be measured on a tester designed based either an open cup or a closed cup method. However, the availability of flash point data are limited and experimental methods to measure flash point are time consuming. For estimations of flash points, modelling based prediction method could serve as an alternative solution.

Many models have been developed for predicting of both pure compounds and mixtures. It has been reported that flash points of pure compounds can be accurately predicted by extrapolating from their physical properties, such as normal boiling points, stoichiometric amount of oxygen required for combustion, vapour pressure and molecular weight etc. In this presentation, the implementation of various models such as Quantitative Structure-Property Relationship and Le Chatelierâ€™s equation based predictive models will be discussed. Additionally, a simulation based on Le Chatelierâ€™s equation to predict flash points of two ethanol-blends (ethanol-biodiesel, ethanol-octane) will be shared. This exercise is in line with the use of bio-ethanol as a renewable fuel with potential to mitigate the consumption of fossil fuels and achieve sustainable development.

458856 Phase Equilibrium of Major Components of Essential Oils in Liquid and Supercritical Carbon Dioxide

Abstract:

The solubility of terpinen-4-ol, alpha-cedrene and alpha-pinene in both liquid and supercritical carbon dioxide were determined using the dynamic solubility method. Experimental conditions were 25°C, 40°C, 50°C and 60°C and pressures corresponding to pure CO_2 densities of 0.2067, 0.3149, 0.4222, 0.5848 and 0.6971 g/mL. Solubility of each component increased with density along each isotherm. Experimental results were correlated by the Peng-Robinson equation of state using conventional mixing rules with two interaction parameters.

456140 Prediction of Liquid Kinematic Viscosities Using the Eos + G^E Mixing Rule with Eyring-Activity Coefficient Models

Abstract:

Research Interests:The prediction of kinematic viscosities are useful for discussing the separation processes.

Teaching Interests:Eyring-activity coefficient models are useful for predicting kinematic viscosities.

Liquid kinematic viscosity is one of thermophysical properties for deigning and/or optimizing chemical processes. A number of predictive methods for determining kinematic viscosities of liquid mixtures have been proposed in the textbook [1]. In the predictive methods, a model based on the Eyringâ€™s absolute rate theory could be useful from a practical point of view [2, 3].

In this paper, we describe the predicted values of liquid kinematic viscosities for ternary systems using binary kinematic viscosity data. Then we describe the predicted values of liquid kinematic viscosities using the cubic equations of state (SRK and Peng-Robinson) + G^Emixing rules (MHV1[4] and Wang-Sandler) with Eyring-activity coefficient (Wilson, NRTL, and Redlich-Kister) and the ASOG-VISCO models [2].

References

1. B.E. Poling, J.M. Prausnitz, and J.P.Oâ€™Connell, The Properties of Gases and Liquids, 5th ed., McGraw-Hill, New York (2001)

2. K. Tochigi, K. Yoshino, and V.K. Rattan, *International Journal of Thermophysics*, Vol.26, No.2 (2005), pp. 413-419.

3. K. Tochigi, T. Okamura, and V.K. Rattan, *Fluid Phase Equilibria*, Vol. 257 (2007), pp. 228-232.

4. K. Kurihara, F. Kato, H. Matsuda, and K. Tochigi, *Netsu Bussei*, Vol.29, No. 4 (2015) pp. 166-172.

459560 Development of a Thermodynamic Model for Confined Fluids

Abstract:

Fluids confined in porous media play a significant role in many engineering applications. Modeling fluids in oil reservoirs, adsorption based separations, and heterogeneous catalysis require the accurate prediction of thermodynamic properties at a wide range of conditions. For confined fluids, this task involves accounting for the fluid-solid interactions induced by proximity to a solid wall. Most engineering models are only able to predict average properties of the system and fail to give information related to the heterogeneity of the confined fluid. Non-uniform behavior in such systems exists due to many effects, including the variation of the fluid properties with distance from the wall, and the presence of different sizes or geometries within the same porous space. Other than these microscopic effects, confined fluids could be subject to effects at a larger scale. For instance, gravity creates a molar distribution of the components in deep oil reservoirs along the depth of the formation. As a result, including the gravitational and confinement effects simultaneously will give a better representation of the properties of the components in these systems.

The objective of this work is to establish a general framework for determining equilibrium properties of fluids in confined media. The thermodynamic model utilizes an equation of state (EoS) to describe both the bulk and confined phases without introducing any additional parameters. On the other hand, the confinement effect could be represented for any solid, regardless of its shape or size, provided that an appropriate model to describe the solid-fluid interactions is available. Another goal is to find equilibrium conditions when several effects are acting on the system, whether they arise from the nature of the confined media or from other sources.

In this model, the variation of properties throughout the system takes place across regions that are defined depending on the effects present. Regions where confinement effects are important are further discretized into layers to capture local distribution within the confinement. For all elements, the volume is specified together with the temperature, and

the total amount of each component. Thus, minimizing the Helmholtz energy determines the number of moles in each element; subsequently other properties are obtained. The Helmholtz energy accounts for internal interactions through EoS; in this work the volume translated Peng Robinson EoS is used. Additionally, the Helmholtz energy function includes an external contribution represented by an adsorption potential to account for fluid-solid interactions. The Steele 10-4-3 potential is used to investigate the confinement of light hydrocarbons in activated carbon. Specifically, the potential is applied to obtain local density and compositional profiles inside the activated carbon silt pores as well as to compute adsorption isotherms for a heterogeneous solid (different pore sizes). The prediction of local behavior is found to be comparable to classical Density Functional Theory (DFT) calculations. When computing the adsorbed amount, good agreement between experimental data and calculations is obtained. Another adsorption potential employed in this work is the Dubinin-Radushkevich-Astakhov (DRA) potential, which is utilized to predict confinement of binary and ternary mixtures of methane, nitrogen, carbon dioxide on activated carbon at various pressures and compositions. The computed isotherms illustrate the adsorption behavior of the studied systems including the preferential adsorption of carbon dioxide over the other two components. These DRA results are found to agree reasonably well with experimental data. Moreover, the ability of the framework to model systems where confinement as well as other effects are present is demonstrated through predicting the molar distribution of methane and carbon dioxide in a porous reservoir where gravitational effects are accounted for.

460602 Understanding Adsorption on the Anatase TiO_2 (101) Surface

Abstract:

Titanium dioxide is one of the most well studied metal oxides due to its wide range of important applications. Of the three structural phases of TiO_2 (rutile, anatase, and brookite), the rutile phase, in particular the (110) surface, has been the most well studied due to its stability and wide availability for study as a prototypical metal oxide; however, the anatase form, which is often the most photocatalytically active form and is adopted by TiO_2 nanoparticles due to its low surface energy, is less well studied. The anatase (101) surface, which comprises the majority of industrial catalysis, is the focus of this particular study. Of particular interest for understanding catalytic behavior of this surface is the study of its fundamental adsorption properties. Adsorption of molecules on the TiO_2 surface often does not readily occur and limits the overall reaction rates and usefulness of catalytic TiO_2. For example, neutral O_2 does not adsorb on (101) anatase TiO_2, limiting the overall rate of photocatalytic organic pollutant degradation, while H_2O forms well-ordered monolayers on this surface. A number of theoretical and experimental studies have been conducted to understand the adsorption of single species on anatase (101) TiO_2, but to our knowledge no single study has conducted a more expansive study of adsorption for a range of molecules to provide a direct comparison of adsorption structures and thermodynamics as well as a fundamental understanding of why some molecules adsorb while others do not. In this work we investigate the adsorption of a wide range of small molecules on the anatase (101) TiO_2 surface using density functional theory (DFT) and develop a fundamental description of adsorption on this important surface.

460981 Phase Transition of Acid Clathrate Hydrate Related to the Proton Conduction: Structural Analysis of Perchloric Acid Hydrate

Abstract:

Clathrate hydrates are nonstoichiometic crystalline compounds which are stabilized by van der Waals interaction between the hydrophobic guest molecules and the hydrogen-bonded water host framework. Unlike classical clathrate hydrates, ionic clathrate hydrates additionally show an ionic interaction between cationic or anionic guest species and counterion-doped host network. A few strong acids form clathrate structures in which anions are encaged into the protonated water cavities. There are two types of protonated host framework. One is hydrogen fluoride-contained host framework such as hexahydrates of HPF_6, $HAsF_6$, or $HSbF_6$ and another is hydrogen fluoride-free host framework such as hydrates of HBF_4 or $HClO_4$. The latter is believed to contain several vacancy sites in the hydrogen bonding lattice for protonation of the host framework. In this work, the neutron powder diffraction patterns of a $DClO_4{\cdot}5.5D_2O$ clathrate were obtained at various temperatures and refined by Rietveld analysis with the direct space method. A position change of vacancy sites was predicted at about 180 K, and this phenomenon is expected to induce a phase transition related to the proton conduction behavior of the perchlorate acid clathrate. The present work provides a better understanding of the proton dynamics of the hydrogen fluoride-free acid clathrate hydrates.

Understanding Segregation Behaviour In AuPt, NiPt And AgAu Bimetallic Nanoparticles Using Distribution Coefficients

Srikanth Divi, Abhijit Chatterjee
Department of Chemical Engineering,
Indian Institute of Technology Bombay, Mumbai, India.

16AIChE Annual Meeting, San Francisco, CA

INTRODUCTION

- Bimetallic nanoparticles (BNPs) often possess peculiar segregation behavior as the particle size, composition, shape and temperature are varied.
- Metal species distribution at the surface is believed to be one of the crucial parameters that influences the activity, selectivity and stability of the catalyst.
- Species distribution depends on a number of factors such as kinetics, thermodynamics, particle size, shape, composition, temperature and the synthesis route.

OBJECTIVE

- New theoretical model to systematically predict the species distribution within the BNP.
- Detailed understanding of thermodynamic segregation behavior in AuPt, NiPt and AuAg BNPs.
- Obtain distribution coefficients as a function of composition, size and temperature.

SIMULATION PROCEDURE

- **Interatomic potential:** Improved embedded atom method (EAM) with mixing parameter θ' (2).

$$E=\sum_i F_i(\rho_i)+\frac{1}{2}\sum_{i,j,j\neq i}\phi_{ij}(r_{ij})$$

$$\phi_{AB}(r)=\frac{1}{2}\left[\theta\phi_{AA}(r)+\frac{1}{\theta}(r)\phi_{BB}(r)\right]$$

- **Simulation method:** Metropolis Monte Carlo 2×10^6 moves
- **Types of moves:** Displacement = 60%, swap move ≈ 40 %
- **BNP shape:** Truncated octahedron
- **BNP sizes:** 711, 2735, and 6895 atoms
- **Compositions:** x_{Pt} = 0.1 to 0.9
- **Temperatures:** 400 K and 800 K
- **Regions:** bulk, {111} facets, {100} facets, edge/vertex
- **Plane analysis method:** To identify the atom phase

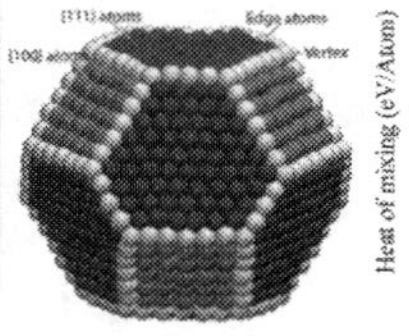

AuPt BNP of 4.3 nm

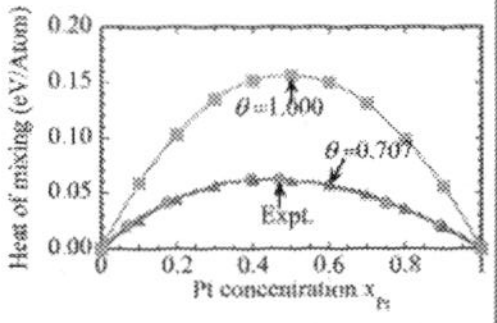

Heat of mixing for bulk AuPt

- For AuPt system, mixing parameter 'θ'' = 0.707 by fitting with heat of mixing experimental data
- Au_xPt_{1-x} BNP has sites in region α, α=Bu, 111, 100 or EV
- No. of atoms in region is denoted by N_α.
- To predict Pt atom x^α_{Pt} in all regions.

$$N^\alpha_{Pt}=N^\alpha x^\alpha_{Pt}$$
$$N^\alpha_{Au}=N^\alpha x^\alpha_{Au}$$
$$x^\alpha_{Au}+x^\alpha_{Pt}=1$$

- Swap move: Pick an atom Pt in α and Au in region β.

DISTRIBUTION COEFFICIENT

- At equilibrium, forward and backward transition rates equal.

Forward move   Backward move

$$Pt[\alpha]+Au[\beta]\leftrightarrow Pt[\beta]+Au[\alpha]$$

- Transition rates for a pair, Pt[α]-Au[β] and Pt[β]-Au[α]:

Rates   Transition probability   Metropolis criterion

$$R_{forward}=2N^\alpha_{Pt}N^\beta_{Au}/N_{atoms}(N_{atoms}-1)*\langle\min(1,\exp(-\Delta U_{\alpha-\beta}/k_BT))\rangle$$

$$R_{backward}=2N^\beta_{Pt}N^\alpha_{Au}/N_{atoms}(N_{atoms}-1)*\langle\min(1,\exp(-\Delta U_{\beta-\alpha}/k_BT))\rangle$$

- Distribution coefficient:

$$\Delta=\frac{x^\alpha_{Pt}/x^\alpha_{Au}}{x^\beta_{Pt}/x^\beta_{Au}}=\frac{\langle\min(1,\exp(-\Delta U_{\beta-\alpha}/k_BT))\rangle}{\langle\min(1,\exp(-\Delta U_{\alpha-\beta}/k_BT))\rangle}$$

- Distribution coefficients valid relations:

$$\Delta_{\alpha-\beta}=\left[\Delta_{\beta-\alpha}\right]^{-1}$$

$$\Delta_{\alpha-\beta}=\Delta_{\alpha-\gamma}\Delta_{\gamma-\beta}$$

- For m different regions present in the BNP, only (m-1) Δ's are independent from the m(m-1) Δ's.

$$N^{111}_{Pt}+N^{100}_{Pt}+N^{EV}_{Pt}+N^{Bu}_{Pt}=N_{atoms}x_{Pt}$$

- The [α-β] pairs chosen are [111-Bu], [100-Bu] and [100-EV].

RESULTS

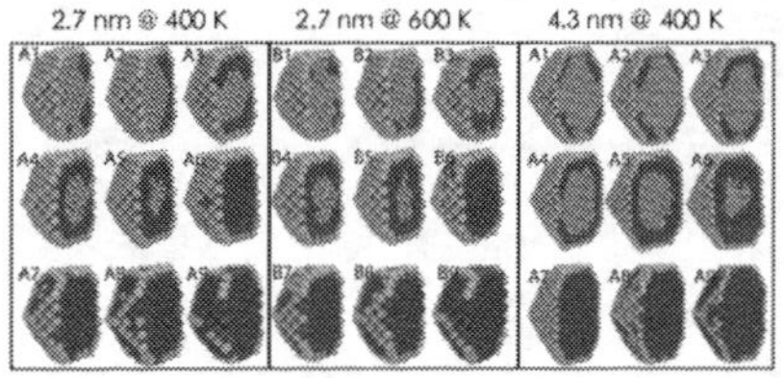

Effect of composition and temperature on 2.7 and 4.3 nm AuPt BNP cross-sectional views

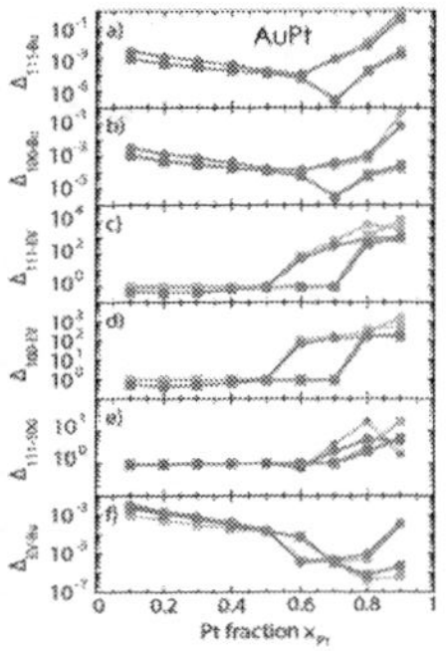

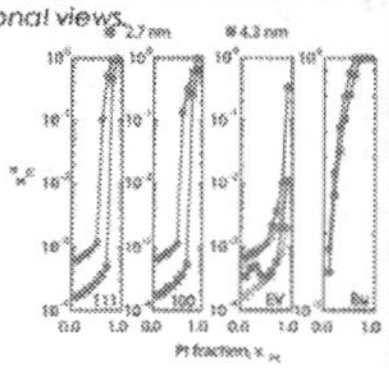

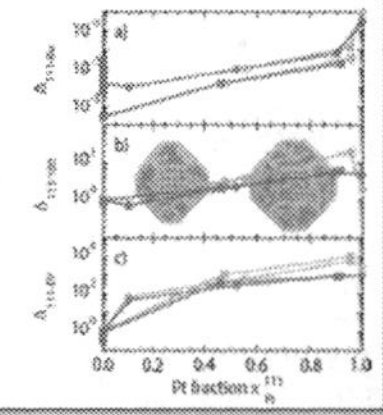

RESULTS (Contd.)

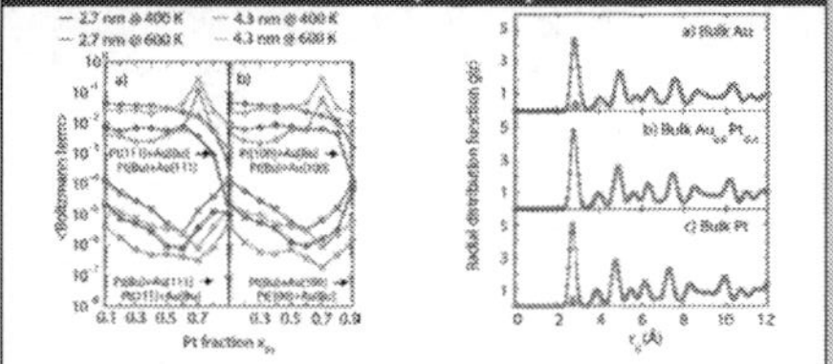

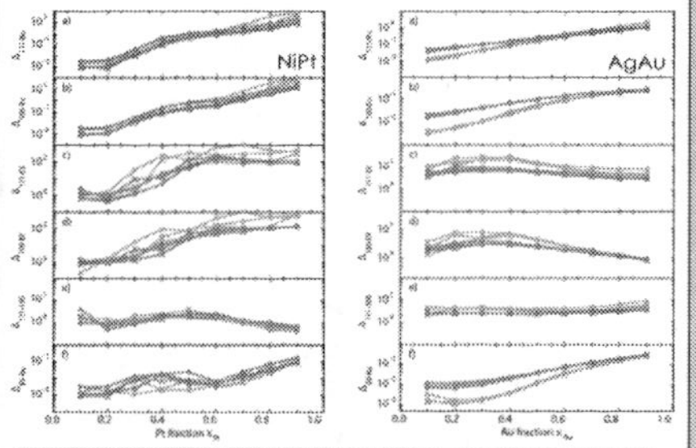

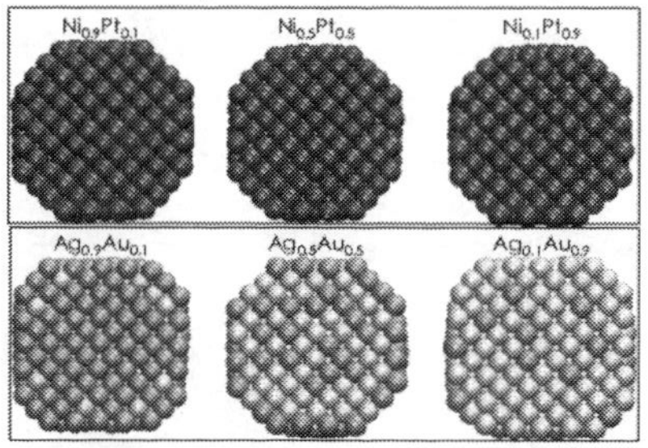

Effect of composition on 2.7 nm BNP cross-sectional views.

CONCLUSION

- A thermodynamic model based on distribution coefficient enables us to systematically predict the segregation behaviour with species distribution in BNP for conditions that were not included in the study.
- Distribution coefficient are found to be independent of size, shape, temperature and in some cases composition.

ACKNOWLEDGEMENTS

Research work funded by Science and Engineering Research Board, Department of Science and Technology Grant No. SB/S3/CE/022/2014 and Indian National Science Academy Grant No. SP/YSP/120/2015/307.

REFERENCES

1. S. Divi, A. Chatterjee, Understanding Segregation Behaviour In AuPt, NiPt And AgAu Bimetallic Nanoparticles Using Distribution Coefficients, J. Phys. Chem. C. (2016), doi:10.1021/acs.jpcc.6b08325.
2. S. Divi, et al., Improved Embedded Atom Method for Bimetallic Alloys (submitted).

http://www.che.iitb.ac.in/online/faculty/abhijit-chatterjee abhijit@che.iitb.ac.in / srikanth@che.iitb.ac.in

461876 Mathematical Modeling of Compressed CO2 Expansion through a Coanda Nozzle

Abstract:

A Coanda nozzle can be used to produce a high velocity mixture of air, gaseous CO_2, and dry ice particles from a supply of liquid CO_2. Such a process is effective (for instance) for residue-free rapid cooling or precision cleaning. A thermodynamic and computational fluid dynamics (CFD) analysis of this flow is presented for the purpose of optimizing the process parameters, which include the temperature of the liquid CO_2supply, the flow rate, and the pressure and nozzle and air configuration.

Temperature, pressure, velocity and mass fraction profile distributions have been modeled on the spray-nozzle system. Flow characteristics have been quantified for dry ice particle sizes of $1.0x10^{-3}$m, $1.0x10^{-4}$m and $1.0x10^{-5}$m. Maximum velocity and residence time as a function of inlet temperature have been simulated and quantified. The analysis suggests that as inlet CO_2temperature increases, the maximum velocity increases and the particle residence time decreases. Decreasing dry ice particle size decreases the residence time in the nozzle channel. Overall, this study provides a basis for optimizing the temperature and velocity profiles, and thus the cooling or cleaning capacity of such Coanda devices.

461952 Modification of Staverman-Guggenheim Combinatorial Entropy Expression

Abstract:

Staverman-Guggenheim[1] (S-G) expression has been widely used to describe the non-ideal entropy of mixing in state of art activity coefficient models, e.g. UNIQUAC[2]/UNIFAC[3], COSMO-SAC[4] etc. It accounts for both size and shape differences of molecules, while Flory-Huggins[5] (F-H) expression only considers size difference and results in an upper limit for entropy of mixing. However, for molecules with large size and shape differences, S-G may return unrealistic results. In this study, a modification is proposed by introducing an exponent on both volume and surface area parameters of S-G. The modified S-G is employed to predict infinite dilute activity coefficients for mixtures containing linear, branched, and cyclic alkanes. It shows significant improvement than original S-G model by comparing to experimental data. The modification is further explained from statistical mechanics point of view. Finally, the modified S-G expression is applied to activity coefficient models like UNIQUAC and NRTL-SAC[6] to show the improvement in correlation and prediction.

1. (a) Staverman, A., The entropy of high polymer solutions. Generalization of formulae. *Recueil des Travaux Chimiques des Pays-Bas* **1950,** *69* (2), 163-174; (b) Guggenheim, E. A., *Mixtures: the theory of the equilibrium properties of some simple classes of mixtures, solutions and alloys*. Clarendon Press: 1952.

2. Abrams, D. S.; Prausnitz, J. M., Statistical thermodynamics of liquid mixtures: A new expression for the excess Gibbs energy of partly or completely miscible systems. *AIChE Journal* **1975,** *21* (1), 116-128.

3. Fredenslund, A.; Jones, R. L.; Prausnitz, J. M., Groupâ€contribution estimation of activity coefficients in nonideal liquid mixtures. *AIChE Journal* **1975,** *21* (6), 1086-1099.

4. Lin, S.-T.; Sandler, S. I., A priori phase equilibrium prediction from a segment contribution solvation model. *Industrial & engineering chemistry research* **2002,** *41* (5), 899-913.

5. (a) Flory, P. J., Thermodynamics of high polymer solutions. *The Journal of Chemical Physics* **1941,** *9* (8), 660-660; (b) Flory, P. J., Thermodynamics of High Polymer Solutions. *The Journal of Chemical Physics* **1942,** *10* (1), 51-61; (c) Huggins, M. L., Thermodynamic Properties of Solutions of Longâ€chain Compounds. *Annals of the New York Academy of Sciences* **1942,** *43* (1), 1-32.

6. Chen, C.-C.; Song, Y., Solubility modeling with a nonrandom two-liquid segment activity coefficient model. *Industrial & engineering chemistry research* **2004,** *43* (26), 8354-8362.

462711 First Principles Design of Active and Durable Catalysts Toward Oxygen Reduction and Evolution Reactions

Abstract:

Design of Active and Durable Catalysts with Non-precious Materials for oxygen reduction and Evolution Reactions: First Principles Prediction and Experimental Validation

Byungchan Han[1], Jeemin Hwang[1]

[1]*Department of Chemcal & Biomolecular Engineering, Yonsei University, Seoul, 03722, Korea*

bchan@yonsei.ac.kr

Using first-principles density functional theory (DFT) calculations and experimental materialization, we design highly active and durable catalysts toward oxygen reduction reaction (ORR) with nonprecious Cu@N-C materials. DFT calculations indicate that encapsulated Cu metal by N-doped carbon shells is a promising electrocatalyst for ORR. To validate the prediction we synthesize three different types of catalysts with various applied processes: (i) hydrothermally treated "Cu@N-C(hydro)", (ii) "Cu@N-C(heat)" heat-treated at T = 1000 °C for 2 h, and (iii) "Cu@N-C(CO_2)" oxidized by CO_2 for 15 min at T = 1000 °C. It is shown that applying the CO_2 treatment can be a key process controlling electronic structures and shell thickness of the materials leading the high ORR catalytic performance.

To alleviate the substantial overpotential problem in water splitting process storing the energy in fuels of O_2 and H_2 we develop hexagonal perovskite oxide with a transition metal of mixed oxidation states. Both computational prediction and experimental measurements consistently show that its performance is better than IrO2.

Reference:

Jeemin Hwang, Byungchan Han et al., JACS, 138, 3541-3547 (2016)

Seung Hyo Noh, Byungchan Han et al., J. Mater. Chem. A, 3, 22031-22034 (2015)

Seung Hyo Noh, Byungchan Han et al., Nano Research, 8(10), 3394-3403 (2015)

462841 Spectroscopic Identification of Hydroxyl Group Positioning in Butanol Hydrates

Abstract:

In this study, the inclusion of large alcohol guest molecules in hydrate structure are investigated through spectroscopic identifications and equilibrium condition measurements. Four clathrate hydrates formed by structural butanol isomers are identified with powder X-ray diffraction (PXRD). The PXRD results showed that 1-butanol, isobutanol, and *tert*-butanol can act as structure II hydrate formers, but 2-butanol cannot be enclathrated in hydrate cages. Raman spectroscopy was used to analyze the guest distributions and inclusion behaviors of large alcohol molecules in these hydrate systems. The existence of a free OH indicates that guest molecules (isobutanol and *tert*-butanol) can be captured in the large cages of structure II hydrates without any hydrogen bonding interaction between the hydroxyl group of the guests and the water-host framework. However, Raman spectra of the binary (1-butanol + CH_4) hydrate did not show the free OH signal, indicating that there could be possible hydrogen bonding interactions between the guests and hosts. We also measured the four-phase equilibrium conditions of the butanols + CH_4 + H_2O systems. The results of this study could provide a better understanding of the complex host â€“ guest behaviors occurring in clathrate hydrate systems.

463709 Solubility of Squalene and Squalene + Palmitic Acid in Supercritical Carbon Dioxide

Abstract:

Amaranth (*Amaranthus hypochondriacus*) is an ancient crop harvested in Mexico. It is highly appreciated for its nutrimental content as protein and carbohydrates source. Amaranth is widely cultivated because its easiness of growing and consumption. Recently Amaranth oil has industrial interest due its squalene content. Amaranth seeds contain 7-8% oil, an important content of palmitic acid represents fatty acids, and squalene as non-saponifiable substance.

Last century, the greatest squalene source was shark liver oil, nowadays there is an arising concern on sea animal protection, new sources of squalene are being searched. Amaranth seed oil may be a profitable alternative.

Supercritical fluid extraction with CO_2 is environmentally benign technology over organic solvents, solubility data are important to develop extraction techniques. Semi-empirical models based on density are common approaches used to correlate experimental solubility in supercritical fluids.

Experiments based on the static-analytical method. Solubility apparatus is able to reach 450 K and 40 MPa. Binary systems; squalene + CO_2 and palmitic acid + CO_2; and ternary system; squalene + palmitic acid + CO_2; were measured at high pressure (up to 30 MPa) and temperature (313 to 333 K).

The self-consistency of the data was tested according to the density-based models proposed by Mendez-Santiago and Teja, and Chrastil, the calculated values agree with an average absolute deviation 14% for first model and lower than 0.4% for Chrastil model.

A SAFT-based Equation of State for Electrolyte Solutions; Focused on Second Order Thermodynamic Properties

Reza Shahriari, Mohammad Reza Dehghani[1]

Thermodynamics Research Laboratory, School of Chemical Engineering,

Iran University of Science and Technology, Narmak, Tehran 16846-13114, Iran

Abstract

In this study, the electrolyte Perturbed Chain Statistical Associating Fluid Theory (ePC-SAFT) equation of state (EoS) has been utilized to predict the second order thermodynamic properties of electrolyte solutions over wide range of pressure and temperature. The main motivation of our work is the need for prediction of derivative properties of electrolyte solutions. The dielectric constant in Born and electrostatic contributions is modified to keep the pressure, temperature and ionic strength dependency. The isobaric heat capacity at various salt concentrations is predicted without any adjustable parameters. The results of this work indicate an acceptable agreement with experimental data.

1. Introduction

The accurate description of second order derivative properties is important for transport properties which are difficult to predict. We need a robust and efficient thermodynamic model to provide good prediction of phase equilibrium calculations and thermo-physical properties over wide range of operational conditions. During last decade, several EoS have been developed based on SAFT theory to account the phase equilibrium calculation of numerous systems. However, the capability of presented model for prediction of second order derivative properties such as heat capacity, speed of sound and etc. is not clear. In the case of electrolyte solutions, several type of electrolyte EoS have focused on phase equilibrium calculations of single or

[1] Corresponding Author: m_dehghani@iust.ac.ir

mixed electrolyte (or solvent) systems, while prediction of second order derivative properties of electrolyte solutions have been spared.

In the case of second order properties estimation, there are some activity coefficient models which correlate partial molal enthalpy and isobaric heat capacity coefficient of electrolyte and nonelectrolyte aqueous solutions[1]. Simonin et. al. [1] used Mean Spherical Approximation-NRTL (MSA-NRTL) model to describe the apparent heat capacity and dilution enthalpy of five 1-1 aqueous electrolyte solutions. They utilized a simple polynomial form to correlate the infinite dilution heat capacity. Then, the apparent molal heat capacity and dilution enthalpy were correlated using five adjustable parameters per each salt. The results showed good agreement with experimental data. K. Thomsen [2] showed that the Extended UNIQUAC model is able to represent the phase behavior and thermal properties of electrolyte solution up to high temperature and salt concentration. This work is one of the well-known activity coefficient models that shows a good result for second order properties such as solution heat capacity and enthalpy over a wide range of electrolyte concentration.

In this work, the modified form of dielectric constant is coupled with ePC-SAFT EoS to consider the effect of temperature, pressure and ion concentration on long rang and solvation terms. Then, the first order derivative properties such as MIAC, AMV and osmotic coefficient are correlated and predicted up to high pressure and temperature. Finally, Gibbs free energy and heat capacity of some electrolyte solutions are predicted.

2. Theory

The PC-SAFT EoS was based on perturbation theory, where the dispersion contribution derived for chain molecules[3]. In the modified PC-SAFT model for electrolyte solutions, the residual Helmholtz free energy consists of five contributions:

$$\frac{A^{res}}{Ik_BT}=\frac{A^{hc}}{Nk_BT}+\frac{A^{disp}}{Nk_BT}+\frac{A^{assoc}}{Nk_BT}+\frac{A^{elec}}{Nk_BT}+\frac{A^{Born}}{Nk_BT} \quad (1)$$

where the superscripts *res, hc, disp, elec, assoc* and *Born* refers to residual, hard chain, dispersion, electrostatic, association and Born contributions, respectively.

The hard-chain term (A^{hc}) and the dispersion term (A^{disp}) are given by[3]:

$$\frac{A^{hc}}{Nk_BT} = \bar{m}\frac{A^{hs}}{Nk_BT} - \sum_i x_i(m_i - 1)\ln g_{ii}^{hs}(\sigma_{ii}) \tag{2}$$

$$\frac{A^{hs}}{Nk_BT} = \frac{1}{\xi_0}\left[\frac{3\xi_1\xi_2}{(1-\xi_3)} + \frac{\xi_2^3}{\xi_3(1-\xi_3)^2} + \left(\frac{\xi_2^3}{\xi_3^2} - \xi_0\right)\ln(1-\xi_3)\right] \tag{3}$$

$$\frac{A^{disp}}{Nk_BT} = -2\pi\rho I_1(\eta,\bar{m})\overline{m^2\epsilon\sigma^3} - \pi\rho\bar{m}C_1I_2(\eta,\bar{m})\overline{m^2\epsilon^2\sigma^3} \tag{4}$$

Chapman et al.[4] used Wertheim's theory to obtain the association contribution, given as follows:

$$\frac{A^{assoc}}{Nk_BT} = \sum_i x_i\left[\sum_{A_i}\left[\ln X^{A_i} - \frac{X^{A_i}}{2}\right] + \frac{M_i}{2}\right] \tag{5}$$

where x_i is the mole fraction of component *i* in the mixture, *M* is the number of associated sites per molecule, and X^{A_i} is the fraction of molecules not bonded at specific interaction site A.

In this work, the free energy due to the electrostatic interactions between ions is considered by Debye Hückel (DH) theory[5].

$$\frac{A^{elec}}{Nk_BT} = -\frac{\kappa}{12\pi k_BT\varepsilon}\sum_i x_i q_i^2 \chi_i \tag{6}$$

The Born equation is used to account the solvation energy as follows [6]:

$$\frac{A^{Born}}{Nk_BT} = -\frac{e^2}{4\pi k_BT\varepsilon_0}\left(1 - \frac{1}{\varepsilon}\right)\sum_i \frac{x_i z_i^2}{\sigma_i} \tag{7}$$

It is clear that the hydrated diameter in Born term is different from hard sphere diameter. We utilized the following empirical relationship between the ion diameter and Born diameter that verified by molecular simulation [7,8]:

$$\sigma_{anion}^{Born} = \sigma_{anion} + 1.7\ \text{Å} \tag{8}$$

$$\sigma_{cation}^{Born} = \sigma_{cation} + 0.2\ \text{Å} \tag{9}$$

Using Born diameters instead of bare ion diameter gives a good result for prediction of hydration Gibbs free energy.

Maribo-Mogensen et al.[9] modified the g-factor using the probability function in Wertheim's association theory. Based on their modification, permittivity of mixture can be obtained using equation (18) [9]:

$$\frac{(2\varepsilon_r + \varepsilon_\infty)(\varepsilon_r - \varepsilon_\infty)}{\varepsilon_r(\varepsilon_\infty + 2)^2} = \frac{\rho}{9\varepsilon_0 k_B T}\sum_I x_i\,\Theta_i g_i \mu_{i,0}^2 \tag{10}$$

Following relations are adopted from previous works [10-13] to calculate the Gibbs free energy of hydration based on EoS:

$$_h\bar{G}^*_{salt}(T,P) = RT\ln\varphi_{salt}\,(T,P,x_{salt}\to 0) + \nu RT\ln(\frac{m^0 d^0 RT}{1000P^0}) \tag{11}$$

where the second term is the conversion from hypothetically ideal gaseous ions being hydrated to the hypothetically ideal $1m$ aqueous solution[11-14]. The reference molality is 1 molar (m^0= 1 mol/kg) and the standard state pressure (ideal gas) is 1 bar (P^0= 1 bar). The d^0 (g/cm^3) is denoted density of pure solvent (in this work water) and ν represent the stoichiometric number of ions in the electrolyte.

First order derivative of eq. 11 respect to temperature leads to partial molar entropy at infinite dilution and second order derivate leads to heat capacity as follow:

$$\Delta_h\bar{S}^*_{salt}(T,P) = -\left(\frac{\partial\Delta_h\bar{G}^*_{salt}}{\partial T}\right)_P \tag{12}$$

$$\Delta_h\bar{H}^*_{salt}(T,P) = \Delta_h\bar{G}^*_{salt}(T,P) + T\,\Delta_h\bar{S}^*_{salt}(T,P) \tag{13}$$

$$\Delta_h\bar{C}^*_{p,salt}(T,P) = -T\left(\frac{\partial^2\Delta_h\bar{G}^*_{salt}}{\partial T^2}\right)_P \tag{14}$$

The infinite dilution partial molar heat capacity of salt is obtained as follows:

$$\bar{C}^0_{p,salt}(T,P) = \Delta_h\bar{C}^*_{p,salt}(T,P) + \nu C_P^{ig}(T) - \nu R\left[1 - 2T\alpha^0 - T^2\left(\frac{\partial\alpha^0}{\partial T}\right)_p\right] \tag{15}$$

where α^0 is the isobaric expansivity of pure water. C_P^{ig} is the heat capacity coefficient of gaseous ions at standard state that is set to 20.785 $\frac{J}{mol.\ K}$ for ions[15].

3. Results

3.1. Pure solvent parameters

In the case of electrolyte solutions, calculation of solvent properties play an important role. The water parameters have been obtained using vapor pressure and saturated liquid density experimental data up to critical temperature; see table 1.

Table 1. PC-SAFT parameters for pure water. Parameters were obtained using vapor pressure and saturated liquid density experimental data [16].

	m	σ	ε/k	ε^{AB}/k	κ^{AB}	ARD(%)	
Scheme						P^{sat}	ρ_{liq}^{sat}
4C	1.0175	3.0348	339.39	1538.43	0.0317	0.246	0.214
Ref. [17]							

Figure 1 shows the calculated vapor pressure and saturated density of pure water compare with experimental data.

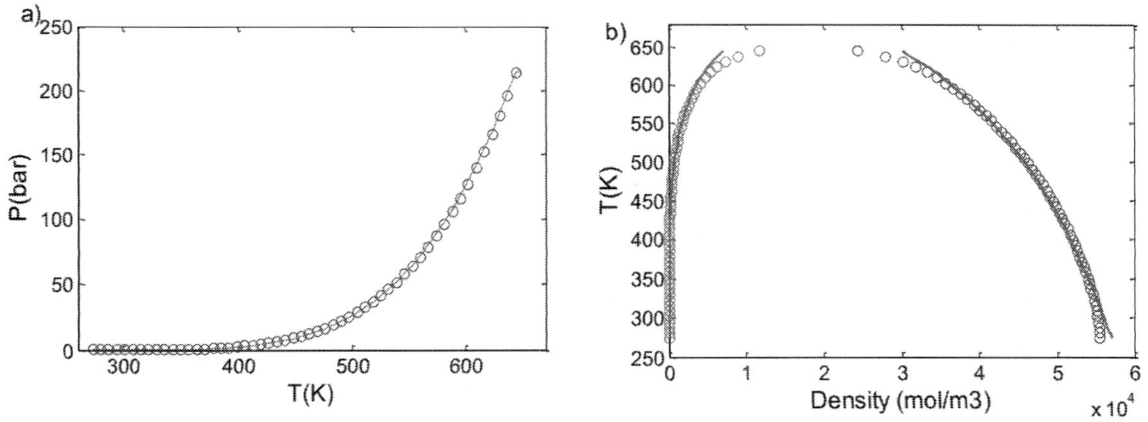

Figure 1. Calculated and experimental vapor pressure and liquid density of water.

3.2. Electrolyte solutions

The model parameters are segment diameter, segment energy and association energy of ions. The segment number for ions set to one based on our previous work [18,19]. The binary interaction parameters (BIP) between all particles are set to zero (k_{ij}=0.0). The MIAC and AMV of twelve aqueous electrolyte solutions at 298.15 K and 1 bar have been simultaneously utilized for parameters optimization. Three parameters contained segment diameter (σ), segment energy (ε/k_B) and association energy (ε^{AB}/k_B) have been adjusted for each ion; see table 2.

Table 2. Estimated ionic parameters using MIAC and AMV data.

Component	m	σ	ε/k	ε^{AB}/k	κ^{AB}	S[1]
Li^{+}	1	2.67824	75.03787	1283.497	0.03	6
Na^{+}	1	2.850063	642.1423	647.3689	0.03	6
K^{+}	1	2.962996	110.0381	130.3039	0.03	6
Ca^{2+}	1	3.731494	1411.569	3197.902	0.03	8
Cl^{-}	1	2.457642	49.4767	281.4294	0.03	5
Br^{-}	1	2.950236	122.7321	234.2157	0.03	5
I^{-}	1	3.553865	172.5888	352.6175	0.03	5
SO_4^{2-}	1	3.5137	1821.552	100.4415	0.03	5

system	ARD (%)=$\frac{100}{N}\sum_i^N \left\lvert\frac{X_i^{exp}-X_i^{calc}}{X_i^{exp}}\right\rvert$		
	$\gamma_{salt}^{\pm}$	AMV	Molality Max. (mol/kg solvent)
LiCl	4.9	5.6	4.5
LiBr	3.8	3.23	4.5
NaCl	3.03	4.01	6.0
NaBr	2.56	1.81	4.0
NaI	3.44	0.821	3.5
KCl	2.27	2.72	5.0
KBr	3.09	2.78	5.5
KI	4.01	1.09	4.5
$CaCl_2$	6.05	3.96	2.0
$CaBr_2$	8.59	1.75	2.0
Na_2SO_4	6.02	2.87	2.0
K_2SO_4	7.36	15.03	1.0
Average ARD(%)	3.80	4.59	

The obtained parameters at 298.15 K and 1 bar were utilized for prediction of first and second order derivative properties up to high temperatures. For the sake of parameter optimization accuracy, a robust and efficient optimization algorithm namely, the Shuffled Complex Evolution (SCE)[20] was utilized. Following objective function was used for parameters optimization:

$$OF = \sum_i \sum_j \left[\left(\frac{\gamma_j^{\pm\, exp} - \gamma_j^{\pm\, calc}}{\gamma_j^{\pm\, exp}} \right)_i + \left(\frac{AMV_j^{exp} - AMV_j^{cal}}{AMV_j^{exp}} \right)_i \right] \quad (16)$$

where i and j are counter over salts and data points, respectively. Results have been shown in figures 2 to 4.

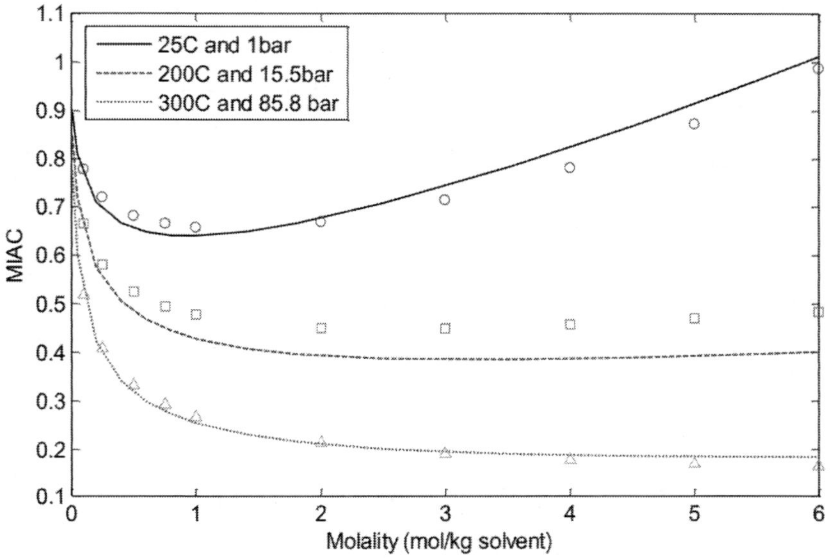

Figure 2. Mean ionic activity coefficient of NaCl aqueous solution at several temperature and pressure[21].

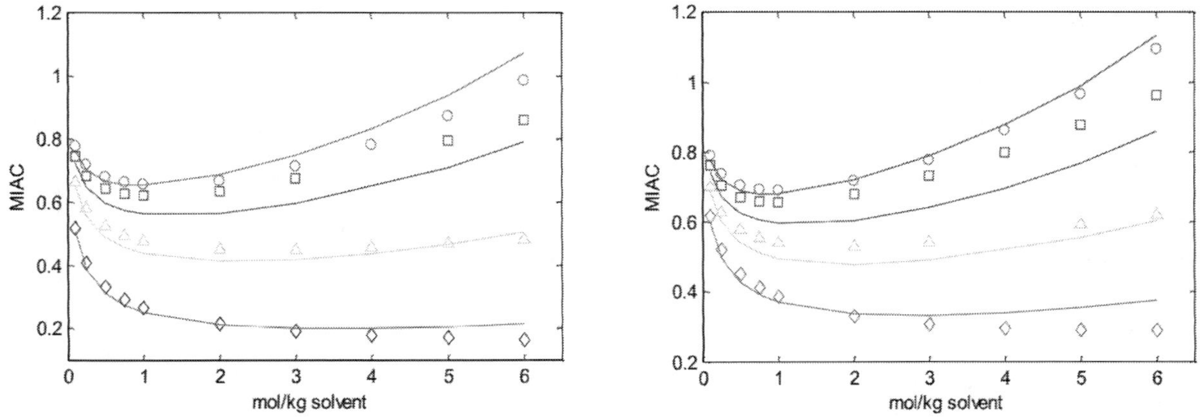

Figure 3. Model calculated (lines) and experimental data (points) MIAC for NaCl. (○) 298.15, (□) 373.15, (Δ) 473.15 and (◊) 573.15. Left figure at 1bar for T<373.15 K and saturated at T>373.15 K. Right figure at 1000 bar [21].

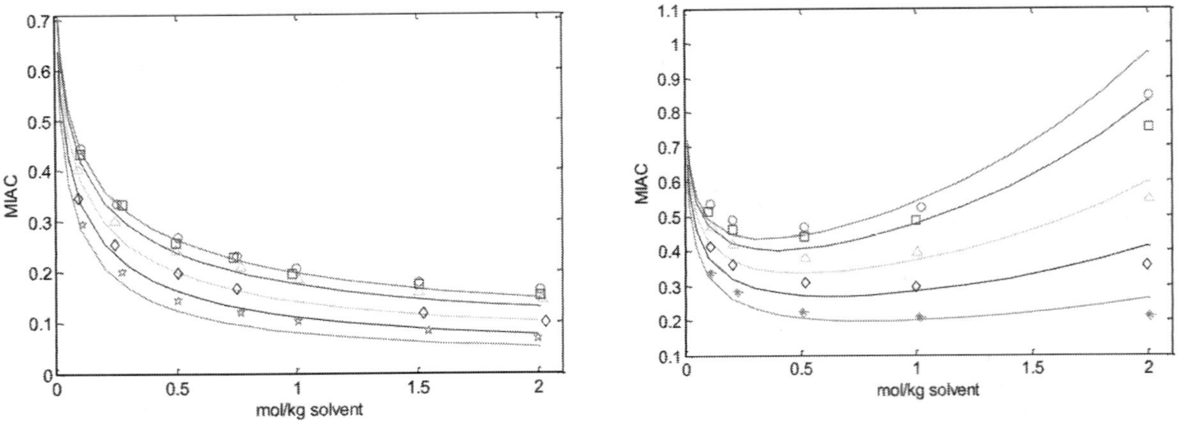

Figure 4. Model calculated (lines) and experimental data (points) MIAC for $CaCl_2$ (right) at 400 bar and Na_2SO_4 (left) at 200 bar. (○) 298.15, (□) 323.15, (Δ) 373.15, (◊) 423.15 and (*) 473.15 K [22,23].

In this study, the Gibbs free energy of hydration has been calculated. Table 3 gives a comparison between predicted enthalpy of hydration and experimental data.

Table 3. Prediction of Gibbs free energy of hydration (KJ/mol) at 1bar and 298.15 K.

Salt	Experiment [24]	This work	SAFT-VR+DE [25]	npmsa + PC-SAFT [26]	Err (%) $=\mathbf{100}\left\|\frac{X^{exp}-X^{calc}}{X^{exp}}\right\|$		
					This work	SAFT-VR+DE	npmsa + PC-SAFT
LiCl	-831.78	-767.59	-693.08	-741.3	7.71	16.67	10.87
LiBr	-805.83	-736.05	-652.21	-715.8	8.65	19.06	11.17
NaCl	-727.6	-755.53	-653.77	-667.9	3.83	10.14	8.20
KCl	-659.81	-710.51	-630.55	-604.1	7.68	4.43	8.44
NaBr	-701.65	-729.99	-616.92	-647.6	4.03	12.07	7.70
KBr	-633.87	-678.97	-593.77	-578.9	7.11	6.32	8.67
NaI	-668.18	-694.58	-568.23	-617.6	3.95	14.95	7.56
KI	-600.40	-649.56	-548.81	-562.6	8.18	8.59	6.29
Average Error					**6.39**	**11.53**	**8.61**

As it is shown, this work give good results compare with other SAFT-based models.

On the other hand, in this work the modified ePC-SAFT model was used to predict heat capacity up to high pressure and temperature without any fitting parameters or additional correlative term. Figure 5 shows a comparison between model prediction and three correlative or semi-empirical models [11,27,28] including, unified theory proposed by Djamali and Cobble [11] (a), and Pitzer based models (b and c).The results show that, SAFT-based EoS that utilized in the present work gives an acceptable prediction compared to other correlations.

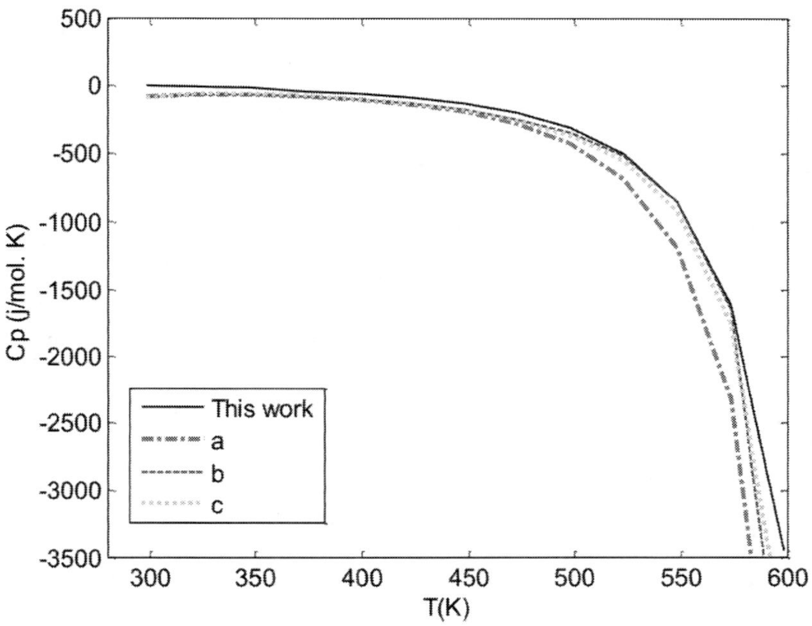

Figure 5. Standard partial molal heat capacity of NaCl aqueous solutions at infinite dilution up to 120 bar and 600 K. a) [11], b) [28], c) [27].

In figures 6, predicted infinite dilution heat capacity of three electrolyte solutions at various temperatures are compared with experimental values.

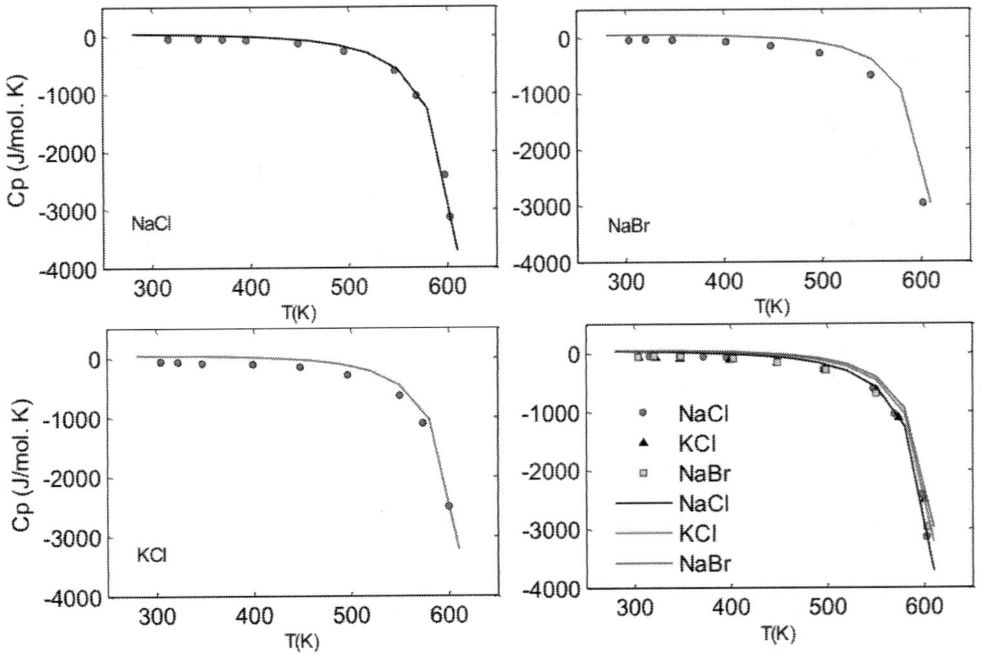

Figure 6. Standard partial molal heat capacity of aqueous electrolyte solutions at infinite dilution at high pressure; NaCl (179 bar), KCl (179 bar) and NaBr (175 bar). Experimental data from [29,30].

The results show that, the standard state heat capacity of salt can be predicted using ion-specific parameters obtained by first order derivative properties (MIAC and AMV). Deviation from experimental data at low temperature relatively is high, especially at ambient condition while it decreases as temperature increases.

In figure 7, Using predicted standard state heat capacity of electrolyte solutions, the isobaric heat capacity of several salts up to 3 molal have been predicted.

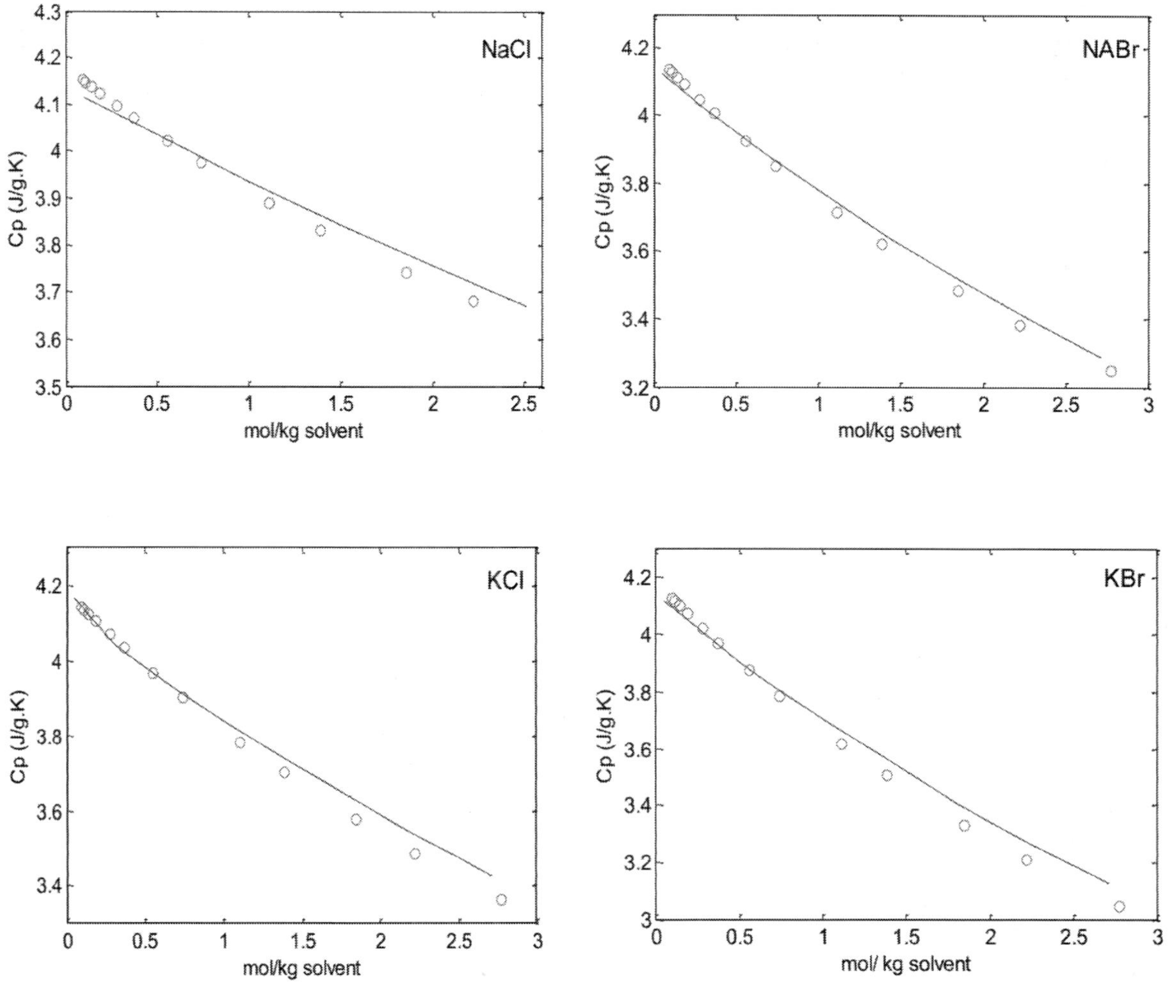

Figure 7. Predicted heat capacity of aqueous electrolyte solutions at 1 bar and 298.15 K [31].

These figures show that, the modified ePC-SAFT results are in good agreement with experimental heat capacity data of electrolyte solutions.

4. Conclusion

The electrolyte PC-SAFT model has been modified and utilized to estimate first and second order derivative properties of electrolyte solution up to high pressure and temperature. The Onsager, Kirkwood, and Fröhlich framework coupled with Wertheim's association theory have been used to represent an appropriate definition for dielectric constant. Our results show that at infinite dilution the solvent properties and solvation contribution play an important role, therefore, second order derivative properties of pure water has been studied in detail. It is shown that, the present model is able to predict infinite dilution heat capacity of electrolyte solution up to high pressure and temperature. In addition, the model capability in prediction of MIAC and AMV at high pressure and temperature was satisfactory. Finally, the isobaric heat capacity of aqueous electrolyte solutions up to 3 molal have been predicted with reasonable deviation compare to experimental data.

Reference

1. Simonin J-P, Bernard O, Papaiconomou N, Kunz W. Description of dilution enthalpies and heat capacities for aqueous solutions within the MSA–NRTL model with ion solvation. Fluid Phase Equilibria. 2008;264:211-219.
2. K.Thomsen. Aqueous electrolytes: model parameters and process simulation. Denmark: Chemical Engineering, Technical University of Denmark; 1997.
3. Gross J, Sadowski G. Perturbed-Chain SAFT: An Equation of State Based on a Perturbation Theory for Chain Molecules. Ind. Eng. Chem. Res. 2001;40:1244-1260.
4. Chapman WG, Gubbins KE, Jackson G, Radosz M. New reference equation of state for associating liquids. Industrial & Engineering Chemistry Research. 1990;29:1709.
5. Debye P, Hückel E. Zur Theorie der Elektrolyte. I.Gefrierpunktserniedrigung und verwandte Erscheinungen. Physik Z. 1923;24:185-206.
6. Born M. Volumen und Hydratations Warme der Ionen. Zeitschrift Zeitschrift für Physik. 1920;1:45–49.
7. Latimer WM, Pitzer KS, Slansky CM. The free energy of hydration of gaseous ions, and the absolute potential of the normal calomel electrode. J. Chem. Phys. 1939;7:108.
8. Rajamani S, Ghosh T, Garde S. Size dependent ion hydration, its asymmetry, and convergence to macroscopic behavior. J. Chem. Phys. 2004;120:4457.
9. M-Mogensen B, Kontogeorgis GM, Thomsen K. Modeling of Dielectric Properties of Complex Fluids with an Equation of State. J. Phys. Chem. B. 2013;117:3389–3397.
10. Sedlbauer J, O'Connell JP, Wood RH. A new equation of state for correlation and prediction of standard molal thermodynamic properties of aqueous species at high temperatures and pressures. Chemical Geology. 2000;163:43-63.
11. Djamali E, Cobble JW. A Unified Theory of the Thermodynamic Properties of Aqueous Electrolytes to Extreme Temperatures and Pressures. J. Phys. Chem. B. 2009;113:2398–2403.
12. Djamali E, Cobble JW. Thermodynamic Properties of Aqueous Polyatomic Ions at Extreme Temperatures and Pressures. J. Phys. Chem. B. 2010;114:3887–3893.
13. Djamali E, Kan AT, Tomson MB. A Priori Prediction of Thermodynamic Properties of Electrolytes in Mixed Aqueous−Organic Solvents to Extreme Temperatures. J. Phys. Chem. B. 2012;116:9033−9042.
14. Chapman WG, Gubbins KE, Jackson G, Radosz M. New Reference Equation of State for Associating Liquids. Industrial & Engineering Chemistry Research. 1990;29:1709-1721.
15. Stull DR, Prophet H. JANAF Thermochemical Tables. In: 37 N-N, ed1971.
16. NIST Chemistry WebBook. National Institute of Standard and Thechnology 2009. http://webbook.nist.gov.
17. Lee B-S, Kim† K-C. Modeling of aqueous electrolyte solutions based on perturbed-chain statistical associating fluid theory incorporated with primitive mean pherical approximation. Korean J. Chem. Eng. 2009;26:1733-1747.
18. Valavi M, Dehghani MR, Shahriari R. Application of modified PHSC model in prediction of phase behavior of single and mixed electrolyte solutions. Fluid Phase Equilibria. 4/25/ 2013;344(0):92-100.
19. Shahriari R, Dehghani MR, Behzadi B. Thermodynamic Modeling of Aqueous Ionic Liquid Solutions Using PC-SAFT Equation of State. Ind. Eng. Chem. Res. 2012;51:10274−10282.
20. Duan QY, Gupta VK, Sorooshian S. Shuffled Complex Evolution Approach for Effective and Efficient Global Minimization. J. Optimi. Theory Appl. 1993;76:501−521.
21. Zaytsev ID, Aseyev GG. Properties of Aqueous Solutions of Electrolytes: CRC Press; 1992.
22. Pabalan RT, Pitzer KS. Heat capacity and other thermodynamic properties of Na2SO4(aq) in hydrothermal solutions and the solubilities ofsodium sulfate minerals in the system sodium-chlorine-sulfate-hydroxylwater to 300 °C. Geochim. Cosmochim. Acta. 1988;52:2393-2404.
23. Holmes HF, Busey RH, Simonson JM, Mesmer RE. CaCl2(aq) at elevated temperatures. Enthalpies of dilution, isopiestic molalities, and thermodynamic properties. J. Chem. Thermodyn. 1994;26:271-298.
24. Robinson RA, Stokes RH. Electrolyte Solutions. 2nd ed. ed. London1959.
25. Das G, Hlushak S, dos Ramos MC, McCabe C. Predicting the thermodynamic properties and dielectric behavior of electrolyte solutions using the SAFT-VR+DE equation of state. AIChE Journal. 2015;61(9):3053-3072.
26. Herzog S, Gross J, Arlt W. Equation of state for aqueous electrolyte systems based on the semirestricted non-primitive mean spherical approximation. Fluid Phase Equilibria. 10/15/ 2010;297(1):23-33.

27. Archer DG. Thermodynamic Properties of the NaCl+H2O System. II. Thermodynamic Properties of NaCl(aq), NaCl·2H2(cr), and Phase Equilibria. J. Phys. Chem. Ref. Data. 1992;21:793–829.
28. Pitzer KS, J. C. Pieper, Busey RH. Thermodynamic Properties of Aqueous Sodium Chloride Solutions. J. Phys. Chem. Ref. Data. 1984;13:1–102.
29. Gates AJ. Thermodynamic of aqueous electrolyte solutions at high temperature and pressures. Newark, Delaware; 1985.
30. Smith-Magowan D, Wood RH. Heat capacity of aqueous sodium chloride from 320 to 600 K measured with a new flow calorimeter. J. Chem. thermodyn. . 1981;13:1047-1073.
31. Rossini FD. APPARENT AND PARTIAL MOLAL HEAT CAPACITIES IN AQUEOUS SOLUTIONS OF 19 UNI-UNIVALENT STRONG ELECTROLYTES. Bureau of standards jornal of research 1931;7.

EULERIAN VOLUME-AVERAGED MODEL AND NUMERICAL SIMULATIONS OF REACTIVE SOLID-LIQUID MIXTURES

Martina Costa Reis, University of Campinas, Campinas, Brazil
Julian Heß, Technische Universität Darmstadt, Darmstadt, Germany
Yongqi Wang, Technische Universität Darmstadt, Darmstadt, Germany

Abstract

Based on the Eulerian spatial averaging theory, a non-equilibrium thermodynamic model for reactive solid-liquid mixtures is presented. Initially, some fundamental properties of average quantities are discussed and, then, averaged balance equations are postulated. Furthermore, a theoretical model for a dilute suspension of calcium carbonate nuclei dispersed in brines is proposed. In order to check the validity of the proposed model, numerical simulations for the gravitational settling of calcium carbonate particles in brines are performed. The obtained results indicate that the suspension behaves as a viscous fluid, where finger-like structures develop at the interface between the solid and liquid phases.

Introduction

Solid-liquid mixtures, also named sols, are colloidal systems in which an aggregate of very fine solid particles is dispersed in a liquid dispersion medium. In contrast with single-phase fluids, the physical principles that govern the thermodynamics of solid-liquid mixtures are definitely more complex due to the presence of multiple deformable moving interfaces, and the possibility of mass, momentum and energy interfacial transfers. Moreover, the existence of fluctuation of variables, which appears as a natural result of turbulences and the motions of the interfaces, is also an inherent complication in solid-liquid mixtures. These factors have motivated the development of models based on averaging methods [1] that provide a continuum approach for multiphase flows, by eliminating local instant fluctuations.

The basic idea of thermodynamic models based on averaging methods is the formulation of interaction laws between the phases and, subsequently, the exploitation of phase averaging operations by which the discrete distribution of phases is smoothed over a representative volume element. At the end, a set of balance equations similar to those of the theory of mixtures is obtained, but capable of providing information about interfacial transport phenomena. Surely, during the formulation of averaged thermodynamic models, unspecified quantities appear in the balance equations, which require the postulation of constitutive equations in order to obtain a closed system of equations that can be numerically solved.

In this work, an Eulerian volume-averaged model is proposed for reactive solid-liquid mixtures, where the formation of the solid phase is due to a precipitation chemical reaction that involves ions dissolved in the liquid phase. Furthermore, a thermodynamic model for dilute suspensions of $CaCO_3$(s) in brines is briefly discussed. In order to check the validity of the proposed thermodynamic model, numerical simulations are performed for the gravitational settling of calcium carbonate particles in brines. The results of the numerical simulations show that finger-like structures similar to those observed in the Rayleigh-Taylor instability appear during the sedimentation.

Principles of Eulerian Averaging Theory

According to the Eulerian spatial averaging theory, a multiphase system is regarded as a collection of various separate phases, each occupying only a part of the space. The total volume occupied by phase k is the union of a large number of isolated sub-regions that are separated by interfaces at which physical properties may undergo a step discontinuity. Hereafter, for the sake of simplicity, the index $k = 1$ will be used to describe quantities of the liquid phase, whereas the index $k = 2$ will be used to denote quantities of the solid phase. In addition, the microscopic scale of the multiphase mixture will be regarded as the scale at which morphological aspects and the kinematics of each phase and of interfaces can be observed.

The governing equations in each sub-region of the system are exactly the same balance equations used to study single-phase flows. Nevertheless, since the solution of these equations is particularly difficult for multiphase flows, the scale at which the system is considered is changed, by averaging the quantities observed at the microscopic scale over a representative averaging element volume $\mathcal{V}$.

Let $\mathcal{V}$ be an averaging element volume, which is made up of the volume of the solid and liquid phases. Each phase k present in the representative averaging element volume contributes for $\mathcal{V}$, such that $\mathcal{V} = \sum_{k=1}^{n} \mathcal{V}_k(t)$. Thus, the average value of a certain field ψ belonging to phase k is defined as

$$\langle \psi_k \rangle (\mathbf{x}, t) = \frac{1}{\mathcal{V}} \int_{\mathcal{V}_k(t)} \psi_k(\mathbf{r}, t)\, d\mathcal{V} = \frac{1}{\mathcal{V}} \int_{\mathcal{V}(t)} \psi(\mathbf{r}, t)\, \gamma_k(\mathbf{r}, t)\, d\mathcal{V}\,, \tag{1}$$

where $\mathbf{x}$ is a spatial coordinate at the macroscopic scale, $\mathbf{r}$ is a spatial coordinate at the microscopic scale, and γ_k is the phase indicator function, a step function defined by

$$\gamma_k(\mathbf{r}, t) = \begin{cases} 1 & \mathbf{r} \in k, \\ 0 & \mathbf{r} \notin k. \end{cases} \tag{2}$$

Similarly, one can also write

$$\langle \psi_k \rangle_k (\mathbf{x}, t) = \frac{1}{\mathcal{V}_k(t)} \int_{\mathcal{V}_k(t)} \psi_k(\mathbf{r}, t)\, d\mathcal{V} = \frac{1}{\mathcal{V}_k(t)} \int_{\mathcal{V}(t)} \psi(\mathbf{r}, t)\, \gamma_k(\mathbf{r}, t)\, d\mathcal{V}\,, \tag{3}$$

$$\overline{\langle \psi_k \rangle} (\mathbf{x}, t) = \frac{\int_{\mathcal{V}_k(t)} \rho_k(\mathbf{r}, t)\, \psi_k(\mathbf{r}, t)\, d\mathcal{V}}{\int_{\mathcal{V}_k(t)} \rho_k(\mathbf{r}, t)\, d\mathcal{V}} = \frac{\langle \rho_k \psi_k \rangle}{\langle \rho_k \rangle}\,, \tag{4}$$

where ρ_k stands for the mass density of phase k, $\langle \psi_k \rangle_k$ is the intrinsic average value of ψ_k, and $\overline{\langle \psi_k \rangle}$ is the mass-weighted average value of ψ_k. Note that in expression (3) a property of phase k is averaged over that phase only, whereas in expression (1) ψ_k is averaged over the entire averaging element volume $\mathcal{V}$. Then, the ratio between expressions (1) and (3) leads to an important relation known as the volume fraction of phase k,

$$\alpha_k(\mathbf{x}, t) = \frac{\mathcal{V}_k(t)}{\mathcal{V}} = \frac{1}{\mathcal{V}} \int_{\mathcal{V}_k(t)} \gamma_k(\mathbf{r}, t)\, d\mathcal{V}\,, \tag{5}$$

where α_k is constrained by $\sum_{k=1}^{n} \alpha_k = 1$, such that $0 \leq \alpha_k \leq 1$.

General spatial averaged balance equation

Since in the Eulerian spatial averaging theory a multiphase system is regarded as a collection of different phases that can exchange mass, linear and angular momenta, energy *etc.* through interfaces, the average fields of a certain phase are not independent on the average fields of the other phases. In fact, each phase of a multiphase system has its own system of balance equations that is similar to those of single-phase flows, but they have interfacial source terms that account for possible interfacial processes too. By defining the average quantities of each phase in terms of parameters observed at the microscopic scale, the general averaged balance equation for a solid-liquid flow [2] is given by,

$$\frac{\partial \langle \psi_k \rangle}{\partial t} - \operatorname{div}\left(\langle \psi_k \rangle \overline{\langle \mathbf{v}_k \rangle} \right) - \operatorname{div}\left\langle \mathbf{\Phi}_k^{total} \right\rangle - \langle \sigma_k \rangle - \langle \pi_k \rangle - \frac{1}{\mathcal{V}} \int_{I_k(t)} \psi_k \left(\mathbf{w} \cdot \mathbf{n}_k - \mathbf{v}_k \cdot \mathbf{n}_k \right) da - \tag{6}$$
$$\frac{1}{\mathcal{V}} \int_{I_k(t)} \mathbf{\Phi}_k \cdot \mathbf{n}_k da = 0\,,$$

where $\mathbf{v}_k$ is the velocity, $\left\langle \mathbf{\Phi}_k^{total} \right\rangle = \langle \mathbf{\Phi}_k \rangle - \left\langle \tilde{\psi}_k \tilde{\mathbf{v}}_k \right\rangle$ is the total flux, such that $\mathbf{\Phi}_k$ denotes the flux of the quantity ψ_k and the symbol $\tilde{\cdot}$ denotes the fluctuation of a certain quantity in relation to its average value, σ_k and π_k are respectively the supply and production of ψ_k, I_k is the interface between the solid and liquid phases, $\mathbf{w}$ is the interface velocity, and $\mathbf{n}_k$ is the outward unit normal vector to the interface of the region occupied by phase k.

Equation (6) includes not only self-interacting terms that express the flux, supply, and production of ψ_k within the volume occupied by phase k in the averaging representative element of volume, but also interfacial transfer terms that describe the transport of ψ_k across the interface I_k. In Table 1, the fields at the microscopic scale for a reactive solid-liquid mixture are given, where $\boldsymbol{T}_k$ is the stress tensor, $\mathbf{b}_k$ is the body force, ε_k is the internal energy, $\mathbf{h}_k$ is the heat flux vector, r_k is the energy supply, η_k is the entropy density, $\mathbf{\Phi}_k$ is the entropy flux, σ_k is the entropy supply, and π_k is the entropy production.

Table 1: Physical quantities at the microscopic scale of a reactive solid-liquid flow.

Quantity	Mass	Linear momentum	Angular momentum	Energy	Entropy
ψ_k	ρ_k	$\rho^k\mathbf{v}^k$	$\mathbf{r}\times\rho_k\mathbf{v}_k$	$\rho_k\varepsilon_k + 1/2\rho_k v_k^2$	$\rho_k\eta_k$
$\boldsymbol{\Phi}_k$	-	$\boldsymbol{T}_k$	$\mathbf{r}\times\boldsymbol{T}_k$	$-\mathbf{h}_k+\boldsymbol{T}_k\mathbf{v}_k$	$-\mathbf{\Phi}_k$
σ_k	-	$\rho_k\mathbf{b}_k$	$\mathbf{r}\times\rho_k\mathbf{b}_k$	$\rho_k r_k + \mathbf{v}_k\cdot\rho_k\mathbf{b}_k$	$\rho_k\sigma_k$
π_k	-	-	-	-	$\rho_k\pi_k$

Eulerian Model for a Dilute Suspension of Calcium Carbonate Particles in Brines

Consider a dilute suspension of calcium carbonate particles in brines, where the diameter of nuclei is of the order of 100 μm. The solid particles are assumed to behave as a deformable material whose granular viscosity and granular bulk viscosity are respectively calculated by the models described by Syamlal and O'Brien [3] and Lun *et. al* [4]. In turn, the liquid phase is assumed to be a Stokes fluid. For the sake of simplicity, the heterogeneous chemical reaction that leads to the formation of $CaCO_3$(s) in the mixture is neglected. Thus, interfacial mass mechanisms can be omitted in the formulation of an Eulerian model for a dilute suspension of calcium carbonate in brines.

Moreover, the interactions between the solid and liquid phases are mostly associated with exchanges of linear momentum, so that temperature plays no significant role in the thermodynamic behavior of the suspension. In accounting for the solid pressure, one also considers that the pressure in the solid phase corresponds to the summation of two effects: one kinetic contribution that arises from the fluctuations of the particles velocities, and other collisional contribution that is due to the collisions between the particles.

The kinematic contribution for the solid pressure can be determined by applying the kinetic theory of gases to a solid-liquid mixture. In this case, an analogy between the random motion of particles caused by interparticle collisions and the thermal motion of molecules in a gas is made, so that the kinetic energy related to the random motion of particles is given by the mean square of particle velocity fluctuations (granular temperature). On the other hand, the collisional contribution is determined by taking into account the granular temperature, the restitution coefficient for interparticle collisions, and a radial distribution function. Basically, the effect of the collisional term on the calculations is to prevent that the calculated particle concentration exceeds the maximum concentration derivable from a given packing limit.

Hence, the Eulerian formulation for a dilute suspension of calcium carbonate in brine is given by

$$\langle\rho_2\rangle_2\frac{\partial\alpha_2}{\partial t} + \langle\rho_2\rangle_2\,\mathrm{div}\left(\overline{\langle\mathbf{v}_2\rangle}\alpha_2\right) = 0\,, \tag{7}$$

$$\langle\rho_2\rangle_2\frac{\partial\left(\overline{\langle\mathbf{v}_2\rangle}\alpha_2\right)}{\partial t} + \langle\rho_2\rangle_2\,\mathrm{div}\left(\alpha_2\overline{\langle\mathbf{v}_2\rangle}\otimes\overline{\langle\mathbf{v}_2\rangle}\right) + \alpha_2\nabla p + \nabla\langle p_2\rangle - \mathrm{div}\,\langle T_2\rangle_2 - \alpha_2\langle\rho_2\rangle_2\overline{\langle\mathbf{g}\rangle} - c_{12}\left(\overline{\langle\mathbf{v}_1\rangle} - \overline{\langle\mathbf{v}_2\rangle}\right) = 0\,, \tag{8}$$

$$\langle\rho_1\rangle_1\frac{\partial\alpha_1}{\partial t} + \langle\rho_1\rangle_1\,\mathrm{div}\left(\overline{\langle\mathbf{v}_1\rangle}\alpha_1\right) = 0\,, \tag{9}$$

$$\langle\rho_1\rangle_1\frac{\partial\left(\overline{\langle\mathbf{v}_1\rangle}\alpha_1\right)}{\partial t} + \langle\rho_1\rangle_1\,\mathrm{div}\left(\alpha_1\overline{\langle\mathbf{v}_1\rangle}\otimes\overline{\langle\mathbf{v}_1\rangle}\right) - \mathrm{div}\,\langle T_1\rangle_1 + \alpha_1\nabla p - \alpha_1\langle\rho_1\rangle_1\overline{\langle\mathbf{g}\rangle} - c_{12}\left(\overline{\langle\mathbf{v}_2\rangle} - \overline{\langle\mathbf{v}_1\rangle}\right) = 0\,, \tag{10}$$

where $\langle T_2 \rangle_2$ is the solid stress tensor that accounts for shear and bulk viscosities, which arise from translation and collision between particles, $\langle T_1 \rangle_1 = \alpha_1 \mu_1 \left(\nabla \overline{\langle \mathbf{v}_1 \rangle} + \nabla \overline{\langle \mathbf{v}_1 \rangle}^T \right)$, such that μ_1 is the shear viscosity of the liquid phase, p is the pressure of the mixture, p_2 is the solid pressure, $\mathbf{g}$ is the gravitational field, and c_{12} is the fluid-solid exchange coefficient given by

$$c_{12} = \frac{\alpha_2 \langle \rho_2 \rangle_2 f}{\tau_2} , \tag{11}$$

such that τ_2 is the particle relaxation time, and f is the drag function.

All definitions of f incorporate a drag coefficient C_D that depends on the relative Reynolds number Re_s and certain physical assumptions. For example, according to the model proposed by Syamlal and O'Brien [3], the drag function is deduced with basis on the assumption that the relation between the gravitational forces and viscous forces is the same for a single particle system and a system of many particles. Then, the coefficients c_{12} and C_D have the following forms,

$$c_{12} = \frac{3}{4} \frac{\alpha_2 \alpha_1 \langle \rho_1 \rangle_1}{\mathbf{v}_{r,2}^2 d_2} C_D \left(\frac{Re_s}{\mathbf{v}_{r,2}} \right) \left| \overline{\langle \mathbf{v}_2 \rangle} - \overline{\langle \mathbf{v}_1 \rangle} \right| , \quad C_D = \left(0.63 + \frac{4.8}{\sqrt{Re_s / \mathbf{v}_{r,2}}} \right) , \tag{12}$$

where d_2 is the diameter of a solid particle, and $\mathbf{v}_{r,2}$ is the relative velocity correlation,

$$\mathbf{v}_{r,2} = 0.5 \left(A - 0.06 Re_s + \sqrt{(0.06 Re_s)^2 + 0.12 Re_s (2B - A) + A^2} \right) , \tag{13}$$

with $A = \alpha_1^{4.14}$, $B = 0.8\alpha_1^{1.28}$ for $\alpha_1 \leq 0.85$, and $B = \alpha_1^{2.65}$ for $\alpha_1 > 0.85$. In turn, the drag force model proposed by Wen and Yu [5] does not account for the influence of particle-particle interactions on the drag force, what makes it suitable for dilute solid-liquid mixtures. In this case, the fluid-solid exchange coefficient and the drag coefficient are given by

$$c_{12} = \frac{3}{4} \frac{\alpha_2 \alpha_1 \langle \rho_1 \rangle_1}{d_2} C_D \alpha_1^{-2.65} \left| \overline{\langle \mathbf{v}_2 \rangle} - \overline{\langle \mathbf{v}_1 \rangle} \right| , \quad C_D = \frac{24}{\alpha_1 Re_s} \left[1 + 0.15 \left(\alpha_1 Re_s \right)^{0.687} \right] . \tag{14}$$

Another drag force model for solid-liquid flows available in the literature is that proposed by Gidaspow [6]. This model consists of the combination of the Wen-Yu model with a correlation function that relates the drag force to the pressure drop in a porous medium. Differently from the Wen-Yu model, the shape of the solid particles is taken into account here. For spherically symmetric particles and dilute solid-liquid mixtures with $\alpha_1 > 0.8$, Gidaspow model reduces to Wen-Yu model. However, for solid-liquid mixtures of spherically symmetric particles with $\alpha_1 \leq 0.8$, c_{12} is given by

$$c_{12} = \frac{150 \alpha_2 (1 - \alpha_1) \mu_1}{\alpha_1 d_2^2} + 1.75 \frac{\alpha_1 \langle \rho_1 \rangle_1 \overline{\langle \mathbf{v}_2 \rangle - \langle \mathbf{v}_1 \rangle}}{d_2} . \tag{15}$$

For dilute suspensions of calcium carbonate, the trajectories of $CaCO_3$ particles do not overlap, and probabilistic considerations about the particles trajectories are irrelevant. Note yet that, in view of the assumption of dilution, frictional interparticle stresses are not taken into account in the averaged balance equation of linear momentum of the solid phase. As a matter of fact, for solid-liquid mixtures whose Bagnold number Ba satisfies the condition $Ba < 40$,

$$Ba = \frac{\langle \rho_2 \rangle_2 d_2 \phi^{0.5} \gamma}{\mu_1} , \tag{16}$$

where ϕ is the packing density of solid particles and γ is the shear rate, one may state that viscous fluid stresses dominate the frictional interparticle forces, and the solid-liquid mixture tends to behave as a viscous fluid. For the suspension here considered, one estimates a value for the Bagnold number that remains far below 40, unless unrealistic values for the shear rates are considered. Therefore, this legitimizes the absence of a frictional term in the solid pressure.

Numerical simulations of the gravitational settling of calcium carbonate particles in brines

The numerical simulations for the gravitational settling of calcium carbonate particles in brines were performed by considering a box of dimensions x x y x z equal to 5 cm x 10 cm x 5 cm. At time $t=$ 0.00 s, a $CaCO_3\,(s)$ layer of 0.20 cm thickness is positioned at the top of the box, which falls under action of the gravitational field. For the sake of simplicity, the effects of coalescence and disintegration of particles were neglected by imposing a restitution coefficient of collision equal to 0.90. No-slip condition was imposed at the walls of the box, and the values for the physical parameters presented in Table 2 are coherent with experimental observations [7, 8]. The drag force was calculated, according to the Syamlal-O'Brien model, Wen-Yu model, and Gidaspow model. Furthermore, the numerical simulations were carried out in ANSYS/FLUENT Release 16.2 [9], by using a phase coupled SIMPLE scheme and a time step size of 0.03 s. First-order schemes for spatial and temporal discretizations were used.

Table 2: Physical constants employed in the numerical simulations.

Physical constant	**Value**
Viscosity of brine	0.001750 kg m^{-1} s^{-1}
Density of brine	1198 kg m^{-3}
Density of $Ca_2CO_3\,(s)$	2800 kg m^{-3}
Diameter of solid particles	100 μm
Restitution coefficient	0.90
Maximum packing limit	0.52

In Figure 1, the volume fraction contour graphs for calcium carbonate at the diagonal plane that connects the vertices (0 cm, 10 cm, 5 cm), (5 cm, 10 cm, 0 cm), and (0 cm, 0 cm, 5 cm) of the box are shown. Then, according to such contour graphs, one initially has a layer of denser particles that overlies a brine solution. Due to gravity, this initial configuration is unstable, and the calcium carbonate nuclei are accelerated downwards, so that pertubations in the interface between the solid and liquid phases start growing. During the settling process, the initial flat interface evolves into fingering structures, which ressemble to the well-known Rayleigh-Taylor instability [10, 11]. As a matter of fact, this situation contrasts with that expected for sedimentation of particles in vacuum, where a slightly planar interface might continuously move downwards until the solid particles cover the bottom of the box as a compact layer [12].

Furthermore, Figure 1 shows the formation of different settling regions. Usually, during the gravitational settling, one observes the formation of three different sedimentation zones. At the top, one initially has a compact layer with the maximum packing limit, where the motion of particles is constrained. In turn, in intermediate zones, one notices a formation of an unstable layer, in which fingering structures develop. In this region, gradients of volume fraction and velocity are observed as a consequence of the fluctuations of density and velocity. Finally, some instants after the beginning of the experiment, a discrete layer will be formed at the bottom of the box, separating the particle-depleted liquid from the suspension. Such a discrete layer evolves with time until an equilibrium state is reached, and no more particles sedimentate.

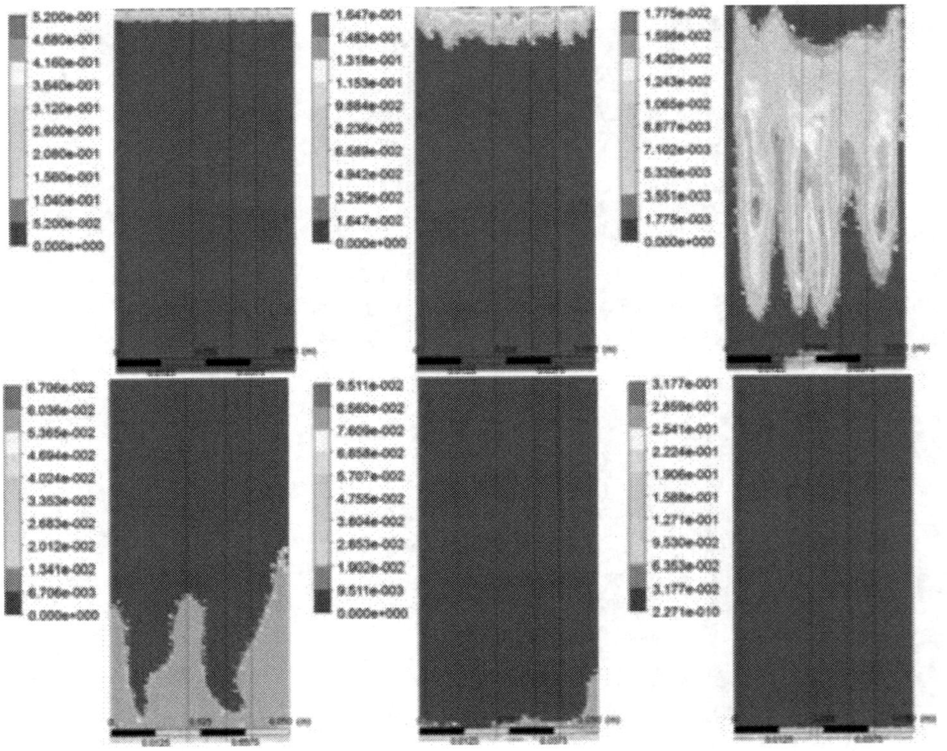

Figure 1: Volume fraction contour graphs for calcium carbonate at a diagonal plane defined by the vertices (0 cm, 10 cm, 5 cm), (5 cm, 10 cm, 0 cm), and (0 cm, 0 cm, 5 cm) of the box. The contour graphs refer to times 0.00, 0.24, 1.56, 3.84, 4.56, and 36 s. Results obtained from the Syamlal-O'Brien model.

For a dilute suspension of calcium carbonate in brine, Figures 1-3 indicate that the solid phase settles with a maximum volume fraction near to 0.31. This suggests that the $CaCO_3$(s) nuclei present some sort of attractive interactions that reduce the average distance between the individual particles, making easier the percolation of brine in the solid phase. As a consequence, at the bottom particles pack with a maximum packing below than that ideally foreseen.

On the other hand, Figures 1-3 also show that the pattern of the fingering structures observed during the sedimentation depend on the drag force model. Figures 1-3 indicate that the fingering structures are more pronounced, when the drag force is calculated in accordance with the Wen-Yu model and Gidaspow model. A possible explanation for this might be related to the fact that the interparticle interactions in the considered suspensions of $CaCO_3$(s) in brine might not be so insignificant, as assumed in the Wen-Yu model and Gidaspow model. In other terms, such interparticle interactions are negligible in comparison to the viscous forces present in the mixture as indicated by the estimation of the Bagnold number, but they should not be neglected when one accounts for the drag force.

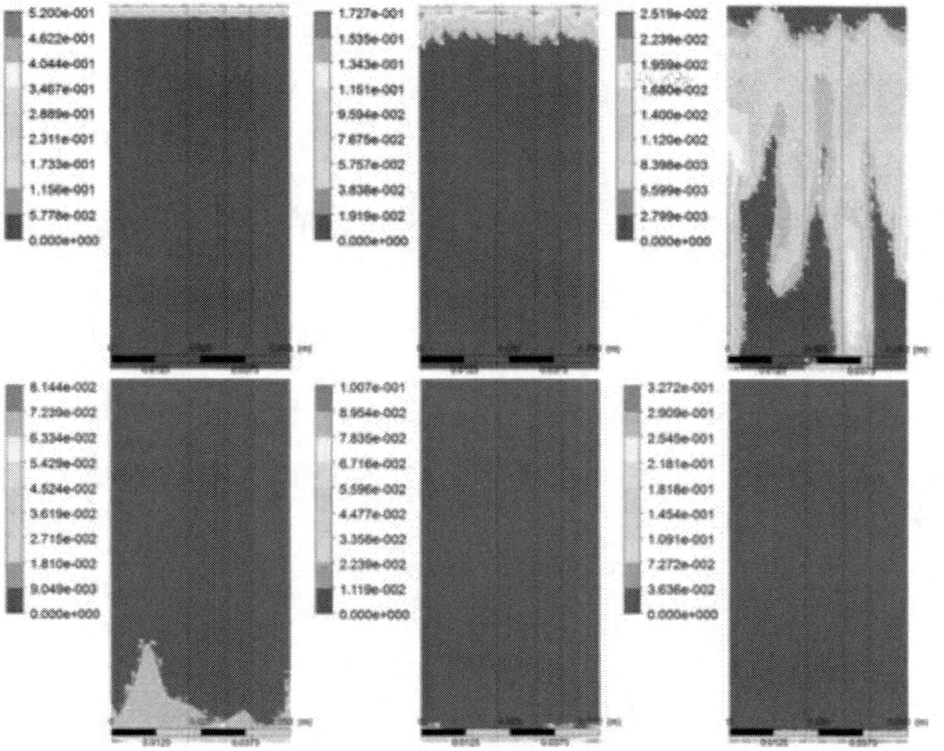

Figure 2: Volume fraction contour graphs for calcium carbonate at a diagonal plane defined by the vertices (0 cm, 10 cm, 5 cm), (5 cm, 10 cm, 0 cm), and (0 cm, 0 cm, 5 cm) of the box. The contour graphs refer to times 0.00, 0.24, 1.56, 3.84, 4.56, and 36 s. Results obtained from the Wen-Yu model.

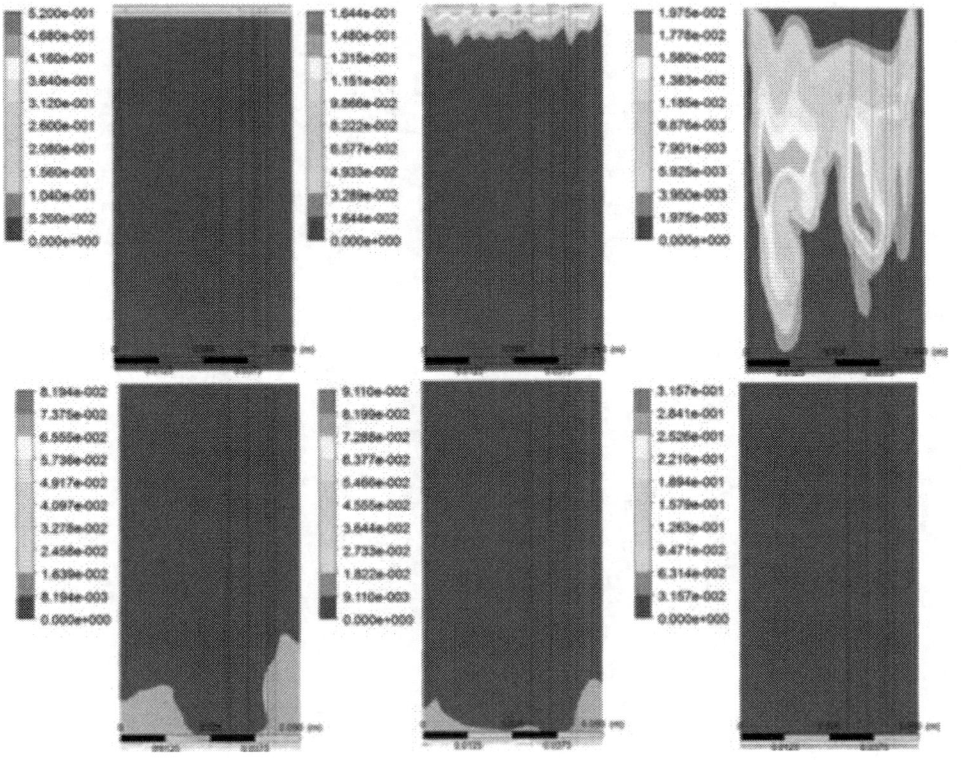

Figure 3: Volume fraction contour graphs for calcium carbonate at a diagonal plane defined by the vertices (0 cm, 10 cm, 5 cm), (5 cm, 10 cm, 0 cm), and (0 cm, 0 cm, 5 cm) of the box. The contour graphs refer to times 0.00, 0.24, 1.56, 3.84, 4.56, and 36 s. Results obtained from the Gidaspow model.

References

[1] Ishii, M. and Hibiki, T. (2006), *Thermo-fluid Dynamics of Two Phase Flows*, New York, NY, Springer.

[2] Whitaker, S. (1973), "The Transport Equations for Multiphase Systems", *Chemical Engineering Science*, January, pp. 139-147.

[3] Syamlal, M. and O'Brien J., T. (1989), "Computer Simulation of Bubbles in a Fluidized Bed", *AIChE Symposium Series*, January, pp. 22-31.

[4] Lun, C.K.K, Savage, S.B., Jeffrey, D.J. and Chepurniy, N. (1984), "Kinetic Theories for Granular Flow: Inelastic Particles in Couette Flow and Slightly Inelastic Particles in a General Flow Field", *Journal of Fluid Mechanics*, March, pp. 223-256.

[5] Wen C., Y.- and Yu H., Y. (1966), "Mechanics of Fluidization", *Chemical Engineering Progress Symposium Series*, pp. 100-111.

[6] Gidaspow, D., Bezburuah, R. and Ding, J. (1992), "Hydrodynamics of Circulating Fluidized Beds. Kinetic Theory Approach", in *Fluidization VII- Proceedings of the 7th Engineering Foundation Conference on Fluidization*, O.E. Potter and D.J. Nicklin, ed. Brisbane, QLD, Engineering Foundation.

[7] Isdale, J.D., Spence, C.M. and Tudhope, J.S. (1972), "Physical Properties of Seawater Solutions: Viscosity", *Desalination*, June, pp. 319-328.

[8] Francke, H. and Thorade, M. (2010), "Density and Viscosity of Brine: An Overview from a Process Engineers Perspective", *Chemie der Erde- Geochemistry*, August, pp. 23-32.

[9] *ANSYS ® Academic Research* (2015), Release 16.2.

[10] Lange, A., Schröter, M., Scherer, M.A., Engel A. and Rehberg, I. (1998), "Fingering Instability in a Water-Sand Mixture", *The European Physical Journal B*, August, pp. 475-484.

[11] Völtz, C., Pesch, W. and Rehberg, I. (2001), "Rayleigh-Taylor Instability in a Sedimenting Suspension", *Physical Review E*, Dezember, pp. 011404-1-011404-7.

[12] Niebling, M.J., Flekkøy, E.G., Måloy, K.J. and Toussaint, R. (2010), "Sedimentation Instabilities: Impact of the fluid compressibility and viscosity", *Physical Review E*, November, pp. 051302-1-051302-9.

464579 Volumetric and Viscometric Properties of a Binary Mixture of a Model Diesel Compound with a Biobased Additive

Abstract:

Introduction and background

As a consequence of ongoing regulations to avoid the depletion of fossil fuels as well as environmental concern about their use, renewable sources have gained increasing attention during the past two decades. Among them are biofuels, having biodiesel an increasing prominence.

One of the grand challenges of the biodiesel industry is to manage the surplus of glycerol generated as a by-product of its manufacturing process, whose price has steeply decreased in the past years [1]. Among the routes to exploit such compound, its chemical valorisation is on the spot as a way to obtain value-added products [2].

Among them are acetals, which derive from the reaction of glycerol with an aldehyde or ketone. Solketal (Sk) is the acetal that has attracted the most attention in terms of research [3], for its synthesis is made through the reaction of glycerol with widely available acetone [4]. Sk has proven its relevance as additive to fuels and biofuels by improving certain performance parameters [5-7].

Data on properties of fuels is important to the design of engines, due to the effect that they have on the combustion quality. Therefore, the determination of volumetric and viscometric properties is of relevance. N-hexadecane (also known as cetane) has typically been employed as a compound to model the behavior of diesel [8], so considering that Sk is a promising additive, this work intends to present the volumetric and viscometric properties of binary mixtures consisting of cetane and solketal.

Methodology

N-hexadecane (purity > 99%) and Solketal (DL-1,2-Isopropylideneglycerol, purity > 98%) by Sigma-Aldrich Co Ltd were used for the the binary mixtures. To prepare samples of varying composition (intervals of 0.1 of molar fraction), a Kern and Sohn ABS 220-4 analytical balance with a 0.0001 g accuracy was used.

Viscosity was measured using a temperature controlled Malvern Bohlin rheometer using a cone (60 mm diameter, 2° angle, 52 µm truncation) and plate geometry. The uncertainties associated to the viscosity and temperature were ±0.001 mPa s and ±0.1 K.

The density measurements were performed in a Krüss K100 tensiometer heated by a temperature controlled Tecam TE-7 Tempette water bath, with an uncertainty in the measurement of ±0.001 mN/m and ±0.01 K, respectively.

Triplicate measurements were performed for each data point for both the properties.

Summary of results and conclusions

The values of density and viscosity of pure n-hexadecane and solketal were validated with already existing data at 293.15, 303.15, 313.15 and 323.15 K for viscosity and 298.15, 303.15, 313.15 and 323.15 K in the case of density. For solketal, whose use is novel, data are still very scarce and referred only to 298.15 K and purities around 95%. The data obtained for the pure n-hexadecane did not differ in more than 3.1% at most for viscosity 0.3% for density.

Fitting of the viscosity (µ) and density (ρ) as a function of temperature were made for the pure components according to equations 1 and 2, respectively:

$\mu = \mu_0 e^{-k/T}$ (1)

$\rho = a + bT$ (2)

The values of the fitting parameters therein included were as follows:

- n-hexadecane: $\mu_0 = 6.24\text{x}10^{-6}$ and k=1846.20 for equation 1; a=0.9747 and $b=-6.84\text{x}10^{-4}$ for equation 2.
- solketal: $\mu_0 = 2.59\text{x}10^{-7}$ and k=3141.09 for equation 1; a=1.3102 and $b=-8.30\text{x}10^{-4}$ for equation 2.

Then, viscosities and densities of the 9 binary mixtures consisting of increasing molar fractions of solketal in n-hexadecane were measured. From the observed data, the excess molar volumes (V^E), viscosity deviations (Δµ) and excess Gibb's free energy of activation for viscous flow (ΔG^E) were computed following equations 3 to 5:

$V^E=(x_1M_1+x_2M_2)/\rho-(x_1M_1/\rho_1+x_2M_2/\rho_{2)}$ (3)

$\Delta\mu=\mu-[x_1\mu_1+(1-x_1)\mu_1]$ (4)

$\Delta G^E=RT[\ln(\mu V)-(x_1\ln(\mu_1V_1)+(1-x_1)\ln(\mu_2V_2)]$ (5)

where x_i and M_i represent the molar fractions and molecular weights of n-hexadecane (1) and solketal (2) in the mixture, V are the molar volumes and R is the ideal gas constant.

Finally, from the excess properties and deviations computed, correlation to a Redlich-Kister type equation was performed.

$P=x_1(1-x_1)\Sigma A_i(1-x_1)^i$ (6)

in which P is the excess property or deviation and A_i are the fitting parameters in the equation.

References

1. Quispe, C.A.G., C.J.R. Coronado, and J.A. Carvalho, Jr., *Glycerol: Production, consumption, prices, characterization and new trends in combustion.* Renewable & Sustainable Energy Reviews, 2013. **27**: p. 475-493.

2. Behr, A., et al., *Improved utilisation of renewable resources: New important derivatives of glycerol.* Green Chemistry, 2008. **10**(1): p. 13-30.

3. Nanda, M.R., et al., *Catalytic conversion of glycerol for sustainable production of solketal as a fuel additive: A review.* Renewable and Sustainable Energy Reviews, 2016. **56**: p. 1022-1031.

4. Esteban, J., M. Ladero, and F. Garcia-Ochoa, *Kinetic modelling of the solventless synthesis of solketal with a sulphonic ion exchange resin.* Chemical Engineering Journal, 2015. **269**: p. 194-202.

5. Garcia, E., et al., *New Class of Acetal Derived from Glycerin as a Biodiesel Fuel Component.* Energy & Fuels, 2008. **22**(6): p. 4274-4280.

6. Giraldo, S.Y., et al., *Síntesis de Aditivos para Biodiesel a partir de Modificaciones Químicas de la Glicerina.* Información tecnológica, 2009. **20**(6): p. 75-84.

7. Mota, C.J.A., et al., *Glycerin Derivatives as Fuel Additives: The Addition of Glycerol/Acetone Ketal (Solketal) in Gasolines.* Energy & Fuels, 2010. **24**: p. 2733-2736.

8. Wang, X., X. Wang, and J. Chen, *Experimental investigations of density and dynamic viscosity of n-hexadecane with three fatty acid methyl esters.* Fuel, 2016. **166**: p. 553-559.

464621 Consideration of Ionic Liquid Dissociation in Thermodynamic Modeling of Ionic Liquids and Mixtures

Abstract:

Ionic liquids (ILs) are considered as "green solvent" for their negligible vapor pressure, and are potential candidates of replacing conventional volatile solvents which are not environment friendly. Some of the applications are CO_2 capturing, separation solvents, refrigerants, novel electrolytes in batteries and photovoltaic cells, etc. Like all electrolytes, properly accounting for the dissociation chemistry of ionic liquids whether in pure or in solvents is essential in developing accurate thermodynamic models to account for the liquid phase non-ideality as function of compositions and temperature. Experimental data show that ionic liquids are partially dissociated [refs]. However, in literature, Ionic liquids are often approximated as all molecular form, completely dissociated, or constant dissociation in solutions for the thermodynamic modeling. Ionic liquids are sometimes treated with different extents of dissociation in the two liquid phases [1]. In this work, we investigate three common ILs, i.e., [emim][$EtSO_4$], [emim][TFO], and [emim][TFA] in water, ethanol, and methanol. Partial dissociation of ionic liquids is explicitly considered together with Symmetric electrolyte Non-Random Two Liquid (eNRTL) [2]. The binary interaction parameters are correlated with experimental data on vapor pressure [1,3-5], infinite dilution activity coefficient [6], excess enthalpy [6], and dissociation extent [1]. Our new approach accurately correlates all available data and the model predicts the ionic liquid dissociation phenomena over the entire concentration and temperature ranges. In next steps, molecular dynamic simulation and experimental measurements are to be conducted to elucidate the nature of the complex formation resulting from the ionic liquids dissociation chemistry.

[1] Simoni LD."Predictive Modeling of Fluid Phase Equilibria for Systems containing Ionic Liquids."University of Notre Dame.2009.(Doctoral dissertation)

[2] Song Y,Chen C-C."Symmetric Electrolyte Nonrandom Two-Liquid Activity Coefficient Model." *Industrial Engineering Chemistry Research*.2009;48:7788-7799.

[3] Orchillés, AV,Miguel PJ,Vercher E,Martínez-Andreu A."Isobaric Vapor-Liquid Equilibria for 1-Propanol+Water+1-Ethyl-3-methylimidazolium Trifluoromethanesulfonate at 100 kPa."*Journal of Chemical & Engineering Data*.2008;53:2426-2431.

[4] Wang JF,Li CX,Wang ZH."Measurement and Prediction of Vapor Pressure of Binary and Ternary Systems Containing 1-Ethyl-3-methylimidazolium Ethyl Sulfate.*Journal of Chemical & Engineering Data*.2007;52:1307-1312.

[5] Calvar N,González B, Gómez E, Domínguez Á."Vapor–Liquid Equilibria for the Ternary System Ethanol + Water + 1-Ethyl-3-methylimidazolium Ethylsulfate and the Corresponding Binary Systems Containing the Ionic Liquid at 101.3 kPa."*Journal of Chemical & Engineering Data*.2008;53:820-825.

[6] Kato R,Gamehling J."Systems with ionic liquids: Measurement of VLE and γ^{∞} data and prediction of their thermodynamic behavior using original UNIFAC, mod. UNIFAC(Do) and COSMO-RS(Ol)."*Journal of Chemical Thermodynamics*.2005;37:603-619.

465343 Accurate and Efficient Thermodynamic Properties from Molecular Simulation By Combining Lammps and Gomc

Abstract:

Many common applications of force fields involve molecular dynamics, but force field development is generally pursued within a Monte Carlo (MC) framework. This distinction can lead to dichotomous mindsets relating to subtle but significant details. For example, most force fields have been developed based on the assumption of spherical symmetry and tail corrections to compute the thermodynamic properties, which is quite convenient for MC simulation. Unfortunately, a fundamental issue arises in the context of molecular dynamics because the discontinuity in the force field undermines the conservation of energy, leading to a range of possible compensating treatments, and poor consistency from one implementation to the next. Furthermore, many common applications involve inhomogeneous fluids where this assumption breaks down. Although Ewald methods have been suggested as a remedy, our experience with these methods has left the issue unresolved.

In this work, we suggest the development of a Transferable Shifted Force Field based on the United Atom perspective (TranSFF-UA), taking the TraPPE-UA model as a starting point. We universally apply a cut and shift of 3Ïƒ, where Ïƒ is the position where the potential initially crosses zero. This seemingly simple change leads to remarkable difficulty when implemented in the commonly available simulation packages that we have considered: LAMMPS, GOMC, HOOMD-BLUE, and Cassandra.

This presentation focuses particularly on the promise and pitfalls of these prospective platforms. At first glance, LAMMPS offers the prospect of greatest consistency by developing the force field entirely within the molecular dynamics framework in which it is to be applied. This necessitates thermodynamic integration, however, which is quite feasible except where one might least suspect a problem: low density and high temperature. LAMMPS shows its greatest strength at low temperature and high density. On adapting for a shifted force, Cassandra and HOOMD-BLUE exhibit inconsistencies that remain unresolved after months (going on years) of interacting with these package developers. GOMC, on the other hand, is easily adapted for the shifted force while providing consistency with LAMMPS and resolving the issue of accuracy at high temperature and low density. At reduced temperatures greater than $T/T_c = 0.7$, the GCMC option of GOMC provides highly efficient generation of phase envelopes and saturation properties. To demonstrate this methodology, vapor pressure, saturated liquid density, and compressed fluid density are analyzed for n-alkanes, branched alkanes, ethers, olefins, naphthenics and aromatic compounds, and water.

This work is supported in part by the Scientific and Technological Research Council of Turkey (TUBITAK) through the grant no. 114M178.

467229 Influence of Ionic Activity on the Osmotic Energy Generation in a Nanotube

Abstract:

In this study, the salinity gradient-driven energy conversion by the reverse electrodialysis (RED) in a nanotube is theoretically investigated with taking into account the activity of ionic species, which is different from the previous studies where the activity was neglected. The influences of salinity gradients on the osmotic current, diffusion potential, transference number of cations, maximum power density, and maximum energy conversion efficiency in a nanotube are comprehensively investigated. Results show that the osmotic current and the maximum power density calculated from the present model with considering the activity of ionic species is smaller than those from the existing model without considering the activity of ionic species. However, the reverse trends are observed for the results of transference number of cations and the maximum energy conversion efficiency. The relative deviations of the maximum power density and energy conversion efficiency between the results with and without considering the activity of ionic species can be on the order as high as ca. 12 %.

467274 Spray Cooling of Hot Steel Plate Using Water Based TiO_2 Nanofluid

Abstract:

Spray cooling of hot steel plate using water based TiO_2 nanofluid

Samarshi Chakraborty [a], Surjya Kanta Pal [b], Sudipto Chakraborty [a]

[a]*Department of Chemical Engineering, Indian Institute of Technology Kharagpur, India.*

[b]*Department of Mechanical Engineering, Indian Institute of Technology Kharagpur, India.*

Abstract

With the progress in the nanotechnology, in past two decade synthesis of nano-sized particle has become well known reality and with ever increasing demand in heat transfer application and thermal engineering sector, high performance cooling is the need of the hour. Engineered coolant such as nanofluids [1] (particle size less than 100 nm) provides much improved thermo-physical properties (thermal conductivity, surface tension) as compared to conventional coolant (i.e. water, ethanol, etc.) therefore can be implemented for various industrial applications such as power production, air conditioning system, electronics, and steel manufacturing process. On the other hand, spray cooling process is a known technique faster heat extraction in numerous industrial applications, such as electronics, metal manufacturing. Cooling rate at run-out table (Temperature range 900°-600° C) helps to achieve desired metal microstructure which in turn leads to improved mechanical properties [2]. The experimental condition maintained during current study is similar to the condition present during the heat treatment of hot steel strip at the run out table. In the present work, TiO_2 nanofluid and spray cooling technique are used to enhance the heat transfer rate of a hot steel plate (initial plate temperature >900°C) in the aforementioned temperature region. TiO_2 nanoparticle used in this study was synthesized using co-precipitation method [3]. Effect of varying nanoparticle concentration on thermo-physical properties (thermal conductivity, surface tension and viscosity) of nanofluid has been studied in details. A stainless steel plate with a dimension of 10 cm (length) × 10 cm (breadth) × 6 mm (thickness) were used throughout the experiment. A full cone spray nozzle (Manufacturer: Lechler 460-843-17-CG) with a spray angle of 45° were used during the experiment. The coolant flow rate and spray height were maintained at 16 lpm and 6 cm, respectively. The transient temperature-time data were measured by three K type thermocouple and recorded via data acquisition system (NIcDAQ-9174 and NI 9211 card) for further processing. An inverse heat conduction solver named INTEMP [4] were used to calculate surface temperature and heat flux from the recorded data. The experimental results reveal that with addition of TiO_2 nanoparticle in water, surface cooling rate increases as compared cooling rate achieved by water. However, with increasing surface cooling rate is observed up to an optimum concentration beyond which it starts to decrease again. Maximum cooling rate of 151.4°C/s was achieved at nanoparticle concentration of 40 ppm which is greater than that achieved by water. Therefore, it can be concluded that TiO_2nanofluid can act as superior coolant in run-out table cooling operation for achieving greater cooling rates.

Table 1: Thermo-physical properties of TiO_2nanofluid at optimum concentration

Coolant	Thermal Conductivity (W/mk)	Surface Tension (mN/m)	Viscosity (mPa-sec)
Pure water	0.57	71.93	0.89
TiO_2nanofluid (40 ppm)	0.76	70.73	1.53

Keywords: Spray cooling; TiO_2nanofluid; Steel.

Reference:

[1] S.U.S. Choi, Enhancing thermal conductivity of fluids with nanoparticles, in: American Society of Mechanical Engineers, Fluids Engineering Division (Publication) FED, 1995, pp. 99-105.

[2] S.D. Cox, S.J. Hardy, D.J. Parker, Influence of runout table operation setup on hot strip quality, subject to initial strip condition: heat transfer issues, Ironmaking & Steelmaking, 28 (2001) 363-372.

[3] R. Saleh, N. Putra, R.E. Wibowo, W.N. Septiadi, S.P. Prakoso, Titanium dioxide nanofluids for heat transfer applications, Experimental Thermal and Fluid Science, 52 (2014) 19-29.

[4] D.M. Trujillo, H.R. Busby, Practical Inverse Analysis in Engineering with 3.5 Disk, CRC Press, Inc., 1997.

468371 Empirical Fundamental Equations of State Correlations Based on Hybrid Datasets

Abstract:

The process engineering industry is in an ever increasing demand for thermodynamic data that cannot be satisfied exclusively with experimental measurements due to cost and time inefficiency. By using fundamental equations of state (EOS), this problem can be circumvented, since EOS extend the range of available experimental data by offering interpolation and extrapolation capability. If the EOS is expressed in terms of a thermodynamic potential, it has the additional ability that every other time-independent thermodynamic property can be obtained as a combination of its derivatives with respect to its natural variables. Unfortunately, EOS exist only for a very limited subset of the pure compounds that are in technological use.

The construction of an EOS normally requires less data than the amount that would be needed to map the entire fluid region of technological relevance. However, in cases of extreme thermodynamic conditions or hazardous fluids, the experimental database is too small anyway. For mixtures, where the range of required data increases drastically with the number of components, the situation is much worse.

Molecular simulation has evolved to a point where it can contribute effectively to thermodynamic data retrieval and complement experimental databases. Its predictive capability is limited only by the molecular interaction model that represents the investigated substance. In contrast to experiments, molecular simulation can straightforwardly target any state point of interest while the associated financial cost and time requirement is only a fraction than that of a corresponding laboratory measurement.

A number of EOS for several substances were constructed using both experimental and simulation data, i.e. with a hybrid database, in the course of this work. Results for Hydrogen Chloride [1], Ethylene Oxide [2],

Hexamethyldisiloxane [3] and Octamethylcyclotetrasiloxane [4] are presented.

[1] *The Journal of Chemical Physics* 139: 041102 (2013)

[2] *Chemical Engineering Science* 121: 87-99 (2014) and 134: 887-890 (2015).

[3] *Fluid Phase Equilibria* 418: 133-151 (2016).

[4] *Journal of Chemical & Engineering Data*, in press (2016).

466535 Physico-Chemical Characterization of Fluorinated Ionic Liquids for Their Use As Blood Substitutes

Abstract:

The fluorination of ionic liquids (ILs) (fully or partially) is of particular interest in areas where perfluorocarbons (PFCs) find relevant applications, such as separation methods and fluorous phase organic synthesis, surfactants in supercritical solvents, substitutes for chlorinated solvents or biomedical applications. In the latter case, PFCs are used as in vivo gas carriers in liquid ventilation or artificial blood substitute formulations [1]. This first generation of products, suitable as oxygen carriers only is now in clinical trials. It has an enormous potential in surgery with extensive blood loss, and different substitutes have already been developed for this purpose.

The nearly null volatility of ionic liquids at atmospheric conditions, their easy recovery and therefore, recyclability, as well as their tunable toxicity completely justify the application of fluorinated ionic liquids (FIL) for improving the available oxygen therapeutics. The aforementioned aspects motivate the development of novel neoteric FILs combining the best of perfluorocarbons properties with those of ILs. However, it is crucial to consider multiple aspects related to their thermophysical behavior in the presence of water and respiratory gases.

In this contribution, the phase, interface and structural behaviour of some of these compounds have been theoretically characterized using the soft-SAFT molecular-based equation of state [2], in combination with the Density Gradient Theory (DGT) [3] for the study of interfacial properties. Based on new density, surface tension and phase equilibrium data [4], several FILs with the imidazolium and pyridinium cations, and the perfluoroalkylsulfonate [$C_4F_9SO_3$] perfluoroalkylcarbonate [$C_4F_9CO_2$] anions have been studied in detail. A molecular model based on the analysis of the distribution charges on the anion has been proposed and optimized to find a reliable set of parameter values describing the main properties of those fluids in good agreement with the data. An analysis of the molecular parameters based on the analogy between the

chemical structures has been performed. Next, the vapor-liquid and liquid-liquid equilibria the respiratory gases (oxygen, nitrogen and carbon dioxide) and water has been reproduced. Preliminary results on the formation of aggregates are also provided. When data were not available, the theoretical approach was used to predict the solubility behaviour of the respiratory gases by transferring the molecular parameters.

Acknowledgements:

F. Llovell and Ana B. Pereiro acknowledge two Short Term Scientific Mission Grants (STSM-CM1206-020315-056728 and STSM-CM1206-010316-071781 STSM), within the framework of the COST Action EXIL - Exchange on Ionic Liquids. Additional support was provided by a TransBio SUDOE collaborative project (TRANSBIO-BCN-GT2-A1-03), the Catalan Government (2014SGR-1582) and the Fundação para a Ciência e Tecnologia (FCT/MEC, Portugal) through the contracts under Investigador FCT 2014 (A. B. Pereiro and J. M. M. Araújo) and through the project PTDC/EQU-FTT/118800/2010.

References:

[1]. J. G. Riess, Chem. Rev. 101 (2001) 2797-2919.

[2]. F. J. Blas, L. F. Vega, Mol. Phys. 92 (1997) 135-150.

[3]. J. W. Cahn, E. J. Hilliard, J. Chem. Phys. 28 (1958) 258–267.

[4]. N.S.M. Vieira, A. Luís, P.M. Reis, P.J. Carvalho, J.A. Lopes-da-Silva, J.M.S.S. Esperança, J.M.M. Araújo, L.P.N. Rebelo, M.G. Freire, A.B. Pereiro, J. Chem. Thermodynamics 97 (2016) 354–361.

469303 Effect of Branching on Phase Behavior of Chain Molecules from a New SAFT Based Model

Abstract:

A new model is proposed to accurately predict the effect of branching on the thermodynamics of chain molecules based on Statistical Associating Fluid Theory (SAFT). Despite numerous experimental and simulation results demonstrating the different phase behavior for isomers, few theoretic models have been developed to predict and explain the effect. Current branch chain models are mostly based on lattice theory, yielding results only qualitatively in agreement with experimental data. Since SAFT has proved extremely successful in characterizing chain molecules, the model proposed in the study makes use of one version of SAFT that assumes Lennard Jones dimer as the reference fluid, and accounts for the branching effect through approximation of Wertheimâ€™s Second Order Perturbation Theory (TPT2). Both the number of branches and position where branches grow are taken into consideration explicitly.

The model is firstly applied to isomeric alkanes and then extended to polyolefins with various branching distribution. Given minimal parameters, the model is applicable to various systems from pure component to mixtures, from melt blends to polymer solutions. Vapor liquid equilibrium (VLE) and liquid liquid equilibrium (LLE) results generated from the model reflect the changes in thermodynamic properties due to addition of branches in good agreement with experimental data. In the future, association will be introduced to facilitate the applicability of the model to a wider category of molecules, specifically those with functional groups.

469305 Comparative Study Between Dissociation Conditions of H_2o + N_2+ Undecane and H_2o + CO_2+ Undecane Mixtures in Hydrate Formation

Abstract:

Gas hydrates are crystalline solid ice-like structures, formed by water and small gas molecules with low molecular weight, which is housed in the cavities formed by the hydrogen bonds at conditions of hydrate formation (high pressure, low temperature, small gas molecules and presence of water). The hydrates are typically stable at low temperatures and high pressures, such as the conditions found in deep waters. In México there are reservoirs with high content of water (20 % or more) like Cantarell where hydrate formation can be simple. In this work, the experimental dissociation conditions for the hydrate formed in the mixtures: $H_2O + C_{11}H_{24} + CO_2$ and $H_2O + C_{11}H_{24} + N_2$ were reported in the temperature range of (265.15-293.15) K and pressures up to 4 MPa for CO_2 and 30 MPa for N_2. The composition expressed as mass fraction of undecane was 0.05 and 0.10. The experimental equipment used in this work is based on the static-synthetic-isochoric method. The combined expanded uncertainties of the dissociation conditions measurements were evaluated to be less than 0.012 MPa, 0.020 K and 0.0020 for pressure, temperature and mass fraction, respectively.

469413 A Comprehensive Molecular Thermodynamic Model for Major Electrolytes in High Salinity Produced Water

Abstract:

Managing produced water in the rapidly developing shale gas industry is a major challenge for protecting the environment [1]. Desalination for reuse of high salinity produced water depends on novel material and process innovations. We present recent advances in the development of a comprehensive thermodynamic model aimed to support heat and mass balance calculations and process simulation of desalination processes with produced water. The major ions of concern for produced water management include Na^+, K^+, Mg^{2+}, Ca^{2+}, Ba^{2+}, Sr^{2+}, Cl^-, HCO_3^- and SO_4^{2-} ions. Based on the electrolyte Non-Random Two Liquid theory (eNRTL) for electrolyte solutions [2], the concentration dependency of the solution nonideality is accounted for with two binary interaction parameters for each of the water–electrolyte interaction pairs and the electrolyte–electrolyte interaction pairs. The temperature dependency of the binary interaction parameters is accounted for with a Gibbs-Helmholtz type expression with three temperature coefficients representing excess Gibbs energy, excess enthalpy, and excess heat capacity contributions. With the binary parameters regressed from thermodynamic data of aqueous single electrolyte binary systems and aqueous two electrolyte ternary systems, the model provides accurate representations and reliable predictions for various thermodynamic properties of quaternary and quinary systems examined thus far for the aqueous Na^+–K^+–Mg^{2+}–Ca^{2+}–Cl^-–SO_4^{2-} hexary system with temperatures from 273.15 K to 473.15 K and concentrations up to salt saturation. The model is being extended to cover Ba^{2+}, Sr^{2+} and HCO_3^- ions and it should become an indispensable scientific tool in the development of novel desalination processes for high salinity produced water in oil and gas productions.

References

[1] D.L. Shaffer, L.H. Arias Chavez, M. Ben-Sasson, S. R.-V. Castrillón, N.Y. Yip, M. Elimelech, Environ. Sci. Technol. 2013, 47:9569-9583.

[2] Y. Song, C.-C. Chen, "Symmetric Electrolyte Nonrandom Two-Liquid Activity Coefficient Model," Ind. Eng. Chem. Res. 2009, 48:7788-7797.

469195 Thermodynamic Modeling of Polyelectrolyte Solutions with Electrolyte Nrtl Model

Abstract:

Thermodynamic modeling of polyelectrolyte solutions with eNRTL model

Yue Yu, Yuan Li, Chau-Chyun Chen

Department of Chemical Engineering

Texas Tech University

Polyelectrolytes are polymers with charged functional groups along the backbone chains. The charged functional groups are electrostatically attached with counterions that carry opposite charges. The charged polymers are attractive to counterions and repulsive to coions. Such behavior makes polyelectrolytes applicable in processes of separation, purification, biomedical systems, and energy storage systems [1-3].

Thermodynamic description of the mobile ions including both counterions and coions in polymer and solution phases draws great interest of research. Early works proposed various excess Gibbs energy models [4-7]. There had been no limiting law for polyelectrolyte solutions like Debye-Hückel theory for simple electrolytes until Manning put forward his counterion condensation theory [8]. Manning defined a "dimensionless linear charge density" of polyelectrolytes as the Manning parameter and stated that counterions would condense on the charged functional groups of polyelectrolytes when charge density was larger than a critical value. Applying Manning's model as the limiting law, Danner and his co-workers incorporated NRTL model for additional short-range interaction contribution and successfully calculated activity and osmotic coefficients [9-10]. Sadowski combined Manning's limiting law and Perturbed-Chain Statistical Association Fluid Theory to describe vapor-liquid equilibrium and polymer swelling [11]. Recently, Manning et al. extended the counterion condensation theory to predict activity coefficients of mobile ions in cross-linked ion exchange membranes in equilibrium with simple electrolyte solutions [12].

This work presents a different approach to model polyelectrolytes. A new concept of an "effective charge number" is introduced for polyelectrolytes to account for the counterion condensation derived from the charge density of the polymer chains. While the real charge number of polyelectrolytes is related to the number of ionic sites or the capacity of attracting counterions, the "effective charge number" represents the electrostatic influence of charged functional groups in the solution. The very successful electrolyte Non-Random Two Liquid (eNRTL) model is then used to describe the excess Gibbs energy of the polyelectrolyte system with the "effective charge number".

eNRTL model has been proven to be applicable to varieties of aqueous and mixed-solvent electrolyte solutions [13]. We show the eNRTL model is capable of systematically correlating and extrapolating thermodynamic properties for all components in polyelectrolyte solutions including osmotic coefficients, solvent activity, mobile ion activity coefficient, Gibbs free energy of the solution, etc. We further show successful modeling results for describing equilibrium of mobile ions in ion exchange polymers and external salt solutions,

1. Jeon, Sung-il, et al. "Desalination via a new membrane capacitive deionization process utilizing flow-electrodes." Energy & Environmental Science 6.5 (2013): 1471-1475.
2. Qiu, Yong, and Kinam Park. "Environment-sensitive hydrogels for drug delivery." Advanced Drug Delivery Reviews 64 (2012): 49-60.
3. Cai, Zhijun, et al. "High performance of lithium-ion polymer battery based on non-aqueous lithiated perfluorinated sulfonic ion-exchange membranes." Energy & Environmental Science 5.2 (2012): 5690-5693.
4. Marcus, R. A. "Calculation of thermodynamic properties of polyelectrolytes." Journal of Chemical Physics 23.6 (1955): 1057-1068.
5. Ise, N. "The mean activity coefficient of polyelectrolytes in aqueous solutions and its related properties." Fortschritte der Hochpolymeren-Forschung. Springer Berlin Heidelberg, 1971. 536-593.
6. Lammertz, Silke, et al. "A model for the Gibbs energy of aqueous solutions of polyelectrolytes." Fluid Phase Equilibria 280.1 (2009): 132-143.
7. Ghalami-Choobar, Bahram, and Tayyebe Nosrati Fallahkar. "Mean activity coefficients measurement of sodium chloride and thermodynamic modeling of sodium chloride and polysodium acrylate aqueous mixtures at T=(298.2 and 303.2) K." Fluid Phase Equilibria 378 (2014): 51-59.
8. Manning, Gerald S. "Limiting laws and counterion condensation in polyelectrolyte solutions I. Colligative properties." The Journal of Chemical Physics 51.3 (1969): 924-933.
9. Nagvekar, Manoj, and R. P. Danner. "An excess Gibbs free energy model for polyelectrolyte solutions." Fluid phase equilibria 53 (1989): 219-227.
10. Nagvekar, Manoj, Funda Tihminlioglu, and Ronald P. Danner. "Colligative properties of polyelectrolyte solutions." Fluid phase equilibria 145.1 (1998): 15-41.
11. Arndt, Markus C., and Gabriele Sadowski. "Thermodynamic model for polyelectrolyte hydrogels." The Journal of Physical Chemistry B 118.35 (2014): 10534-10542.
12. Kamcev, Jovan, et al. "Partitioning of Mobile Ions Between Ion Exchange Polymers and Aqueous Salt Solutions: Importance of Counter-ion Condensation." Physical Chemistry Chemical Physics (2016).
13. Song, Yuhua, and Chau-Chyun Chen. "Symmetric electrolyte nonrandom two-liquid activity coefficient model." Industrial & Engineering Chemistry Research 48.16 (2009): 7788-7797.

469231 Effect of Liquid Hydrocarbons (C_6, C_8, and C_{10}) in Experimental Dissociation Points of Hydrates Formed By H_2o + CO_2

Abstract:

In Mexico, oil production from mature oilfields has been an increase in produced water, for example oil wells of Cantarell. This problem results in the development of alternative processes of separation of water from oil. The formation of gas hydrates can be a very interesting alternative for it. Hydrates are ice-like crystalline solid formed through a combination of water and gas which is housed in the cavities formed by the hydrogen bonds at conditions of hydrate formation (high pressure, low temperature, small gas molecules and presence of water). However, a systematic study of the effect of compounds present in the oil (in this case alkanes) in the hydrate formation with CO_2 must be conducted.

In this work, experimental dissociation conditions of $H_2O + C_6H_{14} + CO_2$, $H_2O + C_8H_{18} + CO_2$ and $H_2O + C_{10}H_{22} + CO_2$ mixtures were reported at temperature range of (263.15-293.15) K and pressures up to 4.5 MPa. The composition expressed as mass fraction of alkane was 0.30. The combined expanded uncertainties of the dissociation conditions were estimated to be less than 0.012 MPa, 0.020 K and 0.0020 for pressure, temperature and mass fraction, respectively

469443 Real-Component Based Molecular Characterization of Petroleum Fluids

Abstract:

Abstract

As light to medium sweet crude oils keep depleting while demand of oil products are increasing, heavier and sourer crude has become a growing component to global oil market. Accompany with that, extreme bulk properties (including high molecular weights, high boiling points, large densities, high viscosities, etc.) are expected. However, current experimental approaches are still limited in characterizing heavy crudes in two aspects: 1) chemical structures of hydrocarbons with carbon number up to merely 12 can be accurately analyzed; 2) distillation cuts can only be obtained up to 750 K of normal boiling point, even with deep vacuum fractionation apparatus[1]. With limited experimental data available, less than 50 wt. % of individual heavy oil or bitumen can be analyzed, which renders simple model extrapolation over the whole crude highly unreliable.

The present work introduces a real-component based molecular characterization method[2] derived from thermodynamically-consistent correlation of available crude oil data on chemical and physical properties. This method not only gives reasonable predictions for properties of crude oil including heavy cuts, but also provides chemical structures and compositions of model constituent components over the whole crude oil, which serves as a cornerstone of high fidelity refining simulation and molecular management. The results from the molecular characterization of three crude oils including Terengganu (API 73.1), HOOPS Blend (API 34.2), and Cold Lake Bitumen (API 8.1) are presented.

Reference

1. Sánchez-Lemus, M. C., et al. (2016). "Physical properties of heavy oil distillation cuts." Fuel 180: 457-472.

2. Chen, C. C. and H. QUE (2013). "Method of characterizing chemical composition of crude oil for petroleum refinery processing". US Patent 20130185044 A1.

470012 Predicting Mosced Parameters for Nonelectrolyte Solids Using Electronic Structure Calculations

Abstract:

Accurate and efficient models to predict the phase-behavior of nonelectrolyte solids in a wide range of solvents are central to the design of novel separation processes. A promising design tool is the MOSCED limiting activity coefficient model which is parameterized for 133 solvents. However, before predictions may be made for a solute of interest, solute MOSCED parameters are required. For novel compounds of interest, the necessary MOSCED parameters are unavailable and sufficient reference data is likely unavailable to regress the necessary parameters.

In this study we explore the use of electronic structure calculations to generate the reference data necessary to obtain solute MOSCED parameters. Specifically, we use solvation free energies computed in a continuum solvent using electronic structure methods. Results will be shown wherein the method is successfully applied to a wide range of compounds of biological, environmental, and industrial interests. We additionally demonstrate how the limiting activity coefficients provided by MOSCED may be used to parameterize an excess Gibbs free energy model and used to make blind equilibrium solubility predictions.

470739 Modeling Phase Equilibria Using Numerical Algebraic Geometry

Abstract:

Accurate prediction of phase equilibrium is a fundamental aspect of applied chemical engineering thermodynamics. However, the resulting equations are often highly nonlinear and difficult to solve. Objective functions may present many local minima and trivial solutions, with multiphase and multicomponent systems being especially challenging. As a result, traditional optimization methods are sometimes wholly insufficient for solving these problems, or require unreasonably high quality initialization. Here, we demonstrate the use of homotopy continuation [and more general numerical algebraic techniques] as a powerful tool in solving these difficult problems. In particular, we outline a parameter free method for calculating phase equilibria from Helmholtz-explicit fundamental equations of state, requiring no use of auxiliary equations. Additionally, we show that polynomial homotopy continuation may be powerfully used to enumerate all existent critical points of multicomponent mixtures. These results serve to demonstrate the power of numerical algebraic geometry across a broad range of classical thermodynamic problems.

470778 Volumetric, Acoustic and Viscometric Properties of Binary Mixtures of (*n*-Butylammonium based ionic liquid + Ethanol) at Different Temperatures

Abstract:

In this work, experimental data of density, speed of sound and viscosity for binary mixtures of {ethanol + *n*-butylammonium methanoate, or + *n*-butylammonium acetate, or + *n*-butylammonium propanoate l, or + n-butylammonium oleate, or + n-butylammonium decanoate, or + n-butylammonium hexanoate ionic liquid} have been determined as a function of composition at T = (293.15 - 313.15) K. The results were used to calculate the excess molar volume, V^E , deviation in isentropic compressibility, Î"k_sÂ , deviation in viscosity, *Î"Î·*, and excess Gibbs free energy of activation of viscous flow, *Î"*G^{*E}, and correlated to the Redlich-Kister polynomial equation. Ionic liquids were synthesized by the method of acid-base neutralization reaction using *n*-butylamine and formic acid, or acetic acid, or propionic acid, or oleic acid, or decanoic acid, or hexanoic acid. *N*-Butylamine (Merck, purity > 99.0%), formic acid (Merck, purity > 98.0%), acetic acid (Merck, purity > 100%), or propionic acid (Sigma-Aldrich, purity > 99.5%), or oleic acid (Sigma-Aldrich or Fluka, purity > 65,0%), decanoic acid (Merck, purity > 98.0%), hexanoic acid (Merck, purity > 98.0%) were used without further purification. Ionic Liquids were characterized by using ^{1}H-NMR, ^{13}C-NMR and FTIR spectroscopy techniques.

The binary mixtures were prepared by weighing the pure components using an electronic balance manufactured by Shimadzu (Model AUY220). The densities and speed of sound of pure components and their solutions were measured using a commercial density and speed of sound analyzer manufactured by Anton Paar (Model DSA5000) The viscosities were measuring using a Stabinger viscometer manufactured by Anton Paar (Model SVM3000/G2). The analysis of ^{1}H-NMR and ^{13}C-NMR were obtained in an Avance III 600 HD Bruker spectrometer operating at 500 MHz for protons and 125 MHz for carbon. The IR spectra were obtained in Fourier transform infrared spectroscopy Nicolet with attenuated total reflectance (FTIR-ATR) (Model 6700).

For all systems studied, the values of excess molar volumes, deviation in isentropic compressibility and deviation in viscosity were negative over the entire composition range. The values of excess Gibbs free energy of activation of viscous flow were positive over the entire composition range. The results obtained were discussed in terms of intermolecular interactions, particularly hydrogen-bonding interactions between like and unlike molecules.

471287 Apparent Molar Volumes of Saccharides in Aqueous Sodium Chloride Mixtures at Different Temperatures

Abstract:

In the present study, density experimental data of aqueous solutions of sodium chloride containing saccharides have been determined at different temperatures and composition. Density of the solutions have been determined by using and vibrating-tube densimeter manufactured by Anton Paar (Model 4500). The saccharides studied were D(+)-lactose, D(+)-glucose, D-fructose and sucrose (molalities saccharides range from 0.095 and 1.7 mol·kg^{-1}) in water and in (0.025; 0.050; 0.075 and 0.1 mol·kg^{-1}) aqueous sodium chloride at temperatures of ***T*** = (283.15, 288.15, 293.15, 298.15 and 303.15) K. The experimental results have been used to calculate the apparent molar volumes, infinite dilution apparent molar volumes and partial molar volumes of transfer of the saccharides from water to aqueous sodium chloride solutions. The results lead to believe that possible hydrophilic-ionic interactions among the hydrophilic groups (-OH, -C=O, -O-) of the saccharides and the ions (Na^+ and Cl^-) of the electrolyte can be predominant in the systems containing D(+)-lactose, D-fructose and sucrose while hydrophobic-ionic interactions among the hydrophobic groups (R=CH) of the solute and the ions (Na^+ and Cl^-) of the co-solute can be predominant in the system containing D(+)-glucose.

470056 Remote Read-out System for Thermal Barcode Based on Phase Change Materials

Abstract:

Covert thermal barcode consists of a series of phase change materials (PCMs) that have high heats of fusion and different melting temperatures. This type of barcode is a promising tool for the protection of identities and item recognition. Paper-based covert thermal barcodes composed of the four types of PCMs: stearic acid, palmitic acid, lauric acid, and icosane were developed with particular sequences. Traditionally, a differential scanning calorimetry (DSC)-based decoding system is used to decode the barcode but is time-consuming and has low-throughput[1]. A remote read-out system using an infrared thermal camera was developed for a fast and easy alternative. The infrared thermal camera is non-contact, high resolution, high-throughput, and repeatable compared with DSC. The camera recorded the temperature variation during the melting and cooling processes of the four types of PCMs. The melting and freezing points were determined through the analysis of temperature variation curves produced. After mathematical smoothing and differencing, rates of temperature variation were obtained and the peaks generated from the sudden change in temperature rate pinpointed the exact melting temperatures. Thus, the PCMs were distinguished by their melting temperatures and three thermal barcodes were successfully decoded. This paper demonstrated that the infrared thermal camera system is a promising method to decode thermal barcodes and simultaneously analyze other thermal properties of materials.

471376 Modeling Sorption and Phase Equilibria Using CP2K Software Suite

Abstract:

CP2K is powerful software suite that allows one to perform electronic strcuture calculation for systems consisting of hundreds of atoms. In this poster, we will present some of the new features of the code that allows one to efficiently sample configurational space for highly associating systems in Monte Carlo simulations. We will also share of some of results that quantify efficacy of recent density functionals in predicting sorption and phase equilibria of molecular systems.

471471 Aluminum Nitride Hydrolysis Enabled By Hydroxyl-Mediated Surface Proton Hopping

Abstract:

The propensity for aluminum nitride (AlN) to hydrolyze has led to its consideration as a redox material for solar thermochemical ammonia (NH_3) synthesis applications where AlN would be intentionally hydrolyzed to produce NH_3 and aluminum oxide (Al_2O_3), which could be subsequently reduced in N_2 to reform AlN and reinitiate the NH_3 synthesis cycle. No quantitative, atomistic mechanism by which AlN, and more generally, metal nitrides react with water to become oxidized and generate NH_3 yet exists. In this work, we used density functional theory (DFT) to examine the reaction mechanisms of the initial stages of AlN hydrolysis, which include: water adsorption and dissociation, hydroxyl-mediated proton diffusion to form NH_3, and desorption of NH_3. We found activation barriers (E_a) for hydrolysis of 79 and 85 kcal/mol for the cases of minimal adsorbed water and additional adsorbed water, respectively, corroborating the high observed temperatures for the onset of steam AlN hydrolysis. We predict AlN hydrolysis to be kinetically limited by the dissociation of strong Al-N bonds required to accumulate protons on surface N atoms to form NH_3. The hydrolysis mechanism we elucidate is enabled by the diffusion of protons across the AlN surface by a hydroxyl-mediated Grotthuss mechanism. A comparison between intrinsic proton diffusion (E_a = 79 kcal/mol) and Grotthuss-like hydroxyl-mediated proton diffusion (E_a = 21 kcal/mol) shows that hydroxyl-mediated proton diffusion is the predominant mechanism in AlN hydrolysis. The large activation barrier for NH_3 generation from AlN (E_a = 79 or 85 kcal/mol depending on water coverage) suggests that in the design of materials for solar thermochemical ammonia synthesis, emphasis should be placed on metal nitrides with less covalent metal-nitrogen bonds and thus more facile N liberation in the form of NH_3.

471502 Molecular Modeling of Nonionic Block Copolymer Micelles

Abstract:

Nonionic block copolymer micelles have been the focus of much interest for both academia and industry over decades. The immiscibility of constituent groups causes micellar aggregates when dissolved in a selective solvent. Over the years a large number of experiments have been done to study temperature and pressure effects onto critical micelle concentration. The development of theory to model micellar solutions is limited by inhomogeneity of micellar structures. Several early successes of modeling temperature effects follow Flory-Hugginsâ€™ theory but ignores pressure effects. We investigate both how temperature and pressure can affect critical micelle concentration by inhomogeneous statistical associating fluid theory(iSAFT). iSAFT is a classical density functional theory based on statistical mechanics that uses physical molecular parameters. We found that iSAFT can capture the pressure and temperature effects onto critical micelle concentration in agreement with experimental data.

Beyond temperature and pressure, microscopic qualities of a block copolymer micelle e.g. polydispersity and chain architecture, can also influence critical micelle concentration and even result in a complex structure. Linear tiblock copolymer in fact has a higher critical micelle concentration than cyclic triblock copolymer. Microscopic entropic effect is significant in self-assembly. Classical thermodynamics, however, fails to take into microscopic effects since it ignores molecular configurations nor variation in microscopic densities. iSAFT is a suitable theoretical model to study local density effect and chain connectivity. We used iSAFT and molecular simulation tool to explain synergistic effects of mixed micellization and polydispersity induced complex micellar structures. The iSAFT theory agrees well with molecular simulation results. With iSAFT, molecular simulation, and physical parameters obtained from bulk fluids, we believe this is a new trend towards quantitative modeling of block copolymer micelles and soft material.

472707 SAFT-Q: A Transferable Parameter Quartic Equation of State Version of the Statistical Associating Fluid Theory for Pure Fluids and Mixtures

Abstract:

Compressibility factor contributions for segment repulsion, covalent bonding, dispersion and association from the early Huang-Radosz (HR) version of the statistical associating fluid theory (SAFT) equation of state [1] are recast in a series of simple terms in packing fraction that results in an overall â€œquarticâ€ in molar volume equation of state. The pure fluid version of this new model contains two universal parameters in the athermal-plus-covalent bonding EOS backbone [2], three dispersion parameters that are directly predictable (transferable) from the SAFT-HR segment number and segment energy and three *pseudo-association strengths* that are also directly predictable (transferable) from the SAFT-HR segment number and energy and the association energy and volume. Theoretically-based mixing rules for the dispersion and association parameters are derived. Results are given for representative substances including single-segment, long-chain non-associating and associating pure fluids. Liquid density, vapor pressure and coexistence curve predictions using the new SAFT-Q EOS with parameters transferred from SAFT-HR are compared with predictions from the conventional SAFT-HR model on which it is based. Vapor-liquid equilibria are predicted for binary systems containing self-associating with non-associating components and with cross-associating components. Extension of the model/approach to other SAFT versions (e.g., soft-SAFT, PC-SAFT and SAFT-VR) is briefly discussed.

REFERENCES

[1] Huang, S. H. and M. Radosz, Equation of State for Small, Large, Polydisperse, and Associating Molecules, *Ind. Eng. Chem. Res.* **29**, 2284-2294 (1990).

[2] Gow, A. S., S. Alkhaldi and S. Demir, Cubic and Quartic Hard-Sphere and Lennard-Jones Chain Equations of State as Foundations for Complex Fluid Modeling, *Fluid Phase Equilibria* **399**, 1-15 (2015).

474952 Directing Colloidal Self-Assembly

Abstract:

Colloidal and nanoparticle self-assembly is a promising approach to the nanomanufacture of advanced functional materials capable of controlling the transport of heat, light, and chemical species. But while thermodynamics dictates the structures that form by self-assembly, the kinetics of colloidal assembly are often trapped into arrested, non-equilibrium states.

In this talk, I will discuss the use of directing electric and magnetic fields to circumvent kinetic bottle necks during colloidal self-assembly. A useful model system has been suspensions of superparamagnetic colloids. In a strong, steady magnetic field, paramagnetic colloids form system-spanning, kinetically arrested networks similar to a gel. From this state, it is possible to phase separate and condense the suspension by toggling the external field [1]. In its evolution towards the equilibrium state, the suspension undergoes a Rayleigh-Plateau instability for a range of field strengths and toggle frequencies [2]. The particles initially chain together to form a percolated network that coarsens diffusively. With time, the surface of the growing domains in the network become unstable. The amplitude of the waves eventually reaches a critical value and the columns pinch off and condense into ellipsoidal structures.

A key advantage of directed self-assembly in toggled fields is the relatively large range of field-strengths, analogous to effective temperatures, that lead to phase separation. Our results demonstrate how kinetic barriers to a colloidal phase transition are subverted through measured, periodic variation of driving forces while retaining the strengths of a "bottom-up" self-assembly process.

[1] Swan, J. W. et al., Proc. Natl. Acad. Sci. U. S. A. 2012, 109, 16023–16028.

[2] Swan, J. W.; Bauer, J. L.; Liu, Y.; Furst, E. M. Soft Matter 2014, 10, 1102–1109.

474863 Dynamic Factors Controlling Bacterial Adhesion to Polymeric and Nanopatterned Surfaces

Abstract:

Bacterial adhesion is generating increasing scientific interest in the context of biofilm formation and increased antibiotic resistance. While the complete elimination of biofilms poses a challenge addressed by surface chemists, the use of biofilms in remediation and other applications drives the creation of material supports on which biofilms can be controlled and maintained. To this end, we discuss here the dynamic aspects of early biofilm formation involving the capture of single cells and their interfacial response on model surfaces, specifically examining the development of electrostatic and hydrogen bonding interactions. We present here a study comparing the capture and relaxation processes of *S. aureus* on biocompatible PEG brushes and hydrogel coatings, polycationic antimicrobial surfaces, and nanopatterned hybrid surfaces containing PEG brush and cationic components. The work benchmarks bacterial behavior against that of silica microspheres on the same surfaces, where the negative silica charge models that on the bacteria and where silica also undergoes hydrogen bonding. Even on surfaces such as hydrogels and brushes that do not capture (arrest) bacteria, evidence for weak interactions are found. For instance in the vicinity of brushes and hydrogels, *S. aureus* undergo associated interfacial motions in flow such as rolling and hopping. While bare silica did not reproduce this behavior because of its greater density of binding sites, a partially PEG-modified silica was shown to behave similarly. By contrast to the motion signatures of adhesion capture on brushy surfaces, we reveal abrupt capture kinetics in flow on cationic surfaces. Beyond the dynamics of capture itself, we find, in the first 30 minutes after capture, evidence for surface-dependent interfacial relaxations that increases the tightness of bacterial adhesion, and we demonstrate how nanoscopic patterning can be adjusted to tune interfacial relaxation and antimicrobial character.

454308 Interfacial Phenomena Undergird Human Eye Health

Abstract:

A healthy human eye relies on the integrity of the tear film to protect the cornea. Interfacial phenomena govern the behavior of the tear-film system. Capillary hydrodynamics sets the thickness of the deposited film (at about 3 μm) and provides the smoothness necessary for precise vision. Meniscus-driven capillary suction pinches film extremities forming McDonald-Brubaker lines that prevent gravity-driven slough off. Spreading of a 100-nm lipid-layer of meibum over the tear film significantly slows evaporation rates that otherwise would consume the entire the tear layer during an interblink. Without capillarity, the human eye cannot function properly. We show here, however, that capillarity also causes dry-eye disease: a burny, itchy, scratchy feeling of dryness and discomfort. Dry eye infects up to 30 % of the global population. It is especially prevalent in the elderly and women, and in arid, windy climates.

The etiology of dry eye arises from clinical observation that during an interblink randomly distributed ruptures occur in the tear film. Under sodium-fluorescein instillation, "black spots" and/or "black streaks" appear in less than a few seconds for people who suffer from dry eye. Rapid tear breakup strongly correlates to dry-eye syndrome. In spite of decades of effort, however, there is no physically consistent explanation for how a 3-μm aqueous film on a water-wetting corneal surface breaks up within a few seconds, nor is there an explanation for how such ruptures give rise to dry-eye symptoms.

We propose local evaporative-driven tear rupture. Our proposal relies on interferometric observations that the 100-nm the meibum coating film is unstable, in agreement with *in-vitro* experiments on duplex oil films spread on water. Once the covering lipid-film ruptures, underlying tear evaporates at much higher rate (near that of pure water). Increased evaporation drives a deepening hole in the tear film relative to surrounding lipid-covered domains. The growing hole is suppressed by curvature-driven healing flow and by osmotic-suction from the cornea due to a local salinity increase. Rupture occurs only when the locally high evaporative flux outweighs the two healing flows. Our one-dimensional stability analysis is in excellent agreement with clinical observations on times for breakup and on environmental roles of wind speed and relative humidity. New *in-vivo* measurements of transient infrared ocular surface temperature along with transient tear breakup area confirm faster evaporation rates in expanding tear-rupture holes.

Connection of tear rupture to dry-eye syndrome is the calculated large increases in salinity underneath the deepening rupture holes. Evaporative-driven tear breakup leads to significant increased salinities at the bottom of the rupture spot (or streak) that we coin salinity "hot spots". Hyperosmolarity causes eye pain and epithelial cell damage. A tear film peppered with salinity hot spots activates corneal afferent nerve receptors (cold) causing pain sensation and eventually dry eye.

Principles of capillarity are integral to functioning of the human eye and can be applied to understand and hopefully ameliorate dry-eye syndrome.

470460 The Sedimentation of Particulate Suspensions Under Orthogonal Shear: Mechanisms at Finite Weissenberg Number

Abstract:

Predicting the settling rates of non-Brownian particles suspended in complex fluids is critical to a number of engineering applications, including oil and gas recovery. A topic of specific importance involves particle suspensions in viscoelastic fluids experiencing an applied shear force in a direction perpendicular to gravity (referred to as orthogonal shear). In viscoelastic fluids, a nonlinear coupling between the orthogonal shear flow and the settling of particles has been observed [1]. Describing the underlying mechanisms and quantifying the effect of this *coupling drag*, which results from orthogonal shear on a sphere settling in a viscoelastic fluid, is the focus of this talk.

This talk looks at the role of fluid elasticity on single particles settling in orthogonal shear using experiments and numerical simulations. Previously, it has been shown that fluid elasticity with orthogonal shear can significantly reduce the settling rate of rigid spheres [2, 3]. New experiments were performed to study this enhanced coupling drag as a function of shear Weissenberg number, sedimentation Weissenberg number, and particle confinement in Boger fluids. This elastic effect was also studied using fully 3D simulations of flow past a rigid sphere, with the FENE-P constitutive model used to describe the polymeric fluid rheology. These simulations show good agreement with the experiments and allow for further insight into the mechanism of elasticity-enhanced drag. Results suggest the coupling drag, as well as the mechanism itself, are dependent on the shear Weissenberg number, sedimentation Weissenberg number, and particle confinement. In the low Weissenberg number regime (<1), these results are compared to existing results and theory. New results at high shear and sedimentation Weissenberg numbers (>1) provide novel insights into the elasticity-induced coupling drag enhancement for particles settling in orthogonal shear.

[1] Van den Brule, B.H.A.A., and Gheissary, G. *J. Non-Newton. Fluid Mech.*, 49 (1993): 123-132.

[2] Padhy, S.et al. *J. Non-Newton. Fluid Mech.*, 197 (2013): 48-60.

[3] Housiadas, K.D., and Tanner, R.I. *J. Non-Newton. Fluid Mech.*, 183-184 (2012): 52-56

460693 Stress-Gradient-Induced Polymer Migration in Microfluidics

Abstract:

We develop a continuum theory for the stress-gradient-induced migration of polymers in presence of solid boundaries and apply this to a model electro-osmotic flow with periodic slip wall velocity. We obtain theoretical results for the steady-state distribution of dilute polymer solutions using a systematic perturbation analysis in Weissenberg number. The theory is also extended to incorporate additional effects due to wall hydrodynamic interactions. By comparing the theoretical results with Brownian Dynamics (BD) and Stochastic Rotation Dynamics (SRD) simulations, we show that our theory can accurately capture the migration phenomena when the polymer coil size is less than half of the length scale over which the velocity gradient changes. Predicting polymer distribution in confined geometries is crucial in lubrication, oil recovery, microelectronics processing, and next generation DNA sequencing technologies.

469401 Simultaneous Flow and Drying in Open Microchannels

Abstract:

Microchannels have applications in microfluidic devices, micromolding, and flexible electronic devices. Liquid flow in open microchannels is controlled by capillarity, which draws the liquid down the channel, and viscous forces, which slow the flow. When a solid coating is desired on the interior of the channel, as in the case of flexible electronic devices, liquid inks that dry and solidify as they flow down the channels are used. In this work, we investigate the balance between flow and drying during capillary-driven flow in open microchannelsâ€"a topic that has received little attention in the literature to-date. Using high speed visualization, we study the influence of drying on the evolution of capillary flow dynamics over time. Specifically, we investigate aqueous polyvinyl alcohol (PVA) solutions imbibed into photolithographically created open microchannels with widths of 10 â€" 200 Âµm and depths of 50 Âµm. As an evaporating liquid travels down a microchannel, we show that the increase in viscosity (specifically the local increase at the contact line) reduces the total travel distance of the liquid and also has a unique influence on the flow dynamics. Specifically, strong departures from the generally accepted Lucas-Washburn theory of capillary flow are revealed. The highly dynamic flow field in the vicinity of the contact line during drying is also investigated using micro-particle image velocimetry. Pinning induced by the high-viscosity contact line is shown to disrupt the well-characterized fountain flow exhibited by liquids in microchannels. Comparisons are made to PVA solutions flowing in a high humidity environment, where evaporation is severely suppressed. This work has important applications in all processes involving the spreading of volatile solutions or suspensions, especially printing processes requiring the patterning of inks in microchannels or other confined volumes.

467397 Plug Flow of Shear-Thinning Liquids in Microchannels

Abstract:

1 Introduction

The demand for sustainable and efficient continuous processing makes micro-fluidic devices, with dimensions in the sub millimetre scale, an attractive option. Two-phase flow chemical operations in microchannels offer significant advantages compared to large-scale processes, such as lower sample consumption, increased safety and high surface-to-volume ratio. However, most studies are limited to Newtonian fluids although fluids with non-Newtonian rheology are very common industrially, including catalytic polymerization reactions, food processing and enhanced oil recovery [1-2]. The present work investigates the dynamics of plug formation of a non-Newtonian shear-thinning aqueous solution and a Newtonian organic fluid in a circular, glass microchannel using a two-colour micro-PIV system. There are, to the authors knowledge, no previous μ-PIV studies in liquid-liquid micro-flows involving non-Newtonian fluids where velocity fields are captured simultaneously in both phases.

2 Experimental setup

Two aqueous glycerol solutions containing xanthan gum (1000 and 2000 ppm) are used as the non-Newtonian fluids while silicone oil (Sigma-Aldrich) is the Newtonian phase. Two oils with different viscosities, 5 and 155 cSt have been considered. The corresponding Newtonian aqueous solution is also used for reference and comparison. All runs are carried out in a circular glass microchannel (Dolomite microfluidics) with inner diameter 200 μm at different combinations of flow rates of the two phases, that varied in the range of 0.01-0.1 ml/min. Using separate syringe pumps (KDS), the two immiscible fluids are introduced in the microchannel via a T-junction. In the two-colour micro-PIV experiments the aqueous phase is seeded with 1 μm carboxylate-modified microspheres FluoSpheres with orange fluorescent colour (540/560 nm) whereas the organic phase is seeded with 1 μm blue polystyrene microspheres particles Fluoro-Max (350/440 nm). The illumination is achieved with a UV double pulsed Nd:YAG laser (Litron Lasers) and the light emitted from the seeded fluids in the test section is led from the microscope to a beam splitter (Andor Technology). Each wavelength of the emitted light is led subsequently to a separate CCD camera. Both cameras are connected to a laser pulse synchronizer (TSI) and a PC. Image processing software (Insight 4G, TSI) is used for extracting information from the acquired video images. The flow is also visualized using a high speed CCD camera and a light source for the illumination of the test section.

3 Results

Under the specific experimental conditions, the organic 5 cSt Newotnian phase always forms the dispersed plugs irrespective of the side branch of the T-junction used to introduce it in the main channel. In the case of the viscous oil 155 cSt, the phase continuity is inversed and the aqueous non-Newotnian phase forms the plugs whereas the organic phase is now the continuous one.

During the plug formation different stages of detachment can be observed [3]. Using the two-colour μ-PIV, averaged velocity profiles were obtained in each phase for different locations of the plug, determined by the position of the plug tip inside the main channel. The velocity fields for the viscous oil 155 cSt (**Figure 1**) reveal that the continuous phase resists the flow of the dispersed phase at the rear of the plug causing a change in the interface curvature (Q_C / Q_D = 2).

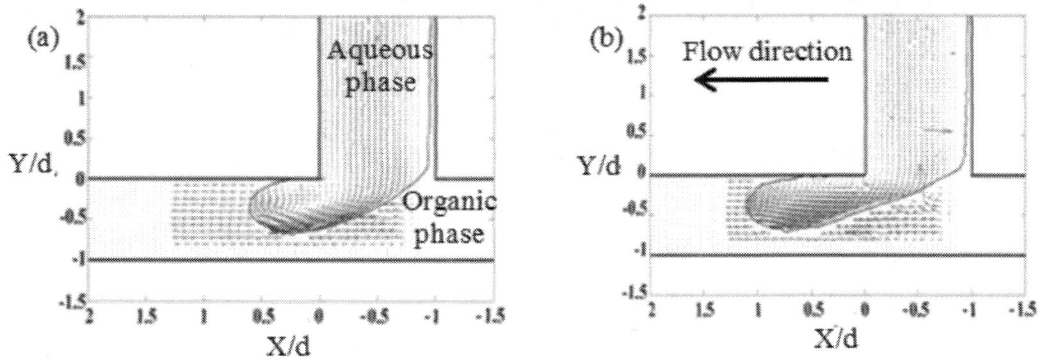

Figure 1: Velocity fields in both phases for plug tip position (a) 0.75 (b) 1.25 (Newtonian case).

In every case a thin filament was produced whose length depends both on the flow rate and the viscosity of the fluids. For the same stage of breakage and same flow rate (Q_C / Q_D = 2), the effect of shear-thinning behaviour is presented in **Figure 2**. The dynamics of the filament breakage change depending on the amount of xanthan gum in the aqueous phase and consequently on rheology of the fluid. For the 2000 ppm non-Newtonian case, the break-up point has moved further downstream and the breakage time is increased. This thin film eventually produces small droplets. The droplet diameter inceases with increasing amount of xanthan gum into the aqueous phase.

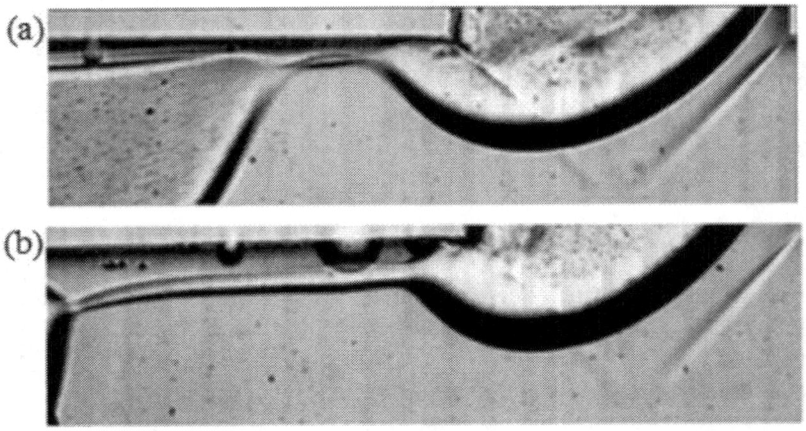

Figure 2: Dynamics of filament breakage for 155 cSt silicone oil and (a) Newtonian aqueous phase (b) non-Newtonian aqueous phase, 2000 ppm.

The experimental results have also been compared with numerical simulations using the CODE BLUE which is a massively-parallel Navier-Stokes solver for both Newtonian and non-Newtonian multiphase flows. The method for the treatment of the fluid interfaces and capillary forces uses a parallelized Front Tracking/Level Set technique which defines the interface both by a discontinuous density and viscosity fields as well as by a local triangular Lagrangian mesh. This structure allows the interface to undergo large deformations including the rupture and/or coalescence of fluid interfaces. Comparisons between experimental and numerical results on the formation of the plug at the channel inlet with the 155 cSt silicone oil can be seen in **Figure 3**.

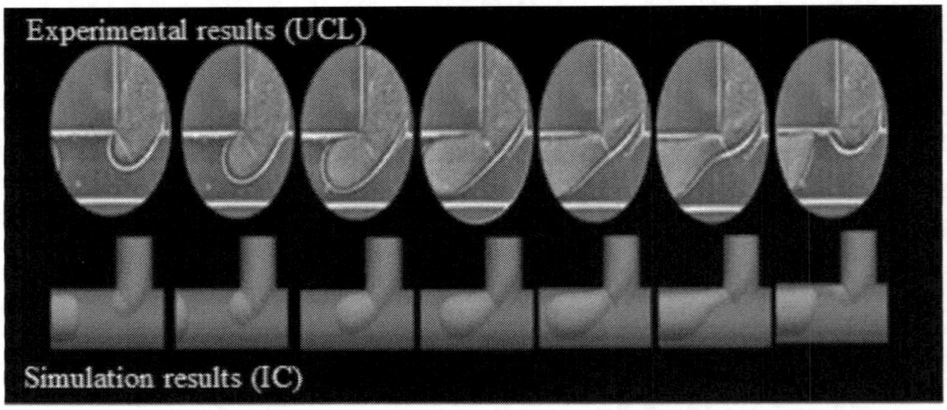

Figure 3: Comparison between experimental results and simulation.

The effect of the non-Newtonian rheology on the characteristics of the fully-formed plug flow in the main channel can be seen in **Figure 4** for the less viscous, 5 cSt, oil. For constant flowrates, the plug shape changes as the amount of xanthan gum increases. The plug acquires a bullet-shape profile with the front part curved due to the viscous forces (**Figure 4b**) while the rear part is similar to the Newtonian case.

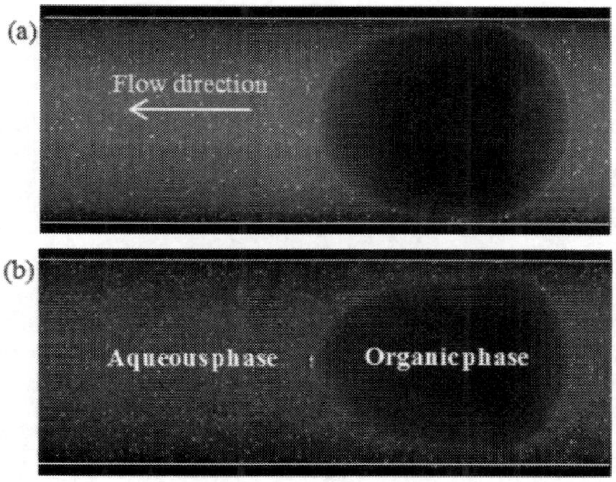

Figure 4: Effect of non-Newtonian behaviour on plug curvature (a) Newtonian fluid (b) non-Newtonian fluid, 2000 ppm (Q_C = 0.07 mL/min and Q_D = 0.01 mL/min).

Acknowledgements

The project was funded by the UK Engineering and Physical Science Research Council (EPSRC) Programme Grant MEMPHIS.

References

[1] Salim, A., Fourar, M., Pironon, J., Sausse, J., 2008. Oil/water two-phase flow in microchannels: flow patterns and pressure drop measurements. *Can. J. Chem. Eng*. 86, 978–988.

[2] Tang, G., Lu, Y., Zhang, S., Wang, F., Tao, W., 2012. Experimental investigation of non-Newtonian liquid flow in microchannels. Journal of non-Newtonian Fluid Mechanics. 174, 21-29.

[3] Garstecki P., Fuerstman MJ., Stone HA., Whitesides GM., 2006. Formation of droplets and bubbles in a microfluidic T-junction scaling and mechanism of break-up. Lab on a Chip 6(3): 437-446.

472166 Single Polymer Dynamics in Large Amplitude Oscillatory Extension (LAOE)

Abstract:

Over the past two decades, advances in fluorescence imaging and particle manipulation have enabled the direct observation of single polymer dynamics in model flows such as shear flow and planar extensional flow. The vast majority of single polymer studies, however, has focused on chain dynamics using simple transient step forcing functions. In order to study single polymer dynamics in non-idealized â€˜modelâ€™ flows, there is a clear need to implement more complicated flow types and transient flow forcing functions. For example, large amplitude oscillatory shear flow (LAOS) has been widely used in recent years to study the linear and nonlinear viscoelasticity of materials in bulk rheology but complementing molecular techniques have not yet been applied.

In this work, we directly probe single polymer dynamics using large amplitude oscillatory extensional flow (LAOE) in precisely controlled microfluidic devices. We are able to generate both small amplitude and large amplitude sinusoidal oscillatory extensional flow in a cross-slot microfluidic device while imaging the conformational dynamics of a single DNA polymers trapped at the stagnation point. Using this experimental technique combined with Brownian dynamics (BD) simulations, we study the single chain dynamics of a Î»-DNA molecule as a function of both the Weissenberg number (*Wi*, flow strength) and Deborah number (*De*, probing frequency). In LAOE, polymer chains undergo repeated cycles of compression, rotation, and stretching between extensional and compressional axes in a highly dynamic and transient manner. To our knowledge, this work is the first demonstrated molecular rheology experiment utilizing two-dimensional time-dependent control over the entire x-y flow plane. Using this approach, we characterize transient chain dynamics by constructing single polymer Lissajous curves, which are defined based on chain conformation and stretching in flow. We further construct a series of Lissajous plots as a function of *Wi* and *De* using experiments and BD simulations. Our results reveal four distinct characteristic shapes of the Lissajous curves, each representing a unique behavioral regime for a single polymer chain. We further study the average chain dynamics in LAOE over an ensemble of polymer chains. Based on the average projected extension, our results show the existence of a critical *Wi* at the coil-stretch transition that is a function of the dimensionless cycle frequency (*De*). In this way, we map out a wide range of the Pipkin space for a single polymer chain ranging from linear steady-state conditions at low *Wi* and *De* to non-linear unsteady states at high *Wi* and *De*. We use a conformational space construction based on experimental data to visualize average chain dynamics in LAOE in Pipkin space both below and above the coil-stretch transition.

460341 Shear Induced Demixing in Large Amplitude Oscillatory Shear Flows

Abstract:

Entangled polymer solutions can exhibit enhanced light scattering and demixing for a broad range of flow geometries in both steady and unsteady flows. The resulting compositional inhomogeneity can lead to dramatic changes in the flow field relative to what would be predicted for the same fluid with a uniform composition. A connection between shear induced demixing and shear-banding-like transitions is well established in simple shear flow, but comparatively little theory has been developed to understand flow induced demixing in more general flow types. Describing inhomogeneous flows with a standard two-fluid model, we have found that shear induced demixing in large amplitude oscillatory shear flow has the same general character as shear induced demixing in simple shear flow. Through a combination of asymptotic methods, we are able to predict stability boundaries and shear banded steady states for a well-entangled polymer solution in a large amplitude oscillatory shear flow. These predictions are confirmed through a direct numerical simulation of the full two-fluid model.

463421 Molecularly Based Criteria for Shear Banding in Transient Flow of Entangled Polymeric Fluids

Abstract:

Dissipative Particle Dynamics (DPD) simulations of polymeric melts in a start-up of shear flow as a function of ramp-time to its steady state value is studied. Here, the molecular findings showing the effect of ramp-time on formation of shear banded structure and chain relaxation behavior is presented. Specifically, it is shown that shear banding emerges at rapid start-up, however homogenous shear prevails when the deformation rate ramp time is sufficiently slow. This finding is in full consistency with prior continuum level linear stability analysis of shear banding in startup of shear flows [1] as well as experimental observations of entangled DNA and polymer solutions [2]. Moreover, it has been revealed for the first time that the ratio of the chain longest orientation relaxation time to that of the time for the imposed deformation rate to reach its steady state value, denoted as MK, plays a central role in determining whether local inhomogeneities are created in the fluid; a prerequisite for shear banding formation. Specifically, if MK is greater than one, local inhomogeneity in the entanglement density in the velocity gradient direction appears and in turn flow transitions to a shear banded structure. Otherwise, homogeneous shear flow prevails. Moreover, local process for the formation of shear banding is examined and is shown that the gradient of average number of entanglements per chain along the velocity gradient direction should reach a critical value for local inhomogeneity and in turn a fluid interface to be formed. Finally, a molecular picture for the interrelation between the chain longest orientation, stress relaxation time, entanglement network dynamics, local inhomogeneities, and shear banding is proposed.

References:

1. R.L. Moorcroft, S.M. Fielding, Physical Review Letters, 110 (2013) 086001.
2. P.E. Boukany, S.-Q. Wang, Macromolecules, 43 (2010) 6950-6952.

472141 Effect of Chain Flexibility on Non-Linear Extensional Response of Linear Polymer Melts

Abstract:

Our work has recently shown the existence of universality in nonlinear flow behavior of polymer melts and polymer solutions [1]. The key conclusion from Wingstrand et al. is that in order for two systems to have the same non-linear flow dynamics, they must have three properties in common, (1) the same number of entanglements, (2) the same number of Kuhn segments between entanglements, and (3) the same monomeric friction between components. By extension, a comprehensive model capable of predicting both linear and nonlinear flow behavior must incorporate these fundamental properties. Since differences in polymer chemistry lead to fundamentally different values of (1)-(2) (and possibly (3)), these results show that the likelihood of two melts having the same nonlinear behavior is improbable. This work determines the dependence of the nonlinear flow behavior on (2) and (3) using a new commercial filament extensional rheometer, VADER 1000 (Rheo Filament ApS). More specifically, we study polystyrene, poly(methyl methacrylate), and poly(tert-butylstyrene) melts with tuned MW to give the same number of entanglements. The samples only differ in the number of Kuhn segments between entanglements (flexibility) and possibly monomeric friction. Small amplitude oscillatory shear is used to confirm the identical linear shear rheology between the three polymer melt samples. Constant strain rate extensional rheology is used to probe the fast flow dynamics of the polymers. The scaled stress growth coefficient as a function of scaled time shows the unique response of each polymer and the trend of increasing number of Kuhn segments per entanglements on strain hardening for various Weissenberg numbers.

[1] Wingstrand, S. L., Alvarez, N. J., Huang, Q. & Hassager, O. Phys. Rev. Lett. 115, 078302 (2015).

456754 Investigating Droplet-Breakup Dynamics for Characterizing Low-Viscosity Elasticity of Dilute Polymer Solutions

Abstract:

The addition of macromolecules to solvent, even in dilute quantities, can significantly alter a fluidâ€™s response in an extensional, or elongational, flow. When high molecular weight polymers are extended in solution, extensional thickening can result, yielding viscosities that are orders of magnitude larger than values found in shear. Examples of applications where weakly viscoelastic fluids are used include inkjet printing, turbulent drag flow reduction, and the atomization of fertilizers and pesticides. Thus, understanding the response of such fluids, particularly in extension, is important for making relevant processing and design decisions. The present research was conducted in an effort to make improvements to and to provide an in-depth assessment of the techniques available for characterizing low-viscosity elasticity. Outcomes of this work include the creation of a new device for extensional characterization and the completion of a study comparing a number of previously described extensional techniques (e.g., ODES-DOS) in an effort to investigate discrepancies and identify limitations in device usage.

Similar to passive breakup of a droplet at a T-junction, a microfluidic device is presented that uses a cross-slot geometry to observe a two-phase droplet breakup process. For viscoelastic fluids, we report that a cylindrical filament forms between droplet segments, which decays exponentially in time. In optically tracking this decay, both transient extensional viscosity and relaxation times were evaluated. For validating and optimizing the device, a range of polyethylene oxide (PEO) and Newtonian solutions were tested. Comparisons of the evolution profiles due to the presence of elasticity were made. The resulting Trouton ratios versus Hencky strains were recorded and compared with the results from other emerging extensional techniques. Experiments were performed to observe the growth of instability, the dynamics of filament thinning, and the presence of any bead formation in the droplet breakup event. These results reveal a complex interplay between viscous, elastic, capillary, and inertial contributions in the extensional devices used.

468218 Brownian Dynamics Simulations of Knotted Polymers Moving through Pores Under a Constant or Periodic External Field

Abstract:

Nanopore sequencing is a technique where DNA moves through a pore and base-pair information is read along the chain as a time-varying electric signal. Currently, one hurdle facing this technique is that DNA passes too quickly through the pore, rendering the signal to be noisy. One relatively unexplored idea is to use entanglements like knots to retard the motion of the chain. If the external field is constant, we find that knots jam at the pore's entrance above a critical tension and halt the polymer's motion. This by itself may not be surprising, but one can use this information to design strategies to control the polymer's speed. For example, if one cycles the external field on and off at the relaxation time scale of the knot, one can control the swelling of the knot at the pore's entrance and ratchet the polymer through the pore. We discuss how knot topology and cycle time affects the polymer's dynamics, and we further discuss what force regimes lead to large fluctuations in the polymer's transit times. We find that cycling the force field can reduce fluctuations near the knot's jamming transition, but can enhance the fluctuations at very high forces since knots get trapped in metastable states during the relaxation process. The study here encompasses a rich variety of physics – some phenomena we observe include slip-stick motion, caging, and molecular individualism.

Decellularization of Porcine Aorta Using Supercritical CO_2

Dominic M. Casali[1] & Michael A. Matthews[1,2]

[1]Department of Chemical Engineering, University of South Carolina, Columbia, SC
[2]Biomedical Engineering Program, University of South Carolina, Columbia, SC

November 15, 2016
AIChE 2016 – San Francisco, CA

UNIVERSITY OF
SOUTH CAROLINA

AIChE
American Institute of Chemical Engineers

1

The Need for Organ Transplants

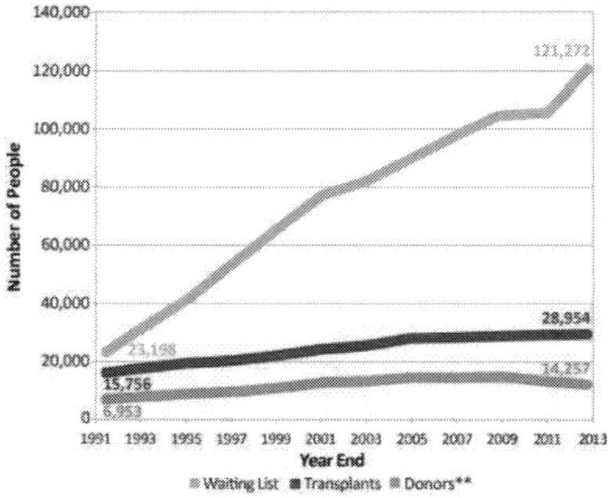

Source: US Dept. of Health & Human Services & Organ Procurement and Transplantation Network

- Over 120,000 people in the US awaiting a transplant
- About 8,000 people die annually while waiting/become too sick
- Average kidney transplant wait time is 3-4 years; others are much longer
- Especially problematic for minorities
  - Fewer potential matches
  - Higher rates of some diseases

2

Tissue Engineering: A Biomaterials Challenge

The Key Question: Where does this come from?

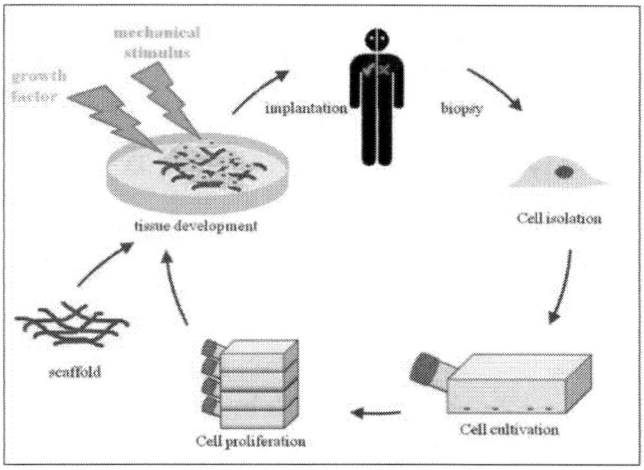

Tissue Engineering Flowchart (Wikimedia Commons)

Scaffold Materials:
- Synthetic Polymers
- Natural Biomaterials

3

Decellularizing Natural Materials

- Removal of cells from an animal tissue; extracellular matrix (ECM) used as a scaffold
- Lower chance of rejection
- Constructive remodeling
- No universal decell. method
- Detergents most common, but:
  - Long treatment times
  - ECM damage
  - Residual cytotoxicity

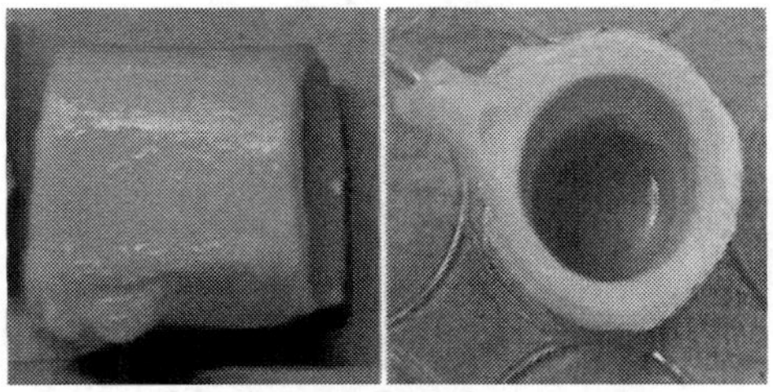

(Left) Untreated and (Right) Decellularized porcine aorta – from Mooney *et al.*, *Cells Tissues Organs*, 2015

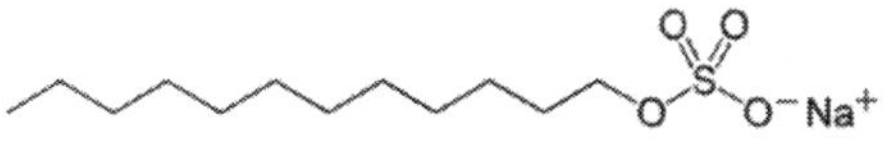

SDS Chemical Structure

4

Decellularization with Supercritical CO_2

- CO_2 is used for removal of other biological molecules:
  - Sterilization
  - Lipid Extraction
  - Endotoxins
  - Allergenic Proteins
- May be faster than common methods (hours vs. days)
- No residual solvents or detergents
- Sawada *et al.* (2008)
  - Only report of $scCO_2$ decellularization
  - Reported tissue dehydration and increased ECM stiffness

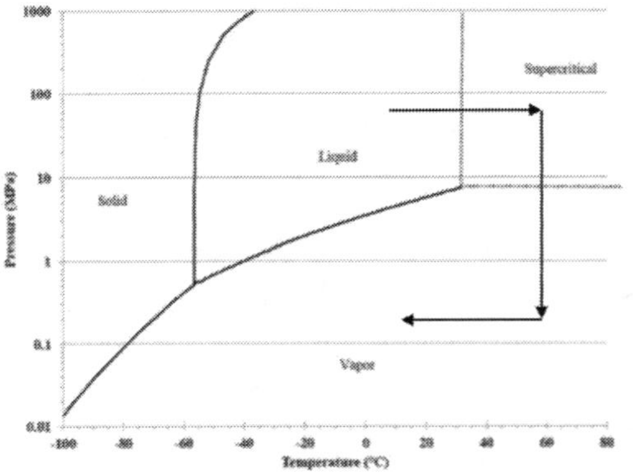

CO_2 Phase Diagram

5

Project Objectives

Overall Goal: Decellularize with $scCO_2$ while maintaining matrix hydration and overall integrity

1. Prevent tissue dehydration
2. Determine extent of decellularization with CO_2
3. Present detergent/CO_2 hybrid method

6

Part 1: Maintaining Hydration State

7

CO_2 Presaturation Apparatus

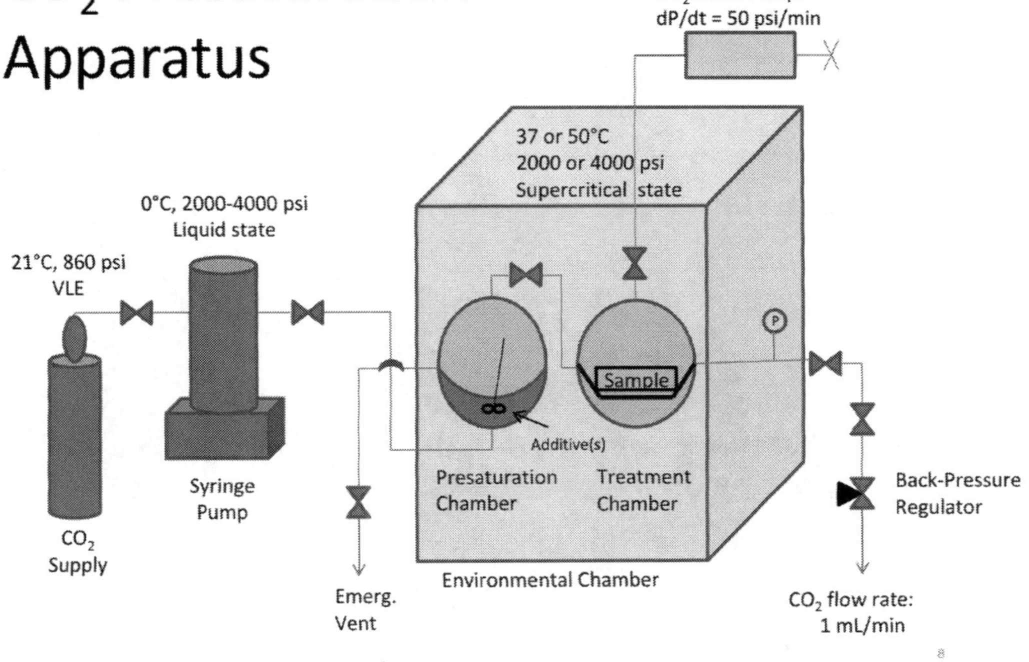

8

CO_2 Treatment of Model Scaffolds

Hydrogel

Model Scaffolds

Porcine Aorta

Dry CO_2

H_2O-Presaturated CO_2

9

Scaffold Dehydration Tests

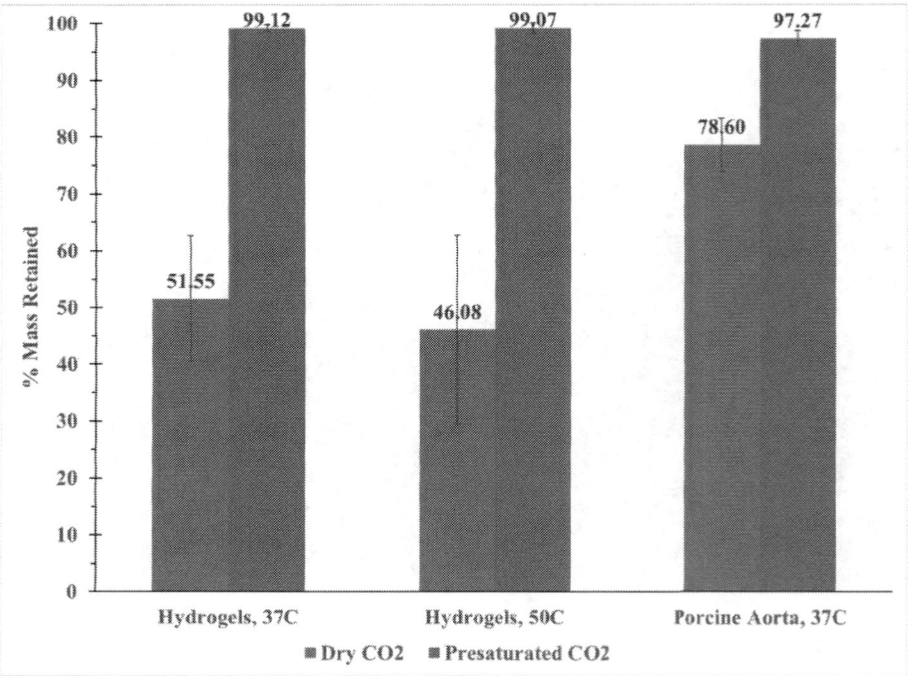

10

Part 1: Conclusions

- Dry CO_2 extracted an appreciable amount of water from both a hydrogel and porcine aorta in one hour of treatment
- Presaturating CO_2 with water almost entirely prevented water extraction during treatment of both biomaterials
- A small amount of volatile material is extracted from porcine aorta even after water presaturation, these are likely other trace volatile components

11

Part 2: CO_2 Treatment & Determining the Extent of Decellularization

12

Decellularization Standards

- No uniform standard exists to confidently verify complete decellularization

- Standards recently proposed by Crapo *et al.* (2011):
  1. No visible nuclei in stained tissue sections – common

  2. Total dsDNA < 0.05 μg/mg dry tissue – sometimes done

  3. All DNA fragments shorter than 200 base pairs – rarely done

13

Treatment Plan

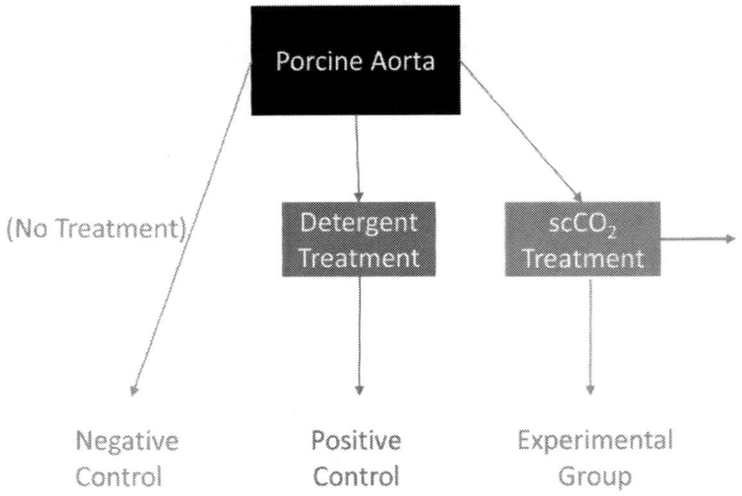

<u>CO_2 Treatments/Additives</u>:
- Dry $scCO_2$
- $scCO_2 + H_2O$
- $scCO_2$ + EtOH
- $scCO_2$ + EtOH + H_2O

<u>Assays</u>:
- H&E Staining
- DNA Quantification

14

Treatment Conditions

- Detergent Treatment (Funamoto *et al.*, 2010):
  - 72 hr treatment
  - 0.1% SDS detergent + DNase, RNase

- CO_2 Treatment with Presaturation:
  - Temperature: 37°C
  - Pressure: 4000 psi (ρ_{CO2} = 0.908 g/mL)
  - Treatment Time: ≈1 hr (normalized to mass)
  - Depressurization Rate: 50 psi/min

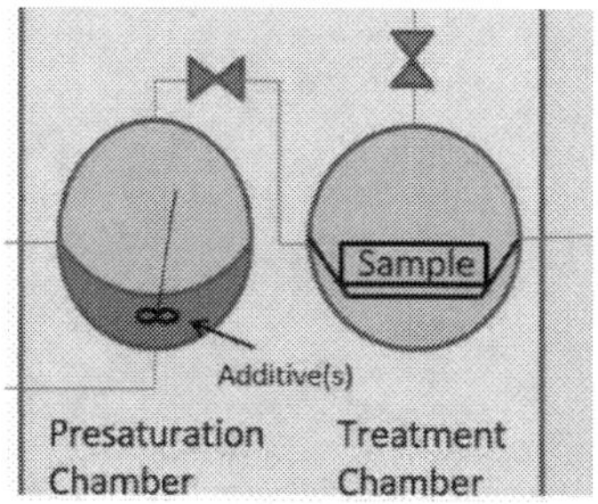

15

Hematoxylin and Eosin (H&E) Stain

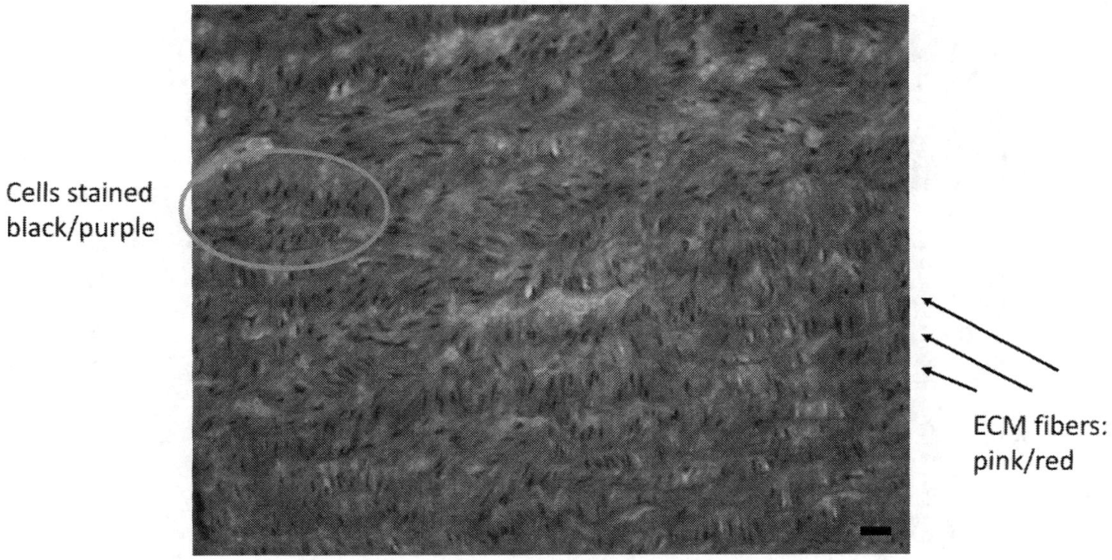

Untreated Aorta

All scale bars = 50 µm

16

H&E Staining – SDS, Dry CO_2

Untreated

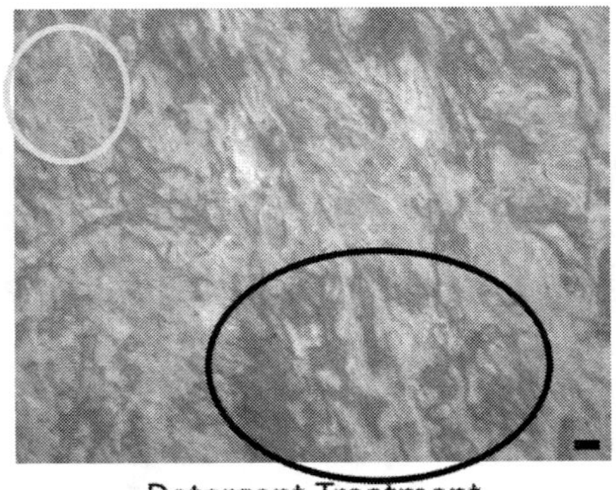

Detergent Treatment

Dry CO_2

All Scale Bars = 50 μm

17

H&E Staining – CO_2 with Additives

Untreated

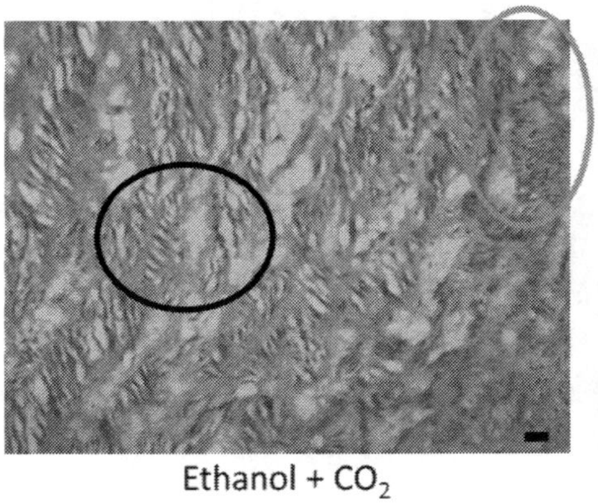

Ethanol + CO_2

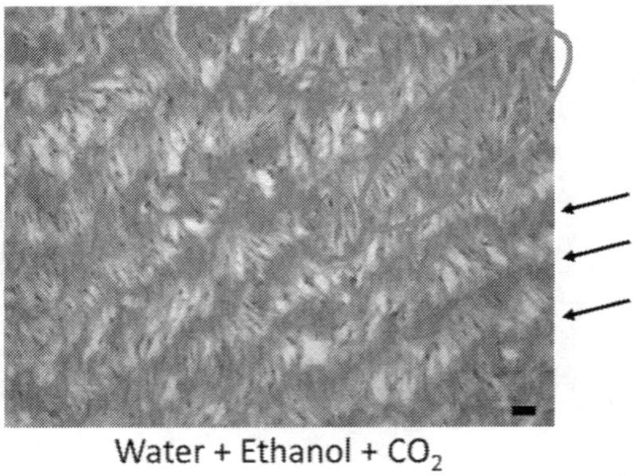

Water + Ethanol + CO_2

All Scale Bars = 50 μm

18

DNA Quantification

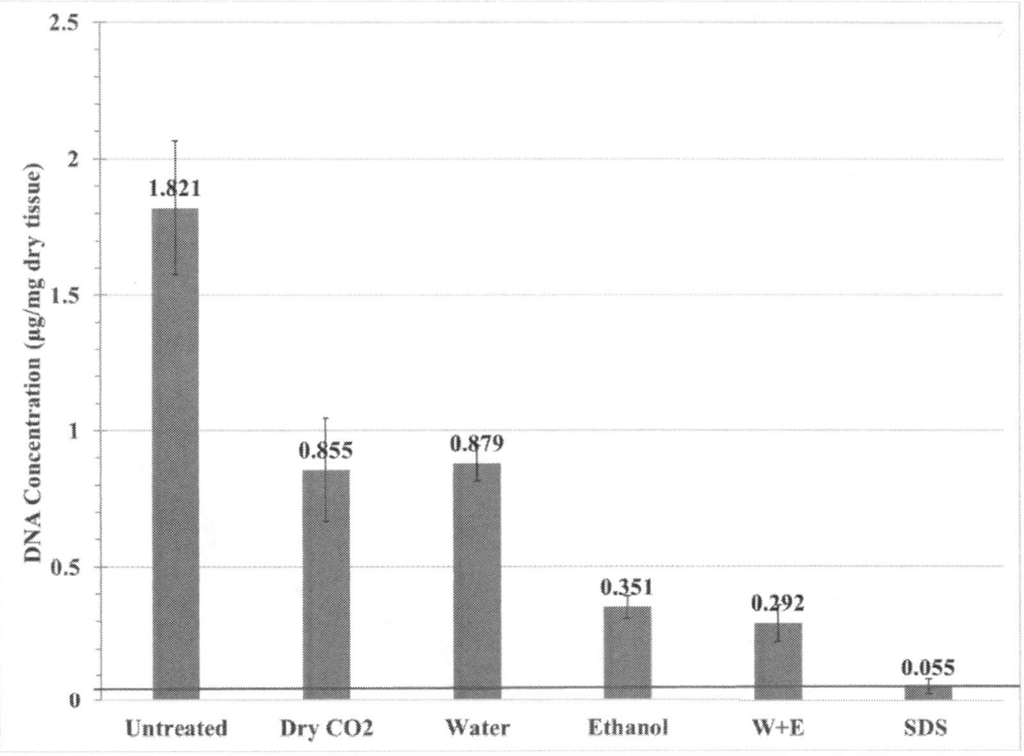

19

Part 2: Conclusions

- CO_2 treatment does not remove all cells from the ECM, even when ethanol is added
- Water helps maintain elastic fibers in the ECM
- CO_2 does not disrupt the cell membrane, while SDS does:
  → What if we treat briefly with SDS before using CO_2?

20

Part 3: SDS/CO_2 Hybrid Treatment

1. 48 hr SDS to rupture cell membranes
2. 1 hr CO_2 with ethanol and water additives to extract freed intracellular components

Possible Advantages:

- Shorter treatment time
- Maintenance of ECM fibers
- Removal of residual detergent

21

H&E Staining – Hybrid Treatment

Untreated

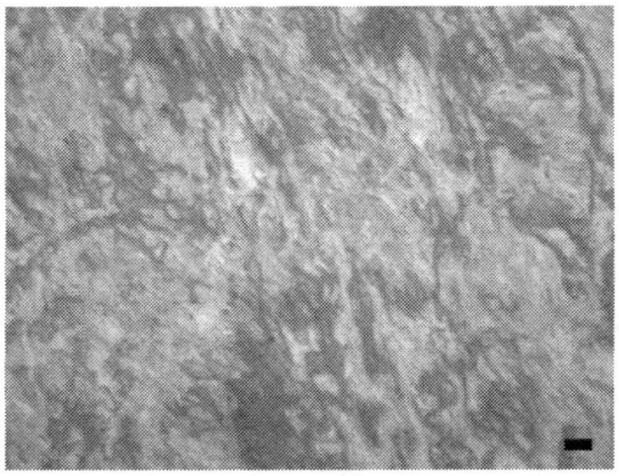

SDS 72 hr

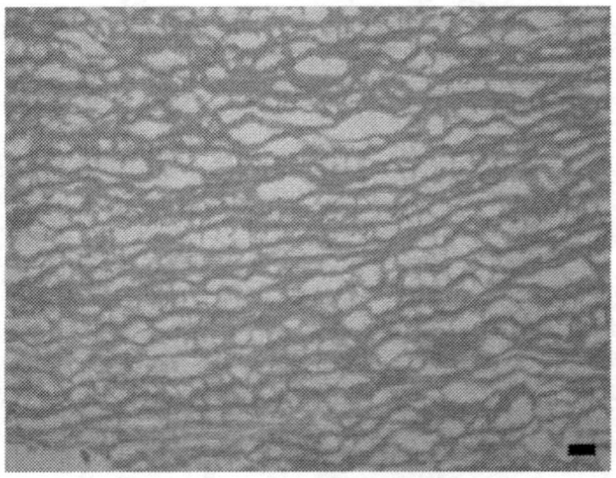

SDS 48 hr + $CO_2/H_2O/EtOH$ 1 hr

22

All Scale Bars = 50 µm

DNA Quantification

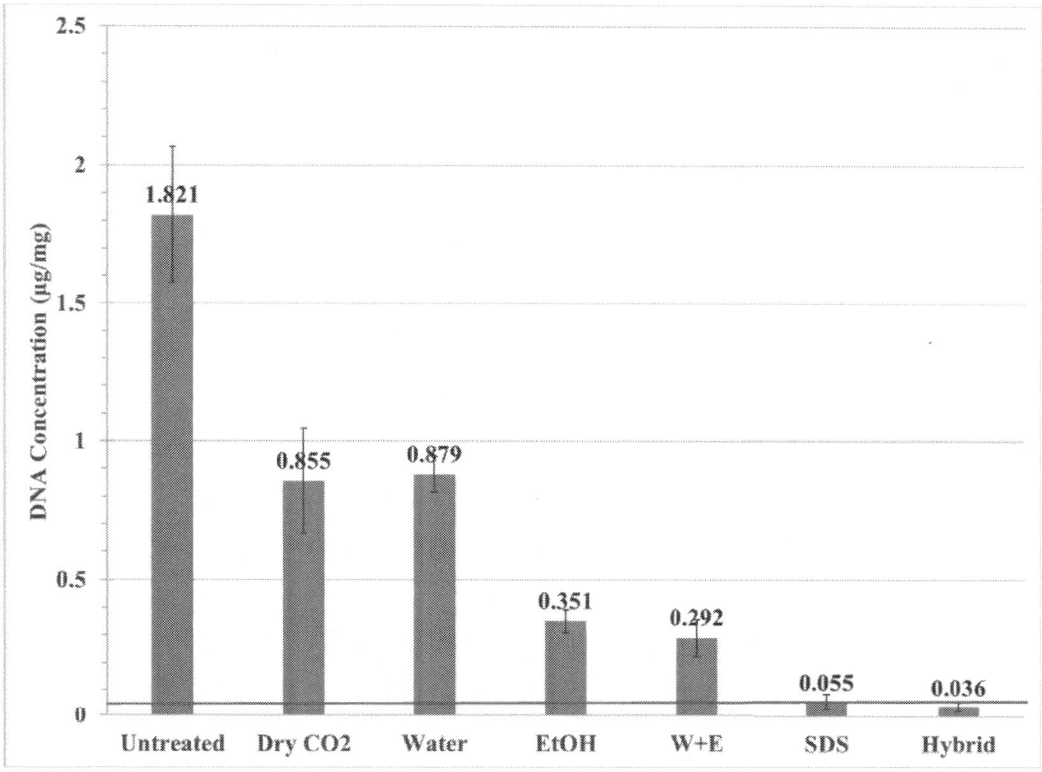

23

Overall Conclusions

- Presaturated CO_2 can be used to maintain water content
- Supercritical CO_2 treatment alone is ineffective for decellularizing porcine aorta
- Preliminary study of a hybrid SDS/CO_2 treatment shows complete decellularization
  - Faster and more effective removal of cellular debris than the control method
  - Residual SDS possibly removed by CO_2

24

Future Work

- Confirm removal of SDS during hybrid treatment
- Optimization of hybrid treatment
  - Reduce SDS and/or CO_2 treatment time
  - Use lower SDS concentration
- Analysis of scaffold physical properties
  - Tensile testing
  - Electron microscopy
- Explore CO_2-based decellularization of other tissues

25

Acknowledgements

- Rachel Handleton

- U. of S.C. Instrument Resource Facility (School of Medicine)
  - Bennie Davidson
  - Sharon Cooper

- U. of S.C. VPR/SPARC Fellowship

26

Decellularization of Porcine Aorta Using Supercritical CO_2

Dominic M. Casali[1] & Michael A. Matthews[1,2]

[1]Department of Chemical Engineering, University of South Carolina, Columbia, SC
[2]Biomedical Engineering Program, University of South Carolina, Columbia, SC

November 15, 2016
AIChE 2016 – San Francisco, CA

UNIVERSITY OF SOUTH CAROLINA

AIChE
American Institute of Chemical Engineers

27

464037 Supercritical Fluid Immobilization of Horseradish Peroxidase on Mesoporous Carbon for Environmental Remediation

Abstract:

Immobilization of enzymes on solid supports is crucial in enzymatic reactions and catalysis as it allows reutilization of soluble enzymes and reduces cost. Covalent immobilization of horseradish peroxidise(HRP) on surface-modified mesoporous activated carbon (SMAC) in combination with supercritical carbon dioxide (scCO2) has been investigated in this study. HRP loading in the scCO2 environment was consistently higher for different HRP: SMAC ratios. The significant increase in the rate and quantity of HRP immobilization is attributed to the favourable transport properties of solutes in scCO2 coupled with greater affinity of HRP for the non-polar supercritical solvent which enhances the diffusivity of enzymes into the porous matrix. The immobilized HRP was successfully employed in removing phenol from water via polymerization of dissolved phenol in the presence of hydrogen peroxide (H2O2). The study shows the advantage of using a mesoporous substrate as the immobilization platform for enzymes in combination with scCO2 as the transport medium.

472322 Supercritical Fractionation of Oligomeric Pyrene Pitches for Kinetic Modeling

Abstract:

Semi-continuous, packed-column supercritical extraction (SCE) with toluene and toluene-NMP (N-methylpyrrolidone) mixtures was used to fractionate various polycyclic aromatic hydrocarbon (PAH) pitches into their constituent oligomers. In particular, the starting model pitch samples were produced by the catalytic polymerization of pure pyrene using AlCl3, or by thermal polymerization, with both being done at temperatures from 300-500 °C. Each sample was reacted for a different period of time in order to vary the oligomeric composition of the final pitch product, such that kinetic data on the pyrene polymerization process could be obtained.

SCE was used to separate each sample into its constituent oligomers, such that an oligmeric mass balance could be performed on the reacted pitch products. The co-solvent NMP provided enhanced solvent power for extracting the higher MW oligomers from the pitch; furthermore, NMP serves as a valuable suppressant of the condensation reactions that can occur between toluene and the pyrene oligomers at the elevated temperatures necessary for SCE, thus keeping product purities high.

465407 Highly Dispersed Metals on Metal Oxide Supports Via Reactive Deposition from Supercritical CO_2

Abstract:

Formation of dispersed metal nanoparticles from precursors deposited on supports from supercritical carbon dioxide has been a topic of interest over the past decade. Materials made by this method have shown promise as active catalysts that employ noble metals efficiently, and they also exhibit potential for adsorptive separations. Examples of such materials include a variety of transition metals deposited on carbon-based supports and metal oxides. For carbon-based materials, including nanotubes and aerogels, the adsorption of organometallic precursors has been shown to follow thermodynamically-limited adsorption isotherms, and this behavior has even been used to control the composition of bimetallic nanoparticles. Our work in adsorption of similar precursors on metal oxide supports indicates a more complex deposition mechanism in these systems.

We have investigated the deposition of platinum and copper acetylacetonates from supercritical CO_2 on both Î³-alumina and anatase catalyst supports. A comparison of these materials with others of similar composition prepared by wet impregnation from methanol led to the observation that metal adsorption from CO_2 is irreversible, while adsorption from methanol can be reversed. A potential explanation for this behavior is that a specific interaction between CO_2 and protonated surface oxides results in surface bicarbonate groups, which in turn undergo ion exchange with the organometallic precursors, leaving the metal ions bound to carbonate sites on the surface. This mechanism should result in highly dispersed metal ions having strong interactions with the surface. During calcination and reduction, the carbonates decompose, leaving metal nanoparticles on the surface. The effect of reduction method and of temperature during the calcination and reduction processes on the both size and morphology of the particles formed will be discussed.

463978 Quantum Dots Synthesis in Continuous Supercritical Micro-Millireactors

Abstract:

The development of nanosciences and nanotechnologies in the 21th century is linked to the progresses made with the nanomaterials preparation methods. Control, reproducibility, scalability and sustainability are the key issues for the design of advanced nanostructured materials. Among the synthesis approaches, the supercritical fluids based flow process presents an efficient alternative for the continuous, controlled, scalable and sustainable synthesis of nanomaterials.

Coupling chemistry and chemical engineering approaches in supercritical fluids conducts to the design of high quality and unique nanostructures. This will be in particular illustrated in this oral presentation through the continuous production of quantum dots (inorganic nanocrystals) made of selenide, sulfide and nitrides (CdSe, core-shell CdSe@CdS, GaN, etc.).

Besides the advantages of microscale flows for the precise control of the obtained nanostructures characteristics, scaling up is also of great interest for realizing the promise of these nanomaterials in devices structures. We will therefore detail briefly some example of scaling (production of CdSe QDs in millifluidic supercritical reactors) demonstrating production rates of up to 1g/day.

472773 CO_2 As an Expanded Solvent for the Fractionation of Lignin

Abstract:

Lignin is an abundant byproduct in both the nascent biofuel and pulp-and-paper industries. Because it is the only abundant, renewable biopolymer with aromaticity, it has been proposed as an alternative for petroleum-derived products for applications ranging from coatings to carbon fibers. However, preliminary work indicates that such applications will require the fractionation of lignin by molecular weight. In this work, CO_2 was added to acetic acid–water mixtures into which the lignin had been dissolved a priori to create expanded solvents. These systems proved to be highly effective in separating the lignin polymer into fractions with well-defined molecular weights. Precipitation of lignin fractions began at CO_2 pressures as low as 7 bar and continued up to the maximum pressure of 55 bar, where the system volume had approximately doubled. The resultant fractions were characterized by GPC, ^{1}H, ^{13}C, and HSQC-NMR, and interesting trends in functionality were obtained, different from those currently being observed by more conventional separation techniques. Finally, the use of CO_2 as an expanded solvent facilitates the separation of the solvent system (acetic acid–water) from the anti-solvent (CO_2) by pressure reduction, eliminating the need for an energy-intensive separation step such as distillation.

463407 Confinement Effect on Chemical Reaction Yield: The Nitric Oxide Dimer Reaction

Abstract:

It is well established that confinement within a nano-porous material, such as an activated carbon, carbon nanotube, or porous oxide can affect reaction yield, reaction rate and even the mechanism of the reaction [1,2]. These effects arise from the strong intermolecular forces between the various reacting species, including the activated state, and the pore walls, but are poorly understood.

Here we report a molecular simulation study of the effects of confinement within a nanoporous carbon material on the equilibrium yield of the nitric oxide dimerization reaction, 2NO=(NO)2. This reaction is chosen for study both because of its importance in atmospheric chemistry and biology, and because experimental data is available for the effect of confinement within carbon materials. Thus Kaneko and coworkers [3] have used magnetic susceptibility measurements to determine the equilibrium yield of the dimer in activated carbon fibers having slit-shaped pores of 0.8 nm width, while Yates and coworkers [4] have used FTIR spectra to find the yield in single-walled carbon nanotubes of diameter 1.35 nm. At the experimental conditions the yield of dimer in the gas phase in equilibrium with the pore phase was less than 1 mol %, whereas the yield within the pores was 100 mol % within the accuracy of the experiment. Monte Carlo simulations reported [5] in 2001 found a large increase in the yield within slit-shaped pores due to the confinement, but the calculated yield was well below the experimental value.

There has also been evidence that nanophases can exhibit high pressures when adsorbed within nanoporous materials due to confinement effects [6,7]. While the pressure in the bulk is a scalar quantity, the pressure inside a pore is a second-order tensor with both normal and tangential components, with respect to the wall. Molecular simulations for simple fluids in pores has shown the presence of high local tangential pressures, even at lower bulk pressures.

In this work, we report a Reactive Monte Carlo study of the nitric oxide dimerization reaction in slit-shaped carbon pores of various widths and over a range of temperatures. We examine several approximations made in earlier molecular simulation studies, in particular the flexibility of the bonds and the force fields used, and show that the latter have a large effect on the calculated yield. In particular, we show that the force fields used for both the monomer-wall and dimer-wall interactions were in error and this was a primary cause of the discrepancy between the simulations and experiment. Using ab initio calculations at the MP2 level we report revised force fields for these interactions, and show that the use of these leads to agreement with experimental results. In addition, we also investigate the relationship between the high in-pore tangential pressures and the higher reaction conversion yields under confinement.

References

[1] C.H. Turner, J.K. Brennan, M. Lisal, W.R. Smith, J.K. Johnson and K.E. Gubbins, "Simulation of Chemical Reaction

Equilibria by the Reaction Ensemble Monte Carlo Method: A Review", Molecular Simulation, 34, 119-146 (2008).

[2] C.H. Turner, J.K. Brennan, J.K. Johnson and K.E. Gubbins, "Effect of Confinement by Porous Materials on Chemical

Reaction Kinetics", Journal of Chemical Physics, 116, 2138-2148 (2002).

[3] K. Kaneko, N. Fukuzaki, K. Kakei, T. Suzuki and S. Ozeki, Langmuir, "Enhancement of NO Dimerization by

Micropore Fields of Activated Carbon Fibers", 5, 960-965 (1989).

[4] O. Byl, P. Kondratyuk and J. T. Yates, "Adsorption and Dimerization of NO inside Single-Walled Carbon Nanotubes –

An Infrared Spectroscopic Study", J. Phys. Chem. B, 2003, 107, 4277-4279.

[5] C. H. Turner, J. K. Johnson and K. E. Gubbins, "Effect of Confinement on Chemical Reaction Equilibria: The Reactions

2NO = (NO)2 and N2+3H2 = 2NH3 in Carbon Micropores", J. Chem. Phys., 114, 1851-1859 (2001).

[6] Yun Long, Jeremy C. Palmer, Benoit Coasne, Małgorzata Śliwinska-Bartkowiak and Keith E. Gubbins, "Pressure

enhancement in carbon nanopores: A major confinement effect", Physical Chemistry Chemical Physics, 13, 17163-17170

(2011).

[7] Yun Long, Jeremy C. Palmer, Benoit Coasne, Małgorzata Śliwinska-Bartkowiak, George Jackson, Erich A. Müller and

Keith E. Gubbins, "On the Molecular Origin of High Pressure Effects in Nanoconfinement: Effects of Surface Chemistry

and Roughness", Journal of Chemical Physics, 139, 144701 (2013)

460605 Self-Assembly of Gemini Surfactants

Abstract:

The self-assembly of molecules into nano-structured materials is a fascinating process because small changes in intermolecular interactions can have a large impact on the final mesoscopic structures. An interesting goal is directed self-assembly where the chemical nature of the molecules drives the assembly into specific nanostructures. Gemini surfactants are composed of two single-tailed surfactants that are connected by a linker. They form a variety of lyotropic liquid crystalline phases depending on the concentration of surfactant, chemical structure of the molecules, and nature of counterions. Using computer simulation we show that both non-electrostatic and electrostatic interactions play an important role in the self-assembly of these systems. We obtain physical insight into the odd-even effect, where the phase behavior depends on whether the linker has an odd or even number of carbon atoms, and the effect of the counterions. We also investigate the structure and dynamics of the Frank-Kasper sigma phase that has been recently discovered in these systems.

475119 Digital Alchemy for Molecular Design of Thermodynamically Stable Structures

Abstract:

We present a generalized thermodynamic ensemble approach to materials design in which particle attributes and their conjugate "alchemical" potentials are included as thermodynamic variables. We show how this approach can be used to find, in computer simulations, molecular and colloidal building blocks optimized for self assembly into target crystal structures.

*with R. Cersonsky, P. Dodd, X. Du, Y. Geng, J. Proctor, and G. van Anders

464472 Understanding Gas Separations Using Polymeric Membranes

Abstract:

Polymeric membranes, which present an efficient solution for emergent technologies, such as CO_2 capture and natural gas purification, are effective at the selective and efficient transport of gases. Here we critically delineate the underpinnings of these separation technologies using a suite of molecular simulation methods. Most of the current understanding of the separation efficiency of glassy polymer membranes is based on an ill-defined quantity, the free-volume. Our new finding, based on computer simulations on coarse grained simulations, is that the local, vibrational dynamics of a dense glassy polymer, as characterized by the long-time plateau value of the mean-squared displacement of the frozen matrix, defines the appropriate dynamic "free-volume" metric. In a related vein, it has been long thought that there is an "upper" bound when permeability (related to the flux) vs. selectivity (product purity) is plotted for a host of glassy membranes and gas solute pairs. This notion has become more tenuous with the knowledge that the upper bound is not fixed, but that it "moves" to improved permeability and better selectivity values with newer classes of polymers being synthesized. We shall use computer simulations to address this issue, specifically the existence of an upper bound and how it is affected by the architecture of the polymer chains in question.

455824 Simulations of Activity Coefficients and Solubilities for Aqueous Electrolytes: Why Are Our Models Off?

Abstract:

Aqueous electrolyte solutions play an important role in chemical engineering separations, and also in geochemical environments and for biophysics. The mean ionic activity coefficients quantify the deviation of salt chemical potential from ideal solution behavior; experimental measurements are available for many salts over broad ranges of concentration and temperature, but there have been practically no prior simulation results of these quantities, because of sampling difficulties for explicit-solvent electrolyte solutions. We have developed a new approach for determination of activity coefficients of aqueous electrolytes [1-2]. Common fixed-point-charge models for water and ions are unable to reproduce simultaneously activity coefficients and solubilities. Polarizable models [3] perform better, but still predict an incorrect temperature dependence of these properties. To complicate matters even more, direct coexistence simulations of salt solubilities systematically overpredict salt solubilities when compared to free-energy calculations using the same molecular models. Ongoing work in the group that aims to pinpoint the origin of these discrepancies will be described.

[1] Z. Mester and A. Z. Panagiotopoulos "Mean ionic activity coefficients in aqueous NaCl solutions from molecular dynamics simulations," *J. Chem. Phys.* **142**: 044507, 10 pp (2015). DOI: http://dx.doi.org/10.1063/1.4906320.

[2] Z. Mester and A. Z. Panagiotopoulos, "Temperature-Dependent Solubilities and Mean Ionic Activity Coefficients of Alkali Halides in Aqueous Solutions from Molecular Dynamics Simulations," *J. Chem. Phys.* **143**: 044505, 10 pp (2015). DOI: http://dx.doi.org/10.1063/1.4926840.

[3] H. Jiang, Z. Mester, O. A. Moultos, I. G. Economou, and A. Z. Panagiotopoulos, "Thermodynamic and Transport Properties of H2O+NaCl from Polarizable Force Fields," *Chem. Theory Comput}* **11**: 3802-3810

(2015). DOI: http://pubs.acs.org/doi/abs/10.1021/acs.jctc.5b00421.

474797 Progress in Electro Acoustic Signal Analysis

Abstract:

We have recently determined a correlation between the acoustic time of flight and intensity and the structure, state of charge, and state of health of closed system electrochemical energy storage. We have shown that regardless of chemistry and geometry, an acoustic time of flight signal can be related to these properties of the cell.

Because a closed cell is effectively a mass redistribution reactor, and in a standard cell the volume is effectively fixed, the distribution of density within a battery must change as a function of state of charge and, along with density, the bulk moduli of the anode and cathode changes as well. The shifts in density and modulus also change the acoustic attenuation in a battery

In this presentation we will review the physical basis of our hypothesis and present progress in 1) correlating structural evolution and failure analysis with acoustic signal evolution, 2) present preliminary inverse models which describe the changes we see in the acoustic/state of charge response and 3) present â€œphysics freeâ€ machine learning results to examine the uniqueness of an acoustic signal for a given battery in a given state of charge and health.

463541 Model-Assisted Development of Microfabricated 3D $Ni(OH)_2$ Electrodes with Rapid Charging Capabilities

Abstract:

Model-assisted Development of Microfabricated 3D $Ni(OH)_2$ Electrodes with Rapid Charging Capabilities

Recently, several studies have focused on the development of materials that realize batteries with improved power densities, and thus, rapid charging and discharging capabilities. Yet, they are primarily based on non-deterministic fabrication techniques which limit the control over critical electrode dimensions affecting the power performance of the battery, such as the electrode surface area and the active material thickness. The study reported herein involves rationally designed and deterministically engineered 3D structures with nearly precise control over the optimized dimensions that address the aforementioned critical factors affecting the high-power capability and ultimately determine the performance of the electrochemical system.

A series of fabrication methods encompassing electrochemical techniques and microfabrication technologies have been utilized for the formation and characterization of the scalable, well-ordered, and high-surface-area 3D architectures that can be potentially used as high-power electrodes in a variety of applications, ranging from autonomous microsystems to macroscale portable electronics. These versatile structures were based on a large number of thin layers of two different materials, i.e., Cu and Ni, the former being sacrificial while the latter was structural, fabricated through a temporary photoresist mold in an alternating sequence via an automated robotic system. The final structure was obtained by the selective removal of the sacrificial Cu layers, leaving behind a large number of separate Ni layers to be coated with the electrochemically active material. Nickel hydroxide ($Ni(OH)_2$), one of the most well-studied active material in secondary (i.e., rechargeable) battery systems along with Li-ion chemistry, was selected as the active material to be deposited onto the 3D multilayer structure. Electrodes with areal capacities as high as 2.43 mAh cm^{-2}

were realized, which were able to deliver more than 50% of their capacities when charged at ultrafast rate of 150C. The resultant electrodes also featured remarkable cycling stability when charged and discharged at high rates for more than 80 cycles.

For the optimization of the characteristic dimensions of the electrodes, a two-dimensional mathematical model has been developed in COMSOL 5.2 by employing fundamental mass transport and reaction kinetics principles. The geometry of a repeating unit of the multilayer electrode structure was studied. Fickâ€™s second law, Nernst-Planck equation, and Butler-Volmerâ€™s equation were used to describe the solid phase diffusion, the electrolyte phase transport, and the electrode/electrolyte interfacial kinetics, respectively. The model was shown to successfully predict the energy/power tradeoff of the fabricated multilayer electrodes. Diffusion coefficient of protons within the $Ni(OH)_2$ active material, a critical parameter that determines the power performance of the $Ni(OH)_2$ electrode, was experimentally measured using cyclic voltammetry. The obtained value of $2{\sim}3*10^{-10}$ cm^2/s is within the range reported from existing literatures and close to the best-fit value ($3.4*10^{-10}$ cm^2/s) obtained from the model. Accordingly, effort was directed towards finding the optimal electrode dimensions that lead to enhanced power performance based on the model. Geometrical parameters that were taken into account included the active material thickness, the interlayer spacing between two individual layers, and the nickel current collector thickness. Projections showed that multilayer structures based on the optimized dimensions exhibited superior performance than the experimentally fabricated non-optimized structures. Depending on the energy and power needs, this model opened up the possibility for the realization of deterministically engineered electrodes for both micro- and macroscale applications.

464527 Electrode-Supported Macroporous Ceramic Membrane Separator for Lithium Ion Batteries

Abstract:

The emergence of lithium-ion batteries as the most popular energy storage system is mainly due to their high energy density and long cycle life. They find application in a wide variety of portable electronic devices as well as electric vehicles and smart grids. However, the relatively frequent occurrences of battery fires and explosions are a major cause of concern. A significant source of this flammability, along with the electrolyte, is the polymeric separator used widely in commercial lithium-ion batteries. Lithium ion batteries with inorganic separators offer advantages of safer operation. An inorganic macroporous membrane separator for lithium-ion battery was prepared by blade coating an α-Al_2O_3 slurry directly on the electrode followed by drying. The improved separator preparation involves a twice-coating process instead of coating the slurry all at once in order to obtain a thin (~40μm) and uniform coat. Unlike state-of-the-art polyolefin separators such as polypropylene (PP) which are selectively wettable with only certain electrolytes, the excellent wettability of α-Al_2O_3 allows the coated separator to function with different electrolytes. The coated α-Al_2O_3 separator has much higher resistance to temperature effects than polyolefin counterparts. The inorganic coated separator retains its dimensional integrity at temperatures as high as 200°C, thus eliminating the possibility of a short circuit during thermal runaway. Lithium ion batteries half-cells and whole cells with coated α-Al_2O_3 separator exhibit electrochemical performance comparable with that with PP separator at room temperature. However, the coated separator shows better cycling performance than PP separator under extreme temperatures when tested over a wide temperature range of -30°C to +60°C. Therefore, the coated α-Al_2O_3 separator is very promising for uses in safe lithium-ion batteries for energy storage applications.

465548 First-Principles Density Functional Theory Modeling of Li Binding: Thermodynamics and Redox Properties of Quinone Derivatives for Lithium-Ion Batteries

Abstract:

First-Principles Density Functional Theory Modeling of Li Binding: Thermodynamics and Redox Properties of Quinone Derivatives for Lithium-Ion Batteries

Ki Chul Kim,[1] Tianyuan Liu,[2] Seung Woo Lee,[2] and Seung Soon Jang[1]

[1]*School of Material Science and Engineering, Georgia Institute of Technology, 771 Ferst Drive NW, Atlanta, GA 30332-0245, USA*

[2]*G. W. Woodruff School of Mechanical Engineering, Georgia Institute of Technology, 801 Ferst Drive NW, Atlanta, GA 30332-0405, USA*

Abstract

The Li-binding thermodynamics and redox potentials of seven different quinone derivatives are investigated to determine their suitability as positive electrode materials for lithium-ion batteries. First, using density functional theory (DFT) calculations on the interactions between the quinone derivatives and Li atoms, we find that the Li atoms primarily bind with the carbonyl groups in the test molecules. Next, we observed that the redox properties of the quinone derivatives can be tuned in the desired direction by systematically modifying their chemical structures using electron-withdrawing functional groups. Further, DFT-based investigations of the redox potentials of the Li-bound quinone derivatives provide insights regarding the changes induced in their redox properties during the discharging process. The redox potential decreases as the number of bound Li atoms is increased. However, we found that the functionalization of the quinone derivatives with carboxylic acids can improve their redox potential as well as their charge capacity. Through this study, we also determined that the cathodic activity of quinone derivatives during the discharging process relies strongly on the solvation effect as well as on the number of carbonyl groups available for further Li binding.

466882 Conductivity Degradation of Polyvinylidene Fluoride Binder during Cycling: Measurements and Simulations for Lithium-Ion Batteries

Abstract:

Battery electrodes are complex multiphase composites which must provide efficient bicontinuous networks for transport of electrons (through the particle phase) and positive lithium ions (through the electrolyte filled pores of the electrode). A crucial but often neglected element of battery electrodes is the binder, typically a mixture of polyvinylidene fluoride (PVDF) and carbon black. The binder has two primary roles â€" to provide mechanical integrity and to improve electrical conduction of the electrodes. Migration of the binder has also been implicated as a potential mechanism of capacity fade in rechargeable lithium ion batteries.

We will present experimental characterization of the polymer binder for battery applications. Mechanical properties of the composite binder will be shown for both dry films and also binder swollen with carbonate electrolytes used in rechargeable lithium batteries. The electrical properties are strongly dependent on the applied stress. The evolution of mechanical and electrical properties of the binder after repeated cycling will be shown.

Then we will examine the impact of cyclic mechanical stresses on the performance of lithium cobalt oxide cathodes. Experimental results will be presented on the evolution of electrical conductivity as a function of mechanical cycling. Mechanisms for electronic conductivity in these complex porous composite electrodes will be investigated using mesoscale simulations using experimentally determined three dimensional structures of battery cathodes. Implications for battery internal resistance and cycling stability will be discussed.

Sandia National Laboratories is a multi-program laboratory managed and operated by Sandia Corporation, a wholly owned subsidiary of Lockheed Martin Corporation, for the U.S. Department of Energyâ€™s National Nuclear Security Administration under contract DE-AC04-94AL85000. SAND2016-4371 A

474683 Polymeric Electrolyte Additives for Suppressing Zinc Dendrite Formation in Rechargeable Batteries with Zinc Anodes

Abstract:

Rechargeable zinc-air batteries are attractive because they utilize non-flammable aqueous electrolytes and offer high energy densities. However, a major roadblock in the development of rechargeable zinc-anodes is the dendritic morphology evolution during zinc electrodeposition (i.e., during battery charging). Zinc dendrites are a major source of capacity loss and early cell failures.

In the present talk, we will demonstrate the use of polymeric additives, such as polyethylenimine, as electrolyte additives for effectively suppressing dendrite formation during zinc electrodeposition. Dendrite suppression by additives will be characterized using in situ optical microscopy. Steady-state and transient electrochemical polarization measurements on a rotating disk electrode, combined with electrochemical quartz crystal microgravimetry will be used to unravel the mechanism by which additives suppresses zinc dendrite growth. Additive injection studies implemented in a rotating disc electrode configuration, coupled with diffusion-adsorption modeling, will be discussed as a reliable technique for precisely measuring the diffusion coefficient and adsorption rate constant of aforementioned polymeric additives. These additives' properties provide insights into the rate-controlling processes responsible for additives-assisted zinc dendrite suppression.

464831 Experimental Investigation of Relaminarization of Turbulence during Natural Circulation Cooling of a Very Hot Channel

Abstract:

Experimental Investigation of Relaminarization of Turbulence during Natural Circulation Cooling of a Very Hot Channel

Apoorva V. Rudra*[a], Narbeh Artoun [b], Sanjoy Banerjee [a, c], Masahiro Kawaji [b, c]

[a]Department of Chemical Engineering, City College of the City University of New York, New York

[b]Department of Mechanical Engineering, City College of the City University of New York, New York

[c] The Energy Institute, City College of the City University of New York, New York

Gaseous coolants have attracted a lot of attention for usage in various industrial applications due to their high thermal efficiency, chemical inertness, and environmental acceptability. Gas coolants are also proposed for Very High Temperature Reactors (VHTR's), which is one of six Generation IV reactors that have been proposed for DOE's Next Generation Nuclear Plant [1]. VHTRs also display benefits of passive safety systems including intra-core natural circulation for heat removal in accident scenarios [2]. Our work addresses the issue of flow laminarization, in which a strongly heated turbulent gas flow, exhibits heat transfer characteristics of laminar flows in the downstream part [3]. Under natural circulation conditions, reductions in the Reynolds number can trigger flow laminarization even if the incoming flow in the riser is above the critical Reynolds number [2]. As the gas is heated, a reduction in the gas density causes the bulk flow to accelerate upward, and the viscosity also increases leading to reduced Reynolds numbers. This large increase in gas viscosity at strongly heated walls leads to relaminarization. In some cases, the buoyant forces can also lead to flow laminarization [4]. On the other hand, the laminar flow can transition to turbulent flow in the downcomer channels as the gas is cooled [5].

In this study, we present the experimental measurements of the coolant flow and heat transfer behavior in a flow channel in a graphite block simulating a prismatic core of a Very High Temperature Gas-Cooled Reactor (VHTR). Natural circulation experiments are conducted using a helium-nitrogen mixture representing a helium-air mixture. A helium analyzer is used to investigate the nitrogen transport from the lower to upper plena as a function of time.

The natural circulation flow rate is measured and the flow rate data is related to the graphite temperatures, operating pressure, and overall concentration of nitrogen in helium. The amount of nitrogen gas injected and thus the overall nitrogen concentration in the gas mixture is systematically changed to determine its effect at different graphite temperature settings in the riser and downcomer. To what extent the local helium concentrations will differ between the lower and upper plena during both steady and transient natural circulation is determined in order to elucidate the air transport process following a postulated accident involving air ingress into the lower plenum. Under these conditions, the gas viscosity increases with temperature within the reactor cooling channels, causing a reduction in the Reynolds number which triggers flow laminarization. The natural circulation flow and heat transfer data along with the viscosity effects are analyzed to identify deterioration in heat transfer due to laminarization.

This experimental work will be supplemented by numerical simulations using CFD models [6] that will help to understand the physics behind the entire process better.

REFERENCES

[1] Dan Gabriel Cacuci (ed.), 2010. Handbook of Nuclear Engineering, DOI 10.1007/978-0-387-98149-9_22, @Springer Science+Business Media LLC.

[2] Valentín, F.I., Artoun, N., Anderson, R. and Kawaji, M., 2015. "Study of Abnormal Heat transfer during forced and natural convection scenarios in a prismatic core of a VHTR: Numerical and Experimental Results," Proc. of 16th International Topical Meeting on Nuclear Reactor Thermal Hydraulics (Nureth-16), Aug. 30-Sept. 4, 2015, Chicago, IL.

[3] Valentín, F.I., Artoun, N., Kawaji, M., and McEligot, D.M., 2015. "Investigation of helium flow laminarization at high temperatures and pressures in a graphite flow channel," Proc. of the 1st Thermal and Fluids Engineering Summer Conference, New York, August 9-12, 2015.

[4] Torii, S., Yang, W., 1997, "Laminarization of turbulent gas flow inside a strongly heated tube," *Int. J. Heat and Mass Transfer*, 40, pp. 3105-3117.

[5] Valentín, F.I., Artoun, N., Kawaji, M., and McEligot, D.M., 2015. "Experimental study of forced convection heat transfer during upward and downward flow of helium at high pressure and high temperature," Proc. of First Thermal and Fluids Engineering Summer Conference, New York, August 9-12, 2015.

[6] Shome, B., 2014. "Numerical Study of turbulent flow in heated circular tube using transitional shear stress transport turbulence model," *Int. J. Thermal Sciences*, Vol 79, pp. 90-102.

468751 Identification of High Energetic Flow Structures in the Separation Chamber of a Uniflow Cyclone with POD

Abstract:

For the optimization of the design criteria of a uniflow cyclone CFD simulation are carried out. However, numerical results require validation, e.g. due the sensitivity of the velocity distribution on the boundary conditions. Therefore reference PIV measurements are used to evaluate the flow field in the separation chamber of the uniflow cyclone [1]. The principal of the measurement technique is to record snapshots of tracer particles in the fluid flow to visualize the movement inside interrogation windows and calculate corresponding vectors [2]. Due to turbulence and local instabilities in the fluid flow pattern the cross-correlation analysis for each snapshot results in varying velocities per vector and snapshot. The goal is to identify the most energetic modes of the flow field to smooth the velocity field and to reach a better comparability with respect to the calculated velocity distribution over the examined region of interest. For the identification of the most corresponding fluid structures, proper orthogonal decomposition (POD) is introduced to extract these statistically coherent structures. The solution of the eigenvalue problem leads to the POD modes (eigenvectors) with the associated eigenvalues. These values represent the contribution of each POD mode to the overall power of the signal. The eigenvalues are ordered downwards. The modes with the highest contribution represent the most energetic structures [3]. This method leads to a better comparability of the CFD calculation and the measured PIV data. Moreover this leads to a more significant prediction of the flow pattern an assists the design criteria decision process.

[1] Kofler, T., Pillei, M., and Kraxner, M., "Influence of the Inlet Vane Geometry On the Uniflow Cyclones Performance (639f)". AIChE 2015 Annual Meeting, 2015.
[2] Raffel, M., ed., Particle image velocimetry: A practical guide, 2nd ed. Experimental fluid mechanics. Berlin: Heidelberg, 2007.
[3] Chen, H., Reuss, D. L., Hung, D. L., and Sick, V., "A practical guide for using proper orthogonal decomposition in engine research," International Journal of Engine Research, vol. 14, no. 4, pp. 307–319,

2013.

470888 Experimental Studies of Solid-Liquid Pipe Flow Using LDV

Abstract:

Mixtures of liquid and solids are present in many engineering and geological processes such as hydraulic conveying, energy conversion and the consolidation of soil sediments. Although they are widespread, our understanding of their fluid mechanics behavior is not complete partly due to a lack of experimental data. This data is needed for the development of engineering models that can optimize industrial operations, reduce energy demands and predict undesirable processes like sedimentation in transport lines.

The present investigation obtained novel experimental data for solid-liquid turbulent dilute flows in the intermediate and inertia dominated regime as characterized by the Stokes number. Velocity profiles for each phase and for the three velocity components were acquired in a pilot scale facility using Laser Doppler Velocimetry (LDV) for flows with particles up to 5 mm in diameter at concentrations from 0.7-2% V/V. In general, the presence of the particles flattened the mean liquid profile and augmented the turbulence intensity in the core region of the pipe. The mean particle profile is much flatter than the fluid with a maximum relative velocity at the center. The particle fluctuation profile is much flatter than the fluid suggesting a fairly uniform solid volume fraction across the pipe. The effect of the Reynolds number is marked with fluctuations being reduced with increasing Reynolds numbers.

457399 The Influence of Hydrodynamic Fluctuations on Turbulent Mixing and Chemical Reactions in a Rotating Channel Flow

Abstract:

Turbulent mixing in a rotating channel flow indicates that the transverse transport of a chemically reactive constituent is mitigated by a coupling between the shear component of the Reynolds stress and the longitudinal component of the mean flux of the reactive constituent. In the region of zero intrinsic vorticity, the dispersion coefficient in the cross flow direction is larger than the dispersion coefficient in the spanwise direction. The dispersion coefficient in the longitudinal direction is relatively small.

Koppula, K.S., A. Bénard, C. A. Petty, 2011, "Turbulent Energy Redistribution in Spanwise Rotating Channel Flows", Ind. Eng. Chem. Res. , 50 (15), 8905-8916.

469792 Computational Study of Turbulent Bubbly Flows

Abstract:

Turbulent bubbly flows are encountered widely in various industrial applications such as nuclear reactor cooling, bubble column reactors, cavitation on ship propellers, and several others. The efficient design of these systems requires the accurate CFD prediction of multiphase flow turbulence. Though the applications involving turbulent bubbly flows are widely encountered, little attention is paid towards developing accurate turbulence CFD models for prediction of these flows. Most works (Ranade, 1997; Behzadi et al., 2004; Lahey, 2005) adopt ad-hoc turbulence models by extending the standard k-ε model by adding a source term for turbulence production due to the presence of the bubbles, without considering the exact form of the turbulent transport equations. Moreover, second order turbulence closure models are rare and not adopted in many works despite the high degree of anisotropy involved in turbulent bubbly flows (Lance and Bataille, 1991). Keeping in mind these weaknesses, in the present study we propose a multiphase Reynolds-stress model for turbulent bubbly flows derived by Reynolds averaging (RA) the two-fluid model (Drew, 1983) following the procedure described in Fox (2014). The RA of the two-fluid model results in many unclosed terms due to the presence of non-linear terms in the two-fluid equations. The unclosed term resulting due to the RA of convection term or in other words Reynolds stress is accounted by deriving an exact form of transport equation for Reynolds stress from the two-fluid model. The values of all the unclosed terms involved in the momentum and the Reynolds stress transport equations are computed by means of fully resolved mesoscale simulations (Zhang, 2002), and the dependency of the unclosed terms on the loading of the bubbles is investigated. The computational domain for mesoscale direct simulations considered in the present study is a three-dimensional column with square cross-section. Periodic boundary conditions are applied to the sides of the column to create a homogenous turbulent flow to isolate and study the gradient independent unclosed terms which account for the exchange of turbulent quantities between the phases. The momentum and turbulence energy budgets computed from the mesoscale simulations are reported. Modeling strategies involved in modeling of the unclosed terms based on the budget are also discussed.

References Behzadi, A., Issa, R.I., Rusche, H., 2004. Modelling of dispersed bubble and droplet flow at high phase fractions. Chem. Eng. Sci. 59, 759–770. Drew, 1983. Mathematical modelling of two-phase flow. Annu. Rev. Fluid Mech. 15, 261–291. Fox, R.O., 2014. On multiphase turbulence models for collisional fluid–particle flows. J. Fluid Mech. 742, 368–424. M. Lance and J. Bataille (1991). Turbulence in the liquid phase of a uniform bubbly air–water flow. J. Fluid Mech., 222, pp 95-118 doi:10.1017/S0022112091001015 Lahey Jr., R.T., 2005. The simulation of multidimensional multiphase flows. Nucl. Eng. Des. 235, 1043–1060. Ranade, V., 1997. Modelling of turbulent flow in a bubble column reactor. Chem. Eng. Res. Des. 75, 14–23. Zhang, Duan Z., and W. Brian VanderHeyden. "The effects of mesoscale structures on the macroscopic momentum equations for two-phase flows." Intl J. Multiphase Flow 28.5 (2002): 805-822.

457716 An Explicit Variable-Density Projection Method for Low-Mach Reacting Flows on Structured Uniform Grids

Abstract:

Low-Mach reacting flows belong to a category of fluid dynamics where acoustic perturbations have negligible influence on the thermodynamics of the flow. They are used to model flares, combustors, reactors, supernovae, and other variable-density flows. However, solving the full equations of gas dynamics in the low-Mach regime is notoriously inefficient due to the presence of acoustic timescales which dictate stability requirements on the timestep size. Pressure projection methods provide a means of excluding these acoustic timescales and allowing one to march in time at the much more attainable convective timescales. In this study, a consistent, explicit projection method is developed for low-Mach reacting flows on structured uniform grids. The approach is based on predicting a dilatation field that is consistent with the thermodynamic state of the flow. The method is fully explicit, first-order in time and requires the solution of a variable coefficient Poisson equation for the pressure. The method has been formally verified and validated against analytical and MMS solutions of variable density flows with mixing and reactions. It has been shown to be robust over a very high range of density ratios up to 1000 and is capable of recovering the incompressible limit.

460097 RANS Modeling of Cluster-Induced and Shear Turbulence in Particle-Laden Channel Flow

Abstract:

A phenomenon often observed in gas-solid flows is the formation of mesoscale clusters of particles due to the relative motion between the solid and fluid phases that is sustained through the dampening of collisional particle motion from interphase momentum coupling inside these clusters. The formation of such sustained clusters, leading to cluster-induced turbulence (CIT), can have a significant impact in industrial processes, particularly in regards to mixing, reaction progress, and heat transfer. One of the challenges in successfully resolving these clusters through Eulerian-Lagrangian simulations in previous work has been the requirement for an overwhelming degree of refinement of the modeled geometry. A less computationally expensive model is needed to model and study CIT in practical applications. This work implements a two-phase Reynolds-Averaged Navier-Stokes (RANS) model to capture the wall-normal flow characteristics in fully developed channel flow with CIT. Both dilute and moderately dense concentrations of particles are considered and compared with high-resolution Eulerian-Lagrangian simulation results. The mean gas velocity is varied to study the relative importance of shear vs. cluster-induced turbulence for fixed mass loading of particles. The inclination of the channel geometry is also varied to examine the effect of the direction of gravity relative to the direction of flow on cluster formation and turbulence statistics.

471492 Why Fibers Are Better Turbulent Drag Reducing Agents Than Polymers

Abstract:

It is typically found in literature that fibers are not as effective as drag reducing agents as polymers, however, for low concentrations, when adding polyelectrolytes (charged polymers) to distilled water, it is found that they are better drag reducing agents than polyelectrolytes in salt water [1]. Since polyelectrolytes in distilled water show rod-like behavior, and polyelectrolytes in salt water show coil-like behavior, this suggests that the common notion that polymers are better at reducing drag needs to be revised.

In this study [2], using hybrid Direct Numerical Simulation with Langevin dynamics, a comparison is performed between polymer and fiber stress tensors in turbulent flow. The stress tensors are found to be similar, suggesting a common drag reducing mechanism in the onset regime for both flexible polymers and rigid fibers. Since fibers do not have an elastic backbone, this must be a viscous effect. Analysis of the viscosity tensor reveals that all terms are negligible, except the off-diagonal shear viscosity associated with rotation. Based on this analysis, we are able to explain why polyelectrolytes in distilled water (i.e fibers) are better drag reducing agents than polyelectrolytes in salt water (polymer coils). Additionally, we identify the rotational orientation time as the unifying time scale setting a new time criterion for drag reduction by both flexible polymers and rigid fibers.

References:
[1] P.S. Virk (1975), "Drag reduction by collapsed and extended polyelectrolytes", Nature, 253, 109-110
[2] A.M.P. Boelens, M. Muthukumar (Forthcoming) "Rotational relaxation time as unifying time scale for polymer and fiber drag reduction", PRE

465918 Turbulent Drag Reduction, Flow Development, and Degradation of Polyox U310 Solutions in a Smooth Pipe

Abstract:

Turbulent drag reduction, flow development, and degradation of Polyox U310 solutions was studied by axial pressure profile measurements in a smooth, segmented, pipe with electro-polished bore of ID 4.58 mm and overall L/D 210. The test pipe assembly, of 7 identical segments, each of L/D 30 with a 0.38 mm ID pressure tap 2.4 L/D from its downstream end, was placed in a single-pass progressive cavity pump-driven flow system fed from a 200 liter tank that held premixed polymer solution. The U310 Polyox, MW ~$10x10^6$, was studied at concentrations C = 0.1 to 10 wppm in deionized water DW solvent.

At any fixed flowrate, individual friction factors for DW solvent were constant to within 1.2% for all tap pairs; and all DW friction factors adhered to the Prandtl-Karman law, $1/\sqrt{f} = 4.0 \log Re\sqrt{f} - 0.4$, within ±0.2 $1/\sqrt{f}$ units for $500 < Re\sqrt{f} < 3000$. Thus the present solvent flows were all fully-developed by the first tap pair 1&2 at mean L/D = 42, in accord with established entrance lengths of 25 – 40 L/D for Newtonian turbulent pipe flow.

Drag reduction by U310 solutions, described by flow enhancements aka "slip" $S' = (1/\sqrt{f_p} - 1/\sqrt{f_n})_{Re\sqrt{f}}$, varied over almost the entire possible range $0 < S' < S'_{mdr}$, from S' ~ 1 for C = 0.1 wppm at low $Re\sqrt{f}$ ~ 500 in the polymeric regime to S' ~ 15 for C = 10 wppm at high $Re\sqrt{f}$ ~ 2500, close to the maximum drag reduction asymptote, $1/\sqrt{f} = 19.0 \log Re\sqrt{f} - 32.4$. Two distinct kinds of S' vs L/D behaviour, described by examples, were discerned: (i) At low C and high $Re\sqrt{f}$, example C = 0.1 wppm and $Re\sqrt{f} > 1000$, S' was highest initially (L/D = 42) and decreased monotonically with increasing L/D to the end (L/D = 192), likely reflecting polymer degradation with increasing downstream distance, and yielding an apparent degradation rate constant kdeg ~ 4 1/s at wall shear stress Tw ~ 100 Pa. (ii) At high C and all $Re\sqrt{f}$, example C = 10 wppm and $500 < Re\sqrt{f} < 2500$, S' was lowest initially (L/D = 42), increased with increasing L/D, and

asymptotically attained a constant S' ~ S'_{mdr} for L/D > 132, which likely represents the entrance length required for fully developed turbulent flow at maximum drag reduction.

455942 Effects of Drag-Reducing Polymers on the Transition and Growth of Turbulent Coherent Structures

Abstract:

Drag-reducing polymers have profound impact on the dynamics of turbulence, not only in the statistically-converged stage but also in its transient development process. Our interest in this area is motivated by two prominent unsolved problems in viscoelastic turbulence. The first is the so-called maximum drag reduction (MDR), an asymptotic upper limit of polymer-induced drag reduction whose mechanism has eluded decades of investigation. Recent progress in the area increasingly points to the laminar-turbulent transition regime as the key to understanding MDR. The second is the transition to turbulence in viscoelastic fluids itself, where complex and sometimes contradictory observations have been made, suggesting the possibility of multiple transition mechanisms. The pivotal solution object controlling the laminar-turbulent transition, the so-called "edge state", is numerically computed. Direct numerical simulation is then used to track the transient development of these marginally turbulent states into fully developed turbulence. The transition process of Newtonian fluids is characterized by a strong breakdown event where turbulent structures sharply intensify across the channel before their eventual decay into the typical coherent structures in the turbulent basin. Polymer additives are found to effectively suppress these high-intensity vortices, which leads to a different transition pathway where a strong breakdown is bypassed.

472903 Evaluating Electrocatalysts for Solar Fuels: Experimental Methods, Performance Metrics, and Catalyst Design Strategies

Abstract:

Electrocatalysis is a key aspect that enables a number of technologies for the renewable, sustainable production of fuels and chemicals. Examples include photovoltaic coupled water electrolysis (PV-electrolysis) and solar photoelectrochemical (PEC) water-splitting, where catalysts are needed for the hydrogen evolution reaction (HER) and the oxygen evolution reaction (OER). Related technologies involve catalysts for CO2 electro-reduction to produce carbon-based fuels and chemials. This presentation will cover several aspects regarding electrocatalysis for these reactions, in particular experimental methods, figures of merit for comparing catalysts, and resulting design strategies for future catalyst development.

474117 Electrochemical Pathways for Sustainable Manufacturing

Abstract:

The chemical industry represents a key component of economic development and prosperity worldwide. However, they confront technology challenges â€“ e.g., reliability of energy supply, lack of energy efficient/transformative manufacturing technologies, waste reduction, and water conservation- that hinder and jeopardize their growth and that could affect their competiveness worldwide. Current chemical industrial production methods have approached their practical performance limits, therefore, new transformational, disruptive, and enabling technologies are needed that will provide solutions to the chemical industry beyond incremental manufacturing improvements. Furthermore, a large and growing number of manufacturers are realizing substantial financial and environmental benefits from sustainable business practices. Sustainable manufacturing is the creation of manufactured products through economically-sound processes that minimize negative environmental impacts while conserving energy and natural resources.

Electrochemistry provides the potential to develop transformational technologies to make more efficient traditional chemical and biological processes. This talk will discuss examples of advanced applications of electrochemical science and technology for the sustainable production of ammonia (chemical), graphene (advanced material), and water remediation.

474722 Challenges and Current Development of Sulfur Cathode in Lithium-Sulfur Battery

Abstract:

Rechargeable lithium-sulfur (Li-S) batteries have exceptional theoretical capacity; however, the practical applications are still elusive to date due to many critical challenges. This presentation summarizes the technical issues of the sulfur cathode and the state of the art strategies in recent years to address these issues from the aspects of lithium polysulfides sequestration, new concepts of sulfur cathode materials, functionality of binders, and the roles of the electrolytes. With regard to conventional sulfur cathodes, liquid-phase reactions involving polysulfides are inevitable. Therefore, the cycle stability is essentially dependent on the affinity of the cathode with lithium polysulfides. Cathode components, including the sulfur host, additive, and binder, should all be active to attract lithium polysulfides. Because polysulfide anions are the electrochemically active species, the sequestration functionality of the cathode components should be specifically designed to attract polysulfide anions. Metal oxides and heteroatoms show promising properties in attracting polysulfides, but the electrochemical â€œinactiveâ€ mass associated with their use needs to be minimized. Polycation binders also seem effective in their ability to attract polysulfide anions due to the electrostatic attraction, but systematic investigation on binders is lacking. In addition to the polysulfide affinity, other properties of polymer binders, including mechanical strength (molecular weight), solubility, and long-term stability in specific electrolytes, are all important to the optimization of the binder system. Moreover, the electrolyte/sulfur weight ratio needs to be watched carefully since almost all studies in open literature used a ratio too high to ever achieve full cell capacity to surpass state of the art Li-ion batteries. It is clear that the development of practical high capacity Li-S batteries still has a long winding road ahead to travel.

457098 Mathematical Modeling of Lithium Ion Cells and Batteries

Abstract:

A physics-based models of lithium ion cells can be used to make predictions of the voltage of the cell as a function of time for charge and discharge given the design parameters and operating conditions of the cell. Such a model enables cell designers to determine the effect of changing design parameters on cell performance before building the cell for a particular application. These models can also be used to design a thermal management system to ensure that the heat generated in the cell is removed before causing overheating and thermal runaway of the cell. These physic-based models of a lithium ion cell can be extended to include multiple spatial dimensions to predict the temperature distribution in a lithium-ion cell for a given set of conditions. A review of the development of physics-based lithium ion cell models will be presented, with particular attention given to predicting temperature distribution in batteries and the effect of volume expansion on battery performance.

461114 Solid-in-Oil (S/O) Nanodispersions for Transdermal Cancer Immunotherapy

Abstract:

Transcutaneous immunization is a novel, non-invasive alternative to conventional immunization by injection. Skin immunocompetence comprised of abundant antigen-presenting cells in the epidermal and dermal layers of the skin can provide an effective tool for transcutaneous immunization, whereas the outermost, hydrophobic layer of skin, the stratum corneum, hinders penetration of antigens into the skin. To realize effective transcutaneous delivery of antigens, we have developed a solid-in-oil (S/O) technique that produces an oil dispersion of hydrophilic biomolecules.

In this study, we applied this technique to a cancer antigen and aimed to develop an effective transcutaneous cancer vaccine. We chose the melanoma, a kind of the skin cancer, to evaluate the cancer immunity. Because a melanoma antigen, TRP-2 peptide, has low solubility in water, we added lysine residues to the N-terminus of the peptide and the newly developed peptide antigen, K-TRP-2, was used as a melanoma antigen. In addition, we tried the use of R-848 as hydrophobic adjuvant to induce stronger immunity. The *in vivo* experiment revealed the effective prevention of tumor and the metastasis was achieved by S/O formulation containing R-848.

473822 Micromotors Generating Self-Propelled Regular Motion By Organic Fuels

Abstract:

Most catalytic micro/nanomotors that have been developed so far use hydrogen peroxide as fuel, while some use hydrazine. These fuels are difficult to apply because they can cause skin irritation, and often form and store disruptive bubbles. In this paper, we demonstrate a novel catalytic Pt micromotor that does not produce bubbles, and is driven by the oxidation of stable, nontoxic primary alcohols and aldehydes with dissolved oxygen. This use of organic oxidation mirrors living systems, and lends this new motor essentially the same characteristics, including decreased motility in low oxygen environments and the direct isothermal conversion of chemical energy into mechanical energy. Interestingly, the motility direction is reversed by replacing the reducing fuels with hydrogen peroxide. Therefore, these micromotors not only provide a novel system in nanotechnology, but also help in further revealing the underlining mechanisms of motility of living organisms.

473467 Pattern Formation of Interfacially Confined Periodically Sequenced Polypeptides

Abstract:

Periodically sequenced peptides can be confined to interfaces and assembled into patterns that present chemical functionalities with exceptional spatial precision. These rationally designed peptides and polypeptides are rapidly becoming useful components in nanostructured materials for applications ranging from drug delivery to energy storage. This presentation will examine several fundamental aspects of self-assembly and pattern formation of well-defined sheet forming peptides confined at interfaces. Our approach involves three steps. (1) We design and synthesize simple periodic peptide sequences, yielding surface-active Î²-strands that self-organize into aggregates to form patterns as a function of the peptide sequence. Rational peptide design allows us to systematically explore the role of hydrophobicity, electrostatics and molecular size on materials properties. (2) We use a set of interfacial characterization tools to examine in intermolecular assembly and supramolecular mechanics of the self-assembled structures. (3) We apply two-dimensional equations of state that define both the phase behavior and the critical surface concentrations of nascent aggregates at the interface. Subsequently, we can apply these parameters to predict the dimensions of pattern formation and to determine the potential of the peptide assemblies as biomimetic materials.

474938 Self-Assembly Under Conditions Mimicking Deep-Sea Hydrothermal Vents

Abstract:

Deep-sea hydrothermal vents are hot spring occurring in the bottom of the ocean. Hot and compressed water, sometimes in the supercritical state (T_c = 374 °C, P_c = 22.1 MPa), gushes out from the sea floor and rapidly cooled to near-freezing temperature by surrounding cold water. In this communication, self-assembly under such unique conditions and its possible implication for the origin of life will be discussed.

Preparation and Properties of Niosomes Prepared with Polyglycerol Fatty Acid Esters Using the Supercritical Carbon Dioxide Reverse Phase Evaporation Method

Shunsuke Yamaguchi [1], *Kenichi Sakai* [2),3)] *and Hideki Sakai* [2),3)]

1). NIKKOL GROUP Cosmos Technical Center Co., Ltd., 3-24-3 Hasune, Itabashiku, Tokyo 174-0046, Japan

2). Department of Pure and Applied Chemistry, Faculty of Science and Technology, Tokyo University of Science, 2641 Yamazaki, Noda, Chiba 278-8510, Japan.

3). Research Institute for Science and Technology, Tokyo University of Science, 2641 Yamazaki, Noda, Chiba 278-8510, Japan.

Keywords: nonionic surfactant, polyglycerol fatty acid ester, supercritical carbon dioxide reverse phase evaporation method, niosome, liposome

Abstract

We have previously reported a new preparation method of liposomes and niosomes (nonionic vesicles) using supercritical carbon dioxide ($scCO_2$) as a solvent (scRPE method). In this study, niosomes were prepared by polyglycerol fatty acid ester (PG ester)-type nonionic surfactants, which are made from naturally-derived materials and have a high safety for a human body and environment, using the scRPE method with ethanol as co-solvent. Decaglycerol distearate (DG2S) and decaglycerol diisostearate (DG2IS) successfully formed niosomes using the scRPE method. On the other hands, decaglycerol monostearate (DG1S) with a high hydrophilic-lipophilic balance (HLB) value gave a spherical micellar solution, and decaglycerol tristearate (DG3S) with a low HLB value gave a gel-like solution. Niosomes of DG2IS had higher trapping efficiency and dispersion stability than DG2S niosomes because the membrane fluidity of DG2IS niosomes was higher than that of DG2S niosomes. The niosomes obtained in the present study would be a useful candidate for applying to cosmetics and drugs because they are formed by nonionic surfactants derived from natural sources and are prepared using the scRPE method without harmful organic solvents.

1. Introduction

Liposomes have phospholipid bilayers with an aqueous phase as their core[1], and have been used as a drug carrier in drug delivery systems, cosmetics, a biomembrane model, a microfield for chemical reactions, and so on[2-8]. One of the most peculiar properties for liposomes lies in retaining water-soluble substances in the inner aqueous phase and oil-soluble substances in the bilayers[9, 10].

Although liposomes are expected to be practically used for wide industrial fields, there are several problems in terms of physical and chemical stability. One of the most useful approaches to improve this problem is a usage of nonionic surfactants in place of phospholipids to form vesicles. They are suitable for industry applications compared with liposomes because of their high physicochemical stability and low cost performance. Furthermore, there are a large number of nonionic surfactants available for the design of nonionic vesicles, niosomes, because of many combinations of hydrophilic groups and lipophilic groups. For instance, there have been many reports to prepare niosomes using polyoxyethylene alkyl ether, sucrose fatty acid ester, polyglycerol fatty acid ester, sorbitan fatty acid ester and polyoxyethylene sorbitan fatty acid ester[11, 12]. However, a variety of preparation methods to obtain liposomes or niosomes require for the addition of harmful organic solvents like chloroform or acetone in their preparation process for dissolving phospholipids or nonionic surfactants[1, 14, 15]. When vesicles are applied to drug carriers and cosmetics, use of organic solvents must be avoided as much as possible. Recently, we have reported about a new environmentally-friendly method, the supercritical carbon dioxide reverse phase evaporation method (scRPE method), for the preparation of liposomes or niosomes in a single step using supercritical CO_2 ($scCO_2$)[16, 17]. However, the niosome formation with the scRPE method has been confirmed only for nonionic surfactants with polyoxyethylene as their hydrophilic groups[18,19]. In pharmaceutical, cosmetic and food fields, consumers have tendency to demand to use products made from more safe and plant-derived materials. Thus, the usage of materials containing petroleum-derived materials like polyoxyethylene-type nonionic surfactants is a tendency to be avoided. However, there were a few reports about niosomes prepared with non-polyoxyethylene-type nonionic surfactants.

In this study, the scRPE method is applied to prepare niosomes with polyglycerol fatty acid ester (PG ester)-type nonionic surfactants. PG ester-type nonionic surfactants are widely used as food additives due to their higher safety. We report in this paper the effects of molecular structure of nonionic surfactants on the niosome formation and their physicochemical properties in terms of trapping efficiency measurements, cryogenic transmission electron microscopy (cryo-TEM), fluorescence anisotropy (FA), differential scanning calorimeter (DSC) and dynamic light scattering (DLS).

2. Experimental

2.1. Materials

Decaglycerol monostearate (DG1S), decaglycerol distearate (DG2S), decaglycerol diisostearate (DG2IS) and decaglycerol tristearate (DG3S) listed in Table 1 provided by Nikko Chemicals (Tokyo, Japan) were used as PG ester-type nonionic surfactants. These materials were derived from natural sources and were used without further purification.

Table 1. Polyglycerol fatty acid ester-type nonionic surfactants used in this study

Chemical structure of nonioinic surfactants	HLB*
Polyglycerol fatty acid ester 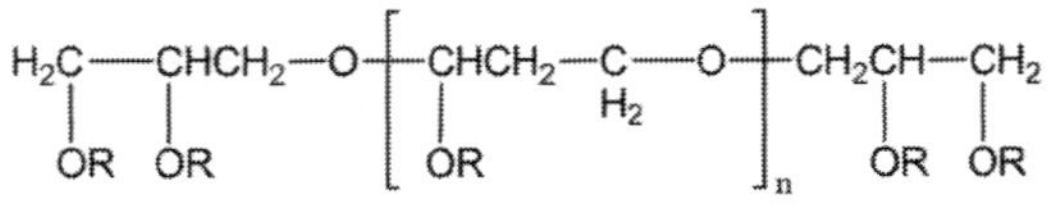 n = ca.10 R = H or stearyl, isostearyl (esterification degree = 1 to 3)	
Decaglycerol monostearate (DG1S)	12.0
Decaglycerol distearate (DG2S)	9.5
Decaglycerol tristearate (DG3S)	7.5
Decaglycerol diisostearate (DG2IS)	10.0

* HLB values were given by suppliers.

2.2. Methods

2.2.1 scRPE Method

Fig. 1 shows a schematic illustration of the experimental apparatus for the scRPE method. After a nonionic surfactant and ethanol at desired amounts were sealed in the cell, CO_2 was introduced into the cell. The cell temperature was then raised to 60°C, while the pressure was kept at desired CO_2 pressure. After a few minutes for equilibration, aqueous solution of 0.2 mol/L glucose, a water-soluble model drug substance, was slowly (0.1 mL/min) introduced using a HPLC pump until total volume of 5 mL and the surfactant concentration of 10 mmol/L. The pressure was then reduced to release CO_2, thereby giving suspensions. The interior of the cell was stirred with a magnetic stirring tip during the preparation process.

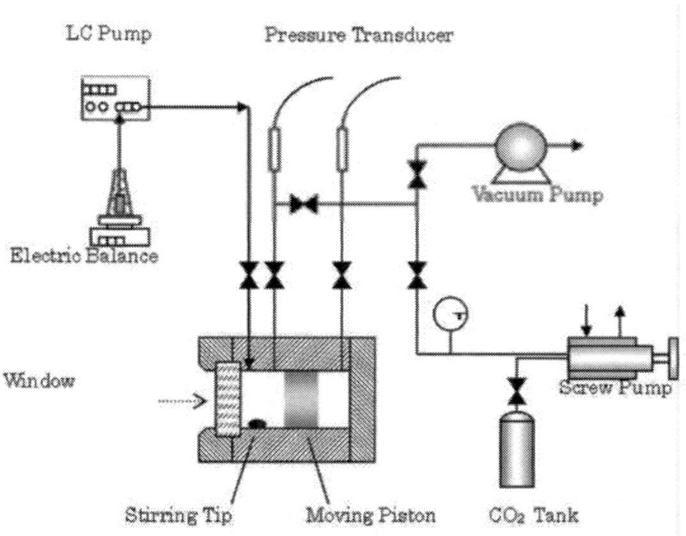

Figure 1. Apparatus for preparation of niosomes with the scRPE method

2.2.2 Evaluation of niosome suspensions

For determination of trapping efficiency of glucose as a water-soluble model drug entrapped in niosomes, the glucose-loaded niosome suspension was dialyzed against water using a cellophane tube (Viscase Scales Co.) to remove unentrapped glucose. The niosomes inside the tube were destroyed by the addition of ethanol. The amount of glucose in the solution was determined by the mutarotase GOD method[20, 21)] using a spectrophotometer (UV-260, Shimadzu Co.).

The hydrodynamic diameter of niosomes was estimated using a NICOMP 380ZLS particle size analyzer equipped with a 5 mW He-Ne laser at a constant detector angle of 90°. The obtained scattering data were fitted using a number-weighted NICOMP-mode analysis to estimate the diffusion coefficient of the surfactant assemblies in the solution. The hydrodynamic diameter was obtained from the diffusion coefficient using the Stokes-Einstein equation.

The formation of niosomes was confirmed by cryogenic transmission electron microscopy (cryo-TEM). A small amount (3–5 μL) of sample solution was placed on the surface of a TEM copper grid covered by a holey carbon film, which was held by a pair of self-locking tweezers mounted on a spring-loaded shaft with a cryo-preparation system (LEICA EM CPC, LEICA microsystems). The sample drop was blotted with a filter paper to form a thin liquid film on the grid (<300 nm), and immediately plunged into liquid ethane cooled by liquid nitrogen (<−170°C). The grid was transferred onto the tip of a cryospecimen holder (CT-3500, Oxford Instruments) under liquid nitrogen. Specimens are kept at temperatures less than −170°C and imaged in a transmission electron microscope (H-7650, Hitachi Science Systems, Ltd.) at an accelerating voltage of 120 kV under low electron dose.

DSC measurements were performed with a differential scanning calorimeter (DSC-60, Shimadzu Co.) at a heating rate of 5 K/min. A stainless steel pan was used as the sample vessel. The reference was distilled water with the same weight as that of the sample.

Examination of the microviscosity of bilayers was made through fluorescence polarization measurements using 1,6-diphenyl-1,3,5-hexatrience (DPH) as a fluorescence probe. DPH was dissolved in tetrahydrofuran, and the resultant solution was added to each of the niosome suspensions to make the molar ratio of DPH to surfactant 1:100. The DPH containing solutions were then incubated at 37°C for 2 h to solubilize the probe and the fluorescence intensity was measured with a fluorescence spectrophotometer (RF-5000, Shimadzu Co.). The excitation and emission wavelengths were 350 and 450nm, respectively. Fluorescence polarization (P) was calculated using the following equation.

$$P = (I_p - I_v)/(I_p + I_v)$$

where I_p and I_v are the fluorescence intensities of the emitted light polarized parallel and vertical to the exciting light, respectively.

3. Results and Discussion

3.1. Solubility of nonionic surfactants in $scCO_2$ and its dependence on ethanol addition

From our previous study[19], it is necessary that nonionic surfactants are completely dissolved in $scCO_2$ and that their hydrophilic and lipophilic balance (HLB) ranges from 9.5 to 9.9 in order to form niosomes using the scRPE method. All nonionic surfactants used in the present study showed a poor solubility in $scCO_2$ at 60°C without co-solvents. Furthermore, in the case of without ethanol, these suspensions did not have trapping efficiency. The addition of ethanol as co-solvent[17] is thus required for the formation of niosomes using these nonionic surfactants. The addition of ethanol is allowed in the fields of cosmetics and pharmaceutical industries if the amount added is proper. Fig. 2 shows the solubility of the nonionic surfactants in $scCO_2$ at 60°C as a function of the amount of ethanol. In this figure, one phase region means complete dissolution of surfactants in $scCO_2$, while two phases region means that a part of the surfactants added cannot dissolve in $scCO_2$. DG1S did not completely dissolve in $scCO_2$ below 200 bar (the upper limit of CO_2 pressure in the present equipment) even if ethanol was added in the surfactant solution. On the other hands, DG2S was soluble in $scCO_2$ at the pressure larger than 130 bar when the concentrations of ethanol were more than 7.5 wt.%. In the case of DG2IS, the complete dissolution was confirmed at ethanol concentrations larger than 2.5 wt.% and at the CO_2 pressure larger than 120 bar. For DG3S system, the dissolution was confirmed at ethanol concentrations larger than 5.0 wt.% and at the CO_2 pressure larger than 105 bar. This result showed that the solubility in $scCO_2$ is increased with increasing the esterification degree of PG ester-type surfactants, because the solubility of $scCO_2$ was close to hexane and $scCO_2$ was easy to dissolve lipophilic surfactants. In the following experiments, on the basis of this result, niosomes were prepared at 60°C, at 7.5 wt.% ethanol concentration, and at 130 bar of CO_2 pressure where DG3S, DG2S and DG2IS were soluble in $scCO_2$.

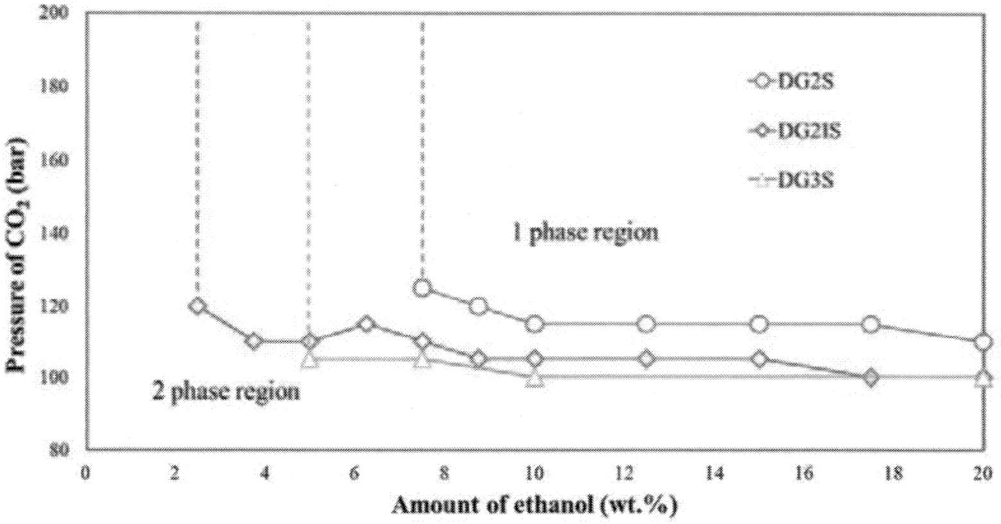

Figure 2. Phase diagram of polyglycerol fatty acid ester-type nonionic surfactants (DG2S, DG2IS and DG3S) in $scCO_2$ as a function of the amount of ethanol at 60°C. DG1S was insoluble in $scCO_2$ at every pressure of CO_2 and ethanol concentration investigated in this study.

3.2. Preparation of niosomes using PG ester-type nonionic surfactants

Preparation of niosomes using PG ester-type nonionic surfactants were conducted by the scRPE method at the proper condition described in 3.1. Table 2 shows the results of visual observation, average particle size and trapping efficiency for the PG ester-type nonionic surfactant suspensions just after preparation. Turbid suspensions were obtained using DG2S and DG2IS, which suggests the formation of niosomes. On the other hands, usage of DG1S and DG3S gave transparent and gel-like solutions, respectively. The average particle size for DG1S was less than 10 nm, which indicated the spherical micelle formation. The aggregates of DG2S and DG2IS had particle sizes of 340 nm and 220 nm, respectively. In addition, the trapping efficiency values for DG2S and

DG2IS were 5 and 20%, respectively, while the DG1S suspension had almost zero % as trapping efficiency.

Table 2. Appearance, average particle sizes, and trapping efficiency for PG ester-type nonionic surfactant suspensions using the scRPE method

	DG1S	DG2S	DG3S	DG2IS
Visual observation			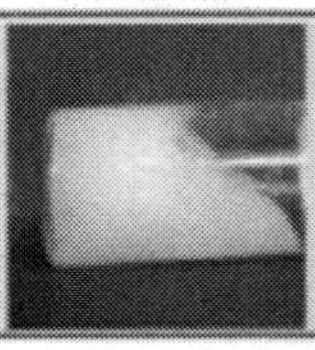	
	Translucent liquid	Turbid liquid	Gel	Turbid liquid
Particle size (nm)	9	340	N.A.	220
Trapping efficiency (%)	0	5	N.A.	20

Figure 3 shows cryo-TEM micrographs for molecular assemblies formed by DG2S and DG2IS. Hollow structures observed in the cryo-TEM micrographs strongly suggest the formation of niosomes for DG2S and DG2IS systems. On the other hands, although DG3S was dissolved in $scCO_2$ by the addition of ethanol as co-solvent, the formation of niosomes was not confirmed and gel-like solution was obtained. Our previous report[19) shows that proper HLB values of nonionic surfactants for the niosome formation range from 9.5 to 9.9. For DG2S and DG2IS, whose HLB values are in or similar to the proper HLB range, niosomes were formed using the scRPE method. Because the HLB value (7.5) of DG3S is smaller than the lower limit of this proper HLB range, lamellar structures were not stably formed for DG3S. In the case of DG1S, the HLB value (12.0) is higher than the upper limit of the proper HLB value for niosome preparation. These findings obtained here showed that high solubility of the PG ester-type nonionic surfactants in $scCO_2$ by the addition of ethanol, as well as usage of nonionic surfactants with the proper HLB value, are required for the formation of niosomes by the scRPE method. These required conditions for the present PG ester-type nonionic surfactants are almost same as those for ethylene oxide-type nonionic surfactants reported in our previous study[19).

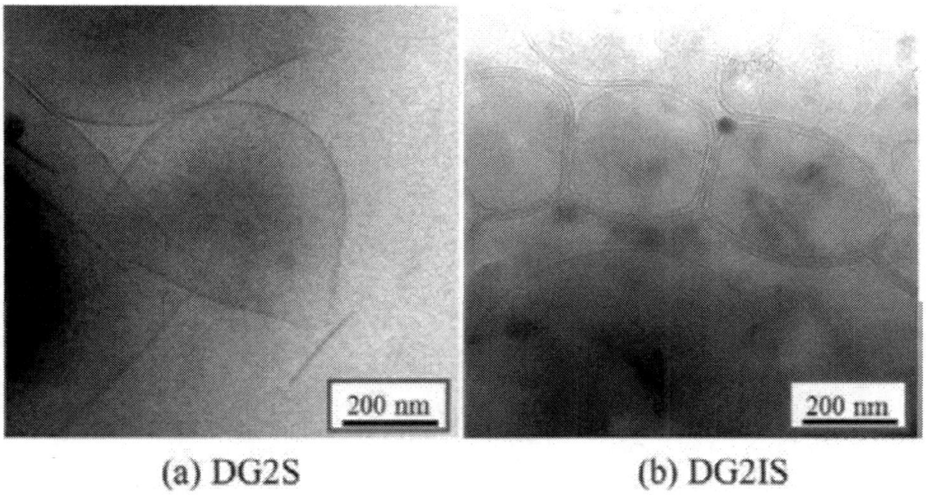

Figure 3. Cryo-TEM micrographs for (a) DG2S and (b) DG2IS niosome suspensions (10mM)

3.3. Comparison between DG2S and DG2IS niosomes

Fig. 4 shows average particle sizes of DG2S and DG2IS niosomes as a function of time after the sample preparation. The particle size of DG2IS niosomes kept an almost constant value at least for 1 week, while that of DG2S niosomes grew gradually from 350 to 1500 nm. This result suggested that DG2IS niosomes had higher dispersion stability than DG2S niosomes. Fig. 5 shows DSC curves for DG2S and DG2IS niosome suspensions. An endothermic peak was observed at 42°C for the DG2S niosome suspension, which indicated a phase transition from gel state (L_β) to liquid crystalline state (L_α). On the other hands, there was no endothermic peak for the DG2IS niosome suspension in the temperature range from 25°C to 70°C. Fig. 6 shows fluorescent anisotropy of DPH incorporated in DG2S and DG2IS niosomes as a function of temperature. The fluorescent anisotropy in DG2S niosomes was gradually decreased from 0.14 to 0.12 up to 40°C, and then rapidly decreased and reached to 0.08 above 45°C. In the case of the DG2IS niosome suspension, the fluorescent anisotropy showed an almost constant value of 0.07 in the temperature range from 5°C to 50°C. These results, as well as the DSC results, showed that the membrane fluidity of DG2IS niosomes was higher than that of DG2S niosomes at 25°C. The cryo-TEM image for DG2S niosomes (Fig.3a) was non-spherical and angular shape having some corners, which also suggested that these niosomes consisted of less flexible bilayers. While niosomes formed by DG2IS with higher membrane fluidity were almost spherical (cryo-TEM image in Fig. 3b). This higher membrane fluidity gives the DG2IS niosomes with higher dispersion stability and higher trapping efficiency.

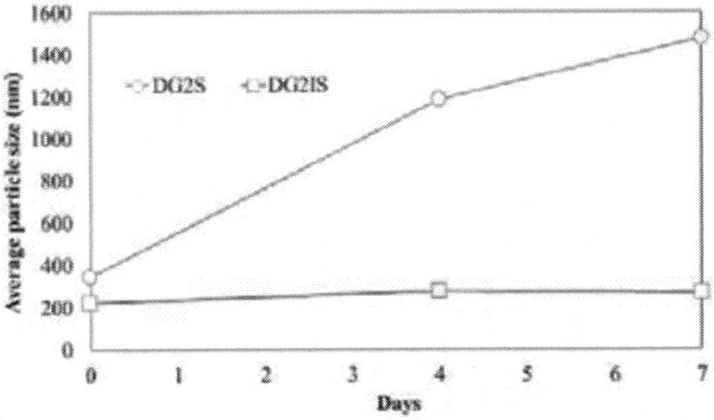

Figure 4. Average particle sizes obtained by DLS for DG2S and DG2IS niosome suspensions as a function of time after the sample preparation

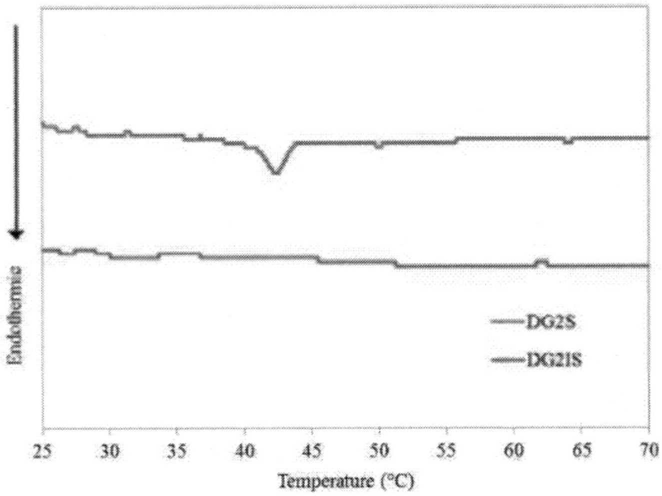

Figure 5. DSC curves for DG2S and DG2IS niosome suspensions.

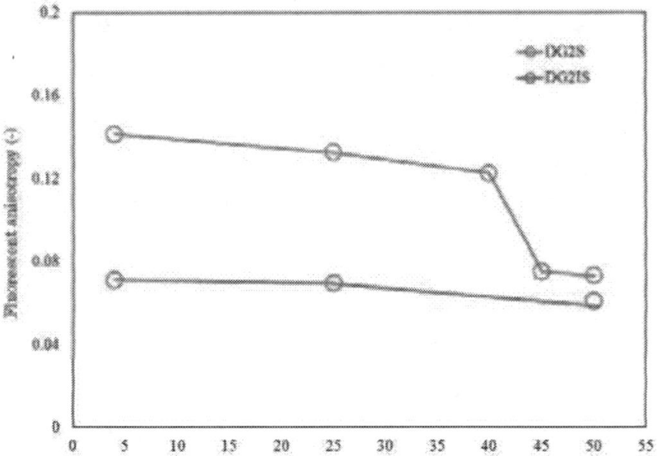

Figure 6. Fluorescent anisotropy for DG2S and DG2IS niosome suspensions as a function of temperature.

4. Conclusions

In this study, we prepared niosomes using PG ester-type nonionic surfactants by the scRPE method. The addition of ethanol as a co-solvent was needed for complete dissolution of the PG ester-type surfactants in $scCO_2$. For DG2S and DG2IS, we succeeded in the formation of niosomes by the scRPE method in the presence of 7.5 % ethanol. These nonionic surfactants have proper HLB values to form lamellar structure in aqueous solutions, which is one of the important factors to obtain niosomes in the scRPE method. Niosomes prepared with DG2IS have higher trapping efficiency and higher dispersion stability than DG2S niosomes because of higher membrane fluidity of DG2IS arisen from branched hydrophobic moiety in the molecule. These findings obtained in this study give us useful information to prepare niosomes using nonionic surfactants derived from naturally sources by the present environment-friendly method.

References

1) Bangham, A. D.; Standish, M. M.; Watkins, J. C. Diffusion of univalent ions across the lamellae of swollen phospholipids, *J. Mol. Biol.*, **13**, 238-252 (1965).

2) Balon, K.; Riebesehl, B. U.; Mueller, B. W. Determination of liposome partitioning of ionizable drugs by titration, *J. Pharm. Sci.*, **88(8)**, 802-806 (1999)

3) Colic, M.; Morse, D. The elusive mechanism of the magnetic `memory' of water, *Colloids Surf., A,* **154 (1-2)**, 167-174 (1999)

4) Montero,M.T.; Hernandez-Borrell, J.; Keough, K.M.W., Fluoroquinolone-biomembrane interactions: monolayer and caloriemtric studies, *Langmuir,* **14 (9)**, 2451-2454 (1998)

5) Komatsu, H.; Okada, S., Increased permeability of phase-separated liposomal membranes with mixtures of ethanol-induced interdigitated and non-interdigitated structures, *Biochim. Biophys. Acta,* **1237 (2)**, 165-175 (1995)

6) Imura T.; Yanagishita H.; Ohira J.; Sakai H.; Abe M.; Kitamoto D., Thermodynamically stable vesicle formation from glycolipid biosurfactant sponge phase, *Colloids and Surfaces B: Biointerfaces*, **43**, 115-121 (2005)

7) Yokouchi y.; Tsunoda Y.; Imura T.; Yamauchi H.; Yokoyama S.; Sakai H.; Abe M., Effect of adsorption of bovine serum albumin on liposomal membrane characteristics, *Colloids and Surfaces B: Biointerfaces*, **20**, 95-103 (2001)

8) Tianshun L.; Rodney J.Y.Ho, Trends and developments in liposome drug delivery systems, *J. Pharm. Sci.*, **90(6)**, 667-680 (2001)

9) Kaneda Y., Virosomes: evolution of the liposome as a targeted drug delivery system, *Adv. Drug Delivery Rev.*, **43 (2-3)**, 197-205 (2000)

10) Marjan, J.; Xie, Z.; Devine, D. V., Liposome-induced activation of the classical complement pathway does not require immunoglobulin, *Biochim. Biophys. Acta,* **1192 (1)**, 35-44 (1994)

11) Manosroi A.; Wongtrakul P.; Manosroi J.; Sakai H.; Sugawara F.; Yuasa M.; Abe M., Characterization of vesicles prepared with various non-ionic surfactants mixed with cholesterol, *Colloids Surf. B: Biointerfaces*, **30**, 129-138 (2003)

12) Nasseri B., Effect of cholesterol and temperature on the elastic properties of niosomal membranes, *Inter. J. Pharm.*, **300**, 95-101 (2005)

13) Uchegbu I.F.; Vyas S.P., Non-ionic surfactant based vesicles (niosomes) in drug delivery, *Inter. J. Pharm.*, **172**, 33-70 (1998)

14) Batzru S.; Korn E.D., Single bilayer liposomes prepared without sonication, *Biochim. Biophys. Acta*, **298**, 1015-1019 (1973)

15) Szoka F. Jr.; Papahadjopoulos D., Procedure for preparation of liposomes with large internal aqueous space and high capture by reverse-phase evaporation, *Proc. Natl. Acad. Sci. U.S.A.*, **75**, 4194-4189 (1978)

16) Otake K.; Imura T.; Sakai H.; Abe M. Development of a New Preparation Method of Liposomes Using Supercritical Carbon Dioxide, *Langmuir*, **17**, 3898-3901 (2001).

17) Otake K.; Shimomura T.; Goto T.; Imura T.; Furuya T.; Yoda S.; Takebayashi Y.; Sakai H.; Abe M., Preparation of Liposomes Using an Improved Supercritical Reverse Phase Evaporation Method, *Langmuir*, **22(6)**, 2543-2550 (2006)

18) Ri K.; Yamaguchi S.; Wongtrakul P.; Hashimoto S.; Otake K.; Ohkubo T.; Sakai H.; Abe M., Preparation and Characterization of Nonionic Surfactant Vesicles Using Supercritical Carbon Dioxide, *Mater. Technol.*, **23(5)**, 340-347 (2005)

19) Yamaguchi S.; Tsuchiya K.; Abe M.; Sakai H., Preparation and Solution Properties of Nonionic Vesicles Using Super Critical Carbon Dioxide Reverse Phase Evaporation Method, *J. Oleo Sci.*, submitted (2015)

20) Miwa, I.; Okuda, J.; Meada, K.; Okuda, G., Mutarotase effect on colorimetric determination of blood glucose with -D-glucose oxidase, *Clin. Chim. Acta,* **37**, 538-540 (1972)

21) Okuda, J.; Miwa, I., Enzymatic micro-determination of D-glucose and its anomer, *Protein, Nucleic Acid, Enzyme,* **17**, 216-224 (1972)

474177 Poloxamer-Based Formulations: Function through Molecular Self-Assembly and Directed Assembly

Abstract:

Amphiphilic polymers of the poly(ethylene oxide)-block-poly(propylene oxide) (PEO-PPO) family, commercially available as Pluronics or Poloxamers and approved for pharmaceutical use [Gu & Alexandridis, J. Pharm. Sci. 2004, 93 (6), 1454-1470; DOI: 10.1002/jps.20021], offer prime examples of self-assembling systems that find diverse applications. The presentation will utilize research findings from our group and others to highlight how the fundamental aspects of Poloxamer block copolymer micellization [Kaizu & Alexandridis, Colloids Surfaces A 2015, 480, 203-213; DOI: 10.1016/j.colsurfa.2014.10.061] and lyotropic liquid crystal formation and nanostructure [Sarkar et al., Macromol. Chem. Phys. 2012, 213 (23), 2514-2528; DOI: 10.1002/macp.201200438] in aqueous media, inform the formulation of Poloxamers applied to emulsions [Kaizu & Alexandridis, J. Colloid Interface Sci. 2016, 466, 138-149; DOI: 10.1016/j.jcis.2015.10.016], colloidal dispersions [Bodratti et al., J. Dispersion Sci. Tech. 2015, 36 (12), 1806-1815; DOI: 10.1080/01932691.2015.1011273], nanoparticle synthesis [Sakai et al., Colloids Surfaces A 2015, 487, 84-91; DOI: 10.1016/j.colsurfa.2015.09.058], and drug delivery [Zhang et al., Nature Commun. 2016, 7, DOI: 10.1038/ncomms11649].

472355 Dynamics of Colloidal Droplets Under Rotating Magnetic Fields

Abstract:

Micron-sized paramagnetic colloids in a AC magnetic field can be used to form 2-D droplets-like structures. By changing the magnetic field strength, we change the interaction energy, leading to the formation of droplets of different phases - such as a fluid or crystal - that grow and coalesce. We characterize these droplets at quasi-equilibrium and quantify their structure to show how the bulk phase differs from the interface, which depends largely on the interaction energy between the particles that make up the cluster. We also determine the stability of these colloidal, clustered, droplets by calculating their energy using a modified dipolar model and measuring the surface excess energy. We look at the fluctuation of the interface and use Fourier analysis to measure interfacial tension and compare the results to the stability found using surface excess energy. We also capture the dynamics of grain boundaries that form as a result of their coalescence with other droplets and compare these dynamics to colloidal droplets of different sizes and phases.

471564 Supramolecular Structural Forces Influence Drainage and Stratification Kinetics in Stratifying Foam Films

Abstract:

Understanding and controlling the drainage kinetics of thin films is an important problem that underlies the stability, lifetime and rheology of foams and emulsions. Foam films containing micelles, colloidal particles or polyelectrolyte-surfactant mixtures exhibit step-wise thinning or stratification, due to the influence of non-DLVO forces, including supramolecular oscillatory structural forces. In this study, we use experiments and theory to investigate the drainage and stratification in vertical and horizontal thin foam films (< 100 nm) formed by aqueous sodium dodecyl sulfate (SDS) solutions. Stratifying thin films exhibit a mosaic pattern in reflected white light microscopy, attributed to the coexistence of domains with various thicknesses, separated by steps. Using Interferometry Digital Imaging Optical Microscopy (IDIOM) protocols developed in the course of this study, we spatially resolve for the first time, the landscape of stratifying freely standing thin films. Further, we determine how the concentration of surfactants and added salt influences the stepwise thinning process for micellar solutions, and how step size can be used for estimating micelle size and interactions. The concentration-dependent aggregation number extracted from our experiments match-up reasonably well with values obtained by other techniques including scattering and fluorescence. Finally, we describe how the interplay between interfacial hydrodynamics and supramolecular structural forces leads to the observed stratification and drainage kinetics and discuss the specific effects of adding salt and increasing surfactant concentration.

460665 Understanding Beer Foam with Dynamic Fluid-Film Interferometry

Abstract:

Foams are commonly found in industrial processes like froth floatation, wastewater treatment, and scrubbers. They are also an important part of consumer products like bread or the foam on a glass of beer. The characteristics of the foam such as the density, mechanical properties, and stability, are dependent on the properties of the surface-active compounds that are present in the liquid.

In the present work, dynamic fluid-film interferometry is utilized to study the time evolution of films captured between a single bubble and the (initially) planar air/solution interface. This technique can be used to track the topological features of the film as a function of time, from which the dynamics of drainage can be quantified. Some of the metrics presented here include the radial extent of the film, the total volume of the film, and the maximum, minimum, and mean film thicknesses. These data are then compared to the dynamic evolution of complete foams formed by sparging air bubbles into a small volume of liquid. Results will be presented for simple surfactant solutions, as well as a range of beers, with the goal of enabling the prediction and/or control of foam characteristics by tuning the properties of the air/solution interface.

461957 Elucidating the Role of Interfacial Rheology on Foam Wetness

Abstract:

Aqueous foams can be found in places ranging from natural systems such as frog nests to applications in the cosmetics, food, and pharmaceutical industries. Foams can be stabilized by the action of surfactants, which are surface-active material that adsorb on gas bubble interfaces. The surfactants slow natural coarsening and the 'drying' of foams over time as the liquid component drains due to gravity. Many studies have examined the effect of different surfactants on foam stability, but less known is their effect on foam density ('wetness'). Previous work from our group has examined both the maximum volume of surfactant solution entrained as a liquid film when a bubble rises to the surface of a surfactant solution and the rate of film drainage for several small-molecule surfactants. The measurements find correlations between the maximum film volume and bulk foam density. In this more recent work we investigate how surfactants with interfacial rheology may affect the volume of entrained fluid, potentially providing a mechanism for modulating bulk foam properties.

To accomplish this, we utilize a model globular protein found in cow's milk, bovine serum albumin (BSA), to study the effect of interfacial rheology on the mechanism of film drainage between air-solution interfaces. Previous studies have established that upon exposure to air-solution interfaces, protein molecules irreversibly adsorb and undergo conformational changes. This can be observed as time-varying interfacial viscoelasticity. We utilize dynamic fluid-film interferometry to analyze the profile of a thin liquid film as it drains from the interstitial space of a single bubble and the bulk interface. Additionally, we will present preliminary results that describe the evolution of the liquid film over time for a chosen BSA concentration and discuss the usefulness of using surface elasticity to predict the foam wetness.

2D modeling of foam flow using an interface tracking method for analysing its stress-strain behavior

Gouthami Sethamaraikkannan, University of Calgary, Canada
Ian Donald Gates, University of Calgary, Canada

Abstract

Bulk foam is the agglomeration of gas bubbles separated from each other by thin liquid films [1]. In this study, we explore a grey-box modeling approach to evaluate effective bulk foam viscosity. The strategy is to model the complex behavior of surface-tension dominated foam flow from first principles, followed by the utilization of an empirical viscosity model. Simulation of foam flow is attempted in two ways. One is by solving the Navier-Stokes equation along with the convection of level set function which indicates the presence of gas, liquid or interface. The other method is the simulation of foam based on the Poiseulle-like flow of liquid phase through the interconnected system of plateau borders. It is observed that the second method is easier to implement in spite of its limitations. Effective viscosity is then empirically determined by different approaches.

1 Introduction

Foams are complex discontinuous systems dominated by surface forces [2]. The volume fraction of gas in a foam is referred to as foam quality. When this fraction is low, bubbles in a foam are spherical and when it is high, bubbles are polyhedral shaped and separated by thin liquid films. Although a foam system is thermodynamically unstable, it has many unique properties such as compressibility, viscoelasticity, nonlinear mechanical behavior etc. Like solids, foams possess finite shear modulus, yet they also flow like a non-Newtonian fluid [3].

Given its unique properties, foams are desirable in various industries including the petroleum recovery and processing industry. Some of the applications in the petroleum industry are oil well drilling, fracking, reservoir injection, foam-assisted SAGD, oil well production, foam fractionation, mobility control etc [4]. The usage of foam often requires accurate prediction of its pressure drop during flow. However, since the dynamic bubble structures in a foam are always changing, rheological characterization is challenging. Many experimental works have produced empirical models correlating foam viscosity with a large number of parameters such as foam quality, texture, foam channel wall interactions, liquid phase properties, temperature, pressure, etc [5]. However these findings have been found to vary largely between studies and are often non reproducible. On the other hand, micromechanical models that can elucidate the relation between foam structure and rheology are difficult to develop due to the complex film-level mechanisms [3, 6].

Keeping these issues in view, in this study we exercise a grey-box technique to determine the effective viscosity of foam. Grey box modeling technique is one which incorporates apriori knowledge about a process into the parameters or variables of a model [7]. The advantage of this approach lies in its ability to consistently reflect the effects of various bulk phase and interfacial properties on dynamic foam flow, while ignoring complex interfacial mechanisms that are difficult to scale up.

2 Foam flow model

The first part of the strategy discussed above is the development of a suitable foam flow model. Owing to the complexity of three-dimensional structures, we restrict ourselves to two-dimensional models. The two modeling approaches that we have tested are interface tracking and drainage through interconnected system of plateau borders.

2.1 Interface tracking

Interface tracking is a technique used in the simulation of multiphase flows with discontinuous interfaces. It works by solving the fundamental Navier-Stokes equation coupled with explicit tracking of the location of interfaces. There are many different interface tracking methods such as volume-of-fluid, level-set, front tracking etc. These have been applied to the study of two-phase flow transients like Rayleigh-Taylor instability, shape and stability of a rising bubble, falling drop etc [8]. Recently, interface tracking methods have also been applied in the field of computer graphics, for the simulation of dynamic foams. Since this method can track foam flow quite accurately depending on the physics made available, it is potentially suitable in the analysis of foam rheology.

2.1.1 Method

Firstly, the Navier-Stokes equation for an incompressible viscous fluid is considered.

$$\frac{\partial \vec{u}}{\partial t} = -(\vec{u}.\bigtriangledown)\vec{u} + \bigtriangledown.(\nu \bigtriangledown \vec{u}) - \frac{\bigtriangledown P}{\rho} + \vec{f} \tag{1}$$

$$\bigtriangledown.\vec{u} = 0 \tag{2}$$

Here $\vec{u}$ is the velocity, P is the pressure, ν is the kinematic viscosity, ρ is the density and $\vec{f}$ is the external force (gravity/buoyancy/surface tension/other external forces).

Coupled with this equation, we track the level set function ϕ defined as,

$$\phi \begin{cases} > 0, & \text{Gas region} \\ < 0, & \text{Liquid region} \\ = 0, & \text{Interface} \end{cases}$$

The convection of this level set function is given by,

$$\frac{\partial \phi}{\partial t} + \vec{u}.\bigtriangledown \phi = 0 \tag{3}$$

In order to generate numerical simulations of the above equations, we employ the CFD module of COMSOL multiphysics. By assuming fixed interface thickness, numerical difficulties are avoided [9]. The following is the level-set equation solved in COMSOL [10].

$$\frac{\partial \phi}{\partial t} + \vec{u}.\bigtriangledown\phi = \gamma\bigtriangledown\cdot\left(\varepsilon_s\bigtriangledown\phi - \phi(1-\phi)\frac{\bigtriangledown\phi}{\mid\bigtriangledown\phi\mid}\right) \tag{4}$$

Parameters ε and γ handle numerical stabilization and reinitialization. Finally, the surface tension force in equation 1 is given by,

$$\vec{f_{st}} = -\sigma\kappa\vec{n}\delta(\phi) \tag{5}$$

where,

$$\vec{n} = \frac{\bigtriangledown\phi}{\mid\bigtriangledown\phi\mid}, \kappa = \bigtriangledown.\frac{\bigtriangledown\phi}{\mid\bigtriangledown\phi\mid}, \delta = 6\mid\bigtriangledown\phi\mid\mid\phi(1+\phi)\mid \tag{6}$$

2.2 Flow through interconnected system of plateau borders

The other method that we have applied for the modeling of foam flow is flow through the interconnected system of plateau borders. The complex geometry of a foam can be classified into three distinct regions. (a) Thin liquid films or lamella where two bubble surfaces meet, (b) Plateau borders or channels which are junctions of liquid lying between three bubbles (b) Nodes, which are the surfaces where four plateau borders meet. Since film thickness in a foam are typically very low, most of the liquid is assumed to be lying within the plateau borders and nodes [11].

Ignoring film rupture and ostwald ripening, fluid flow between bubble walls is determined by the surface mobility factor M. When M is low, flow is Poiseulle-like and when it is high, flow is plug-like [12]. The standard foam drainage equation developed by Goldfarb et al [13]. applies to Poiseulle flow of the liquid phase under gravity in dry foams. Non-dimensionalizing this equation, we get [14]

$$\frac{\partial\varepsilon^*}{\partial t^*} + \frac{\partial\varepsilon^{*2}}{\partial z^*} = \frac{\partial}{\partial z^*}\left(\sqrt{\varepsilon^*}\frac{\partial\varepsilon^*}{\partial z}\right) \tag{7}$$

where ε is the liquid volume fraction, $\varepsilon^* = \varepsilon/\varepsilon_0$, $t^* = t/t_0$ and $z^* = z/z_0$. $z_0 = H$, (the maximum height of foam), $t_0 = \frac{12\alpha_2 \mu \rho g H^3}{\delta_\varepsilon \sigma^2}$ and $\varepsilon_0 = \left(\frac{\sigma}{2\rho g H L}\right)^2 \delta_\varepsilon$, μ is bulk viscosity, ρ is bulk density, σ is surface tension coefficient, L is the length of the plateau border and α_2, δ_ε are dimensionless parameters. Extending this equation to 2D drainage in a foam channel network with heterogeneous cross section radii, the drainage equation becomes [14]

$$\frac{\partial \varepsilon^*}{\partial t^*} + \frac{\partial \varepsilon^{*2}}{\partial z^*} = \frac{\partial}{\partial z^*}\left(\sqrt{\varepsilon^*}\frac{\partial \varepsilon^*}{\partial z}\right) + \frac{1}{n^2}\frac{\partial}{\partial x^*}\left(\sqrt{\varepsilon^*}\frac{\partial \varepsilon^*}{\partial x}\right) \tag{8}$$

where $x^* = x/x_0$, $x_0 = W$ (the width of the foam column) and $n = W/H$. Assuming that flow is fully developed and there is no liquid entering the top or the sides of the foam, we have the following boundary conditions

$$\varepsilon^*(z^*, x^*, t^*) = 0, \quad z^* < 0 \tag{9a}$$

$$\varepsilon^*(z^*, x^*, t^*) = 0, \quad (x^* < 0) \quad \cup \quad (x^* > 1) \tag{9b}$$

$$\frac{\partial}{\partial t^*}\varepsilon^*(z^*, x^*, t^*) = 0, \quad z^* = 1 \tag{9c}$$

Along with appropriate initial conditions, explicitly discretized forms of equations 8 and 9 are solved to describe foam flow.

3 Effective foam viscosity

After we have obtained foam flow behaviour from the two first principles modeling approaches described above, foam flow information is utilized in empirical models for the evaluation of effective foam viscosity. The different models that we have tested are presented in the results section seen below.

4 Simulation results

4.1 Interface tracking

While it seemed appropriate to use interface tracking to encapsulate complex flow behaviour, evaluation of the model was very expensive computationally. For instance, seen below are the results of interface tracking model on a single soap bubble with an air-water interface. Simulations were performed on a PC with 2.8 GHz CPU and 32 GB RAM.

(a) Soap bubble with large interface thickness falling down and collapsing

(b) Soap bubble with smaller interface thickness floating upwards, bubble size decreases.

Figure 1: Interface tracking of a single soap bubble

While a bubble with larger interface thickness (figure 1a) was simulated in 1 hour, bubble with a smaller interface thickness (which is typical) was simulated in 30 hours (figure 1b). Since in a foam there are atleast thousands of bubbles, simulation time will be extremely high.

4.2 Flow through interconnected system of plateau borders

Simulation of foam dynamics based on flow of liquid phase was much easier to implement since flow equations were explicitly discretized. However, in order to avoid numerical oscillations or instability, $\Delta t^*/\Delta z^*$ and $\Delta t/\Delta x^*$ ratios were kept well below 10^{-4}. Figure 2 shows the results of this model in a network with homogeneous channel cross-section radii.

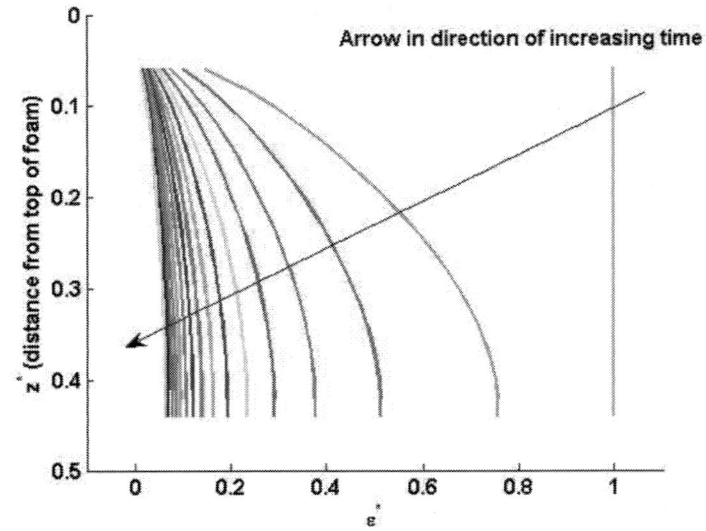

Figure 2: Liquid volume fraction in a homogeneous foam channel network.

4.3 Effective viscosity based on foam drainage model

Since the above models derived above establish the effects of gravity, capillary and viscous forces on a foam at a very low computational cost, we utilize the resulting numerical solution of the liquid volume fraction property, $\varepsilon(z,x,t)$ to empirically determine effective viscosity.

4.3.1 Approach 1: Volume-weighted averaging

From $\varepsilon(z,x,t)$ one can determine $\varepsilon(t)$, which is the total liquid volume fraction at any instant t. The effective foam density ρ^f and foam viscosity μ^f can be obtained by volumetric averaging as,

$$\rho^f = \rho_l \varepsilon(t) + \rho_g (1 - \varepsilon(t)) \tag{10}$$

$$\mu^f = \mu_l \varepsilon(t) + \mu_g (1 - \varepsilon(t)) \tag{11}$$

where ρ_l, μ_l are the density and viscosity of liquid, while ρ_g, μ_g are the density, viscosity of gas.

4.3.2 Approach 2: Fully developed flow with no-slip

Assuming fully developed flow of foam with no-slip condition on pipe walls, effective viscosity can be evaluated analogous to the Poiseulle law as,

$$\Delta P = \frac{K \mu^f u^f H}{R_{pipe}^2} \tag{12}$$

where H and R_{pipe} are the length and radius of pipe. $u = Q/\pi R_{pipe}^2$ where Q is the voulmetric flow rate and K is an empirical constant. ΔP is the sum of stresses due to gravity and capillary forces in this case. Assuming an empirical model for the capillary stresses in terms of $\varepsilon(t)$,

$$\Delta P = \rho^f g + \gamma_0 \left(\frac{\varepsilon_0}{\varepsilon}\right)^n \tag{13}$$

where ε_0 is the initial liquid volume fraction and γ_0, n are empirical constants.

4.3.3 Results

For the evaluation of effective foam viscosity, values of physical parameters used in simulations are shown in Table 1. First, the volume-weighted averaging approach was tested on a foam channel network with homogeneous cross section

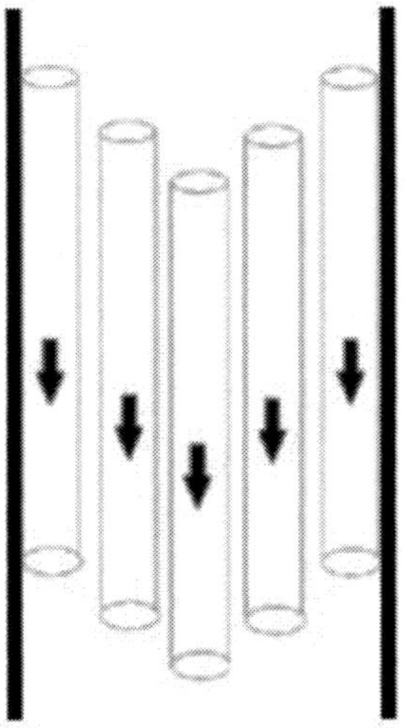

Figure 3: Foam flow in a pipe. Each channel (through which the liquid phase flows) is now moving at different velocities due to no-slip condition at walls. Flow is assumed to be fully developed.

radii and foam column height (same as foam column width) equal to 1 cm. Over a period of 8 minutes, liquid volume fraction decreases from 0.09 to 0.004. Figure 4 shows the resulting foam viscosity. It is seen that foam viscosity approaches gas viscosity over time.

On the other hand, if we consider the second approach, the effective viscosity values are different. Figure 5 shows effective foam viscosity for the same foam column described above, at different values of the empirical constants. This clearly reflects that foam viscosity continues to increase with the increase in foam quality (i.e., decrease in liquid volume fraction). This exponential behaviour of foam behaviour was also observed in the experimental work on shaving foam by Wu et al [15].

Table 1: Parameter values used in simulations

Parameter	Description	Unit	Value
ρ_l	Liquid density	kg/m^3	1000
ρ_g	Gas density	kg/m^3	1.225
μ_l	Liquid viscosity	kg/ms	10^{-3}
μ_g	Gas viscosity	kg/ms	2×10^{-5}
σ	Gas-Liquid interfacial tension	N/m	0.015
H	Height of foam column	m	0.01
W	Width of foam column	m	0.01
L	Average length of plateau border	m	10^{-4}
δ_ε	dimensionless parameter	-	0.171
α_2	dimensionless parameter	-	50

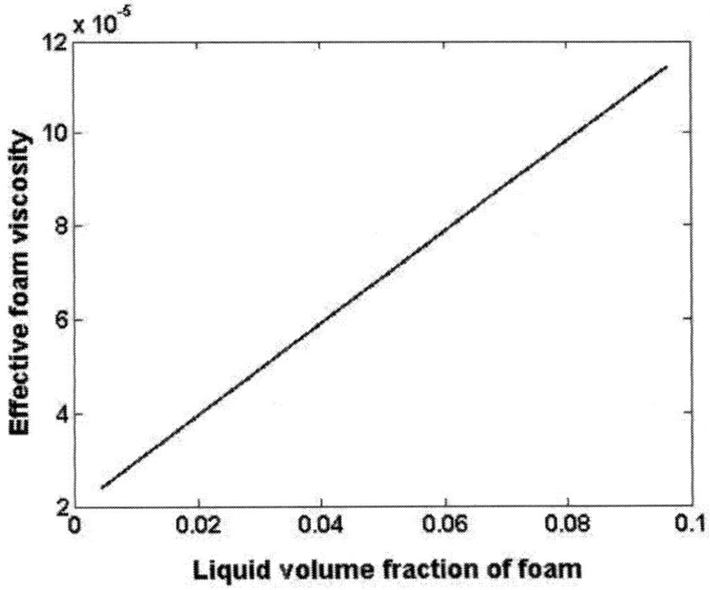

Figure 4: Volumetric-averaged foam viscosity for a foam column with homogeneous cross section radii and $H = W = 1cm$

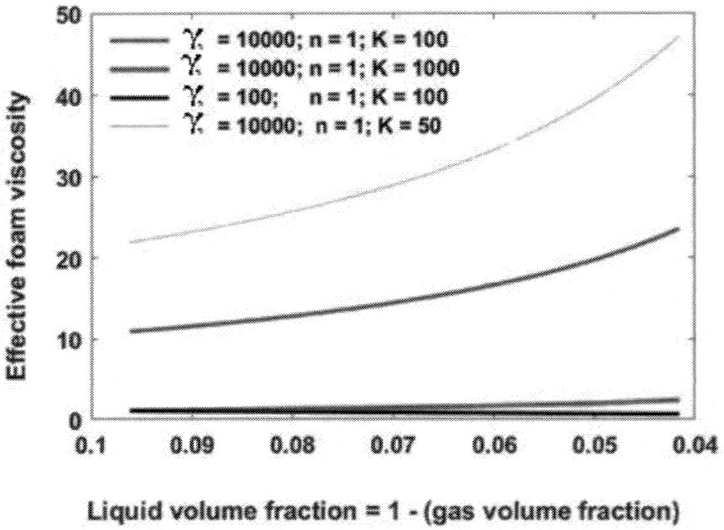

Figure 5: Foam viscosity derived based on Poiseulle-foam analogy, for a foam column with homogeneous cross section radii and $H = W = 1cm$

5 Shaving foam

Figure 6 shows the image of shaving foam observed through diascopic light in a confocal microscope. Using these images from different times, we can extract bubblesize distributions for use in the foam flow model. However, the foam on a microscope slide is under stress from microscope slide and slide cover on the top and bottom. Since the empirical model for foam foam assumes fully developed foam flow, the model is better suited at pipe dimensions that are greater than bubble size. In this regard, we have procured a dynamic foam analyser from KRUSS (model DFA100) to obtain dynamic bubble size distributions in a foam. Using this equipment, we can validate our model and make pressure drop predictions for different types of foam.

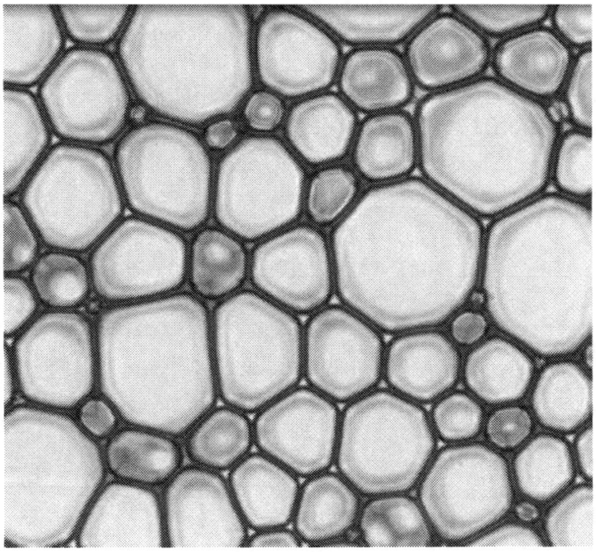

Figure 6: Bubbles in a shaving foam

6 Conclusion

In this study we have developed a novel method for the determination of foam rheology behavior. The effects of parameters affecting internal dynamics is obtained through fundamental models based on first principles. The resulting information is then incorporated in suitable empirical models for determining effective foam viscosity. Fundamental models elucidating bubble dynamics were resolved through interface tracking methods and modeling flow of liquid phase through the interconnected system of plateau borders. While the former method was found to be highly computationally expensive, the latter method had very low simulation times. In order to determine effective viscosity, two approaches were used. The method of volume-weighted averaging indicates a low viscosity at a high foam quality which is contrary to observed behaviour. On the other hand, foam viscosity derived by analogy to fully developed Poiseulle flow of foam in a pipe generated an exponential trend with high viscosity values at high foam quality.

References

[1] W.R. Rossen. Foams in enhanced oil recovery. *Foams-Theory, Measurement, and Applications), R.K. Prud'Homme and S.A. Khan ed*, pages 413–464, 1996.

[2] Wen Zheng, Jun-Hai Yong, and Jean-Claude Paul. Simulation of bubbles. *Graphical Models*, 71(6):229–239, 2009.

[3] Andrew M Kraynik. Foam flows. *Annual Review of Fluid Mechanics*, 20(1):325–357, 1988.

[4] Laurier L Schramm. Fundamentals and applications in the petroleum industry. *Adv. Chem*, 231:3–24, 1992.

[5] Luis Fernando Bonilla, Subhash N Shah, et al. Experimental investigation on the rheology of foams. In *SPE/CERI Gas Technology Symposium*. Society of Petroleum Engineers, 2000.

[6] HM Princen and AD Kiss. Rheology of foams and highly concentrated emulsions: III. static shear modulus. *Journal of Colloid and Interface Science*, 112(2):427–437, 1986.

[7] Björn Sohlberg. Grey box modelling for model predictive control of a heating process. *Journal of Process Control*, 13(3):225–238, 2003.

[8] Gregor Cerne, Stojan Petelin, and Iztok Tiselj. Coupling of the interface tracking and the two-fluid models for the simulation of incompressible two-phase flow. *Journal of computational physics*, 171(2):776–804, 2001.

[9] H Grzybowski and R Mosdorf. Modelling of two-phase flow in a minichannel using level-set method. In *Journal of Physics: Conference Series*, volume 530, page 012049. IOP Publishing, 2014.

[10] Mark A. Finlayson. *COMSOL Multiphysics User's Guide*.

[11] Stephan A Koehler, Sascha Hilgenfeldt, and Howard A Stone. A generalized view of foam drainage: experiment and theory. *Langmuir*, 16(15):6327–6341, 2000.

[12] M Safouane, A Saint-Jalmes, V Bergeron, and D Langevin. Viscosity effects in foam drainage: Newtonian and non-newtonian foaming fluids. *The European Physical Journal E*, 19(2):195–202, 2006.

[13] I Goldfarb, I, B Kahn, K, and Schreiber I. Liquid flow in foams. *Fluid Dynamics*, 23:244–249, 1988.

[14] G Senthamaraikkannan and I Gates. Analyzing the effects of bulk media parameters on foam stability with 2D drainage models. *(submitted)*, 2016.

[15] Maw-Sheng Wu, Maureen E Sullivan, and Diana J Yee. The viscosity of a foam (air in water emulsion). *Colloids and surfaces*, 12:375–380, 1984.

472505 Monitor and Optimize Emulsion Droplet Size with Real-Time *in Situ* measurements

Abstract:

Emulsions play a critical role in many unit operations and affect product quality in many industries, including the production of liquid formulations in personal care, pharmaceutical and agrochemical products, as well as in the production of many foods, flavorings, and polymer products. Designing a robust process that delivers the required droplet size distribution in a suspension or emulsion is a common challenge for scientists and engineers who are tasked with optimizing product quality and process efficiency.

Characterizing emulsion droplet size effectively allows scientists and engineers to more effectively solve process bottlenecks and improve product quality and consistency. Traditionally, scientists and engineers have relied on off-line particle size analyzers, to perform emulsion characterization. Off-line analysis methods typically require sample preparation and sample dilution, which in the case of emulsions, often causes changes in the droplet system that make it difficult to relate the off-line results to real-time, dynamic changes in the droplet system that can impact process efficiency and product quality.

The emergence of PAT technologies for particle and droplet system characterization allow scientists and engineers to track the rate and degree of change of particles and droplets in suspensions and emulsions as they naturally exists in process without the need for sampling or sample preparation. In-process measurement of particles and droplets can provide real time information about how droplets behave under changing process conditions, which can directly relate Critical Process Parameters such as stir/shear rates, surfactant selection and dosage, and temperature; to Critical Quality Attributes including droplet or particle size and product stability, for better holistic achievement of Quality by Design.

This presentation will review case studies demonstrating the use of a Design of Experiments in combination with common in-situ PAT tools, including FBRM and PVM technologies, to develop detailed process understanding and enable process optimization of emulsion-based products.

464907 Arrested Coalescence of Viscoelastic Droplets: Restructuring Dynamics

Abstract:

Microstructured viscoelastic droplets in oil-in-water emulsions can arrest during coalescence and acquire non-spherical shapes depending on their size, concentration and connectivity. The rheology and flow of these droplet networks has a strong effect on the texture and rheology of various foods, cosmetics, paints, and pharmaceutical products. Previous work examined monodisperse binary droplet interactions to develop a physical model of arrest and final droplet shape. This work considers the arrest of polydisperse droplets and structures formed from a larger number of droplets in order to develop a model of full emulsions. A new mechanism of arrested droplet restructuring is identified and shown to form denser networks, but only under specific conditions. Image analysis mapping distributions of strain through droplet networks is used as a way to determine the significance of arrest in an emulsion but also to better understand the interactions of interfacial and elastic driving forces. This work allows the design of anisotropic emulsion microstructures and prediction of the overall performance of formulations.

465820 Stability of Drop and Bubble Collisions: Coalescence Maps

Abstract:

Predicting the outcome of droplet and bubble collisions is an invaluable tool for the optimisation of a wide range of industrial processes, including emulsion and foam stability in food, and also mineral and pharmaceutical processing. In particular, the stability of both drop and bubble collisions is of particular importance in lab-on-a-chip (LOC) devices, where the continual interaction of drop or bubble pairs is necessary for effective operation. Previously, our group has developed and applied experimental and theoretical methods to investigate the dynamic interactions between drop or bubble pairs using Atomic Force Microscopy (AFM). Here we apply our methods to quantitatively describe the outcome of both drop and bubble collisions under conditions of direct relevance to LOC devices.

Using an experimentally validated drop interaction model, which accounts for interface deformation, surface forces and thin film drainage, we are able to predict the outcome of droplet or bubble collisions over a wide range of important parameters directly relevant to LOC devices. The parameters of interest include drop or bubble size, approach velocity and interfacial tension (determined by the presence of surfactants within the system). Our results are presented in the form of â€˜coalescence mapsâ€™, a practical and simple depiction of collision stability. Our characterisation of both drop and bubble collisions uncovers rich and complex behaviour, depending primarily upon the influence of different types of surface forces present.

471852 Nanoemulsion Formation: Controlling and Predicting Droplet Size

Abstract:

Nanoemulsions are liquid-liquid dispersions with droplet sizes on the order of 100 nm. They are used in diverse areas like drug delivery, pharmaceutics and development of smart food drinks, and possess exceptional properties such as high surface area, robust stability and tunable rheology. However, there is little knowledge in literature about the factors dictating the droplet size of nanoemulsions. In this work, we highlight that the droplet size correlations for macroemulsions (droplet size ~ 100 um) cannot be extrapolated to nanoemulsions. Hence, we derive a new criterion for predicting droplet size of nanoemulsions that takes into account the physics at small length scales. We validate our scaling theory with a wide range of experimental droplet size data. Further, building on the proposed scaling theory, we propose a new breakage model for predicting the frequency of successful droplet breakup event, the rate-controlling step in nanoemulsion formation. A population balance analysis with the proposed breakage model provides insights about the kinetics of droplet size change and successfully captures the effect of droplet viscosity, unlike the earlier breakage models. The predictions of the proposed breakage model are supported by the experimental droplet size data. Lastly, we also obtain a parity plot for the kinetic droplet size data prepared with different oil viscosities and different power density input. The proposed scaling theory and models are useful for rational design of nanoemulsions.

460735 Fluorocarbon Oil Reinforced Triple Emulsion Drops

Abstract:

There is a significant interest in developing a general class of microcapsules that enable encapsulation of any kind of cargo with high loading efficiency, while simultaneously preventing premature leakage until exposure to some external stimulus such as mechanical force. However, there has been little progress toward the production of microcapsules that fulfills these demands. In this work, we present triple emulsion drops with an ultra-thin fluorocarbon oil layer that separates the innermost cargo from the polymeric shell to provide possible solution to these challenges. The combined favorable effects of omniphobicity and the chemically inert property of fluorocarbon oil enable encapsulation of a broad range of polar and non-polar cargoes as well as prevention of premature leakage of small actives in microcapsules. We use the unprecedented flow control of multi-phasic fluids in microfluidics to produce these multi-layered emulsion drops. They have an ultra-thin intermediate layer that separates the innermost cargo phase from the polymeric shell phase, leading to high loading efficiency of actives in these microcapsules. We demonstrate the effectiveness of these microcapsules by successfully encapsulating small molecules and active biomolecules that would otherwise be very challenging.

464440 Ionic Memristors and Nonlinear Ionic Circuits for Future Massively Multiplexed Liquid Biopsy Electrochemical Biosensor Arrays

Abstract:

Recent discoveries of a multitude of protein, microRNA, ctDNA and other molecular biomarkers for cancer suggest that early screening and dynamic therapy management would soon be possible to significantly increase cancer survival rate. However, the screening tests for these biomarkers are only effective if they are applied frequently and outside laboratories ---- a turn-key liquid biopsy technology needs to be developed to detect and quantify a large number of molecular biomarker targets from a small volume of blood sample. I will review my current progress in developing such a platform and describe some future plans/challenges.

I have developed micro/nanofluidic circuit components that allow precise measurement and control of ion and biomolecule transport in physiological buffers. I intend to synthesize these fluidic circuit units into large-scale fluidic-based ionic circuits, including multiplexed active electrochemical sensor arrays, for cancer liquid biopsy. The ionic circuits require a robust flip-flop latch element to digitize and memorize the ionic signals and to enable robust sequential operations. They also require a high-flux switching/amplifying element to individually address each sensing/memory unit and to sustain the long-range ionic signal propagation. I have focused on the development of these two elements in my PhD work.

A bistable ionic memristor based on a reversible redox reaction on a silicon microelectrode is the basis of my ionic latch element [Sun et al, Small, 11, 5206-5213 (2015)]. The memory of ions transporting across the silicon/solution interface is due to the electrochemical formation of an interfacial nanoscale oxide film on the silicon microelectrode under anodic polarization. While under cathodic bias, the hydride formation decomposes the thin oxide layer. Combining these two processes we produce a reversible and history dependent interfacial resistance of the silicon microelectrode, which gives this ionic memristor distinct on-off current states. The electrochemical oxide film growth and decomposition together with their contribution to the resistive switching are studied by X-ray photoelectron spectroscopy. By designing an optimal latching operation protocol and appropriate impedance matching with the surrounding microfluidic ionic channel, we are able to realize a robust flip-flop memory action with a more than 10 on/off ratio and a 1s short switching time. The ion current signal can hence be sampled and digitized into the ionic memristor electrochemically.

To realize the switching/amplifying element, a high-flux ionic transistor and amplifier based on non-equilibrium ion transport through an ion-selective nanoporous membrane is then designed which allows a small normal gating voltage signal to control a large draining ionic cross current. It can allow for individual addressing of the ionic memristor and help synthesize multiplexed electrochemical sensors with ionic memristors into large-scale electrochemical circuits [Sun et al, LabChip,16, 1171-1177 (2016)]. Because of the charge selectivity, asymmetric concentration polarization occurs across such a membrane upon the application of a normal electric field. By coupling the ion-selective membrane with a tangential microfluidic channel adjacent to the membrane, I utilized coion flux balance to produce a depletion front in the microchannel that is sensitive to the gating voltage, thus creating an ionic transistor with significant conductance amplification. The optimal membrane ionic transistor we have designed based on this non-equilibrium flux balance principle is able to achieve a transconductance of more than 12 μS, which is larger than previous reported for ionic transistors by a factor of 40. It corresponds to a voltage amplification factor of 10.

I will report my preliminary result on controlling a single ionic memristor by the ionic transistor and using the ion current amplifier to improve a single nucleic acid electrochemical membrane sensor [Slouka et al, Annu Rev Anal Chem, 7, 317 (2014)]. The ionic transistor will be shown to enhance the electrochemical membrane sensor by orders of magnitude with fM sensitivity. It can also mediate the communication between the membrane sensor and the ionic memristor to selectively register the sensing result to the ionic latch or to read out the sensing result. I will then discuss my plan to scale up the control-memory electrochemical circuit architecture to massively multiplexed smart electrochemical sensor arrays with a parallel operation paradigm. Advantages and challenges of this future biopsy platform for microRNA profiling against the current lab-bound micro-array and rtPCR technologies will also be discussed [Egatz-Gomez et al, Biomicrofluidics, 10, 032902 (2016)]. More details of the ion current amplification design for improving electrochemical molecular sensing will be presented in Session T9002 and Session T3003.

469714 High Energy Density Carbides and Nitrides Based Supercapacitors

Abstract:

High-surface-area early transition-metal carbides and nitrides possess very high capacitances and hold promise for use in high energy density supercapacitors as a consequence of their high surface areas, pseudocapacitive nature and high electronic conductivities. These materials are also stable in a variety of aqueous and non-aqueous electrolytes. Most of the work on carbides and nitrides for supercapacitors has been done in aqueous media. Vanadium nitride (VN) and molybdenum nitride (Mo_2N) have been reported to possess specific capacitances as high as 1340 Fg^{-1} in 1.1 V aqueous KOH and 346 Fg^{-1} in 0.8 V aqueous H_2SO_4 electrolytes, respectively.

Further development of early transition-metal carbides and nitrides for use in supercapacitors would benefit from a detailed understanding of their charge storage mechanisms. This work aims to characterize the charge storage mechanisms for high-surface-area early transition-metal (Ti, V, Nb, Mo, and W) carbides and nitrides in aqueous electrolytes. We have identified the active ions involved in the charge storage mechanisms (OH^- and H^+), located the active sites for charge storage, quantified the extent of pseudocapacitance and the amount of inserted active ions, and proposed the pseudocapacitive charge storage mechanisms for these materials.

Given findings regarding the storage mechanism, we hypothesized that protic, non-aqueous electrolytes would enable increased energy densities, given the higher operating voltages afforded by non-aqueous electrolytes. We were able to expand the operating voltage windows for Ti, V, Nb, Mo and W carbides and nitrides (i.e. from 1.1 V to 2.0 V for TiN) using protic ionic liquid electrolytes instead of aqueous electrolytes. For most materials, a three- to four-fold increase in energy density was observed while maintaining similar power density.

471621 Revealing the Dynamics of Electrochemical Interfaces in Lithium Batteries

Abstract:

Electrochemical interfaces, spanning across a few nanometers, are the place active species meet, multiple physics impinge, and lots of uncertainties hide. Exploring such uncertainties will facilitate the rational design of high-performance energy conversion and storage materials and systems. This presentation summarizes the candidateâ€™s original contributions in three different areas. **(i)** In porous electrodes of carbon-coated $LiFePO_4$ nanoparticles, the dynamic process consists of lithium ion insertion from the liquid electrolyte to the interstitial vacancies in the solid particles and electron transfer from the carbon coating to the Fe site of the solid particle. Besides the usual factors like concentration and temperature, the kinetics is largely affected by the phase transformation process inside the solid particles, which results in a particle-by-particle reaction pathway throughout the porous electrode. Modeling the dynamics at both the single-particle [1] and many-particleÂ [2]Â scales enables the extraction of local reaction rates, which can be accurately explained by the Marcus theory [3]. **(ii)** In lithium metal anodes, whisker-like electrodeposits always develop by mechanisms that still remain elusive. Experimental results in novel capillary cells reveal that the insulating solid-electrolyte interphase (SEI) layer on lithium metal is the reason of the root-growing whiskers, which interweave with each other to form a mossy structure and can be blocked by nanoporous ceramic separators [4]. Only at the diffusion limitation, would tip-growing dendritic lithium start to grow and penetrate the nanopores to short the cell [5]. The safety of rechargeable lithium metal batteries can therefore be increase by using concentrated liquid electrolyte and nanoprous ceramic separators. **(iii)**Prototypes of energy-dense lithium-bromine/oxygen fuel cell using a liquid-solid-liquid hybrid electrolyte [6], and the degradation of the key component, the glass ceramic solid electrolyte, will also be discussed [7].

References

1. Peng Bai, Daniel A. Cogswell, Martin Z. Bazant. Suppression of phase separation in LiFePO4 nanoparticles during battery discharge. Nano Letters, 11(11): 4890-4896 (2011).
2. Peng Bai, Guangyu Tian. Statistical kinetics of phase-transforming nanoparticles in LiFePO4 porous electrodes. Electrochimica Acta, 89: 644-651 (2013).
3. Peng Bai, Martin Z. Bazant. Charge transfer kinetics at the solid-solid interface of porous electrodes. Nature Communications, 5:3585 (2014).
4. Peng Bai, Ju Li, Fikile Brushett and Martin Z. Bazant. Lithium growth mechanisms in liquid electrolytes. In revision.
5. Peng Bai, Miao Wang, Fikile Brushett and Martin Z. Bazant. Interactions between lithium growths and nanoporous ceramic separators: blockage versus penetration. To be submitted.
6. Peng Bai, Venkat Viswanathan and Martin Z. Bazant. A dual-mode rechargeable lithium-bromine/oxygen fuel cell. Journal of Materials Chemistry A, 3: 14165-14172 (2015)
7. Peng Bai and Martin Z. Bazant. Performance and degradation of a lithium-bromine rechargeable fuel cell using highly concentrated catholytes. Electrochimica Acta, 202: 216-223 (2016).

464245 Electrocatalytic Water Splitting at Ultrathin (Oxy)Hydroxide Films/Precious Metal Interfaces

Abstract:

Design of catalysts with simultaneously enhanced activity, stability and material efficiency for the electrochemical water splitting is of paramount importance for the next generation of alkaline electrochemical devices. Ultrathin (oxy)hydroxide films on precious metal substrates possess impressive activities for these reactions, but the filmsâ€™ structure and stability are unknown, and the catalytic mechanism is still unclear. Because of the limited resolution of many in-situ characterization techniques at complex liquid/solid interfaces, first principles-based modeling is strongly positioned to provide atomistic insights into the activity and stability of these fascinating systems.

In this talk, we will first introduce the methodologies we have recently developed towards highly accurate prediction of Pourbaix diagram of transition metal (oxy)hydroxides. Next, using monolayer Ni (oxy)hydroxide films as an example, we will describe aÂ simple scheme to study the structures and the stability of these films on precious metal surfaces. We will show how the ultrathin films can be dramatically stabilized with respect to the corresponding bulk analogues. Next, using hydrogen evolution reaction as an example, we will demonstrate how these techniques can be applied to understand the steady state, the active phases, and the catalytic mechanism of bi-functional interfaces. We will then demonstrate the extension of the present understanding to real world catalysts, i.e. precious metal nanoparticles supported on ultrathin transition metal (oxy)hydroxide films. Finally, we will show this understanding can be used to design new bi-functional catalysts with improved performances.

[1] J. Greeley, I. E. L. Stephens, A. S. Bondarenko, T. P. Johansson, H. A. Hansen, T. F. Jaramillo, J. Rossmeisl, I. Chorkendorff, J. K. Nørskov, *Nat. Chem.* **2009**, *1*, 552-556.

[2] Subbaraman, R.; Stamenkovic, V.; Markovic, N. et al, Science 2011, 334, 1256.

[3] Subbaraman, R.; Greeley, J.; Stamenkovic, V.; Markovic, N. et al, *Nat Mater* **2012**, *11*, 550

[4] Z. Zeng, M. K. Y. Chan, Z.-J. Zhao, J. Kubal, D. Fan, J. Greeley, *J. Phys. Chem. C* **2015**, *119*, 18177

453363 the Fabrication and Use of Bipolar Membrane in Hybrid PEM/AEM Fuel Cells

Abstract:

The Fabrication and Use of Bipolar Membrane in Hybrid PEM/AEM Fuel Cells

John M. Ahlfield[*†], Lisha Liu[‡], and Paul A. Kohl[†]

[†]School of Chemical and Biomolecular Engineering

[‡]School of Materials Science and Engineering

Georgia Institute of Technology, Atlanta, GA

Bipolar membranes fuel cells utilizing both anion and cation conductive materials have several advantages compared to their purely acidic or alkaline counterparts. The potential advantages include improved water management through self-hydration and facile electrode kinetics. Material transport properties play an important role in determining viability of membrane and ionomer materials. In addition, the cation-anion junction itself is a critical element in determining device performance because it must be sufficiently conductive to ionic species and mechanically stable to withstand the internal pressure from water formation. A series of electrochemical devices using different bipolar membranes have been fabricated for use in direct methanol and hydrogen fuel cells. This study examines both single- and two-membrane systems containing cation-anion junctions. These fuel cells were characterized by performance metrics and electrochemical impedance spectroscopy to determine specific areas for improvement in the bipolar devices. Operation under varying humidity was studied in order to understand water management necessary for bipolar

fuel cells. These results will be used in the future optimization of bipolar devices.

Financial support from US Office of the Deputy Assistant Secretary of the Army for Defense Exports and Cooperation (DASA-DE&C) is gratefully acknowledged.

457587 High Efficiency, Low Cost Electrochemical Ammonia Production: Challenges and Opportunities

Abstract:

A majority of ammonia (NH_3) production today occurs via the Haber-Bosch process which involves the heterogeneous reaction of nitrogen (N_2) obtained from air, and hydrogen (H_2) obtained from fossil fuels. The process occurs at high pressure (150–300 atm) and high temperature (400°–500°C) over an iron-based catalyst. It is one of the most impactful developments in human history, but it comes at a price. NH_3 is a top 10 energy consuming chemical in the United States (U.S.) and one of the most energy intensive chemicals to manufacture in the world. In addition to consuming 1-2% of world-wide energy, the production of NH_3 also generates 350 million metric tons of carbon dioxide (CO_2) emissions yearly, which is ~1% of total greenhouse gas emissions world-wide. Additional emissions are incurred because of the need to transport the NH_3 from large centralized plants. As such, novel, scalable NH_3 manufacturing techniques could have an extremely large impact on total energy consumption and emissions.

One alternative approach for NH_3 manufacturing is to use a green feedstock as the proton source for the reaction. Additionally, electricity could be used to drive the reaction, decreasing the need for high pressure and heat, and reducing energy demand. This electrochemically driven process would be compatible with intermittent operation and enable utilization (and monetization) of renewable electricity without the need for transmission capacity expansion. Further reduction of emissions would be realized through the reduced need for ammonia transport. Since electrochemical technology based on flow cells is highly scalable, products could support a range of production capacities. As megawatt (MW)-scale electrolysis systems are already becoming a reality at companies such as Proton OnSite, localized ammonia production at relevant scales for agriculture is not hard to envision. There is also a natural synergy in using distributed wind power for fertilizer production. In the Plains and Upper Midwest, excess wind production capacity, transmission limitations, and high regional demand for nitrogen based fertilizers combine to create excellent economic drivers for this technology. Additionally, there are many other industrial uses for ammonia besides agricultural fertilizers. Ammonia is used to synthesize a variety of chemicals including urea, nitric acid and pharmaceutical compounds. It is also important in emissions capture as well as refrigeration, and potentially could be used in a fuel cell for electricity generation. This flexibility in use makes ammonia an attractive renewable energy storage option.

A variety of electrochemical systems have been investigated in the literature. A majority of them either require high temperatures, or an acidic environment for functionality. High temperatures make the process less practical for consumer use, or rapid intermittent operation with renewable energy. In addition, the acidic environments require costly materials of construction compared to a basic environment and severely limit the options for catalyst materials, potentially eliminating many highly active and selective catalysts from the design. While proton exchange membranes (PEMs) have shown extremely long lifetimes and fast ion transport in other electrochemical applications, ammonia is a weak base, and it is expected to readily react with acidic membranes to reduce proton conductivity and speculatively, membrane lifetime. An anion exchange membrane (AEM)-based technology is ideal for ammonia synthesis because the membranes are not expected to react with ammonia as readily, they enable low-cost materials of construction, and they allow the utilization of a wider array of low-cost catalysts.

Proton OnSite and their partners have made several advancements in area of AEM-based electrochemical ammonia production. Through their work, they have gathered insight on catalyst materials, membrane/ionomer materials, testing protocols and system integration. However significant challenges still exist in product analysis, materials screening and selection, and suitable testing protocols that eliminate sources of contamination, and precisely control reaction conditions. While Proton OnSite has designed a second generation test rig to address some of these challenges, there are many opportunities for creative materials design, testing development, integration of materials, and fundamental reaction understanding in the area electrochemical ammonia generation.

469219 Improving Performance in Alkaline Membrane Fuel Cells through Enhanced Water Management

Abstract:

Water management in anion exchange membrane fuel cells (AEMFCs) is more complex than proton exchange membrane fuel cells (PEMFCs). In the PEM system, water is generated at the cathode and otherwise only serves as the membrane proton transport medium. In the hydroxide AEMFC, water is generated at the anode, consumed at the cathode, and is greatly needed for transport of the larger hydroxide ion. The challenge this introduces is the need to provide adequate water to maintain the membrane humidity without flooding the catalyst or gas diffusion layers.[1] One method being used to achieve the necessary high membrane water content is utilizing hydrophilic gas diffusion layers without microporous layers.[2] Others have used gas diffusion layers with a hydrophobic coated microporous layer while raising the temperature of the humidifiers well above the cell temperature.[3] Gas stream dew points above 100% relative humidity and back pressure are common approaches in increasing the membrane water content. However there is instability in the cell performance and conditions that generate high performance levels are not easily repeatable. The problem is the line between proper membrane hydration and flooded catalysts layers is very thin, and non-existent in some cases.

In this study, the influence of the membrane, ionomer and gas diffusion layer as well as the flow rate and dew points of the anode and cathode gases on AEMFC performance were explored. Tokuyama A201 membranes and AS-4 ionomer were investigated alongside quaternary-ammonium-functionalized radiation-grafted ETFE alkaline anion-exchange membranes and ionomers.[4] Using a hydrophilic gas diffusion layer without a microporous layer increased membrane hydration, but as expected increased the potential for flooding. Manipulating the dew points led to the counter-intuitive discovery that the cell performs better with the humidity higher at the anode than the cathode, despite water generation and electro-osmotic drag towards that electrode. In fact, removing too much water from the anode caused instability in the cell, while increasing the water at the anode decreased the membrane resistivity. Back diffusion likely plays an important role in membrane hydration and hydroxide transport through the membrane. A very high flow rate (1.0 L/min) also increased cell performance, despite being several orders of magnitude above the stoichiometric need.

Combining all areas of improvement resulted in a very high performing AEMFC with a maximum current density of 2.2 A/cm^2 (at 0.1 V) and max power density of 670 mW/cm^2 (880 mW/cm^2 iR-corrected) with a membrane resistivity of 75 mOhms*cm^2 (Figure 1a). Only a minor drop in the current was observed using air at the cathode with the same 1.0 L/min flow rate as oxygen, giving max current density of 1.7 A/cm^2 and a max power density of 580 mW/cm^2 with a resistivity of 74 mOhms*cm^2 (Figure 1b). This near identical behavior confirms that the amount of reactant present supplied by the higher flow rate is not necessary, but the volumetric flow rate is needed for water management. It is likely that the pressure drop along the single pass cell hardware allows the gas to "jump the bar" only at very high flowrates, which results in better water removal and limits cell flooding in the cell.

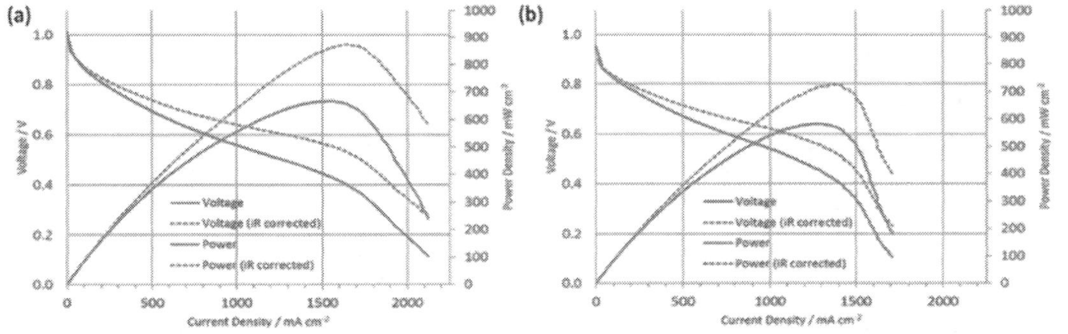

Figure 1: 10 mV/s linear sweep polarization curve and power curve with (a) oxygen and (b) air at the cathode.

References:

1. T. D. Myles, A. M. Kiss, K. N. Grew, A. A. Peracchio, G. J. Nelson and W. K. Chiu, *J.Electrochem.Soc.*, **158**, 7 (2011).

2. R. B. Kaspar, M. P. Letterio, J. A. Wittkopf, K. Gong, S. Gu and Y. Yan, *J.Electrochem.Soc.*, **162**, 6 (2015).

3. M. Mamlouk, J. Horsfall, C. Williams and K. Scott, *Int J Hydrogen Energy.*, **37**, 16 (2012).

4. S. D. Poynton, R. C. Slade, T. J. Omasta, W. E. Mustain, R. Escudero-Cid, P. Ocón and J. R. Varcoe, *Journal of Materials Chemistry A.*, **2**, 14 (2014).

468120 Surface Transport in Charged Porous Media

Abstract:

Surface transport occurs in a variety of physical processes of interest in many fields: engineering, physics, geology, colloidal science, and biology. Many times, these phenomena are not well understood and, therefore, not considered in the treatment of transport processes. There are few studies of surface transport in charged porous media when an electric field is generated by movement of ions in response to a voltage applied to an ideal conductor.

A new model that describes ion transport by bulk and surface transport inside charged porous carbon electrodes is proposed to simulate the CDI process. A general treatment of bulk and surface transport processes is presented in terms of the method of volumeÂ averaging. In addition to the spatial-averaging theorem, the analysis requires the surface-averaging theorem plus a closure scheme. The closure problem is formulatedÂ and transport parameters are calculated using Changâ€™s unit cells.Â The influence of surface migration terms is also discussed. The derived transport equations lead to well-known results in the case of non-charged materials.

462195 Self-Diffusion of Heptane inside Aggregates of Porous Alumina Crystallites By Pulsed Field Gradient NMR

Abstract:

Porous alumina has piqued the interest of the research community due to its relevance for industrial applications where it can be used as a catalyst, catalyst support, adsorbent, and template for the fabrication of nanomaterials. Specifically, the ability to fabricate alumina with a high surface area (up to 700 m^2/g) and well-defined pore structure where the majority of pores fall into the mesopore range makes it highly attractive in these applications. This has led to various studies focused on synthesizing this material under conditions which control the pore structure and size. Such modifications in the pore structure, specifically variations in porosity, have been shown to directly influence molecular transport properties. Until now, diffusion studies of porous alumina have mostly been performed using macroscopic techniques.

Here we report microscopic studies of n-heptane diffusion in porous alumina samples exhibiting a hierarchy of pore sizes that span the range from micropore to macropore sizes. Diffusion measurements were performed using ^1H and ^{13}C pulsed field gradient (PFG) NMR utilizing a high field (17.6 T) and high field gradients (up to 30 T/m) to obtain length scales of displacements in the range of about 1-20 mm; much smaller than previously reported studies. The measured heptane diffusivities inside the crystal aggregates were found to be several times smaller than the diffusivity in the bulk liquid heptane. The diffusivities inside the aggregates were correlated with the aggregate structural characteristics, which include pore volume, pore size, and packing density of the individual crystals in the aggregates.

466354 Mutual Diffusion of Binary Liquid Mixtures Containing Methanol, Ethanol, Acetone, Benzene, Cyclohexane, Toluene and Carbon Tetrachloride

Abstract:

Nowadays, modern rate-based methods, which involve mass and energy transfer models, are employed to solve complex macroscopic modelling and simulation issues in the chemical engineering field. These non-equilibrium methods require not only diffusion data, but also other transport coefficients like shear viscosity and thermal conductivity for pure components as well as mixtures. Thus, there is a growing need for accurate transport properties and better methods for their prediction. Owing to the rapid development of computing power, molecular modelling and simulation has emerged as an alternative for such predictions, especially when dealing with hazardous substances or challenging thermodynamic conditions.

We present the results of a comprehensive study on the prediction of mutual diffusion coefficients in binary mixtures on the basis of molecular dynamics simulations and the Green-Kubo formalism. The diffusion coefficients of all 20 binary liquid mixtures that can be formed out of methanol, ethanol, acetone, benzene, cyclohexane, toluene and carbon tetrachloride without a miscibility gap at ambient conditions of temperature and pressure are assessed in the entire composition range. The considered mixtures show a varying mixing behavior from almost ideal to strongly non-ideal. Radial distribution functions are analyzed to gain an understanding of the liquid structure influencing the diffusion processes. It is shown that cluster formation in mixtures containing one alcoholic component has a significant impact on the diffusion process. The estimation of the thermodynamic factor from experimental vapor-liquid equilibrium data is investigated, considering three excess Gibbs energy models, i.e. Wilson, NRTL and UNIQUAC. It is found that the Wilson model yields the thermodynamic factor that best suits the simulation results for the prediction of the Fick diffusion coefficient. Four semi-empirical methods for the prediction of the self-diffusion coefficients and nine

predictive equations for the Fick diffusion coefficient are assessed and it is found that methods based on local composition models are more reliable. Finally, the shear viscosity and thermal conductivity are predicted and in most cases favorably compared with experimental literature values.

463394 Transport of Chemotactic Bacteria in Granular Media Containing a Distribution of Non-Aqueous Phase Liquid Contaminants

Abstract:

Chemotaxis describes the transport of a population of self-propelled bacteria towards or away from regions of increasing chemical concentration. Bacterial cells can be characterized as colloidal particles given their typical size distribution; however the added motility and directional bias that chemotactic cells exhibit often yield interesting particle transport phenomena that can be positively exploited. One particular area of application in which bacterial chemotaxis plays an important role is that of bioremediation of non-aqueous phase liquid (NAPL) contaminants. NAPL contaminants, typically introduced into groundwater systems via oil spills and industrial waste streams, are difficult to remediate due to their low aqueous solubility and the physically heterogeneous nature of subsurface environments. As a result, NAPL contaminants can be trapped in the subsurface for long periods of time while they slowly dissolve into surrounding groundwater. Bioremediation of groundwater occurs via pollutant-degrading bacteria, many of which exhibit chemotaxis. That is, the bacteria can detect the presence of organic contaminants (which serve as a carbon source for them) and preferentially swim towards regions of higher chemical concentration because it is advantageous for them to do so. The movement of chemotactic bacteria towards the contaminant source, irrespective of whether it is trapped in the interstices of the soil matrix, leads to increased pollutant accessibility and biodegradation.

This experimental study employed a laboratory-scale continuous flow sand column to better understand chemotactic transport in response to a uniform distribution of entrapped NAPL. The presence of NAPL within the column created localized concentration gradients. 2,2,4,4,6,8,8-heptamethylnonane (HMN) containing dissolved naphthalene was used as a model NAPL. An equal concentration of *Pseudomonas putida* G7, which is chemotactic to naphthalene, and *Pseudomonas putida* G7 Y1, a non-chemotactic mutant strain, was introduced simultaneously into the sand column and effluent bacterial concentrations were measured with time. To aid differentiation between bacterial strains, each cell type was stained with a fluorescent label prior to injection into the column. Experiments with sand only and

sand/HMN-NAPL with no dissolved naphthalene were also conducted to serve as controls. We tested the hypothesis that due to chemotaxis, an increase in the retention of *P. putida* G7 will be observed within the sand column containing the HMN-naphthalene-NAPL (as compared to the control cases). This increase in chemotactic bacterial retention within the polluted porous media is expected to lead to an increase in contaminant biodegradation. Moment analysis parameters, such as percent recovery and mean travel time, were obtained from species breakthrough curves. Additionally, a one-dimensional advection-dispersion equation was fitted to species breakthrough curves and the influence of chemotaxis on apparent model parameters was analyzed. Initial experiments in which solid naphthalene crystals were evenly distributed throughout the sand column resulted in a 43% decrease in percent recovery of *P. putida* G7 from the column, compared to the control case with no naphthalene present, thus supporting our hypothesis. In contrast, transport of the non-chemotactic mutant strain was not influenced by the presence of naphthalene. The results of this experimental study are important for understanding how bacterial chemotaxis may influence the accessibility of contaminants present within NAPL contaminated aquifers.

470428 Size- and Morphology-Controlled Synthesis of Zinc Oxide Nanoparticles in a Two-Phase Millifluidic Reactor

Abstract:

We present a novel technique for the continuous synthesis of nanoparticles using a two-phase millifluidic reactor. Conducting the synthesis reaction inside droplets of controlled size offers several advantages such as eliminating temperature and concentration gradients inside the reactor as well as preventing reactor fouling. Additionally, by locating the synthesis reagents initially in separate phases the rate of reagent addition can be controlled and the need for having a reliable means of droplet merging in order to combine the synthesis reagents is eliminated. The technique is demonstrated by synthesizing zinc oxide nanoparticles inside aqueous droplets that initially contain zinc acetate. Sodium hydroxide then diffuses into the droplet phase in a controlled manner, producing zinc oxide nanoparticles with a narrow particle size distribution as compared to a batch reactor. Spherical as well as plate-like shapes of zinc oxide nanoparticles are seen for the ranges of parameters studied. The size and morphology of the nanoparticles is shown to be easily controlled by varying the droplet size, reagent concentrations, reaction temperature, and residence time. The effect of adding surfactants to either the bulk or droplet phase on the zinc oxide size and shape is also presented.

AUTHOR INDEX

AUTHOR INDEX

AUTHOR INDEX

AUTHOR INDEX

AUTHOR INDEX

AUTHOR INDEX

AUTHOR INDEX

AUTHOR INDEX

AUTHOR INDEX

AUTHOR INDEX

AUTHOR INDEX

AUTHOR INDEX

AUTHOR INDEX

AUTHOR INDEX

AUTHOR INDEX

AUTHOR INDEX

AUTHOR INDEX

AUTHOR INDEX

AIChE
120 Wall Street, FL 23
New York, NY 10005-4020

ISBN 978-1-5108-3434-7

9 781510 834347

Engineering Sciences and Fundamentals 2016

Core Programming Area at the 2016 AIChE Annual Meeting

San Francisco, California, USA
13 - 18 November 2016

Volume 2 of 2

AIChE
The Global Home of Chemical Engineers

Engineering Sciences and Fundamentals 2016

Core Programming Area at the 2016 AIChE Annual Meeting

San Francisco, California, USA
13 - 18 November 2016

Volume 2 of 2

ISBN: 978-1-5108-3434-7

Printed from e-media with permission by:

Curran Associates, Inc.
57 Morehouse Lane
Red Hook, NY 12571

CURRAN ASSOCIATES INC. proceedings.com

Printed by Curran Associates, Inc. (2017)

For permission requests, please contact AIChE
at the address below.

AIChE
120 Wall Street, FL 23
New York, NY 10005-4020

Phone: (800) 242-4363
Fax: (203) 775-5177

www.aiche.org

Additional copies of this publication are available from:

Curran Associates, Inc.
57 Morehouse Lane
Red Hook, NY 12571 USA
Phone: 845-758-0400
Fax: 845-758-2633
Email: curran@proceedings.com
Web: www.proceedings.com

TABLE OF CONTENTS

VOLUME 1

VOLUME 2

Analysis of high-speed rotating flow in polar $(r - \theta)$ coordinate

S. Pradhan

Department of Chemical Engineering, Indian Institute of Science, Bangalore 560 012, India

The generalized analytical model for the radial boundary layer in a high-speed rotating cylinder is formulated for studying the secondary radial flow due to insertion of mass, momentum and energy into the rotating cylinder in the polar $(r-\theta)$ plane. The analytical solution includes the sixth order differential equation for the radial boundary layer at the cylindrical surface in terms of master potential (χ), which is derived from the equations of motion in a polar $(r-\theta)$ plane. The linearization approximation is used, where the equations of motion are truncated at linear order in the velocity and pressure disturbances to the base flow, which is a solid-body rotation. Additional assumptions in the analytical model include constant temperature in the base state (isothermal condition), and high Reynolds number, but there is no limitation on the stratification parameter. In this limit, the gas flow is restricted to a boundary layer of thickness $(\mathrm{Re}^{-1/3}R)$ at the wall of the cylinder. Here, the stratification parameter $\mathcal{A} = \sqrt{m\Omega^2R^2/(2k_BT)}$. This quantity$(\mathcal{A})$ is the ratio of the peripheral speed (ΩR) to the most probable molecular speed $\sqrt{2k_BT/m}$, the Reynolds number $\mathrm{Re} = (\rho_w\Omega R^2)/\mu$, where m is the molecular mass, Ω and R are the rotational speed and radius of the cylinder, k_B is the Boltzmann constant, T is the gas temperature, ρ_w is the gas density at wall, and μ is the gas viscosity. For the case of mass/momentum/energy insertion into the flow, the separation-of-variables procedure is used, and the appropriate homogeneous boundary conditions are specified so that the linear operators in the meridional and radial directions are self-adjoint. The discrete eigenvalues and eigenfunctions of the linear operators (sixth-order and second-order in the radial and meridional directions for the generalized analytical equation) are obtained. The solutions for the secondary radial flows are determined in terms of these eigenvalues and eigenfunctions. The major advantage of the present formulation is that it is not restricted to the asymptotic limit of high stratification parameter, and the shock type solutions (Rankine-Hugoniot relations) are obtained simultaneously. These solutions are compared with direct simulation Monte Carlo (DSMC) simulations and found excellent agreement (with a difference of less than 10%) between the predictions of the analytical model and the DSMC simulations, provided the boundary conditions in the analytical model are accurately specified. In a high speed rotating field we examine the mass flow rate through the stationary intake tube. The simulations show that the scaled mass flow rate $m/(H\Omega)$ increases progressively from zero, at the stagnation condition, to 0.024 as the equilibrium back pressure is reduced. Here, H is the total gas holdup. The slow down of the circumferential velocity of the bulk of the rotating fluid due to the presence of stationary intake tube is studied for stratification parameter in the range $0.707 - 3.535$, and found significant slow down (between 8 to 28%), which induces the secondary radial flow towards the axis, and it further excites the secondary axial flow, which could be very important for the centrifugal gas separation processes. An important finding is that the stagnation pressure (no mass flow through the intake tube) is strongly affected by the wall gap, as well as with stratification parameter indicating a strong coupling between the local temperature, density, pressure and velocity fields.

Key words: High speed rotating flow, DSMC Simulation, Rarefied gas flow.

1. Introduction

Centrifugal gas separation processes effect separation by utilizing the difference in the mole fraction in a high speed rotating cylinder caused by the difference in molecular mass, and consequently the centrifual force density. These have been widely used in isotope separation because chemical separation methods cannot be used to separate isotopes of the same chemical species (Cohen (1951), Avery & Davies (1973), Von Halle, Wood & Lowry (1978), Brouwers (1978), Soubbaramayer (1979), Hoglund , Shacter & Von Halle (1979), Wood & Morton (1980), Sakurai (1981), Olander (1981), Wood & Sanders (1983), Wood, Jordan & Gunzburger (1984), Copper & Morton (1988), Wood & Babarsky (1992), Wood (1995), Wood, Mason & Soubbaramayer (1996), Omnes (2007)). Radioactive isotopes, including zinc, cadmium, iridium, selenium, tellurium, germanium and molybdenum, are used in medical and industrial applications. In medical applications, radio-isotopes are used both for diagnostics such as gallium scans for locations of inflammation and infection, indium for leukocyte imaging and molybdenum/technetium scans for high bone growth, as well as for therapy including cancer therapy using isotopes of iodine and iridium. An important industrial application is the separation of uranium isotopes for nuclear power generation. Other important applications include non-destructive testing using gamma cameras, and use of zinc isotopes to reduce corrosion in reactor cooling systems.

The principle of isotope separation in high speed centrifugal field in long rotating cylinders is based on the difference in the centrifugal force on different isotopes due to their difference in mass. Due to this, the concentration of the heavier isotope is higher then the mean concentration at the wall, while the lighter isotope is more concentrated near the center. In a closed rotating cylinder with no inflow/outflow, the gas rotated like a solid body, and the concentration separation is easily predicted on the basis of hydrostatics alone. However, in an operating centrifuge, there is an inflow and outflow of gases, and these require stationary and rotating inserts. Secondary flows could also be generated due to temperature gradients along the side wall of the rotating cylinder as well as temperature difference between the end caps. The achievable separation is primarily determined by the secondary axial and radial flows. Therefore, the accurate modeling of these secondary flows and its numerical analysis based on molecular simulations is of critical importance in predicting the centrifuge performance.

Centrifugal separation in a high-speed rotating field is a widely used method for isotope separation (Fraenkal (1959); Barcilon & Pedlosky (1967); Hunter (1967); Salant (1968); Homsy & Hudson (1969); Greenspan (1968); Bark & Bark (1976); Hittinger, Holt & Soubbaramayer (1981); Hittinger, Holt, Soubbaramayer & Cortet (1983); Gunzburger & Wood (1982); Jones (1982); Jung (1983); Kai (1983); Gunzburger, Wood & Jordan (1984); Heuser, Ribando & Wood (1986); Copper & Morton (1988); Babarsky & Wood (1990, 1992); Bourn, Peterson & Wood (1999); Doneddu, Robin & Wood (2000); Babarsky, Herbst & Wood (2002); Maslen & Moore (1956); Stewartson (1957); Sozou (1959); Thompson & Chen (1980); Viecelli (1983)). The small difference in the molecular masses of the isotopes provides the only economically feasible method for effecting separation. The centrifugal force results in a small difference in concentration of the two isotopes, with a higher concentration of the heavier isotope near the wall of the cylinder, and a higher concentration of the lighter isotope near the axis. In order to separate the isotopes, stationary tubes and baffles are inserted into the rotating cylinder. The different types of external forcing (wall thermal drive, end-cap thermal drive, feed drive, mechanical drive) cause a axial and radial secondary flow in the cylinder. This secondary flow has a strong effect on the efficiency of separation.

Here, we have studied the secondary radial flow due to insertion of mass, momentum and energy into the rotating cylinder in the polar $(r - \theta)$ plane. The presence of a stationary insert (intake tube) within the flow domain causes a disturbance to the base flow (solid body rotation). A detached bow shock is formed in front of the stationary insert (intake tube), and behind the detached bow shock, secondary radial flow is induced towards the axis, and it excites the secondary axial flow, which could be very important for the centrifugal gas separation processes. The inward radial flow also produces a vortex column near the stationary insert. A mechanically driven gas centrifuge utilizes stationary intake tube (which is used to collect the separated gases), to generate the secondary radial flow, and it further excites the secondary axial flow. For this reason, it is important to analyze the secondary radial flow due to insertion of mass, momentum and energy into the rotating cylinder in the polar $(r-\theta)$ plane to assess the performance of a mechanically driven gas centrifuge.

There are two objectives of the present analysis of the rarefied gas flow in a rotating cylinder in the polar $(r - \theta)$ plane. The first is to remove the restriction of high stratification parameter, and to generalize the solutions to low rotation speeds where the stratification parameter may be $O(1)$. Secondly, we would like to compare the predictions with molecular simulations based on the direct simulation Monte Carlo (DSMC) method for rarefied gas flows in a polar $(r-\theta)$ plane, in order to quantify the errors resulting from the approximations made in the analytical solution. Here, we obtained the generalized analytical solution in a polar $(r - \theta)$ plane and compared with molecular simulations based on direct simulation Monte Carlo (DSMC) method (Bird (1963, 1994)) due to insertion of mass, momentum and energy into the rotating cylinder in the polar $(r - \theta)$ plane.

A complete flow simulation involves the solution of the compressible mass, momentum and energy equations for the gas velocity in the cylinder, along with the ideal gas equation of state. There have been previous numerical simulations of axisymmetric flows in high speed rotating cylinder (Lahargue & Soubbaramayer (1978); Soubbaramayer (1979); Harada (1980); Hittinger, Holt & Soubbaramayer (1981); Hittinger, Holt, Soubbaramayer & Cortet (1983); Homsy & Hudson (1969); Hunter (1967); Kai (1983); Matsuda, Tamura & Sawada (1989); Mikami (1981); Sakurai (1981); Omnes (2007); Olander (1981); Brouwers (1978); Hoglund , Shacter & Von Halle (1979); Ribando (1984); Soubbaramayer & Billet (1980)). Apart from the complicated formulation of the differential equations for five variables (density, three components of velocity and temperature), there are two important conceptual difficulties in the numerical solution of the continuum equations. The flow in the cylinder is usually at high Mach number, where the gas is concentrated at the wall of the cylinder in a 'radial boundary layer' of thickness $\mathrm{Re}^{-1/3}R$ much smaller than the cylinder diameter. While the continuum approximation is valid in the radial boundary layer, the density in the central region is sufficiently small that the flow is in the free-molecular regime. There is no simple way to determine the boundary between the continuum and free-molecular regimes, and to prescribe boundary conditions at this boundary. Another important issue is the modeling of the wall slip in the simulations. While the no-slip conditions are applicable to low Mach number flows, there is usually substantial slip at the wall which needs to be modeled in continuum simulations.

The analysis of flow field and concentration field in a high speed rotating cylinder is a challenging job which faces the following difficulties. In a high speed rotating cylinder the flow velocities are very large due to the high rotation rates, required to generate the necessary concentration separation between the species with nearly equal molecular masses. The stratification parameter, $\mathcal{A} = \sqrt{m\Omega^2 R^2/(2k_B T)}$, which is the ratio of the peripheral speed (ΩR) to the most probable molecular speed $\sqrt{2k_B T/m}$, is typically much larger than 1. This generates two different complications in the simulation procedure. In contrast to most low stratification parameter flows commonly encountered, the high stratification parameter flows are 'compressible'. The simulation of high-speed rotating compressible flows is challenging due to the presence of curved detached bow shock waves involving sharp and nearly discontinuous density variations. There is a large variation in the total gas density across the high speed rotating cylinder, and the density at the wall is typically $(5-6)$ orders of magnitude larger than that at the center. Associated with the large variation in the density is a similar variation in the 'mean free path', which is the average distance travel by a molecule between two successive collisions. The mean free path also varies by $(5-6)$ orders of magnitude over the radius of the cylinder. The mean free path is typically small at the wall of the cylinder, but could be quite large at the center. The Knudsen number, which is the ratio of the mean free path and the macroscopic scale (cylinder radius), is small at the wall, but is greater than 1 at the center. Continuum Navier-Stokes equations cannot be applied when the Knudsen number is larger than 1, because these equations are derived assuming that the mean free path is small compared to the macroscopic scales, thus facilitating the use of a gradient expansion.

There are uncertainties in the specification of boundary conditions at the bounding surfaces. Usually, in continuum simulations, the velocity at a solid surface is equal to the velocity of the surface (no-slip condition), and the temperature of the fluid at the surface is equal to the surface temperature. However, these boundary conditions cannot be used in rarefied gas dynamics, where the mean free path is comparable to the system size, and there could be slip at the bounding surfaces. General expressions are not available for the slip boundary condition, since the slip depends on the type of gas and solid surface.

In addition to the outer wall of the cylinder, it is also necessary to specify boundary conditions at the axis. For incompressible flows, symmetry boundary conditions are specified at the axis of the cylinder. However, in high speed rotating flows, the density at the center is very small and the mean free path is comparable to the macroscopic scale (cylinder radius). Due to this, it is not possible to extend the continuum description to the center of the cylinder. Boundary conditions have to be applied at an artificial 'inner boundary' at which the density is close to zero. There is no clearly defined way to specify the location of this boundary, but one has to show that the results of the simulation are insensitive to the inner boundary location. In addition, procedures have to be devised to specify velocity and temperature at this boundary.

In practical applications, there are rotating and stationary inserts in the cylinder for the inflow/outflow of gases. These inserts have to be incorporated as solid boundaries in a simulation, resulting in a complex simulation domain. This increases the complexity of the simulation grid required, and the number of grid points at which the hydrodynamic fields are evaluated.

All of the above difficulties in a continuum simulation can be overcome if we do a simulation at the molecular level, where the motion of the molecules and their interactions with the bounding surfaces are directly incorporated. For dilute gases, the most efficient simulation technique is the direct simulation Monte Carlo (DSMC) technique (Bird (1963, 1994); Vincenti & Kruger (1967); Cercignani & Lampis (1971); Chen &

Boyd (1996); Cercignani (2000); Chapman & Cowling (1970); Wagner (1992); Rieffel (1988); Garcia (2000); Nanbu (1980); Macrossan (2001); Fan & Shen (2001); Quanhua & Boyd (2002)). A discrete simulation technique does not become invalid when the mean free path is larger than the system size, because the simulation does not use averaged equations. So even very low density regions can be simulations. In a discrete simulation technique, the interaction of molecules with each other and with the solid surfaces are specified, and there is no need to specify boundary conditions for the hydrodynamic fields. So complications with artificial boundaries do not arise. Though DSMC technique overcomes several of the difficulties of a continuum simulation, the computational power and time required for the simulation is much higher than that for a continuum simulation. This is due to the requirement for simulating, directly, a simplified version of the molecular motion. This requires the evaluation of the complete velocity distribution function of the molecule. and temperature fields.

In direct simulation Monte Carlo (DSMC) simulation procedure, the Boltzmann equation is simulated using a probabilistic Monte Carlo procedure to obtain the velocity distribution function, and the density, velocity and temperature are obtained by taking moments of the distribution function. Here, one 'simulated' molecule represents a large number of real molecules, and the collision probability is chosen in such a manner that the collision frequency, mean free path and the temperature of the simulated molecules is equal to those for the real molecules. This procedure can be used for the simulation of both the rarefied gas regime and the continuum regime, and this has been used previously for centrifugal flows (Roberts (1985); Roberts & Hausman (1988); Roblin & Doneddu (2000); Pradhan & Kumaran (2011); Kumaran & Pradhan (2014)). However, DSMC simualtions are computationally intensive, due to the requirement of having a large number of simulated molecules in each differential volume in order to reduce statistical fluctuations.

Two simulation techniques are discussed here in the order of increasing computational complexity and decreasing analytical complexity. While the analysis of the analytical model is complicated, this method can be used for obtaining quick and approximate computational results. In contrast, the direct simulation Monte Carlo (DSMC) method is conceptually simple, since it simulates the motion of molecules with specified intermolecular interactions. However, since the number of molecules to be simulated is very large, this is computationally very expensive. This can be deployed to obtain accurate computational results in narrow regions of parameter space identified by more approximate method. Thus, the simulation methods in this hierarchy complement each other and provide a comprehensive framework for simulating strongly rotating flows.

In our previous study for the axial flow at the radial boundary layer and radial flow near the end caps (Pradhan & Kumaran (2011), Kumaran & Pradhan (2014)) in an axisymmetric $(r - z)$ plane has revealed several novel features of high speed rotating flows. The generalized Onsager and Carrier-Maslen models (Pradhan & Kumaran (2011)) were formulated in the axisymmetric $(r - z)$ plane where the approximation of high stratification parameter was relaxed. In this, a generalized sixth order operator was derived, which is applicable even when the stratification parameter is $O(1)$, though the assumption of high Reynolds number and high aspect ratio were made. This operator is also self-adjoint with real eigenvalues and orthogonal eigenfunctions, though these have to be determined numerically. For the secondary flow, it was found that the predictions of the generalized Onsager model were in very good agreement with DSMC results even when the secondary flow is as large as the 20% of the base solid-body rotation, the stratification parameter is as low as 0.707, the Reynolds number is as low as 100 and the cylinder length is only two times the diameter. However, these models still assume that the molecular masses of the two species are equal while calculating the secondary flows, and that the mole fractions are a constant. Next, the mass difference between the two species in a binary gas mixture was incorporated into the generalized Onsager and Carrier-Maslen models (Kumaran & Pradhan (2014)) to make it applicable to systems where the difference in molecular masses is not much smaller than the average molecular mass. In the base state, the flow is considered to be a solid body rotation with constant temperature, but the variation in the mole fraction is taken into account. Due to this, there is a variation in the viscosity and thermal conductivity in the radial direction. In addition, the total pressure is written as the sum of partial pressures of the two species, and is not approximated as an exponential function of the distance from the wall. The dilute gas equation of state is used for the pressure, while the Chapman-Enskog expressions (Chapman & Cowling (1970)) for hard sphere gases are used for the thermal conductivity and viscosity. In the analysis of the secondary flow, the equation for the master potential is expanded in an asymptotic series in the parameter $\epsilon = (\Delta m/m_{av})$, where Δm is the difference in the molecular masses of the two species, and m_{av} is a suitably defined average molecular mass. In the leading approximation where terms of $O(\epsilon)$ or smaller are neglected. The leading order equation and boundary conditions for the master potential are nearly identical to the sixth order differential equation of the generalized Onsager model (Pradhan & Kumaran (2011)); the important difference being that the density is not assumed to decrease exponentially from the wall, but the actual species densities are used. The the mean velocity and temperature obtained by solving the leading order equations are used in the species balance equations for the mole fractions of the two species, and the corrections for the number density and mass density of the two species due to the secondary flow are obtained. These appear as inhomogeneous terms in the equation for the $O(\epsilon)$ correction for the master potential; this equation is solved with appropriate boundary conditions to obtain the $O(\epsilon)$ correction to the master potential and flow velocity. This is then used in the $O(\epsilon)$ contribution to the species balance equation, and the hierarchal procedure is repeated until the $O(\epsilon^2)$ correction to the secondary flow velocity and mole fraction are calculated. A similar hierarchial procedure is used for the Carrier-Maslen model for the end-cap boundary layers. The results of the Onsager hierarchy, up to $O(\epsilon^2)$, are compared with the results of DSMC simulations for the axisymmetric flow of a two-species hard sphere molecular gas. The constitutive relations for mass diffusion, viscosity and heat conduction predicted by the DSMC simulations are compared with the Chapman-Enskog expressions (Chapman & Cowling (1970)) in order to validate the constitutive relations used in the analysis. After this, the results of the $O(\epsilon^2)$ Onsager hierarchy for the secondary flow are compared with DSMC simulations for three types of forcing, a linear wall temperature profile, inflow/outflow of gas along the axis, and mass/momentum input due to inserts within the flow (Pradhan & Kumaran (2011), Kumaran & Pradhan (2014)).

The condition which has not yet been relaxed in the analytical model is the condition on the linearization of the equations. It will be too complicated to include the non-linear terms in the conservation equations in the analytical model, and it will likely not be possible to obtain analytically tractable equations.

There have been previous studies on non-axisymmetric rotating compressible flows in a cylindrical geometry (Matsuda & Nakagawa (1983), Matsuda, & Hashimotko (1976), Wood & Babarsky (1992), Babarsky & Wood (1990)). Matsuda & Nakagawa (1983) studied the flow in a pie-shaped cylinder of infinite length with walls at fixed angles in the meridional direction, and reported the presence of 'buyoancy' boundary layer of thickness $\mathrm{Re}^{-1/2}R$ at the walls perpendicular to the meridional direction. Wood & Babarsky (1992) and Babarsky & Wood (1990) extended the work of Matsuda & Nakagawa (1983) to a finite-length, pie-shaped cylinder about the vertex. They presented an eigenfunction expansion based on the spectral decomposition of the resulting differential operator. The present work provides an opportunity for validating both the analytical model as well as the DSMC method for high-speed rotating flows in the polar $(r - \theta)$ plane. The basic difference of flow dynamics between axisymmetric $(r - z)$ and a polar $(r - \theta)$ configuration is that, in $(r - z)$ plane secondary flow is subsonic, whereas in $(r - \theta)$ plane the azimuthal flow is supersonic and encounter a curved detached bow shock wave in front of the stationary inserts and generates a sub-sonic inward radial flow.

Babarsky, Herbst & Wood (2002); Bourn, Peterson & Wood (1999) introduced the properties of the temperature potential and its relationship to the original problem of axisymmetric flow in a rapidly rotating centrifuge based on Onsager master potential and provide a firm mathematical foundation for the solutions of these two problems. They have shown that the temperature potential plays an analogus role as Onsager master potential, and the Onsager minimum entropy principle for axisymmetric flow have been derived in order to place the discussion of temperature potential in context. The governing equations and accompanying boundary conditions have been derived, and the relationship of these new eigensolutions to the original Onsager eigensolution is explained in terms of supersymmetry. The authors state that one of the advantages of the temperature-potential method is that it is easier to study the non-axisymmetric flows, since the nonaxisymmetric temperature operator can be constructed by considering it to be a small perturbation of the self-adjoint operator associated with the axisymmetric temperature potential.

Two-dimensional continuum simulations have also been carried out in the $(r - \theta)$ plane for the rotating flow past a fixed object. Here, one of the objectives is to analyse the high-speed shock wave that forms when a rotating flow encounters a stationary object. Hittinger, Holt & Soubbaramayer (1981) and Hittinger, Holt, Soubbaramayer & Cortet (1983) obtained a bow shock in front of a stationary object in their simulations. Two-dimensional numerical simulation of flow past a stationary object have been done by Mikami (1981) and Mikami (1985), and three dimensional inlet-flows has been simulated by Aoki & Suzuki (1985). In the simulations, they assumed a uniform upstream flow, and were able to resolve the shock waves that result. The centrifugally driven thermal convection near the cooled rotor cover of a centrifuge studied by Mikami (1973). In this analysis, the boundary layer equations are reduced to ordinary differential equations by the Lees-Levy transformation, and solved numerically by applying the quasi-linearization method. Sample solutions for radial and azimuthal velocity profiles are presented. Kai (1983) studied the drag on the gas due to a stationary cylindrical object, and then equated it with a viscous drag due to an axisymmetric equivalent disk. A modified version of the Newton method was adopted to get the full solution in static state for nonlinear equations.

The secondary flows generated inside a high speed rotating cylinder have been studied by numerical solutions of the compressible Navier-Stokes equations by many authors in two-dimensional axisymmetric $(r - z)$ configuration, in the polar $(r - \theta)$ configuration, and in three dimensional $(r - \theta - z)$ configuration. These equations are solved, subject to boundary conditions, to determine the density, velocity and temperature fields. Nakayama & Torii (1974) analysed the Stewartson layers by assuming that the height is larger than the thickness of the $\mathrm{Re}^{-\frac{1}{4}}R$ and $\mathrm{Re}^{-\frac{1}{3}}R$ layers, by expanding the equations in terms of the ratio of thickness to the height. Bark & Bark (1976) determined the exact solution for the $\mathrm{Re}^{-\frac{1}{4}}R$ layer and asymptotic solutions for the $\mathrm{Re}^{-\frac{1}{3}}R$ layer, as well as numerical solutions, assuming that the cylinder walls are thermally conducting, and the wall temperature is a constant. The work of Matsuda (1976a) was one of the first to stress the importance of the thermal boundary condition on the flow field, particularly if the walls are thermally insulated. The combined effect of the compressibility of the fluid and the thermal insulation of the walls produces a flow distinctly different from that of Boussinesq fluids, and from that of compressible fluids confined in a cylinder with thermally conducting walls. Matsuda & Hashimoto (1978) divided the flow field in r-z plane into several regions, the inner inviscid core, the Stewartson $\mathrm{Re}^{-\frac{1}{4}}R$ layer along the sidewall, its extension to the top and bottom Ekman layers with thickness $\mathrm{Re}^{-\frac{1}{2}}R$, the Stewartson $\mathrm{Re}^{-\frac{1}{3}}R$ layer and its extension to the Ekman layers. In a frame rotating with the angular velocity Ω there is no flow in the inner core on account of the Taylor-Proudman theorem. Along the sidewall, however, there must be an azimuthal flow in order to satisfy the side-wall conditions. This azimuthal flow does not satisfy the boundary conditions at the top and bottom end plates, leading to the formation of the top and bottom Ekman layers with thickness $\mathrm{Re}^{-\frac{1}{2}}R$, which in turn induce a meridional flow of order $\mathrm{Re}^{-\frac{1}{2}}R$ in the $\mathrm{Re}^{-\frac{1}{4}}R$ layer owing to Ekman suction. This flow then circulates through the $\mathrm{Re}^{-\frac{1}{3}}R$ layer and in the Ekman extensions.

Three dimensional Navier-Stokes simulations of flows affected by an inserted stationary object, modeled as a stationary disk extended in the meridional direction, has been performed by Volosciuk (1981) and Walz, Volosciuk & Schutz (1983). Numerical investigations have also been made of thermally, mechanically and thermo-mechanically driven flows of a weakly compressible gas in a cylinder assuming an axisymmetric continuum flow (Harada (1980)). The finite difference method was devloped for these computations based on the Euler upwind scheme. It was found that an almost cylindrical Couette flow was established in an annulus, despite the end effects of the plates which generate machanically driven circulations. The two mechanically driven circulations interact with the thermally driven circulations. The authors also investigated the axisymmetric continuum flow of a strongly compressible rotating gas driven thermally and mechanically in an annulus using finite difference method. The DuFort-Frankel upwind scheme was used for efficient computations of the strongly compressible rotating flows, and it was found that the thermally driven flow becomes weaker as the compressibility is increased. The viscous dissipation increases the gas temperature of the mechanically driven flow despite cooling by the expansion work.

Sakurai (1981) analyzed the axisymmetric steady centrifugally driven thermal convection of a compressible fluid in a rapidly rotating circular. They clarified the importance of the effect of the flow-induced volume change of a fluid particle, and propose a method of solution. The effect of inserted objects was analysed in terms of linearized thin-wing theory by Sakurai (1981). Their analysis showed that the drag force on one inserted object was not affected by other upstream objects, or by the wake of the same object due to fluid circulation, though the objects do interact with respect to their effect on the flow.

Many previous studies have obtained velocity profiles in rotating flows using the compressible Navier-Stokes equations. There have been studies of high-speed stratified flows past objects (Hausman & Roberts (1984), Roberts (1985), Roberts & Hausman (1988), Roberts & Hausman (1988)). The present work provides an opportunity for validating the analytical model as well as the DSMC method for the high-speed rotating flows in a polar $(r-\theta)$ plane. The molecular simulation procedure is more computationally intensive than the continuum Navier-Stokes equations, but it also has two important advantages. The Navier-Stokes equations are written for the macroscopic density and velocity fields, and these are applicable only when the mean free path in the gas is small compared to the macroscopic scale (radius of the cylinder or boundary layer thickness). There are many situations in high-speed rotating flows where the gas density in the central region of the cylinder is sufficiently small that the mean free path is comparable to the cylinder radius. In these cases, the Navier-Stokes equations are not valid near the center, and it is necessary to specify artificial 'boundary conditions' at a surface within the flow. In this context, Roblin & Doneddu (2000) has been performed DSMC simulations to study the feed gas expansion from the central pipe, and the residual gas contained in the casing in which the cylinder rotates. These simulations were used to provide more realistic boundary conditions for continuum fluid flow calculations. The possible dependence of the solutions on these artificial boundary conditions poses difficulties in the interpretation of results. No such difficulties exist in the molecular simulations, since it is only necessary to specify particle-wall interaction parameters (Pradhan & Kumaran (2011); Kumaran & Pradhan (2014)).

The another advantage of the molecular simulations is that the velocity slip at the wall resulting from the particle-wall interaction law can be obtained directly from the simulations (Pradhan & Kumaran (2011), Kumaran & Pradhan (2014)), whereas the slip velocity has to be modeled in the continuum numerical solutions. In the present simulations, we also find that there is substantial 'temperature slip' in the DSMC simulations, that is, the gas temperature at the wall is not equal to the wall temperature. Both velocity and temperature slip at the walls are measured in the DSMC simulations, and these are then incorporated into the analytical solutions, in order to make a quantitative comparison of the results of the analysis and simulations. The comparisons are carried out for stratification parameter in the rage $0.707 - 3.535$ and for Reynolds numbers as low as 100, in order to quantitatively estimate the accuracy of the analytical model.

In molecular simulations, the direct simulation Monte Carlo (DSMC) technique is most suited for the simulation of rarefied gases. In this technique, the Boltzmann equation is solved using a probabilistic solution procedure. The positions and velocities of simulated molecules in the flow domain are evolved in time. The macroscopic properties (density, velocity, temperature) are determined by averaging over the molecular properties in cells. The positions and velocities between collisions are evolved deterministically using Newton's laws, while a stochastic Monte Carlo technique is used for collisions between the nearby molecules. Collisions are only permitted between molecules in the same sub-cell to ensure spatial localization. The number of simulated molecules is small compared to the number of real molecules in the system - typically, one simulated molecule represents $(10^{14} - 10^{18})$ real molecules. The advantage of the procedure is that the collision frequency and molecular diameter of the simulated molecules can be adjusted to obtain the same collision frequency and mean free path as the real molecules. Despite this reduction, the number of molecules per cell has to be large for accurate statistical averaging. Typically, $(50 - 100)$ simulated molecules are required per cell for reliable averages to be obtained. Due to this, the total number of molecules becomes very large (about 100 times the number of cells), and the procedure is computationally very intensive. The advantages of this procedure is that properties have to be specified only at the molecular level. It is not necessary to specify continuum parameters such as the equation of state, the viscosity, thermal conductivity and diffusivity. The properties specified are the inter-molecular interaction parameters such as the molecular diameter, molecular mass, and collision cross section, and the molecule-wall collision parameters. Molecular models range from the simplest hard-sphere model, where the molecules are treated as rigid hard spheres, to the soft-sphere model and to more sophisticated models which incorporate the internal energy and the anisotropy in molecular interactions. Similar parameters need to be specified for particle-wall interactions.

The DSMC simulations are computationally very intensive, both with regard to the computational speed and memory requirements. In two dimensions, a proper discretization requires of the order of $(10^4 - 10^5)$ cells, and consequently the number of molecules is $(10^6 - 10^7)$. In the computation, if we assign 100 bytes for the position and velocity of each molecule, the access memory required is $(10^8 - 10^9)$. Thus, 1 GB is the minimum memory required for carrying out simulations even in two dimensions; this is certainly feasible. In three dimensions, the number of cells increases by a further factor of $(10^2 - 10^3)$ in the third dimension, and a consequent increase in the memory requirement of the order of 100 GB to 1 TB. With regard to the speed of computation, it is necessary to generate a large number of random numbers at each collision in the stochastic collision model. This increases significantly the number of computations required at each time step. As computational power increases, discrete simulations are likely to become more and more attractive in comparison to other techniques.

Apart from the large computational requirements, this procedure does not have most of the disadvantages present in continuum models. It is not necessary to specify the equation of state and transport properties, since these emerge naturally by statistical averaging over the molecular properties. There is no difficulty in simulating high Mach number or high Knudsen number flows because a molecular description is being used. Velocity and temperature slip at the walls are accurately captured once the particle-wall interaction parameters are specified. There is no need to make a transition from a continuum to free-molecule regime because the description is at the molecular level. Consequently, there is also no need to specify inner boundary conditions.

The implementation of sources of mass, momentum and energy within the flow is technically non-trivial in the DSMC simulation. In order to mimic a Dirac delta function in the simulations, mass, momentum and energy are inserted/removed over a small source region of radius 10^{-3} times the cylinder radius at the designated location. When quantitative comparison has to be made with analytical solutions for specified rates of input/output of mass, momentum and energy, care has to be taken in the simulations due to the coupling between mass, momentum and energy sources. When molecules are added in simulations, there is the simultaneous addition of the momentum and energy of these molecules as well. In order to add mass alone, it is necessary to reduce the momentum and energy of all molecules in the source region so that total momentum and energy are conserved. In a similar manner, momentum is added by increasing the mean velocity of the molecules in the source region, while addition of energy requires increasing the fluctuating velocity of the molecules while keeping the total mean velocity a constant. Another complication is that addition of energy at a location near the wall in the simulation causes an increase in the local fluctuating velocities of molecules, which heats up the wall. This violates the constant temperature boundary condition that is used in the analytical solution, and so it is necessary to extract heat locally at the wall in the simulation to ensure that the wall temperature is maintained a constant (Pradhan & Kumaran (2011), Kumaran & Pradhan (2014)).

Here, we formulate a generalized analytical model in the polar $(r - \theta)$ plane in a high speed rotating cylinder, where the restriction on high stratification parameter is removed. In order to set the context for the numerical comparison, we revisit the approximations in the analytical model presented here in the polar $(r - \theta)$ plane in a high speed rotating cylinder of radius R. The starting point of the analytical model is the Navier-Stokes, mass, momentum and energy conservation equations in the polar $(r - \theta)$ coordinate, where r and θ are the radial and azimuthal coordinates. The base state is a solid-body rotation, in which there is an exponential decay of the density from the wall of the cylinder proportional to $\rho_w \exp\left(-\mathcal{A}^2(1 - (r^2/R^2))\right)$ due to the centrifugal force. Here

$\mathcal{A} = \sqrt{m\Omega^2 R^2/(2k_B T)}$ is the stratification parameter, m is the molecular mass, Ω and R are the angular velocity and radius of the cylinder, k_B is the Boltzmann constant, T is the absolute temperature, and ρ_w is the gas density at the wall. The equations are linearized in the amplitudes of the secondary radial flow velocity, density and temperature about a base state which is a solid body rotation. A scaled radial coordinate is defined by $\xi = \mathcal{A}^2 \left(1 - (r^2/R^2)\right)$, and it is assumed that the Reynolds number $(\rho_w \Omega R^2/\mu)$ is large. However, there is no restriction on the stratification parameter, and since we have used the conservative form of the compressible mass, momentum and energy conservation equations, in deriving the generalized analytical model, obviously, the shock type solutions (Rankine-Hugoniot relations) are automatically satisfied. In this limit, the equations are reduced to a sixth-order differential equation in ξ for the 'master potential' χ, which is related to the stream function ψ by the relation $\psi = (1/r)(\partial\chi/\partial r)$. After the approximations, the linear operator in the generalized analytical model is dependent explicitely on the stratification parameter, though it is still a self-adjoint operator with discrete eigenvalues and orthogonal eigenfunctions if the boundary conditions are properly specified. In addition, the equations are solved on a finite domain, $0 \leq \xi \leq \mathcal{A}^2$.

The effect of a stationary insert in the flow domain is of three types. The first is the mass source or sink due to the addition or removal of gas in a particular region in the flow domain. The second effect is due to the force exerted on the gas by a stationary inserts, which is modelled locally as a momentum source or sink. The time rate of addition/removal of momentum is equal to the applied force. The third effect is a localised energy source, which increases or decreases the temperature of the gas. All the three effects are incorporated in the present molecular simulation (direct simulation Monte Carlo (DSMC) method) as follows. The addition or removal of mass is accomplished by either adding or removing molecules at specified locations in the simulation cell at specified rates. The pressure is specified at inflow and outflow boundaries by connecting a pressure bath, which is a large volume of gas maintained at a constant pressure. Forces are exerted by imposing impulses on the molecules at specified locations, in such a manner that the rate of change of impulse is equal to the applied force. Energy is added by increasing the mean square velocities of the molecules at locations where heating takes place.

In §2, the mass, momentum and energy conservation equations in the polar $(r-\theta)$ plane in a high speed rotating cylinder are presented and different dimensionless numbers are defined. The equations for perturbations to the base state (solid body rotation) are then used to get the equation for the master potential in §3. In §4, the radial boundary layer equations are solved to obtain the generalized analytical equation in the polar $(r-\theta)$ plane. In §5, a brief summary of the molecular simulation based on direct simulation Monte Carlo (DSMC) method is presented. In §6, flow fields due to the insertion of mass, momentum and energy into the rotating cylinder are studied for a single-component gas for stratification parameter ($\mathcal{A}$) in the range $(0.707 - 3.535)$, and analytical results are compared with direct simulation Monte Carlo (DSMC) simulations. In §7, DSMC simulations are carried out for a binary gas mixture and results are compared with single-component gas. The important conclusions of the analysis are summarized in §8.

2. Formulation

The configuration consists of a right cicular cylinder of radius R and angle $(\theta = 2\pi)$ rotating at a constant angular velocity Ω. A cylindrical coordinate system with origin at the centre of the cylinder $(r = 0)$, with radial, and meridional coordinates r, and θ respectively, is used. Throughout this analysis, we assume that the flow is in the $(r-\theta)$

plane, so there is no dependence on the axial coordinate z. Gravitational acceleration is neglected, since it has a negligible effect on high-speed gas flows. The compressible mass, momentum and energy conservation equations in cylindrical polar $(r-\theta)$ coordinate system are provided in the appendix for reference. There are six variables in these equations, the density ρ, the three components of the velocity v_r, v_θ, v_z, the pressure p and the temperature T. In equations A 1 to A 5, S_m is the mass source density (mass input per unit volume per unit time), f_r, f_θ and f_z are the force densities in r, θ and z directions respectively, and $\mathcal{S}_e$ is the energy source density (energy input per unit volume per unit time).

The pressure is related to the density by the ideal gas equation of state,

$$\rho = \frac{pm}{k_B T}, \tag{2.1}$$

where m is the mass of a molecule and k_B is the Boltzmann constant.

If the temperature on the walls of the cylinder is uniform, and there is no inflow or outflow of gas, a steady solution for the equations of motion is one where the gas undergoes solid-body rotation with angular velocity Ω at constant temperature. The temperature $\bar{T}$ and the components of the velocity $\bar{\mathbf{v}}$ in the base state in this steady state are,

$$\bar{v}_\theta = \Omega r, \quad \bar{v}_z = 0, \quad \bar{v}_r = 0, \quad \bar{T} = T_0, \tag{2.2}$$

where T_0 is the constant temperature in the steady state. The pressure is obtained by solving the radial momentum equation,

$$\frac{d\bar{p}}{dr} = \bar{\rho}\Omega^2 r = \frac{\bar{p}m\Omega^2 r}{k_B T_0}, \tag{2.3}$$

where $\bar{\rho}$ is the density at steady state, together with the ideal gas equation of state,

$$\bar{\rho} = \frac{\bar{p}m}{k_B T_0}, \tag{2.4}$$

where m is the mass of a molecule and k_B is the Boltzmann constant. The above equation is easily solved to obtain

$$\frac{\bar{p}}{\bar{p}_w} = \frac{\bar{\rho}}{\bar{\rho}_w} = \exp\left(-\mathcal{A}^2\left(1 - \frac{r^2}{R^2}\right)\right), \tag{2.5}$$

where $\bar{p}_w$ and $\bar{\rho}_w$ are the pressure and density at the wall, and $\mathcal{A}$ is the stratification parameter. The total mass of gas in the cylinder is related to ρ_w by,

$$\text{Mass of gas} = \frac{\pi R^2 Z \bar{\rho}_w (1 - \exp(-\mathcal{A}^2))}{\mathcal{A}^2}, \tag{2.6}$$

where Z is the total height of the cylinder.

It is useful to discuss the different dimensionless numbers used in the analysis. The stratification parameter obtained from the equation for the density is,

$$\mathcal{A} = \sqrt{\frac{m\Omega^2 R^2}{2k_B T_0}} \propto \frac{R\Omega}{c_s}, \tag{2.7}$$

where $(R\Omega)$ is the characteristic flow velocity, and $c_s = \sqrt{\gamma k_B T_0/m}$ is the speed of sound. The Reynolds number, used a little later, is defined by

$$\text{Re} = \frac{\rho_w \Omega R^2}{\bar{\mu}}, \tag{2.8}$$

where $\bar{\mu}$ is the viscosity of the fluid in the unperturbed state. Note that the viscosity of a dilute gas depends only on the temperature, and not on the density. Therefore, the viscosity is a constant in the base state where the temperature is uniform. The viscosity of a dilute gas scales as,

$$\bar{\mu} \sim \frac{\sqrt{mk_B T_0}}{d^2}, \tag{2.9}$$

where m is the molecular mass and d is the molecular diameter. Using the above expression for the viscosity, it can easily be verified that the Reynolds number can be rewritten as,

$$\mathrm{Re} \sim \frac{\mathcal{A}}{\mathrm{Kn}}, \tag{2.10}$$

where the Knudsen number is defined by

$$\mathrm{Kn} = (m/\sqrt{2}\pi\rho_w d^2 R) = (\lambda/R), \tag{2.11}$$

where $\lambda = (m/\sqrt{2}\pi\rho_w d^2)$ is the mean free path at the wall. The other dimensionless groups used are the Prandtl number and the ratio of specific heats,

$$\mathrm{Pr} = \frac{C_p\bar{\mu}}{K}, \tag{2.12}$$

$$\gamma = \frac{C_p}{C_v}, \tag{2.13}$$

which are both $O(1)$ in the ideal gas limit.

3. Equations of motion in polar $(r - \theta)$ coordinate

The disturbance to the base state due to insertion of small mass, momentum and energy into the rotating cylinder, is incorporated by perturbing the base velocity and pressure profile as follows:

$$\rho = \bar{\rho} + \hat{\rho}, \tag{3.1}$$

$$\begin{aligned} p &= \bar{p} + \hat{p}, \\ T &= \bar{T} + \hat{T}, \\ v_r &= \hat{v}_r, \\ v_\theta &= r\Omega + \hat{v}_\theta, \\ v_z &= \hat{v}_z. \end{aligned} \tag{3.2}$$

The above forms for the dynamical variables are inserted into the mass, momentum and energy conservation equations, and linearized in the perturbation (hatted) quantities. Steady state is assumed, so that all the time derivatives are zero. After subtracting out the corresponding base state equations, we obtain the following qeuations for the perturbations:

$$\frac{1}{r}\frac{\partial}{\partial r}(r\bar{\rho}\hat{v}_r) + \frac{1}{r}\frac{\partial}{\partial \theta}(\bar{\rho}\hat{v}_\theta + \hat{\rho}\Omega r) = S_m, \tag{3.3}$$

$$-\hat{\rho}\Omega^2 r - 2\bar{\rho}\Omega\hat{v}_\theta + \bar{\rho}\Omega\frac{\partial \hat{v}_r}{\partial \theta} = -\frac{\partial \hat{p}}{\partial r} + \bar{\mu}\left(\frac{\partial}{\partial r}\left(\frac{1}{r}\frac{\partial}{\partial r}(r\hat{v}_r)\right) + \frac{1}{r^2}\frac{\partial^2 \hat{v}_r}{\partial \theta^2} - \frac{2}{r^2}\frac{\partial \hat{v}_\theta}{\partial \theta}\right) + \left(f_r + \frac{\bar{\mu}}{3}\frac{\partial S_m}{\partial r}\right), \tag{3.4}$$

$$2\bar{\rho}\Omega\hat{v}_r+\bar{\rho}\Omega\frac{\partial\hat{v}_\theta}{\partial\theta} = -\frac{1}{r}\frac{\partial\hat{p}}{\partial\theta}+\bar{\mu}\left(\frac{\partial}{\partial r}\left(\frac{1}{r}\frac{\partial}{\partial r}(r\hat{v}_\theta)\right)+\frac{1}{r^2}\frac{\partial^2\hat{v}_\theta}{\partial\theta^2}+\frac{2}{r^2}\frac{\partial\hat{v}_r}{\partial\theta}\right)+\left(f_\theta+\frac{1}{r}\frac{\bar{\mu}}{3}\frac{\partial S_m}{\partial\theta}\right), \tag{3.5}$$

$$\bar{\rho}\Omega\frac{\partial\hat{v}_z}{\partial\theta} = \bar{\mu}\left(\frac{1}{r}\frac{\partial}{\partial r}\left(r\frac{\partial\hat{v}_z}{\partial r}\right)+\frac{1}{r^2}\frac{\partial^2\hat{v}_z}{\partial\theta^2}\right)+f_z, \tag{3.6}$$

$$-\bar{\rho}\Omega^2 r\hat{v}_r+\bar{\rho}\Omega C_p\frac{\partial\hat{T}}{\partial\theta} = -\bar{p}S_m+\bar{K}\left(\frac{1}{r}\frac{\partial}{\partial r}\left(r\frac{\partial\hat{T}}{\partial r}\right)+\frac{1}{r^2}\frac{\partial^2\hat{T}}{\partial\theta^2}\right)+S_e. \tag{3.7}$$

Here, $\bar{\mu}$ and $\bar{K}$ are the viscosity and thermal conductivity of the gas in the base state, where the temperature is T_0. Note that in the ideal gas limit, the viscosity and thermal conductivity are functions of the temperature alone, and they do not depend on the density. In equation 3.4, the term on the left is the contribution to the r momemtum equation due to the centrifugal force, while in equation 3.5, the left side represents the contribution to the θ momentum equation due to the Coriolis force. In the energy equation 3.7, the term on the left is the pressure work done due to the compression or expansion of the gas, and we have used the equation of state 2.4 and the mass conservation equation 3.3 to simplify this term. The perturbation to the equation of state is

$$\hat{\rho} = \left(\frac{m}{k_B T_0}\right)\hat{p}-\bar{\rho}\left(\frac{\hat{T}}{T_0}\right). \tag{3.8}$$

First, we eliminate the density in the r momentum equation, 3.4, using the equation of state, equation 3.8, to obtain

$$\begin{aligned}&\bar{\rho}\Omega\frac{\partial\hat{v}_r}{\partial\theta}-\frac{\Omega^2 rm\hat{p}}{k_B T_0}+\frac{\Omega^2 r\bar{\rho}\hat{T}}{T_0}-2\bar{\rho}\Omega\hat{v}_\theta+\frac{\partial\hat{p}}{\partial r}\\ &=\bar{\mu}\left(\frac{\partial}{\partial r}\left(\frac{1}{r}\frac{\partial}{\partial r}(r\hat{v}_r)\right)+\frac{1}{r^2}\frac{\partial^2\hat{v}_r}{\partial\theta^2}-\frac{2}{r^2}\frac{\partial\hat{v}_\theta}{\partial\theta}\right)+\left(f_r+\frac{\bar{\mu}}{3}\frac{\partial S_m}{\partial r}\right).\end{aligned} \tag{3.9}$$

The identity,

$$\begin{aligned}\bar{\rho}\frac{\partial}{\partial r}\left(\frac{\hat{p}}{\bar{\rho}}\right) &= \frac{\partial\hat{p}}{\partial r}-\frac{\hat{p}}{\bar{\rho}}\frac{\partial\bar{\rho}}{\partial r}\\ &= \frac{\partial\hat{p}}{\partial r}-\frac{\hat{p}m\Omega^2 r}{k_B T_0},\end{aligned} \tag{3.10}$$

can be inserted into the momentum conservation equation, 3.9, to obtain

$$\begin{aligned}&\frac{\Omega}{r}\frac{\partial\hat{v}_r}{\partial\theta}+\frac{1}{r}\frac{\partial}{\partial r}\left(\frac{\hat{p}}{\bar{\rho}}\right)-\frac{2\Omega\hat{v}_\theta}{r}-\frac{\Omega^2\hat{T}}{T_0}\\ &=\frac{\bar{\mu}}{\bar{\rho}r}\left(\frac{\partial}{\partial r}\left(\frac{1}{r}\frac{\partial}{\partial r}(r\hat{v}_r)\right)+\frac{1}{r^2}\frac{\partial^2\hat{v}_r}{\partial\theta^2}-\frac{2}{r^2}\frac{\partial\hat{v}_\theta}{\partial\theta}\right)+\frac{1}{\bar{\rho}r}\left(f_r+\frac{\bar{\mu}}{3}\frac{\partial S_m}{\partial r}\right).\end{aligned} \tag{3.11}$$

If we take the derivative of the above equation with respect to the θ coordinate, we obtain

$$\begin{aligned}&\frac{\Omega}{r}\frac{\partial^2\hat{v}_r}{\partial\theta^2}+\frac{1}{r}\frac{\partial}{\partial r}\left(\frac{1}{\bar{\rho}}\frac{\partial\hat{p}}{\partial\theta}\right)-\Omega\frac{d}{d\theta}\left(\frac{2\hat{v}_\theta}{r}-\frac{\Omega\hat{T}}{T_0}\right)\\ &=\frac{\bar{\mu}}{\bar{\rho}r}\frac{\partial}{\partial\theta}\left(\frac{\partial}{\partial r}\left(\frac{1}{r}\frac{\partial}{\partial r}(r\hat{v}_r)\right)+\frac{1}{r^2}\frac{\partial^2\hat{v}_r}{\partial\theta^2}-\frac{2}{r^2}\frac{\partial\hat{v}_\theta}{\partial\theta}\right)+\frac{1}{\bar{\rho}r}\frac{\partial}{\partial\theta}\left(f_r+\frac{\bar{\mu}}{3}\frac{\partial S_m}{\partial r}\right).\end{aligned} \tag{3.12}$$

Equation 3.5 is used to substitute for $(\partial \hat{p}/\partial \theta)$ in the above equation, to obtain

$$\begin{aligned}
&\frac{\Omega}{r}\frac{\partial^2 \hat{v}_r}{\partial \theta^2} + \frac{1}{r}\frac{\partial}{\partial r}\left(\frac{r}{\bar{\rho}}\left(-2\bar{\rho}\Omega\hat{v}_r - \bar{\rho}\Omega\frac{\partial \hat{v}_\theta}{\partial \theta} + \bar{\mu}\left(\frac{\partial}{\partial r}\left(\frac{1}{r}\frac{\partial}{\partial r}(r\hat{v}_\theta)\right)\right.\right.\right. \\
&\left.\left.\left. + \frac{1}{r^2}\frac{\partial^2 \hat{v}_\theta}{\partial \theta^2} + \frac{2}{r^2}\frac{\partial \hat{v}_r}{\partial \theta} + \left(f_\theta + \frac{1}{r}\frac{\bar{\mu}}{3}\frac{\partial S_m}{\partial \theta}\right) - \Omega\frac{d}{d\theta}\left(\frac{2\hat{v}_\theta}{r} - \frac{\Omega\hat{T}}{T_0}\right)\right)\right)\right) = \\
&\frac{\bar{\mu}}{\bar{\rho}r}\frac{\partial}{\partial \theta}\left(\frac{\partial}{\partial r}\left(\frac{1}{r}\frac{\partial}{\partial r}(r\hat{v}_r)\right) + \frac{1}{r^2}\frac{\partial^2 \hat{v}_r}{\partial \theta^2} - \frac{2}{r^2}\frac{\partial \hat{v}_\theta}{\partial \theta}\right) \\
&+ \frac{1}{\bar{\rho}r}\frac{\partial}{\partial \theta}\left(f_r + \frac{\bar{\mu}}{3}\frac{\partial S_m}{\partial r}\right).
\end{aligned} \tag{3.13}$$

Divided by $\bar{\mu}$ and after simplification it gives,

$$\begin{aligned}
&\frac{\Omega}{\bar{\mu}r}\frac{\partial^2 \hat{v}_r}{\partial \theta^2} + \frac{1}{\bar{\mu}r}\frac{\partial}{\partial r}\left(-2r\Omega\hat{v}_r - r\Omega\frac{\partial \hat{v}_\theta}{\partial \theta}\right) \\
&+ \frac{1}{r}\frac{\partial}{\partial r}\left(\frac{r}{\bar{\rho}}\left(\frac{\partial}{\partial r}\left(\frac{1}{r}\frac{\partial}{\partial r}(r\hat{v}_\theta)\right)\right)\right) + \frac{1}{r}\frac{\partial}{\partial r}\left(\frac{1}{\bar{\rho}r}\frac{\partial^2 \hat{v}_\theta}{\partial \theta^2}\right) \\
&+ \frac{1}{r}\frac{\partial}{\partial r}\left(\frac{2}{\bar{\rho}r}\frac{\partial \hat{v}_r}{\partial \theta}\right) - \frac{1}{\bar{\rho}r}\frac{\partial}{\partial \theta}\left(\frac{\partial}{\partial r}\left(\frac{1}{r}\frac{\partial}{\partial r}(r\hat{v}_r)\right)\right) \\
&+ \frac{1}{\bar{\rho}r^3}\left(-\frac{\partial^3 \hat{v}_r}{\partial \theta^3} + 2\frac{\partial^2 \hat{v}_\theta}{\partial \theta^2}\right) - \frac{\Omega}{\bar{\mu}}\frac{d}{d\theta}\left(\frac{2\hat{v}_\theta}{r} - \frac{\Omega\hat{T}}{T_0}\right) = \\
&\frac{1}{\bar{\mu}}\left(\frac{1}{\bar{\rho}r}\frac{\partial}{\partial \theta}\left(f_r + \frac{\bar{\mu}}{3}\frac{\partial S_m}{\partial r}\right) - \frac{1}{r}\frac{\partial}{\partial r}\left(\frac{r}{\bar{\rho}}\left(f_\theta + \frac{1}{r}\frac{\bar{\mu}}{3}\frac{\partial S_m}{\partial \theta}\right)\right)\right).
\end{aligned} \tag{3.14}$$

Here, the viscosity $\bar{\mu}$ in the base state is independent of position, because the temperature is uniform and the viscosity depends only on temperature.

The velocity components $\hat{v}_r$ and $\hat{v}_\theta$ can be expressed in terms of a stream function using the mass conservation equation 3.3,

$$r\bar{\rho}\hat{v}_r = -\frac{\partial \psi}{\partial \theta} + \int_0^r S_m\, r dr, \tag{3.15}$$

$$\bar{\rho}\hat{v}_\theta = \frac{\partial \psi}{\partial r} - \hat{\rho}\Omega r, \tag{3.16}$$

where the difference in the stream function between two radial locations gives the angular mass flux in the region between these two locations,

$$\psi(a,\theta) - \psi(r,\theta) = \int_r^a (\bar{\rho}\hat{v}_\theta + \hat{\rho}\Omega r)\, dr. \tag{3.17}$$

Expressing in terms of the stream function, equation 3.14 is

$$\begin{aligned}
&\frac{\Omega}{\bar{\mu}r}\frac{\partial^2}{\partial \theta^2}\left(\frac{1}{\bar{\rho}r}\left(-\frac{\partial \psi}{\partial \theta} + \int_0^r S_m\, r dr\right)\right) + \frac{1}{\bar{\mu}r}\frac{\partial}{\partial r}\left(\frac{-2\Omega}{\bar{\rho}}\left(-\frac{\partial \psi}{\partial \theta} + \int_0^r S_m\, r dr\right)\right) \\
&-r\Omega\frac{\partial}{\partial \theta}\left(\frac{1}{\bar{\rho}}\left(\frac{\partial \psi}{\partial r} - \hat{\rho}\Omega r\right)\right) + \frac{1}{r}\frac{\partial}{\partial r}\left(\frac{r}{\bar{\rho}}\left(\frac{\partial}{\partial r}\left(\frac{1}{r}\frac{\partial}{\partial r}\left(\frac{r}{\bar{\rho}}\left(\frac{\partial \psi}{\partial r} - \hat{\rho}\Omega r\right)\right)\right)\right)\right) \\
&+\frac{1}{r}\frac{\partial}{\partial r}\left(\frac{1}{\bar{\rho}r}\frac{\partial^2}{\partial \theta^2}\left(\frac{1}{\bar{\rho}}\left(\frac{\partial \psi}{\partial r} - \hat{\rho}\Omega r\right)\right)\right) + \frac{1}{r}\frac{\partial}{\partial r}\left(\frac{2}{\bar{\rho}^2 r^2}\frac{\partial}{\partial \theta}\left(-\frac{\partial \psi}{\partial \theta} + \int_0^r S_m\, r dr\right)\right) \\
&-\frac{1}{\bar{\rho}r}\frac{\partial}{\partial \theta}\left(\frac{\partial}{\partial r}\left(\frac{1}{r}\frac{\partial}{\partial r}\left(\frac{1}{\bar{\rho}}\left(-\frac{\partial \psi}{\partial \theta} + \int_0^r S_m\, r dr\right)\right)\right)\right) - \frac{1}{\bar{\rho}^2 r^4}\frac{\partial^3}{\partial \theta^3}\left(-\frac{\partial \psi}{\partial \theta} + \int_0^r S_m\, r dr\right)
\end{aligned}$$

$$+\frac{2}{\bar{\rho}^2 r^3}\frac{\partial^2}{\partial\theta^2}\left(\frac{\partial\psi}{\partial r}-\hat{\rho}\Omega r\right)-\frac{\Omega}{\bar{\mu}}\frac{d}{d\theta}\left(\frac{2}{\bar{\rho}r}\left(\frac{\partial\psi}{\partial r}-\hat{\rho}\Omega r\right)-\frac{\Omega\hat{T}}{T_0}\right)=$$
$$\frac{1}{\bar{\mu}}\left(\frac{1}{\bar{\rho}r}\frac{\partial}{\partial\theta}\left(f_r+\frac{\bar{\mu}}{3}\frac{\partial S_m}{\partial r}\right)-\frac{1}{r}\frac{\partial}{\partial r}\left(\frac{r}{\bar{\rho}}\left(f_\theta+\frac{1}{r}\frac{\bar{\mu}}{3}\frac{\partial S_m}{\partial\theta}\right)\right)\right). \quad (3.18)$$

The stream function ψ is expressed in terms of the 'master potential' χ,

$$\psi=\frac{1}{r}\frac{\partial\chi}{\partial r}. \quad (3.19)$$

This is inserted into the momentum conservation equation 3.13 to obtain

$$\frac{\Omega}{\bar{\mu}r}\frac{\partial^2}{\partial\theta^2}\left(\frac{1}{\bar{\rho}r}\left(-\frac{\partial}{\partial\theta}\left(\frac{1}{r}\frac{\partial\chi}{\partial r}\right)+\int_0^r S_m\,r dr\right)\right)+\frac{1}{\bar{\mu}r}\frac{\partial}{\partial r}\left(\frac{-2\Omega}{\bar{\rho}}\left(-\frac{\partial}{\partial\theta}\left(\frac{1}{r}\frac{\partial\chi}{\partial r}\right)+\int_0^r S_m\,r dr\right)\right)$$
$$-r\Omega\frac{\partial}{\partial\theta}\left(\frac{1}{\bar{\rho}}\left(\frac{\partial}{\partial r}\left(\frac{1}{r}\frac{\partial\chi}{\partial r}\right)-\hat{\rho}\Omega r\right)\right)+\frac{1}{r}\frac{\partial}{\partial r}\left(\frac{r}{\bar{\rho}}\left(\frac{\partial}{\partial r}\left(\frac{1}{r}\frac{\partial}{\partial r}\left(\frac{r}{\bar{\rho}}\left(\frac{\partial}{\partial r}\left(\frac{1}{r}\frac{\partial\chi}{\partial r}\right)-\hat{\rho}\Omega r\right)\right)\right)\right)\right)$$
$$+\frac{1}{r}\frac{\partial}{\partial r}\left(\frac{1}{\bar{\rho}r}\frac{\partial^2}{\partial\theta^2}\left(\frac{1}{\bar{\rho}}\left(\frac{\partial}{\partial r}\left(\frac{1}{r}\frac{\partial\chi}{\partial r}\right)-\hat{\rho}\Omega r\right)\right)\right)+\frac{1}{r}\frac{\partial}{\partial r}\left(\frac{2}{\bar{\rho}^2 r^2}\frac{\partial}{\partial\theta}\left(-\frac{\partial}{\partial\theta}\left(\frac{1}{r}\frac{\partial\chi}{\partial r}\right)+\int_0^r S_m\,r dr\right)\right)$$
$$-\frac{1}{\bar{\rho}r}\frac{\partial}{\partial\theta}\left(\frac{\partial}{\partial r}\left(\frac{1}{r}\frac{\partial}{\partial r}\left(\frac{1}{\bar{\rho}}\left(-\frac{\partial}{\partial\theta}\left(\frac{1}{r}\frac{\partial\chi}{\partial r}\right)+\int_0^r S_m\,r dr\right)\right)\right)\right)-\frac{1}{\bar{\rho}^2 r^4}\frac{\partial^3}{\partial\theta^3}\left(-\frac{\partial}{\partial\theta}\left(\frac{1}{r}\frac{\partial\chi}{\partial r}\right)+\int_0^r S_m\,r dr\right)$$
$$+\frac{2}{\bar{\rho}^2 r^3}\frac{\partial^2}{\partial\theta^2}\left(\frac{\partial}{\partial r}\left(\frac{1}{r}\frac{\partial\chi}{\partial r}\right)-\hat{\rho}\Omega r\right)-\frac{\Omega}{\bar{\mu}}\frac{d}{d\theta}\left(\frac{2}{\bar{\rho}r}\left(\frac{\partial}{\partial r}\left(\frac{1}{r}\frac{\partial\chi}{\partial r}\right)-\hat{\rho}\Omega r\right)\right)-\frac{\Omega^2}{\bar{\mu}}\frac{\partial}{\partial}\left(\frac{\hat{T}}{T_0}\right)=$$
$$\frac{1}{\bar{\mu}}\left(\frac{1}{\bar{\rho}r}\frac{\partial}{\partial\theta}\left(f_r+\frac{\bar{\mu}}{3}\frac{\partial S_m}{\partial r}\right)-\frac{1}{r}\frac{\partial}{\partial r}\left(\frac{r}{\bar{\rho}}\left(f_\theta+\frac{1}{r}\frac{\bar{\mu}}{3}\frac{\partial S_m}{\partial\theta}\right)\right)\right). \quad (3.20)$$

4. Radial boundary layer

When the boundary layer is in the radial coordinate, the radial derivatives of $\hat{v}_\theta$ and $\hat{T}$ on the right sides of equations 3.5 and 3.7 are balanced by the left sides. The temperature perturbation $\hat{T}$ in 3.20 can be expressed in terms of the master potential from the energy equation, 3.7. The energy equation is

$$-\bar{\rho}\Omega^2 r\hat{v}_r+\bar{\rho}\Omega C_p\frac{\partial\hat{T}}{\partial\theta}=-\bar{p}S_m+\bar{K}\left(\frac{1}{r}\frac{\partial}{\partial r}\left(r\frac{\partial\hat{T}}{\partial r}\right)+\frac{\partial^2\hat{T}}{\partial z^2}\right)+S_e. \quad (4.1)$$

Expressing the radial velocity $\hat{v}_r$ in terms of the master potential using 3.15 and 3.19, we obtain

$$\frac{\Omega^2}{r}\frac{\partial}{\partial r}\left(\frac{\partial\chi}{\partial\theta}\right)-\Omega^2\int_{r_0}^r S_m\,r dr+\bar{\rho}C_p\Omega\frac{\partial\hat{T}}{\partial\theta}=\bar{K}\left(\frac{1}{r}\frac{\partial}{\partial r}\left(r\frac{\partial\hat{T}}{\partial r}\right)+\frac{1}{r^2}\frac{\partial^2\hat{T}}{\partial\theta^2}\right)+(S_e-\bar{p}S_m). \quad (4.2)$$

The above equation can be multiplied by r, integrated once with respect to the radial coordinate with lower limit r_0 and upper limit r, to obtain

$$\Omega^2\left(\frac{\partial\chi}{\partial\theta}\right)-\Omega^2\int_{r_0}^r r\left(\int_{r_0}^r S_m\,r dr\right)dr+\int_{r_0}^r r\,dr\,\bar{\rho}C_p\Omega\left(\frac{\partial\hat{T}}{\partial\theta}\right)=$$
$$\bar{K}\left(r\frac{\partial\hat{T}}{\partial r}\right)\Bigg|_r-\bar{K}\,r\frac{\partial\hat{T}}{\partial r}\Bigg|_{r=r_0}+\int_{r_0}^r\frac{dr}{r}\left(\frac{\partial^2\hat{T}}{\partial\theta^2}\right)$$

$$+\int_{r_0}^{r} (S_e - \bar{p} S_m)\, r\, dr. \tag{4.3}$$

This is then multiplied by $(\Omega^2/\bar{\mu} T_0 r^2)$, and differentiated once with respect to θ, to obtain

$$\frac{\Omega^4}{\bar{K}\bar{\mu}T_0 r^2}\left(\frac{\partial^2 \chi}{\partial\theta^2}\right) - \frac{\Omega^4}{\bar{K}\bar{\mu}T_0 r^2}\frac{\partial}{\partial\theta}\int_{r_0}^{r} r\left(\int_{r_0}^{r}\frac{\partial S_m}{\partial\theta} r dr\right) dr + R_T + C_T$$
$$= \frac{\Omega^2}{\bar{\mu}}\frac{1}{r}\frac{\partial}{\partial r}\frac{\partial}{\partial\theta}\left(\frac{\hat{T}}{T_0}\right) + \frac{\Omega^2}{\bar{K}\bar{\mu}T_0 r^2}\frac{\partial}{\partial\theta}\int_{r_0}^{r}(S_e - \bar{p}S_m)\, r\, dr, \tag{4.4}$$

where

$$R_T = \frac{\Omega^2}{\bar{K}\bar{\mu}T_0 r^2}\int_{r_0}^{r}\frac{dr}{r}\left(\frac{\partial^3 \hat{T}}{\partial\theta^3}\right) + \frac{\Omega^2}{\bar{K}\bar{\mu}T_0 r^2}\int_{r_0}^{r} r\, dr\, \bar{\rho} C_p \Omega\left(\frac{\partial^2 \hat{T}}{\partial\theta^2}\right), \tag{4.5}$$

and the term appearing due to the constant of integration in equation 4.3 is,

$$C_T = \frac{\Omega^2}{\bar{\mu}T_0 r^2}\frac{\partial}{\partial\theta}\left(r\frac{\partial\hat{T}}{\partial r}\right)_{r=r_0} - \frac{\Omega^4}{\bar{K}\bar{\mu}T_0 r^2}\left(\frac{\partial^2\chi}{\partial\theta^2}\right)_{r=r_0}, \tag{4.6}$$

where r_0 is a reference location. For the flow in a cylinder, $r_0 = 0$ (at the centre).

We take the derivative $(1/r)(\partial/\partial r)$ of the equation for the master potential, equation 3.20, and then insert the equation 4.4, to obtain

$$\frac{\Omega}{\bar{\mu}r}\frac{1}{r}\frac{\partial}{\partial r}\left(\frac{\partial^2}{\partial\theta^2}\left(\frac{1}{\bar{\rho}r}\left(-\frac{\partial}{\partial\theta}\left(\frac{1}{r}\frac{\partial\chi}{\partial r}\right) + \int_0^r S_m\, r dr\right)\right)\right)$$
$$+\frac{1}{\bar{\mu}}\frac{1}{r}\frac{\partial}{\partial r}\left(\frac{1}{r}\frac{\partial}{\partial r}\left(\frac{-2\Omega}{\bar{\rho}}\left(-\frac{\partial}{\partial\theta}\left(\frac{1}{r}\frac{\partial\chi}{\partial r}\right) + \int_0^r S_m\, r dr\right)\right)\right)$$
$$-r\Omega\frac{1}{r}\frac{\partial}{\partial r}\left(\frac{\partial}{\partial\theta}\left(\frac{1}{\bar{\rho}}\left(\frac{\partial}{\partial r}\left(\frac{1}{r}\frac{\partial\chi}{\partial r}\right)\right)\right)\right)$$
$$-\hat{\rho}\Omega r + \frac{1}{r}\frac{\partial}{\partial r}\left(\frac{1}{r}\frac{\partial}{\partial r}\left(\frac{r}{\bar{\rho}}\left(\frac{\partial}{\partial r}\left(\frac{1}{r}\frac{\partial}{\partial r}\left(\frac{r}{\bar{\rho}}\left(\frac{\partial}{\partial r}\left(\frac{1}{r}\frac{\partial\chi}{\partial r}\right) - \hat{\rho}\Omega r\right)\right)\right)\right)\right)\right)$$
$$+\frac{1}{r}\frac{\partial}{\partial r}\left(\frac{1}{r}\frac{\partial}{\partial r}\left(\frac{1}{\bar{\rho}r}\frac{\partial^2}{\partial\theta^2}\left(\frac{1}{\bar{\rho}}\left(\frac{\partial}{\partial r}\left(\frac{1}{r}\frac{\partial\chi}{\partial r}\right) - \hat{\rho}\Omega r\right)\right)\right)\right)$$
$$+\frac{1}{r}\frac{\partial}{\partial r}\left(\frac{1}{r}\frac{\partial}{\partial r}\left(\frac{2}{\bar{\rho}^2 r^2}\frac{\partial}{\partial\theta}\left(-\frac{\partial}{\partial\theta}\left(\frac{1}{r}\frac{\partial\chi}{\partial r}\right) + \int_0^r S_m\, r dr\right)\right)\right)$$
$$-\frac{1}{r}\frac{\partial}{\partial r}\left(\frac{1}{\bar{\rho}r}\frac{\partial}{\partial\theta}\left(\frac{\partial}{\partial r}\left(\frac{1}{r}\frac{\partial}{\partial r}\left(\frac{1}{\bar{\rho}}\left(-\frac{\partial}{\partial\theta}\left(\frac{1}{r}\frac{\partial\chi}{\partial r}\right) + \int_0^r S_m\, r dr\right)\right)\right)\right)\right)$$
$$-\frac{1}{r}\frac{\partial}{\partial r}\left(\frac{1}{\bar{\rho}^2 r^4}\frac{\partial^3}{\partial\theta^3}\left(-\frac{\partial}{\partial\theta}\left(\frac{1}{r}\frac{\partial\chi}{\partial r}\right) + \int_0^r S_m\, r dr\right)\right)$$
$$+\frac{1}{r}\frac{\partial}{\partial r}\left(\frac{2}{\bar{\rho}^2 r^3}\frac{\partial^2}{\partial\theta^2}\left(\frac{\partial}{\partial r}\left(\frac{1}{r}\frac{\partial\chi}{\partial r}\right) - \hat{\rho}\Omega r\right)\right)$$
$$-\frac{1}{r}\frac{\partial}{\partial r}\left(\frac{\Omega}{\bar{\mu}}\frac{d}{d\theta}\left(\frac{2}{\bar{\rho}r}\left(\frac{\partial}{\partial r}\left(\frac{1}{r}\frac{\partial\chi}{\partial r}\right) - \hat{\rho}\Omega r\right)\right)\right)$$
$$-\frac{\Omega^4}{\bar{K}\bar{\mu}T_0 r^2}\left(\frac{\partial^2\chi}{\partial\theta^2}\right) + R_T + C_T$$
$$= \frac{\Omega^2}{\bar{K}\bar{\mu}T_0 r^2}\frac{\partial}{\partial\theta}\int_{r_0}^{r}(S_e - \bar{p}S_m)\, r\, dr$$

$$+\frac{1}{\bar{\mu}}\frac{1}{r}\frac{\partial}{\partial r}\left(\frac{1}{\bar{\rho}r}\frac{\partial}{\partial\theta}\left(f_r+\frac{\bar{\mu}}{3}\frac{\partial S_m}{\partial r}\right)\right)$$
$$-\frac{1}{\bar{\mu}}\frac{1}{r}\frac{\partial}{\partial r}\left(\frac{1}{r}\frac{\partial}{\partial r}\left(\frac{r}{\bar{\rho}}\left(f_\theta+\frac{1}{r}\frac{\bar{\mu}}{3}\frac{\partial S_m}{\partial\theta}\right)\right)\right). \quad (4.7)$$

The radial and azimuthal coordinates are expressed in terms of the scaled radial ξ, and azimuthal coordinates θ^*, defined by

$$\xi=\mathcal{A}^2\left(1-\frac{r^2}{R^2}\right), \quad (4.8)$$

$$\theta^*=\frac{\theta}{2\pi}. \quad (4.9)$$

With these transformation, it can easily be verified that

$$\frac{1}{r}\frac{\partial}{\partial r}=-\frac{2\mathcal{A}^2}{R^2}\frac{\partial}{\partial\xi}, \quad (4.10)$$

$$\frac{\partial}{\partial\theta}=\frac{1}{2\pi}\frac{\partial}{\partial\theta^*}, \quad (4.11)$$

where $\delta=(R/\sqrt{2}\mathcal{A})$ is the boundary layer thickness in the radial coordinate. When the equation for the master potential is expressed in terms of these scaled co-ordinates, and the base density profile

$$\bar{\rho}=\rho_w\exp\left(-\xi\right) \quad (4.12)$$

is used, we obtain

$$\frac{64\mathcal{A}^{12}}{R^8\rho_w^2}\frac{\partial^2}{\partial\xi^2}\left(e^\xi\left(1-\frac{\xi}{\mathcal{A}^2}\right)\frac{\partial^2}{\partial\xi^2}\left(e^\xi\left(1-\frac{\xi}{\mathcal{A}^2}\right)\frac{\partial^2\chi}{\partial\xi^2}\right)\right)$$
$$+\frac{\Omega\mathcal{A}^4}{2\pi^3\bar{\mu}\rho_w R^5}\frac{1}{\left(1-\frac{\xi}{\mathcal{A}^2}\right)^{1/2}}\frac{\partial}{\partial\xi}\left(\frac{\partial^2}{\partial\theta^{*2}}\left(\frac{e^\xi}{\left(1-\frac{\xi}{\mathcal{A}^2}\right)^{1/2}}\frac{\partial}{\partial\theta^*}\frac{\partial\chi}{\partial\xi}\right)\right)$$
$$-\frac{8\Omega\mathcal{A}^6}{\pi\bar{\mu}R^6\rho_w}\frac{\partial^2}{\partial\xi^2}\left(e^\xi\left(\frac{\partial}{\partial\theta^*}\frac{\partial\chi}{\partial\xi}\right)\right)$$
$$+\frac{4\mathcal{A}^6\Omega}{\pi\rho_w R^4}\frac{\partial}{\partial\xi}\left(1-\frac{\xi}{\mathcal{A}^2}\right)^{1/2}\frac{\partial}{\partial\theta^*}\left(e^\xi\left(1-\frac{\xi}{\mathcal{A}^2}\right)^{1/2}\frac{\partial^2\chi}{\partial\xi^2}\right)$$
$$+\frac{4\mathcal{A}^8}{\pi^2\rho_w^2R^8}\frac{\partial^2}{\partial\xi^2}\left(\frac{e^\xi}{\left(1-\frac{\xi}{\mathcal{A}^2}\right)^{1/2}}\left(\frac{\partial^2}{\partial\theta^{*2}}e^\xi\left(1-\frac{\xi}{\mathcal{A}^2}\right)^{1/2}\frac{\partial^2\chi}{\partial\xi^2}\right)\right)$$
$$+\frac{4\mathcal{A}^6}{\pi^2\rho_w^2R^8}\frac{\partial^2}{\partial\xi^2}\left(\frac{e^\xi}{\left(1-\frac{\xi}{\mathcal{A}^2}\right)}\left(\frac{\partial}{\partial\theta^*}\frac{\partial^2\chi}{\partial\theta^*\partial\xi}\right)\right)$$
$$+\frac{4\mathcal{A}^8}{\pi^2\rho_w^2R^8}\frac{\partial}{\partial\xi}\left(\frac{e^\xi}{\left(1-\frac{\xi}{\mathcal{A}^2}\right)^{1/2}}\left(\frac{\partial}{\partial\theta^*}\left(1-\frac{\xi}{\mathcal{A}^2}\right)\frac{\partial^2}{\partial\xi^2}\left(e^\xi\frac{\partial^2\chi}{\partial\theta^*\partial\xi}\right)\right)\right)$$

$$+\frac{\mathcal{A}^4}{2\pi^3\rho_w^2R^8}\frac{\partial}{\partial\xi}\left(\frac{e^{\xi}}{\left(1-\frac{\xi}{\mathcal{A}^2}\right)^2}\left(\frac{\partial^3}{\partial\theta^{*3}}\left(\frac{\partial\chi}{\partial\xi}\right)\right)\right)$$

$$+\frac{4\mathcal{A}^6}{\pi^2\rho_w^2R^8}\frac{\partial}{\partial\xi}\left(\frac{e^{2\xi}}{\left(1-\frac{\xi}{\mathcal{A}^2}\right)^{3/2}}\left(\frac{\partial^2}{\partial\theta^{*2}}\left(1-\frac{\xi}{\mathcal{A}^2}\right)^{1/2}\frac{\partial^2\chi}{\partial\xi^2}\right)\right)$$

$$+\frac{8\mathcal{A}^6\Omega}{\pi\rho_w\bar{\mu}R^6}\frac{\partial}{\partial\xi}\left(\frac{\partial}{\partial\theta^*}\left(e^{\xi}\frac{\partial^2\chi}{\partial\xi^2}\right)\right)$$

$$+\frac{\Omega^4}{2\pi\bar{K}T_0\left(1-\frac{\xi}{\mathcal{A}^2}\right)}\left(\frac{\partial^2\chi}{\partial\theta^{*2}}\right)$$

$$+R_T+C_T=S, \tag{4.13}$$

where S represents the source term in the equation for the master potential obtained from the mass, momentum and energy sources.

The above equation can be reduced to dimensionless form by dividing by the prefactor of the first term on the left, and defining a scaled master potential. The scaling for master potential (χ) is best inferred by balancing the first term on the left and the term on the right in equation 4.2 which provides $\chi^*=(\chi\,/\,\bar{\mu}\,R^2\,2\pi)$. Using this scaled master potential, we obtain

$$\frac{\partial^2}{\partial\xi^2}\left(e^{\xi}\left(1-\frac{\xi}{\mathcal{A}^2}\right)\frac{\partial^2}{\partial\xi^2}\left(e^{\xi}\left(1-\frac{\xi}{\mathcal{A}^2}\right)\frac{\partial^2\chi^*}{\partial\xi^2}\right)\right)$$

$$+\frac{1}{128\pi^3\mathcal{A}^8\mathrm{Re}}\frac{1}{\left(1-\frac{\xi}{\mathcal{A}^2}\right)^{1/2}}\frac{\partial}{\partial\xi}\left(\frac{\partial^2}{\partial\theta^{*2}}\left(\frac{e^{\xi}}{\left(1-\frac{\xi}{\mathcal{A}^2}\right)^{1/2}}\frac{\partial}{\partial\theta^*}\frac{\partial\chi^*}{\partial\xi}\right)\right)$$

$$-\frac{1}{8\pi\mathcal{A}^6\mathrm{Re}}\frac{\partial^2}{\partial\xi^2}\left(e^{\xi}\left(\frac{\partial}{\partial\theta^*}\frac{\partial\chi^*}{\partial\xi}\right)\right)$$

$$+\frac{1}{16\pi\mathcal{A}^6\mathrm{Re}}\frac{\partial}{\partial\xi}\left(1-\frac{\xi}{\mathcal{A}^2}\right)^{1/2}\frac{\partial}{\partial\theta^*}\left(e^{\xi}\left(1-\frac{\xi}{\mathcal{A}^2}\right)^{1/2}\frac{\partial^2\chi^*}{\partial\xi^2}\right)$$

$$+\frac{1}{16\pi\mathcal{A}^4\mathrm{Re}}\frac{\partial^2}{\partial\xi^2}\left(\frac{e^{\xi}}{\left(1-\frac{\xi}{\mathcal{A}^2}\right)^{1/2}}\left(\frac{\partial^2}{\partial\theta^{*2}}e^{\xi}\left(1-\frac{\xi}{\mathcal{A}^2}\right)^{1/2}\frac{\partial^2\chi^*}{\partial\xi^2}\right)\right)$$

$$+\frac{1}{16\pi^2\mathcal{A}^6\mathrm{Re}}\frac{\partial^2}{\partial\xi^2}\left(\frac{e^{\xi}}{\left(1-\frac{\xi}{\mathcal{A}^2}\right)}\left(\frac{\partial}{\partial\theta^*}\frac{\partial^2\chi^*}{\partial\theta^*\partial\xi}\right)\right)$$

$$+\frac{1}{16\pi^2\mathcal{A}^4\mathrm{Re}}\frac{\partial}{\partial\xi}\left(\frac{e^{\xi}}{\left(1-\frac{\xi}{\mathcal{A}^2}\right)^{1/2}}\left(\frac{\partial}{\partial\theta^*}\left(1-\frac{\xi}{\mathcal{A}^2}\right)\frac{\partial^2}{\partial\xi^2}\left(e^{\xi}\frac{\partial^2\chi^*}{\partial\theta^*\partial\xi}\right)\right)\right)$$

$$+\frac{1}{128\pi^3\mathcal{A}^8\mathrm{Re}}\frac{\partial}{\partial\xi}\left(\frac{e^{\xi}}{\left(1-\frac{\xi}{\mathcal{A}^2}\right)^2}\left(\frac{\partial^3}{\partial\theta^{*3}}\left(\frac{\partial\chi^*}{\partial\xi}\right)\right)\right)$$

$$+\frac{1}{16\pi^2\mathcal{A}^6\mathrm{Re}}\frac{\partial}{\partial\xi}\left(\frac{e^{2\xi}}{\left(1-\frac{\xi}{\mathcal{A}^2}\right)^{3/2}}\left(\frac{\partial^2}{\partial\theta^{*2}}\left(1-\frac{\xi}{\mathcal{A}^2}\right)^{1/2}\frac{\partial^2\chi^*}{\partial\xi^2}\right)\right)$$
$$+\frac{1}{8\pi\mathcal{A}^6\mathrm{Re}}\frac{\partial}{\partial\xi}\left(\frac{\partial}{\partial\theta^*}\left(e^{\xi}\frac{\partial^2\chi^*}{\partial\xi^2}\right)\right)$$
$$+\frac{2\mathcal{A}^2\mathrm{Pr}\left(1-\gamma^{-1}\right)}{\left(1-\frac{\xi}{\mathcal{A}^2}\right)}\frac{\mathrm{Re}^2}{64\left(2\pi\right)^2\mathcal{A}^{12}}\left(\frac{\partial^2\chi^*}{\partial\theta^{*2}}\right)$$
$$+R^*_{\hat{T}}+C^*_{\hat{T}}=S^*, \quad (4.14)$$

where the Reynolds number, stratification parameter and the Prandtl number are given in equations 2.8, 2.7 and 2.12 respectively, and

$$R^*_{\hat{T}}=-\left(\frac{1}{2\pi}\right)^4\frac{1}{\mathcal{A}^{14}}\left(\left(1-\xi/\mathcal{A}^2\right)^{-1}\int_{\xi_0}^{\xi}\frac{d\xi}{\left(1-\xi/\mathcal{A}^2\right)}\left(\frac{\partial^3\hat{T}^*}{\partial\theta^{*3}}\right)\right)-$$
$$\left(\frac{1}{8\pi^2}\right)\frac{1}{\mathcal{A}^{14}}\left(\left(1-\xi/\mathcal{A}^2\right)^{-1}\int_{\xi_0}^{\xi}e^{-\xi}\,d\xi\left(\frac{\partial^2\hat{T}^*}{\partial\theta^{*2}}\right)\right), \quad (4.15)$$

$$C^*_{\hat{T}}=-\left(\frac{1}{2\pi}\right)\frac{1}{\mathcal{A}^{10}}\left(\left(1-\xi/\mathcal{A}^2\right)^{-1}\frac{\partial}{\partial\theta^*}\left(\left(1-\xi/\mathcal{A}^2\right)\left.\frac{\partial\hat{T}^*}{\partial\xi}\right|_{\xi=\xi_0}\right)\right)+$$
$$\left(\frac{1}{2\pi}\right)\frac{1}{\mathcal{A}^{12}}\left(\left(1-\xi/\mathcal{A}^2\right)^{-1}\left.\frac{\partial^2\chi^*}{\partial\theta^{*2}}\right|_{\xi=\xi_0}\right), \quad (4.16)$$

where $\hat{v}^*_\theta=\hat{v}_\theta/\left(\Omega R\right)$ and $\hat{T}^*=\hat{T}/T_0$ and S^* is the scaled source term, where

$$S^*(\xi,\theta^*)=\frac{\Omega}{\bar{\mu}R\rho_w}\frac{1}{4\pi^2}\left(\frac{1}{\left(1-\frac{\xi}{\mathcal{A}^2}\right)^{\frac{1}{2}}}\frac{\partial}{\partial\xi}\left(\frac{\partial^2}{\partial\theta^{*2}}\left(e^{\xi}\int_{\xi_0}^{\xi}S_m d\xi^*\right)\right)\right.$$
$$+\frac{4\mathcal{A}^2\Omega}{\bar{\mu}R^2\rho_w}\left(\frac{\partial^2}{\partial\xi^2}\left(e^{\xi}\int_{\xi_0}^{\xi}S_m d\xi^*\right)\right)$$
$$-\frac{4\mathcal{A}^2}{R^4\rho_w^2}\frac{1}{2\pi}\left(\frac{\partial^2}{\partial\xi^2}\left(\frac{e^{2\xi}}{\left(1-\frac{\xi}{\mathcal{A}^2}\right)}\int_{\xi_0}^{\xi}\frac{\partial S_m\left(\xi^{**},\theta^*\right)}{\partial\theta^*}d\xi^*\right)\right)$$
$$+\frac{4\mathcal{A}^4}{R^6\rho_w^2}\frac{1}{2\pi}\left(\frac{e^{\xi}}{\left(1-\frac{\xi}{\mathcal{A}^2}\right)^{\frac{1}{2}}}\left(\frac{\partial}{\partial\theta^*}\left(1-\frac{\xi}{\mathcal{A}^2}\right)^{\frac{1}{2}}\frac{\partial^2}{\partial\xi^2}\left(e^{\xi}\int_{\xi_0}^{\xi}S_m d\xi^*\right)\right)\right)$$
$$\left.-\frac{1}{R^4\rho_w^2}\frac{1}{8\pi^3}\left(\frac{\partial}{\partial\xi}\left(\frac{e^{2\xi}}{\left(1-\frac{\xi}{\mathcal{A}^2}\right)^2}\left(\frac{\partial^3}{\partial\theta^{*3}}\int_{\xi_0}^{\xi}S_m d\xi^*\right)\right)\right)\right)$$

$$
\begin{aligned}
&-\left(\frac{2\mathrm{Re}^2\mathcal{A}^2\mathrm{Pr}(1-\gamma^{-1})}{(1-\xi/\mathcal{A}^2)}\right)\frac{1}{256\pi^2\mathcal{A}^{16}}\frac{R\Omega}{\rho_w}\times \\
&\int_{\xi_0}^{\xi}\left[\int_{\xi_0}^{\xi^*}\frac{\partial \mathcal{S}_m(\xi^{**},\theta^*)}{\partial\theta^*}d\xi^{**}\right]d\xi^* \\
&+\frac{\Omega^2}{2\mathcal{A}^2\bar{K}\bar{\mu}T_0}\frac{1}{2\pi}\frac{1}{\left(1-\frac{\xi}{\mathcal{A}^2}\right)}\int_{\xi_0}^{\xi}\frac{\partial\left(S_e-p_we^{-\xi}S_m\right)}{\partial\theta^*}d\xi^* \\
&-\frac{2\mathcal{A}^2}{\bar{\mu}R^3\rho_w}\frac{1}{2\pi}\left(\frac{\partial}{\partial\xi}\left(\frac{e^{\xi}}{\left(1-\frac{\xi}{\mathcal{A}^2}\right)^{\frac{1}{2}}}\frac{\partial}{\partial\theta^*}\left(f_r-\frac{2\bar{\mu}}{3}\frac{\mathcal{A}^2}{R}\left(1-\frac{\xi}{\mathcal{A}^2}\right)^{\frac{1}{2}}\frac{\partial S_m}{\partial\xi}\right)\right)\right) \\
&-\frac{4\mathcal{A}^4}{\bar{\mu}R^3\rho_w}\left(\frac{\partial^2}{\partial\xi^2}\left(\left(1-\frac{\xi}{\mathcal{A}^2}\right)^{\frac{1}{2}}e^{\xi}\left(f_\theta+\frac{1}{R\left(1-\frac{\xi}{\mathcal{A}^2}\right)^{\frac{1}{2}}}\frac{\bar{\mu}}{3}\frac{1}{2\pi}\frac{\partial S_m}{\partial\theta^*}\right)\right)\right). \qquad (4.17)
\end{aligned}
$$

The equation 4.17 provides the conditions under which the source strengths are significant, that is, for $\mathcal{S}_m \sim (R^4\rho_w^2/\mathcal{A}^2)$, $f_r \sim (\bar{\mu}R^3\rho_w/\mathcal{A}^2)$, $f_\theta \sim (\bar{\mu}R^3\rho_w/\mathcal{A}^4)$ and $S_e \sim (\bar{K}T_0\bar{\mu}\mathcal{A}^2/\Omega^2)$.

The following terms in equation 4.14 are neglected to formulate the 'analytical equation' in the polar $(r-\theta)$ plane.

(a) The second term on the left side of equation 4.13 is neglected, because the parameter $\left(1/\left(128\pi^3\mathcal{A}^8\mathrm{Re}\right)\right) \ll 1$. Clearly, this is valid when the Reynolds number (Re) and stratification parameter ($\mathcal{A}$) is large.

(b) The third term on the left side of equation 4.13 is neglected, because the parameter $\left(1/\left(8\pi\mathcal{A}^6\mathrm{Re}\right)\right) \ll 1$. Clearly, this is valid when the Reynolds number (Re) and stratification parameter ($\mathcal{A}$) is large.

(c) The fourth term on the left side of equation 4.13 is neglected, because the parameter $\left(1/\left(16\pi\mathcal{A}^6\mathrm{Re}\right)\right) \ll 1$. Clearly, this is valid when the Reynolds number (Re) and stratification parameter ($\mathcal{A}$) is large.

(d) The fifth term on the left side of equation 4.13 is neglected, because the parameter $\left(1/\left(16\pi\mathcal{A}^4\mathrm{Re}\right)\right) \ll 1$. Clearly, this is valid when the Reynolds number (Re) and stratification parameter ($\mathcal{A}$) is large.

(e) The sixth term on the left side of equation 4.13 is neglected, because the parameter $\left(1/\left(16\pi^2\mathcal{A}^6\mathrm{Re}\right)\right) \ll 1$. Clearly, this is valid when the Reynolds number (Re) and stratification parameter ($\mathcal{A}$) is large. ,

(f) The seventh term on the left side of equation 4.13 is neglected, because the parameter $\left(1/\left(16\pi^2\mathcal{A}^4\mathrm{Re}\right)\right) \ll 1$. Clearly, this is valid when the Reynolds number (Re) and stratification parameter ($\mathcal{A}$) is large.

(g) The eighth term on the left side of equation 4.13 is neglected, because the parameter $\left(1/\left(128\pi^3\mathcal{A}^8\mathrm{Re}\right)\right) \ll 1$. Clearly, this is valid when the Reynolds number (Re) and stratification parameter ($\mathcal{A}$) is large.

(h) The ninth term on the left side of equation 4.13 is neglected, because the parameter $\left(1/\left(16\pi^2\mathcal{A}^6\mathrm{Re}\right)\right) \ll 1$. Clearly, this is valid when the Reynolds number (Re) and stratification parameter ($\mathcal{A}$) is large.

(i) The tenth term on the left side of equation 4.13 is neglected, because the parameter $\left(1/\left(8\pi\mathcal{A}^6\mathrm{Re}\right)\right) \ll 1$. Clearly, this is valid when the Reynolds number (Re) and stratification parameter ($\mathcal{A}$) is large.

(j) The terms R_T^* and C_T^* on the left side of equation 4.13 is also neglected. The value of these terms depends on the location of the boundary surface $\xi = \xi_0$. In the present case, this boundary is at the center of the cylinder. Symmetry requires that the $(\partial T/\partial \theta) = 0$, and $(\partial T/\partial r) = 0$. Therefore, R_T^* and C_T^* are identically zero in this case.

(k) Within the first term on the left of equation 4.13, a term proportional to $(\xi/\mathcal{A}^2)$ can be neglected by assuming that $(1-\xi/\mathcal{A}^2) \approx 1$. Clearly, this is valid when the stratification parameter $\mathcal{A} \gg 1$. Similarly, within the eleventh term on the left, a term proportional to $(\xi/\mathcal{A}^2)$ can be neglected by assuming that $(1-\xi/\mathcal{A}^2) \approx 1$. Clearly, this is valid when the stratification parameter $\mathcal{A} \gg 1$.

Therefore, in the limit of $\mathrm{Re} \gg 1$ and $\mathcal{A} \gg 1$, the equations 4.14 can be reduced to

$$\frac{\partial^2}{\partial \xi^2}\left(e^{\xi}\frac{\partial^2}{\partial \xi^2}\left(e^{\xi}\frac{\partial^2 \chi^*}{\partial \xi^2}\right)\right) + \left(2\mathcal{A}^2 \mathrm{Pr}\left(1-\gamma^{-1}\right)\right)\frac{\mathrm{Re}^2}{64\left(2\pi\right)^2 \mathcal{A}^{12}}\frac{\partial^2 \chi^*}{\partial \theta^{*2}} = S^*. \tag{4.18}$$

We refer the equation (4.18) as the 'analytical equation' in the polar $(r-\theta)$ plane. Clearly, this is valid when the stratification parameter $\mathcal{A} \gg 1$ and Reynolds number $\mathrm{Re} \gg 1$.

The limitation $\mathcal{A} \gg 1$ imposed by condition 11 above can be relaxed if we formulate a 'generalized analytical equation' in the polar $(r-\theta)$ plane,

$$L_6^{\mathcal{A}}(\chi^*) + \frac{\mathrm{Re}^2}{64\left(2\pi\right)^2 \mathcal{A}^{12}}\frac{\partial^2 \chi^*}{\partial \theta^{*2}} = S^*, \tag{4.19}$$

where the linear differential operator $L_6^{\mathcal{A}}$ in the ξ co-ordinate is defined by

$$L_6^{\mathcal{A}} = \left(\frac{2\mathcal{A}^2\mathrm{Pr}(1-\gamma^{-1})}{(1-\xi/\mathcal{A}^2)}\right)^{-1} \times \left(\frac{\partial^2}{\partial \xi^2}\left(e^{\xi}\left(1-\frac{\xi}{\mathcal{A}^2}\right)\frac{\partial^2}{\partial \xi^2}\left(e^{\xi}\left(1-\frac{\xi}{\mathcal{A}^2}\right)\frac{\partial^2}{\partial \xi^2}\right)\right)\right). \tag{4.20}$$

The operator has been written with the superscript $^{\mathcal{A}}$ to indicate that there is an explicit dependence on the stratification parameter. This operator also depends on the Prandtl number Pr and the ratio of specific heats, but these are gas properties and not flow properties, and are both $O(1)$ for an ideal gas. Therefore, the the only flow dependence in the operator $L_6^{\mathcal{A}}$ enters through the stratification parameter $(\mathcal{A})$, while the Prandtl number and ideal gas constant are determined by the nature of the gas. Here, we will assume an ideal gas, for which $\mathrm{Pr} = 0.95$ and $\gamma = 1.066$.

It is easily verified that this operator is also self-adjoint, when the inner product is defined by

$$(A_1, B_1) = \int_0^{\xi_0} d\xi w(\xi) A_1(\xi) B_1(\xi), \tag{4.21}$$

where the weighting function $w(\xi)$ is given by

$$w(\xi) = \left(\frac{2\mathcal{A}^2\mathrm{Pr}\left(1-\gamma^{-1}\right)}{(1-\xi/\mathcal{A}^2)}\right), \tag{4.22}$$

and the boundary conditions are appropriately specified, as discussed next. Therefore, the eigenvalues of this operator are real, and the eigenfunctions are orthogonal. We refer to equation 4.19 as the 'generalized analytical equation' in the polar $(r-\theta)$ plane, where the scaled source term (S^*) was defined in equation 4.17.

Since the equation 4.19 is a sixth order differential equation, it is necessary to specify three boundary conditions at each boundary. Here we will investigate the secondary radial flow due to mass/momentum/energy sources within the domain, with homogeneous boundary conditions (constant temperature). As shown below, the solution for the master potential with internal sources can be obtained by the method of separation of variables, since the differential operators in both the axial and meridional directions are self-adjoint. The following expressions for the velocity components are useful for specifying the boundary conditions,

$$\hat{v}_r = \frac{2R\Omega\mathcal{A}^2}{\mathrm{Re}}\left(1-\frac{\xi}{\mathcal{A}^2}\right)^{-1/2} e^{\xi}\frac{\partial^2\chi^*}{\partial\xi\partial\theta^*}, \tag{4.23}$$

$$\hat{v}_\theta = \frac{8\pi R\Omega\mathcal{A}^4}{\mathrm{Re}}\left(1-\frac{\xi}{\mathcal{A}^2}\right)^{1/2} e^{\xi}\frac{\partial^2\chi^*}{\partial\xi^2}. \tag{4.24}$$

The homogeneous boundary conditions for 4.19 are as follows. At the wall $((r = R); (r^* = r/R = 1); (\xi = 0))$, the boundary conditions are the no-penetration condition in the radial direction, the no-slip in the meridional direction and the zero temperature perturbation condition. The homogeneous radial and meridional velocity boundary conditions, $\hat{v}_r = 0$ and $\hat{v}_\theta = 0$ at $((r = R); (r/R = 1); (\xi = 0))$, reduce to,

$$\frac{\partial\chi^*}{\partial\xi} = 0, \tag{4.25}$$

$$\frac{\partial^2\chi^*}{\partial\xi^2} = 0, \tag{4.26}$$

while a combination of the zero meridional velocity and temperature conditions is formulated from equation 3.20 for the master potential,

$$\frac{1}{r}\frac{\partial}{\partial r}\left(\frac{r}{\bar{\rho}}\left(\frac{\partial}{\partial r}\left(\frac{1}{r}\frac{\partial}{\partial r}\left(\frac{r}{\bar{\rho}}\left(\frac{\partial}{\partial r}\left(\frac{1}{r}\frac{\partial\chi}{\partial r}\right)\right)\right)\right)\right)\right) = 0, \tag{4.27}$$

which is expressed in terms of scaled coordinates as

$$\frac{\partial}{\partial\xi}\left(e^{\xi}\left(1-\frac{\xi}{\mathcal{A}^2}\right)\frac{\partial^2}{\partial\xi^2}\left(e^{\xi}\left(1-\frac{\xi}{\mathcal{A}^2}\right)\frac{\partial^2\chi^*}{\partial\xi^2}\right)\right) = 0 \tag{4.28}$$

or

$$L_5^{\mathcal{A}}(\chi^*) = 0, \tag{4.29}$$

where the operator $L_5^{\mathcal{A}}$ is

$$L_5^{\mathcal{A}} = \frac{\partial}{\partial\xi}\left(e^{\xi}\left(1-\frac{\xi}{\mathcal{A}^2}\right)\frac{\partial^2}{\partial\xi^2}\left(e^{\xi}\left(1-\frac{\xi}{\mathcal{A}^2}\right)\frac{\partial^2}{\partial\xi^2}\right)\right). \tag{4.30}$$

At the center of the cylinder, $(\xi = \xi_0 = \mathcal{A}^2)$ (which corresponds to $((r = 0); (r^* = r/R = 0))$), the symmetry conditions are zero radial velocity $\hat{v}_r = 0$, zero azimuthal shear stress $\tau_{r\theta} = 0$. The zero radial velocity boundary condition at the centre requires that $(\partial\psi/\partial\theta) = 0$, or

$$\frac{\partial\chi^*}{\partial\xi} = 0. \tag{4.31}$$

Symmetry requires that the azimuthal shear stress $\tau_{r\theta}$ has to be zero at the centre of the cylinder. Therefore,

$$r\frac{\partial}{\partial r}\left(\frac{\hat{v}_\theta}{r}\right) = 0, \tag{4.32}$$

or

$$r\frac{\partial}{\partial r}\left(\frac{1}{\bar{\rho}r}\frac{\partial\psi}{\partial r}\right) = 0, \tag{4.33}$$

which is expressed in terms of scaled coordinates as

$$\frac{\partial}{\partial\xi}\left(e^{\xi}\left(1-\frac{\xi}{A^2}\right)\frac{\partial^2\chi^*}{\partial\xi^2}\right) = 0, \tag{4.34}$$

or

$$L_3^A\,(\chi^*) = 0, \tag{4.35}$$

where

$$L_3^A = \frac{\partial}{\partial\xi}\left(e^{\xi}\left(1-\frac{\xi}{A^2}\right)\frac{\partial^2}{\partial\xi^2}\right). \tag{4.36}$$

In addition, since the velocity and temperature perturbations are related to the derivatives of the master potential, the master potential itself can be set equal to zero, without loss of generality, at the axis of the cylinder. We choose to set the master potential equal to zero at the axis. Then, from the boundary condition 4.31, the master potential is equal to zero at the axis $((r=0);\ (r^* = r/R = 0),\ (\xi = \xi_0))$, that is,

$$\chi^* = 0. \tag{4.37}$$

In the θ direction, the boundary conditions are as follows,

$$\chi^*(\theta^*=0) = \chi^*(\theta^*=1), \tag{4.38}$$

To obtain the solution, we use the separation-of-variables procedure. First, we define the eigenvalues and eigenfunctions of the operators L_6^A and $(\partial^2/\partial\theta^{*2})$. For the L_6^A operator, the eigen-relation can be written as,

$$L_6^A F_n(\xi) = \alpha_n F_n(\xi), \tag{4.39}$$

where α_n are the discrete eigenvalues. The adjoint of the linear operator L_6^A, with boundary conditions equations 4.25 to 4.37, can be calculated as,

$$\begin{aligned}
&\int_0^{\xi_0} d\xi A_1(\xi)L_6^A(B_1(\xi))\\
&= \Bigg(A_1L_5B_1 - (d_\xi A_1)L_4B_1 + (d_\xi^2A_1)\left(e^{\xi}\left(1-\frac{\xi}{A^2}\right)\right)L_3B_1\\
&\quad -(d_\xi^2B_1)\left(e^{\xi}\left(1-\frac{\xi}{A^2}\right)\right)L_3A_1 + (d_\xi A_1)L_4B_1 - B_1L_5A_1\Bigg)\Bigg|_{\xi=0}^{\xi=\xi_0}\\
&\quad + \int_0^{\xi_0} d\xi B_1(\xi)L_6^A(A_1(\xi))
\end{aligned} \tag{4.40}$$

where

$$L_5^A A_1 = \frac{\partial}{\partial\xi}\left(e^{\xi}\left(1-\frac{\xi}{A^2}\right)\frac{\partial^2}{\partial\xi^2}\left(e^{\xi}\left(1-\frac{\xi}{A^2}\right)\frac{\partial^2A_1}{\partial\xi^2}\right)\right), \tag{4.41}$$

$$L_4^A A_1 = e^{\xi}\left(1-\frac{\xi}{A^2}\right)\frac{\partial^2}{\partial\xi^2}\left(e^{\xi}\left(1-\frac{\xi}{A^2}\right)\frac{\partial^2A_1}{\partial\xi^2}\right), \tag{4.42}$$

Eigenvalue	$\mathcal{A} = 0.707$	$\mathcal{A} = 2.121$	$\mathcal{A} = 3.535$
α_1	1.163426781	1.018965341	0.988653417
α_2	3.894532109	2.101305067	1.990673421
α_3	5.384321785	3.102328976	2.979863209
α_4	6.550432198	3.757192418	3.590768210
α_5	7.522923167	4.314588795	4.100844231
α_6	8.408542189	4.748457210	4.539599803
α_7	9.269542087	5.106905623	4.978805681
α_8	9.980945328	5.411974328	5.267641095
α_9	10.521989056	5.608567231	5.428856018
α_{10}	10.798215649	5.700896743	5.560698274

TABLE 1. First 10 eigenvalues of $L_6^{\mathcal{A}}$ operator at stratification parameter = 0.707, 2.121, 3.535.

$$L_3^{\mathcal{A}} A_1 = \frac{\partial}{\partial \xi} \left(e^{\xi} \left(1 - \frac{\xi}{\mathcal{A}^2} \right) \frac{\partial^2 A_1}{\partial \xi^2} \right). \tag{4.43}$$

It is clear, from equation 4.40, that the operator $L_6^{\mathcal{A}}$, with boundary conditions (equations 4.25 to 4.37) is a self-adjoint operator, since all the terms on the right side obtained by partial integration are zero. Therefore, the eigenvalues α_n in equation 4.39 are real, and the eigenfunctions $F_n(\xi)$ are orthogonal. We define the eigen-functions so that they are orthonormal:

$$\int_0^{\xi_0} d\xi F_n(\xi) F_m(\xi) = \delta_{mn}. \tag{4.44}$$

The eigenvalues and eigenfunctions are determined numerically using fourth-order Runge-Kutta integration and the shooting technique. The first ten eigenvalues (α_n) of the operator $L_6^{\mathcal{A}}$ at $\mathcal{A} = 0.707$, 2.121, and 3.535 are given in table 1.

The eigen equation for the θ direction is,

$$\frac{\partial^2 G_n(\theta^*)}{\partial \theta^{*2}} = \beta_n G_n(\theta^*), \tag{4.45}$$

with boundary condition 4.38. For this case, the eigen-values are given by $\beta_n = (n\pi)^2$, and the orthonormal eigen-functions are,

$$G_n(\theta^*) = \sqrt{2} \sin(n\pi\theta^*). \tag{4.46}$$

The master potential can be expressed in terms of the eigenfunctions of the $L_6^{\mathcal{A}}$ and the $(\partial^2/\partial\theta^{*2})$ operators,

$$\chi^* = \sum_{n=1}^{\infty} \sum_{m=1}^{\infty} A_{nm} F_n(\xi) G_m(\theta^*). \tag{4.47}$$

The above expressions are substituted into the differential equation A 4, and the orthogonality condition is used, to obtain the coefficients A_{mn},

$$A_{nm} = \frac{1}{\alpha_n + B^* \beta_m} \int_0^{\xi_0} d\xi \int_0^1 d\theta^* S^*(\xi, \theta^*) F_n(\xi) G_m(\theta^*). \tag{4.48}$$

In the case of delta function sources,

$$S^*(\xi, \theta^*) = \sum_{p=1}^{P} S_p^* \delta(\xi - \xi_p, \theta^* - \theta_p^*), \tag{4.49}$$

the coefficients A_{nm} reduce to,

$$A_{nm} = \frac{1}{\alpha_n + B^* \beta_m} \sum_{p=1}^{P} S_p^* F_n(\xi_p) G_m(\theta_p^*). \tag{4.50}$$

Here, P is the total number of sources. In the case of a the complicated source expression 4.17, which involves derivatives and integrals, if the mass, momentum and energy sources are of the form,

$$\begin{pmatrix} \mathcal{S}_m \\ f_r \\ f_\theta \\ f_z \\ \mathcal{S}_e \end{pmatrix} = \sum_{p=1}^{P} \begin{pmatrix} \mathcal{S}_{mp} \\ f_{rp} \\ f_{\theta p} \\ f_{zp} \\ \mathcal{S}_{ep} \end{pmatrix} \delta(\xi - \xi_p)\delta(\theta^* - \theta_p^*), \tag{4.51}$$

the expression for the coefficients can be determined using integration by parts.

5. DSMC simulation

The direct simulation Monte Carlo method (DSMC) is a numerical method for modeling rarefied gas flows. Here, the Boltzmann equation for the probability distribution function for gas molecules is solved using a probabilistic (Monte Carlo) simulation technique. The Boltzmann equation is more fundamental than the Navier-Stokes equations, since it describes the evolution of the velocity distribution function of the gas molecules. The Navier-Stokes equations are obtained by taking the first, second and third moments of the Boltzmann equation with respect to the molecular velocities. While taking moments, the assumption of 'local equilibrium' (velocity distribution function close to the Maxwell-Boltzmann distribution) is made, thereby restricting the Navier-Stokes equations to near-equilibrium flows. Since there is no assumption of local equilibrium in the Boltzmann equation, it is valid even for far-from equilibrium flows, provided the molecular chaos approximation is satisfied.

The DSMC method uses 'simulated molecules' which represent a large number of real gas molecules. A basic assumption is that the inter-molecular collisions and molecular motion between collisions can be decoupled over time scales comparable to the time between collisions. The translation of molecules between collisions is deterministically modeled, while the inter-molecular collisions and molecular-surfaces are incorporated using probabilistic models. The domain is divided into a large number of cells, and properties are evaluated in these cells. The cells are further divided into 'sub-cells' for collision localization.

In the DSMC simulations, inter-molecular collisions are executed at random from all molecules within 'sub-cells' within the simulation cell, whose linear dimension is much smaller than the mean free path. The probability of collisions between a pair of molecules within a sub-cell is selected equal to that for a dilute gas with the same temperature as that in the cell. In this manner, the temperature (mean square velocity), mean free path and collision frequency for the simulated molecules are equal to those for the real molecules, provided the sub-cell dimension is much smaller than the mean free path.

The most common molecular model is the hard-sphere model, but other models such as the variable hard sphere and variable soft sphere models are also used in order to accurately reproduce the density and temperature dependence of the viscosity. The hydrodynamic fields (density, velocity, temperature) are obtained by averaging over the molecular distribution functions in the DSMC simulations.

At the boundaries, it is necessary to prescribe interaction laws between the molecules and the solid surfaces. Accurate specification of the interaction laws requires a detailed knowledge of the molecule-surface interaction. However, there are standard interaction laws based on the surface velocity and temperature which are sufficient for predicting the hydrodynamic fields accurately. The most commonly used is the 'diffuse reflection' conditions, where a molecule incident on the surface gets reflected with a Gaussian velocity distribution with the same mean velocity and temperature as that of the surface. Another commonly used condition is the specular reflection condition, that the velocity normal to the surface is reversed, while that tangential to the surface is unchanged. It is important to note that these boundary conditions are different from the no-slip conditions in the continuum models. In high Knudsen number flows, the diffuse and specular reflection conditions do predict a slip velocity at the surface. Therefore, these simulations could be used even in situations where there is a slip at the surface.

From kinetic theory of dilute gases Chapman & Cowling (1970), the viscosity of a hard-sphere gas in the dilute limit is given by,

$$\mu = \frac{5\sqrt{mk_B T}}{16\sqrt{\pi}d^2}, \tag{5.1}$$

where k_B is the Boltzmann constant, while the speed of sound c_s is given by the relation

$$mnc_s^2 = \gamma p = \gamma n k_B T, \tag{5.2}$$

where m is the molecular mass, n is the number density, and p is the pressure. From this the stratification parameter $\mathcal{A}$, Reynolds number Re, cylinder radius R, molecular diameter d and volume fraction of the molecules (ϕ) are related by the following equation:

$$\mathrm{Re} = \frac{96}{5} \times \frac{\sqrt{2}}{\sqrt{\pi}} \mathcal{A} \phi \frac{R}{d}, \tag{5.3}$$

where the volume fraction of the molecules is given by

$$\phi = \frac{\pi}{6} d^3 n, \tag{5.4}$$

where n is the number density of the molecules. The important point to note, from 5.3, is that the Reynolds number is proportional to the Mach number times $\phi(R/d)$, where R is the macroscopic scale. Therefore if we are keeping the volume fraction (density) a constant, increasing (decreasing) the Reynolds number at constant Mach number corresponds to increasing (decreasing) the system size R in relation to the molecular diameter.

The details of the DSMC simulation procedure are discussed in Bird (1963, 1981, 1987, 1994), and so we do not repeat the discussion here, but restrict attention to the parameters in the simulations. The number of cells along the radial direction is taken as 400 with clustering towards the rotating sidewall as well as near the stationary insert, while 3200 cells are taken along the azimuthal direction with clustering near the stationary insert. The cell width is about one-third of the local mean free path. The number of subcells per cell is taken as 4 for both radial and azimuthal directions. The number of simulated particles per cell is taken as 100. In our simulations, the number of real molecules per simulated particles varied between 10^{16} and 10^{18} so that the statistical fluctuations not becomes large.

The fluid is assumed to be an ideal gas with the molecular mass $58.442636 \times 10^{-26}$ kg/molecule, molecular diameter $= 5.5 \times 10^{-10}$ m, viscosity index 0.933, specific heat ratio $\gamma = 1.066$, and Prandtl number Pr= 0.95. It is assumed that the gas molecules undergo diffuse reflection from all the solid surfaces with the thermal accommodation coefficient equal to 1. The time step in the simulation is chosen to be small compared to the time between collisions. The mean collision time is calculated as, $t = (\lambda / c_{av})$, where λ is the mean free path and c_{av} is the average molecular velocity. The minimum value of mean free path at the rotating sidewall, calculated from the gas density at the wall, is $\lambda = 2.316742 \times 10^{-7}$ m. The average molecular velocity is calculated as $\sqrt{k_B T/m} =$ 84.17717 m/sec. This gives the mean collision time (t) $= 2.752221 \times 10^{-9}$ sec. The time step used in the simulations is 5.0×10^{-10} sec, which is less than mean collision time.

Two-dimensional polar $(r - \theta)$ configuration is considered in the present simulations. The flow field is studied by placing the stationary insert within the flow domain which causes the disturbance to the base flow (solid body rotation). Due to the presence of stationary insert in the high speed rotating field, there is a shock wave formation and subsequently generation of secondary radial flow. The effect of a stationary insert in the flow domain is of three types. The first is the mass source or sink due to the addition or removal of gas in a particular region in the flow domain. The second effect is due to the force exerted on the gas by a stationary inserts, which is modelled locally as a momentum source or sink. The time rate of addition/removal of momentum is equal to the applied force. The third effect is a localized energy source, which increases or decreases the temperature of the gas. All the three effects are incorporated in the molecular simulation as follows. The addition or removal of mass is accomplished by either adding or removing molecules at specified locations in the simulation cell at specified rates. The pressure is specified at inflow and outflow boundaries by connecting a pressure bath, which is a large volume of gas maintained at a constant pressure. Forces are exerted by imposing impulses on the molecules at specified locations, in such a manner that the rate of change of impulse is equal to the applied force. Energy is added by increasing the mean square velocities of the molecules at locations where heating takes place.

In the present analysis, we have varied the stratification parameter in the range (0.707–3.535), Reynolds number in the range $(10^2 - 10^6)$, dimensionless scaled wall gap $d^* = (R - r)/R$ in the range $(0.01 - 0.15)$. We place the sources/sinks of mass/momentum/energy at two locations, one outside the radial boundary layer at $((r^* = 0.86); (\xi_s^* = 0.25))$, and the second inside the radial boundary layer at $((r^* = 0.99); (\xi_s^* = 0.01))$, where $\xi^* = \xi/\mathcal{A}^2$. The computational domain and coordinate system is shown in figure 1.

6. Results (single-component gas)

First, we consider the single-component gas with the molecular mass $58.442636 \times 10^{-26}$ kg/molecule, molecular diameter $= 5.5 \times 10^{-10}$ m, viscosity index 0.933, specific heat ratio $\gamma = 1.066$, and Prandtl number Pr= 0.95. It is assumed that the gas molecules undergo diffuse reflection from all the solid surfaces with the thermal accommodation coefficient equal to 1.

Before proceeding to simulate the flow field in the two-dimensional polar $(r - \theta)$ coordinate by placing a stationary intake tube within the flow domain, we first study the simplified flow field by placing a small circular source domain of radius 10^{-3} times the cylinder radius R for adding/removing mass, momentum and energy. In the case of mass source, a molecule is added at random from the source domain at fixed time intervals determined by the desired rate of mass input. The velocity of the added molecules is chosen from a random Gaussian distribution with the same mean velocity and temper-

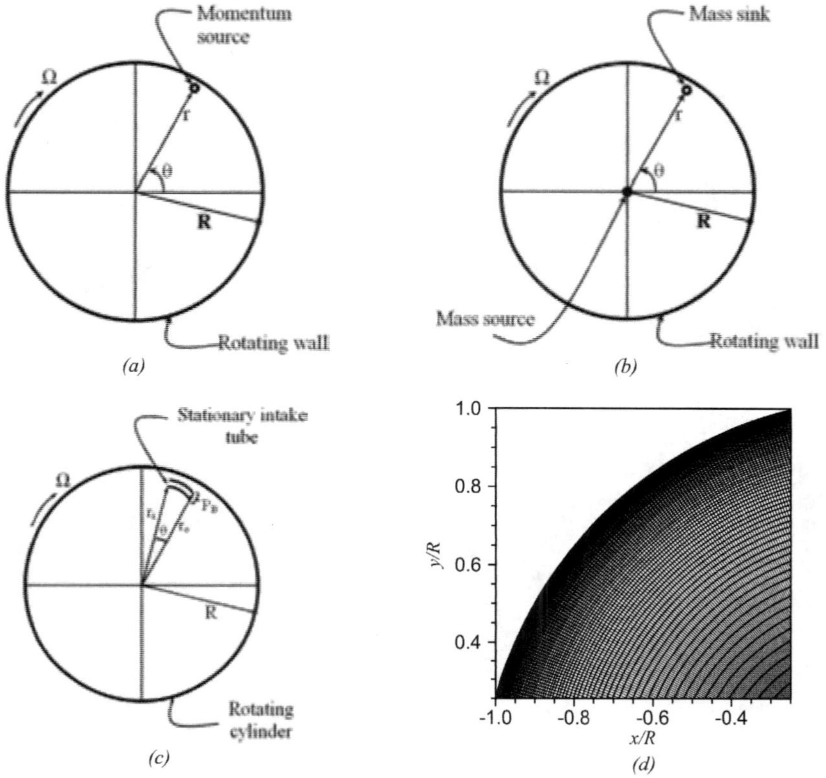

FIGURE 1. Computational domain and coordinate system in the two-dimensional polar $(r-\theta)$ coordinate for momentum source (a); mass source/sink (b); stationary intake tube (c); and the cell arrangement in the flow domain (sectional zoomed view) (d). In sub-figure (a), the momentum source is placed in two different radial locations, one outside the radial boundary layer $((r^* = 0.86)$; $(\xi_s^* = 0.25))$, and other inside the radial boundary layer $((r^* = 0.99)$; $(\xi_s^* = 0.01))$. In sub-figur (b), the mass source is placed at the centre $((r^* = 0)$, and mass sink is placed in two different radial locations, one outside the radial boundary layer $((r^* = 0.86)$; $(\xi_s^* = 0.25))$, and other inside the radial boundary layer $((r^* = 0.99)$; $(\xi_s^* = 0.01))$. In sub-figure (c), the stationary intake tube is placed inside the rotating field with inner radius r_i and outer radius r_o. The intake tube is extended in the θ direction from 60° to 75°. The back pressure is applied at the rear-end of the intake tube at $\theta = 60^o$. Here, $\xi = \mathcal{A}^2\left(1-(r^2/R^2)\right) = \mathcal{A}^2\left(1-r^{*2}\right)$, and $r^* = \sqrt{(1-(\xi/\mathcal{A}^2))} = \sqrt{(1-\xi^*)}$, where $r^* = r/R$, $\xi^* = \xi/\mathcal{A}^2$, and the angle θ is expressed in degree.

ature as that of the gas in the domain. When a molecule is added, the mean velocity and the fluctuating velocity of all other molecules in the domain are scaled by a constant factor so that the total momentum and energy of all the molecules in the source domain remain unchanged due to the addition of a molecule. In the case of a mass sink, molecules are removed at random from the domain at time intervals fixed by the desired rate of mass output. When a molecule is removed, the mean and the fluctuating velocity of all other molecules in the source domain are scaled by a constant factor so that the total momentum and energy in the domain remain unchanged.

Forces (momentum sources) exerted on the gas due to stationary inserts are modelled by imposing impulses (increasing the momentum) of all the molecules in the direction of a force at each collision step. The magnitude of the impulse is determined such that

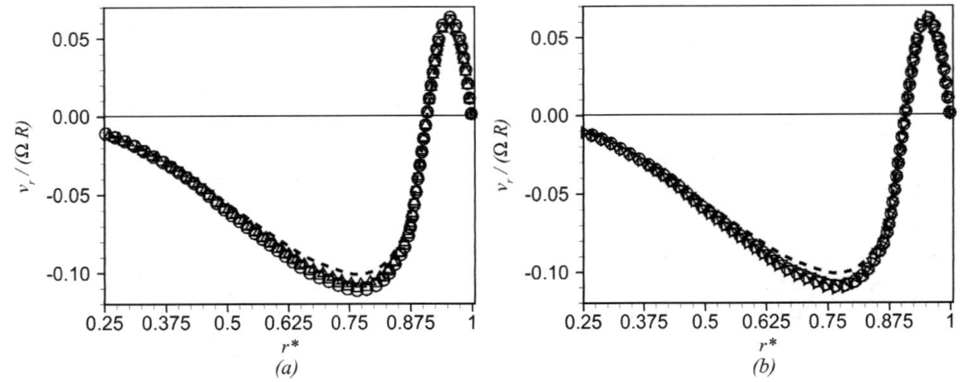

FIGURE 2. Radial velocity, scaled by (ΩR)), as a function of r^* for $\mathcal{A} = 3.535$; and $\mathrm{Re} = 10^2$. Flow induced by the momentum source inside the radial boundary layer at $((r^* = 0.99)$; $(\xi_s^* = 0.01))$. Sub-figure (a) for different time steps (t) in seconds. The symbols show the DSMC results for sub-figure (a) : $(t = 5 \times 10^{-10}$ (○), $t = 1 \times 10^{-9}$ (□), $t = 1.5 \times 10^{-9}$ (△)). Sub-figure (b) for different numbers of simulated molecules per cell (n_s). The symbols show the DSMC results for sub-figure (b) : $(n_s = 100$ (▷), $n_s = 80$ (○), $n_s = 60$ (▽)). The lines show the generalized analytical model (4.19) with homogeneous (zero temperature perturbation) boundary condition (−−). Here, $\xi = \mathcal{A}^2\left(1-(r^2/R^2)\right) = \mathcal{A}^2\left(1-r^{*2}\right)$, and $r^* = \sqrt{(1-(\xi/\mathcal{A}^2))} = \sqrt{(1-\xi^*)}$, where $r^* = r/R$, $\xi^* = \xi/\mathcal{A}^2$.

the ratio of the impulse and the time between collisions is equal to the applied force. A localised energy source, which increases or decreases the temperature of the gas, is imposed by scaling the fluctuating velocity of all the molecules in the source domain. The scaling of the velocity is determined such that the increase in energy divided by the time between collisions is equal to the desired energy input rate.

The flow geometry and coordinate system are shown in figure 1 (a), for the momentum source, and in figure 1 (b), for mass source/sink. In our calculation, we place the sources/sinks at two locations, one outside the radial boundary layer at $((r^* = 0.86)$; $(\xi_s^* = 0.25))$, and the second inside the radial boundary layer at $((r^* = 0.99)$; $(\xi_s^* = 0.01))$, where $r^* = r/R$; and $\xi_s^* = \xi_s/\mathcal{A}^2$. Here, we choose to express the results in terms of the scaled radial coordinate r^*, where $r^* = r/R$ for notational simplicity. Note that $\xi = \mathcal{A}^2\left(1-(r^2/R^2)\right) = \mathcal{A}^2\left(1-r^{*2}\right)$. Therefore, $r^* = \sqrt{(1-(\xi/\mathcal{A}^2))} = \sqrt{(1-\xi^*}$, where $r^* = r/R$, $\xi^* = \xi/\mathcal{A}^2$.

In order to check the convergence of the simulations, we have varied both the time step and the number of simulated molecules per cell, keeping the number of real molecules a constant. In all cases, the time step is decreased by a factor of 5 and the number of simulated molecules increased by a factor of 2, and verify that there is less than 1.5 % change in the resulting secondary radial flow profiles. The comparisons for a momentum source for $\mathrm{Re} = 10^2$, $\mathcal{A} = 3.535$, and source strength $F^* = 0.2$ is shown in figure 2. Different time steps $(5 \times 10^{-10}$, 1×10^{-9}, 1.5×10^{-9} sec) and different numbers of simulated molecules per cell (100, 80, 60) are used to test the convergence. It is found that the results are insensitive to the time step and the number of simulated molecules per cell. This clearly shows the convergence of the simulation results. Similar results are obtained for mass source/sink within the flow, but we do not provide the details here.

6.1. *Momentum source (single-component gas)*

The secondary radial flow generated by a momentum source, as shown in figure 1 (a), has been studied at different Reynolds numbers and stratification parameters. Results for the stratification parameters $\mathcal{A} = 3.535, 2.121, 0.707$, Reynolds number $\mathrm{Re} = 10^2$ are shown in figure 5. This because the smallest Reynolds number provides the most stringent test. The momentum source is placed in two different radial locations, one outside the radial boundary layer at $((r^* = 0.86); (\xi_s^* = 0.25))$, and other inside the radial boundary layer at $((r^* = 0.99); (\xi_s^* = 0.01))$.

The influx of momentum/energy near the wall is imposed by increasing the mean/fluctuating velocity of the molecules at the location. When the location is near the wall, the molecules with augmented momentum/energy from the source collide with the wall and locally increase the temperature. This results in a non-zero temperature perturabtion at the wall. The dimensionless temperature slip (T_s^*) is defined by

$$T_s^* = \frac{T - T_a}{T}, \tag{6.1}$$

where T is the gas temperature at the wall obtained from DSMC simulation, and T_a is the constant gas temperature at the wall assumed in the analytical solution.

We have examined, in detail, the temperature slip at the rotating sidewall near the source in the DSMC simulations. The variation of temperature slip for momentum source and mass source/sink is shown in figure 3 (a, b), for the momentum source with configuration shown in figure 1 (a), and for the mass source/sink configuration shown in figure 1 (b). We find that the temperature slip (T_s^*) decreases as the Reynolds number is increased and temperature slip (T_s^*) increases as the stratification parameter is increased for these cases, and is at a maximum for $\mathrm{Re} = 10^2$, and stratification parameter $\mathcal{A} = 3.535$, shown in figure 3 (a, b). As expected, the temperature slip is larger when the source is placed inside the boundary layer $((r^* = 0.99); (\xi_s^* = 0.01))$, owing to the larger number density and the higher frequency of molecular collisions. There is little variation in the temperature slip with Reynolds number, but there is a systemmatic increase in the temperature slip as the stratification parameter is increased. The largest scaled temperature slip (T_s^*) is close to 0.72 for the momentum source in figure 3 (a), and is close to 0.65 for the mass source/sink in figure 3 (b).

The temperature slip violates the homogeneous (zero temperature perturbation) boundary condition used in the analytical solution, and so results in a difference in the magnitudes of the secondary flow predicted. It is observed that there is a significant difference between the analytical and simulation results. In order to mimic the homogeneous boundary condition for the gas temperature assumed in the analytical solutions, it is necessary to remove energy from the wall. This is done by reducing the wall temperature as explained in the simulation section. Once the homogeneous boundary conditions are satisfied in the simulations, it is observed that there is quantitative agreement (with a difference of less than 10 %) between the analytical and simulation results as shown in figure 4.

The analytical solutions are compared with the results of DSMC simulations, where we have taken the precautions discussed above to ensure that homogeneous boundary conditions are satisfied. The analytical solutions are obtained by considering homogeneous boundary condition with internal source. The radial velocity, scaled by (ΩR), is plotted as a function of normalized radial coordinate $(r^* = r/R)$ for stratification parameter $\mathcal{A} = 3.535, 2.121, 0.707$, Reynolds number $\mathrm{Re} = 10^2$, for isothermal sidewall (with energy removal from the wall) as shown in figure 4. We find that the secondary radial flow is weaker when stratification parameter is reduced from 3.535 to 0.707.

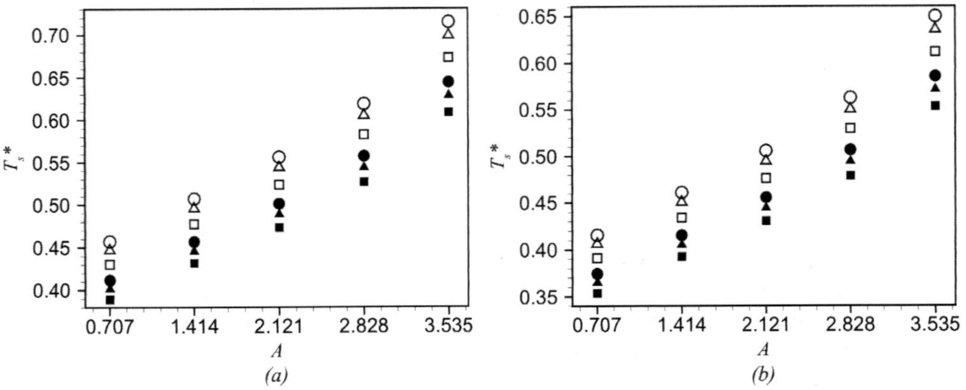

FIGURE 3. The variation of dimensionless scaled temperature slip (T_s^*) as a function of stratification parameter ($\mathcal{A}$) with source strength $F^* = 0.2$. Sub-figure (a) for momentum source; and sub-figure (b) for mass source/sink. The symbols show the DSMC results for the sources inside the radial boundary layer at (($r^* = 0.99$); ($\xi_s^* = 0.01$)) for : (Re $= 10^6$ (□), Re $= 10^4$ (△), Re $= 10^2$ (○)); and for the sources outside the radial boundary layer at (($r^* = 0.86$); ($\xi_s^* = 0.25$)) for : (Re $= 10^6$ (■), Re $= 10^4$ (▲), Re $= 10^2$ (•)). Here, $r^* = r/R$, $\xi^* = \xi/\mathcal{A}^2$.

The results show quantitative agreement between the DSMC simulations and the analytical results of the generalized analytical model (equation 4.19), provided the care is take to ensure that the temperature is a constant at the sidewall. There is agreement to within 10 % even when the Reynolds number is as low as 100, the stratification parameter is as low as 0.707. The results for higher Reynolds numbers and stratification parameters, not shown here, are in even better agreement with the simulation results. This shows that the 'generalized analytical model' (equation 4.19) captures the secondary radial flows quantitatively even at stratification parameter as low as 0.707, provided the boundary conditions in the analytical model are accurately specified. The predictions of the 'generalized analytical model' (equation 4.19) is also significantly better than that of the 'analytical model' (equation 4.18), which is restricted to thin boundary layers in the limit of high stratification parameter. We have also carried out further tests, whose results are not shown here for conciseness. When constant temperature boundary conditions are not enforced in the simulations, we find that the difference between analytical and simulation results is more than one and half times that when the constant temperature boundary conditions are enforced.

It should be noted that energy removal is necessary because the energy dissipation term (double dot product of the viscous stress and rate of deformation tensors) is second-order in the velocity perturbation. This is not included in the linearization approximation that we have used, as can be seen from the energy equation. Therefore, the viscous heating is not captured by the generalized analytical model (equation 4.19), and it is necessary to explicitly remove energy to maintain a constant temperature at the wall.

6.2. *Mass source/sink (single-component gas)*

The secondary radial flow generated by a mass source/sink, as shown in figure 1 (b), has also been studied at different Reynolds numbers and stratification parameters. Results for the stratification parameters $\mathcal{A} = 3.535, 2.121, 0.707$, Re $= 10^2$, with $F^* = 0.2$, are shown in figure 5. This because the smallest Reynolds number provides the most

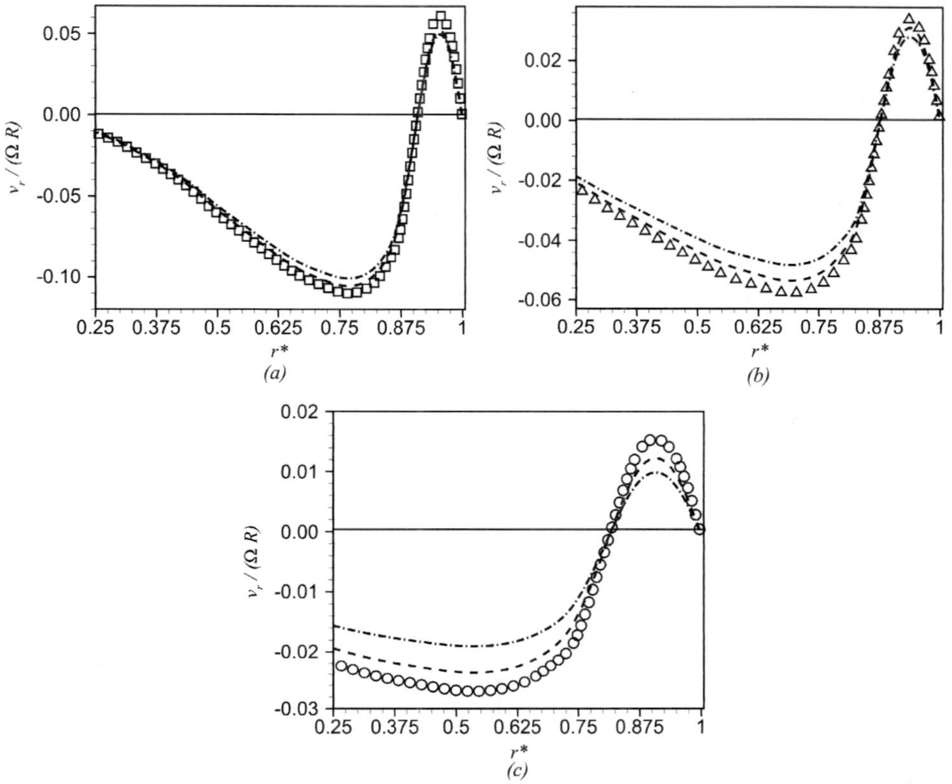

FIGURE 4. Radial velocity, scaled by (ΩR)), as a function of r^* for $\mathcal{A} = 3.535$ (a); $\mathcal{A} = 2.121$ (b); $\mathcal{A} = 0.707$ (c), for Re $= 10^2$. Flow induced by momentum source inside the radial boundary layer at $((r^* = 0.99)$; $(\xi_s^* = 0.01))$ with source strength $= 0.2$. The symbols show the DSMC results for isothermal sidewall (with energy removal from wall) for : ($\mathcal{A} = 3.535$ (□), $\mathcal{A} = 2.121$ (△), $\mathcal{A} = 0.707$ (○)). The lines show the generalized analytical model (4.19) with homogeneous (zero temperature perturbation) boundary condition (——), and analytical model (4.18) with homogeneous (zero temperature perturbation) boundary condition (· — ·). Here, $\xi = \mathcal{A}^2\ (1 - (r^2/R^2))$ $= \mathcal{A}^2\ (1 - r^{*2})$, and $r^* = \sqrt{(1 - (\xi/\mathcal{A}^2))} = \sqrt{(1 - \xi^*)}$, where $r^* = r/R$, $\xi^* = \xi/\mathcal{A}^2$.

stringent test. Here, the dimensionless scaled source strength is defined by

$$F^* = \frac{F}{H\Omega}, \tag{6.2}$$

where F is the mass flow rate, H is the total gas holdup, and Ω is the angular velocity. The scaled source strength (F^*) is a measure of the perturbation to the base flow due to the mass flow rate for mass source/sink type of forcing. Here, we have carried out simulations for $F^* = 0.2$. The actual mass flow rate (F) is varied in such a way that the desired scaled source strength is obtained at different stratification parameter, and Reynolds number. The mass sink is placed in two different radial locations, one inside $((r^* = 0.99)$; $(\xi_s^* = 0.01))$ and other outside $((r^* = 0.86)$; $(\xi_s^* = 0.25))$ the radial boundary layer. Mass is introduced in the flow domain by adding molecules to the system at $r^* = 0$ through the circle of radius $10^{-3}R$ which is connected to a pressure bath. Similarly, the mass is removed by removing molecules through a circle of the same radius $10^{-3}R$ at $((r^* = 0.86)$; $(\xi_s^* = 0.25))$ (outside the radial boundary layer) and at $((r^* = 0.99)$; $(\xi_s^*$

$= 0.01$)) (inside the radial boundary layer) which is also connected to a prerssure bath to get the same desired scaled mass flow rate $F^* = 0.2$.

The analytical solutions are compared with the results of DSMC simulations, where we have taken the precautions to ensure that the homogeneous boundary conditions are satisfied. The radial velocity, scaled by (ΩR), is plotted as a function of normalized radial coordinate ($r^* = r/R$) for the stratification parameters $\mathcal{A} = 3.535$, 2.121, 0.707, Re $= 10^2$, with $F^* = 0.2$, for isothermal sidewall (with energy removal from the wall) as shown in figure 5. We find that the secondary radial flow is weaker when stratification parameter is reduced from 3.535 to 0.707 in this case as well.

The results show quantitative agreement between the DSMC simulations and the analytical results of the 'generalized analytical model' (equation 4.19), provided the care is take to ensure that the temperature is a constant at the sidewall. There is agreement to within 10 % even when the Reynolds number is as low as 100, the stratification parameter is as low as 0.707. The results for higher Reynolds numbers and stratification parameters, not shown here, are in even better agreement with the simulation results. This shows that the 'generalized analytical model' (equation 4.19) captures the secondary radial flows quantitatively even at stratification parameter as low as 0.707, provided the boundary conditions in the analytical model are accurately specified. The predictions of the 'generalized analytical model' (equation 4.19) is also significantly better than that of the 'analytical model' (equation 4.18).

6.3. *Stationary intake tube (single-component gas)*

Next, we have studied the flow field in polar $(r - \theta)$ plane by placing a stationary intake tube within the flow domain. The effect of an insert in the flow domain is of three types. The first is the mass source/sink due to the addition or removal of gas in a particular region in the flow domain. The second effect is due to the force exerted on the gas by a stationary inserts, which is modelled locally as a momentum source or sink. The time rate of addition/removal of momentum is equal to the applied force. The third effect is a localized energy source, which increases or decreases the temperature of the gas. All the three effects are incorporated in the molecular simulation as follows. The addition or removal of mass is accomplished by either adding or removing molecules at specified locations in the simulation cell at specified rates. The pressure is specified at inflow and outflow boundaries by connecting a pressure bath, which is a large volume of gas maintained at a constant pressure. Forces are exerted by imposing impulses on the molecules at specified locations, in such a manner that the rate of change of impulse is equal to the applied force. Energy is added by increasing the mean square velocities of the molecules at locations where heating takes place.

The computational domain and coordinate system is shown in figure 1 (c). The density and temperature contours close to the stationary intake tube (zoomed view) for $\mathcal{A} = 3.535$, Re $= 10^2$, for the forward flow (equilibrium back pressure is less than the stagnation pressure) is shown in figure 6 (a), and figure 7 (a); and for the reverse flow (equilibrium back pressure is more than the stagnation pressure) is shown in figure 6 (b), and figure 7 (b). A detached bow shock wave is clearly observed in the simulations.

We have also examined, in detail, the temperature slip at the rotating sidewall in the DSMC simulations. The stationary intake tube inside the rotating field acts as a source/sink of mass/momentum/energy. When the location is near the wall, the molecules with augmented momentum/energy from the source collide with the wall and locally increase the temperature. This results in a non-zero temperature perturabtion at the wall, as shown by the sybmols (∘)in figure 8 (a), for $\mathcal{A} = 3.535$, Re $= 10^2$. At the wall nearest the source, there is an increase of close to 91 % in the temperature at the wall.

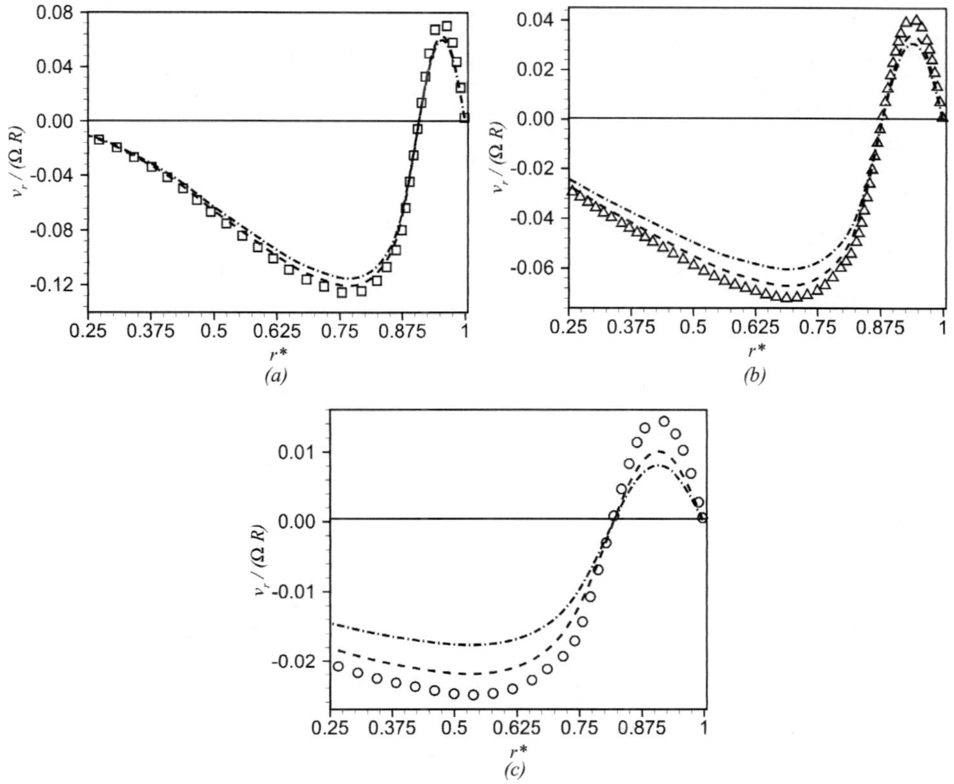

FIGURE 5. Radial velocity, scaled by (ΩR)), as a function of r^* for $\mathcal{A} = 3.535$ (a); $\mathcal{A} = 2.121$ (b); $\mathcal{A} = 0.707$ (c), for Re $= 10^2$, with source strength $F^* = 0.2$. Mass is introduced in the flow domain by adding molecules to the system at $r^* = 0$ through the circle of radius $10^{-3}R$. The mass sink is placed inside the radial boundary layer at (($r^* = 0.99$); ($\xi_s^* = 0.01$)). The mass is removed by removing molecules through a circle of the same radius $10^{-3}R$ at the radial location (($r^* = 0.99$); ($\xi_s^* = 0.01$)). The symbols show the DSMC results for isothermal sidewall (with energy removal from wall) for : ($\mathcal{A} = 3.535$ (□), $\mathcal{A} = 2.121$ (△), $\mathcal{A} = 0.707$ (○)). The lines show the generalized analytical model (4.19) with homogeneous (zero temperature perturbation) boundary condition (−−), and analytical model (4.18) with homogeneous (zero temperature perturbation) boundary condition (· − ·). Here, $\xi = \mathcal{A}^2\left(1 - (r^2/R^2)\right) = \mathcal{A}^2\left(1 - r^{*2}\right)$, and $r^* = \sqrt{(1 - (\xi/\mathcal{A}^2))} = \sqrt{(1 - \xi^*)}$, where $r^* = r/R$, $\xi^* = \xi/\mathcal{A}^2$.

This results in a substantial increase in the secondary radial flow as shown in figure 9. We find that the magnitude of the secondary radial flow velocity is about 1.3 − 1.75 times when constant temperature boundary conditions are not enforced. We find that the dimensionless temperature slip (T_s^*) increases as the stratification parameter is increased, as shown in figure 8 (b). It is also observed that the temperature slip is more in front of the intake tube ($\theta = 75^o$) compared to at the rear-end ($\theta = 60^o$.)

Azimuthal velocity in the flow domain, scaled by (ΩR), is plotted along the radial coordinate (r^*) for the stratification parameters $\mathcal{A} = 0.707$, 2.121, 3.535, and Re $= 10^2$, at $\theta = 75^o$ (in front of the intake tube) and $\theta = 60^o$ (rear-end of the intake tube), as shown in figure 10. It is clearly observed that there is a substantial slow down of the circumferential velocity (or equivalent angular momentum loss) $v_{\theta+(slow\ down)}$ of the bulk

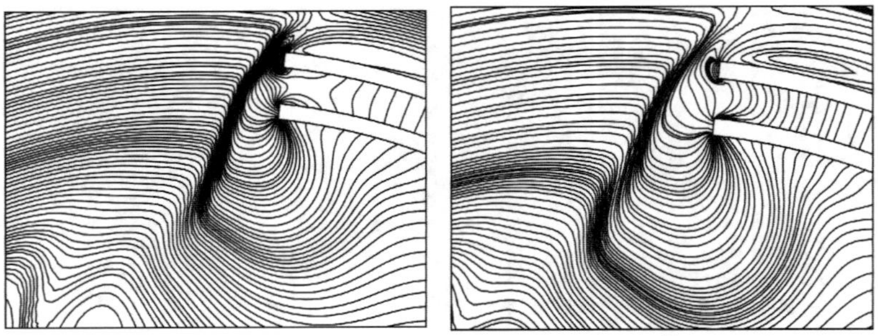

FIGURE 6. Density contours close to the stationary intake tube (zoomed view) for $\mathcal{A} = 3.535$, $\mathrm{Re} = 10^2$, for the forward flow (equilibrium back pressure is less than the stagnation pressure) (a); and for the reverse flow (equilibrium back pressure is more than the stagnation pressure) (b). A detached bow shock is formed in front of the stationary insert (intake tube), and behind the detached bow shock, secondary radial flow is induced towards the axis, and it excites the secondary axial flow.

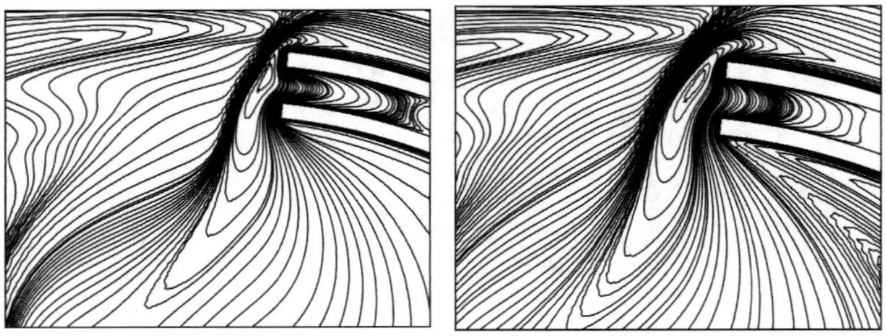

FIGURE 7. Temperature contours close to the stationary intake tube (zoomed view) for $\mathcal{A} = 3.535$, $\mathrm{Re} = 10^2$, for the forward flow (when the equilibrium back pressure is less than the stagnation pressure) (a); and for the reverse flow (when the equilibrium back pressure is more than the stagnation pressure (b). A detached bow shock wave is clearly observed.

of the rotating fluid due to the presence of stationary insert (intake tube) within the high speed rotating field.

The slow down of the circumferential velocity $v_{\theta(slow\,down)}$ of the bulk of the rotating fluid in the polar $(r-\theta)$ plane can be calculated as

$$v_{\theta(slow\,down)} = \frac{1}{2\pi R}\int_0^{2\pi}\left(\int_0^R \left(v_{\theta(sbr)} - v_\theta\right) dr\right) d\theta. \tag{6.3}$$

Using the scaled variables, $r^* = r/R$, $\theta^* = \theta/(2\pi)$, and $v_\theta^* = v_\theta/(R\Omega)$, we obtain

$$v^*_{\theta(slow\,down)} = \frac{v_{\theta(slow\,down)}}{R\Omega} = \int_0^1\left(\int_0^1 \left(v^*_{\theta(sbr)} - v^*_\theta\right) dr^*\right) d\theta^*. \tag{6.4}$$

Here, $v_{\theta(sbr)}$ is the circumferential velocity in the base state (solid body rotation). The slow down of the circumferential velocity $v_{\theta(slow\,down)}$ of the bulk of the rotating fluid for stratification parameters $\mathcal{A} = 0.707$, 2.121, 3.535, at $\mathrm{Re} = 10^2$, is given in table 2. This clearly indicates that the $v_{\theta(slow\,down)}$ increases from 8 % to 28 % as the stratification parameter $\mathcal{A}$ is increased from 0.707 to 3.535, at $\mathrm{Re} = 10^2$, which induces the secondary

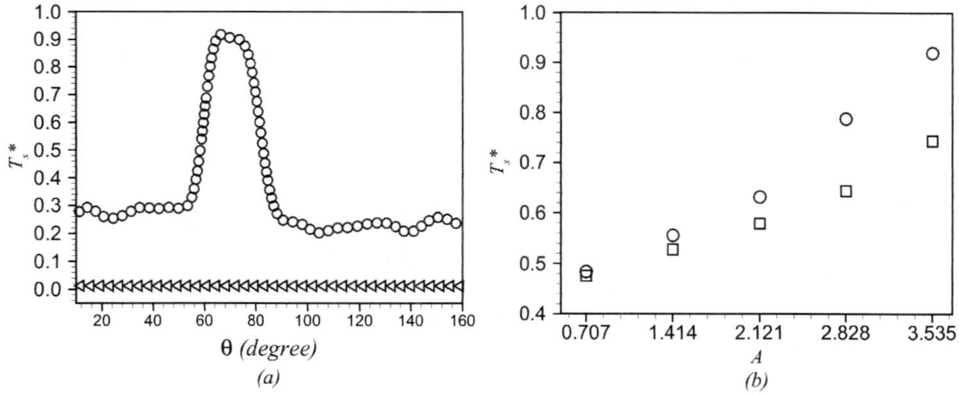

FIGURE 8. Sub-figure (a) shows the variation of dimensionless scaled temperature slip (T_s^*) along θ at the rotating wall for $\mathcal{A}$ = 3.535, and Re = 10^2, with energy removal ($\cdot - \cdot$), without energy removal (∘); sub-figure (b) shows the variation of dimensionless scaled temperature slip (T_s^*) as a function of stratification parameter ($\mathcal{A}$) for Re = 10^2. The symbols show the DSMC results without energy removal in sub-figure (b) for : ($\theta = 75^o$ (∘) : in front of the intake tube, $\theta = 60^0$ (□) : rear-end of the intake tube).

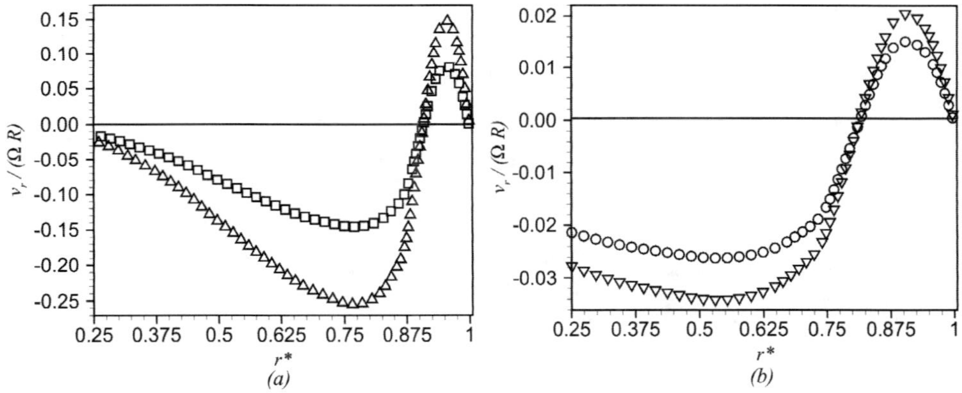

FIGURE 9. Radial velocity, scaled by (ΩR)), as a function of r^* at $\theta = 75^o$ for $\mathcal{A}$ = 3.535 (a); and $\mathcal{A}$ = 0.707 (b), for Re = 10^2. The symbols show the DSMC results for sub-figure (a) : (with energy removal (□), without energy removal (△)). The symbols show the DSMC results for sub-figure (b) : (with energy removal (∘), without energy removal (▽)).

radial flow towards the axis, and it further excites the secondary axial flow, which could be very important for the centrifugal gas separation processes.

The mass flow rate through the stationary intake tube, scaled by ($H\Omega$), is studied for different back pressures, as shown in figure 11 for $\mathcal{A}$ = 2.121; Re = 10^2, and d^* = 0.125. Here, H is the gas holdup in kg, Ω is the angular velocity sec^{-1}, d is the wall gap $d = (R - r)$, and $d^* = d/R$. The simulations show that the scaled mass flow rate $m/(H\Omega)$ increases progressively from zero, at the stagnation condition P_S/P_W = 1.6489, for $\mathcal{A}$ = 2.121, Re = 10^2, d^* = 0.125 to 0.024 when the back pressure is P_B/P_W = 0.4692, for $\mathcal{A}$ = 2.121, Re = 10^2, d^* = 0.125.

The stagnation pressure, scaled by the wall pressure (P_W) is plotted as a function of

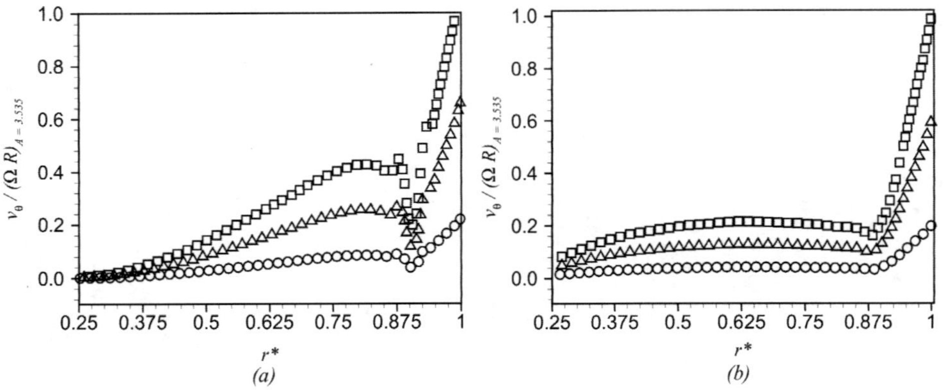

FIGURE 10. Azimuthal velocity, scaled by $((\Omega R))$, as a function of r^* at two different locations of θ. Sub-figure (a), in front of the intake tube: ($\theta = 75^o$); and sub-figure (b), at the rear-end of the intake tube: ($\theta = 60^o$) for $\mathcal{A} = 0.707$, 2.121, 3.535; for Re $= 10^2$. The symbols show the DSMC results for : ($\mathcal{A} = 3.535$ (□), $\mathcal{A} = 2.121$ (△), $\mathcal{A} = 0.707$ (○)).

Stratification parameter	$v_{\theta(slow\ down)}$ (%)
$\mathcal{A} = 0.707$	8
$\mathcal{A} = 2.121$	17
$\mathcal{A} = 3.535$	28

TABLE 2. The slow down of the circumferential velocity $v_{\theta(slow\ down)}$ of the bulk of the rotating fluid for stratification parameters $\mathcal{A} = 0.707$, 2.121, 3.535, at Re $= 10^2$

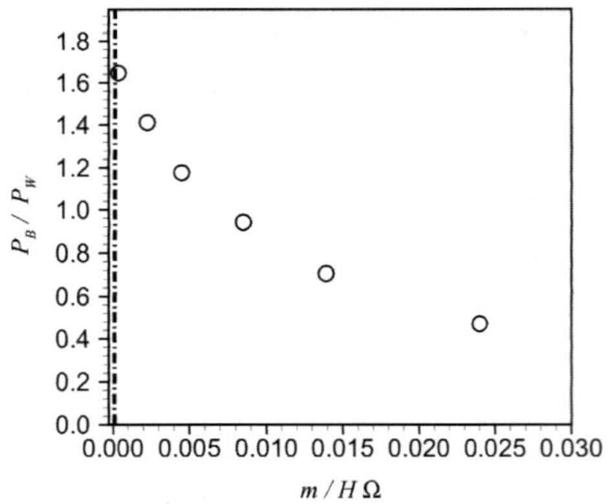

FIGURE 11. Mass flow rate through the intake tube, scaled by $(H\Omega)$, as a function of equilibrium back pressure, scaled by the wall pressure (P_W), for $\mathcal{A} = 2.121$; Re $= 10^2$, $d^* = 0.125$. DSMC results are shown by symbols (○). The line ($\cdot - \cdot$) corresponds to the zero mass flow rate to get the stagnation pressure. Here, d is the wall gap $d = (R - r)$, and $d^* = d/R$.

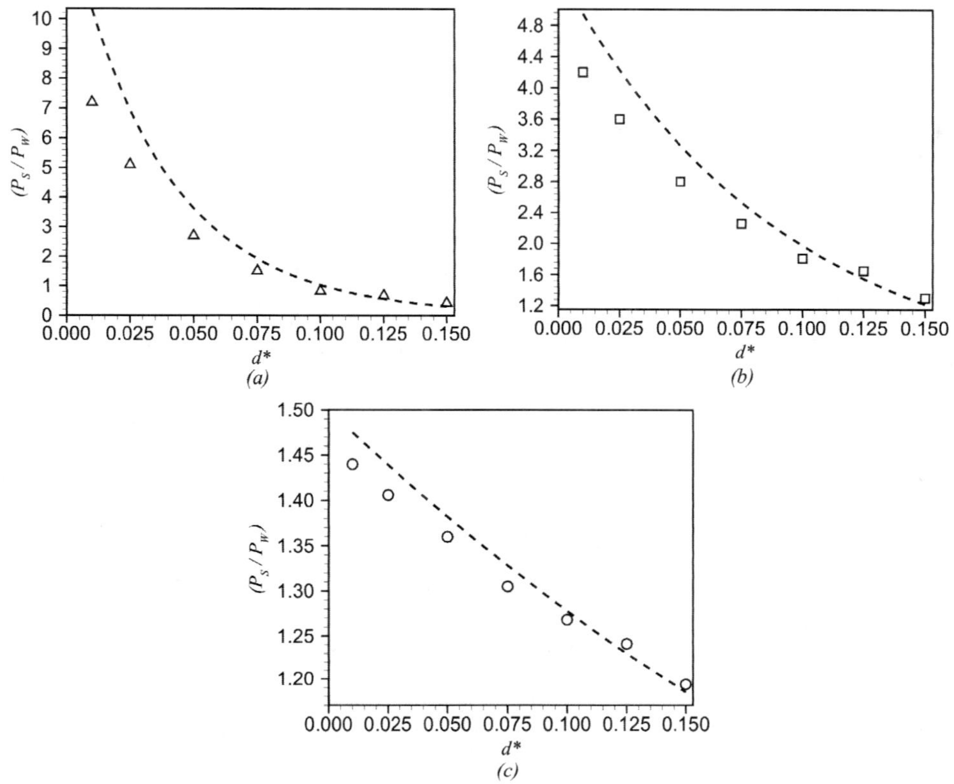

FIGURE 12. The stagnation pressure, scaled by wall pressure(P_W)), as a function of scaled wall gap ($d^* = d/R$), for $\mathcal{A} = 3.535$ (a); $\mathcal{A} = 2.121$ (b); $\mathcal{A} = 0.707$ (c), for Re $= 10^2$. The symbols show the DSMC results for : ($\mathcal{A} = 3.535$ ($\triangle$) (a), $\mathcal{A} = 2.121$ ($\square$) (b), $\mathcal{A} = 0.707$ ($\circ$)) (c). The lines ($--$) show the results for the base state (solid body rotation) (Equation 6.5).

dimensionless wall gap (d^*) for the stratification parameters ($\mathcal{A}$) in the range (0.707 - 3.535), for Re $= 10^2$, as shown in figure 12. Here, $d^* = d/R$.

It is clearly observed that the stagnation pressure increases with stratification parameter ($\mathcal{A}$), and decreases as the wall gap is increased. In the base state (solid body rotation), the stagnation pressure is expressed as

$$P_S = \left(1 + \mathcal{A}^2 \frac{r^2}{R^2}\right) \times \exp\left(-\mathcal{A}^2 \left(1 - \frac{r^2}{R^2}\right)\right), \qquad (6.5)$$

where the wall gap $d = (R - r)$.

7. Binary gas mixture

Next, we have carried out the DSMC simulations in two-dimensional polar ($r-\theta$) plane for a binary gas mixture with large mass difference ($\epsilon_a = (2\Delta m/(m_1 + m_2)) = 0.5$), and we compare the simulation results with those of single component gas. Here, Δm is the difference in the molecular masses of the two species $\Delta m = (m_1 - m_2)$. The fluid is an ideal gas mixture in which both species have equal molecular diameter $d = 5.5 \times 10^{-10}$ m, average molecular mass $m_{av} = (m_1 + m_2)/2 = 58.4426 \times 10^{-26}$ kg/molecule, specific heat ratio $\gamma = 1.066$, and Prandtl number Pr$= 0.95$. It is assumed that the gas molecules

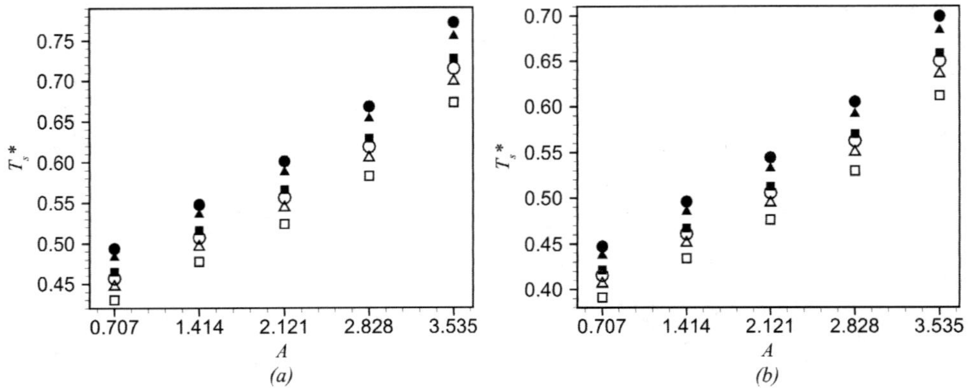

FIGURE 13. The variation of dimensionless scaled temperature slip (T_s^*) as a function of stratification parameter ($\mathcal{A}$). Sub-figure (a) for momentum source; and sub-figure (b) for mass source/sink. The symbols show the DSMC results for the sources inside the radial boundary layer at (($r^* = 0.99$); ($\xi_s^* = 0.01$)) for single-component gas : (Re = 10^6 (□), Re = 10^4 (△), Re = 10^2 (○)); and the DSMC results for a binary gas mixture with $\epsilon_a = 0.5$ for : (Re = 10^6 (■), Re = 10^4 (▲), Re = 10^2 (•)). Here, $r^* = r/R$; $\xi_s^* = \xi_s/\mathcal{A}^2$, and $\epsilon_a = (2\Delta m/(m_1 + m_2))$.

undergo diffuse reflection from all the solid surfaces with the thermal accommodation coefficient equal to 1.

The stratification parameter $\mathcal{A}$ is varied in the range 0.707-3.535, and Reynolds number Re in the range $10^2 - 10^6$. The flow geometry and coordinate system are shown in figures 1.

7.1. *Momentum source (binary gas mixture)*

We find that there is substantial temperature slip at the rotating sidewall near the source, and slip increases as the mass difference between two species in a binary gas mixture increases. Here, we have presented the results for $\epsilon_a = (2\Delta m/(m_1+m_2)) = 0.5$; for Reynolds number (Re) in the range ($10^2 - 10^6$); and stratification parameters ($\mathcal{A}$) in the range (0.707 - 3.535) as shown in figure 13.

We have examined, in detail, the temperature slip at the rotating sidewall near the source in the DSMC simulations. The variation of temperature slip for momentum source and mass source/sink is shown in figure 13 (a, b), for the momentum source with configuration shown in figure 1 (a), and for the mass source/sink configuration shown in figure 1 (b). We find that the temperature slip (T_s^*) decreases as the Reynolds number is increased and temperature slip (T_s^*) increases as the stratification parameter is increased for these cases, and is at a maximum for Re = 10^2, and stratification parameter $\mathcal{A}$ = 3.535, shown in figure 13 (a, b). We also find that temperature slip increases as the mass difference between two species increases in a binary gas mixture. Here, we have shown the results for $\epsilon_a = (2\Delta m/(m_1+m_2)) = 0.5$; and Reynolds number Re in the range $10^2 - 10^6$; and stratification parameters $\mathcal{A}$ in the range $0.707 - 3.535$ as shown in figure 13.

Radial velocity in the flow domain, scaled by (ΩR), is plotted along the radial coordinate (r^*) as shown in figure 14. Here, we have shown the results only for the case where energy is removed at the wall so that the homogeneous temperature boundary condition is maintained, and the momentum source is placed inside the radial boundary layer at (($r^* = 0.99$); ($\xi_s^* = 0.01$)). We find that the secondary radial flow is weaker when stratification parameter is reduced. The peak value of the secondary radial flow for a binary

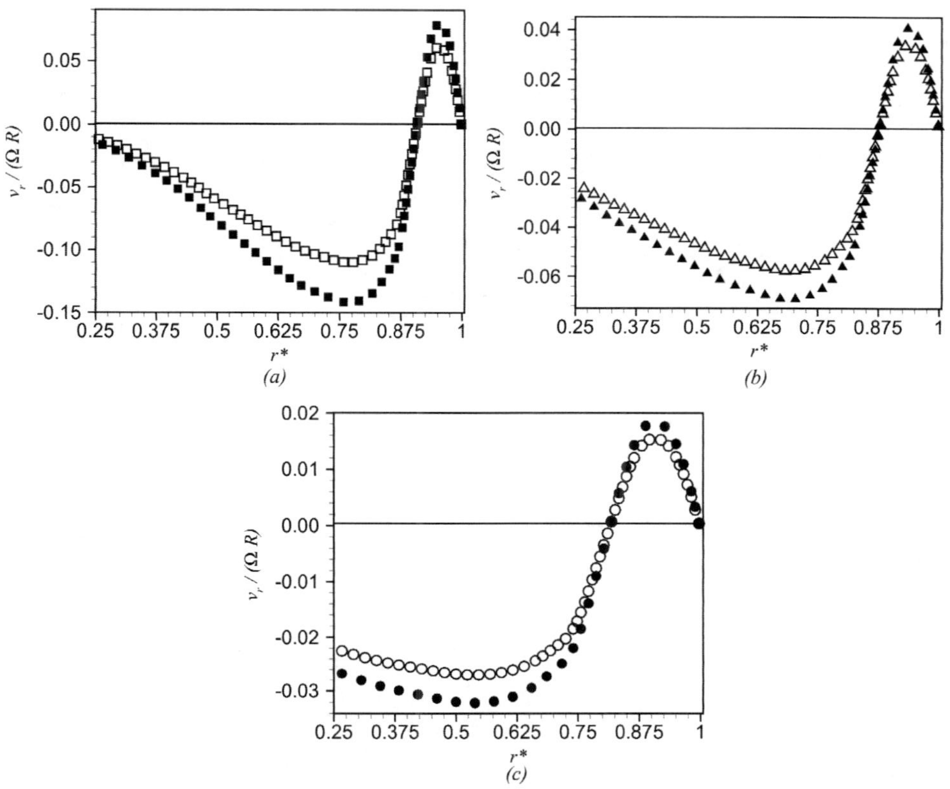

FIGURE 14. Radial velocity, scaled by (ΩR)), as a function of r^* for $\mathcal{A} = 3.535$ (a); $\mathcal{A} = 2.121$ (b); $\mathcal{A} = 0.707$ (c), for Re $= 10^2$. Flow induced by momentum source inside the radial boundary layer at $((r^* = 0.99)$; $(\xi_s^* = 0.01))$. The symbols show the DSMC results for isothermal sidewall for single-component gas : ($\mathcal{A} = 3.535$ (□), $\mathcal{A} = 2.121$ (△), $\mathcal{A} = 0.707$ (∘)); and the DSMC results for a binary gas mixture with $\epsilon_a = 0.5$ for : ($\mathcal{A} = 3.535$ (■), $\mathcal{A} = 2.121$ (▲), $\mathcal{A} = 0.707$ (•)). Here, $r^* = r/R$; and $\xi_s^* = \xi_s/\mathcal{A}^2$.

gas mixture with $\epsilon_a = 0.5$ increases by (19 - 30)% compared to the single-component gas, for the stratification parameter $\mathcal{A}$ in the range $0.707 - 3.535$ at Re $= 10^2$.

7.2. *Mass source/sink (binary gas mixture)*

Next, we study the secondary radial flow for the mass source/sink type of forcing. Substantial temperature slip at the rotating sidewall near the source is obtained in this case as well. Here, we have presented the results for $\epsilon_a = (2\Delta m/(m_1 + m_2)) = 0.5$; for Reynolds number (Re) in the range $(10^2 - 10^6)$; and stratification parameters ($\mathcal{A}$) in the range (0.707 - 3.535) as shown in figure 13. We find that the temperature slip increases as the mass difference between two species in a binary gas mixture is increased.

The radial velocity in the flow domain, scaled by (ΩR), is plotted along the scaled radial coordinate (r^*) for the isothermal sidewall as shown in figure 15. The mass source/sink is placed inside the radial boundary layer at $((r^* = 0.99)$; $(\xi_s^* = 0.01))$. We find that the secondary radial flow is weaker when stratification parameter is reduced in this case as well. The peak value of the secondary radial flow for a binary gas mixture with $\epsilon_a = 0.5$ increases by (20 - 25)% compared to the single-component gas, for the stratification parameter $\mathcal{A}$ in the range $0.707 - 3.535$ at Re $= 10^2$.

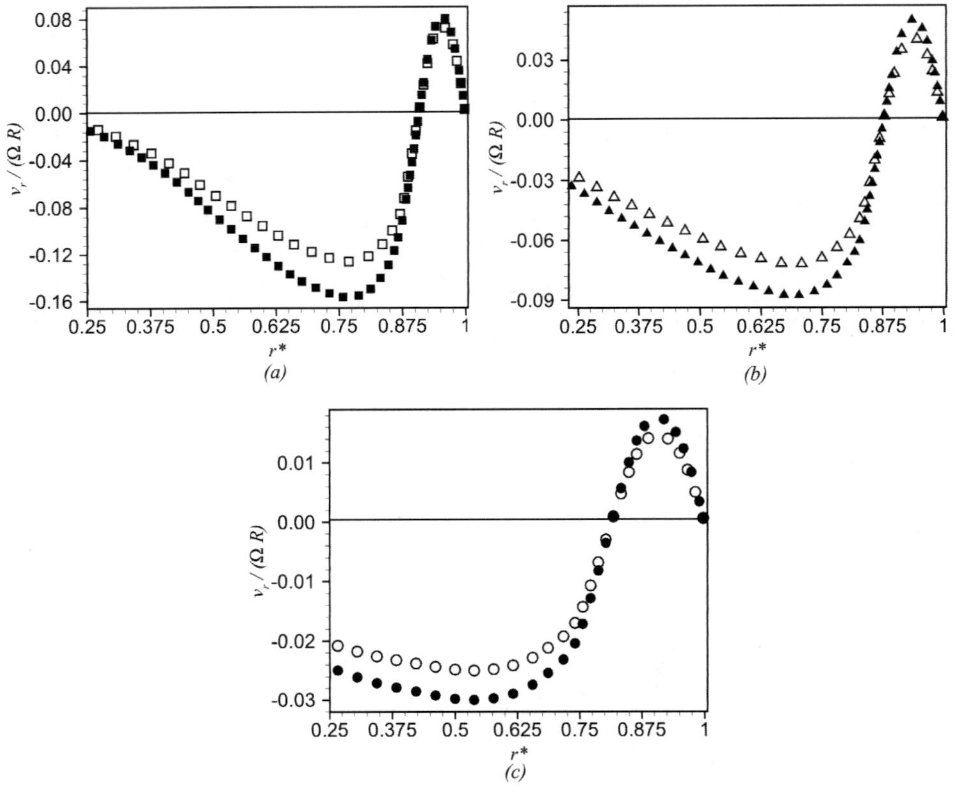

FIGURE 15. Radial velocity, scaled by (ΩR)), as a function of r^* for $\mathcal{A} = 3.535$ (a); $\mathcal{A} = 2.121$ (b); $\mathcal{A} = 0.707$ (c), for Re $= 10^2$, with source strength $F^* = 0.2$. Flow induced by a mass source at ($r^* = 0$) and a mass sink inside the radial boundary layer at (($r^* = 0.99$); ($\xi_s^* = 0.01$)). The symbols show the DSMC results for isothermal sidewall for single-component gas : ($\mathcal{A} = 3.535$ (□), $\mathcal{A} = 2.121$ (△), $\mathcal{A} = 0.707$ (○)); and the DSMC results for a binary gas mixture with $\epsilon_a = 0.5$ for : ($\mathcal{A} = 3.535$ (■), $\mathcal{A} = 2.121$ (▲), $\mathcal{A} = 0.707$ (•)). Here, $r^* = r/R$; and $\xi_s^* = \xi_s/\mathcal{A}^2$.

7.3. *Stationary intake tube (binary gas mixture)*

We have further carried out the DSMC simulations for a binary gas mixture by placing a stationary intake tube within the flow domain. First, we calculate the temperature slip at the rotating sidewall for a binary gas mixture with $\epsilon_a = 0.5$ as shown in figure 16. We find that the temperature slip increases as the mass difference between two species in a binary gas mixture is increased.

The radial velocity in the flow domain, scaled by (ΩR), is plotted along the radial coordinate (r^*) for the isothermal sidewall as shown in figure 17. We find that the secondary radial flow is weaker when stratification parameter is reduced, and the peak value of the secondary radial flow for a binary gas mixture with $\epsilon_a = 0.5$ increases by (19 - 30)% compared to the single-component gas, for the stratification parameter $\mathcal{A}$ in the range $0.707 - 3.535$ at Re $= 10^2$.

8. Conclusions

The objective of the present work is to carry out a comprehensive study of the comparison of the results obtained from analytical equation (4.18), and generalized analytical

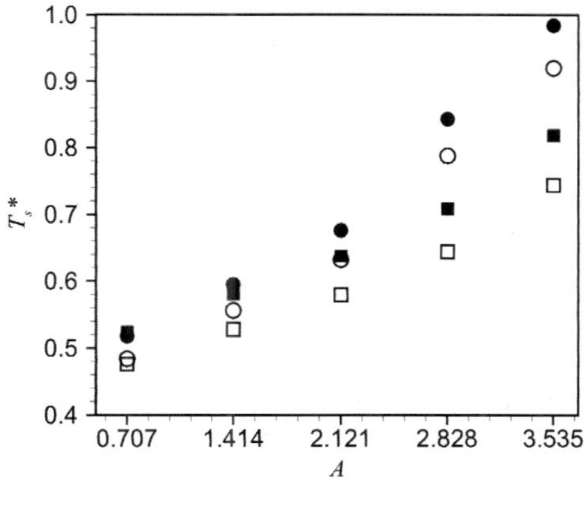

FIGURE 16. The variation of dimensionless scaled temperature slip (T_s^*) as a function of stratification parameter ($\mathcal{A}$) for Re $= 10^2$. The symbols show the DSMC results without energy removal for single-component gas : ($\theta = 75^o$ (∘), $\theta = 60^o$ (□)); and the DSMC results for a binary gas mixture with $\epsilon_a = 0.5$ for : ($\theta = 75^o$ (•), $\theta = 60^o$ (■)).

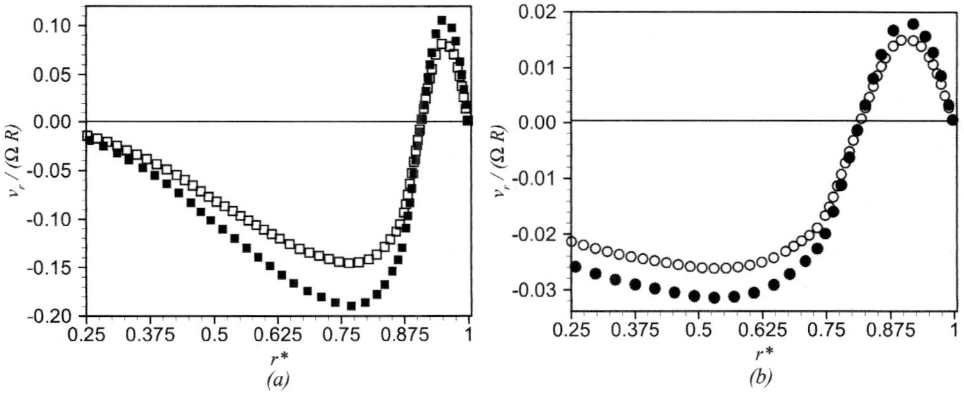

FIGURE 17. Radial velocity, scaled by (ΩR)), as a function of r^* for $\mathcal{A} = 3.535$ (a); and $\mathcal{A} = 0.707$ (b), for Re $= 10^2$. The symbols show the DSMC results for sub-figure (a) : (single-component gas (□), binary gas mixture with $\epsilon_a = 0.5$ (■); The symbols show the DSMC results for sub-figure (b) : (single-component gas (∘), binary gas mixture with $\epsilon_a = 0.5$ (•).

equation (4.19), derived in §4, in the polar ($r - \theta$) plane, with molecular simulations based on the direct simulation Monte Carlo (DSMC) method, for the secondary radial flow in a rotating cylinder. Several situations have been considered, including delta-function momentum source, and mass source/sink within the domain of the flow. Simulations have been carried out for Reynolds number as low as 100, stratification parameter as low as 0.707, and with source strength $= 0.2$.

Here, we have studied the secondary radial flow due to insertion of mass, momentum and energy into the rotating cylinder in the polar ($r - \theta$) plane. The presence of a stationary insert (intake tube) within the flow domain causes a disturbance to the base flow (solid body rotation). A detached bow shock is formed in front of the stationary insert (intake tube), and behind the detached bow shock, secondary radial flow is induced towards the axis, and it excites the secondary axial flow, which could be very important for the centrifugal gas separation processes. The inward radial flow also produces a vortex column near the stationary insert. A mechanically driven gas centrifuge utilizes stationary intake tube (which is used to collect the separated gases), to generate the secondary radial flow, and it further excites the secondary axial flow.

The generalized analytical model for the radial boundary layer (4.19) in a high-speed rotating cylinder is formulated for studying the gas flow field due to insertion of mass, momentum and energy into the rotating cylinder in the polar $(r-\theta)$ plane. The analytical solution includes the sixth order differential equation for the radial boundary layer at the cylindrical surface in terms of master potential (χ), which is derived from the equations of motion in a polar $(r-\theta)$ plane. The linearization approximation is used, where the equations of motion are truncated at linear order in the velocity and pressure disturbances to the base flow, which is a solid-body rotation. Additional assumptions in the analytical model include constant temperature in the base state (isothermal condition), and high Reynolds number, but there is no limitation on the stratification parameter. In this limit, the gas flow is restricted to a boundary layer of thickness $(\mathrm{Re}^{-1/3}R)$ at the wall of the cylinder.

For the case of mass/momentum/energy insertion into the flow, the separation-of-variables procedure is used, and the appropriate homogeneous boundary conditions are specified so that the linear operators in the meridional and radial directions are self-adjoint. The discrete eigenvalues and eigenfunctions of the linear operators (sixth-order and second-order in the radial and meridional directions for the generalized analytical equation) are obtained. The solutions for the secondary radial flows are determined in terms of these eigenvalues and eigenfunctions.

The major advantage of the present formulation is that it is not restricted to the asymptotic limit of high stratification parameter. Consequently, the model is not restricted to just boundary layer flows at the curved surface of the cylinder, but is even applicable to situations where the flow domain extends all the way to the centre of the cylinder. Therefore, the model can be used for much slower flows, and the shock type solutions (Rankine-Hugoniot relations) are obtained simultaneously.

The analytical solutions are compared with direct simulation Monte Carlo (DSMC) simulations for two types of forcing, the momentum source, and mass source/sink for stratification parameter in the range $0.707-3.535$, at Reynolds number as low as 100, and source strength as high as 0.2. Even for this strong forcing, we find that the generalized analytical solutions are in excellent agreement (error within 10%) with the results of DSMC simulations even when the stratification parameter is as low as 0.707, and the Reynolds number is as low as 100, provided precautions are taken to ensure that the homogeneous boundary conditions assumed in the generalized analytical equation (4.19) are accurately reproduced in the simulations. This shows that the generalized analytical model has excellent predictive capabilities even at relatively low Mach and Reynolds numbers, provided the care is taken in the specification of boundary conditions. The generalized analytical equation (4.19) also represent a significant better agreement with the results of DSMC simulations, over the analytical equation (4.18), which is restricted to thin boundary layers in the limit of high stratification parameter. It should be noted that energy removal is necessary because the energy dissipation term (double dot product of the viscous stress and rate of deformation tensors) is second-order in the velocity perturbation. This is not included in the linearization approximation that we have used, as can be seen from the energy equation. Therefore, the viscous heating is not captured by the generalized analytical model, and it is necessary to explicitly remove energy to maintain a constant temperature at the wall.

In a high speed rotating field we examine the mass flow rate through the stationary intake tube. The simulations show that the scaled mass flow rate $m/(H\Omega)$ increases progressively from zero, at the stagnation condition $P_S/P_W = 1.6489$, $\mathcal{A} = 2.121$, Re $= 10^2$, $d^* = 0.125$, to, 0.024 when the back pressure is reduced to $P_B/P_W = 0.4692$. The slow down of the circumferential velocity of the bulk of the rotating fluid due to the presence of stationary intake tube within the rotating flow field is studied for stratification parameter in the range $0.707 - 3.535$, and found significant slow down (between 8 to 28%), which induces the secondary radial flow towards the axis, and it further excites the secondary axial flow, which could be very important for the centrifugal gas separation processes. An important finding is that the stagnation pressure (no mass flow through the intake tube) is strongly affected by the wall gap, as well as with stratification parameter indicating a strong coupling between the local temperature, density, pressure and velocity fields.

We have also carried out the DSMC simulations for a binary gas mixture with large mass difference ($\epsilon_a = (2\Delta m/(m_1+m_2)) = 0.5$), and compare the simulation results with those of single-component gas incorporating the homogeneous boundary conditions. An important point to note is that in all cases, internal momentum source, mass source/sink, and stationary intake tube, the peak value of the secondary radial flow for a binary gas mixture with $\epsilon_a = 0.5$ increases by (19 - 30)% compared to the single-component gas, as stratification parameter $\mathcal{A}$ is increased from 0.707 to 3.535, for Re $= 10^2$.

The present work provides an opportunity for validating the analytical model as well as the DSMC method for the high-speed rotating flow in the polar $(r - \theta)$ coordinate. DSMC simulation technique does not become invalid when the mean free path is comparable to the system size (cylinder radius), because the simulation does not use averaged equations. There are many situations in high-speed rotating flows where the gas density in the central region of the cylinder is sufficiently small that the mean free path is comparable to the cylinder radius. In these cases, the Navier-Stokes equations are not valid near the centre, and it is necessary to specify 'artificial boundary conditions' at a fixed inner surface within the flow. The possible dependence of the solutions on these 'artificial boundary conditions' poses difficulties in the interpretation of results. No such difficulties exist in the molecular simulations, since it is only necessary to specify particle-wall interaction parameters. Molecular simulations also have the distinct advantage that the wall slip emerges naturally from the particlewall collision laws, and it does not have to be modelled. Though DSMC simulation technique overcomes several of the difficulties of a continuum simulation, the computational power and time required for the simulation is much higher than that for a continuum simulation. This is due to the requirement for simulating, directly, a simplified version of the molecular motion. This requires the evaluation of the complete velocity distribution function of the molecule, and temperature fields.

Appendix A. Conservation equations in cylindrical coordinates

The compressible mass, momentum and energy conservation equations in cylindrical polar $(r - \theta)$ coordinate system are

$$\frac{\partial \rho}{\partial t} + \nabla.(\rho \mathbf{v}) = S_m, \tag{A 1}$$

$$\rho\left(\frac{\partial v_r}{\partial t} + v_r\frac{\partial v_r}{\partial r} - \frac{v_\theta^2}{r} + \frac{v_\theta}{r}\frac{\partial v_r}{\partial \theta}\right) = -\frac{\partial p}{\partial r} + \left[\frac{1}{r}\frac{\partial}{\partial r}(\tau_{rr}r) - \frac{\tau_{\theta\theta}}{r} + \frac{1}{r}\frac{\partial(\tau_{\theta r})}{\partial \theta}\right] + f_r, \tag{A 2}$$

$$\rho\left(\frac{\partial v_\theta}{\partial t}+v_r\frac{\partial v_\theta}{\partial r}+\frac{v_r v_\theta}{r}+\frac{v_\theta}{r}\frac{\partial v_\theta}{\partial \theta}\right)$$
$$=-\frac{1}{r}\frac{\partial p}{\partial \theta}+\left[\frac{1}{r^2}\frac{\partial}{\partial r}\left(r^2\tau_{r\theta}\right)+\frac{1}{r}\frac{\partial(\tau_{\theta\theta})}{\partial \theta}\right]+f_\theta, \tag{A 3}$$

$$\rho\left(\frac{\partial v_z}{\partial t}+v_r\frac{\partial v_z}{\partial r}+\frac{v_\theta}{r}\frac{\partial v_z}{\partial \theta}\right)$$
$$=\left[\frac{1}{r}\frac{\partial}{\partial r}\left(r\tau_{rz}\right)+\frac{1}{r}\frac{\partial(\tau_{\theta z})}{\partial \theta}\right]+f_z, \tag{A 4}$$

$$\rho C_v\left(\frac{\partial T}{\partial t}+v_r\frac{\partial T}{\partial r}+\frac{v_\theta}{r}\frac{\partial T}{\partial \theta}\right)$$
$$=-p(\nabla.\mathbf{v})+\tau_{rr}\frac{\partial v_r}{\partial r}+\tau_{\theta\theta}\left(\frac{1}{r}\frac{\partial v_\theta}{\partial \theta}+\frac{v_r}{r}\right)+\tau_{r\theta}\left(\frac{1}{r}\frac{\partial v_r}{\partial \theta}+r\frac{\partial}{\partial r}\left(\frac{v_\theta}{r}\right)\right)+\tau_{\theta z}\left(\frac{1}{r}\frac{\partial v_z}{\partial \theta}\right)+\tau_{rz}\left(\frac{\partial v_z}{\partial r}\right)$$
$$+\frac{1}{r}\frac{\partial}{\partial r}\left(rK\frac{\partial T}{\partial r}\right)+\frac{1}{r}\frac{\partial}{\partial \theta}\left(\frac{K}{r}\frac{\partial T}{\partial \theta}\right)+S_e, \tag{A 5}$$

where

$$\nabla.\mathbf{v}=\frac{1}{r}\frac{\partial}{\partial r}\left(rv_r\right)+\frac{1}{r}\frac{\partial v_\theta}{\partial \theta}, \tag{A 6}$$

$$\tau_{rr}=\mu\left(2\frac{\partial v_r}{\partial r}-\frac{2}{3}\nabla.\mathbf{v}\right), \tag{A 7}$$

$$\tau_{\theta\theta}=\mu\left(2\left(\frac{1}{r}\frac{\partial v_\theta}{\partial \theta}+\frac{v_r}{r}\right)-\frac{2}{3}\nabla.\mathbf{v}\right), \tag{A 8}$$

$$\tau_{zz}=\mu\left(-\frac{2}{3}\nabla.\mathbf{v}\right), \tag{A 9}$$

$$\tau_{r\theta}=\mu\left(r\frac{\partial}{\partial r}\left(\frac{v_\theta}{r}\right)+\frac{1}{r}\frac{\partial v_r}{\partial \theta}\right), \tag{A 10}$$

$$\tau_{rz}=\mu\left(\frac{\partial v_z}{\partial r}\right), \tag{A 11}$$

$$\tau_{\theta z}=\frac{\mu}{r}\left(\frac{\partial v_z}{\partial \theta}\right). \tag{A 12}$$

Equation (A 1) is the mass conservation equation, equations (A 2), (A 3) and (A 4) are the momentum conservation equations in the r, θ and z directions, and (A 5) is the energy conservation equation.

REFERENCES

ANDERSON, J. D. 2005 Fundamentals of Aerodynamics. *McGraw-Hill.*

AOKI, E. & SUZUKI, M. 1985 Numerical studies on gas flow around scoop of a gas centrifuge. *In Proc. Sixth Workshop on Gases in Strong Rotation, Tokyo (ed. Y. Takashima)*, 11.

AVERY, D. G. & DAVIES, E. 1973 Uranium enrichment by gas centrifuge *Mills and Boon Ltd. London* **1**, 1.

BABARSKY, R. J. & WOOD, H. G. 1990 Approximate eigensolutions for non-axisymmetric rotating compressible flows. *Computer Meth. Appl. Mech. and Engg.* **81**, 317.

BABARSKY, R. J. & WOOD, H. G. 1992 Analysis of the flow field in a gas centrifuge by means of a temperature potential. *In Proc. of the Third International Workshop on Separation Phenomena in Liquids and Gases.(ed. Wood, H. G.)* , 27.

BABARSKY, R. J. & HERBST, W. I. & WOOD, H. G. 2002 A new variational approach to gas flow in a rotating system. *Physics of fluids.* **14(10)**, 3624.

BARCILON, V. & PEDLOSKY, J. 1967 On the steady motions produced by a stable stratification in a rapidly rotating fluid. *J. Fluid Mech.* **29**, 673.

BARK, F. H. & BARK,T. H. 1976 On vertical boundary layers in a rapidly rotating gas. *J. Fluid Mech.* **78**, 749.

BIRD, G. A. 1963 Approach to translational equilibrium in a rigid sphere gas. *Phys. Fluids.* **6**, 1518.

BIRD, G. A. 1981 Monte-Carlo simulation in an engineering context. *In S.S. Fischer, editor, Rarefied Gas Dynamics, proceedings of the 12th international symposium - part-1.*, 239.

BIRD, G. A. 1987 Direct simulation of high-vorticity gas flows. *Phys. Fluids.* **30**, 364.

BIRD, G. A. 1994 Molecular gas dynamics and the direct simulation of gas flows. *Clarendon Press. Oxford.*

BOURN, R. & PETERSON, T. D. & WOOD, H. G. 1999 Solution of the pancake model for flow in a gas centrifuge by means of a temperature potential. *Comput. Meth. Appl. Mech. Engng.* **178**, 183.

BROUWERS, J. J. H. 1978 On compressible flow in a gas centrifuge and its effect on maximum separative power. *Nucl. Technology.* **39**, 311.

CERCIGNANI, C. & LAMPIS, M. 1971 Kinetic models for gas surface interactions. *Transport Theory Stat. Physics.* **1**, 101.

CERCIGNANI, C. 2000 Rarefied gas dynamics. From basic concepts to actual calculations. *Cambridge University Press.*

CHAPMAN, S. & COWLING, T. G. 1970 The Mathematical Theory of Non-Uniform Gases. *Cambridge University Press.*

CHEN, G. & BOYD, I. D. 1996 Statistical error analysis for the Direct Simulation Monte Carlo technique. *Journal of Computational Physics.* **126**, 434.

COHEN, K. 1951 The theory of isotope separation as applied to the large scale production of U235. *McGraw-Hill.*

COPPER, G. R. & MORTON, J. B. 1988 Driven Wave Motions in a Rotating Gas. *J. Fluid Struct* **2**, 453.

DONEDDU, F. & ROBIN, P. & WOOD, H. G. 2000 Optimization Studies for Gas Centrifuges. *Separation Science and Technology.* **35**, 1207.

FAN, J. & SHEN, C. 2001 Statistical simulation of low-speed rarefied gas flows. *J. Comp. Phys.* **167**, 393.

FRAENKAL, L. E. 1959 A Cylindrical sound pulse in a rotating gas. *J. Fluid Mech.* **36**, 637.

GARCIA. 2000 Time step truncation error in direct simulation Monte Carlo. *Physics of fluids* **12**, 2621.

GREENSPAN, H. P. 1968 The theory of Rotating Fluids. *Cambridge University Press* **1**, 1.

GUNZBURGER, M. D. & WOOD, H. G. 1982 A finite element method for the Onsager pancake equation. *Comp. Meth. Appl. Mech. Engg.* **31**, 43.

GUNZBURGER, M. D. & WOOD, H. G.& JORDAN, J. A. 1984 A finite element method for gas centrifuge problem. *SIAM J. Sci. Stat. Comp.* **5**.

HARADA, I. 1980 A Numerical Study of Wikely Compressible Rotating Flows in a Gas Centrifuge. *Nucleas Sc. and Engg.* **73**, 225.

HARADA, I. 1980 Computation of Strongly Compressible Rotating Flows. *J. Computational Physics.* **38**, 335.

HAUSMAN, M. A. & ROBERTS, W. W. 1984 Three dimensional stratified gas flows past an obstacle. *J. Comp. Phys.* **55**, 347.

HEUSER, G. E. & RIBANDO, R. J. & WOOD, H. G. 1986 A numerical simulation of internal waves in a rotating fluid. *Comput. Methods Appl. Mech. Engrg.* **57**, 207.

HITTINGER, M. & HOLT, M. & SOUBBARAMAYER, B. 1981 Numerical solution of the flow field near a gas centrifuge scoop. *In Proc. Fourth Workshop on Gases in Strong Rotation, Oxford (ed. E. Ratz)* , 22.

HITTINGER, M. & HOLT, M. & SOUBBARAMAYER, B. & CORTET, C. 1983 Simulation of gas flow in front of the scoop of a gas centrifuge. *In Proc. Fifth Workshop on Gases in Strong Rotation, Charlottesrille (ed. H. G. Wood)* , 515.

HOGLUND, R. L. & SHACTER, J. & VON HALLE, E. 1979 Diffusion separation methods. *Encyclopedia of chemical technology.* **7**, 3rd ed.

HOMSY, G. M. & HUDSON, J. L. 1967 Centrifugally driven thermal convection in a rotating cylinder. *J. Fluid Mech.* **35**, 33.

HUNTER, C. 1967 The axisymmetric flow in a rotating annulus due to horizontally applied temperature gradient. *J. Fluid Mech.* **27**, 753.

JONES, C. A. 1982 On flow between counter-rotating cylinders. *J. Fluid Mech.* **120**, 433.

JUNG, E. 1983 An analytic solution of the linearized flow equations using the method of eigenvalues. *In Proc. Fifth Workshop on Gases in Strong Rotation, Charlottesrille (ed. H. G. Wood)* , 247.

KAI, T. 1983 Analysis of Separation Performance of Gas Centrifuge. *In Proc. Fifth Workshop on Gases in Strong Rotation, Charlottesrille (ed. H. G. Wood)* , 1.

KUMARAN, V. & PRADHAN, S. 2014 The generalized Onsager model for a binary gas mixture. *J. Fluid Mech.* **753**, 307.

LAHARGUE, J. P. & SOUBBARAMAYER 1978 A numerical model for the investigation of the flow and isotope concentration field in an ultracentrifuge. *Comp. Meth. Appl. Mech. Engg.* **15**, 259.

MACROSSAN, M. N. 2001 DSMC : A fast simulation method for rarefied flow. *J. Comp. Phys.* **173**, 600.

MASLEN, S. H. & MOORE, F. K. 1956 On strong transverse waves without shocks in a circular cylinder. *J. Aeronautical Sciences* **23**, 583.

MATSUDA, T. & HASHIMOTKO, K. 1976 Thermally, mechanically or externally driven flow in a gas centrifuge with insulated horizontal end plates. *J. Fluid Mech.* **78**, 337.

MATSUDA, T. 1976 A New proposal of gas centrifuge with desirable counter-current. *J. Nucl.Sci. Tech.* **13**, 98.

MATSUDA, T. & HASHIMOTO, K. 1978 The Structure of the Stewartson layers in a gas centrifuge. Part 1: Insulated end plates. *J. Fluid Mech.* **85**, 433.

MATSUDA, T. & NAKAGAWA, K. 1983 A new type of boundary layer in a rapidly rotating gas. *J. Fluid Mech.* **126**, 431.

MATSUDA, T. & TAMURA, N. & SAWADA, K. 1989 Three-dimensional numerical simulation of flows past scoops in a gas centrifuge. *J. Fluid Mech.* **201**, 203.

MIKAMI, H. 1973 Thermally induced flow in a gas centrifuge. *J.Nucl.Sci.Tech.* **10**, 580.

MIKAMI, H. 1981 Rotating supersonic flow about scoop inlet using an unsteady implicit technique. *In Proc. Fourth Workshop on Gases in Strong Rotation, Oxford (ed. E. Ratz)*, 94.

MIKAMI, H. 1985 Two dimensional numerical calculation of flow fields about scoop inlet by the piecewise linear method. *In Proc. Sixth Workshop on Gases in Strong Rotation, Tokyo (ed. Y.Takashima)*, 67.

NAKAYAMA, W. & TORII, T. 1974 Numerical analysis of separative power of isotope centrifuges. *J. Nucl. Sci. Technology.* **11**, 495.

NANBU. 1980 Direct simulation scheme derived from the Boltzmann equation I. Multicomponent gases. *J. Phys. Soc. of Japan.* **45**, 2042.

OLANDER, D. R. 1981 The theory of uranium enrichment by the gas centrifuge. *Progress in Nuclear Energy.* **8**, 1.

OMNES, P. 2007 Numerical and physical comparisons of two models of a gas centrifuge. *Computers Fluids* **36**, 1028.

PRADHAN, S. & KUMARAN, V. 2011 The generalized Onsager model for the secondary flow in a high-speed rotating cylinder. *J. Fluid Mech.* **686**, 109.

QUANHUA, S. & BOYD, I. D. 2002 A Direct Simulation Method for Subsonic Microscale Gas Flows . *J. Computational Physics.* **179**, 400.

RIBANDO, R. J. 1984 A Finite difference solution of Onsager's model for flow in a gas centrifuge. *Computers Fluids* **12 (3)**, 235.

RIEFFEL, M. A. 1999 A method for estimating the computational requirements of DSMC simulations. *J. Comp. Phys.* **149**, 95.

ROBERTS, W. W. 1985 Three dimensional stratified gas flows past impact probes and scoops : N-body Monte Carlo calculations. *In Proc. Sixth Workshop on Gases in Strong Rotation, Tokyo (ed. Y. Takashima.)*, 115.

ROBERTS, W. W. & HAUSMAN, M. A. 1988 Hypersonic, stratified gas flows past an obstacle : Direct Simulation Monte Carlo calculations. *J. Comp. Phys.* **77**, 283.

ROBLIN, P. & DONEDDU, F. 2000 Direct Monte-Carlo Simulations In a Gas Centrifuge. *In Proc. 22nd International Symposium on Rarefied Gas Dynamics. (ed. T. J. Bartel & M. C. Gallis.)*, 169.

SAKURAI, T. 1981 Linearized thin-wing theory of gas-centrifuge scoops. *J. Fluid Mech.* **103**, 257.

SALANT, R. F. 1968 Syymmetric normal modes in a uniformly rotating fluid. *J. Accoustical Society of America.* **43**, 1302.

STEWARTSON, K. 1957 On almost rigid rotations. *J. Fluid Mech.* **3**, 17.

SOUBBARAMAYER. 1979 Centrifugation. *In Uranium Enrichment (Ed. S. Villani, Springer)* **1**, 1.

SOUBBARAMAYER & BILLET, J. 1980 A numerical method for optimizing the gas flow field in a centrifuge. *Comp. Meth. Appl. Mech. Engg.* **24**, 165.

SOZOU, C. 1959 Adiabatic transverse modes in a uniformly rotating fluid. *J. Fluid Mech.* **36**, 605.

THOMPSON, J. J. & CHEN, Y. P. 1980 Eigen vector solution of the linearized equations of thermal convection in a closed gas centrifuge. Part-1 : Problem formulation and solution by collocation. *Nuclear Engineering and Design* **58**, 97.

VIECELLI, J. A. 1983 Exponential difference operator approximation for the sixth order Onsager equation. *J. Comp. Phys.* **50**, 162.

VINCENTI, W. G. & KRUGER, C. H. 1967 Introduction to Physical Gas Dynumics. *Wiley, New York.*

VOLOSCIUK, K. 1981 Application of the vortex transport equations to the calculation of 2 and 3 dimensional extraction chamber flows. *In Proc. Fourth Workshop on Gases in Strong Rotation, Oxford (ed. E.Ratz)*, 604.

VON HALLE E. & WOOD, H. G., LOWRY, R. A. 1978 The effect of vacuum core boundary conditions on separation in the gas centrifuge. *Nucl. Technology.* **62**, 325.

WAGNER, W. 1992 A convergence proof for Bird's Direct Simulation Monte Carlo method for the Boltzmann equation. *Journal of Statistical Physics.* **66**, 1011.

WALZ, A. & VOLOSCIUK, K. & SCHUTZ, H. 1983 Numerical investigations of the flow field near a model of a scoop using the vortex transport equations. *In Proc. Fifth Workshop on Gases in Strong Rotation, Charlottesrille (ed. H.G. Wood)*, 425.

WOOD, H. G. & MORTON, J. B. 1980 Onsager's pancake approximation for the fluid dynamics of a gas centrifuge. *J. Fluid Mech.* **101**, 1.

WOOD, H. G. & SANDERS, G. 1983 Rotating compressible flows with internal sources and sinks. *J. Fluid Mech.* **127**, 299.

WOOD, H. G. & JORDAN, J. A. & GUNZBURGER, M. D. 1984 The effect of curvature on the flow field in rapidly rotating gas centrifuges. *J. Fluid Mech.* **140**, 373.

WOOD, H. G. & BABARSKY, R. J. 1992 Analysis of a rapidly rotating gas in a pie-shaped cylinder. *J. Fluid Mech.* **239**, 249.

WOOD, H. G. 1995 Analysis of feed effects on a single-stage gas centrifuge cascade. *Separation Science and Technology.* **30(13)**, 2631.

WOOD, H. G. & MASON, T. C. & SOUBBARAMAYER. 1996 Multi-Isotope separation in a gas centrifuge using Onsager's Pancake model. *Separation Science and Technology.* **31(9)**, 1185.

472728 Nanoscale Dynamics of Silica-Based Nanofluids for Asphaltene Removal and Wettability Alteration

Abstract:

Nanoscale dynamics of silica-based nanofluids for asphaltene removal and wettability alteration

Tatiana Montoya, Nashaat N. Nassar*, Gerardo Vitale

[1]Department of Chemical and Petroleum Engineering, University of Calgary, 2500 University Drive NW, Calgary, Alberta, Canada.

Corresponding author e-mail: nassar@ucalgary.ca

Abstract

Asphaltenes form the solubility class that is considered the heaviest, most aromatic and most surface-active fraction of crude oil. Because of their complex chemical structures, polarizability and amphiphilic behavior, asphaltenes exhibit a self–associating feature that promotes aggregation and subsequently increases the viscosity of crude oil. Due to these characteristics, asphaltenes can cause different problems at different stages of crude oil production, any of which cause a loss of productivity. Thus, understanding the adsorption and deposition of asphaltenes onto solid surfaces has practical significance for monitoring the fluid-property variations that are common during oil production. Therefore, the deposition mechanism of asphaltene onto sand and alumina surfaces and the factors influencing it are investigated in this study. This is because asphaltene deposition is important in many different domains of the oil industry, such as alteration of reservoir rock wettability because its potential to be adsorbed on reservoir rock, plugging of flow lines because of asphaltene deposit build-up, and refinery catalyst deactivation because of asphaltene adsorption at active sites, causing significant production losses. Accordingly, appropriate mitigation techniques, for surface exposed to asphaltene or operating conditions, can be identified.

It has been demonstrated that the use of nanoparticles may improve the mobility of oil and alter the formation wettability. This is because nanoparticles may be used as adsorbents and catalysts in the oil industry for enhancing in situ upgrading and recovery processes. Therefore, this study aims at inhibiting of formation damage caused by asphaltene deposition and subsequently enhance oil recovery using silica-based nanofluid flooding approach, with varying nanoparticle chemical surface.

The potential flow of silica-based nanofluids at concentration of 100 ppm with varying acidity in displacing the asphaltene layers adsorbed on silica surfaces is investigated in an attempt to explore the applicability of these nanoparticles as a chemical modifier to alter the reservoir wettability. The quartz crystal microbalance with dissipation (QCM-D) was used to determine the adsorption of the nanoparticles on silica or alumina which has a thin layer of asphaltenes (preadsorbed), and/or its removal. The operating variables studied were nanoparticles surface acidity, nanoparticle size, flowrate and temperature. The role of silica-base nanofluid on the wettability alteration and morphology mitigation of the asphaltene-contaminated (coated) silica surface was determined by contact angle measurement and atomic force microscope (AFM) imaging. Dynamic light scattering (DLS) was also used to determine the role of the modified chemical surface on aggregation. Thus, effluent samples were taken at different flowing times for DLS measurements. The measurements were conducted before and after the nanofluids flow. The results demonstrated the possibility of using basic nanoparticles to modify the wettability, through removing the hydrocarbon that adsorbed/deposited on inorganic solids (clays), in order to improve heavy oil recovery.

Keywords: Nanofluids, quarz crystal microbalance (QCM-D), Asphaltenes.

467729 Impact of Non-Ideality Revealed in Numerical Simulations of Mixing of Hydrocarbons and Supercritical Water

Abstract:

Upgrading heavy crude oil using near-critical or supercritical water (NCW/SCW) shows advantages such as high yielding of light oils, less coke formation, removal of sulfur content and heavy metals, etc. We study the thermodynamic and mixing process of crude oil and NCW/SCW, using representative hydrocarbons, e.g., n-C10, n-C30, benzene, naphthalene, and 1-decylnaphthalene. GCA-EoS, the first equation of state that takes into account association using a SAFT-like group contribution term, is used to predict the thermodynamic properties and phase equilibrium compositions. Chung et al. correlation method is used to estimate the viscosity and thermal conductivity of the supercritical mixture, and the Tracer Liu-Silva-Macedo equation is used to calculate the binary diffusion coefficients. The mixing process of a hydrocarbon droplet (D=0.5mm) in a bulk of NCW/SCW at 24 MPa is simulated for different components and at different temperations. Our results reveal the importance of the non-ideal diffusion in this mixing process of supercritical fluids: Above the upper critical solution temperature, the diffusion of water into hydrocarbons is significantly delayed. The characteristics of alkanes and aromatics when mixing with NCW/SCW are summarized based on our numerical study.

466027 Scaleout of Pipeline Drag Reduction By Wall-Injected Polyox U310 Solutions to Marine Boundary Layers

Abstract:

Turbulent drag reduction by wall-injected Polyox U310 solutions was studied by simultaneous axial pressure profile and flow visualization measurements at Reynolds numbers from 40000 to 102000 in a transparent test pipe of ID 15.8 mm and L/D 430. Friction factors for solvent, deionized water, adhered to the Prandtl-Karman law for Newtonian fully-developed turbulent flow in smooth pipes, $1/\sqrt{f} = 4.0 \log Re\sqrt{f} - 0.4$, to within 0.5 units of $1/\sqrt{f}$. The U310 Polyox, of MW ~ $10x10^6$, was injected as a concentrated solution C_{inj} = 100 wppm through 6 equi-spaced 2.3 mm ID circumferential ports at friction-normalized injection velocities $V_{inj}^+ \sim 1 << $ pipeline mean velocity $V^+ \sim 25$. Downstream polymer concentrations C_{line} varied from 0.05 to 2 wppm, exhibiting eventual fractional drag reductions DR_{line} from ~ 0.03 to nearly the asymptotic maximum drag reduction DR_{MDR} ~ 0.75 possible at the present Re. Polymer dissolution into the flow was assessed from the initial increase in fractional drag reduction towards the eventual downstream DR_{line}, which provided a characteristic development distance $(L/D)_{dr}$ ~ 60, and also from flash photography which provided a characteristic distance $(L/D)_{pv}$ ~ 100 where red-dyed strands of the injected solution disappeared. Finally, an attempt was made to "scale-out" the pipeline drag reduction observations to a flat plate boundary layer by treating the latter as an equivalent pipe flow with $U_{CL} = U_o$, the free-stream velocity, $R^+ = du_T/v$, with d thickness, u_T friction velocity and v kinematic viscosity at the boundary layer trailing edge L. Drag reduction from the present pipe flow results, measured by flow enhancement or slip $S'_{line} = (1/\sqrt{f_p} - 1/\sqrt{f_n})_{Re\sqrt{f}}$, was translated to the classical HMS Highburton Polyox W301 ejection trials of Canham et. al (1971). This scale-out suggested that the present Polyox U310 might have provided their reported drag reductions at roughly 1/50 th their polymer expenditure rate per unit of hull wetted area.

466209 SNG Technology Via Clathrate Hydrates for Large Scale Storage of Natural Gas

Abstract:

Increasing global demand and widespread applications of natural gas (NG) necessitate the development of more effective and efficient methods for NG storage and transportation. Conventional methods of NG storage are either energy intensive - LNG (Liquefied Natural Gas) or pose high safety risk as in CNG (Compressed Natural Gas). SNG (Solidified natural gas) via clathrate hydrates is advantageous due to its characteristic advantages that include high volumetric capacity, non-explosive nature, environmentally benign and being relatively less energy intensive compared to conventional methods of NG storage. However, a significant challenge that impedes the commercial development of SNG technology is the slow rate of hydrate formation. For the first time, we report an increased rate of mixed methane hydrate formation in a simple unstirred tank reactor configuration (UTR) at moderate temperature and pressure conditions employing tetrahydrofuran (THF) as a promoter. THF was observed to function as a thermodynamic as well as an excellent kinetic promoter for hydrate formation at 283.2 K and moderate pressures (3.0-7.2 MPa). Substantial methane gas uptake of 3.52 (±0.13) kmol/m^3 of water is possible even at reduced experimental pressure of 3.0 MPa and 283.2 K in 2 h employing the unstirred reactor configuration. At higher temperature of 293.2 K, similar methane gas uptake and formation kinetics (as that observed at 283.2 K) is possible with the addition of just 100 ppm sodium dodecyl sulphate (SDS) surfactant due to the synergism between THF and SDS. Further, we demonstrate multi-scale experimental validation of our method to a volumetric sample scale-up factor of 120 and internal reactor diameter scale-up factor of 10. It is noted that substantial decrease in the severity of the storage conditions can be achieved by the use of THF as a promoter. SNG technology has a potential for a large scale commercial deployment.

471344 A Water-Immiscible Heavy Guest Molecule for Fast Methane Hydrate Formation

Abstract:

Clathrate hydrates have an ice-like appearance but a three dimensional structure having hydrogen bonded water cavities in which small guest molecules such as methane, hydrogen, carbon dioxide and light organic molecules are encaged. They can play an important role in CO_2 separation and also be a safe and economical medium for natural gas storage. But to apply gas hydrate techniques to industrial processes, we have to overcome the two barriers: (1) Gas hydrates are usually formed at the interface between the bulk gas and water phases. Once the solid hydrates cover the interface, gas hydrate formation decreases significantly because the hydrate layer becomes a barrier for mass transfer, and thus hydrate conversion is very low. (2) The gas hydrate formation kinetics is extremely unfavorable. This talk addresses the circumvention of the mass transfer and kinetic limits by introducing a water-immiscible heavy guest molecule of idomethane to enhance methane hydrate formation. We will demonstrate the key idea for maintaining liquid-liquid interfaces available for continuous methane gas access to the hydrate growth front and achieving a complete conversion to methane-idomethane binary hydrates within 2 minutes in an unstirred reactor system.

470972 Molecular Dynamics Studies of Structure II Hydrate Nucleation Using Advanced Sampling Techniques

Abstract:

Clathrate hydrates are solid, non-stoichiometric combinations of guest molecules entrapped in a solid water lattice. The guest molecules are generally light gases, such as methane or carbon dioxide. Clathrate hydrates are potentially an important energy resource; some estimates suggest there is more energy trapped in hydrate form than in all other fossil fuel sources on earth. Hydrates also represent a major flow assurance hazard in the petroleum industry. Currently active research on applications of clathrate hydrates ranges from energy and flow assurance to gas separations and transportation. Many such applications of clathrate hydrates require a firm understanding of the mechanism of hydrate formation.

The formation of a new hydrate phase occurs from a supersaturated liquid solution, through a process called nucleation. In nucleation, the formation of the new phase must overcome an activation barrier. The nucleation process is inherently a molecular-level phenomenon, occurring on length- and time-scales that are difficult to probe with experimental techniques. The length and time-scales of hydrate nucleation appear well-suited to be studied through molecular simulation. However, in most cases the associated activation barrier is large. Therefore, the nucleation is a rare event – the waiting time between nucleation events is orders-of-magnitude longer than the nucleation events themselves. Correspondingly, in straightforward simulations most of the computational power is expended waiting for the nucleation event to happen, rather than simulating the nucleation events of interest. In order to study hydrate nucleation with molecular dynamics simulations, we use an advanced sampling technique known as forward flux sampling (FFS).

We study the nucleation of sII clathrate hydrates from a supersaturated liquid solution with FFS. The water and guest molecules are described by a coarse-grained model which uses monatomic water and was developed by Jacobson et al.[1] In this presentation, we report results from large-scale FFS calculations comprising over 190,000 individual molecular dynamics trajectories. A committor probability analysis is used in conjunction with FFS to elucidate the nucleation mechanism. We estimate the reaction coordinate with a least squares estimation technique developed for FFS by Borrero and Escobedo.[2] Elements of the labile cluster hypothesis, local structuring hypothesis, and blob-mechanism are observed at different stages of the nucleation. We discuss the ability of several proposed order parameters to capture nucleation of sII hydrates. Our results highlight the various factors at play in clathrate hydrate nucleation.

(1) Jacobson, L. C.; Hujo, W.; Molinero, V. ***J. Phys. Chem. B*** **2010**, ***114*** (43), 13796–13807.

(2) Borrero, E. E.; Escobedo, F. A. ***J. Chem. Phys.*** **2007**, ***127*** (16), 164101.

467021 Estimation of Direct Transition Mechanism for Molecular Diffusion in Type *I* Gas Hydrates Using Density Functional Theory

Abstract:

Remarkable research efforts have been performed concerning the study of hydrates and organic clathrates using different theoretical approaches, such as molecular equations of state, molecular simulations including both Molecular Dynamics and Monte Carlo techniques, and different ab initio Quantum Mechanics approaches[1-3]. In this work, CO_2 and CH_4 type I hydrates have been studied using electronic Density Functional Theory (DFT)[4]and the Quantum Theory of Atoms in Molecules (QTAIM). Several mechanisms for transport of gases inside hydrates have been proposed so far in literature. Previous results have pointed out the apparent imposibility of some guests passing directly through faces connecting adjacent cages without destroying the water structure. Both types of cells included in the structure of type I hydrate were modeled as isolated double semi-flexible atomic systems. Interaction potentials of guest molecules with the enclathrating cell, when moving between neighbour cells were calculated using B3LYP/6-311+g(d,p) DFT approximation. Our calculations show that direct transitions are feasible through hexagonal and pentagonal faces without compromising the overall structure integrity in opposition to other results previously reported in literature. This stability has been explored using the QTAIM theory and reveals that even in the case that some bond may break during the transition, all of them are recovered, because the face distorsion is absorbed locally by the hydrogen bond network. The validity of the theory level selected has been stated, and the high anisotropy of the guest-cell interaction potential for the molecules analysed is shown, which may be considered in the formulation of hydrate thermodynamic models as equations of state, and also for the description of transport properties.

References:

[1] A. K. Sum, C. A. Koh and E. D. Sloan, Ind. Eng. Chem. Res, 48, 7457, 2009.

[2] Á. Vidal-Vidal, M. Pérez-Rodríguez, J.-P. Torré and M. M. Piñeiro, Phys. Chem. Chem. Phys., 17, 6963, 2015.

[3] Á. Vidal-Vidal, M. Pérez-Rodríguez and M. M. Piñeiro, RSC Advances, 6, 1966, 2016.

[3] Gaussian 09 Revision D.01, Gaussian Inc. Wallingford CT, 2009.

Acknowledgements

The authors acknowledge CESGA (www.cesga.es) in Santiago de Compostela, Spain, for providing access to computing facilities, and Ministerio de Economía y Competitividad (grant n. FIS2015-68910-P), also in Spain, for financial support.

473336 Measurements and Modeling of Gas Hydrate Formation from High Salinity Systems

Abstract:

In this study, measurements and modeling of gas hydrate phase equilibrium have been performed for aqueous solutions with high salt concentration (near-saturated and saturated solutions) to comprehensively study the interplay of hydrate formation and salt precipitation. As no data exist for hydrate phase equilibria, including the triple point for CH_4 [g] – Solution [l] – Hydrate [s] and quadruple one for CH_4 [g] – NaCl [s] – Solution [l] – Hydrate [s], measurements have been performed with undersaturated and saturated solutions of NaCl, $CaCl_2$, KCl, and $MgCl_2$, as single salts and mixtures, using an isochoric method for pressures up to 206 MPa. For the modeling part, we have developed a simple correlation to determine the hydrate suppression temperature as a function of the salt concentration. This correlation can predict the hydrate equilibrium conditions to within about 1 ^{o}C. Moreover, the approach used for the correlation can be readily extended to other salt systems.

461436 Performance of Biodegradable Kinetic Inhibitors on Natural Gas Hydrate Formation

Abstract:

Clathrate hydrates of natural gas are solid crystalline inclusions that form when suitably sized hydrocarbon components into a lattice structures formed by water molecules at adequately high pressures and low temperatures. The formation of gas hydrates is particularly critical in subsea production, transport, and processing facilities where conditions are more suitable. Several alternative strategies exist for managing hydrate formation and growth whereas use of low dosage kinetic inhibitors is one method that has gained attention over the recent two decades as the required active concentration is below 1.0 wt. % . Although kinetic hydrate inhibitors require much lower dosages than thermodynamic inhibitors their use may be restricted due to poor biodegradability. The purpose of this study is to examine and compare the performance of two commercially available synthetic kinetic hydrate inhibitors within the biodegradable class, tested through well-established screening methods.

466837 Towards a Consistent Thermodynamic Theory for Gas-Hydrates

Abstract:

One of the main criticism of van der Waals-Platteeuw theory, as currently applied for gas hydrates, is the large number of parameters that are required to predict the phase equilibria. These parameters are obtained through regression with experimental data on gas hydrate phase equilibria. As a result, these adjustable carry little physical meaning[1]. In this work, we demonstrate a method to substantially reduce the number of fitted parameters to two per guest molecule while maintaining sufficient accuracy in the prediction of the theromodynamic properties of gas hydrates. This has been possible due to availability of accurate cage occupancy data[2]; development of accurate molecular models for water[3] and guest molecules; and development of methods to rigorously compute the partition function of gas hydrates[4]. We demonstrate our method on gas hydrates formed from methane, ethane and carbon dioxide as well as from their binary mixtures.

References

1. A. A. Chialvo, M. Houssa and P. T. Cummings, *J. Phys. Chem. B*, **106**, 442-451, (2002)

2. J. Qin and W. F. Kuhs, *AIChE Journal*, **59**, 2155-2167, (2013)

3. C. Vega and J. L. F. Abascal, *PCCP*, **13**, 19663-19688, (2011)

4. S. Ravipati and S. N. Punnathanam, *J. Phys. Chem. C*, **119**, 12365-12377, (2015)

471365 Exploring the Free Energy Landscape of Gas Hydrate Nucleation By Forward Flux Sampling

Abstract:

Although the importance of gas hydrate has been widely recognized in diverse areas such as energy recovery, flow assurance, global climate change, and gas storage/transportation, a quantitate knowledge of hydrate nucleation mechanism is still missing. The impediment to acquiring this understanding lies mainly in its stochastic nature and ultra-fine scales, making it challenging to explore hydrate nucleation experimentally. Molecular simulation, which is an ideal tool for unveiling process at the molecular level, can only access a very limited volume of the system, thus requiring a long induction time to observe a spontaneous nucleation event under a realistic thermodynamic condition.

Here we present an advanced computational strategy based on forward flux sampling (FFS)[1] to overcome the limit of the direct molecular simulation. Through this strategy, we pursue a quantitative, "first principle" description on the key factors contributing to the crystallization of hydrate. To facilitate the application of FFS in studying methane hydrate formation, we developed an effective order parameter based on the topological analysis of the tetrahedral network[2]. The order parameter capitalizes the signature of hydrate structure, i.e., polyhedral cages, and is capable of efficiently distinguishing hydrate from ice and liquid water while allowing the formation of different hydrate phases, i.e., sI, sII, and amorphous. The quality of the developed order parameter was subsequently verified by the rigorous pB histogram analysis, i.e., committor probability analysis, which indeed showed that the order parameter describes well the reaction coordinates of hydrate nucleation.

Employing the developed method, we explicitly explore hydrate nucleation on its nucleation rate, free energy landscape, and structural evolution, for different guest species, and under conditions where spontaneous hydrate nucleation becomes too slow to occur in direct simulation.

First, analyzing the obtained large ensemble of hydrate nucleation trajectories, we show hydrate formation, on average, is facilitated by a gradual transition from amorphous to crystalline structure, which provides the direct support to the two-step nucleation mechanism of gas hydrate. Importantly, the analysis also reveals the diversity of hydrate nucleation behaviors. In particular, there also exist nucleation pathways where hydrate crystallizes directly into crystalline phase, without necessarily going through the amorphous stage. The diversity of hydrate nucleation pathways highlights the complexity of hydrate nucleation, and calls for further study for hydrate nucleation mechanism under various conditions.

Second, we obtained the free energy landscape along the nucleation pathway through combining the FFS and backward fluxing sampling (BFS)[3]. Since the stationary distribution as a function of the order parameter is calculated independent of nucleation theory, the obtained free energy profile can be used to test the applicability of the classical nucleation theory (CNT). Interestingly, the calculated free energy profile for homogeneous ice nucleation was found to fit perfectly against CNT, a result supporting our previous studies[4,5] demonstrating the classical nature of ice nucleation. In comparison, the calculated free energy profile for hydrate nucleation shows noticeable deviation from the CNT, particularly at small cluster sizes. The difference may not be surprising, and can be attributed to the non-classical nucleation pathways of hydrate and the irregular shape of hydrate nucleus. Nevertheless, the differences aside, the CNT was found to describe the free energy landscape of hydrate nucleation reasonably well, despite its non-classical nature.

Acknowledgement:

The work is supported by NSF through award CBET-1264438.

References:

(1) Allen, R. J.; Frenkel, D.; Wolde, Ten, P. R. Simulating Rare Events in Equilibrium or Nonequilibrium Stochastic Systems. ***J Chem Phys* 2006**, ***124*** (2), 024102.

(2) Bi, Y.; Li, T. Probing Methane Hydrate Nucleation Through the Forward Flux Sampling Method. ***J Phys Chem B* 2014**, ***118*** (47), 13324–13332.

(3) Valeriani, C.; Allen, R. J.; Morelli, M. J.; Frenkel, D.; Rein ten Wolde, P. Computing Stationary Distributions in Equilibrium and Nonequilibrium Systems with Forward Flux Sampling. ***J Chem Phys* 2007**, ***127*** (11), 114109.

(4) Li, T.; Donadio, D.; Russo, G.; Galli, G. Homogeneous Ice Nucleation From Supercooled Water. ***Phys Chem Chem Phys* 2011**, ***13*** (44), 19807–19813.

(5) Cabriolu, R.; Li, T. Ice Nucleation on Carbon Surface Supports the Classical Theory for Heterogeneous Nucleation. ***Physical Review E* 2015**, ***91***, 052402.

457040 Nonlinear Velocity Profiles of Sheared Active Bacterial Suspensions

Abstract:

Bacterial suspensions under shear exhibit interesting rheology including a remarkable "superfluid-like" behavior with vanishing viscosities. Hydrodynamic theories have shown that such a bacterial "superfluid" is associated with nonlinear shear banding flows. However, although the bulk rheology of bacterial suspensions has been experimentally studied, velocity profiles within sheared bacterial suspensions have not been systematically explored. Here, we experimentally investigate the microscopic dynamics of *E. coli* suspensions under planar oscillatory shear. Using high-speed confocal rheometry, we measure the velocity profile of sheared bacterial suspensions between two shear plates. We find that at low shear rates, high-concentration active bacterial suspensions exhibit an unusual nonlinear velocity profile in the "superfluid" state. Such a nonlinear profile gradually transitions into a conventional linear velocity profile with increasing shear rates. We demonstrate that the nonlinear shear flow is a direct consequence of the collective motion of swarming bacteria. The phase diagram of sheared bacterial suspensions is systematically measured as functions of shear rates and bacterial concentrations. We quantitatively interpret the observed phase diagram based on the competition between the shear stress of fluid flows and the active stress of swarming bacteria. Our study provides new insights into rheology of bacterial suspensions and sheds light on shear-induced dynamics of active fluids.

462309 Effects of Elasticity and Hydrodynamic Interactions on Locomotion in a Coarse-Grained Model of Monotrichous Bacteria

Abstract:

The locomotion of singly-flagellated (monotrichous) bacteria in viscous fluid is a quintessential example of movement at low Reynolds number. Observations of such organisms have shown that the hook protein connecting cell body and flagellum may buckle when the flagellum generates a large enough propulsive force. This local instability is sufficient to produce complex trajectories even for relatively rigid flagella. We first model a swimmer as rigid cell body and rigid flagellum (with simple toy models and a fully detailed model) connected by a flexible hook, then discretize the flagellum as a series of rods connected by flexible springs to incorporate elasticity. In the first case, the hook always buckles above a critical flexibility, and the resulting misaligned swimmer undergoes complex trajectories. In the second case, the flagellum itself cannot maintain its equilibrium helical shape above a critical flexibility, leading to significantly less propulsion than the rigid linked swimmer. The inclusion of hydrodynamic interactions affects the stability boundaries in all cases. These considerations may explain the biology and composition of particular organisms, and provide insight into optimal designs and materials for artificial microswimmers.

455288 Effects of Director Field on Dynamics of Bacteria in Liquid Crystals

Abstract:

Orientationally ordered liquid crystal environment is known to align the swimming direction of rod-like bacteria such as *B. subtilis* parallel to the local director. We present experimental studies of *B. subtilis* in the nematic cells with the director field that is either (i) perpendicular to the direction of swimming or (ii) contains topological defect. Perpendicular alignment (i) forces a bacterium to adopt one of the two dynamic states: (a) spinning around the longitudinal axis but not moving in the plane of the cell; (b) swimming perpendicularly to the director with hairpin-like reversals of the direction. The tumbling behavior of the bacteria adopts two modes not seen in isotropic media, namely, (1) a direct 180 degree reversal of swimming direction and (2) a random change of swimming direction through a transient spinning state. Pre-imposed topological defects (ii) allow one to control the trajectory of bacterial motion, in particular, to rectify polarity of circular orbiting and to trigger pumping.

The work is supported by NSF DMR-1507637 and DMS-1434185.

472119 Locomotion of *C. Elegans*: 2D and 3D Maneuvers in Undulatory Burrowing and Swimming

Abstract:

A free-living nematode *Caenorhabditis elegans* is a powerful genetic model, essential for studies ranging from behavior to neural control to aging; since the nematode locomotion is the key observable, it is used to evaluate the impact of mutations and pharmacological interventions. In its natural environment *C. elegans* efficiently moves in 3D complex media, i.e., it burrows and swims in decomposing organic matter and water. The available quantitative investigations of the locomotion of *C. elegans* have so far been limited to two dimensional motion. Recently [Phys. Fluids 25, 081902 (2013)], we provided a quantitative analysis of turning maneuvers of crawling and swimming nematodes on flat surfaces and in 2D fluid layers. We now follow with the first 3D description of how *C. elegans* navigates complex 3D environments. We show that by superposing body torsion and 2D undulations, a burrowing or swimming nematode can rotate the undulation plane. A combination of these roll maneuvers and 2D turns associated with a purposeful variation of undulation-wave parameters allows the nematode to explore the 3D space. As a conclusion, we analyze 3D chemotaxis of nematodes burrowing in gel and swimming in water to demonstrate an important application of our nematode maneuverability model.

456077 Reverse Marangoni Surfing

Abstract:

We theoretically study the surfing motion of chemically and thermally active particles located at a flat liquid-gas interface that sits above a liquid layer of finite depth. The particles' activity creates and maintains a surface tension gradient resulting in the auto-surfing. It is intuitively perceived that Marangoni surfers propel towards the direction with a higher surface tension. Remarkably, we find that the surfers may propel in the lower surface tension direction depending on their geometry and proximity to the bottom of the liquid layer. In particular, our analytical calculations for Stokes flow and diffusion-dominated scalar (i.e. chemical concentration and temperature) fields indicate that spherical particles undergo reverse Marangoni propulsion under confinement whereas disk-shaped surfers always move in the expected direction. We extend our results by proposing an approximate formula for the propulsion speed of oblate spheroidal particles based on the speeds of spheres and disks.

471549 A New Class of Colloidal Swimmers Based on Magnetically Actuated Assemblies of Metallo-Dielectric Microcubes

Abstract:

We present results and analyze the origins of motility of a new class of externally powered active clusters made by magnetic assembly and actuation of metallo-dielectric microcubes. We have previously shown how metallo-dielectric Janus/patchy spheres and microcubes acquire complex polarization patterns in magnetic fields, leading to multidirectional interactions and assembly into microclusters and microchains. The field-induced polarization of the ferromagnetic facets leads to dipole-dipole and field-dipole interactions and reconfiguration of the neighboring microcubes, which is directed by their conformational restrictions. In effect, the external field energy enables programmed cluster dynamics. The reversible magnetically controlled folding and unfolding of these clusters can be used to create novel active structures. In this talk, we show how pre-assembled microcube clusters form the basis of a new class of microswimmers. These assembled micro-cube clusters are able to achieve net motion if (i) the stroke pattern is made non-symmetric in time (by adjusting magnetic field sequence) and (ii) the fluid medium (i.e., polymeric solutions) possesses non-Newtonian rheological properties such as shear-thinning viscosity. The magnetically powered assemblies can be moved dynamically back and forth by inverting the time-asymmetric stroke sequence (e.g., rapid opening and slow closing *vs.* slow opening and rapid closing) or made to stay in place by applying time-symmetric fields. Surprisingly, the direction of their motion is dependent on the cluster sequence and the length of the re-folding chain. We believe that this behavior is connected to polymeric fluidâ€™s extensional viscosities, which is known to be dependent on strain. The results will be interpreted on the basis of a model of â €œcoupled scallopsâ€ configuration projected upon the metallo-polymer clusters. We will discuss the potential of making more complex dynamic structures that could self-propel in various complex fluids such as liquid crystals.

470672 Origins of Concentration Gradients for Diffusiophoresis

Abstract:

Fluid and particle transport that is driven by gradients of pressure, gravity, or electro-magnetic potential is well-known and studied in many fields. A subtler type of transport, called diffusiophoresis, occurs in a gradient of chemical concentration, either electrolyte or non-electrolyte. Although this mechanism of particle transport is well-known, the diffusiophoresis mechanism is often considered to be an esoteric laboratory phenomenon. However, in this talk it is shown that concentration gradients can develop in a surprisingly wide variety of physical phenomena – imposed gradients, asymmetric reactions, dissolution, crystallization, evaporation, mixing, sedimentation, and others – so that diffusiophoresis is in fact a very common transport mechanism in complex fluids, in both natural and artificial systems. It is anticipated that in georeservoir extractions, physiological systems, drying operations, laboratory and industrial separations, crystallization operations, membrane processes, and many other situations, diffusiophoresis is already occurring – often without being recognized – and that opportunities exist for designing this transport to great advantage. In this talk methods are shown for identifying when diffusiophoresis might be important, and for analyzing transport in complex fluid systems, from both a modeling and experimental standpoint.

469938 Boundary Guidance Due to Hydrodynamics Interactions in the Navigation of Diffusiophoretically Self Propelled Colloid

Abstract:

Self-propelled, chemically-powered colloidal locomotors are swimmers designed to transverse small scale landscapes in a range of applications involving micropumping, sensing and cargo transport. Applications can require that the locomotors navigate precisely prescribed pathways, in which case onboard steering mechanisms are incorporated into the engine design. We demonstrate, through numerical simulation how boundaries can simply and passively guide locomotors to move along their surfaces. As a model system we choose an engine design in which a spherical Janus colloid coated with a symmetrical catalyst cap converts fuel in a continuous liquid phase into a product solute on one side of the colloid. The solute is repelled from the colloid through a short range interaction in a thin layer around the particle which is much smaller than the swimmer radius. The concentration gradient which develops because of the asymmetric production on the active Janus side creates a slip velocity in the layer which propels the swimmer in the direction opposite to the cap (self-diffusiophoresis). Previous research has demonstrated that diffusiophoretically-driven Janus swimmers reaching a planar wall can, for a range of active cap areas, rotate to a configuration in which the active side is partially inclined from the wall and the locomotor skims at a constant separation distance. In this configuration, the net diffusiophoretic propulsive torque balances the hydrodynamic torque resulting from the wall proximity, and the net diffusiophoretic force normal to the wall is equal to zero. The propulsive force along the surface balances the hydrodynamic resistance, allowing the colloid to skim. This represents the simplest form of passive guidance.

Here we investigate the more complicated example of boundary guidance in which the swimmer skimming along a planar wall encounters a second wall which intersects the first. We use a boundary integral method to compute the resulting colloid trajectory as a function of the size of the active cap area and the intersection angle. We find that torques exerted on the swimmer arising from the hydrodynamic interaction with the wall and the diffusiophoretic propulsive force cause the colloid to rotate from its stable skimming configuration. We develop criteria, as a function of the intersection angle and cap size , to describe when the colloid resumes skimming along the intersecting boundary, or rebounds from the wall breaking the boundary guidance.

471553 Collective Dynamics of Catalytically Self-Propelled Particles

Abstract:

Colloidal motors capable of autonomously traversing liquids exhibit remarkable nonequilibrium phenomena including clustering and phase separation. Regarding motions of micron and sub-micron motors, the nature of viscous medium and the role of Brownian motions are substantial and hence, it is crucial to perceive the microhydrodynamics beside the underlying driving force for propulsion in probing the collective dynamics of the system. An increasing number of experiments on catalytically-driven (active) colloidal particles have shown that the interactions of chemically active particles are more complicated than usual interaction of two nonreactive (passive) particles. Indeed, each chemically active particle changes the distribution of product (or reactant) solutes which, in turn, alters the motion of the other particles. In this case, the motions of active particles are influenced not only by hydrodynamic interactions but also via diffusive interactions of the solute distribution generated (or consumed) at the surface of the active areas of each colloid with the boundary of other colloids.

In this research, we first present an analytical solution based on a continuum approach to address the pair interaction of two partially active colloids with arbitrary orientations. Colloids' translational and angular velocities at Stokes flow regime are obtained using Reynolds Reciprocal Theorem (RRT) based on an asymptotic approach in which the net interaction creates a slip-velocity at the surface which actuates the motion. Our analysis indicates two possible scenarios for pair trajectories of catalytic self-propelled particles: either the particles approach, come into contact and assemble or they interact and move away from each other (escape). For motions of the colloids, it is found that the direction of particle rotations is the key factor in determining the escape or assembly scenario. Next, we extend this work to account for many body interactions of a suspension of self-propelled particles by Stokesian Dynamics (SD) simulations. The phoretic interactions between the colloidal motors are short-range and hence can be taken into account in a pairwise additive fashion and the hydrodynamic and Brownian forces are taken into account according to the standard SD. We believe the proposed numerical platform can shed light in our challenging search for a

fundamental understanding of reductionist systems in active matter.

472299 Diffusion of Ellipsoids in Active Fluids

Abstract:

Enhanced diffusion of passive tracers immersed in active fluids is a universal feature of active fluids and has been studied extensively in recent years. In analogy to microrheology used in equilibrium complex fluids, it is reasonable to assume that the unusual dynamics of passive tracers reveal intrinsic properties of active fluids. Nevertheless, experiments, simulations and theories have all shown that the translational dynamics of spherical tracers in active fluids are qualitatively similar, independent of whether the active fluids are pushers or pullersâ€"the two fundamental classes of active fluids. Is it possible to distinguish between pushers and pullers by simply imaging the dynamics of passive tracers? Here, we experimentally studied the diffusion of isolated ellipsoids in both pusher and puller active fluids using suspensions of E. coli and algae as our model systems. Due to the asymmetric nature, ellipsoids possess a rotational degree of freedom, which has not been studied in the context of active fluids. Surprisingly, we found that although the translational and rotational diffusions of ellipsoids show the same enhancement in both pushers and pullers similar to that of spherical tracers, the coupling between the two degrees of freedom exhibits completely opposite trends in the two classes of active fluids. We showed that these opposite translation-rotation couplings arise from a sharp difference in the anisotropy of the body-frame diffusion of ellipsoids. An ellipsoid diffuses fastest along its major axis when immersed in pullers, whereas it diffuses slowest along the major axis in pushers. Based on these experimental findings, we have developed a simple mean-field hydrodynamic model, which qualitatively explains the observed behaviors. Our study provides not only a new insight into organizing principles of active fluids, but also a convenient tool for probing the fundamental classes of active fluids. We acknowledge support from ACS PRF (54168-DNI9) and the Packard Foundation.

466570 Emergence of Force Chains in Granular Materials

Abstract:

An important but unresolved question on the mechanics of granular materials is the relation between the stress and the microstructure. Several experiments on two dimensional disks have reported the presence of anisotropic quasi-linear structures, called "force chains", which have been shown to carry a large fraction of the stress. Several studies have tried to quantify these structures using conventional mechanical definitions and network theory-based measures, but the problem largely remains unsolved. We report here the results of our computational studies on static and sheared collections of grains.Â Our main advance is the definition of an order parameter Qâ€ which measures long-range correlations based on the linearity of contacts. The value of Q is limited to the range [0, 1], the upper limit representing particles aligned in a straight line, and the lower limit representing a self-avoiding walk that cannot take turns greater than 90 degrees. The size (number of particles) of the clusters N(Q) shows a sharp transition at a critical value of Q_c of the order parameter, similar to phenomena observed in phase transitionâ€™s. The average inter-particle normal force of clusters corresponding to a particular value of Q shows a sharp peak exactly at Q_c, which is reminiscent of the divergence of certain quantities in phase transitions as a critical point is approached. We hypothesize that the force chains observed experimentally and numerically are the linear-most set of percolating clusters, as corroborated by the percolation probabilities of these clusters. All our observations hold even for randomly generated particle configurations, where $Q_{c\hat{A}}$ shows strong dependence on the coordination number. Thus, it appears that self-organization is not necessary for the emergence of force chains, rather these chain-like structures are a generic property of random geometric networks. Finally, we show that the cluster size distribution does not depend on the number of particles in the system, indicating that the long-range correlations are not system-size dependent.

461003 Transition in a Dense Granular Flow Down an Inclined Plane

Abstract:

A granular flow down an inclined plane exhibits a discontinuous transition from a disordered, random state to an ordered state consisting of layers of particles with in-layer hexagonal order, when there is a small change in the roughness of the base. The rough base was created using a random arrangement of frozen particles at the base, and the roughness was varied by varying the ratio of the frozen and moving particle diameters. The transition base roughness depends on the interaction model used for the particle interactions, but is independent of the layer height and angle of inclination. In the ordered state, there are layering of particles into distinct layers sliding on each other, and ordering of particles in the layers. The volume fraction in the ordered state is higher than that in the disordered state, but the resistance to flow is much lower and there is an increase in the flow rate by two orders of magnitude when there is a disorder-order transition. The velocity profile in both the ordered and disordered states follow the Bagnold law, but the Bagnold coefficients are very different. In addition to base roughness created by frozen particles, the effect of a different forms of base roughness, which are sinusoidal perturbations of varying amplitude and wavelength, is examined. The results indicate that the results for the roughness are not a linear superposition of sinusoidal modes of different amplitude, and the response to an external perturbation also do not exhibit the exponential decay at low amplitudes expected from linear response theories.

467558 DNS Simulation of Liquid Bridges Between Particles

Abstract:

Liquid transfer within granular materials can result in alteration of the physical properties of the material due to the formation of liquid bridges between individual particles. Cohesive and viscous forces lead to complex and still not well understood granular behaviour. This can have a significant impact on various industrial processes such as coating, drying or granulation. Understanding how liquid is transferred in the micro scale between individual particles has been very challenging both experimentally and computationally [1].

The objective of the current work is the use of Computational Fluid Dynamics in order to:

1. Investigate the capillary forces due to the formation of a liquid bridge between a pair of particles.

2. Investigate the Viscous forces due to high wet particle collision velocities.

3. Develop a liquid transfer model applicable in larger scale particle simulations. Specifically we examine the liquid bridge rupture point and redistribution of the liquid on the involved particles, depending on liquid bridge volume, particle size ratio, liquid viscosity, surface tension and collison velocity.

The abovementioned microscale invastigation can be used as a tool for the development of closed form expressions that can be easily used in a macroscale DEM-CFD simulation of a fluidized bed agglomeration process.

REFERENCES:

[1] C.D. Willett, A.J. Simon, M.J. Adams, J.P.K. Seville, Pendular capillary bridges, In: in: A. Salman, M. Hounslow, J.P.K. Seville (Eds.), Granulation, Handbook of PowderTechnology, vol. 11, Elsevier, Amsterdam, 2007, pp. 1317â €“1350.

ACKNOWELEDGMENTS:

This work was supported by the TMAPPP Marie Curie Initial Training Network, funded through the People Programme (Marie Curie Actions) of the European Union's Seventh Framework Programme under grant agreement No. ITN607453.

463677 Fingering Instability of a Suspension Film Spreading on a Spinning Disk

Abstract:

We have experimentally investigated the spreading of a suspension drop when rotated atop a spinning disc using flow visualization techniques. The suspension is created from 53–106 micron glass beads suspended in a low viscosity, partially wetting Newtonian liquid having same density as the glass beads. The suspension drop (of differing particle volume fractions) is placed centrally on a horizontal disc and the disc is then rotated at a desired speed. The spreading behavior is captured using a high speed camera and the acquired images are analyzed to find the edges of the spreading drop. For all particle volume fractions ($_p$), the suspension drop spreads upto a critical radius, before the contact line (drop edge) develops instabilities which further grow into fingers. The dependence on $_p$ is, however, found to be quite remarkable. The critical radius for the onset of instability shows an increase with increase in the particle fraction ($_p$) before decreasing slightly at the highest value of $_p$ studied, while the instability wavelength (λ) exhibits a non-monotonic dependence. The value of λ is close to that for a partially wetting liquid at lower $_p$, it decreases with increasing $_p$ to a minimum before increasing again at largest $_p$. The non-monotonic trends observed for λ are discussed in light of the linear stability analysis of thin film equations derived for suspensions by Cook et al., [1] and Balmforth et al., [2].

References:

[1] B. P. Cook, O. Alexandrov, and A. L. Bertozzi, "Linear stability of particle-laden thin films," Eur. Phys. J. Special Topics 166, 77 (2009).

[2] N. Balmforth, S. Ghadge, and T. Myers, "Surface tension driven fingering of a viscoplastic film," J. Non-Newtonian Fluid Mech. 142, 143 (2007).

469264 A Rheological Signature of Frictional Interactions in Shear Thickening Suspensions

Abstract:

We explore the relative contributions from hydrodynamic lubrication and frictional contact forces to colloidal shear thickening by measuring both the viscosity and first normal stress difference in suspensions of silica spheres over a wide range of volume fractions. At moderate volume fractions we find $N_1<0$, consistent with hydrodynamic models, however at higher volume fractions and shear stresses these models break down and we instead observe dilation ($N_1>0$), indicating frictional contact networks. Remarkably, there is no signature of this transition in the viscosity, instead this change in the sign of N_1 occurs while the shear thickening remains continuous.

These results suggest a scenario where shear thickening is driven primarily by the formation of frictional contacts, with hydrodynamic forces playing a supporting role at lower concentrations. Motivated by this picture, we introduce a simple model which combines these frictional and hydrodynamic contributions and accurately fits the measured viscosity over a wide range of particle volume fraction and shear stress.

471988 A New Conformation Tensor Based Macroscopic Model for Emulsions at Finite Reynolds Numbers

Abstract:

Multiphase flows are commonly encountered in numerous industrial processes. This work develops a new, thermodynamically consistent, conformation tensor based model that describes signature rheological features of emulsion droplets seen at finite Reynold number flows. The rheological behavior of emulsion droplets at Re<<1 has been well described previously using a contravariant conformation tensor, **C**, in the Maffetone and Minale (1998) model. This model describes the dynamics of droplets in dilute emulsions through a conformation tensor of constant determinant (det(**C**)=1) and employs the Gordon-Schowalter time derivative. Recently, there has been renewed interest in describing the rheological behavior of emulsions at finite Reynolds number. Li and Sarkar (2005) have carried out simulations of single droplets in shear flows at finite Reynolds number. Unlike zero Reynolds flows where the droplets are observed to orient in the flow direction, in the presence of inertia, the droplets orient increasingly in the velocity gradient direction. Furthermore, negative first normal stress differences and positive second normal stress differences are seen. The observation of negative first normal stress differences is also characterized by orientation of the major axis of the droplets at angles, θ, greater than 45^0 above the flow direction. The findings of Li and Sarkar (2005) have also been demonstrated using analytic calculations by Raja et al. (2010) who associate the emergence of negative first normal stresses to a critical Ohnesorge number. This dimensionless group describes the competing effect of viscous forces on one hand and inertia and interfacial forces on the other. The effect of this dimensionless group has not been described by a macroscopic model to the best our knowledge.

In this work, we describe the systematic development of an extended Maffetone-Minale model using the General Equation for Reversible Irreversible Coupling (GENERIC) revealing a new dissipative term in the time derivative leading to a new codeformational time derivative incorporating two parameters. In addition, the extended Maffetone Minale model also includes a stress tensor, defined self consistently from the

GENERIC framework. The new time derivative includes an additional parameter beyond the non-affine parameter,ξ , typically associated with the Johnson-Segalman model for polymers. For specific values of the new parameter, ζ, the macroscopic model allows us to predict negative first normal stresses. Futhermore, the droplet morphology associated with the negative first normal stresses is oriented increasingly in the velocity gradient direction with $\theta>45^0$. Therefore, we are now able to parameterize the macroscopic model in terms of the Ohnesorge number allowing us to describe the effects of inertia on the droplet dynamics in the extended Maffetone Minale model and describe the results of Li and Sarkar (2005). Finally, by examining the asymptotic solutions of the extended Maffetone-Minale model, and comparing against analytic solutions by Raja et al. (2010) we are able to elucidate the role of the Ohnesorge number in controlling the onset of negative normal stresses associated with inertia.

References

1. Maffettone, P.L. and Minale, M., 1998. Equation of change for ellipsoidal drops in viscous flow. *Journal of Non-Newtonian Fluid Mechanics, 78*(2), pp.227-241.
2. Ait-Kadi, A., Ramazani, A., Grmela, M. and Zhou, C., 1999. "Volume preserving" rheological models for polymer melts and solutions using the GENERIC formalism. *Journal of Rheology (1978-present), 43*(1), pp.51-72.
3. Li, X. and Sarkar, K., 2005. Effects of inertia on the rheology of a dilute emulsion of drops in shear. *Journal of Rheology, 49*(6), pp.1377-1394.
4. Raja, R.V., Subramanian, G. and Koch, D.L., 2010. Inertial effects on the rheology of a dilute emulsion. *Journal of Fluid Mechanics, 646*, pp.255-296.
5. Beris, A.N. and Edwards, B.J., 1994. Thermodynamics of flowing systems: with internal microstructure.
6. Grmela, M. and Öttinger, H.C., 1997. Dynamics and thermodynamics of complex fluids. I. Development of a general formalism. *Physical Review E,56*(6), p.6620.

471347 Direct Tracking of Particles and Quantification of Margination in Blood Flow

Abstract:

There has been a growing interest in using micro- and nano-particles as drug carriers, especially for cancer treatment (1). However, little is known about the rheology of particles in blood flow, which must be understood if the drug-carrying particles are to be administered intravenously. Decuzzi *et al.* and Graham *et al.* (2, 3) showed in their modeling that particles can undergo a "margination" phenomenon wherein the particles trend toward the periphery of blood vessels. The implication is a higher chance for the particles to diffuse into the tumor through the leaky vasculature typically found near tumor sites. Experimentally, most existing studies use particle adhesion to channel walls during flow as a mechanism for quantifying the margination propensity of particles. However, adhesion is also affected by other factors, such as hydrodynamic drag and ligand density, further complicating the interpretations of experimental results (4). In this presentation, we will present our findings on directly tracking the spatial distribution and velocity of particles flowing through microfluidic devices. A margination parameter, M, is further defined as the total number of particles found within the cell-free layers normalized by the total number of particles that passed through the channel. The effects of particle size, suspending medium, and apparent shear rate on margination propensity were quantified using this parameter.

References

1. Davis, ME. Fighting cancer with nanoparticle medicinesâ€•The nanoscale matters. MRS Bulletin. 2012;37:828-835.

2. Decuzzi P, Lee S, Bushan B, Ferrari M. A theoretical model for the margination of particles within blood vessels. Annals Biomed. Eng. 2005;33(2):179-190.

3. Kumar A, Graham MD. Margination and segregation in confined flows of blood and other multicomponent suspensions. Soft Matter. 2012;8(41):10536â€“48.

4. Carboni E, Tschudi K, Nam J, Liu X, Ma A. Particle margination and its implications on intravenous anticancer drug delivery. AAPS PharmSciTech. 2014;15(3):762-71

469420 The Partial Drift Volume Induced By a Translating Spherical Bubble at Finite Reynolds Number

Abstract:

The Lagrangian displacement, or drift, of fluid elements caused by moving bodies is important to pool boiling, dispersion in fixed beds, and oceanic bio-mixing. Previous theoretical calculations of drift have been conducted at the limits of inviscid flow and Stokes flow. Here, the drift caused by the translation of a bubble through a quiescent fluid is quantified at finite Reynolds number (*Re*). We consider a spherical, non-deformable bubble of radius *a* translating at a constant speed *U* through an unbounded Newtonian fluid. The bubble starts at the origin and moves in a straight line; axisymmetry of the flow about the direction of translation is assumed. The flow around the bubble is obtained for $0.01 < Re < 100$ by numerically integrating the Navier-Stokes equations using a spectral element method. Then, the distortion of a circular material sheet of fluid elements of radius $R \gg a$ centered on and perpendicular to the bubble's axis of translation is computed. This is accomplished by integrating the trajectories of individual fluid elements on the material sheet over time. The sheet is positioned at an initial distance *L* from the bubble at time $t = 0$. The volume swept out by this material sheet after a time $t = t_f$ defines the partial drift volume V_D. We quantify V_D with respect to the parameters *Re*, *L*, *R*, and t_f. At finite *Re*, the displacement of fluid elements on the sheet, and hence V_D, grow unbounded as $t_f \to \infty$. This is the result of the slow decay of the velocity field in the wake of the bubble, which acts as a point source of momentum at large distances. Specifically, we find that $V_D \sim \ln(t_f)$ as $t_f \to \infty$. However, the coefficient of this scaling depends on R/L. We also find that increasing *Re* decreases V_D relative to that of the bubble. Notably, $V_D \sim Re^{-0.5}$ for $Re \gg 1$, which is the same as the scaling for the drag on the bubble at large *Re*. Physically, fluid elements are entrained by the viscous wake behind the bubble, which decreases in extent as *Re* increases.

466653 Motion of a Spherical Particle Near a Porous Boundary

Abstract:

The flow of suspensions near porous boundaries occurs in a variety of industrial processes, such as shear-enhanced filtration, and in living systems, such as blood flowing through capillaries. Some of the important questions that arise are how the permeability of the porous walls and the concomitant velocity slip affect the dynamics of individual particles, and the effective rheology of the suspension. In this study, we consider the motion of a single rigid sphere near a plane permeable wall; the Stokes equations are solved by obtaining the velocity and pressure fields as eigenfunction expansions in the bispherical co-ordinate system. The boundary conditions at the porous wall are Darcyâ €™s flux condition normal to the surface and Saffmanâ€™s slip condition in the tangential direction. In addition, we impose the condition of zero net flux across the porous wall, and obtain the non-zero pressure on the other side of the porous wall. The drag and torque on the particle moving near a porous wall are computed and compared with those obtained for an impermeable wall. The effect of various parameters such as distance of the particle from the wall, permeability, slip coefficient and wall thickness are studied. Finally, we comment on the implications of our results on recent measurements of the suspension stress near porous boundaries.

456086 A Reciprocal Theorem for Convective Heat and Mass Transfer from a Particle in Stokes and Potential Flows

Abstract:

In the study of convective heat and mass transfer from a particle, key quantities of interest are usually the average rate of transfer and mean distribution of the scalar (i.e. temperature or concentration) at the particle surface. Calculating these quantities using conventional equations requires the detailed knowledge of the scalar field, which is available predominantly for problems involving uniform scalar and flux boundary conditions. Here, we derive a reciprocal relation between two diffusing scalars that are advected by oppositely driven Stokes or potential flows whose streamline configurations are identical. This relation leads to alternative expressions for the aforementioned average quantities based on the solution of the scalar field for uniform surface conditions. We exemplify our results via two applications: (i) heat transfer from a sphere with non-uniform boundary conditions in Stokes flow at small Peclet numbers, and (ii) extension of Brenner's theorem for the invariance of heat transfer rate to flow reversal.

455915 Interfacial Tensions of Industrial Fluids from a Molecular-Based Square Gradient Theory

Abstract:

Several correlations exist which, based on semi-empirical corresponding states principia or otherwise, allow the calculation of the interfacial tension (IFT) of industrially relevant fluids. However, their application is typically restricted to the chemical family used to fix the constants involved. In this work we combine a molecular thermodynamic theory and molecular simulations to obtain faithful description of the tensiometry of molecular models of fluids and a mapping of it to experimental data. Specifically, this work combines a theoretical approach based on the statistical associating fluid theory (SAFT) SAFT-VR Mie EoS with the square gradient theory (SGT) and molecular dynamics simulations (MD). This approach is based on the description of the interfacial properties for short flexible chains composed of 2, 3, 4, 5, and 6 freely-jointed tangent spheres through a Mie λ - 6 (λ= 8, 10, 12, 20) potential. From the MD results, a simple, flexible and accurate expression for the correlation of the influence parameter in SGT is obtained. This expression provides a route to calculate the influence parameters for pure chain fluids by only using the molecular characteristics of the model fluid (m_s, ε, σ, λ). By combining this approach with previous mappings of the Mie potential to pure fluids (Mejía, A. Herdes, C. and Müller EA. Force fields for coarse-grained molecular simulations from a corresponding states correlation. Ind Eng Chem Res. 2014; 53:4131–4141) one can effectively predict the bulk and interfacial properties of pure fluids from the knowledge of only three widely available properties: the critical temperature, the acentric factor, and a liquid density. A key aspect of the methodology is the internal consistency of the molecular model, that is, both the theory and the simulations are based on the same set of unique force field descriptors. We provide a predictive tool to determine IFTs for a wide range of molecules including hydrocarbons, fluorocarbons, polar molecules, among others. The proposed methodology is tested against comparable existing correlations in the literature, proving to be vastly superior, exhibiting an average absolute deviation of 2.2%.

456536 Nonequilibrium Molecular Dynamics Simulations to Understand Thermal Rectification and the Role of Interfacial Resistances Â in Solid-Liquid and Liquid-Liquid Systems

Abstract:

Thermal transport can be manipulated using designed composite structures since these modify the high frequency harmonic interactions between adjacent vibrating atoms. The method of nonequilibrium molecular dynamics has been used to demonstrate thermal rectification in confined liquids by manipulating the solid-liquid and liquid-liquid interfaces as well as asymmetric temperature dependence of thermal conductivity. Our studies have shown that while previously it was thought that rectification in general will only be possible in solids , these manipulations allow rectification to be observed in a wide range of systems. We were able to demonstrate rectification in polar liquids in contact with hydrophobic/hydrophilic surfaces, and liquids such as benzene and water that have asymmetric temperature dependence of thermal conductivity. These investigations have also allowed us to improve our current understanding of the behavior of Kapitza resistance (thermal resistance in the interfacial boundary), and how it can be used to help control heat transfer rate in nano-systems.

457325 Modeling Water+Hydrocarbons and Water+Oxygenates Phase Equilibrium with the GC-PPC-SAFT Equation of State

Abstract:

The use of lignocellulosic biomass derived from agricultural and forest residues is currently considered as a promising alternative for the production of so-called second-generation biofuels. One of the challenges in the design of processes for converting these feedstocks into valuable products is the description of phase equilibrium: due to the highly oxygenated nature of these products and the presence of large amounts of water, both vapour-liquid and liquid-liquid equilibria can occur. Moreover, the large variety of compounds involved in these processes and the lack of experimental data make the use of predictive thermodynamic models very interesting.

The present work is focused on improving the GC-PPC-SAFT parameterization of pure water and its mixture with hydrocarbons and oxygenates. A previous work with this model [1] has led to succesful modelling of the phase equilibrium of these mixtures, but large density deviations were observed for pure water. Here, we used the approach proposed by Held et al. [2], consisting in modifying the temperature dependence of the hard sphere diameter of water in order to improve the global phase behaviour of this compound in a wide temperature range. The deviations in water vapour pressure and especially liquid density are now significantly reduced compared to the previous model [1], to as low as 2.2% and 0.6% ARD respectively, over a temperature range from 0ºC to 550ºC. Four new parameters are introduced.

Mixtures of water with alkanes, aromatics, alcohols, aldehydes, ketones, ethers and esters are then investigated using the group-contribution concept for predicting their LLE and VLE. Results in terms of mutual solubilities are found to be similar or better than our previous works on these mixtures [3]. The VLLE of some water-alcohol heteroazeotropic mixtures are also investigated. A comparison is also made between our purely predictive approach and the one using fitted binary interaction parameters (k_{ij}). The results so far reflect that after employing our new modifications, the model moves closer to real behaviour

and prediction accuracy is enhanced for water mixtures.

References

[1] N. H. Dong, J.C. de-Hemptinne, R. Lugo, J. P. Passarello, P. Tobaly. "Modeling Liquid-Liquid and Vapor-Liquid enquilbria of Binary systems containing water with an Alkane, an aromatic hydrocarbon, an alcohol or a gas, using GC-PPC-SAFT" Industrial Engineering and Chemical Research, 2011, 50: 7467-7483.

[2] C. Held, L.F. Cameretti, G. Sadowski "Modeling of electrolyte solutions Part 1. Fully dissociated electrolytes" Fluid Phase Equilibria, 2008, 270: 87-96.

[3] T.B. Nguyen, J.C. de Hemptinne, B. Creton, G.M. Kontogeorgis. " Improving GC-PPC-SAFT equation of state for LLE of hydrocarbons and oxygenated compounds with water" Fluid Phase Equilibria, 2014, 372: 113-125.

463542 A Novel Approach of Modeling Orientational Hydrogen Bonding Interactions in Associating Fluids with the COSMO-SAC Activity Coefficient Model

Abstract:

Based on solid theoretical background in statistical thermodynamics and quantum mechanics, the COSMO-SAC model is a powerful predictive thermodynamic model which can provide reliable thermodynamic properties and phase behavior of mixture fluids. The model has been applied in many problems such as vapor-liquid equilibrium, liquid-liquid equilibrium, drug solubility, partition coefficient ionic liquid screening, etc over the last decades. The core knowledge in COSMO-SAC model is how to evaluate the interaction between molecules on the basis of surface segments whose properties such as position and charge density can be determined from quantum mechanical calculation. In recent years, some refinements such as including the temperature dependence on electronic interactions and differentiating the different types of hydrogen bonding interactions are introduced to improve the accuracy of COSMO-SAC model (Mu, Rarey et al. 2007). In this work, we propose a new approach which allows for consideration of spatial orientation of hydrogen bonds in the COSMO-SAC model. Instead of using a cutoff charge or a predefined hydrogen bonding atoms, we adopt the concept of molecular orbital, Valence Shell Electron Pair Repulsion theory (PETRUCCI, R.H 2002), to identify the hydrogen bonding surfaces (Fig. 1). The inclusion of molecular orbital information allows for specific the hydrogen bond interactions within certain spatial orientations and results in a more accurate description regarding the vapor-liquid equilibrium (VLE) of mixtures containing different hydrogen bonding specie. We examined 35 associating fluids (whose include water, alcohols, amines and amides) in total 598 mixtures (7583 data points, temperature range from 263.15K to 548.15 K). The prediction accuracy is found to be 6.75% (AARD-P%) and 2.67% (AAD-y%) in pressure and vapor composition, respectively. The results are more accurate than the original model (7.16% and 2.83%) (Hsieh, Sandler et al. 2010), while the number of universal parameters used are reduced from 10 to 9.

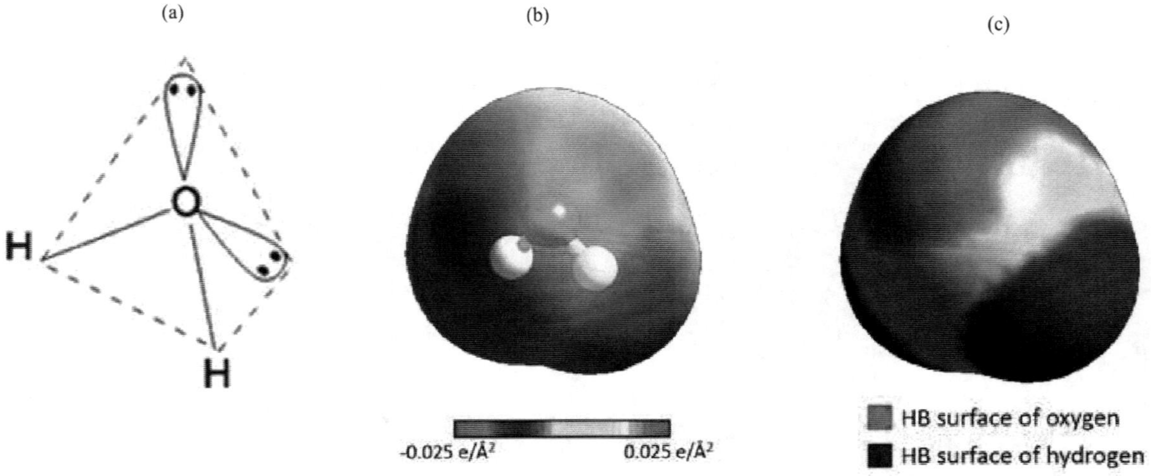

Figure 1. (a) The structure of water molecule and its lone pair describing by VSEPR theory. (b) The screening charge distribution on water surface. (c) The spatial allowance in hydrogen bond restricted by the lone pair within a cut-off radius.

References

1. Hsieh, C.-M., S. I. Sandler and S.-T. Lin (2010). "Improvements of COSMO-SAC for vapor¡Vliquid and liquid¡Vliquid equilibrium predictions." Fluid Phase Equilibria 297(1): 90-97.

2. Mu, T. C., J. Rarey and J. Gmehling (2007). "Performance of COSMO-RS with sigma profiles from different model chemistries." Industrial & Engineering Chemistry Research 46(20): 6612-6629.

3. PETRUCCI, R.H., HARWOOD, W.S. and HERRING, F.G. (2002). General ChemistryPrinciples and Modern Applications, 8th edn. Prentice Hall, Upeer Saddle River, NJ.

468270 Automatized Determination of Fundamental Eos Based on Molecular Simulations in the Cloud

Abstract:

Thermodynamic data are essential for the design and optimization of chemical engineering processes. Typically, the chemical industry is relying on experimental investigations to generate such data. Depending on the type of fluid, this undertaking is difficult to conduct or in the case of hazardous substances, e.g. being explosive, toxic or mutagenic, this might even be practically impossible. As an attractive alternative route, molecular modelling and simulation on the basis of classical force fields has been proven to be an adequate prediction method due to the steady development effort of the last decades. Along with its powerful predictive capabilities, large datasets of thermophysical data can be rapidly produced with considerably less effort than laboratory measurements.

It has recently been shown that molecular simulation data are useful for the construction of empirical fundamental equation of state (EOS) correlations that are explicit in terms of a thermodynamic potential, which is most often the Helmholtz energy [1]. Thermodynamic potentials are one of the most essential building blocks of thermodynamics, because every other time-independent thermodynamic property can be obtained as a combination of the partial derivatives of the thermodynamic potential with respect to its independent variables. Unfortunately, the exact mathematical expression of a thermodynamic potential, with the exception of very simple systems (e.g. ideal gas), is unknown. The construction of an EOS therefore necessarily means that the mathematical form of such a correlation has to be fitted to available thermodynamic data. Once the parameterized correlation is available, it provides thermodynamic data in a consistent manner. Naturally, the more thermodynamically independent properties are considered during the construction of the EOS, the better the representation of other properties is. Therefore, in order to create such a correlation, a large number of independent thermodynamic data points has to be at disposal. This scenario is in fact the application field for the statistical mechanical formalism proposed by Lustig [2,3]. Using his formalism, any derivative of the Helmholtz energy, which intentionally belongs to the same thermodynamic potential in which the EOS is explicit, can be obtained at a given state point by a single canonical ensemble molecular simulation. Since any thermodynamic property is just a combination of these derivatives, the lack of thermodynamically independent data is not a limiting factor anymore, and EOS can be created purely on the basis of molecular simulation data.

However, setting up molecular simulations and fitting an EOS using simulation data requires considerable expertise. The goal of this work is to automatize and optimize this workflow by combining the molecular simulation tool *ms*2 [4] with a simple EOS fitting algorithm, and providing an interface for the user that requires minimal knowledge about the background operations, making EOS development more appealing to the chemical industry. To be specific, this was done here by using a cloud-based approach employing a simple graphical user interface for creating the necessary input for simulations. In the background, a job manager efficiently allocates the molecular simulation runs to the available high performance computing (HPC) resources. Once the simulation results are ready, the EOS fitting software takes over control and provides the empirical EOS correlation along with the quality of representation of the simulation data considered. The capabilities of this approach are tested for ethylene oxide and phosgene [5], two particularly hazardous substances of industrial relevance with scarce experimental databases. **References**

[1] Rutkai, G.; Thol, M.; Lustig, R.; Span, R.; Vrabec, J. *The Journal of Chemical Physics* 2013, 139, 041102.

[2] Lustig, R. *Molecular Simulation* 2011, 37, 457–465.

[3] Lustig, R. *Molecular Physics* 2012, 110, 3041–3052.

[4] Glass, C. W.; Reiser, S.; Rutkai, G.; Deublein, S.; Köster, A.; Guevara-Carrion, G.;Wafai, A.; Horsch, M.; Bernreuther, M.; Windmann, T.; Hasse, H.; Vrabec, J. *Computer Physics Communications* 2014, 185, 3302–3306.

[5] Rutkai, G.; Vrabec, J. *Journal of Chemical & Engineering Data* 2015, 60, 2895–2905.

455518 Phase Behavior and Density Measurements of (CO_2 + Toluene) at Temperatures from (298 to 448) K and Pressures up to 65 Mpa

Abstract:

Introduction

Carbon Capture and Storage (CCS) is a crucial technology in the drive to reduce anthropogenic CO_2 emissions with the aim of keeping the global temperature rise below 2 °C above pre-industrial levels.[1] According to an IEA report in 2015, in all sectors the amount of CO_2 captured and stored needs to be about 6 billion tonne per year until 2050 to achieve the 2 °C target.[2] CCS technologies can be applied in many industrial processes such as natural gas, hydrogen, steel and cement production, power generation and so on. Early commercial projects have been achieved by combining CCS with enhanced oil recovery (EOR) technology.

The understanding of the thermophysical properties for CO_2 + hydrocarbons systems at reservoir conditions is very important for the correct design and optimization of carbon storage and EOR. The available saturated-phase densities, high-pressure liquid densities and phase behaviour data for these mixtures are limited, especially at reservoir conditions.[3] These thermophysical properties are essential to determine the amount of CO_2 that can be stored in depleted oil reservoirs as well as to understand the convective transport through the reservoir. Such experimental data are also important in a wide range of industrial applications. Furthermore, the thermodynamic models used for reservoir and/or process simulation are optimized and/or validated with these experimental data. In this work, new density and phase behaviour measurements of (CO_2 + toluene) systems have been made at temperatures from (298 to 450) K and pressures up to 70 MPa. The measurements have been carried out in specially-developed apparatus, previously reported by our group.[4,5,6] New empirical models have also been developed to represent the densities for this particular mixture. Finally, the new density and vapour-liquid-equilibrium data are compared with the predictions of the SAFT-γ Mie[7] and Peng-Robison equations of state. A key feature of the present study is that it incorporates both phase-equilibrium and phase-density measurements over extremely wide ranges of temperature and pressure, permitting the thermodynamic models to be tested in a rigorous way.

Results and Modelling

Dew- and bubble-point measurements for (CO_2 + toluene) were carried out at T/K = 298.15, 323.15, 373.15 and 423.15 and at pressures up to 20 MPa by means of a synthetic method described previosuly.[4] Critical points were also determined by direct visual observation. These vapour-liquid-equilibria results were found to be in good agreement with data existing in the literature.

Saturated-phase density measurements for (CO_2 + toluene) were performed along six isotherms from 298 K to 450 K and at pressures up to the critical. On most of the isotherms investigated, the bubble-point densities decreased, and the dew-point densities increased, monotonically with increasing pressure. However, at $T = 298$ K and $T = 323$ K, the bubble-point densities initially increased with increasing pressure until a maximum was reached after which they deceased monotonically to the critical point. The experimental dew- and bubble-point data were fitted on each isotherm, with absolute average deviations of about 2 $kg{\cdot}m^{-3}$, using an empirical function incorporating an appropriate non-analytic term. This permitted the critical pressure to be estimated precisely and the results were found to be in close agreement with previous data.[8]

The high-pressure compressed-liquid densities were measured with a vibrating U-tube densimeter as a function of pressure along nine isotherms at T/K = 283, 298, 323, 348, 373, 398, 423, 448 and 473 at pressures up to 70 MPa. The carbon dioxide mole fractions in the binary mixtures studied were 0.0, 0.2, 0.4, 0.6 and 0.8 and the overall uncertainty of the density measurements was ≤ 1 $kg{\cdot}m^{-1}$ at 95 % confidence. In order to maintain the homogeneity of the mixture, only pressures above the bubble curve were considered. An empirical model was used to correlate the high-pressure liquid densities along isotherms. To verify the consistency of the data, we estimated the bubble pressures at each composition by combining the high-pressure liquid-phase and the saturated-phase density data. Bubble points obtained from these densities measurements matched the ones determined directly from the VLE measurements.

The new density and vapour liquid equilibrium data were compared with the predictions of the SAFT-γ Mie equation of state, which implements a group-contribution approach and the Mie potential to represent segment-segment interactions. The results showed that the SAFT-g Mie model predicted the phase behaviour, saturated-phase densities and high-pressure compressed-liquid density for this mixture very well. For comparison, we also tested the using Peng-Robison equation of state with the classical mixing rules and one binary parameter. As expected, this simple model represented the vapour liquid equilibrium quite well but gave inferior results for the saturated-phase and compressed-liquid densities.

Reference

1. International Climate change conference in Paris, December, 2015.
2. IEA, 2015, Energy Technology Perspectives, International Energy Agency, Paris, France
3. S. W. Løvseth, H. G. J. Stang, A. Austegard, S. F. Westman, R. Span, R. Wegge, Energy Procedia, 86 (2016) 469.
4. S. Z. Al Ghafri, G. C. Maitland, J.P. M. Trusler, Fluid Phase Equilibria 365 (2014) 20.
5. E. C. Efika, R. Hoballah, X. Li, E. F. May, M. Nania, Y. Sanchez-Vicente, J.P. M. Trusler J. Chem. Thermodynamics 93 (2016) 347.
6. S. Z. Al Ghafri, G. C. Maitland, and J. P. M. Trusler, J. Chem. Eng. Data, 2013, 58, 402−412
7. S. Dufal, V. Papaioannou, M. Sadeqzadeh, T. Pogiatzis, A. Chremos, C. S. Adjiman, G. Jackson, and A. Galindo J. Chem. Eng. Data, 59 (2014) 3272.
8. J. W. Ziegler, T. L. Chester, D. P. Innis, S. H. Page and J. G. Dorsey, J.W. Ziegler, Anal. Chem. 67 (1995) 456.

Acknowledgment

We gratefully acknowledge the funding of QCCSRC provided jointly by Qatar Petroleum, Shell, and the Qatar Science and Technology Park, and their permission to publish this research.

474306 Direct Measurements of Cholesterol s Impact on Lipid Hydration and Packing in Phosphatidylcholine Lipid Bilayers

Abstract:

The role of cholesterol in physiological and model biological membranes has been studied extensively. A particularly active research area is the importance of cholesterol in modulating lipid rafts domains, which are thought to be involved in a myriad of cellular processes from protein sorting to signal transduction. Fundamental biophysical studies on model systems have begun to tease out the thermodynamics and molecular level interactions between cholesterol and lipids. In this work, high resolution x-ray diffraction from single supported bilayers and monolayers at the air water interface of phosphatidylcholine-cholesterol mixtures clearly elucidates cholesterol-lipid complex formation, phase separation, and suggests the formation of a cholesterol superlattice with a specific stoichiometric ratio that occurs in lipid bilayers but not in the lipid monolayers. Importantly, convincing evidence will be presented that cholesterol alters the hydration of saturated phosphatidylcholine lipid bilayers, resulting in significant changes in lipid packing and orientation.

466188 Growth Behavior of Amyloid Fibrils on Membrane Interfaces of Lipid Membranes

Abstract:

Amyloid β protein (Aβ) that is a small peptide (4.5 kDa) forms amyloid fibrils and is a possible causative factor for Alzheimer's disease. Morphology of fibrils is one of key factors in determining its cytotoxicity[1]. Liposome is a closed lipid bilayer membrane composed of phospholipids and provides the variety of morphologies of Aβ fibrils[2]. The fibril growth behavior was likely to depend on the interfacial properties of liposome such as the charge, hydrophobicity, or membrane fluidity. The initial step where Aβ binds to and accmulates on liposome membranes has been clarified, to some extent [3,4]. Meanwhile, the relationship between the initial step for the growth of amyloid fibrils and their morphologies has been still unclear. In this study, we investigated the relationship between the accumulation step of Aβ on lipid membranes and the growth process, in the molecular aspects.<span">In the observation with a time-of-flight secondary imaging mass spectroscopy (ToF-SIMS), an invasion of Aβ into 1,2-oleoyl-*sn*-glycero-3-pospocholine (DOPC) membrane with high fluidity and a deposit of Aβ on 1,2-palmitoyl-*sn*-glycero-3-pospocholine (DPPC) membrane with low fluidity were confirmed. The accumulation pattern of Aβ obtained by ToF-SIMS were consistent with the result obtained from other spectroscopic methods. Morphology of fibrils on lipid membranes was also characteristic: fibrils formed on DPPC membranes were short, straight and rigid whereas fibrils on DOPC membrane were branched. This growth behavior indicated the environment of lipid membranes being obstacle for elongation of fibrils. Therefore, the position of Aβ accumulation in/on lipid membranes appeared to determine the extent that the environment of lipid membranes such as headgroup mobility or laerall difusion of lipid molecules influeced the elongation process of fibrils.

1) R. Roychaudhuri et al., *J. Biol Chem*, **284**, 4749-4753 (2009)

2) T. Shimanouchi et al., *AIChE J.*, **58**, 3625-3632 (2012)

3) K. Matsuzaki et al., *Biochim. Biophys. Acta* **1768**, 1935-1942(2007)

4) M. Terakawa-Suzuki et al., *J. Biol. Chem.*, **290**, 815-826 (2015)

474337 Influence of Gold Nanoparticle Surface Chemistry and Diameter upon Alzheimer's Disease Amyloid-β Protein Aggregation

Abstract:

Deposits of aggregated amyloid-β protein (Aβ) are a pathological hallmark of Alzheimer's disease (AD). Thus, one therapeutic strategy is to eliminate these deposits by halting Aβ aggregation. While a variety of possible aggregation inhibitors have been explored, only nanoparticles (NPs) exhibit promise at low substoichiometric ratios. With tunable size, shape, and surface properties, NPs present an ideal platform for rationally designed Aβ aggregation inhibitors. In this study, we characterized the inhibitory capabilities of gold nanospheres exhibiting different surface coatings and diameters. Both diameter and surface chemistry were found to modulate the extent of aggregation, while NP electric charge influenced aggregate morphology. Notably, 8 nm and 18 nm poly(acrylic acid) NPs abrogated Aβ aggregation at a substoichiometric ratio of 1:2,000,000. Theoretical calculations suggest that this stoichiometry could arise from altered solution conditions near the NP surface. Insights provided by this study will inform future rational design of effective NP-based therapeutics for AD.

466428 Chiral Recognition of Small Molecules at Boundary Edge in Heterogeneous Liposome Membranes

Abstract:

In order to investigate the interaction of hydrophilic molecules with liposomal membranes, we employed 1-(4-(trimethylamino)phenyl)-6-phenyl-1,3,5-hexatriene and 1,2-dioleoyl-*sn*-glycero-3-phosphoethanolamine-*N*-(5-dimethylamino-1-naphthalenesulfonyl) as fluorescent probes to monitor the surface regions of the membrane, and the results for various liposomes were plotted in correlation diagrams. According to the formation of a variety of phase states, different tendencies of decreasing surface hydrophobicity were observed in the liposomes that were modified with high concentrations of cholesterol or in the liposomes that were composed of ternary components. These liposomes, with hydrophobic surfaces, also showed preferential adsorption of l-histidine (l-His) or S-Ibuprofen, and the hydrophobicity of the liposomal membrane at the surface increased during their adsorption regardless of the initial liposomal properties. Furthermore, we revealed that accelerated adsorption of l-His and preferential binding was induced in ternary liposomes forming boundaries between two separate phases.

466552 Enzyme-Reactive Self-Assembling Peptides for Biomacromolecular Functionalization

Abstract:

Supramolecular nanostructures comprised of self-assembling peptides are a promising soft material in bionanotechnology. They can form nanocarriers of small drugs/biomolecules for pharmaceuticals, nanofibrous networks as an artificial extracellular matrix for regenerative medicine, and so forth. Enzyme-responsiveness of these materials has recently been attracted attention because it enables the efficient and highly specific control of materials functions under physiological conditions. However most of the published work utilized enzymes to fabricate or decompose the assembly, and enzymatic decoration of the nanostructures has not been focused.

Herein we developed novel peptide assemblies that can be further functionalized with biomacromolecules using enzyme catalysis. The enzyme used in our study is microbial transglutaminase, MTG. MTG catalyzes the acyl transfer reaction between Î³-carboxyamide of glutamine (Q) residues and Îµ-amino group of lysine (K) residues or primary amines. MTG-recognizable self-assembling peptides were designed by introducing aromatic group and hydrophobic amino acid sequence at the N-terminal of Q. As a biomacromolecule, enhanced green fluorescent protein (EGFP) was employed, of which C-terminal was attached with the MTG-recognizable lysine-containing sequence, K-tag. We synthesized several self-assembling peptides and investigated the influence of the assembled structures on the MTG reactivity. The enzymatic functionalization of EGFP on the peptides structures was also examined.

466766 Module Library Approach for Designing the High Performance Artificial Cellulosome on Nanomaterials

Abstract:

An economical conversion from inedible cellulosic biomass materials to water-soluble sugar units is expected to produce food, feed, chemicals and fuels. Microbial cellulolytic enzymes, called cellulase is an enzyme which hydrolyzes cellulose to water-soluble sugar units with low energy and low environmental loads, and more than 1000 cellulases are characterized until now. About 33% of these cellulases have carbohydrate-binding domain (CBD), fused to catalytic domain (CD), and the CBDs promotes effective degradation of cellulose by increasing substantial substrate concentration around CD. Furthermore, in some rumen bacteria, CDs with different functions are clustered on a giant scaffold protein containing a CBD, via cohesin-dockerin interaction, called cellulosome, to efficiently degrade cellulose. We have proposed a new design of artificial cellulosome, called hybrid nano-cellulosome (HyNaC), by clustering recombinant biotinylated CDs and CBDs on streptavidin-modified CdSe quantum dot via biotin-avidin interaction.[1] This HyNaC structure showed drastic enhancement of degradation for water insoluble cellulose.[2,3]

In this study, we selected optimal CD and CBD from the library containing 50 kinds of CDs and 20 kinds of CBDs for screening high performance HyNaC. The HyNaC structure format produced more reducing sugar than the corresponding native module structure, and a replacement of the kind of CBD contributed considerably to the enhancement of the degradation of amorphous cellulose on HyNaC, whereas, chimera module enzyme with each CBD did not show. In addition, mixture experiment of HyNaC with different CBD shows the synergistic effect between CBDs observed on HyNaC with most active CD of E2 from *Thermobifida fusca.* Finally, to understand the mechanism of the enhanced degradation on HyNaC by variety of CBD, we investigated the binding affinity and the saturated binding amount of CBD library to amorphous and crystalline cellulose and extracted a rule to enhance the cellulolytic activity of HyNaC based on CBD.

[1] Kim et al. *Small*, 7, 656 (2011)

[2] Kim et al. *Catalysis Science & Technology*, 2, 499 (2012)

[3] Nakazawa et al. *ACS catalysis*, 3, 1342 (2013)

468572 Simulating Airflow in Human and Nonhuman Primate Airways with Applications to Aerosolized Drug Delivery Animal Testing

Abstract:

Assessing the human health benefits and risks from the inhalation of aerosolized medications is often performed by extrapolating experimental data taken using nonhuman primates animal studies to humans. Published results from these studies are often measurements of biological dose-response data that are corrected using coarse-grained allometric scaling to make predictions for humans. In this work we analyze how drug-delivery differs between species from a transport perspective. We use computational fluid dynamics alongside extensive in-vivo animal inhalation experiments to understand the dynamics of airflow in nonhuman primate lungs, and report on differences in aerosol transport and deposition mechanisms between humans and primates.

In the respiratory tract, inhaled aerosols must navigate the airways of the body and be deposited in targeted ways to maximize drug efficiency. Accurate analysis of drug efficacy and dosage requires an understanding of where these particles go upon inhalation. We report deposition fractions in airways across a range of particle sizes, breathing rates, and in numerous lung geometries. These computational results are validated with PET/CT experimental results. We compare these results with our previous report to show that differences in length scales between species produce fundamentally different time-dependent flow behavior and aerosol deposition mechanisms. Particle filtering in the nasal passages are reported for both species. In addition to patient-specific lungs, we compare and contrast the flow through statistically averaged bronchial trees of both humans and rhesus monkeys.

Additionally, we report on the success of novel simulation techniques that allow us to simulate deposition during a full inhalation/exhalation cycle. We employ advanced boundary conditions constructed using reduced-order modeling of the lower generations of the bifurcating airways and alveoli.

469054 Aerosol Emission during Human Speech

Abstract:

The physical pathways governing airborne disease transmission between humans are poorly understood. The traditional emphasis has been on coughing and sneezing, which are dramatic expiratory events that yield easily visible droplets. Recent research, however, suggests that normal speech can release even larger quantities of aerosols that are too small to see with the naked eye, but are nonetheless large enough to carry a variety of pathogens (e.g., influenza A). This observation raises an important question: what types of speech emit the most aerosols? Here we show that the concentration of aerosols emitted during healthy human speech is positively correlated with both the amplitude (loudness) and fundamental frequency (pitch) of the vocalization. Experimental measurements with an aerodynamic particle sizer (APS) indicate that speaking in a loud voice (~95 decibels) yields up to fifty times mores aerosols than in a quiet voice (~75 decibels), and that sounds associated with certain phonemes (e.g., [a] or [o]) yield more aerosols than others. We interpret these results in terms of the egressive airflow rate associated with each phoneme and the corresponding fundamental frequency, which is known to vary significantly with gender and age. The results suggest that individual speech patterns, including choice of language spoken, could affect the probability of airborne disease transmission.

471342 Investigating the Effect of Spatially Varying Wall Shear Stress on Lymphatic Endothelial Cell Alignment and Transcriptional Regulation

Abstract:

The lymphatic system is an essential though sometimes overlooked part of the cardiovascular system. The best-known role of the lymphatic system is to drain interstitial tissues. In addition, both recent and historical evidence demonstrate a central role for lymphatic vessels in lipid absorption, immune surveillance, and in regulating salt-dependent blood pressure. Failure to establish adequate tissue drainage results in lymphedema, which is painful and debilitating, and is a common outcome of cancer treatment. At present there is no cure for lymphedema.

Lymphatic valves are necessary for unidirectional lymph flow and thus for the physiological function of the lymphatic circulatory system. The inability to treat valvular dysfunction constitutes a critical bottleneck in the treatment of lymphatic diseases including lymphedema. Unfortunately, adults have only a limited ability to generate new lymphatic valves. There is likewise little progress in developing lymphatic valves, or any other cardiovascular valve, *in vitro*.

In preliminary results, we have built a novel 2D *in vitro* assay that reproduces key aspects of the fluid flow environment near valves, such as flow through a constriction, and discovered that lymphatic endothelial cells seem to recapitulate the first steps in valve formation when exposed to this flow environment. We show that human lymphatic microvascular endothelial cells (hLMVECs) align perpendicular to the flow direction at the region of the maximum wall shear stress (WSS) (**Fig. 1**), i.e. at the maximum constriction site, and exhibit a highly nuclear localization of Prox1, a transcription factor required for valve formation (**Fig. 2**). Further experiments demonstrated that chemical inhibition of the G-Protein Coupled Receptor, Sphingosine 1-Phosphate Receptor 1 (S1PR1), which we have previously showed to be important in hLMVEC response to fluid flow, with antagonist 5 μM W146 blocks both the perpendicular alignment of hLMVECs with respect to flow (**Fig. 1B**) and the upregulation of Prox1 (**Fig. 2B**).

Ongoing work aims to build an *in vitro* 3D assay to guide lymphatic valve formation, and to further characterize the interaction of spatiotemporal flow-based stimuli with signaling pathways that are known to be required for valvulogenesis. If successful, this approach will allow us to precisely define the molecular and biomechanical factors that are required for valve formation. Moreover, it will lay the groundwork for valve growth *in vitro*, which may impact the understanding and treatment not only of lymphatic dysfunction, but also valvulopathies of the heart and venous systems.

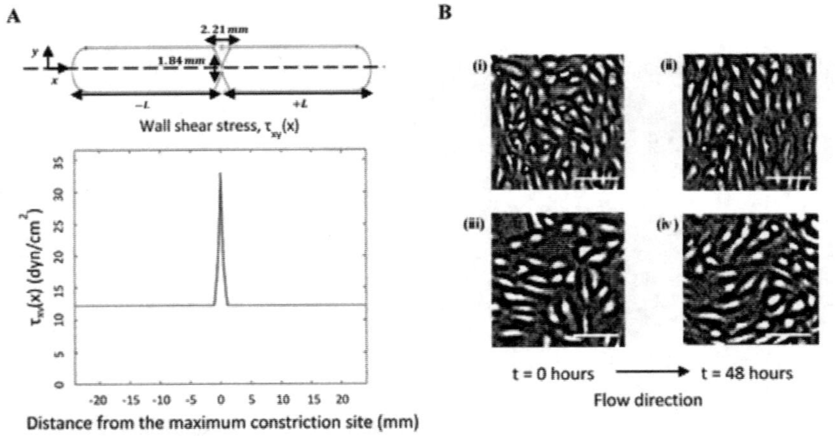

Figure 1. (A) WSS was calculated using 2-D lubrication theory. WSS profile is shown for the constricted channel at the $y = 0$ midline for a chosen minimum width of 1.84 mm, with maximum WSS of 33 dyn/cm^2. The distance along the channel is plotted on the x-axis (flow direction) with reference to the maximum constriction site at $x = 0$. (B) Cells within the constriction (33 dyn/cm^2): (i), (ii) *wild-type* hLMVECs at 0 and 48 hours, respectively, and (iii) and (iv) S1PR1 inhibited hLMVECs at 0 and 48 hours, respectively. hLMVECs collectively align perpendicular to the flow direction (3C-(ii)), and they lose this perpendicular orientation upon S1PR1 inhibition (3C-(iv)). Scale bar: 100 μm.

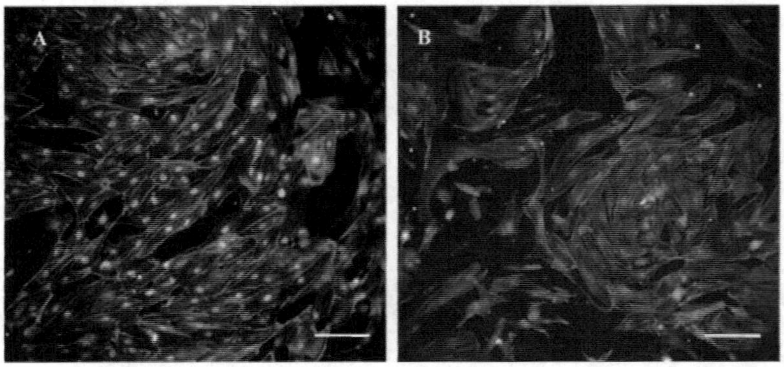

Figure 2. Immunofluorescence images of hLMVECs subjected to flow through a constriction (Fig. 1). Prox1 (green) and actin (red) at 20X magnification: (A) *wild-type* hLMVECs, and (B) S1PR1 inhibited hLMVECs (W146; 5 μM). hLMVECs exhibit a highly nuclear localization of Prox1, and this nuclear localization appears to be decreased in the presence of S1PR1 inhibition (B). Scale bar: 100 μm.

461851 Sub-Cellular Modeling of Platelet Transport and Adhesion in Microcirculation

Abstract:

Abstract

Platelet transport and its adhesive function are the major contributing factors to the formation of thrombi in the circulation. We perform high-fidelity mesoscopic simulations of blood flow in microchannels with or without constriction, where the wall shear rates within the constrictions reach levels as high as 8, 000 $s-1$, similar to those encountered in moderate atherosclerotic plaques. Both red blood cells and platelets are resolved at sub- cellular resolution using the Dissipative Particle Dynamics (DPD) method, and stochastic bond formation/dissociation are included to account for platelet adhesive dynamics at the site of injury. We find that higher levels of constriction and wall shear rates lead to significantly enhanced margination of platelets, which may explain the experimental observations of enhanced post-stenosis platelet aggregation. Platelet adhesion inside the constrictions is mostly mediated through GPIba receptors on platelet's surface and von Willebrand Factor multimers. We study the mechanics of adhesion under different shear rate and flow conditions as well.

470205 Quantifying Protein Viscosity As a Method for Determining Biopharmaceutical Degradation

Abstract:

Quantifying Protein Viscosity as a Method for Determining Biopharmaceutical Degradation

Katherine N. Clayton[1], Dong Hoon Lee[1], Steven T. Wereley[1], Tamara L. Kinzer-Ursem[2]

[1]School of Mechanical Engineering, Purdue University

[2]Weldon School of Biomedical Engineering, Purdue University

Patients requiring protein-based biopharmaceutical prescriptions run the risk of inactivation of their biologic due to heating or short-shelf life. However, the maintenance of protein folding and activity levels of these biopharmaceuticals is essential for the safety and efficacy of patient treatment. Developing quick and efficient methods for determining protein degradation could be of use in current laboratory work flows. Here, we present Particle Scattering Diffusometry (PSD), a viscosity measurement technique that uses low sample volumes (< 4mL) that can be integrated into micro- or nanofluidic systems. Based on the fundamental principles of diffusion, particles (~200 nm) undergoing Brownian motion are imaged under fluorescence microscopy. The diffusion coefficient of the particles is calculated by correlating successive particle images (at time *Dt*) to one another (cross-correlation, s_c) and the particle image with itself (autocorrelation, s_a) at a magnification (*M*):

$$D = \frac{s_c^2 - s_a^2}{16M^2\Delta t}$$

The viscosity of the solution is then calculated from the Stokes-Einstein equation.

In this work we use PSD to measure amounts of protein degradation of a common biologic used in diabetes treatment as a function of protein viscosity. We address insulin stability by comparing degraded and intact insulin samples between concentrations of 1 to 10 mg/ml. Particles introduced to the solution are imaged with PSD. Solution viscosity is calculated from the measured diffusion coefficient of the particles. We find that the viscosity of denatured insulin exponentially increases with concentration, whereas intact insulin maintains a similar viscosity across a range of concentrations. Additionally, we combined denatured and intact insulin to quantify changes in the viscosity when a certain fraction of the insulin is denatured. Not unexpectedly, we find that there is a decrease in the viscosity of insulin with increasing concentrations of intact insulin. Importantly, we are able to detect small changes in solution viscosity when low amounts of denatured insulin is present within the samples. By achieving viscosity measurements of denatured insulin presence in sample, PSD can be used to test for biopharmaceutical sample stability for quality control in field testing.

In summary, we have established a rapid (~8 seconds) and sensitive technology for detecting biomarkers in very low sample volumes (microliters) using instrumentation and tools that are common to most laboratory settings (fluorescent microscope, CCD camera, and computer). Based on the success of using PSD for biopharmaceutical viscosity analysis, we expect to translate these techniques for use in a wide range of protein-based biopharmaceuticals. In this presentation we will to discuss the development of the PSD technique, our current results and challenges associated with PSD measurements, as well as future work to integrate this method into basic laboratory settings.

464648 Ellipsoidal Microhydrodynamics: New Velocity Representations for Stokes Flow in Ellipsoidal Geometries

Abstract:

The ellipsoid and its degenerate version (the spheroid) is a most convenient starting point for the study of nonspherical effects in many physical systems including fluid mechanics. The classical work for the steady translational motion of an ellipsoid in Stokes flow dates to Oberbeck's (1876) solution, but more recently, a new velocity representation for the ellipsoidal geometry has been discovered thanks to an eigenfunction analysis of the Stokes double layer operator, with special linkages to Brenner's (1960s) papers on Stokes flow past an ellipsoid. The power of the new approach will be illustrated with examples from biofluid dynamics and colloidal hydrodynamics.

462590 Modulation of Self-Assembly in Biological Motor Complexes

Abstract:

The flagellum in bacteria is a sensitive detector of perturbations in environmental viscosities. The driver for such sensing of mechanical signals is the flagellar motor, which tailors motility to suit the environment. Mechanosensitive torque-generating elements that rotate the motor detect mechanical signals, and reversibly remodel their assembly in response to changes in the viscous loads. Remodeling likely plays a role in triggering signaling pathways that control cell-morphology and motility. To determine the mechanism of remodeling and its role in signaling, we tested remodeling over varying viscous loads. We subjected wild-type strains and strains that lacked downstream-transducers of mechanical signals to large mechanical stimuli. From the initial speed-response (following the stimulus), we estimated the amount of torque-generated per stator-unit in the presence and absence of the effectors. Subsequent measurements revealed the absence of feedback from downstream effectors on motor remodeling. Although crucial for cell-survival and adaptation, proteins involved in downstream-signaling are unlikely to regulate motor-mechanosensing.

466894 Collective Motion in Viscoelastic Fluids

Abstract:

Swimming microorganisms and their collective motion have received growing attention for their importance in pathology, reproduction and ecology. The suspension of self-propelled microorganisms exhibits many turbulent-like behaviors, such as locally correlated motions, enhanced particle diffusion, and enhanced fluid mixing. Even though several studies have been conducted on the collective motion in a Newtonian fluid, the natural habitat of microorganisms is often non-Newtonian. The non-Newtonian fluid rheological properties, such as viscoelasticity of saliva, mucus and biofilm, significantly affect the swimming properties and hydrodynamic interaction of microorganisms. In this work, we use direct numerical simulation to investigate the collective motion of rod-like swimmers in viscoelastic fluids. We demonstrate that the fluid elasticity has different effects on a suspension of microswimmers compared to an isolated swimmer. It has a stronger effect on a suspension of pushers compared to pullers. The induced polymer stresses enhance the lateral attraction of pushers and break down the large scale fluid structures.

464941 Population Dynamics of Chemotactic Bacteria in Response to Multiple Chemical Stimuli

Abstract:

Chemotactic bacteria sense and respond to temporal and spatial gradients of chemical cues in their surroundings. This phenomenon plays a critical role in many microbial processes such as groundwater bioremediation, microbially-enhanced oil recovery, nitrogen fixation in legumes, and pathogenesis of disease. Chemical heterogeneity in these natural systems may produce numerous competing signals from various directions. Predicting the migration behavior of bacterial populations under such conditions is necessary for designing effective treatment strategies.

In this study we observed the response of *Escherichia coli* to a mixture of chemoattractant (methylaspartate) and a chemorepellant (nickel ion) in various spatial configurations. The goal was to predict the overall response based on transport parameters that quantified the chemotactic response to individual chemicals. We used a combination of experimental assays and mathematical modeling. A microfluidic device designed to generate steady, linear concentration profiles was used to evaluate the chemotaxis transport properties for each chemoeffector separately. Then bacterial concentration profiles for various mixture configurations were compared to model predictions. We initially considered a modeling approach using an additive response in which the macroscopic expressions for the chemotactic velocities were simply added together assuming a negative velocity for the chemorepellant. Because the chemosensory mechanism is well-documented for *E. coli*, we also considered a molecular-level modeling approach that incorporated several key steps in the pathway in order to deduce the mechanism by which bacteria process multiple signal inputs to generate an integrated response. Finally, we explored the separation of a mixed microbial culture based on the differential response to a single chemoeffector.

471977 Transport of Filamentous Viruses As Reporters in Lateral Flow Assays

Abstract:

Lateral-flow assays (LFAs) are a simple and rapid point-of-care format to detect biological analytes such as proteins, viruses, and bacteria. In LFAs, reporter particles bearing antibodies to a given analyte are transported by capillary action through a fibrous affinity membrane; particles bind to the analytes during flow and are in turn captured on the membrane, thereby generating a visual signal. While LFAs are particularly well-suited for use in economically challenged areas with limited laboratory infrastructure, their lack of sensitivity limits their use in scenarios requiring early detection. Previously, we showed that filamentous viruses can be employed as viral nanoparticle reporters in a quantitative LFA whose limit of detection is up to two orders of magnitude less than that of existing methods, but the reasons underlying this enhanced sensitivity remain poorly understood.

Here, we test the idea that the transport properties of filamentous viruses in LFA membranes gives rise to the increased sensitivity of viral nanoparticle LFAs. High-aspect-ratio rodlike M13 bacteriophage (M13) displaying biotin on the M13 tail protein p3 are conjugated with fluorescent labels and suspended in solutions of high-molecular-weight polymers, which mimic viscoelastic biofluids. We functionalize glass fiber LFA membranes with antibodies to M13, which primarily capture M13 on the p8 coat proteins on the lateral phage surface, and with avidin, which captures M13 at the biotinylated tail. Using microscopy and image analysis, we characterize the flow and binding of viruses bearing side- or tip-binding recognition elements in solutions of various polymer concentration. The number of M13 captured on the membrane is greater for side-binding than for tip-binding, as expected from the number of recognition elements, and in both cases depends on the flux of M13. We show that most binding events require the rodlike viruses to reorient prior to capture on the membrane. This result is consistent with the idea that the elongated M13 shape couples with the local flow field in an open and disordered fibrous LFA membrane to increase capture and hence LFA sensitivity.

452495 Polyoxometalate Application in Redox Flow Battery

Abstract:

The impact of renewable energy has increased over the past decade. Based on a recent U.S. Energy Information Administration (EIA) report, renewable energy sources supplied 13% of the energy consumed in the US. The need for improved methods of renewable energy storage has also increased. Redox flow batteries (RFBs), having both high energy capacity and a flexible layout, are a promising option to help solve the storage problem. Polyoxometalates (POMs) are attractive active species for use in RFBs, owing to their versatile, tunable properties and multielectron redox reactions.

The object of this research is to select POMs which can be used in non-aqueous RFBs in a systematic way and to demonstrate their performance for energy storage. Electrochemical characteristics of different keggin POMs ($XM_mO_{40}^{n-}$) in acetonitrile were examined. The 1A and 2A groups of counter-cation exchanged POMs – Li_3, Na_3, $Mg_{1.5}$ and $Ca_{1.5}$ salts – were studied. Heteroatom-exchanged POMs including $XMo_{12}O_{40}^{3-/4-}$ and $XW_{12}O_{40}^{3-/4-}$ forms ($Li_3PMo_{12}O_{40}$, $Li_3AsPMo_{12}O_{40}$, $Li_4SiMo_{12}O_{40}$, $Li_3PW_{12}O_{40}$, $Li_4SiW_{12}O_{40}$.) were also tested. Lastly, two series of framework-substituted POMs, $Li_4PMo_{11}VO_{40}$, $Li_5PMo_{10}V_2O_{40}$, and $Li_6PMo_9V_3O_{40}$ and $Li_3PMo_{12}O_{40}$, $Li_3PW_{12}O_{40}$, $Li_3PMo_{11}WO_{40}$ and $Li_3Mo_6W_6O_{40}$, were examined.

The 1A and 2A groups of counter-cation exchanged POMs generally showed 3 redox couples with slightly different peaks and peak positions. The effect of 1A or 2A cation exchange resulted in minor differences in observed POM redox potentials. The effect of heteroatom exchange on the POM redox potentials was observed to vary with the negative charge density of the P-, As-, and Si-substituted Keggin ions, however the number of accessible redox couples was not affected. In contrast, framework metal atom variation

affected the redox potentials and the numbers of redox couples. However, the mechanism of how these metal substitutions into the Keggin structure alters the POM electrochemical properties is still unknown.

The most promising POM materials of those studied were used in charge/discharge tests in acetonitrile, with $Li_3PMo_{12}O_{40}$ chosen as the model. 90% coulombic efficiency of the $Li_3PMo_{12}O_{40}$ charge/discharge test was found, suggesting both high stability and reversibility of the POMs under these conditions. Our results suggest that POMs in non-aqueous media warrant further study as active species for the development of RFBs with high energy density and high stability.

460844 Insights into the Solvation of Vanadium Ions in the Vanadium Redox Flow Battery Electrolyte Using Molecular Dynamics and Metadynamics

Abstract:

Insights into the solvation of vanadium ions in the vanadium redox flow battery electrolyte using molecular dynamics and metadynamics

Sukriti Gupta,[a] Nyunt Wai,[a] Tuti M. Lim,[b] and Samir H. Mushrif [a,c*]

Email: shmushrif@ntu.edu.sg

[a]Energy Research Institute and Interdisciplinary Graduate School, Nanyang Technological University (NTU), Singapore [b]School of Civil and Environmental Engineering, NTU, Singapore [c]School of Chemical and Biomedical Engineering, NTU, Singapore.

Abstract

Vanadium Redox Flow Batteries (VRFB) can be integrated with renewable energy technologies like Solar or Wind energy to store the energy generated by these systems and to supply it according to requirements. The energy storage capacity of VRFB depends on the volume of the electrolyte and the amount of ions dissolved in it, whereas its power depends on the electrode surface area and cell stack. [1] Thus, the main advantage of using a redox flow battery over any other rechargeable battery system is that the energy storage capacity can be increased to any extent by increasing the size of the electrolyte storage tanks. In an attempt to commercialize VRFB, efforts are being made to reduce the

size of electrolyte storage tanks, by increasing the solubility of vanadium ions in these electrolytes. Additionally, these electrolytes are stable only in the temperature range of 10 to 40°C. Different approaches, like changing the sulphuric acid concentration, using a mixture of hydrochloric and sulphuric acids, or using various chemicals as additives have been tried. [2] Increasing the sulphuric acid concentration stabilizes VO_2^+ ions better, but the remaining V^{2+}, V^{3+} and VO^{2+} ions become relatively unstable. [3] Different additives alter the solubility of vanadium ions to different extents. However, fundamental insights into the interactions that govern the solubility and stability of vanadium ions in the electrolyte system are lacking and hence systematic screening and selection of additives is not feasible. Hence, classical molecular dynamics simulations, coupled with metadynamics were performed to investigate the local molecular solvation structure of vanadium ions in the electrolyte system. Free energy landscapes corresponding to the migration of chloride, sulphate and bi-sulphate ions were computed. Force-field parameters for vanadium ions were recently developed in our group and were implemented in these simulations [4]. These simulations allow us to estimate (i) the relative stability of counter ions within the local solvation shell of vanadium ions and (ii) the kinetics associated with the migration of counter ions and with the reorganization of water solvation shell around vanadium.

We observed that it is kinetically possible for sulphate, bisulphate and chloride ions to enter the 1st solvation shell of all four vanadium ions at room temperature and they are thermodynamically stable there. Sulphate ions coordinate with the vanadium ion via two or three of its oxygens by replacing the first solvation shell water molecules. This distorts the octahedral structure of VO_2^+, thus increasing chances for interaction with other vanadium ions, possibly leading to the formation polymers and dimers. Bisulphate ion however could only share one of its oxygen with vanadium ion. As the chloride ion is small and can only replace one water molecule from the 1st solvation shell of vanadium ions, it does not disturb the octahedral structure of vanadium ions. Chloride ions in the electrolyte can compete with sulphate ions, to enter the 1st solvation shell.

References

[1] F. Grossmith, Efficient Vanadium Redox Flow Cell, Journal of Electrochemical Society. (1987) 2950–2953.

[2] A. Parasuraman, T.M. Lim, C. Menictas, M. Skyllas-Kazacos, Review of material research and development for vanadium redox flow battery applications, Electrochimica Acta. 101 (2013) 27–40.

[3] F. Rahman, M. Skyllas-Kazacos, Vanadium redox battery: Positive half-cell electrolyte studies, Journal of Power Sources. 189 (2009) 1212–1219.

[4] S. Gupta, N. Wai, T.M. Lim, S.H. Mushrif, Force-field parameters for vanadium ions (+2, +3, +4, +5) to investigate their interactions within the vanadium redox flow battery electrolyte solution, Journal of Molecular Liquids. 215 (2016) 596–602.

465085 Membraneless Electrolyzers for Water Electrolysis

Abstract:

Hydrogen (H_2) is a carbon- free energy carrier with applications in the transportation, electricity, and industrial sectors. The vast majority of today's H_2 is produced by steam methane reforming, but the production of H_2 from water electrolysis offers a carbon-free route when the electricity is supplied from solar or wind.[1,2] Unfortunately, the price of H_2 produced by water electrolysis (≈$5 / kg H_2) remains above the U.S. DOE target of $2-4 /kg H_2. In this work, we explore simple and scalable membraneless electrolyzers based on angled mesh flow-through electrodes as a means of decreasing the electrolyzer capital costs and thereby bring down the price of producing H_2 from water electrolysis. The electrolyzer capital cost is especially important in instances of low capacity factor and time of use pricing, which are expected to become more prevalent as variable solar and wind generators capture a larger share of the electricity market.

Conventional polymer electrolyte membrane (PEM) electrolyzers rely on membranes to conduct H^+ ions between electrodes while separating the H_2 and O_2 product species, but these membranes can be costly, prone to failure, limit the choice of solution pH, and require a fairly complex MEA-based architecture.[3,4] Recently, we have demonstrated 3D printed membraneless electrolyzers based on flowing electrolyte through two angled mesh flow-through electrodes, which results in efficient separation of the H_2 and O_2 product gases with minimal product crossover.[3] Ongoing research builds off of this initial study by using *in situ* high speed video analysis to quantitatively measure current density distributions along angled mesh electrodes under various operating conditions. Experimental observations are compared with modeled current density distributions, providing a useful framework for further optimizing device performance and better understanding the hydrodynamics associated with gas-evolving flow-through electrodes.

References

[1.] B. Kroposki, *et al.*, *National Renewable Energy Laboratory - Technical report* (2006).

[2.] Millet, P., *et al.*, *International Journal of Hydrogen Energy* 35, (2010): 5043–5052.

[3.] G.D. O'Neil, C. Christian, D. Brown, J.T. Davis, D.E. Brown, D.V. Esposito, *J. Electrochemical Society*, 163, (2016): (In Press).

461792 Formation of Percolating Carbon Networks in Battery Processing and Their Effects on Electrode Performance

Abstract:

The increasing demand for energy storage has led to continual improvements in materials for battery electrodes. However, the overall processing of these materials into devices and the need for composite electrodes has remained essentially the same. Currently, the optimal processing conditions are determined through time-, labor-, and capital-intensive trial and error. Determining the fundamental particle and electrochemical physics that result in high-performance composite electrodes is so far an untapped field of potential. Our work strives to fully understand from a rheological perspective the interactions of components in electrode slurries. These insights will allow the manufacturing process of new materials to be accelerated, resulting in faster, more efficient implementation of new materials into effective electrodes.

The mixing, coating, drying and calendaring of active material, conductive additive, and polymer binder in solvent affect the final battery microstructure. Although it is well-known that the electrode slurry microstructure impacts the final battery performance, there is little agreement on what constitutes a favorable microstructure [1,2]. Slurries may be broadly divided into two rheological categories, fluids and gels. In a fluid-like system, suspended particles typically interact via hydrodynamic forces. In a colloidal gel, particles form a percolating network that extends across the volume of the system through surface or depletion interactions.

In this work, we study a model system of lithium nickel manganese cobalt oxide (NMC), carbon black, polyvinyldifluoride (PVDF), and 1-methyl-2-pyrrolidinone (NMP) to determine the effects of slurry microstructure on battery performance. Slurries are characterized using small-amplitude oscillatory shear (SAOS) in which a constant strain is applied at increasing angular frequencies and the viscous and elastic components of the stress are determined as a function of frequency (frequency sweep). Figure 1 compares the frequency sweeps for a fluid system and a gel system formed with micron-sized and nano-sized active material, while maintaining constant PVDF molecular weight, carbon black, and mixing parameters. A fluid system is depicted by G″ larger than G′. Gel systems are typically observed when G′ dominates.

Suspended micron-sized active material does not form a gel in the presence of polymer binder. However, the addition of a critical volume fraction of free nano-sized carbon additive causes the suspension to gel. We find that the critical gelation volume fraction of free carbon black (40nm) is equal to 0.5 vol%. The typical quantity used in commercial battery slurry recipes is 2 vol% [3]. The amount of free carbon is highly dependent on the commonly used dry-mixing step, where active material and conductive additive are blended together. It has been shown that dry –mixing causes conductive additive to coat the active material reducing the amount of free carbon in the suspension. In other words, in the absence of dry-mixing, all the carbon is above the critical concentration and readily forms a gel, i.e. percolating network. However, depending on the dry-mixing conditions, enough carbon is coated on the active material such that the amount of free carbon is too low to form a colloidal gel. By varying the dry mixing step of slurry preparation, a wide range of slurry microstructures can be achieved without varying the total carbon concentration. Contrary, nano-sized active material readily forms a gel due to its size and large volume fraction in the electrode slurry.

Slurries with characterized microstructures are coated, dried, and calendered into electrodes in order to relate slurry microstructure to electronic conductivity, life cycle and rate capability of the processed electrodes. Measurements of the electronic conductivity and battery performance show that gel slurries form electrodes with higher initial conductivity and lower porosity. Electrodes from gel slurries also cycle longer and have higher capacities than fluid slurries. We hypothesize that the continuous carbon network is necessary to provide sufficient electronic conductivity. Future work will investigate the dependence of performance and electrode tortuosity on slurry microstructure and the effects of drying parameters on battery performance.

References

[1]W. Bauer, D. Noetzel, Ceram. Int. 40 (2014) 4591.

[2]G.-W. Lee, J.H. Ryu, W. Han, K.H. Ahn, S.M. Journal of Power Sources 195 (2010) 6049–6054.

[3] T. Marks, S. Trussler, A. J. Smith, D. Xiong and J. R. Dahn, J. Electrochem. Soc. 2011, Volume 158, Issue 1, Pages A51-A57.

459699 Multi-Scale Model of Electrochemical Double Layer Capacitors: Predicting Performance Limitations

Abstract:

Mobile energy storage applications such as electric cars often require quick changes between the charging and discharging. For such operation mode, conventional batteries are not very suitable due to their internal resistance originating from Faradaic reactions. On the other hand, Electrochemical Double Layer Capacitors (EDLCs) store energy by the physical separation of charges on their large surface area. However, even for EDLCs, the utilizable capacitance is a function of the time-scale at which they are operated and this is strongly dependent on the construction and physical parameters of the device. The mathematical modeling is a useful for the identification of factors influencing the EDLC performance and the evaluation of their importance.
We simulated the dynamic behavior of an EDLC using a spatially resolved model based on the porous electrode theory. The established model of transport on the scale of one cell [1] was extended with an additional dimension describing the transport into the carbon particles, effectively separating the spatial scales of different pore sizes. The dynamic response of the model was obtained by simulating the most common electrochemical techniques: the Cyclic Voltammetry, Electrochemical Impedance Spectroscopy and Galvanostatic Cycling.
Our results show a large influence of the construction and physical parameters, such as the electrode thickness (L_e), separator thickness (L_s) and electrolyte conductivity ($\hat{I}^o$) on the performance of EDLCs. In agreement with literature experimental data, the time constant (inverse of the frequency at which the capacitance drops to half of the maximum value) was an increasing function of L_e and L_s and a decreasing function of $\hat{I}^o$. The effect of the potential dependence of capacitance (described by the Gouy-Chapman-Stern and Bikerman theories) was shown to be important only for cells that are close to a complete discharge. The main limitation was found to be on the scale of the whole cell (two electrodes and separator), while the transport into the particles became a limiting factor only if the particle size was unrealistically large. Using a simple RC circuit model, the results were generalized into a simplified relation allowing

for a quick evaluation of performance for the design of new devices.
This work provides an insight into the performance limitation of EDLCs and identifies the critical parameters in this respect. The multi-scale approach that was developed in this work allows to compare the importance of phenomena occurring at pores of different size.

References

[1] Verbrugge, M. W.; Liu, P. Microstructural Analysis and Mathematical Modeling of Electric Double-Layer Supercapacitors. Journal of the Electrochemical Society 2005, 152, D79

470093 Novel Columnar Liquid Crystals Functionalized Reduced Graphene Oxide Based Electrodes for Photocapacitor Applications

Abstract:

Columnar liquid crystals (CLCs) are known to be photosensitive materials given their Ï€-stacked column arrangements formed by the self-assembly of aromatic molecules which permits to alternate electron donor-acceptor molecules. By taking advantages of such CLCs arrangement properties, it is then possible to harvest and store solar energy through the development of electrochemical capacitors. Our research group has focused on growing these CLCs stacks over a reduced graphene oxide-based electrode for charge storage enhancement and capacitor device fabrication. This approach, and in comparison with single-molecule light harvesting, offers broad light absorption (which enhances the efficiency) and a simpler fabrication path. Also, based on the utilized approach, molecule stability and material self-healing capabilities are among the obtained benefits which, ultimately, addresses several drawbacks of opto-electronic materials. Preliminary results have shown that these CLCs materials exhibit faradaic charge transfer photoactive behavior given differences in capacitance measurements observed under â€œdarkâ€ and illuminated conditions. Materials physicochemical properties have been studied by X-Ray diffraction (XRD), Fourier transform infrared (FTIR), and scanning electron microscope (SEM). Finally, electrochemical studies were conducted by means of cyclic voltammetry (CV) in a 3-electrode configuration.

457793 On the Origin of the Interfacial Phase Transition Causing the Stabilization of Water/Oil Emulsions By Asphaltenes: Accounting for Excluded Area Effects through a Lattice Gas Equation of State

Abstract:

Stabilization of water in oil emulsions by asphaltenes (a solubility class of crude oil comprising polyaromatic and heterocyclic molecules) has for a long time been attributed to the formation of a cohesive â€œgelâ€ at the interface, as revealed by the formation of wrinkles upon contraction of water droplets aged in asphaltenes solutions. This interpretation however suffers from many shortcomings, the most obvious one being the timescale discrepancy between emulsion stabilization (fast) and transition from fluid to solid rheological properties of adsorbed asphaltenes layers (slow). An alternative paradigm recently emerged from the systematic study of surface pressure-interfacial area and surface pressure-dilatational elasticity curves which proved independent from adsorption conditions and in particular adsorption time. This demonstrated that asphaltenes follow an equation of state, which could be approximated by a Langmuir model. The corresponding surface excess coverage (the reciprocal of molecular area) was found particularly consistent with the observed conformation of asphaltenes at the water surface (with the aromatic core flat on water) and their average core size. From this starting point, droplet contraction experiments could be reanalyzed quantitatively: deviations of the surface pressure isotherm from the Langmuir equation of state and of the droplet from the Laplacian shape appeared to occur when surface coverage approached maximum packing. In other words transition to a solid like interface would be driven by steric effects as it is the case for the jamming of particle laden interfaces. The parallel with particles was further extended to the analysis of emulsion stability. The critical mass of asphaltenes per unit interfacial area for preventing droplets coalescence was actually found to be very close to the ratio of molar weight to molar area. This indicates that asphaltenes block coalescence when approaching packing at water surface just like particles. As in Pickering emulsions, this condition can be reached much faster in an emulsion which is stirred and undergoes limited coalescence than in a rheometer where interfacial area remains constant and mass transport is diffusion limited. On a theoretical point of view it is however troublesome that molecules following a Langmuir equation of state would exhibit a fluid to solid transition due to steric interactions like particles do. The localized adsorption assumption of the Langmuir model should actually rule out any hard core repulsion between adsorbed molecules as it imposes that their size is smaller than that of adsorption sites. Reversely this restriction with respect to the relative size of adsorbates and adsorption sites should preclude the use of the Langmuir equation of state for molecules that are bigger than water molecules (3 to 4 times for asphaltenes). These inconsistencies are largely alleviated when considering lattice gas equations of state in which an adsorbate covers several adsorption sites. Such models actually predict a phase transition at high coverage due to excluded area effects. On the other hand when plotted similarly to experimental data, lattice gas equations of state (below their phase transition) can fairly be approximated by the Langmuir equation of state. In particular the limiting elasticity versus surface pressure relationship seems to be primarily dependent upon molecular area. This means that molecular area can be first evaluated from experimental data with the Langmuir equation of state, then used to calculate the number of water molecules covered by each adsorbed molecule. Proper lattice gas models can then be identified from the literature and compared with more details to experimental data. For asphaltenes the selection is further refined based upon the geometry of their aromatic cores which are roughly hexagonal. The corresponding equation of state (hexagons covering three sites on a triangular lattice) is extensively described in the literature and is found to match experimental observations up to a surprising level of detail. In particular the predicted phase transition domain exactly matches the onset and final steep increase of the error to the Laplacian shape during droplet contraction. Furthermore, the predicted relationship between limiting elasticity and surface pressure exhibits the same inflexion point as experimental data. At low coverage both first and second derivative of elasticity with respect to pressure are positive, but at high coverage the second derivative becomes null: elasticity is somewhat proportional to pressure. This peculiar feature is verified for all lattice gas equations of state tested so far, with a proportionality constant ranging from 2 for small adsorbates to 5 for large ones. Interestingly similar observations were repeatedly reported for dilatational rheology experiments on surfactants (with a proportionality constant circa 2-3) and proteins (with a proportionality constant circa 4-5).

471339 Aging Oil-Water Interfaces with Asphaltene Adsorption: Interface Rheology and Heterogeneity

Abstract:

Water is often used in the production and processing of oil, but the water and oil must ultimately be separated. Such separations can be very challenging, because various surface-active materials naturally exist in crude oil that stabilize water-oil interfaces. Polycyclic aromatic hydrocarbons called asphaltenes are a broad class of surface-active compounds that adsorb at water-oil interfaces and stabilize such emulsions.

We use ferromagnetic microbuttons as interfacial rheological probes to probe the evolution of oil/water interfaces as asphaltenes adsorb, and the effect of chemical additives like ethylcellulose (EC) on the evolution of the mechanical properties of the interface. Oil/water interfaces progressively stiffen as asphaltenes adsorb; this process, however, can be prevented or reversed with the addition of EC.

To probe the mechanism behind these observations, we visualize the deformation (strain) field of the oil/water interface in response to the stresses imposed by the microbutton. Asphaltene-adsorbed oil/water interfaces show significant mechanical heterogeneity, with pronounced stiff and weak regions. EC affects these heterogeneous regions differently, suggesting various hypotheses for its action on asphaltene-stabilized interfaces.

Our study reveals the rich properties of water/asphaltene/oil interfaces, and highlights new tools to probe mechanically heterogeneous interfaces as they evolve in response to their local chemical environments.

453866 Novel Colloid Chemistry Approach to Resolving Emulsions for Fouling Control in Steam Crackers

Abstract:

Steam cracking is the principal technology used to produce olefins such as ethylene, propylene and butadiene. Cracked gas from the furnaces is cooled in a quench water tower where steam and pyrolysis gasoline condense. Water is an effective heat transfer medium, but mixing of hydrocarbons and water can lead to the formation of stable emulsions. If the subsequent gasoline-water separation is incomplete hydrocarbons entrained in the water phase make their way into downstream units. This potentially leads to severe fouling issues, notably in the dilution steam system or saturators where the steam used in the cracking process is generated.

The study described in this paper is unique in that it addresses emulsion formation in quench water towers in steam crackers from a different angle, probing the emulsifying properties of different interfacially active fractions, and analyzing the correlation between emulsion stability and the physico-chemical properties of the interfacial films formed by these fractions.

Freeze-drying according to a protocol developed in-house was used to extract potentially emulsion stabilizing species present in quench water sampled from a steam cracker. Such species were subsequently fractionated into three solubility classes: water soluble, soluble in polar organic solvents and soluble in apolar organic solvents. Liquid chromatography/mass spectrometry analyses indicated differences in the chemical composition of these fractions, with oligomeric series and species having a high oxygen content detected solely in the fractions soluble in organic solvents.

Bottle test results demonstrated that the interfacial material soluble in polar organic solvents was the most effective in stabilizing water-in-gasoline (W/O) and gasoline-in-water (O/W) emulsions. The type of emulsion formed with the fraction soluble in organic solvents was a function of the energy imparted to the system. O/W emulsions required lower energy to form, but also destabilized more rapidly, compared to W/O type emulsions. These results suggest that water droplets in gasoline were stabilized by relatively large molecules that did not promptly desorb from the interface after sorption had taken place. Conversely, gasoline droplets in water were likely stabilized by molecules that required limited energy to adsorb to and desorb from the gasoline-water interface. Further investigation of these aspects will be the object of future work.

The water soluble fraction decreased the stability of emulsions stabilized by the fraction soluble in organic solvents, and its demulsifying properties increased from pH 8.5 to pH 5.5. In the absence of the water soluble fraction emulsion stability was largely independent of pH, suggesting that electrostatic forces played a marginal role when emulsions were stabilized by the fraction soluble in organic solvents.

Scanning Electron Microscopy (SEM)/Energy Dispersive X-Ray Spectroscopy (EDX) was used to probe the morphology and elemental composition of the films extracted from the gasoline-water interface and deposited onto silicon wafers using a modified Langmuir-Blodgett method. SEM images highlighted the presence of particles at the gasoline-water interface, suggesting Pickering stabilization mechanisms. SEM/EDX data reveal a correlation between the composition of interfacial films and emulsion stability. Interfacial films contained a high density of particles containing carbon, sulfur and/or iron when emulsions were most stable, and a low density of such particles otherwise. It is speculated that sulfur or iron particles, rendered amphiphilic due to contact with organic material, were particularly effective in stabilizing the interface. The hypothesis that sulfur or iron compounds acted as aggregation nuclei for organic species with emulsifying properties cannot be discounted. Additional efforts will made to analyze sulfur and iron species as part of our future research.

Interfacial tension measurements were conducted using the pendant drop method. The interfacial tension decrease with time for gasoline-water interfaces was relatively low and similar for all fractions, in agreement with the hypothesis that the material adsorbing at the gasoline-water interface was particle-like.

The compressional visco-elastic moduli of interfacial films and the force required to break them were determined using a pulsating drop rheometer and a modified Wilhelmy plate method, respectively. Measurements were conducted at pH 5.5 and 8.5, with either one or all solubility fractions added to the system. In pulsating drop measurements a sessile gasoline droplet was formed in water and its volume was varied sinusoidally, while continuously monitoring the change in the interfacial tension caused by the droplet deformation. In the modified Wilhelmy plate method, films were aged for 30 minutes, after which they were stretched in a controlled manner using a Wilhelmy plate till film rupture occurred, while continuously monitoring the force exerted on the plate. The compressional visco-elastic moduli and the force required to break the films were similar with all fractions, and do not account for the differences in their stabilizing properties.

The zeta potential of gasoline droplets dispersed in water was also found to be similar for all fractions, and does not explain their effect on emulsion stability. Zeta potential measurements further revealed that gasoline droplets in water were neutrally charged at pH 5.5, at which emulsions were stable with the fraction soluble in organic solvents. This result supports the hypothesis that electrostatic repulsion was not a relevant mechanism in stabilizing emulsions in the presence of the fraction soluble in organic solvents.

In summary, this study shows that interfacially active species in a stable gasoline-water emulsion from a steam cracker are comprised of three major fractions: 1) two fractions soluble in organic solvents (polar and apolar), which contain stabilizing carbon, iron and sulfur compounds, and 2) a water soluble fraction with pH-dependent demulsifying properties. The stabilizing fractions soluble in organic solvents are complex mixtures of compounds, which play different roles in stabilizing W/O and O/W emulsions.

It is believed that the ongoing pioneering work presented in this paper will ultimately open up novel avenues for chemical resolution of problematic emulsions in steam crackers.

453994 Mechanisms of Pickering Foam Stabilization during Bioreactor Operation

Abstract:

Pickering foams have garnered increasing attention because of their characteristically long decay times. However, their persistence can present a significant challenge when foam formation is undesired in the course of industrial processes such as bioreactor operation. In the later stages of cell culture proliferation in bioreactors, Pickering stabilization of the foam head occurs, causing the traditional remediation strategy of antifoam addition to be less effective. The correlation between foam stability and process age is strongly tied to the accumulation of biomass, specifically an increase in the number and the distribution of size and shape of biological particles. We present results and analysis of the stabilization effects and breakdown behavior of foams stabilized by these biological particles. Through the use of optical microscopy we observe the entrapment of cells in thin foam films and at Plateau borders. In order to elucidate the mechanisms of foam stabilization in this system, we study samples extracted at different time points in bioreactor cell culture processes each with corresponding differences in the cell concentration and viability as well as differences in the concentration of cell debris fragments and macromolecules. We compare data for interfacial adsorption, entrapment in the thin liquid films and Plateau borders, and accumulation in real bulk foams. We explore how each class of biological particles and surface-active agents contributes to foam formation and foam stabilization on their own and through interaction between the molecular and particle species. Additionally, we discuss the role that antifoam addition and deactivation has on the lifetime of these biological particle-based Pickering foams.

468306 Formation of Multi-Nanoemulsions for Colloidal Synthesis

Abstract:

In recent years, complex emulsions â€“ i.e., droplets with internal structure â€“ have generated great research interest due to their potential applications in materials, foods, cosmetics, pharmaceuticals, and chemical separations. Microfluidic methods have already demonstrated the ability to create micron- and larger scale complex emulsions with breathtaking sophistication and control, as well as compartmentalize encapsulation of molecules within them. However, scaling the size of such droplets to the nanoscale has been extremely challenging due to limitations on the devices and energies required to produce nanoscale droplets, i.e. nanoemulsions. Here, we report the ability to fabricate complex water-in-oil nanoemulsions of various morphologies, and use them as templates for forming complex structured nanoparticles. To produce complex morphologies, we combine high-energy emulsification methods with co-surfactant pairs possessing highly asymmetric molecular geometry. The former aids the generation of nanoscale droplets, and the latter influences their morphology through frustrated interfacial curvature, resulting in the reproducible generation of nanoscale core-shell and multi-core shell morphologies. The size, stability, internal morphology and chemical compartmentalization of these complex nanoemulsions have been quantified using a combination of scattering, optical microscopy and cryogenic-transmission electron microscopy techniques. We show that complex droplet morphologies are retained upon the addition of various material pre-cursors, and that the droplets are stable over the time scales required for material chemistry, thereby enabling their use as templates for complex nanoparticles. By devising and understanding new processes to make complex nanoemulsions, and particles within them, we hope to demonstrate their use as a powerful and facile tool for generating unique nanostructures with many potential applications in nanotechnology.

459877 Novel Surfactants for Stabilizing CO_2-in-Oil Emulsions and Natural Gas-in-Oil Foams for Dry Hydraulic Fracturing

Abstract:

Liquid CO_2 is an attractive fluid for many hydraulic fracturing applications because it contains no water (i.e., it is "dry" does not damage water-sensitive sandstone or shale formations) and it is easily removed from the wellbore after the fracturing process is complete. Despite these attributes, very few "dry" fracks with liquid CO_2 are conducted. The foremost technical reason that pure CO_2 remains an ineffective dry fracking fluid is its very low viscosity at typical fracturing pressures. Based on models of rock mechanics, a fracture formed with pure CO_2 is significantly smaller than fractures formed with "slickwater". Further, the low viscosity of CO_2 hinders its ability to effectively transport high concentrations of large proppant particles into the fractures. There are several companies that combine liquid CO_2 with an aqueous surfactant solution to generate CO_2-in-water emulsions that have a high apparent viscosity. However, the presence of water in these CO_2-in-water foams renders them "wet". We propose a "dry foam" composed of bubbles of liquid CO_2 within films of an oil phase. Several research teams have previously explored CO_2-in-oil emulsions that were stabilized with conventional surfactants composed of hydrophilic and hydrophobic segments, often with the addition of water or ethylene glycol to the mixture. However the dissolution of these surfactants in the oil often results in gelation of the oil that makes it difficult to pump, and upon mixing of the oil + surfactant mixture with CO_2 the resultant emulsions are often unstable. Therefore we have designed and synthesized novel non-fluorous surfactants for stabilizing CO_2-in-oil emulsions. In accordance with Bancroft's rule, these surfactants are more oil-soluble than CO_2-soluble. Further, each surfactant has two hydrophobic segments; one that is CO_2-philic and oil-phobic, while the other segment is oil-philic and CO_2-phobic. Finally, the surfactants have also been designed to not gel the oil. High pressure CO_2-in-mineral oil emulsion stability measurements and high pressure CO_2-mineral oil IFT results will be presented that demonstrate the remarkable efficacy of these novel surfactants relative to conventional surfactants.

Preliminary results for novel surfactants designed to stabilize dry natural gas-in-oil foams will also be presented. Such foams may be attractive for dry fracturing because the foam would have a much higher apparent viscosity than pure methane, natural gas could be obtained from nearby producing wells, and the flowback hydrocarbons (natural gas and oil) would mix with the produced gas and oil.

460970 Water-Oil Janus Emulsions: Microfluidic Synthesis and Morphology Design

Abstract:

Janus emulsions and particles have been widely studied because of their asymmetry and corresponding applications in different fields. There are trials using in situ microfluidic methods to prepare water-oil Janus emulsions, However, the expansion of the range was still limited. Furthermore, even the theory of crucial factors for the formation of Janus morphologies has been described, convenient and systematic experiments for adjusting the morphologies have not been done.

In this work we propose a method to design the morphology of waterâ€"oil Janus emulsions. By using a double-bore capillary microfluidic device, we in-situ form the water-oil Janus droplets with the flow ratio of water and oil phases in wide range. Then we built a model to guide our experiments to realize the morphology design. The model is based on the lowest interfacial energy theory and the geometry. Moreover, by combining the lowest interfacial energy theory and our fluids properties, we could adjust the Janus morphology by simply adjusting the mass fraction of the surfactants in continuous phase and the flow rates ratio. According to our model guidance, we customize the emulsion morphologies to designed ideas with chosen surfactant mass fraction and flow ratio. Moreover, we achieve the methodology of designing Janus morphologies and first in-situ prepare water-oil Janus droplets with controlled structure. We successfully prepared Janus emulsions with gradual morphology changes, which would be meaningful in fields that have a high demand for morphology designing of amphiphilic Janus particles.

The materials we used were deionized water as the inner water phase, ethoxylated trimethylolpropane triacrylate monomer (ETPTA) as the inner oil phase, and liquid paraffin with various mass fractions of Span 80 (surfactant) as the outer oil phase. In experiments, we adjust the mass fraction of Span 80 to adjust the interfacial tension between these liquids. Another independent variable is the flow ratio of ETPTA to water phase. With clear understanding of the

lowest energy theory and the geometry of Janus, we find the mass fraction of surfactant and the flow ratio are the only two variables that influence the Janus morphology. As long as they are fixed, the morphology is determined. Here in our work, we build Matlab code to predict the morphologies. When inputting the interfacial tension of liquids(which is decided by the mass fraction of surfactant) and the flow ratio of ETPTA to water, the Janus morphology could be simulated and shown as a figure. Thus we could know the Janus morphology before we do experiments. The experiment conditions(the mass fraction and the flow ratio) according to the targeted Janus morphology.

In conclusion, by combining the geometry theory, numerical modelling and experimental properties we obtained a succinct method to design the Janus morphologies and prepare specific Janus morphologies by adjusting the flow rates ratio and the mass fractions of the surfactant in the outer phase. With the clear understanding of the formation mechanism and the succinct and useful microfluidic technology and working systems, we achieved a method to design Janus morphologies with high efficiency, that would be meaningful in fields like catalysis with expensive materials.

464810 Multiscale Design of Emulsions with Vegetable Oils of Promising Amazonian Species

Abstract:

Colombia has been classified worldwide as a mega-diverse country, because hosts approximately 10% of life forms reported. It can be considered an advantage, in comparison with other countries, for development of new products and services derived for biodiversity use. Therefore, it is possible to deduce that there is a need to carry out more research projects aimed towards efficient design of products, including natural ingredients in their formulation (bioprospection). The application of innovative techniques of bioprospection can lead to the finding of new biological species that can be used as the primary source to obtain new products.

Oils and fats as raw materials has a great importance for several industries such as food, animal food, toiletries and pharmaceutical industry. Nonetheless only oils derived from African palm, soybean, some oilseed, and animal fats have been considered economically important. Some endemic palms are potential species for oil extraction, but logistical difficulties of amazon region and strict regulations for Colombia´s biodiversity use have caused that these species do not have an economical role in oils and fats chain. Hence oils of endemic palms are under – used, and factors such as their metabolic profile, nutritional potential and phytotherapeutic potential are ignored.

Here, we propose a rational methodology of bioprospection over some fruits from reported species of the Arecaceae family, using a multiscale design as a strategy to achieve a systematic design of stable emulsions, mainly for the cosmetic industry. The species *Euterpe precatoria, Oenocarpus bataua, and Mauritia flexuosa* are abundant in Colombia's amazon region, and have a significant amount of ethereal extract; therefore, it is possible to take advantage on this characteristic considering to exploit them with a sustainable approach. For this, an expeller press process has been used. It allows to find 8% - 30% w/w yields depending on the specie, thus, comparing the values of extraction with petroleum ether, it is possible to conclude that pressing process allows an oil recovery of 67 – 77% w/w. Each oil was evaluated by quantifying the quality indices, lipid and triglycerides profile analysis. It was found that the content of unsaturated fatty acids is 74% - 83% for the three species being triolein the compound of higher relative concentration.

Since many finished cosmetic products are made from colloidal mixtures, in order to do a complete biosprospection research, we focused on finding a successful strategy for emulsion design. The use of emulsions in cosmetic products formulation allows to give specific characteristics to final product through mechanisms as encapsulation of active compounds, modification of rheological or optical properties or adjustment of organoleptic attributes. Therefore, emulsions with each one of the extracted oils are made using a biosurfactant as emulsifyin agent. Biosurfactants are a structurally quite diverse group of compounds synthesized by microorganisms, that can be be located at the interface between two liquid substances of different polarity, this feature gives the molecule abilities as act surfactant reducing surface tension, modifying the rheological behavior and making possible micellar organization (Nitschke & Costa, 2007; Ranasalva et al, 2014). The used biosurfactant was the protein membrane A of *Escherichia coli* (OmpA) which shows high homology with emulsifier protein (AlnA) from *Acinetobacter radioresistens* (Torens et al, 2002). OmpA´s amphiphilic structure is due its hydrophobic β-barrel domain and hydrophilic periplasmic domain. Thus, OmpA displays a significant role as a potential biosurfactant since it increases the stability of dodecane – water emulsions (Aguilera et al., 2014).

Finally, emulsions of different oils were made using 15 – 18% w/w of oil and 0,001 – 0,05% w/w of biosurfactant. To determine the stability of the emulsion, the microscopic and macroscopic scales have been evaluated (Pradilla, Vargas, & Alvarez, 2015). Firstly, using methods of light backscattering in order to obtain the variation of the droplet size over time. Subsequently, rheology properties were specified to decide the best application of the emulsion being studied. It was found that emulsions of each oil have their own rheological behavior and destabilization pattern.

In conclusion, the use of biosurfactants and non – timber natural forest products as raw matherials in cosmetic products can be an effective strategy for sustainable use of Colombian biodiversity. For this, it necessary to understand that the emulsion formulation is a critical step of product design because it has a direct impact in product properties and quality.

References

Aguilera Segura, S.; Macías, A. P., Carrero Pinto, D., Vargas, W. L., Vives-Florez, M. J., Castro Barrera, H.E., Álvarez, O.A. and Gonzalez Barrios, A. F. (2014). *Escherichia coli's* OmpA as Biosurfactant for Cosmetic Industry: Stability Analysis and Experimental Validation Based on Molecular Simulations. In L. F. Castillo, M. Cristancho, G. Isaza, A. Pinzón, & J. M. C. Rodríguez (Eds.), *Advances in Computational Biology SE - 38* (Vol. 232, pp. 265–271). Springer International Publishing.

Dirección Nacional de Planeación. (2010). *Cadena produtiva de oleaginosas, aceites y grasas*. Retrieved from https://colaboracion.dnp.gov.co/CDT/Desarrollo Empresarial/Oleaginosas.pdf [02/02/16]

Dirección Nacional de Planeación. Departamento Nacional de Planeación. Consejo Nacional de Política Económica y Social. Documento CONPES 3697. Política para el Desarrollo Comercial de la Biotecnología a partir del uso sostenible de la biodiversidad (2011). Bogotá D.C. Colombia.

Nitschke, M., & Costa, S. 2007. Biosurfactants in food industry. Food Science & Technology 18: 252-259.

Ranasalva, N.; Sunil, R. & Poovarasaan, G. 2014. Importance of biosurfactant in food industry. IOSR Journal of Agriculture and Veterinary Science, 7 (5): 6 – 9.

Pradilla, D., Vargas, W., & Alvarez, O. (2015). The application of a multi-scale approach to the manufacture of concentrated and highly concentrated emulsions. *Chemical Engineering Research and Design*, *95*, 162–172.

Torens, A.; Orr, E.; Paitan, Y.; Ron, E.Z. and Rosenberg, E. (2002). The Active Component of Bioemulsifier Alasan from *Acinetobacter radioresistens* KA53 Is an OmpA – like Protein. Journal of Bacteriology. 184 (1): 165 – 170.

465094 Effect of Solid Particles on Interfacial Rheology and Transient Stability of Water-in-Oil Emulsions

Abstract:

Solid stabilized emulsions commonly occur in many industries, especially in the energy industry. Fundamental insight on the transient stability of concentrated emulsions stabilized using solid particles is critical to manage these processes. This work focuses on quantifying the relationship between the stability of emulsions and their associated interfacial properties. In this work, nuclear magnetic resonance (NMR) is utilized to characterize transient behavior of concentrated emulsions. A series of experiments are presented for model oils and well-characterized surfactants and solids in order to elucidate the impact of stabilizer type on dynamic emulsion behavior. A particular focus of the presentation is on the impact of silica and carbon nanotube stabilizers on dynamic emulsion behavior. Additional experiments were performed to investigate the effect of salinity on the transient stability of emulsions. Dynamic oilâ^'water interfacial tension and interfacial rheological measurements using the oscillatory pendant drop method were performed to characterize the interfacial rheological properties of concentrated water-in-oil emulsions stabilized using solid particles and carbon nanotubes.

451689 Microstructural Origins of Gel-like Rheology in Solid-Stabilized Emulsions

Abstract:

Pickering-Ramsden emulsions rely on solid particles to stabilize fluid-fluid interfaces and are commonly found in many applications including pharmaceuticals, petroleum refining and various consumer products. Significant to many current and emerging technological applications of Pickering-Ramsden emulsions are their resulting microstructures and rheological behavior, which can influence processing, functionality, and stability of end-use products. Depending on the processing and physicochemical conditions, these multiphase mixtures can exhibit rich rheological behavior including liquid-like viscoelasticity, creep deformation, gelation, and yielding. The purpose of this study is to investigate the microstructural mechanisms that lead to gelation in solid-stabilized emulsions, and how the emergent morphology governs the mechanical behavior of the resultant mixture. We utilize two fluid systems to create four solid-stabilized emulsion gels with differing internal microstructures. These include bicontinuous fluid domains separated by a jammed colloidal monolayer (bijels); tenuous, percolating droplet networks bridged together by colloidal particles (Pickering emulsion gels); regular, concentrated Pickering-Ramsden emulsions; and tightly packed arrangements of faceted solid-stabilized droplets (biliquid foams). The microstructures and mechanical properties of the mixtures are characterized by 3D confocal imaging and oscillatory rheology, and the correlations between the two are investigated. We observe nontrivial dependence of the zero-shear elastic modulus and yield stress on colloid volume fraction in these systems, including non-monotonic dependence and the emergence of a maximum strength in bridging systems. Our observations will be discussed in the context of the emulsion microstructure. In fact, access to these four classes of solid-stabilized emulsions with markedly different morphologies suitably positions us to investigate these questions in a systematic fashion. Our findings have important implications for a wide range of different industries that utilize mixtures of fine particles and immiscible fluids to create new food formulations, cosmetic products, and composite materials.

458227 Experimental Verification of Structural Transition and Dissociation Enthalpy Change during CH_4 - CO_2 Replacement That Occurs in Various Gas Hydrate Structures

Abstract:

This study investigated the CH_4-CO_2 replacement that occurs in various gas hydrate structures (sI, sII, and sH), primarily focusing on thermodynamic stability, structural transition, and dissociation enthalpy change for its dual function of CH_4 recovery and CO_2 sequestration. To verify the influence of CO_2 injection on the thermodynamic stability of various gas hydrate structures, phase equilibria of the initial CH_4 hydrate (sI), CH_4 + C_3H_8 hydrate (sII), and CH_4 + 2,2-dimethylbutane (neohexane, NH) hydrate (sH) were compared with those of corresponding gas hydrates replaced with CO_2. The shift in the phase equilibrium conditions after replacement implied that the substantial extent of the replacement was achievable after the replacement reaction proceeded. In addition, to examine a possible structural transition and to elucidate cage-dependent guest distributions, the initial and replaced hydrates were analyzed via carbon-13 nuclear magnetic resonance (^{13}C NMR). The ^{13}C NMR spectroscopic results confirmed that the CH_4 - CO_2 replacement occurred in the sI-isostructural system and the CH_4 + C_3H_8 - CO_2 replacement occurred in the sII-isostructural system, whereas the CH_4 + NH - CO_2 replacement underwent the structural transformation from sH to sI. To reveal heat generated or absorbed during the replacement process and its influence on the thermal properties of the replaced hydrates, changes in heat flow and dissociation enthalpies (*Î”*H_d) were measured using a high-pressure micro-differential scanning calorimeter (HP *Î¼*-DSC). During the CH_4 - CO_2 replacement and the CH_4 + C_3H_8 - CO_2 replacement, no significant changes in heat flow emanating from endothermic or exothermic reactions were observed, indicating that the isostructural replacement occurred without significant hydrate dissociation or reformation. However, a significant endothermic peak followed by a large exothermic peak was observed in the CH_4 + NH - CO_2 replacement, indicating that the initial sH hydrate dissociation was followed by the formation of sI hydrate as the replacement reaction proceeded. The DH_d value of the replaced hydrate in the structure-transitional system was significantly lower than that of the initial CH_4 + NH hydrate and even slightly lower than that of pure CO_2 hydrate. However, the DH_d values of the replaced hydrates in the isostructural replacement did not change as remarkably as in the structure-transitional replacement. The overall experimental results provide further insights into the cage-specific occupation of external gas molecules and thermodynamic stability for the real replacement occurring in natural gas hydrate reservoirs for CH_4 recovery and CO_2 sequestration.

INVESTIGATION OF HYDRATE SHELL FORMATION ON WATER-CONTAINING PARTICLES

Yutaek Seo, Hyunho Kim, Department of Naval Architecture and Ocean Engineering, Seoul National University, Seoul, South Korea
Jong-Tae Chung, YoungGeun Kim, KOGAS Research Institute, South Korea
Colin D. Wood, CSIRO Manufacturing, Clayton, Australia

Introduction

Gas hydrates are nonstoichiometric crystalline compounds, consisting of host water and guest hydrocarbon molecules such as methane, ethane, and propane. They are formed through hydrogen bonding between water molecules enclathrating guest molecules at high pressure and low temperature. These conditions are found in subsea oil and gas flowlines as the production of offshore oil and gas fields has been moving into deeper and remote regions. Industry practice has been relying on the injection of vast quantities of alcohol- or glycol-based thermodynamic hydrate inhibitor (THI) to prevent hydrate formation in subsea flowlines. However, handling large amount of THI in remote offshore fields significantly increases the capital and operational expenditure, thus adversely affecting the economics of field development. As such hydrate prevention strategies are now moving toward hydrate risk management by either delaying the formation or preventing the agglomeration of hydrate particles.

There are several options available to industry for avoiding hydrate blockages including thermal management of the hydrocarbon fluids via insulation of flowlines and/or injection of thermodynamic hydrate inhibitors (THIs) such as methanol or mono-ethylene glycol (MEG). The concentration of THI in the aqueous phase must be sufficiently high to avoid hydrate formation. This is calculated from the flowline pressure and temperature, however, the injection rate of THI increases depending on the volume of the aqueous phase. The handling of THI in offshore fields significantly increases the operating expenditure due to the large injection volume, and it was estimated that the cost of methanol injected into a flowline in offshore oil fields producing 50,000 bbl/day water can reach 50% of the operating costs.[14] Although this is the conventional strategy it is now facing challenges such as limited space on floating structures and bulk logistics, as the exploration and production of offshore oil and gas fields moves to deeper and remote regions. As such hydrate prevention strategies are now moving toward hydrate risk management, where the hydrates are allowed to form, but the formation is delayed or the agglomeration is prevented before blocking flowlines.

In our previous work[23], we investigated the potential to use polyacrylamide-based hydrogel particles for simultaneously inhibiting the formation of gas hydrates and preventing their agglomeration. The particles were synthesized specifically for the application and swell to a controlled degree in water whilst remaining discrete. Stable torque was observed for the polymer hydrogels during the hydrate formation while there were sharp increases in torque before the stirrer stops for a water and decane mixture. The experimental results showed that the hydrogel particles were well dispersed in the decane phase even after the hydrate formation and do not deposit under stirring because the particles remain discrete due to the polymer network. As such polymer hydrogel particles have potential to prevent hydrate bed formation and deposition without using surfactants. Moreover the hydrogel particles can be further optimized to include thermodynamic or kinetic hydrate inhibitors. In the present study, polymer hydrogel particles were synthesized that incorporate MEG or a KHI, and their hydrate inhibition performance was tested by measuring the hydrate onset time, initial growth rate, hydrate fraction, and torque changes. Control experiments for MEG and KHI solutions (without polymer hydrogels) were carried out to investigate the effect of the polymer hydrogels. We report the results from this extensive set of experiments and suggest a mechanism for hydrate particles deposition in the presence of both hydrate inhibitor and the hydrogel particle. The obtained results suggests that the polymer hydrogel can be utilized as a versatile base material for hydrate inhibitors to be coupled with either thermodynamic hydrate inhibitors or kinetic hydrate inhibitors.

Experiments

Materials and synthesis of hydrogel particles

The distilled water used for hydrate experiments was purchased from OCI and decane was from Sigma-Aldrich. The simulated natural gas (CH_4: 90mol%, C_2H_6: 6 mol%, C_3H_8: 3 mol%, and C_4H_{10}: 1 mol%) was provided by Special gas (Korea). All of the chemicals for the hydrogel polymer synthesis were purchased from Sigma-Aldrich and were used as received including: polyacrylamide-*co*-acrylic acid partial sodium salt (PAM-*co*-AA), M_w 520 000, M_n 150 000, typical acrylamide level 80%; *N*-(3-(dimethylamino)propyl)-*N'*-ethylcarbodiimide hydrochloride (EDC, commercial grade); *N*-hydroxysuccinimide (NHS, 98%); and 1,2-diamino ethane (EDA >99%). Tris(2-aminoethyl)amine (TREN);, technical grade; heptane (HPLC grade, >99.5%).

The synthesis of the hydrogel particles has been described previously and involves the carbodiimide-mediated cross-linking (CMC) reaction of PAM-*co*-AA in an inverse suspension. For MEG swollen hydrogels, the hydrogel particles containing water were dried by precipitating the polymer in ethanol to remove the water. The resulting dried polymer was swollen in a 20% w/w% solution of MEG and the final polymer concentration was 13%.

After formation of the hydrogel the water was removed by freeze drying for approximately 4 hrs. These polymers were ground using a mortar and pestal. To assist the grounding, dry ice was applied to enhance the brittleness as needed. The dried particles were swelled by suspending in heptane and adding water in a dropwise manner while stirring vigorously to ensure the particles were separated in the jar. The heptane was removed by decanting and the particles were left to air dry resulting in a bead-like gel. Microscopy images of the MEG hydrogel are shown in Figure 1. The images prove that the hydrogel particles are stable even after the cycles of hydrate formation and dissociation.

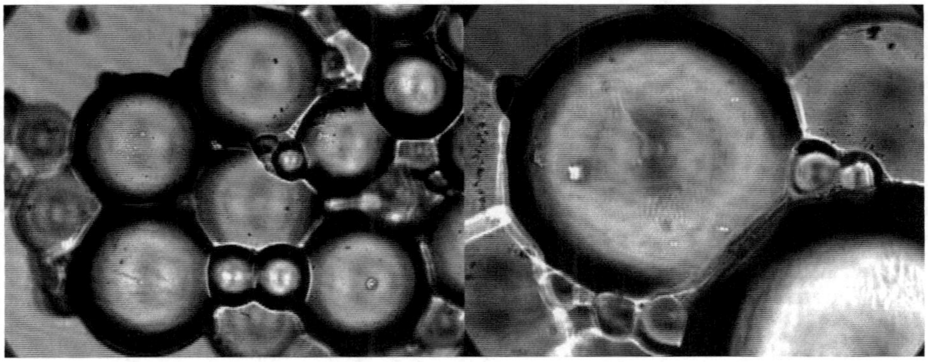

Figure 1. Microscopy images of MEG hydrogel after hydrate formation and dissociation. (left): 10x, (right):20x

High pressure autoclave experiments

The experiment was conducted using a high pressure autoclave made by 316 SUS and equipped with anchor type impeller. The autoclave body was immersed in a bath circulator during the experiments (Jeio tech, RW 2040G) to control the temperature of the inner fluids. The temperature and pressure was measured using a platinum resistance thermocouple (0.15 °C accuracy) and pressure transducer (0.1 bar accuracy in the range of 0-200 bar), respectively. The anchor type impeller was mounted on a solid shaft and was driven by a motor (BLDC 90). The stirring speed was 600 rpm for all experiments to obtain Reynolds number of 32,000. A torque sensor (TRD-50KC) with a platinum-coated connector was used to indicate resistance-to-flow due to the formation of hydrates (0.3% accuracy). Total 80 mL of liquid phase was loaded into the autoclave that had a 360 mL internal volume. Decane and distilled water were used to represent liquid hydrocarbon phase and aqueous phase for bulk water experiment, which is a control without hydrogel polymer in aqueous phase. The water cut was 60% for all experiments in this work. For studying hydrate formation in MEG-PAM hydrogel particles, the bulk distilled water was replaced by MEG-PAM hydrogel particles and mixed with decane.

Pressure, temperature, and torque were continuously monitored during the formation and dissociation cycles of hydrates as seen in Figure 2. The reactor was pressurized to 120 bar at 24 °C and the system was stirred at 600 rpm to saturate liquid phase with gas for 2 hours. Then the temperature was decreased at a constant rate to achieve the target temperature by cooling rate of 0.25 °C per min from 24 °C to 4 °C. After reaching 4 °C, the system was maintained at 4 °C for ten hours, and then the temperature was increased to 28 °C for three hours to dissociate the hydrate.

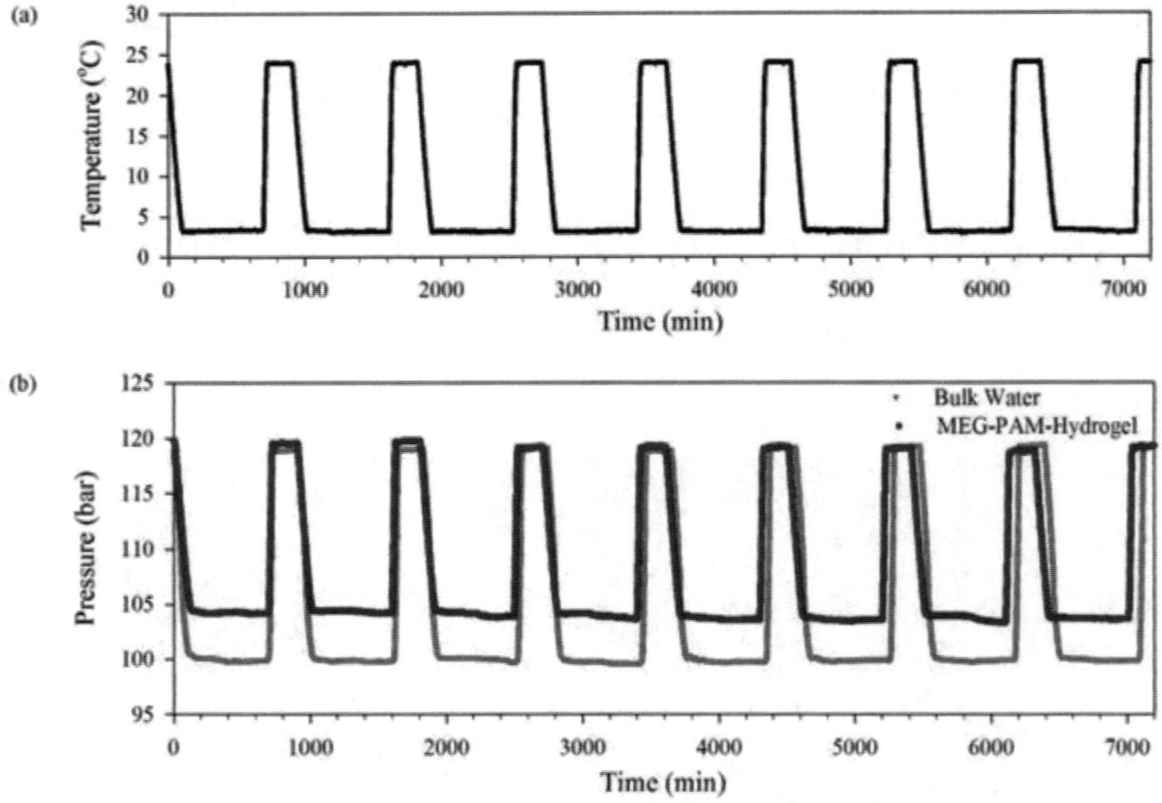

Figure 2. (a) Temperature and (b) pressure changes of bulk water and MEG-PAM hydrogel systems during the eight cycles of hydrate formation and dissociation.

Figure 2 (a) shows the temperature cycles for the entire cycles while Figure 2 (b) shows the pressure changes for both bulk water system (red) and MEG-PAM hydrogel system (blue). A total of eight cycles of hydrate formation and dissociation were repeated for both bulk water and MEG-PAM hydrogel, respectively. Hydrate formation was characterized by decreases in pressure due to consumption of the gas during the formation process. As seen in Figure 2, the MEG-PAM hydrogels showed different hydrate growth rate when comparing to that of bulk water. The details of hydrate onset time, water conversion rate, and hydrate volume fraction were estimated from the amount of gas consumed during the hydrate formation.

Results and Discussion

The hydrate formation characteristics for the hydrogel particles including the onset time, subcooling temperature, hydrate fraction, and torque changes were compared to those of bulk water (without hydrogel). Eight experiments were performed to investigate the hydrate formation characteristics for bulk water and MEG-PAM hydrogel particles (total 16 experiments). The hydrate volume fractions, Φ_{hyd} in the liquid phase were calculated based on the gas consumption, which was determined from the pressure difference between the experimental pressure and the calculated pressure with the assumption of no hydrate formation.[12, 13] The obtained hydrate growth profiles for each bulk water and MEG-PAM hydrogel particles are shown in Figure S1 and Figure S2, respectively. The average of hydrate volume fraction at corresponding time were calculated and represented together with the experimental results until 100 min after the hydrate formation (Figure S1 and Figure S2). The standard deviation was 1.8% with the maximum deviation of 3.76% for bulk water until 100min while the standard deviation was 0.85% with the maximum deviation of 1.61% for MEG-PAM hydrogel particles. The averaged hydrate growth profiles are used to investigate the hydrate formation kinetics for both bulk water and MEG-PAM hydrogels.

Figure 3 (a) shows the hydrate volume fraction increase during the hydrate formation in bulk water and MEG-PAM hydrogel particles. The water cut was maintained at 60% for all the experiments and time zero indicates the moment of hydrate onset. As can be seen in Figure 3 (a), for bulk water, initially the hydrate growth is fast and slows just after the hydrate onset. However, the rate of hydrate formation increases again and hydrates grow up to a hydrate fraction of 0.3, where the rate becomes slow again before reaching the maximum value of 0.4. However, for MEG-PAM hydrogel particles, a steady growth of hydrate is observed until 10 min after the onset, then it slowly reached the maximum value of 0.08. The final hydrate fraction at 600 min after the onset is presented in Table 1 and Table 2 for bulk water and MEG-PAM hydrogel particles, respectively.

Recent studies on the hydrate plug formation using a flow loop and autoclaves suggested that the hydrate fraction in the liquid phase is an important parameter to describe the flow ability of hydrate particles. [14,15] Hydrate particles are segregated heterogeneously followed by deposition and bedding on the wall with increasing hydrate fraction. To describe this process, Akhfash et al. suggested the disruption of the water-oil interface due to hydrate particle growth followed by full dispersion of the water phase in continuous oil phase leading rapid hydrate growth. [16] This hydrate particle growth mechanism was developed based on the hydrate volume fraction profile with time calculated form the consumed amount of gas during the hydrate formation. For fully dispersed water particles, initial hydrate growth rate drops below the prediction 15 min after hydrate nucleation, where the model assume the water particles would be fully dispersed in oil phase and the surface area for hydrate growth would remain constant. An increase in mass transfer resistance and a decrease in the interfacial water-oil surface area would cause this sudden decrease, however the growth eventually accelerates to a rate faster that the prediction, which can be attributed to the generation of additional surface area between the water and oil phase. Thus a sufficient amount of hydrate is formed resulting in hydrate plug formation leading large spikes and fluctuations in the torque. As seen in Figure 3 (a), the obtained hydrate growth profile for bulk water was similar to those in literature, suggesting the mechanism of hydrate plug formation in literature would be applicable to the hydrate formation process in this work.

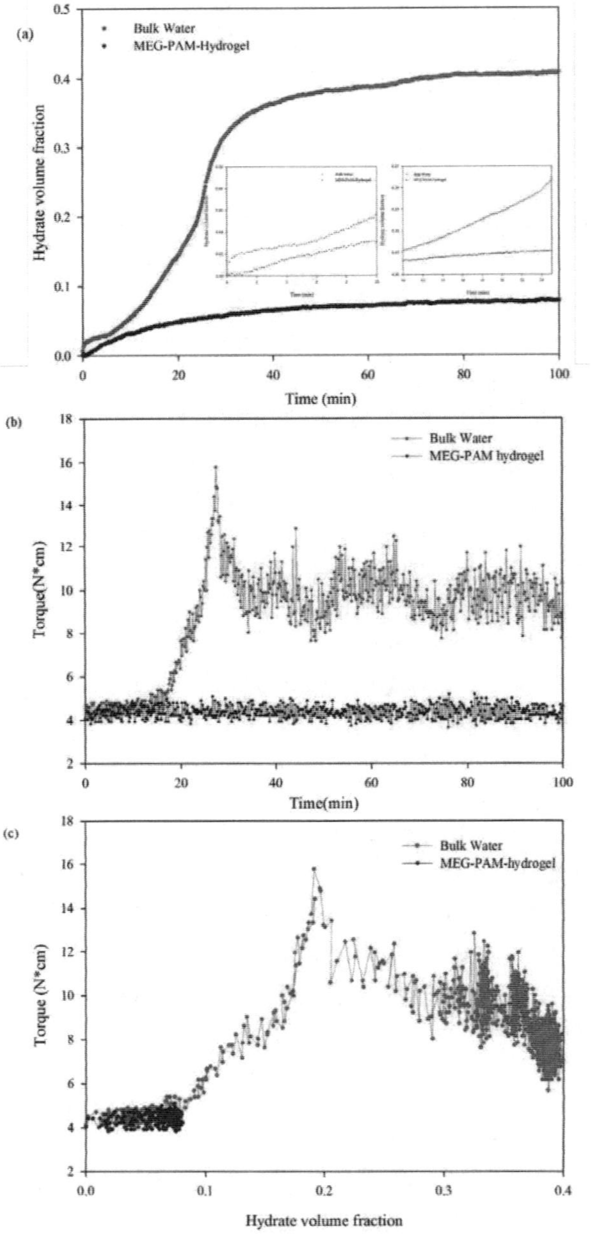

Figure 3. (a) Average hydrate volume fraction (Inset figures are the enlarged area of 0-10 min and 10-25 min.), (b) torque change, and (c) relation between hydrate volume fraction and torque of bulk water and MEG-PAM-Hydrogel systems during single cycle.

Unlike bulk water, there was no acceleration in hydrate growth for MEG-PAM hydrogel particles, as shown in Figure 3 (a). Fast initial growth was followed by steady decrease in the growth rate, which is well predicted by the hydrate growth model suggested by other researchers [2,16] when assuming a constant surface area and the onset of diffusion limitations through the hydrate shell thickened for certain period of time. In the onset figure in Figure 3 (a), the growth rates for bulk water and MEG-PAM hydrogel were compared during the initial stage of hydrate formation. The growth rates for bulk water and MEG-PAM hydrogel systems were initially 0.005/min and 0.003/min from 0 to 10 min; however, the rates changed when hydrate growth proceeded further. The growth rate of bulk water from 10 to 25 min was increased to 0.011/ min suggesting an acceleration of the hydrate formation with the increased rate almost three times greater than initial value. On the other hand, the rate for MEG-PAM hydrogel decreased to 0.0015/min, which was only half of the initial value. With the increased growth rate for bulk water during 10 to 25 min, the hydrate volume fraction increased up to 0.217, then reached final fraction of 0.4 at 100 min. However, the hydrate volume fraction for MEG-PAM hydrogel only reached 0.05 at 25 min, then 0.08 at 100 min.

These results suggest that the mechanism of hydrate growth is different between bulk water and MEG-PAM hydrogel particles. Without the polymer hydrogel, bulk water was dispersed as droplets in the decane and these water droplets are covered by a hydrate shell upon hydrate onset on the surface of the droplets. If the hydrate shell is annealed enough, the cohesion between hydrate particles can be minimized. However as seen in Figure 3 (b), 15 min after the hydrate onset, the torque substantially increases and reaches to the maximum value of 16 N cm, where the hydrate fraction is about 0.2. It takes less than 10 min for the torque to increase and to reach the maximum. The cohesion and sintering between hydrate particles could be becoming significant when the hydrate fraction increases more than 0.1. The torque spike is followed by severe fluctuation until the hydrate fraction reaches 0.4, as seen in Figure 3 (c). The maximum torque followed by severe fluctuation is an indication of hydrate plug formation observed in flow loop and autoclave studies. [14, 17] Therefore, as seen in Figure 3, hydrate plug formation is confirmed for bulk water when hydrate volume fraction increases more than 0.2 from the torque spike and fluctuation. For eight cycles of hydrate formation and dissociation, as shown in Figure 4, hydrate formation and growth always induce the torque spikes followed by fluctuation. It clearly confirms that the hydrate plug will form upon hydrate onset and growth due to cohesive interaction between hydrate shell-covered water droplets.

Compared to fast plug formation in bulk water, MEG-PAM hydrogels do not show any evidence of the plug formation. Figure 3 (a) shows the hydrate growth on the surface of MEG-PAM hydrogels steadily decreases with time and the hydrate fraction in liquid phase reaches only 0.08. The torque remains stable during the hydrate formation, suggesting negligible resistance-to-flow due to hydrate shell formation on the surface of MEG-PAM hydrogel particles. The water conversion to hydrate was only 20.9% for the hydrogel particles, thus it is expected that MEG-PAM hydrogel particles are covered by hydrate shell and the cohesion between the particles is minimized. This is possibly due to the polymer hydrogel absorbs water inside hydrogel core and prevents its breakout through the hydrate shell. As can be seen in Figure 3 (a) and (b), the torque remains stable during the early stage of hydrate growth, it then also remain stable at lower torque values when the hydrate volume fraction becomes stable.

The average hydrate onset time was 20.9 min and the average subcooling temperature was 4.9 °C for bulk water. The addition of MEG in hydrogel particles increased the hydrate onset time to 60.0 min as well as the subcooling temperature to 8.8 °C. These results indicate the onset of hydrate crystals was delayed significantly due to the shift of the hydrate equilibrium condition by adding 20 wt% MEG into the aqueous phase, reducing the thermal driving force for hydrate formation. Our previous work suggested the addition of hydrogel slightly increased the hydrate onset time and subcooling temperature, however in this work, by incorporating MEG into hydrogels, longer hydrate onset time and higher subcooling temperatures were achieved. 74.5 vol.% of the water was converted, representing a hydrate volume fraction of 0.5 in the liquid phase. On the other hand, for MEG-PAM hydrogels, only 20.9 % of water was converted to hydrate resulting in a lower hydrate volume fraction of 0.15 in the liquid phase. While the 20.9 % of water was lost to hydrate shell, MEG still remained inside water core without loss. The concentration of MEG in the hydrogel particles slightly increases to 24 wt% considering the loss of water into hydrate shell, suggesting that the MEG concentration inside hydrogel core would limit the hydrate inward growth, indicating self-inhibition by MEG.

These results suggest that the hydrate shell formation changes the mechanical properties or reduces the liquid bridge of the hydrogel particles, which in turn prevents the hydrate from agglomerating. In the case of an offshore flowline the formation of hydrate shell on the surface of water droplets dispersed in a hydrocarbon phase results in agglomeration of the droplets and eventually a hydrate plug will be formed. The formation of a hydrate shell on the surface of MEG-PAM hydrogel did not induce an increase in torque, suggesting the particles are well dispersed in the hydrocarbon phase preventing the cohesive interaction and sintering of hydrate particles. Thus by adding MEG and PAM-*co*-AA hydrogel polymer into aqueous phase, a new concept for hydrate inhibition can be developed as it involved thermodynamic, kinetic, and anti-agglomeration inhibition performance.

References

1. Sloan, E. D.; Koh, C. A. (1991) "Clathte Hydrates of Natural Gases", 3rd ed.; *CRC Press (Taylor and Francis Group)*: Boca Raton, FL, **2008**.
2. Seo, Y.; Shin, K.; Kim, H.; Wood, C. D.; Tian, W.; Kozielski, K. A. Preventing Gas Hydrate Agglomeration with Polymer Hydrogels, *Energy Fuels,* **2014**, *28*, 4409-4420.
3. Cha, M.; Shin, K.; Kim, J.; Chang, D.; Seo, Y.; Lee, H.; Kang, S.-P. Thermodynamic and kinetic hydrate inhibition performance of aqueous ethylene glycol solutions for natural gas. *Chem. Eng. Sci.* **2013**, 99, 184−190.
4. Cha, M.; Shin, K.; Seo, Y.; Shin, J. Y.; Kang, S. -P. Catastrophic growth of gas hydrates in the presence of kinetic hydrate inhibitors. *J. Phys. Chem. A* **2013**, 117, 13988−13995.
5. Sohn, Y.; Kim, J.; Shin, K.; Chang, D.; Seo, Y.; Aman, Z. M.; May, E. F. Hydrate plug formation risk with varying water cut and inhibitor concentrations. *Chem. Eng. Sci.* **2015**, 126, 711-718.

459582 *Natural Gas Hydrates in Marine Locations: Effect of Stimulation Temperature on Gas and Water Production*

Abstract:

Natural gas (NG) is the cleanest burning fossil fuel, generating the least amount of CO_2 per unit energy produced. Following the growing energy demand and the depletion of conventional energy resources, there is a need to harness energy from the unconventional sources, such as shale gas and natural gas hydrates (NGH). The total carbon content sequestered in NGH is of the order of 10,000 Gt C, more than twice of the conventional fossil fuels (gas, oil and coal) combined. In this work, we introduce a new approach to determine the fractional conversion of methane under an excess water environment to mimic marine locations, which is found to be around 81.5%. We further investigated the gas and water production profiles from the hydrate bearing sediments. The dissociation of methane hydrate was done under a constant bottom-hole pressure of 4.5 MPa subjecting to different thermal stimulation ranging from 278.7 K to 285.2 K. It was found that a low temperature driving force would result in an extremely low dissociation rate shedding light on impact of global warming on the stability of hydrates in nature, and a minimum temperature of 280.7 K (corresponding to 2.1 K temperature driving force) is required to achieve 90% dissociation within 10 hours. The importance of water management in developing efficient methods to effectively recover energy from hydrate bearing sediment will be presented.

465792 Enhanced Kinetic Inhibition of Gas Hydrate Formation Using the Mixture of Poly(N-vinylcaprolactam) and Ionic Liquids

Abstract:

To investigate the synergetic effect by the mixture of ionic liquids (ILs), consisting of various cations and tetrafluoroborate anion, and well-know polymeric inhibitor, poly(N-vinylcaprolactam, PVCap) tests of kinetic inhibition on natural gas hydrate formation were performed. Results shows that the PVCap-IL mixtures revealed significantly enhanced kinetic hydrate inhibition performance. In particular, the mixture of PVCap and 1-hexyl-1-methylpyrrolidinium tetrafluroborate (HMP-BF4) showed the best hydrate inhibition performance even at higher pressure condition. It exhibited the highest hydrate nucleation inhibiting performance when it was at individual used, additional study was conducted using the mixtures of the two at various concentration combinations. The combination of 1:0.5 via weight basis between PVCap and HMP-BF4 showed the longest induction time. It is presumed that the enhanced synergetic effect of the ILs on the natural gas hydrate inhibition may arise from the prevention of the small cage formation by the ILs, which induces the inhibition of metastable structure I hydrate formation.

466445 Mechanism for the Replacement of CH4 in Methane Hydrates with CO2 in the Solid Phase

Abstract:

Methane hydrates is a kind of nonstoichiometric crystalline compound composed of water and methane at low temperatures and high pressures, and is regarded as a potential energy resource for the future. In addition to energy gas production, reducing global warming gas is also a concerned issue. One intriguing idea for the simultaneous recovery of energy and sequestration of global warming gas is proposed by the transformation of methane hydrate into carbon dioxide (CO_2) hydrate without melting the network of hydrogen-bonded water molecules. Some experiments have shown that methane hydrate can be changed into CO_2 hydrate by injecting liquid CO_2 into methane hydrate powders, and some theories has been proposed that there should be some porous vacancies formed by water molecules (water vacancy) in the hydrate structure to facilitate the replacement reaction. However, the exact mechanism is still unclear.

Molecular dynamics (MD) simulation has been a useful tool to unveil the molecular level details of gas hydrate. In this work, we use MD simulation to exam the moving or exchanging behavior of methane and CO_2 molecular in the crystalline hydrate with and without the presence of water vacancies. Our simulation results show that water vacancies are high energy sites in the hydrate structure. For the system with low concentration of water vacancy, we found that the vacancy propagates within the hydrate structure; however, its propagation does not stimulate the movement of methane or CO_2 molecule between cages. For the system with high concentration of water vacancy, the initially separated vacancies are found to aggregate into a special defect structure, centering around small (5^{12}) cages and resulting in broken surrounding large cages ($5^{12}6^2$). The movements of methane or CO_2 molecular are found to take place only in such aggregated defect structures. To confirm the validity of our simulation, we also check the diffusion coefficient of methane and carbon dioxide molecular, which are found to be in good agreement with experiment.

468587 Phase Equilibrium Measurements for Semi-Clathrate Hydrates of Water+Decane+Carbon Dioxide+Tbab Systems

Abstract:

Regardless of developments in the energy sector related to the search of renewable energies, it is a fact that fossil fuels still dominate the market, so its use is very important. In recent years, the oil industry record an increase of produced water during the extraction of these resources in oil production wells, so it is necessary to develop a low cost effective process for separating water from oil. During the extraction process the oil-gas-water phases are present. A novel alternative that can be considered for the separation of these phases is the gas hydrates formation. Gas hydrates are crystalline compounds similar to ice, composed of water molecules that form a clathrate structure (host molecule) and a molecule of a gas (guest molecule), which is encapsulated within the structure. Hydrate formation is due to a physical process, so there is no chemical interaction between the host and guest molecules at high pressures and low temperatures. However, given the energy costs, the separation of produced water must be carried out at near ambient conditions. This is possible in hydrate formation by adding a thermodynamic promoter, shifting the phase equilibrium at ambient conditions.

In this work, we reported the dissociation conditions for semi-clathrate hydrates of Water+Decane+Carbon Dioxide+tetra-n-butylammonium bromide (TBAB) systems. The dissociation conditions have been determinate using an apparatus based in an isochoric technique and designed to operate at pressure up to 40 MPa and at temperature range of (258.15-373.15) K. The experimental measurements were performed at temperature range of (275-282) K and pressures up to 4 MPa. The combined expanded uncertainties of the dissociation conditions were evaluated to be less than 0.012 MPa, 0.020 K and 0.0020 for pressure, temperature and mass fraction, respectively.

16AIChE
Annual Meeting, San Francisco, CA

2016 AIChE® Annual Meeting

Thermodynamic Study on Proline Inhibition of Methane Hydrate Formation

Bo Chen, Yangyang Li, Qiang Sun*, Xuqiang Guo*

State Key Laboratory of Heavy Oil Processing

中国石油大学(北京)
China University of Petroleum - Beijing (CUPB)

2016 AIChE® Annual Meeting

Contents

1.Introduction

- The major thermodynamic hydrate inhibitors.
- A. Alcohols is the most popular This in industry. Methanol and ethylene glycol.
- B. Inorganic salt is also used as This. Such as NaCl,KCl,$CaCl_2$.

1.Introduction

- Biological hydrate inhibitors

Antifreeze proteins

Amino acids

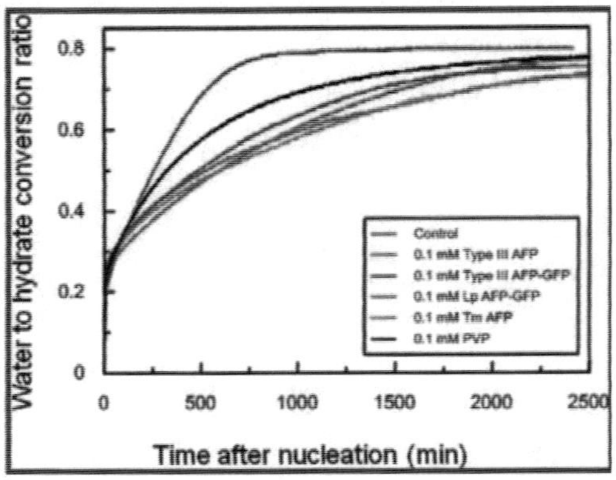

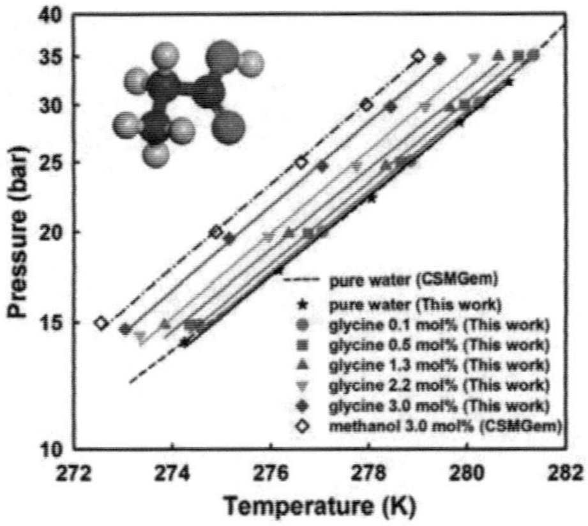

1.Introduction

Biological hydrate inhibitors

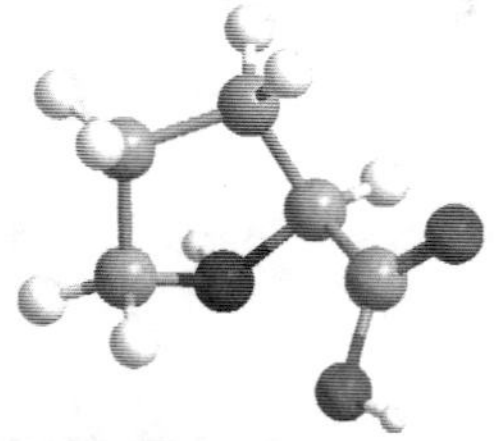

Proline

Osmolye

Proline is one of the most powerful osmolyes and synthesized and accumulate in cells as a response to high salt and very cold environments.

Unique structure

Proline has a unique structure among the amino acids; the amino nitrogen binds with the side chain, leading to formation of a pyrolidine ring with the α-carbon.

High solubility

Proline has the highest solubility of amino acids, . water solubility of proline is 162.3g/100ml(at 25℃).

2016 AIChE® Annual Meeting

2. Experiment

Apparatus

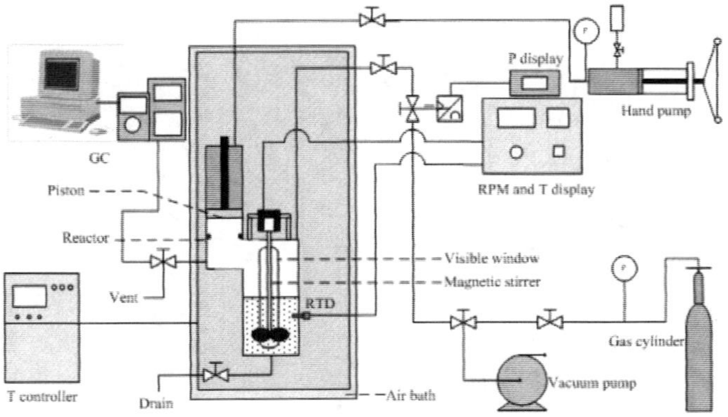

2. Experiment

❖ Experiment

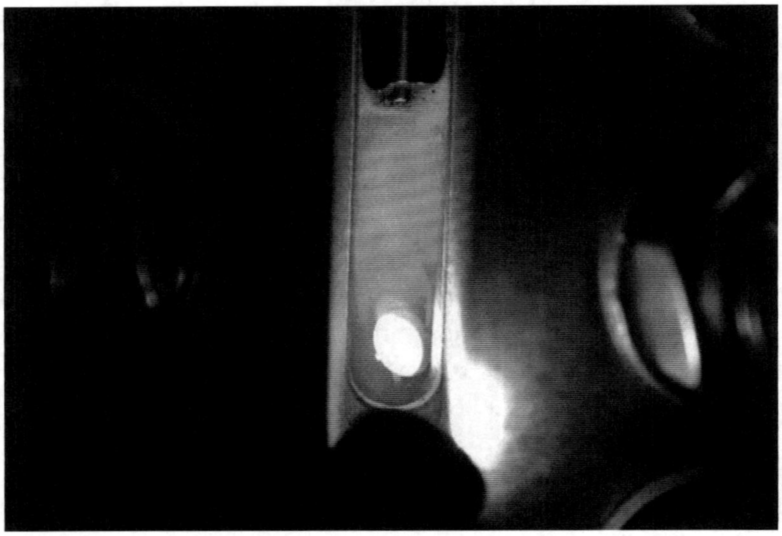

2016 AIChE® Annual Meeting

3.Working

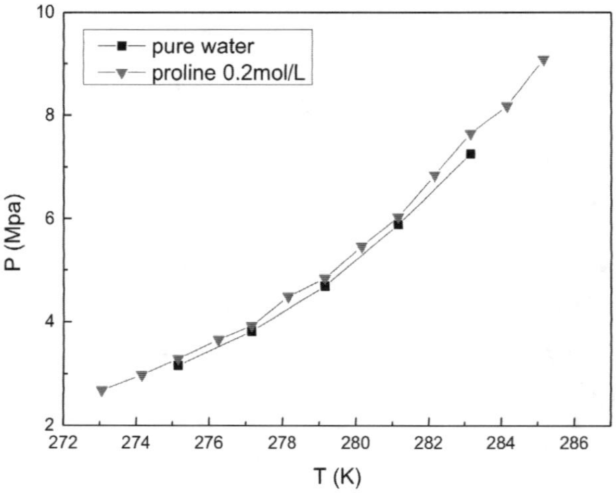

3.Working

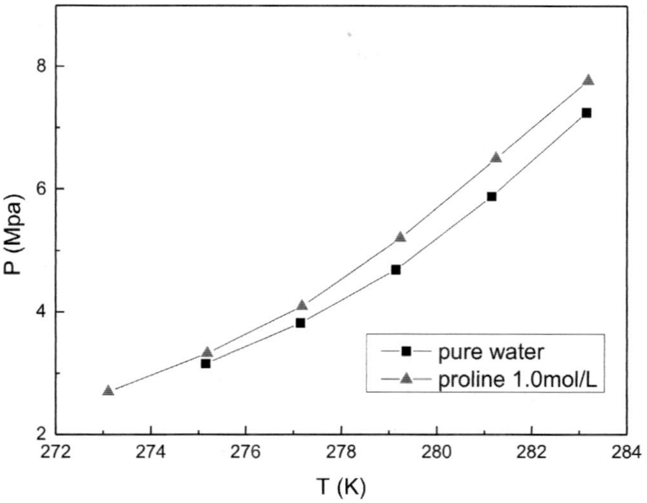

2016 AIChE® Annual Meeting

3.Working

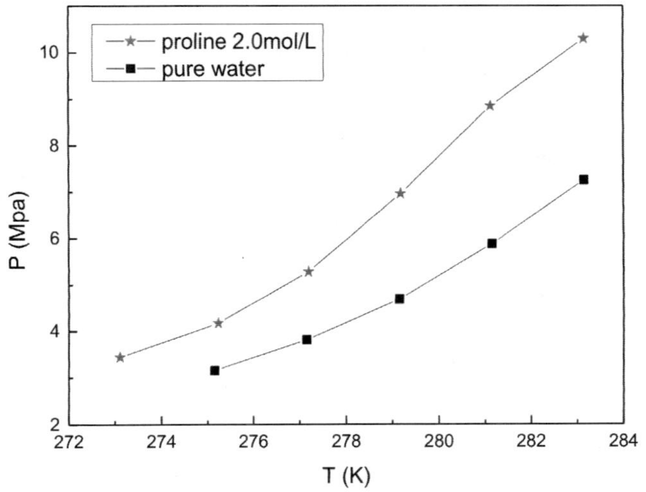

3.Working

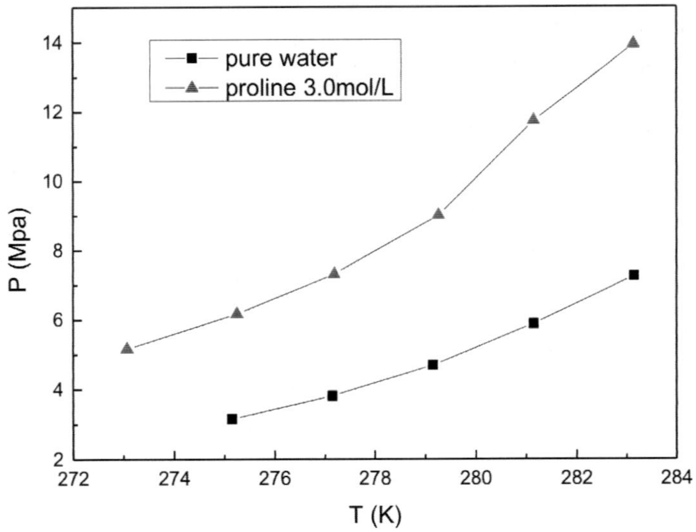

2016 AIChE® Annual Meeting

3.Working

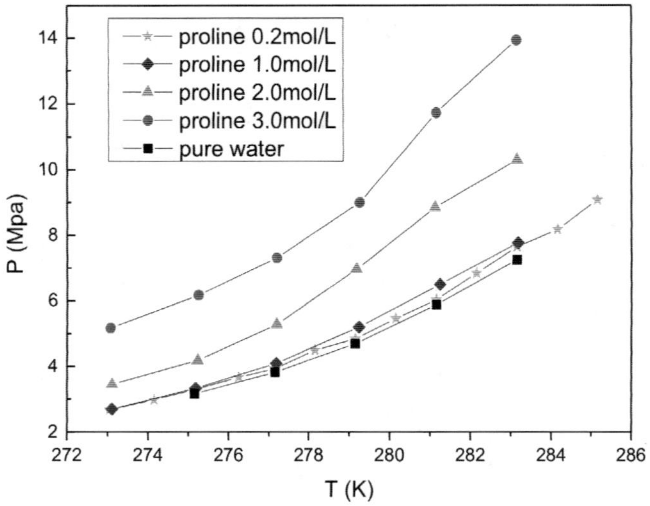

3.Working

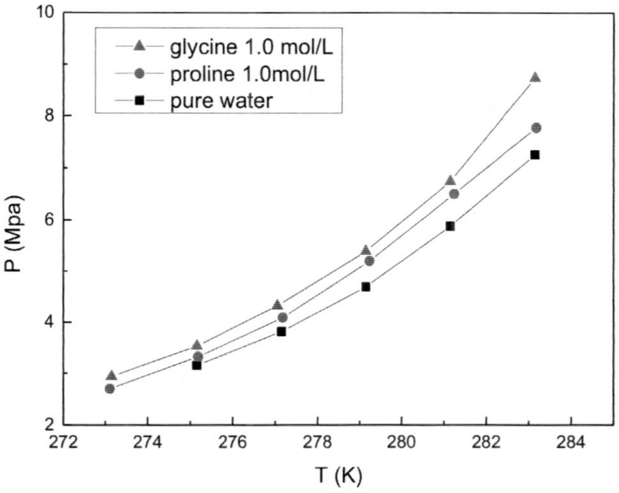

2016 AIChE® Annual Meeting

3.Working

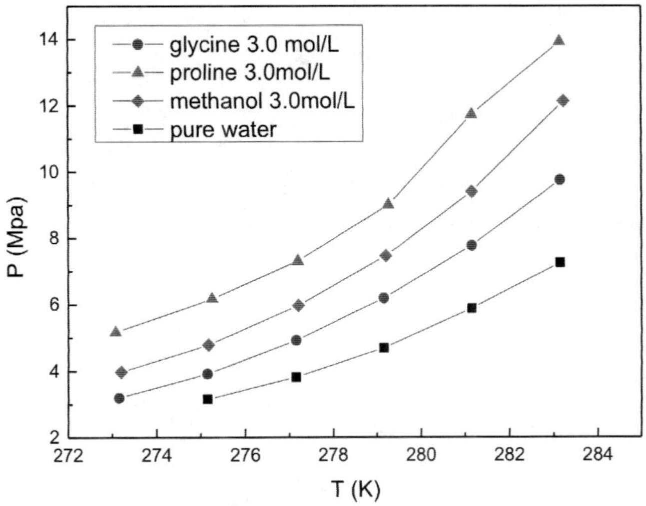

4. Conclusions

- Proline is a powerful thermodynamic hydrate inhibitor as a Biological hydrate inhibitors.
- Because of the best solubility as an amino acid, proline is most powerful THI in amino acids.

Thank You !

2015313018@student.cup.edu.cn

State Key Laboratory of Heavy Oil Processing

中国石油大学(北京)
China University of Petroleum - Beijing (CUPB)

469722 Fundamental Investigation into the Thermal Decomposition of Methane Hydrates and the Impact of CO_2 Sequestration

Abstract:

Fundamental investigation into the thermal decomposition of methane hydrates and the impact of CO_2 sequestration

Fundamentals of the hydrate formation and dissociation in a large-scale laboratory reactor were studied using a thermal stimulation and thermal stimulation combined with CO_2 sequestration method. Methane hydrate production can be coupled with simultaneous CO_2 sequestration by converting injected CO_2 into hydrate phase. Carbon sequestration makes this process carbon neutral to some extent. A Large-Scale Hydrate Vessel (LSHV) of internal volume 59.3 liters was used to duplicate the permafrost conditions in which hydrates form naturally. Methane hydrate formations were achieved and the formation of hydrates in LSHV was confirmed by comparing the Pressure-Temperature data with the pure CH_4 hydrate equilibrium curve. A testing matrix was developed and executed to study the effect of heating rate (20 W & 100 W) and CO_2 injection rate (155 ml/min and 1000 ml/min) on methane recovery and CO_2 sequestration at a constant initial hydrate saturation (50 % by pore volume). In addition to these tests, 2 baseline tests were conducted where no CO_2 was injected during the dissociation for comparison purposes. Results from those tests show that at 100 W heating rate the number of moles of CH_4 recovered increased by addition of CO_2 at 155 ml/min whereas, it decreased the CH_4 production by adding CO_2 at 1000 ml/min; relative to baseline tests where no CO_2 was injected. Similar results were obtained for 20 W heating tests. The number of moles of CO_2 sequestered increased from a value of 13 moles at 155 ml/min to 40 moles at 1000 ml/min. The motivation for this work comes from the need to utilize unconventional natural gas resources. Natural gas extraction from the hydrates can also enable sequestration of carbon dioxide helping reduce the carbon footprint and contribute to reducing overall greenhouse gas emissions. Currently, the conventional gas extraction and recovery methods are cheaper than the gas hydrate recovery methods and it is expected that this will change as, more efficient methods are being developed in gas hydrate field. It is required to focus the research on a lab scale experiments to provide economically feasible guidelines for large-scale production tests.

465045 Charge Transport Model to Predict Intrinsic Reliability for Dielectric Materials in Integrated Circuits

Abstract:

Dielectric breakdown is the formation of a conductive path in an insulating material, and it is observed in nature (lightning strikes) and man-made equipment (spark plugs, microelectronic devices). In the microelectronics industry, dielectric breakdown is an increasing concern for reliability engineers as device dimensions shrink in order to increase computing power and speed [1]. However, the ability to understand and to predict the breakdown behavior of dielectrics in integrated circuits is becoming more complex as new materials are introduced and the manufacturing process constantly changes. Due to time constraints, engineers employ a strategy to test a limited sample population at several high voltages in order to wear out the dielectric over time and cause failure, also known as time-dependent-dielectric-breakdown [2]. In order to extrapolate test results to low voltages, empirical models are used to fit the experimental data and then predict device lifetime at operating conditions. There is dispute over which empirical model provides the most accurate prediction, and it is most likely that a single model cannot be used to predict failure for every type of dielectric and process involved in an integrated circuit. As a result, a charge transport model was developed to provide a more comprehensive understanding of dielectric failure [3], in order to tie together the physical mechanism of failure to the dielectric material (e.g. dielectric constant, k) and the processing conditions.

The charge transport model is based on electron conduction and defect/trap formation in the dielectric, and uses Poisson's equation to calculate the potential distribution, continuity equations for mobile and trapped electrons, and a defect reaction equation [3]. The mobile carrier concentration injected into the dielectric depends on the probability of the electrons to overcome the energy barrier from the Fermi level of the metal electrode to the conduction band of the dielectric, known as Schottky emission [4]. Failure originates from the interaction of high energy electrons with the dielectric matrix. The electrons gain energy due to the applied electric field, and lose energy to the matrix through collisions with defects or other constituents of the dielectric [5]. The energy released during the collisions becomes available to generate additional defects, and the mobile electrons become trapped within the dielectric by these defects. As the defect concentration increases over time, the probability for trapped electrons to tunnel from trap-to-trap through the dielectric increases. This trap-assisted tunneling is quickly accelerated once a critical defect concentration is reached [6]. The emptying of traps due to tunneling electrons enhances the local field at the cathode. This causes an increase in the injection of electrons, and further increases defect generation. This is known as positive feedback, or feed-forward, failure [7], and is the cause for the abrupt increase in current seen at failure. The charge transport model is one-dimensional and is solved by numerical methods using COMSOL® software. All of the parameters used in the model have some theoretical basis, are material properties of the dielectric, or are experimental conditions.

Current and failure data was obtained from several sources in literature for low-k SiCOH [8, 9] and high-k SiN [10] dielectrics. Both materials are commonly found in integrated circuits. Similar tests were also conducted at GLOBALFOUNDRIES on low-k SiCOH materials. Model parameters were set based on the dielectric material (dielectric constant, thickness, etc.) and experimental conditions (temperature). The boundary condition for the voltage at the cathode was altered to match the test type (constant voltage or voltage ramp) for each data set. The model reproduces the current as a function of time for constant voltage tests, capturing the current decay at early stress times and the subsequent current increase due to defect generation. Current vs. voltage curves are also reproduced for voltage ramp tests. For both test types, the rapid increase in current observed at failure is reproduced by the model. Experimentally, a large jump in current is defined as the breakdown point, and this method was also used in our simulations to determine the time-to-failure as a function of the applied electric field for high voltage testing. The model is then used predictively to determine device failure times at low voltages, or near operating conditions. These low-voltage predictions are compared to the most commonly used empirical models for dielectric reliability. Finally, simulations are also run altering key parameters, such as the dielectric constant or thickness, in order to hypothesize how these parameters can affect device reliability.

References

[1] Chen, F., C. Graas, M. Shinosky, K. Zhao, S. Narasimha, X.H. Liu, and C. Tian. 2015. "Breakdown Data Generation and in-Die Deconvolution Methodology to Address BEOL and MOL Dielectric Breakdown Challenges." *Microelectronics Reliability* 55 (12, Part B): 2727–47.

[2] McPherson, J. W. 2012. "Time Dependent Dielectric Breakdown Physics – Models Revisited." *Microelectronics Reliability* 52 (9–10): 1753–60.

[3] Ogden, S.P., J. Borja, J.L. Plawsky, T.-M. Lu, K.B. Yeap, and W.N. Gill. 2015. "Charge Transport Model to Predict Intrinsic Reliability for Dielectric Materials." *Journal of Applied Physics* 118 (12): 124102.

[4] Yeap, K. B., T. Shen, G. W. Zhang, S. F. Yap, B. Holt, A. Gondal, S. Choi, S. L. Liew, W. Yao, and P. Justison. 2015. "Impact of Electrode Surface Modulation on Time-Dependent Dielectric Breakdown." In *Reliability Physics Symposium (IRPS), 2015 IEEE International*, 2A.1.1–2A.1.5.

[5] Canright, G. S., and G. D. Mahan. 1987. "Time-Dependent Theory of Hot Electrons Using the Discrete Boltzmann Equation." *Physical Review B* 36 (2): 1025–31.

[6] Lombardo, S., J. H. Stathis, and B. P. Linder. 2003. "Breakdown Transients in Ultrathin Gate Oxides: Transition in the Degradation Rate." *Physical Review Letters* 90 (16): 167601.

[7] Chen, I. C., S. Holland, and C. Hut. 1985. "A Quantitative Physical Model for Time-Dependent Breakdown in SiO2." In *Reliability Physics Symposium, 1985. 23rd Annual*, 24–31.

[8] Chery, E., X. Federspiel, D. Roy, F. Volpi, and J. -M. Chaix. 2013. "Identification of the (√E + 1/E)-Dependence of Porous Low-K Time Dependent Dielectric Breakdown Using over One Year Long Package Level Tests." *Microelectronic Engineering*, Insulating Films on Semiconductors 2013, 109 (September): 90–93.

[9] Croes, K., C. Wu, D. Kocaay, Y. Li, Ph Roussel, J. Bömmels, and Zs. Tőkei. 2015. "Current Understanding of BEOL TDDB Lifetime Models." *ECS Journal of Solid State Science and Technology* 4 (1): N3094–97.

[10] Allers, K. -H. 2004. "Prediction of Dielectric Reliability from I–V Characteristics: Poole–Frenkel Conduction Mechanism Leading to √E Model for Silicon Nitride MIM Capacitor." *Microelectronics Reliability* 44 (3): 411–23.

461932 Fluid Dynamics Modeling for a Novel High Throughput Pipette Viscometer

Abstract:

High throughput research methods enable combinatorial approaches for rapid exploration of wide parameter spaces when developing new chemistries and materials. Robotic automation of the processes facilitates the high rate of data gathering, including the resulting physical and chemical properties of the new materials. In the development of fluidic materials, viscosity measurements are often a critical screening step, providing data necessary to understand both performance and manufacturing attributes related to flow behavior. Examples of such materials include agrochemicals, paints, paper coatings, pressure sensitive adhesives, elastomers, thermosets, silicones, water soluble polymers and cellulosic systems, food thickeners, de-icers, cooling emulsions, polishing suspensions, skin creams, shampoos, cleaning solutions, liquid detergents, and many more consumer products.

This presentation will describe a novel high throughput system for measuring viscosity of fluids under different conditions of shear rate, temperature, etc. This method utilizes the transient flow of complex fluids in robot-controlled pipettes, and is easily incorporated into the high throughput sequencing when transferring fluid from one set of vessels to another using microtiter plate formats. The high throughput advantages include small sample volumes (less than 1ml) and high sample rates (greater than 100 samples per hour) through parallel operation. The viscosity determination is based on mass and pressure measurements during both aspiration and dispensing of the fluid. The viscosity screening tool has been successfully utilized for a multitude of complex fluids including oils, paints, solvents, and detergents.

Numerical methods involving computational fluid dynamics (CFD), as well as analytical methods, are used to understand and characterize the capabilities and limitations of this viscosity measurement technique. The numerical tools are an essential component of the viscosity determination, especially when working with complex, non-Newtonian fluids. The transient CFD approach uses a moving-mesh method to capture the specified piston displacement and a volume-of-fluid method to track the fluid-air interface in the pipette. The fluid rheology is specified and the air is treated as a compressible ideal gas. Capillary and gravitational forces are included in the momentum equation. The pressure profiles in the pipette, along with the liquid aspiration rate and volume, are monitored as a function of time. The analytical method involves an interdependent mass and pressure balance at each time step. It can be set up using a spreadsheet, with individual time steps represented on each row of the spreadsheet.

The results obtained from the CFD and spreadsheet methods are in almost perfect agreement with each other for constant viscosity fluids. They also closely match the experimental data involving fluids with standard viscosities. The advantage of the spreadsheet method is its computational speed, whereas the CFD approach allows the analysis of complex rheologies. This capability is essential in evaluating the experimental data for non-Newtonian fluids.

The generalized Onsager model and DSMC simulations of high-speed rotating flow with swirling feed

S. Pradhan

Department of Chemical Engineering, Indian Institute of Science, Bangalore 560 012, India

The generalized Onsager model for the radial boundary layer and of the generalized Carrier-Maslen model for the axial boundary layer at the end-caps in a high-speed rotating cylinder ((S. Pradhan & V. Kumaran, J. Fluid Mech., 2011, vol. 686, pp. 109-159); (V. Kumaran & S. Pradhan, J. Fluid Mech., 2014, vol. 753, pp. 307-359)), are extended to incorporate the angular momentum of the feed gas for a swirling feed for single component gas and binary gas mixture. For a single component gas, the analytical solutions are obtained for the sixth-order generalized Onsager equations for the master potential, and for the fourth-order generalized Carrier-Maslen equation for the velocity potential. In both cases, the equations are linearized in the perturbation to the base flow, which is a solid-body rotation. The equations are restricted to the limit of high Reynolds number and (length/radius) ratio, but there is no limitation on the stratification parameter. The linear operators in the generalized Onsager and generalized Carrier-Maslen equations with swirling feed are still self-adjoint, and so the eigenfunctions form a complete orthogonal basis set. However, the differential operator depends explicitly on the stratification parameter, and so it is necessary to evaluate the eigenvalues and eigenfunctions numerically. For the case of mass/momentum/energy insertion into the flow, the separation of variables procedure is used, and the appropriate homogeneous boundary conditions are specified so that the linear operators in the axial and radial directions are self-adjoint. The discrete eigenvalues and eigenfunctions of the linear operators (sixth and second order in the radial and axial directions for the generalized Onsager equation, and fourth and second order in the axial and radial directions for the generalized Carrier-Maslen equation) are determined. The solutions for the secondary flows with swirling feed are determined in terms of these eigenvalues and eigenfunctions. These solutions are compared with direct simulation Monte Carlo (DSMC) simulations and found excellent agreement (with a difference of less than 15%) between the predictions of the analytical model and the DSMC simulations, provided the boundary conditions in the analytical model are accurately specified. The major advantages of the swirling feed, associated with the angular momentum of the feed gas, is that, it results in a reduction of the angular momentum loss of the rotating gas due to feed injection near the feed point by $(3-21)\%$, reduces the axial spreading of the feed gas by $(4-24)\%$, minimizes the formation of small secondary vortices near the feed zone, and increases the overall axial mass flux by $(16-27)\%$, and thereby enhance the efficiency of the feed drive for the centrifugal gas separation process.

Key words: High speed rotating flow, Swirling feed, DSMC Simulations, Rarefied gas flow.

---

1. Introduction

Centrifugal gas separation processes effect separation by utilizing the difference in the mole fraction in a high speed rotating cylinder caused by the difference in molecular mass, and consequently the centrifual force density. These have been widely used in isotope separation because chemical separation methods cannot be used to separate isotopes of the same chemical species (Cohen (1951), Avery & Davies (1973), Von Halle, Wood & Lowry (1978), Brouwers (1978), Soubbaramayer (1979), Hoglund , Shacter & Von Halle (1979), Wood & Morton (1980), Sakurai (1981), Olander (1981), Wood & Sanders (1983), Wood, Jordan & Gunzburger (1984), Copper & Morton (1988), Wood & Babarsky (1992), Wood (1995), Wood, Mason & Soubbaramayer (1996), Omnes (2007), Pradhan & Kumaran (2011), Kumaran & Pradhan (2014)). Radioactive isotopes, including zinc, cadmium, iridium, selenium, tellurium, germanium and molybdenum, are used in medical and industrial applications. In medical applications, radio-isotopes are used both for diagnostics such as gallium scans for locations of inflammation and infection, indium for leukocyte imaging and molybdenum/technetium scans for high bone growth, as well as for therapy including cancer therapy using isotopes of iodine and iridium. Nuclear diagnostic imaging has an important role in the identification and management of conditions such as heart disease, brain disorder, lung and kidney functions and a broad range of cancers. The high sensitivity and specificity of nuclear diagnostic imaging techniques offer the important advantages of being able to identify diseases at an early stage, to track disease progression, to allow for accurate disease staging and to provide predictive information about likely success of alternative therapy options. One of the important diagnostic techniques is gamma imaging. Gamma cameras are used to detect diseases of various organs including heart, brain, bone, lung and the thyroid using radio-isotope of gallium, indium, and iodine. Natural titanium is being used to encapsulate radioactive seeds for brachytherapy, in particular for prostate cancer. Depleted molybdenum is used in the experimental process of transmutation of the long-lived radio-isotopes, which are present in spent nuclear fuel. Cadmium isotope is employed as a precursor for the important radio pharmaceutical isotope indium, which has many varied medical applications. The cadmium isotope is placed in a cyclotron, and thereby is exposed to proton bombardment. The cadmium captures a proton then releases two neutrons such that indium is formed. Radioactive indium is used for several purposes such as a detector (of heart transplant rejection, of osteomyelitis and with gallium of soft tissue infection); imaging (of abdominal infections, of white blood cells, and of tumours), antibody labelling, cellular dosimetry, myocardial scans, and finally for the treatment of leukaemia. Cadmium isotopes are also being used in helium-cadmium lasers. Isotopically pure cadmium boost the output of the laser and also increase its coherence length. Iridium isotope is used in industrial applications for the non-destructive testing of constructions. Furthermore, iridium isotopes are being used for many medical applications such as brachytherapy for the treatment of prostate, brain, breast and gynaecological cancers. Xenon isotope is used as a precursor for radioactive iodine, which is an important thryroid diagnostic radiopharmaceutical isotope, and also used for diagnostic analysis of brain, kidney and neurological diseases. Iodine isotope is also used in diagnostic imaging and detection of osteoporosis, and in radiotherapy treatment of prostate cancer through seed implantation. An important industrial application is the separation of uranium isotopes for nuclear power generation. Other important application include the use of zinc isotopes to reduce corrosion in reactor cooling systems.

Centrifugal separation in a high-speed rotating flow field is a widely used method for isotope separation (Cohen (1951); Maslen & Moore (1956); Stewartson (1957);

Sozou (1959); Barcilon & Pedlosky (1967); Hunter (1967); Greenspan (1968); Homsy & Hudson (1969); Avery & Davies (1973); Mikami (1973); Nakayama & Torii (1974); Bark & Bark (1976); Matsuda, & Hashimotko (1976); Matsuda (1976a); Matsuda & Hashimoto (1978); Von Halle, Wood & Lowry (1978); Brouwers (1978); Hoglund , Shacter & Von Halle (1979); Soubbaramayer (1979); Thompson & Chen (1980); Wood & Morton (1980); Hittinger, Holt & Soubbaramayer (1981); Sakurai (1981); Hittinger, Holt & Soubbaramayer (1981); Mikami (1981); Volosciuk (1981); Gunzburger & Wood (1982); Viecelli (1983); Hittinger, Holt, Soubbaramayer & Cortet (1983); Jung (1983); Kai (1983); Walz, Volosciuk & Schutz (1983); Wood & Sanders (1983); Hausman & Roberts (1984); Gunzburger, Wood & Jordan (1984); Ribando (1984); Mikami (1985); Heuser, Ribando & Wood (1986); Copper & Morton (1988); Matsuda, Tamura & Sawada (1989); Babarsky & Wood (1990, 1992); Wood (1995); Wood, Mason & Soubbaramayer (1996); Bourn, Peterson & Wood (1999); Doneddu, Robin & Wood (2000); Babarsky, Herbst & Wood (2002); Omnes (2007); Pradhan & Kumaran (2011); Kumaran & Pradhan (2014)). The small difference in the molecular masses of the isotopes provides the only economically feasible method for effecting separation. The centrifugal force results in a small difference in concentration of the two isotopes, with a higher concentration of the heavier isotope near the wall of the cylinder, and a higher concentration of the lighter isotope near the axis. In order to separate the two isotopes, stationary tubes and baffles are inserted into the rotating cylinder. The inserts (identified as mechanical drive), inflow/outflow of the feed gas into the rotating cylinder (identified as feed drive), as well as temperature gradient along the rotating sidewall (identified as wall-thermal drive), and the temperature difference between the end-caps (identified as end-cap thermal drive), cause a secondary flow in the rotating cylinder, which could be very important for the centrifugal gas separation process.

The in-flow of the feed gas into the rotating cylinder associated with angular momentum (which we refer as swirling feed), results in a reduction of the angular momentum loss of the rotating gas due to feed injection near the feed points, as well as it reduces the axial spreading of the feed gas, minimizes the formation of small secondary vortices near the feed zone, and increases the overall axial mass flux, and thereby enhance the efficiency of the centrifugal gas separation process. Therefore, it is important to study the effect of the swirling feed on secondary flow field (identified as feed drive) for the centrifugal gas separation process based on analytical model and molecular simulations, the direct simulation Monte Carlo (DSMC) method, to assess the performance of a gas centrifuge.

The Onsager model was developed for calculating the secondary flows generated by applied forces in a rotating fluid in a cylinder. The secondary flow velocities are small compared to the velocity of the base flow, which is a solid body rotation about the cylinder axis. The linearization approximation is used for the secondary flow, where the equations of motion are truncated at linear order in the velocity and pressure disturbances to the base flow. Additional assumptions in the Onsager model include high aspect ratio (length of the cylinder is large compared to the diameter), constant temperature in the base state (isothermal condition), high Reynolds number and high stratification parameter $\mathcal{A}$, where the Reynolds number $\mathrm{Re} = (\rho_w \Omega R^2)/\mu$, the stratification parameter $\mathcal{A} = \sqrt{(m\Omega^2 R^2)/(2k_B T)}$, where R is the cylinder radius, ρ_w is the gas density at the wall, μ is the gas viscosity (which is independent of density for rarefied gases at constant temperature), Ω is the angular velocity, T is the absolute temperature, and k_B is the Boltzmann constant. The analysis is often carried out in terms of the inverse of the Reynolds number, which is referrred to as the Ekman number, but here we choose to express the results in terms of the Reynolds number for notational simplicity. In this limit, the gas flow is restricted to a boundary layer of thickness ($\mathrm{Re}^{-1/3}R$) at the wall of the cylinder. The flow at the end-caps is also restricted to a thin boundary layer of thickness $\mathrm{Re}^{-1/2}R$ in the axial coordinate, and analytical solutions for the end-cap flow were first derived by Carrier and Maslen (Olander (1981)).

The present analysis of rarefied gas flow in a rotating cylinder with swirling feed has two objectives. The first is to generalize the solutions to low rotation speeds where the stratification parameter may be $O(1)$, incorporating the angular momentum of the feed gas for a swirling feed. Secondly, we would like to compare the predictions with molecular simulations based on the direct simulation Monte Carlo (DSMC) method for rarefied gas flows with swirling feed, in order to quantify the errors resulting from the approximations at different aspect ratios, Reynolds numbers and stratification parameters. In this study, we have obtained the analytical and numerical solutions for the secondary flows generated at both, the cylinder sidewall and at the end-caps due to external gas in- and out-flow at the axis of the cylinder. Next, the effect of sources of mass, momentum and energy within the flow domain are also analyzed with applied wall temperature profile. Our studies show that the generalized Onsager model with swirling feed (for the radial boundary layer at the cylinder sidewall) and the generalized Carrier-Maslen model with swirling feed (for the axial boundary layer at the end-caps) are in agreement with numerical solutions to within 15% for stratification parameter as low as 0.707, Reynolds number as low as 100 and for cylinders with length/diameter ratio as low as 2, and the secondary flow velocity is as high as 0.2 times the maximum base flow velocity, the ratio $2\Delta m/(m_1 + m_2)$ is as high as 0.5, and the scaled angular momentum of the feed gas $F_\theta^* = F_\theta/(\theta_{mi}\Omega)$ is as high as 0.2, provided the wall-slip in both the flow velocity and temperature are incorporated in the analytical solutions.

A complete flow simulation involves the solution of the compressible mas, momentum and energy equations for the gas velocity in the cylinder, along with the ideal gas equation of state. There have been previous numerical simulations of axisymmetric flows in high speed rotating cylinder (Lahargue & Soubbaramayer (1978); Soubbaramayer & Billet (1980)). Apart from the complicated formulation of the differential equations for five variables (density, three components of velocity and temperature), there are two important conceptual difficulties in the numerical solution of the continuum equations. The flow in the cylinder is usually at high stratification parameter, where the gas is concentrated at the wall of the cylinder in a 'radial boundary layer' of thickness $\mathrm{Re}^{-1/3}R$ much smaller than the cylinder diameter. While the continuum approximation is valid in the radial boundary layer, the density in the central region is sufficiently small that the flow is in the free-molecular regime. There is no simple way to determine the boundary between the continuum and free-molecular regimes, and to prescribe boundary conditions at this boundary. Another important issue is the modeling of the wall slip in the simulations. While the no-slip conditions are applicable to low stratification parameter flows, there is usually substantial slip velocity at the wall at high stratification parameter which needs to be modeled in continuum simulations.

An alternative is to use a molecular simulation procedure (Bird (1994, 1963, 1981, 1987); Nanbu (1980); Quanhua & Boyd (2002); Vincenti & Kruger (1967); Chen & Boyd (1996); Garcia (2000); Chapman & Cowling (1970); Cercignani & Lampis (1971); Wagner (1992); Rieffel (1988); Cercignani (2000); Macrossan (2001)). In the direct simulation Monte Carlo (DSMC) procedure, the Boltzmann equation is simulated using a probabilistic Monte Carlo procedure to obtain the velocity distribution function, and the density, velocity and temperature are obtained by taking moments of the distribution function. Here, one 'simulated' molecule represents a large number of real molecules, and the collision probability is chosen in such a manner that the collision frequency, mean free path and the temperature of the simulated molecules is equal to those for the real molecules. This procedure can be used for the simulation of both the rarefied gas regime and the continuum regime, and this has been used previously for centrifugal flows (Roberts (1985); Roberts & Hausman (1988); Roblin & Doneddu (2000); Pradhan & Kumaran (2011); Kumaran & Pradhan (2014)). However, these simualtions are computationally intensive, due to the requirement of having a large number of simulated molecules in each differential volume in order to reduce statistical fluctuations.

A differential equation for the 'master potential' for the secondary flow in a rotating cylinder was first presented in an unpublished report by Onsager in 1965 (Wood & Morton (1980)). This equation was a sixth-order differential equation in the radial coordinate, and was valid for high rotational velocities where the gas density decreases exponentially from the wall of the cylinder due to the centrifugal force (the ideal gas equation is used to relate the pressure to the density). The model, which has subsequently been referred to as the 'Onsager pancake model' predicts that the flow is restricted to a thin boundary layer at the wall of thickness $\mathrm{Re}^{-1/3}R$ smaller than the cylinder radius. The flow at the end-caps of the cylinder was analysed by Carrier & Maslen (1962) (Olander (1981)), who derived a fourth-order differential equation for the axial variation of the stream function at the end-caps. This model predicts that the flow is restricted to a boundary layer of thickness $\mathrm{Re}^{-1/2}R$ at the end-caps.

The basic derivation of the Onsager model was presented by Wood & Morton (1980), and the formulation of the boundary conditions was also discussed. The equations for the master potential were derived in a very general form and include the effects of sources of mass, momentum and energy. It was shown that the complete solution for the flow in the radial boundary layer can be obtained as the sum of three types of fundamental solutions, the zero eigenvalue mode for which the axial eigenvalues are zero and there is no variation in the axial direction, the end-driven mode where the eigenvalues are real and there is an exponential decay from the end-caps, and the lateral modes where the eigenvalues are imaginary and the solutions decay sinusoidally in the axial direction. The solutions for the Ekman boundary layer at the end-caps were also presented, with some numerical results. The Onsager model equations derived by Wood & Morton (1980) have been approximately solved by a variety of techniques like eigenfunction-expansion methods (Wood & Morton (1980); Pradhan & Kumaran (2011)), finite-difference methods (Viecelli (1983)), and finite-element methods (Gunzburger & Wood (1982) and Gunzburger, Wood & Jordan (1984)).

The main assumptions taken into account in deriving the Onsager equation are that the configuration is axisymmetric and the flow field is a small perturbation about an isothermal solid-body rotation. The rotational speed of the cylinder is high enough that the bulk of the fluid is present in the 'Stewartson layer' of thickness $\mathrm{Re}^{-1/3}R$ smaller than the cylinder radius at the outer wall. These assumption enables us to linearize the equations of viscous compressible flow, neglect terms that are small in the limit of high Reynolds number and stratification parameter, and obtain a sixth order equation in the radial direction for the 'master-potential', which is related to the density, velocity and temperature fields. The curvature of the outer surface is usually neglected since the boundary layer thickness is small compared to the cylinder radius, but Wood, Jordan & Gunzburger (1984) also included the effect of curvature of the outer wall of the cylinder, to obtain the Onsager-Maslen equation.

The Onsager model is not valid at the end-caps of the cylinder where there is a radial flow in a thin axial boundary layer. Here, procedures similar to the Onsager model, but with the assumption that the 'axial boundary layer' is much smaller than the cylinder radius, have been used to derive the 'Carrier-Maslen' model (Olander (1981)), which is a fourth order differential equation for the stream function. This has been solved to obtain the flow profiles at the end-caps.

In our previous study for the axial flow in the radial boundary layer and radial flow near the end-caps (Pradhan & Kumaran (2011), Kumaran & Pradhan (2014)) in an axisymmetric $(r - z)$ plane has revealed several novel features of high speed rotating flows. The generalized Onsager and Carrier-Maslen models (Pradhan & Kumaran (2011)) were formulated in the axisymmetric $(r - z)$ plane where the approximation of high stratification parameter was relaxed. In this, a generalized sixth order operator was derived, which is applicable even when the stratification parameter is $O(1)$, though the assumption of high Reynolds number and high aspect ratio were made. This operator is also self-adjoint with real eigenvalues and orthogonal eigenfunctions, though these have to be determined numerically. For the secondary flow, it was found that the predictions of the generalized Onsager model were in very good agreement with DSMC results even when the secondary flow is as large as the 20% of the base solid-body rotation, the stratification parameter is as low as 0.707, the Reynolds number is as low as 100 and the cylinder length is only two times the diameter. However, these models still assume that the molecular masses of the two species are equal while calculating the secondary flows, and that the mole fractions are a constant. Next, the mass difference between the two species in a binary gas mixture was incorporated into the generalized Onsager and generalized Carrier-Maslen models (Kumaran & Pradhan (2014)) to make it applicable to systems where the difference in molecular masses is not much smaller than the average molecular mass. In the base state, the flow is considered to be a solid body rotation with constant temperature, but the variation in the mole fraction is taken into account. Due to this, there is a variation in the viscosity and thermal conductivity in the radial direction. In addition, the total pressure is written as the sum of partial pressures of the two species, and is not approximated as an exponential function of the distance from the wall. The dilute gas equation of state is used for the pressure, while the Chapman-Enskog expressions (Chapman & Cowling (1970)) for hard sphere gases are used for the thermal conductivity and viscosity. In the analysis of the secondary flow, the equation for the master potential is expanded in an asymptotic series in the parameter $\epsilon = (\Delta m/m_{av})$, where Δm is the difference in the molecular masses of the two species, and m_{av} is a suitably defined average molecular mass. In the leading approximation where terms of $O(\epsilon)$ or smaller are neglected. The leading order equation and boundary conditions for the master potential are nearly identical to the sixth order differential equation of the generalized Onsager model (Pradhan & Kumaran (2011)); the important difference being that the density is not assumed to decrease exponentially from the wall, but the actual species densities are used. The mean velocity and temperature obtained by solving the leading order equations are used in the species balance equations for the mole fractions of the two species, and the corrections for the number density and mass density of the two species due to the secondary flow are obtained. These appear as inhomogeneous terms in the equation for the $O(\epsilon)$ correction for the master potential; this equation is solved with appropriate boundary conditions to obtain the $O(\epsilon)$ correction to the master potential and flow velocity. This is then used in the $O(\epsilon)$ contribution to the species balance equation, and the hierarchal procedure is repeated until the $O(\epsilon^2)$ correction to the secondary flow velocity and mole fraction are calculated. A similar hierarchial procedure is used for the Carrier-Maslen model for the end-cap boundary layers. The results of the Onsager hierarchy, up to $O(\epsilon^2)$, are compared with the results of DSMC simulations for the axisymmetric flow of a two-species ideal gas. The constitutive relations for mass diffusion, viscosity and heat conduction predicted by the DSMC simulations are compared with the Chapman-Enskog expressions (Chapman & Cowling (1970)) in order to validate the constitutive relations used in the analysis. After this, the results of the $O(\epsilon^2)$ Onsager hierarchy for the secondary flow are compared with DSMC simulations for three types of forcing, a linear wall temperature profile, inflow/outflow of gas along the axis, and mass/momentum input due to inserts within the flow (Pradhan & Kumaran (2011), Kumaran & Pradhan (2014)).

Although Onsager pancake model was originally developed to study the steady, axisymmetric flows in a rapidly rotating cylinder, it has later been applied to non-axisymmetric flows. Matsuda & Nakagawa (1983) studied the flow in a pie-shaped cylinder of infinite length with walls at fixed angles in the meridional direction, and reported the presence of 'buyoancy' boundary layer of thickness $\text{Re}^{-1/2}R$ at the walls perpendicular to the meridional direction. Wood & Babarsky (1992) and Babarsky & Wood (1990) extended the work of Matsuda & Nakagawa (1983) to a finite-length, pie-shaped cylinder about the vertex. They presented an eigenfunction expansion based on the spectral decomposition of the resulting differential operator.

Our previous study on high-speed rotating flows in the polar $(r - \theta)$ plane (Pradhan (2016)) has revealed several novel features of high speed rotating flows. The generalized analytical model for the radial boundary layer was formulated for studying the secondary radial flow due to insertion of mass, momentum and energy into the rotating cylinder. The analytical solution includes the sixth order differential equation for the radial boundary layer at the cylindrical surface in terms of master potential (χ), which is derived from the equations of motion in a polar $(r - \theta)$ plane. The linearization approximation is used, where the equations of motion are truncated at linear order in the velocity and pressure disturbances to the base flow, which is a solid-body rotation. Additional assumptions in the analytical model include constant temperature in the base state (isothermal condition), and high Reynolds number, but there is no limitation on the stratification parameter. In this limit, the gas flow is restricted to a boundary layer of thickness $(\text{Re}^{-1/3}R)$ at the wall of the cylinder. For the case of mass/momentum/energy insertion into the flow, the separation-of-variables procedure is used, and the appropriate homogeneous boundary conditions are specified so that the linear operators in the meridional and radial directions are self-adjoint. The discrete eigenvalues and eigenfunctions of the linear operators (sixth-order and second-order in the radial and meridional directions for the generalized analytical equation) are obtained. The solutions for the secondary radial flows are determined in terms of these eigenvalues and eigenfunctions. The major advantage of this formulation is that it is not restricted to the asymptotic limit of high stratification parameter, and since we have used the conservative form of the compressible mass, momentum and energy conservation equations in deriving the generalized analytical model, obviously, the shock type solutions (Rankine-Hugoniot relations) are automatically satisfied. These solutions are compared with direct simulation Monte Carlo (DSMC) simulations. The comparison shows that it is necessary to take certain precautions when making a quantitative comparison between analysis and simulations. The commonly used 'diffuse reflection' boundary conditions at solid walls in DSMC simulations result in a non-zero slip. These have to be incorporated in the analysis in order to make quantitative predictions. In the case of mass/momentum/energy sources within the flow, it is necessary to ensure that the homogeneous boundary conditions (zero temperature perturbation) are accurately satisfied in the simulations. When these precautions are taken, there is quantitative agreement (with a difference of less than 10%) between the predictions of the analytical model and the DSMC simulations, even when stratification parameter is as low as 0.707, and the Reynolds number is as low as 100. The predictions of the generalized analytical model is also significantly better than that of the analytical model, which is restricted to thin boundary layers in the limit of high stratification parameter. In a high speed rotating field we examine the mass flow rate through the stationary intake tube. The simulations show that the scaled mass flow rate $m/(H\Omega)$ increases progressively from zero, at the stagnation condition, to 0.024 as the equilibrium back pressure is reduced. Here, H is the total gas holdup. The slow down of the circumferential velocity of the bulk of the rotating fluid due to the presence of stationary intake tube is studied for stratification parameter in the range $0.707 - 3.535$, and found significant slow down (between 8 to 28%), which induces the secondary radial flow towards the axis, and it further excites the secondary axial flow, which could be very important for the centrifugal gas separation processes. An important finding is that the stagnation pressure (no mass flow through the intake tube) is strongly affected by the wall gap, as well as with stratification parameter indicating a strong coupling between the local temperature, density, pressure and velocity fields. We have further carried out the DSMC simulations for a binary gas mixture with large mass difference ($\epsilon_a = (2\Delta m/(m_1 + m_2)) = 0.5$), and compare the simulation results with those of single-component gas for homogeneous temperature boundary conditions. Here, $\Delta m = (m_1 - m_2)$ is the difference in the molecular masses of the two species. An important point to note is that in all cases, internal momentum source, mass source/sink, and stationary intake tube, the peak value of the secondary radial flow for a binary gas mixture with $\epsilon_a = 0.5$ increases by (19 - 30)% compared to the single-component gas, as stratification parameter $\mathcal{A}$ is increased from 0.707 to 3.535, for Re $= 10^2$.

There are two broad classes of theoretical approaches to high-speed rotating flows in a cylinder. The first is based on the Onsager and Carrier-Maslen models, where the linearization approximation is used to derive equations for the master potential and the velocity potential, which are then solved analytically. The second is based on the linearization of the compressible Navier-Stokes equations. The latter is more exact, in the sense that there are fewer approximations made in deriving the final equations, and also because shock-type solutions can be obtained in the plane perpendicular to the axis of rotation. However, continuum equations are unsuited to flows such as that in a high-speed rotating cylinder, where there is large variation in the Knudsen number, and the mean free path is comparable to the macroscopic scale in some regions. It is necessary to specify boundary conditions at an artificial boundary where there is transition from the continuum to the free-molecular regime.

The Onsager and Carrier-Maslen equations have the advantage that analytical solutions can be obtained using the eigenfunction expansion procedure, resulting in a simpler formulation with better physical understanding. Moreover, the structure of the analytical solution ensures that there is no need to specify artificial boundary conditions, and the continuum equations for the master potential can be solved all the way to the centre of the cylinder. Of course, there is the question of whether the continuum equations accurately reproduce the flow dynamics in the region where the Knudsen number is not small. This leads to the second part of our study, where we compare the results of the generalized Onsager and generalized Carrier-Maslen models with molecular simulations based on the DSMC technique.

Here, we study the generalized Onsager and Carrier-Maslen equations with swirling feed associated with the angular momentum of the feed gas, where the restriction on high stratification parameter is removed. This involves retaining terms in the equation for the master potential which were neglected due to the high stratification parameter assumptions, while still neglecting terms due to the assumption of high (length/radius) ratio and high Reynolds number. First, we show how the equation for the master potential is obtained from the linearized Navier-Stokes equations, and evaluate the magnitudes of all the terms. This order of-magnitude analysis is used to justify the simplification to the generalized Onsager equation for the axial flow at the curved surface of the cylinder, and the generalized Carrier-Maslen model for the flow along the end-caps. This serves to clarify the approximations involved in deriving the Onsager and Carrier-Maslen equations (Pradhan & Kumaran (2011)).

The resulting sixth-order operator for the master potential in the generalized Onsager model, and the fourth-order operator for the velocity potential in the generalized Carrier-Maslen model, are still self-adjoint, and therefore the eigenvalues of this operator are real, the eigenfunctions are orthogonal, and they form a complete function space. However, the eigenvalues and the eigenfunctions are now dependent on the stratification parameter, and are evaluated either by a Frobenius expansion or by numerical integration. These are then used to obtain solutions for the secondary axial flow and radial end-cap flow due to a temperature gradient along the curved surface of the cylinder, and where it is due to gas feed along the axis. We also examine the secondary flows due to mass, momentum and energy inputs/outputs within the flow domain, by solving the inhomogeneous generalized Onsager equation using the method of separation of variables.

The analytical results are compared with molecular simulations based on direct simulation Monte Carlo (DSMC) method (Bird (1994), Bird (1963), Cercignani & Lampis (1971), Bird (1981), Bird (1987), Chen & Boyd (1996)). The present work provides an opportunity for validating both the generalized Onsager and generalized Carrier-Maslen model as well as the DSMC method for high-speed rotating flows with swirling feed. The molecular simulation procedure is more computationally intensive than the continuum Navier-Stokes equations, but it also has two important advantages. The Navier-Stokes equations are written for the macroscopic density and velocity fields, and these are applicable only when the mean free path in the gas is small compared to the macroscopic scale (radius of the cylinder or boundary layer thickness). There are many situations in high-speed rotating flows where the gas density in the central region of the cylinder is sufficiently small that the mean free path is comparable to the cylinder radius. In these cases, the Navier-Stokes equations are not valid near the center, and it is necessary to specify artificial 'boundary conditions' at a surface within the flow. In this context, Roblin & Doneddu (2000) has been performed DSMC simulations to study the feed gas expansion from the central pipe, and the residual gas contained in the casing in which the cylinder rotates. These simulations were used to provide more realistic boundary conditions for continuum fluid flow calculations. In the second case, they use DSMC simulations to determine the mechanical power of the rotating cylinder and the heat flux dissipated on the rotating cylinder and the fixed cylinder of the casing. The possible dependence of the solutions on these artificial boundary conditions poses difficulties in the interpretation of results. No such difficulties exist in the molecular simulations, since it is only necessary to specify particle-wall interaction parameters.

Another advantage of molecular simulations is that the velocity slip at the wall resulting from the particle-wall interaction law can be obtained directly from the simulations, whereas the slip velocity has to be modeled in the continuum numerical solutions. In the present simulations, we also find that there is substantial 'temperature slip' in the DSMC simulations, that is, the gas temperature at the wall is not equal to the wall temperature. Both velocity and temperature slip at the walls are measured in the DSMC simulations, and these are then incorporated into the analytical solutions, in order to make a quantitative comparison of the results of the analysis and simulations. The comparisons are carried out for relatively large secondary flows, where the scaled temperature gradient (for thermally driven flow) or the scaled secondary flow rate (for flows due to input/output of gas in the cylinder) is as large as 0.2, and where the aspect ratio (ratio of length to diameter of cylinder) is as small as 2. In addition, we consider stratification parameter as low as 0.707 and Reynolds numbers as low as 100, in order to quantitatively estimate the accuracy of the generalized Onsager model.

There are technical intricacies in the implementation of sources of mass, momentum and energy within the flow in the DSMC simulations. In order to mimic a Dirac delta function in the simulations, mass, momentum and energy are inserted/removed over a small source domain of radius 10^{-3} times the cylinder radius at the designated location. When quantitative comparison has to be made with analytical solutions for specified rates of input/output of mass, momentum and energy, care has to be taken in the simulations due to the coupling between mass, momentum and energy sources. When molecules are added in simulations, there is the simultaneous addition of the momentum and energy of these molecules. In order to add mass alone, it is necessary to reduce the momentum and energy of all molecules in the source domain so that total momentum and energy are conserved. In a similar manner, momentum is added by increasing the mean velocity of the molecules in the source domain, while addition of energy requires increasing the fluctuating velocity of the molecules while keeping the total mean velocity constant. Another complication is that addition of energy at a location near the wall in the simulation causes an increase in the local fluctuating velocities of molecules, which heats up the wall. This violates the constant temperature boundary condition that is used in the analytical solution, and so it is necessary to extract heat locally at the wall in the simulation to ensure that the wall temperature is kept constant.

In §2, the linearized mass, momentum and energy conservation equations for an axisymmetric flow in a cylindrical coordinate system are presented and different dimensionless numbers are defined. The equations for perturbations to the base state (solid body rotation) are then used to get the equation for the master potential in §3. In §4, the radial boundary layer equations are solved to obtain the generalized Onsager equation incorporating the angular momentum of the feed gas in a swirling feed, while the equation for the stream function for the axial boundary layer is used to obtain the generalized Carrier-Maslen equation for the axial boundary layer in §5. In §6, a brief summary on molecular simulation based on the direct simulation Monte Carlo (DSMC) method is presented. In §7, secondary flow due to the combination of inflow/outflow of gas along the axis, the applied linear wall temperature profile, the internal sources/sinks of mass, momentum, and energy within the flow domain with swirling feed is studied for a single component gas and analytical results are compared with those of DSMC simulations. In §8, secondary flow field is analyzed for a binary gas mixture and analytical results are compared with those of DSMC simulations. The important conclusions of the analysis are summarized in §9.

2. Formulation

The configuration consists of a right circular cylinder of radius R and height Z rotating at a constant angular velocity Ω. A cylindrical coordinate system with origin at the bottom of the cylinder along the axis of rotation, with radial, meridional and axial coordinates r, θ, and z respectively, is used. Throughout this analysis, we assume that the flow is axisymmetric, so there is no dependence on the meridional coordinate θ. Gravitational acceleration is neglected, since it has a negligible effect for high speed gas flows. The compressible mass, momentum and energy conservation equations in the cylindrical coordinate system are provided in the appendix for reference. There are six variables in these equations, the density ρ, the three components of the velocity v_r, v_θ, v_z, the pressure p and the temperature T. In equations (A 1) to (A 5), $\mathcal{S}_m$ is the mass source density (mass input per unit volume per unit time), f_r, f_θ and f_z are the force densities in r, θ and z directions respectively, S_e is the energy source density (energy input per unit volume per unit time), and $f_{\theta,feed}$ is the force densities in θ direction, associated with the swirling feed gas, is explicitly added in the θ momentum equation (3.4).

The pressure is related to the density by the ideal gas equation of state,

$$\rho = \frac{pm}{k_B T}, \tag{2.1}$$

where m is the mass of a molecule and k_B is the Boltzmann constant.

If the temperature on the walls of the cylinder is uniform, and there is no inflow or outflow of gas, a steady solution for the equations of motion is one where the gas undergoes solid-body rotation with angular velocity Ω at constant temperature. The temperature $\bar{T}$ and the components of the velocity $\bar{\mathbf{v}}$ in the base state in this steady state are

$$\bar{v}_\theta = \Omega r, \ \ \bar{v}_z = 0, \ \ \bar{v}_r = 0, \ \ \bar{T} = T_0, \tag{2.2}$$

where T_0 is the constant temperature. The pressure is obtained by solving the radial momentum equation,

$$\frac{d\bar{p}}{dr} = \bar{\rho}\Omega^2 r = \frac{\bar{p} m \Omega^2 r}{k_B T_0}, \tag{2.3}$$

where $\bar{\rho}$ is the density at steady state, together with the ideal gas equation of state,

$$\bar{\rho} = \frac{\bar{p} m}{k_B T_0}, \tag{2.4}$$

where m is the mass of a molecule and k_B is the Boltzmann constant. The above equation is easily solved to obtain

$$\frac{\bar{p}}{\bar{p}_w} = \frac{\bar{\rho}}{\bar{\rho}_w} = \exp\left(-\mathcal{A}^2\left(1 - \frac{r^2}{R^2}\right)\right), \tag{2.5}$$

where $\bar{p}_w$ and $\bar{\rho}_w$ are the pressure and density at the wall in the unperturbed state, and $\mathcal{A}$ is the stratification parameter defined below. The total mass of gas in the cylinder is related to ρ_w by,

$$\text{Mass of gas} = \frac{\pi R^2 Z \bar{\rho}_w (1 - \exp(-\mathcal{A}^2))}{\mathcal{A}^2}, \tag{2.6}$$

where Z is the total height of the cylinder.

It is useful to discuss the different dimensionless numbers used in the analysis. The stratification parameter obtained from the equation for the density is

$$\mathcal{A} = \sqrt{\frac{m\Omega^2 R^2}{2 k_B T_0}} \propto \frac{R\Omega}{c_s}, \tag{2.7}$$

where $R\Omega$ is the characteristic flow velocity, and $c_s = \sqrt{\gamma k_B T_0 / m}$ is the speed of sound. The Reynolds number, used a little later, is defined as

$$\text{Re} = \frac{\rho_w \Omega R^2}{\bar{\mu}}, \tag{2.8}$$

where $\bar{\mu}$ is the viscosity of the fluid in the unperturbed state. The viscosity of a dilute gas scales as

$$\bar{\mu} \sim \frac{\sqrt{m k_B T_0}}{d^2}, \tag{2.9}$$

where m is the molecular mass and d is the molecular diameter. Using the above expression for the viscosity, it can easily be verified that the Reynolds number can be rewritten as

$$\text{Re} \sim \frac{\mathcal{A}}{\text{Kn}}, \tag{2.10}$$

where the Knudsen number is defined by

$$\text{Kn} = (m/\sqrt{2}\pi\rho_w d^2 R) = (\lambda/R), \tag{2.11}$$

where $\lambda = (m/\sqrt{2}\pi\rho_w d^2)$ is the mean free path at the wall. The other dimensionless groups used are the Prandtl number and the ratio of specific heats,

$$\text{Pr} = \frac{C_p\bar{\mu}}{\bar{K}}, \tag{2.12}$$

$$\gamma = \frac{C_p}{C_v}, \tag{2.13}$$

which are both $O(1)$ for an ideal gas.

3. Equations of motion in axisymmetric coordinates

The disturbance to the base state due to a temperature gradient at the walls, or due to a small mass flux through the cylinder, is incorporated by perturbing the base velocity and pressure profile as follows:

$$\begin{aligned} \rho &= \bar{\rho} + \hat{\rho}, \\ p &= \bar{p} + \hat{p}, \\ T &= \bar{T} + \hat{T}, \\ v_r &= \hat{v}_r, \\ v_\theta &= r\Omega + \hat{v}_\theta, \\ v_z &= \hat{v}_z. \end{aligned} \tag{3.1}$$

The above forms for the dynamical variables are inserted into the mass, momentum and energy conservation equations, and linearized in the perturbation (hatted) quantities. Steady state is assumed, so that all the time derivatives are zero. The complete equations, which are linearized, are given in Appendix-A. After subtracting the corresponding base state equations, we obtain the following equations for the perturbations:

$$\frac{1}{r}\frac{\partial}{\partial r}(r\bar{\rho}\hat{v}_r) + \frac{\partial}{\partial z}(\bar{\rho}\hat{v}_z) = \mathcal{S}_m, \tag{3.2}$$

$$-\hat{\rho}\Omega^2 r - 2\bar{\rho}\Omega\hat{v}_\theta = -\frac{\partial\hat{p}}{\partial r} + \bar{\mu}\left(\frac{\partial}{\partial r}\left(\frac{1}{r}\frac{\partial}{\partial r}(r\hat{v}_r)\right) + \frac{\partial^2\hat{v}_r}{\partial z^2}\right) + \left(f_r + \frac{\bar{\mu}}{3}\frac{\partial S_m}{\partial r}\right), \tag{3.3}$$

$$2\bar{\rho}\Omega\hat{v}_r = \bar{\mu}\left(\frac{\partial}{\partial r}\left(\frac{1}{r}\frac{\partial}{\partial r}(r\hat{v}_\theta)\right) + \frac{\partial^2\hat{v}_\theta}{\partial z^2}\right) + (f_\theta + f_{\theta,feed}), \tag{3.4}$$

$$0 = -\frac{\partial\hat{p}}{\partial z} + \bar{\mu}\left(\frac{1}{r}\frac{\partial}{\partial r}\left(r\frac{\partial\hat{v}_z}{\partial r}\right) + \frac{\partial^2\hat{v}_z}{\partial z^2}\right) + \left(f_z + \frac{\bar{\mu}}{3}\frac{\partial S_m}{\partial z}\right), \tag{3.5}$$

$$-\bar{\rho}\Omega^2 r\hat{v}_r = -\bar{p}S_m + \bar{K}\left(\frac{1}{r}\frac{\partial}{\partial r}\left(r\frac{\partial\hat{T}}{\partial r}\right) + \frac{\partial^2\hat{T}}{\partial z^2}\right) + \mathcal{S}_e. \tag{3.6}$$

Here, $f_{\theta,feed}$ is the force densities in θ direction, associated with the swirling feed gas, is explicitly added in the θ momentum equation (3.4). $\bar{\mu}$ and $\bar{K}$ are the viscosity and thermal conductivity of the gas in the base state, where the temperature is T_0. Note that in the ideal gas limit, the viscosity and thermal conductivity are functions of the temperature alone, and they do not depend on the density. In equation (3.3), the term on the left is the contribution to the r momentum equation due to the centrifugal force, while in equation (3.4), the left side represents the contribution to the θ momentum equation due to the Coriolis force. In the energy equation (3.6), the term on the left is the pressure work done due to the compression or expansion of the gas, and we have used the equation of state (2.4) and the mass conservation equation (3.2) to simplify this term. The perturbation to the equation of state is,

$$\hat{\rho} = \left(\frac{m}{k_B T_0}\right)\hat{p} - \bar{\rho}\left(\frac{\hat{T}}{T_0}\right). \tag{3.7}$$

Following our previous study (Pradhan & Kumaran (2011)), and expressing the velocity components $\hat{v}_r$ and $\hat{v}_z$ in terms of a stream function using the mass conservation equation 3.2,

$$r\bar{\rho}\hat{v}_r = -\frac{\partial\psi}{\partial z} + \int_{r_0}^{r} S_m\, r dr, \tag{3.8}$$

$$r\bar{\rho}\hat{v}_z = \frac{\partial\psi}{\partial r}, \tag{3.9}$$

where the difference in the stream function between two radial locations gives the axial mass flux in the region between these two locations,

$$\psi(a,z) - \psi(r,z) = \int_{r}^{a} \bar{\rho}\hat{v}_z\, r\, dr., \tag{3.10}$$

we obtain

$$\begin{aligned}\frac{1}{r}\frac{\partial}{\partial r}\left(\frac{1}{\bar{\rho}r}\frac{\partial}{\partial r}\left(r\frac{\partial}{\partial r}\right) + \frac{1}{\bar{\rho}}\frac{\partial^2}{\partial z^2}\right)\left(\frac{1}{\bar{\rho}r}\frac{\partial\psi}{\partial r}\right) + \left(\frac{1}{\bar{\rho}r}\frac{\partial}{\partial r}\left(\frac{1}{r}\frac{\partial}{\partial r}\left(r\frac{\partial}{\partial z}\right)\right) + \frac{1}{\bar{\rho}r}\frac{\partial^3}{\partial z^3}\right)&\\ \times\left(\frac{1}{\bar{\rho}r}\left(\frac{\partial\psi}{\partial z} - \int_{r_0}^{r} S_m\, r dr\right)\right) - \frac{\Omega}{\bar{\mu}}\frac{\partial}{\partial z}\left(\frac{2\hat{v}_\theta}{r} - \frac{\Omega\hat{T}}{T_0}\right)&\\ = \frac{1}{\bar{\mu}\bar{\rho}r}\frac{\partial}{\partial z}\left(f_r + \frac{\bar{\mu}}{3}\frac{\partial S_m}{\partial r}\right) - \frac{1}{\bar{\mu}r}\frac{\partial}{\partial r}\left(\frac{1}{\bar{\rho}}\left(f_z + \frac{\bar{\mu}}{3}\frac{\partial S_m}{\partial z}\right)\right).&\end{aligned} \tag{3.11}$$

The stream function ψ is expressed in terms of the 'master potential' χ:

$$\psi = \frac{1}{r}\frac{\partial\chi}{\partial r}. \tag{3.12}$$

This is inserted into the equation 3.13 to obtain,

$$\begin{aligned}\frac{1}{r}\frac{\partial}{\partial r}\left(\frac{1}{\bar{\rho}r}\frac{\partial}{\partial r}\left(r\frac{\partial}{\partial r}\right) + \frac{1}{\bar{\rho}}\frac{\partial^2}{\partial z^2}\right)\left(\frac{1}{\bar{\rho}r}\frac{\partial}{\partial r}\left(\frac{1}{r}\frac{\partial\chi}{\partial r}\right)\right)&\\ + \left(\frac{1}{\bar{\rho}r}\frac{\partial}{\partial r}\left(\frac{1}{r}\frac{\partial}{\partial r}\left(r\frac{\partial}{\partial z}\right)\right) + \frac{1}{\bar{\rho}r}\frac{\partial^3}{\partial z^3}\right)&\\ \times\left(\frac{1}{\bar{\rho}r^2}\frac{\partial}{\partial r}\left(\frac{\partial\chi}{\partial z}\right) - \frac{1}{\bar{\rho}r}\int_{r_0}^{r} S_m\, r dr\right) - \Omega\frac{\partial}{\partial z}\left(\frac{2\hat{v}_\theta}{r} - \frac{\Omega\hat{T}}{T_0}\right)&\\ = \frac{1}{\bar{\mu}\bar{\rho}r}\frac{\partial}{\partial z}\left(f_r + \frac{\bar{\mu}}{3}\frac{\partial S_m}{\partial r}\right) - \frac{1}{\bar{\mu}r}\frac{\partial}{\partial r}\left(\frac{1}{\bar{\rho}}\left(f_z + \frac{\bar{\mu}}{3}\frac{\partial S_m}{\partial z}\right)\right).&\end{aligned} \tag{3.13}$$

4. Axial flow in radial boundary layer

Further development of the equation for the master potential depends on whether the boundary layer considered is in the radial or the axial direction. When the boundary layer is in the radial coordinate, the radial derivatives of $\hat{v}_\theta$ and $\hat{T}$ on the right sides of equations 3.4 and 3.6 are balanced by the left sides. Following our previous study (Pradhan & Kumaran (2011)), we obtain

$$\frac{\partial^2}{\partial\xi^2}\left(e^\xi\frac{\partial}{\partial\xi}\left(\left(1-\frac{\xi}{\mathcal{A}^2}\right)\frac{\partial}{\partial\xi}\left(e^\xi\frac{\partial^2\chi^*}{\partial\xi^2}\right)\right)\right)+\frac{1}{4\mathcal{A}^4}\frac{R^2}{Z^2}\frac{\partial^2}{\partial\xi^2}\left(e^{2\xi}\frac{\partial^2}{\partial\xi^2}\frac{\partial^2\chi^*}{\partial\eta^2}\right)+$$
$$\frac{1}{4\mathcal{A}^4}\frac{R^2}{Z^2}\frac{\partial}{\partial\xi}\left(e^\xi\frac{\partial^2}{\partial\xi^2}\left(e^\xi\frac{\partial}{\partial\xi}\frac{\partial^2\chi^*}{\partial\eta^2}\right)\right)+\frac{1}{16\mathcal{A}^8}\frac{R^4}{Z^4}\frac{\partial}{\partial\xi}\left(\frac{e^{2\xi}}{(1-(\xi/\mathcal{A}^2))}\frac{\partial}{\partial\xi}\frac{\partial^4\chi^*}{\partial\eta^4}\right)$$
$$+\left(\frac{2\mathrm{Re}^2\mathcal{A}^2\mathrm{Pr}(1-\gamma^{-1})}{(1-\xi/\mathcal{A}^2)}+\frac{4\mathrm{Re}^2}{(1-\xi/\mathcal{A}^2)^2}\right)\frac{1}{64\mathcal{A}^{12}}\frac{R^2}{Z^2}\frac{\partial^2\chi^*}{\partial\eta^2}$$
$$+(R^*_\theta+R^*_T)+(C^*_\theta+C^*_T)=S^*, \tag{4.1}$$

where the Reynolds number, stratification parameter and the Prandtl number are given in equations 2.8, 2.7 and 2.12 respectively, and the dimensionless scaled variables are

$$\xi=\mathcal{A}^2\left(1-\frac{r^2}{R^2}\right), \tag{4.2}$$

$$\eta=\frac{z}{Z}, \tag{4.3}$$

$$\chi^*=\frac{\chi}{\bar{\mu}\,R^2\,Z}, \tag{4.4}$$

$$R^*_\theta=\frac{1}{64\mathcal{A}^{14}}\frac{R^4}{Z^4}\left(\left(1-\xi/\mathcal{A}^2\right)^{-2}\int_{\xi_0}^{\xi}d\xi\,\left(1-\xi/\mathcal{A}^2\right)^{1/2}\left(\frac{\partial^3\hat{v}^*_\theta}{\partial\eta^3}\right)\right), \tag{4.5}$$

$$C^*_\theta=\frac{1}{16\mathcal{A}^{10}}\frac{R^2}{Z^2}\left(\left(1-\xi/\mathcal{A}^2\right)^{-2}\left(\frac{\partial\hat{v}^*_\theta}{\partial\xi}\bigg|_{\xi=\xi_0}\right)\right), \tag{4.6}$$

$$R^*_{\hat{T}}=-\frac{1}{128\mathcal{A}^{14}}\frac{R^4}{Z^4}\left(\left(1-\xi/\mathcal{A}^2\right)^{-1}\int_{\xi_0}^{\xi}d\xi\left(\frac{\partial^3\hat{T}^*}{\partial\eta^3}\right)\right), \tag{4.7}$$

$$C^*_{\hat{T}}=-\frac{1}{32\mathcal{A}^{10}}\frac{R^2}{Z^2}\left(\left(1-\xi/\mathcal{A}^2\right)^{-1}\frac{\partial}{\partial\eta}\frac{\partial\hat{T}^*}{\partial\xi}\bigg|_{\xi=\xi_0}\right), \tag{4.8}$$

where $\hat{v}^*_\theta=(\hat{v}_\theta/\Omega R)$ and $\hat{T}^*=\left(\hat{T}/T_0\right)$ and S^* is the scaled source term, expressed as,

$$\mathcal{S}^*(\xi,\eta)=\frac{\mathrm{Re}}{32\mathcal{A}^{10}}\frac{R}{Z}\frac{R^2}{\mathcal{A}^4\bar{K}T_0}\left(1-\xi/\mathcal{A}^2\right)^{-1}\int_{\xi_0}^{\xi}\frac{\partial(\mathcal{S}_e-p_we^{-\xi}\mathcal{S}_m)}{\partial\eta}\,d\xi^*$$
$$-\frac{\mathrm{Re}}{16\mathcal{A}^{10}}\frac{R}{Z}\frac{R}{\mathcal{A}^4\Omega\bar{\mu}}\left(1-\xi/\mathcal{A}^2\right)^{-2}\int_{\xi_0}^{\xi}\frac{\partial\left(f_\theta+f_{\theta,feed}\right)}{\partial\eta}\left(1-\frac{\xi^*}{\mathcal{A}^2}\right)^{\frac{1}{2}}d\xi^*$$
$$-\left(\frac{2\mathrm{Re}^2\mathcal{A}^2\mathrm{Pr}(1-\gamma^{-1})}{(1-\xi/\mathcal{A}^2)}+\frac{4\mathrm{Re}^2}{\left(1-\xi/\mathcal{A}^2\right)^2}\right)\frac{1}{256\mathcal{A}^{16}}\frac{R}{\rho_w\Omega Z}\times$$

$$\int_{\xi_0}^{\xi}\left[\int_{\xi_0}^{\xi^*}\frac{\partial S_m(\xi^{**},\eta)}{\partial\eta}d\xi^{**}\right]d\xi^*$$
$$+\frac{R^4}{16\mathcal{A}^8Z^2\bar{\mu}}\left(\frac{\partial}{\partial\xi}\left(e^{\xi}\frac{\partial}{\partial\xi}\left(\frac{\partial}{\partial\xi}\left(\left(1-\frac{\xi}{\mathcal{A}^2}\right)^{\frac{1}{2}}\frac{\partial}{\partial\eta}\right)\right)\right)\right)$$
$$+\frac{R^2}{Z^2}\frac{\partial}{\partial\xi}\left(\frac{e^{\xi}}{(1-\xi/\mathcal{A}^2)^{\frac{1}{2}}}\frac{\partial^3}{\partial\eta^3}\right)\Bigg)\Bigg)$$
$$\times\left[\frac{e^{\xi}}{(1-\xi/\mathcal{A}^2)^{\frac{1}{2}}}\int_{\xi_0}^{\xi}S_m\,d\xi^*\right]$$
$$-\frac{\mathrm{Re}}{32\mathcal{A}^{10}}\left[\frac{1}{\rho_w\Omega^2 Z}\frac{\partial}{\partial\xi}\left(\frac{1}{(1-\xi/\mathcal{A}^2)^{\frac{1}{2}}}\frac{\partial\left(e^{\xi}\left(f_r-\frac{2}{3}\bar{\mu}\frac{\mathcal{A}^2}{R}\left(1-\frac{\xi}{\mathcal{A}^2}\right)^{\frac{1}{2}}\frac{\partial S_m}{\partial\xi}\right)\right)}{\partial\eta}\right)\right.$$
$$\left.+\frac{1}{\rho_w\Omega^2 R}\frac{\partial^2\left(e^{\xi}\left(f_z+\frac{\bar{\mu}}{3Z}\frac{\partial S_m}{\partial\eta}\right)\right)}{\partial\xi^2}\right]. \tag{4.9}$$

In the above equation, the contribution to S^* due to the fourth term on the right side of the above equation (fourth and fifth lines of equation (4.9)) is $O(\mathrm{Re}^{-1}(R/Z))$ smaller than that due to the third term on the right side (third line in equation 4.9). Therefore, this can be neglected in the leading approximation. The magnitude of the other terms do depend on the source strengths, and so these cannot be neglected in general. However, equation (4.9) provides the conditions under which the source strengths are significant, that is, for $\mathcal{S}_m \sim (\rho_w\Omega Z/R)$, $f_r \sim (\rho_w\Omega^2 R/\mathrm{Re})$, $(f_\theta+f_{\theta,feed}) \sim (\Omega\bar{\mu}Z/R^2\mathrm{Re})$, $f_z \sim (\rho_w\Omega^2 R/\mathrm{Re})$ and $S_e \sim (\bar{K}T_0 Z/R^3\mathrm{Re})$.

Following our previous study (Pradhan & Kumaran (2011)), the limitation $\mathcal{A} \gg 1$ can be relaxed if we formulate a generalized Onsager equation,

$$L_6^{\mathcal{A}}(\chi^*)+\frac{\mathrm{Re}^2}{64\mathcal{A}^{12}}\frac{R^2}{Z^2}\frac{\partial^2\chi^*}{\partial\eta^2}=S^*, \tag{4.10}$$

where the scaled source term (S^*) was defined in equation (4.9), and the linear differential operator $L_6^{\mathcal{A}}$ in the ξ coordinate is defined by

$$L_6^{\mathcal{A}}=\left(\frac{2\mathcal{A}^2\mathrm{Pr}(1-\gamma^{-1})}{(1-\xi/\mathcal{A}^2)}+\frac{4}{(1-\xi/\mathcal{A}^2)^2}\right)^{-1}\times\left(\frac{\partial^2}{\partial\xi^2}\left(e^{\xi}\frac{\partial}{\partial\xi}\left(\left(1-\frac{\xi}{\mathcal{A}^2}\right)\frac{\partial}{\partial\xi}\left(e^{\xi}\frac{\partial^2}{\partial\xi^2}\right)\right)\right)\right). \tag{4.11}$$

The operator has been written with the superscript $^{\mathcal{A}}$ to indicate that there is an explicit dependence on the stratification parameter. This operator also depends on the Prandtl number Pr and the ratio of specific heats, but these are gas properties and not flow properties, and are both $O(1)$ for an ideal gas. Therefore, the only flow dependence in the operator $L_6^{\mathcal{A}}$ enters through the stratification parameter, while the Prandtl number and ideal gas constant are determined by the nature of the gas. Here, we will assume an ideal gas, for which $\mathrm{Pr} = 0.95$ and $\gamma = 1.066$.

It is easily verified that this operator is also self-adjoint, when the inner product is defined by

$$(A_1, B_1)=\int_0^{\xi_0}d\xi w(\xi)A_1(\xi)B_1(\xi), \tag{4.12}$$

where the weighting function $w(\xi)$ is given by

$$w(\xi) = \left(\frac{2\mathcal{A}^2 \mathrm{Pr}(1-\gamma^{-1})}{(1-\xi/\mathcal{A}^2)} + \frac{4}{(1-\xi/\mathcal{A}^2)^2} \right), \tag{4.13}$$

when the boundary conditions are appropriately specified as discussed next. Therefore, the eigenvalues of this operator are real, and the eigenfunctions are orthogonal. We refer to equation (4.10) as the 'generalized' Onsager equation. Since the equation (4.1) is a sixth order differential equation, it is necessary to specify three boundary conditions at each boundary. Here we will consider the homogeneous form of the generalized Onsager equation and we will investigate the secondary flow due to wall thermal gradients, external inflow/outflow of gas into the rotating cylinder. For the flow in a cylinder where $r_0 = 0$, the boundary conditions at the center are as follows. The following expressions for the velocity components and the axial flow rate are useful for specifying the boundary conditions,

$$\hat{v}_r = \frac{2R\Omega\mathcal{A}^2}{\mathrm{Re}} \left(1 - \frac{\xi}{\mathcal{A}^2}\right)^{-1/2} e^{\xi} \frac{\partial^2 \chi^*}{\partial\xi\partial\eta}, \tag{4.14}$$

$$\hat{v}_z = \frac{4Z\Omega\mathcal{A}^4}{\mathrm{Re}} e^{\xi} \frac{\partial^2 \chi^*}{\partial\xi^2}, \tag{4.15}$$

$$\begin{aligned} F_z &= 2\pi \int_0^R r\, dr\, \bar{\rho}\hat{v}_z \\ &= \frac{4\pi\rho_w R^2 \Omega Z \mathcal{A}^2}{\mathrm{Re}} \left. \frac{\partial \chi^*}{\partial \xi} \right|_0^{\xi}, \end{aligned} \tag{4.16}$$

where F_z is the total flow rate in the axial direction. Since the velocity potential is related only to derivatives of the master potential χ, adding a constant to χ does not change the velocity fields. Therefore, without loss of generality, we can impose at the center,

$$\chi^* = 0. \tag{4.17}$$

The zero radial velocity boundary condition at the center requires that $(\partial\psi/\partial z) = 0$, or,

$$\frac{\partial \chi^*}{\partial \xi} = 0. \tag{4.18}$$

Symmetry requires that the radial derivative of the axial velocity has to be zero at the center of the cylinder. Therefore,

$$\frac{\partial \hat{v}_z}{\partial r} = 0. \tag{4.19}$$

Now,

$$\bar{\rho}\hat{v}_z = \frac{1}{r}\frac{\partial}{\partial r}\left(\frac{1}{r}\frac{\partial \chi}{\partial r}\right), \tag{4.20}$$

or,

$$\bar{\rho}_w \, \hat{v}_z = \left(\frac{4\mathcal{A}^4}{R^4}\right)\left(e^{\xi}\frac{\partial^2 \chi}{\partial \xi^2}\right). \tag{4.21}$$

Taking the derivative with respect to ξ gives

$$\bar{\rho}_w \left(\frac{\partial \hat{v}_z}{\partial \xi}\right) = \left(\frac{4\mathcal{A}^4}{R^4}\right)\left(\frac{\partial}{\partial \xi}\left(e^{\xi}\frac{\partial^2 \chi}{\partial \xi^2}\right)\right). \tag{4.22}$$

From the condition that the radial derivative of the axial velocity is zero, we obtain,

$$\frac{\partial}{\partial \xi}\left(e^{\xi}\frac{\partial^2 \chi^*}{\partial \xi^2}\right) = 0, \tag{4.23}$$

or,

$$L_3^{\mathcal{A}}(\chi^*) = 0, \tag{4.24}$$

where,

$$L_3^{\mathcal{A}}(\chi^*) = \left(1 - \frac{\xi}{\mathcal{A}^2}\right)\frac{\partial}{\partial \xi}\left(e^{\xi}\frac{\partial^2 \chi^*}{\partial \xi^2}\right). \tag{4.25}$$

At the wall $r = R(\xi = 0)$, the condition for the total axial mass flow rate (equation 4.16) can be used to obtain a boundary condition for the master potential χ^*. Since $(\partial \chi^*/\partial \xi) = 0$ at the center, the total flow rate condition reduces to,

$$\frac{\partial \chi^*}{\partial \xi} = F^*, \tag{4.26}$$

where $F^* = (F\mathrm{Re}/4\pi\mathcal{A}^2 R^2 \Omega Z \rho_w)$ is the scaled axial flow rate. The axial velocity condition at the wall can be written as,

$$\frac{\partial^2 \chi^*}{\partial \xi^2} = V_z^*, \tag{4.27}$$

where $V_z^* = (\hat{v}_z \mathrm{Re}/4\mathcal{A}^4 \Omega Z)$ is the scaled axial velocity at the wall. It might be expected that the axial velocity should be zero at the wall from the no-slip condition, but in the DSMC simulations, we find that the axial velocity is not zero at the wall, and there is a slip velocity. This slip velocity V_z^* is included in the boundary condition (4.27).

The third boundary condition at the outer wall emerges from the temperature at the outer surface of the cylinder. This boundary condition is formulated from equation (3.13) for the master potential, in which we neglect the axial gradients of the master potential in comparison to the radial derivatives. Since the velocity perturbation $\hat{v}_\theta$ is zero at the cylindrical surface, the boundary condition reduces to

$$\frac{1}{r}\frac{\partial}{\partial r}\left(\frac{1}{\bar{\rho} r}\frac{\partial}{\partial r}\left(r\frac{\partial}{\partial r}\right)\right)\left(\frac{1}{r}\frac{\partial}{\partial r}\left(\frac{1}{\bar{\rho} r}\frac{\partial \chi}{\partial r}\right)\right) = -\frac{\Omega^2}{\bar{\mu} T_0}\left(\frac{\partial \hat{T}}{\partial z}\right). \tag{4.28}$$

The temperature boundary condition at the outer wall (from equation 4.28) is expressed in terms of scaled coordinates as

$$L_5^{\mathcal{A}}(\chi^*) = -\frac{\mathrm{Re}^2}{32\mathcal{A}^{10}}\frac{R^2}{Z^2}\left(\frac{1}{T_0}\frac{\partial \hat{T}}{\partial \eta}\right), \tag{4.29}$$

where the operator $L_5^{\mathcal{A}}$ is

$$L_5^{\mathcal{A}} = \frac{\partial}{\partial \xi}\left(e^{\xi}\frac{\partial}{\partial \xi}\left(\left(1 - \frac{\xi}{\mathcal{A}^2}\right)\frac{\partial}{\partial \xi}\left(e^{\xi}\frac{\partial^2}{\partial \xi^2}\right)\right)\right). \tag{4.30}$$

Note that in general, equation (4.29) provides the boundary condition for a linear combination of the perturbation in $\hat{v}_\theta$ and $\hat{T}$, and not their individual boundary conditions. Since $\hat{v}_\theta$ is zero at the outer surface, equation (4.29) reduced to a boundary condition for $\hat{T}$.

Analytical solutions are obtained for the generalized Onsager equation 4.10:

$$\chi^* = (C_0 + D_0\eta)\left(2\mathcal{A}^2(1 - 2\mathcal{A}^2)e^{-2\xi} + 4\mathcal{A}^4(2 + 2\mathcal{A}^2 - 2\xi)\mathrm{Ei}(2\mathcal{A}^2 - 2\xi)e^{-2\mathcal{A}^2}\right.$$

$$
\begin{aligned}
&-8\mathcal{A}^4 e^{-(\xi+\mathcal{A}^2)}\mathrm{Ei}(\mathcal{A}^2-\xi)\Big) \\
&+(C_1+D_1\eta)\Big(-2\mathcal{A}^2 e^{-2\xi}+2\mathcal{A}^2(2+2\mathcal{A}^2-2\xi)e^{-2\mathcal{A}^2}\mathrm{Ei}(2\mathcal{A}^2-2\xi) \\
&-4\mathcal{A}^2 e^{-(\xi+\mathcal{A}^2)}\mathrm{Ei}(\mathcal{A}^2-\xi)\Big) \\
&+(C_2+D_2\eta)\Big(-4\mathcal{A}^2 e^{-\xi}+2\mathcal{A}^2(2+2\mathcal{A}^2-2\xi)e^{-\mathcal{A}^2}\mathrm{Ei}(\mathcal{A}^2-\xi)\Big) \\
&+(C_3+D_3\eta)e^{-\xi}+(C_4+D_4\eta)\xi+(C_5+D_5\eta),
\end{aligned} \tag{4.31}
$$

where $Ei(x)$ is the exponential integral

$$
Ei(x)=\int_{-\infty}^{x} dt\frac{e^t}{t}. \tag{4.32}
$$

The coefficients in the above equation have to be determined numerically from the boundary conditions (4.17-4.24) at $\xi=\mathcal{A}^2$, and (4.26-4.29) at $\xi=0$.

Next, we consider the secondary flow due to mass/momentum/energy sources within the domain, with homogeneous boundary conditions (constant temperature with no net inflow/outflow). Since equation 4.10 is linear in the master potential, any solution can be written as the sum of two parts, one due to secondary flow driven by wall inhomogeneities and the second due to internal sources. As shown below, the solution for the master potential with internal sources can be obtained by the method of separation of variables, since the differential operators in both the axial and radial directions are self-adjoint. The homogeneous boundary conditions for the equation 4.10 are as follows.

At the wall ($\xi=0$), the boundary conditions are the no-penetration condition in the radial direction, the no-slip in the axial and meridional direction and the zero temperature perturbation condition. The homogeneous radial and axial velocity boundary conditions, $\hat{v}_r=0$ and $\hat{v}_z=0$ at $r=R$ ($\xi=0$), reduce to

$$
\frac{\partial \chi^*}{\partial \xi}=0, \tag{4.33}
$$

$$
\frac{\partial^2 \chi^*}{\partial \xi^2}=0, \tag{4.34}
$$

while a combination of the zero meridional velocity and temperature conditions is given by

$$
L_5^{\mathcal{A}}(\chi^*)=0, \tag{4.35}
$$

where the operator $L_5^{\mathcal{A}}$ is given by

$$
L_5^{\mathcal{A}}=\frac{\partial}{\partial \xi}\left(e^{\xi}\frac{\partial}{\partial \xi}\left(\left(1-\frac{\xi}{\mathcal{A}^2}\right)\frac{\partial}{\partial \xi}\left(e^{\xi}\frac{\partial^2}{\partial \xi^2}\right)\right)\right). \tag{4.36}
$$

At the center of the cylinder, $\xi=\xi_0=(\mathcal{A}^2)$ (which corresponds to $r=0$ along the axis of the cylinder), the symmetry conditions are zero radial velocity $\hat{v}_r=0$, zero axial shear stress ($\frac{\partial \hat{v}_z}{\partial r}=0$), which can be written as

$$
\frac{\partial \chi^*}{\partial \xi}=0, \tag{4.37}
$$

$$
L_3^{\mathcal{A}}(\chi^*)=0, \tag{4.38}
$$

where the operator L_3^A is given by

$$L_3^A(\chi^*) = \left(1 - \frac{\xi}{A^2}\right) \frac{\partial}{\partial \xi} \left(e^{\xi} \frac{\partial^2 \chi^*}{\partial \xi^2}\right). \tag{4.39}$$

In addition, since the velocity and temperature perturbations are related to the derivatives of the master potential, the master potential itself can be set equal to zero, without loss of generality, at any point along the axis of the cylinder. We choose to set the master potential equal to zero at a point along the axis. Then, from the boundary condition (4.37), the master potential is equal to zero at all points on the axis $\xi = \xi_0$, that is,

$$\chi^* = 0. \tag{4.40}$$

In the η direction, the boundary conditions are the zero normal velocity conditions at the end-caps,

$$\chi^* = 0 \quad \text{at} \quad \eta = 0, \tag{4.41}$$

$$\chi^* = 0 \quad \text{at} \quad \eta = 1. \tag{4.42}$$

It is not possible to apply tangential velocity and heat flux conditions in the η direction, since the governing equation (4.10) is a second-order differential equation in η.

To obtain the solution, we use the separation-of-variables procedure. First, we define the eigenvalues and eigenfunctions of the operators L_6^A and $(\partial^2/\partial\eta^2)$. For the L_6^A operator, the eigen relation can be written as

$$L_6^A F_n(\xi) = \alpha_n F_n(\xi), \tag{4.43}$$

where α_n are the discrete eigenvalues. The adjoint of the linear operator L_6^A, with boundary conditions equations (4.33) to (4.40), can be calculated as

$$\begin{aligned}
&\int_0^{\xi_0} d\xi w(\xi) A_1(\xi) L_6^A(B_1(\xi)) \\
&= \Bigg(A_1 L_5^A B_1 - (d_\xi A_1) L_4^A B_1 + (d_\xi^2 A_1)\left(e^{\xi}\left(1 - \frac{\xi}{A^2}\right)\right) L_3^A B_1 \\
&\quad - (d_\xi^2 B_1)\left(e^{\xi}\left(1 - \frac{\xi}{A^2}\right)\right) L_3^A A_1 + (d_\xi A_1) L_4^A B_1 - B_1 L_5^A A_1\Bigg)\Bigg|_{\xi=0}^{\xi=\xi_0} \\
&\quad + \int_0^{\xi_0} d\xi w(\xi) B_1(\xi) L_6^A(A_1(\xi)),
\end{aligned} \tag{4.44}$$

where

$$L_5^A = \frac{\partial}{\partial \xi}\left(e^{\xi} \frac{\partial}{\partial \xi}\left(\left(1 - \frac{\xi}{A^2}\right) \frac{\partial}{\partial \xi}\left(e^{\xi} \frac{\partial^2 A_1}{\partial \xi^2}\right)\right)\right), \tag{4.45}$$

$$L_4^A A_1 = e^{\xi} \frac{\partial}{\partial \xi}\left(\left(1 - \frac{\xi}{A^2}\right) \frac{\partial}{\partial \xi}\left(e^{\xi} \frac{\partial^2 A_1}{\partial \xi^2}\right)\right), \tag{4.46}$$

$$L_3^A A_1 = \left(1 - \frac{\xi}{A^2}\right) \frac{\partial}{\partial \xi}\left(e^{\xi} \frac{\partial^2 A_1}{\partial \xi^2}\right). \tag{4.47}$$

It is clear, from equation (4.44), that the operator L_6^A, with boundary conditions equations (4.33) to (4.40) is a self-adjoint operator, since all the terms on the right side obtained by partial integration are zero. Therefore, the eigenvalues α_n in equation (4.43) are real, and the eigenfunctions $F_n(\xi)$ are orthogonal. We define the eigenfunctions so

Eigenvalue	$\mathcal{A} = 0.707$	$\mathcal{A} = 2.121$	$\mathcal{A} = 3.535$
α_1	2.041214592	1.482059289	1.221756348
α_2	6.280779784	3.809995603	3.552378653
α_3	8.795687234	5.229456790	4.898674393
α_4	10.610548978	6.185634981	5.877902475
α_5	12.086534131	6.960542317	6.661652340
α_6	13.256783429	7.580054231	7.298259826
α_7	14.260934275	8.113994373	7.820965234
α_8	15.146965314	8.570763421	8.296539810
α_9	15.920934879	8.976178439	8.711321412
α_{10}	16.147321968	9.151874328	9.031678329

TABLE 1. First 10 eigenvalues of $L_6^{\mathcal{A}}$ operator at stratification parameter $\mathcal{A} = 0.707$, 2.121, 3.535.

that they are orthonormal:

$$\int_0^{\xi_0} d\xi F_n(\xi) F_m(\xi) = \delta_{mn}. \tag{4.48}$$

The eigenvalues and eigenfunctions are determined numerically using fourth-order Runge-Kutta integration and the shooting technique. The first ten eigenvalues (α_n) of the operator $L_6^{\mathcal{A}}$ at $\mathcal{A} = 0.707$, 2.121, 3.535 are given in table 1.

The eigen equation for the η direction is

$$\frac{\partial^2 G_n(\eta)}{\partial \eta^2} = \beta_n G_n(\eta), \tag{4.49}$$

with boundary conditions (4.41) and (4.42). For this case, the eigen-values are given by $\beta_n = (n\pi)^2$, and the orthonormal eigenfunctions are

$$G_n(\eta) = \sqrt{2} \sin(n\pi\eta). \tag{4.50}$$

The master potential can be expressed in terms of the eigenfunctions of the $L_6^{\mathcal{A}}$ and the $(\partial^2/\partial\eta^2)$ operators:

$$\chi^* = \sum_{n=1}^{\infty} \sum_{m=1}^{\infty} A_{nm} F_n(\xi) G_m(\eta). \tag{4.51}$$

The above expressions are substituted into the differential equation 4.10, and the orthogonality condition is used, to obtain the coefficients A_{mn}:

$$A_{nm} = \frac{1}{\alpha_n + ((\mathrm{Re}/8\mathcal{A}^6)(R/Z))^2 \beta_m} \int_0^{\xi_0} d\xi \int_0^1 d\eta S^*(\xi, \eta) F_n(\xi) G_m(\eta). \tag{4.52}$$

In the case of delta function sources,

$$\mathcal{S}^*(\xi, \eta) = \sum_{p=1}^{P} \mathcal{S}_p^* \delta(\xi - \xi_p, \eta - \eta_p), \tag{4.53}$$

the coefficients A_{nm} reduce to,

$$A_{nm} = \frac{1}{\alpha_n + ((\mathrm{Re}/8\mathcal{A}^6)(R/Z))^2\beta_m} \sum_{p=1}^{P} S_p^* F_n(\xi_p) G_m(\eta_p). \tag{4.54}$$

Here, P is the total number of sources. In the case of a the complicated source expression (4.9), which involves derivatives and integrals, if the mass, momentum and energy sources are of the form

$$\begin{pmatrix} \mathcal{S}_m \\ f_r \\ f_\theta + f_{\theta,feed} \\ f_z \\ \mathcal{S}_e \end{pmatrix} = \sum_{p=1}^{P} \begin{pmatrix} \mathcal{S}_{mp} \\ f_{rp} \\ f_{\theta p} + f_{\theta p,feed} \\ f_{zp} \\ \mathcal{S}_{ep} \end{pmatrix} \delta(\xi - \xi_p)\delta(\eta - \eta_p), \tag{4.55}$$

the expression for the coefficients can be determined using integration by parts:

$$\begin{aligned}
&(\alpha_n + ((\mathrm{Re}/8\mathcal{A}^6)(R/Z))^2\beta_m)A_{nm} \\
&= \frac{\mathrm{Re}}{32\mathcal{A}^{10}} \frac{R}{Z} \frac{R^2(\mathcal{S}_{ep} - p_w e^{-\xi_c}\mathcal{S}_m)}{4\mathcal{A}^4 \bar{K} T_0} \left(- \left.\frac{\partial G_m}{\partial \eta}\right|_{\eta=\eta_c} \right) \int_{\xi_c}^{0} d\xi \frac{F_n(\xi)}{(1-\xi/\mathcal{A}^2)} \\
&- \frac{\mathrm{Re}}{16\mathcal{A}^{10}} \frac{R}{Z} \frac{(f_{\theta p} + f_{\theta p,feed})R}{4\mathcal{A}^4 \Omega \bar{\mu}} \left(- \left.\frac{\partial G_m}{\partial \eta}\right|_{\eta=\eta_c} \right) \left(1 - \frac{\xi_c^*}{\mathcal{A}^2}\right)^{\frac{1}{2}} \int_{\xi_c}^{0} d\xi \frac{F_n(\xi)}{(1-\xi/\mathcal{A}^2)^2} \\
&- \left[\left(\frac{2\mathrm{Re}^2 \mathcal{A}^2 \mathrm{Pr}(1-\gamma^{-1})}{(1-\xi/\mathcal{A}^2)} + \frac{4\mathrm{Re}^2}{(1-\xi/\mathcal{A}^2)^2} \right) \right. \\
&\left. \frac{\mathcal{S}_{mp} R}{256\mathcal{A}^{16}\rho_w \Omega Z} \left(- \left.\frac{\partial G_m}{\partial \eta}\right|_{\eta=\eta_c} \right) \int_{\xi_c}^{0} d\xi F_n(\xi)(\xi - \xi_c) \right] \\
&- \frac{\mathrm{Re}}{32\mathcal{A}^{10}} \left[\frac{\left(f_{rp} - \frac{2}{3R}\bar{\mu}\mathcal{A}^2 \left(1 - \xi_c/\mathcal{A}^2\right)^{1/2} \frac{\partial \mathcal{S}_{mp}}{\partial \xi} \right)}{\rho_w \Omega^2 Z} \left(\left.\frac{\partial G_m}{\partial \eta}\right|_{\eta=\eta_c} \right) \right. \\
&\left. \left(\frac{e^{\xi_c}}{1 - \xi_c/\mathcal{A}^2)^{1/2}} \left.\frac{\partial F_n}{\partial \xi}\right|_{\xi=\xi_c} \right) + \frac{\left(f_{zp} + \frac{\bar{\mu}}{3Z} \frac{\partial \mathcal{S}_{mp}}{\partial \eta} \right)}{\rho_w \Omega^2 R} \left(G_m(\eta_p) e^{\xi_p} \left.\frac{\partial^2 F_n(\xi)}{\partial \xi^2}\right|_{\xi=\xi_p} \right) \right].
\end{aligned} \tag{4.57}$$

Here, the fourth term on the right side of equation (4.9) has been neglected, since this is small compared to the third term for high Reynolds number and aspect ratio.

5. Radial flow at end-caps

Discussion is restricted to an radial flow at the end-caps where there are no sources or sinks within the boundary layer which turns out to be $O(\mathrm{Re}^{-1/2})$ smaller than the cylinder diameter. For the radial flow, the axial derivatives of $\hat{v}_\theta$ and $\hat{T}$ on the right sides of equations (3.4) and (3.6) are balanced by the left sides. Following our previous study (Pradhan & Kumaran (2011)), we obtain

$$\frac{16\mathcal{A}^8}{\mathrm{Re}^2} \frac{\partial}{\partial \xi} \left(e^{\xi} \frac{\partial}{\partial \xi} \left(1 - \frac{\xi}{\mathcal{A}^2} \right) \frac{\partial}{\partial \xi} \left(e^{\xi} \frac{\partial \psi^*}{\partial \xi} \right) \right) + \frac{4\mathcal{A}^4}{\mathrm{Re}^2} \frac{R^2}{Z^2} \frac{\partial}{\partial \xi} \left(e^{2\xi} \frac{\partial}{\partial \xi} \frac{\partial^2 \psi^*}{\partial \eta^{*2}} \right)$$

$$+\frac{4\mathcal{A}^4}{\mathrm{Re}^2}\frac{R^2}{Z^2}e^{\xi}\frac{\partial^2}{\partial\xi^2}\left(e^{\xi}\frac{\partial^2\psi^*}{\partial\eta^{*2}}\right)+\frac{e^{2\xi}}{(1-(\xi/\mathcal{A}^2))}\frac{\partial^4\psi^*}{\partial\eta^{*4}}$$
$$+\left(\frac{4}{(1-\xi/\mathcal{A}^2)}+2\mathcal{A}^2\mathrm{Pr}(1-\gamma^{-1})\right)\psi^*$$
$$+\frac{R^4}{Z^4\mathrm{Re}^2}(S_\theta^*+S_T^*+C_\theta^*+C_T^*)=0, \quad (5.1)$$

where,

$$\psi^*=\frac{\psi}{\bar{\mu}Z}, \quad (5.2)$$

$$\eta^*=\mathrm{Re}^{1/2}(Z/R)\eta, \quad (5.3)$$

$$S_\theta^*=\frac{8\mathcal{A}^4\mathrm{Re}^2Z^4}{R^4}\int_{\eta_0}^{\eta}d\eta\left(\frac{\partial^2}{\partial\xi^2}((1-\xi/\mathcal{A}^2)^{1/2}\hat{v}_\theta^*)\right), \quad (5.4)$$

$$C_\theta^*=-\frac{2\mathrm{Re}^2Z^2}{R^2}\frac{1}{(1-\xi/\mathcal{A}^2)^{1/2}}\left.\frac{\partial\hat{v}_\theta^*}{\partial\eta}\right|_{\eta=\eta_0}, \quad (5.5)$$

$$S_T^*=-\frac{4\mathcal{A}^4\mathrm{Re}^2Z^4}{R^4}\frac{\partial}{\partial\xi}\left((1-\xi/\mathcal{A}^2)\frac{\partial}{\partial\xi}\left(\int_{\eta_0}^{\eta}d\eta\hat{T}^*\right)\right), \quad (5.6)$$

$$C_T^*=\frac{\mathrm{Re}^2Z^2}{R^2}\left.\frac{\partial\hat{T}^*}{\partial\eta}\right|_{\eta=\eta_0}, \quad (5.7)$$

$$\hat{v}_\theta^*=\frac{\hat{v}_\theta}{\Omega R}, \quad (5.8)$$

$$\hat{T}^*=\frac{\hat{T}}{T_0}. \quad (5.9)$$

In the generalized Carrier-Maslen equation, the above equation is approximated as (Pradhan & Kumaran (2011))

$$\frac{e^{2\xi}}{(1-(\xi/\mathcal{A}^2))}\frac{\partial^4\psi^*}{\partial\eta^{*4}}+\left(\frac{4}{(1-\xi/\mathcal{A}^2)}+2\mathcal{A}^2\mathrm{Pr}(1-\gamma^{-1})\right)\psi^*=0. \quad (5.10)$$

The solution for the homogeneous form of the generalized Carrier-Maslen equation is easily obtained in terms of the variable $\eta^\dagger$,

$$\eta^\dagger=\eta^*\left\{\left(e^{(-2\xi)}\left(4+2\mathcal{A}^2\mathrm{Pr}(1-\gamma^{-1})(1-\xi/\mathcal{A}^2)\right)\right)\right\}^{\frac{1}{4}}. \quad (5.11)$$

The general solution for the scaled stream function in terms of this scaled coordinate is,

$$\psi^*=C_1e^{((1+\imath)\eta^\dagger/\sqrt{2})}+C_2e^{(-(1+\imath)\eta^\dagger/\sqrt{2})}+C_3e^{((1-\imath)\eta^\dagger/\sqrt{2})}+C_4e^{(-(1-\imath)\eta^\dagger/\sqrt{2})}. \quad (5.12)$$

Since the velocities decay to zero as $z\to\infty$, we can assume without loss of generality that $\psi^*(r,z\to\infty)=0$. Consequently, the constants C_1 and C_3 in equation (5.12) are both zero.

The velocity components can be expressed in terms of the stream function as,

$$\hat{v}_r=-\frac{1}{\rho r}\frac{\partial\psi}{\partial z}$$
$$=\frac{\bar{\mu}Z}{\bar{\rho}r}\frac{\mathrm{Re}^{1/2}}{R}\frac{\partial\psi^*}{\partial\eta^\dagger}\frac{\partial\eta^\dagger}{\partial\eta^*} \quad (5.13)$$

$$= \frac{\bar{\mu} Z}{\bar{\rho} r}\frac{\mathrm{Re}^{1/2}}{R}\left\{\left(e^{(-2\xi)}\left(4 + 2\mathcal{A}^2\mathrm{Pr}(1-\gamma^{-1})(1-\xi/\mathcal{A}^2)\right)\right)\right\}^{\frac{1}{4}}\frac{\partial\psi^*}{\partial\eta^\dagger}. \tag{5.14}$$

and

$$\begin{aligned}\hat{v}_z &= \frac{1}{\bar{\rho} r}\frac{\partial\psi}{\partial r}\\ &= -\frac{2\mathcal{A}^2\bar{\mu}Z}{\bar{\rho}R^2}\frac{\partial\psi^*}{\partial\xi}\\ &= -\frac{2\mathcal{A}^2\bar{\mu}Z}{\bar{\rho}R^2}\frac{\partial\psi^*}{\partial\eta^\dagger}\frac{\partial\eta^\dagger}{\partial\xi}.\end{aligned} \tag{5.15}$$

Equations (5.14) and (5.15) show that both the normal and tangential velocities are proportional to $(\partial\psi^*/\partial\eta^\dagger)$. Therefore, a zero normal velocity (no penetration) condition also implies that the tangential velocity is zero within the context of the generalized Carrier-Maslen approximation; it is necessary to include higher order corrections to specify the tangential and normal velocities independently. However, this does not turn out to be a restriction in the present case, because the molecular simulations show that the tangential velocity at the end-caps is numerically small, and the zero normal velocity approximation is a good one for the radial flow at the end-caps. The zero normal velocity condition serves to fix one of the constants in the solution (5.12). The final constant is fixed from the flux condition,

$$\begin{aligned}\psi(r, z=\infty) - \psi(r, z=0) &= \int_0^\infty r dz \bar{\rho} v_r\\ &= \frac{F}{2\pi},\end{aligned} \tag{5.16}$$

where F is the total mass flow rate. This can be expressed in terms of the dimensionless variables as

$$\psi^*(r, z=0) = -\frac{F}{2\pi\bar{\mu}Z}. \tag{5.17}$$

Here, we have assumed that $\psi^*(r, z\to\infty) = 0$ without loss of generality. With this, the final solution for the stream function in terms of the flow rate is

$$\psi^* = \frac{F}{2\pi\bar{\mu}Z}\left(\frac{1+\imath}{2}e^{(-(1+\imath)\eta^\dagger/\sqrt{2})} + \frac{1-\imath}{2}e^{(-(1-\imath)\eta^\dagger/\sqrt{2})}\right). \tag{5.18}$$

6. DSMC simulation

The direct simulation Monte Carlo method (DSMC) is a numerical method for modeling rarefied gas flows. Here, the Boltzmann equation for the probability distribution function for gas molecules is solved using a probabilistic (Monte Carlo) simulation technique. The Boltzmann equation is more fundamental than the Navier-Stokes equations, since it describes the evolution of the velocity distribution function of the gas molecules. The Navier-Stokes equations are obtained by taking the first, second and third moments of the Boltzmann equation with respect to the molecular velocities. While taking moments, the assumption of 'local equilibrium' (velocity distribution function close to the Maxwell-Boltzmann distribution) is made, thereby restricting the Navier-Stokes equations to near-equilibrium flows. Since there is no assumption of local equilibrium in the Boltzmann equation, it is valid even for far-from equilibrium flows, provided the molecular chaos approximation is satisfied.

The DSMC method uses 'simulated molecules' which represent a large number of real gas molecules. A basic assumption is that the inter-molecular collisions and molecular motion between collisions can be decoupled over time scales comparable to the time between collisions. The translation of molecules between collisions is deterministically modeled, while the inter-molecular collisions and molecular-surfaces are incorporated using probabilistic models. The domain is divided into a large number of cells, and properties are evaluated in these cells. The cells are further divided into 'sub-cells' for collision localization.

In the DSMC simulations, inter-molecular collisions are executed at random from all molecules within 'sub-cells' within the simulation cell, whose linear dimension is much smaller than the mean free path. The probability of collisions between a pair of molecules within a sub-cell is selected equal to that for a dilute gas with the same temperature as that in the cell. In this manner, the temperature (mean square velocity), mean free path and collision frequency for the simulated molecules are equal to those for the real molecules, provided the sub-cell dimension is much smaller than the mean free path.

The most common model is the hard-sphere model, but other models such as the variable hard sphere and variable soft sphere models are also used in order to accurately reproduce the density and temperature dependence of the viscosity. The hydrodynamic fields (density, velocity, temperature) are obtained by averaging over the molecular distribution functions in the DSMC simulations.

At the boundaries, it is necessary to prescribe interaction laws between the molecules and the solid surfaces. Accurate specification of the interaction laws requires a detailed knowledge of the molecule-surface interaction. However, there are standard interaction laws based on the surface velocity and temperature which are sufficient for predicting the hydrodynamic fields accurately. The most commonly used is the 'diffuse reflection' conditions, where a molecule incident on the surface gets reflected with a Gaussian velocity distribution with the same mean velocity and temperature as that of the surface. Another commonly used condition is the specular reflection condition, that the velocity normal to the surface is reversed, while that tangential to the surface is unchanged. It is important to note that these boundary conditions are different from the no-slip conditions in the continuum models. In high Knudsen number flows, the diffuse and specular reflection conditions do predict a slip velocity at the surface. Therefore, these simulations could be used even in situations where there is a slip at the surface.

From kinetic theory of dilute gases (Chapman & Cowling (1970)), the viscosity of a hard-sphere gas in the dilute limit is given by

$$\mu = \frac{5\sqrt{mk_BT}}{16\sqrt{\pi}d^2}, \tag{6.1}$$

where k_B is the Boltzmann constant, while the speed of sound c_s is given by the relation

$$mnc_s^2 = \gamma p = \gamma nk_BT, \tag{6.2}$$

where m is the molecular mass, n is the number density, and p is the gas pressure. From this the stratification parameter $\mathcal{A}$, Reynolds number Re, cylinder radius R, molecular diameter d and volume fraction of the molecules ϕ are related by the following equation:

$$\mathrm{Re} = \frac{96}{5} \times \frac{\sqrt{2}}{\sqrt{\pi}} \mathcal{A}\,\phi\,\frac{R}{d}, \tag{6.3}$$

where the volume fraction of the molecules is given by

$$\phi = \frac{\pi}{6}\,d^3\,n, \tag{6.4}$$

where n is the number density of the molecules. The important point to note, from 6.3, is that the Reynolds number is proportional to the stratification parameter times $\phi(R/d)$, where R is the macroscopic scale. Therefore if we are keeping the volume fraction (density) a constant, increasing (decreasing) the Reynolds number at constant stratification parameter corresponds to increasing (decreasing) the system size R in relation to the molecular diameter.

The details of the DSMC simulation procedure are discussed in Bird (1963) and Bird (1994), and so we do not repeat the discussion here, but restrict attention to the parameters in the simulations. The number of cells along the radial direction is 400 with clustering towards the rotating sidewall, while there are 1600, 3200 and 6400 intervals along the axial direction with clustering towards the top and bottom end-caps for aspect ratio $Z/D = 2$, 4, and 8 respectively. The cell width is no greater than one-third of the local mean free path. The number of subcells per cell is taken as 4 for both radial and axial directions. The number of simulated particles per cell is taken as 100. In our simulations, the number of real molecules per simulated particles varied between 10^{16} and 10^{18}, so that the statistical fluctuations not becomes large.

The fluid is assumed to be an ideal gas with the molecular mass $58.442636 \times 10^{-26}$ kg/molecule, molecular diameter $= 5.5 \times 10^{-10}$ m, specific heat ratio $\gamma = 1.066$, and Prandtl number Pr$= 0.95$. It is assumed that the gas molecules undergo diffuse reflection from all the solid surfaces with the thermal accommodation coefficient equal to 1. The time step in the simulation is chosen to be small compared to the time between collisions. The mean collision time is calculated as, $t = (\lambda / c_{av})$, where λ is the mean free path and c_{av} is the average molecular velocity. The minimum value of mean free path at the rotating sidewall, calculated from the gas density at the wall, is $\lambda = 2.316742 \times 10^{-7}$ m. The average molecular velocity is calculated as $\sqrt{k_B T/m} = 84.17717$ m/sec. This gives the mean collision time (t) $= 2.752221 \times 10^{-9}$ sec. The time step used in the simulations is 1.0×10^{-10} sec, which is less than mean collision time.

Simulations were carried out for stratification parameters in the range $0.707 - 3.535$, and Reynolds numbers in the range $10^2 - 10^6$. Three different aspect ratios, $Z/D = 2$, 4 and 8, are considered in the analysis (where Z is the height and D is the diameter of the cylinder). Only the comparisons for $Z/D = 2$ are shown here. This because the smallest aspect ratio provides the most stringent test. The dimensionless scaled angular momentum of the feed gas $F_\theta^* = F_\theta/(\theta_{mi}\Omega)$ is taken as 0.2. There is better agreement between simulations and theory as the aspect ratio increases, and so results are reported only for the smallest aspect ratio $(Z/D) = 2$. In addition, there is better agreement between simulations and theory as the Reynolds number increases, and so results are reported only for the lowest Reynolds number, that is Re $= 10^2$. There are three types of forcing for the secondary flow,

(*a*) For inflow/outflow of gas along the axis, the feed gas with scaled angular momentum $F_\theta^* = F_\theta/(\theta_{mi}\Omega) = 0.2$, is introduced at the axis of the cylinder at the axial location $\eta = (z/Z) = 0.5$, and $r = 0$, and outflow of the gas at two axial locations along the axis, one near the bottom end cap at $\eta = (z/Z) = 0.15$ and other near top end cap at $\eta = (z/Z) = 0.85$.

(*b*) Inflow of feed gas with scaled angular momentum $F_\theta^* = F_\theta/(\theta_{mi}\Omega) = 0.2$, is introduced at the axis of the cylinder at the axial location $\eta = (z/Z) = 0.5$, and $r = 0$, and outflow of the gas at two axial locations, one near the bottom end cap at $(\eta = (z/Z) = 0.15)$ and other near top end cap $(\eta = (z/Z) = 0.85)$, at two different radial locations, one inside the radial boundary layer at $\xi^* = \xi/\mathcal{A}^2 = 0.01$, and other outside the radial boundary layer at $\xi^* = 0.25$, with azimuthal-momentum source inside the flow domain at $\eta =$ 0.15, and 0.85, and at two different radial locations, one inside the radial boundary layer at $\xi^* = \xi/\mathcal{A}^2 = 0.01$, and other outside the radial boundary layer at $\xi^* = 0.25$, with a linear wall temperature profile at the rotating sidewall.

(c) Inflow of feed gas with angular momentum $F_\theta^* = F_\theta/(\theta_{mi}\Omega) = 0.2$, is introduced at the axis of the cylinder at the axial location $\eta = (z/Z) = 0.5$, and $r = 0$, and outflow of the gas at two axial locations, one near the bottom end cap at ($\eta = (z/Z) = 0.15$) and other near top end cap ($\eta = (z/Z) = 0.85$), at two different radial locations, one inside the radial boundary layer at $\xi^* = \xi/\mathcal{A}^2 = 0.01$, and other outside the radial boundary layer at $\xi^* = 0.25$, with azimuthal-momentum source inside the flow domain at $\eta = 0.15$, and 0.85, and at two different radial locations, one inside the radial boundary layer at $\xi^* = \xi/\mathcal{A}^2 = 0.01$, and other outside the radial boundary layer at $\xi^* = 0.25$, with a linear wall temperature profile at the rotating sidewall, and with rotating insert.

(a) In the case of a wall thermal drive, a linear temperature profile is imposed at the rotating sidewall. The wall temperature influences the gas temperature through the diffuse reflection boundary condition, where the velocity of the reflected molecules is chosen from a Maxwell-Boltzmann distribution with mean velocity and temperature equal to those of the wall. The non-dimensional parameter $F_T = (Z/\bar{T})(dT/dz)$ is used to quantify the temperature gradient, where $\bar{T}$ is the wall temperature averaged over the height.

(b) For inflow/outflow of gas along the axis, the gas is introduced at the axis of the cylinder at $\eta = 0.5$ and $r = 0$. The gas flows radially outward due to the centrifugal acceleration, and then splits at the wall into two streams, one going upward along the upper half and the other going downward along the lower half. At the two end caps, there is a radially inward flow towards the center, and the gas is taken out at two locations along the axis, one near the bottom end cap ($\eta = (z/Z) = 0.15$) and other near the top end cap ($\eta = (z/Z) = 0.85$). Here, Z is the total cylinder height and z is the axial distance from the bottom end cap.
In this case, the dimensionless scaled feed rate is defined by

$$F^* = \frac{F}{H\Omega}, \tag{6.5}$$

and dimensionless scaled angular momentum of the feed gas is defined by

$$F_\theta^* = \frac{F_\theta}{\theta_{mi}\Omega}, \tag{6.6}$$

where F is the feed rate, H is the total gas holdup, Ω is the angular velocity, and F_θ is the angular momentum of the feed gas in the θ direction, and θ_{mi} is the moment of inertia of the rotating cylinder. The dimensionless scaled feed rate F^* provides the magnitude of the perturbation to the base flow due to external feed. Simulations have been carried out for $F^* = 0.2$, with $F_\theta^* = 0.2$. The actual feed rate (F) is varied in such a way that the desired scaled feed rate F^* is obtained at different stratification parameter, Reynolds numbers and aspect ratios. Similarly, the actual angular momentum of the feed gas F_θ is varied in such a way that the desired scaled angular momentum F_θ^* is obtained at different stratification parameter, Reynolds numbers and aspect ratios.

(c) The internal sources of mass, momentum and energy are incorporated in the DSMC simulations as follows. A small circular source domain of radius 10^{-3} times the cylinder radius (R) is selected for adding/removing mass, momentum and energy to approximate a delta function. In the case of mass source, a molecule is added at random into the source domain at fixed time intervals determined by the desired rate of mass input. The velocity of the added molecules is chosen from a random Gaussian distribution with the same mean velocity and temperature as that of the gas in the domain. When a molecule is added, the mean velocity and the fluctuating velocity of all other molecules in the domain are scaled by a constant factor so that the total momentum and energy of all the molecules in the source domain remain unchanged due to the addition of a molecule. In the case of a mass sink, molecules are removed at random from the domain at time intervals fixed by the desired rate of mass output. When a molecule is removed, the mean and the fluctuating velocity of all other molecules in the source domain are scaled by a constant factor so that the total momentum and energy in the domain remain unchanged. The simulations are carried out for two configurations, one with source/sink outside the radial boundary layer at ($\xi^* = 0.25$), and the second inside the radial boundary layer at ($\xi^* = 0.01$) respectively. Here, $\xi^* = \xi/\mathcal{A}^2$.

(*d*) Forces (momentum sources) exerted on the gas due to stationary inserts are modeled by imposing impulses (increasing the momentum) of all the molecules in the direction of a force at each collision step. The magnitude of the impulse is determined such that the ratio of the impulse and the time between collisions is equal to the applied force. A localized energy source, which increases or decreases the temperature of the gas, is imposed by scaling the fluctuating velocity of all the molecules in the source domain. The scaling of the velocity is determined such that the increase in energy divided by the time between collisions is equal to the desired energy input rate.

7. Results (single-component gas)

First, we consider the single component ideal gas with the molecular mass $58.442636 \times 10^{-26}$ kg/molecule, molecular diameter $= 5.5 \times 10^{-10}$ m, specific heat ratio $\gamma = 1.066$, and Prandtl number Pr$= 0.95$. It is assumed that the gas molecules undergo diffuse reflection from all the solid surfaces with the thermal accommodation coefficient equal to 1.

7.1. *Inflow/outflow of gas along the axis with swirling feed (single-component gas)*

The flow geometry and coordinate system for inflow/outflow of gas along the axis is shown in figure 1 (a) and the characteristic streamline plot in figure 1 (b). The gas is introduced at the axis of the cylinder at the axial location $\eta = 0.5$ and $r = 0$. The gas flows radially outward due to the centrifugal acceleration, and then splits at the wall into two streams, one going upward along the upper half and the other going downward along the lower half. At the two end caps, there is a radially inward flow towards the center, and the gas is taken out at two locations along the axis, one near the bottom end cap at $\eta = (z/Z) = 0.15$ and other near top end cap at ($\eta = (z/Z) = 0.85$). Here, Z is the total cylinder height and z is the axial distance from the bottom end cap.

Here, we have carried out the simulations for $F^* = 0.2$, with $F^*_\theta = 0.2$. The actual feed rate (F) is varied in such a way that the desired scaled feed rate is obtained at different stratification parameter, Reynolds numbers and aspect ratios.

Substantial axial velocity slip at the rotating sidewall is obtained in the simulations in this case. The axial velocity at the rotating sidewall, non-dimensionalized by $(F^* \, (R\Omega))$, is plotted as a function of stratification parameter $\mathcal{A}$ at the axial location $(\eta) = 0.675$ as shown in figure 2. It is observed that the ratio of the slip velocity and the maximum secondary flow velocity $(v_z/(R\Omega F^*))$, actually decreases as the stratification parameter is increased. It is important to incorporate the axial velocity and temperature slip in order to obtain quantitative agreement between the analytical results and simulations.

The master potential profiles due to inflow/outflow of gas along the axis as a function of ξ^* at $\eta = 0.325$ and 0.675, are shown in figures 3, for $\mathcal{A} = 3.535, 2.121, 1.414, 0.707$; $\mathrm{Re} = 10^2$, and for aspect ratio (Z/D) $= 2$; $F^* = 0.2$, and $F^*_\theta = 0.2$.

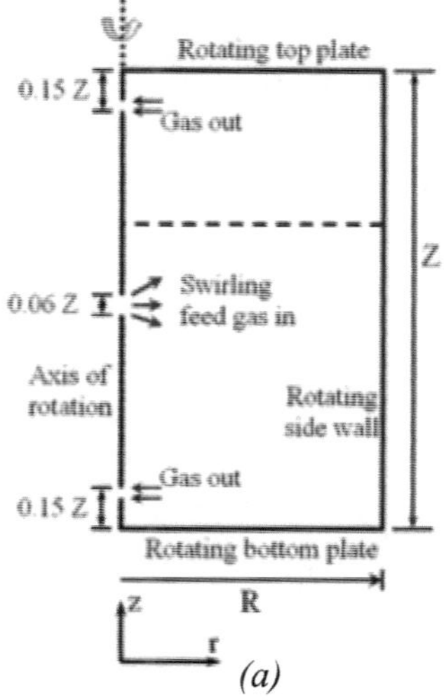

(a)

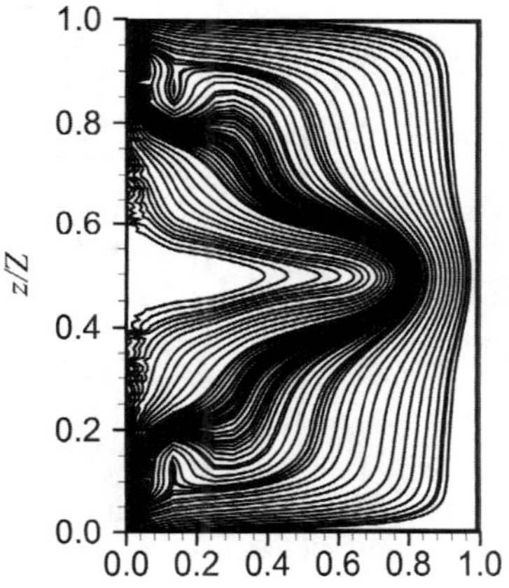

(b)

FIGURE 1. Computational domain and coordinate system (a); and characteristic streamline plot from DSMC simulations (b), at $\mathcal{A} = 3.535$, Re $= 10^4$, $F^* = 0.2$, aspect ratio $(Z/D) = 4$, and $F_\theta^* = 0.2$, for inflow/outflow of gas along the axis of the cylinder with swirling feed.

The analytical solutions are compared with the results of DSMC simulations with swirling feed $F_\theta^* = 0.2$, for stratification parameter $\mathcal{A}$ in the range $0.707 - 3.535$, F^* as high as 0.2, aspect ratio (Z/D) as low as 2, and Reynolds number Re as low as 10^2, as shown in figure 4. The smallest aspect ratio, smallest Reynolds number, and highest secondary flow velocity provides the most stringent test. When the slip velocity and temperature are correctly incorporated in the analytical solution, the results are in quantitative agreement with simulations to within 15% even when the stratification parameter is as low as 0.707 and the Reynolds number is as low as 100 and the aspect ratio (length/diameter) of the cylinder is as low as 2, and the secondary flow velocity is as high as 0.2 times the maximum base flow velocity, and the dimensionless scaled angular momentum of the feed gas $F_\theta^* = F_\theta/(\theta_{mi}\Omega)$ as high as 0.2.

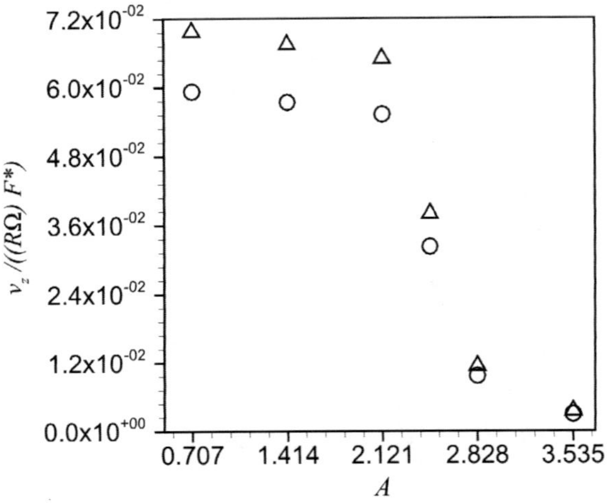

FIGURE 2. Axial slip velocity at the curved surface, scaled by $((R\Omega)F^*)$, as a function of $\mathcal{A}$, for inflow/outflow of gas along the axis at $\eta = 0.675$, for aspect ratio $(Z/D) = 2$, $\mathrm{Re} = 10^2$, and $F^* = 0.2$. The symbols show the DSMC results for : ($F_\theta^* = 0$ ($\circ$), $F_\theta^* = 0.2$ ($\triangle$)). Here, $F^* = F/(H\Omega)$, and $F_\theta^* = F_\theta/(\theta_{mi}\Omega)$.

7.2. *Radial flow at end-caps with swirling feed (single-component gas)*

The analytical solutions, obtained by solving the generalized Carrier-Maslen equation are compared with the results of DSMC simulations incorporating the angular momentum of the feed gas in a swirling feed. As discussed below, it is necessary to specify the total mass flow rate at a fixed radial location in order to obtain the analytical solution. This mass flux is a constant for the inflow/outflow of gas along the axis, because there is an upward flow along the wall in the upper half, and an inward flow along the end-caps and removal of gas at the axis of the cylinder. The wall slip velocity, which was an important factor in our analysis of the axial flow, turns out not to be significant for the end-caps. The maximum scaled slip velocity (scaled by $R\Omega$) obtained in our simulation with swirling feed is 6.2×10^{-4}.

The boundary conditions in the generalized Carrier-Maslen solution are specified as follows. The structure of the generalized Carrier-Maslen solution is such that the leading-order tangential and normal velocities are proportional to each other. Therefore, the no-penetration (zero normal velocity) condition results in the tangential velocity at the surface also being zero. The solution for the streamfunction is then completely determined from the total radial flow rate at the end-caps,

$$\psi^*(r, z = 0) = -\frac{F}{2\pi\bar{\mu}Z}, \tag{7.1}$$

with

$$\psi(r, z \to \infty) = 0, \tag{7.2}$$

where ψ^* is the scaled stream function $(\psi/(\bar{\mu}Z))$, and F is the mass flow rate. With the above two conditions, as well as the zero normal velocity condition, the analytical solution for the radial flow at the end-caps is completely specified.

The radial velocity at the end-caps, non-dimensionalised by $(R\Omega)$, is plotted as a function of scaled axial coordinate (η^*) for $\mathcal{A} = 3.535$, 2.121, and 0.707, in figure 5, figure 6,

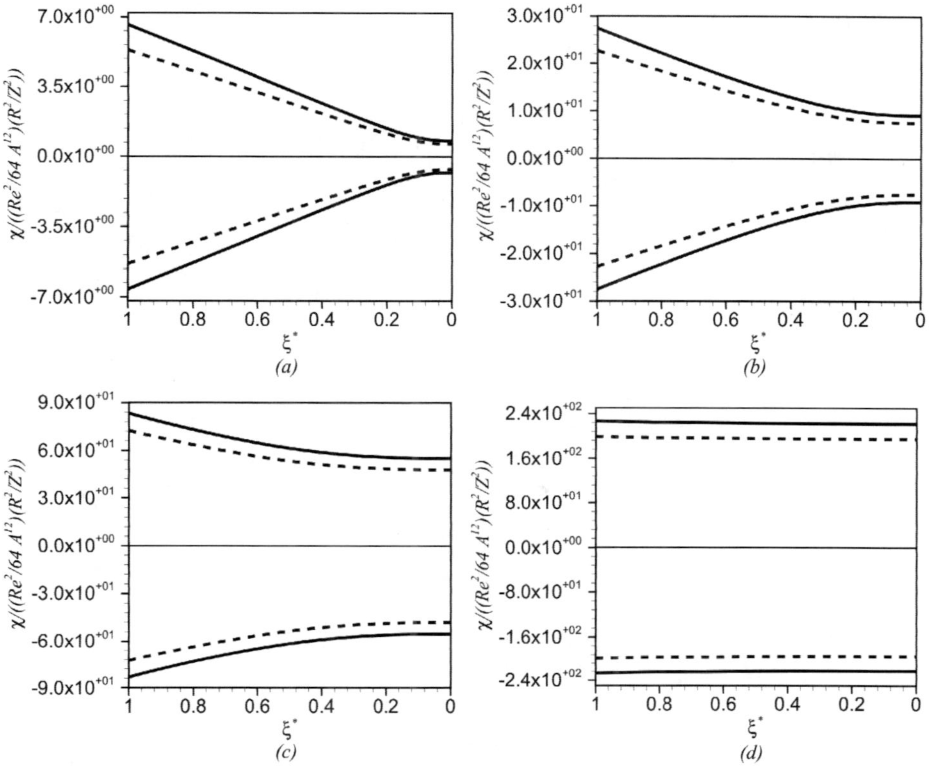

FIGURE 3. Master potential scaled by $((\mathrm{Re}^2/64\mathcal{A}^{12})(R^2/Z^2))$, as a function of (ξ^*) at $\eta = 0.325$ and 0.675, Re $= 10^2$, aspect ratio (Z/D) $= 2$, $F^* = 0.2$, for $\mathcal{A} = 3.535$ (a); $\mathcal{A} = 2.121$ (b); $\mathcal{A} = 1.414$ (c); and $\mathcal{A} = 0.707$ (d), for the secondary flow due to inflow/outflow of gas along the axis. The lines show the analytical solutions with wall slip for : ($F_\theta^* = 0$ (−−); $F_\theta^* = 0.2$ (—)). Here, $F^* = F/(H\Omega)$, and $F_\theta^* = F_\theta/(\theta_{mi}\Omega)$.

and figure 7 respectively. The simulation results show that the slip in the tangential velocity at the end-caps is numerically small. In the simulations, it was found that the radial velocity increases as the stratification parameter is increased.

Figure 5, figure 6, and figure 7 shows that the highly confined boundary layer flow with a small boundary layer thickness close to the cylindrical wall at $\xi^* = 0.1$ expands out due to a flow with lower velocity and higher boundary layer thickness at $\xi^* = 0.5$. Note that for the lowest Reynolds number of 100, the width of the flow extends up to $\eta^* = 8$, or up to 0.8 times the cylinder radius. Even for such a large boundary layer thickness, there is quantitative agreement between the analytical results for the generalized Carrier-Maslen model and the DSMC simulations even for $\mathcal{A} = 0.707$.

7.3. *Inflow of feed gas along the axis, the applied linear wall temperature profile with internal sources/sinks of mass, momentum, and energy within the flow domain, and with swirling feed (single-component gas)*

Next, we study the secondary flow field due to the combination of inflow of feed gas along the axis, the applied linear wall temperature profile along the rotating sidewall, with internal sources/sinks of mass, momentum, and energy within the flow domain, with swirling feed, as shown in figure 8 (a), and the characteristic streamlines and velocity

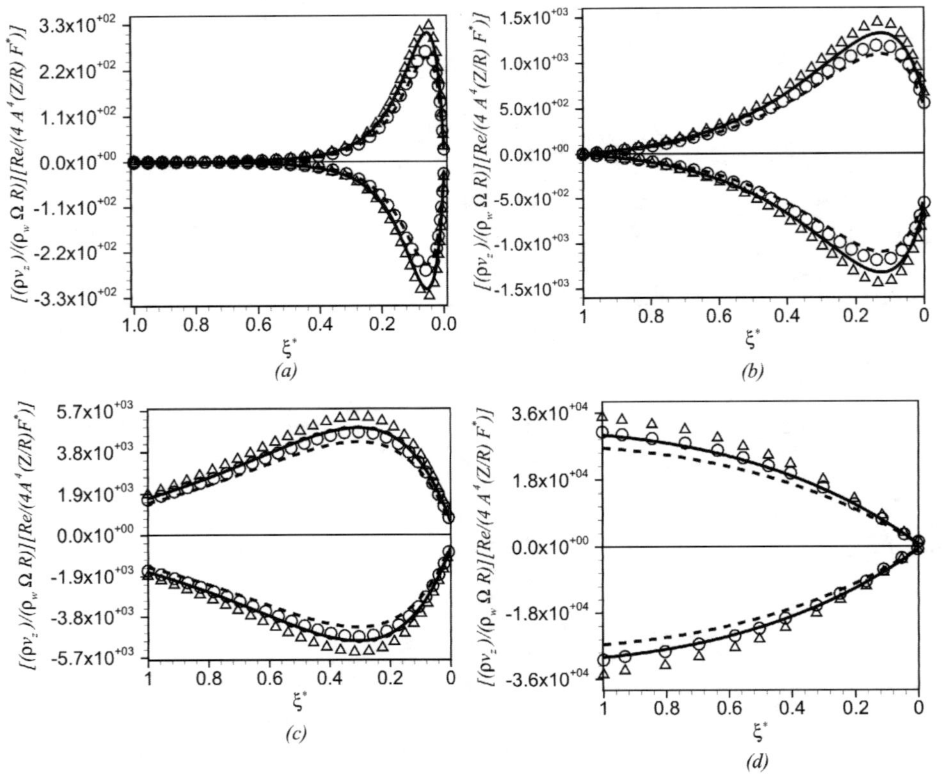

FIGURE 4. Axial mass flux, scaled by $(\rho_w \Omega R)$, $(4\mathcal{A}^4(Z/R)F^*/\mathrm{Re})$, as a function of (ξ^*) at $\eta = 0.325$ and 0.675, for aspect ratio $(Z/D) = 2$, $\mathrm{Re} = 10^2$, $F^* = 0.2$, and stratification parameter $\mathcal{A} = 3.535$ (a); $\mathcal{A} = 2.121$ (b); $\mathcal{A} = 1.414$ (c); and $\mathcal{A} = 0.707$ (d). The symbols show the DSMC results for : ($F_\theta^* = 0$ (○), $F_\theta^* = 0.2$ (△)). The lines show the generalized Onsager model with slip for : ($F_\theta^* = 0$ (−−); $F_\theta^* = 0.2$ (—)). Here, $F^* = F/(H\Omega)$, and $F_\theta^* = F_\theta/(\theta_{mi}\Omega)$.

vectors in figure 8 (b, c). In this case, a linear temperature profile is applied along the rotating sidewall. This results in a generation of secondary axial flow along the wall. When the gas reaches the top end cap, there is an outward radial flow towards the rotating sidewall along the top end cap, and a simultaneous downward flow along the rotating wall of the cylinder. The circulation is completed by an inward radial flow towards the axis along the bottom end cap. Simulations were carried out for stratification parameters $\mathcal{A}$ in the range $0.707 - 3.535$, and Reynolds numbers in the range $10^2 - 10^6$. Three different aspect ratios, $Z/D = 2$, 4 and 8, are considered in the analysis (where Z is the height and D is the diameter of the cylinder). Only the comparisons for $Z/D = 2$ are shown here. This because the smallest aspect ratio provides the most stringent test. For comparison, the scaled wall temperature gradient is defined by

$$F_T = \frac{Z}{T_0}\frac{\partial T}{\partial z}, \tag{7.3}$$

where T_0 is the average temperature. The value of F_T provides the magnitude of the perturbation to the base flow due to the wall temperature gradient. The value of F_T used in the present analysis is 0.2.

In the simulations, the temperature difference applied along the rotating sidewall is

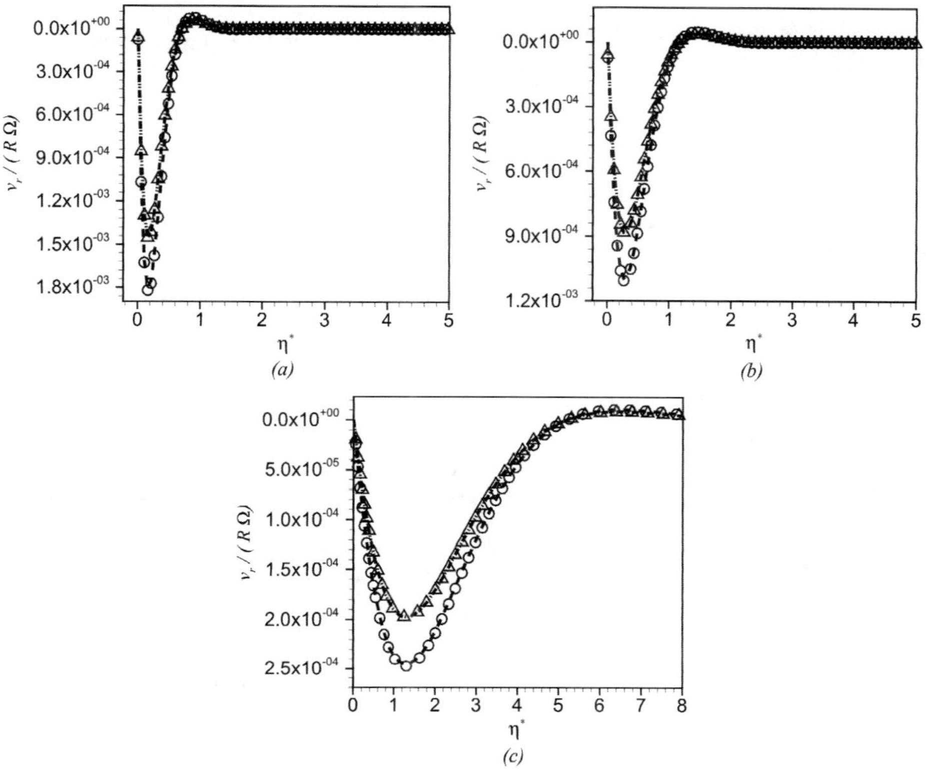

FIGURE 5. Radial velocity, scaled by $(R\Omega)$, as a function of scaled axial coordinate (η^*), at $\xi^* = 0.1$ (a); $\xi^* = 0.2$ (b); and $\xi^* = 0.5$ (c), for $\mathcal{A} = 3.535$, aspect ratio $(Z/D) = 2$, Re $= 100$, $F^* = 0.2$. The symbols show the DSMC results for : ($F^*_\theta = 0$ ($\triangle$), $F^*_\theta = 0.2$ ($\circ$)); and the lines show the results for the generalized Carrier-Maslen model with no slip for : ($F^*_\theta = 0$ ($\cdot\cdot - \cdot\cdot$), $F^*_\theta = 0.2$ ($--$)).

independent of the aspect ratio, so that the temperature gradient is inversely proportional to the aspect ratio. Therefore, the system with the smallest aspect ratio has the highest temperature gradient. Although the imposed temperature along the wall varies linearly with height, our results showed that the gas temperature at the wall is different, by a maximum of 20 %, from the wall temperature. It should be noted that the temperature of the gas is not fixed at the wall in the simulations; rather, the gas temperature is set by the diffuse reflection condition, with the reflected molecules having the same temperature as the wall. Consequently, the average temperature of the gas molecules after collision is equal to the wall temperature, but the average temperature of molecules before collision is not. This results in a significant 'temperature slip' at the wall,. Since the analytical solutions were obtained for a linear variation in the temperature gradient with height, we use the linear approximation for the temperature gradient. The temperature slip has very little variation with Reynolds number Re, though it does increase with stratification parameter $\mathcal{A}$.

In the simulations, it was verified that the temperature slip is a non-continuum effect consistent with the temperature variation in the radial direction, as follows. The average temperature of a molecule reflected from the wall is equal to the wall temperature, T_w. The average temperature of a molecule hitting the wall depends on the temperature at the

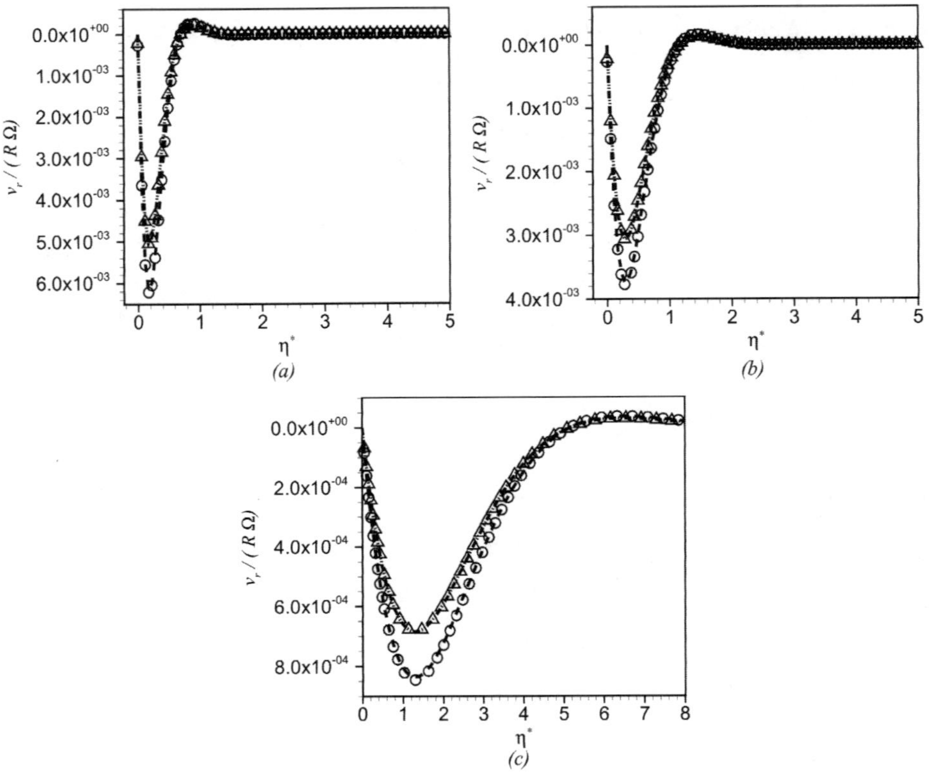

FIGURE 6. Radial velocity, scaled by $(R\Omega)$, as a function of scaled axial coordinate (η^*), at $\xi^* = 0.1$ (a); $\xi^* = 0.2$ (b); and $\xi^* = 0.5$ (c), for $\mathcal{A} = 2.121$, aspect ratio $(Z/D) = 2$, Re $= 100$, $F^* = 0.2$. The symbols show the DSMC results for : ($F^*_\theta = 0$ ($\triangle$), $F^*_\theta = 0.2$ ($\circ$)); and the lines show the results for the generalized Carrier-Maslen model with no slip for : ($F^*_\theta = 0$ $(\cdot\cdot - \cdot\cdot)$, $F^*_\theta = 0.2$ $(--)$).

radial position where the molecule had its most recent collision. In order to calculate this, we determine the probability distribution $P_o(r)$, which is defined such that $P_o(r)dr$ is the probability that the molecule hitting the wall had its most recent collision in the interval dr about the location r. The probability $P_o(r)$ is normalized so that $\int_0^R dr P_o(r) = 1$. The average temperature of the molecules incident on the wall is then

$$T_i = \int_0^R dr P_o(r) T(r). \tag{7.4}$$

For consistency, we require that the gas temperature at the wall, T_{gw}, is the average temperature of the molecules incident and reflected, that is,

$$T_{gw} = (T_i + T_w)/2, \tag{7.5}$$

where T_w is the wall temperature. In the numerical simulations, we have independently verified that T_{gw} calculated from equations 7.4 and 7.5 is equal to that actually measured in the simulations at the wall.

The internal sources of mass, momentum and energy are incorporated in the DSMC simulations as follows. A small circular source domain of radius 10^{-3} times the cylinder radius (R) is selected for adding/removing mass, momentum and energy. In the case of

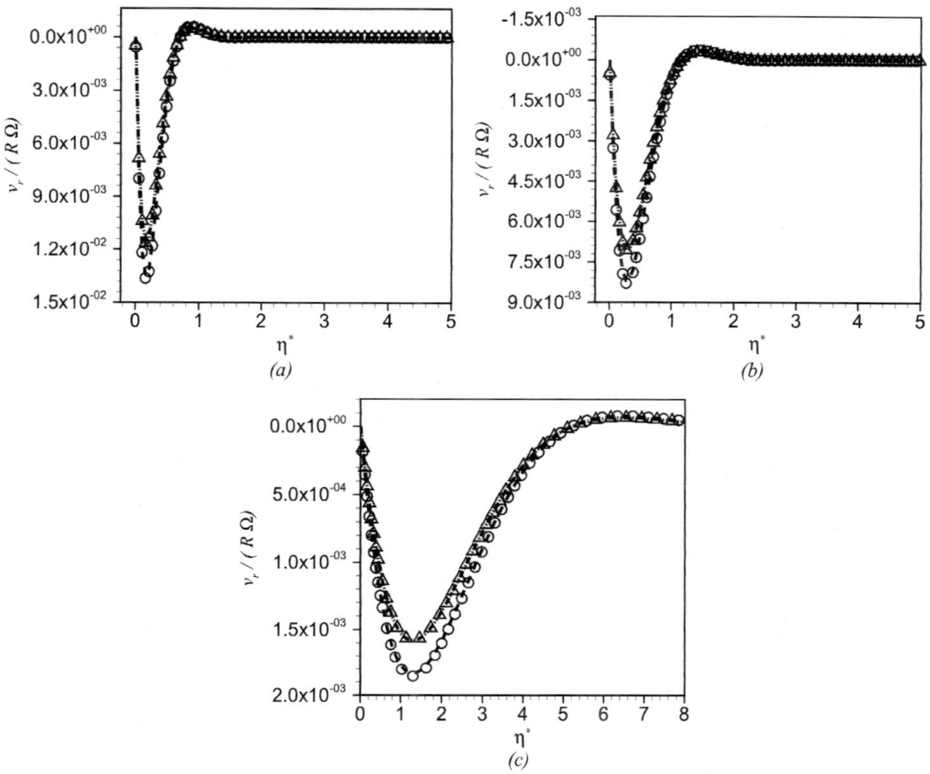

FIGURE 7. Radial velocity, scaled by $(R\Omega)$, as a function of scaled axial coordinate (η^*), at $\xi^* = 0.1$ (a); $\xi^* = 0.2$ (b); and $\xi^* = 0.5$ (c), for $\mathcal{A} = 0.707$, aspect ratio $(Z/D) = 2$, Re $= 100$, $F^* = 0.2$. The symbols show the DSMC results for : ($F^*_\theta = 0$ ($\triangle$), $F^*_\theta = 0.2$ ($\circ$)); and the lines show the results for the generalized Carrier-Maslen model with no slip for : ($F^*_\theta = 0$ $(\cdot\cdot - \cdot\cdot)$, $F^*_\theta = 0.2$ $(--)$).

mass source, a molecule is added at random from the source domain at fixed time intervals determined by the desired rate of mass input. The velocity of the added molecules is chosen from a random Gaussian distribution with the same mean velocity and temperature as that of the gas in the domain. When a molecule is added, the mean velocity and the fluctuating velocity of all other molecules in the domain are scaled by a constant factor so that the total momentum and energy of all the molecules in the source domain remain unchanged due to the addition of a molecule. In the case of a mass sink, molecules are removed at random from the domain at time intervals fixed by the desired rate of mass output. When a molecule is removed, the mean and the fluctuating velocity of all other molecules in the source domain are scaled by a constant factor so that the total momentum and energy in the domain remain unchanged.

Forces (momentum sources) exerted on the gas due to stationary inserts are modelled by imposing impulses (increasing the momentum) of all the molecules in the direction of a force at each collision step. The magnitude of the impulse is determined such that the ratio of the impulse and the time between collisions is equal to the applied force. A localized energy source, which increases or decreases the temperature of the gas, is imposed by scaling the fluctuating velocity of all the molecules in the source domain.

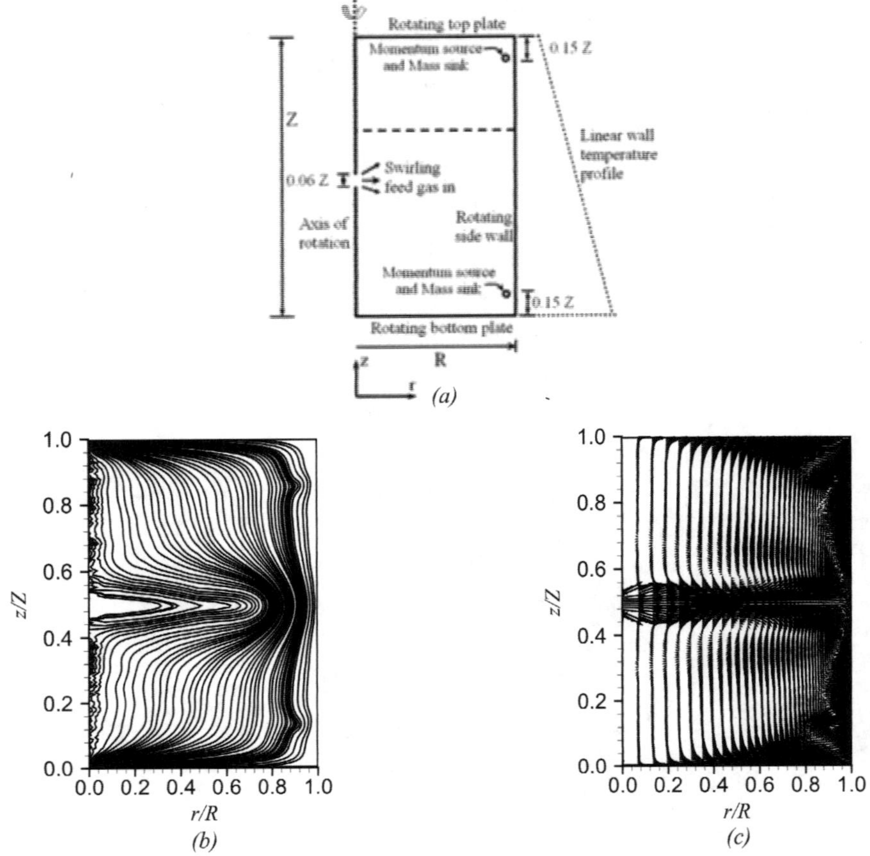

FIGURE 8. Computational domain and coordinate system (a); the characteristic streamline plot (b); and the velocity vectors (c), from DSMC simulations at $\mathcal{A} = 3.535$, Re $= 10^4$, $F^* = 0.2$, $F_T = 0.2$, $F_\theta^* = 0.2$, and aspect ratio $(Z/D) = 4$, for the secondary flow field due to the combination of inflow of feed gas along the axis, the applied linear wall temperature profile along the rotating sidewall, with internal sources/sinks of mass, momentum, and energy within the flow domain, with swirling feed.

The scaling of the velocity is determined such that the increase in energy divided by the time between collisions is equal to the desired energy input rate.

In case of internal mass source/sink, the mass is introduced in the flow domain by adding molecules to the system through the circle of radius $10^{-3}\,R$ which is connected to a prerssure bath so that the desired scaled mass flow rate $F^* = 0.2$. Similarly the mass is removed by removing molecules from the circle which is also connected to a prerssure bath to get the same desired scaled mass flow rate $F^* = 0.2$.

In our calculation, we place the internal sources/sinks at two radial locations, one outside the radial boundary layer at $(\xi_s^* = 0.25)$, and the second inside the radial boundary layer at $(\xi_s^* = 0.01)$, where $\xi_s^* = \xi_s/\mathcal{A}^2$.

We have also examined, in detail, the velocity slip at the rotating sidewall in the DSMC simulations due to the combination of inflow of feed gas along the axis, the applied linear wall temperature profile along the rotating sidewall, with internal sources/sinks of mass, momentum, and energy within the flow domain, with swirling feed. The variation of scaled

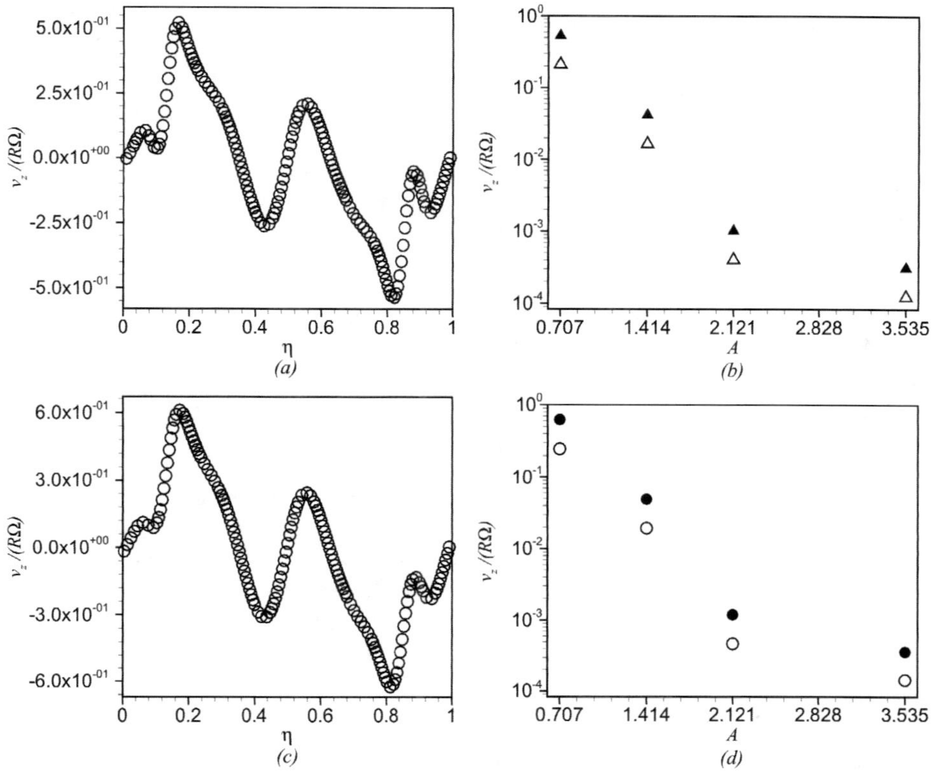

FIGURE 9. Axial slip velocity at the rotating sidewall, scaled by $(R\Omega)$, as a function of η, for $\mathcal{A} = 0.707$, Re $= 10^2$, aspect ratio $Z/D = 2$, $F^* = 0.2$, $F_T = 0.2$, $\xi_s^* = 0.01$, and $F_\theta^* = 0$ (a); $F_\theta^* = 0.2$ (c). The scaled slip velocity $v_z/(R\Omega)$ at the rotating sidewall as a function of stratification parameter $\mathcal{A}$, at $\eta = 0.81$ (maximum slip), for Re $= 10^2$, aspect ratio $Z/D = 2$, $F^* = 0.2$, $F_T = 0.2$, and $F_\theta^* = 0$ (b); $F_\theta^* = 0.2$ (d). The sub-figures (a, b) for without swirling feed $F_\theta^* = 0$; and sub-figures (c, d) with swirling feed $F_\theta^* = 0.2$. The symbols show the DSMC results for sub-figures (b), without swirling feed $F_\theta^* = 0$: ($\xi_s^* = 0.25$ ($\triangle$), $\xi_s^* = 0.01$ ($\blacktriangle$)); and the symbols show the DSMC results for sub-figures (d), with swirling feed $F_\theta^* = 0.2$: ($\xi_s^* = 0.25$ ($\circ$), $\xi_s^* = 0.01$ ($\bullet$)).

velocity slip $v_z/(R\Omega)$, is shown in figure 9 (a, b). We find that the scaled slip velocity increases as the Reynolds number is increased and as the stratification parameter is decreased, and is at a maximum for Re $= 10^6$ and stratification parameter $\mathcal{A} = 0.707$, shown in figure 9 (b). The maximum slip velocity is close to the location of the sources. The variation of the maximum slip velocity with Reynolds number and stratification parameter is shown in figure 9 (b), for two radial locations, $\xi_s^* = 0.01$ inside the radial boundary layer and $\xi_s^* = 0.25$ outside the radial boundary layer. As expected, the slip velocity is larger when the source is placed inside the radial boundary layer at $\xi_s^* = 0.01$, owing to the larger number density and the higher frequency of molecular collisions. There is very little variation in the scaled slip velocity with Reynolds number, but there is a systematic decrease in the scaled slip velocity as the stratification parameter is increased. In all cases, the largest scaled slip velocity is 0.625 times $OmegaR$ when the scaled secondary flow rate is 0.2 times that of the primary rotation rate.

The influx of momentum/energy near the wall is imposed by increasing the mean/fluctuating velocity of the molecules at the location. When the location is near the wall, the molecules with augmented momentum/energy from the source collide with the wall and locally increase the temperature. This results in a non-zero temperature perturabtion at the wall, as shown in figure 10 (a, c), for $\mathcal{A} = 3.535$, Re $= 10^2$, aspect ratio $Z/D = 2$, $F^* = 0.2$, $F_T = 0.2$, for swirling feed and without swirling feed. Here, the dimensionless temperature slip T_s^* is defined by

$$T_s^* = \frac{T - T_a}{T}, \tag{7.6}$$

where T is the gas temperature at the wall obtained from DSMC simulation, and T_a is the constant gas temperature at the wall assumed in the analytical solution. The dimensionless temperature slip T_s^* increases as the Reynolds number is decreased, and temperature slip T_s^* increases as the stratification parameter is increased, as shown in figure 10 (b, d), for without and with swirling feed, and the temperature slip T_s^*, as shown in figure 10 (a, c), for swirling feed and without swirling feed, is the the largest for the Reynolds number Re $= 10^2$ and stratification parameter $\mathcal{A} = 3.535$, the ranges of parameter studied in our simulations.

The temperature slip violates the homogeneous (zero temperature perturbation) boundary condition used in the analytical solution, and so results in a difference in the magnitudes of the secondary flow predicted. When constant temperature boundary conditions are not enforced in the simulations, not shown here, we find that the difference between analytical and simulation results is more than one and half times that when the constant temperature boundary conditions are enforced.

In order to mimic the homogeneous boundary condition for the gas temperature assumed in the analysis, it is necessary to remove energy from the wall. This is done by reducing the wall temperature as explained in the simulation section, and the gas temperature at the wall obtained in this manner is shown by the dashed lines in figure 10 (a, c). Once the homogeneous temperature boundary conditions and velocity slip are satisfied, it is observed that there is quantitative agreement between the analytical and simulation results as shown in figure 11.

It should be noted that energy removal is necessary because the energy dissipation term (double dot product of the viscous stress and rate of deformation tensors) is second-order in the velocity perturbation. This is not included in the linearization approximation that we have used, as can be seen from the energy equation 3.6. Therefore, the viscous heating is not captured by the generalized Onsager model, and it is necessary to explicitly remove energy to maintain a constant temperature at the wall.

The analytical solutions are compared with the results of DSMC simulations, where we have taken the precautions to ensure that the homogeneous boundary conditions are satisfied, and incorporating the velocity slip at the wall. The analytical solutions are obtained by considering (a) homogeneous boundary condition with internal source, and (b) inhomogeneous velocity slip boundary condition with no internal source. The final solution is obtained by superposition of (a) and (b).

The axial mass flux, scaled by $((\rho_w \Omega R)$, $(4\mathcal{A}^4(Z/R)(F_T, F^*)/\text{Re}))$, is plotted as a function of the normalized radial scale height ξ^* for different stratification parameters and for Reynolds number Re $= 10^2$, $F^* = 0.2$, $F_T = 0.2$, for the isothermal sidewall (with energy removal from wall) as shown in figure 11, when the internal sources/sinks are placed within the radial boundary layer at $\xi_s^* = 0.01$, and outside the radial boundary layer at $\xi_s^* = 0.25$, for swirling feed ($F_\theta^* = 0.2$), and without swirling feed ($F_\theta^* = 0$). The flux is scaled in such a manner that the analytical solution is independent of the Reynolds number and aspect ratio, and depends only on the stratification parameter. We have shown the results for the case where energy is removed at the wall so that the

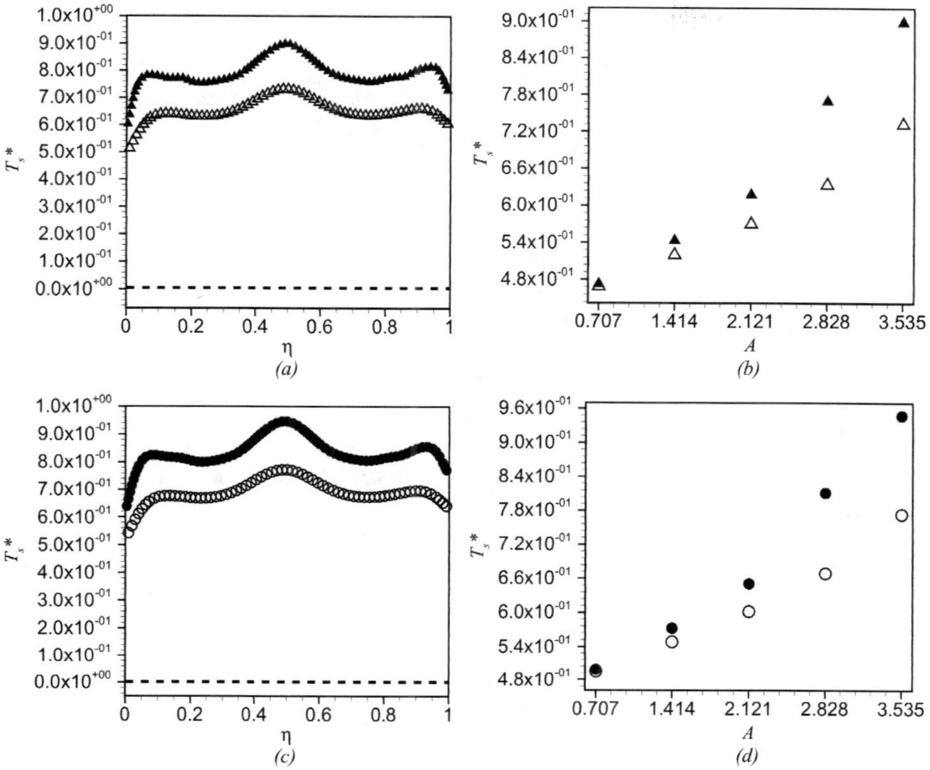

FIGURE 10. The scaled temperature slip (T_s^*) at the rotating sidewall, as a function of η, for $\mathcal{A}$ = 3.535, Re = 10^2, aspect ratio Z/D = 2, F^* = 0.2, F_T = 0.2, and F_θ^* = 0 (a); F_θ^* = 0.2 (c). The scaled temperature slip (T_s^*) at the rotating sidewall as a function of stratification parameter $\mathcal{A}$, for Re = 10^2, aspect ratio Z/D = 2, F^* = 0.2, F_T = 0.2, and F_θ^* = 0 (b); F_θ^* = 0.2 (d). The sub-figures (a, b) for without swirling feed F_θ^* = 0; and sub-figures (c, d) with swirling feed F_θ^* = 0.2. The symbols show the DSMC results for sub-figures (a) : (ξ_s^* = 0.25 ($\triangle$), ξ_s^* = 0.01 ($\blacktriangle$)); The symbols show the DSMC results for sub-figures (c) : (ξ_s^* = 0.25 ($\circ$), ξ_s^* = 0.01 ($\bullet$)). The symbols show the DSMC results for sub-figures (b), without swirling feed F_θ^* = 0 : (ξ_s^* = 0.25 ($\triangle$), ξ_s^* = 0.01 ($\blacktriangle$)); and the symbols show the DSMC results for sub-figures (d), with swirling feed F_θ^* = 0.2 : (ξ_s^* = 0.25 ($\circ$), ξ_s^* = 0.01 ($\bullet$)). In sub-figures (a, c) the line (−−) shows the result with energy removal.

homogeneous temperature boundary condition is maintained. The results are shown at the axial locaiton η = 0.675. We find that the resultant scaled axial mass flux is higher with swirling feed (F_θ^* = 0.2), in comparison to without swirling feed (F_θ^* = 0), owing to the reduction of the angular momentum loss of the rotating gas due to feed injection near the feed point, and further reduction of the axial spreading of the feed gas. It is also observed that the magnitude of the axial mass flux is larger when the sources are placed within the radial boundary layer at ξ_s^* = 0.01, in comparison to the case where the sources are outside the radial boundary layer at ξ_s^* = 0.25, owing to the larger number density and the higher frequency of molecular collisions.

There is quantitative agreement (with a difference of less than 15 %) between the predictions of the analytical model and the simulations even when the stratification parameter is 0.707, the Reynolds number is 100 and the cylinder length is twice the

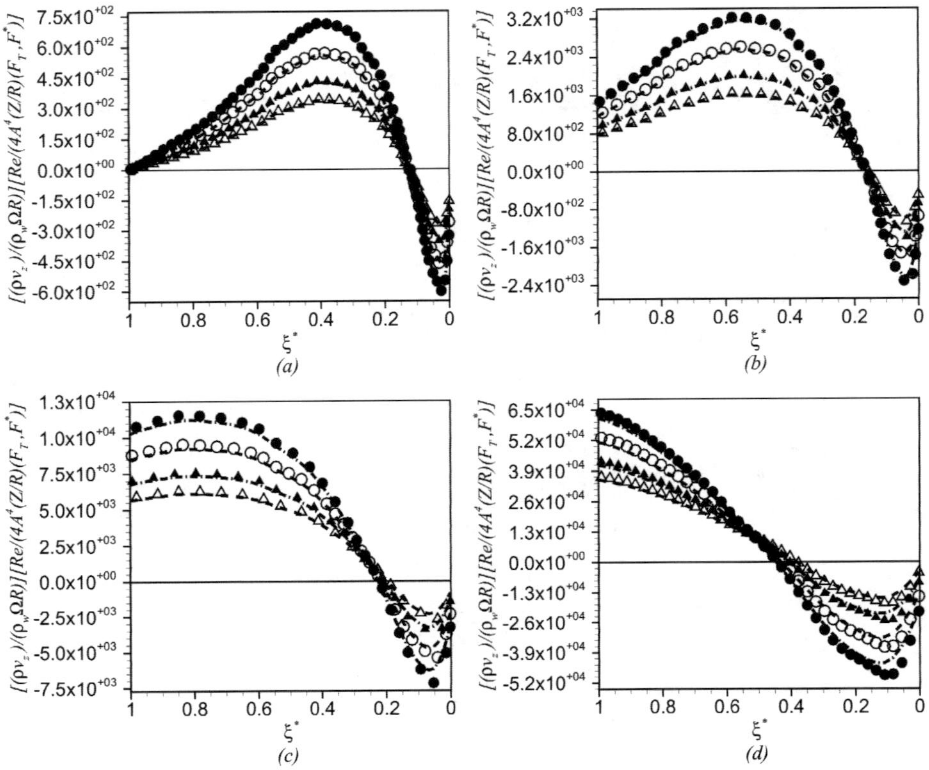

FIGURE 11. Axial mass flux, scaled by $(\rho_w \Omega R)(4\mathcal{A}^4(Z/R)(F_T, F^*)/\mathrm{Re})$, as a function of (ξ^*) at $\eta = 0.675$ for aspect ratio $(Z/D) = 2$, $F^* = 0.2$, $F_T = 0.2$, $\mathrm{Re} = 10^2$, and stratification parameter $\mathcal{A} = 3.535$ (a); $\mathcal{A} = 2.121$ (b); $\mathcal{A} = 1.414$ (c); and $\mathcal{A} = 0.707$ (d). The symbols show the DSMC results without swirling feed $F_\theta^* = 0$ for : ($\xi_s^* = 0.25$ (△); $\xi_s^* = 0.01$ (○)), and with swirling feed $F_\theta^* = 0.2$ for : ($\xi_s^* = 0.25$ (▲); $\xi_s^* = 0.01$ (●)). The lines show the generalized Onsager model with slip, and without swirling feed $F_\theta^* = 0$ (−−); and generalized Onsager model with slip, and with swirling feed $F_\theta^* = 0.2$ (· − ·).

diameter. We find better agreement for higher aspect ratios, higher Reynolds numbers and stratification parameters, though the results are not shown here.

7.4. *Inflow of feed gas along the axis, the applied linear wall temperature profile with internal sources/sinks of mass, momentum, and energy within the flow domain, with rotating insert, and with swirling feed (single-component gas)*

Next, we study the secondary flow generated by a combination of inflow of feed gas along the axis, the applied linear wall temperature profile along the rotating sidewall, with internal sources/sinks of mass, momentum, and energy within the flow domain, with rotating insert within the flow domain, and with swirling feed, as shown in figure 12 (a), and the characteristic streamlines and velocity vectors in figure 12 (b, c).

Substantial temperature and velocity slip at the rotating sidewall is obtained in the simulations in this case as well, as shown in figure 13, and figure 14 respectively, for swirling feed ($F_\theta^* = 0.2$), and without swirling feed ($F_\theta^* = 0$). We find that the temperature slip has very little variation with Reynolds number Re, though it does increase with stratification parameter $\mathcal{A}$. We also find that the scaled slip velocity $v_z/(R\Omega)$ increases as the stratification parameter is decreased, and is at a maximum for stratification parame-

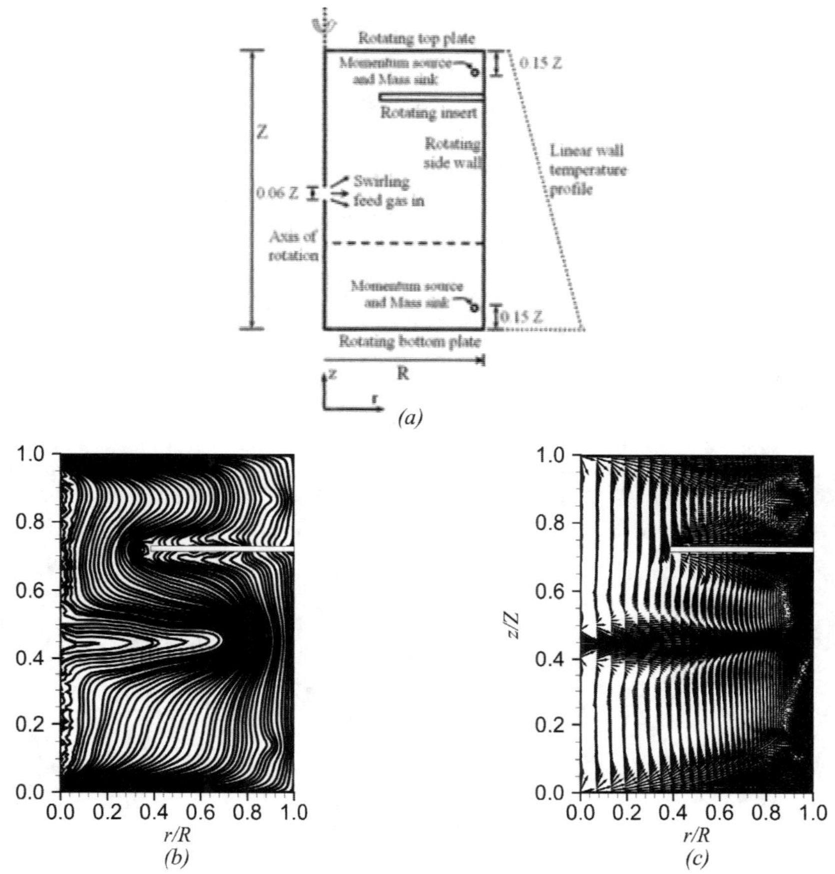

FIGURE 12. Computational domain and coordinate system (a); the characteristic streamline plot (b); and the velocity vectors (c), from DSMC simulations at $\mathcal{A} = 3.535$, Re $= 10^4$, $F^* = 0.2$, $F_T = 0.2$, $F^*_\theta = 0.2$, and aspect ratio $(Z/D) = 4$, for the secondary flow field due to the combination of inflow of feed gas along the axis, the applied linear wall temperature profile along the rotating sidewall, with internal sources/sinks of mass, momentum, and energy within the flow domain, with rotating insert within the flow domain, with swirling feed.

ter $\mathcal{A} = 0.707$, as shown in figure 14. The maximum slip velocity is obtained close to the location of the sources. The variation of the maximum slip velocity with stratification parameter for Re $= 10^2$ is shown in figure 14.

The axial mass flux, scaled by $((\rho_w \Omega R), (4\mathcal{A}^4(Z/R)(F_T, F^*)/\text{Re}))$, is plotted as a function of the normalized radial scale height ξ^* for different stratification parameters and for Reynolds number Re $= 10^2$, $F^* = 0.2$, $F_T = 0.2$, for the isothermal sidewall (with energy removal from wall) as shown in figure 15, when the internal sources/sinks are placed within the radial boundary layer at $\xi^*_s = 0.01$, and outside the radial boundary layer at $\xi^*_s = 0.25$, and for swirling feed ($F^*_\theta = 0.2$), and without swirling feed ($F^*_\theta = 0$). The flux is scaled in such a manner that the analytical solution is independent of the Reynolds number and aspect ratio, and depends only on the stratification parameter. We have shown the results for the case where energy is removed at the wall so that the homogeneous temperature boundary condition is maintained. The results are shown at the axial locaiton $\eta = 0.325$. We find that the resultant scaled axial mass flux increased

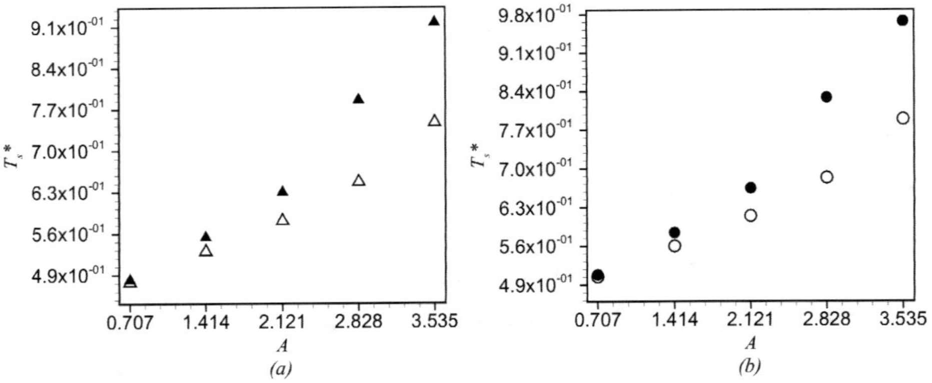

FIGURE 13. The scaled temperature slip T_s^* at the rotating sidewall, as a function of stratification parameter $\mathcal{A}$, for Re $= 10^2$, aspect ratio $Z/D = 2$, $F^* = 0.2$, $F_T = 0.2$. The sub-figure (a) for without swirling feed $F_\theta^* = 0$; and sub-figures (b) with swirling feed $F_\theta^* = 0.2$. The symbols show the DSMC results for sub-figures (a) : ($\xi_s^* = 0.25$ (△), $\xi_s^* = 0.01$ (▲)); and the symbols show the DSMC results for sub-figures (b) : ($\xi_s^* = 0.25$ (○), $\xi_s^* = 0.01$ (●)).

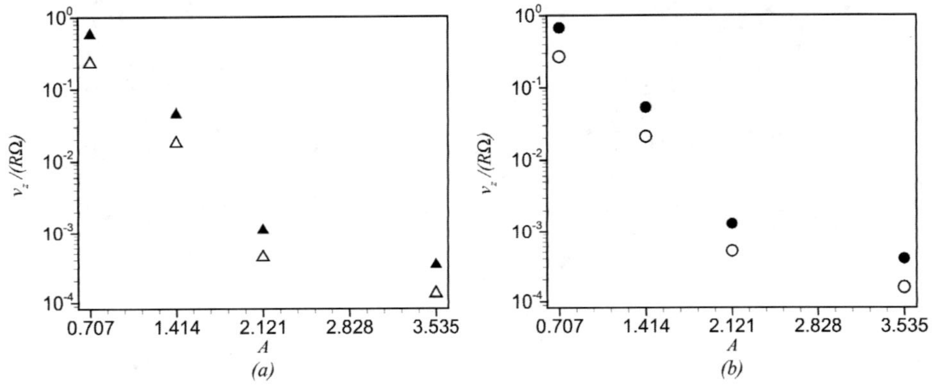

FIGURE 14. The scaled slip velocity $v_z/(R\Omega)$ at the rotating sidewall, as a function of stratification parameter $\mathcal{A}$, for Re $= 10^2$, aspect ratio $Z/D = 2$, $F^* = 0.2$, $F_T = 0.2$. The sub-figures (a) for without swirling feed $F_\theta^* = 0$; and sub-figures (b) for with swirling feed $F_\theta^* = 0.2$. The symbols show the DSMC results for sub-figures (a) : ($\xi_s^* = 0.25$ (△), $\xi_s^* = 0.01$ (▲)); and the symbols show the DSMC results for sub-figures (b) : ($\xi_s^* = 0.25$ (○), $\xi_s^* = 0.01$ (●)).

by $(16 - 27)\%$, with swirling feed ($F_\theta^* = 0.2$), in comparison to without swirling feed ($F_\theta^* = 0$), owing to the reduction of the angular momentum loss of the rotating gas due to feed injection near the feed point, and further reduction of the axial spreading of the feed gas. It is also observed that the magnitude of the axial mass flux is larger when the sources are placed within the radial boundary layer at $\xi_s^* = 0.01$, in comparison to the case where the sources are outside the radial boundary layer at $\xi_s^* = 0.25$, owing to the larger number density and the higher frequency of molecular collisions. It is further observed that the magnitude of the axial mass flux is larger in presence of a rotating insert in this case (figure 12 (a)) compared to the previous case (figure 8 (a)). This is because of the shielding of the internal sources/sinks of mass, momentum, and energy, placed near the top end cap at the axial location $\eta = 0.85$ by the rotating insert, that helps to seize the secondary axial flow generated in the opposite direction by the internal sources/sinks of mass, momentum, and energy at $\eta = 0.85$, with respect to the secondary

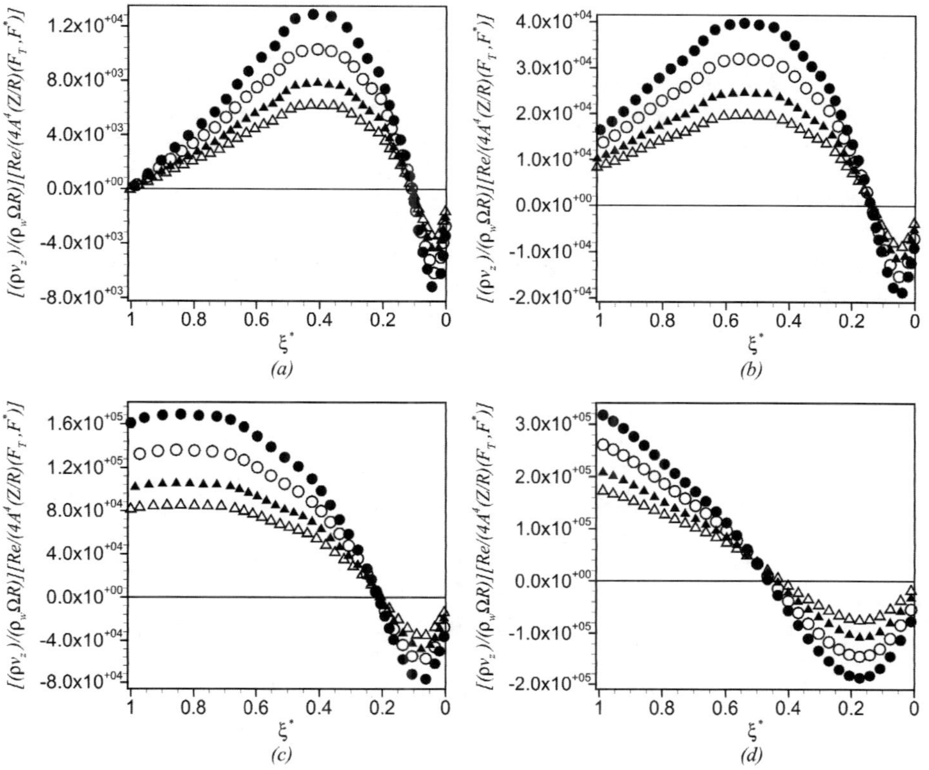

FIGURE 15. Axial mass flux, scaled by $(\rho_w \, \Omega R)(4\mathcal{A}^4(Z/R)(F_T, F^*)/\mathrm{Re})$, as a function of (ξ^*) at $\eta = 0.325$ for aspect ratio $(Z/D) = 2$, $F^* = 0.2$, $F_T = 0.2$, $\mathrm{Re} = 10^2$, and stratification parameter $\mathcal{A} = 3.535$ (a); $\mathcal{A} = 2.121$ (b); $\mathcal{A} = 1.414$ (c); and $\mathcal{A} = 0.707$ (d). The symbols show the DSMC results without swirling feed $F_\theta^* = 0$ for : ($\xi_s^* = 0.25$ ($\triangle$); $\xi_s^* = 0.01$ ($\circ$), and with swirling feed $F_\theta^* = 0.2$ for : ($\xi_s^* = 0.25$ ($\blacktriangle$); $\xi_s^* = 0.01$ ($\bullet$). Here, $F^* = F/(H\Omega)$, and $F_\theta^* = F_\theta/(\theta_{mi}\Omega)$.

axial flow generated by the internal sources/sinks of mass, momentum, and energy at $\eta = 0.15$, and in this way an effective axial mass flux is produced, which could be very important for the centrifugal gas separation process.

8. Binary gas mixture

Next, we study the secondary flow field for a binary gas mixture incorporating the angular momentum in the feed gas, which we refer as 'swirling feed'. In the base state, the flow is considered to be a solid body rotation with constant temperature, but the variation in the mole fraction is taken into account. Due to this, there is a variation in the viscosity and thermal conductivity in the radial direction. In addition, the total pressure is written as the sum of partial pressures of the two species, and is not approximated as an exponential function of the distance from the wall. The dilute gas equation of state is used for the pressure, while the Chapman-Enskog expressions (Chapman & Cowling (1970), Kumaran & Pradhan (2014)) for hard sphere gases are used for the thermal conductivity and viscosity.

Following our previous study (Kumaran & Pradhan (2014)), in the analysis of the secondary flow for a binary gas mixture, the equation for the master potential is expanded in an asymptotic series in the parameter $\epsilon = (\Delta m/m_{av})$, where Δm is the difference in the molecular masses of the two species, and m_{av} is a suitably defined average molecular mass. In the leading approximation the terms of $O(\epsilon)$ or smaller are neglected. The leading order equation and boundary conditions for the master potential are nearly identical to the sixth order differential equation of the generalized Onsager model (Pradhan & Kumaran (2011)); the important difference being that the density is not assumed to decrease exponentially from the wall, but the actual species densities are used. The mean velocity and temperature obtained by solving the leading order equations are used in the species balance equations for the mole fractions of the two species, and the corrections for the number density and mass density of the two species due to the secondary flow are obtained. These appear as inhomogeneous terms in the equation for the $O(\epsilon)$ correction for the master potential; this equation is solved with appropriate boundary conditions to obtain the $O(\epsilon)$ correction to the master potential and flow velocity. This is then used in the $O(\epsilon)$ contribution to the species balance equation, and the hierarchal procedure is repeated until the $O(\epsilon^2)$ correction to the secondary flow velocity and mole fraction are calculated. A similar hierarchial procedure is used for the Carrier-Maslen model for the end-cap boundary layers. The results of the Onsager hierarchy, up to $O(\epsilon^2)$, are compared with the results of DSMC simulations for the axisymmetric flow of a two-species ideal gas with swirling feed, by incorporating the angular momentum in the feed gas. The results of the $O(\epsilon^2)$ Onsager hierarchy for the secondary flow are compared with DSMC simulations for three types of forcing, (1) inflow/outflow of gas along the axis, as shown in figure 1 (a), (2) combination of inflow of feed gas along the axis, the applied linear wall temperature profile along the rotating sidewall, with internal sources/sinks of mass, momentum, and energy within the flow domain, with swirling feed, as shown in figure 8 (a), and (3) combination of inflow of feed gas along the axis, the applied linear wall temperature profile along the rotating sidewall, with internal sources/sinks of mass, momentum, and energy within the flow domain, with rotating insert, and with swirling feed, as shown in figure 12 (a).

The initial gas mixture contains 50% mole-fraction of each component with $\epsilon_a = (2\Delta m/(m_1+m_2)) = 0.5$. The fluid is an ideal gas mixture in which both species have equal molecular diameter $d = 5.5 \times 10^{-10}$ m, average molecular mass $m_{av} = (m_1 + m_2)/2 = 58.4426 \times 10^{-26}$ kg/molecule, specific heat ratio $\gamma = 1.066$, and Prandtl number Pr= 0.95. It is assumed that the gas molecules undergo diffuse reflection from all the solid surfaces with the thermal accommodation coefficient equal to 1.

The stratification parameter $\mathcal{A}$ is varied in the range $0.707 - 3.535$, and Reynolds number in the range $10^2 - 10^6$ with aspect ratio $Z/D = 2, 4, 8$. There is better agreement between simulations and theory as the aspect ratio increases, and so results are reported only for the smallest aspect ratio $(Z/D) = 2$. In addition, there is better agreement between simulations and theory as the Reynolds number increases, and so results are reported only for the lowest Reynolds number, Re $= 10^2$. In the present analysis we have considered the value of $\epsilon_a = (2\Delta m/(m_1+m_2)) = 0.5$. The flow geometry and coordinate system for different types of forcing for the secondary flow are shown in figure 1 (a), figure 8 (a), figure 12 (a).

8.1. *Inflow/outflow of gas along the axis with swirling feed (binary gas mixture)*

The flow geometry and coordinate system are shown in figure 1 (a) for the secondary flow generated due to inflow/outflow of gas along the axis. Substantial axial velocity slip at the rotating sidewall is obtained in the simulations in this case as well. The axial velocity at the rotating sidewall, non-dimensionalized by $(R\Omega)$, and scaled by F^*, is

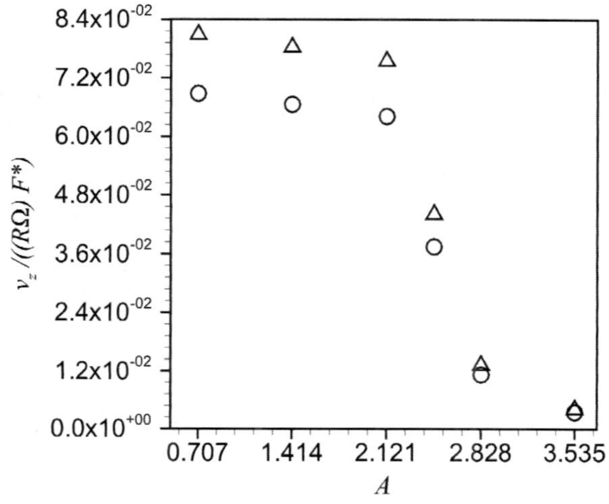

FIGURE 16. Axial slip velocity at the rotating sidewall, scaled by $((R\Omega)F^*)$, as a function of $\mathcal{A}$, for inflow/outflow of gas along the axis, at $\eta = 0.675$, for aspect ratio $Z/D = 2$, $\mathrm{Re} = 10^2$, and $F^* = 0.2$, for a binary gas mixture with $\epsilon_a = 0.5$. The symbols show the DSMC results for : $(F_\theta^* = 0$ ($\circ$), $F_\theta^* = 0.2$ ($\triangle$)). Here, $F^* = F/(H\Omega)$, and $F_\theta^* = F_\theta/(\theta_{mi}\Omega)$.

plotted as a function of stratification parameter $\mathcal{A}$ at the axial location $(\eta) = 0.675$ for inflow/outflow of gas along the axis, as shown in figure 16. It is observed that the ratio of the slip velocity and the maximum secondary flow velocity $(v_z/(R\Omega F^*))$, actually decreases as the stratification parameter is increased. The slip velocity for a binary gas mixture with $\epsilon_a = 0.5$ is observed to be larger than that for a single-component gas, as can be seen from figures 2, and 16, owing to the higher frequency of molecular collisions between dissimilar molecules.

The master potential profiles for inflow/outflow of gas along the axis as a function of ξ^* at $\eta = 0.325$ and 0.675, are shown in figures 17, for $\mathcal{A} = 3.535$, 2.121, 1.414, 0.707; for $\mathrm{Re} = 10^2$, aspect ratio $Z/D = 2$; $F^* = 0.2$, and $F_\theta^* = 0.2$, for a binary gas mixture with $\epsilon_a = 0.5$.

The analytical solutions are compared with the results of DSMC simulations for a binary gas mixture with $\epsilon_a = 0.5$, and with swirling feed as shown in figure 18. The axial mass flux, scaled by $(\rho_w \Omega R)$, $(4\mathcal{A}^4(Z/R)F^*/\mathrm{Re})$, so that the analytical solution is independent of Reynolds number and aspect ratio but depends only on the stratification parameter, is plotted as a function of normalized radial scale height ξ^* for stratification parameter in the range $0.707 - 3.535$, and at Reynolds number $\mathrm{Re} = 10^2$ for isothermal sidewall (with energy removal from wall) as shown in figure 18. It should be noted that energy removal is necessary because the energy dissipation term (double dot product of the viscous stress and rate of deformation tensors) is second-order in the velocity perturbation. This is not included in the linearization approximation that we have used, as can be seen from the energy equation 3.6. Therefore, the viscous heating is not captured by the generalized Onsager model, and it is necessary to explicitly remove energy to maintain a constant temperature at the wall.

We find that the $O(\epsilon)$ correction to the master potential of the generalized Onsager model is comparable to the leading order solution for the lowest Reynolds number $\mathrm{Re} = 10^2$, and stratification parameter $\mathcal{A} = 0.707$, for $\epsilon_a = 0.5$. However, the second correction is significantly smaller than the first correction, implying convergence of the series. The $O(\epsilon)$ and $O(\epsilon^2)$ solutions for the axial mass flux do consistently converge towards to the

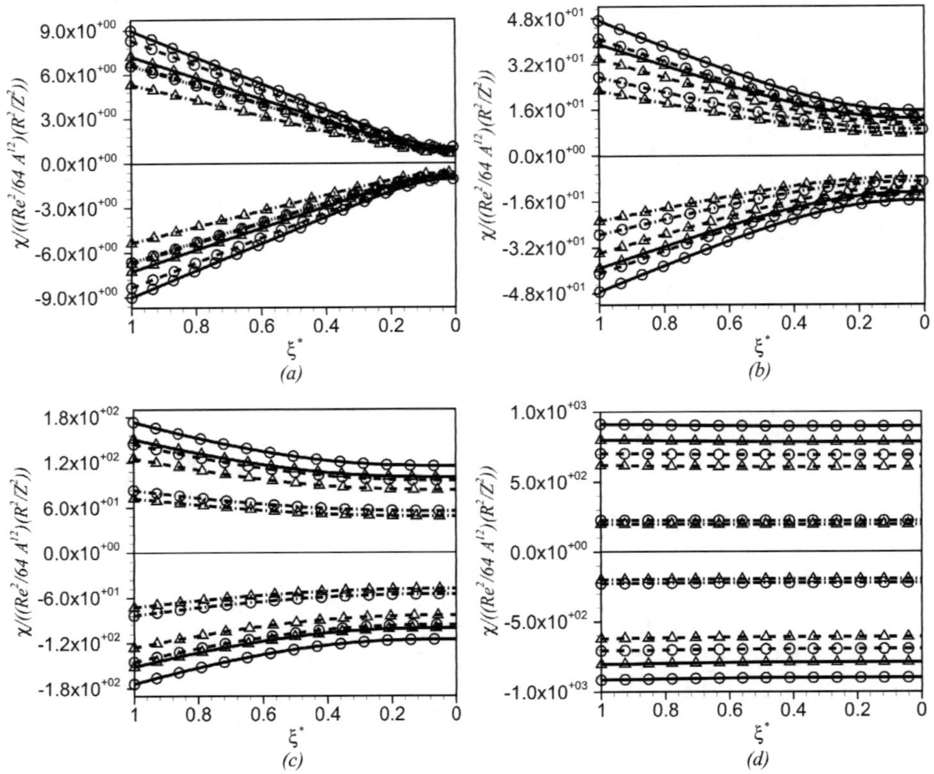

FIGURE 17. Master potential, scaled by $((\mathrm{Re}^2/64\mathcal{A}^{12})(R^2/Z^2))$, as a function of (ξ^*) at $\eta =$ 0.325 and 0.675, aspect ratio $Z/D = 2$, $F_* = 0.2$, for $\mathcal{A} = 3.535$ (a); $\mathcal{A} = 2.121$ (b); $\mathcal{A} = 1.414$ (c); and $\mathcal{A} = 0.707$ (d), for the secondary flow due to inflow/outflow of gas along the axis, for a binary gas mixture with $\epsilon_a = 0.5$. The lines show the analytical solutions with wall slip correct to : $(O(1) : (\cdot - \cdot)$; $O(\epsilon) : (--)$; $O(\epsilon^2) : (—))$, and the symbols show the results for : $(F_\theta^* = 0$ $(\triangle)$; $F_\theta^* = 0.2$ $(\circ))$. Here, $F^* = F/(H\Omega)$, and $F_\theta^* = F_\theta/(\theta_{mi}\Omega)$.

DSMC results. There is a difference of $10 - 15\%$ between the $O(\epsilon^2)$ solution and the DSMC result for the lowest values of the stratification parameter $\mathcal{A} = 0.707$ considered here, but the difference is negligible for higher values of the stratification parameter $\mathcal{A} = 3.535$, provided the slip velocity and homogeneous temperature boundary conditions are correctly incorporated in the analytical solution. It is observed that the magnitude of the axial mass flux is larger with swirling feed ($F_\theta^* = 0.2$), in comparison to the without swirling feed ($F_\theta^* = 0$), owing to the associated angular momentum of the feed gas which reduces the angular momentum loss of the rotating gas due to feed injection near the feed point, and further reduces the axial spreading of the feed gas.

8.2. *Radial flow at end-caps with swirling feed (binary gas mixture)*

The Carrier-Maslen solutions, incorporating the $O(\epsilon^2)$ correction due to the mass difference, are compared with the results of DSMC simulations with swirling feed for the radial flow at the end-caps for the configuration where there is inflow and outflow along the axis of the cylinder. It is necessary to specify the total mass flow rate at a fixed radial location in order to specify the boundary condition. The mass flux is specified as one half of the rate of inflow along the axis, since the inflow at the axis separates symmetrically

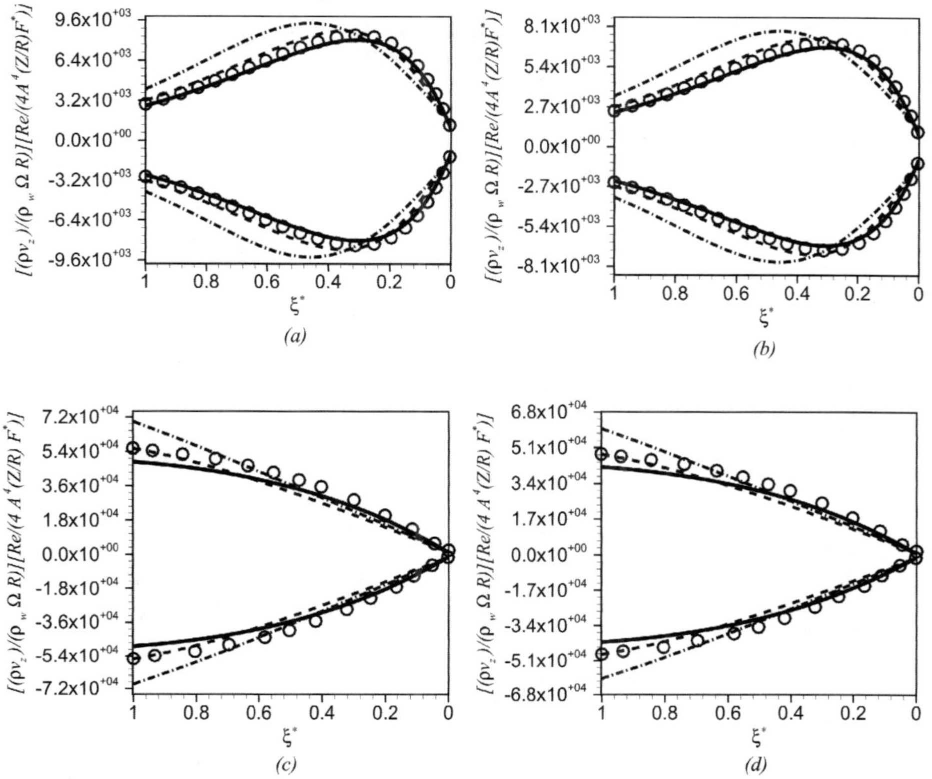

FIGURE 18. Axial mass flux, scaled by $(\rho_w \Omega R)$, $(4\mathcal{A}^4(Z/R)F^*/\text{Re})$, as a function of (ξ^*) at $\eta =$ 0.325 and 0.675, for aspect ratio $(Z/D) = 2$, Reynolds number $\text{Re} = 10^2$, $F^* = 0.2$, for a binary gas mixture with $\epsilon_a = 0.5$, and stratification parameter $\mathcal{A} = 1.414$ (a, b); and $\mathcal{A} = 0.707$ (c, d). The sub-figures (a, c) show the results with swirling feed $F^*_\theta = 0.2$, and the sub-figures (b, d) show the results without swirling feed $F^*_\theta = 0$. The symbol ($\circ$) show the DSMC results, and the lines show the analytical solutions correct to $O(1)$: $(\cdot - \cdot)$, $O(\epsilon)$: $(--)$, and $O(\epsilon^2)$: (—).

at the cylindrical wall and one half of the inflow goes upwards/downwards and inwards along the end-caps. The wall slip for the radial flow at the end-caps is found to be much smaller than that for the axial flow at the rotating sidewall.

The radial velocity at the end-caps, non-dimensionalised by $(R\Omega)$, is plotted as a function of scaled axial coordinate η^* at $\xi^* = 0.1$, 0.2 and 0.5 in figures 19, and 20. Progressing inwards from the wall of the cylinder, it is observed that there is an expansion of the boundary layer by a factor of 5 when the distance from the cylinder wall is increased from $\xi^* = 0.1$ to $\xi^* = 0.5$, while the mass flux remains a constant. The prediction for the radial velocity correct to $O(\epsilon^2)$ is in quantitative agreement with the DSMC solutions. The leading order result (neglecting the mass difference between the two species) does capture the trends, but the magnitude is lower than the DSMC results.

An interesting feature is that the relative difference between the solutions correct to $O(1)$ and $O(\epsilon^2)$ for the Carrier-Maslen solutions is much smaller than that for the generalized Onsager model. There are two reasons for this. Firstly, the $O(\epsilon)$ corrections to the Carrier-Maslen equations could be incorporated for $\psi^{(0)}$, since these were known functions of the number and mass density in the base flow. Secondly, the inhomogeneous term for $\psi^{(2)}$, contains a factor of $1/\text{Re}$, which is small in the high-Reynolds-number limit. For

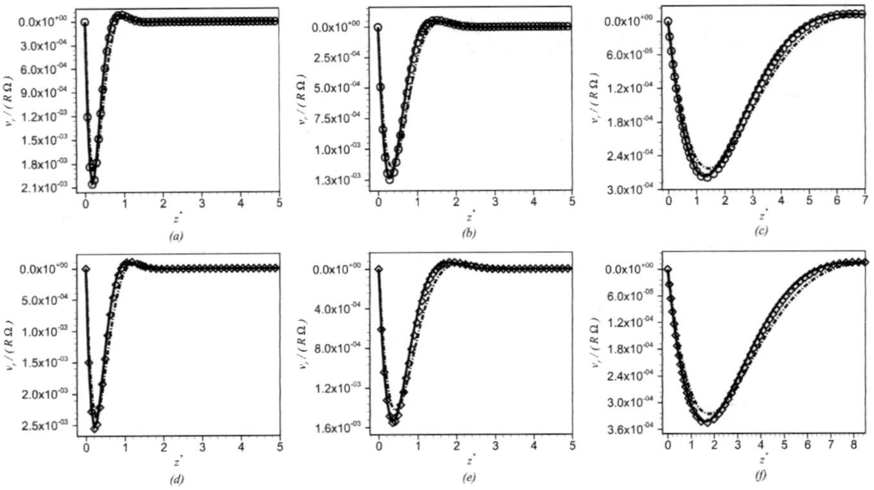

FIGURE 19. Radial velocity, scaled by $(R\Omega)$, as a function of scaled axial coordinate (z^*), at $(\xi^*) = 0.1$ (a, d); $(\xi^*) = 0.2$ (b, e); and $(\xi^*) = 0.5$ (c, f), for $\mathcal{A} = 3.535$, aspect ratio (Z/D) = 2, for a binary gas mixture with $\epsilon_a = 0.5$, Re = 10^2, $F^* = 0.2$. The sub-figures (a, b, c) show the results without swirling feed $F^*_\theta = 0$; and the sub-figures (d, e, f) show the results with swirling feed with $F^*_\theta = 0.2$. The symbols show the DSMC results for : ($F^*_\theta = 0$ ($\circ$); $F^*_\theta = 0.2$ ($\Diamond$)). The lines show the analytical solutions correct to $O(1)$: $(\cdot - \cdot)$, and $O(\epsilon^2)$: (—). Here, $z^* = z/\delta z$, where δz is the axial boundary layer thickness = $(R\,\mathrm{Re}^{-1/2})$.

these two reasons, the leading approximation for the generalized Carrier-Maslen equations gives a significantly more accurate solution in comparison to that for the generalized Onsager equation.

It is observed that the magnitude of the radial velocity at the end-caps is larger with swirling feed ($F^*_\theta = 0.2$), in comparison to the without swirling feed ($F^*_\theta = 0$), owing to the associated angular momentum of the feed gas, which reduces the angular momentum loss of the rotating gas due to feed injection near the feed point, and further reducces the axial spreading of the feed gas.

8.3. *Inflow of feed gas along the axis, the applied linear wall temperature profile with internal sources/sinks of mass, momentum, and energy within the flow domain, and with swirling feed (binary gas mixture)*

Next, we study the secondary flow field due to the combination of inflow of feed gas along the axis, the applied linear wall temperature profile along the rotating sidewall, with internal sources/sinks of mass, momentum, and energy within the flow domain, with swirling feed, for a binary gas mixture with $\epsilon_a = 0.5$. The flow geometry and coordinate system are shown in figure 8 (a). Substantial temperature slip at the rotating sidewall is obtained in the simulations in this case as well, as shown in figure 21. The temperature slip for a binary gas mixture with $\epsilon_a = 0.5$ is found to be larger than that for a single-component gas. We find that the temperature slip has very little variation with Reynolds number Re, though it does increase with stratification parameter $\mathcal{A}$.

Substantial velocity slip at the rotating sidewall is obtained in the simulations in this case as well as shown in figure 22. The axial velocity at the rotating sidewall, non-dimensionalized by $(R\Omega)$, is plotted as a function of stratification parameter $\mathcal{A}$ at the axial location $(\eta) = 0.675$ as shown in figure 22. It is observed that the slip velocity increases as the Reynolds number and (F_T, F^*) are increased, and the ratio of the slip velocity

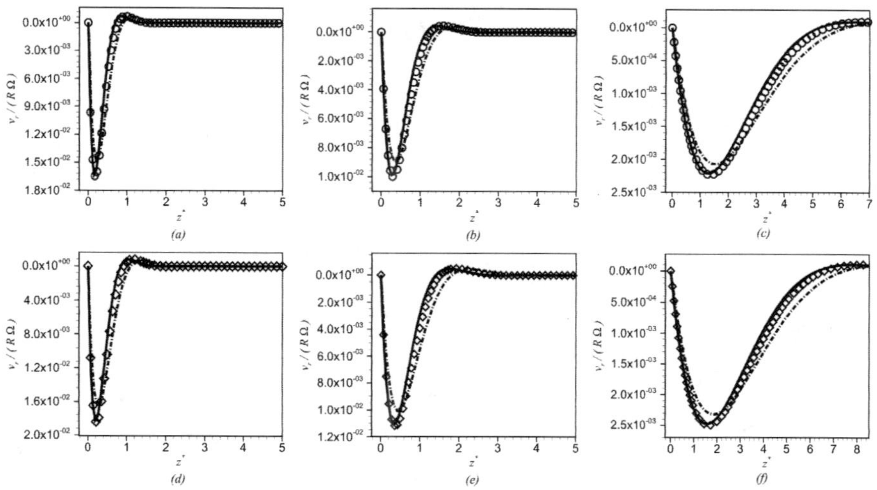

FIGURE 20. Radial velocity, scaled by $(R\Omega)$, as a function of scaled axial coordinate (z^*), at $(\xi^*) = 0.1$ (a, d); $(\xi^*) = 0.2$ (b, e); and $(\xi^*) = 0.5$ (c, f), for $\mathcal{A} = 0.707$, aspect ratio (Z/D) = 2, for a binary gas mixture with $\epsilon_a = 0.5$, $\mathrm{Re} = 10^2$, $F^* = 0.2$. The sub-figures (a, b, c) show the results without swirling feed $F^*_\theta = 0$; and the sub-figures (d, e, f) show the results with swirling feed with $F^*_\theta = 0.2$. The symbols show the DSMC results for : ($F^*_\theta = 0$ (○); $F^*_\theta = 0.2$ (◇)). The lines show the analytical solutions correct to $O(1)$: (· − ·), and $O(\epsilon^2)$: (—). Here, $z^* = z/\delta z$, where δz is the axial boundary layer thickness $= (R\,\mathrm{Re}^{-1/2})$.

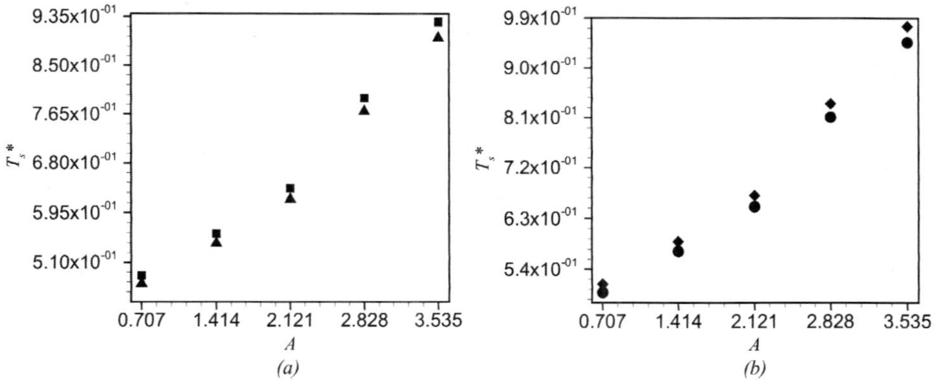

FIGURE 21. The scaled temperature slip T^*_s at the rotating sidewall, as a function of stratification parameter $\mathcal{A}$, for $\mathrm{Re} = 10^2$, aspect ratio $Z/D = 2$, $F^* = 0.2$, $F_T = 0.2$. The sub-figure (a) for without swirling feed $F^*_\theta = 0$; and sub-figures (b) with swirling feed $F^*_\theta = 0.2$. The symbols show the DSMC results for sub-figures (a) for $\xi^*_s = 0.01$: ($\epsilon_a = 0$ (▲), $\epsilon_a = 0.5$ (■)). The symbols show the DSMC results for sub-figures (b) for $\xi^*_s = 0.01$: ($\epsilon_a = 0$ (•), $\epsilon_a = 0.5$ (♦)).

and the peripheral velocity $v_z/(R\Omega)$, actually decreases as the stratification parameter is increased. The slip velocity for a binary gas mixture with $\epsilon_a = 0.5$ is observed to be larger than that for a single-component gas, as shown in figure 22.

The axial mass flux, scaled by $((\rho_w \Omega R), (4\mathcal{A}^4(Z/R)(F_T, F^*)/\mathrm{Re}))$, is plotted as a function of the normalized radial scale height ξ^* for stratification parameter $\mathcal{A}$ in the range 0.707 − 3.535, Reynolds number $\mathrm{Re} = 10^2$, aspect ratio $(Z/D) = 2$, $F^* = 0.2$, $F_T = 0.2$, for the isothermal sidewall (with energy removal from wall) as shown in figures

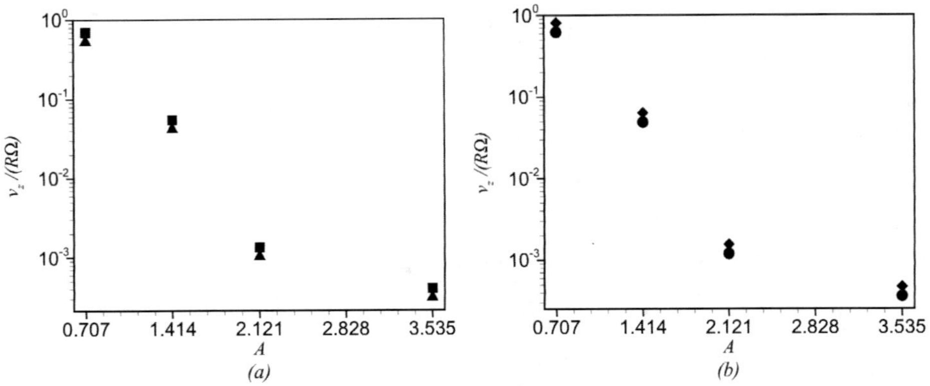

FIGURE 22. The scaled slip velocity $v_z/(R\Omega)$ at the rotating sidewall, as a function of stratification parameter $\mathcal{A}$, for Re $= 10^2$, aspect ratio $Z/D = 2$, $F^* = 0.2$, $F_T = 0.2$. The sub-figure (a) for without swirling feed $F_\theta^* = 0$; and sub-figures (b) with swirling feed $F_\theta^* = 0.2$. The symbols show the DSMC results for sub-figures (a) for $\xi_s^* = 0.01$: ($\epsilon_a = 0$ (▲), $\epsilon_a = 0.5$ (■)). The symbols show the DSMC results for sub-figures (b) for $\xi_s^* = 0.01$: ($\epsilon_a = 0$ (•), $\epsilon_a = 0.5$ (♦)).

23, 24, 25, and 26, when the internal sources/sinks are placed within the radial boundary layer at $\xi_s^* = 0.01$, and outside the radial boundary layer at $\xi_s^* = 0.25$. The flux is scaled in such a manner that the analytical solution is independent of the Reynolds number and aspect ratio, and depends only on the stratification parameter. We have shown the results for the case where energy is removed at the wall so that the homogeneous temperature boundary condition is maintained. The results are shown at the axial locaiton $\eta = 0.675$. In all cases, the solutions correct to $O(\epsilon^2)$ are in much better agreement with DSMC results than the $O(1)$ solutions which do not incorporate the mass difference. There is a maximum difference of about 15% between the analytical and DSMC results for the lowest stratification parameter 0.707, Reynolds number Re $= 10^2$, and aspect ratio Z/D $= 2$, but there is much better agreement between analytical and DSMC results for higher values of the stratification parameter, Reynolds number and aspect ratio.

It is observed that the magnitude of the axial mass flux is larger with swirling feed ($F_\theta^* = 0.2$), in comparison to the without swirling feed ($F_\theta^* = 0$), owing to the associated angular momentum of the feed gas which reduces the angular momentum loss of the rotating gas due to feed injection near the feed point, and further reduces the axial spreading of the feed gas. It is also observed that the magnitude of the axial mass flux is larger when the sources are placed within the radial boundary layer at $\xi_s^* = 0.01$, in comparison to the case where the sources are outside the radial boundary layer at $\xi_s^* = 0.25$, owing to the larger number density and the higher frequency of molecular collisions.

8.4. *Inflow of feed gas along the axis, the applied linear wall temperature profile with internal sources/sinks of mass, momentum, and energy within the flow domain, with rotating insert, and with swirling feed (binary gas mixture)*

Next, we consider the secondary flow generated by a combination of inflow of feed gas along the axis, the applied linear wall temperature profile along the rotating sidewall, with internal sources/sinks of mass, momentum, and energy within the flow domain, with rotating insert within the flow domain, with swirling feed, for a binary gas mixture with $\epsilon_a = 0.5$, as shown in figure 12 (a). Substantial temperature slip at the rotating sidewall is obtained in the simulations in this case as well, as shown in figure 27. The

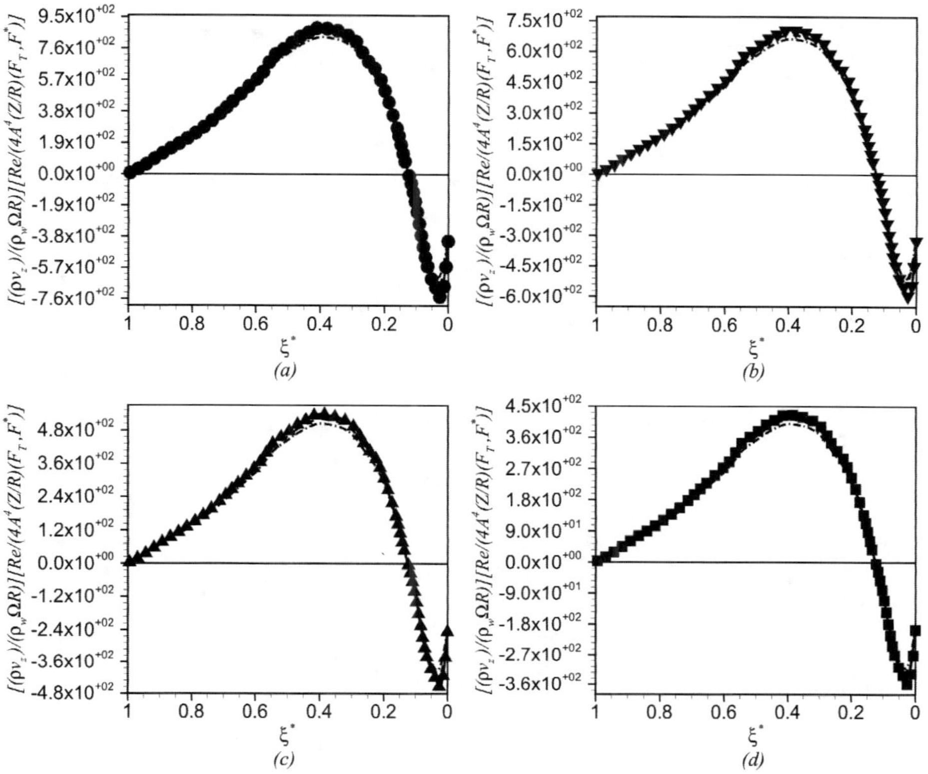

FIGURE 23. Axial mass flux, scaled by $(\rho_w \Omega R)(4\mathcal{A}^4(Z/R)(F_T, F^*)/\mathrm{Re})$, as a function of (ξ^*) at $\eta = 0.675$ for aspect ratio $(Z/D) = 2$, $F^* = 0.2$, $F_T = 0.2$, $\mathrm{Re} = 10^2$, for a binary gas mixture with $\epsilon_a = 0.5$, and stratification parameter $\mathcal{A} = 3.535$. The sub-figures (a, c) for swirling feed $(F_\theta^* = 0.2)$; and sub-figures (b, d) without swirling feed $(F_\theta^* = 0)$. The sub-figures (a, b) when the source is within the radial boundary layer at $\xi_s^* = 0.01$; and sub-figures (c, d) outside the radial boundary layer at $\xi_s^* = 0.25$. The symbols show the DSMC results with swirling feed $F_\theta^* = 0.2$ for : ($\xi_s^* = 0.01$ (●), $\xi_s^* = 0.25$ (▲)); and without swirling feed $F_\theta^* = 0$ for : ($\xi_s^* = 0.01$ (▼), $\xi_s^* = 0.25$ (■)). The lines show the analytical solutions with wall slip for : correct to $O(1)$ (· − ·), $O(\epsilon)$ (−−), and $O(\epsilon^2)$ (—).

temperature slip for a binary gas mixture with $\epsilon_a = 0.5$ is observed to be larger than that for a single-component gas. We find that the temperature slip has very little variation with Reynolds number Re, though it does increase with stratification parameter $\mathcal{A}$.

Substantial velocity slip at the rotating sidewall is obtained in the simulations in this case as well as shown in figure 28. The axial velocity at the rotating sidewall, non-dimensionalized by $(R\Omega)$, is plotted as a function of stratification parameter $\mathcal{A}$ at the axial location $(\eta) = 0.325$ as shown in figure 22. It is observed that the slip velocity increases as the Reynolds number and (F_T, F^*) are increased, and the ratio of the slip velocity and the peripheral velocity $v_z/(R\Omega)$, actually decreases as the stratification parameter is increased. The slip velocity for a binary gas mixture with $\epsilon_a = 0.5$ is observed to be larger than that for a single-component gas, as shown in figure 28.

The axial mass flux, scaled by $((\rho_w \Omega R), (4\mathcal{A}^4(Z/R)(F_T, F^*)/\mathrm{Re}))$, is plotted as a function of the normalized radial scale height ξ^* for stratification parameter $\mathcal{A}$ in the range $0.707 - 3.535$, Reynolds number $\mathrm{Re} = 10^2$, aspect ratio $(Z/D) = 2$, $F^* = 0.2$, $F_T = 0.2$, for the isothermal sidewall (with energy removal from wall), as shown in figure 29,

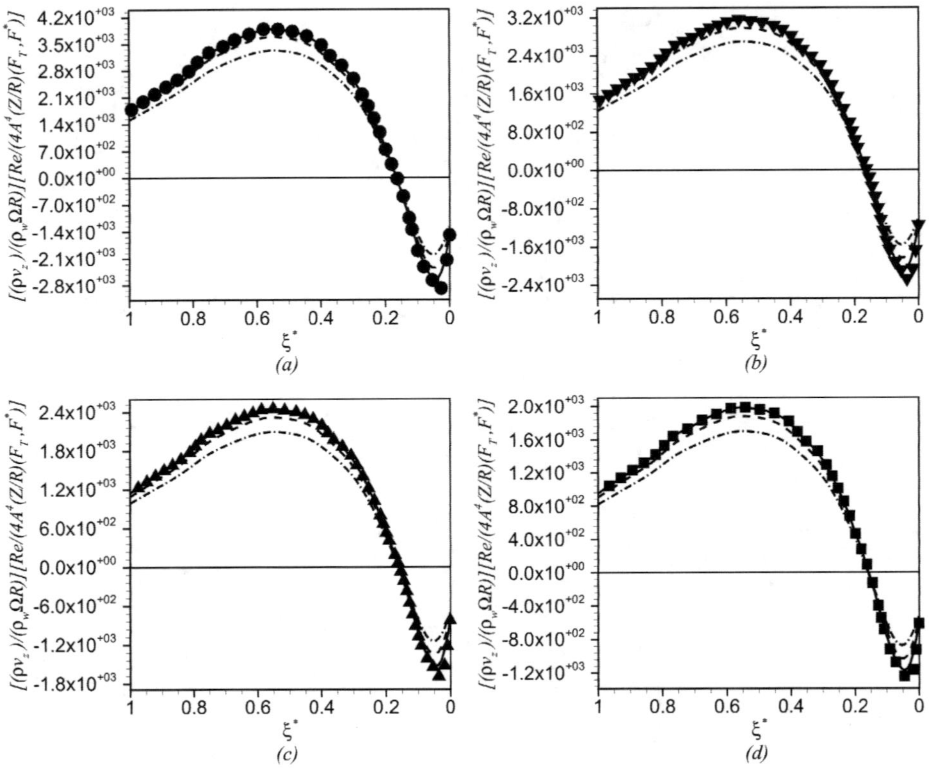

FIGURE 24. Axial mass flux, scaled by $(\rho_w \Omega R)(4\mathcal{A}^4(Z/R)(F_T, F^*)/\mathrm{Re})$, as a function of (ξ^*) at $\eta = 0.675$ for aspect ratio $(Z/D) = 2$, $F^* = 0.2$, $F_T = 0.2$, $\mathrm{Re} = 10^2$, for a binary gas mixture with $\epsilon_a = 0.5$, and stratification parameter $\mathcal{A} = 2.121$. The sub-figures (a, c) for swirling feed $(F_\theta^* = 0.2)$; and sub-figures (b, d) without swirling feed $(F_\theta^* = 0)$. The sub-figures (a, b) when the source is within the radial boundary layer at $\xi_s^* = 0.01$; and sub-figures (c, d) outside the radial boundary layer at $\xi_s^* = 0.25$. The symbols show the DSMC results with swirling feed $F_\theta^* = 0.2$ for : ($\xi_s^* = 0.01$ (•), $\xi_s^* = 0.25$ (▲)); and without swirling feed $F_\theta^* = 0$ for : ($\xi_s^* = 0.01$ (▼), $\xi_s^* = 0.25$ (■)). The lines show the analytical solutions with wall slip for : correct to $O(1)$ (· – ·), $O(\epsilon)$ (– –), and $O(\epsilon^2)$ (—).

when the internal sources/sinks are placed within the radial boundary layer at $\xi_s^* = 0.01$, and outside the radial boundary layer at $\xi_s^* = 0.25$. The flux is scaled in such a manner that the analytical solution is independent of the Reynolds number and aspect ratio, and depends only on the stratification parameter. We have shown the results for the case where energy is removed at the wall so that the homogeneous temperature boundary condition is maintained. The results are shown at the axial locaiton $\eta = 0.325$.

It is observed that the magnitude of the axial mass flux is larger with swirling feed $(F_\theta^* = 0.2)$, in comparison to the without swirling feed $(F_\theta^* = 0)$, owing to the associated angular momentum of the feed gas which reduces the angular momentum loss of the rotating gas due to feed injection near the feed point, and further reduces the axial spreading of the feed gas. It is also observed that the magnitude of the axial mass flux is larger when the sources are placed within the radial boundary layer at $\xi_s^* = 0.01$, in comparison to the case where the sources are outside the radial boundary layer at $\xi_s^* = 0.25$, owing to the larger number density and the higher frequency of molecular collisions.

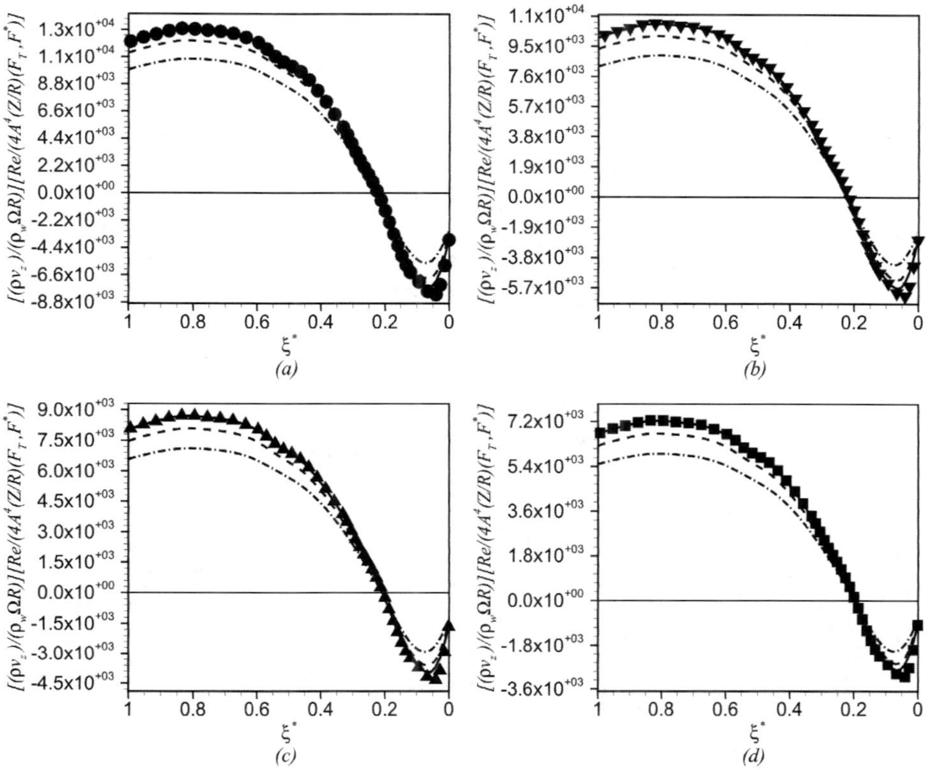

FIGURE 25. Axial mass flux, scaled by $(\rho_w \Omega R)(4\mathcal{A}^4(Z/R)(F_T, F^*)/\text{Re})$, as a function of (ξ^*) at $\eta = 0.675$ for aspect ratio $(Z/D) = 2$, $F^* = 0.2$, $F_T = 0.2$, $\text{Re} = 10^2$, for a binary gas mixture with $\epsilon_a = 0.5$, and stratification parameter $\mathcal{A} = 1.414$. The sub-figures (a, c) for swirling feed $(F_\theta^* = 0.2)$; and sub-figures (b, d) without swirling feed $(F_\theta^* = 0)$. The sub-figures (a, b) when the source is within the radial boundary layer at $\xi_s^* = 0.01$; and sub-figures (c, d) outside the radial boundary layer at $\xi_s^* = 0.25$. The symbols show the DSMC results with swirling feed $F_\theta^* = 0.2$ for : ($\xi_s^* = 0.01$ (●), $\xi_s^* = 0.25$ (▲)); and without swirling feed $F_\theta^* = 0$ for : ($\xi_s^* = 0.01$ (▼), $\xi_s^* = 0.25$ (■)). The lines show the analytical solutions with wall slip for : correct to $O(1)$ $(\cdot - \cdot)$, $O(\epsilon)$ $(--)$, and $O(\epsilon^2)$ (—).

8.5. *Axial spreading of the feed gas*

Next, we study the axial spreading of the feed gas due to feed injection as shown in figure 30 (a, b). It is found that there is significant reduction $(4 - 24)\%$ of the axial spreading of the feed gas near the feed point with swirling feed $(F_\theta^* = 0.2)$, in comparison to the without swirling feed $(F_\theta^* = 0)$, owing to the associated angular momentum of the feed gas, and thereby increases the efficiency of the feed drive for the centrifugal gas separation process.

The axial spreading of the feed gas for stratification parameter $\mathcal{A} = 3.535$, 2.121, 1.414, 0.707; at $\text{Re} = 10^2$, aspect ratio $(Z/D) = 4$, $F^* = 0.2$, $F_T = 0.2$, and $\epsilon_a = 0.5$, for swirling feed $(F_\theta^* = 0.2)$ and without swirling feed $(F_\theta^* = 0)$ are given in table 2.

8.6. *Circumferential velocity loss of the rotating gas due to feed injection*

We further study the 'circumferential velocity loss' of the rotating gas due to feed injection near the feed point for swirling feed $(F_\theta^* = 0.2)$, and without swirling feed $(F_\theta^* = 0)$ as

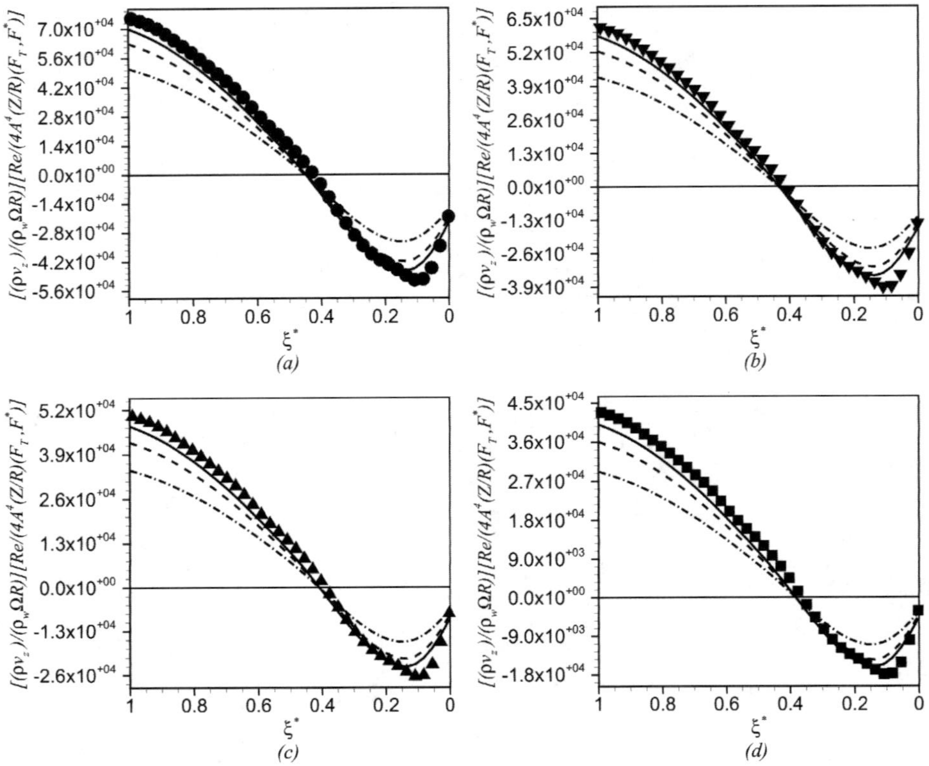

FIGURE 26. Axial mass flux, scaled by $(\rho_w \Omega R)(4\mathcal{A}^4(Z/R)(F_T, F^*)/\mathrm{Re})$, as a function of (ξ^*) at $\eta = 0.675$ for aspect ratio $(Z/D) = 2$, $F^* = 0.2$, $F_T = 0.2$, $\mathrm{Re} = 10^2$, for a binary gas mixture with $\epsilon_a = 0.5$, and stratification parameter $\mathcal{A} = 0.707$. The sub-figures (a, c) for swirling feed $(F_\theta^* = 0.2)$; and sub-figures (b, d) without swirling feed $(F_\theta^* = 0)$. The sub-figures (a, b) when the source is within the radial boundary layer at $\xi_s^* = 0.01$; and sub-figures (c, d) outside the radial boundary layer at $\xi_s^* = 0.25$. The symbols show the DSMC results with swirling feed $F_\theta^* = 0.2$ for : ($\xi_s^* = 0.01$ (•), $\xi_s^* = 0.25$ (▲)); and without swirling feed $F_\theta^* = 0$ for : ($\xi_s^* = 0.01$ (▼), $\xi_s^* = 0.25$ (■)). The lines show the analytical solutions with wall slip for : correct to $O(1)$ $(\cdot - \cdot)$, $O(\epsilon)$ (−−), and $O(\epsilon^2)$ (—).

Stratification parameter	$\eta_{(axial-spreading),sf}$ (%)	$\eta_{(axial-spreading),nsf}$ (%)
$\mathcal{A} = 0.707$	3	7
$\mathcal{A} = 1.414$	4	10
$\mathcal{A} = 2.121$	6	16
$\mathcal{A} = 3.535$	9	33

TABLE 2. The axial spreading of the feed gas for stratification parameter $\mathcal{A} = 3.535$, 2.121, 1.414, 0.707; at $\mathrm{Re} = 10^2$, aspect ratio $(Z/D) = 4$, $F^* = 0.2$, $F_T = 0.2$, and $\epsilon_a = 0.5$, for swirling feed $(F_\theta^* = 0.2)$ and without swirling feed $(F_\theta^* = 0)$.

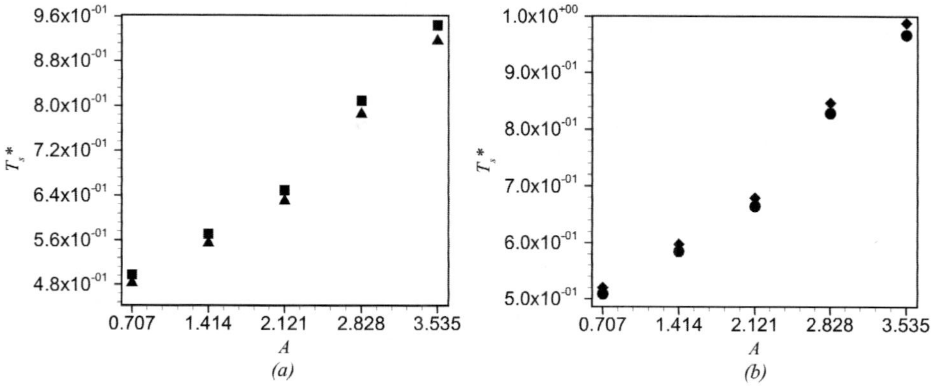

FIGURE 27. The scaled temperature slip T_s^* at the rotating sidewall, as a function of stratification parameter $\mathcal{A}$, for Re $= 10^2$, aspect ratio $Z/D = 2$, $F^* = 0.2$, $F_T = 0.2$. The sub-figure (a) for without swirling feed $F_\theta^* = 0$; and sub-figures (b) with swirling feed $F_\theta^* = 0.2$. The symbols show the DSMC results for sub-figures (a) for $\xi_s^* = 0.01$: ($\epsilon_a = 0$ (▲), $\epsilon_a = 0.5$ (■)). The symbols show the DSMC results for sub-figures (b) for $\xi_s^* = 0.01$: ($\epsilon_a = 0$ (•), $\epsilon_a = 0.5$ (♦)).

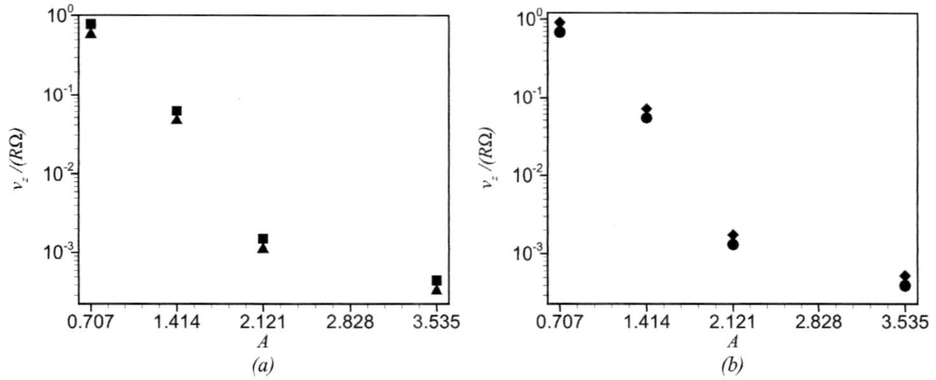

FIGURE 28. The scaled slip velocity $v_z/(R\Omega)$ at the rotating sidewall, as a function of stratification parameter $\mathcal{A}$, for Re $= 10^2$, aspect ratio $Z/D = 2$, $F^* = 0.2$, $F_T = 0.2$. The sub-figure (a) for without swirling feed $F_\theta^* = 0$; and sub-figures (b) with swirling feed $F_\theta^* = 0.2$. The symbols show the DSMC results for sub-figures (a) for $\xi_s^* = 0.01$: ($\epsilon_a = 0$ (▲), $\epsilon_a = 0.5$ (■)). The symbols show the DSMC results for sub-figures (b) for $\xi_s^* = 0.01$: ($\epsilon_a = 0$ (•), $\epsilon_a = 0.5$ (♦)).

shown in figure 31. The reduction of the 'circumferential velocity loss' of the rotating gas due to swirling feed injection can be calculated as

$$v_{(\theta-loss),\,reduction} = \frac{1}{RZ}\int_0^Z \left(\int_0^R \left(v_{\theta,sf} - v_{\theta,nsf}\right) dr \right) dz, \qquad (8.1)$$

where, $v_{\theta,sf}$, and $v_{\theta,nsf}$, are the circumferential velocities of the rotating gas with swirling feed and without swirling feed respectively. Using the scaled variables, $r^* = r/R$, $\eta = z/Z$, and $v_\theta^* = v_\theta/(R\Omega)$, we obtain

$$v^*_{(\theta-loss),\,reduction} = \frac{v_{(\theta-loss),\,reduction}}{\Omega R} = \int_0^1 \left(\int_0^1 \left(v^*_{\theta,sf} - v^*_{\theta,nsf}\right) dr^* \right) d\eta. \qquad (8.2)$$

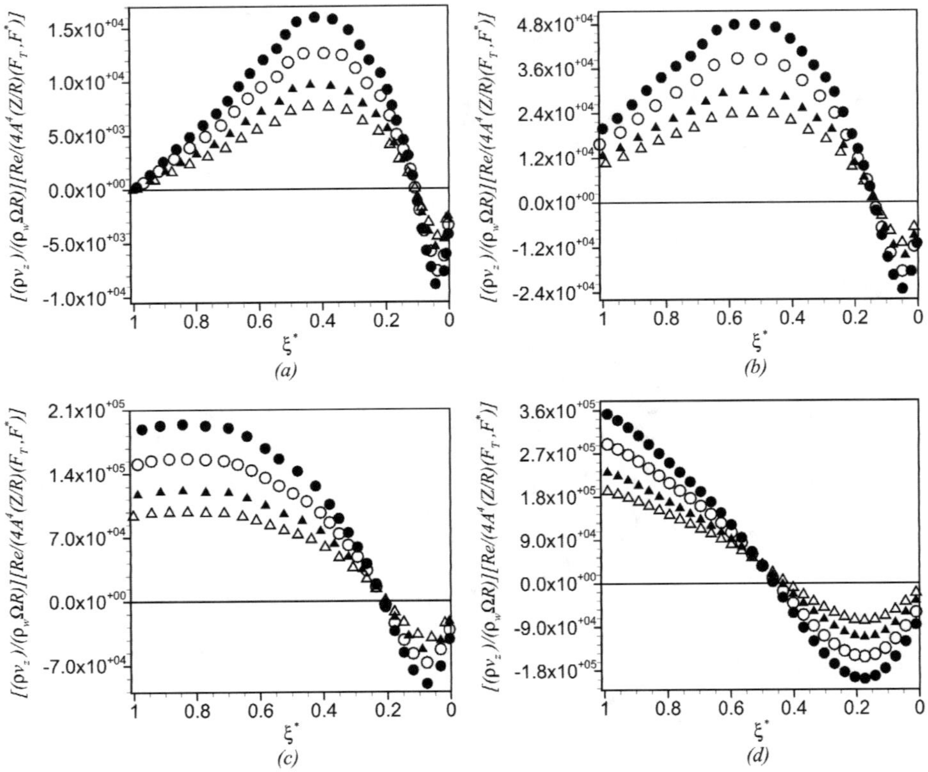

FIGURE 29. Axial mass flux, scaled by $(\rho_w\,\Omega R)(4\mathcal{A}^4(Z/R)(F_T, F^*)/\mathrm{Re})$, as a function of (ξ^*) at $\eta = 0.325$ for aspect ratio $(Z/D) = 2$, $F^* = 0.2$, $F_T = 0.2$, $\mathrm{Re} = 10^2$, for a binary gas mixture with $\epsilon_a = 0.5$, and stratification parameter $\mathcal{A} = 3.535$ (a); $\mathcal{A} = 2.121$ (b); $\mathcal{A} = 1.414$ (c); and $\mathcal{A} = 0.707$ (d). The symbols show the DSMC results without swirling feed $F^*_\theta = 0$ for : ($\xi^*_s = 0.25$ (△); $\xi^*_s = 0.01$ (○), and with swirling feed $F^*_\theta = 0.2$ for : ($\xi^*_s = 0.25$ (▲); $\xi^*_s = 0.01$ (●).

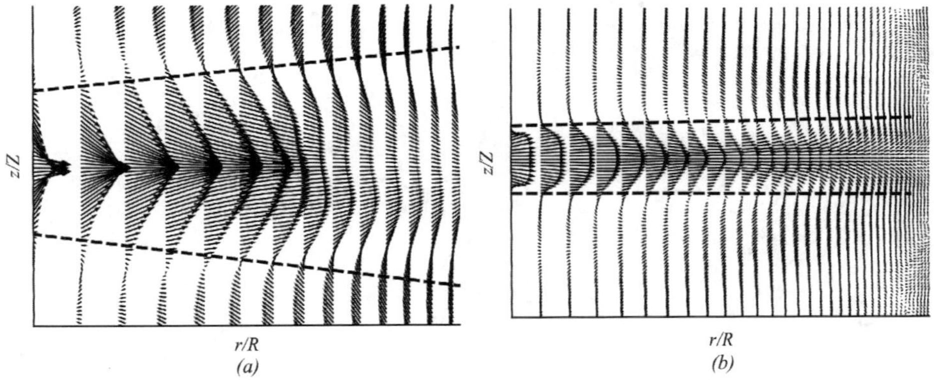

FIGURE 30. Axial spreading of the feed gas (zoomed view) near the feed point without-swirling feed ($F^*_\theta = 0$) (a); with swirling feed ($F^*_\theta = 0.2$) (b), for stratification parameter $\mathcal{A} = 3.535$, aspect ratio $(Z/D) = 4$, $\mathrm{Re} = 10^2$, $F^* = 0.2$, $F_T = 0.2$, and $\epsilon_a = 0.5$.

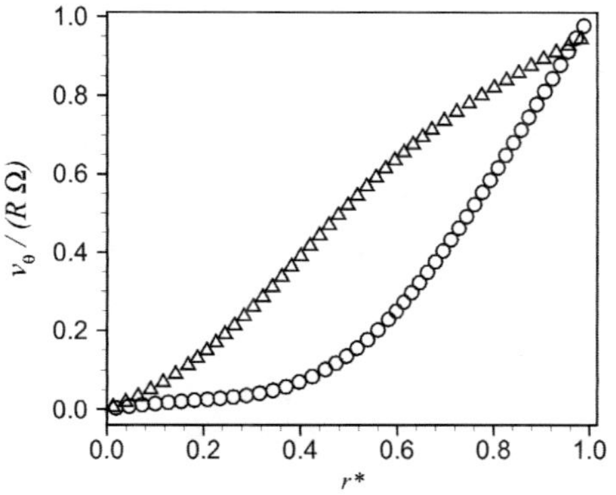

FIGURE 31. The circumferential velocity of the rotating gas, scaled by $(R\Omega)$ as a function of scaled radial coordinate $r^* = r/R$, near the feed point, for swirling feed $F_\theta^* = 0.2$ ($\triangle$), and without swirling feed $F_\theta^* = 0$ ($\circ$), for stratification parameter $\mathcal{A} = 3.535$, aspect ratio $(Z/D) = 4$, $\mathrm{Re} = 10^2$, $F^* = 0.2$, $F_T = 0.2$, and $\epsilon_a = 0.5$. Here, $F^* = F/(H\Omega)$, and $F_\theta^* = F_\theta/(\theta_{mi}\Omega)$.

Stratification parameter	$v_{(\theta-loss),sf}$ (%)	$v_{(\theta-loss),nsf}$ (%)
$\mathcal{A} = 0.707$	2	5
$\mathcal{A} = 1.414$	7	12
$\mathcal{A} = 2.121$	11	20
$\mathcal{A} = 3.535$	13	34

TABLE 3. Circumferential velocity loss of the rotating gas due to in-flow of the feed gas into the rotating cylinder at $\eta = 0.5$, $r = 0$, for stratification parameter $\mathcal{A} = 3.535$, 2.121, 1.414, 0.707; at $\mathrm{Re} = 10^2$, aspect ratio $(Z/D) = 4$, $F^* = 0.2$, $F_T = 0.2$, and $\epsilon_a = 0.5$, for swirling feed (F_θ^* = 0.2) and without swirling feed ($F_\theta^* = 0$).

It is found that with swirling feed ($F_\theta^* = 0.2$), the 'circumferential velocity loss' (or in terms of angular momentum) of the rotating gas due to feed injection reduces significantly (by $3 - 21\%$), owing to the associated angular momentum of the swirling feed gas, and thereby increases the efficiency of the feed drive for the centrifugal gas separation process.

The circumferential velocity loss (or in terms of angular momentum) of the rotating gas due to in-flow of the feed gas into the rotating cylinder at $\eta = 0.5$, $r = 0$, for stratification parameter $\mathcal{A} = 3.535$, 2.121, 1.414, 0.707; at $\mathrm{Re} = 10^2$, aspect ratio $(Z/D) = 4$, $F^* = 0.2$, $F_T = 0.2$, and $\epsilon_a = 0.5$, for swirling feed ($F_\theta^* = 0.2$) and without swirling feed (F_θ^* = 0) are given in table 3.

9. Conclusions

In this study, the generalized Onsager model for the radial boundary layer and of the generalized Carrier-Maslen model for the axial boundary layer at the end-caps in a high-speed rotating cylinder are extended to incorporate the angular momentum of the feed gas for a swirling feed in a high speed rotating cylinder for a single component gas and binary gas mixture. For a single component gas, the analytical solutions are obtained for the sixth-order generalized Onsager equations for the master potential, and for the fourth-order generalized Carrier-Maslen equation for the velocity potential, incorporating the angular momentum of the feed gas for a swirling feed. In both cases, the equations are linearized in the perturbation to the base flow, which is a solid-body rotation. The equations are restricted to the limit of high Reynolds number and (length/radius) ratio, but there is no limitation on the stratification parameter. The linear operators in the generalized Onsager and generalized Carrier-Maslen equations are still self-adjoint, and so the eigenfunctions form a complete orthogonal basis set. However, the differential operator depends explicitly on the stratification parameter, and so it is necessary to evaluate the eigenvalues and eigenfunctions numerically.

For the case of mass/momentum/energy insertion into the flow, the separation-ofvariables procedure is used, and the appropriate homogeneous boundary conditions are specified so that the linear operators in the axial and radial directions are self-adjoint. The discrete eigenvalues and eigenfunctions of the linear operators are obtained. The solutions for the secondary flows for a swirling feed, incorporating the angular momentum of the feed gas, is determined in terms of these eigenvalues and eigenfunctions.

The major advantage of the present formulation, incorporating the angular momentum of the feed gas in a swirling feed, is that it is not restricted to the asymptotic limit of high stratification parameter, even though we have assumed that the Reynolds number and aspect ratio are high. Consequently, the model is not restricted to just boundary layer flows at the curved surface of the cylinder, but is even applicable to situations where the flow domain extends all the way to the centre of the cylinder. Therefore, the model can be used for much slower flows than the original Onsager model, and this provides a way to bridge the existing gap between the Onsager model for very high-speed (high-stratification-parameter) flows, and incompressible flows where the incompressible Navier-Stokes equations can be used. However, the modified generalized Carrier-Maslen model is applicable only to boundary layer flows at the end-caps, though even here, there is now no restriction that the stratification parameter should be much greater than 1.

We compare the analytical results with those of DSMC simulations, incorporating the angular momentum of the feed gas in a swirling feed. The comparison shows that it is necessary to take certain precautions when making a quantitative comparison between analysis and simulations. It is necessary to ensure that the wall velocity slip and 'temperature slip' (difference between gas temperature at the wall and wall temperature) are accurately incorporated in the analytical solutions. In the case of mass/momentum/energy insertion into the flow, it is necessary to accurately enforce the homogeneous boundary conditions at the walls in the simulations, since the analysis assumes homogeneous boundary conditions. When these precautions are taken, we find that there is quantitative agreement (with a difference of less than 15 %) between the predictions of the analytical model and the simulations even when the secondary flow is as large as the 20 % of the base solid-body rotation, the stratification parameter is as low as 0.707, the Reynolds number is as low as 100 and the cylinder length is only twice the diameter, and the scaled angular momentum of the feed gas in a swirling feed is as high as 0.2.

Even for this strong forcing, we find that the analytical solutions for a swirling feed are in excellent agreement (error within 15 %) with the results of DSMC simulations even when the stratification parameter is as low as 0.707, the Reynolds number is as low as 100, and for cylinders with length/diameter ratio as low as 2, and the scaled angular momentum of the feed gas in a swirling feed is as high as 0.2, provided precautions are taken to ensure that the homogeneous boundary conditions assumed in the analytical solutions are accurately reproduced in the simulations. The linear superposition of solutions obtained for inhomogeneous boundary conditions and for internal sources of mass, momentum and energy are also found to be in good agreement with simulations. This shows that the the generalized Onsager and generalized Carrier-Maslen models with swirling feed have excellent predictive capabilities even at relatively low Reynolds number and stratification parameter, and for cylinders of relatively low aspect ratio, provided that care is taken in the specification of boundary conditions. It should be noted that energy removal is necessary because the energy dissipation term (double dot product of the viscous stress and rate of deformation tensors) is second-order in the velocity perturbation. This is not included in the linearization approximation that we have used, as can be seen from the energy equation (3.6). Therefore, the viscous heating is not captured by the generalized Onsager model, and it is necessary to explicitly remove energy to maintain a constant temperature at the wall.

Next, we study the generalized Onsager and generalized Carrier-Maslen model for a binary gas mixture with swirling feed, incorporating the angular momentum of the feed gas in a high speed rotating cylinder. In the base state, the flow is considered to be a solid body rotation with constant temperature, but the variation in the mole fraction is taken into account. For the secondary flow, the mass, momentum and energy equations in axisymmetric coordinates are expanded in an asymptotic series in a parameter $\epsilon = (\Delta m/m_{av})$, where Δm is the difference in the molecular masses of the two species, and the average molecular mass m_{av} is defined as $m_{av} = ((\rho_{w1}m_1 + \rho_{w2}m_2)/\rho_w)$, where ρ_{w1} and ρ_{w2} are the mass densities of the two species at the wall, and $\rho_w = \rho_{w1} + \rho_{w2}$. The equation for the master potential and the boundary conditions are derived correct to $O(\epsilon^2)$. The leading order equation for the master potential contains a self-adjoint sixth order operator in the radial direction. This is solved, subject to boundary conditions, to obtain the $O(1)$ approximation for the secondary flow with swirling feed. The $O(\epsilon)$ and $O(\epsilon^2)$ equations contain inhomogeneous terms which depend on the lower order solutions, and these are solved in a hierarchal manner to obtain the $O(\epsilon)$ and $O(\epsilon^2)$ corrections to the master potential. A similar hierarchial procedure is used for the Carrier-Maslen model for a swirling feed, incorporating the angular momentum of the feed gas, for the end-cap boundary layers.

The results of the Onsager hierarchy, up to $O(\epsilon^2)$, are compared with the results of DSMC simulations for a binary ideal gas mixture for secondary flow due to a inflow/outflow of gas along the axis, mass/momentum/energy sources in the flow, with applied linear wall temperature profile, incorporating the angular momentum of the feed gas for a swirling feed. There is excellent agreement between the solutions for the secondary flow correct to $O(\epsilon^2)$ and the simulations, to within 15 %, even when the stratification parameter is as low as 0.707, the Reynolds number is as low as 100 and the aspect ratio (length/diameter) of the cylinder is as low as 2, and the secondary flow velocity is as high as 0.2 times the maximum base flow velocity, and the ratio of the mass difference and the average mass $(2\Delta m/(m_1 + m_2))$ is as high as 0.5, and the scaled angular momentum of the feed gas $F_\theta^* = F_\theta/(\theta_{mi}\Omega)$ is as high as 0.2.

The major advantages of the swirling feed, associated with the angular momentum of the feed gas, is that it results in a reduction of the angular momentum loss of the rotating gas due to feed injection near the feed point by $(3 - 21)\%$, reduces the axial spreading of the feed gas by $(4 - 24)\%$, minimizes the formation of small secondary vortices near the feed zone, and increases the overall axial mass flux by $(16 - 27)\%$, and thereby enhance the efficiency of the feed drive for the centrifugal gas separation process.

Appendix A. Conservation equations in cylindrical coordinates

The mass momentum and energy conservation equations for an axisymmetric flow in a cylindrical coordinate system are,

$$\frac{\partial \rho}{\partial t} + \nabla.(\rho \mathbf{v}) = \mathcal{S}_m, \tag{A 1}$$

$$\rho\left(\frac{\partial v_r}{\partial t} + v_r\frac{\partial v_r}{\partial r} - \frac{v_\theta^2}{r} + v_z\frac{\partial v_r}{\partial z}\right) = -\frac{\partial p}{\partial r} + \left[\frac{1}{r}\frac{\partial (r\tau_{rr})}{\partial r} - \frac{\tau_{\theta\theta}}{r} + \frac{\partial \tau_{rz}}{\partial z}\right] + f_r, \tag{A 2}$$

$$\rho\left(\frac{\partial v_\theta}{\partial t} + v_r\frac{\partial v_\theta}{\partial r} + \frac{v_r v_\theta}{r} + v_z\frac{\partial v_\theta}{\partial z}\right) = \left[\frac{1}{r^2}\frac{\partial (r^2\tau_{r\theta})}{\partial r} + \frac{\partial \tau_{\theta z}}{\partial z}\right] + (f_\theta + f_{\theta,feed}), \tag{A 3}$$

$$\rho\left(\frac{\partial v_z}{\partial t} + v_r\frac{\partial v_z}{\partial r} + v_z\frac{\partial v_z}{\partial z}\right) = -\frac{\partial p}{\partial z} + \left[\frac{1}{r}\frac{\partial (r\tau_{rz})}{\partial r} + \frac{\partial \tau_{zz}}{\partial z}\right] + f_z, \tag{A 4}$$

$$\begin{aligned}\rho C_v\left(\frac{\partial T}{\partial t} + v_r\frac{\partial T}{\partial r} + v_z\frac{\partial T}{\partial z}\right) &= -p(\nabla.\mathbf{v}) + \tau_{rr}\frac{\partial v_r}{\partial r} + \tau_{\theta\theta}\frac{v_r}{r} + \tau_{zz}\frac{\partial v_z}{\partial z} \\ &+\tau_{r\theta}\left(r\frac{\partial}{\partial r}\left(\frac{v_\theta}{r}\right)\right) + \tau_{\theta z}\left(\frac{\partial v_\theta}{\partial z}\right) + \tau_{rz}\left(\frac{\partial v_z}{\partial r} + \frac{\partial v_r}{\partial z}\right) \\ &+\frac{1}{r}\frac{\partial}{\partial r}\left(rK\frac{\partial T}{\partial r}\right) + \frac{\partial}{\partial z}\left(K\frac{\partial T}{\partial z}\right) + \mathcal{S}_e,\end{aligned} \tag{A 5}$$

where

$$\nabla.\mathbf{v} = \frac{1}{r}\frac{\partial (r v_r)}{\partial r} + \frac{\partial v_z}{\partial z}, \tag{A 6}$$

$$\tau_{rr} = \mu\left(2\frac{\partial v_r}{\partial r} - \frac{2}{3}\nabla.\mathbf{v}\right), \tag{A 7}$$

$$\tau_{\theta\theta} = \mu\left(2\frac{v_r}{r} - \frac{2}{3}\nabla.\mathbf{v}\right), \tag{A 8}$$

$$\tau_{zz} = \mu\left(2\frac{\partial v_z}{\partial z} - \frac{2}{3}\nabla.\mathbf{v}\right), \tag{A 9}$$

$$\tau_{r\theta} = \mu\left(r\frac{\partial}{\partial r}\left(\frac{v_\theta}{r}\right)\right), \tag{A 10}$$

$$\tau_{rz} = \mu\left(\frac{\partial v_r}{\partial z} + \frac{\partial v_z}{\partial r}\right), \tag{A 11}$$

$$\tau_{\theta z} = \mu\left(\frac{\partial v_\theta}{\partial z}\right). \tag{A 12}$$

Equation (A 1) is the mass conservation equation, equations (A 2), (A 3) and (A 4) are the momentum conservation equations in the r, θ and z directions, and equation (A 5) is the energy conservation equation.

REFERENCES

AVERY, D. G. & DAVIES, E. 1973 Uranium enrichment by gas centrifuge *Mills and Boon Ltd. London* **1**, 1.

BABARSKY, R. J. & WOOD, H. G. 1990 Approximate eigensolutions for non-axisymmetric rotating compressible flows. *Computer Meth. Appl. Mech. and Engg.* **81**, 317.

BABARSKY, R. J. & WOOD, H. G. 1992 Analysis of the flow field in a gas centrifuge by means of a temperature potential. *In Proc. of the Third International Workshop on Separation Phenomena in Liquids and Gases.(ed. Wood, H. G.)* , 27.

BABARSKY, R. J. & HERBST, W. I. & WOOD, H. G. 2002 A new variational approach to gas flow in a rotating system. *Physics of fluids.* **14(10)**, 3624.

BARCILON, V. & PEDLOSKY, J. 1967 On the steady motions produced by a stable stratification in a rapidly rotating fluid. *J. Fluid Mech.* **29**, 673.

BARK, F. H. & BARK,T. H. 1976 On vertical boundary layers in a rapidly rotating gas. *J. Fluid Mech.* **78**, 749.

BIRD, G. A. 1963 Approach to translational equilibrium in a rigid sphere gas. *Phys. Fluids.* **6**, 1518.

BIRD, G. A. 1981 Monte-Carlo simulation in an engineering context. *In S.S. Fischer, editor, Rarefied Gas Dynamics, proceedings of the 12th international symposium - part-1.*, 239.

BIRD, G. A. 1987 Direct simulation of high-vorticity gas flows. *Phys. Fluids.* **30**, 364.

BIRD, G. A. 1994 Molecular gas dynamics and the direct simulation of gas flows. *Clarendon Press. Oxford.*

BOURN, R. & PETERSON, T. D. & WOOD, H. G. 1999 Solution of the pancake model for flow in a gas centrifuge by means of a temperature potential. *Comput. Meth. Appl. Mech. Engng.* **178**, 183.

BROUWERS, J. J. H. 1978 On compressible flow in a gas centrifuge and its effect on maximum separative power. *Nucl. Technology.* **39**, 311.

CERCIGNANI, C. & LAMPIS, M. 1971 Kinetic models for gas surface interactions. *Transport Theory Stat. Physics.* **1**, 101.

CERCIGNANI, C. 2000 Rarefied gas dynamics. From basic concepts to actual calculations. *Cambridge University Press.*

CHAPMAN, S. & COWLING, T. G. 1970 The Mathematical Theory of Non-Uniform Gases. *Cambridge University Press.*

CHEN, G. & BOYD, I. D. 1996 Statistical error analysis for the Direct Simulation Monte Carlo technique. *Journal of Computational Physics.* **126**, 434.

COHEN, K. 1951 The theory of isotope separation as applied to the large scale production of U235. *McGraw-Hill.*

COPPER, G. R. & MORTON, J. B. 1988 Driven Wave Motions in a Rotating Gas. *J. Fluid Struct* **2**, 453.

DONEDDU, F. & ROBIN, P. & WOOD, H. G. 2000 Optimization Studies for Gas Centrifuges. *Separation Science and Technology.* **35**, 1207.

FAN, J. & SHEN, C. 2001 Statistical simulation of low-speed rarefied gas flows. *J. Comp. Phys.* **167**, 393.

FRAENKAL, L. E. 1959 A Cylindrical sound pulse in a rotating gas. *J. Fluid Mech.* **36**, 637.

GARCIA. 2000 Time step truncation error in direct simulation Monte Carlo. *Physics of fluids* **12**, 2621.

GREENSPAN, H. P. 1968 The theory of Rotating Fluids. *Cambridge University Press* **1**, 1.

GUNZBURGER, M. D. & WOOD, H. G. 1982 A finite element method for the Onsager pancake equation. *Comp. Meth. Appl. Mech. Engg.* **31**, 43.

GUNZBURGER, M. D. & WOOD, H. G.& JORDAN, J. A. 1984 A finite element method for gas centrifuge problem. *SIAM J. Sci. Stat. Comp.* **5.**

HARADA, I. 1980 A Numerical Study of Wikely Compressible Rotating Flows in a Gas Centrifuge. *Nucleas Sc. and Engg.* **73**, 225.

HARADA, I. 1980 Computation of Strongly Compressible Rotating Flows. *J. Computational Physics.* **38**, 335.

HAUSMAN, M. A. & ROBERTS, W. W. 1984 Three dimensional stratified gas flows past an obstacle. *J. Comp. Phys.* **55**, 347.

HEUSER, G. E. & RIBANDO, R. J. & WOOD, H. G. 1986 A numerical simulation of internal waves in a rotating fluid. *Comput. Methods Appl. Mech. Engrg.* **57**, 207.

HITTINGER, M. & HOLT, M. & SOUBBARAMAYER, B. 1981 Numerical solution of the flow field near a gas centrifuge scoop. *In Proc. Fourth Workshop on Gases in Strong Rotation, Oxford (ed. E. Ratz)*, 22.

HITTINGER, M. & HOLT, M. & SOUBBARAMAYER, B. & CORTET, C. 1983 Simulation of gas flow in front of the scoop of a gas centrifuge. *In Proc. Fifth Workshop on Gases in Strong Rotation, Charlottesrille (ed. H. G. Wood)*, 515.

HOGLUND, R. L. & SHACTER, J. & VON HALLE, E. 1979 Diffusion separation methods. *Encyclopedia of chemical technology.* **7**, 3rd ed.

HOMSY, G. M. & HUDSON, J. L. 1967 Centrifugally driven thermal convection in a rotating cylinder. *J. Fluid Mech.* **35**, 33.

HUNTER, C. 1967 The axisymmetric flow in a rotating annulus due to horizontally applied temperature gradient. *J. Fluid Mech.* **27**, 753.

JONES, C. A. 1982 On flow between counter-rotating cylinders. *J. Fluid Mech.* **120**, 433.

JUNG, E. 1983 An analytic solution of the linearized flow equations using the method of eigenvalues. *In Proc. Fifth Workshop on Gases in Strong Rotation, Charlottesrille (ed. H. G. Wood)*, 247.

KAI, T. 1983 Analysis of Separation Performance of Gas Centrifuge. *In Proc. Fifth Workshop on Gases in Strong Rotation, Charlottesrille (ed. H. G. Wood)*, 1.

KUMARAN, V. & PRADHAN, S. 2014 The generalized Onsager model for a binary gas mixture. *J. Fluid Mech.* **753**, 307.

LAHARGUE, J. P. & SOUBBARAMAYER 1978 A numerical model for the investigation of the flow and isotope concentration field in an ultracentrifuge. *Comp. Meth. Appl. Mech. Engg.* **15**, 259.

MACROSSAN, M. N. 2001 DSMC : A fast simulation method for rarefied flow. *J. Comp. Phys.* **173**, 600.

MASLEN, S. H. & MOORE, F. K. 1956 On strong transverse waves without shocks in a circular cylinder. *J. Aeronautical Sciences* **23**, 583.

MATSUDA, T. & HASHIMOTKO, K. 1976 Thermally, mechanically or externally driven flow in a gas centrifuge with insulated horizontal end plates. *J. Fluid Mech.* **78**, 337.

MATSUDA, T. 1976 A New proposal of gas centrifuge with desirable counter-current. *J. Nucl.Sci. Tech.* **13**, 98.

MATSUDA, T. & HASHIMOTO, K. 1978 The Structure of the Stewartson layers in a gas centrifuge. Part 1: Insulated end plates. *J. Fluid Mech.* **85**, 433.

MATSUDA, T. & NAKAGAWA, K. 1983 A new type of boundary layer in a rapidly rotating gas. *J. Fluid Mech.* **126**, 431.

MATSUDA, T. & TAMURA, N. & SAWADA, K. 1989 Three-dimensional numerical simulation of flows past scoops in a gas centrifuge. *J. Fluid Mech.* **201**, 203.

MIKAMI, H. 1973 Thermally induced flow in a gas centrifuge. *J.Nucl.Sci.Tech.* **10**, 580.

MIKAMI, H. 1981 Rotating supersonic flow about scoop inlet using an unsteady implicit technique. *In Proc. Fourth Workshop on Gases in Strong Rotation, Oxford (ed. E. Ratz)*, 94.

MIKAMI, H. 1985 Two dimensional numerical calculation of flow fields about scoop inlet by the piecewise linear method. *In Proc. Sixth Workshop on Gases in Strong Rotation, Tokyo (ed. Y. Takashima)*, 67.

NAKAYAMA, W. & TORII, T. 1974 Numerical analysis of separative power of isotope centrifuges. *J. Nucl. Sci. Technology.* **11**, 495.

NANBU. 1980 Direct simulation scheme derived from the Boltzmann equation I. Multicomponent gases. *J. Phys. Soc. of Japan.* **45**, 2042.

OLANDER, D. R. 1981 The theory of uranium enrichment by the gas centrifuge. *Progress in Nuclear Energy.* **8**, 1.

OMNES, P. 2007 Numerical and physical comparisons of two models of a gas centrifuge. *Computers Fluids* **36**, 1028.

PRADHAN, S. & KUMARAN, V. 2011 The generalized Onsager model for the secondary flow in a high-speed rotating cylinder. *J. Fluid Mech.* **686**, 109.

PRADHAN, S. 2016 The generalized analytical model and DSMC simulations of high-speed rotating flow in polar $(r - \theta)$ coordinate. *Submitted to J. Fluid Mech.*

QUANHUA, S. & BOYD, I. D. 2002 A Direct Simulation Method for Subsonic Microscale Gas Flows . *J. Computational Physics.* **179**, 400.

RIBANDO, R. J. 1984 A Finite difference solution of Onsager's model for flow in a gas centrifuge. *Computers Fluids* **12 (3)**, 235.

RIEFFEL, M. A. 1999 A method for estimating the computational requirements of DSMC simulations. *J. Comp. Phys.* **149**, 95.

ROBERTS, W. W. 1985 Three dimensional stratified gas flows past impact probes and scoops : N-body Monte Carlo calculations. *In Proc. Sixth Workshop on Gases in Strong Rotation, Tokyo (ed. Y. Takashima.)*, 115.

ROBERTS, W. W. & HAUSMAN, M. A. 1988 Hypersonic, stratified gas flows past an obstacle : Direct Simulation Monte Carlo calculations. *J. Comp. Phys.* **77**, 283.

ROBLIN, P. & DONEDDU, F. 2000 Direct Monte-Carlo Simulations In a Gas Centrifuge. *In Proc. 22nd International Symposium on Rarefied Gas Dynamics. (ed. T. J. Bartel & M. C. Gallis.)*, 169.

SAKURAI, T. 1981 Linearized thin-wing theory of gas-centrifuge scoops. *J. Fluid Mech.* **103**, 257.

SALANT, R. F. 1968 Syymmetric normal modes in a uniformly rotating fluid. *J. Accoustical Society of America.* **43**, 1302.

STEWARTSON, K. 1957 On almost rigid rotations. *J. Fluid Mech.* **3**, 17.

SOUBBARAMAYER. 1979 Centrifugation. *In Uranium Enrichment (Ed. S. Villani, Springer)* **1**, 1.

SOUBBARAMAYER & BILLET, J. 1980 A numerical method for optimizing the gas flow field in a centrifuge. *Comp. Meth. Appl. Mech. Engg.* **24**, 165.

SOZOU, C. 1959 Adiabatic transverse modes in a uniformly rotating fluid. *J. Fluid Mech.* **36**, 605.

THOMPSON, J. J. & CHEN, Y. P. 1980 Eigen vector solution of the linearized equations of thermal convection in a closed gas centrifuge. Part-1 : Problem formulation and solution by collocation. *Nuclear Engineering and Design* **58**, 97.

VIECELLI, J. A. 1983 Exponential difference operator approximation for the sixth order Onsager equation. *J. Comp. Phys.* **50**, 162.

VINCENTI, W. G. & KRUGER, C. H. 1967 Introduction to Physical Gas Dynumics. *Wiley, New York.*

VOLOSCIUK, K. 1981 Application of the vortex transport equations to the calculation of 2 and 3 dimensional extraction chamber flows. *In Proc. Fourth Workshop on Gases in Strong Rotation, Oxford (ed. E.Ratz)*, 604.

VON HALLE E. & WOOD, H. G., LOWRY, R. A. 1978 The effect of vacuum core boundary conditions on separation in the gas centrifuge. *Nucl. Technology.* **62**, 325.

WAGNER, W. 1992 A convergence proof for Bird's Direct Simulation Monte Carlo method for the Boltzmann equation. *Journal of Statistical Physics.* **66**, 1011.

WALZ, A. & VOLOSCIUK, K. & SCHUTZ, H. 1983 Numerical investigations of the flow field near a model of a scoop using the vortex transport equations. *In Proc. Fifth Workshop on Gases in Strong Rotation,Charlottesrille (ed. H.G.Wood)*, 425.

WOOD, H. G. & MORTON, J. B. 1980 Onsager's pancake approximation for the fluid dynamics of a gas centrifuge. *J. Fluid Mech.* **101**, 1.

WOOD, H. G. & SANDERS, G. 1983 Rotating compressible flows with internal sources and sinks. *J. Fluid Mech.* **127**, 299.

WOOD, H. G. & JORDAN, J. A. & GUNZBURGER, M. D. 1984 The effect of curvature on the flow field in rapidly rotating gas centrifuges. *J. Fluid Mech.* **140**, 373.

WOOD, H. G. & BABARSKY, R. J. 1992 Analysis of a rapidly rotating gas in a pie-shaped cylinder. *J. Fluid Mech.* **239**, 249.

WOOD, H. G. 1995 Analysis of feed effects on a single-stage gas centrifuge cascade. *Separation Science and Technology.* **30(13)**, 2631.

WOOD, H. G. & MASON, T. C. & SOUBBARAMAYER. 1996 Multi-Isotope separation in a gas centrifuge using Onsager's Pancake model. *Separation Science and Technology.* **31(9)**, 1185.

459541 A Numerical Model of Chemically Reacting Flow through Additively Manufactured 3D Structures

Abstract:

There is currently a ubiquitous reliance on randomly packed powders in chemical reactors, separation columns, and other devices that manipulate fluid flow and reactive species transport. Replacing randomly arranged materials by deterministically fabricated devices with optimized geometries achieves major performance and efficiency improvements. Notable examples can be found in gas chromatography, microfluidic medical devices, and recently emerging "3D battery" structures. We explore additive manufacturing as a means to achieve such deterministic structures, and their potential benefits for exchange chromatography columns. Exchange chromatography is of scientific interest because the second-order chemical reaction between phases results in a theoretically infinitely sharp boundary that does not broaden as it progresses, although broadening mechanisms do occur in practice, and are easily studied. In our case, we examine the exchange of hydrogen and deuterium with a hypothetical 3D-printed metal hydride column in a coupled reaction and diffusion-convection numerical model. We aim to design, as a proof of principle, a hydrogen-deuterium separation column that achieves more sharply defined separations at a lower pressure drop than a similarly sized packed-powder column. Our results suggest that a front can be obtained in a hypothetical 3D structure, and that it is sharper than the one obtained in a 1D structure for the same pressure drop. Other preliminary results suggest that 1D arrays with defective pores result in broad fronts in the total flow due to varying flow rates between the tubes, whereas the 3D structure will allow mixing between inhomogeneous regions, maintaining a sharp front.

Sandia National Laboratories is a multi-program laboratory managed and operated by Sandia Corporation, a wholly owned subsidiary of Lockheed Martin Corporation, for the U.S. Department of Energy's National Nuclear Security Administration under contract DE-AC04-94AL85000.

472816 Computational Fluid Dynamics Modeling of Momentum and Heat Transfer in Nanofluids

Abstract:

Nanofluids are a class of heat transport fluids created by suspending nano-scaled metallic or nonmetallic particles into a base fluid. Some experimental investigations have revealed that the nanofluids have remarkably higher thermal conductivities than those of conventional pure fluids and are more suited for practical application than the existing techniques of heat transfer enhancement using millimeter and/or micrometer-sized particles in fluids. Use of nanoparticles reduces pressure drop, system wear, and overall mass of the system leading to a reduction in costs over existing enhancement techniques. The enhancement of the heat transfer coefficient is determined experimentally using a CuO based nanofluids by dispersing CuO nanoparticles (40nm) with different particles loadings (0.25%, 1% and 2%wt) into water. The heat transfer coefficient enhancement ratio at different concentrations of CuO in water were calculated at 40 and 70°C. The experimental results illustrated that various factors including flow rate (Re number), particle concentration and temperature are all capable of impacting the enhancement ratio. To further explain the impact of these variable on the hydrodynamic and thermal parameters of a nanofluid, we developed a CFD model using an Eulerian-Lagrangian approach to study the nature of both the laminar and turbulent flow fields of the fluid phase as well as kinematic and dynamic motion of the dispersed nanoparticles. The main goal is to provide additional information about the fluid and particle dynamic to explain the observed behavior in the experimentally observed trends of the heat transfer coefficient enhancement of CuO/water relative to both nanoparticle concentrations and fluid flowrates. Our results indicate that heat transfer enhancement significantly depends on particle motion within the system and is highly dependent upon the position of nanoparticles relative to the tube wall.

The simulation results revealed that in some cases under the laminar flow regime, particles are able to move through thermal boundary layer into the bulk flow of the fluid. When compared with experimental results, it was found that heat transfer coefficient was highest in these cases. Based on both the experimental and theoretical results under

turbulent flow, particles add little to the mixing and turbulence within the fluid resulting in negligible if any enhancements. Based on our results, laminar flow should be chosen so that the heat transfer coefficient can be effectively enhanced in these systems as the highest enhancements in this work were observed under these conditions. However, it is very important that the nanoparticles pass thermal boundary layer to provide the highest heat transfer enhancement. To ensure that this is the case, simulations of the particle trajectories should be performed based on the exact geometry of the system to ensure that the correct particle concentrations and fluid temperature are used.

481186 The Soret Effect in Polyaromatic Hydrocarbons: Thermal Separation From Alkanes

Abstract:

Thermodiffusion in liquid mixtures may explain some counter-intuitive but naturally occurring phenomena such as hydrocarbon reservoirs with heavier component(s) stratified on top of lighter ones. However, beyond benchmark systems, systematic measurements of thermodiffusion in binary organic mixtures are lacking. We use an Optical Beam Deflection apparatus to simultaneously probe Fickian and thermal diffusion in binary solution mixtures of polycyclic aromatic hydrocarbons dissolved in alkanes, and measure both Fickian diffusion D and the Soret coefficient S_T, and then obtain the thermodiffusion coefficient D_T. In a series of nine binary mixtures, we vary both the size of the aromatic compound from two to four rings, as well as the length of the alkane chain, from 6 to 16 carbons. To probe the effect of increasing ring size, we include a 6-ringed aromatic compound, coronene, and toluene as a solvent, due to the insolubility of coronene in alkanes. Our results suggest that Fickian diffusion increases with the inverse of solvent viscosity and also with decreasing molecular weight of the solute. While both of these trends match our intuition, the behavior of S_T and D_T are more complicated. We find that S_T and D_T increase with the solute molecular weight when the solvent is held fixed, and that the impact of solute ring size is higher in shorter chain alkane solvents.

476849 Next-Generation High-Efficiency Hybrid Solar to Hydrogen Conversion with Integrated Storage

Abstract:

By 2050 solar technologies are projected to deliver the majority of the world's electricity. Solar energy can be used to generate both heat and electrical power. Most solar panels are designed for only one of these purposes, so an electrical photovoltaic (PV) panel is typically less than 20% efficient.

It is well known that PV cells experience a deterioration in performance (efficiency) when they are operated at high temperatures, and that this leads to particularly high losses when the solar resource is at its highest. For example, a drop in PV cell efficiency in the range 10-20% can be expected, depending on the PV technology, when operating PV cells at temperatures of ~65 °C (or, 40 °C higher than room temperature), which is easily experienced in hot climates. Hybrid PV-thermal (PV/T) solar collector technology combines PV modules with the contacting flow of a cooling fluid (gas or liquid) in a number of configurations and geometries, and offers some advantages when space is at a premium and there is demand for both heat and power. Moreover, a major motivation in PV/T technology is to cool the PV cells and therefore to increase their electrical efficiency, while delivering a potentially useful thermal output (hot fluid stream). PV/T collectors are a highly efficient technology, capable of achieving system efficiencies (electrical plus thermal) in excess of 70%.

By far the most common use of the thermal-energy output from PV/T systems (in fact most solar-thermal collector technologies) is to provide hot water at 50-60 °C for households or commercial use, however, a wide range of opportunities arise at higher temperatures whence additional power-generation cycles (e.g., with organic Rankine cycles, thermoelectric generators, amongst other) or thermal-cooling technologies (e.g., with desiccant, ad/absorption refrigeration cycles, amongst other) can be combined with PV/T modules. These additional options become viable at temperatures typically above ~80 °C, and importantly, become increasingly efficient at progressively higher temperatures.

Operating PV/T panels efficiently at high temperatures for this purpose is a significant scientific and engineering challenge, since the two modes of heat-transfer loss that result directly in the drop in performance, namely convection and radiation, are both exacerbated at higher temperatures. The reduction of convective losses can be addressed by evacuation of the panel, while radiation losses can be decreased by using selective coatings. This talk will discuss recent advances in evacuated PV/T technology with optimized surfaces and selective coatings combined with a secondary power generation sub-system based on organic Rankine cycles (ORCs) for high-efficiency conversion of solar energy to electrical power and consequently hydrogen via electrolysis. Projections at this stage suggest an improvement of the order of 20-30% relative to PV-only systems, for the same area.

In this solar-to-electricity-to-hydrogen system, energy storage options include thermal storage, electrical storage and hydrogen storage. Of these, thermal storage is relatively low cost and can be performed over a range of scales with sensible heat storage (e.g., hot water tanks), phase-change materials (PCMs) covering a range of organic or inorganic substances (e.g., molten salt tanks), or via thermo-chemical energy storage which utilizes the energy alternately absorbed and released in reversible chemical reactions and has the highest theoretical storage capacity of these solutions. Electrical-energy storage on the other hand is highly dependent on scale. At small scales of application a variety of battery technologies exist or are currently under development, which are highly suitable solutions. At large-scales, on the other hand, the costs and potential environmental impact of the widespread use of batteries (arising from the use of hazardous or scarce substances) allow serious competition from mechanical or thermo-mechanical systems, which become the technology of choice. Relevant conventional electricity-storage solutions include pumped hydro (PHS) and compressed air energy storage (CAES).

A significant effort has been placed in recent years on the development of additional electricity storage technologies, driven by the increasing penetration into the grid of renewable energy technologies for electricity generation, many of which have an unpredictably intermittent nature. Of increasing interest are the so-called cryogenic 'liquid-air' (or, liquid-nitrogen) energy storage (LAES) system and a novel storage technology known as Pumped Thermal Electricity Storage (PTES). LAES uses excess electricity to operate a refrigeration cycle with which to liquefy air, which is then stored in a tank at atmospheric pressure. When the electricity is needed, the liquid air is evaporated and used to drive a turbine. PTES, on the other hand, uses excess electricity to operate a heat pump cycle with which to convert the electrical energy into thermal energy. The thermal energy, both hot and cold relative to atmosphere, is then stored in two large reservoirs (hot and cold). The reservoirs contain a material with a high heat capacity, and are able to store the energy much more compactly than PHS. When required, the thermal energy can be converted back into electricity by running the heat pump in reverse, as a heat engine. The projected round-trip efficiencies of both LAES and PTES are lower than PHS, but these systems have a number of potential benefits, including low capital cost, no geographical constraints (i.e., elevations, caverns, etc.) in terms of deployment, and no associated environmental impact. This talk will present recent advances in the development of these technologies, compare the various options with each other and with alternative thermal and hydrogen storage in the context of solar-energy conversion and storage, and finally discuss the next-steps that would enable their successful deployment. In isolation, thermal-energy storage emerges as the technological option with the lowest cost and highest technical maturity, whereas at larger scales electricity storage can benefit as part of a wider storage, transmission and distribution network that includes additional intermittent generation from renewables.

481381 Understanding and Eliminating Interaction of Cells in an Electrochemical Stack

Abstract:

Typically, both the fuel cells and electrolysis cells are stacked in series by means of impermeable graphite or metallic bipolar plates. The open-circuit voltage of most types of fuel cells is approximately 1 V. To support sufficient voltage and power required for commercial applications, cells are assembled in a stack. By connecting multiple cells in series, the voltage of a stack is multiplied, generating the required high power output. Similarly, electrolysis cells are stacked to uptake the high voltages and power, and reach a demanding hydrogen production; the hydrogen generation-rate of a single cell is limited to electrodesâ€™ active area.

In an ideal stack, the design and operating conditions for all cells should be identical throughout the stack. Overall stack performance would then be the summation of the equal stand-alone performances of cells. However, in a real stack, the cells are not perfectly identical and they might differ either in design or operating conditions or both for a variety of reasons: imperfections during manufacturing, a non-uniform inlet flow distribution, degradation of one or more cells and external boundary conditions, such as heat transfer with the surroundings. Though such non-uniformities are encountered at a single cell level, the performance differences are translated, very often not to a single cell only, but to a group of cells located in neighborhood. In other words, one poorly operating or perturbed cell could affect the performance of other cells in the stack. Such a coupling of the cells is the focus of investigation here as it could be detrimental for the performance and lifetime of the stack.

As described above, the individual cells in a stack are separated by impermeable bipolar plates. Thus, we can classify the transport phenomena in their stacks according to the region of influence. Mass, momentum and species transport are confined to individual cells, whereas charge and heat transfer take place globally across the cells such that the cells become coupled and influence each other. Since the cells are coupled via heat and charge transport, abnormal behavior

of one cell can propagate to the neighboring cells. Two adjacent cells can be said to be (1) electrically decoupled when their local current density distributions do not influence and perturb each other, and (2) thermally decoupled if there is negligible heat transfer between the two cells. We analyze charge and heat transport across the cells in a stack for the above two decoupling conditions. In short, electrical decoupling of the cells in a stack is found to depend on the geometrical and material properties of the bipolar plate and the operating parameters. Similarly, the thermal decoupling is governed by the design and operating parameters of coolant flow fields. We finally identify the design guidelines that ensure decoupling of the cells via both the transport modes. The guidelines could be of use for stack designers when selecting the bipolar plate material and dimensions and coolant flow field conditions, so as to minimize the spread of disturbances in a cell to neighboring cells.

Acknowledgements: The research is supported by the National Research Foundation, Singapore under its Competitive Research Program (Award No. NRF-CRP8-2011-04).

454078 Advancement of Hydrogen Storage Technologies

Abstract:

Hydrogen is seen as one of the major storage options for renewable energy. However, its storage still faces a number of challenges. Safe, dense and efficient hydrogen storage has to overcome a number of thermodynamic constraints. A big issue, for example, is the compromise between simple hydrogen release with a small energy demand on the one side and loss-free storage at ambient conditions with a high density on the other side. The interplay between enthalpy and entropy change during hydrogen uptake plays an important role in this respect.

In this contribution different steps in the development of hydrogen storage technologies, such as Liquid Organic Hydrogen Carrier (LOHC), will be presented. First, a study on the identification of potentials for the improvement of hydrogen storage technologies is conducted. This outlines the impact of different measures and gives insight into the question what should be improved.

Based on these findings, a systematic development of carrier materials and technologies will be demonstrated. A comparison of different hydrogen storage options is discussed.

The role of process simulation in this context will be outlined. In the last step a special emphasis will be put on the issue of reliability and uncertainty of the respective process models.

458291 Thermodynamic and Economic Evaluation of Hydrogen Transport Technologies

Abstract:

Due to its high gravimetric energy density, hydrogen is considered as an interesting energy carrying substance for a future energy system based on renewable resources. In this context, energy distribution and thereby efficient, cheap and secure transport of hydrogen is an important aspect.

Today, hydrogen transport is almost exclusively realized by compressed or liquefied gas. Transportation as compressed gas affords high pressures, generating a certain security risk while the amount of transported hydrogen is still low due to the low storage density. Liquefied gas can reach substantially higher storage densities and thereby capacities but suffers from large energy demands for liquefaction and boil off-losses during storage. Several alternative storage technologies came into the focus of research to overcome these limitations. In the context of hydrogen transport, especially carrier-based hydrogen storage technologies like metal hydrides or liquid organic hydrogen carriers (LOHCs) seem promising as these technologies enable a dense storage of hydrogen at ambient conditions.

In this contribution, a thermodynamic and economic evaluation of several hydrogen transport technologies is presented. The analysis commences with the determination of the energy demands for hydrogen conditioning (e.g. compression, liquefaction), the actual transport and the provision of hydrogen gas at the point of use (e.g. hydrogen liberation from carrier). For all technologies considered in the analysis the energy demand for the actual transport is comparatively low. The conditioning (for pressurized and liquefied gas) and provision step (for carrier-based transport) have a far higher energy demand. Options to reduce the energy demands are pointed out. The cost calculation includes investment, maintenance, staff, energy, and material costs and shows the potential advantage of chemical carrier technologies over the established transport technologies.

459518 State-of-the-Art of the Liquid Organic Hydrogen Carrier (LOHC) Technology

Abstract:

LOHC are a safe and dense solution of storing hydrogen at ambient conditions. The basis is a reversible chemical reaction of hydrogen with the LOHC. This technology is studied worldwide since 5 years.

Despite the fact that several component pairs (loaded and unloaded form) are able to be a LOHC, the basic process design is known.

The lecture highlights technical applications of this technology. This progress will push the development of a hydrogen-based energy system.

481365 On using CO_2 and Renewable Energy for Autonomous Cities

Abstract:

An innovative concept of district heating using carbon dioxide is proposed to supply energy services in cities. The principle is to use CO2 as a district heating fluid distributed in liquid and gas form at a temperature of 15-20 °C. Heating needs are then supplied by condensing CO2 while cooling needs are supplied by evaporation, the temperature levels being adjusted using heat pumps. The energy balance is closed by exchanging heat between the district heating fluid and the environment accessing therefore attractive sources. The advantage of such system is the supply of the heating/cooling services at the temperature it requires and the possible exchange between heating and cooling demands. The use of the district heating system allows indeed to connect the most attractive environmental sources (geothermal, lake, river, waste water, industrial waste heat,. . .) with the users in order to maximise the efficiency of the system. Using this type of technology, it has been demonstrated that heating and cooling the city center of a European city can be reduced by a factor 5 to 6 (i.e. 80% energy savings) while having a pay back period of around 7 years . The concept can be extended to integrate the renewable energy resources in particular solar photovoltaics (PV). One of the difficulty of the PV systems is the excess of electricity available in the summer that needs to be stored to supply services in the winter. The idea here is to electro-chemically convert excess of electricity into methane by combining electrolysis of water and the Sabatier reaction that converts CO2 and H2 into methane. This can be done in co-electrolysis concepts
using solid oxide electrochemical membranes. Methane can then be stored in liquid form. In the winter time, the methane is converted in decentralised solid oxide fuel cells into electricity and heat while separating the CO2. The CO2 is collected in the district CO2 network at 50 bar and stored in the liquid form for the summer season when it will again be converted into methane. The produced electricity is used to supply the needs of the households, to drive the heat-pumps of the CO2 based district network and even electrical cars. It is therefore possible to develop cities that are supplied only by renewable sources. Furthermore, the methanation concept can also be applied to convert the waste biomass produced in the city into methane.

461623 Dielectrophoretic Assembly of Nanowires in Shear Flows

Abstract:

Fluid flow in conjunction with an electric field has been shown to be a promising approach to precisely assemble nanowires or carbon nanotubes over pre-patterned electrode sites on a substrate [1]. When the flow rate of the suspension of nanowires and the strength of the electric field are tuned correctly, individual nanowires are deposited across each electrode, aligned with the electric field. While the proof of concept is complete, several open questions remain for the practical application of this technique for controlled assembly of nanowires. What is the quantitative relationship between flow and dielectrophoretic forces that lead to assembly? What is the effect of Brownian motion? What is the smallest nanowire that can be assembled in this manner?

To address these questions, we present simulations and analysis to clarify the interplay among dielectrophoresis, hydrodynamics and Brownian motion that determines whether nanowires will successfully assemble for a given set of experimental parameters, including flow rate, potential differences across the electrodes, size of electrodes and nanowires, and electrical properties of the nanowires. It is found that within a finite distance from the substrate, there exists a depletion layer largely void of nanowires (due to hydrodynamics, steric effects and nanowire anisotropy) and a capture zone (due to dielectrophoresis competing with hydrodynamics). Nanowires are captured when the capture width exceeds the depletion width; otherwise they are carried away with the flow. Successful assembly is aided by strong electric fields and low flowrates. Scaling arguments are presented to identify two dimensionless numbers that are ratios of hydrodynamic, Brownian and dielectrophoretic forces and mediate particle dynamics and predict flowrate-voltage combinations for successful assembly. With decreasing nanowire length, increasing diffusion distributes particles more broadly, thereby aiding assembly by moving some nanowires into the region where the electric field is strong enough for capture. However, for very short nanowires, the strength of diffusion eventually becomes limiting and retards assembly. Diffusion in conjunction with the practical necessity to avoid dielectric breakdown of solvent is shown to impose a minimum nanowire length for which assembly is feasible. Depending upon

material properties and field frequency, this is about 20 nm for semiconducting nanowires.

1. Freer, E. M.; Grachev, O.; Duan, X.; Martin, S.; Stumbo, D. P. High-yield self-limiting single-nanowire assembly with dielectrophoresis. *Nat Nano* **5**, 525-530, 2010.

457486 Microstructure of Rigid Rod Suspensions

Abstract:

Nonspherical particles suspended in a continuous liquid phase have a tendency to align because of particle-fluid interactions. In the absence of fluid dynamic couples, spontaneous self-alignment can occur due to excluded volume contraints on rotary Brownian motion. This phenomenon prevents a return-to-isotropy from an anisotropic state. Low-order moments of the rotary Smoluchowski equation are used to explore the influence of the rotary Péclet number on the shear viscosity of the suspension. An algebraic closure for the fourth-order orientation operator in terms of the second-order orientation operator is used to predict the orientation states of the suspension. The mapping satisfies the six-fold symmetry properties the six-fold projection properties of the exact orientation tetradic. The instantaneous orientation states predicted by the theory are realizable for all Peclet numbers and for all excluded volume potentials.

YoChan Kim, André Bénard, and Charles A. Petty, Microstructure and Rheology of Rigid Suspensions, Ind. Eng. Chem. Res., 2015, 54, 4497-4504.

462016 Dynamics of Rigidly Foldable Sheets in Shear Flow

Abstract:

We study the dynamics of piecewise rigid sheets containing predefined crease lines in shear flow. The crease lines act like hinge joints along which the sheet may fold rigidly, i.e. without bending any other crease line. We choose the crease lines such that they tessellate the sheet into a two-dimensional array of parallelograms. When all the hinges are fully open the sheet is planar, whereas when all are closed the sheet folds over itself to form a compact flat structure. Such a tessellation is known as a `Miura`-pattern in the origami community. Due to rigidity constraints the folded state of a `Miura`-sheet can be described using a single fold angle.

We model a hinged sheet using the framework of constrained multibody systems in the absence of inertia. The hydrodynamic drag on each parallelogram is calculated based on an inscribed elliptic disk. We present results highlighting the various modes of coupled tumbling and folding motion of the sheet. Furthermore, when the joints are associated with a bending potential, we study the buckling behavior of an intially planar sheet in shear flow.

470725 In Search of Physical Meaning: Defining Transient Parameters for Nonlinear Viscoelasticity

Abstract:

Many soft materials are subjected to rheologically nonlinear deformations and flows in their native biological environment, during processing, or in their intended end use. Control over the nonlinear rheology of soft materials is therefore crucial for product formation and use, and it is important that there be a physically-motivated way of understanding responses to well-characterized deformations. In this work, a complete set of model-independent viscoelastic functions for understanding responses to transient nonlinear rheological tests is presented, using large-amplitude oscillatory shear strain as a model nonlinear protocol. The derivation makes no assumptions about symmetries, and is therefore applicable to the responses to any input, allowing researchers to unambiguously define time-dependent moduli, viscosities, compliances, fluidities, and normal stress coefficients. A legend for interpreting the dynamic trajectories in modulus space is provided, along with explicit definitions of the rates at which the moduli change. These provide a quantitative mechanism to identify when, and by how much, a material response stiffens, softens, thickens, or thins while being deformed. In addition to providing analytical expressions for the moduli, the derivation *requires* the definition of a conceptually new term. This means there exist three, not two, time-dependent nonlinear viscoelastic functions by which any response can be fully described. The third function accounts for nonlinear properties such as yield stresses and the shifting of the strain equilibrium. This complete analysis scheme is unique in making a distinction between the strains in the lab and material frames. The quantitative sequence of physical processes analysis, which is fully developed in this work, allows for comprehensive physical interpretations of responses to transient deformations of any kind to be made, including the steady alternance responses to LAOS, time-dependent oscillatory shear startup responses, and thixotropic and anti-thixotropic responses.

470385 Numerical Study of the Shear Rheology of Suspensions in Viscoelastic Fluids at Low Volume Fraction

Abstract:

The rheology of viscoelastic fluids with particulate fillers is of both industrial and academic interest due to the wide use of such materials in manufacturing, such as injection molding and 3D printing, and technologies such as oil recovery. However, existing models to describe the steady shear rheology of viscoelastic suspensions fail to adequately predict experimental results even at low volume fractions. This reflects an insufficient fundamental understanding of the mechanism by which particles alter the bulk stress of viscoelastic fluids. Furthermore, the nonlinearities of the suspending fluid present a challenge for deriving analytical results. As such, fully 3-D simulations that can resolve particle-scale hydrodynamics can provide useful insights into such systems and inform more predictive models. We present steady shear simulations of freely suspended rigid spheres at low volume fractions in a viscoelastic fluid described by the Giesekus model. We use a fully parallel code based on an unstructured finite volume formulation for incompressible flow. The Giesekus model is implemented using a log-conformation formulation, which allows us to probe the high Weissenberg number regime (large deformation rate and high elasticity). The code can accommodate boundary-fitted meshes, which allow high resolution of particle boundaries, as well as an immersed-boundary formulation that can efficiently handle multi-body dynamic simulations. Using these tools we study both dilute and semi-dilute suspensions, making comparisons to theoretical results and experimental measurements. Single particle results are used to validate the theories in the dilute limit and explore the effect of high shear rates. We show that the competing effect between the particle extra stress and polymer stresses induced in the fluid by particle disturbances determines the particle contribution to the bulk rheology. At high shear rates, the particle-induced polymer stress dominates. We also present multi-particle simulations at low volume fractions to show how hydrodynamic interactions between particles affect the microstructure and resulting bulk stresses.

473703 Numerical Analysis of DNA-Functionalized Colloidal Particle Deposition in a Channel Flow

Abstract:

Aggregating particulate flows are at the heart of an extremely broad range of phenomena in science, medicine and engineering, including biofilm formation in pipes and cellular aggregation in blood flow. These phenomena also are extremely popular targets for computer simulations, and impressive algorithmic developments over the years have enabled sophisticated, multiscale simulations of aggregating particulate flows in a variety of settings [e.g., 1,2]. Although various numerical models for particulate flow have been shown to capture many qualitative phenomena correctly, in most cases their predictive ability is not well established. Often, this is due to the sheer complexity of the various coupled phenomena at play.

One example in which this complexity is highly evident is the aggregation of platelets flowing over an injury in a blood vessel [1]. Here, each platelet represents a complex 'particle' with time-dependent shape, activity (or 'stickiness') that is also able to communicate with its local environment via an intricate sequence of biochemical pathways. The aggregation of these particles over the injury site is a multifaceted problem that requires the consideration of fluid mechanics, biochemical reaction networks, and particle assembly. While microfluidic models of platelet aggregation have greatly simplified the analysis of thrombosis [3], the complexity of the platelet biology is still present. Here, we consider a simplified microfluidic model system that removes much of this complexity, while retaining some of the essential physics. In particular, DNA-functionalized spherical particles [4] are used to simulate activated platelets, which interact with, and stick to, a patch on the microfluidic channel that presents a brush of complementary DNA single-strands.

The experimental model is simulated using several types of fluid-structure simulation (e.g., immersed boundary method and Brownian dynamics) in order to assess the roles of inter-particle DNA-mediated and hydrodynamic interactions, Brownian fluctuations, and various geometric and flow factors. The interplay between these factors and the resulting aggregation behavior is studied statistically with the aim of generating a reduced-order representation of the aggregation process.

[1] M. H. Flamm et al., Blood **120**, 190 (2012).

[2]. J. K. W. Chesnutt and J. S. Marshall, Comput Fluids **38**, 1782 (2009).

[3] T. V. Colace, G. W. Tormoen, O. J. T. McCarty, and S. L. Diamond, Annu Rev Biomed Eng. **15**, 283 (2013).

[4] M. T. Casey, R. T. Scarlett, W. B. Rogers, I. Jenkins, T. Sinno, and J. C. Crocker, Nature Communications 3, 1209 (1-8) (2012).

472891 Stress-Driven Colloidal Crystal Reassembly through Darcy Flow during Vibration Assisted Convective Deposition

Abstract:

The realization of structural diversity in colloidal crystals obtained by self-assembly techniques remains constrained by thermodynamic considerations and current limits on our ability to alter structure over large scales using imposed fields and confinement. Following previous work, a convective-based procedure to fabricate multi-layer colloidal crystal films with extensive square-like symmetry is enabled by periodic substrate motion imposed during the continuous assembly. The generation of extensive single crystalline (100) fcc domains as large as 15 mm^2 and covering nearly 40% of the colloidal crystalline film is possible by simply tuning coating conditions and multi-layer film thickness. Preferential orientation of the square-packed domains with respect to the direction of deposition is attributed to domain generation based upon a shear-related mechanism. We propose the orientation of these crystals is a result of the stress caused by Darcy flow through the crystal. Crystal orientation changes the relative resistance, causing domains to preferentially anneal into structures with prefered orientation. Further evidence is the periodic defects found in the (100) upward-facing domains that cannot be fabricated by particle-wise assembly.

468141 The Role of Hydrodynamic Interactions in Colloidal Dispersionswith Short-Ranged Attraction and Long-Ranged Repulsion

Abstract:

Colloidal gels are formed during arrested phase separation. Sub-micron, mutually attractive particles aggregate to form a system-spanning network with high interfacial area, far from equilibrium. Models for microstructural evolution during colloidal gelation have often struggled to match experimental results with long standing questions regarding the role of hydrodynamic interactions. In the present work, we demonstrate simulations of gelation in a dispersion of colloids interacting pair-wise via a short-ranged attraction and long-ranged repulsion with and without hydrodynamic interactions between the suspended particles. The disparities between these simulations are striking and mirror the experimental-theoretical mismatch in the literature. The hydrodynamic simulations agree with experimental observations, however.

We explore a simple model of the competing transport processes in gelation that anticipates these disparities, and conclude that hydrodynamic forces are essential. Near the gel boundary, there exists a competition between compaction of individual aggregates, which suppresses gelation and coagulation of aggregates, which enhances it. The time scale for compaction is mildly slowed by hydrodynamic interactions, while the time scale for coagulation is greatly accelerated by collective motion of particles within an aggregate. This enhancement to coagulation leads to a shift in the gel boundary to lower strengths of attraction and lower particle concentrations when compared to models that neglect hydrodynamic interactions.When aggregation is reaction limited, as is the case for particles interacting via a short-ranged attraction and a long-ranged repulsion, the diffusive dynamics of particle clusters are key to establishing the correct aggregation rate. Long-ranged hydrodynamic interactions between particles result in aggregates that diffuse anisotropically, which promote the growth of percolated networks as opposed to condensed domains.

This result necessitates a fundamental rethinking of how both microscopic and macroscopic models for gelation kinetics in colloids are developed. Only simulations that correctly account for hydrodynamic interactions will provide a realistic description of the stresses in a deforming network and correctly reproduce phenomena observed in experiments such as vorticity alignment under shear flow. We have now developed such a minimal model that captures aggregation kinetics accurately.

470742 Time Resolved Study of Shear-Induced Microstructures of Concentrated Silica Dispersions By SAXS

Abstract:

Concentrated colloidal dispersions sometimes show non-equilibrium structure formation during shear. These microstructural behaviors affect the macroscopic rheological response. However, there is still a lack of detailed understanding of the link between rheology and microstructure formation, especially regarding the onset of shear thinning and shear thickened/jammed structure. Here, we made simultaneous measurements of rheology and shear-induced structure by small angle x-ray scattering (SAXS) at Sector 8-ID-I of Advanced Photon Source. Concentrated silica dispersions were loaded in Couette cell. The x-ray scattering was measured on velocity-vorticity plane. Under moderate oscillatory shear, particles are rearranged into hexagonal packed layered structures stacking along gradient direction. This rearrangement resulted in reducing the resistance against flow, leading to shear thinning. Upon faster shear, the ordered structure breaks down. But we found that this order-to-disorder transition occurs in a different shear stress regime than the typical shear thickening state, and its occurrence is highly dependent upon oscillatory frequency, particle size and its distribution.

460197 The Motion of Magnetic Oblate Spheroids Suspended in Newtonian Fluids Under Magnetic Fields

Abstract:

Anisotropic structures are found in many natural materials. People are committed to developing materials that contain aligned anisotropic particles. In this study, we used a rotating magnetic field to align magnetic microdisks. Both rotational and translational motions of particles and different structures of particles are observed. We have developed an analytic solution that describes the alignment dynamics of microdisks. The analytic solution covers the entire possible frequency range of a rotating magnetic field. We also have developed asymptotic solutions at both the high-frequency limit and the low-frequency limit of the rotating magnetic field, which can be easily implemented in industrial applications. Experimental results confirm the theoretical models. We have simulated the translational motion of the microdisks and the formation of different structures that result from induced magnetic fields within the particles. With the knowledge of the dynamics of microdisks suspended in fluids under magnetic fields, we can create composites with anisotropic structures. The three-axis electromagnetic coil system used in this study can align the particles to the desired orientation, and the particles can be frozen inside a solid matrix using a UV-curable polymer.

461181 Modeling of CO_2 Solubility in Electrolyte Solutions and Brines Using Statistical Associating Fluid Theory

Abstract:

Understanding the behavior of CO_2, especially its solubility, in electrolyte solutions and brines is of great importance to the design of a CO_2 geologic sequestration project. Since the properties of CO_2 bearing electrolyte solutions are strongly affected by the long-rang Coulomb interactions of ions, an accurate description of ion-ion interactions is important to the modeling of CO_2-H_2O-electrolyte system. In this work, CO_2 solubilities in single and mixed electrolyte solutions as well as synthetic geologic formation brines were correlated or predicted with statistical associating fluid theory (SAFT). [1, 2] In the proposed SAFT equation of state, the ion-ion interactions were represented by an improved mean spherical approximation in the primitive model using a parameter *K* to correct the excess energies (â€œKMSAâ€ for short). [3] The universal correction factor *K* was obtained by adjusting the internal and excess energies of mean spherical approximation to Monte Carlo simulation data. With the SAFT-KMSA model, we studied a range of aqueous electrolyte solutions including Na^+, K^+, Ca^{2+}, Mg^{2+}, Cl^-, Br^- and SO_4^{2-} from 298.15 K to 473.15 K. The cross interactions for CO_2-H_2O and CO_2-ion dispersion interactions were adjusted with binary interaction parameters k_{ij}, which were fitted to CO_2 solubility data in water and single electrolyte solutions. Without additional parameters to fit, the proposed model was conveniently extended to predict CO_2 solubilities in mixed electrolyte solutions, including NaCl+KCl, KCl+$CaCl_2$, NaCl+KCl+$CaCl_2$, and in synthetic geologic formation brines. The proposed model shows good predictive power for CO_2 solubility in mixed electrolyte solutions and brines.

[1] Jiang, H.; Panagiotopoulos, A. Z.; Economou, I. G. *Geochim. Cosmochim. Acta*. **2016**, 176, 185.

[2] Tan, S. P.; Ji, X.; Adidharma, H.; Radosz, M. *J. Phys. Chem. B*. **2006**, 110, 16694.

[3] Jiang, H.; Adidharma, H. *Mol. Simul*. **2015**, 41, 727.

461444 Molecular Simulations of Ammonium-Based Protic Ionic Liquids

Abstract:

The structure and transport properties of a series of trialkylammonium triflate protic ionic liquids (PILs) and associated solutions involving neutral species were studied using molecular dynamics (MD). It is well know that the use of classical force fields where the ions carry net unit charges can lead to overstructured liquids and poor estimation of dynamical and transport properties, in particular too-high viscosities and too-low diffusion constants and ionic conductivities. The application of a uniform charge scaling factor can effectively address this issue, and is now commonly used. In this work we have determined the optimal scaling factors for several trialkylammonium triflate PILs simulated with the force field of Lopes and Padua (*J. Phys. Chem. B* **2004**, 108, 16893) by comparison with recent experimental data from Yasuda *et al.* (*J. Chem. Eng. Data* **2013**, 58, 2724). Simulations performed with optimized charge scaling showed generally good agreement with experimental data, and in all cases were much improved over the original force field. While diffusion constants could be very well-described over a large temperature range, even with optimized charge scaling the viscosity tended to be overestimated at low temperatures, and the ionic conductivity underestimated at most temperatures. We hypothesize that this is due to poor reproduction of the temperature dependence of the density; while the density near room temperature is well-reproduced, the thermal expansion coefficients of the simulated PILs are systematically low, leading to over-high densities at higher temperatures. The optimal scaling factors were found to vary significantly even between PILs where the cation differed only by a single methyl group. In the second part of this work, the solvation and dynamics of the neutral products of proton transfer reactions (trialkylamines and triflic acid) were studied in their parent PILs, in order to better understand how acid-base equilibria in PILs affects their properties.

465525 Molecular Dynamics Simulations of Mixtures of Refrigerants and Deep Eutectic Solvents

Abstract:

Molecular dynamics simulations of mixtures of refrigerants and deep eutectic solvents

Rubaiyet Abedin, John C. Flake and Francisco R. Hung

Cain Department of Chemical Engineering, Louisiana State University, Baton Rouge, LA 70803

Heating and cooling buildings in the U.S. consumes an enormous amount of energy (>10 quadrillion BTU), and is responsible for adding ~1 billion metric tons of CO_2 in the earth's atmosphere every year. Much of this energy is used as electricity in vapor-compression systems; however, this technology is mature and only evolutionary improvements are expected in the near future. Remarkably, a few studies have shown that several common ionic liquids (ILs) can be combined with standard fluorocarbon refrigerants for use in absorption refrigeration systems that use waste heat at relatively low temperatures (~100 °C). Nevertheless, there is limited understanding (and data) on the VLE behavior of these systems, and only one working example of an absorption system using this type of mixture. Furthermore, deep eutectic solvents (DESs), a relatively new class of solvents, share many of the properties of ILs while being considerable cheaper and mostly nontoxic. A fundamental understanding of how the chemical structure of the different species affects the solubility of fluorocarbons in a DES is crucial to design mixtures suitable for use in absorption refrigeration systems that use solar energy or waste heat. In this work, we performed molecular dynamics simulations of mixtures of a conventional fluorocarbon refrigerant, 1,1,1,2-tetrafluoroethane (R134a) with three deep eutectic solvents (1:2 choline chloride/urea, 1:2 choline chloride/glycerol and 1:2 choline chloride/ethylene glycol). We report and discuss a number of properties for these systems, including Henry's law constants of R134a in the three DESs as a function of temperature, radial distribution functions, and diffusion coefficients.

466418 Phase Equilibria in Ionic Liquid- Alkanol Solutions

Abstract:

Ionic liquids (IL) can offer an enhanced potential in improvement of existing processes due to the fact that their properties can be specifically customized for a particular application by the corresponding combination of cations and anions. To provide preliminary screening of the great number of possible combinations, a-priori predictive tools such as Conductor-like Screening Model for Realistic Solvation (COSMO-RS) are of indisputable value.

To apply any molecular model, the right representation of a molecule is necessary. Till now, it is still not clear in what form IL molecules are present in solution and thus how to treat them in the molecular calculations: as a single neutral compound (molecular form/ion pair), as an equimolar mixture of cations and anions (dissociated form/single ions) or as a combination of both forms. Moreover, it has recently been shown that in diluted aqueous and methanol solutions ILs can also build ionic clusters, consisting of several cations and anions. So, IL may be present in solutions in any of the above-mentioned forms. The main objective of this work is to study the influence of different representations of IL on liquid-liquid (LLE) and solid-liquid (SLE) phase equilibria in the frame of COSMO-RS model and focuses on the IL-alcohol systems.

In the present study, the aggregation behavior of ionic liquids (1-alkyl-3-methylimidazolium chlorides) are experimentally studied in diluted alcohol (from methanol to butanol) solutions via mass spectrometry (QTrap, AB Sciex). Formation of positively charged clusters consisting of up to eight cations and seven anions as well as negatively charged clusters containing up to nine anions and eight cations have been observed. Cluster solvation energies calculated using COSMO-RS reproduce the experimentally obtained trends.

The COSMO-RS predictions of SLE and LLE are carried out in binary mixtures of 1-alkyl-3-methylimidazolium chlorides as well as with imidazolium-based cations and other anions (e.g., hexafluorophosphate,

bis(trifluoromethylsulfonyl)amide, tetrafluoroborate) with alkanols (from methanol to octanol). The calculations are based on different representation of IL in the liquid phase (ion pairs, single ions, clusters) and compared to experimental data available in literature.

The experimental and modeling results are analyzed as a function of the length of the side chain in IL cation and polarity of the alcohol. The obtained trends are evaluated to explore the effect of different representations of IL and to gain preliminary understanding of how to model IL molecules in solution: as completely dissociated to single ions, as neutral molecules, as charged clusters or as a mixture of all those.

MODELING GAS SEPARATION OF GREENHOUSE GASES WITH IONIC LIQUIDS AND DEEP EUTECTIC SOLVENT

Joel Orduña-Lloret, IQS School of Engineering, Universitat Ramon Llull, Barcelona, Spain
Lourdes F. Vega, The Petroleum Institute, Abu Dhabi. United Arabs Emirates
Fèlix Llovell, IQS School of Engineering, Universitat Ramon Llull, Barcelona, Spain

Abstract

This work illustrates several practical examples based on the solubility of greenhouse gases on selected ionic liquids (ILs) and Deep Eutectic Solvents (DESs), in order to evaluate its feasibility as a possible greener media for gas separation. A coarse-grained molecular model, within the framework of the soft-SAFT equation of state is presented for each molecule based on structural information. Particular attention is paid on the hydrogen-bonding effects on these molecules when building an accurate molecular model to describe the main gas-solvent interactions. The absorption of relevant gases is accurately describe and compared with experimental data using a minimum amount of information, trying to assess the most appropriate solvent and the optimal operating conditions to maximize each gas separation.

Introduction

In the search for new types of green solvents, ionic liquids (ILs) and deep eutectic solvents (DESs) have come to the fore. The popularity of ILs grew rapidly due to their low melting points, negligible vapour pressure and wide liquid range, among other properties. However, the popular conception that they are all intrinsically green is not true. Some ILs are as toxic as some organic solvents, depending on their chemical structure. In addition, their synthesis is typically expensive compared to conventional volatile organic compounds. Recently, DESs have appeared as potential solvents as they share many of the IL properties but coming in at lower cost. They are mixtures of different substances in particular proportions that melt and freeze at a lower temperature than those of the individual components or of mixtures of the components in other proportions. This property is a consequence of hydrogen bonding between the components. DESs are non-volatile, biodegradable and do not react with water, becoming potential candidate for a wide variety of applications. During the past decade, these compounds have drawn academic and commercial attention due to their unusual chemical and physical properties that make them valuable alternatives to traditional compounds for different applications, such as separation and extraction processes.1 One of the applications that is raising more interest is their possible use as solvents to carry gas separation in the context of carbon capture. It is well known that conventional technology of using chemical absorption through an aqueous solution of amine has important drawbacks, including loss of solvent, corrosion of facility, and high energy demand regeneration of the absorbent.[2] ILs and DESs are a possible kind of novel media that, with an appropriate selection, could result in a non-contaminated target gas especially attractive for the separation of these gases.

At the current moment, there is still a need for a further understanding of the dependence of the gas behavior on ILs and, specially, on DESs. Molecular modeling techniques and, in particular, molecular-based equations of state (EoSs) provide an attractive option to screen ILs properties and give quick answers to guide the experiments, becoming a crucial tool for process design.[3] In the framework of EoSs, the soft-SAFT EoS[4] provides a physically grounded theoretical approach to describe the thermophysical properties and phase behavior of complex molecules and their mixtures in good agreement with the experimental data. Previous contributions have already been successfully applied to model the solubility of certain gases on ILs.[5-13] The key of this success is based on the fact that the equation explicitly accounts for hydrogen bonding effects in a specific term coming from Wertheim's first-order perturbation theory.[14] Consequently, the purpose of this work is to exploit the possibilities of soft-SAFT as a powerful tool to assess gas separation processes by exploring new ILs and describing, for the first time, some promising DESs as alternatives with a lower cost.

Methodology

The soft Statistical Associating Fluid Theory (SAFT)[4] is a well-established variant of the original SAFT equation of state.[15] Based on Wertheim's first-order perturbation theory[14], it provides a framework in which the effects of molecular shape and intermolecular interactions on the thermodynamic properties of a system are explicitly accounted for. The equation calculates the total free energy of the system, splitting it into a sum of different contributions that consider several molecular effects. For associating and polar molecules, the soft-SAFT approach calculates the Helmholtz free energy of the system as a sum of an ideal contribution A^{ideal}, a reference term A^{ref}, for the attractive and repulsive forces between the segments that form the molecules, a chain contribution A^{chain}, which accounts for the connectivity of the segments in the molecules, a contribution due to site-site intermolecular association A^{assoc}, and a polar term, A^{polar}:

$$\frac{A^{total}}{Nk_BT}-\frac{A^{ideal}}{Nk_BT}=\frac{A^{res}}{Nk_BT}=\frac{A^{ref}}{Nk_BT}+\frac{A^{chain}}{Nk_BT}+\frac{A^{assoc}}{Nk_BT}+\frac{A^{polar}}{Nk_BT}, \tag{1}$$

where N is the number of molecules, k_B is the Boltzmann constant and T is the temperature. A^{res} is the residual Helmholtz free energy density of the system. The reference term of the soft-SAFT EoS is given by a Lennard-Jones (LJ) spherical fluid, considering repulsive and attractive interactions of the monomers in a single contribution. This intermolecular potential includes the segment diameter of the monomers σ_{ii} and the dispersive energy between segments ε_{ii}/k_B. The equation of Johnson et al,[16] adjusted to molecular simulations of Lennard-Jones monomers, is used here to calculate the free energy of this term.

The chain and association terms come from Wertheim's theory and are formally identical in all SAFT equations:

$$\frac{A^{chain}}{Nk_BT}=\sum_i x_i(1-m_i)\ln g_{LJ} \tag{2}$$

$$\frac{A^{assoc}}{Nk_BT}=\sum_i x_i\sum_a\left(\ln X_i^a-\frac{X_i^a}{2}\right)+\frac{M_i}{2} \tag{3}$$

The chain term includes the chain length parameter m_i, while the association term involves two additional parameters defining the volume (κ^{HB}) and energy (ε^{HB}/k_B) of association. The polar term is calculated using an expansion of the Helmholtz energy in terms of a perturbed quadrupole-quadrupole potential with the Padé approximation,[17] and it includes the quadrupole moment Q of a molecule.

The reference term extension to mixtures is done by applying the van der Waals one-fluid theory with the Lorentz-Berthelot mixing rules for the size σ_{ii} and energy ε_{ii} parameters:

$$\sigma_{ij} = \frac{\sigma_{ii} + \sigma_{jj}}{2} \tag{4}$$

$$\varepsilon_{ij} = \xi\left(\varepsilon_{ii}\varepsilon_{jj}\right)^{1/2} \tag{5}$$

ξ is an adjustable binary parameter to adjust the cross-dispersive energy of the systems.

Molecular models

The success of a molecular-based equation of state relies on the development of a simple but accurate molecular model, able to capture the main physical features of the compound through a coarse-grained approach. Hence, it is necessary to develop a model for each one of the compounds involved in this work (gases, ILs and DESs) within the soft-SAFT framework.

Carbon dioxide (CO_2) and nitrous oxide (N_2O) are modeled as a LJ chain in which explicit quadrupolar interactions are taken into account. Consequently, in addition to the molecular parameters described above, they also require the quadrupole moment, Q, and x_p, the fraction of segments in the chain that contains the quadrupole. x_p is fixed to 1/3 for CO_2 and N_2O to mimic molecules as three segments with a quadrupole in one of them. The effective quadrupole moments for these compounds, obtained from fitting, was within the range of the bibliographic values. The final set of molecular parameters are taken from references 12 and 18. Hydrogen sulfide (H_2S) is modeled as an associating molecule with 3 interaction sites: 2 for the sites of type *H* for the hydrogen atoms and 1 site of type *e* for the electronegativity of the sulfur, with only *e-H* interactions allowed. Finally, sulfur dioxide (SO_2) is represented by a LJ chain with two associating sites of different nature (positive-negative) to mimic in an effective manner the nature of the dipole moment of the molecule. Details about the choice of these models and the final molecular parameters of H_2S and SO_2 can be found in reference 8.

The modeling of ILs with soft-SAFT is based on the assumption that dispersion forces, specific steric interactions and, possibly, the formation of short-lived ion pairs reduce the ionic character of ionic liquids, as observed in molecular simulations.[19] Based on this principle, they are modeled as a single chain, where the cation and the anion are together. In order to account for the highly directional interactions of shorter range, a number of association sites have to be specified. These association schemes are based on the guidance of COSMO-RS quantum calculations and previous experience. In this particular work, a few selected imidazolium ILs have been described to evaluate its performance for gas separation purposes. The [C_4mim][$N(CN)_2$] IL is modeled as an associating molecule with two sites, one A site representing the N−atom interactions with the cation, and one B site representing the delocalized charge due to the two extra N−atoms on the anion. On the other hand, the preliminary COSMO-RS calculations for the [C_4mim][SCN] IL present two near charged density peaks related to the delocalization of the anion negative charge due to its resonance structure. Due to its nearness, a preliminary study was made proposing either one or two association sites. As analogous results were found, the simplest one-site model was proposed. The [C_4mim][Ac] pair was modeled considering two associating sites, one A site which would mimic the O−atom interactions with the cation, and one B site representing the delocalized charge due to the second O−atom on the anion. Finally, [C_4mim][$MeSO_4$] is described with two associating sites that describe the specific cation-anion interactions due to the charges and the asymmetry: one site A, representing the interaction of the sulfur with the cation, and one site B, representing the anion delocalized charge, due to the presence of O-atoms. In all cases, only AB interactions between different ionic liquids molecules are permitted.

Concerning DESs, the same pseudo-pure component approach has been used to model these compounds, with an average molar mass calculated based on the composition of the DES constituents. One of the main issues in proposing a molecular model for DESs is the lack of information related to the physical interactions occurring in this solvent. A recent contribution on the Choline Chloride:Urea (1:2) DES, based on molecular dynamics simulations, indicates the presence of a "soup" of multiple hydrogen bonding interactions.[20] In a first simple approach, these interactions are accounted by the addition of four association sites to this compound, two of positive nature (representing the OH from Choline and NH from Urea) and two of negative nature (representing the Cl^- from Chloride and O=C from Urea). Finally, the Tetraalkylammonium Chloride:Lactic Acid (TXACl:LA) DESs have also been studied in this work. To our knowledge, there is no available information on the interactions dominating its physicochemical behavior. As a first trial, we have followed the pseudo-pure component approach suggested by Kroon and co-workers with PC-SAFT,[21] and have modeled those compounds with two association sites (one positive A and one negative B) to consider the self-association between the hydrogen bond donors and the hydrogen bonding acceptors of the DES.

The molecular parameters for all the gases are fitted to phase equilibrium liquid densities and vapor pressures. On the other hand, due to their very low vapor pressures, the molecular parameters of ILs and DESs are fitted to temperature-density data at atmospheric pressure. The molecular parameters for the ILs are taken from a previous contribution,[12] while those for DESs are presented here for the first time. In all cases, we have chosen the most meaningful set of parameters from a physical point of view, allowing the establishment of trends with the molecular weight of each family of ILs and DESs.

RESULTS

Ionic Liquids

Two main examples are highlighted in this section. In this first example, several [C_4mim] ILs have been tested in order to find the most appropriate anion for the difficult CO_2/N_2O separation. The solubility of these gases in ILs with the anions $[N(CN)_2]^-$, $[SCN]^-$ and $[Ac]^-$ are studied and compared among them, searching for the best system for separation purposes. The adequacy of the proposed molecular models and parameters values for each IL is confirmed by the description of the solubility of CO_2 and N_2O. A fully predictive approach (no binary parameter used) is found for the N_2O systems at several temperatures. CO_2 solubility is also predicted for [C_4mim][$N(CN)_2$], while a temperature independent binary interaction parameter close to unity (ξ=0.98) is used for [C_4mim][SCN]. For [C_4mim][Ac], the particular phase equilibrium of its mixture with CO_2 due to a chemisorption process, is addressed through the representation of this chemical reaction by a cross-associating scheme involving the addition of two association sites in the CO_2 only able to cross-associate with the IL (κ^{HB}=950Å^3, ε^{HB}/k_B=5500K). Excellent agreement is found between the modeling and the experimental data,[22] as it can be seen in **Figure 1** for the [C_4mim][Ac] case.

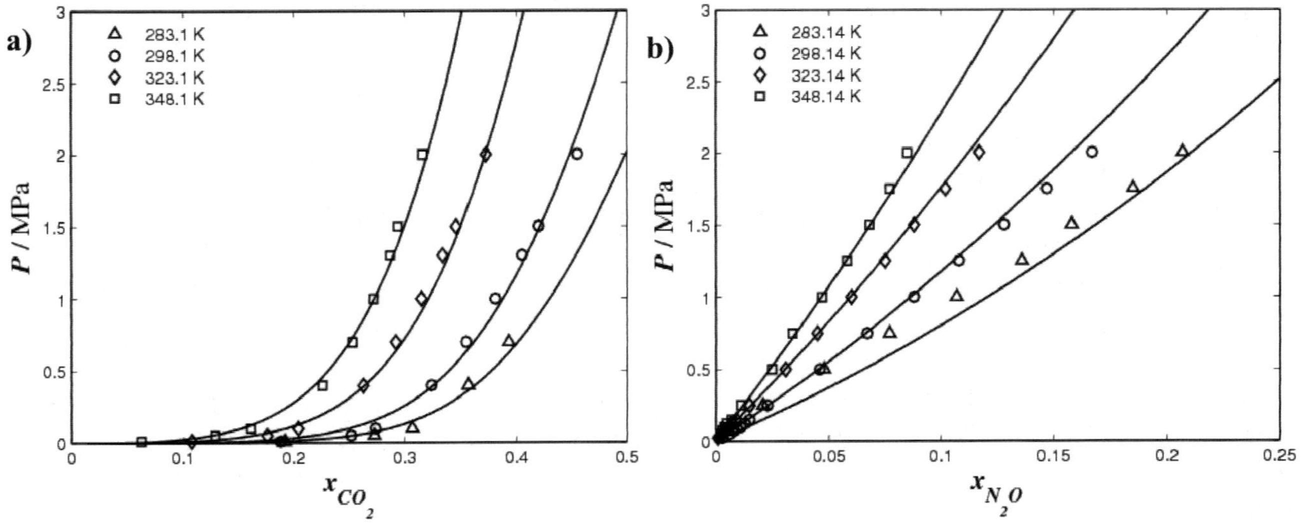

Figure 1. *Pressure-composition diagram of the binary system* ***a)*** *CO_2 + [C_4mim][Ac] at 283.10 K (triangles), 298.10 K (circles), 323.10 K (diamonds), and 343.10 K (squares); and* ***b)*** *N_2O+ [C_4mim][Ac] at 283.14 K (triangles), 298.11 K (circles), 323.14 K (diamonds), and 347.71 K (squares). Symbols are experimental data[22] and the solid lines represent the soft-SAFT EoS results.*

The Henry's constants and the calculated selectivity resulting from the soft-SAFT calculations are shown in **Table 1**. From these results, it is concluded that [C_4mim][Ac] is the only appropriate IL to assess this separation.

Table 1. Henry´s law constants (MPa) and calculated selectivity for CO_2 and N_2O in the selected ILs at 298K.

	H_{CO2}*	H_{N2O}*	$S_{CO2/N2O}$*
[C_4mim][N(CN)$_2$]	8.86	9.11	1.03
[C_4mim][SCN]	12.0	9.66	0.80
[C_4mim][Ac]	0.027	10.9	400

Another interesting example is described next. In this case, the butylmethylimidazolium methylsulfate [C_4mim][$MeSO_4$] IL has been studied to assess the CO_2/SO_2 and CO_2/H_2S gas separations. In this case, the solubility of these gases on the selected IL have been reproduced in agreement with the experimental data using one binary parameter.[13] Once the binary mixtures are accurately described, the ternary diagrams of the mixtures SO_2 + CO_2 + [C_4mim][$MeSO_4$] and H_2S + CO_2 + [C_4mim][$MeSO_4$] have been calculated in a predictive manner at 298.15K in order to obtain additional insight about the best conditions for the separation of these acid gases from CO_2, and to better determine the effect of the immiscibility found in the CO_2 binary mixture. As it can be seen in **Figure 2a** for the SO_2 case, large areas of immiscibility have been found, preventing the separation using elevated concentrations of the ionic liquid at moderate to high pressures.

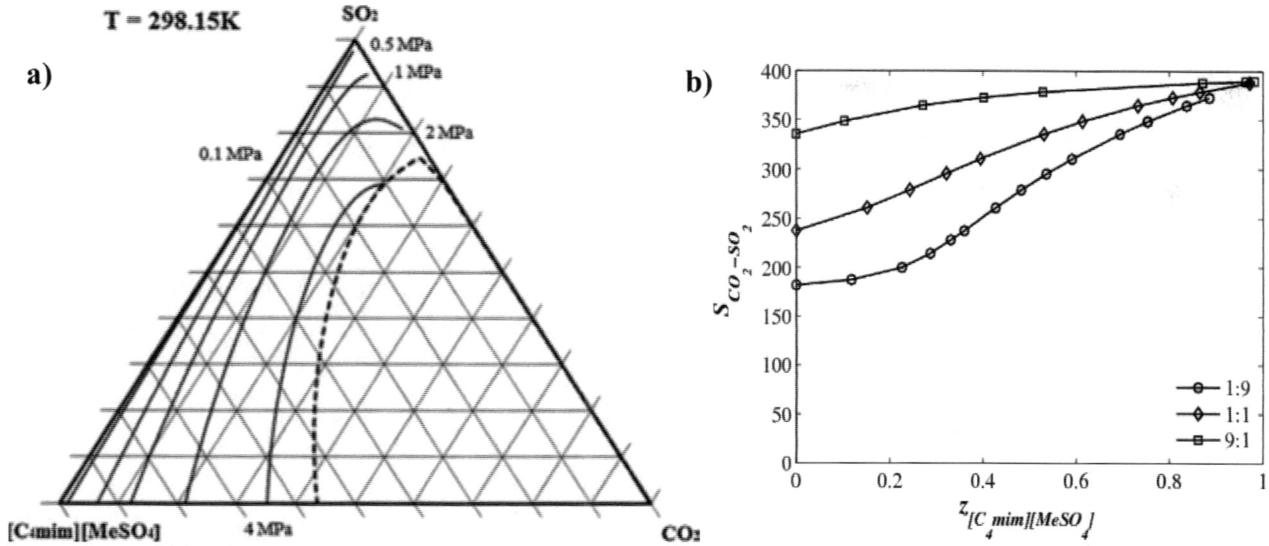

Figure 2. ***a)*** *Isothermal ternary phase diagram at T=298.15 K predicted with soft-SAFT for the system CO_2-SO_2-[C_4mim][$MeSO_4$]. The solid curves represent the calculation at different isobars at 0.1 MPa, 0.5 MPa, 1 MPa 2 MPa and 4 MPa, while the dashed curve represents the VLLE immiscibility curve. b) Selectivity of the CO_2/SO_2 separation as a function of the feed[C_4mim][$MeSO_4$] mole % at three different CO_2/SO_2 feed ratios of 1:9 (circles), 1:1 (diamonds) and 9:1 (squares) at T = 298.15 K and P = 0.1 MPa.*

The selectivity at 298.15K and P_{atm} has been calculated as a function of three different feed ratios, as it is shown in **Figure 2b** for the SO_2 case. The results have shown that the conditions studied are suitable for a very favorable separation of CO_2 and SO_2 using [C_4mim][$MeSO_4$], even at very low concentrations of the IL, as very high selectivity values (over 200) are achieved. For the H_2S case, moderate improvements are obtained, with a selectivity value around 11, independently of the IL feed ratio, suggesting that there is still some room for improvement by the search of a more suitable solvent.

Deep Eutectic Solvents

As mentioned in the molecular models section, a pseudo-pure component approach has been proposed to describe some selected DESs. The results of the fitting to liquid density at atmospheric pressure have provided the set of molecular parameters given in Table 2.

Table 2. Molecular parameters optimized with SAFT for selected Deep Eutectic Solvents.

Compound	M_w (g/mol)	m	σ (Å)	ε/k_B (K)	ε^{HB}/k_B (K)	κ^{HB} (Å^3)	AAD_D (%)
ChCl: Urea (1:2)	95.44	3.090	3.464	355.1	3478	2000	0.0078
TMACl:LA (1:2)	96.59	3.146	3.479	390.0	3200	1390	0.0456
TEACl:LA (1:2)	115.29	3.503	3.594	390.0	3200	1390	0.0459
TBACl:LA(1:2)	152.69	4.030	3.870	390.0	3200	1390	0.0173
TBACl:LA(1:3)	137.04	4.408	3.596	390.0	3200	1390	0.0290

In an effort to follow a more transferable methodology, the dispersive energy and the association parameters of al TXACl:LA DESs were kept constant, while both the segment chain *m* and the segment diameter σ increase with the molecular weight and can be easily correlated. This information is important in order to determine the parameters of other members of this family for which experimental data are not available.

Once a set of molecular parameters has been found for each DES, its accuracy is tested by modeling the solubility of different gases in them. For the ChCl:Urea (1:2) DES, the description of the solubility of CO_2 and SO_2 has been modeled using a temperature-independent binary parameter, finding good agreement with the experimental data,[23,24] as it is shown in **Figure 3**. The validation of the model will allow further calculations at other conditions where experimental data are not available.

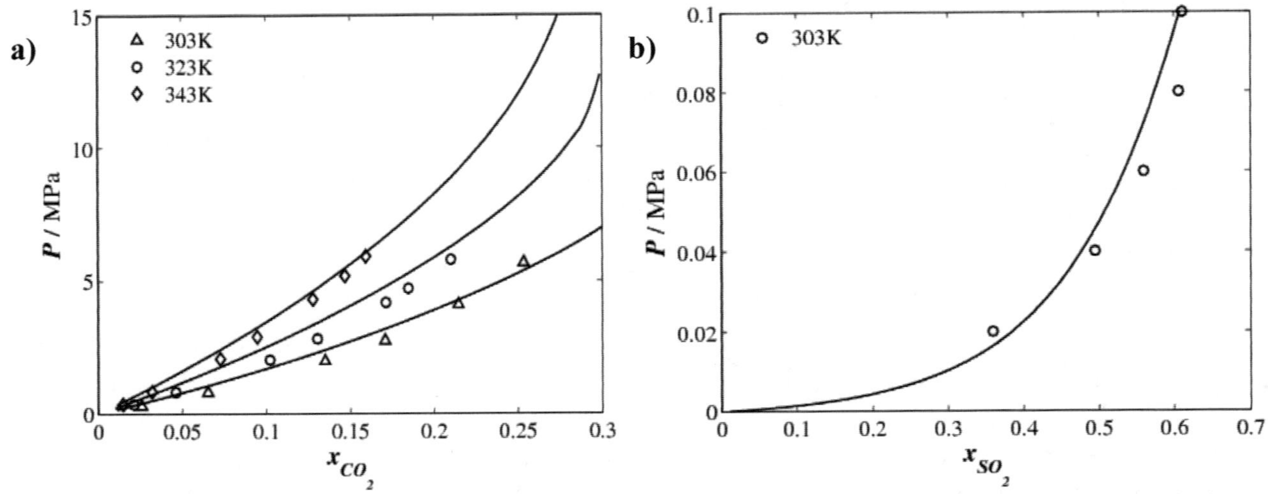

Figure 3. *Solubility in Choline Chloride:Urea (1:2) of* ***a)*** *CO_2 at 303 K (triangles), 323 K (circles) and 343 K (diamonds) using ξ=1.05.* ***b)*** *SO_2 at 303 K (circles) using ξ=1.21. Symbols are experimental data[23,24], while the lines are the soft-SAFT calculations*

Concerning the TXACl:LA DESs, the validity of the proposed sets of parameters has been checked by determining how accurately the solubility of CO_2 in those compounds is described. As done for ILs and for ChCl:Urea(1:2), the binary parameter ξ has been fitted to experimental data[21] in order to obtain quantitative agreement. For each mixture, only the intermediate temperature isotherm is used to find the optimum value of ξ, while the other two isotherms are predicted by transferring this value. This procedure gave excellent results, as it can be seen in the example shown in **Figure 4a** for the TEACl:LA (1:2) case. Although all the optimal values were around the unity, this fitting revealed a linear ξ decrease as far as the molecular weight of the compound was increased. The transferability of the molecular model was tested by only fitting the three TXACl:LA DESs having a (1:2) ratio, and predicting the binary parameter of TBACl:LA (1:3) from the correlation obtained with the molecular weight. The predicted value (ξ=0.976) allows an excellent description of the CO_2 solubility at the three isotherms (see **Figure 4b**), reinforcing the accuracy of the simple, but reliable molecular model proposed with soft-SAFT.

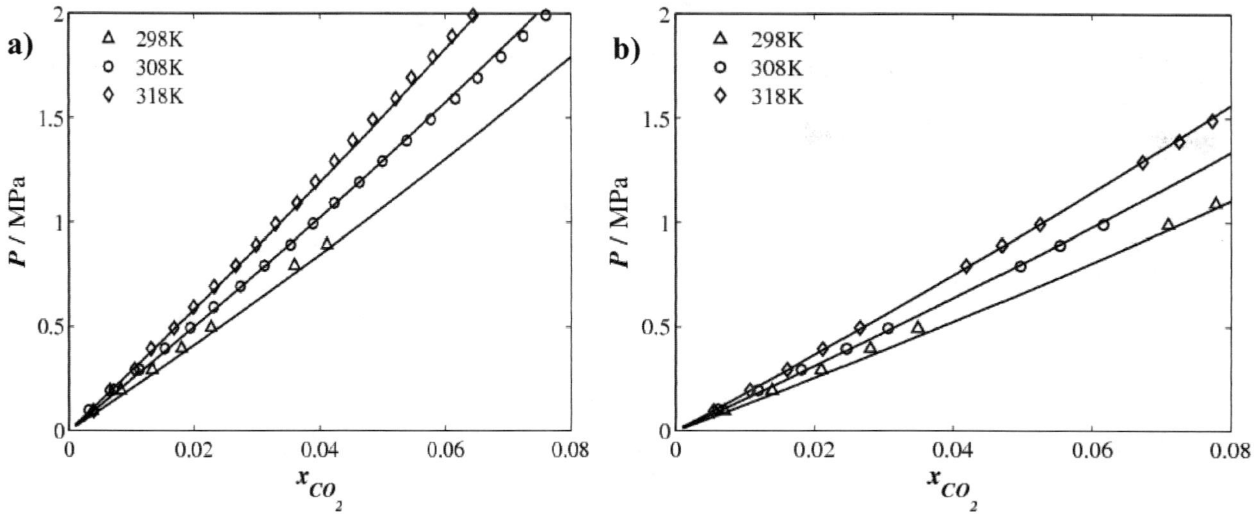

Figure 4. *Solubility of CO_2 at 298 K (triangles), 308 K (circles) and 318 K (diamonds) in **a)** TEACl:LA (1:2) ξ=0.995 and **b)** TBACl:LA (1:3) with predicted ξ=0.976. Symbols are experimental data[21] and lines are the soft-SAFT calculations.*

CONCLUSIONS

The use of imidazolium-based ionic liquids and deep eutectic solvent for the study of the separation of gases and contaminants has been addressed using the soft-SAFT molecular-based equation of state. A relatively simple molecular model, where the cation and the anion are considered as a chain for ILs, and the two compounds forming the eutectic mixture are considered as an individual for DESs has been proposed for the ionic liquids. The molecular parameters of ILs and DESs have been fitted to temperature-density data at atmospheric pressure, finding optimal sets with very good agreement with the experimental data. Then, the solubility of CO_2, SO_2, H_2S and N_2O on selected solvents has been tested. In all cases, very good agreement is found in a wide range of temperatures and pressures, using a constant temperature independent binary parameter, allowing predictions at conditions where experimental data is not available. In the practical examples highlighted here, it is confirmed that [C$_4$mim][Ac] selectivity toward CO_2 is several orders of magnitude higher than for N_2O, showing that this IL can be used for a practical separation of CO_2/N_2O mixtures. The soft-SAFT model allows the calculation of the selectivity in a predictive manner at any temperature. Also, it is seen that [C$_4$mim][MeSO$_4$] is a good option to separate CO_2 from SO_2, while better alternatives can be found for the H_2S separation. The prediction of the ternary diagram behaviour from soft-SAFT allows a direct calculation of the selectivity at any feed ratio condition. Finally, the preliminary results of the modelling of DESs are in very good agreement with the experimental data at all temperatures for both CO_2 and SO_2. In particular, the TXACl:LA family is described with a transferable set of parameters, which allow the description of the CO_2 solubility for other eutectic ratios in a predictive manner. From these results, it is expected to screen the optimal condition for gas separation without the necessity of carrying out a vast amount of experiments.

This work highlights how molecular modeling tools can assess the choice of the right solvent in a particular application, something very useful when dealing with ILs. Further research is expected to present more complex and realistic models of Deep Eutectic Solvents (DES), by modeling them as two independent species and considering the main hydrogen-bonding interactions that occur in the system, based on recent molecular simulation results.[20]

ACKNOWLEDGEMENTS

Financial support from the Catalan government under project 2014-SGR1582 is gratefully acknowledged.

REFERENCES

1. Aparicio, S., Atilhan, M., Karadas, F. (2010), Review on the use of Ionic Liquids (ILs) as Alternative Fluids for CO_2 Capture and Natural Gas sweetening, *Ind. Eng. Chem. Res.* 49 pp. 9580-9595.
2. Olajire, A. (2010) CO_2 capture and separation technologies for end-of-pipe applications – A review. *Energy*, 45 pp. 2610–2628.
3. Vega, L. F., Vilaseca, O., Llovell, F., Andreu, J. S. (2010), Modeling ionic liquids and the solubility of gases in them: Recent advances and perspectives, *Fluid Phase Equilib.* 294, pp. 15-30.
4. Blas, F. J., Vega, L. F. (1997), Thermodynamic behaviour of homonuclear and heteronuclear Lennard-Jones chains with association sites from simulation and theory, *Mol. Phys.* 92, pp. 135-150.
5. Andreu, J.S., Vega, L.F. (2007), Capturing the solubility behavior of CO_2 in ionic liquids by a simple model, *J. Phys. Chem. C,* 111, pp. 16028-16034.
6. Llovell, F., Valente, E., Vilaseca, O., Vega, L. F. (2011), Modeling Complex Associating Mixtures with [C_n-mim][Tf_2N] Ionic Liquids: Predictions from the Soft-SAFT Equation, *J. Phys. Chem. B.* 115, pp. 4387-4398.
7. Llovell, F., Vilaseca, O. and Vega, L.F. (2012), Thermodynamic Modeling of Imidazolium-Based Ionic Liquids with the $[PF_6]^-$ Anion for Separation Purposes, *Sep. Sci. & Tech.* 47, pp. 399-410.
8. Llovell, F., Marcos, R.M., Mac Dowell, N., Vega, L.F. (2012), Modeling the Absorption of Weak Electrolytes and Acid Gases with Ionic Liquids Using the Soft-SAFT Approach, *J. Phys. Chem. B.* 116, pp. 7709−7718.
9. Vega, L.F., Llovell, F., Blas, F.J. (2009), Capturing the Solubility Minima of n-Alkanes in Water by Soft-SAFT, *J. Phys. Chem. B* 113, pp. 7621-7630
10. Llovell, F., Oliveira. M.B., Coutinho. J.A.P., Vega, L.F. (2015), Modeling the [NTf_2] Pyridinium Ionic Liquids Family and Their Mixtures with the soft Statistical Associating Fluid Theory Equation of State, *J. Phys. Chem. B.* 116, pp. 9089-9100.
11. Pereira, L.M.C., Oliveira, M.B., Dias, A.M.A., Llovell, F., Vega, L.F., Carvalho, P.J., Coutinho, J.A.P. (2013), High pressure solubility of CH_4, N_2O and N_2 in 1-butyl-3-methylimidazolium dicyanamide: Solubilities, selectivities and soft-SAFT modeling, *Int. J. Greenh. Gas Cont.* 19, pp. 299–309.
12. Pereira, L.M.C., Oliveira, M.B., Llovell, F., Vega, L.F., Coutinho, J.A.P. (2014) Assessing the N_2O/CO_2 high pressure separation using ionic liquids with the soft-SAFT EoS *J. Supercrit. Fluids* 92, pp. 231–241.
13. Llovell, F., Oliveira. M.B., Coutinho. J.A.P., Vega, L.F. (2015), Solubility of greenhouse and acid gases on the [C_4mim][$MeSO_4$] ionic liquid for gas separation and CO_2 conversion. *Catal. Today* 255, pp. 87–96
14. Wertheim, M.S. (1984), Fluids with highly directional attractive forces. 2. thermodynamic perturbation-theory and integral-equations, *J. Stat. Phys.* 35 pp. 35-47.

15. Chapman, W.G., Gubbins, K.E., Jackson, G., Radosz, M. (1989), SAFT: Equation-of-state solution model for associating fluids, *Fluid Phase Equilib.,* 52, pp. 31-38.
16. Johnson, J.K., Zollweg, J.A., Gubbins, K.E. (1993), The Lennard–Jones equation of state revisited, *Mol. Phys.* 78, pp. 591-618.
17. Stell, G., Rasaiah, J. C., Narang, H. (1974), Thermodynamic perturbation theory for simple polar fluids. II, *Mol. Phys.* 27, pp. 1393-1414.
18. Dias, A.M.A., Carrier, H., Daridon, J.L., Pàmies, J.C., Vega, L.F., Coutinho, J.A.P., Marrucho, I.M. (2006), Vapor–liquid equilibrium of carbon dioxide–perfluoroalkane mixtures: experimental data and SAFT modeling, *Ind. & Eng. Chem. Res.* 45, pp. 2341-2350.
19. Weiss, V.C., Heggen, B., Müller-Plathe, F. (2010), Critical Parameters and Surface Tension of the Room Temperature Ionic Liquid [bmim][PF6]: A Corresponding-States Analysis of Experimental and New Simulation Data. *J. Phys. Chem. C* 114, pp. 3599-3608.
20. Ashworth, C.R., Matthews, R. P., Welton, T., Hunt. P.A. (2016), Doubly ionic hydrogen bond interactions within the choline chloride–urea deep eutectic solvent, *Phys. Chem. Chem. Phys.* 18, pp. 18145-18160.
21. Zubeir, L.F., Held, C., Sadowski, G., Kroon, M.C. (2016), PC-SAFT modeling of CO_2 solubilities in Deep Eutectic Solvents*, J. Phys. Chem. B,* 120, pp. 2300-2310.
22. Shiflett, M.B., Elliott, B.A., Niehaus, A.M.S., Yokozeki, A. (2012) Separation of N_2O and CO_2 using room-temperature ionic liquid [bmim][Ac], *Sep. Sci. & Tech.* 47, pp. 411–421.
23. Leron, R.B., Caparanga, A.R., Li, M.H. (2013), Carbon dioxide solubility in a deep eutectic solvent based on choline chloride and urea at T = 303.15–343.15 K and moderate pressures*, J. Taiwan Inst. Chem. Eng.* 44, pp. 879-885.
24. Sun, S., Niu, Y., Xu, Q., Sun, Z., Wei, X. (2016), Efficient SO_2 Absorptions by Four Kinds of Deep Eutectic Solvents Based on Choline Chloride, *Ind. Eng. Chem. Res*. 54**,** 8019−8024.

468604 Modeling Aqueous Alkanolamine + Carbon Dioxide Mixtures Using the SAFT-VR Mie and SAFT-γ Mie Equations of State

Abstract:

An accurate description of the thermodynamic properties of fluids is useful for the design of better chemical processes. The predictive capabilities of two molecular-based equations of state are assessed: SAFT-VR Mie [1], employing a homonuclear model where each molecule is described by its own specific parameters, and SAFT-γ Mie [2], a group contribution (GC) version of SAFT in which molecules are represented by a set of functional groups, each with their own set of parameters. In common with other SAFT approaches, the molecules are described as chains of spherical segments. The segment-segment interactions are described with a Mie potential, which has been shown to provide a good prediction of the fluid-phase equilibria (including the near-critical region) and properties that are second-derivatives of the Helmholtz free energy (for example heat capacities), by allowing for varying degrees of softness/hardness of the interactions [3]. In this study we analyse the predictions obtained when modelling mixtures of aqueous alkanolamines and carbon dioxide (CO_2) that are of relevance in CO_2 capture studies. This involves the prediction of thermodynamic properties that are not considered in the development of the molecular models, and in the case of the GC approach the predictions for molecules for which experimental data are not used in the model development. The thermodynamic models are currently being used to describe process models for carbon capture and separation [4,5].

References

[1] Lafitte, Thomas, et al. "Accurate statistical associating fluid theory for chain molecules formed from Mie segments." *The Journal of chemical physics*, 139.15 (2013): 154504.

[2] Papaioannou, Vasileios, et al. "Group contribution methodology based on the statistical associating fluid theory for heteronuclear molecules formed from Mie segments." *The Journal of chemical physics*, 140.5 (2014): 054107.

[3] Dufal, Simon, et al. "Developing intermolecular potential models for use with the SAFT VR Mie equation of state." *AIChE Journal*, 61.9 (2015): 2891-2912.

[4] Brand, Charles V., et al. "On the use of molecular-based thermodynamic models to assess the performance of solvents for CO_2 capture processes: monoethanolamine solutions." *Faraday Discussions* (2016).

[5] Burger, Jakob, et al. "A hierarchical method to integrated solvent and process design of physical CO_2 absorption using the SAFT γ Mie approach." *AIChE Journal*, 61.10 (2015): 3249-3269.

470321 Comprehensive Thermodynamic Modeling of Complex Mixed-Solvent Electrolyte Systems: An Investigation on Water-Hydrogen Chloride-Methanol Ternary System

Abstract:

Comprehensive Thermodynamic Modeling of Complex Mixed-Solvent Electrolyte Systems: An Investigation on Water-Hydrogen Chloride-Methanol Ternary System

Sina Hassanjani Saravi and Chau-Chyun Chen

Department of Chemical Engineering, Texas Tech University

The binary system of HCl + Water which was comprehensively presented in our previous thermodynamic modeling work [1] is extended to cover the ternary system of HCl + water + methanol. Prior studies on the HCl + water + methanol ternary system, a complex mixed-solvent electrolyte system, are rare. Lu et al. [2] reported experimental VLLE data (TPxy) for this system in atmospheric pressure and temperature range from 70 to 95 °C. Furthermore they calculated mean ionic activity coefficient based on Pitzer model up to 8 molal. Deyhimi et al. [3] determined mean molal activity coefficients for HCl in the mixture based on potentiometric data of a galvanic cell and modeled the system by use of Pitzer and Pitzer – Simonson – Clegg ion-interaction equations with HCl concentration limited to 5.75 molality at 298.15 K.

In this study, a comprehensive thermodynamic model is presented based on symmetric electrolyte Non-Random Two-Liquid (eNRTL) activity coefficient model. The model includes phase equilibrium properties such as vapor-liquid equilibrium and liquid-liquid equilibrium, calorimetric properties such as liquid enthalpy and heat capacity, and speciation properties such as molecular and ionic species concentrations. With use of binary interaction parameters accounting for molecule-molecule binary interactions and molecule-electrolyte interactions, the model accurately correlates and calculates all thermodynamic properties

including mean ionic activity coefficient, osmotic coefficient, vapor pressure, boiling point, dew point, excess enthalpy and heat capacity. Hence, a reliable extrapolation can be performed to predict various properties over the entire range of concentrations. The VLE and LLE predictions along with calorimetric predictions for heat capacity and excess enthalpy of the solution are calculated and reported in this work. Speciation results which show the ionization of the solution species are depicted. Finally the ternary phase diagram which shows the two-phase region for this ternary system is illustrated.

References:

1. S.H, Saravi, C.-C., Chen "Thermodynamic modeling of hydrochloric acid-water binary system with electrolyte Non-Random Two-Liquid (eNRTL) model," *Manuscript in preparation*
2. X. Lu, Y. Wang, J. Shi, "The thermodynamics of concentrated nonaqueous strong electrolyte solutions -vapor - liquid equilibrium of hydrochloride - methanol mixture," *Journal of Huagong Xuebao,* 41 (1991) 695-701
3. F. Deyhimi, Z. Karimzadeh, M. Abedi, "Pitzer and Pitzer–Simonson–Clegg ion-interaction modeling approaches: ternary HCl + methanol + water electrolyte system," *Journal of Molecular Liquids,* 150 (2009) 62–67

451450 ThermophysicalÂ Properties of Ionic Liquids - *Ilthermo* Archival Database: 3. Engineering Applications

Abstract:

Critically evaluated experimental data are essential to carry out innovative chemical process design, or to improve material and energy efficiencies of existing chemical processes. Developed under the auspices of IUPAC (Project 2003-020-2-100), *ILThermo* (http://ilthermo.boulder.nist.gov/) is a data archive of experimental thermophysical properties (including 120 thermodynamic, transport and thermochemical properties) of ionic liquids and mixtures that contain them. Version 1.0 was announced at COIL-1 [1]. Detailed capabilities were described [2]. Recently, Version 2.0 [3] was released by NIST's Thermodynamics Research Center. *ILThermo* is a web-based open-access database that provides data and metadata from published experimental studies of ionic liquids, including numerical values of thermophysical properties, chemical structures, measurement methods, sample purity, critically evaluated standard uncertainties of property values, as well as other significant measurement details. Version 2.0 is keyed on chemical structure of each chemical species that is curated, which brings inherent advantages to users of the database. Best practices are described for search and retrieval of thermophysical properties and phase behavior data from *ILThermo* for use in engineering applications.

References

[1] J. W. Magee, J. A. Widegren, M. Frenkel, Q. Dong, C. Muzny, R. D. Chirico and V. V. Diky "Comprehensive Data Retrieval System for Ionic Liquids," 1st International Congress on Ionic Liquids, Salzburg, Austria, June 19-22, 2005.

[2] Q. Dong, C. D. Muzny, A. Kazakov, V. Diky, J. W. Magee, J. A. Widegren, R. D. Chirico, K. N. Marsh, and M. Frenkel, *J. Chem. Eng. Data*, 2007, 52, 1151-1159.

[3] A. Kazakov, J.W. Magee, R.D. Chirico, V. Diky, C.D. Muzny, K. Kroenlein and M. Frenkel "NIST Standard Reference Database 147: NIST Ionic Liquids Database - (*ILThermo*)", Version 2.0, National Institute of Standards and Technology, Gaithersburg MD, 20899, http://ilthermo.boulder.nist.gov .

473743 Clickable Janus Particles

Abstract:

The rising interest in the use of particles as solid surfactants in multiphasic systems has been motivated by prospects to increase emulsion stability and facilitate surfactant removal and recycling. While many systems have utilized homogeneous particles to form stable Pickering emulsions, amphiphilic Janus particles offer enhanced attachment energies and the potential to tailor the functionality to specific fluid phases. Current methods for the fabrication of Janus particles in bulk solutions rely on phase separation mechanisms which are highly dependent on the interaction energies between precursors, requiring unique synthesis conditions. In order for Janus particles to be viable solid surfactants on a large scale, it is necessary to develop versatile methods of fabrication that enable the chemical composition to be tailored to specific applications. Inspired by recent advances in click chemistry for particle synthesis, we developed clickable Janus particles containing alkyne pendant groups that can participate in thiol-yne click reactions to transform the chemical functionality and surface charge. The integration of alkyne pendant groups allows us to take advantage of the suite of thiols already commercially available for synthesis of a wide range of functional Janus particles. We demonstrate functionalizing clickable Janus particles to be amphiphilic by introducing either carboxyl, hydroxyl, or amine moieties. Click chemistry not only allows us to tailor the chemical functionality and surface charge of Janus particles, but also enables us to control the extent of functionalization which we show influences the particle morphology and emulsion stability. Furthermore, multi-functional Janus particles can be easily synthesized by performing sequential click reactions. The use of thiol-yne click chemistry to functionalize clickable Janus particles offers a fast-reacting, scalable method to design functional Janus particles for potential use as solid surfactants.

473308 Synthesis of Optically Complex, Porous and Anisometric Polymeric Microparticles By Templating from Liquid Crystalline Droplets

Abstract:

We report that aqueous dispersions of micrometer-sized liquid crystal (LC) droplets provide the basis of a general and facile methodology for the templated synthesis of spherical and non-spherical polymeric microparticles with complex internal structure and porosity. Specifically, we prepared nematic droplets of reactive/non-reactive mesogens with distinct internal configurations using a range of approaches, photo-polymerized the reactive mesogen 4-(3-acryloyoxypropyloxy) benzoic acid 2-methyl-1,4-phenylene ester (RM257) within the LC droplets, and then extracted the non-reactive mesogens to yield polymeric particles. We found that LC droplets exhibiting bipolar, radial, axial or preradial configurations template the formation of lemon-shaped, spherical, spherocylindrical or tear-shaped polymeric microparticles, respectively. Each type of microparticle exhibited distinct optical signatures indicating the presence of an internal LC-templated, anisotropic polymer network. In addition, by using a microfluidic system to generate monodisperse LC droplets containing 10% - 40% wt/wt of RM257, we formed lemon-shaped microparticles with tailored aspect ratios ranging from 2.4 to 1.2. We show also the mass density of spherical microparticles templated from radial LC droplets can be tuned to range from 0.2 to 0.6 g/cm^3, revealing the introduction of porosity (confirmed by electron microscopy) with a volume-average pore diameter of 39 ± 16 nm (obtained from nitrogen sorption isotherms).

464876 Soft Polyhedral Colloids

Abstract:

Controled production of anisotropic, polyhedral, colloids is a key need in development of products and processes based on self-assembly for numerous applications. The faceting of the particles, and the regulated truncation of the facets, is a control variable in the particle-particle interactions that dictate order, disorder, and the final structures formed in suspensions. A common difficulty in controlling the production of such particles is the kinetics and polydispersity encountered during crystallization of solid particles. We show a flexible method for producing soft polyhedral colloids with a wide range of polyhedral shapes and variable degrees of truncation. The particles originate from an emulsion precursor that enables better control over size polydispersity and the soft nature of the particles provides flexible control over the growth rates that dictate specific shape formation.

465154 Measuring the Interactions of Anisotropic Particles Using Total Internal Reflection Microscopy

Abstract:

Controlled self-assembly in colloidal systems has benefited from a set of surface force measurement techniques optimised for systems of spherical particles and other symmetric geometries. With growing interest in the assembly of anisotropic particles (*e.g.* carbon nanotubes settling at an interface, or the alignment of rod-like zinc oxide crystals during solar cell fabrication), there is a strong incentive to develop surface force measurement techniques optimised for systems exhibiting either geometric or compositional anisotropy. One promising approach is the extension of total internal reflection microscopy (TIRM), an established direct force measurement technique capable of providing three-dimensional particle position with nanometre resolution at sub-millisecond timescales. Conventional TIRM is used to measure the hydrodynamics and order-k_{BT} interaction potentials of spherical micro or nanoparticles levitated above a surface, and is based on observing the inelastic scattering of evanescent fields; for spheres, this scattering is well-described by generalised Lorentz-Mie theory. An extension of TIRM for anisotropic particles requires a description of evanescent scattering to include up to three additional spatial dimensions (*i.e.* pitch, tilt and yaw) to account for non-degenerate rotational degrees of freedom. Bridging this gap in complexity has required the construction of finite-element models to solve Maxwellâ€™s field equations directly, enabling the quantification of scattering in systems of particles with arbitrary geometry and composition. The insights gained from this modeling work have guided the design and assembly of an experimental apparatus that extends the utility of TIRM to observe systems of diffusing anisotropic particles while retaining the advantages of high spatiotemporal resolution. We will discuss our development of anisotropic TIRM, describe the way in which evanescent scattering models have been used to enable and complement experimental analysis, and present some innovations in particle sizing related to this new method.

471559 Colloidal Molecules Assembled from Binary Spheres Under an AC Electric Field

Abstract:

Colloidal particles are often used as analogues for simple atoms and molecules, however these systems typically lack the complexities present in their counterparts. Here we report the assembly of colloidal molecules from a binary mixture of spheres (1, 1.6, 2, and 4 μm) under an alternating current electric field. The assembled oligomers typically consist of a large sphere in the center and a number of smaller satellite spheres surrounding the large one. In deionized water, the number of satellite particles (i.e., the coordination number) increases with increasing size ratio of the constituent particles. For a given binary mixture, the oligomer coordination number decreases with increasing frequency of the applied field. These trends can be correctly captured by computing the electrostatic energy of the systems based on induced dipolar interactions. By suspending the particles in polyvinylpyrrolidone (PVP) aqueous solution, we can further tune the bond length of the oligomers independently from their coordination numbers. The addition of PVP also allows us to fix the assembled colloidal molecules permanently so that they stay intact even after the electric field is turned off. Our method provides a robust way to produce a family of colloidal molecules with controllable geometry.

464671 3D Position and Orientation Control of Single and Assembled Colloidal Superellipsoids in Electric Fields

Abstract:

The ability to assemble anisotropic nano- and micro- colloidal particles into hierarchically ordered structures that are also potentially reconfigurable provides the basis for exotic materials and controllable devices in emerging technologies. However, current capabilities are limited by the numbers and types of non-trivial microstructures that can be reliably produced and manipulated using colloidal assembly. These limitations are due to fundamental problems with designing, controlling, and optimizing the thermodynamics and kinetics of assembly processes for complex colloidal components. Our approach to this problem is to use inhomogeneous AC electric fields to manipulate the position, orientation, and assembly of superellipsoidal colloids having up to three different semi-principal axes. We begin by first developing rigorous potentials to capture the position and orientation dependent DLVO particle-substrate and particle-field interactions. Microscopy results are used to validate these theoretical expressions by showing measured and modeled potential energy landscapes for up to 12 unique states for single tri-axial particles (including all combinations of position and orientation). We then show how assemblies of many particles can be used to obtain a variety of microstructures of superellipsoidal colloids, which can be controllably reconfigured between different states. Finally, we report results for the kinetics of assembly and reconfiguration, which we are using to inform the development of optimal control theories for manipulating such systems. Ultimately, with the ability to measure, model, and tune anisotropic colloidal interactions and dynamics in inhomogeneous electric fields, we demonstrate how time varying fields can be used to manipulate non-equilibrium pathways for the assembly, disassembly, reconfiguration, and repair of hierarchically ordered anisotropic colloidal microstructures.

462874 Assembly and Actuation in Concentrated Suspensions of Janus Ellipsoids

Abstract:

The emergence of anisotropic colloidal particles as building blocks for self-assembly has opened up a route to achieve bottom-top assembly of diverse functional structures with complex symmetries. For instance, Janus ellipsoidal particles that posses both shape and interaction anisotropy self- assemble into fiber-like structures in which the degree of positional order is controlled by the electrolyte concentration (*Nature Materials* **14** 117-124 (2015). These structures may also be reversibly actuated by application of an external electric field. Here, we have studied the equilibrium self-assembly of dense suspensions of Janus ellipsoidal particles via confocal laser-scanning microscopy (CLSM). The work examines the role of the particle concentration on the kinetics of the assembly process and the phases that can be obtained. At particular electrolyte concentrations (~1mM), where the screened electrostatic repulsion between the particles facilitates the assembly process, we have observed a domain structure of the assembled fibers, with local nematic order within each domain. To exploit the reconfigurability of these structures in applications where actuation by external stimuli is desirable, we investigate the actuation capability in a highly populated network of the assembled fibers. Depending on the frequency of the applied field, the fibers align with the external field, the domain boundaries disappear, and field-induced interactions between neighboring fibers are observed. Our findings are relevant to generating shape-memory colloidal fibers suitable for applications where rapid, on demand reconfiguration is needed.

472619 Clustering and Dynamics in Anisotropic Colloidal Glasses and Gels Containing Weakly Adsorbing Polymers

Abstract:

We report on large-scale structures and dynamics in an anisotropic colloidal system with weakly adsorbing polymers. Specifically, we utilize rheology, ultra-small angle neutron and x-ray scattering (USANS and USAXS), dynamic light scattering (DLS) and x-ray photon correlation spectroscopy (XPCS) to explore disk-shaped colloids, Laponite RD, in solution with polyethylene oxide (PEO). Aqueous dispersions of laponite form a repulsive colloidal glass. Nanometer to micron-sized aggregates form with the addition of PEO chains. DLS results indicate that, for lower laponite concentrations, these aggregates are stable in size and display diffusive behavior, suggesting that the fluids and solids we obtain are a liquid and glass of clusters, respectively. Through rheology, we observe the variations in the viscoelastic moduli (G' and G'') which can be correlated with the changes in the low-q power law exponent extracted from an x-ray and neutron scattering experiment.

472597 Dynamics of a Magnetic Colloidal Chain in a Confined Spherical Cavity

Abstract:

Study of confinement effects of magnetically driven colloidal chains under confinement provides insight of understanding some fundamental polymer dynamics such as chain entanglement or the cellular processes such as DNA replication or viral transfection. We present a bead-spring model through coarse-grained Brownian Dynamics simulation to describe the confinement dynamics inside a spherical cavity. Radial contraction is employed as the confining procedure, and a possible rotating magnetic field will be introduced to actuate dynamics in the system. Our aim is to study the conformational properties of confined semiflexible colloidal chain with respect to the contraction rate and level of confinement (i.e. weak or strong confinement regimes), and discuss the change of dynamics along the chain when a short-ranged wall interaction is present.

471535 The Surface Pressure and Microstructure of Carbon Nanotubes at an Air-Water Interface

Abstract:

Interface-driven assembly of particles has been demonstrated by a number of authors (1). The corresponding physics is intriguing and fairly well documented for micron-size particles with a moderate aspect ratio (<10) (2). Carbon nanotubes (CNTs) are rolled cylinders of graphene sheets with an aspect ratio exceeding 50 and their interfacial behavior remains largely unexplored. In the present work, we investigated the surface pressure and interfacial assembly of both functionalized and non-functionalized CNTs at an air-water interface (3). Firstly, our findings indicated that both types of CNTs adsorbed strongly at the air-water interface, leading to an increase in surface pressure. Volmer equation of states (4) has been applied to model and understand the surface pressure data. Secondly, the CNT-decorated interface was studied both *in situ*optically and with electron microscopy after performing Langmuir-Schaefer deposition. Both types of CNTs showed some degree of orientational ordering upon compression. To quantify the degree of orientation for each type of CNTs, the scalar order parameters were calculated from the electron micrographs. The findings may have implications on the development of more stable emulsions and the production of CNT thin films with controllable packing density and orientational ordering. This work is supported by NSF CAREER award #1253613.

References

1. Pieranski, P., Two-dimensional interfacial colloidal crystals. Phys. Rev. Lett., 45(7), 569-572 (1980).
2. Lewandowski, E.P. *et al.*, Orientation and self-assembly of cylindrical particles by anisotropic capillary interactions. Langmuir, 26(19), 15142-15154 (2010).
3. Vora, S. R. *et al.*, Surface pressure and microstructure of carbon nanotubes at an airâ€"water interface. Langmuir, 31(16), 4663 - 4672 (2015).
4. Fainerman, V.B. *et al.*, Surface-pressure isotherms of monolayers formed by microsize and nanosize particles. Langmuir, 22(4), 1701-1705 (2006).

456863 Electrophoresis in Dense Colloidal Suspensions: Role of Particle Concentration and Debye Layer Thickness

Abstract:

The electrophoretic mobility of colloidal particles is a property that controls their kinetics of self-assembly in an external electric field. To assemble a colloidal crystal, the applied field induces a phase transition by driving particle densification in the vicinity of the electrode with opposite polarity. The deposition results in a spatially varying particle volume fraction in the direction of the applied field. Therefore, the electrophoretic velocity, which is a decreasing function of the colloid volume fraction, also varies along an axis perpendicular to the electrode surface, and this retardation can significantly affect crystallization rates. To better understand the impact of particle concentration on the electrophoretic velocity, we have studied the electrophoresis phenomenon in suspensions of sterically stabilized poly(methyl methacrylate) colloids in a nonpolar solvent at different volume fractions using confocal laser scanning microscopy. A number of complex phenomena are observed, including significant effects of device geometry and electrolyte concentration. These measurements show that the spatiotemporal evolution of the electrophoretic deposition velocity in response to the changing colloid concentration in the device can yield mobilities that vary by as much as a factor of ten. These results are compared to available theory of the volume fraction dependent electrophoretic mobility. The work furthermore sheds light on the complexity of electrophoretic colloidal deposition under concentrated conditions, a situation that approximates a number of important processes in the coatings industry.

467223 New Methods for Rapid Calculation of Thermal Forces in Coarse Grained Simulations of Complex Fluids

Abstract:

Coarse-grained simulations of soft materials and complex fluids such as colloidal, surfactant, and polymer solutions require rapid calculation of the hydrodynamic forces exerted on suspended, microscopic constituents (particles). Because of the small length scales associated with colloids, surfactants, and polymers, low Reynolds number flows dictate these forces. Inertia of the fluid and of the constituents is negligible on time scales that are relevant for rheological measurements. Many coarse-grained simulation methods for modeling soft materials construct an explicit model of the solvent, determining both the position and momentum of the solvent and suspended particles as a function of time. These include Lattice-Boltzmann, Dissipative Particle Dynamics, Stochastic Rotation Dynamics, Multi-particle Collision Dynamics, and others. Always, degrees of freedom associated with the solvent are neglected, and a stochastic force satisfying the fluctuation dissipation theorem must be exerted on all components of the dispersion.

The explicit solvent models were developed in part because of difficulties performing rapid calculations with implicit solvent models, such as Brownian Dynamics, in which the hydrodynamic interactions between particles are modeled deterministically and depend only on the relative configuration of particles. The chief advantage of explicit solvent models comes from sampling of the stochastic forces, which is no more expensive computationally than generating a string of random numbers. For implicit solvent models, the stochastic forces must be drawn from a normal distribution whose covariance is a complicated function of the particle configuration. For a system of interacting N particles, drawing a single sample requires $O(N^3)$ operations, if numerically exact techniques from linear algebra are employed. So-called â€œfastâ€ methods can approximate the sampling with roughly $O(N^m \log N)$ computational complexity, where m is a coefficient greater than one which depends on the configuration of the particles. Typical values of m

range from 1.25 to 1.5. This puts a serious limit on application of implicit solvent methods to large-scale simulation.

In the presented work, we will demonstrate an optimal and spectrally accurate, fast method for calculation of thermal forces in implicit solvent simulations of soft materials such as Brownian Dynamics. The computational complexity of this approach is $O(N (\log N)^{d/(d+3)})$, where d is the fractal dimension of the particulate structures being modeled. Remarkably, this new approach adapts to the structure of the material under study by properly balancing the computational effort spent evaluating near-field and far-field contributions to the hydrodynamic interactions among the suspended particles. We term this algorithm optimal because in the periodic geometries needed to model viscometric flows of soft materials, log-linear computational scaling is the best available. This presented approach for drawing samples achieves this same asymptotic scaling. Applications of this approach to modeling colloidal gelation, solutions of macromolecules as well as other heterogeneous colloidal media far from equilibrium will be discussed.

469706 Effect of Ion Sterics on Diffusiophoresis in Concentrated Electrolytes

Abstract:

Diffusiophoretic migration of a charged colloidal particle in an electrolyte gradient has been known since the pioneering work of Derjaguin et al. in 1947. Recent interest in diffusiophoresis has been motivated by new microfluidic measurement techniques, and applications in membrane filtration, mineral replacement reactions, and colloidal self-propulsion, for example. Here, we revisit the fundamental problem of predicting the diffusiophoretic velocity of a spherical, uniformly charged colloid. All existing work in this regard utilizes the Poisson-Nernst-Planck (PNP) model for point-like, non-interacting ions. However, it is well known that the PNP model can lead to impossibly large ion concentrations near even moderately charged surfaces. Thus, we consider modified PNP equations that account for steric repulsion between finite-sized ions, via the Bikerman, Carnahan-Starling, and BMCSL equations of state. We compute the diffusiophoretic velocity using an (experimentally relevant) asymptotic scheme for thin Debye layers. Ion steric effects have a dramatic influence: e.g, while the PNP model predicts a reversal in the diffusiophoretic velocity in certain electrolytes (e.g. KCl) with increasing particle zeta potential â€" at precisely the zeta potentials at which the validity of the PNP model is questionable â€" we show that ion sterics essentially eliminates this reversal. This finding is of particular relevance to diffusiophoretic migration in high salinity solutions, where the effects of ion sterics are expected to be most prominent.

469486 Gravitational Collapse of Reversible Colloidal Gels Via Large-Scale Dynamic Simulation

Abstract:

We investigate the phenomenon of gravitational collapse in colloidal gels via dynamic simulation.Â In moderately concentrated gels formed via arrested phase separation, rupture and re-formation of bonds of strength O(kT) permit ongoing structural rearrangements that lead to temporal evolution â€“ aging â€“ of gel structure and rheology. The reversible nature of the bonds also allows the gel to transition from solid-like behavior to liquid-like behavior under external forcing, and back to solid-like behavior when forcing is removed. But such gels have also been shown to be susceptible to sudden and catastrophic collapse of the entire structural network, after which the gel sediments into a dense layer, eliminating any intended functionality of the network scaffold. Although the phenomenon is well studied in the experimental literature, the microscopic mechanism underlying the collapse is not understood. To study this behavior, we conduct large-scale dynamic simulation to model structural and rheological evolution in colloidal gels subjected to gravitational stress, examining the detailed micro-mechanics in three temporal regimes: slow, pre-collapse evolution; collapse and rapid sedimentation; and long-time compaction. A range of attraction strengths and gel ages, and their effect on the critical force that triggers collapse, are studied.

469254 Dynamics and Rotational Regimes of Semi-Flexible Colloidal Chains

Abstract:

The interaction between flexible filaments and a surrounding fluid has been a topic of study for many years, due to its relevance to both naturally occurring and industrially relevant phenomena. However, the connection between a fiberâ €™s elastic properties, the external driving forces, and the eventual dynamics of the fiber is not well understood. Fibers in the semi-flexible regime, such as microtubules and carbon nanotubes, are of particular interest, as the competition of elastic, hydrodynamic, and external forces appear to cause deviations from the scalings expected of rigid and flexible fibers. We thus investigate the dynamical behavior of semi-flexible fibers experimentally, through the use of a tunable model system consisting of linked paramagnetic particle chains; when chains are driven by an external rotating magnetic field, which exerts a torque profile analogous to that of shear flow, we observe multiple regimes of dynamical behavior. By pairing experiments, computational models, and theoretical calculations, we find that the dynamics of the system depend on the dimensionless Mason and magnetoelastic numbers, which we then use to identify and classify the regimes of different rotational modes.

472000 *Catalytic Motors Near Curved Surfaces: Guided and Trapped Motion*

Abstract:

Chemically self-propelled colloidal locomotors are active particles designed to translate along micro- or nanoscale landscapes to deliver various applications, such as micropumping, sensing, drug delivery and cargo transport. In spite of significant enhancements in design and efficiency of catalytic motors, the precise navigation and onboard steering of the colloidal motors still remains as a challenging task. This fact relies on the relative importance of the viscous nature of the medium and the Brownian motion of the particle. Here, we examine the self-diffusiophoresis of a small Janus colloid near a larger stationary solid spherical surface. In the context of self-diffusiophoresis, the colloid itself generates a flow as a result of non-uniform distribution of reactant (or product) solutes due to a chemical reaction taking place asymmetrically on its surface. The dynamics of the Janus particle near a curved no-slip stationary wall that is impenetrable to the solute is found by analytical solution of the solute conservation in conjunction with the application of the Reynolds Reciprocal Theorem of the Stokes flow. It is found that the motions of active particles are influenced not only by hydrodynamic interactions but also via interactions of the solute distribution generated (or consumed) at the surface of the active area of each colloid with the boundary. For deterministic motions, i.e. infinite Péclet numbers, it is observed that the trajectory of the motor depends on two parameters: the colloid active coverage and its relative orientation angle with respect to the wall. For axisymmetric motions, when the active area of the colloid is facing the wall, the motor is repelled by the accumulation of solute in the gap between the swimmer and the wall. For the cases where the cap is opposite to the wall, the swimmer moves towards the wall due to the repulsive interaction of the accumulated solute on the active cap side. In asymmetric incidents on the other hand, the trajectories are more complex as the motor can also rotate due to both hydrodynamic and phoretic interactions with the curved wall. We discuss all possible scenarios for the trajectories of a swimmer approaching obliquely to a curved stationary wall and elucidate the role of hydrodynamic and phoretic interactions on the dynamics of the motor.

469442 The Effect of Suspended Colloidal Particles on the Stability of High-Density Particles Against Sedimentation

Abstract:

The stability against sedimentation of 1 wt% TiO_2 particles, with diameters of 300±200 nm and density of 4.2 g/cm^3, in 0.5 wt% (10.8 mM) or 2.0 wt% (43.4 mM) aqueous vesicular dispersions of the cationic surfactant DDAB (didodecyldimethyl-ammonium bromide) was examined; see Yang, Y.-J., D. S. Corti, and E. I. Franses, Langmuir, 31, 8802-8808, 2015. Whereas at the low DDAB concentrations the particles settled completely by 1 cm over two months, at the high DDAB concentrations the particles remained suspended for over 18 months. A variety of methods, including cryo-TEM (transmission electron microscopy), dynamic light scattering, and electrophoretic mobility measurements, were used to probe the microstructure, size, surface potential, and mobility of vesicles [*Yang et al.*, 2015]. The hollow vesicles were shown to have diameters of about 400-600 nm. At 0.5 wt% DDAB, the vesicles were quite mobile, with an estimated volume fraction of less than 0.2. At 2.0 wt% DDAB, the vesicles were quite immobile, probably owing to their high volume fraction of over 0.5, making them to arrange into a close-packed formation. The close-packed vesicles provide strong resistance to the sedimentation of the dense particles, but the dispersions are nonetheless highly shear-thinning, and free flowing, with moderate bulk viscosities.

The mechanism of the stabilization of the dense particles against sedimentation with the use of close-packed vesicles was also examined in general by using Brownian dynamics simulations (BDS) for the first time. BDS were performed for mixtures of two types of colloidal spherical particles interacting via DLVO potentials. For one type, the Peclet number, $Pe = \pi d^3 \Delta\rho g L / 6 k_B T$, Pe is high; Pe is defined as the ratio of the sedimentation flux, $v_{sed}c$, to the macroscopic diffusion flux, Dc/L, over the scale L of the sample, where d is the particle diameter, $\Delta\rho$ is the particle-dispersion medium density difference, c is the particle concentration, D is the diffusion coefficient, v_{sed} is the sedimentation velocity, g is the gravitational acceleration, k_B is Boltzmann's constant, and T is the absolute temperature. These particles, of diameter d_1 and volume fraction Φ_1, will settle rapidly on their own. For the second type of particles, Pe is zero, to represent non-settling "rigid vesicles". The effects of the particle size d_2 and volume fraction Φ_2 of the "vesicles" on the sedimentation stability of the high-Pe particles were examined. Stability maps of d_2 and Φ_2 ranges were generated, to provide general guidelines on preventing, or slowing down, the settling of the high-Pe particles. In addition, the local microviscosities around the dense particles and the viscosities of the vesicular dispersions were computed from the simulations as a function of the relevant shear rates or shear stresses. The resulting viscosities are consistent with the sedimentation observations and the measured shear-dependent viscosities of the TiO_2-DDAB suspensions. Hence, the close-packed vesicular dispersions provide a high local resistance to the dense particles gravity-induced flow, which corresponds to low local shear stresses, but a much smaller resistance to the suspension bulk flow, which corresponds to high shear stresses. The experiments and the BDS results indicate that close-packed vesicular dispersions provided a general method for long-term stabilization against sedimentation of high-Peclet number particles of a wide range of diameters and for a wide range of volume fractions.

472007 Deformation of Fractal Clusters Under Mixed Shear Flows

Abstract:

In this work Stokesian Dynamics was used to investigate the deformation of fractal colloidal clusters under a flow field characterized by a special mixed linear shear flow, in which the ratio between the strain rate and the vorticity was adjusted. Interparticle interactions are accounted for using DLVO theory as well as tangential forces to allow particle bonds to sustain bending moments. Different deformation patterns were obtained for clusters under different strain/vorticity ratios. Clusters were shown to deform to highly elongated and nearly linear structures when the strain rate is higher than the vorticity, while they undergo periodic deformations when the vorticity is higher than strain rate. The frequency of the periodic deformation can be analyzed through Fourier transform and can be fitted by an ellipsoid model. In the case where the strain rate is equal to vorticity, the mixed shear becomes a simple shear. The deformation of clusters under simple shear was found to be independent of the shear stress if a cluster size and its fractal dimension are fixed. The strength of DLVO interactions does not show a direct influence on cluster deformation, but can change the role played by tangential forces among particles, thus indirectly affecting the cluster deformation. In the final part of the work, the influence of the deformation of clusters the fluid viscosity is investigated. A semi empirical relationship between the cluster fractal dimension and the system stresslet, which determines the cluster contribution to the suspension viscosity, was derived and verified by simulations.

465340 The Rheology and Microstructure of an Aging Thermoreversible Colloidal Gel

Abstract:

The properties of colloidal gels are known to change with gel age, but the processes and mechanisms by which aging occurs is not well understood and limits our ability to predict macroscopic behavior in these systems. Colloidal gels are disordered systems that are trapped in a dynamically arrested, non-equilibrium state. It is generally accepted that the aging of colloidal gels is governed by their potential energy landscape of which the thermodynamic equilibrium is the global minimum; however, the system cannot reach equilibrium on an experimental timescale [1]. Instead, the gel samples local potential energy minima resulting in aging behavior. Due to its fundamental significance and industrial applicability, we investigate the microstructural basis of aging in a well-studied, model adhesive hard sphere (AHS) system, consisting of silica nanoparticles grafted with an octadecyl brush dispersed in tetradecane [2, 3].

In this work, we quantitatively relate rheological aging to structural aging by simultaneously measuring the bulk properties and gel microstructure using rheometry and small angle neutron scattering (Rheo-SANS), respectively [4]. Specifically, we develop a quantitative and predictive relationship between the macroscopic properties and the underlying microstructure (*i.e.*the effective strength of attraction) of an aging colloidal gel, which is independent of the gel's thermal and shear history. Analysis with mode coupling theory is consistent with local particle rearrangements as the mechanism of aging, which lead to monotonically increasing interaction strengths in a continuously evolving gel and strongly support aging as a trajectory in the free energy landscape. The analyses and conclusions of this study may be 1) industrially relevant to products that age on commercial timescales, such as paints and pharmaceuticals, 2) applicable to other dynamically arrested systems, such as metallic glasses, and 3) used in the design of new materials.

References:

[1] Royall, C. P., Williams, S. R., Ohtsuka, T., and Tanaka, H., "Direct observation of a local structural mechanism for dynamic arrest," Nature Materials **7**, 556-561 (2008).

[2] Eberle, A. P. R., Castaneda-Priego, R., Kim, J. M., and Wagner, N. J., "Dynamical Arrest, Percolation, Gelation, and Glass Formation in Model Nanoparticle Dispersions with Thermoreversible Adhesive Interactions," Langmuir **28**, 1866-1878 (2012).

[3] Eberle, A. P. R., Wagner, N. J., Akgun, B., and Satija, S. K., "Temperature-Dependent Nanostructure of an End-Tethered Octadecane Brush in Tetradecane and Nanoparticle Phase Behavior," Langmuir **26**, 3003-3007 (2010).

[4] **Gordon, M.B.**, Kloxin, C.J., and Wagner, N.J., "Rheology and Microstructure of an Aging Thermoreversible Colloidal Gel," Journal of Rheology. Submitted.

471710 Influence of Attractive Forces and Particle Size Distribution on Shear Thickening of Suspensions

Abstract:

The shear thickening of suspensions includes a remarkable phenomenon termed "discontinuous shear thickening" in which the viscosity of a suspension can increase several orders of magnitude as the shear rate is increased over a small range. This increase may have dramatic impact on applications. The transition from a lubricated to a "frictional" rheology due to the contribution of contact forces has been identified as one mechanism for shear thickening in suspensions. Recent simulations have been successful in quantitatively reproducing experimentally-observed, continuous and discontinuous shear thickening for both Brownian and non-Brownian suspensions [1-2]. A related experimental observation [3] is that attractive forces can lead to large low-shear viscosity and potentially yield stress, with the result that shear-thinning occurs and shear thickening is obscured. Here, we will show that this behavior is captured by our simulation tool including hydrodynamic effects, contact and friction, as well as attractive colloidal forces. We study the impact of various cohesion strengths on shear thickening curves at various suspension concentrations. We also investigate the non-trivial effect of particle size distribution, which may manifest itself through differences in inter-particle interactions (repulsive or attractive) and effective maximum packing increase.

[1] Seto, Ryohei, et al. "Discontinuous shear thickening of frictional hard-sphere suspensions." *Physical Review Letters* 111.21 (2013): 218301.

[2] Mari, Romain, et al. "Discontinuous shear thickening in Brownian suspensions by dynamic simulation." *Proceedings of the National Academy of Sciences* 112.50 (2015): 15326-15330.

[3] Gopalakrishnan, V., and C. F. Zukoski. "Effect of attractions on shear thickening in dense suspensions." *Journal of Rheology (1978-present)* 48.6 (2004): 1321-1344.

479748 Material Design for High-Performance Electrocatalyst for the Reduction of Oxygen

Abstract:

Requirements of high activity, stability and low cost post new challenges for the design of advanced electrocatalysts. In this talk, I will mainly present recent approaches to the design of core-shell bimetallic nanostructures as low-platinum group metal (PGM) catalysts for electrochemical reduction of oxygen. In all cases, design and processing of catalysts with optimal surface atomic structures and favored dynamics under reactive environments are important. In situ, variable-temperature environmental transmission electron microscopy (TEM) techniques can be quite useful in this regard for the realization of designed structures, such as preferred catalyst surfaces both in atomic arrangement and composition. Such information helps in the control of catalyst with high-level of accuracy. The focus will be on the solution phase design and post-synthesis treatment of nanoparticle catalysts based on both theoretical and experimental understanding; *in-situ* environmental TEM (ETEM) in studying the dynamics of structural behavior of electrocatalyst under reactive environment; and structure-property relationship of bimetallic catalysts towards oxygen reduction reaction (ORR).

479749 Highly Stable and Active Carbon Supports for Pt Cathode Catalysts in Polymer Electrolyte Fuel Cells

Abstract:

Proton electrolyte fuel cells (PEFCs) have been widely considered one of environmentally friendly and highly efficient electric power sources for a wide variety of power applications ranging from portable and stationary power supplies to transportation. However, some of the most critical barriers to their commercialisation have been the intrinsic high cost and insufficient long-term performance durability required for practical fuel cell systems. Among others, the spherical carbon black particle support used for Pt catalyst in MEAs have a number of intrinsic structural disadvantage. It is partially responsible for current insufficient performance and durability of MEAs. They include (1) inefficient O_2 transport due to significant tortuosity in carbon black-based cathode, (2) non-uniform distribution of ionomer due to significant microporosity of carbon blacks, (3) catastrophic corrosion during start-up/shut-down events due to low corrosion resistance, (4) low electrical conductivity, and (5) weak binding sites with Pt nanoparticles.

University at Buffalo, SUNY has developed a new type of high-surface areas (up to 850 m^2/g) and large size nitrogen-diped graphene tubes (NGT) in a highly controlled manner and at high yields. The diameters of these graphene tubes can be tuned from 50 to 500 nm, which is much larger than conventiona carbon nanotubes (20 nm). It should be noted that the graphene tube is low cost using inexpensive dicyandiamide as precursors ($1-2/kg) and the synthesis method is easily scaled up. Apart from the obvious advantage of high electronic conductivity and enhanced corrosion resistance, the large tube size provided better Pt nanoparticle dispersion and more facil mass transfer including O_2 and H_2O. Furthermore, the novel ORR active nitrogen-doped graphene tubes are able to not only offer remarkable support effects by geometrically and electronically modifying the deposited Pt nanoparticles, but also provide a large amount of complementary nonprecious ORR active sites, thereby significantly improving the overall activity and durability of the Pt catalysts. Thus, relative to conventional carbon blacks, the highly graphitized and active N-GT is an ideal

support for low-PGM catalysts for high-performance MEAs with enhanced mass transport and durability along with ideal interfaces with ionomer.

465068 Metal Oxides As Protective Layers on Carbons and Electrocatalyst Supports

Abstract:

Carbon has long been used as an electrode material in electrochemical processes because it is electrically conductive and easily available at a cheap price. Unfortunately, carbon is not thermodynamically stable in aqueous systems due to its low oxidation potential. Electrochemical corrosion of carbon can be a serious problem in many kinetically accelerated processes, for example, in low temperature fuel cells, rechargeable batteries, and supercapacitors. It has become a great challenge to discover new materials, both electrically conductive and chemically resistant to corrosion, to replace carbon.

In this talk, a brief review of metal oxides as electrocatalyst support will be given first, focused especially on fundamental electrochemical aspects in oxygen reduction reaction (ORR). Following that I will present our studies on preparing high surface area metal oxides in the form of nanoscale particles and nanocoatings on, for example, carbons. We have developed techniques to make titania and niobia nanocoatings on different carbons, including carbon black and carbon nanotubes. To make the oxides electrically conductive, carbon activation and elemental doping were used and found to be effective. Carbon activation was also found to lower phase transformation temperatures, making it possible to transform the metal oxides to the right phase without sintering and thus retaining the nanoscale morphology of the coatings. When Pt catalyst nanoparticles were deposited on the oxides, they were more uniformly dispersed and stabilized due to metal-support interactions. The Pt catalysts also showed better activity.

470323 Novel Metal Oxide As Highly Acid Stable Electrocatalyst for Oxygen Evolution Reaction in Water Splitting

Abstract:

Electrochemical process is among the most important systems in chemical engineering field for addressing the energy and environmental science problems, such as sustainable energy generation and storage. Water splitting is one process that focuses on renewable and environmental friendly approach to the generation of hydrogen fuel. The key reaction, oxygen evolution reaction (OER), has a sluggish kinetic and thus huge over-potential, so there is an urgent need for solving this problem for efficient electrolyzer. Recent results show that RuO_2 and IrO_2 are still the most useful OER catalysts due to their performance in acidic system; although their durability still does not match the requirement. In this presentation, I will discuss a new class of metal oxide can be used as highly active and durable electrocatalyst for OER. Sol-gel synthesized oxide catalyst with improved intermediate kinetic initiation was analyzed and compared with commercial RuO_2. The OER activity was analyzed by cyclic voltammetry (CV) with potential range from 1.1 to 1.6 V and scan rate of 10 mV/s. The result showed that the newly developed OER catalyst had much higher current density in the range and gave smaller values of onset potential as well as Tafel slope compare to RuO_2 catalyst. For durability analysis, corresponding 10,000 cycles of CV were performed. The newly-developed catalyst also maintained exceptional high current density and low Tafel slope. This new discovery of the OER catalysts may help to improve the efficiency of hydrogen generation based on water splitting and provide an aspect on potential materials of electrocatalysts.

455758 Transition-Metal Doped Carbon Electrocatalysts for Oxygen Reduction

Abstract:

Active and durable non-precious metal catalysts (NPMCs) are highly demanded to replace the rare and expensive Pt for the oxygen reduction reaction (ORR) in polymer electroltye fuel cells and metal-air batteris. Compared to extensively studied ORR catalysis in alkaline media, development of high-performance NPMCs to replace Pt in acidic electrolytes remains grand challenges. Great progress has been made over the last decade to develop transition metal-derived nitrogen-doped carbons (M-N-C, M: Fe or Co), the most promising class of NPMC for ORR electrocatalysis. Among studied catalysts, the Fe-N-C formulation holds the greatest promise for the ORR in acid. Here, we report a new highly active and stable Fe-N-C catalyst featured with well-dispersed atomic Fe in porous carbon matrix, which was prepared through one single thermal conversion from Fe-doped ZIF-8, a metal-organic framework (MOF) containing Zn^{2+} and well-defined $Fe\text{-}N_4$ coordination. Unlike other Fe-N-C catalyst preparation, no additional tedious post-treatments such as acid leaching and the second heating treatment are required in this work. Notably, an O_2-free environment for preparing the Fe-doped ZIF-8 precursor is found to be crucial for yielding uniform Fe distribution into highly porous N-doped carbon matrix. The resulting new Fe-N-C catalyst exhibited exceptionally improved ORR activity with a very high half-wave potential (0.82 V *vs.* RHE) and sufficient potential cycling stability in acid. Opposite to previous observation, the highly active Fe-N-C catalyst is in the absence of any graphitized nanocarbons, which would lead to a new discussion in the field for understanding the role of carbon during the ORR electrocatalysis.

465561 Probing the Synergy in Nickel-Molybdenum Hydrogen Evolution Catalysts

Abstract:

Electrochemical production of hydrogen offers an alternative to fossil fuel intensive methane reforming. On a commercial scale, water electrolysis is done with either proton exchange membrane based systems (acidic) or in alkaline electrolyzers. In alkaline environments, nickel and nickel alloys are widely used. While nickel-molybdenum is known to be highly active, the reason for this high activity is not understood, nor are the potential degradation mechanisms. To investigate the synergy that exists in this system, we fabricated well defined nickel electrodes with preformed nanocrystals and explored the catalytic activity with and without the influence of molybdenum. First, we looked at the promoting effect of molybdenum substrates. We observe a synergistic increase in catalytic activity as well as a strong wetting effect of the nickel on molybdenum. When we compare the activity of different size nanocrystals on molybdenum, we see that large particles are more active on a mass basis than small particles. This does not scale with either the surface area or interface length, suppors the existence of a bulk interaction between nickel and molybdenum. To expand on these observations, we directly investigated the role of molybdate deposition on nickel by performing chronoamperometry experiments with molybdate added to the electrolyte. The molybdate experiments show that a mixed molybdenum-nickel phase can achieve even higher activtity. Over the course of the molybdate deposition experiments, we observe a slow increase in the activity followed by an eventual decline. By performing ex-situ SEM and auger depth profiling on electrodes at various timepoints, we are able to identify the most active morphology of nickel molybdenum. Our observations support the existence of a bifunctional mechanism, in which one species acts to stabilize water as a proton is abstracted and another site acts to combine these protons into hydrogen gas. While nickel can perform this alone, it is more effective in the presence of molybdenum.

473109 High Pressure Fluid Phase Behavior of Two Essential Oil Systems with CO_2

Abstract:

From numerous studies on binary and ternary mixtures with carbon dioxide as the near-critical component, it became apparent that the fluid phase behavior of these systems can be quite complex. For instance, with many solutes carbon dioxide shows a second liquid phase. In processing essential oils, we are dealing with complex mixtures of organic constituents of which many of them may show fluid multiphase behavior with carbon dioxide, of course, depending on temperature, pressure and composition of the system. Therefore, it has to be expected that also essential oils may show such a complex multiphase behavior. This study reports on phase equilibrium experiments of two essential oils, namely eucalyptus and alecrim oil in the presence of carbon dioxide as the near-critical solvent. It has been established that both oils show liquid-liquid immiscibility with carbon dioxide with all characteristics of type-III fluid phase behavior in the classification of Van Konynenburg and Scott. Although both essential oils have a completely different molecular composition, it was found that only minor differences occurred in their fluid phase behavior.

A New Pure-Component Equation of State Designed for Accurate Reproduction of Phase-Equilibrium, Caloric and Volumetric Properties with Emphasis on Supercritical-Property Prediction

Prof. Jean-Noël Jaubert, Dr. Romain Privat, Dr. Silvia Lasala, Yohann Le Guennec, Université de Lorraine, Ecole Nationale Supérieure des Industries Chimiques, Laboratoire Réactions et Génie des Procédés (UMR CNRS 7274), 1 rue Grandville, 54000 Nancy, France.

Introduction

Numerous modifications to the Van der Waals model have been presented over the years with the aim of representing with increasing accuracy the thermodynamics of complex systems. As a matter of facts, the most of those do not incorporate a substantial difference in the functional combination of attractive and repulsive forces, with respect to the original formulation introduced by Van der Waals. Although the analytical expressions of the repulsive and attractive terms proposed in literature do not correctly quantify the actual repulsive and attractive contribution to pressure, their sum results in a quantitative representation of fluid properties being sufficiently accurate to make their combination the "cornerstone of the generalized Van der Waals theory". From the Van der Waals proposal, even the most successful two-parameter cubic equations of state (EoS) still express their attractive and repulsive term by introducing a parameter a, which is a measure of the attractive forces ("energy parameter") between molecules, and the parameter b, which is a measure of the size ("intrinsic volume" or "co-volume") of the molecules. The pure-component a energy parameter is directly proportional to the temperature-dependent alpha function $\alpha(T)$ which is a measure of how the a parameter deviates from its critical value. A wide variety of α-functions has been proposed over the past years. Those contain parameters which are adjusted according to different criteria: either over a whole set of components or for each specie ("generalized" versus "component-specific" α-function); either considering the whole temperature domain or specifically in the subcritical or supercritical regions ("overall" versus "domain-specific" α-function). Furthermore, among the mostly applied formulations, two different functional forms are identifiable: polynomial and exponential forms.

In this work, the popular component-specific Twu (1988) alpha function (denoted Twu88), which exhibits an exponential form has been selected to model Vapor-Liquid Equilibrium (VLE) behavior of mixtures containing supercritical components. A key outcome of our work is that the use of the Twu88 α-function together with some sets of component-dependent parameters has resulted in the inaccurate VLE modeling of such mixtures, in spite of the use of advanced EoS/g^E mixing rules. A detailed examination of this issue has shown that the application of such sets of parameters entails the generation of abnormal deviations of the α-functions, when extrapolating at supercritical conditions. In parallel, the use of the simple generalized and polynomial Soave alpha function incorporated in the same advanced EoS/g^E mixing rules as previous, has led to very accurate results. In the basis of these counterintuitive results, it was attempted to understand how the use of an EoS coupled with a generalized alpha function (with no adjustable parameter) could induce better modeling results than an EoS coupled with a more flexible alpha function involving 2 adjustable parameters. It has been shown that the calibration of α-functions requires the imposition of constraints to the parameters under regression in order to guarantee consistent predictions of the derived thermodynamic properties (residual enthalpy, entropy, heat capacity ...) at all temperatures (sub- or super-critical). On the basis of such conclusions, the Twu88 alpha-function has been re-parameterized for a large series of pure components (more than 900). In the light of the newly regressed parameters, a generalized Twu-type α function, performing better than Soave's α function, has been proposed. The use of this new generalized Twu alpha function has been used to model the VLE data of binary systems containing a supercritical component and has led to improved modeling results with respect to the use of the simple Soave alpha-function.

The functional form of α-functions

One of the most influential contribution in the prediction of vapor pressures of non-polar and slightly polar pure compounds has been the introduction of the Soave α-function [1]. Over the years, further modifications have been introduced to this functional form, mainly aimed at improving the correlation of vapor pressures. However, most of the α-functions presented in literature have been developed without evaluating their predictive capability in the supercritical domain. It is nevertheless well acknowledged that one of the main targets, at the basis of the modifications to the Soave function, should consist in the improvement of the behavior in the supercritical domain, in addition to that of increasing its accuracy in correlating vapor pressures of any fluid. The only attempt found in the literature, to improve the behavior of α-functions in the supercritical region, consisted in optimizing the parameters of the function, separately, in the subcritical and supercritical domain. However, no modification has been found that introduces a novel functional form being able to improve the predictions in both the subcritical and supercritical regions, without distinguishing between parameters specifically applied in one of the two domains. To this modeling gap it should be associated the lack, in the literature, of specific theoretically based guidelines for defining an α-function that could guarantee consistent predictions of the derived thermodynamic properties, in the whole temperature range. With that respect, the purpose of this section is to present requirements that an α-function should fulfill.

As for the original Van der Waals equation of state, in contemporary cubic EoS the critical attractive parameter (a_c), the covolume b and the critical molar volume, v_c^{EoS}, can be determined by applying the so-called critical constraints recalled in Equ. (1).

$$\begin{cases} P_{c,exp} - P(T_{c,exp}, v_c^{EoS}) = 0 \\ \left(\dfrac{\partial P}{\partial v}\right)_T \Bigg|_{\substack{T=T_{c,exp} \\ v=v_c^{EoS}}} = 0 \\ \left(\dfrac{\partial^2 P}{\partial v^2}\right)_T \Bigg|_{\substack{T=T_{c,exp} \\ v=v_c^{EoS}}} = 0 \end{cases} \quad (1)$$

where $T_{c,exp}$ and $P_{c,exp}$ are the experimental critical temperature and pressure and where the critical molar volume is determined from the considered equation of state: $v_c^{EoS} = v(T_{c,exp}; P_{c,exp})$. The resulting critical energy parameter and covolume are usually expressed in terms of the EoS-dependent coefficients Ω_a and Ω_b:

$$\begin{cases} a_c = a(T_{c,exp}) = \Omega_a R^2 T_{c,exp}^2 / P_{c,exp} \\ b = \Omega_b R T_{c,exp} / P_{c,exp} \end{cases} \quad (2)$$

Differently from the covolume, necessarily temperature independent, at a temperature different from T_c the energy parameter, *a(T)*, is written in terms of its value at the critical point, a_c, and of the α-function:

$$a(T) = a_c \alpha(T) \tag{3}$$

It follows that $\alpha(T)$ is a non-dimensional factor which becomes unity at the critical temperature:

$$\alpha(T_c) = 1 \tag{4}$$

Moreover, to avoid non-physical negative values of the attractive term of cubic equations of state, an α-function should be a positive function at all temperatures:

$$\text{For all T, } \alpha(T) > 0 \tag{5}$$

At low temperatures, the average molecular kinetic energy is drastically reduced, resulting in a system where molecules are particularly prone to interact, upon collision. This explains the reason for which the EoS energy parameter should become increasingly important when temperature tends to zero. A decreasing, convex α-function, as the temperature approaches zero satisfies such requirements so that:

$$\begin{cases} \lim\limits_{T \to 0^+} \dfrac{d\alpha}{dT} < 0 \\ \lim\limits_{T \to 0^+} \dfrac{d^2\alpha}{dT^2} > 0 \end{cases} \tag{6}$$

The behavior of the attractive parameter at infinite temperature is a question that remains open. For some authors the attractive term is zero at infinite temperature, for others, it takes a non-null finite value. If α-functions are obtained by forcing the EoS to reproduce properly selected experimental data, Colina et al. [2] clearly show that at the infinite temperature limit α-functions should exhibit negative values. As highlighted by themselves, finite negative values of the attractive term, at the infinite-temperature limit, would result from the necessity of counterbalancing the defects of the repulsive term, which at high temperature dominates over the negligible attractive one. Colina et al. also give proof that by assuming that: (i) at infinite temperature, the cubic EoS reduces to a hard-body equation of state and that (ii) the Van der Waals term RT/(v-b) correctly represents hard-body repulsions, the limiting value of α-function at infinite temperature should obviously be zero. As discussed by Abott and Prausnitz [3], there is however no theoretical evidence that the attractive part of cubic EoS cancels out at infinite temperature. Similarly, Sandler [4] merely alleges that at very high temperatures only the hard core part of the potential is important rather than the weaker (or soft) dispersive energies. Thus, both Abbott and Prausnitz [3] and Sandler [4] leave some uncertainty on whether the α-function at infinite temperature should assume a zero rather than finite value. Since no hard theoretical evidences have been found on this issue, we do not conclude about the theoretical limit of the α-function at infinite temperatures. In order to develop a new α-function we however practically need to quantify the limit of the function at infinite temperature. For such a practical reason, being aware of the irrelevance of the selected value attributed to this limit, we consider it equal to zero. We will also assume that the α-function is concave when T becomes infinite.

We can thus summarize our observations by:

$$\begin{cases} \alpha(T_c) = 1 \\ \forall\ T,\ \alpha(T) > 0 \\ \lim\limits_{T\to 0^+} \dfrac{d\alpha}{dT} < 0 \text{ and } \lim\limits_{T\to 0^+} \dfrac{d^2\alpha}{dT^2} > 0 \\ \lim\limits_{T\to +\infty} \dfrac{d\alpha}{dT} = 0 \text{ and } \lim\limits_{T\to +\infty} \dfrac{d^2\alpha}{dT^2} > 0 \end{cases} \tag{7}$$

In order to understand how do the derivatives of an α-function behave at intermediate temperatures, it is possible to consider the expression of the residual molar heat capacity at constant volume. Starting for the general form of a cubic EoS:

$$P = \frac{RT}{v-b} - \frac{a_c \times \alpha(T)}{(v-r_1 b)(v-r_2 b)} \tag{8}$$

One gets:

$$c_v^{res}(T,v) = -\frac{T a_c}{(r_1 - r_2)b} \ln\left(\frac{v - r_1 b}{v - r_2 b}\right)\left(\frac{d^2\alpha}{dT^2}\right) \tag{9}$$

From equation (9), it results that the presence of inflection points on an α-function at a specific temperature T* entails a zero $c_v^{res}(T^*)$:

$$\left.\left(\frac{d^2\alpha}{dT^2}\right)\right|_{T=T^*} = 0 \ \Rightarrow\ c_v^{res}(T^*,v) = 0 \ \ \text{(for all v)} \tag{10}$$

As a result of the non-dependency from pressure of $d^2\alpha/dT^2$, at the temperature T* the fluid would be characterized by a unique heat capacity value (that of the corresponding perfect gas since $c_v^{res}(T^*)=0$) for any value of the molar volume, i.e. whatever the pressure would be $[c_v^{res}(T^*,v) = c_v^{perfect\ gas}(T^*)]$. In addition to this already unphysical condition, a further consequence has to be pointed out. Due to conditions in (7), we can state that the α-function is necessarily convex when T approaches zero or infinite. Consequently, the presence of one inflection point, on the curve representing the α-function, would involve the occurrence of, at least, a second inflection point at temperature T**. This would imply the existence of two temperatures, T* and T**, where the system would behave as a perfect gas, in terms of heat capacities ($c_v^{res}(T^*) = c_v^{res}(T^{**}) = 0$) and in correspondence of which all the isobars in a (c_v,T) plane would intersect. Such an absurd conclusion demonstrates that: (i) the α(T) must be a monotonically decreasing function of the temperature and (ii) $d^2\alpha/dT^2$ must be strictly positive at all temperatures.

To sum up, an α-function must verify:

$$\begin{cases} \alpha(T_c) = 1 \\ \left.\begin{array}{l} \alpha(T) > 0 \\ \dfrac{d\alpha}{dT} < 0 \\ \dfrac{d^2\alpha}{dT^2} > 0 \end{array}\right] \quad \text{for all T} \\ \lim\limits_{T \to +\infty} \alpha(T) = 0 \end{cases} \tag{11}$$

In this work, the popular component-specific Twu alpha function (denoted Twu88) [5], which exhibits an exponential form has been selected to model VLE behavior of mixtures containing supercritical components. Such an α-function, in which L and M are two adjustable parameters writes:

$$\alpha_{88} = T_r^{\,2(M-1)} \exp\left[L\left(1 - T_r^{\,2M}\right)\right] \tag{12}$$

L and M were fitted in order to check the constraints in Equ. (11). By doing so, inaccurate VLE modeling of such mixtures was obtained, in spite of the use of advanced EoS/gE mixing rules. A detailed examination of this issue has shown that the application of the constraints described in Equ. (11) entails the generation of abnormal deviations of the α-function, when extrapolating at supercritical conditions.

It was however possible to fix this issue by imposing to the third derivative of α with respect to temperature to be negative. We can thus conclude that a necessary condition to get accurate results in the supercritical region with a cubic EoS is that:

$$\frac{d^3\alpha}{dT^3} < 0 \text{ for all T} \tag{13}$$

On the basis of such conclusion, the Twu88 alpha-function has been re-parameterized for a large series of pure components (more than 900). The parameters L and M were determined in order to check the mathematical constraints described in Equs. (11) and (13).

In the light of the newly regressed parameters, a generalized Twu-type α function, has been proposed.

Results and conclusion

As stated in the previous section, parameters L and M could be correlated as a function of the acentric factor in order to propose a new generalized Twu-type α function. For the Peng-Robinson EoS [6-10], the following correlations were obtained [11-12]:

$$\begin{cases} L_{PR} = 0.1208\omega^2 + 0.6066\omega + 0.0848 \\ M_{PR} = 0.1727\omega^2 - 0.2635\omega + 0.8873 \end{cases} \tag{14}$$

With such parameters, it was possible to predict with a very high accuracy the properties of many pure components. On a database of 900 pure components, the vapor pressures, the enthalpies of vaporization and the heat capacities were predicted with an accuracy of 1%, 2%

and 4% respectively. Such deviations are much smaller than those observed with a Soave-type alpha-function, demonstrating the importance of respecting the constraints detailed in Equs. (11) and (13). As a next step, such an α-function will be tested over mixtures through advanced mixing rules which combine the PR EoS with the residual part of an activity coefficient model. A special attention will be devoted to type III phase diagrams in the classification scheme of Van Konynenburg and Scott [13] since it is a great challenge to accurately correlate such systems.

Nomenclature:

$\alpha(T)$: alpha function of a cubic equation of state

c_v^{res} : residual molar heat capacity at constant volume

P: pressure

$P_{c,exp}$: experimental critical pressure

T: temperature

$T_{c,exp}$: experimental critical temperature

v: molar volume

References:

[1] Soave G., Equilibrium constants from a modified Redlich-Kwong equation of state. *Chem. Eng. Sci.* **1972**, 27(6):1197–1203.

[2] Colina C., Santos J., Olivera-Fuentes C., High-temperature behavior of the cohesion parameter of cubic equations of state. *High Temp.-High Press.* **1977**, 29(5): 525–532.

[3] Abbott M.M., Prausnitz J.M., Generalized van der waals theory: a classical perspective. *Fluid Phase Equilib.* **1987**, 37:29–62.

[4] Sandler S.I., The generalized van der Waals partition function as a basis for excess free energy models. *J. Supercrit. Fluids.* **2010**, 55(2):496–502.

[5] Twu C.H., A Modified Redlich-Kwong Equation of State for Highly Polar, supercritical Systems. Proceedings of the International Symposium on Thermodynamics in Chemical Engineering and Industry, **1988**, p. 148.

[6] Peng D.Y., Robinson D.B., A new two-constant equation of state. *Ind. Eng. Chem. Fundam.* **1976**, 15(1):59–64.

[7] Jaubert J.N., Mutelet F., VLE predictions with the Peng–Robinson equation of state and temperature–dependent k_{ij} calculated through a group contribution method. *Fluid Phase Equilib.* **2004**, 224:285–304.

[8] Vitu S., Privat R., Jaubert J.N., Mutelet F., Predicting the phase equilibria of CO_2 + hydrocarbon systems with the PPR78 model (PR EoS and k_{ij} calculated through a group contribution method). *J. Supercrit. Fluids.* **2008**, 45:1–26.

[9] Jaubert J.N., Privat R., Mutelet F., Predicting the phase equilibria of synthetic petroleum fluids with the PPR78 approach. *AIChE J.* **2010**, 56:3225–3235.

[10] Jaubert J.N., Privat R., Relationship between the binary interaction parameters (k_{ij}) of the Peng–Robinson and those of the Soave–Redlich–Kwong equations of state. Application to the definition of the PR2SRK model. *Fluid Phase Equilib.* **2010**, 295:26–37.

[11] Le Guennec Y., Lasala S., Privat R., Jaubert J.N., A consistency test for α-functions of cubic equations of state. *Fluid Phase Equilib.* **2016**, 427:513–538.
[12] Le Guennec Y., Privat R., Jaubert J.N., Development of the *translated-consistent tc*-PR and *tc*-RK cubic equations of state for a safe and accurate prediction of volumetric, energetic and saturation properties of pure compounds in the sub- and super-critical domains. *Fluid Phase Equilib.* **2016**, 429:301–312.
[13] Privat R., Jaubert J.N., Classification of global fluid–phase equilibrium behaviors in binary systems. *Chem. Eng. Res. Des.* **2013**, 91:1807–1839.

464209 New Procedure for Enhancing SAFT Molecular Parameters Transferability: The Role of Derivative Properties

Abstract:

The design and optimization of industrial processes, involving the correct selection of suitable solvents, the most advantageous unit operations, and an adequate separation sequence, relies upon the accurate prediction of the behavior the system at different thermodynamic conditions. Besides the phase equilibria description, knowledge of derivative properties, such as heats of vaporization, isochoric and isobaric heat capacities, isothermal compressibility and thermal expansion coefficients, are essential to fulfill the conditions of an optimized industrial process or product.

It has been shown that second order derivatives are more sensitive to errors than the first order derivatives of the thermodynamic function. Consequently, a correct description of first order derivatives, such as phase equilibria calculations, do not assure an accurate prediction of second order derivatives of the same thermodynamic function (heat capacities, speed of sound, etc.) for the same compound or mixture. In spite of the importance of derivative properties and their sensitivity to errors, most modeling efforts have been devoted to the description of standard phase equilibria, while the works on derivative properties are more limited.

We will present a new robust procedure, involving vapor-liquid equilibria (VLE) and derivative properties regression, in a step forward obtaining reliable and transferable parameters from molecular-based equations of state. The method deals with the introduction and testing of coupling factors ranging from 0 (only *Cp*) to 1 (only VLE), to change the weight of one set of properties over the other in the fitting procedure; this allows assessing the role of derivative properties in the robustness of the parameters. The technique is illustrated with the soft-SAFT equation by calculating a large number of thermodynamic properties of different compounds, and in a wide range of thermodynamic conditions, for compounds from the *n*-alkanes, *n*-perfluorocarbons, 1-alkanols and glycols families, as representative of different types of molecular interactions. The most relevant thermodynamic properties to be included in the parameters fitting, the weight they should have in the regression procedure and the influence of association will be discussed in detail.

The method presented here is robust, simple, straightforward to implement and not equation dependent; it can be transferred to any other equation, providing robust molecular parameters for global equations.

Partial financial support for this work has been provided by the Catalan government under project 2014-SGR1582.

458552 Description and Prediction of Polymer Phase Behavior with Lattice Cluster Theory: An Homage to One of the Great Experimentalists

Abstract:

When chemical engineers are faced with mixtures containing polymers, many difficulties arise, which are not present for other systems. For example, polymers are often poly-disperse with respect to molar mass, chemical composition and branching. Concerning the latter, they can have many different architectures starting from linear over short-chain branched to hyper-branched polymers. Aside from being hard to even characterize, these difficulties make modeling and even more prediction of phase behavior a very challenging task at least. The phase behavior of polymers is not only hard to model, but also experimentally very challenging to determine and it needs a tremendous effort to arrive at reliable experimental data, which is again crucial for developing and testing thermodynamic models of polymers. One of the biggest names in the area of experimental investigation of polymer phase equilibria is Theo de Loos. We try to highlight here on the example of our works on lattice cluster theory, how large the impact was that he has made on our research and our quest for understanding the polymer phase behavior. For this purpose, we go through the different equilibria, i.e. solid-liquid, liquid-liquid, vapor-liquid, solid-liquid-liquid, and vapor-liquid-liquid we have described or predicted over the past years using lattice cluster theory and discuss the challenges associated with polymer phase equilibria, the applicability and limits of lattice cluster theory as well as possible extensions of the theory and combinations thereof with other theories. Not much to our surprise, in almost all of our publications on these topics we have used at least some experimental data of Theo de Loos, showing just how important reliable experimental data is to the application of lattice cluster theory as an engineering tool and to its further development.

465180 Phase Equilibria of Mixtures Containing Alternative Solvents for Green Chemistry

Abstract:

Solvent selection is very important in the design and development of the reaction or separation and purification processes. Previously, conventional organic solvents were used for these solvents. However, these organic solvents are generally volatile organic compounds (VOCs), and they have environmental drawbacks such as being environmentally polluting [1]. Therefore, it is very important to find alternative environmentally friendly solvents that allow an effective reaction or separation.

Our group has studied ionic liquids (ILs) and fluorous solvents as green solvents for alternative solvents of VOCs. ILs have unique properties, such as high polarity, negligible vapor pressure, and conductive properties. ILs have been considered as entrainers for the extractive distillation or the liquid-liquid extraction [2,3]. On the other hand, fluorous solvents such as perfluoroalkane and perfluoropoly-ether have attracted a great deal of interest, because of their use in fluorous phase organic synthesis, in particular in fluorous catalysis (fluorous biphasic chemistry) [4]. When fluorous solvents are used instead of VOCs, the products can be easily separated in the biphasic system. The fluorous solvents can also be easily recycled. Therefore, the fluorous solvents can be used as a recyclable alternative to organic reaction solvents. A reliable knowledge of the phase equilibria containing IL or fluorous solvent is important for the design and development of the separation and purification processes. This paper reports our experimental results of phase equilibria of mixgtures containing green alternative solvents.

Firstly, we focused on liquid–liquid extraction as a separation method for essential oils. We investigated limonene and linalool as the citrus essential oil, and 1-alkyl-3-methyl-imidazolium bis(trifluoromethanesulfonyl)imide ($[C_nMIM]^+[TFSI]^-$) (n = 4 and 6) and 1-butyl-3-methyl-imidazolium acetate ($[BMIM]^+[Ac]^-$) as IL. The liquid-liquid equilibrium (LLE) data of the binary mixtures

$[C_nMIM]^+[TFSI]^-$ + linalool, and ternary mixture limonene + linalool+ $[BMIM]^+[Ac]^-$were determined by cloud point method.

Next, we investigated Galden® HT-135 (1,1,2,3,3,3-hexafluorooxidized polymd), a polyether-type perfluorinated solvent, as a fluorous solvent with high boiling point. The boiling point of Galden® HT-135 is 408 K [1,5], and therefore, its volatility is lower than perfluorohexane (boiling point: 329.75 K [6]). In addition, Galden® HT-135 is cheaper than perfluorohexane. So, Galden® HT-135 has been used as a reaction solvent for organic synthesis by some groups [5,7,8]. The LLE of binary mixtures Galden HT-135®with selected alkanes and alcohols were measured up to the upper critical solution temperature (UCST). The experimental LLE behavior was compered to peviously reported those of binary mixtures perfluoroalkane + alkane [9, 10].

References

[1] F. M. Kerton, Alternative Solvents for Green Chemistry, RSC Publishing, Cambridge, UK, 2009.

[2] K. N, Marsh, J. A, Boxall, R. Lichtenthaler, Fluid Phase Equilib. 219 (2004) 93-98.

[3] M. Krummen, P. Wasserscheid, J. Gmehling, J. Chem. Eng. Data 47 (2002) 1411-1417.

[4] I. T. Horváth, J. Rábai, Science 266 (1994) 72-75.

[5] I. Ryu, H. Matsubara, H. Nakamura, D.P. Curran, Chem. Rec. 8 (2008) 351–363.

[6] B. E. Poling, J. M. Prausnitz, J. P. O'Connell, The Properties of Gases and Liquids, 5th ed., McGraw-Hill, New York, 2001.

[7] H. Matsubara, M. Tsukida, S. Yasuda, I. Ryu, J. Fluorine Chem. 129 (2008) 951–954.

[8] X. Hao, A. Yoshida, J. Nishikido, Tetrahedron Lett. 46 (2005) 2697–2700.

[9] H. Matsuda, A. Kitabatake, M. Kosuge, K. Kurihara, K. Tochigi, K. Ochi, Fluid Phase Equilib. 297 (2010) 187–191.

[10] H. Matsuda, Y. Hirota, K. Kurihara, K. Tochigi, K. Ochi, Fluid Phase Equilib. 357 (2013) 71–75.

466955 The Symbiosis of Measuring and Predicting Property Data

Abstract:

Property data can be determined from measurements at every single state point of interest such as temperature and pressure. While measurements are the most precise method to determine property data, the experimental effort can be tremendous. Experimental effort can be reduced by correlating property data with models like equations of state while maintaining very good accuracy. In case of the PCP-SAFT equation of state, at least 3 experiments must be performed per pure component. Predicting property data can eliminate the need for experiments altogether, but the accuracy is not sufficient for many applications. Thus, we can choose between measurement, correlation and prediction trading-off effort against accuracy. We propose combining information from several approaches to increase accuracy at low effort. The PCP-SAFT equation of state is used as model. As a starting point, PCP-SAFT parameters are predicted using methods developed in our group (J. Phys. Chem. B 112, 5693-5701). For this purpose, molecular descriptors are obtained from quantum mechanics. The predicted PCP-SAFT parameters are then refined using information from 1) the predictive model COSMO-RS or 2) a single measurement point of the vapor pressure. The new information is used to refit the PCP-SAFT parameter set using a log-likelihood function combining all available information. For a database containing 26 components, the combination of the predictive models for PCP-SAFT and COSMO-RS halves the errors to ~20% error in vapor pressure from about ~40 error for each individual prediction method. Combining predictive PCP-SAFT with a single vapor pressure experiment leads to almost the accuracy of pure correlation (<5% error in vapor pressure). Model-based Optimal Experimental Design can be used to determine the experiment which should be conducted for the largest gain of accuracy in our proposed combination method. The proposed combination of predictive models with selected experiments thus allows to greatly improve accuracy of property data at low effort.

462430 Bubble Pressure, Enthalpy of Mixing and Excess Volume Measurement and Prediction for n-Alkane + Aromatic and Naphthenic Hydrocarbon Binary Mixtures

Abstract:

Experimental phase behavior data are scarce for long chain n-alkane + aromatic and naphthenic mixtures. In two recent works [1, 2] we showed that commonly used cubic equations of state systematically predict both qualitatively and quantitatively incorrect phase behavior unless negative binary interaction parameter (k_{ij}) values are used. We also showed that the PC-SAFT equation of state accurately computes the phase behavior, with binary interaction parameters set to zero, and therefore can be used to estimate k_{ij} values for cubic equations of state in the absence of experimental phase equilibrium data. However, the investigation was restricted to the correctness of phase behavior type [3], bubble pressures and critical points for a limited number of mixtures for which data were available at the time of publication. In this follow up study, a more extensive set of high-precision bubble-pressure, excess volume and enthalpy of mixing measurements are presented for representative binary mixtures of n-alkanes + aromatic and naphthenic compounds. The selected mixtures comprise binaries of n-alkanes, from n-C_{16} to C_{36}, combined with alkyl-benzene, alkyl-naphthtalene, and their naphthenic analogues such as alkyl-cyclohexane. The experimental data are compared with predicted and regressed outcomes obtained using the Peng-Robinson, Soave-Redlich-Kwong and the PC-SAFT equations of state where interaction parameter values are regressed from bubble pressure data. Outcomes are evaluated. Expected applications of this work include improved phase behavior model accuracy applicable to hydrocarbon production, transport and refining environments, and improved corresponding state based transport property prediction, particularly for viscosity, that require accurate phase behaviors and phase compositions as inputs.

[1] S. Ahitan, M. A. Satyro, J. M. Shaw, "Systematic Misprediction of n-Alkane + Aromatic and Naphthenic Hydrocarbon Phase Behavior Using Common Equation of State", J. Chem. Eng. Data 2015.

DOI:10.1021/acs.jced.5b00539

[2] S. Ahitan, J. M. Shaw, "Quantitative Comparison Between Predicted and Experimental Binary n-Alkane + Benzene Phase Behaviors Using Cubic and PC-SAFT EOS", Submitted to Fluid Phase Equilibria 2016.

[3] P. H. van Konynenburg, R. L. Scott, "Critical lines and phase-equilibria in binary van der Waals mixtures", *Philos. Trans. R. Soc.,* A1980, 298, 495-540.

472859 Surface Tension-Driven Flows for Convective Deposition

Abstract:

Convective deposition, traditionally driven by evaporation of a solvent from a thin film that results in well-ordered particle arrays from suspension, is performed under various conditions that result in surface tension gradients. This includes the resulting gradient in temperature when heating the substrate, concentration graditent of added surfactant, and concentration gradient when using a binary solvent of ethanol and water. In each case, these surface tension gradients can result in flow profiles and deposition modes that vary greatly from traditional convective deposition. Resulting ultra-fast deposition of well-ordered particle monolayers and streaking instabilities will be explained in terms of these surface tension gradients, breaking previous paradigms surounding convective deposition.

465413 Elastic Deformation of Soft Supported Films of Finite Thickness during the Drainage of a Viscous Fluid

Abstract:

Elastic deformation of soft supported films of finite thickness during the drainage of a viscous fluid

Joelle Frechette and Yumo Wang, Chemical and Biomolecular Engineering, Johns Hopkins University, Baltimore MD 21218

Elastohydrodynamic deformation can cause lift and reduce friction during sliding and alter the rheological properties of soft colloidal particles. It also modifies the shape of approaching surfaces, a determining factor for the adhesion dynamics to wet or flooded surfaces. When studying elastohydrodynamics in soft matter it is a challenge to measure simultaneously the hydrodynamic forces and the deformation, both necessary to understand how contact is reached and the coupling between deformation of the solid and viscous dissipation in the fluid.

In this presentation the characterization of the spatiotemporal deformation of an elastic film during the radial drainage of fluid from a narrowing gap will be discussed. As predicted, we will show that the elastic deformation of a thick film takes the form of a dimple and prevents full contact to be reached.[1] We reveal the importance of measuring absolute surface separation when working with soft materials by showing the effects of viscoelasticity of the solid in the fluid film profiles, and point out that elastic compliance leads to lower hydrodynamic forces at a given time while deformation leads to stronger forces at a given fluid film thickness.With thin supported elastic films the stress caused by fluid drainage becomes increasingly transferred to the underlying rigid substrate and the dimple formation is suppressed, which allows the surfaces to reach full contact (See top of Figure 1). For intermediate film thicknesses we observe shapes that are more complex when close to contact (see bottom of Figure 1). These shapes originate from lubrication pressure and its corresponding deformation distribution on a curved, layered compliant system. We extend existing models for elastohydrodynamic deformation to the more general case of the compliant material being part of a multi-layer system instead of a half-space. When taking the finite thickness of the compliant film into account we predict the same complex morphology as the ones observed in experiments. Our work is the first reported and validated prediction that such complex surface profiles exist in elastic solid and we draw analogies to the cases of "wimple" or "pimple" that have been observed in droplets.

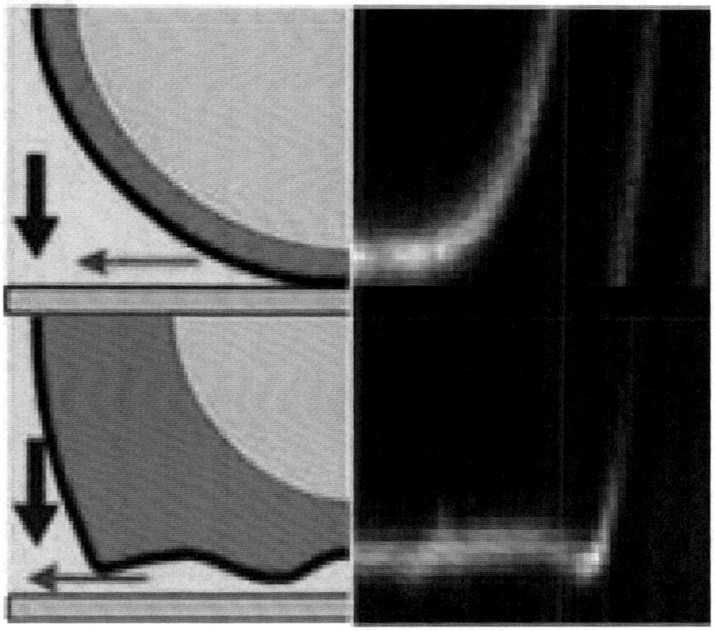

FIG. 1. (Left) Schematic of the elastohydrodynamic problem highlighted on effects of film thickness. Grey core: rigid substrate; Red layer: compliant coating. Up: thin film; down: intermediate thickness film. (Right) experimental deformation profile of the compliant coatings caused by lubrication forces with corresponding film thicknesses.

Reference:

1. Wang, Y., C. Dhong, and J. Frechette, *Out-of-Contact Elastohydrodynamic Deformation due to Lubrication Forces.* Physical review letters, **2015**. 115(24): p. 248302.

450080 Dynamic Wetting Failure in Surfactant Solutions

Abstract:

The influence of insoluble surfactants on dynamic wetting failure during displacement of Newtonian fluids in a rectangular channel is studied in this work. A hydrodynamic model for steady Stokes flows of dilute surfactant solutions is developed and evaluated using three approaches: (i) a one-dimensional (1D) lubrication-type approach, (ii) a novel hybrid of a 1D description of the receding phase and a 2D description of the advancing phase, and (iii) an asymptotic theory of Cox (J. Fluid Mech. 168 (1986) 195). Steady-state solution families in the form of macroscopic contact angles as a function of the capillary number are determined and limit points are identified. When air is the receding fluid, Marangoni stresses are found to increase the receding-phase pressure gradients near the contact line by thinning the air film without significantly changing the capillary-pressure gradients there. As a consequence, the limit points shift to lower capillary numbers and the onset of wetting failure is promoted. The model predictions are then used to interpret decades-old experimental observations concerning the influence of surfactants on air entrainment (Burley & Kennedy, Chem. Eng. Sci. 31 (1976) 901). In addition to being a computationally efficient alternative for the rectangular geometries considered here, the hybrid modelling approach developed in this paper could also be applied to more complicated geometries where a thin air layer is present near a contact line.

465874 Rinsing Flow Instabilities on Rotating Silicon Wafers â€“ Effect of Viscosity and Wettability

Abstract:

Cleaning of silicon wafers is a crucial step in manufacturing silicon circuits for microelectronic applications. The thrust to develop an efficient and environment-friendly way to remove small contaminants has led to the procedure of rinsing a rotating circular wafer by an impinging jet of liquid (water). The wafer is usually pre-wet by a liquid through spin-coating before the rinsing fluid is impinged on it. A common concern in this procedure is the occurrence of fingering instabilities. They eventually form rivulets that are acted on by the centrifugal forces and spread faster than the radially expanding wave-front. This results in stagnation zones in between rivulets that are potential locations of particle build-up. Therefore, it is preferred that these instabilities are suppressed to ensure efficient rinsing.

Earlier work by our group involved rinsing experiments on pre-wet wafers and the observation that pre-wetting wafers with the same liquid significantly reduced instabilities. The influence of surface tension of the pre-wet, coating fluid on the instabilities was also explored. The wave-front shape is quantified by a parameter called the excess perimeter which is defined as the ratio of the actual perimeter of the wave-front to the circumference of the circle of the same area.

In the present work, we study the effect of the viscosity of the coating fluid and the influence of the Saffman-Taylor instabilities for this specific system. Our study also analyzes the influence of substrate wettability by comparing the rinsing wave-front development on hydrophilic and hydrophobic substrates. The flow profile for a hydrophobic, pre-wet hydrophilic and dry hydrophilic substrates show very different behaviors. The flow rate of the rinsing water and the rotation rate of the silicon wafer were varied to determine their impact. This investigation of the dependence of flow rate, rotation rate, fluid viscosity and substrate wettability strives to determine the exact conditions for instabilities in this system. This work seeks to gain significant physical insight into the fluid dynamics problem of an impinging jet on a rotating substrate where the surface tension, viscosity, centrifugal and Coriolis forces are all present. They all play a key role in shaping the wave-front propagation and associated instabilities that need to be minimized for an effective industrial process development.

464839 Flow in the Thin Film Created By Normal and Oblique Turbulent Liquid Jets Impinging on a Vertical Wall: Application to Cleaning

Abstract:

Liquid jets are used for various applications including cleaning, heat transfer, and cutting. When a coherent liquid jet impinges on a vertical wall it forms a thin film, spreading radially away from the point of impingement, until a point where the outward momentum is balanced by surface tension. At this point, the liquid film changes its thickness abruptly giving a jump. A model for the jump location, based on Nusseltâ€™s film theory, was presented by Wilson et al. (Chem. Eng. Sci, 2012, 68, 449-460). In this study, the flow field created by a turbulent liquid jet and the location of the film jump are studied and a new model is presented. The liquid film passes through three zones, namely the laminar boundary layer, the laminar film and the turbulent film zone. The location of the laminar to turbulent transition is described theoretically. In addition, the analysis explains why the location of film jump is observed to be insensitive to the nature of the substrate at high flow rates. The model is compared with new, as well as previously reported, data for the film jump location and film thickness when a jet impinges perpendicularly and at an oblique angle. The average velocity of the liquid was estimated from the initial growth of the film. Good agreement is obtained between the measurements and the model, which has no adjustable parameters. The results are then applied to the cleaning of two representative soil types; (1) Polyvinyl acetate layer which interacts with the liquid in the film, and (2) hydrophobic, non-interacting petroleum jelly.

471514 Nanoscopic Thickness Transitions and Domain Growth Dynamics in Stratifying, Micellar Foam Films

Abstract:

The stability, rheology and applications of foams, emulsions and colloidal sols depend on the hydrodynamics and thermodynamics of thin liquid films that separate bubbles, drops and particles respectively. Thin liquid films containing micelles, colloidal particles, liquid crystals or polyelectrolyteâ€“surfactant mixtures exhibit step-wise thinning or stratification, often attributed to the layer-by-layer removal of the aforementioned supramolecular structures. Stratification proceeds through emergence and growth of thinner circular domains within a thicker film, and the domain expansion dynamics are the primary focus of this study. Domain and associated thickness variation in thin foam films (h < 100 nm) made from sodium dodecyl sulfate (SDS) micellar solutions are examined using a porous plate cell with a novel technique we call Interferometry Digital Imaging Optical Microscopy (IDIOM). We show that the domain expansion dynamics exhibit two distinct growth regimes: a transition from a constant diffusivity to a constant velocity regime is realized when a topological instability occurs at the contact line between the growing thinner isolated domain and the surrounding thicker film. We show a similar transition is also realized after a section of the expanding domain coalesces with the Plateau border. Though several studies have focused on the expansion dynamics of isolated domains that exhibit a diffusion-like scaling, the change in expansion kinetics observed after domains contact with the Plateau border has not been reported and analyzed before. We have developed a theoretical model that incorporates both hydrodynamic and thermodynamic effects for capturing domain expansion kinetics in stratifying foam films. Ultimately, a better understanding of stratification and drainage kinetics, and the role played by supramolecular structural forces, will enable design and creation of foams that drain more rapidly (are fragile and unstable) or more slowly (are stable and long lasting) as desired for applications like food and beverages, oil recovery, pharmaceuticals and cosmetics.

460958 Ultra-Fast Microfluidic Mixing By Soft-Wall Turbulence

Abstract:

Experiments in a soft-walled microchannel show that there is a laminar-turbulence transition at a Reynolds number significantly lower than that for a rigid channel due to the fluid-wall interactions. The transition Reynolds number could be as low as 226 for a channel with smallest dimension 35 microns, in contrast to the transition Reynolds number of 1200 for a rigid channel. The flow after transition shows many of the characteristics of a turbulent flow, including the large maximum in the streamwise velocity fluctuations close to the wall, and the non-zero Reynolds stress. There are also significant differences with wall bounded turbulence --- the fluctuating velocities and the Reynolds stress do not decrease to zero at the soft wall. The turbulence production appears to be a maximum at the wall itself indicating that the turbulence generating mechanism involves wall motion; this is very different from the generation of turbulence due to bursting of eddies in the near-wall region in hard wall bounded turbulence. After transition, the mixing time across a channel of width 0.5 mm is smaller, by a factor of 10^5 , than that for a laminar flow, and complete mixing is achieved within a channel length of 2 cm. The increased mixing rate comes at very little energy cost, because the pressure drop is comparable to that required in current microfluidic devices, and it increases continuously and modestly at transition. This is because the channel length required to achieve complete mixing, 2 cm, is much smaller than that used in microfluidic devices that employ diffusive mixing; in addition, the deformation of the soft wall decreases the resistance to flow. Thus, soft-wall turbulence provides a simple and powerful strategy for ultra-fast mixing in a microfluidic applications.

464395 Bubble Shape Hysteresis: The Effect of Compressibility

Abstract:

The maximum bubble pressure method is a classical procedure to obtain the surface tension at liquid/air interfaces. Since surface tension is an important consideration for the stability of foams, this technique, which is valued because of its simplicity, is widely used. The name of the method is derived from the observation that a gas, when emerging from a capillary, undergoes a rapid expansion from a small, spherical cap within the capillary, to a much larger bubble. Associated with this transition, is a maximum in the pressure within the gas, which can be shown to be simply related to the surface tension. In this paper, we describe a more complete analysis of this shape transition that quantifies the effect of compressibility, previously not taken into account. This analysis is complemented by experimental measurements where the bubble volume was measured as a function of internal pressure for bubbles that are both emerging from capillaries and are being withdrawn into them. A principal finding of this work is the existence of a hysteresis in this relationship, depending on the geometry of the experimental system.

Experimentally, we observe that, as a result of compressibility, there are two regimes of bubble expansion and contraction. In the first regime, the bubble expands and contracts reversibly, with the bubble retracing the expansion profile obtained during contraction. However, in the second regime, expansion and contraction of the bubble is no longer reversible, thus exhibiting a hysteresis. In this second regime, the bubble undergoes a non-equilibrium jump in size between expansion and contraction.

Theoretically we show that the experimental observations are a consequence of compressibility influencing the equilibrium states of bubbles as a function of two non-dimensional parameters, **A** and **P**. **A** characterizes the relative magnitude of the volume of a hemispherical bubble to the dead volume in the system, and **P** characterizes the relative magnitude of the maximum capillary pressure to the ambient static pressure. We show that (at a given value of **P**) for values of **A** above a critical value, the bubble expands and contracts reversibly as in the first regime seen in our experiments. However for values of **A** below a critical value, the bubble shapes become hysteretic. Using a phase diagram of **A** and **P**, we show the two regimes of bubble formation (with and without hysteresis). Understanding bubble-shape hysteresis and its dependence on compressibility can have important consequences to the design of dilatational interfacial rheology experiments where bubbles are continuously compressed and expanded.

Three-Dimensional Numerical Simulation of Spherical Drops in Liquid-Liquid and in Liquid-Gas Systems with Emphasis on the Influence of the State of Internal Motion on Drag

Christopher A. Edelmann, Patrick LeClercq, and Berthold Noll

DLR Stuttgart, Germany

Introduction

Drops have been studied for a long time in various areas and a vast amount of literature on this topic exists. An introduction as well as a review of work until 1978 can be found in the book by Clift et al. [1]. Important for the present work is the experimental investigation by Thorsen et al. [9] who studied the terminal velocity of liquid drops falling through water. They found that for some substances a sudden increase of terminal velocity occurs at a certain drop diameter and attributed this phenomena to "different states or levels of internal circulation".
Feng and Michaelides [3] simulated cases for viscosity ratios between 0 and ∞ and $Re < 1000$ with a two-dimensional, axisymmetric method. It was found that the internal motion is rather insensitive to the internal Reynolds number Re_i and resembles a Hill's vortex (Hill [4]). For certain cases with high density of the inner fluid, a secondary circular internal motion emerges in the rear part. Furthermore, Feng and Michaelides [3] state that the density ratio (κ) of inner to outer density has nearly no influence on drag. Sugioka and Komori [8] simulated water drops in air with three-dimensional simulations. At $Re = 300$ they found that even for a uniform air flow, the outer as well as the inner flow field is not axisymmetric. Engberg and Kenig [2] investigated the observation that in the experiments by Wegener et al. [10], in some cases, two different rise velocities were found for the same drop diameter. Even though their main objective was the deformation of the drops, they found a non-axisymmetric solution for nearly spherical drops and draw the conclusion that "the widely-used assumption of axial symmetry has to be carefully checked".

Model and Governing Equations

A fixed spherical drop with diameter d, placed in a uniform flow with inflow velocity U_o is considered. The flow field inside and outside of the drop is governed by the unsteady incompressible Navier-Stokes equations for non-reacting Newtonian fluids with constant viscosity and density, without external forces. They can be written in vector notation as:

$$0 = \nabla \cdot \mathbf{v} \quad (1)$$

$$\rho \frac{D\mathbf{v}}{Dt} = \mu \nabla^2 \mathbf{v} - \nabla p \quad (2)$$

where $\mathbf{v}$ is the velocity vector, ρ is the density, μ is the viscosity and p is the pressure. The flow fields inside and outside of the drop are separated by the drop surface, where a discrete jump in viscosity and density occurs. The ratios between the inner and the outer fluids are:

$$\lambda = \frac{\mu_i}{\mu_o} \quad (3)$$

$$\kappa = \frac{\rho_i}{\rho_o} \quad (4)$$

where the subscripts o and i denote the outer and the inner fluid, respectively. At the fluid-fluid interface, the total velocity and the shear stress are continuous. As the position of the interface is fixed and there is no interfacial mass transfer, the normal velocities of both phases at the interface are zero. Furthermore, both the tangential velocities and the shear stresses match.
The outer and inner Reynolds numbers are defined as

$$Re = Re_o = \frac{U_o d \rho_o}{\mu_o} \quad (5)$$

$$Re_i = \frac{U_i d \rho_i}{\mu_i} \quad (6)$$

where U_i is the maximum absolute velocity inside the drop ($|\mathbf{v}_i|_{max}$), which is for all cases equal to the maximum absolute velocity at the drop surface. For the present study it is defined that $d \cdot \rho_0 / \mu_0 = 1$, which means that $Re = Re_0 = U_0$.
As the drop is fixed and the inflow is uniform, the drag coefficient consists of two parts. A pressure (or form) part (denoted by the subscript p) and a friction (or viscous) component (denoted by the subscript f):

$$C_D = C_{D,p} + C_{D,f} = 8 \frac{F_{D,p} + F_{D,f}}{\pi \rho_o U_o^2 d^2} \quad (7)$$

where C_D is the drag coefficient and F_D is the drag force. The Strouhal number is defined as:

$$St = \frac{fd}{U_o} \quad (8)$$

where f denotes the frequency.

Numerical method

The three-dimensional simulations of the flow inside and outside the drop were performed with the German Aerospace Center (DLR) inhouse code THETA. THETA possesses various incompressible finite-volume solvers (and one compressible) and works on unstructured grids. It is well validated and used for a variety of computational fluid dynamic (CFD) problems. Recently a 4th-order as well as a 2nd-order low dissipation scheme for the spatial discretization were implemented by Löwe et al. [6].
For the present simulations, the convective and the diffusive fluxes were discretized using central differences. For the velocity in the momentum fluxes the 4th-order low-dissipation low-dispersion scheme according to Löwe et al. [6] was used. The Crank-Nicolson scheme was selected for the time-discretization. For the splitting of the pressure and the velocity calculations, the incremental variant of the projection method is applied. Convergence in time was carefully checked by monitoring the residuals of the pressure and velocity as well as the drag forces.
The drop is situated in the middle of a three-dimensional, cubic computational domain. At the

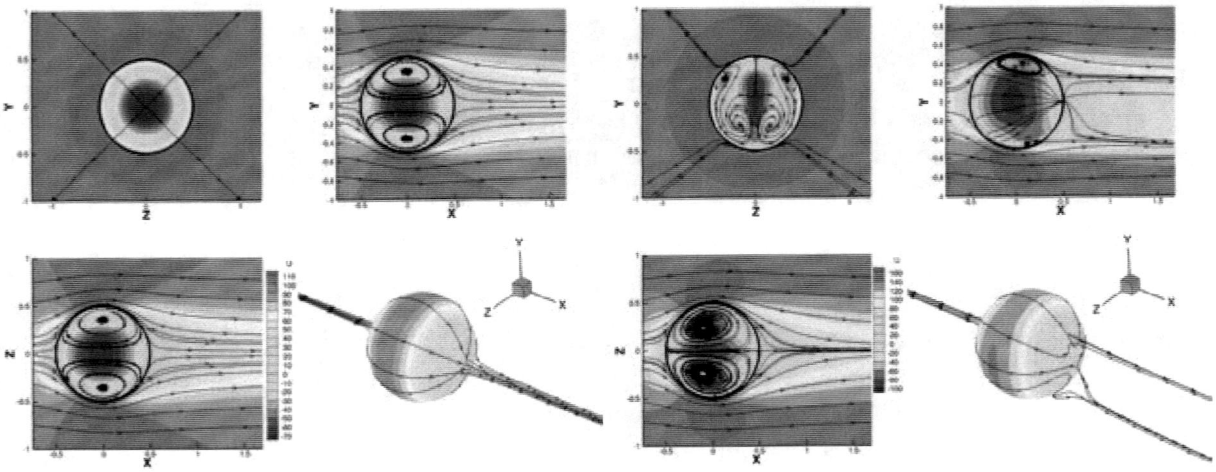

Figure 1: Streamlines inside and outside of a fluid sphere and contours of streamwise velocity for $Re = 100$, $\lambda = 0.5$, $\kappa = 1.5$.

Figure 2: Streamlines inside and outside of a fluid sphere and contours of streamwise velocity for $Re = 150$, $\lambda = 0.5$, $\kappa = 1.5$.

inflow boundary plane, a uniform velocity U_o is prescribed, whereas at the outflow a condition is used where the pressure is fixed and the velocities are extrapolated. At the side planes, symmetry conditions are applied. The simulations are run until a steady or periodic solution is found. The computational mesh is highly refined near the drop surface, with the same height for the first elements (Δ) on both sides from the interface. The height for the first grid spacing from the fluid-fluid interface was chosen such that the velocity boundary layer is sufficiently resolved. The boundary layer thickness was estimated by $\delta \sim Re^{-1/2}d$. Even for Reynolds numbers up to 10000, Δ/d is well below the estimated boundary layer thickness of about 0.01. The domain size is $40d$ in all three spatial dimensions. The size was chosen based on a literature review (see e.g. Johnson and Patel [5] or Sugioka and Komori [8]) and validated to be sufficient for $Re \geq 5$.

Results

Two different basic regimes of fluid properties are studied. First, a liquid drop in a liquid ambient phase and second a liquid drop in gas. For the first case, a viscosity ratio of $\lambda = 0.5$ and a density ratio $\kappa = 1.5$ were chosen. These parameters resemble ethyl bromide drops in water at 25°C and 1 bar, a system researched by Thorsen et al. [9]. The second case is based on n-heptane drops in air at 30°C which results in approximately $\lambda = 20$ and $\kappa = 600$. To study the influence of a varying density ratio and thus inner Reynolds number, another case at $\lambda = 20$ and $\kappa = 1200$ is investigated.

Flow Fields

The flow field of the steady solution for a case at $Re = 100$, $\lambda = 0.5$, $\kappa = 1.5$ is shown in figure 1. It can be seen that the flow field is cylindrical symmetric around the x-axis. The outer flow remains attached to the drop and inside of the drop the flow field resembles a Hill's vortex.

Based on the aforementioned case, the inflow velocity is increased such that the outer Reynolds number is 150. As illustrated in figure 2, the flow field is very different from the case at $Re = 100$.

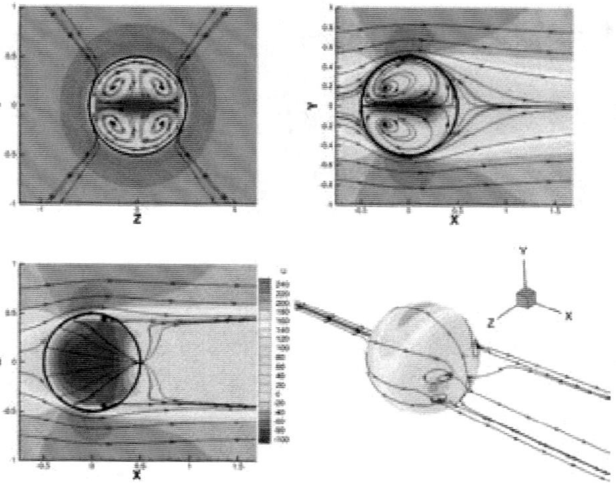

Figure 3: Streamlines inside and outside of a fluid sphere and contours of streamwise velocity for $Re = 200$, $\lambda = 0.5$, $\kappa = 1.5$.

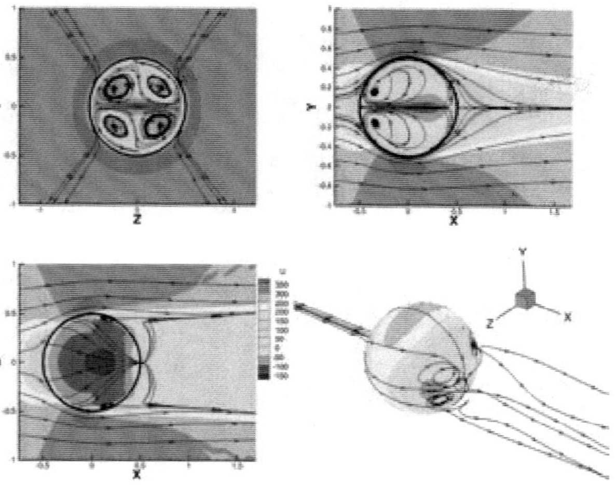

Figure 4: Streamlines inside and outside of a fluid sphere and contours of streamwise velocity for $Re = 300$, $\lambda = 0.5$, $\kappa = 1.5$. A snapshot is shown as the flow field is quasi-steady.

The cylindrical symmetry is broken but now a plane of symmetry is found. The outer flow is still attached in the x-z-plane but a detachment is detected in the x-y-plane. Inside the drop, streamlines with complicated, three-dimensional path are obtained. Four vortical structures are observed in the y-z-plane. In the x-z-plane, the streamlines resemble a Hill's vortex but the streamlines are no longer closed curves. In the x-y-plane, the flow in the rear part of the drop at the surface is reversed compared to the above described case at $Re = 100$. At both the upper and lower separation point (of the outer flow) a vortex is found inside the sphere. It should be noted that this flow field solution cannot be obtained with a two-dimensional calculation as performed by e.g. Feng and Michaelides [3]. However, it seems that the flow field reported by Engberg and Kenig [2] for a $d = 3$mm toluene droplet in water (the slow solution) is very close to the solution described above.

The Reynolds number of the outer flow is increased to $Re = 200$. The solution, illustrated in figure 3 is in some aspects different to the aforementioned case at $Re = 150$. Now, the steady flow field is two-fold axisymmetric around the x-axis and furthermore, two symmetry planes are observed. Similar to the case at $Re = 150$, the flow remains attached in one plane[1], the x-y-plane in this case, and detaches in the other plane (the x-z-plane). The upper separation point in the x-z moved upstream and two very small recirculation zones are observed.

The Reynolds number is further increased to $Re = 300$. No steady solution can be found in this case. However, high frequency oscillations ($St \approx 0.33$) with low amplitude remain and the flow field is only quasi-steady. Furthermore, as it can be seen from figure 4, no plane of symmetry is observed, but the two-fold axisymmety around the x-axis is retained. In other aspects, the flow field resembles the one at $Re = 200$ (see figure 3).

When the outer Reynolds number is further increased to 400, the asymptotic solution at large times is highly unsteady. Different modes of internal circulation can be observed from time to time, even though the flow field is superimposed by disturbances. The governing oscillation of this mode swapping is of relatively low frequency with a Strouhal number of approximately

[1]It is assumed that the orientation of the axes is arbitrary. Due to numerical reasons, the flow field orients itself to the axes. See e.g. Johnson and Patel [5] for a similar case.

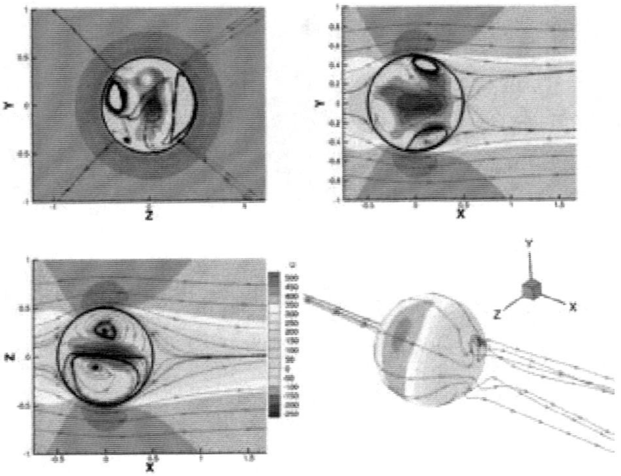

Figure 5: Streamlines inside and outside of a fluid sphere and contours of streamwise velocity for $Re = 400$, $\lambda = 0.5$, $\kappa = 1.5$. A snapshot is shown as the flow field is unsteady.

0.055. A snapshot of the flow field at the last timestep simulated is illustrated in figure 5. It can be seen that the flow field resembles at that point in time the solution shown in figure 4 for the case at $Re = 300$.

In the liquid-gas regime, a case at $Re = 100$ with $\lambda = 20$ and $\kappa = 600$ is chosen as the starting point. In this case, the steady inner flow field resembles a Hill's vortex, as depicted in figure 6. In contrast to the liquid-liquid regime, a trailing vortex can be seen in the outer flow field. The trailing vortex itself is detached from the sphere because of the intact Hill's vortex like solution inside the drop.

Instead of increasing the inflow velocity, the density of the fluid inside the sphere is increased. A change in density of the inner fluid changes κ and thus the inner Reynolds number, without altering the outer Reynolds number. The flow field of a case at $Re = 100$, $\lambda = 20$, $\kappa = 1200$ is shown in figure 7. Similar to the case at $Re = 300$ in the liquid-liquid regime (see figure 4), a unsteady, but quasi-steady solution exhibiting a two-fold symmetry around the x-axis is obtained. The trailing vortex partly attaches to the drop (in the x-z-plane). However in other aspects, like the size of the trailing vortex or the location of the flow separation from the sphere, only minimal changes are observed.

Both the Reynolds number and the density ratio where varied to find a steady internal flow mode beside the Hill's vortex like solution. However, it seems that in the liquid-gas regime, compared to the liquid-liquid regime, the two additional alternative modes of internal circulation observed in the liquid-liquid regime do not exist.

At higher internal Reynolds numbers, a highly unsteady flow field is found, similar to the one in the liquid-liquid regime shown in figure 5.

Drag

It is observed that in the liquid-liquid regime, the flow field is altered by the change of the inner flow mode. Thus, it is expected, that this influences the drag as well. Therefore, distinct differences in drag of own results compared to previous two-dimensional numerical simulations should be found. Furthermore, as mentioned above, ethyl bromide drops in water were studied

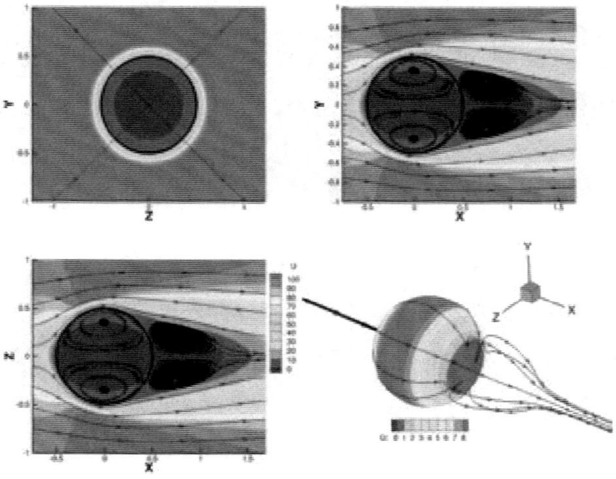

Figure 6: Streamlines inside and outside of a fluid sphere and contours of streamwise velocity for $Re = 100$, $\lambda = 20$, $\kappa = 600$.

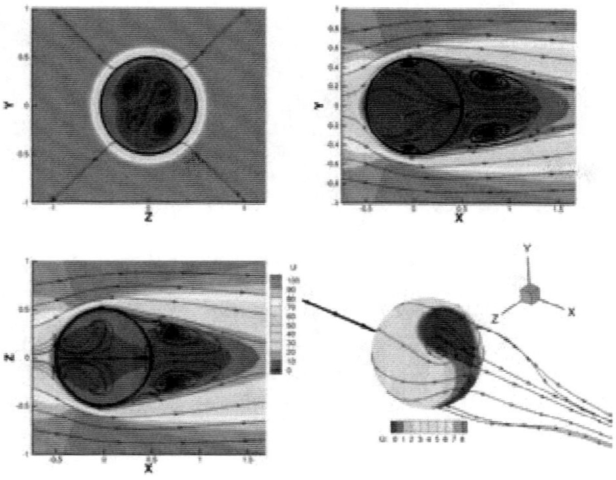

Figure 7: Streamlines inside and outside of a fluid sphere and contours of streamwise velocity for $Re = 100$, $\lambda = 20$, $\kappa = 1200$. A snapshot is shown as the flow field is quasi-steady.

experimentally by Thorsen et al. [9]. At least for the steady cases, a direct comparison with the experimental results is possible.

In figure 8, drag coefficients versus Reynolds number from various sources are plotted. It can be seen that the correlation function for viscous spheres is a bit off the numerical results by Feng and Michaelides [3] but follows the general trend. The function for rigid spheres gives much higher drag coefficients and illustrates the extreme case of an immobile surface. At $Re = 50$ and $Re = 100$ the present results are very close to the numerical results from Feng and Michaelides [3], whereas the results deviate at higher Reynolds numbers. This is due to the different internal flow modes, as Feng and Michaelides [3] only conducted two-dimensional axisymmetric simulations. If for the present cases with alternative modes, the density ratio is reduced such that a Hill's vortex like internal solution is obtained, the resulting drag coefficient is very close to the results by Feng and Michaelides [3]. However, the most important information from figure 8 is that the present simulation results are in good agreement with the experiments by Thorsen et al. [9]. This supports the idea that the present results are of practical importance. However, the unsteady cases are worth a closer look, as Thorsen et al. [9] observed no oscillations for drops up to d=0.204 cm ($Re = 520$). For the quasi-steady cases, the disturbance amplitudes are very low. Furthermore, the simulated Strouhal number of 0.33 is not too close to the characteristic Strouhal number for ethyl bromide of 0.44 given by Thorsen et al. [9]. Thus it seems reasonable that the oscillations remain at a very low amplitude which can be hardly detected in an experiment. However, at $Re = 400$ and $Re = 500$ strong oscillations were found in the present study. E.g. at $Re = 400$ the oscillation amplitude of the drag coefficient is roughly 25%. In this case, coupled interactions with the outer flow field and the liquid-liquid interface will take place, which cannot be captured by the present method. Thus, it must be doubted that the flow field shown in figure 5 is close to the physical flow field of an ethyl bromide drop in water of this size.

In the liquid-gas regime, it is expected that the drag is only minimally altered by the internal flow. This is first of all due to the very low velocities at the drop surface because of the high viscosity ratio. In table 1 drag coefficients of cases at $Re = 100$ are compared. The drag coefficient for viscous spheres at $\lambda = 20$ with $\kappa = 600$ and $\kappa = 1200$ are shown, as well as the drag coefficient for a rigid sphere (indicated by $\lambda = \infty$). As predicted, all three cases have

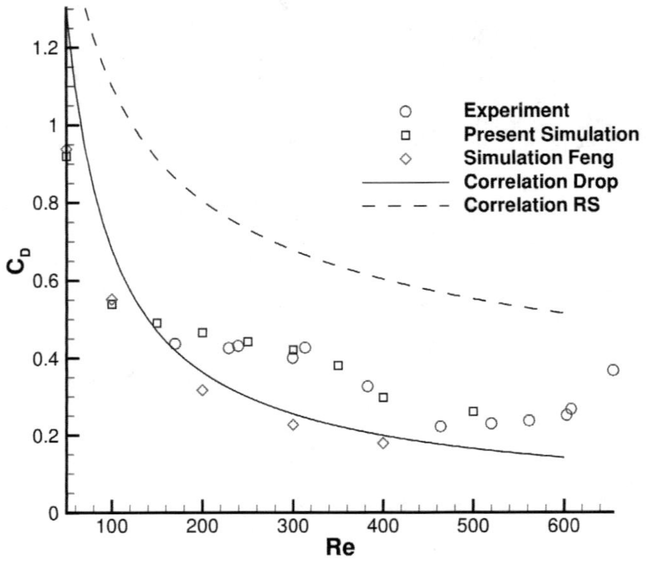

Figure 8: Drag coefficient versus Reynolds number for ethyl bromide drops in water. Experiment: Data by Thorsen et al. [9]. Present Simulation: Results at $\lambda = 0.5$, $\kappa = 1.5$. Simulation Feng: Two-dimensional axisymmetric simulations at $\lambda = 0.5$, $\kappa = 0.5$ from Feng and Michaelides [3]. Correlation Drop: Correlation for viscous spheres from Feng and Michaelides [3]. Correlation RS: Correlation for rigid spheres by Putnam [7].

very similar drag coefficients. Even the Hill's vortex like case deviates by only $\approx 4\%$ from the rigid sphere result. In accordance with previous studies (e.g. Feng and Michaelides [3]) only a secondary influence of the internal motion on drag is found at $\lambda = 20$.

Conclusions

In the present work, the results of detailed, three-dimensional numerical simulations of spherical drops in a uniform flow were presented. Liquid drops surrounded by a liquid as well as liquid drops in a gaseous ambient phase were considered. It seems that the assumption of two-dimensional axisymmetric flow fields inside and outside of droplets is not justified in many cases. With increasing Reynolds number, different types of internal flow modes are found, which do

Table 1: Drag coefficients for viscous spheres in the liquid-gas regime and for a rigid sphere. The rigid sphere is indicated by $\lambda = \infty$.

Re	λ	κ	C_D
100	20	600	1.045
100	20	1200	1.076
100	∞	–	1.089

not resemble the well known Hill's vortex solution. In the liquid-liquid regime, a major influence of the type of internal circulation on the drag is obtained, whereas the influence in liquid-gas systems is negligible.

References

[1] R Clift, J Grace, and M Weber. *Bubbles, drops, and particles.* Academic Press, New York, 1978.

[2] Roland F. Engberg and Eugeny Y. Kenig. An investigation of the influence of initial deformation on fluid dynamics of toluene droplets in water. *International Journal of Multiphase Flow*, 76:144 – 157, 2015.

[3] Zhi-Gang Feng and Efstathios E. Michaelides. Drag coefficients of viscous spheres at intermediate and high reynolds numbers. *Journal of Fluids Engineering*, 123(4):841–849, May 2001.

[4] M. J. M. Hill. On a spherical vortex. *Philosophical Transactions of the Royal Society of London A: Mathematical, Physical and Engineering Sciences*, 185:213–245, 1894.

[5] T. A. Johnson and V. C. Patel. Flow past a sphere up to a reynolds number of 300. *Journal of Fluid Mechanics*, 378:19–70, 1 1999.

[6] Johannes Löwe, Axel Probst, Tobias Knopp, and Roland Kessler. A low-dissipation low-dispersion second-order scheme for unstructured finite-volume flow solvers. In *AIAA SciTech.* American Institute of Aeronautics and Astronautics, jan 2015.

[7] A Putnam. Integratable form of droplet drag coefficient. *Ars Journal*, 31(10):1467–1468, 1961.

[8] Ken-Ichi Sugioka and Satoru Komori. Drag and lift forces acting on a spherical water droplet in homogeneous linear shear air flow. *Journal of Fluid Mechanics*, 570:155–175, 1 2007.

[9] G. Thorsen, R.M. Stordalen, and S.G. Terjesen. On the terminal velocity of circulating and oscillating liquid drops. *Chemical Engineering Science*, 23(5):413 – 426, 1968.

[10] Mirco Wegener, Matthias Kraume, and Anja R. Paschedag. Terminal and transient drop rise velocity of single toluene droplets in water. *AIChE Journal*, 56(1):2–10, 2010.

468699 Shear-Induced Alignment of Janus Particle Lamellar Structures

Abstract:

Control over the alignment of colloidal structures plays a crucial role in advanced reconfigurable materials. In this work, we study the alignment of Janus particle lamellar structures under shear flow via Brownian dynamics simulations. Lamellar alignment (orientation relative to flow direction) is measured as a function of the particle volume fraction, strength of the anisotropic interaction potential, and the Péclet number (Pe)—the ratio of viscous shear and Brownian forces. In conditions where lamellar structures are formed, three orientation regimes are observed: (1) Random orientation for Pe $<<$ 0.1, (2) parallel orientation—lamellae with their normals parallel to the direction of the velocity gradient—in the range of 0.1 $<$ Pe $<$ 10 and (3) perpendicular orientation—lamellae with their normals parallel to the vorticity direction—for Pe $>$ 10. Our results suggest that the change of parallel to perpendicular orientation is caused by hydrodynamic torque on the particles overcoming Brownian rotary motion and torques resulting from interparticle interactions. We compare these results with analogous alignment obtained in smectic liquid crystalline and block copolymers under shear, in order to extend the implications of our study to other layer-forming fluids. This initial study of shear-induced alignment on lamellar structures formed by Janus colloidal particles also opens the door for future applications where a reversible actuator for structure orientation is required.

463584 Hydrodynamic Simulations of Constant Stress and Pressure Rheology of Dense Colloidal Suspensions

Abstract:

We study the constant stress and pressure rheology of polydisperse colloidal suspensions with volume fractions up to the jamming density via a new implementation of Stokesian dynamics, the Spectral Ewald Accelerated Stokesian Dynamics (SEASD). The non-Brownian limit of the simulations agrees quantitatively with the experiments of Boyer et al. [Phys. Rev. Lett. 107, 188301 (2011)] with only a shift in the maximum volume fraction. Consistent with non-hydrodynamic simulations, the suspension shear viscosity diverges algebraically with the distance from the arrest volume fraction along constant pressure contours. The low volume fraction continuous shear thickening observed at constant volume fraction no longer occurs at fixed pressure. Structurally, in the high imposed pressure limit, the suspensions develop string-like ordered structures that disappear with further stress increase. Using a Péclet number based on the long-time self-diffusivity, we discover a universal collapse of the interaction contribution of the friction coefficient, suggesting that a non-equilibrium Stokes-Einstein-Sutherland relation for the long-time self-diffusivity and the suspension shear viscosity holds with an effective temperature proportional to the suspension osmotic pressure for all imposed stresses and pressures. With the same Péclet number, the structural signature of the suspensions also collapses to a master curve. That similar collapses are also found in non-hydrodynamic simulations suggests that the hydrodynamic interactions only quantitatively affect the behavior of dense suspensions near flow-arrest transitions.

468958 Hydrodynamically Interacting Colloids inside a Spherical Cavity As a Model for Intracellular Transport: Hydrodynamic Entrainment and Active Motion

Abstract:

We study diffusion and rheology of hydrodynamically interacting particles confined by a spherical cavity via dynamic simulation, as a model for intracellular and other micro-confined biophysical transport. Previous models of 3D confined transport either focused on the motion of a single confined particle or studied collections of particles while neglecting hydrodynamic interactions. Such approximations are motivated by the challenge of modeling both far-field many-body and near-field lubrication interactions and how such interactions are influenced by the presence of a confining cavity. However, biophysical and other confined suspension systems are crowded, watery environments populated with a multitude of hydrodynamically interacting microscopic particles. In the present work, we utilize our framework for spherically confined suspensions to study the effect of 3D confinement on particle *entrainment* as well as on the active motion of a probe particle. Recent studies have shown that in unbound suspensions, increasing the suspension volume fraction does not lead to a change in the algebraic decay of hydrodynamic entrainment, that is, crowding does not lead to hydrodynamic screening. In contrast, studies involving a dilute system of particles near a single planar wall have observed that the presence of the boundary can lead to a change in the algebraic decay of hydrodynamic entrainment. In this work, we show how the presence of the spherical cavity impacts entrained motion in 3D confined colloidal suspensions. We calculate the hydrodynamic entrainment between particle pairs when these are immersed in an intervening medium of freely diffusing bath particles. This allows us to determine how hydrodynamic entrainment changes as a function of particle position, volume fractions, and particle-to-cavity size ratio. In addition to hydrodynamic entrainment, active motion is an important transport mechanism in many biophysical systems. We study the effect of confinement on active motion by using as a model system the motion of a single probe particle driven through the 3D confined suspension by an external force. By tracking the motion of the particle, as well as the evolution of the suspension stress rheological properties such as the micro-viscosity, osmotic pressure, and normal stresses are determined as a function of probe forcing in the linear response regime.

455090 Modeling the Transient Shear Flow of a Carbon Black Soft Colloidal System Using a Scalar Structural Parameter Thixotropic Model

Abstract:

Recent work using small angle neutron scattering under flow has identified the existence of a microstructure that is dependent on flow conditions [1,2] in soft colloidal systems. On the other hand, over the last twenty years or so, a large number of empirically based thixotropic rheological models has been developed that involve, albeit phenomenologically, a single scalar structural parameter, lambda [3-6]. These models involve the combination of a lambda-dependent decomposition of the shear stress to an elastic and a viscous contribution with a relaxation-based evolution equation for lambda. The rich behavior of the soft colloidal systems is primarily due to the interaction of the structure and the hydrodynamic force and this can be particularly enhanced applied through in transient flow, and/or large amplitude oscillatory shear flow.

Applying large amplitude oscillatory shear (LAOS) to complex fluids induces nonlinear rheological responses, that, with proper modeling, can be used to sensitively probe the underlying microstructure and its dynamics. We demonstrate this for several concentrated colloidal suspensions using a newly developed semi-empirical, thixotropic master-equation developed around a scalar internal structural parameter, and the best features of several published thixotropic constitutive equations [3,4,5,6]. We examine two soft colloidal systems, all involving collective phenomena of yield stress and thixotropy. The systems are a 3.23vol% carbon black in napthenic oil [7,8,9], a thermo-reversible, adhesive hard sphere system, 30 nm silica particles in n-tetradecane [2].

The new master equation uses the structure parameter with a constitutive equation to model the elastic and viscous components of rheological responses to shear rate. Both steady state and time-dependent shear experimental data are fit using a recently developed robust numerical method that stochastically and effectively searches for the global optimum of a suitably forced least squares residual in the allowed parameter space [9]. With the best fit model

parameters large amplitude oscillatory shear is predicted and compared to LAOS experimental data. The quality of the model is evaluated by comparing the difference between prediction and experimental LAOS data that are system dependent, and predicting strain rate frequency superposition [10]. While the master equation has been found to be able to qualitatively fit the experimental data, important quantitative are still were observed. The advantages and disadvantages of the new rheological model limitations of are going to be discussed focusing on the elucidation of the limitations of generic single scalar internal structural parameter models. These results and model comparisons are used to identify areas for improvement in thixotropic suspension modeling. Limitations of single scalar structural modeling will be shown.

Reference

[1] C.R. Lopez-Barron, A.K. Gurnon, A.P.R. Eberle, L. Porcar, and N.J. Wagner. Physical Review 89, 042301 (2014), 1-11.

[2] J. M. Kim. PhD thesis, University of Delaware, 2013.

[3] J. Mewis and N. J. Wagner, Colloidal Suspension Rheology, Cambridge Univ. Press, 2012.

[4] P. de Souza Mendes and R. Thompson. Rheol. Acta (2013) 52:673-694.

[5] A. Mujumbdar, A. N. Beris and A. B. Metzner. J. Non-Newtonian Fluid Mech. 102 (2002) 157-178.

[6] C.J. Dimitriou, R.H. Ewoldt and G.H. McKinley. J. Rheol. 571(1), 27-70.

[7] K. Dullaert and J. Mewis. Rheol. Acta (2005) 45: 23-32.

[8] K. Dullaert and J. Mewis. JNNFM 139 (2006) 21-30.

[9] M.J. Armstrong, PhD Thesis, University of Delaware (2015).

[10] H.M. Wyss, K. Mikayzaki, J. Mattsson, A. Hu, D.R. Reichman and D.A. Weitz. PRL 98, 238303 (2007).

467432 Microrheological Characterization of the Dynamic Transition of Hydrogenated Castor Oil Colloidal Gels Due to Osmotic Pressure Gradients

Abstract:

Nature provides colloidal materials with wide ranges of structure and function leading to materials with diverse rheology and applicability. Although these materials are used in industrial applications, their potential is not fully realized due to a lack of in depth understanding of the structure-function relationship on the microscale as they undergo dynamic processes. Uncovering the heterogeneous, dynamic structure during critical transitions of colloidal gels will enable these materials to be used for more advanced applications, including rheological modification of composite materials and dynamically adaptable materials that tolerate harsh. We characterize the dynamic transitions (sol-gel and gel-sol) of hydrogenated castor oil (HCO), which transitions due to an osmotic pressure gradient. This scaffold consists of HCO colloids, a surfactant (linear alkylbenzene sulfonate) and water. Multiple particle tracking microrheology (MPT) measures these phase transitions. In MPT, we embed fluorescent probe particles into the sample and measure their Brownian motion, which is related to rheological properties using the Generalized Stokes-Einstein Relation. Probe movement is tracked during expansion and contraction of the gel until a steady state has been reached. MPT data are analyzed using time-cure superposition identifying critical transition times and critical relaxation exponents that determine the amount of energy stored by the material. HCO gels have critical relaxation exponents of n_{exp} = 0.77±0.08, for expansion, and n_{con} = 0.90±0.15, for contraction. During expansion and contraction HCO gels evolve heterogeneously, this heterogeneity is measured spatially and rheologically. van Hove correlation functions, characterize the one-dimensional probability distribution function of particles. Heterogeneity of the gel during the phase change is quantified by comparing variances of single particle van Hove correlation functions using an F-test with a 95% confidence interval determining if individual particles are probing statistically different microenvironments within the gel. These phase changes have rheological heterogeneous microenvironments that are homogeneously

distributed throughout the field of view. Although HCO gels do evolve heterogeneously, this work determines that these heterogeneities do not significantly change traditional MPT measurements but the analysis techniques developed provide additional information on the unique heterogeneous scaffold microenvironments. This creates a toolbox that can be widely applied to other scaffolds during dynamic transitions.

466633 Microrheology of Monoclonal and Bispecific Antibodies

Abstract:

A key challenge encountered in the current development of therapeutic protein solutions is the need to measure their viscosity to identify stable, syringeable formulations in a large composition space. Commercially available techniques such as capillary viscometry and rotational rheometers are frequently used, but require relatively large sample volumes. This sample size requirement restricts the number of rheological measurements in the early development stage when only small amounts of protein are typically available.

This talk will focus on characterizing the viscosity of protein therapeutics over a wide range of compositions with a minimal amount of material. Microrheology techniques are powerful methods to study scarce biomaterials [1]. We use multiple particle tracking (MPT) to provide a high throughput sample processing platform for protein therapeutics. This work examines the microviscosity of three IgG monoclonal antibodies and two bispecific monoclonal antibodies assembled from IgG half-antibodies over a large range of concentrations (1â€"180 mg/mL). The protein solutions exhibit Newtonian fluid behavior over a frequency range of 0.05 to 30 s^{-1}, and do not show evidence of further microstructure development due to protein-protein association. An analysis of covariance demonstrates the high accuracy of small volume microrheology measurements. Based on the relative error between measured and tabulated viscosities, the uncertainty of viscosities derived from particle tracking is less than 2% of the true value.

[1] K.M. Schultz & E.M. Furst. Soft Matter 8, 6198-6205, 2012.

464901 Microstructure and Yielding of Microfiber Gels

Abstract:

Complex fluids have been studied and developed throughout most sectors of industry and medicine because of their unique rheological properties. Suspending colloidal particles with attractive interactions to form a network is a common way to impart a yield stress to a fluid. Large aspect ratio cellulose microfibers are able to a form poroelastic network at low volume fractions via aggregation and entanglement, forming a gel without significantly modifying viscosity. The gels have a small but useful yield stress and a better ability to suspend particles than non-interacting higher volume fraction glasses because the sparse fiber networks can significantly restructure at small strains. Yielding behavior can thus strongly depend on the fluid microstructure.

In our study, a new active microrheology technique is developed and applied to study deformation and yielding of aqueous cellulose fiber gels. Confocal imaging shows how gel yield stress relates to structural deformation rate because of localized network restructuring. Such response is advantageous to applications like surface coatings, nasal sprays, cosmetics, and foods. Understanding the mechanism of rate- and length-scale dependent yielding, and relating microstructure changes to bulk rheology, will enhance our ability to formulate, model, and design complex fluids with novel performance.

469523 Transient Yield in Reversible Colloidal Gels: A Micro-Mechanical Perspective

Abstract:

We study the nonlinear rheology of colloidal gels via large-scale dynamic simulation, with a view toward understanding the micro-mechanical origins of the transition from solid-like to liquid-like behavior during flow startup and its connection to energy storage and viscous dissipation. Such materials often exhibit an overshoot in the stress during flow startup, but the underlying microstructural origins of this behavior remain unclear. To understand this overshoot behavior, a fixed strain rate is imposed on a reversible colloidal gel, where thermal fluctuations enable quiescent gel aging. It has been suggested flow occurs only after clusters first break free from the network and then disintegrate, leading to two stress peaks that vary with age, flow strength, volume fraction, bond strength, and pre-strain history. However, our detailed studies of the microstructural evolution during startup challenge this view. We present a new model of stress development, relaxation, and microstructural evolution in reversible colloidal gels in which the ongoing age-coarsening process plays a central role.

462663 Role of Particle Roughness in Shear Thickening Colloidal Suspensions

Abstract:

The connection between roughness and friction is not straightforward despite decades of study. Here, we examine the effect of surface roughness on suspensions that undergo continuous shear thickening (CST) and discontinuous shear thickening (DST), with special emphasis on both viscosity and normal stresses. We provide direct evidence that increasing the surface roughness in model colloidal suspensions causes the CST-DST-dilatancy transition to occur at lower volume fractions (Ï•) and shear rates. Increasing the root-mean-squared roughness beyond the hydrodynamic lubrication length scale shifts the divergence in viscosity from random close packing for frictionless spheres (Ï•max ~ 0.64) to Ï•max ~ 0.54, in accordance with theory developed for frictional grains. Our results show that the forces required to push rough colloids into close contact are an order of magnitude lower than that of smooth colloids. The data also show that lubrication is dominant in concentrated suspensions of smooth colloids, but friction becomes the primary mechanism for DST when particles are in close contact at very high shear rates, Ï•, and surface roughness.

457869 Axial Dispersion of Brownian Colloids in Microfluidic Channels

Abstract:

Particles in non-uniform flow fields undergo an enhanced axial dispersion compared to diffusion in the absence of flow. Qualitatively, axial dispersion is enhanced by flow because particles diffuse across streamlines and advect at rates different from the average velocity, resulting in a net spreading relative to the mean. Taylor and Aris first calculated the asymptotic form of this dispersion coefficient for point-like tracer particles in a cylindrical tube under the assumptions that the particles diffuse isotropically and explore all streamlines of the flow field uniformly. Colloidal dispersions, especially under confined flow, exhibit markedly different behavior due to their finite size and particle-particle and particle-wall hydrodynamic interactions.

We present a complete theoretical framework for the axial dispersion of a Brownian colloidal suspension confined in a parallel plate channel, extending the Taylor-Aris treatment to particles with diameters comparable to the channel width. The theoretical model incorporates the effects of confinement on the colloid distribution, corrections to the velocity profile due to the effects of colloid concentration on the suspension viscosity, and position-dependent anisotropic diffusivities. We test the theoretical model using explicit-solvent molecular dynamics simulations that fully incorporate hydrodynamic correlations and thermal fluctuations, and obtain good quantitative agreement between theory and simulations. We find that the non-uniform colloid structuring that arises in confinement due to excluded volume between the colloids and channel walls significantly impacts the transport properties of the suspension. The developed model should prove useful in many applications involving the axial dispersion of colloids, including extracting diffusion coefficients from microfluidic experiments and modeling the transport of colloids in geological fractures.

470155 Bubble Meets Droplet: Particle-Assisted Reconfiguration of Wetting and Dewetting

Abstract:

Wetting in colloidal multiphase systems plays a major role in many industrial processes, including encapsulation, enhanced oil recovery, water purification, ore processing, and food technology. The wetting of a bubble or a liquid droplet by a second fluid in a third immiscible fluid medium can result in three possible wetting morphologies: non-wetting, partial engulfment, or complete engulfment. A given fluid system, however, does not always adopt the desired wetting configuration; hence there is interest in methods to tune wetting configuration on demand. Particles with suitable wettability can be strongly adsorbed to the interface of immiscible fluids. Although it was recognized more than a century ago that particles can act as stabilizers in colloidal multiphase systems, prior to this work it was still unknown whether particles can also act as wetting or progressive de-wetting modifiers. We demonstrate that wetting configuration can be dynamically tuned by selection of appropriate colloidal particles, as predicted by an effective spreading coefficient. The wetting and engulfment of an air bubble by an oil droplet in a water medium was used as a model system relevant in a wide variety of industrial processes. We demonstrate that particles can be used both to promote "bubble wetting", thereby inducing complete bubble engulfment by an oil drop, as well as to trigger "bubble de-wetting", *i.e.* substantially reducing the oil-bubble contact area. This study yields a new on demand strategy to predict and control the wetting configuration in colloidal multiphase systems, with potential benefits in a wide range of research fields, industrial processes, and commercial products.

465208 *In Situ* Mechanical Testing of Bijel Fibers

Abstract:

Bicontinuous interfacially jammed emulsions (bijels) are a new class of soft materials [1] with promising applications in various areas ranging from healthcare, food, energy and reaction engineering due to their unique structural, mechanical and transport properties. We recently presented a new method to prepare bijels via solvent transfer-induced phase separation (STRIPS), which enables continuous fabrication of hierarchically structured bijel fibers using a microfluidic platform [2]. Understanding the mechanical properties of these bijel fibers is crucial for their implementation in new technologies. Here, we introduce a novel microfluidic *in situ* technique to measure the yield strength of bijel fibers, and potentially the mechanics of other fibrous materials. The yield strength of bijel fibers depends on the various factors such as the age of the bijel material, the nanoparticle concentration and the strength of particle-particle attraction. We show that such fibers with controlled aspect ratios can be fabricated by varying the flow rates in a co-axial microfluidic device and explain the formation process with a simple force-balance model. These findings broaden the potential for the use of STRIPS bijels in applications with different mechanical demands. Moreover, our *in situ* mechanical testing method could potentially enable the mechanical characterization of various soft fibrous materials such as hydrogels and protein complexes.

[1] Herzig, E. M., et al. (2007). Nature materials, 6(12), 966-971.

[2] Haase, M. F., Stebe, K. J., & Lee, D. (2015). *Advanced Materials*, *27*(44), 7065-7071.

462502 Structural Effects on Adsorption Kinetics of Phosphatidylcholine Vesicles at an Air-Water Interface

Abstract:

The dynamic surface tension of several purified phospholipids onto a clean air-water interface was studied via drop volume tensiometry in order to gain insight into the mechanism of adsorption. Long chain phospholipids spontaneously organize into vesicles when dispersed in water due to their high hydrophobic character. Upon creation of a clean air-water interface, these vesicles will diffuse to a region just below the surface and then kinetically adsorb to the interface. We are able to gain insight into this adsorption process via dynamic surface tension experiments. We have demonstrated that the adsorption of purified phospholipid vesicles to an air-water interface is kinetically limited, rather than diffusion limited for liquid-phase vesicles. The mechanistic process that takes places during the kinetically controlled adsorption step, however, remains unresolved. We have examined the effects of vesicle size, temperature, and chain length on dynamic surface tension as a means of developing an appropriate kinetic model. Dynamic surface tension profiles are then coupled with dynamic light scattering results for comparison to model predictions.

472558 Tau Misfolding and Assembly at Lipid Membrane Interfaces

Abstract:

Interaction between the microtubule associated, intrinsically disordered, tau protein with the lipid membrane has been shown to induce tau structural compaction and membrane disruption, suggesting possible membrane-mediated pathways for tau aggregation and toxicity in neurodegenerative tauopathies, including Alzheimerâ€™s disease (AD) and familial frontotemporal dementia. In this study, we examined the effects of hyperphosphorylation, an early event in AD pathogenesis, and tau domain composition on tau-membrane interactions using the psudo-phosphorylation mutant hTau40/3Epi and the repeat domain of tau K18, which lacks the projection domain, respectively. Complementary X-ray and neutron scattering techniques were used to resolve Ã…-level structural details of tau association with model lipid monolayers at the air/water interface and quartz supported lipid bilayers. Results show that neither hyperphosphorylation nor truncation of the tau projection domain significantly affected the tauâ€™s intrinsic surface activity or membrane interaction. The mutant and K18 were highly surface active and exhibited selective, yet strong interactions with anionic lipid membranes. Furthermore, binding of the tau proteins to lipid membranes induced tau structural compaction of and disrupted membrane morphology, lipid packing, and structural integrity. Results from this study thus show that membrane-based pathways that induce tau aggregation and lead to toxicity are insensitive towards details of the amino acid composition of the tau protein, and could serve as a general mechanism by which tau misfolds and aggregate into pathogenic aggregates.

465223 MD Simulation Study of Direct Permeation of Nanoparticle Across Cell Membrane Under External Electric Field

Abstract:

Nanoparticles (NPs) have been attracted much attention for biomedical and pharmaceutical applications. In most of the applications, NPs are required to translocate across the cell membrane and to reach the cell cytosol. Experimental studies have reported that by applying an electric field NPs can directly permeate across the cell membrane without confinement of NPs by endocytic vesicles, while damage to the cell can often be a concern. Understanding of the mechanism underlying the direct permeation of NPs under an external electric field can greatly contribute to realize a technology for the direct delivery of NPs. Here we investigated the permeation of a cationic gold NP across a phospholipid bilayer under an external electric field using a coarse-grained molecular dynamics simulation. When an external electric field that is equal to the membrane breakdown intensity was applied, a typical NP delivery by electroporation was shown: the cationic gold NP directly permeated across a lipid bilayer without membrane wrapping of the NP, while a persistent transmembrane pore was formed. However, when a specific range of electric field that is lower than the membrane breakdown intensity was applied, a unique permeation pathway was exhibited: the generated transmembrane pore immediately resealed after the direct permeation of NP. Furthermore, we found that affinity of the NP for the membrane surface is a key for the self-resealing of the pore. Our finding suggests that by applying an electric field in a proper range NPs can be directly delivered into the cell with less cellular damage.

470515 Curvature Capillary Repulsion

Abstract:

On a curved fluid interface, microparticles interact with the curvature field in a manner akin to a charged multipole in an applied external electrostatic field. The interaction is driven by capillarity where particles move to minimize the excess area associated with the distortions that they make in the interface. Here, we present capillary curvature repulsion. We study microparticles migration on a curved host interface with zero mean curvature, which is created by pinning an oil-water interface around a micropost. We identify curvature fields in which microparticles migrate away from high curvature sites. By a combination of theory and experiment, we show how these interactions occur and their relationship to particle and post geometry. By tuning repulsion and attraction, we define sites far from bounding surfaces for particle assembly at equilibrium on interfaces, and explore collective behaviors of particles near these sites. Understanding of effects of curvature on the behavior of microparticles at the interfaces provides implications to create reconfigurable hierarchical structures.

464182 On Precipitation during the Coalescence of Reactive Droplets

Abstract:

The coalescence of sessile drops of completely miscible liquids can be affected by a Marangoni flow, induced by the surface tension difference of the two liquids [1,2]. Although the surface tension is strongly related to solute's concentration, the coalescence of drops containing reactive liquids was not investigated so far. However, the coalescence behavior of reactive droplets is encountered in a wide range of applications such as liquid–liquid extraction, ink jet printing, nanoparticle synthesis, etc. Moreover, controlling reagent mixing and transport phenomena during the droplet coalescence is of major importance in the new precipitation in emulsion process we are currently developing [3,4].

We consider the coalescence behavior of sessile drops containing aqueous solutions of cerium nitrate and oxalic acid. Upon contact, these compounds form insoluble cerium oxalate.

$$Ce(NO_3)_3 + 3C_2H_2O_4 + 10H_2O \rightarrow Ce_2(C_2O_4)_3.10H_2O + 6HNO_3$$

We observed that in a given and limited range of component concentrations and surface tension differences between both liquids (the latter was controlled using diol/water binary solvents), the precipitating cerium oxalate surprisingly forms periodic patterns of fringes with different light-scattering properties, that allow easy detection by strongly contrasted images (Figure 1).

SEM observation, performed after complete drying of the sample, reveals individual needle-like particles of typical length 25µm in the dark, strongly light-scattering zone; whereas the latter appear aggregated in large "microflowers" (*i.e.*65 µm) in the light areas. Additional experiments using confocal microscopy indicate that the two types of particle result from different precipitation mechanisms. However, XRD shows that the particles exhibit a monoclinic structure typical of cerium oxalate decahydrate, regardless of their

shape.

The relation between the various control parameters (initial contact angles, reactant concentrations, viscosity, surface tension difference), the stripe property (spacing of the fringes), and the Marangoni flow was further analyzed by optical microscopy and image processing. The results highlight a strong coupling between the transport processes and the fast chemical reaction [5]. On the other hand, for given "hydrodynamic" parameters (*e.g.* initial contact angle and surface tension difference), the occurrence of the reaction appeared to modify the coalescence behavior, compared to the non-reactive cases investigated by Karpitschka and Riegler [1,2]. Coalescence of reacting droplets has therefore proved to be a valuable tool to further study crystal nucleation and growth mechanisms.

In this aim, the solvent influence on the precipitates is currently examined. Water, binary water/diol and ternary water/oil/hydrotrope mixtures are considered as solvent [6]. Still triggering the reaction by the coalescence of sessile droplets, first results have highlighted that mesoscale solubilization in structured solvents [7], changes the particle morphology to a more compact mesocrystal shape described by [8].

[1] S. Karpitschka and H. Riegler, Langmuir, 26, **2010**, 11823-11829.

[2] S. Karpitschka and H. Riegler, J. Fluid Mech, 743, **2014**, R1.

[3] S. Charton *et al.*, Chem. Eng. Res. Des., 91, **2013**, 660-669.

[4] M. Ollivier *et al.*, Proceedings of ATALANTE2016, Montpellier.

[5] M. Jehannin *et al.*, Langmuir, 31, **2015**, 11484-11490.

[6] M. Jehannin *et al.*, *submitted to*Physical Chemistry Letters.

[7] W. Kunz *et al.*, Current Opinion in Colloids and Interface Science, **2016**, *in press.*

[8] H. Cölfen and M. Antonietti, *Mesocrystals and Nonclassical Crystallization*, **2008**, John Wiley & Sons, Ltd

Figure 1: Periodic patterns appearing during drop coalescence (optical imaging via top view camera) and (after drying) by scanning electron microscopy. Black domains in the top picture correspond to needlelike crystals, while microflowers of micron-size aggregated needles are observed in the light domains. Fringe formation is observed regardless of the Marangoni flow direction.

Keywords: Coalescence; Marangoni flow; Precipitation Mechanisms

455912 Kinetic Influence of Polyvinylpyrrolidone in the Shape-Control Mechanism of Ag Nanocrystal Synthesis

Abstract:

In the shape-controlled synthesis of colloidal Ag nanocrystals, structure-directing agents, particularly polyvinylpyrrolidone (PVP), are known to be a key ingredient in making nanostructures with well-defined shapes. Although many Ag nanocrystals have been successfully synthesized using PVP, the mechanism by which PVP actuates shape control remains elusive. Surface-adsorbed structure-directing agents can influence both the surface free energies of the nanocrystalâ€™s facets and the kinetics of the Ag atom deposition. To probe the kinetic influence of the adsorbed PVP film, we use umbrella sampling molecular dynamics simulations to calculate the potential of mean force (PMF) and diffusion coefficient profiles of Ag atom deposition onto Ag(100) and Ag(111) in ethylene glycol solution with PVP. The computed PMF and diffusion coefficient profiles are used to calculate the mean-first passage times and implement extensive Brownian dynamics simulations, which allows the kinetic effectsÂ to be quantitatively evaluated. Our results show that PVP films can regulate the flux of Ag atoms to be greater towards Ag(111) than Ag(100). Under kinetic control, {100}-faceted nanocrystals will be formed when the Ag flux is greater towards Ag(111). The predicted kinetic Wulff shapes are in agreement with the analogous experimental system, which supports our hypothesis that the kinetic influences of PVP dominate the mechanism of shape control for the seed-mediated growth of large nanocrystals.

456293 Microfluidic Mass Transfer Process Intensification for Micron Droplet Generation with Dilute Surfactant Concentration

Abstract:

Microfluidic emulsification always involves fresh interface formation during droplet generation, which induces surfactant mass transfer and thus dynamic interfacial tensions (IFTs). As surfactant transfer onto freshly created interface often limits decrease of IFTs and droplet sizes, excessive amounts of surfactant would usually be added to speed up surfactant transfer and eliminate dynamic effects. Here we have systematically investigated the dynamic effects of surfactants in our newly designed capillary embedded T-junction microfluidic device. Within the asymmetric devices, hydrodynamic forces would intensify surfactant molecules concentrating to the point of droplet break-up, leading to extremely low dynamic IFTs temporarily (~0mN/m). Micron droplets could thus be generated with dilute surfactant concentrations (dynamic effects would be eliminated with surfactant concentrations of SDS 0.05wt% 1/6 CMC, Tween20 0.20wt% 4/7 CMC, F68 0.20wt% 2/3 CMC). The presented results provide us with more in-depth understanding of dynamic effects of surfactants as well as a novel method for surfactant mass transfer intensification and thus micron droplet generation with dilute surfactant concentrations.

474101 Novel Methodology to Measure the Contact Angle Between a Liquid and a Powder Bed By Using Droplet Penetration Technique

Abstract:

Determination of the contact angle between a liquid and a solid porous material is not trivial but crucial for many engineering applications including pharmaceutical, agricultural, printing, etc. and it can be also used to characterize the solid particle surface energy by measuring the contact angle with multiple liquids. In this work, we use the droplet penetration technique which has the advantages over the more traditional "Washburn" column technique in that the amount of material necessary to perform the experiment and the time it takes are orders of magnitude smaller.

The novel model is based on the comparison between the non-dimensional volume of liquid penetrated at the same non-dimensional times when using two different liquids: the test liquid (water or Diiodomethane) and a reference liquid (PDMS), which completely wets the granular material. This approach takes advantage of the fact that under certain assumptions, easily verified in most cases, the non-dimensional dynamics of a droplet penetrating into a porous material doesn't depend on any length scale except for the radius of the area of contact between the droplet and the material. As a result, when this scale is constant, the non-dimensional dynamics is universal. The advantage of this approach is that the resulting contact angle does not depend on the actual sorption dynamics: whether unidimensional as usually assumed, with the front propagating only in the axial direction, or bi-dimensional, when there is also front propagation in the radial direction. As long as a large part of the penetration process occurs at constant contact area, condition that is amply confirmed in our systems, we found a simple analytical expression to determine the contact angle based only on computing the ratio between non-dimensional penetration times of the test and reference liquids.

We show the universality of this approach by performing experiments with droplet of different size, different liquids and different powders. We also perform the analysis for different times (and penetrated volume) and compare the results with those obtained using the more traditional approach "Washburn" column technique.

469600 Improvements in the Anion Exchange Membrane Transport of Carbonate and Bicarbonate for Low-Temperature CO_2 Capture and Energy Conversion

Abstract:

In recent years, anion exchange membrane (AEM) devices have become a topic of significantly increased interest. Unlike liquid electrolyte alkaline fuel cells, which precipitate potassium carbonate salts when exposed to carbon dioxide, the lack of mobile cations in anion exchange membranes prevents the rapid cell death that results from salting. Though carbonate and bicarbonate transport in anion exchange membranes has typically been poor, recent improvements have facilitated better mobility and opened the door for many different types of carbonate based devices.[1] Electrochemical carbon dioxide separators[2, 3], room temperature carbonate fuel cells[4], and methane activation reactors[5] are all promising applications of carbonate based anion exchange membranes, but with typical current densities reported in the single mA/cm^2 range, improvements in carbonate transport are necessary to improve performance.

In a typical carbonate-based device, oxygen is reduced with carbon dioxide at the cathode, generally through an indirect pathway first to hydroxide, then (bi)carbonate. This pathway results in a mixed anion composition throughout the AEM where three very different anions can carry the charge across the membrane. Bicarbonate and carbonate (especially bicarbonate) have lower conductivities than hydroxide, but this cannot alone explain the 2 to 3 order of magnitude performance loss from a carbonate-free cell (typically 100s of mA/cm^2 to 1000s of mA/cm^2). In a carbon dioxide pump, the oxygen evolution reaction results in evolved carbon dioxide, while in a carbonate fuel cell hydrogen oxidation results in carbon dioxide without evolving oxygen. The cell potential and resulting half reactions influence the mixed transport and the amount of carbonate, bicarbonate, and hydroxide anions transported. In both carbon dioxide pumps and carbonate fuel cells a relationship is observed between the evolution of carbon dioxide at the anode and the flow of electrons in the cell. This relationship between current and evolved CO_2 in both pump mode and fuel cell mode is indicative of the anion ratio and the exchange mechanisms occurring at the anode the cathode.

In this study, a peak current density of 400 mA/cm^2 was achieved using a carbonate-exchange membrane fuel cell (Figure 1a) and 100 mA/cm^2 was reached in CO_2 pumping mode (Figure 1b). The difference in the current is a reflection of the difference in the dominating anion in the membrane as well as the difference in anode reactivity of hydrogen oxidation vs. carbonate electrolysis. This performance was achieved with radiation grafted anion exchange membranes and ionomers.[6] It was observed that performance was very sensitive to AEM humidification; that there is a delicate balance between maintaining a high water content in the membrane to facilitate anion transport and preventing catalyst and GDL flooding. The order of magnitude increase in the carbonate/bicarbonate transport observed in this work significantly increases the viability of membrane based carbon dioxide pumps, fuel cells, and fuel conversion devices. However, our results also point to advances that are needed in both catalysts and membranes to increase the faradaic efficiency in high current density carbonate devices.

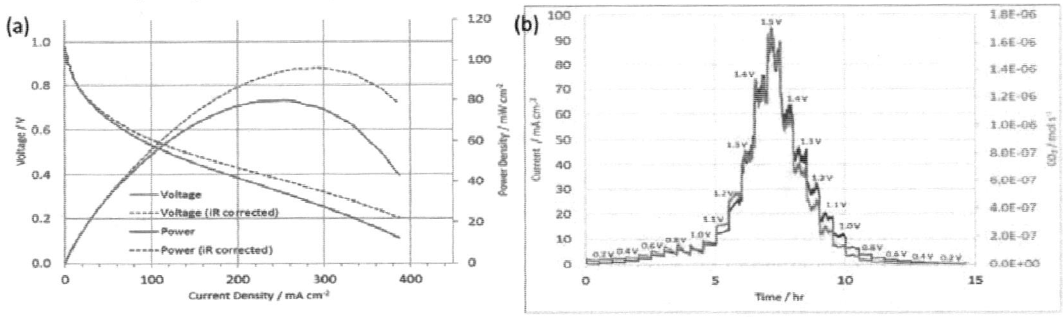

Figure 1: (a) Polarization curve and power curve for a carbonate based fuel cell from 10 mV/s linear sweep voltammetry. (b) 30 minute voltage holds taken from 0.1 V up to 1.5V and back down to 0.1 V in 0.1 V steps. The carbon dioxide was measured in the anode effluent by NDIR.

References:

1. J. R. Varcoe, P. Atanassov, D. R. Dekel, A. M. Herring, M. A. Hickner, P. A. Kohl, A. R. Kucernak, W. E. Mustain, K. Nijmeijer, K. Scott, T. Xu and L. Zhuang, Anion-exchange membranes in electrochemical energy systems. *Energy & environmental science.,* 7, 10 (2014).

2. W. A. Rigdon, T. J. Omasta, C. A. Lewis, L. Zhu, M. A. Hickner, S. D. Poynton, J. R. Varcoe, J. N. Renner, K. E. Ayers and W. E. Mustain, Carbonate dynamics and opportunities with low temperature, AEM-based electrochemical CO2 separators. *Journal of Electrochemical Energy Conversion and Storage.,* February SI (2017). DOI: 10.1115/1.4033411.

3. J. Landon and J. R. Kitchin, Electrochemical concentration of carbon dioxide from an oxygen/carbon dioxide containing gas stream. *J.Electrochem.Soc.,* **157**, 8 (2010).

4. C. M. Lang, K. Kim and P. A. Kohl, High-energy density, room-temperature carbonate fuel cell. *Electrochemical and solid-state letters.,* **9**, 12 (2006).

5. N. Spinner and W. E. Mustain, Electrochemical methane activation and conversion to oxygenates at room temperature. *J. Electrochem. Soc.,* **160**, 11 (2013).

6. S. D. Poynton, R. C. Slade, T. J. Omasta, W. E. Mustain, R. Escudero-Cid, P. Ocón and J. R. Varcoe, Preparation of radiation-grafted powders for use as anion exchange ionomers in alkaline polymer electrolyte fuel cells. *Journal of Materials Chemistry A.,* **2**, 14 (2014).

472742 A New Approach to Measure Internal Resistance of Microbial Fuel Cells

Abstract:

A new approach of measuring the internal resistance of microbial fuel cells have been investigated. Our approach measures internal resistance applying a voltage scan that is dictated by the energy generating ability of the microorganisms. The voltage scan is employed by the charging a capacitor. It reflects true characteristic value of the internal resistance because time scale of the measurement is consistent with the time scale of microbial growth. The method is tested through experimental determination of the internal resistance of a microbial fuel cell inoculated with *Shewanella oneidensis*, MR-1. The results are analyzed using an equivalent circuit model and compared with internal resistance measured using current-step method. The results demonstrate that capacitor charging method can be an alternative electrochemical technique to measure internal resistance that considers time scale of microbial growth.

465023 Ab Inio Insights into the Electrochemical Double Layer

Abstract:

An ongoing challenge in computational electrochemistry is the accurate determination of electrochemical barriers due to the large cost associated with using adequately-sized cells to eliminate any artifacts contributing to the final result. A recent publication from our group provides a simple method to obtain these barriers more efficiently by employing a capacitor model.[1] DFT calculations using this capacitor model have demonstrated, contrary to accepted knowledge, that the charge of a hydronium (H_3O^+) ion in the outer Helmholtz plane is not +1, but closer to +0.6. Thus, it is the goal of the current project to understand whether this is a physical phenomenon and the implications thereof on how we fundamentally treat electrochemical barriers.

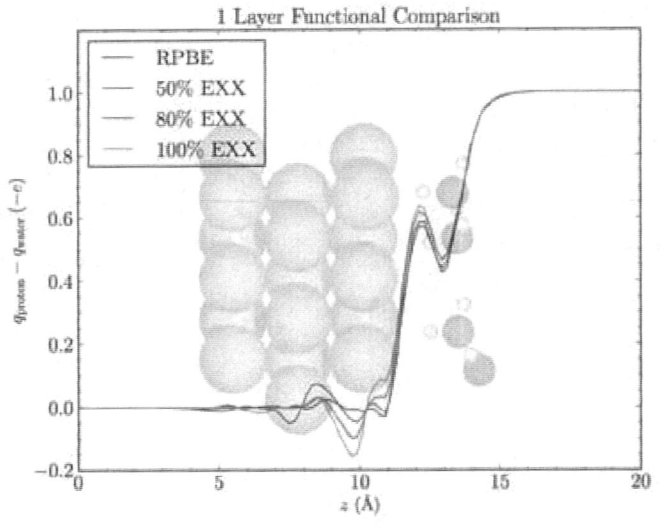

Figure 1. Cumulative charge difference between the protonated system (Pt slab + water + H) and the system with one fewer H (Pt slab + water) as a function of the z-coordinate of the unit cell, calculated with various functionals. Including various amounts of exact exchange only increases the electron density very slightly on the metal compared to GGA-level DFT. At the metal surface on the side of the solvent, there is a difference of 0.6 electrons, implying that the solvated H_3O^+ has a +0.6 charge instead of +1 in the first Helmholtz layer.

(1) Chan, K.; Norskov, J. K. Electrochemical Barriers Made Simple. *J. Phys. Chem. Lett.* **2015**, *6* (14), 2663-2668.

455959 In-Situ Formation of Polyethylene Glycol (PEG) Hydrogel Membrane in Microfluidic Fuel Cells

Abstract:

Membranless microfluidic fuel cells operated under two co-laminar flows often face serious fuel cross-over problems, especially when the flow rates are close to zero. In this study, we show that polyethylene glycol (PEG) solutions can be cross-linked in situ in microfluidic channels to form a PEG hydrogel membrane. This membrane prevents mixing of two incompatible electrolyte solutions while allowing diffusion of certain fuel molecules and ions through the membrane. By using PEG monomers of different molecular weight and UV exposure time, the selectivity of the membrane can be tailored with high accuracy. As a proof of principle, we develop a glucose fuel cell with two incompatible electrolytes (acid and base) which do not mix together even at zero flow rate. This glucose fuel cell is able to generate an open circuit voltage up to 400 mV from 5 mM of glucose. This device will be potentially useful for biomedical devices which use glucose as a power source.

469705 Identifying Layered Double Hydroxides for Electrochemical Anion Intercalation

Abstract:

Identifying Layered Double Hydroxides for Electrochemical Anion Intercalation

Matthias J. Young, Nicholas M. Bedford, Tatyana Kiryutina, Taylor J. Woehl

Applied Chemicals and Materials Division, Materials Measurement Laboratory, National Institute of Standards and Technology, Boulder, CO

Electrochemical anion intercalation is a developing field which promises to enable rapid electrochemical desalination and open a new frontier for anion-based battery technology. Layered double hydroxides (LDHs) are common anion exchange materials with atomic structures similar to naturally-occuring clays, and have garnered recent interest in the electrochemical community due to their catalytic and corrosion-resistant properties. In this work, we examine whether the interlayer anions in LDHs are capable of reversibly incorporating under applied electrical bias. We have employed atomistic materials simulations to model the electrochemical properties of LDH structures with multivalent metal centers. From this modeling we identified candidate compositions of the LDHs to be experimentally evaluated for anion intercalation. LDH nanoplates were synthesized using coprecipitation and evaluated as electrode materials for anion intercalation using *in situ* electrochemical quartz crystal microbalance and *in situ* synchrotron characterization techniques.

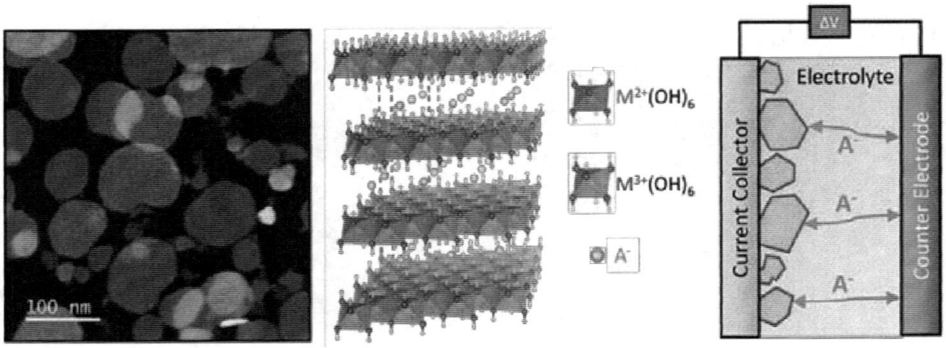

470709 Ion Selectivity in the Nanoporous Electrodes of Electrical Double Layer Capacitors with Room-Temperature Ionic Liquids

Abstract:

Ion selectivity in nanoporous materials plays an important role in the performance of electrical double layer capacitors (EDLCs) containing room-temperature ionic liquids (RTIL) but remains poorly understood due to difficulties in both materials characterization and monitoring of ion partition at small scales. In this work, we study the partitioning of RTIL mixtures in the nanopores of carbon derived carbide (CDC), an amorphous porous material commonly used in EDLCs. Based on a course-grained model for RTILs, the classical density functional theory (CDFT) allow us to identify key parameters controlling ion selectivity and to optimize EDLC performance by tuning the electrolyte composition. The theoretical predictions on the effect electrolyte composition on capacitance find good agreement with voltammetry measurements for EDLCs consisting of CDC and a ternary ion mixture containing 1-ethyl-3-methylimidazolium (EMIM) cation, bis(trifluoromethane)sulfonimide (TFSI) and tetrafluoroborate (BF_4) anions.

469796 Charge Storage Mechanisms of High-Surface-Area Carbides and Nitrides

Abstract:

High-surface-area early transition-metal carbides and nitrides possess very high capacitances and hold promise for use in high energy density supercapacitors as a consequence of their high surface areas, pseudocapacitive nature and high electronic conductivities. These materials are also stable in a variety of aqueous and non-aqueous electrolytes. Most of the work on carbides and nitrides for supercapacitors has been done in aqueous media. Vanadium nitride (VN) and molybdenum nitride (Mo_2N) have been reported to possess specific capacitances as high as 1340 Fg^{-1} in 1.1 V aqueous KOH and 346 Fg^{-1} in 0.8 V aqueous H_2SO_4 electrolytes, respectively.

Further development of early transition-metal carbides and nitrides for use in supercapacitors would benefit from a detailed understanding of their charge storage mechanisms. This work aims to characterize the charge storage mechanisms for high-surface-area early transition-metal (Ti, V, Nb, Mo, and W) carbides and nitrides in aqueous electrolytes. We have identified the active ions involved in the charge storage mechanisms (OH^- and H^+), located the active sites for charge storage, quantified the extent of pseudocapacitance and the amount of inserted active ions, and proposed the pseudocapacitive charge storage mechanisms for these materials.

Given findings regarding the storage mechanism, we hypothesized that protic, non-aqueous electrolytes would enable increased energy densities, given the higher operating voltages afforded by non-aqueous electrolytes. We were able to expand the operating voltage windows for Ti, V, Nb, Mo and W carbides and nitrides (i.e. from 1.1 V to 2.0 V for TiN) using protic ionic liquid electrolytes instead of aqueous electrolytes. For most materials, a three- to four-fold increase in energy density was observed while maintaining similar power density.

472147 Pseudocapacitive Hydrogen and Hydroxide Storage in High-Surface-Area Carbides and Nitrides

Abstract:

High-surface-area early transition-metal carbides and nitrides possess very high capacitances and hold promise for use in high energy density supercapacitors as a consequence of their high surface areas, pseudocapacitive nature and high electronic conductivities. These materials are also stable in a variety of aqueous and non-aqueous electrolytes. Most of the work on carbides and nitrides for supercapacitors has been done in aqueous media. Vanadium nitride (VN) and molybdenum nitride (Mo_2N) have been reported to possess specific capacitances as high as 1340 Fg^{-1} in 1.1 V aqueous KOH and 346 Fg^{-1} in 0.8 V aqueous H_2SO_4 electrolytes, respectively.

Further development of early transition-metal carbides and nitrides for use in supercapacitors would benefit from a detailed understanding of their charge storage mechanisms. This work aims to characterize the charge storage mechanisms for high-surface-area early transition-metal (Ti, V, Nb, Mo, and W) carbides and nitrides in aqueous electrolytes. We have identified the active ions involved in the charge storage mechanisms (OH^- and H^+), located the active sites for charge storage, quantified the extent of pseudocapacitance and the amount of inserted active ions, and proposed the pseudocapacitive charge storage mechanisms for these materials.

Given findings regarding the storage mechanism, we hypothesized that protic, non-aqueous electrolytes would enable increased energy densities, given the higher operating voltages afforded by non-aqueous electrolytes. We were able to expand the operating voltage windows for Ti, V, Nb, Mo and W carbides and nitrides (i.e. from 1.1 V to 2.0 V for TiN) using protic ionic liquid electrolytes instead of aqueous electrolytes. For most materials, a three- to four-fold increase in energy density was observed while maintaining similar power density.

454812 Pendant Drops and Liquid Jets in Miscible Environments

Abstract:

The spreading of liquids is a classical problem in interfacial fluid mechanics and, historically, the examination has been limited to immiscible systems. We have reported previously on our studies and observations of the spreading of sessile drops in miscible environments, which have distinctly different shape evolution and power law dynamics from sessile drops that spread in immiscible environments. We have extended this work to include the shape evolution of pendant drops and liquid jets existing in a miscible environment. By examining pendant drops and liquid jets, the need to account for surface energies arising from a solid-fluid interface, as in the sessile drop problem, is eliminated.

As time evolves, diffusion across the miscible liquid-liquid boundary proceeds due to the chemical potential difference between the two initially distinct, homogeneous phases. Diffusion, in turn, imparts a time-dependence to the properties of the liquids in the diffusive region – notably the density, viscosity, and interfacial tension – that influence the shape evolution. It was found for sessile drops in a miscible environment that gravitational forces dominate the spreading process, and is expected for pendant drops and liquid jets as well.

A series of liquid pairs have been studied (corn syrup-water, glycerol-water, glycerol-ethanol, tricresyl phosphate-ethanol). Various volumes of droplets, and diameters and flow rates of liquid jets have been considered.

Figure 1: Image sequence taken in time of a corn syrup pendant drop immersed in water. A strand emanates from the apex of the drop and continues to flow as the entire drop descends and elongates.

Particle tracking velocimetry has been performed to identify the internal flow pattern of the pendant drops and liquid jets.

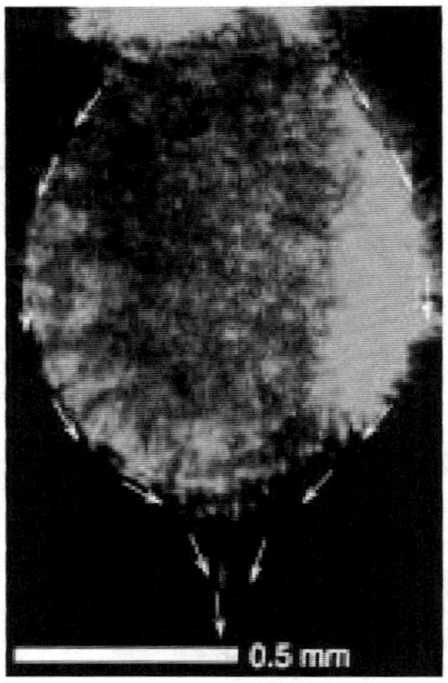

Figure 2: Image from a particle tracking velocimetry experiment of a corn syrup pendant drop immersed in water. The corn syrup contains 6 μm microspheres at a concentration of 10-3 g/ml, which scatter incident light. The arrows indicate velocity vectors obtained from particle movement. Motion was largely restricted to the corn syrup-water interface. Particles within the pendant drop and away from the liquid-liquid interface move with the drop as it descends and elongates; within the reference frame of the drop, these interior particles do not move significantly.

473864 Reversible and Irreversible Deformation of Particle-Coated Droplets

Abstract:

We report results from an experimental study of controlled deformation of particle-coated droplets (a.k.a. liquid marbles), which typically do not wet solid surfaces.

By applying gentle compression, we derive critical conditions at which rupture occurs, i.e. when the coating of particles cracks open and allows wetting to occur.

Interestingly, before this critical threshold, if the droplets are compressed and then released, the droplets can undergo an irreversible deformation whereby they do not regain their original shape, as one might expect.

By comparing shapes prior to and after compression, we quantify the deformation for a range of particle sizes.

In addition to previous studies on liquid marbles, our observations add another layer of hidden complexity and non-linearity to these seemingly elemental structures

472894 Controlling Interfacial Composition, Coverage and Mechanics for Stable Capsule Formation

Abstract:

Nanoparticle-surfactant mixtures can form stable fluid interfaces with large dilatational modulus. Surfactants modify the nanoparticle hydrophobicity and therefore tune the wettability of the nanoparticle to nearby interfaces. Adsorption timescales of the nanoparticle-surfactant complexes to interfaces are controlled by diffusion. These diffusion timescales are comparable to residence times in microfluidic devices, allowing the interfacial coverage on monodisperse droplets and bubbles to be controlled. For confined bubbles translating in a long microchannel, sufficiently long residence times lead to the formation of extremely stable gas-filled capsules. In this talk we vary the dispersion composition to vary the particle hydrophobicity and adsorption to interfaces. We examine the role of dispersion composition on capsule formation, and independently probe the interfacial mechanics at similar surface coverages and compositions using a custom built microtensiometer.

463228 Coalescence Inhibition through Asphaltene Adsorption

Abstract:

Asphaltenes are complex polar components of heavy oils and are only marginally stable in organic solvents. Because of this marginal solubility, they have a propensity to adsorb onto oil/water interfaces to form viscoelastic layers. This can result in a strong stabilization of water-in-oil emulsions that cause difficulties in refining operations. In this paper we examine the coalescence process of water droplets in the presence of asphaltenes using a newly developed instrument where the thickness of thin, draining films can be measured in space and time. Using this instrument, thin film thickness, internal droplet pressure and coalescence times can be measured. These results are further compared against measurements of the interfacial shear viscoelasticity of asphaltene-laden interfaces. In addition to asphaltene concentration in a model hydrocarbon (toluene is used), the aging times of the interface was varied. As expected, both asphaltene concentration and aging time are found to strongly affect coalescence dynamics. In addition to reporting on water droplet coalescence dynamics, we found that contact of water against asphaltene/toluene solutions induced the production of a remarkable â€œmicroemulsionâ€ where micron-sized water droplets, stabilized by asphaltenes, spontaneously appeared.

468483 Continuous Droplet-Based Viscometer

Abstract:

Continuous viscosity measurements have a wide range of applications from industrial chemical production to medical diagnosis. In this work, we have developed a simple droplet-based (i.e., water-in-oil) continuous viscometer capable of measuring viscosity changes in less than 10 seconds and consuming a total sample volume of less than 1 µl per hour. The viscometer employs a flow-focusing geometry and generates droplets under constant pressure. The length of the droplets (L_d) is highly correlated to the aqueous-phase viscosity (μ_{aq}) at high ratios of aqueous-inlet to oil-inlet pressure (AIP/OIP), yielding a linear relationship between μ_{aq} and $1/(L_d - L_c)$ where $L_c$is the minimal obtainable droplet length.

Theoretical analysis was conducted to predict the droplet size and verify the experimentally found linear relationship. The theoretical analysis is based on the mechanism for flow-rate controlled breakup proposed by Garstecki *et al.* in a T-junction geometry. They found that the droplet length is linearly proportional to the ratio of the dispersed-phase to continuous-phase flow rate. We combined this relationship with a pressure balance between the applied pressure and the pressure drop from the inlet to the outlet of the device to derive the flow rate ratio. The resulting equations can successfully predict droplet lengths and verify the linear relationship between μ_{aq} and $1/(L_d - L_c)$. The equations can also be applied to optimize the device geometry (i.e., channel widths, depths and lengths).

468633 On the Effect of Surfactants on Drop Coalescence with Liquid/Liquid Interfaces

Abstract:

1. Introduction

Drop coalescence is commonly encountered in a wide range of industrial applications in food processing and in petrochemical and pharmaceutical fields. In such cases, the surface of the drop is often coated with surfactants, whose properties largely affect the coalescence behaviour. In current oil extraction methods, water is injected into the well to displace the residual oil in rock gaps and enhance crude oil recovery. During injection, oil-water dispersions form. In some cases, surfactants are also added to facilitate the formation of these dispersions. Previous studies have shown that the oil recovery efficiency is affected by this addition of surfactants. As it is difficult to trace the distribution of the surfactants along the droplet surface, the effect of surfactants on the coalescence process is still an active research area. In this work, advanced laser velocity field measurements were conducted, using the Particle Image Velocimetry (PIV) technique, to investigate the effect of surfactants on the coalescence of single aqueous drop with an aqueous/organic interface.

2. Experimental Set-up

The aqueous phase was 80% w/w glycerol/water mixture, while the organic phase was a low viscosity kerosene-like Exxsol D80 oil. The concentration of the aqueous mixture was chosen to match the refractive index of the organic phase, necessary to minimize optical diffractions to produce more accurate laser based flow measurements. Various concentrations of Span 80 were introduced to the organic phase to study the effect of a surfactant on the coalescence mechanism. The properties of the fluids are shown in Table 1.

Table 1: Properties of fluids used.

	φ (w/w)	ρ (kg/m3)	μ (mPa.s)	γ (mN/m)
Water/glycerol	20%	1210	54.0	--
Span80 / Exxsol D80	0	804	1.75	26.73
	$2 \cdot 10^{-5}$	804	1.75	23.43
	$1 \cdot 10^{-4}$	804	1.75	13
	$2 \cdot 10^{-4}$	804	1.75	7.8
	$5 \cdot 10^{-4}$	804	1.75	2.16

The experiments were conducted in a transparent acrylic tank of 50 mm square section width and 150 mm in height, as shown in Fig 1. The aqueous/organic interface was placed at 40 mm from the bottom of the cell while the organic/air was placed at 100 mm. Droplets of water/glycerol mixture were generated by injecting the liquid into the Exxsol D80 through a cylindrical stainless steel nozzle of 80 mm length and 2 mm inner diameter. The nozzle was fixed at approximately 20 mm above the aqueous/organic interface to minimise any variation on the droplet impact locations. For the investigation of the velocity fields with PIV, the droplets were seeded with 1 micron Rhodamine tracer particles. To clearly observe the interface between the aqueous and organic phase, a small amount of Rhodamine 6G dye was introduced to the aqueous phase close to the interface. These seeding particles were removed with a suction pump after each experimental run.

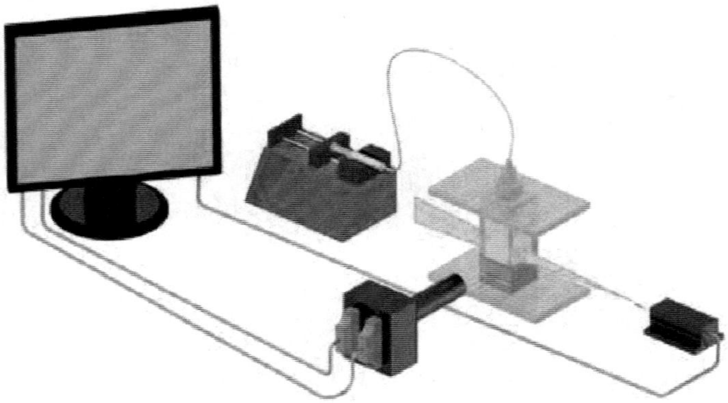

Fig. 1. Experimental set-up.

3. Results and Discussion

In Fig. 2, the variation of the relative height h_s of the droplet top point over time t for various surfactant concentrations is plotted. Prior to the rupture of the interface, the increasing surfactant concentration reduced the interfacial tension and as a result the deformation of the interface was increased. During the rupture process, for the pure system or low surfactant concentration, $\varphi = 0$, $\varphi = 2 \cdot 10^{-5}$ and $\varphi = 1 \cdot 10^{-4}$, the upper part of the droplet surface oscillated at t = 50 ms, t = 55 ms and t = 85 ms, respectively. While for the higher concentration systems ($\varphi = 2 \cdot 10^{-4}$ and $\varphi = 5 \cdot 10^{-4}$) the top point of droplet approached to the equilibrium interface smoothly and slowly, but at much longer times than at the lower concentrations. It is worth mentioning that the top point in these cases was not reaching heights below the final interface level as was the case with the lower surfactant systems.

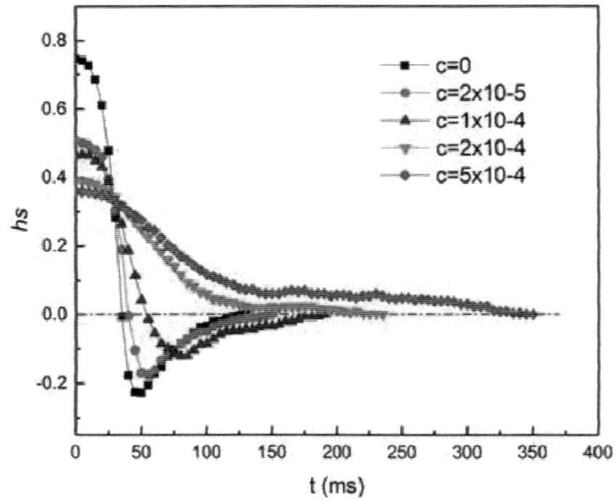

Fig. 2. The variation of the relative height h_s of the top point of the droplet

The velocity fields inside the droplet for different surfactant concentrations and different time steps are presented in

Figs. 3 and 4. Post interface rupture, two counter-rotating vortices were observed at the bottom of the droplet. With increasing time, the two counter-rotating vortices started to attach to the upper part of the droplet while their intensities were gradually increased until a transition point was reached, after which the intensity decreased. At this initial stage the fluid dynamics inside the droplets were dominated by the vortices present, while after the transition point the flow was dominated by the downward axial fluid motion. Also, it was observed that for increasing surfactant concentration the intensity of the two counter rotating vortices was decreased. The location of the two vortices inside the droplet was also tracked and compared for different surfactant concentrations and kinetic energy distributions for different surfactant concentrations were computed.

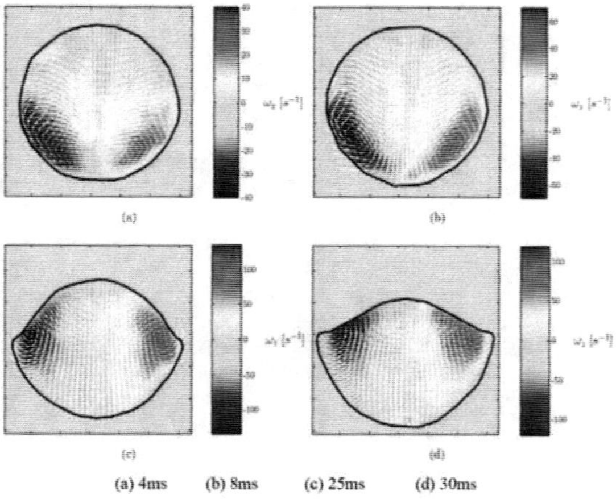

Fig. 3. Velocity fields and contour of the vorticity evolution for $\varphi = 1\cdot10^{-4}$

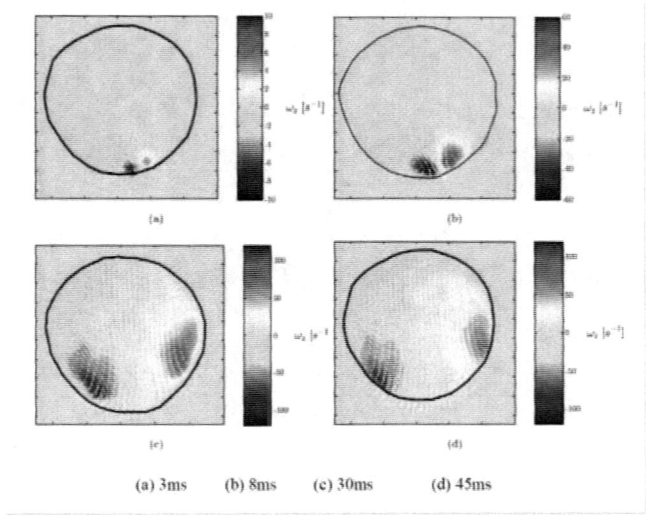

Fig. 4. Velocity fields and contour of the vorticity evolution for $\varphi = 2\cdot10^{-4}$

Acknowledgements

The project was funded by the UK Engineering and Physical Science Research Council (EPSRC) Programme Grant MEMPHIS. T. Dong would also like to thank CSC and UCL for his PhD studentship.

464703 The Effect of a Yield Stress on the Drainage of the Thin Film Between Two Colliding Newtonian Drops

Abstract:

A long-standing problem in oil sands extraction is the presence of 2-3% of residual water in the form of very fine drops (about 1 micron in size) in the diluted bitumen product. A strategy that has been evaluated many times in the past to remove this water is to coalesce the small water droplets with themselves or with bigger water droplets to produce larger drops that can be separated by conventional methods such as inclined settling and centrifugation. Unfortunately, these studies have indicated that these water-in-bitumen emulsions are extremely difficult to break[1]. In the literature, the prevalent explanation for this stability is that molecules in bitumen called asphaltenes adsorb at the oil-water interface and render it rigid, thus strongly increasing the time required for coalescence. Recently, an alternative explanation has emerged to explain this stability[2], which suggests that bitumen is a fluid with a low yield stress due to the capability of the asphaltenes to associate with themselves. During the process of hydrodynamic drainage, if the stresses within the thin film fall below the yield stress of the asphaltene network, the film will stop draining, *before* non-hydrodynamic effects come into effect.

The objective of this paper is to assess the validity of the yield-stress explanation for water-in-bitumen emulsion stability. A survey of the literature reveals that a theoretical understanding of the drainage of the thin film of Bingham fluid between two colliding drops is still lacking, with the exception of two studies that make ad-hoc assumptions about the film shape[3,4]. In this work, we examine this problem for low capillary numbers via a combination of scaling analysis and detailed numerical simulations based on the lubrication analysis[5,6]. One obvious trend is that the introduction of a yield stress in the suspending fluid retards the drainage process relative to a Newtonian fluid of the same viscosity. But there are at least other three notable features of the film drainage process of Bingham fluids. First, the presence of yield stress prolongs the spherical regime of the drainage process and delays transition into the dimpled regime. Second, the

drainage time shows a minimum with respect to the force applied to push the drops towards each other. Third, drainage may be arrested completely below a critical height due to the yield stress. This critical height scales as the square of the yield stress, the cube of the drop radius and inversely as the square of the interfacial tension. Counterintuitively, the critical height independent of the force colliding the two drops! This and other distinguishing characteristics of the drainage process will be elucidated in the presentation. We will conclude the presentation with an evaluation of the 'yield stress' explanation for the stability of water-in-bitumen emulsions, outlining the regime of parameters for which this explanation may be valid.

References

1. F. Rao and Q. Liu, "Froth Treatment in Athabasca Oil Sands Bitumen Recovery Process: A Review", Energy Fuels **27**, 7199–7207 (2013).

Liquid Emulsion Films" , Langmuir **30**, 3024–3033 (2014).

3. S. Hartland and S. A. Jeelani, "Drainage in Thin Planar Non-Newtonian Fluid Films," Can. J. Chem. Eng. **65**, 384-390 (1987).

4. S. Hartland and S. A. Jeelani, "Drainage of Thin Dimpled Non-Newtonian Fluid Films," J. Phys. Chem. **90**, 6054-6059 (1986).

5. A. Saboni, C. Gourdon and A. K. Chesters, "Drainage and Rupture of partially mobile films during coalescence in liquid-liquid systems under a constant interaction force," J. Colloid Int. Sci. **175**, 27-35 (1995).

6. P. Santoro and M. Lowenberg, "Coalescence of drops with tangentially mobile interfaces: effect of ambient flow," Ann. N.Y. Acad. Sci. **1161**, 277-291 (2009).

456555 Theory for Thin-Film Evolution Between a Sphere and a Planar Free Surface

Abstract:

The spatiotemporal evolution of thin liquid films is ubiquitous in the mechanics of emulsions. The so-called "thin film equations," i.e. the Navier-Stokes equations simplified in the thin-film limit, have been applied in numerous technological problems in order to explain the detailed flow physics within a liquid with one or more free surfaces. It is well known that excess interfacial stresses at one or more free surfaces, due to the presence of surface-active species, can have a major influence on the dynamics of thin films. Recent interferometry experiments involving a bubble being pressed against a flat interface have shown that 1) the temporal scaling law for thin-film drainage can have a different exponent depending on the type of surfactant used, and 2) an interesting flow instability can occur in a thin film squeezed between two fluid-fluid interfaces when surfactant is present.

In this talk, the theoretical problem of a sphere passing through an initially planar, fluid-fluid interface is discussed, with the particular aim of explaining observations 1 and 2 above. The theory is built up under the auspice of the thin-film approximation, with suitable homage paid to prior theoretical descriptions. The deformation is computed for axisymmetric films for all time. The importance of three length scales in the problem, namely the capillary length, the sphere radius, and the characteristic thickness of the film, are illuminated in light of the new theory. Four regimes of the motion are characterized, namely the approach, penetration, drainage, and eventual rupture of the thin film. The implications for the mechanics of concentrated emulsions and foams are discussed.

471900 Numerical Calculations of the Rheology and Droplet Behavior of Emulsions in the Presence of Surfactants

Abstract:

Surfactants are found in many natural and industrial processes. The addition of surfactants to an emulsion reduces the interfacial tension of the surface of the drops, and affects the drop deformation and the resulting rheology. Prior studies on surfactant-covered drops have focused primarily on the drop deformation and modes of drop breakup in shear and extensional flows, with little focus on the rheology of the system. In this work, we numerically study the rheology of emulsions in the presence of surfactants in a broad range of kinematics by solving the boundary-integral (BI) equation for the evolution of the drop surface, coupled with the convection-diffusion equation for the distribution of the surfactant on the surface of the drops. The general approach to constitutive modeling of non-Newtonian liquids based on the Oldroyd equation with variable coefficients, which was developed in our prior work (Martin et al. 2014), is applied to this system. This approach involves numerically calculating the rheological functions for two canonical steady flows--simple shear and planar extension at arbitrary flow intensities--and using these functions to find the values of the five Oldroyd parameters as functions of an instantaneous invariant of the flow, chosen as the excess energy dissipation rate due to the presence of drops. This constitutive modeling scheme, which was originally applied to predict the rheology in dilute emulsions of surfactant-free deformable drops (Martin et al. 2014), is extended in this work to account for the presence of surfactant. The resulting constitutive model is applied to various test cases, different from simple shear and planar extension, to evaluate the effectiveness of the model in predicting the emulsion stress for an arbitrary history of deformation of a material element, both Lagrangian steady and unsteady, and to compare the results to the precise BI calculations. Our model achieves excellent stress predictions for planar-mixed flows and uniaxial extension, with more deviation from the exact results for uniaxial compression at large capillary numbers. Another type of kinematics considered is the flow in the ball bearing between two eccentric rotating spheres, where the rheological response of a small surfactant-covered drop is evaluated along a material trajectory. The results confirm the validity of our general scheme of constitutive modeling based on the generalized Oldroyd equation, with variable coefficients fitted to two steady base flows, for a variety of complex, non-Newtonian fluids.

As well as the application of the generalized Oldroyd model to the case of emulsions with surfactant, the behavior of the droplets and the resulting surfactant distribution are evaluated based on three different surface equations of state: the linear, Langmuir, and Frumkin equations. The chosen equation of state provides a relationship between the surfactant concentration on the surface of the drop to the interfacial tension. This work expands on previous work of Pawar and Stebe (1996) and Bazhlekov et al. (2006) to determine how the selected equation of state influences the drop deformation and the resulting rheology of the emulsion in the presence of surfactants.

1. Martin, R., A. Zinchenko, and R. Davis, "A generalized Oldroyd's model for non-Newtonian liquids with applications to a dilute emulsion of deformable drops," *J. Rheol.* **58**: 759-778 (2014).

2. Pawar, Y., and K. Stebe, "Marangoni effects on drop deformation in an extensional flow: The role of surfactant physical chemistry. I. Insoluble surfactants," *Phys. Fluids* **8**: 1738-1751 (1996).

3. Bazhlekov, I. B., P. D. Anderson, and H. E. H. Meijer, "Numerical investigation of the effect of insoluble surfactants on drop deformation and breakup in simple shear flow," *J. Colloid and Interface Sci.* **298**: 369-394 (2006).

472568 Multianalyte Electrochemical Sensors on a Monolith Electrode By Optically Probing the Electrical Double Layer

Abstract:

Electrochemical sensors are arguably the most successful chemical sensing devices primarily because of their high sensitivity at low cost. The key advantage of electrochemical devices is the active signal due to chemical reaction allowing specific recognition of the chemical with a low incidence of false positives. However, only one redox reaction per electrode is possible to measure. An opto-electrochemical method, called Scanning Electrometer for Electrical Double-layer (SEED), to quantitatively detect multiple, individual redox reactions on a monolith electrode will be described. SEED optically measure the charging/discharging of the electrical double layer to quantify the local redox current. By scanning a laser beam, local redox current density distribution can be quantitatively mapped to detect multiple analytes on a monolith (working) electrode. SEED is responsive to (redox) reactions of 0.1 atto-mole of molecules. The local redox current is measured as the change in the charge in the electrical double-layer which significantly changes due to redox reaction. SEED signal enhances as the sensing area reduces in size. Characteristics of SEED and application to chemical and biochemical sensing, and genomics will be described.

473889 Role of Defects in Enhancing the Electrochemical Properties of Transition Metal Oxide

Abstract:

Transition metal oxides are fascinating class of materials that are valued for their catalytic, ferroelectric, superconducting, and optical properties, and hence they find uses in diverse applications such as heterogeneous catalysts, sensors, solar cells, and high temperature superconductors. Such rich properties of metal oxide are partly enabled by their complex surface chemistry and the presence of lattice vacancies that significantly perturb the electronic structure of the material. Hence, an in-depth understanding of the origin and nature of these defect states is important for further expanding the scope of oxide-based systems for various applications. The focus of this presentation is on elucidating the nature of interaction of various chemicals with defects in metal oxide. The chemical-structure-property correlations developed in this work can enable rational engineering of a new type of nanoscale, low temperature catalysts having tailored functionality and selectivity that can significantly reduce the energy cost and carbon footprint of many important chemical reactions.

452861 The Degradation of Additives during the Electrodeposition of Copper for Metamaterial Fabrication

Abstract:

Metamaterials, a class of artificial optically-active materials with exceptional properties based on their patterned structures, typically use metallic and dielectric repeating features to provide a negative index of refraction. In the fabrication of metamaterials utilized in this study, copper is selected for the metallic part of metamaterial due to its high thermal and electrical conductivity, excellent mechanical properties and relative ease of acquisition.

Electrodeposition from a highly-acidic copper electrolyte is the most attractive approach for fabricating the conductive component of the metamaterial in this study due to its ease of control and high deposition rate. However, since the presence of special high-aspect ratio patterns in the metamaterial requires bottom-up super-filling and leveling of micron and sub-micron scale volumes; additives such as chloride ions, 3-mercapto-1-propanesulfonate (MPSA), polyethylene glycol (PEG), and polyvinylpyrrolidone (PVP) are introduced into the electrodeposition process in order to provide excellent conformality and adhesion. Most of the present study focuses on investigating the optimal parameters of electrodeposition of copper such as the effects of additive aging, the degradation of MPSA and the electrodeposition parameters of deposition voltage and its corresponding current density.

Electrochemical methods such as cyclic voltammetry, chronopotentiometry and controlled potential coulometry probe the mechanism of both the specific effects of additives and the interplay between additives. In parallel, COMSOL finite element modeling (FEM) software is used to simulate the 2D electrodeposition of copper to investigate the copper super-filling, mass transfer and the electrochemical behavior of the electrodeposition system.

Atomic force microscopy (AFM), four-point probe sheet-resistivity measurements, optical microscopy, optical profilometry and X-ray diffraction (XRD) are used as characterization tools to map the surface topography, sheet resistances, surface roughness and crystal structure of the deposited films obtained under different experimental conditions. Nuclear magnetic resonance (NMR) is conducted to analyze the degradation of the additives during the electrodeposition process. Combined with the electrochemical methods, optimized parameters for electrodeposition of copper are obtained and desired fabrication methods for metamaterials are achieved in this study.

470927 Field Effect Control of Electrochemical Reaction Kinetics at Back-Gated, UltrathinÂ Semiconductor Electrodes

Abstract:

An electrochemical reaction refers to a process of chemical changes caused by the electrical current, or a process of generating electricity by chemical reactions, in which direction and rate of the reaction can be controlled by electrode potential (*i.e.* energy of holes and electrons in the electrode). Because an electrochemical reaction involves with charge transfer across an electrode/electrolyte interface (though it may not be the only process), the reaction kinetics is determined to a large extent by the electronic structures (*i.e.* energy level distribution and electronic occupation) of the electrode material and the redox species in the electrolyte. In this talk, we propose a new approach exploiting the *field effect*, in addition to electrode potential, to modulate the electronic structures and electrochemical reaction kinetics at electrode/electrolyte interfaces.

As a model system, we prepared working electrodes with â€˜gate-insulator-semiconductorâ€™ structure, which is similar to that of field effect transistors (FETs). Herein, we employed ultrathin semiconductor layers so that electrochemistry at the semiconductor surface are effectively modulated by a voltage bias applied to the gate. On those *gate-tunable* electrodes, we observed continuous, *in-situ* modulation of outer-sphere electron transfer kinetics at the semiconductor/electrolyte interface with voltage biases applied to the gate. For example, the reduction potential of 2,3,5,6-tetrabromo-1,4-benzoquinone (TBBQ) on a 5 nm thick ZnO electrode could be shifted by ~0.4 V. With further control experiments, we found that the observed gate-controlled redox reaction kinetics is essentially attributed to band alignment shift at the electrode/electrolyte interface.

462103 Electrochemical Synthesis of Organic Nanorods on Gold Nanoparticles Seeds

Abstract:

This talk is focused on nanoparticle-directed synthesis of tetrathiafulvalene charge-transfer salt (TTFBr) and potassium tetracyanoplatinate (KCP) organic nanorods using electrochemical deposition. Gold nanoparticles (AuNPs) acting here as nucleation seeds are deposited on highly oriented pyrolytic graphite (HOPG). The nanorods, either TTFBr or KCP, are deposited subsequently on the AuNPs-decorated HOPG for the study of seed-mediated crystallization process at the nanoscale. Both the seed and the organic crystal are deposited using electrochemical chemistry. The deposition processes are monitored by cyclic voltammetry and in-situ atomic force microscope. The field emission scanning electron microscope is also employed to characterize the seed and crystal nanostructures. The TTFBr and KCP crystals deposited on AuNPs-decorated HOPG show preferential nucleation on the AuNP in comparison with crystals electrochemically deposited on the planar HOPG. The size of the crystals is controlled predominantly by the solution concentration and the seed size. Nanorods as small as 7 nm in height are nucleated on AuNPs of 20 nm in height. We hypothesize that the local curvature of the seed particle, that limits the cross sectional dimension and introduces an interfacial strain, contributes to the nanoconfinement effect observed. Currently we are also applying this seed-directed synthesis method to fabricate organic nanowire sensors on gold lithographical patterns. The prototype sensor shows potential to compete with commercial sensors in detecting toxic and flammable gases. This research contributes a better understanding of crystallization at the nanoscale and a possible low-cost technology for chemical sensing applications.

462273 Interfacial Structure and Capacitance of Li-Doped Ionic Liquid Electrolytes from Molecular Simulations

Abstract:

Ionic liquids have been proposed as candidate electrolytes for high-energy density, rechargeable batteries, supercapacitors, and hybrid energy storage devices. Though Li-salt is often present in these systems, its influence on interfacial properties is largely uncharacterized. We, thereby, present an extensive computational analysis, supported by experimental comparisons, of the properties of a representative set of these electrolytes at an ideal carbon interface as a function of Li-salt doping and voltage. We have performed polarizable molecular (MD) dynamics simulations, using the APPLE&P force field, to evaluate electric double layer (EDL) capacitance and distribution of Li^+ in the EDL. Differential capacitance exhibits the characteristic â€œcamelâ€ profile and is insensitive to Li-doping. Li^+ localizes in the second molecular layer of the EDL, which is a result of confinement from free energy barriers associated with ion layering. Joint MD/electronic structure computations show the electrochemical window of the electrolytes to be a weak function of Li-doping. Estimates of supercapacitor specific energy are made using the computed window and capacitance. The magnitude and trends in specific energy are in good agreement with experiment.

465711 Reactivity of an Alucone Thin Film Coating on a Lithiated Si Anode of Li-Ion Batteries

Abstract:

Lithium-ion batteries are one of the most common types of rechargeable batteries to power small, portable electronic devices. Current research efforts are focused on achieving the energy density required for electric cars and high-end portable electronics. Silicon anodes in Li-ion batteries yield higher theoretical capacities than graphite, the most common anode material; nevertheless, the significant volume changes occurring in the silicon anode during lithiation and de-lithiation limits its ability to retain capacity for an extended number of charge/discharge cycles. A solid electrolyte interphase (SEI) layer that naturally forms on the silicon surface from the reduction of the electrolyte components can improve the mechanical stability of the anode at the expense of hindering Li transport. In contrast, surface modification of the silicon anode may be a viable strategy to accommodate volumetric changes, preventing cracking and pulverization of the electrode, while maintaining or improving ionic and electric conductivity. Aluminum alcoxide (alucone) films to improve the stability of the anode have been successfully deposited and characterized on lithiated silicon surfaces, and reportedly become electronically conductive after Li atoms saturate the favorable sites of the film. In the present work, we use density functional theory and ab initio molecular dynamics simulations to study the reactivity of the film and the anode surface in contact with a typical organic solvent, ethylene carbonate. The solvent molecules are found to be able to penetrate the film and undergo reduction inside the film and at the film-anode and film-electrolyte interfaces via one- and two-electron mechanisms. Such reaction mechanisms are examined along with the structure, stability and suitability of the film to improve the anode, as well as the implications for the design of enhanced electrode materials for Li-ion batteries.

471645 Microstructure of Room-Temperature Ionic Liquids at Charged Surfaces Revealed By Integrated Modeling and Experimental Approaches

Abstract:

The interfaces of room-temperature ionic liquids (RTILs) at charged surfaces are of great importance for many applications, such as electrical energy storage and lubrication. Many nanoscale techniques (e.g., atomic force microscopy, AFM) are developed to reveal the microstructure of electrical double layers (EDLs) formed by RTILs at charged solid surface, and with the help of molecular dynamics (MD) simulation one can obtain detailed information of EDLs at molecular level and have a better understanding of the interfacial phenomena.

In this talk, firstly we report direct observation of structure and properties of the topological defects formed in ordered ionic liquid layers at carbon interfaces. The observation of layer structure at structural defects such as step edges reinforced by MD simulations define spatial resolution of the method. We have observed the internal structure of topological and structural defects on the ionic-liquid surface interfaces using 3D force mapping. The structural defects were found to result only in short range changes in ionic liquid ordering, with the changes in 3D force fields consistent with rigid shift of the layers with narrow disordered region between the two, consistent with MD simulations. At the same time, formation of serendipitous dislocation-type topological defects were observed on planar surfaces. The decrease in the ordering was observed within the defect, consistent with behavior expected for classical liquid crystals.

Secondly, further MD work was done to investigate the influence of the thickness of the electrode surface step on the microstructure of interfacial ILs. A strong correlation was observed between the interfacial IL structure and the step thickness in electrode surface as well as the ion size. Specifically, when the step thickness is commensurate with ion size, the interfacial layering of cation/anion is more evident; whereas, the layering tends to be less defined when the step thickness is close to the half of ion size. Furthermore, two-dimensional microstructure of ion layers exhibits different patterns and alignments of counterion/co-ion lattice at neutral and charged electrodes.

Thirdly, beyond carbon surfaces, we then explore the ILs at charged mica surfaces. By comparing the measured force profiles from AFM to ion density profiles from MD simulations, we were able to determine which ion of the ionic liquid the AFM tips are sensitive to during force measurements, and this is determined by ion larger ion volume and/or weight. By varying the anion or cation of the ionic liquids we are able to move from a regime where we detect the position of the anions (e.g. [emim][Tf_2N]) to one where we detect the position of the cations (e.g. [emim][BF_4]). Using AFM data alone we cannot identify which ion we are sensitive to, and through a direct comparison of experiment data with molecular dynamics simulation, we are able to gain a better understanding of the mechanisms leading to the measured force profiles, therefore making future interpretation of force-curves at ionic-liquid solid interfaces easier. We will also show the interfacial structure of humid ILs at charged mica surfaces, in particular, the water accumulation will differ from that at carbon surfaces.

470229 Mathematical Models for the Impedance Response of Subcutaneous Glucose Sensors

Abstract:

Since 1974, the artificial pancreas has been acknowledged by the greater scientific community to be a possible way to manage type 1 diabetes [1]. In 2008, a case study showed that a subcutaneous continuous glucose monitor sensor could be implanted for 28 days in a patient with type 1 diabetes [2]. However, commercially available subcutaneous continuous glucose monitors are generally changed every seven days to avoid unreliable results. The object of this work is to explore the potential of impedance spectroscopy to monitor the state of health of subcutaneous continuous glucose monitor sensors.

A mathematical model has been developed for the impedance response of a glucose oxidase enzyme-based electrochemical biosensor. The model accounts for a glucose limiting membrane GLM, which controls the amount of glucose participating in the enzymatic reaction. The glucose oxidase was assumed to be immobilized within a thin film adjacent to the electrode. In the glucose oxidase layer, a process of enzymatic catalysis transforms the glucose into peroxide, which can be detected electrochemically. This system may be considered to be a special case of the coupled homogeneous and heterogeneous reactions addressed by Levich [3].

The model development required two steps. The nonlinear coupled differential equations governing this system were solved under the assumption of a steady state. The steady-state concentrations resulting from the steady-state simulation were used in the solution of the linearized set of differential equations describing the sinusoidal steady state. The enzymatic catalysis was treated in terms of two homogeneous reactions, one consuming the glucose oxidase and forming gluconic acid, and the other regenerating the glucose oxidase and forming the peroxide.

References

1. A. M. Albisser, B. S. Leibel, T. G. Ewart, Z. Davidovac, C. K. Botz, and W. Zingg, Diabetes 23 (1974), 389â€“396.
2. J. J. Chamberlain and D. Small, Clin. Diabetes, 26 (2008), 138â€“139.
3. V. G. Levich, Physicochemical Hydrodynamics, Prentice Hall, Englewood Cliffs, NJ, (1962), p. 345-357.

470435 The Effect of Alkali and Quaternary Ammonium Cations on the Electrochemical Interface Studied By *in-Situ* Infrared Spectroscopy

Abstract:

Recent computational studies have suggested that the adsorption of alkali metal cations may have significant impact on electrochemical processes, particularly those which occur at low potentials such as the hydrogen evolution (HER) and oxidation (HOR) reactions. It has also been demonstrated that the reactivity and selectivity of other important electrochemical processes such as CO_2 reduction are influenced by the presence of alkali cations in the electrolyte. In this study we employ surface-enhanced infrared absorption spectroscopy to study the adsorption of alkali and various quaternary ammonium cations using CO as a probe molecule on Pt electrodes. Linearly adsorbed CO is displaced at low electrode potentials (below the potential of zero charge) by electrostatically bound cations and forced to a bridge-bonded configuration. It is found that larger alkali cations such as K^+ have stronger surface interactions with adsorbed CO than smaller alkali cations. Quaternary ammonium cations show the least pronounced effect.

453073 Using Polymers As Solvents for Active Pharmaceutical Ingredients

Abstract:

Active pharmaceutical ingredients (APIs) are often complex compounds exhibiting a very low aqueous solubility. Therefore, they only slowly dissolve in the body when administered as crystalline solids, which leads to a very low bioavailability. For this reason, about 80% of the promising APIs currently under development will never make it into a medicine. In contrast, amorphous API forms exhibit higher solubility as well as a higher dissolution rates than the crystalline API forms. However, they a thermodynamically less stable and will recrystallize upon storage or in solution.

One possibility to stabilize amorphous APIs is to molecularly dissolve them in an amorphous polymer which acts as solvent and carrier matrix. However, the API loading in pharmaceutical formulations usually exceeds the API solubility in the polymer. Therefore, these formulations are usually not thermodynamically stable, but amorphous phase separation and even recrystallization of the API might occur during storage. This depends on the thermodynamic phase behaviour and is to a great extend influenced by the kind of API and polymer, by temperature, and by relative humidity.

Due to this complexity, these API/polymer formulations are so far usually found by trial-and-error procedures. Thermodynamic understanding and modeling of the underlying phenomena, however, is a valuable tool to improve and to intensify this process.

The talk will give an overview about the thermodynamic phase behaviour of API/polymer systems and in particular about the influence of temperature and relative humidity. The phenomena will be discussed on the basis of experimental data for various systems. Finally, it will be shown that thermodynamic modelling today allows for reliable correlations and even predictions of the phase behaviour of these formulations. It thus can drastically reduce

the experimental effort for developing the optimal API formulation and processing.

455686 Molecular Thermodynamics for Branched Networks and Bicontinuous Phases in Water+Oil+Surfactant Microemulsions

Abstract:

Predicting the macroscopic behavior of microemulsions is important for fabrication of cosmetics, food and pharmaceuticals, production of oil, etc. Macroscopic properties of microemulsion systems strongly depend on their mesoscale structure. The phenomenological approach of Helfrich and Safran leads to the well-known state diagram of a microemulsion delineating the zones of stable droplets that have spherical or cylindrical shape, the lamellar structures, and the emulsification failure, all in terms of the elasticity moduli and the spontaneous curvature of the surfactant film.

In this work we complement this state diagram of a microemulsion by locating the stability zones for the mesoscale structures that have elements of a toroidal shape: perforated bilayers, sponge phases, and branched wormlike aggregates. We discuss the relation between the surfactantâ€™s molecular properties and the elasticity constants that determine the behavior of the microemulsion using simple mean field theories. We discuss the mechanism of stabilization of perforations and branches between the aggregates and the effect of surfactant structure: the length and flexibility of the hydrocarbon tail, the size and charge of the head. For classical ionic surfactants and diblock copolymers containing weak polyelectrolyte subchain, these theories help to describe the response of the microemulsion to a number of important factors, including the acidity and salinity of the medium and the specific chemistry of added salt.

We thank the Russian Science Foundation (project # 16-13-10042) for financial support.

455336 Properties of Amphiphilic Dendritic Polymers in Solutions, at the Air/Water Interface and on Solid Substrates

Abstract:

Properties of amphiphilic dendritic polymers in solutions, at the air/water interface and on solid substrates

J. Gregorowicz,[a] A. Brzozowska,[a,] J. Paczesny,[a] M. Luszczyk,[a] K. Nikiforov,[a] V. Sashuk,[a] P. Parzuchowski,[b] G. Rokicki,[b] [a] *Institute of Physical Chemistry, Polish Academy of Sciences, Kasprzaka 44/52, 01-224 Warsaw, Poland*

[b] *Faculty of Chemistry, Warsaw University of Technology, Noakowskiego 3, 00-664 Warsaw, Poland*

Dendritic polymers with a tree-like architecture are considered promising candidates for functional supramolecular materials. These nanoscale structures have a variety of prospective applications, e.g. adhesives, functional coatings, chemical sensors, drug delivery systems, separation processes and biomimetic systems. To achieve this, a comprehensive understanding of structure – property relationships is required. Dendritic macromolecules are represented by two main classes of structures: dendrimers and hyperbranched polymers. Dendrimers are macromolecules with highly regular branching, having a three-like structure that emerge from a central core. The synthesis of dendrimners is well controlled and as a result dendrimer macromolecules have very well defined structure reflecting complete branching and regularity. As opposed to dendrimers, random hyperbranched polymers are in fact mixtures of species differing in molecular weight and the degree of branching. Nevertheless, hyperbranched polymers have more possibilities of potential applications, since they are less expensive and simple to produce. There are still ongoing discussions as to the extent to which dendrimers and hyperbranched polymers possess similar properties.

In this contribution, behaviour of amphiphilic hyperbranched polyesters and poly(propylene imine) dendrimers at the air/water interface, on solid substrates and in solutions were investigated. All macromolecules used in the experiments have alkyl chains as terminal functionalities.

The phase equilibria in binary systems formed by the hyperbranched polyester H3200 (synthesized from Boltorn H30 through the esterification of OH-end groups with a mixture of eicosanoic (arachidic) and docosanoic (behemic) acids leading to C20/C22 (3:1) alkane chains as end groups) and alcohols are reported. Both liquid – liquid and solid – liquid equilibria were investigated. The influence of the presence of water on the phase behaviour of selected H3200 solutions is elaborated and the properties of the solid phase precipitated from alcohol solution were investigated. It was shown that a small amount of water present in these systems has a significant impact on the extent of the liquid – liquid phase split. At temperatures below 325 K alcohol solutions of H3200 form gels over a wide concentration range of the polymer. It seems that in these systems gelation results from arrested phase instability.

Despite defected structure and dispersity HBPs with terminal hydroxyl groups substituted with alkyl chains form well-organized monolayers at the air/water interface and on solid substrates. The monolayers were studied using surface pressure (static and dynamic modes) and Kelvin potential measurements, Brewster angle microscopy, relaxation measurements and compression/decompression experiments. It was shown that the HBPs molecules arranged into monolayers have polyester interior in contact with the water subphase and the alkyl chains are all pointing toward the air, forming a parallel-packed hydrophobic layer. The structure of this hydrophobic layer changes from soft liquid – like to stiff solid – like depending on the length of the terminal functionalities and temperature. The HBPs monolayers were transferred onto silicon and gold substrates using the Langmuir – Blodgett technique. Detailed structural analysis of monolayers deposited at solid substrates reveled that alkyl tails remained arranged in an up-right orientation with dense liquid-crystalline ordering.

Poly(propylene imine) (PPI) dendrimers of third and fifth generations with terminal amine groups substituted with alkyl chains also form well-organized monolayers at the air/water interface and on solid substrates. It was shown that the dendrimer molecules arranged into monolayers have core amine part in contact with the water subphase and the alkyl chains are all pointing toward the air, forming a parallel-packed hydrophobic layer. Also here, the structure of this hydrophobic layer changes from soft liquid – like to stiff solid – like depending on the length of the terminal alkyl chains and temperature. In comparison to the HBPs dendrimes macromolecules form more stiff and less temperature sensitive monolayers at the air/water interface. HBPs macromolecules in the monolayer interact mainly through alkyl chains. In case of PPI dendrimers interaction of terminal alkyl chains is strengthen by hydrogen bonds between secondary amine groups. This effect makes the monolayers stiffer and more resistant to temperature changes.

463691 Comparative Study Between Simultaneous Determinations of Dynamic Viscosity and Density of Ethanol Â Using a Straight and Coil Capillary Viscometers at Temperatures Between (313-353) K and Pressures up to 30 Mpa

Abstract:

In this work, a comparative study between viscosity and density determinations of ethanol using two different capillaries viscometers (straight and coil) was performed. The main objective of this analysis is to determine the influence that have the length of the capillary tube and the existing centripetal force in the coil tube in the determination of liquid viscosities and densities of pure fluids. For this reason, ethanol was used as the test fluid, determining the deviation between the data obtained with both tubes and the international literature. Moreover, for both cases the experimental uncertainty was estimated. The results show that, for the operating conditions used, the coil capillary viscometer has the smallest deviation from literature data (for viscosity: 0.45% coil capillary and 0.6% straight capillary; for density: 0.04% coil capillary and 0.051% straight capillary). This can be explained because there is a long length difference between the two tubes (5.5 m and 0.76 m for coil and straight tube, respectively). In the estimation of relative combined expanded uncertainty, there is no noticeable difference between the two tubes (for viscosity: 0.72% and 0.73% for coil and straight capillary, respectively; for density: 0.07% and 0.071% for coil and straight capillary). It is noteworthy that the effect of centripetal force in the calculation of the uncertainty was evaluated to be ± 0.12%. All measurements were carried out at temperature up to 353 K and pressures up to 30 MPa

464778 Fluid Phase Equilbria at High Pressures

Abstract:

The high-pressure phase equilibria of fluids [1] has important consequences, ranging from practical applications to frontier sciences. A well-known example of the former is the removal of caffeine from coffee via supercritical extraction whereas modelling the chemical behavior of Saturn and Jupiter is a more exotic application. The diversity of phenomena expands as the number of components in increases. So-called â€œislands of immiscibilityâ€ are observed when three of more component are involved resulting from the complex interplay of mixing and de-mixing processes. The phenomena are ultimately governed by fundamental differences in the interactions between molecules and experimental phase behavior data have provided the impetus for both the development of theory [2] and molecular simulation [3] techniques. This work highlights progress in understanding phase equilibria at high prressures, emphasising new developments [4] that may challenge our understanding of phenomena at high pressures.

[1] R. J. Sadus, High Pressure Phase Behaviour of Multicomponent Fluid Mixtures, Elsevier, Amsterdam, 1992.

[2] Y. S. Wei and R.J. Sadus, AIChE J. 46, 169 (2000).

[3] R. J. Sadus, Molecular Simulation of Fluids: Theory, Algorithms and Object-Orientation, Elsevier, Amsterdam, 1999.

[4] T. M. Yigzawe and R. J. Sadus, J. Chem. Phys. 138, 194502 (2013).

453543 Spray Drying of Pharmaceuticals- Solubility and Diffusivity in the Acetone-Hpmcas System

Abstract:

In the pharmaceutical industry, one option to improve dissolution and kinetic solubility of poorly soluble active pharmaceutical ingredients (API) is to incorporate them into a soluble polymer matrix using spray drying. A key player in this process is, HPMCAS (hypromellose acetate succinate), a cellulose based polymer which has a high glass transition temperature and is used for trapping the API in its amorphous state. Throughout this process, there are different limitations that effect the rate of solvent removal from spray dried amorphous dispersions. The most limiting factor is the rate at which the diffusion occurs in the solvent-polymer system. Unlike liquid-liquid diffusion coefficients, liquid-polymer diffusion coefficients typically change over four orders of magnitude as the dispersion is dried. Without a basic understanding of this process, development and scale-up are empirical and costly. The solubility and diffusivity in the acetone-HPMCAS system have been experimentally determined using three experimental techniques: inverse gas chromatography, gravimetric sorption and differential pressure decay. Below the effective glass transition temperature, the mutual diffusion behavior changes. The Vrentas-Duda free-volume theory has been extended to accurately capture this change in diffusive behavior in both infinite and finite concentration experiments.

464424 Computing Equation of State Parameters of Gases from Monte Carlo Simulations

Abstract:

Phase equilibria calculations are extremely important for designing and operating industrial processes. Equation of states (EoSs) are widely used for this purpose, but often accurate experimental data is required to calibrate the EoS parameters. Alternatively, molecular simulations can be used to compute phase equilibria from the knowledge of molecular properties. Monte Carlo (MC) simulations in ensembles with a fixed chemical potential or fugacity, for example the grand-canonical or the osmotic ensemble, are often used to compute phase equilibria. Chemical potentials can be computed either with an equation of state (EoS) or from molecular simulations. The accuracy of the computed chemical potentials depends on the quality of the (critical) parameters used in the EoS and the applied force field in the simulations. We investigated the consistency of both approaches for computing fugacities of the industrially relevant gases CO2, CH4, CO, H2, N2, and H2S. The critical temperature (Tc), pressure (Pc), and acentric factors (Ï ‰) of these gases are computed from MC simulations in the Gibbs ensemble. The effect of cutoff radius and tail corrections on the computed values of Tc, Pc, and Ï‰ is investigated. In addition, MC simulations in the Gibbs ensemble are used to compute the VLE of the 15 possible binary systems comprising the gases CO2, CH4, CO, H2, N2, and H2S, and the ternary systems CO2/CH4/H2S and CO2/CO/H2. Binary interaction parameters (kij) of these natural/synthesis gas mixtures are obtained by fitting the Peng-Robinson (PR) EoS to the binary VLE data from the MC simulations. The computed properties from the MC simulations are compared with the PR EoS, the GERG EoS, and experimental results. The MC results show that including tail corrections in the simulations is crucial to obtain accurate critical properties. The force fields used for the gases can reproduce the fugacities of the gases within 5% of the experimental data. The dew-point curves of all the 15 binaries were predicted correctly by the MC simulations, but the bubble-point curves for the systems H2/CO, CH4/H2, H2S/N2, and H2S/CO significantly deviate from the experiments. The results show that molecular simulations can be used to compute the VLE of natural gas mixtures, but it is crucial to apply the correct force field specifications to obtain good agreement with experiments. We have

submitted a paper to the special issue of Fluid Phase Equilibria to honor Prof. Theo de Loos.

452702 Low-Salinity Waterflooding of Carbonate Reservoirs: Bulk and Surface Aqueous Speciation of Calcite

Abstract:

Low-salinity waterflooding (LSW) refers to improved oil recovery achieved by waterflooding a reservoir with brine of lower salinity than that of the *in-situ* field water. The process is alluring because of simplicity, favorable economics, and large potential, especially in carbonate reservoirs which constitute over 60 % of world oil resources. Success of LSW relies heavily injected brine to alter reservoir surface chemistry of rock-brine-crude oil interfaces. Application of LSW to carbonate reservoirs, however, is especially challenging because of high brine salinity and, more importantly, because of high reactivity of the rock minerals. Here, we tackle the complex physicochemical processes in chemically active carbonates flooded with diluted brine that is saturated with atmospheric CO_2. Later work focuses on the important role of crude oil in LSW.

We prove that calcite rapidly equilibrates with aqueous solution within very short distances of order mm, even at the high flow rates in fractures of over 100 ft/day. Thus, injected-brine concentrations rapidly equilibrate with calcite rock and are not those of the flooding brine. We present new multi-reaction chemical-equilibrium calculations to describe rock-equilibrated brine compositions. In particular, injected-brine pH is altered substantially because of buffering due to rock dissolution. It is difficult to achieve alkalinities much different than pH ~ 8.3 quite independent of injected-brine pH. We also establish, for the first time, that carbon-dioxide content of a carbonate reservoir, originating from CO_2–rich crude oil and gas, plays a dominant role in setting aqueous pH and speciation and not that injected.

A simple ion-complexing equilibrium model is presented to predict calcite surface charge as a function of reservoir aqueous-solution composition. Our new multispecies ion-exchange reaction model illustrates that the surface charge of aqueous calcite may be either positive or negative depending on aqueous speciation. There is no single point of zero charge; all dissolved aqueous species are charge determining. We find excellent agreement with measured zeta potentials of calcite mineral.

Our evaluation of calcite bulk and surface equilibria draws several important inferences about proposed LSW oil-recovery mechanisms. Diffuse double layer expansion (DLE) is not possible unless brine ionic strength is below 0.01 molar. Because of rapid rock/brine equilibration, the dissolution mechanism for releasing adhered oil is eliminated. Also, fines mobilization and concomitant oil release cannot occur because there are few loose fines and clays in a majority of carbonates. Except at prohibitively high injected alkalinity, LSW cannot be an ultralow interfacial-tension alkaline flood because carbonate dissolution exhausts all injected base near the wellbore and lowers pH to that set by the rock and by formation CO_2.

Our work stresses the importance of rock/water equilibration under reservoir conditions when considering LSW. A simple, but effective formulation is now available for scoping calculations of LSW performance.

455362 Utilizing Solubilized Solvent for Surfactant Based Enhanced Oil Recovery

Abstract:

In enhanced oil recovery, the production is influenced by the ability of injected surfactants to adsorb and to modify interfacial tension at different interfaces. It has previously been demonstrated that surfactant and microemulsion additives enhance the production to a different extent but the mechanisms behind the differences in their action are not fully understood.

Here, we report salient new results on the adsorption from microemulsion at solid-liquid, and liquid-liquid interfaces. The simulation involves an aqueous solution in the presence of an oil (heptane) phase. The solution consists of nonionic surfactant dodecylhepta(oxy-ethylene)ether or $C_{12}E_7$, and a solubilized terpene solvent. We found that the presence of solvent inside the micelles causes the mechanism of adsorption behavior to deviate from those expected for adsorption from micellar surfactant solutions. In the case of solubilized terpene, the swollen micelles adsorb on the surface as one entity. The delivery of a surfactant to the interface and the associated reduction of the interfacial tension is influenced by the change in interaction potential between the surface and surfactant aggregate, and it is controlled by the solvent concentration.

Another significant finding is that both surfactant and terpene solvent can contribute to the two-phase flow in a nanopore containing oil and water. There exists a slip at the oil-water interface which inhibits the momentum transfer from the water phase to the oil phase. The surfactant acts like a linker that diminishes the slip at the oil-water interface, hence, allowing the momentum transfer from the water phase to the oil phase more effectively. The solvent, originally solubilized in a microemulsion droplet, penetrates the thin film of oil. Furthermore, those solvent molecules displace up to 14% the oil molecules in the adsorbed phase and transform that portion of oil into free oil. The resulting mixture of oil and solvent has different properties from the oil alone, indicating a primary difference between the mechanism of action between surfactant and a combination of surfactant and solvent.

The results are important for our understanding of microemulsion behavior under confinement and its application to organic rich shale oil recovery.

464536 Wettability at the Pore Level: A New Set of Spectacles for Higher Spatial Resolution

Abstract:

The understanding of pore wetting is an important factor for the success of water flooding in Enhanced Oil Recovery processes. Till date a conclusive study about the fundamental chemical understanding of all the factors affecting the wettability of a reservoir remains elusive. This is primarily due to the absence of techniques with higher spatial resolution to observe the wettability at pore scale and also due to the complex nature of intrinsic properties of the rock. Due to the lack of higher resolution techniques, pore wetting has been classically estimated using macroscopic method i.e., the contact angle method. This technique may give a misled interpretation of the sample wettability when aged in liquid environments such as DI water and Decane.

In this study we have investigated the wettability on an idealized flat surface of a mineral which is representative of a carbonate reservoir, i.e., atomically flat idealized calcite crystal. We first observed the wettability of calcite at the macroscopic level using Static Contact Angle (SCA) and then observed the wettability at Nanoscopic level. This study was performed to observe if the macroscopic results are a true measure of sample wettability.

Freshly cleaved calcite undergoes a wettability transition from being super hydrophilic to hydrophobic nature on being exposed to the atmosphere for 120 hours. The SCAs changed from being 0° when freshly cleaved on both the planes of calcite, before stabilizing at ~82.8° for Plane 1 and ~ 90.3° for Plane 2 after 120 hours. When observed with a higher spatial resolution using an Atomic Force Microscope (AFM), it was observed when calcite is freshly cleaved in atmosphere the dangling bonds of calcite are balanced by hydrolyzed water. The transition in wettability is due to the growth of these hydrate layers which were confirmed by AFM imaging and FTIR spectrum. The quantitative analyses at the nanoscale i.e., the force of adhesions were in good agreement with the macroscopic results.

When calcite was aged in Deionized water the SCAs changed from 0° when freshly cleaved to ~ 15.4° for Plane 1 and ~18.4° for Plane 2 after aging for 240 hours. The calcite crystal undergoes dissolution with time which was observed with AFM, FTIR and ICP mass spectroscopy. The macroscopic studies were carried out by drying the sample with Nitrogen and then measuring the SCA. However, when probed under an AFM it was observed that there is always a layer of water present on top of the sample surface. When the droplet from contact angle goniometer is put on top of the sample surface, it will interact with water layers present at the nanoscale and therefore alter the contact angle results; thereby putting the macroscopic wettability results into question. These observations also hint that macroscopic measurements alone are not sufficient to understand the cause for the observed time dependent wettability transition of calcite and call for the use of higher spatial resolution techniques.

We believe that our work gives a nanoscopic insight into the wettability of calcite when aged in different mediums. Our work has shown that SCA results of calcite when aged in liquid environments like DI water cannot be taken to be the true measurement of the sample wettability unlike in the case when calcite was aged in ambient atmosphere.

467234 Diffusiophoresis of Colloidal Particles in the Limit of Very High Ionic Strength

Abstract:

Diffusiophoresis is the deterministic migration of a colloidal particle through an aqueous solution, driven by a gradient in concentration of a solute which experiences a long-range interaction with the colloidal particle. The prefix "diffusio-" refers to the need for the range of interaction between particle and solute to be large enough that any stress exerted along the particle surface by the concentration gradient can result in flow of the fluid relative to the particle.

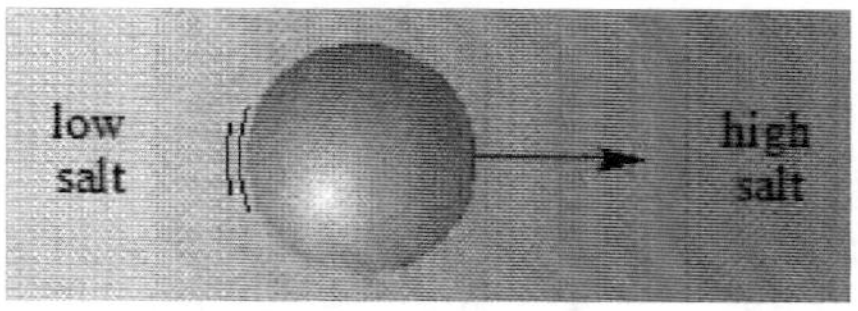

For charged colloidal particles and electrolytes, the particle-solute interaction is electrostatic and the range equals the Debye screening length, which is inversely proportional to the square-root of ionic strength. In the limit of very high ionic strengths, diffusiophoresis and electrophoresis are expected to vanish.

We measured the rate of deposition of 320 nm anionic latex particles onto the face of a Nuclepore membrane having 100 nm pores, which separates a stirred solution having 4 M of NaCl from another stirred solution having 0.4 M of NaCl plus 5wt% latex. The weight of deposit grew linearly with time over the 20 minutes of the experiment and corresponds to a particle speed of about 3 micron/sec. A diffusion potential of 7 mV was also measured with the high-salt side being more positive.

The observed dependence of the deposition speed and diffusion potential on the ratio of salt concentrations (see Fig. 1) is consistent with a mechanism in which the negatively charged latex particles undergo electrophoresis in the electric field induced by the salt gradient. Given that the Debye length is only 0.15 nm (less than one water molecule) at 4 M, it is astonishing that the particle moves at all. Indeed the measured electrophoretic mobility of the latex decreases monotonically to zero at about 5 M of NaCl (see Fig. 2).

One likely reason for our astonishment is that the equation used for calculating Debye length for a given salt concentration (as well as nearly all the models for electrophoresis and diffusiophoresis) is based on the assumption that the electrolyte solutions are thermodynamically ideal. But at these high salt concentration, steric repulsion between ions of finite size makes the solution highly nonideal.

The observed persistence of diffusiophoretic transport of particles in high salinity gradients bodes well for the application of functionalized nanoparticles in enhanced oil recovery during which medium-salinity flood waters contact connate brines in reservoirs.*

Screen Clipping

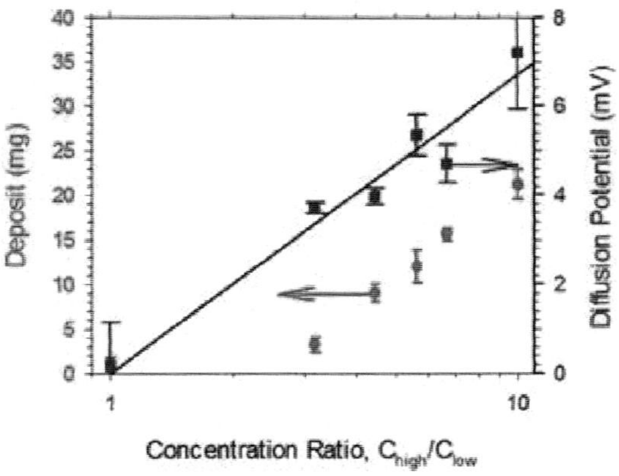

Fig. 1

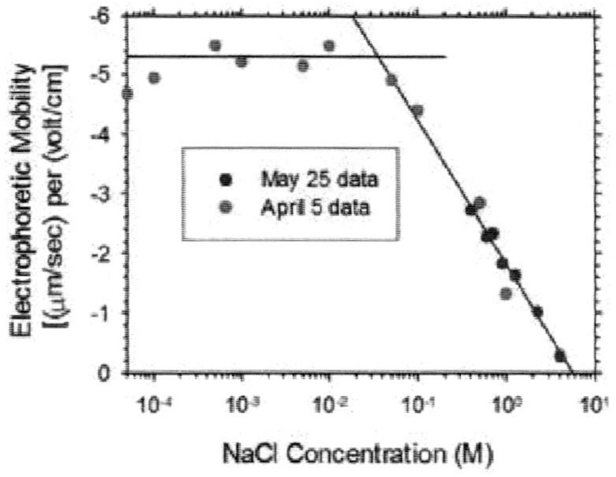

Fig. 2

* A. Kar *et al.*, *ACS Nano* **9**, 746-753 (2015).

469777 Surfactant-Oil Interactions on Surfaces

Abstract:

Oil spills can cause detrimental effects on ecosystems as a significant amount of released oil is adhered to surfaces and can remain there for long time. The interactions between oil components and surfaces (inorganic and organic) may be greatly affected by surface active compounds (natural or synthetic). Our work addresses the fundamentals of oil-solid interactions by studying the adsorption/desorption behavior of oil (fresh or weathered) on surfaces as affected by the presence of surfactants. In this study, we utilize Quartz Crystal Microbalance with Dissipation (QCM-D) and Atomic Force Microscopy (AFM) to investigate the adsorption and desorption of anionic and nonionic synthetic surfactants, of biosurfactants, as well as of model oil components on gold and silica surfaces, as representative metal and mineral surfaces, respectively. QCM-D can monitor of the dynamics of adsorption and desorption in real-time with high sensitivity. The adsorption behaviors of the surfactants below and above critical micelle concentration (cmc) are compared. The results are discussed in the context of interactions between oil components with mineral surfaces. A fundamental understanding of such interactions is critical for both oil spill remediation and oil extraction.

470424 Eco-Friendly Sacrificial Amphiphiles As Chemical Herders for Oil Spill Remediaion

Abstract:

Currently, the two principal principal methods for responding to a maritime oil spill are the use of floating barrier booms to encircle and contract the spill, so that it may be removed by mechanical skimming or in-situ burning, and the aerial spraying of surfactants onto the spill to cause the oil slick to emulsify into small droplets that are carried away from the site or are broken down by microbes. Taken together they are an inadequate toobox: Booms are difficult to deploy, and dispersants do not really remove the oil from the marine biota, and the effect of currently utilized dispersants on marine life is under dispute.

The object of this proposal is to develop eco-friendly chemical herders for removing maritime oil spills. In this method, surfactant is aerially sprayed around the spill, to lower the large surface tension of the air/sea surface uncontaminated by the oil. The reduction in the surface tension causes the floating oil layer to retract and thicken to the point where it can be ignited, and the spill burned away. The most effective surfactant herders that have emerged from laboratory and field testing are silioxane polymers, which do not easily or rapidly degrade in the marine environment. The overall goal is to develop a new class of eco-friendly herders which are based on using a hydrophobic molecule found in abundance in the marine biota, phytol, as the surfactant tail. Effective surfactant herders are made by covalently attaching phytol to cationic and nonionic polar groups. The attachment is through an ester linkage which is located near the double bond of the phytol, allowing the polar groups to be hydrolyzed on adsorption of the surfactant to the air/sea interface to regenerate the phytol. Phytol-derivatized herders can penetrate into the biofilms which form on the sea surface, so that the air/sea tension can be effectively lowered on adsorption, and the sacrificial nature of the polar attachment, with regeneration of phytol, allows long-lasting herder action as the phytol remains on the surface.

We report results on two molecules derivitized with pyridinium and imidazolium cationic groups. In small pan scale experiments, these herders retracted oil slicks by up to ~500%, 700% and 2500% at 5 °C, 20 °C and 35 °C respectively, during the first 10 minutes of experiment, which is equivalent to the current best chemical herders in practice. We also demonstrate that the hydrophobic group of the phytol molecule, with single methylene group branches, which interlock with each other, forming a stable Langmuir monolayer with low tension and sustained herding action.

471550 Interfacial Properties of Skins at the Crude Oil – Water Interface: Effect of Salt Concentration, Temperature and Surface Active Chemicals

Abstract:

Brine chemistry in waterflooding of hydrocarbon reservoirs has gained much attention recently. The composition of injected brine may affect significantly the oil recovery. The low salinity brines and small amount of surface active chemicals may result in high oil recovery. A number of recent studies have focused on determining the mechanisms involved; however full understanding is still lacking. Most authors advocate fluid-rock interactions. Formation of molecular structures at the fluid-fluid interface may be also an important mechanism. Crude oil-water interface shows a viscoelastic character from adsorption of organic compounds which may lead to formation of skins. We observe interfacial skins at low salinity at the crude oil-water interface. Some authors report formation of skins at the interface of the aqueous phase and the oil phase composed of asphaltenes dissolved in toluene.

In this work we present a systematic analysis of the interfacial viscoelasticity and interfacial tension of oil-water interfaces as a function of temperature, and small amount of pentanol added to the aqueous phase. The concentration of pentanol in the aqueous phase is 0.5 wt%. We vary the concentration of NaCl in our experiments.

Viscoelastic measurements are made using a shear rheometer with a DuNoüy ring at a small strain amplitude (0.1%) and frequency of oscillations (0.5 rad/sec). Interfacial tension measurements are made using a Krüss K12 processor tensiometer adapted with a duNoüy ring.

Interfacial viscoelasticity develops relatively fast in oil-water interfaces, and in all cases, stabilizes at 48 hours. At room temperature, the interface is found to be more elastic at low salt concentrations and become more viscous at higher salt concentrations. By increasing the temperature, the interfacial storage

modulus (G) and loss modulus (G'') both increase, while the elastic character behavior remains independent of temperature. In another set of experiments, when we add small amount of pentanol (0.5 wt%) to water, both moduli decrease and the interface becomes more viscous.

A skin is observed at the interface below 0.01 M NaCl while there is no interface skin at higher salt concentrations. We do observe the same with the alcohol at low salt concentrations. Interfacial tension measurements reveal a steep decrease at room temperature when the salt concentration is higher than 0.01 M NaCl. Interestingly, the effect of salt on interfacial tension is in line with interfacial viscoelasticity data.

Increase in interfacial viscoelasticity can be explained in terms of electrostatic interactions in the double electrical layer at the water-oil interface. The screening of the charge leads to adsorption of the polar components in the oil onto the interface. At high salt concentrations, the screening is decreased reducing the adsorption of polar components onto the interface. Our study of fluid-fluid interfaces may shed light on a key mechanism in improved oil recovery by optimizing brine chemistry of waterflooding performance.

471995 Studies of Low Salinity Waterflooding Via Glass Micromodels with Triangular Pore-Throat Architectures

Abstract:

Glass micromodels have been extensively used to simulate and investigate crude oil, brine, and surface interactions due to their wetting properties, rigidity, and ability to precisely emulate a reservoirâ€™s pore topology. However, most micromodels are fabricated with two dimensional patterning, implying that feature depths are constant with varying width, which sub-optimally describes a porous structure. We have successfully fabricated micromodels with arbitrary triangular cross sections via femtosecond pulsed laser cutting. As such, we have achieved arbitrary geometric control over the device fabrication and thus a much more accurate recapitulation of a geological porous media. With this fabrication technique, we are now able to study pore-level depth-dependent multiphase flow phenomena. Additionally, a newly developed thermal glass bonding technique enables quick and effortless fabrication of glass micromodels at the bench scale. This platform was employed to study the low salinity effect by simulating waterflooding processes over a range of salinities for both connate water and injected brine. Real-time monitoring of the crude oil behavior during aging, combined with a comparison of the brine chemistry before and after waterflooding provides an insight into realistic interactions occurring among crude oil, brine, and rock. Consequently, a comprehensive mechanistic explanation of low salinity effects can be pursued in a hypothesis-directed manner.

472431 Visualization Foam Destruction in Porous Media Micromodels

Abstract:

A microfluidic model of porous media with permeability contrast was designed to study foam stability in the presence of oil. The smart rheology of foam can help mitigate reservoir heterogeneity by separating into dry- and wet-phases such that the flow resistance in high-perm region is much higher than that in low-perm region. Foam bubbles are able to block the high-perm region and divert fluids into low permeable regions. One hypothesis is that the phase separation of foam is due to the difference in capillary entry pressure. Here we will describe the behavior of foam in the presence of various oils in the micromodels. We systematically compared the effect of crude oil with paraffin oil on foam film stability. We found out that crude oil is more detrimental to foam because of its complex composition. It is also noted that foam stability in presence of crude oil can be greatly enhanced by blending anionic surfactant with zwitterionic surfactant like betaines.

472852 Modeling Foam Flooding in Varied Dimensions: The Transport of CO2 Soluble Surfactant for Foam Mobility Control

Abstract:

CO_2 flood is one of the most successful and promising enhanced oil recovery technologies. However the displacement is limited by viscous fingering, gravity segregation and reservoir heterogeneity. Foaming the CO_2 and brine with a tailored surfactant can simultaneously address these three problems. In this study, we use a texture-implicit local-equilibrium model called STARS to investigate the effect of partition coefficient for CO_2 soluble surfactants on foam transport in both 1-D and 2-D porous media. We inject these surfactants with the CO_2 phase to enhance the injectivity of Surfactant-Alternating-Gas (SAG) process and to improve the sweep efficiency of foam process. All test results, from different cases studied, have demonstrated that when surfactant partitions approximately equally between gaseous phase and aqueous phase, foam favors oil displacement in terms of apparent viscosity (1-D), foam propagation speed (1-D), mobility reduction factor (2-D) and sweep efficiency (2-D).

465440 Molecular Dynamics Simulation Study on Dicationic Ionic Liquid Electrolytes in Supercapacitors

Abstract:

Room temperature ionic liquids(RTILs)-based supercapacitors has attracted increasing research interests due to the outstanding electrochemical properties of RTILs, such as high electrochemical stability and wider electrochemical windows, endowing them with high energy density. Monocationic ionicliquids(MILs) are the most frequently investigated RTIL electrolytes. Recently, dicationic ionic liquids (DILs) composed of the dication carrying two unit charges and associating with two anions are winning attentions due to their high thermal stability and low volatility. Thus, DILs can be excellent electrolyte candidates for supercapacitors working at extreme environmental conditions. In this work, we investigate the capacitive performance of DIL electrolytes in graphene electrode-based supercapacitors by molecular dynamics simulation. A variety of factors influencing the DIL performance in supercapacitors including alkyl chain length, anion type, the symmetry of dications as well as solvents are taken into account. Moreover, comparison of the interfacial behaviors of MILs and DILs in electric double layer(EDLs) of supercapacitors help reveal the hidden correlation between RTIL structures and their capacitive performance.

469739 Impurities in a Room-Temperature Ionic Liquid for Supercapacitors: Friend or Foe?

Abstract:

Recent years have witnessed growing interests in both the fundamentals and applications of electric double layer capacitors (EDLCs), also known as supercapacitors. A number of strategies have been explored to optimize the device performance in terms of both the energy and power densities. Because the properties of electric double layers (EDLs) are sensitive to the ion distributions in the close vicinity of the electrode surfaces, the supercapacitor performance can be easily altered in the presence of impurity. In this work, we study the impurity effects on EDLC performance using a coarse-grained model of room-temperature ionic liquids (RTILs) and the classical density functional theory (CDFT). The theoretical investigation is focused on the generic features of impurity molecules, such as the molecular size, polarity, self-interaction energy, and binding affinities with the electrode surface and surrounding ions. We identify key parameters controlling both positive and detrimental influences of impurity on ionic distribution as well as the energy storage in porous electrodes. The theoretical predictions provide insights for the impurity behavior at the electrolyte/electrode interface and, more important, useful guidance for formulation of additives to boost supercapacitor performance.

462263 Influence of Molecular Simulation Model Accuracy on the Interfacial Properties of an Ionic Liquid

Abstract:

Increasing the energy storage capability of ionic liquid supercapacitors will require better understanding of ion-electrode interactions. We have probed the influence of these interactions on the structure and differential capacitance of an ionic liquid ([EMIM][BF_4]) at an ideal graphite interface as a function of model accuracy. Of note, differential capacitance is determined through newly derived and validated fluctuation formulas. In terms of model accuracy, we test electrostatic techniques, electrode charging techniques, and electrolyte interatomic potentials. For electrostatic summations, we employ high cost, high fidelity techniques as well as less expensive, approximate techniques for summation in slab geometry. For electrode charging, uniform, constant-charge and environmentally responsive, constant-potential conditions are employed. For the ionic liquid, constant charge and atomically polarizable models are employed. We comment on the role of model accuracy on the structure and energetics of the electric double layer as well as on the magnitude and shape of differential capacitance.

467007 Elucidating the Cytotoxic Mechanism of Imidazolium-Based Ionic Liquids Via Molecular Simulations

Abstract:

Immense interest in the use of ionic liquids (ILs) for a variety of industrially relevant processes has provided an urgent need to better understand their molecular interactions with biological systems. Depending on their chemical structure, ILs can widely range in toxicological classification. Better understanding of the molecular interactions between ILs and biological systems can help facilitate their rational design. Using both atomistic and coarse-grained molecular dynamics (MD) simulations, we attempt to elucidate the physical mechanism for IL-induced disruption of a model cell membrane. In our previous work, IL-induced morphological deformations to a lipid membrane was experimentally demonstrated as one mode of IL toxicity on unicellular organisms. We investigate the effects of ionic liquids 1-n-alkyl-3-methylimidazolium chloride ([$C_n$mim]Cl), with *n* ranging from 4-12, on several lipid bilayer morphologies including the free standing planar bilayer, bilayer disk, double bilayer, and small vesicle. Our results suggest that the disruption of the lipid bilayer is strongly dependent on the asymmetric insertion of ILs from one side of the lipid bilayer surface.

470702 Time-Dependent Density Functional Theory for Ion Transport in Room-Temperature Ionic Liquids

Abstract:

Understanding the charging kinetics of electric double layers (EDLs) is of fundamental significance importance for the design and development of novel electrochemical Nano-devices such as supercapacitors and field-effect transistors. In this work, we study the dynamic behavior of room-temperature ionic liquids (RTILs) using a classical time-dependent density functional theory (TD-DFT) which that accounts for the molecular excluded volume effects and electrostatic correlations to study the dynamic process of charging and discharging of EDLs in ionic liquid. Our theoretical results for the charging and discharging of EDLs show that the counterintuitive charging kinetics in ionic liquids is closely affiliated with the local segregation of counterions and co-ions near the electrode surface, leading to formation of layer-by-layer ionic densities. We also find that the charging and discharging are symmetrical processes, the surface charge density for the charging and the discharging both follow the exponential decay. While an equivalent circuit (EC) model predicts a monotonic increase of the surface charge upon application of an electrode voltage, TDDFT predicts three types of charging behavior depending on the system geometry and the surface electrical potential.

465224 Measuring Ion Dynamics at Ionic Liquid-Fluid Interfaces

Abstract:

The dynamics of ion and molecular transport at ionic liquid-fluid interfaces are important in a variety of applications, including ion exchange (IL-IL), liquid-liquid extraction (IL-solution), and water sorption (IL-H_2O). The self-assembled nanostructure and concentrated ion effects present in ILs complicates ion mobility, and often precludes the use of standard analytical electrolyte transport models. Here, we report a novel method for characterizing transport in ILs by measuring the spatio-temporal evolution of IL concentration and structural fields after IL-fluid contact using microfluidic Fabry-Perot interferometry and birefringence visualization. In these experiments, two fluids are contacted within a semireflective microfluidic device. Light passed through the device forms Fringes of Equal Chromatic Order (FECO), which are determined by the local refractive index near the interface. Tracking changes in FECO with time allows for the calculation of changes in the refractive index (and subsequently the local concentration), while birefringence allows for tracking development of liquid crystalline nanostructure. In this work, we apply the method to study the contact of methylimidazolium-based ILs with other ILs that form different structural mesophases and with water. Measuring the dynamic concentration profiles provides insight into the concentration-dependent macroscopic diffusivity of ions within binary IL mixtures, and also gives rapid mapping of phase behavior relative to other methods. We anticipate that the method will provide measures of ion diffusivity and structure that are complementary to alternative techniques (e.g. conductivity, pulsed-field NMR, QENS), and help elucidate the influence of mesophase structure and concentrated ion effects on ion transport.

469741 Solution Thermodynamics of Amphiphilic Polymers in Ionic Liquids

Abstract:

Ionic liquids (ILs), organic salts that are fluid at ambient conditions, are a novel class of compounds with a combinatorially great chemical diversity and unique properties. The very low volatility and high thermal and chemical stability that many ILs exhibit, render them promising as solvents. We consider amphiphilic polymer + ionic liquid + water systems, where we investigate using isothermal titration calorimetry the thermodynamics (dilution/dissolution, association) and underlying interactions between polymer and ionic liquid and also between ionic liquid and water, and the amphiphilic polymer solution properties at different solvent conditions (ionic liquid - water ratios). This research informs applications of ionic liquids in products formulated with functional polymers.

455198 Ionic Liquids in Microdroplet System: Droplet Formation and Interfacial Mass Transfer Study

Abstract:

Ionic liquids (ILs), which are characterized as liquids composed of anions and cations exclusively, are attracting more and more attention due to their superior features and versatile applications in catalysis, electrochemistry, separation and so on. In this work, we proposed to accommodate ILs in microdroplet system to utilize them more efficiently and controllably. Firstly, droplet formation with ILs as one of the two phases were studied experimentally. Flow regime classification and droplet size scaling laws were investigated in detail and compared with those in regular fluid systems thoroughly. The high viscosity of ILs turned out to have significant influence on this process. Further, when water/IL droplets were formed where water served as the dispersed phase and IL ($[C_4mim][PF_6]$) served as the continuous phase, interfacial mass transfer in this system was studied using micro-LIF (laser-induced fluorescence) technique. Not only were qualitative details analyzed based on the visualization results, but the mass transfer coefficient were also calculated quantitatively. The results showed that the droplet system could enhance the interfacial mass transfer from three aspects: increased driving force, larger interfacial area and higher mass transfer coefficient. This work demonstrated the great potential of combining ILs and droplet-based microfluidics to enhance the mass transfer in IL-involved multiphase systems and promote the whole process efficiency.

458649 The Unique Simultaneously Oleophobic/Hydrophilic Behavior of a Nanometer-Thick Ionic Liquid on the Silica Surface

Abstract:

The unique simultaneously oleophobic/hydrophilic behavior of a nanometer-thick ionic liquid on the silica surface

Victor Manrique and Lei Li

Department of Chemical & Petroleum Engineering

University of Pittsburgh

Pittsburgh, PA 15261

Ionic liquids (ILs) have attracted extensive interests due to their exceptional physicochemical properties in the past two decades. Many promising applications of ILs, e.g., catalysis, electrochemistry, energy storage and lubrication, involve IL-solid interface. Therefore, it is critical to understand the molecular-level structure of ILs at the IL/solid interface. Previously, it has been demonstrated that contact angle testing is an effective approach to uncover the conformation of nanometer-thick films on the solid substrate. In the current study, nanometer-thick ionic liquid has been dip-coated on the silica substrate. Contact angles tests showed that IL/silica sample has unique simultaneously oleophobic/hydrophilic behavior. Based on ellipsometry, AFM and XPS characterizations, the molecular-level conformation of ILs has been determined. The effect of the chemical structure of both ILs and solid substrates has been discussed.

462076 Evaluating Performance of PYR13FSI As an Electrolyte with Lifsi Salt

Abstract:

One of the principal components of battery systems is the electrolyte. The electrolyte passes current through ions between the two electrodes. Important characteristics of the electrolyte include its conductivity and stability. The conductivity of the electrolyte determines the ohmic loss of a cell where higher conductivity leads to lower losses. The cathodic stability of the electrolyte determines the composition and thickness of a layer that forms on the negative electrode surface because of side reactions between the active material and the electrolyte. This layer is called the solid electrolyte interphase or SEI. The formation of an SEI layer leads to capacity fade in the cell because of the loss of active material. However, the formation of a stable SEI layer is essential to preventing continuous capacity fade in a cell. The SEI layer, when formed, reduces the rate of further side reactions while ideally still allowing ions to pass.

Recently, ionic liquids have been explored as a possible new electrolyte. Ionic liquids have shown promise in increasing stability of conventional battery systems. Ionic liquids are considered a safer, more environmentally friendly solution than current conventional electrolytes. N-Propyl-N-Methylpyrrolidinium bis(fluorosulfonyl)imide (PYR13FSI) has been identified among ionic liquids as showing promise for performing well as an electrolyte. $LiFePO_4$ cells using LiTFSI and LiFSI salts in PYR13FSI have been cycled. The cycling behaviors along with postmortem SEM images of electrode surfaces are compared. Mixtures of this ionic liquid electrolyte with a conventional electrolyte (EC:DEC) have also been examined.

458067 A Comprehensive Approach to Produce Industrially Relevant Sulfur Cathode for Lithium-Sulfur Batteries

Abstract:

Lithium-sulfur (Li-S) batteries have attracted great attention over the world because of their high theoretical specific energy. However, due to its low conductivity and the shuttling of lithium polysulfides, the performance of Li-S batteries are hindered by the low utilization and rapid capacity fading. In this work, we exam a comprehensive approach to produce a high capacity sulfur cathode by systematic studies of the carbon host structures, multi-functional polymer binders, and the use of electrolyte.

Porous carbon sheet structure was carefully synthesized as the host for sulfur by using graphene oxide as a shape-directing template and conductive filler. These graphene oxide sheets were uniformly coated with carbon on both sides, which provided enough space for sulfur and polysulfide, also shortened the ion transfer distance, ensuring a good performance of the carbon-sulfur composite. Using different CO_2 activation time, the resultant carbon materials have systematic variation of surface area, total pore volume and pore size distribution. The electrochemical characterization clearly shows that a porous carbon sheet structure with balanced surface area (total pore volume) and pore size distribution is essential to achieve high overall capacity of the sulfur-carbon composite.

In addition to the sulfur-carbon composite, we also propose and investigate ionomers with positively charged backbones (polycations) as binder materials to improve the polysulfide sequestration. For the first time, our electrochemical quart crystal microbalance experiments demonstrated that polycation binder significantly improved the sequestration of the lithium polysulfides during the lithiation process, and thus resulting improved capacity and capacity retention.

Finally we demonstrate that the electrolyte/sulfur weight ratio needs to be watched closely since it may be the decisive parameter to determine the achievable practical specific capacity of Li-S batteries. With minimized electrolyte/sulfur ratio, we demonstrate high capacity of a sulfur cathode with high areal loading.

463413 Effect of Salt Concentration on the Reactivity and Dynamics of the Electrolyte at the Li Anode of Lithium/Sulfur Batteries

Abstract:

Due to the increasing number of commercial applications that require high performance energy storage devices such as electrical vehicles, the need for improvements in conventional lithium-ion batteries is arising very quickly. In the screening of alternative chemistries, the Lithium/Sulfur (Li/S) technology appears as one of the most promising candidates due to its high energy density, and reduced cost compared to traditional Li-ion systems. However, a number of technical challenges have prevented Li/S technology from becoming feasible for commercialization. One of them is the need for an electrolyte that exhibits stability even near the lithium-metal anode, and, in the case of decomposition, the formation of a solid-electrolyte interface (SEI) with beneficial properties for the cell is desired. Recent studies have shown that the nature and composition of the salt in the electrolyte may improve the battery performance. Therefore, a comprehensive understanding of the electrolyte stability far and near the Li-anode as well as the reaction mechanisms that can take place may help us to elucidate the improvements in performance of such system and provide some guidelines for enhanced electrolytes. In this work, we use density functional theory (DFT) and *ab initio* molecular dynamics (AIMD) methods in order to investigate the effect of salt concentration on electrolyte mixtures composed by 1, 2-dimethoxyethane (DME) and salt. Here, two different salts: LiFSI (Lithium bis(fluorosulfonyl)) and LiTFSI (Lithium bis(trifluoromethanesulfonyl)imide) are studied at both high (4M) and low (1M) concentrations near and relatively far from the anode. The reaction pathways, structure of fragments, distribution of charge, and structure of the SEI are examined and characterized in detail. Finally, some insights on the formation of an electrolyte coordinated network due to high salt concentration are provided.

464885 Countering the Polysulfide Shuttle Reaction with Next Generation Cathodes and Electrolytes

Abstract:

Batteries with large amount of energy storage are needed to meet the requirements of modern technology and electronics like electric vehicles or cell phones. Current lithium ion batteries are unable to meet these modern energy demands; new battery technology is required. The lithium-sulfur (Li-S) battery is promising technology that has the theoretical energy capacity to meet societal needs. However, several large issues plague Li-S battery performance such as the polysulfide shuttle reaction which results from migration to the anode side of soluble Li polysulfide species formed at the cathode surface. This parasitic reaction can be countered by using next generation carbon composite electrodes and advanced electrolytes that strongly adsorb soluble long chained lithium polysulfides. However, finding these materials can be difficult due to complex interactions and chemistry. The molecular processes are not well understood.

Quantum scale density functional theory (DFT) and ab-initio molecular dynamics (AIMD) were used to investigate and screen different materials and electrolytes in order to identify cathode and electrolyte combinations that improve battery performance. Promising electrolytes and cathode materials can be determined by computational simulations Chalcogenides and materials used for hydrodesulphurization like MoS_2, MnO_2, and Al_2O_3 were investigated due to their strong interactions with sulfur. However, retaining polysulfides at the cathode may also be dependent on the nature of the electrolyte. Electrolytes consisting of Dimethoxyethane, Dioxolane, and fluorinated ethers in the presence of lithium salts were also investigated both independently and in the presence of the cathode. We gained a better fundamental understanding of interfacial phenomena and established the basis for a rational design that may improve overall Li-S battery performance by reducing the shuttle redox reaction.

460686 Nanostructured Transition Metal Nitrides As Promising Cathode Materials for Lithium Sulfur Batteries

Abstract:

The Lithium Sulfur (Li-S) battery system is considered as one of the most promising options for electric vehicles to achieve a higher driving range, due to their higher energy density compared to conventional lithium ion batteries. However, there are some challenges facing Liâˆ'S battery commercialization, such as: low active material utilization due to the insulating nature of sulfur, high self-discharge products due to redox shuttle reactions, and high rate of capacity fade due to the irreversible deposition of lithium sulfide (Li_2S, Li_2S_2) on the cathode and Li anode surface. Much of the research has focused on enhancing carbon materialsâ€™ electrical conductivity and ability to absorb and accommodate polysulfides in order to enhance Li/S battery cycleability and capacity performance.

In this research, we investigated different transitional metal nitrides such as: Tungsten nitride (WN), Molybdenum Nitride (Mo_2N) and Vanadium Nitride (VN) as novel cathode materials for Li/dissolved polysulfide batteries. Transition metal nitrides are well known materials for super-capacitors and lithium-ion batteries due to their high reversible insertion and extraction of ionic species and the capability of storing lithium by the intercalation mechanism. In order to obtain a better understanding of the electrochemical performance and the mechanisms underlying polysulfides redox reactions of all cathode materials, X-ray diffraction (XRD), scanning electron microscopy (SEM), X-ray photoelectron spectroscopy (XPS) analysis, in addition to electrochemical spectra (EIS) and cyclic voltammetry (CV) measurement have been applied. Among these metal nitrides, WN exhibited the most promising capacity performance as it can adsorb lithium polysulfides effectively and transfer electrons in a facile manner. As a result, the active material and surface area loss were reduced and the capacity and capacity retention of the cell were enhanced.

460670 Vertically Ordered Titanium Nitride Nanotube As a Highly Efficient Cathode Material for Lithium Sulfur Batteries

Abstract:

Abstract

Lithiumâ€"sulfur (Li-S) batteries could potentially revolutionize the rechargeable battery market due to their low cost and high energy density. However, low active material utilization due to poor sulfur conductivity, high rate of capacity fade due to the dissolution of lithium polysulfide intermediates in the liquid electrolyte, and active material loss due to the irreversible deposition of insoluble lithium sulfide (Li_2S_n, nâ‰¤2) on the electrodes surface are the main challenges facing further Li-S battery development.

Our recent study on TiN nanoparticles as a cathode material for Li-S battery using X-ray photoelectron spectroscopy (XPS), revealed that sulfur interacts with TiN nanoparticles during the discharge process. The resultant S-Ti-N retains the sulfur species on the cathode surface which can also adsorb higher order polysulfide species and undergo reduction to lower order polysulfides by transfer of electrons from the TiN electrode. In our current research, we used vertically aligned titanium nitride (VATN) nano-tubes as a cathode material for a Li-S battery with lithium/dissolved polysulfide configuration. VATN enhanced Li-S battery performance by increasing the active material utilization, and decreasing the volumetric expansion and redox shuttle reaction. In addition, the effect of nanotube size and length on Li-S battery cell capacity performance and retention were investigated. Electrochemical performance and the surface chemical environment of this new cathode material for the lithium polysulfide conversion reactions was studied by cyclic voltammetry, AC impedance spectroscopy and XPS analysis. The results may be of great significance to electrode material design for more efficient Li/S batteries.

473106 Non-Invasive in OperandoÂ Imaging ofÂ Lithium-SulfurÂ Battery Using Light Microscopy

Abstract:

Lithium-sulfur battery could increase the energy density of current Li-ion batteries by 300-500%, but their overall performance (cycle life, Coulombic efficiency, areal capacity, reproducibility) is still not ready for large-scale application. Mechanistic understanding of Li-S battery could provide guidance to overcome the intrinsic materials challenges, but are limited mainly due to the instability of sulfur species under electron or X-ray irradiation, and when removed from their native environment. A non-invasive in operando techniques is ideal for studying Li-S batteries. Light microscopy has been heavily used in biological science but has been paid little attention in materials science, mainly due to its lower resolution than electron microscopy. However, because of its ambient working condition, low sample damage, and different contrast mechanism, optical microscopy is ideal for studying complex and delicate materials system under real time, real environment, without radiation damage. Using optical techniques and specifically designed transparent cells, we have discovered new metastable species in operating Li-S batteries, and provided direct evidence on the debated reaction mechanism. The newly elucidated mechanism is correlated to the performance of the battery as well. This light microscopy technique could also be powerful in studying other high-energy batteries with delicate/complex electrochemistry.

470757 Diffusiophoretic Effects on Coffee Ring Patterns

Abstract:

Coffee rings are a natural phenomena ubiquitous in the evaporative drying of colloidal suspensions. They are thought to be associated with only evaporation-induced capillary convection. In applications related to spray painting, coating and ink-jet printing, non-equilibrium convection-driven particle aggregation causes peripheral ring structures. The ring structure can be suppressed using higher aspect ratio particles or Marangoni convection. On the other hand, various applications require formation of strong coffee rings, while other applications want to avoid the ring structure altogether. Despite several technological advances to control the coffee ring effect, modulation from well-defined ring to completely dispersed has not been achieved successfully. Here we show both experimentally and theoretically that diffusiophoresis occurs during the evaporation of saline droplets and is efficient to concentrate or disperse the coffee rings according to the salt generated intrinsic electric fields. The coffee ring structure can be controlled by choosing various salts, whose ions have different relative diffusivities, by adjusting the particle zeta potential, or by changing the rate of evaporation. In this talk the essential physics of the process is shown, an application to particle separation is shown, and the incorporation of active colloidal motors in the coffee ring structures is explored.

458128 Microdynamics and Elasticity of Fractal Gels with Embedded Active Colloids

Abstract:

Colloidal gels are a state of soft matter with attractively bonded particles in a network; these networks display rheological properties such as linear elasticity at low-frequency and a yield stress. Controlling these properties is important in industries where gels are used, such as pharmaceuticals, agricultural formulations, and consumer products. We present a method to control the microscopic dynamics of colloidal gels through active motion. We study how the introduction of active particles affects the dynamics of fractal colloidal gels comprised of passive, Brownian particles. Platinum coated Janus particles are embedded within the gel network and activated by the addition of hydrogen peroxide through a porous hydrogel membrane. We visualize gel structure and dynamics through two channel confocal microscopy. Through image analysis, we determine the positions of particles within the gel and compute their mean squared displacement (MSD). We find that the MSD of particles within the gel increases after the embedded Janus particles are activated. We analyze the MSD enhancement as a function of hydrogen peroxide concentration and the ratio of active to passive colloids. The active to passive gel MSD ratio increases as we vary the peroxide concentration from 0% to 12% at a fixed active to passive colloid ratio of 0.05. At a fixed peroxide concentration of 5%, the gel MSD increases as we increase the amount of active particles within the gel. We explore the implications of the enhanced microscopic dynamics for the macroscopic linear elasticity of the active gels.

472809 Investigation of the Motion of Patchy Particle Swimmers Near Fluid/Fluid and Solid/Liquid Interfaces

Abstract:

Recently, active colloids have been the subject of growing interest among scientists, due to their vast variety of applications including drug delivery, particle separation and Lab on a chip particle transport. Active colloids known as self-propelled particles swim in a hydrogen peroxide solution by decomposition of H_2O_2 into oxygen and H_2O. The asymmetric distribution of the product molecules around the particle leads to the particleâ€™s autonomous motion. Many studies have focused on the motion and behavior of theses particles both in bulk and near interfaces, for example, some of these studies have examined the directionality of the motion of Janus particles and others have examined how the particles align with respect to the wall.

In this context, we have previously probed the motion of patchy particles with varying sizes near a solid/liquid interface. We found evidence that the patchy particlesâ€™ rotational dynamics and their orientation with respect to the wall impact their swimming velocity and have attributed it to a preferential rotation of the particle around its swimming axis. [1]

Here, we present our approach to visualize the cap orientation of particles partially coated with platinum with respect to the fluid/fluid and solid/liquid interfaces. Further, we will show statistically relevant data on the orientation, fluctuation, and translation of colloidal patchy particles and their self-propulsion behavior in the presence of fluid/fluid and solid/liquid interfaces. The orientation and rotational motion of patchy particles will be analyzed and compared with respect to the type of interface they are interacting with.

[1] Z. Jalilvand, A. B. Pawar, and I. Kretzschmar, Patchy particle swimmers: a controlled self- diffusiophoretic motion, In Preparation, 2016.

472994 Acitve Toroids

Abstract:

The study of multicellular sheets is of great interest because of their important role in tissue development, morphogenesis, and wound healing. In confluent layer of epithelia at sufficiently high densities, cells coordinate their behavior and can collectively respond to physical and chemical stimuli. For example, heterogeneity of tissue tension within a planar 2D monolayer has been shown to correlate with the collective cell migration pattern as well as with the local increase in cell proliferation. However, much less is known about how cells respond to out-of-plane curvatures that are on the order of 1 to 100 cell lengths. Here, we investigate the proliferation and structure of MDCK on toroidal hydrogel substrates; unlike spherical or planar substrates, toroidal substrates have regions of both positive and negative Gaussian curvature. We can control the range of curvature via tuning the size and aspect ratio of the hydrogel torus, allowing us to explore the relation between substrate curvature, cell arrangement, and cell proliferation.

468362 Simultaneous Aggregation and Height Bifurcation of Colloidal Particles Near Electrodes in Oscillatory Electric Fields

Abstract:

Micrometer-scale particles suspended in aqueous NaCl solutions aggregate laterally near the electrode upon application of a low-frequency (100 Hz) field, but the same particles suspended in NaOH solutions are instead observed to separate laterally. The underlying mechanism for the electrolyte dependence remains obscure. Recent work by Woehl et al. (PRX, 2015) revealed that, contrary to previous reports, particles suspended in NaOH solutions indeed aggregate under some conditions while simultaneously exhibiting a distinct bifurcation in average height above the electrode. Here we elaborate on this observation by demonstrating the existence of a critical frequency (25 Hz) below which particles in NaOH aggregate laterally and above which they separate. Furthermore, we establish that the height of the upper layer of particles scales as the inverse square root of the applied frequency ($h \sim \omega^{-1/2}$), with particles reaching stable heights as high as 30 μm off the electrode at sufficiently low frequencies. The results indicate that the current demarcation of electrolytes as either aggregating or separating is misleading, and that the key role of the electrolyte instead is to set the magnitude of a critical frequency at which particles transition between the two behaviors.

472296 Sticking and Slipping of Active Bacteria on Silanized Glass Surfaces

Abstract:

Bacterial biofilms foul a wide range of engineered surfaces, from pipelines to membranes to biomedical implants, and lead to deleterious costs for industry and for human health. Designing strategies to reduce bacterial fouling requires fundamental understanding of mechanisms by which active and motile bacteria attach to surfaces. Here, we correlate the near-surface motion of *Escherichia coli* bacteria with transient adhesion and deposition onto silanized glass surfaces. We image bacteria as they flow through a linear channel and attach to the surface using confocal microscopy, and track their position and orientation over time using image processing algorithms. The deposition rate of bacteria on surfaces of varying silane chemistry and chain length does not correlate with surface energy or zeta potential. Instead, we find that a metric based on the degree of near-surface flagella-driven motion is inversely correlated with the rate at which bacteria deposit on these surfaces. On very smooth surfaces these bacteria exhibit mobile adhesion, in which surface-associated cells slip along the surface in the direction of flow, but this behavior cannot be described by colloidal models that predict detachment from hydrodynamic and/or adhesive interactions. We find that *E. coli* bacteria with and without flagella exhibit mobile adhesion, indicating that this behavior is not driven by these appendages, but cells that express fimbriae do not exhibit mobile adhesion. These experiments suggest that surface-associated motion enabled by surface appendages alters bacterial adhesion to surfaces in the earliest stages of biofilm formation.

471940 A Quantitative Theory of Magnetic Janus Particle Assembly

Abstract:

Under a magnetic field, Janus particles with magnetic caps (magJP) assemble into 2D chains. These chains give the fluid they are immersed in interesting rheological properties. More specifically, the longer the chains, the greater the viscosity. The rate of the magJP assembly is dependent on the dipole strength of the particles and is studied in this project. Predicting and controlling the magJP assembly rate will enable the engineering of materials that have desirable properties.

We have carried out experiments that measure the assembly rate of the magJP chains. The particles are synthesized, confined to a 2D environment, placed in a magnetic field, and tracked as they assemble. Particle tracking software has been developed for the project, and has been used to collect information from the videos including: chain concentrations vs time, particle trajectories, and dipole-dipole interaction energy. The results are fit to the Smoluchowski coagulation equations to obtain rate constants for chain aggregation.

The talk will discuss the relevance of these rate equations. A theory is proposed for prediction of the magJP rate constants when varying: concentration, dipole strength, dipole shift, viscosity, and particle size. The theory is shown to predict the magJP rate constants and future experiments are proposed to further test the theory.

460835 Active Janus Motors in Acoustic Confinement

Abstract:

We analyze the active random motion of self-propelled Janus motors confined in an external, near-Harmonic acoustic trap. The external trap behaves as an "osmotic barrier" that confines the swimmers inside the trapping region, analogous to semipermeable membranes that confine passive Brownian particles inside a boundary. We discover that active particles exposed to a weak trap have a familiar Boltzmann-like probability distribution – analogous to that of passive, equilibrium Brownian suspensions – except with the thermal energy k_BT replaced by an intrinsic kinetic 'activity' scale of active matter. However, this equivalence breaks down for stronger traps, where the trap size becomes a macroscopic length scale that confines the run length of the microswimmers. From the swimmers' restricted motion inside the trap, we calculate the unique mechanical 'swim pressure' generated by active systems originating from the force required to confine them by boundaries. Finally, we analyze the crossover from ballistic to diffusive behavior of active matter by tracking the explosion process of a dense active crystal.

452947 Shape-Directed Micro-Rotors Powered By Ultrasound

Abstract:

Shape, symmetry and other geometric features are widely used for programming self-assembly of passive colloids. These parameters can also be used as powerful strategies for engineering motion of active colloids and controlling their interactions and collective dynamics. Unlike self-propulsion directed by anisotropic surface (*e.g.*, catalytic, magnetic and light-driven Janus particles, acoustic bimetallic nanorods), shape-directed motion of active colloids is far less explored. A recent experimental study reported that gold (Au) nanorods (length 2μm, diameter 330nm) with convex and concave tips propel linearly toward their concave end at a speed of 200μm/s in megahertz (MHz) acoustic field. Other shapes and symmetry of individual particles can lead to more complex forms of locomotion than simple linear motion. In this talk, we present shape-directed rotational motion of Au micro-gears (spinners) driven by ultrasound. Both clockwise (CW) and counterclockwise (CCW) rotation can be achieved depending on the chirality of the spinners. We investigate the role of asymmetry, size and number of fins on the rotational speed of individual spinners. Shape-directed rotation can also be harnessed in other external stimuli such as chemical fuel and electric field. Shape- and symmetry-dependent actuation offers new opportunities for designing active colloids with complex forms of locomotion for enhanced functionality and studying their collective dynamics.

469471 Influence of Electrolyte Concentration on the Aggregation Ofcolloidal Particles Near Electrodes in Oscillatory Fields

Abstract:

Micron-scale particles suspended in various aqueous electrolytes have been widely observed to aggregate near electrodes in response to oscillatory electric fields, a phenomenon believed to result from electrically induced flows around the particles. Previous work has focused on elucidating the effects of the applied field strength, frequency, and electrolyte type on the aggregation rate of particles, with less attention paid to the ionic strength. Here we demonstrate that an applied field causes micron-scale particles in aqueous NaCl to rapidly aggregate over a wide range of ionic strengths, but with significant differences in aggregation morphology. Optical microscopy observations reveal that at higher ionic strengths (1 mM) particles arrange as hexagonally closed-packed (HCP) crystals, but at lower ionic strengths (0.05 mM) the particles arrange in randomly closed-packed (RCP) structures. We interpret this behavior in terms of two complementary effects: an increased particle diffusivity at lower ionic strengths due to increased particle height over the electrode and the existence of a deep secondary minimum in the particle pair interaction potential at higher ionic strength that traps particles in close proximity to one another. The results suggest that electrically induced crystallization will readily occur only over a narrow range of ionic strengths.

451997 Influence of Cobalt and Sodium Doping on MnO/CNT Composite Anode Materials for Li-Ion Batteries

Abstract:

Influence of Cobalt and Sodium Doping on MnO/CNT Composite Anode Materials for Li-Ion Batteries

A. Palmieri, R. Kashfi, S. Yazdani, M. Pettes, and W. E. Mustain (University of Connecticut)

Abstract Text:

Li ion batteries (LIBs) have been the leading technology in the portable electronics market since their discovery by Sony in the 1990s. Commercially, their setup is constituted by a graphite anode, a lithium metal oxide cathode and $LiPF_6$ dissolved in a mixture of carbonates as electrolyte, characterized by a maximum energy specific density 150 Wh/Kg [1]. However, the needs of newer and cleaner forms of energy is pushing lithium ion batteries towards higher scale applications, such as the electric grid or automotive, requiring the energy and power density to be at least doubled in the near future.

Materials advances at both sides of the cell have been actively investigated for several years. The cathodic side of the cell presents very few choices of materials that are suitable to be used in LIBs ($LiCoO_2$, $LiMnO_2$, $LiFeO_4$), all with a theoretical capacity lower than 300 mAh/g, and even lower practically achievable capacity.

At the anode, there is a much wider choice of active materials. Silicon has been one of the most widely investigated options in recent years because of its very high theoretical capacity (>3000 mAh/g) and availability. However, pure silicon materials show very rapid deep capacity fade due to the large volumetric expansion occurring during the alloying/dealloying electrochemical reaction (more than 250% of the initial volume). To mitigate this, several

materials and operational adjustments have been made and modern silicon composites are only able to achieve a capacity ca. 1000 mAh/g, only 1/3 of the theoretical capacity [2]. Moreover, reference [3] showed that, due to the cathodic limitation of the cell, there is a "capacity penalty" when the anode material capacity is too much larger than the cathode material, and the anode capacity should only be around 700 mAh/g (until new cathode materials are discovered) to fully realize improvements to the overall cell performance.

This opens up the periodic table to a wide number of available chemistries including metal oxides [4] which have been the focus of many electrochemical studies so far because of their higher capacity relative to graphite and environmental friendliness. Among the metal oxides, manganese oxides are attractive because they are inexpensive and highly available [5]. Unfortunately, to date few studies have been shown with acceptable capacity and retention have been reported, particularly under realistic operating conditions.

In this work, our team will discuss the synthesis of a novel bi-metallic Mn/Co oxide active material imbedded into a low loading multiwalled carbon nanotubes matrix, as shown in Figure 1. We will show the impact of cobalt doping on half-cell performance. Different Cobalt dopant contents (0%,5%,10%,15% and 20%) were prepared and their electrochemical performance was tested, including capacity retention (Figure 2) and rate capability up to 1600 mA/g. Lastly, Na was used as a doping as well, increasing the performance of the manganese active material, which can be tied to enhancement in the electronic conductivity.

References.

1. M.M. Thackeray, C. Wolverton and E.D. Isaacs, *Energy and Environmental Science*, **5**, (2012), 7854.
2. T. Osaka, H. Nara, T. Momma and T. Yokoshima, *J. Mater. Chem. A*, **2,** (2014) 883.
3. M. Karulkar, *J. Power Sources*, **273** (2015) 1194-1201.
4. N. Spinner, L. Zhang and W.E. Mustain, *J. Mat. Chem. A*, **2**(6) (2014) 1627.
5. L. Li, A.R.O. Raji and J.M. Tour, *Advanced Materials*, **25**, (2013), 6298.

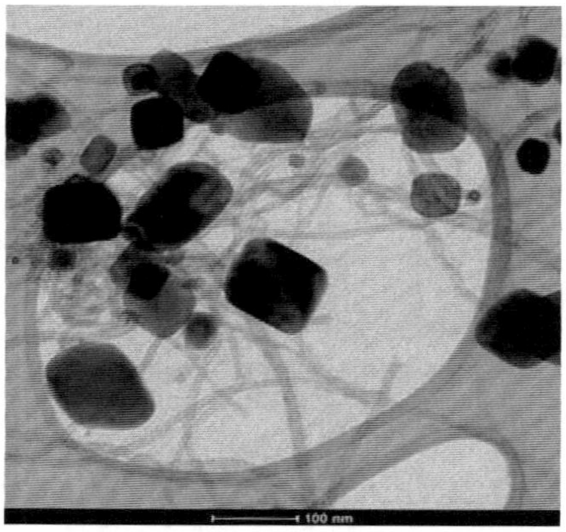

Figure 1: Mn/Co oxide nanoparticles supported on multiwalled carbon nanotubes

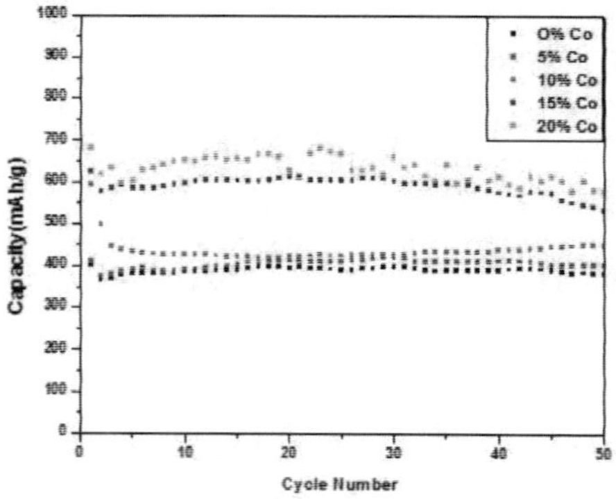

Figure 2: Mn/Co oxide different weight percentages capacity retention

472034 Engineered Silicon Graphene Oxide Anodes for Lithium Ion Batteries

Abstract:

Advances in technologies for lithium-ion batteries (LIBs) have been so spectacular in recent years that they have become one of the most popular sources for portable computing and telecommunication equipment. In the last five years there has been a lot of work on the electrochemical properties of electrode Materials for Lithium Ion Batteries (LIBs) using Graphene (G) and Graphene Oxide (GO) with silicon, however, these works suffer from several drawbacks associated with the breakage of the silicon layer. We present a new technology that avoids breakage of the silicon layer at the SEI interface and characterize these new materials. We incorporate silicon as the high capacity electrochemically active material which has the potential to greatly increase the capacity of these materials. We synthesized anodes using reduced graphene oxide as the matrix material for silicon nanoparticles. These composites are thermally reduced to convert GO to a graphitic structure capable of providing high electrical conductivity as well as high lithium ion conductivity. We solidify the composite to effectively trap the electrochemically active silicon nanoparticles within the matrix, preventing segregation of the active components from the conductive matrix. We present the results of both thermally reduced and chemically reduced graphene oxide based anodes. Our results show anodes cycled at very high charge and discharge rates of 3A/g are possible. We vary the percent of silicon inside the graphene oxide structures, and we report capacities starting at 2000 mA-hr/g.

459354 High Throughput Approach to Accelerate Electrolyte Discovery for Post Li-Ion Batteries

Abstract:

Transforming transportation and electricity grid with high performance, low cost energy storage require development of post Li-ion technology and innovations in electrodes and electrolytes, alike. A non-aqueous multivalent (e.g., Mg^{2+}, Ca^{2+} and Zn^{2+}) metal cell is one of the potential candidates for a post-lithium-ion battery. The theoretical volumetric capacity of a metal anode coupled with the lack of dendrite formation at a multivalent metal anode provide an attractive opportunity in energy storage.[1] The ability of Mg batteries to provide much higher volumetric capacity, particularly on the anode side where the Mg metal can theoretically provide 3833 mA h/cc as compared to the Li counterpart graphite (800 mA h/cc) at a lower cost makes the technology an attractive candidate for future batteries. On the other hand Zn metal anodes coupled with a reversible intercalation cathode chemistry have a number of promising features: (1) highly efficient ($\geq$ 99%) reversible Zn deposition on Zn metal anode in high performance nonaqueous Zn electrolytes (e.g., high anodic stability (maximum 3.8 V) and ionic conductivity) (2) relatively lower activation barrier energy for migration of Zn^{2+} ions in a variety of cathode materials (e.g., $FePO_4$, NiO_2, and V_2O_5) (3) similar ionic radius compared with Li^+ and Mg^{2+} ions (4) much higher volumetric capacity compared to other multivalent metals, such as Mg and Ca. However, development and commercialization of multivalent batteries require not only improved electrode discovery and development but also novel electrolytes, which are compatible with both the multivalent metal as well as the multivalent cathode. Thus a fundamental understanding of molecular level properties of these electrolytes is required to improve the electrochemical stability and the charge transfer properties. An automatic High-throughput computational infrastructure has been constructed for the electrolyte genome project supported by the US Joint Center for Energy Storage Research (JCESR) coupled with experimental analysis[2]. In this work, we present a multi-scale modelling approach for multivalent salts in various solvents. We uncover a novel effect between concentration dependent ion pair formation and anion stability at reducing potential, e.g., at the metal anode[1]. We elucidate systematic correlations between molecular level interactions and composite electrolyte properties, such as electrochemical stability, solvation structure, and dynamics. We find that Mg electrolytes are highly prone to ion pair formation, even at modest concentrations, for a wide range of solvents with different dielectric constants, which have implications for dynamics as well as charge transfer. Specifically, we observe that, at Mg metal potentials, the ion pair undergoes partial reduction at the Mg cation center ($Mg^{2+} \rightarrow Mg^+$), which competes with the charge transfer mechanism and can activate the anion to render it susceptible to decomposition. Specifically, $TFSI^-$ exhibits a significant bond weakening while paired with the transient, partially reduced Mg^+. We uncover the origin of anodic stability for a range of nonaqueous zinc electrolytes. By examination of electrochemical, structural, and transport properties of nonaqueous zinc electrolytes with varying concentrations, it is demonstrated that the acetonitrile $-Zn(TFSI)_2$, acetonitrile$-Zn(CF_3SO_3)_2$, and propylene carbonate$-Zn(TFSI)_2$ electrolytes can not only support highly reversible Zn deposition behavior on a Zn metal anode ($\geq$ 99% of Coulombic efficiency) but also provide high anodic stability (up to 3.8 V vs Zn/Zn^{2+}).[3] The combination of multi-scale modeling with experimental techniques provides unprecedented insight into the origin of the electrochemical, structural, and transport properties in multivalent electrolytes.

References

[1] J. Muldoon, C. B. Bucur and T. Gregory, *Chem. Rev.* 2014, **114**, 11683

[2] N. N. Rajput, X. Qu, N. Sa, A. K. Burrell, K.A. Persson, "The Coupling between Stability and Ion Pair Formation in Magnesium Electrolytes from First-Principles Quantum Mechanics and Classical Molecular Dynamics", Journal of the American Chemical Society, 2015.

[3] X. Qu, A. Jain, N. N. Rajput,Y.Zhang, S.P.Png, M. Brafman, L.Cheng, E. Maginn, L.A. Curtiss, K.A. Persson, "Electrolyte Genome Project: A Big Data Approach in Battery Materials Discovery", Computational Materials Science, 2015.

[4] S. Han, N.N. Rajput, X. Qu, B. Pan, M. He, M. Ferrandon, C.Liao, K.Persson, A. Burrell, ACS Applied Materials & Interfaces, 2016, 8(5), pp 3021-3031

470201 Electrolyte Salt Considerations Toward a High-Capacity, Reversible Li-Air (Li-O_2) Battery

Abstract:

The nonaqueous lithium-oxygen (Li-O_2) electrochemistry has garnered significant attention due to its high theoretical specific energy compared to the state-of-the-art lithium-ion battery. Commonly consisting of a lithium negative electrode, an oxygen-saturated porous carbon positive electrode, and an anhydrous ether-based electrolyte, the nonaqueous Li-O_2 battery operates via the electrochemical formation (discharge) and decomposition (charge) of lithium peroxide, Li_2O_2. Unfortunately, lithium peroxide (Li_2O_2) is an insulator and is insoluble in ether-based electrolytes, causing charge transport issues at even small depths of discharge. Additionally, Li_2O_2 participates in irreversible parasitic reactions with both the ether-based electrolyte and carbon positive electrode, inhibiting the reversibility of the desired oxygen electrochemistry.[1]

In this work, we present recent insights into electrolyte considerations toward mitigating these challenges. We report on our recent results that appropriately selecting the salt anion in the electrolyte solution can circumvent passivation related to Li_2O_2 deposition, resulting in a fourfold increase in Li-O_2 cell capacity without adversely affecting electrochemical stability.[2] This improvement is a result of enhanced Li^+ stability in solution imparted by anions with high Lewis basicity. Enhanced Li^+ stability in solution induces solubility of the lithium superoxide (LiO_2) intermediate to Li_2O_2 formation, leading to Li_2O_2 formation in aggregated toroid structures, leaving carbon sites open for further reaction. Our anion selection was confirmed to have no effect on electrochemical stability via differential electrochemical mass spectrometry and Li_2O_2 titrations. This anion selection strategy could be useful in any electrochemical system where adjusting intermediate stability in solution could induce desired mechanisms of product formation, and it corroborates the development of a molten salt electrolyte of entirely $LiNO_3$ and KNO_3 toward high Li_2O_2 solubility and the absence of organic solvent decomposition.[3]

References:

(1) McCloskey, B. D.; Burke, C. M.; Nichols, J. E.; Renfrew, S. E. *Chem Commun* **2015**, *51*, 12701.

(2) Burke, C. M.; Pande, V.; Khetan, A.; Viswanathan, V.; McCloskey, B. D. *Proceedings of the National Academy of Sciences* **2015**, *112*, 9293.

(3) Giordani, V.; Tozier, D.; Tan, H.; Burke, C. M.; Gallant, B. M.; Uddin, J.; Greer, J. R.; McCloskey, B. D.; Chase, G. V.; Addison, D. *Journal of the American Chemical Society* **2016**, *138*, 2656.

466735 Computational and Experimental Investigation of Stable Polymer-Based Electrolytes for Li-O_2 Batteries

Abstract:

To develop solid-state polymer Li-O_2 batteries with long cycle life and high roundtrip efficiency, polymer-based electrolytes that are stable against oxygen and oxygen reduction products in a basic and oxidizing environment are needed. We developed computational and experimental screening tests to probe the stability of various organic molecules. The screening results and the structures of these organic molecules can be used to assist the structural design of stable polymer-based electrolytes. Four descriptors were computed by Density functional theory and used to screen the stability of the electrolyte *a priori*. We investigated four possible *chemical* reaction mechanisms, namely, hydrogen abstraction, deprotonation, nucleophilic attack, and base-induced elimination, by computing the bond dissociation energy (BDE), deprotonation energy (~pKa), and the energetics of nucleophilic/elimination reaction pathway(s). Chemically stable molecules should have high BDE and deprotonation free energy for every hydrogen/proton and high stability against nucleophilic attack or elimination. In addition, the oxidation and reduction potentials were computed to assess the *electrochemical* stability of the electrolyte components. Promising candidates should be stable against oxidation in the voltage range where the battery operates. In particular, we seek electrolyte components with oxidation potentials larger than 4.5 V vs Li.

The aforementioned computational screening test was performed on a library of small molecules with representative functional groups. The calculations were validated against experiments. In particular, the oxidation potentials of a range of lithium sulfonamide salts (side-chain candidates in the polymer design) were computed and compared to the electrochemical stability windows determined experimentally. Our comprehensive computational screening methods allow us to identify stable functional groups and select promising candidates for the polymer electrolyte backbone as well as side-chain groups. A number of chemically and electrochemically stable small molecules were identified, and their molecular structures provide valuable insight into the design of stable polymer electrolytes.

465999 Mechanistic Insights into the Sodium-Oxygen Battery Cathode Electrochemistry

Abstract:

The nonaqueous lithium-oxygen (Li-O_2) battery has received increasing attention over the past decade, owing largely to its high theoretical specific energy density compared to present state-of-the-art battery technologies. Prior study has demonstrated that the primary discharge product, lithium peroxide (Li_2O_2) is produced via a two-electron oxygen reduction reaction at the cathode.[1] Though this reaction has been shown to occur in reasonably stable nonaqueous aprotic electrolytes, the efficiency of O_2 conversion to Li_2O_2 on discharge and the evolution of O_2 on charge is less than ideal.[1]

A similar system, the nonaqueous sodium-oxygen (Na-O_2) battery, has been reported as a possible alternative to the Li-O_2 battery.[2] Despite its lower theoretical specific energy, Na-O_2 batteries have been reported to provide improved cyclability and energy efficiency compared to their Li counterparts.[2-4] Sodium's natural abundance and lower cost also make large-scale implementation an attractive possibility.[4] Nevertheless, many questions remain about the fundamental electrochemical processes occurring in a Na-O_2 cell. For example, since the Li- and Na-O_2 systems appear to be analogous, one would expect the primary discharge product in aprotic electrolytes to be sodium peroxide (Na_2O_2) produced in a two-electron battery reaction. However, in batteries employing an electrolyte with an ethereal solvent, a one-electron reaction producing sodium superoxide (NaO_2) has been widely reported.[2-6] Since Na_2O_2 formation is slightly thermodynamically favored over NaO_2, this result is particularly surprising. Furthermore, prior work suggests that NaO_2 experiences fewer parasitic side reactions than Li_2O_2, leading to favorable overpotentials on charge and a higher cycle efficiency.[3] However, subsequent studies have shown that Na-O_2 battery performance can be highly sensitive to operating conditions and contaminants, leading to changes in capacity, discharge product morphology, and overall cell performance.[4-6]

In this work, we present unpublished research concerning the operation of nonaqueous Na-O_2 cells while varying key parameters, including the sodium ion concentration in the electrolyte, the oxygen pressure in the headspace, and the current density employed on discharge and charge. We characterize the dependence of cell performance on these conditions, including the cell capacity, discharge and charge overpotentials, and the "sudden death" phenomenon, a precipitous decrease in potential on discharge and increase in potential on charge that limits cell capacity and efficiency. The one-electron oxygen reduction and evolution reactions at the cathode are verified quantitatively by pressure decay/rise measurements and differential electrochemical mass spectrometry (DEMS). Further qualitative study is performed using scanning electron microscopy to examine the morphology and deposition of the discharge product. Through this study, we elucidate the limitations of Na-O_2 cell capacity and the mechanistic origins of the "sudden death" phenomenon.

Acknowledgements:
This material is based upon work supported by the National Science Foundation Graduate Research Fellowship Program under Grant No. DGE-1106400.

References:

1. McCloskey, B. D.; Burke, C. M.; Nichols, J. E.; Renfrew, S. E. *Chem. Commun.* **2015**, *51* (64), 12701-12715.
2. Hartmann, P.; Bender, C. L.; Vračar, M.; Dürr, A. K.; Garsuch, A.; Janek, J.; Adelhelm, P. *Nat. Mater.* **2013,** *12* (3), 228-232.
3. McCloskey, B. D.; Garcia, J. M.; Luntz, A. C. *J. Phys. Chem. Lett.* **2014,** *5* (7), 1230-1235.
4. Adelhelm, P.; Hartmann, P.; Bender, C. L.; Busche, M.; Eufinger, C.; Janek, J. *Beilstein J. Nanotechnol.* **2015,** *6*, 1016-1055.
5. Knudsen, K. B.; Nichols, J. E.; Vegge, T.; Luntz, A. C.; McCloskey, B. D.; Hjelm, J. *J. Phys. Chem. C* [Just Accepted], DOI: 10.1021/acs.jpcc.6b02788. Published online: May 2, 2016.
6. Xia, C.; Black, R.; Fernandes, R.; Adams, B.; Nazar, L. F. *Nat. Chem.* **2015,** *7* (6), 496-501.

470514 Anode Performance and Safety Evaluation of Potassium-Ion Batteries

Abstract:

With the high cost and scarcity of lithium, alternative metal-ion battery systems need to be developed for scalable, inexpensive and large scale energy storage devices. Potassium-ion batteries (KIBs) have recently shown promise due to their capability to reversibly intercalate in graphite, the standard lithium-ion anode, for a specific capacity of 279 mAh/g. Recently, we have explored various carbonaceous and inorganic intercalation materials as anodes in KIBs with impressive rate performance and cycling stability. Electrochemical and material characterization enabled understanding of how the bulky ion and resulting volumetric changes upon cycling affects the semi-graphitic/graphitic anode material, as well as the electrochemical mechanisms during potassiation. In addition, to evaluate their applicability, we have investigated the safety of these materials in KIBs, in terms of thermal runaway range, degradation reaction identification, and enthalpy of reactions caused by degradation of secondary electrolyte interphase leading to exothermic heat generation.

471622 Nucleation of Entropic Crystals

Abstract:

Many hard polyhedra crystallize into ordered structures, driven by directional entropic forces. Here, we use Monte Carlo simulations to investigate the nucleation behavior of polyhedra that all crystallize in the same structure - the face-centered cubic close sphere packing. We describe how the nucleation of anisotropic shapes, with varying amounts of faceting and multiple symmetries, differs from the behavior of spheres. We quantify the nucleation behavior by measuring crystallization probabilities and nucleation energy barriers, as well as cluster size distributions and local ordering in the dense fluid. Our results elucidate the impact that faceting and particle symmetry have on the nucleation behavior in entropically stabilized crystal structures.

459248 Two-Step Crystal Nucleation Is Selected Because of Lower Surface Free Energy Barrier

Abstract:

Nucleation of crystals from solution underlies multitudinous natural and engineered processes. The formation of a crystal nucleus in a supersaturated environment creates an energy-rich interface. According to classical theory, the associated surface free energy barrier is overcome by random fluctuations leading to direct assembly of monomers into a crystalline embryo. Discrepancies between classical predictions and experimental observations suggested a two-step mechanism of crystal nucleation, whereby nuclei are hosted by disordered solute-rich precursors. A fundamental controversy of two-step nucleation is that whereas precursor populations often occupy a minor volume, nucleation confined to precursors is faster than direct nucleation. Here we demonstrate that nucleation of insulin crystals, which occurs with a highly regulated rate in mammalian pancreases as a part of insulin biosynthesis, is assisted by protein-rich clusters of radius 160 â€“ 200 nm that occupy ca. 10^{-6} of the solution volume. We show that the surface free energy encountered by newly nucleated crystals is ca. 80-fold lower than the independently determined surface free energy of the crystal-solution interface; the latter value dictates the barrier for direct crystal nucleation in the solution. This comparison indicates that two-step nucleation is faster than the direct pathway and is selected as a phase transformation route due to a lower surface free energy barrier and suggests that nucleation of crystals can be controlled by manipulating the properties of the disordered precursors.

455080 Modeling the Step Velocity of Non-Centrosymmetric Growth Units and Accounting for Stable/Unstable Layers

Abstract:

The velocities of step edges on crystal faces are anisotropic. This must be accurately captured for a mechanistic model of crystal growth to be successful, since both a spiral growth regime and a birth-and-spread two-dimensional nucleation and growth regime depend on step velocities. The key components to the step velocity are the density of kinks along the step edge (the kink density) and the net rate of growth unit incorporation into kinks (the kink rate).

Non-centrosymmetric growth units represent the majority of industrially relevant molecules, but introduce significant complexity into this problem. The pattern of different growth units along the step edge produces multiple kink sites, and so the simple centrosymmetric expressions for kink density [1] and kink rate [2] are no longer applicable. Kuvadia and Doherty [3] introduced a framework for the mechanistic modeling of crystals with non-centrosymmetric growth units: they considered hypothetical transformations to form an analogous Boltzmann distribution for the kink density and determined an expression for the kink rate for a general number of growth units. Importantly, they also introduced the concept of stable/unstable step edge layers, which are unique to non-centrosymmetric growth units.

We present several improvements and generalizations of this framework that enable a more universal application and provide the ability to better capture the effect of stable/unstable layers. First, we use surface energies around each step edge site to form the Boltzmann-distributed kink density, which enables calculations of specific site densities. This also ensures the required collapse to the centrosymmetric case under appropriate limits. Second, we consider that specific kink sites are unable to move at their native kink rate velocity indefinitely; they can become limited following collision events with other kink sites. This effect can be captured by considering characteristic timescales for possible collisions and a characteristic timescale (based on kink annihilation) for edge reorganization. Each kink is allowed to advance at its native velocity

for a fraction of the reorganization timescale and is then appropriately limited. Physically, this corresponds to the exclusion of overhang growth along the step surface. To calculate the step velocity, the individual contributions from each type of kink on the surface are summed (i.e., specific kink densities and specific native kink rates, which may become limited).

We have tested this expression for the case of a 2-layer step with alternating layers of 'A' and 'B' growth units against kinetic Monte Carlo simulations, varying the edge energy of B. The predicted step velocities are in excellent agreement with simulations and can capture the trend that as the B edge energy is increased, double kinks become dominant on the step edge surface. The resulting models enable accurate morphology prediction for a wide class of molecular solutes.

References

1. Frenkel, J. On the surface motion of particles in crystals and the natural roughness of crystalline faces. *J. Phys. (Moscow)* 1945, *9*, 392-398.
2. Voronkov, V. The movement of an elementary step by means of the formation of one-dimensional nuclei. *Sov. Phys. Crystallogr.* 1970, *15*, 8-13.
3. Kuvadia, Z. B.; Doherty, M. F. Spiral Growth Model for Faceted Crystals of Non-Centrosymmetric Organic Molecules Grown from Solution. *Cryst. Growth Des.* 2011, *11*, 2780-2802.

471664 The Use of Ionic Liquids for API Purifications and Its Effect on Nucleation Kinetics

Abstract:

The Use of Ionic Liquids for API purifications and its effect on nucleation kinetics

Samir A. Kulkarni[a,b]*, Cameron C. Weber[a,c], and Allan S. Myerson[a]

[a] Department of Chemical Engineering, Massachusetts Institute of Technology, 77 Massachusetts Avenue, Cambridge, Massachusetts 02139, USA

[b] SSCI, A Division of Albany Molecular Research, Inc. (AMRI), 3065 Kent Avenue West Lafayette, IN 47906-1076

[c] Department of Chemistry, Imperial College London, London SW7 2AZ, UK

email: samir.kulkarni@amriglobal.com; cameron.weber@imperial.ac.uk; myerson@mit.edu

Keywords: Crystallization, Nucleation, Ionic Liquids

The source of this interest arises from the favorable properties some Ionic Liquids can exhibit, such as low flammability, low vapor pressures, tunable physicochemical properties through appropriate ion selection, large liquidus ranges and good thermal stability. The tunability of Ionic Liquids has been one of their most widely touted and exploited benefits as solvents, as the ability to manipulate solution interactions can lead to the design of Ionic Liquids that are good solvents for chosen target compounds. This has been explored with regards to the solubility of pharmaceutical substances, and their crystallization. These approaches

typically involve the use of antisolvent crystallization or a combination of liquid-liquid extraction and antisolvent crystallization to recover the target [1]. Surprisingly, studies into the use of Ionic Liquids as solvents for the purification by crystallization of organic solids, particularly pharmaceuticals, remains relatively limited despite the industrial significance of this methodology. In order to address these concerns we have considered the use of cooling crystallization in an Ionic Liquid as a purification methodology to reduce the complexity of the resultant solution. The Ionic Liquid possesses a relatively high viscosity over a wide range of temperature and the high initial viscosity can act as kinetic barrier to both nucleation and subsequent crystallization [2].

To ascertain the generality of our approach and to probe the limits of miscibility and thermal stability, we have examined a wide range of different active pharmaceutical compounds (APIs) as model systems for the cooling crystallization approach [3]. To demonstrate proof of principle and examine the efficacy of this purification approach, the cooling crystallization of acetaminophen with various impurities has been conducted and compared with results obtained for antisolvent processes and with cooling crystallizations conducted in water. The results in all cases indicated that impurity inclusion was comparable or improved compared to antisolvent crystallizations from either water, organic solvent systems or Ionic Liquids with substantially improved yields. Such a straightforward crystallization approach taking advantage of the low vapor pressure and wide ranging miscibility of ionic liquids with various solvents suggests a route towards a general purification procedure for many small molecule APIs in Ionic Liquids. The ionic liquid displayed a diverse effect on nucleation kinetics, as ionic liquid was more effective than ethanol-water system to reduce nucleation rate and to enhance growth time. Specific ionic liquids thus provide an additional tool in nucleation control in solution crystallization.

References:

1. CC Weber, AJ Kunov-Kruse, RD Rogers, AS Myerson, Manipulation of ionic liquid anion–solute–antisolvent interactions for the purification of acetaminophen, Chem. Commun., 2015,51, 4294-4297.

2. Kulkarni SA, Kadam SS, Meekes H, Stankiewicz AI, ter Horst JH (2013) Crystal Nucleation Kinetics from Induction Times and Metastable Zone Widths. Crystal Growth & Design 13 (6):2435-2440.

3. Cameron C Weber, Samir A Kulkarni, Andreas J Kunov-Kruse, Robin D Rogers, Allan S Myerson The Use of Cooling Crystallization in an Ionic Liquid System for the Purification of Pharmaceuticals. Cryst. Growth Des., 2015, 15 (10), pp 4946–4951.

458624 Homogeneous Bubble Nucleation in Liquid Carbon Dioxide from a Hybrid of Molecular Dynamics Simulation and Density Gradient Theory

Abstract:

Bubble nucleation is the first stage in cavitation. This phenomenon is well known in industrial processes and generally unwanted. It may occur, if a liquid is brought into the metastable range of the vapor-liquid equilibrium. It is usually described by classical nucleation theory (CNT), which relies on extreme simplifications, e.g. assuming that the usually nanoscopic bubbles have the same properties as the saturated bulk vapor and that their surface tension is the same as that of the planar interface[1–3]. In many cases CNT fails drastically[4–7], while in other cases, it gives fair predictions. This behavior is not fully understood, but it is argued sometimes that in nucleation different phenomena combine and sometimes errors cancel out[8,9].

Another method used for studies of nucleation in literature[10,11] is density functional theory (DFT) or simplifications thereof like the density gradient theory (DGT) first applied by Cahn and Hilliard to droplet nucleation in demixing incompressible liquids[12]. DFT and DGT have the advantage of correct limiting behavior approaching both binodal and spinodal. They do also inherently consider the size dependence of surface tension. Using full DFT, however, the free energy of the heterogeneous system must be known as a functional of density, which is most often not the case. Also, both theories only give information on the thermodynamic barrier, but not on the kinetics of bubble nucleation. This is usually addressed by using expressions for the kinetic pre-factor obtained in the framework of CNT. This may be dangerous, since CNT is usually developed for describing the whole process so that using only the pre-exponential factor of CNT may lead to poor results. We propose here a different method with a clean separation between kinetics and thermodynamics.

In order to properly introduce kinetics into the DGT based nucleation model, we hybridize it with molecular simulation. As an example, CO_2 is studied. All our computations are based on the molecular model of Merker et al.[13] of CO_2. The thermodynamic and interfacial properties of this model are used to fit the parameters of the PC-SAFT[14] equation of state (EOS), as well as the influence parameter from DGT. This way, a thermodynamic link between molecular simulation and thermodynamic model is achieved, which is further confirmed by comparison to molecular simulation data in the metastable range. In a next step, the homogeneous bubble nucleation of the molecular model is simulated for two temperatures using molecular dynamics simulation following an approach first introduced by Yasuoka and Matsumoto[15] for droplet nucleation and later on applied by Diemand et al.[16] for bubble nucleation. The kinetic pre-factor of the proposed hybrid theory is then fit to the direct simulation data at two temperatures, where we assume an Arrhenius type dependence on temperature. This way, our hybrid method describes bubble nucleation in the whole metastable range from triple point to critical point without relying on any further assumptions. The results compare favorably to our direct simulations of the nucleation rates and to available experimental data for CO_2[17,18], while CNT shows the wrong temperature dependence and is off several orders of magnitude in terms of nucleation rates.

(1) Blander, M.; Katz, J. L. The Thermodynamics of Cluster Formation in Nucleation Theory. *J Stat Phys* **1972**, *4* (1), 55–59.

(2) Vehkamäki, H. *Classical Nucleation Theory in Multicomponent Systems*; Springer: Heidelberg, 2006.

(3) Blander, M.; Katz, J. L. Bubble Nucleation in Liquids. *AIChE J.* **1975**, *21* (5), 833–848.

(4) Baidakov, V. G.; Bobrov, K. S. Spontaneous Cavitation in a Lennard-Jones Liquid at Negative Pressures. *The Journal of Chemical Physics* **2014**, *140* (18), 184506.

(5) Gonzalez, M. A.; Abascal, J. L. F.; Valeriani, C.; Bresme, F. Bubble Nucleation in Simple and Molecular Liquids via the Largest Spherical Cavity Method. *The Journal of Chemical Physics* **2015**, *142* (15), 154903.

(6) Römer, F.; Kraska, T. Homogeneous Nucleation and Growth in Supersaturated Zinc Vapor Investigated by Molecular Dynamics Simulation. *The Journal of Chemical Physics* **2007**, *127* (23), 234509.

(7) Sinha, S.; Bhabhe, A.; Laksmono, H.; Wölk, J.; Strey, R.; Wyslouzil, B. Argon Nucleation in a Cryogenic Supersonic Nozzle. *The Journal of Chemical Physics* **2010**, *132* (6), 64304.

(8) Römer, F.; Fischer, B.; Kraska, T. Investigation of The Nucleation and Growth of Methanol Clusters from Supersaturated Vapor by Molecular Dynamics Simulations. *Soft Materials* **2012**, *10* (1–3), 130–152.

(9) Wedekind, J.; Hyvärinen, A.-P.; Brus, D.; Reguera, D. Unraveling the ``Pressure Effect'' in Nucleation. *Phys. Rev. Lett.* **2008**, *101* (12), 125703.

(10) Laaksonen, A.; Talanquer, V.; Oxtoby, D. W. Nucleation: Measurements, Theory, and Atmospheric Applications. *Annual Review of Physical Chemistry* **1995**, *46* (1), 489–524.

(11) Oxtoby, D. W. Density Functional Methods in the Statistical Mechanics of Materials. *Annual Review of Materials Research* **2002**, *32* (1), 39–52.

(12) Cahn, J. W.; Hilliard, J. E. Free Energy of a Nonuniform System. III. Nucleation in a Two Component Incompressible Fluid. *The Journal of Chemical Physics* **1959**, *31* (3), 688–699.

(13) Merker, T.; Engin, C.; Vrabec, J.; Hasse, H. Molecular Model for Carbon Dioxide Optimized to Vapor-Liquid Equilibria. *The Journal of Chemical Physics* **2010**, *132* (23), 234512.

(14) Gross, J.; Sadowski, G. Perturbed-Chain SAFT: An Equation of State Based on a Perturbation Theory for Chain Molecules. *Ind. Eng. Chem. Res.* **2001**, *40* (4), 1244–1260.

(15) Yasuoka, K.; Matsumoto, M. Molecular Dynamics of Homogeneous Nucleation in the Vapor Phase. I. Lennard-Jones Fluid. *The Journal of Chemical Physics* **1998**, *109* (19), 8451–8462.

(16) Diemand, J.; Angélil, R.; Tanaka, K. K.; Tanaka, H. Direct Simulations of Homogeneous Bubble Nucleation: Agreement with Classical Nucleation Theory and No Local Hot Spots. *Phys. Rev. E* **2014**, *90* (5), 52407.

(17) Straub, J. In *Proceedings of the 3rd International Conference on Chemical Thermodynamics and Physico-Chemical Techniques at High Temperatures, Baden, Austria*; Butterworth: London, 1973; Vol. 2.

(18) Huang, J. S.; Goldburg, W. I.; Moldover, M. R. Observation of Anomalously Large Supercooling in Carbon Dioxide. *Phys. Rev. Lett.* **1975**, *34* (11), 639–642.

466555 Simulation of Homogeneous Nucleation of Co-Crystals from a Binary Hard Sphere Fluid Mixture

Abstract:

Co-crystal formation from fluid-mixtures is quite common in a large number of systems. The simplest systems that shows co-crystal (also called substitutionally ordered solids) formation are binary hard sphere mixtures. In this work we study the nucleation of AB_2 type solid compounds using Monte Carlo molecular simulations in a binary hard sphere mixtures with size ratio of 0.55. The conditions chosen for the study lie in the region where nucleation of AB_2 type solid competes with that of pure A solid with FCC (face-centered-cubic) structure. The fluid phase compositon is kept equal to that of the AB_2 type solid. The nucleation free-energy barriers are computed using the seeding technique of Sanz *et al.* [*J. Am. Chem. Soc.* **135**, 15008 (2013)]. Our computed free energy barriers for homogeneous nucleation show that the AB_2 type solid is always favored even under conditions where the pure A solid is more stable. This is primarily due to similarity in the composition of fluid-phase and the AB_2 type solid which in turn leads to much lower interfacial tension between the crystal nucleus and the fluid phase. This system is an example of how the fluid phase composition affect the structure of the nucleating solid phase during crystallization and has relevance to solvent-induced polymorphism seen in molecular crystals.

473323 Molecular Self-Assembly in Liquid Crystals

Abstract:

Topological defects present in liquid crystals (LCs) have been widely used to organize colloidal dispersions and template polymerizations, leading to a range of assemblies, elastomers and gels with complex mechanical and optical properties. However, little is understood about molecular-level assembly processes within defects, although within the nanoscopic cores of singular topological defects the ordering of mesogens is known to be diminished relative to the bulk LC. This presentation will report that nanoscopic environments defined by LC topological defects can selectively trigger processes of molecular self-assembly. By using fluorescence microscopy, cryogenic transmission electron microscopy and super-resolution optical microscopy, key signatures of molecular self-assembly of amphiphilic molecules in topological defects are observed - including cooperativity, reversibility, and controlled growth of the molecular assemblies. By using amphiphiles that can be photocrosslinked, we also demonstrate preservation of molecular assemblies templated by defects, including nanoscopic "o-rings" synthesized from "Saturn-ring" disclinations. Our results reveal that topological defects in LCs are a versatile class of three-dimensional, dynamic and reconfigurable templates that can direct processes of molecular self-assembly in a manner that is strongly analogous to other classes of macromolecular templates (e.g., polymer—surfactant complexes).

469136 Using Advanced Rheological and Neutron Scattering Techniques to Determine Signatures of Branching in Wormlike Micelles (WLMs)

Abstract:

The molecular design of soft materials with optimal flow properties is important in applications ranging from polymer processing to drug delivery, where materials undergo both steady and dynamic nonlinear deformations during processing and/or use [1]. Self-assembled wormlike micelles (WLMs) are of particular scientific and technological interest due to their ability to branch, break, and reform under shear [2]. The unique properties of WLMs can lead to nonlinear flow phenomena and instabilities such as shear banding [1]. As measuring chain branching is a long-standing scientific challenge, branched WLMs are also used as a model system for studying branched polymers. Indeed, nonlinear shear and large amplitude oscillatory shear (LAOS) measurements have shown to be useful in the study of chain branching [1], but a quantitative understanding of how topological branching in WLMs affects the shear rheology and dynamics is not well understood. Understanding how the flow behavior and molecular topology are coupled under these nonlinear flows requires advancements in experimental methods that combine time- and spatially-resolved small angle neutron scattering measurements of structure with rheometry (flow-SANS) for various types of nonlinear deformations [3-5].

We use multiple rheological and neutron techniques to explore the relationship between branching, microstructure, dynamics and nonlinear responses using a model WLM series where we have exquisite control over the branching [5-7]. The rheology, shear-induced microstructures, and dynamics of the WLMs are examined to determine rheological and structural signatures of branching. The degree of branching in the mixed cationic/anionic surfactant (CTAT/SDBS) solutions is controlled via the addition of the hydrotropic salt sodium tosylate [5-7] at two surfactant concentrations: 1.5% wt and 4% wt. The phase behavior has been mapped extensively and we have explored the degree of branching and network formation using static

SANS, cryo-TEM and rheo-optical methods [5-7]. Neutron spin echo (NSE) is performed to determine characteristic differences in the solution dynamics. Rheological signatures of branching are determined with various nonlinear deformations, including steady shear, shear startup, orthogonal superposition (OSP), and large amplitude oscillatory shear (LAOS) on a dimensionless basis of shear rate (Wi) and frequency (De). To develop flow-structural relationships, we employ flow-SANS and nonlinear rheological techniques [3, 4] in conjunction with newly developed methods of time-resolved data analysis that improve experimental resolution by orders of magnitude [8]. The shear-induced ordering of the micelles is spatially and temporally characterized under steady shear, shear startup, and large amplitude oscillatory shear (LAOS) in various shear planes.

Micellar branching leads to deviations from Maxwellian behavior in the LVE rheology, and can alter or eliminate steady shear banding [6]. Steady shear and steady shear startup rheological results indicate that branching inhibits shear banding at both surfactant concentrations, where faster transients and high power law indices are observed with increasing branching. As observed in branched polymers [9], the stress overshoot behavior immediately following shear startup is reflective of the topology. Branched structures mitigate this overshoot, leading to shear thinning as opposed to shear banding. Qualitatively similar trends are seen in the LAOS rheology. In conditions where stress overshoots are observed, the overshoot is mitigated by the introduction of branching. Finally, the orthogonal dynamic moduli (G' and G") are characterized under steady shear with OSP rheology. The normalized orthogonal crossover modulus, G_c, plateau modulus, G_0, and relaxation time ($t_R = 1/w_c$) as a function of Wi decrease more rapidly with branching, indicating a break down of network-like structures.

Neutron spin echo (NSE) measurements are used to identify characteristic relaxation processes in the branched WLM solutions. The stretch exponent, β, derived from fits to the dynamic structure factor is an indicator of solution morphology, where a value of $\beta = 3/4$ is expected for a wormlike chain and a value of $\beta = 2/3$ is predicted for a flexible membrane [10]. The stretch exponent is similar between systems at high q-values, where the basic cylindrical morphology of the solutions is probed. However, systematic differences in the stretch exponent are seen with decreasing q-values, where branching may play a significant role in the solution dynamics. While $\beta = 3/4$ is observed for solutions with low branching at these low q-positions, the stretch exponent approaches $\beta = 2/3$ in the very highly branched solution, indicative of a branched network morphology. While the relaxation rate of the WLMs is similar between solutions at high q-values, deviations also occur at lower q-values, allowing characteristic differences in branching to be quantified.

The shear-induced ordering and stability of the solutions is finally characterized in various shear planes under steady shear, shear startup and LAOS using flow-SANS methods. Local segmental orientation and alignment (A_f) in the flow-gradient plane is found to be a complex function of the branching level, radial position, and deformation type. Results in the 1-2 plane confirm shear banding in solutions with lower branching levels for all deformation types. The shear banding instability is also mitigated by branching for all deformation types [6,7]. Time-resolved startup measurements offer additional structural signatures of shear banding, where a A_f long transient is observed for shear banding solutions that is also dependent on branching level. Little to no transience is observed in shear thinning conditions, or for the shear thinning, highly branched solutions. The time- and spatially-dependent alignment factor can also be used to identify the shear band formation and interface location.

The combination of nonlinear rheology and advanced neutron scattering measurements of the microstructure and dynamics shows distinct differences in the rheological response, flow-induced microstructure, and solution dynamics of WLMs with branching. Importantly, highly branched WLMs offer enhanced solution stability, as shear banding and related instabilities are inhibited for all deformation types at comparable dimensionless frequencies and shear rates. This research employs advanced neutron techniques to determine characteristic differences in the flow-induced microstructure, topology and dynamics of branched WLMs, and is part of a broader effort to characterize branching in chemical polymers and self-assembled systems.

References:

[1] Hyun, K., et al., Progress in Polymer Science **36**(12), 2011.

[2] Lerouge, S. and J. F. Berret, Polymer Characterization **230**, 2009.

[3] Gurnon, A.K., et al., Soft Matter **10**(16), 2014.

[4] Lopez-Barron, C.R., et al., Physical Review E **89**(4), 2014.

[5] Schubert, B., N.J. Wagner, and E.W. Kaler, Langmuir **19**(10), 2003.

[6] Calabrese, M.A., et al., Journal of Rheology, **59**(5), 2015.

[7] Calabrese, M. A., et al., under review Journal of Rheology, 2016.

[8] Calabrese, M. A., N. J. Wagner, and S. A. Rogers, Soft Matter **12**, 2016.

[9] Snijkers, F., et al., Journal of Rheology, **57**(4), 2013.

[10] Zilman, A. and R. Granek, Physical Review Letters **77**, 1996.

464640 Ionically Crosslinked Polymer Networks for the Multiple-Month Release of Small Molecules

Abstract:

Self-assembled hydrogels are widely studied as materials for sustained release applications. Yet, unless their payloads are able to bind or bond to the network, or are loaded into colloids (e.g., nanoparticles or liposomes) that are dispersed within the gels toÂ moderate theÂ release rates, most self-assembled hydrogels are too permeable to deliver small hydrophilic molecules over long (multiple-week or month) times. Here, we report on how gel-like coacervates, which self-assemble via ionotropic gelation of poly(allylamine hydrochloride) (PAH) with either pyrophosphate (PPi) or tripolyphosphate (TPP), can be used to address this limitation. Rheological analysis of these gel-like networks reveals them to have very high ($G'_{\text{â˘ž}}$Â > $10^{5\text{Â}}$ Pa) plateau moduli, which are indicative of small (nanometer-scale) hydrodynamic mesh sizes. These small mesh sizes make PAH/PPi and PAH/TPP complexes effective barriers to solute diffusion, and enable them to sustain small hydrophilic molecule release over many months. Isothermal titration calorimetry (ITC) analysis shows that this highly-sustained release occurs even without payload/network binding, which confirms that the slow release is primarily caused by the small mesh size (and not payload-specific binding interactions). The release rates achieved with these materials can also be easily tuned by varying the encapsulated payload content and gel thickness. This ability to obtain tunable release regardless of the payload/network binding suggests that PAH/PPi and PAH/TPP networks can serve as simple sustained release platforms for a wide range of small molecules.

471713 Photo-Controlled Bcl-2 siRNA and Paclitaxel Co-Delivery Via Azobenzene Based Catanionic Vesicles

Abstract:

The lack of targeted delivery and triggered release during traditional chemotherapy causes high toxicity in the human body and extends the treatment. The usage of nanocarriers for chemotherapeutic drugs aids to increase cancer treatment efficiency. Additionally, a potential therapeutic genetic material, siRNA, can be used to deactivate targeted mRNAs that cause inhibition of disease-associated protein production. The main advantage is using nanocarrier to design certain size nanoparticles that have the tendency of accumulation in tumor cells more than in normal cells. Thus, we have developed a novel method to deliver both siRNA and a chemotherapeutic drug to cancer cells via photo-assisted catanionic vesicles. Spontaneous catanionic self-assemblies are formed upon the interaction of an azobenzene based cationic surfactant and a conventional anionic surfactant. As a result of the photo-responsive property of azobenzene moieties, the photoresponsive mixture microstructure can be switched with light illumination from vesicles to either micelles, lamellar structures, or free surfactant monomers. Specifically, the unique photo-assisted transition of *vesicles* to *free monomers* will resolve two major obstacles in cancer treatment such as passive targeting of chemotherapeutics through cancerous cells and triggered release with light illumination. The photoresponsive catanionic vesicles are self-assembled nanocapsules that can be used for encapsulation and controlled release of substances by UV light exposure. In this work, the effect of hydrophobic tail of surfactants on particle size, charge and surface characteristics will be measured by small-angle neutron scattering, dynamic light scattering, zeta potential meter and cryo-transmission electron microscopy to create an ideal delivery vector. Furthermore, the encapsulation efficiency and transfection rates of Bcl-2 siRNA and paclitaxel co-delivery through MDA-MB-231 human breast cancer cells will be explored. Therefore, azobenzene based catanionic vesicles will be used as a co-delivery tool to develop siRNA based therapeutics with a hydrophobic anticancer drug. It is expected to improve tumor therapeutic efficacy by the usage of the novel photo-assisted catanionic delivery agent.

471252 Succession of Alkane Conformational Motifs Bound within Hydrophobic Nano-Capsule Assemblies

Abstract:

As an approach to computationally predict the preferred conformation of molecules confined within nano-spaces, we report here on a strategy that accurately reproduces the binding motif of *n*-alkanes within dimeric octa-acid capsular complexes. Previously, NMR analysis of these complexes revealed a succession of guest packing motifs from *extended*, to *helical*, to *hairpin*, to *spinning top* structures with increasing chain length. Here, we report a molecular simulation study of alkane conformational preferences within these host-guest assemblies to uncover the factors stabilizing distinct conformers. The simulated alkane conformers follow the trends inferred from ^{1}H-NMR experiments, while guest proton chemical shifts evaluated from Gauge Invariant Atomic Orbital calculations provide further evidence our simulations capture guest packing within these assemblies. Analysis of chain length and dihedral distributions indicates that packing under confinement to minimize non-polar guest and host interior contact with water largely drives the transitions. Mean intramolecular distance maps and transfer free energy differences suggest the transition between the *extended* and *helical* motifs is a continuous process accompanied by increases in the guest *gauche* population to allow accommodation within the complex, while breaks observed between the *helical*/*hairpin* and *hairpin*/*spinning top* motifs are indicative of true transitions. Our results represent the first bridging of empirical and simulation data for flexible guests encapsulated within confined nano-spaces, and constitute an effective strategy by which guest packing motifs within artificial or natural compartments can be rationalized and/or predicted *a priori*.

467051 Self-Assembly of Nanoparticles into Biomimetic Capsid-like Nanoshells

Abstract:

Nanoshells self-assembled from biomacromolecules are common in biology but not among inorganic nanomaterials due to difficulties with replication of spherical tiling without molecular precision of folded biomolecules. Here, we show that semiconductor nanoparticles can spontaneously organize into porous nanoshells without specific lock-and-key interactions. The association of stabilizer-free CdS nanoparticles into self-limited spherical capsules is the result of scale-modified electrostatic and dispersion forces. They cannot be described by DLVO theory whereas molecular dynamics simulations with atomistic description of nanoparticles and solvent reveal the conditions necessary for the emergence of nanoshells. Morphology of the simulated assemblies formed under different conditions matched nearly perfectly the TEM tomography data. This study bridges the gap between biological and inorganic self-assembling nanosystems and conceptualizes a new pathway to spontaneous compartmentalization for a wide range of natural and man-made inorganic nanoparticles.

467212 Interactions Between Antimalarials and Hematin Crystal Surface Sites Determine the Mode of Growth Inhibition

Abstract:

Dynamic events at solid-liquid interfaces mediate the kinetics of crystallization wherein the molecular interactions between the crystal surface and solvent or adsorbates can alter anisotropic growth rates. These fundamental processes mediate natural and synthetic crystallization pathways, giving rise to materials with optical, mechanical, and electrical properties for a variety of applications. Our studies employ a model crystal system, hematin, which has a wide range of potential applications. In malaria pathophysiology, hematin crystallization ensures the survival of the parasite *Plasmodium falciparum* during its asexual life cycle.[1-3] Survival within its human host involves the parasite entering erythrocytes (red blood cells), catabolizing hemoglobin and releasing free heme (Fe^{2+}), which undergoes rapid oxidation to form toxic hematin (Fe^{3+}) within its digestive vacuole (DV). The parasite utilizes the process of heme detoxification by sequestering the toxin into innocuous crystalline hemozoin within the complex DV environment. Current antimalarial compounds are thought to prevent the growth of hematin crystals, referred to as hemozoin (*in vivo*) or beta-hematin (*in vitro*). Identification of antimalarial drug action on this phase transformation could provide a foundation for drug design to overcome parasite drug resistance. Here, we will discuss our effort to understand the molecular mechanism(s) of hematin crystallization and the modes of antimalarial drug action in physiologically-relevant growth media.

In situ atomic force microscopy (AFM) reveals that hematin crystallization occurs by a classical mechanism of layer-by-layer growth involving the addition of individual molecules to specific sites presented on the crystal surface. We show that hematin crystals form in a two-phase medium, which we designed to mimic the parasiteâ€™s DV comprising a complex system with a primarily acidic aqueous phase and lipid subphase.[2] We quantified the rate of layer generation and the velocities of anisotropic step advancement as a function of supersaturation in the absence and presence of antimalarials.[1] We identified that current antimalarials act in unique modes of inhibition. Our current studies aim to elucidate the relationship between the functional moieties present on current antimalarials and free hematin as well as the solid-liquid interface presented on predominant {100} crystal surfaces. In parallel, we studied the prominent functional groups of antimalarial compounds using a combination of *in situ* AFM studies and chemical force microscopy (CFM) measurements. Identifying key factors governing molecular recognition of inhibitors to specific crystal surface sites is crucial to the development of rational design platforms for screening new drug compounds.

References:

[1] Olafson, K.N., Ketchum, M.A., Rimer, J.D., Vekilov, P.G., Mechanisms of Hematin Crystallization and Inhibition by the Antimalarial Drug Chloroquine, ***Proc. Natl. Acad. Sci.*** 16 (2015) 4946-4951

[2] Olafson, K.N., Rimer, J.D., Vekilov, P.G., **Growth of Large Hematin Crystals in Biomimetic Solutions,** *Cryst. Growth Des.* 14 (2014) 2123-2127

[3] Ketchum, M.A., Olafson, K.N., Rimer, J.D., Vekilov, P.G., **Hematin Crystallization from Aqueous and Organic Solvents,** *J. Chem. Phys.* 139 (2013) 121911(1-9)

466783 pH Driven Reorientation of Cytochrome *c* on Silica Nanoparticles: Effect of Surface Charge Distribution and Dipole Moment

Abstract:

In the past decade, nanoscience has found numerous biomedical applications, such as targeted molecular delivery, inâ€‘vivo imaging, sensing, and gene therapy. In almost all these applications, the active nanosized particles come in direct contact with biomolecular environments containing various proteins. Hence, developing a fundamental understanding on how proteins interact and self-assemble at nanoparticle surfaces is of eminent importance. Here we report an experimental study of the pH dependent spatial orientation of adsorbed cytochrome *c* onto silica nanoparticles (diameter 7 to 20Â nm). We use small angle neutron scattering (SANS) to determine the equilibrium assembled state and spatial orientation of adsorbed cytochrome c in the pH range 2-11. We find that the protein orientation on nanoparticleâ€™s surface is strongly correlated with its surface charge distribution and net dipole moment. Our comprehensive analysis of SANS data and protein surface charge distribution calculations indicates that the protein binds to the silica surface via its largest positive surface patch. This study provides a new way of analyzing and understanding protein-nanoparticle interactions, which may further contribute to the development of new approaches for fabricating surface-engineered materials for biomedical applications.

455953 Destabilization of Proteins in Extreme Conditions: Molecular Simulations of the Denaturation of Trp-Cage Miniprotein

Abstract:

Proteins are stable in their compact, folded structures only in a narrow range of physiological conditions. Deviation from these conditions, such as changes in temperature and salt concentrations, can destabilize the folded structures. Investigating the mechanism(s) of protein denaturation induced by these extreme conditions has a fundamental importance in understanding of the evolution of organisms with extreme habitats and formulating effective long-term preservation strategies for biological therapeutics. Using replica-exchange molecular dynamics simulations, we examine the mechanisms of cold- and salt-induced denaturation of a fully-atomistic model of the Trp-cage miniprotein, and we make direct comparison with the unfolding behavior at high temperatures. In contrast to the complete unraveling of the secondary and tertiary structures observed during heat denaturation, the cold-denatured structure of the Trp-cage is compact and partially folded with stable alpha-helix, despite the loss of salt bridges. We also find that the salt-denatured structures exhibit a strong resemblance to the cold-unfolded Trp-cages. The mechanism of cold- and salt-induced protein denaturation is examined by analyzing the structural change, thermodynamic quantities, and protein-solvent interaction.

470416 Understanding the Structural Differences Between Psychrophilic and Thermophilic Enzymes: A Molecular Dynamics Study

Abstract:

Industrial use of enzymes has grown at a tremendous rate. The global market for industrial enzymes is predicted to reach $7 billion by 2018. This is not surprising, given enzymatic catalysis results in high yields, low cost and is environmentally friendly compared to many conventional chemical catalysts. A wide range of industrial markets, such as food industry, textile industry, pharmaceuticals, deployable biosensors and production of biofuels are exploring enzymatic catalysis. These diverse markets demand more robust enzymes which can tolerate environments ranging from low to high temperatures. However, most enzymes are inactive and denature at such extreme conditions.

Extremophile proteins â€" that include thermophilic and psychrophilic proteins are being investigated as enzymes for extreme temperatures. Thermophilic and psychrophilic proteins refer to enzymes that are active at high and low temperatures respectively but not at ambient temperatures. Thermophilic enzymes are more stable at high temperatures while psychrophilic enzymes tend to have shorter lifetimes and tend to be less stable at ambient temperatures. The underlying reasons for the differences in the range of the temperatures that these enzymes are active is not yet understood. Studies have suggested that the active site flexibility of psychrophilic enzymes provides their high activity at low temperatures. In contrast, the loss of flexibility helps thermophilic enzymes retain stability at high temperatures. We test this hypothesis through large-scale all-atom molecular dynamics (MD) simulations. Specifically, we perform simulations of *G. thermocatenulatus* (GTL) â€" a thermophilic enzyme and *P. lipolyticum* (M37L) â€" a psychrophilic enzyme in a range of temperatures. We investigate the differences in the global and local flexibility, specifically near the active site in both enzymes. In our talk, we will present these results and comment on their implications for designing enzymes with a broader range of stability and activity.

464675 Self-Assembled Peptides with RGD Motifs As Scaffolds for Tissue Engineering

Abstract:

Self-assembled peptides gain increasing interest as biocompatible and biodegradable scaffolds for tissue engineering (1). Rationally designed self-assembling building blocks that carry cell attachment motifs such as RGD are especially attractive. We have been using a combination of theoretical and experimental approaches towards such rational designs (2-5). We have been especially focusing on modular designs that consist of a central ultrashort amphiphilic motif derived from the adenovirus fiber shaft (2-4). This central motif is combined with the RGD motif and cysteine residues that allow further functionalization possibilities, such as conjugation of growth factors or attachment to surfaces. We employed replica exchange MD simulations and free energy calculations using CHARMM (6) to computationally investigate the self-assembly properties of possible designed peptide sequences. Our simulations provide insights into the amyloidogenic Î²-sheet rich self-assembling conformations of the peptides. We performed a detailed analysis on the highly-ordered and well- aligned Î²-sheet containing conformations and provided insights into the structural arrangement of the peptidesâ€™ functional groups. The designer peptides experimentally self-assemble into fibers that are structurally characterized with Transmission Electron Microscopy, Scanning Electron Microscopy and X-ray fiber diffraction. Furthermore, they support cell attachment and proliferation of model cell lines. Such short self-assembling peptides that are amenable to computational design offer open-ended possibilities towards multifunctional tissue engineering scaffolds of the future.

References

1. Loo Y, Goktas M, Tekinay AB, Guler MO, Hauser CAE, Mitraki A. (2015) Self-assembled proteins and peptides as scaffolds for tissue regeneration. Advanced Healthcare Materials 16: 2557-86.

2. Tamamis, P., Kasotakis, E., Mitraki, A., and Archontis, G. (2009) Amyloid-like self-assembly of peptide sequences from the adenovirus fiber shaft: insights from molecular dynamics simulations. J. Phys. Chem. B. 113: 15639-15647.

3. Tamamis, P., Kasotakis, E., Archontis, G., Mitraki, A. (2014) Combination of Theoretical and Experimental Approaches for the Design and Study of Fibril-forming Peptides. In Protein Design: Methods and Applications. Methods Mol. Biol. 1216: 53-70.

4. Tamamis, P., Archontis, G. (2011) Amyloid-like Self-Assembly of a Dodecapeptide Sequence from the Adenovirus Fiber Shaft: Perspectives from Molecular Dynamics Simulations. Journal of Non-Crystalline Solids, 357 (2): 717-722.

5. Tamamis, P., Terzaki, K., Kassinopoulos, M., Mastrogiannis, L., Mossou, E., Forsyth, V. T., Mitchell, E. P., Mitraki, A., Archontis, G. (2014) Self-Assembly of an Aspartate-Rich Sequence from the Adenovirus Fiber Shaft: Insights from Molecular Dynamics Simulations and Experiments. Journal of Physical Chemistry B, 118 (7): 1765-1774.

6. B.R. Brooks, C.L. Brooks, III, A.D. MacKerell, Jr., L. Nilsson, R.J. Petrella, B. Roux, Y. Won, G. Archontis, C. Bartels, S. Boresch, A. Caflisch, L. Caves, Q. Cui, A.R. Dinner, M. Feig, S. Fischer, J. Gao, M. Hodoscek, W. Im, K. Kuczera, T. Lazaridis, J. Ma, V. Ovchinnikov, E. Paci, R.W. Pastor, C.B. Post, J.Z. Pu, M. Schaefer, B. Tidor, R. M. Venable, H. L. Woodcock, X. Wu, W. Yang, D.M. York, and M. Karplus. (2009) CHARMM: The Biomolecular Simulation Program J Comput Chem. 30(10): 1545â€“1614.

472328 Membrane Guided Self-Assembly of Dimers and Trimers of Tight Junction Proteins

Abstract:

Tight junctions (TJ) constituted by the claudin family of transmembrane proteins determine solute permeability in the paracellular space of various tissues. Multimeric aggregation of the claudin tight junction strands is difficult to probe using experimental methods. However, understanding how the membrane environment guides the claudin-claudin interactions into TJ assembly has far-reaching significance to a range of biochemical, mechanistic and physiological processes. In this study, we have employed molecular simulations to probe the membrane-driven self-assembly of *cis* interactions in seven prototypic classic claudins. Hydrophobic mismatch between the transmembrane domains and the bulk lipid medium initiates polymerization that leads to dimeric, trimeric, and higher order assemblies. There are both conformational and morphological differences between individual claudins. Specifically in the case of claudin-2, we observed dimeric conformations that are potentially the cation channel forming dimer, which is important for cation absorption in the kidneys. Furthermore, trimeric assemblies were observed in claudin-2, -4 and -19 that match with the receptor configurations for trimeric *Clostridium perfringens* enterotoxin (CPE), revealing the potential mechanism for pathogenesis. These important findings observed in the long-timescale self-assembly simulations provide unprecedented molecular level clues about the tight junction architecture.

459217 Insights into Self-Assembly Mechanisms of Intrinsically Disordered Proteins from Atomistic Simulations

Abstract:

Intrinsically disordered proteins (IDPs), which lack well-defined stable structures yet still perform key biological functions, comprise more than one-third of the eukaryotic proteome. Well-ordered fibril structures or aggregation of many IDPs are associated with the pathology of a number of debilitating human diseases including Alzheimer's, Parkinson's, Huntington's, amyotrophic lateral sclerosis, type II diabetes, etc. Moreover, certain types of low sequence-complexity IDPs form non-membrane bound intracellular assemblages that co-exist with the cytoplasm showing liquid-like properties, i.e. deform under shear, fuse with one another and take a spherical shape. In this work, we study IDPs, such as TAR DNA-binding protein 43 (TDP-43) or FUsed in Sarcoma (FUS) or Islet Amyloid PolyPeptide (IAPP), forming which are the major constituents of membrane-free assemblages and/or well-ordered pathological aggregates. Structural properties of these IDPs from molecular simulations shed light on intra- and inter-molecular interactions leading self-assembly/aggregation of these proteins.

470463 How Macromolecular Solvation Influences Its Conformations

Abstract:

Under ambient conditions, globular proteins sacrifice their conformational entropy, and fold into well-defined â €œnativeâ€ states. They do so because folding leads to favorable intra-molecular interactions, and importantly, to more favorable (or less unfavorable) interactions with the solvent. Consequently, proteins can be denatured under a variety of conditions, by either making the folded protein-solvent interactions less favorable, or unfolded protein-solvent interactions more favorable. Such conditions include lower temperatures, higher pressures, proximity to hydrophobic interfaces, and the presence of co-solutes that act as denaturants. Using hydrophobic polymers as model globular proteins, here we study how the hydration environment of a polymer influences its collapse. We find that near ambient conditions, typical hydrophobic macromolecules are situated at the edge of a collective collapse transition, making them sensitive to perturbations such as the application of pressure or the presence of co-solutes.

462113 Viscous Solvents Enable a Model Prebiotic Nucleic Acid Replication Cycle

Abstract:

Increased control of molecular self-assembly is important for the further development of nucleic acid-based technologies. While nucleic acid technology research has been carried out almost exclusively in aqueous buffer, alternative solvents could increase our ability to direct nucleic acid assembly. Deep eutectic solvents (DES), closely related to ionic liquids, are characterized by significantly lower melting points than their separate components. Recently, DES have been shown to preserve nucleic acid base pairing and duplex formation, offering a relatively unexplored medium for directing nucleic acid assembly.

In this work, we show that DES provide a highly effective means for tuning nucleic acid thermodynamics and kinetics. The power of alternative solvents to alter a nucleic acid assembly pathway is illustrated by our ability to circumvent the â€œstrand inhibition problemâ€ by thermally cycling nucleic acids in a DES. Strand inhibition describes the propensity of a long nucleic acid polymer to form a duplex with its complementary strand rather than to serve as a template for assembly of oligonucleotides. This longstanding problem in the field of prebiotic chemistry has limited demonstration of a prebiotic nucleic acid replication mechanism.

We address the problem of nucleic acid strand inhibition by employing highly viscous solvents to control the annealing rates of long templates and shorter oligonucleotides. In a viscous environment, nucleic acid mobility is slowed in a length-dependent manner, so that long nucleic acid polymers (such as template and copy strands) are kinetically trapped. Meanwhile, short oligonucleotides can diffuse quickly and bind to their complementary targets on the templates. Template duplex formation is further slowed by the formation of intramolecular secondary structure on the template single strands. Based on these thermodynamic and kinetic effects, we show that thermal cycling in a viscous environment can be used to copy a gene-length region (over 300 nucleotides) of a longer mixed sequence template, using both DNA and RNA systems.

468232 Tension-Dependent Free Energies of Nucleosome Unwrapping

Abstract:

Nucleosomes form the basic unit of compaction within eukaryotic genomes and their locations represent an important, yet poorly understood, mechanism of genetic regulation. Quantifying the strength of interactions within the nucleosome is a central problem in biophysics and is critical to understanding how nucleosome positions influence gene expression. By comparing to single-molecule experiments, we demonstrate that a coarse-grained molecular model of the nucleosome can reproduce key aspects of nucleosome unwrapping. Using detailed simulations of DNA and histone proteins, we calculate the tension-dependent free energy surface corresponding to the unwrapping process. The model reproduces quantitatively the forces required to unwrap the nucleosome, and reveals the role played by electrostatic interactions during this process. We then demonstrate that histone modifications and DNA sequence can have significant effects on the energies of nucleosome formation. Most notably, we show that histone tails are crucial for stabilizing the outer turn of nucleosomal DNA.

467109 An Optimization-Based Approach for the Design of Self-Assembled DNA Tiles

Abstract:

In structural DNA nanotechnology, DNA strands exceed their traditional role in gene expression and are being used as programmable building blocks for nanoscale structures. This self-assembly is driven by hybridization following the Watson-Crick rule.[1-3] Depending on the design of the DNA strands, structures with a rich set of nanoscale features can be self-assembled. A special class of DNA structures are the so-called DNA tiles,[4] which are self-assembled from a relatively small number of DNA strands. Experimental methods to self-assemble DNA tiles for a given design of DNA strands are well developed. However, a challenge is to find the optimal design of the individual DNA strands to self-assemble a desired DNA tile.

In general, the design of DNA strands for self-assembly of tiles involves two steps, 1) structural design and, 2) sequence generation. For structural design, the aim is to select optimal locations for so-called crossovers, which involves bridging of two DNA double helices (see Figure 1). Each crossover connects two double helices by a pair of shared single strands DNA.[1,3] For sequence generation, an optimal sequence of nucleotides for each DNA strand needs to be selected to enable the formation of the selected crossovers. Generally speaking, efficient algorithms for sequence generation exist. However, for the structural design, existing methods typically only provide a modeling framework for prediction and visualization of a stable structure for a given structural design (i.e., for solving the 'forward' problem).[6-9] In contrast, finding a structural design that allows for self-assembly of a desired final structure with high stability (i.e., solving the 'inverse' problem) remains challenging.

The objective of this work is to investigate an optimization-based approach for the optimal structural design of self-assembled DNA tiles. A semi-empirical model from literature [8] is used to predict the potential energy of a self-assembled DNA tile.

First, minimization of the potential energy for a given design (i.e., locations of crossovers) is studied for three different types of tiles.[5,10,11] The optimization problem was implemented in Matlab (2014a, Mathworks, Inc.) using the solver *fmincon* and in General Algebraic Modeling System (GAMS) Release 24.4.6. (GAMS Development Corporation, Washington, DC, USA, 2015) using different non-linear programming (NLP) solvers. Figure 2 illustrates the structural configurations of various self-assembled DNA tiles that were found after minimization of the potential energy. Different local minima were identified depending on the randomly chosen initial guesses. In all cases, the structure with the lowest potential energy corresponded to the intended structure reported in literature. Furthermore, in all cases, a significant part of the initial guesses converged to local minima. The structural configurations corresponding to those local minima could be significantly different from the intended structure (Figure 1), which illustrates the practical importance of avoiding local. Both a multi-start optimization method and deterministic global optimization using GAMS/BARON were investigated to identify the global minimum.

Second, self-assembly of the DNA tensegrity triangle [5,12](Figure 1) is chosen as the case to investigate model-based optimization of the structural design. This tile has been synthesized experimentally[5] and has practical relevance as template for protein crystallization. The optimization problem aims to minimize the potential energy of the tile, to maximize stability, using the orientation and position of the helices and the locations of the crossovers as degrees of freedom. Since the length of the DNA strands is typically short, decisions regarding the locations of crossovers are discrete decisions (represented by integer variables). Therefore, the optimization problem is formulated as a mixed-integer nonlinear programming (MINLP) problem. The optimization problem was implemented in GAMS using the solver SBB, which is based on branch-and-bound. The solution obtained with the MINLP formulation was compared to the solution obtained by exhaustive enumeration of the integer variables and solving an NLP problem at each node. The lowest potential energy could be obtained by using 27 nucleotides between each crossover for an equilateral triangle design, which matches experimental findings[5] and was also found using branch-and-bound with a 10-fold smaller CPU time compared to exhaustive enumeration. Finally, the design problem was also solved with the number of nucleotides between crossovers on each side of the triangle as a decision variable. The proposed MINLP optimization method was again able to identify the same mixed-integer solution compared to exhaustive enumeration at a much smaller CPU time (200-fold), which demonstrates the strength of a branch-and-bound algorithm for structural design of self-assembled DNA tiles.

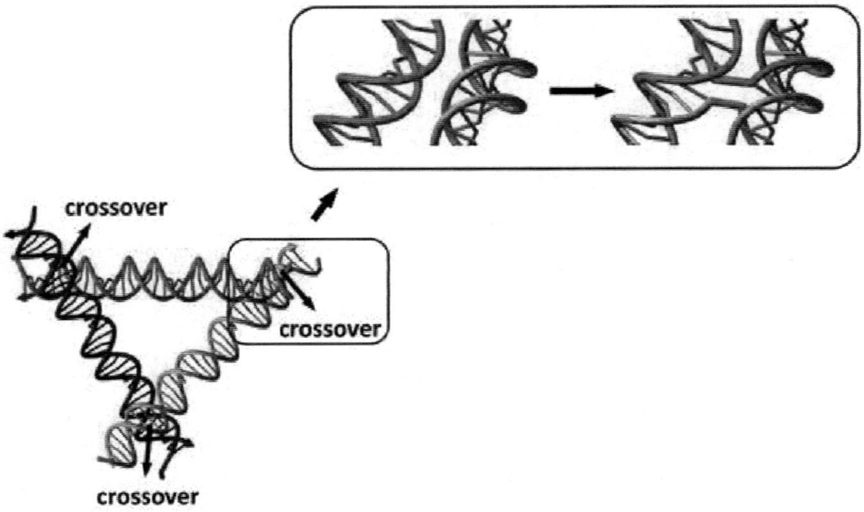

Figure 1: An example of a self-assembled DNA tile (tensegrity triangle) and detailed view of the structure of crossovers that interlink DNA helices.

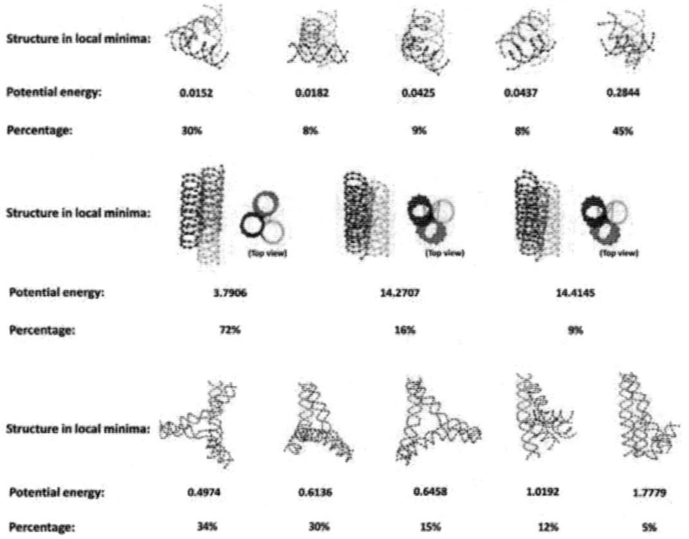

Figure 2: Identified local minima when minimizing the potential-energy of three different types of DNA tiles with fixed locations of crossovers. The percentage of randomly chosen initial guesses that led to each structure is indicated below the potential energy of the structure

Reference

[1] Seeman, N. C. An overview of structural DNA nanotechnology. *Mol. Biotechnol.* **37,**246-257 (2007).

[2] Seeman, N. C. DNA in a material world. *Nature* **421,**427-431 (2003).

[3] Seeman, N. C. DNA nanotechnology: novel DNA constructions. *Annu. Rev. Biophys. Biomol. Struct.* **27,**225-248 (1998).

[4] Park, S. H. *et al.* Programmable DNA self-assemblies for nanoscale organization of ligands and proteins. *Nano Lett.* **5,**729-733 (2005).

[5] Zheng, J. *et al.* From molecular to macroscopic via the rational design of a self-assembled 3D DNA

crystal. *Nature* **461,**74-77 (2009).

[6] Birac, J. J., Sherman, W. B., Kopatsch, J., Constantinou, P. E. & Seeman, N. C. Architecture with GIDEON, a program for design in structural DNA nanotechnology. *J. Mol. Graph. Model.* **25,**470-480 (2006).

[7] Castro, C. E. *et al.* A primer to scaffolded DNA origami. *Nat. Methods* **8,**221-229 (2011).

[8] Zhu, J., Wei, B., Yuan, Y. & Mi, Y. UNIQUIMER 3D, a software system for structural DNA nanotechnology design, analysis and evaluation. *Nucleic Acids Res.* **37,**2164-2175 (2009).

[9] Pan, K. *et al.* Lattice-free prediction of three-dimensional structure of programmed DNA assemblies. *Nat. Commun.* **5,**(2014).

[10] Wei, B. & Mi, Y. A new triple crossover triangle (TXT) motif for DNA self-assembly. *Biomacromolecules* **6,**2528-2532 (2005).

[11] He, Y. *et al.* Self-Assembly of Hexagonal DNA Two-Dimensional (2D) Arrays. *J. Am. Chem. Soc.* 12202-12203 (2005).

[12] Liu, D., Wang, M., Deng, Z., Walulu, R. & Mao, C. Tensegrity: construction of rigid DNA triangles with flexible four-arm DNA junctions. *J. Am. Chem. Soc.* **126,** 2324-2325 (2004).

472842 Linear Independence and Optimality Concepts Revisited in Thermodynamics and Kinetics

Abstract:

In this work, the concept of reaction linear independence is placed in a rigorous mathematical framework. It is shown that different conclusions may be reached regarding reaction independence depending on whether such independence is assessed in an equilibrium sense or in a kinetic sense. Mathematical techniques are proposed for the identification of independent reactions at equilibrium conditions. In the process of developing these results, necessary conditions for optimality are developed to assess reaction and phase equilibrium using a Gibbs Free Energy Minimization formulation. Case studies regarding natural gas reforming based hydrogen production utilizing a Gibbs equilibrium reaction model are presented. These case studies are used to quantify the carbon formation boundary of the reaction.

456846 Designing an Extraterrestrial Submarine for Titan. I. Nitrogen Solubility Modeling in the Titan Hydrocarbon Seas

Abstract:

An extraterrestrial submarine is currently being designed to explore the depths of the hydrocarbon rich seas on the Saturn moon Titan under the NASA Innovative Advanced Concepts phase 2 study. Cassini flyby data compiled over the past decade indicates that Titan's northern pole sustains stable seas of variable concentrations of ethane, methane, and nitrogen at a surface temperature of 93K and 152 kPa pressure. Meanwhile the atmosphere near the seas is primarily gaseous nitrogen, with small amounts of gaseous methane. The average density of the Titan seas is half that of Earth while Titan gravity is 1/6 that of Earth. Other than Earth, Titan is the only other known body in the solar system that sustains stable, accessible seas. The driving force for exploration of Titan is both geological and astrobiological in nature, to determine if hydrocarbon based life is possible on Titan and to improve our understanding of the history and evolution of hydrocarbons in the solar system.

A one year exploration mission is baselined to study the seas of Titan. Multiple vehicle concepts were studied, but a submarine proved to be the most robust option. To meet science exploration objectives, the submarine must operate autonomously, study atmosphere/sea exchange, interact with the seabed at pressures up to 10 atm, traverse large distances, hover at the surface and at any depth within the seas, and be capable of tolerating wide variations in Titan properties. Results of the Phase 1 study are available in the literature [1], along with detailed information on the trade studies and preliminary design of the power, thermal, and ballast control subsystems for the Saturn Titan submarine [2]. In short, a 200kg, 1000W Stirling Radioisotope Generator system was baselined for power, yielding 3800 W waste heat into the seas. Aerogel insulation, a pump coolant loop system, and cold plates were baselined for thermal control to maintain internal science and communication electronics at ambient temperature. A high pressure gaseous neon bottle and pressure-fed system with mechanical separation was used to control the amount of liquid inside the ballast tanks to control the vertical ascent and descent of the sub. Mechanical separation was required to avoid solubility issues of neon dissolving in the ethane/methane seas and to avoid impurities when recovering the pressurant.

Critical to the design and operation of the submarine is knowledge of the amount of dissolved nitrogen gas within the cryogenic liquid ethane and methane seas. Solubility is defined as a property of a solute to dissolve in a particular solvent. Solubility is a special case of vapor/liquid equilibrium (VLE) in a subcritical mixture. Estimates taken during the initial trade studies showed that GN_2 had a relatively high solubility of 4% in pure liquid ethane at the surface; solubility of air in water is negligible (<0.1% at 1.03 MPa) during terrestrial submarine operation. The solubility of GN_2 in liquid methane is expected to be even higher. It is to be determined if heat leak rates from the power source into the liquid are high enough to cause effervescence. However, dissolved nitrogen gas that comes out of solution will cause several problems for the submarine, in particular, there are three concerns:

1. In a quiescent case, GN_2 bubbles could interfere with sensitive science measurements while the submarine is hovering.
2. In a moving case, GN_2 bubbles that form along the craft could coalesce at the aft end of the submarine and cause cavitation in the propellers.
3. GN_2 that come out of solution in the liquid inside the ballast tanks could hinder pressure control and thus affect vertical control of the submarine. Accurate solubility models are therefore needed to mitigate these concerns.

This presentation presents brief details on the Phase 1 design of the submarine, along with thermal design concerns to be addressed in the Phase 2 design. Then, a review of the literature is presented to gather relevant VLE data for binary nitrogen/methane and nitrogen/ethane mixtures. It is shown that there is a large volume of VLE data for nitrogen/methane and nitrogen/ethane mixtures at higher pressures and temperatures typical of those incurred during transport and refinement of liquefied natural gas here on Earth. However, VLE data for nitrogen/methane/ethane mixtures are poorly represented in the literature, especially near Titan sea conditions near the freezing point. Analysis of the data does in fact confirm that nitrogen solubility is higher in liquid methane, and that solubility increases with both increasing pressure and decreasing temperature. For pure liquid methane, at 152 kPa and 93K, the solubility of GN_2 in liquid methane is about 17.5%.

This work was funded through the NASA Innovative Advanced Concepts (NIAC) Phase 2 Titan Submarine Project.

References

[1] Oleson, S., Lorenz, R., and Paul, M. "Phase I Final Report: Titan Submarine" *NASA-TM-2015-218831* July, 2015.

[2] Hartwig, J.W., Colozza, A., Lorenz, R.D., Oleson, S., Paul, M., and Walsh, J. "Exploring the Depths of Kraken Mare – Power, Thermal Analysis, and Ballast Control for the Saturn Titan Submarine" *Cryogenics* 74, 31 – 46. 2016.

468106 Testing DFT Predictions at Finite Temperatures: Anomalous Phase Behavior and Long-Range Superstructures of CoPt

Abstract:

High-throughput ab-initio studies of phase stability in multi-component alloy systems have garnered a great deal of research interest in the past decade. However, significantly less attention has been paid to the sometimes startling results of these studies, such as the implied wealth of undiscovered stable phases present in many 3*d*-5*d* transition metal binaries. This shortcoming is especially well-demonstrated in the case of Co-Pt, which has a long experimental history demonstrating three stable mixed phases: $L1_0$ CoPt and $L1_2$ Co_3Pt and $CoPt_3$; in contrast, results from density functional theory suggest a set of (yet-unobserved) long-period commensurate phases resembling the β_2 superstructure are present at Pt-rich compositions. We have performed an in-depth ab-initio study of the Co-Pt binary, and fit our results to a cluster expansion Hamiltonian appropriate for use in Monte Carlo simulations. Using semi-grand canonical Monte Carlo, we demonstrate that density functional theory implies high-temperature phase behavior wholly incompatible with experimental results[1]. Our case study demonstrates that effective Hamiltonian and statistical mechanic techniques provide a logical "next step" in ab-initio studies where simulation and experiment disagree.

[1] E. Decolvenaere, M.J. Gordon, A. Van der Ven, Phys. Rev. B 92, 085119 (2015)

457265 Demonstration of the Equivalence of Ring-Additivity and Group-Contribution Methods for Properties of Unsubstituted Polycyclic Aromatic Hydrocarbons (PAH)

Abstract:

Abstract:

In the present work, a new ring-additivity method for the estimation of pure-component properties of unsubstituted benzenoid PAH (polycyclic aromatic hydrocarbons containing only 6-membered rings, also called PAH6) is shown to be equivalent to all other known group-additivity/group-contribution estimation methods against which direct comparisons can be made.

There are only 12 ring groups for PAH6. The original notation of Gutman and Cyvin [1] uses the letter L, A, or P followed by the number of rings with which it is fused (from 1 to 6). For the cases of being fused with 1, 5, and 6 rings, there is no ambiguity, so they are only labeled with L; for 2, 3, or 4 rings, there are three possible configurations. Their simplified numbering scheme [2] only uses the numbers 1 through 12. In previous work [3], a nomenclature was developed using H to denote a 6-sided central ring, followed by the number of rings fused to it; for the cases of 2-4 rings fused to the central ring, this is followed by a dash then a number indicating the location of the rings. The notation is designed to facilitate computer implementation of the ring additivity method.

There is are a great many group-contribution (GC) methods for determining pure-component properties. The majority of those for critical properties only use two groups ACH and AC (or an atom-additivity method which can be converted to these groups, such as in Wilson and Jasperson), such as those of Joback, Forman and Thodos, Somayajulu, Constantinou and Gani, and Nannoolal et al. [3,4]; some of these also have contributions reflecting the existence, size, and type of rings. For a PAH6 C_xH_y, # ACH groups = y,

#AC groups = x-y. The number of rings n = 1 + (#AC/2).

Avaullée et al. [5] split AC groups into those at the junction of 2 rings (on the periphery = ACp) or 3 rings (in the interior = ACi). The relations are #ACp = y-6 and #ACi = x-2y+6 [6]. An additional group (M1) reduces to n (= number of rings). The remaining relevant structural group (M2) can only be obtained by considering the structure of the entire PAH molecule. Marrero and Gani [7] use four third-order AROMFUSED group contributions, which include 5 of the possible 12 PAH6 ring groups. The present approach subsumes their first-order ACH and AC groups directly into the ring groups, combining both the first- and third-order group contributions.

Each of the 12 ring groups is the following sum of the contributions of the ACH, ACp, and ACi groups: #ACH + #ACp/2 + #ACi/3, plus additional groups specific to the estimation technique. Each of the 12 groups then is the equivalent to anywhere from 2 to 5 ACH and AC groups. In this manner, the ring groups are directly equivalent to all of the above-cited GC methods.

The sets of groups which have been developed to estimate thermochemical properties of PAH (ΔH, S, Cp) are significantly more detailed than those in the methods used above. The method of Benson (exemplified here by Stein et al. [8]) splits the AC groups into three groups based on the number of AC groups to which they are bound, which ranges from one to three: here referred to as AC1, AC2, and AC3, respectively. Herndon et al. [9] further divide AC3 into peripheral and internal groups, here called AC3p and AC3i, respectively. (AC1 and AC2 are both peripheral AC groups.) Moiseeva et al. [10] further divide the AC3i groups into those which are connected to either 0 or 1 ACp groups (AC3i1) or 2 or 3 ACp groups (AC3i2). Steele et al. [11] (as described by Cohen [12]) suggested that the ACH group be split into three subgroups based on the number of AC groups to which it is bound (0, 1, or 2). While this level of refinement of the ACH groups has never been implemented, the PAH6 ring groups can easily accommodate this revision.

For the PAH6 ring groups to accommodate any of the additional levels of complexity in splitting the AC groups into groups besides ACp and ACi, the types and locations of rings fused to a given ring group must be considered. Otherwise, some of the atom locations have ambiguous as group assignments as follows: (1) AC1 or AC2, (2) AC2 or AC3p, or (3) AC3i1 or AC3i2. This refinement has been done with little increase in computational difficulty, again leading to a a one-to-one correspondence between the new ring-additivity method and these more detailed group-contribution methods.

On the surface, it might seem to be adding an unnecessary level of complexity to group contribution methods by switching to groups consisting of entire rings, especially given the decades-long use of the approaches for PAH6. However, GC methods have had limited success in predicting properties for PAH containing both 5- and 6-membered rings (PAH5/6), especially for those with curvature induced by internal 5-membered rings [13,14].

A ring-additivity method for PAH5/6 (those containing both 5- and 6-membered rings) and fullerenes (closed-cage all-carbon molecules containing both 5- and 6-membered rings) might prove more successful in predicting the properties of those molecules than previous GC methods, since the ring groups would include much more information about the types of curvature produced by the proximity of atoms at the junction of two 6-membered rings and one 5-membered ring.

However, even if only those PAH5/6 and fullerene structures are considered which obey the "isolated pentagon rule" (IPR), 50 new ring-additivity groups will need to be added to the 12 PAH6 ring groups to describe these compounds [3]. It was therefore thought prudent to first explore the feasibility of using the simpler PAH6 ring additivity method. The current work demonstrates how theoretically sound and readily applicable this method is to benzenoid PAH. Even though the equivalence between the new PAH6 ring additivity method has been demonstrated, further testing is in progress to see if new insights can be gained by its implementation. Development of the PAH5/6 ring additivity method is also underway.

Furthermore, the current PAH6 ring groups can serve as an excellent starting point for developing groups for substituted PAH6, since the rings will carry information about the bonding environment at the location of the substituents. Further work along these lines is warranted.

References:

[1] Gutman, I.; Cyvin, S.J. (1989) "Introduction to the Theory of Benzenoid Hydrocarbons", Springer-Verlag, New York.

[2] Gutman, I.; Furtula, B.; Radenković, S. (2004) "Relation between Pauling and Coulson Bond Orders in Benzenoid Hydrocarbons, *Z.Naturforsch.*, **59a**, 699-704.

[3] Pope, C. (2015) "Estimation of Diffusion Coefficients of Polycyclic Aromatic Hydrocarbons (PAH) and Fullerenes", poster presented at the American Institute of Chemical Engineers Annual Meeting, November 2015, Salt Lake City, UT. See also references cited therein.

[4] Pope, C.J. "Estimation of Normal Boiling Point, Critical Properties, and Lennard-Jones ---Parameters for Polycyclic Aromatic Hydrocarbons and Fullerenes", poster presented at the American Institute of Chemical Engineers Annual Meeting, November 2013, San Francisco, CA. See also references cited therein.

[5] Avaullée, L.; Trassy, L; Neau, E.; Jaubert, J.N. (1997) "Thermodynamic modeling for petroleum fluids I. Equation of state and group contribution for the estimation of thermodynamic parameters of heavy hydrocarbons", *Fluid Phase Equil.*, **139**:155-170.

[6] Dias, J.R. (1987) "Handbook of Polycyclic Hydrocarbons Part A: Benzenoid Hydrocarbons", Elsevier, New York, pp. 122-123.

[7] Marrero, J.; Gani, R. (2001) "Group-contribution based estimation of pure component properties", *Fluid Phase Equil.*, **183–184**:183–208.

[8] Stein, S.E.; Golden, D.M.; Benson, S.W. (1977) "Predictive Scheme for Thermochemical Properties of Polycyclic Aromatic Hydrocarbons", *J.Phys.Chem.*, **81**:314-317.

[9] Herndon, W. C.; Nowak, P. C.; Connor, D. A.; Lin, P. (1992) "Empirical Model Calculations for Thermodynamic and Structural Properties of Condensed Polycyclic Aromatic Hydrocarbons", *J.Am.Chem.Soc.*, **114**:41–47.

[10] Moiseeva, N.F.; Dorofeeva, O.V.; Jorish, V.S. (1989) "Development of Benson Group Additivity Method for Estimation of Ideal Gas Thermodynamic Properties of Polycyclic Aromatic Hydrocarbons", *Thermochim.Acta*, **153**:77–85.

[11] Steele, W.V.; Chirico, R.D.; Nguyen,A.; Hossenlopp, I.A.; Smith, N.K. (1990) "Determination of Ideal-Gas Enthalpies of Formation for Key Compounds", *AIChE Symp.Ser.*, **86**:138–154.

[12] Cohen, N. (1996) "Revised Group Additivity Values for Enthalpies of Formation (at 298 K) of Carbon-Hydrogen and Carbon-Hydrogen-Oxygen Compounds", *J.Phys.Chem.Ref.Data*, **25**:1411–1481.

[13] Pope, C.J.; Howard, J.B. (1995) "Thermochemical Properties of Curved PAH and Fullerenes: A Group Additivity Method Compared with MM3(92) and MOPAC Predictions", *J.Phys.Chem.*, **99**:4306–4316.

[14] Allison, T.C.; Burgess, D.R., Jr. (2015) "First-Principles Prediction of Enthalpies of Formation for Polycyclic Aromatic Hydrocarbons and Derivatives", *J.Phys.Chem.A*, **119**:11329-11365.

465884 Experimental Measurements and Predictions of the Properties of Refrigerant Fluids Using Ab-Initio, SAFT-VR Mie and COSMO Calculations

Abstract:

On account of the constraints imposed by the European and International legislations, the refrigerant industry must constantly find alternative refrigerant fluids that have lower impacts on the global warming of Earth and Ozone layer. The hydrofluoroolefins (HFO) have been proposed to replace the currently used hydro fluorocarbons (HFC), as they have a much lower global warming potential. In order to select the optimal mixture composition for the design and operation of a refrigeration process, it is necessary to know the phase diagram and thermodynamic properties of the mixtures. In particular, the vapor-liquid equilibria (VLE) and the location of azeotropes must be accurately known. Besides, the design of chemical processes and heat engines requires a good knowledge of both the ideal gas and the residual thermodynamic properties of the refrigerants. The aim of this work is to propose various thermodynamic approaches to predict these properties.

First, we performed ab initio calculations on isolated molecules to determine the ideal gas properties of refrigerant fluids. Ideal enthalpies of formation and heat capacities can be accurately predicted by using either DFT [1] or some specific semi-empirical methods. We then used several activity coefficient models based on the COSMO approach, to predict the VLE of refrigerant mixtures. VLE and azeotropes can be well predicted by the COSMO-RS model [2]. While the first versions of COSMO-SAC are very inaccurate for such systems, it is possible to obtain good predictions with the COSMO-SAC-dsp model [3] by readjusting some universal parameters. New experimental measurements of vapor-liquid equilibria on mixtures involving HFO have been performed with a static analytic equipment, to complete the literature data for mixtures of HFO with other refrigerants.

We then propose a modified version of the SAFT-VR-Mie [4] equation of state (EoS) that includes a polar contribution to explicitly take electrostatic interaction into account. The polar SAFT-VR-Mie EoS is used to describe the VLE and some derivative thermodynamic properties of refrigerant mixtures. A new association model that takes the formation of cyclic clusters is proposed for hydrogen fluoride (HF), which is one major compound of the refrigerant production process. The new models can be used in process simulation software, thanks to the cape-open and Simulis Thermodynamics® interfaces.

References

[1] A. Osmont et al., Combustion and Flame, 151, 262 (2007)

[2] F. Eckert, A. Klamt, Fluid Phase Equilib., 172, 73 (2010)

[3] C.-H. Hsieh et al., Fluid Phase Equilib., 367, 709 (2014)

[4] T. Lafitte et al., J. Chem. Phys., 139,154504 (2013).

Acknowledgements

The results presented in this work are part of the ongoing project "PREDIREF". We are grateful to the National French Research Agency (ANR-13-CDII-0008) for providing financial support to this project.

471474 Exploring the Bulk Phase Properties and Structural Properties for CO_2 and so$_2$ Via First Principles Calculations

Abstract:

To capture the underlying chemistry and physics of a system, it is necessary to accurately describe the intermolecular interactions such as repulsion, polarization, hydrogen bonding, and van der Waals interactions. Among these interactions, van der Waals (dispersion) interactions are weak in nature as compared to covalent bonds and hydrogen bonding, but govern thermophysical properties of condensed phase systems. CO_2 and SO_2 molecules have numerous industrial applications and predicting their condensed phase and microscopic properties by first principles calculations can be very helpful in the designing of the processes. The latest density functionals can predict with reasonable accuracy the minima and binding energy for dimers or small cluster of molecules, however, their performance for condensed phase is somewhat uncertain. The aim of the present work is to assess the accuracy of Kohn-Sham density functional theory (DFT) by using different functionals that can account for predicting the bulk phase properties and structural properties for CO_2 and SO_2 via first principles Monte Carlo simulations. In this work, we have investigated the accuracy of dispersion corrected GGA (BLYP, PBE), hybrid (PBE0) meta-hybrid GGA (MO6-2X) functional and nonlocal van der Waals functionals.

472216 First-Principles Phase Diagrams: Iron at Earth's Inner Core Conditions with Full Inclusion of Anharmonic and Finite-Size Effects

Abstract:

Phase diagrams provide an essential source of information about phase composition and stability at different thermodynamic states, which are crucial to know for many applications. However, its evaluation at extreme temperature and pressure conditions can be challenging for experimentalists. At such states, computational methods can work as a rigorous alternative. Since such assessments involve evaluation of finite temperature free energies, the computation cost can easily exceed the available resources. Therefore, advances are greatly needed to make such assessments possible with minimal computational effort.

Building the phase diagram of iron at conditions relevant to Earth's inner core (IC) is an example of such challenging and scientifically interesting problem, and a prototype for other applications. The IC is widely believed to be in a solid state composed, mainly, of iron [1] at state of about 350 GPa and 6000 K. However, the exact crystal structure (which is important to understand the complex seismic data observed in Earth's IC and hence understand how Earth was evolved) is not fully understood/determined to the date (both experimentally and theoretically).

*Ab initio*simulation methods (e.g. DFT), provide reliable and rigorous tools that can help predicting iron crystal structure at such conditions. Unfortunately, free energy calculations based on DFT are computationally expensive — especially with the large number of Fe electrons (around 16 electrons) that contribute significantly at such high pressure. Therefore, a wide range of levels of theories have been used to approximate the free energy; e.g., static energy [2], quasiharmonic approximation [3], and self-consistent phonon calculations [4].

In this work [5], we utilize a novel machinery developed by our group (called "harmonically-mapped averaging (HMA)" method [6,7]) to tackle this long standing problem using DFT method as implemented in VASP. The speed up gained by using HMA method to measure thermodynamic properties is at least an order of magnitude faster than "direct" methods. This enabled measuring of the full (harmonic and anharmonic) Gibbs free energy of Iron at several states. In addition, the finite-size effects in free energy were estimated based on a harmonic-based technique described here [8]. Applying these techniques to the three known iron candidate structures (hcp, fcc, and bcc), we were able, to the first time, to build a completely *ab initio*-based phase diagram of iron at Earth's IC conditions in the thermodynamic limit.

[1] F. Birch, *J. Geophys. Res.* **57**, 227 (1952).
[2] P. Soderlind, J. Moriarty, and J. Wills, *Phys. Rev. B* **53**, 14063 (1996).
[3] L. Stixrude, *Phys. Rev. Lett.* **108**, 055505 (2012).
[4] W. Luo, B. Johansson, O. Eriksson, S. Arapan, P. Souvatzis, M. I. Katsnelson, and R. Ahuja, *Proc. Natl. Acad. Sci.* **107**,9962 (2010).
[5] S. G. Moustafa, A. J. Schultz, E. Zurek, and D. A. Kofke, *In preparation*, (2016).
[6] S. G. Moustafa, A. J. Schultz, D. A. Kofke, *Phys. Rev. E* 2015, **92**, 043303.
[7] A. J. Schultz, S. G. Moustafa, L. Weisong, S. J. Weinstein, and D. A. Kofke, *J. Chem. Theory Comput.*, 2016, **12** (4), pp 1491-1498.
[8] T. B. Tan, A. J. Schultz, and D. A. Kofke, *J. Chem. Phys.* **133**, 134104 (2010).

463879 Resolving Non-Specific and Specific Adhesive Interactions of Catechols at Solid/Liquid Interfaces at the Molecular Scale

Abstract:

The adhesive system of mussels evolved into a powerful and adaptive system with affinity to a wide range of surfaces. It is widely known that thereby 3,4-dihydroxyphenylalanine (Dopa) plays a central role. However underlying binding energies and binding mechanims remain unknown at the single molecular scale. Here, we use single molecule force spectroscopy [1, 2] to estimate binding energies of single catechols with a large range of opposing chemical functionalities. Our data demonstrates significant interactions of Dopa with all functionalities, yet most interactions fall within the medium-strong range of 10-20 k_BT. Only bi-dentate binding to TiO_2 surfaces exhibits a higher binding energy of 29 k_BT. Our data also demonstrates at the single molecule level that oxidized Dopa and amines exhibit interaction energies in the range of covalent bonds, confirming the important role of Dopa for cross-linking in the bulk mussel adhesive. We anticipate that our approach and data will further advance the understanding of biologic and technologic adhesives.

[1] S. Raman, T. Utzig, T. Baimpos, B. R. Shrestha, M. Valtiner, *Nature Communications* **2014**, *5*, 7.

[2] T. Utzig, S. Raman, M. Valtiner, *Langmuir* **2015**, *31*, 2722-2729.

461173 Using Replica Exchange Molecular Dynamics Simulations to Examine Sequence Specificity in the Adsorption of Single-Stranded DNA (ssDNA) on Chiral Single-Walled Carbon Nanotubes (SWCNTs)

Abstract:

Single walled-carbon nanotubes (SWCNTs) possess unique properties that make them attractive for a number of applications. However, the difficulty in obtaining uniform samples in terms of chirality inhibits their widespread use. Various separation techniques address this problem through use of surfactants in order to suspend the tubes in aqueous media and then applying other approaches such as aqueous two phase extraction (ATPE) or ion-exchange chromatography (IEX) to sort the SWCNTs by their physiochemical properties. Single-stranded DNA (ssDNA) is a very effective dispersant that shows sequence-specific binding behavior. Previous experimental work has shown that this specificity can be harnessed to tune the separation in favor of particular nanotube chiralities. Currently, the nature of this specificity is not yet well understood and optimal ssDNA/SWCNT bindings must be searched by trial and error. In this study, we use replica exchange molecular dynamics (REMD) simulations to study and compare the binding of experimentally determined optimal and unfavorable binding pairs. We intend to compare the effect and interplay between a number of proposed binding factors including inter- and intra-strand interactions of the ssDNA, the ratio of Kuhn length to tube diameter, and the ssDNA equilibrium conformation relative to the chiral angle of the SWCNT. From these analyses our goal is to provide better insight into the sequence/chirality specific binding mechanism, and eventually to develop a model that allows for the prediction of other pairs leading to the efficient sorting of a desired nanotube chirality.

472061 Adsorption Characterization of Biomolecular Growth Modifiers of Calcium Oxalate Monohydrate Crystals

Abstract:

Controlling crystallization processes can facilitate the tailoring of materials properties by modulating growth, composition, and structure. Crystal growth modifiers range from small organic molecules to macromolecules such as polymers and proteins. In nature, ensembles of biomolecular modifiers are known to mediate biomineralization, the assembly of organic-inorganic materials (e.g. bones, teeth, exoskeletons). Adsorption of biomolecules at crystal interfaces can either hinder or promote the addition of growth units and subsequently produce unique crystal habits through enhanced anisotropic growth. Understanding the fundamental principles underlying such interactions can inform the design and control of crystal habits, and forms the central objective of this work. We exemplify the study of such systems using bovine serum albumin (BSA) and its constitutive peptides along with calcium oxalate monohydrate (COM) crystals as a model system. In our previous work we have demonstrated BSA to be an effective growth modifier of COM crystals. Here we extend that work to further investigate the thermodynamic and molecular aspects of BSA-COM interactions using COM particles for adsorption isotherm experiments. Using isothermal titration calorimetry (ITC), we extract enthalpic and entropic contributions to equilibrium free energy of BSA adsorption on COM. To understand the specific and putative contributions from fragments of solvated BSA interfaces, as well as buried hydrophobic fragments, we study short peptides derived from a BSA-Trypsin digest. COM binders are identified from the digest library and their efficacies as modifiers are evaluated at the bulk scale by quantifying their effects on crystal habit. The binding thermodynamics and affinities of the lead peptides are compared to the results obtained from the native BSA protein. A key outcome of this work will be to establish rational principles for designing peptides as crystal growth modifiers from known proteins that can be applied to other proteins and crystallization processes. This has implications in understanding the fundamentals of modifier-crystal interactions, as well as in tuning structure and property of both biogenic and synthetic materials for various applications.

457811 Single Particle Tracking in Cushioned, Blebbed Supported Lipid Bilayers Enables Studies of Transmembrane Protein Diffusion

Abstract:

Supported Lipid Bilayers (SLBâ€™s) are effective models for studying some biomembrane phenomena. A thin layer of water between the substrate and the bilayer engenders 2D fluidity and enables studies of lipid diffusion and peripheral membrane protein diffusion. However, the water layer is not thick enough to prevent friction between most transmembrane proteins and the substrate. Because of this, it is difficult to study the diffusive properties of proteins that protrude significantly from the membrane. Here, we study several related cushioning strategies that are easy to construct and support the mobility of most transmembrane proteins. All cushions make use of PEGylated lipids to lift the bilayer away from the substrate. The concentration and length of the PEGylated lipids can be varied to maximize mobility for a protein of choice. The PEGylated lipids can also be biotinylated to allow for a double cushion strategy. In this approach, streptavidin is first used to passivate the substrate and will form bonds with the PEGylated lipids to anchor them in place. The bilayer can be formed by vesicle fusion, allowing us to incorporate membrane proteins from cell blebs without using detergents or other artifactual methods. The efficacy of the different cushioning strategies was assessed through single particle tracking (SPT) of fluorescently tagged DHHC20, a ~40 kDa acyltransferase with 4 transmembrane domains. We will discuss the biological implications of these results and the applications of this platform to studying cytoskeleton-mediated confinement of plasma membrane components. We will also briefly discuss a complementary method for carrying out Brownian dynamics simulations to model protein diffusion. The results of these simulations will be used to put forth a model for the mechanism of cytoskeletal confinement based on hydrodynamic interactions with immobilized membrane proteins.

458093 Protein Release from Contact Lenses Monitored By Interfacial Viscoelasticity

Abstract:

Contact lenses are the most prescribed biomaterial with over 125 million wearers who are often plagued by deposition of proteins and lipids in the eye [1]. Protein adsorption, especially denatured protein, increases the risk of microbial cell attachment to the lens and inflammatory diseases such as giant papillary conjunctivitis; their effect on tear film stability have yet to be studied [2]. This study demonstrates the use of an optical platform to monitor drainage of fluid on contact lenses, interfacial rheology, and two-photon microscopy to analyze both the uptake and release of lysozyme in contact lenses. Lysozyme was chosen as the protein of interest here because of its high concentration in the tear film (20-30% of the total tear film protein), relatively small size, and high isoelectric point (11.4), making it positively charged in the tear film (pH 7.4) [2]. Our research aims to understand how lysozyme deposition and desorption affects interfacial properties of a fluid-air interface for conventional and silicone hydrogel contact lenses, which is of importance for a stable tear film.

We found that after exposure to lysozyme, a conventional hydrogel (1 Day Acuvue, Johnson & Johnson) and a silicone hydrogel (Acuvue Oasys, Johnson & Johnson) have markedly different thin film stability profiles, as evidenced by the critical film thickness for breakup in phosphate-buffered saline drainage experiments. This change in film stability can be attributed to the elevation in interfacial rheology during lysozyme desorption from lenses. While the bulk rheology did not significantly change during lysozyme desorption, and surface pressure measurements were not sensitive to protein network formation at the interface, interfacial rheology measurements were consistent with the thin film stability observed in experiments. In tandem with thin film drainage measurements, interfacial rheology can be used to determine the effects of other protein and lipid deposits on tear film stability and, fundamentally, the effects of individual tear film components on tear film stability.

References

[1] J. E. Key. *Eye Contact Lens*. **33**, 343 (2007).

[2] D. Luensmann, et al., *Cont Lens Anterior Eye*. **35**,53 (2012).

[3] L. N. Subbaraman, L. Jones. *J Biomed Sci Polym*. **21**, 343 (2010).

472273 Protein Self-Assembly at the Blood-Brain Barrier Tight Junction Interface

Abstract:

The macromolecular interface of the blood brain barrier (BBB) is constituted by the self-assembly of claudin-5 membrane proteins into tight junctions. Tight junctions which appear as an anatomizing network of strands between two endothelial cells, have semi-permeable molecular pores formed by claudin-5 adhesion interactions. These pores are impermeable to a vast majority of drugs including those that are vital for neurodegenerative diseases. Understanding the complex architecture of the tight junction molecular interface is critical to the design of novel targeted therapeutic agents targeting neurodegenerative diseases. Here we have elucidated the self-assembly of claudin-5 into tight junction strands and further investigated the influence of membrane micro-environment in driving claudin-5 assembly. We employed coarse grained molecular dynamics simulation to characterize the self-assembly of claudin-5 into the tight junctions. Altering the membrane composition especially with cholesterol and sphingolipids affects the orientation and consequently the conformations of claudin-5 interactions. Using umbrella sampling and steered molecular dynamics we assessed the potential mean of force of the dimeric interactions, which provided distinctions on the stability of each dimers. These molecular insights are vital to the advancement of tight junction research.

472461 Targeted Mutagenesis and Combinatorial Library Screening Enables Control of Protein Orientation on Surfaces and Increased Activity of Adsorbed Proteins

Abstract:

While, non-specific adsorption is widely used for immobilizing proteins on solid surfaces, the random nature of protein adsorption may reduce the activity of immobilized proteins due to occlusion of the active site. We hypothesized that the orientation a protein assumes on a given surface can be controlled by systematically introducing mutations into a region distant from its active site, thereby retaining activity of the immobilized protein. To test this hypothesis, we generated a combinatorial protein library by randomizing targeted residues in a binding protein derived from highly stable, non-immunoglobulin Sso7d scaffold; mutations were targeted in a region that is distant from the binding site. This library was screened to isolate binders that retain binding to its cognate target (chicken immunoglobulin Y, cIgY) as well as exhibit adsorption on unmodified silica. A single mutant â€“ Sso7d-2B5 â€“ was selected for further characterization. Sso7d-2B5 retained binding to cIgY with an apparent dissociation constant similar to that of the parent protein; both mutant and parent proteins saturated the surface of silica with similar densities. Strikingly, however, silica beads coated with Sso7d-2B5 could achieve up to seven-fold higher capture of cIgY than beads coated with the parent protein. These results strongly suggest that mutations introduced in Sso7d-2B5 alter its orientation relative to the parent protein, when adsorbed on silica surfaces. Our approach also provides a generalizable strategy for introducing mutations in proteins so as to improve their activity upon immobilization, and has direct relevance to development of protein-based biosensors and biocatalysts.

464301 Controlling Protein Unfolding at the Solution-Solid Interface By Modifying the Nanoscale Environment of the Surface

Abstract:

Many biotechnologies require proteins in their native state at solid-liquid interfaces (*e.g.* biocatalysis, biosensing, chromatography, biomaterials, *etcâ€¦*). Unfortunately proteins often adopt non-native conformations and reduced activity on surfaces. While this phenomena has been of great interest to numerous fields, and therefore the focus of many studies, the specific mechanisms of surface-mediated protein denaturation remain elusive. Traditional biophysical techniques are difficult to apply to interfacial systems due to limited signal from adsorbates relative to protein in the contacting solution. Furthermore, the information gathered from traditional approaches are difficult to interpret due to ensemble averaging of heterogeneous, dynamic systems. We studied the mechanism of lysozyme (T4L) unfolding on fused silica (FS) using single-molecule methods that provided direct insight into the cause of denaturation. The structure of T4L was monitored with FoÌˆrster resonance energy transfer (FRET) while simultaneously tracking the adsorption, diffusion, and desorption of individual molecules at the solidâˆ'solution interface. Single-molecule trajectories were collected for hundreds of thousands of adsorbed T4L molecules in a high throughput manner. Analysis of these trajectories suggested that the unfolding of T4L on FS was mediated by surface diffusion and occurred on isolated nanoscale sites. The frequency of these denaturing sites was controlled by changing the surface coverage, increasing protein concentration passivated denaturing sites. Since the addition of soluble factors is not always appropriate, we are investigating the effects of protein nanoscale environments in order to control the structure and diffusion of adsorbed protein. This includes nanoscale patterning of surface topography, which we propose as a method to hinder interfacial diffusion and stabilize the folded state of adsorbed protein through confinement.

473399 Role of the Air-Solution Interface on the Mechanism of Subvisible Particle Formation in an IgG1 Mab Solution

Abstract:

Mechanical agitation of monoclonal antibody (mAb) solutions often leads to protein particle formation. Using a Langmuir trough, controlled interfacial stresses were applied to various IgG1 mAb formulations, and the resulting protein particles formed at the interface and in the bulk solution were characterized using atomic force microscopy and flow digital imaging. Bulk protein particle formation after being subjected to interfacial stresses was also compared to mAb solutions agitated in glass vials and unstressed controls. mAb samples subjected to 750-1000 interfacial compression-expansion cycles in six hours, contained high particles numbers in bulk solution, and displayed similar particulation trends when agitated in vials. At compression rates of 50 cycles in six hours, however, particle levels in mAb solutions were comparable to unstressed controls, despite protein aggregates being present at the air-solution interface. Our results together suggest that while the air-solution interface serves as a nucleation site for initiating protein aggregation, the number of protein particles measured in bulk mAb solutions depend on the total number of compression cycles that proteins at the air-solution interface are subjected to within a fixed time.

472521 Elucidation of Self-Assembly of Silk Fibroin Proteins into Functional Biomaterials

Abstract:

After billions of years of evolution, nature has learned what works, what is appropriate, and what lasts. We aim to understand related phenomena and associated principles responsible for nature's building blocks and then to selectively adapt ideas from nature in order to engineer novel materials with unique and superior properties. In our research laboratory, mussels, silk, and lotus leaves are three key natural systems from which we get inspiration, particularly regarding their chemistry, composition, and structure. In this presentation, the structure-property relationships of Bombyx mori silk fibroin (SF) proteins will be discussed. Materials fabricated from SF proteins have been known to exhibit excellent eco/biocompatibility, biodegradability, and tunable mechanical properties. Although many applications of silk rely on thin films and sub-micron feature sizes, mechanisms and dynamics on the formation of specific structures (e.g. microneedles, spun fibers, or thin films) at the microscopic level have not been well addressed. In particular, there is a limited amount of information on how such structures self-assemble from their constituent molecules (i.e. SF proteins) and how the SF proteins interact with each other and behave in solution under nanoconfined environments. This presentation summarizes our recent studies and findings in this context. We have studied how SF protein polymers aggregate or bundle to form continuous fibers when they are pressed together and/or sheared against each other, using the Surface Forces Apparatus (SFA), as well as various mechanical characterization and microscopy imaging techniques. The presentation also involves the discussion and comparison of the complex dynamics of force profiles for SF proteins with other protein-based phenomena. We anticipate that the study of dynamics and self-assembly SF protein polymers is not only of fundamental research value, but also of practical significance for better designing and fabricating novel and reliable SF-based materials in nanoscale.

462114 Soluto-Inertial Phenomena: Designing Long Range, Long-Lasting, Surface-Specific Colloidal Interactions

Abstract:

Equilibrium interactions between particles in aqueous suspensions are limited to distances less than 1 μm. Here, we describe a versatile concept to design and engineer non-equilibrium interactions whose magnitude and direction depends on the surface chemistry of the suspended particles, and whose range may extend over hundreds of microns and last thousands of seconds. The mechanism described here relies on diffusiophoresis, in which suspended particles migrate in response to gradients in solution. Three ingredients are involved: a soluto-inertial ÔbeaconÕ designed to emit a steady flux of solute over long time scales; suspended particles that migrate in response to the solute flux, and the solute itself, which mediates the interaction. We demonstrate soluto-inertial interactions that extend for nearly half a millimeter, last for tens of minutes, and which are attractive or repulsive, depending on the surface chemistry of the suspended particles. Experiments agree quantitatively with scaling arguments and numerical computations, confirming the basic phenomenon, revealing design strategies, and suggesting a broad set of new possibilities for the manipulation and control of suspended particles.

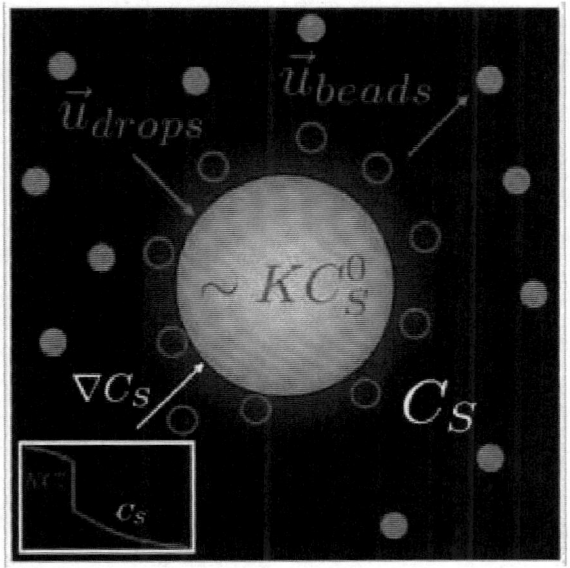

Figure 1: Long-range soluto-inertial interactions. A soluto-inertial beacon (gray), initially loaded with a high solute concentration, is placed in a solute- free suspension. A solute out-flux is established during equilibration, driving nearby suspended particles into diffusiophoretic migration. The magnitude and direction of migration depends on interactions between the particle surface and the solute, depicted here by particles of different surface chemistries (orange and green) that migrate either up or down the solute gradient. Inset: schematic radial profile of solute concentration inside and outside of the beacon.

472233 Microfluidic Synthesis of Janus Particles with Controlled Size and Hydrophilic/Hydrophobic Domains for Stable Pickering Emulsion Formation

Abstract:

Stable Pickering emulsions are important to systems where controlled confinement of an oil or water phase are critical to its applications. For example, micro-capsules encased by particles can serve as rigid vehicles for drugs or precious food ingredients. These coated emulsions can harbor oil in water (o/w) or even water in oil (w/o) phases. Nonetheless, the stability of the Pickering emulsions depend on the wetting properties of the particle. For this reason amphiphilic "Janus" particles (both hydrophilic and hydrophobic portions on same particle) have proven to be major assets in forming stable Pickering emulsions. The key to forming either o/w or w/o emulsions depends on the ratio of lengths of the hydrophobic to hydrophilic parts.

Polyethylene Glycol diacrylate (PEGDA), ethoxilated trimethylopropane triacrylate (ETPTA – monomer to be polymerized), 2-hydroxy-2-methyl-propiophenone (HMP – photo initiator), and water were mixed together to form the hydrophilic portion of the inner-phase. 99 wt% trimethylopropane triacrylate (TMPTA) and 1 wt% 2-hydroxy-2-methyl-propiophenone were added together to form the hydrophobic portion of the inner-phase. HMP would serve as the photo-initiator for polymerization under UV light. 95 wt% silicone oil (viscosity of 20 centi stokes) and 5 wt% Dow Corning Resin 749, as the surfactant served as the continuous phase. Janus particles were formed using flow of these solutions in a glass capillary microfluidic device. Flowrates were varied to control particle sizes (125- 400 microns) and hydrophilic to hydrophobic domain volume ratios (0.14 – 7). At a fixed surfactant concentration, variations of the ratio of inner water to oil flow rates lead to the formation of amphiphilic embedded particles of different sizes and hydrophobic portions. Using UV light, these droplets were cross-linked via photo-polymerization to form fairly monodispersed particles. It was demonstrated that Janus particles created with a hydrophilic to hydrophobic volume ratio of 7 favored an o/w emulsion while particles with a volume ratio of 0.14 favored w/o emulsion. The size of

the emulsions were controlled by changing the size of the Janus particles. An effort was also made to discuss the emulsion final form in terms of the interfacial tension of the two particle domains with respect to the dispersed and or continuous phase.

462844 Behavior of Metallodielectric Janus Particles Under Direct Current Electric Fields: Electrokinetics and Assembly

Abstract:

Anisotropic colloids (e.g. ellipsoids, dumbbells, and dimers) have shown interesting, field-induced dynamics that can be harnessed in active systems and directed assembly. Janus particles (JPs), in particular, have been an important model system for active motion due to asymmetric flows caused by induced charge electroosmosis and/or electrohydrodynamics. To date, most studies have focused on the behavior of these particles in alternating current (AC) fields at high frequency. An alternative method is to apply direct current (DC) fields, which has the advantage of creating gradients in osmotic pressure that result in rapid concentration of colloids. In this work, we explore the behavior of metallodielectric JPs under DC electric fields. The fields are applied in cells that consist of two ITO slides separated by a dielectric spacer. The behavior of particles is strongly dependent on the magnitude of the electric field. When small positive potentials are applied at the bottom electrode, negatively charged JPs move downwards in agreement with electrophoretic theory. In contrast, when large potentials are applied, JPs rapidly move upwards, at several microns per second, until reaching levitation heights of tens of microns. In this regime, an upward force dominates over the electrophoretic force. The upward motion is observed regardless of field polarity, as long as the field strength is large. If the concentration of particles is large enough, collective dynamics emerge that result in particles forming levitated clusters. Such field-induced associations potentially offer new possibilities for directed assembly and active systems.

462892 Non-Equilibrium Colloidal Interactions and Chain Folding Dynamics in Time-Dependent Magnetic Fields

Abstract:

The ability to engineer colloidal assembly processes is of interest to produce ordered structures for meta-materials with exotic properties. Based on this motivation, much research has been conducted on colloidal crystallization at near equilibrium conditions. However, such systems are prone to kinetic barriers and traps as they transition from high free energy to low free energy states. The use of rotating magnetic fields is one approach that has been shown to produce two-dimensional crystals via non-equilibrium processes. Recent results from our group have shown that time averaged hydrodynamic and dipolar interactions in rotating magnetic fields significantly affect particle pairs and multi-particle configurations during these non-equilibrium, steady-state assembly processes. To better understand and characterize these interactions, we conduct optical microscopy and simulated computer experiments of two-particles and multiple particles in a rotating magnetic field at several magnitudes and frequencies. By fitting two-particle trajectories to the Fokker-Planck (FP) equation, we can calculate the effective free energy between colloids as well as the relative diffusivity. This approach can then be applied to a multiple particle system by fitting the FP equation to order parameters that describe how a chain folds and the particles rearrange to form a perfect crystal. The use of order parameters quantifies both statistical mechanical (pseudo free energy) and fluid mechanical (hydrodynamic) contributions. With the ability to measure and tune kT-scale colloidal interactions and quantitatively model how such interactions are connected to dynamically changing microstructures, we demonstrate the potential for control of the assembly, disassembly, and repair of colloidal crystals under steady state, time-dependent fields.

469285 Characterizing Phase Separation in Colloid-Polymer Dispersions through Solvent Extraction in Microfluidic Droplet Arrays

Abstract:

Nanoliter droplets formed using microfluidic techniques have proven useful as components in tunable emulsions and as uniform microreactors. When observed for extended periods, there is significant transport of solvent from the droplet phases to the continuous fluid and material of the microfluidic device. This transport is utilized to control to composition of multicomponent colloidal dispersions during solvent removal. A specific example of solvent removal into a region of phase separation of a polymer-colloid suspension will be discussed. The phase boundaries are determined optically and the properties of the resulting concentrated suspensions investigated. The advantage of this approach is the potential to characterize drying of colloidal suspensions without the complexity of three phase contact lines and solid-liquid interfaces.

469928 Dynamic Simulation of Aging in a Hard-Sphere Colloidal Glass after Volume-Fraction Jumps

Abstract:

In this work we develop a computational framework for studying glassy dynamics in colloidal glass-formers from a micromechanical perspective. Many molecular liquids transition from a liquid to a crystalline state upon cooling. If the cooling is sufficiently rapid, however, the material can undergo a glass transition, whereby it solidifies without crystallization, forming a glass in which the amorphous structure is retained but the dynamics become extremely slow (Angell, 1995). Similar behavior has been observed in colloidal fluids, whereby they form a colloidal glass upon a "concentration quench" (Pusey and van Megen, 1986). Much analysis in the colloidal-glass literature assumes that the behavior of colloidal systems at concentrations near the jamming transition is similar to that of a molecular glass near its glass transition, hence making the simpler colloid a good model for glass-like materials. Recent experimental studies by McKenna and co-workers (Di et al, 2011; Di, Peng, and McKenna, 2014) reveal that non-equilibrium aging and structural recovery of colloidal systems differ qualitatively from molecular glassy dynamics when interrogated for the McKenna-Kovacs signatures (Kovacs, 1963): the intrinsic isotherm, the asymmetry of approach to equilibrium, and the memory effect. The reasons for such differences have not yet been elucidated beyond the broad hypothesis that colloids may not be "true" glass-forming systems despite many shared dynamical and mechanical signatures. To bridge this gap, we study via large-scale dynamic simulation the structural relaxation, dynamics and rheology of a hard-sphere colloidal glass after various volume-fraction jumps. In this talk, we focus on the computational aspects of executing such jumps and the impact of quenching methods on the post-jump particle dynamics. Three methods are utilized to model a volume-fraction jump: container shrinkage (similar to droplet evaporation); fluid removal through a sink (mimicking fluid aspiration); and a jump in particle size (to model thermo-reversible particle swelling). During and following each jump, the positions, velocities, and particle-phase stress are tracked and utilized to characterize relaxation time scales, heterogeneous pressure, and structural changes. In addition, we implemented the protocols of the McKenna-Kovacs signature experiments to study the approach to the intransient state. For

all such tests, the quench (concentration jump) was executed via each of the three methods, and the results compared to provide insight not just into implications of computational test protocol but also the effects of boundaries and quench rate on system behavior. A comparison to the creep experiments of soft colloids (Peng and McKenna, 2016) show qualitative agreement.

470349 Programming DNA-Mediated Colloidal Self-Assembly to Control 3D Binary Superlattice Formation

Abstract:

DNA-functionalized colloids are programmable materials that can self-assemble into a variety of crystal structures. Selective hybridization of conjugate base pairs of DNA molecules plays a fundamental role in tailoring the effective colloidal interactions as a useful tool to guide the formation of desired super-lattices. In binary systems, one promising advantage is the ability to independently tune the cross (A-B) and like (A-A or B-B) particle interactions via tethering two different mixtures of suitable strand sequences on different particles, known as multi-flavoring approach. This useful approach also allows controlling one of the like-like pair-interactions (A-A) independently of the other one (B-B). In this work, we conduct computer simulations to investigate the self-assembly of DNA-functionalized colloids into various 3-D binary crystals by programming cross- and like-particle interactions. The molecular dynamics simulations show that the variation of the strength of cross pair-potential relative to like interaction provides selectivity between body-centered cubic (BCC) and close-packed (FCC and HCP) structures. Moreover, changing the two like-interaction strengths (A-A and B-B) relative to one another, controls the composition of the crystals that form. In addition, the transformation kinetics and the thermodynamic stability of the formed structures will be studied.

457804 Clusters of Polyhedra in Spherical Confinement

Abstract:

Dense particle packing in a confining volume remains largely unexplored, despite its relevance to fields of study as diverse as plasmonics, industrial packaging and transport, colloidal molecule design, blood clotting, and information storage. We report simulation results for dense clusters of the Platonic solids in spherical confinement, for up to N = 60 constituent particles. We discuss the resultant interplay between isotropic geometrical confinement and anisotropic particle shape, classify and compare the symmetry point groups of our clusters, and make an interesting connection to optimal spherical codes. Our results showcase the structural diversity and experimental utility of families of solutions to the problem of packing in confinement.

473317 Straining Soft Colloids in Aqueous Nematic Liquid Crystals

Abstract:

Liquid crystals (LCs), due to their long-range molecular ordering, are anisotropic, elastic fluids. This presentation will report that elastic stresses imparted by nematic LCs can dynamically shape soft colloids and tune their physical properties. Specifically, we use giant unilamellar vesicles (GUVs) as soft colloids, and explore the interplay of mechanical strain when the GUVs are confined within aqueous chromonic LC phases. Accompanying thermal quenching from isotropic to LC phases, we observe the elasticity of the LC phases to transform initially spherical GUVs (diameters 2-50 μm) into two distinct populations of GUVs with spindle-like shapes and aspect ratios as large as 10. Large GUVs are strained to a small extent ($R/r < 1.54$, where R and r are the major and minor radii, respectively) consistent with an LC elasticity-induced expansion of lipid membrane surface area of up to 3% and conservation of the internal GUV volume. Small GUVs, in contrast, form highly elongated spindles ($1.54 < R/r < 10$) that arise from an efflux of LC from the GUVs during the shape transformation, consistent with LC-induced straining of the membrane leading to transient membrane pore formation. A thermodynamic analysis of both populations of GUVs reveals the final shapes adopted by these soft colloids are dominated by a competition between the LC elasticity and an energy (~0.01 mN/m) associated with the GUV-LC interface. Overall, these results provide new insight into the coupling of strain in soft materials and suggest new designs of LC-based responsive and reconfigurable colloidal dispersions.

456738 Growth and Agglomeration Rates Regulate Colloid Fractal Dimension

Abstract:

The formation and growth of crystals in suspensions are central in a great deal of applications, ranging from imaging, to photovoltaics, to drug formulation in the pharmaceutical industry.[1, 2] Typically, when modeling crystal formation, a quite complex number of phenomena need to be accounted for, such as nucleation, growth, aggregation, and breakage.[1, 3] Deterministic models require typically the fractal dimension (d_f) of the clusters as a key input parameter for correctly describing aggregation. The fractal dimension describes the spatial organization of the aggregates and is typically comprised between 1 (linear aggregates) and 3 (compact assemblies). In the most general sense, d_f is a function of the aggregation regime considered,[4] but in the frame of crystal agglomeration it is assumed to be equal to 3, given that crystals grow and potential voids of the forming clusters are rapidly filled. While this condition is often fulfilled, situations exist where aggregation prevails over growth, as it happens in the undesired formation of crystals in flow reactors, often resulting in clogging.[5] In this frame, the scope of the present work is i) to develop a Mont Carlo (MC) method able to describe the simultaneous growth and aggregation of particles, ii) to evaluate the d_f of the growing/aggregating particles in different conditions.

In particular, a growth mechanism has been implemented in a MC code able to describe diffusion-limited cluster aggregation (DLCA) and reaction-limited cluster aggregation (RLCA). The clusters fractal dimension is evaluated as a function of parameters such as the ratio of growth and aggregation rates, the occupied volume fraction and the particles sticking probability. The present model represents an interesting tool able to provide the fractal dimension to be employed in deterministic models to properly describe the simultaneous growth and agglomeration processes.

[1] Ochsenbein DR, Schorsch S, Vetter T, Mazzotti M, Morari M. Growth Rate Estimation of beta L-Glutamic Acid

from Online Measurements of Multidimensional Particle Size Distributions and Concentration. Ind Eng Chem Res. 2014;53:9136-48.

[2] van Embden J, Chesman ASR, Jasieniak JJ. The Heat-Up Synthesis of Colloidal Nanocrystals. Chem Mater. 2015;27:2246-85.

[3] Lindenberg C, Scholl J, Vicum L, Mazzotti M, Brozio J. L-glutamic acid precipitation: Agglomeration effects. Cryst Growth Des. 2008;8:224-37.

[4] Friedlander SK. Smoke, dust, and haze fundamentals of aerosol dynamics. 2nd ed. New York: Oxford University Press; 2000.

[5] Hartman RL, Naber JR, Zaborenko N, Buchwald SL, Jensen KF. Overcoming the Challenges of Solid Bridging and Constriction during Pd-Catalyzed C-N Bond Formation in Microreactors. Org Process Res Dev. 2010;14:1347-57.

467506 Hematin Crystallization Mechanisms Suggest How Antimalarial Drugs Operate

Abstract:

Hematin is a byproduct of heme detoxification in malaria parasites. Current antimalarial compounds are thought to inhibit the growth of hematin crystals, referred to as hemozoin (*in vivo*) or b-hematin (*in vitro*). Identification of antimalarial drug action on this phase transformation could provide a foundation for drug design to overcome parasite resistance to antimalarials. Continued resurgence of antimalarial drug resistance drives research effort to better characterize drug-crystal interactions from a fundamental perspective. To this end, we aim to better understand the molecular mechanism(s) of hematin crystallization and antimalarial drug action within physiologically-relevant environments as the basis for elucidating growth sites on hematin crystal surfaces and design more effective growth inhibitors that selectively bind to these sites.

We employ *in situ* atomic force microscopy (AFM) in parallel with bulk solution and crystallization studies as a platform to study the mechanisms of crystal growth and inhibitor-crystal interactions in biomimetic growth solutions. Time-resolved *in situ* AFM measurements reveal that hematin crystallization occurs by the classical pathway of single molecule incorporation, specifically involving two-dimensional nucleation of layers. We quantified the rate of layer generation and the velocities of anisotropic step advancement as a function of supersaturation in the absence and presence of antimalarials. We identified unique modes of binding for current antimalarials as inhibition pathways: step pinning, kink blocking, and step bunching. In parallel, we studied the prominent functional groups that comprise antimalarial compounds by *in situ* AFM. Identifying fundamental principles governing the molecular recognition of antimalarial drugs to specific sites could lead to the rational design of new compounds with improved efficiency to overcome parasite resistance to current antimalarials.

461662 Atomistic Simulation of the Aqueous-Phase Synthesis of Au Nanoparticles

Abstract:

The catalytic, electronic, and optical properties of metal nanoparticles strongly depend on their three-dimensional atomic structure. In particular, Au nanoparticles have demonstrated a wide range of properties, which can be tuned for different catalytic and biomedical applications. In this work, we have adopted a kinetic Monte Carlo (KMC) simulation approach for modeling the atomistic growth behavior of Au nanoparticles, in order to provide a direct link to the experimentally-observed growth behavior. Such a model can provide information about the impact of individual system events on the final atomic-scale structural features of the Au nanoparticles, and this can help guide the experimental synthesis process. The advantage of our approach is that, compared to traditional molecular dynamics simulations or electronic structure calculations, we can track the atomistic nanoparticle structural evolution on time scales that approach the actual experiments. This has allowed us to perform several different comparisons against experimental benchmarks, and it has helped transition our KMC simulations from a hypothetical toy model into a more experimentally-relevant test-bed. In addition, while previous analytical models have been used for predicting nanoparticle growth, the KMC approach can track atomistic-scale details and account for neighbor-neighbor interactions, structural defects, and structural heterogeneity. These aspects can be particularly influential during nanoparticle growth, leading to a wide variety of interesting nanoparticle geometries.

Here, we report the effects of synthesis temperature and initial precursor concentration on the Au nanoparticle growth behavior, and we find that our model can be trained (via automated feedback) to adequately reproduce the experimental growth curves at the same conditions. The fitting procedure results in reasonable parameter values, including activation energy barriers that are consistent with related experimental measurements. In addition, since our KMC simulations preserve the atomistic details of our growing Au nanoparticle, this allows for detailed structural analyses. Although our nanoparticles are roughly spherical, the maximum/minimum dimensions deviate from the average by approximately 12.5%, which is consistent with the corresponding experiments. Also, our surface texture analysis highlights the changes in the surface structure as a function of time. While the nanoparticles show similar surface structures throughout the growth process, there can be some significant differences during the initial growth at different synthesis conditions. To the best of our knowledge, this is one of the first attempts at modeling the atomistic growth behavior of realistic metal nanoparticles in solution. In the future, we intend to perform additional benchmarking of more atomistic growth features, including an extension of our model to capture anisotropic nanoparticle growth.

470151 A Mechanistic Growth Model for Organic Salt Morphology Prediction

Abstract:

There is growing interest in the pharmaceutical sector to develop strategies for digital drug product design, starting with the molecular structure of a target molecule and creating good initial estimates for all the solid form characteristics, including crystal morphology. For morphology prediction, a detailed understanding of crystal growth is vital. In the past several decades various growth models have been proposed to predict the crystal morphology of organic molecules, which laid the foundation to design schemes to manipulate and tailor the crystal morphology in accordance with the application requirements. Recent developments include [1-2].

However, the prediction of growth rates for organic salts still remains blank, due to their complicated crystal structures and complex atomic interactions. For example, organic salts usually consist of nonstoichiometric ions as the growth units, which makes it difficult to find the periodic bond chains (PBCs) within the crystal lattice; moreover, the short-range intermolecular interactions, and the long-range electrostatic interactions in organic salts may need to be considered; the kink sites in organic salts are complicated due to the existence of positive and negative ions, etc. With these considerations in mind, we propose a mechanistic growth model for organic salt morphology prediction, which identifies periodic bond chains (PBCs) using the concept of building units, and accurately captures the short-range/long-range electrostatic interactions and partial charges. We illustrate the method on ammonium acetate grown from the vapor phase. The electrostatic interaction energies in the kink sites were calculated using a space partitioning method that is computationally efficient. This algorithm for studying the solid-state interactions and building units is based on a mechanistic spiral growth model, which may be useful for predicting the morphology of a wide variety of organic salts.

References:

1. Li J, Tilbury CJ, Kim SH, Doherty MF, A design aid for crystal growth engineering. Progress in Materials Science 82, 1-38 (2016).

2. Li J, Tilbury CJ, Joswiak MN, Peters B, Doherty MF, Rate expressions for kink attachment and detachment during crystal growth. Crystal Growth & Design DOI:10.1021/acs.cgd.6b00292 (2016).

453851 Growing Macroscopic Hydrates Using a Bioinspired Approach and Investigating Dehydration Induced Polymorphism

Abstract:

Growing of macroscopic single crystals of organic materials is not necessarily a straight forward procedure. Carbamazepine dihydrate is one of these cases. Traditional approaches of crystallization of this, particular, compound using mixtures of alcohols with water lead to crystals with high aspect ratios (needle shaped or high aspect ratio plates)[1]. These crystals are not suitable for wettability or XPS measurements on individual facets[2]. In this work, a bioinspired crystallization method is employed for the growth of macroscopic carbamazepine dihydrate crystals.

In a work on the interactions of surfactants with carbamazepine in aqueous solutions, investigators have reported the formation of prismatic carbamazepine dihydrate crystals in the micron scale[3]. From the surfactants employed, it seems that the carbonyl groups of sodium taurocholate influence the crystal habit. Unfortunately, sodium taurocholate proved not to be adequate for the growth of crystals of larger sizes, that can be used as seeds for the growth of macroscopic crystals. Also, other studies suggest that the growth of large protein crystals is favored by biomimetic solutions, imitating the design of the digestive vacuole of malarian parasites such as *Plasmodium falciparum*, comprising of an aqueous and an organic phase[4,5].

A two phase system with cyclohexanone in the organic phase and water in the other phase was developed. An adequate amount of carbamazepine was added in the solution under stirring and the mixture was heated using a water bath. When all the material was dissolved, stirring stopped and a temperature gradient of $1^{o}C*d^{-1}$ was applied. After a few days, depending on the amount of carbamazepine added, crystals started to appear inside precursor droplets formed on the interface between the two phases. When the temperature of the water bath reached 4°C the crystals were harvested. They were quite large with the (0 2 0) facet to be the dominant one, having sizes up to 2x1 cm. The harvested crystals were examined using XRD and found to be carbamazepine dihydrate crystals. Similar experiments were performed with heavy alcohols and butanone, but the crystals obtained were not that promising, even though they were in all cases dihydrate crystals, in terms of crystal habit. The shift from the needle like shape to a more prismatic one can be thus attributed to the carbonyl chemistry.

Molecular dynamics simulations suggest that in such systems droplets of the aqueous phase are formed in the organic phase[6]. This droplets were tracked using Dynamic Light Scattering and found to be about 500 nm in diameter. Nonetheless, these droplets do not seem to act as nucleation sites. Only the main water cyclohexanone interface operates as a nucleation site. This preferential nucleation can be explained on the ground of heterogeneous nucleation. The shape factor for nucleation on an interface should result to a much higher decrease on the Gibbs free energy of nucleation rather than the shape factor for nucleation on a droplet[7,8]. Furthermore, molecular dynamics simulations propose the formation of pockets of enhanced water concentration on the interface. These pockets is speculated to promote even further nucleation.

Drying experiments performed on these macroscopic crystals. From SEM images obtained, it was shown that dehydration is driven by well-ordered cracks formed on the surface of the dominant facet. Two types of cracks appear on the (0 2 0) facet: 1) macroscopic cracks vertical to the (0 2 0) facet with crack spacing a few microns big and 2) cracks with smaller spacing inclined, both clockwise and counterclockwise, with respect to the primary cracks of type 1. These two types of cracks correspond to the following main cleavage planes: (h 0 0), (-1 ±1 1) and (0 ±1 1). Furthermore, macroscopic crystals were sliced with razor blade and the examination of their core revealed an additional cleavage plane, (0 k 0). These findings are important, as previous studies reported only the (h 0 0) cleavage plane. It should, also, be mentioned that the crack appearance is independent of the application of vacuum upon drying.

The anhydrous form nucleates, during dehydration, on the surface of the dihydrate crystals in the form of small islands. The dehydration mechanism for macroscopic crystals results solely to the triclinic polymorph of anhydrous carbamazepine. The triclinic polymorph was depicted both with XRD and SEM. The SEM images are quite interesting showing the needle shaped triclinic crystals growing vertically on the surface of the dihydrate. Comparing the results of the dehydration induced polymorphism for macroscopic dihydrate crystals and micron-sized crystals it can be deduced that the crystal size and habit influence the final polymorph. It seems that for cases where the need for heat dissipation is greater (i.e. in macroscopic crystals), the triclinic polymorph nucleating during drying is higher. This is because the needle-shaped triclinic crystals have a large surface area to volume ratio providing better heat dissipation. This concomitant nucleation scheme has been reported computationally via Monte Carlo simulations[9,10]. However, the suggested drying mechanism can be better understood, mechanistically, on the same ground Machon described phase transformations on nanoparticles using the one dimensional Ginzburg-Landau equation[11, 12].

This study expands the current knowledge on nucleation of organic materials using bioinspired approaches. It is quite important to notice the nucleation of dihydrate and the absence of any other polymorphs. This shows that for such systems, the chemical potential associated with the aqueous phase and with the small amount of water dissolved in cyclohexanone drive the nucleation. It is, also, of particular importance that the drying mechanism is revealed; suggesting that the heat dissipation mechanism drives the dehydration induced polymorphism. The image analysis revealed also the nucleation mechanism of the anhydrous polymorphs resembling to the anisotropic nucleation model proposed in the past[13,14].

Disclosure:

Imperial College London and AbbVie jointly participated in study design, research, data collection, analysis and interpretation of data, writing, reviewing, and approving the publication. Eftychios Hadjittofis is a graduate student at Imperial College London; Jerry Y. Y. Heng is a professor at Imperial College London. They all have no additional conflicts of interest to report. Geoff G. Z. Zhang is an employee of AbbVie and may own AbbVie stock.

References:

1. Khoo JY, Shah UV, Schaepertoens M, Williams DR, Heng JYY. Process-induced phase transformation of carbamazepine dihydrate to its polymorphic anhydrates. *Powder Technol.* 2013;236:114-121.

2. Ho R, Hinder SJ, Watts JF, Dilworth SE, Williams DR, Heng JYY. Determination of surface heterogeneity of d-mannitol by sessile drop contact angle and finite concentration inverse gas chromatography. *Int J Pharm.* 2010;387(1–2):79-86.

3. Rodriguez-Hornedo N, Murphy D. Surfactant-facilitated crystallization of dihydrate carbamazepine during dissolution of anhydrous polymorph. *Journal of Pharmaceutical Sciences.* 2004;93(2):449-460.

4. Olafson KN, Rimer JD, Vekilov PG. Growth of large hematin crystals in biomimetic solutions. *Crystal Growth and Design.* 2014;14(5):2123-2127.

5. Vekilov PG, Rimer JD, Olafson KN, Ketchum MA. Lipid or aqueous medium for hematin crystallization? *Crystal Engineering Communications.* 2015;17:7790-7780.

6. Wick CD, Chang TM. Computational observation of pockets of enhanced water concentration at the 1 Octanol/water interface. *The Journal of Physical Chemistry B.* 2014;118(7785-7791).

7. Fletcher NH. Size effect in heterogeneous nucleation. *Journal of Chemical Physics.* 1958;29:572-576.

8. Johnson WC, White CL, Marth PE, et al. Influence of crystallography on aspects of solid-solid nucleation theory. *Metallurgical transactions A.* 1975;6(4):911-919.

9. Sanders DP, Larralde H, Leyraz F. Competitive nucleation and the ostwald rule in a generalized potts model with multiple metastable phases. *Physical Review B.* 2007;75:132101.

10. Hedges LO, Whitelam S. Limit of validity of ostwald's rule of stages in a statistical mechanical model of crystallization. *Journal of Chemical Physics.* 2011;135:164902.

11. Machon D, Piot L, Hapiuk D, et al. Thermodynamics of nanoparticles: Experimental protocol based on a comprehensive ginzburg-landau interpretation. *Nano Letters.* 2014;14(1):269-276.

12. Machon D, Melinion P. Size-dependent pressure-induced amorphization: A thermodynamic panorama. *Physical Chemistry Chemical Physics.* 2014;17(2):903

13. Bianchet MA, Rigotti G, Blesa MA. Surface nucleation and anisotropic growth model for solid state reactions. *Solid State Ionics.* 1990;42(1):21-27.

14. Bianchet MA, Rigotti G, Blesa MA. Surface nucleation and anisotropic growth model for the dehydration of 11-aminoundecanoic acid dihydrate. *Solid State Ionics.* 1995;78(1–2):49-53.

468938 A Microfluidic Droplet Platform for Investigating Nucleation Kinetics of Hydrocarbons for Application to Wax Crystallization

Abstract:

Crystallization of wax present in crude oil is a major issue for its storage, processing and transportation. As a result, there is a significant interest in identifying inhibitors that can mitigate the nucleation of wax crystals in crude oil. To address this need, droplet-based microfluidics is used to quantify the nucleation kinetics of n-alkanes - a simple model system for wax. In nucleation studies, it is useful to have control of the droplet volume as it dictates the number of impurities (like dust) present per droplet that can act as nucleation site. Therefore, a valve-based flow-focusing device is used to produce droplets of controllable volume. A downstream storage chamber is included to incubate the produced droplets so that nucleation studies can be conducted. The modular design of the microfluidic system provides flexibility, allowing adjustment to specific experimental requirements. Hexadecane-in-water emulsions are produced using Tween 20 and sodium dodecyl sulfate as surfactants. The effect of droplet size and surfactant type on nucleation rate of n-hexadecane is investigated for the first time in a droplet-based microfluidic platform. The results show that the degree of nucleation in hexadecane could be controlled by selecting an appropriate surfactant and controlling the droplet size. The experimental data is used to test current models of nucleation kinetics. This work will enable establishment of a microfluidic platform for future studies of nucleation in paraffins and their mixtures with inhibitors.

463932 Sonocrystallization of L-Asparagine Monohydrate: Modeling and Optimization.

Abstract:

Producing crystals with target attributes consistently is a great challenge in any crystallization process due to stochastic nature of the nucleation event and the limited number of manipulated variables. The use of power ultrasound (defined as frequencies from 20 to 100 kHz) provides an interesting approach to induce nucleation at low supersaturation in a controlled way [1]. Several studies have demonstrated the benefits of applying ultrasound during a cooling or anti-solvent crystallization process such as faster and uniform primary nucleation, higher yield, reduced agglomeration, narrow crystal size distribution, control of polymorphism, and generation of internal seed [2]-[3].

The development of an effective mathematical model describing the crystallization dynamics is a crucial issue towards finding the optimal process performance and to control the crystal product properties. Here we develop and validate a mathematical model based on population balance framework for the ultrasound-assisted unseeded batch cooling crystallization of L-asparagine monohydrate, an important amino acid, from its aqueous solution. The population balance model considers the relevant fundamental events of the crystallization process, such as nucleation, growth, and breakage phenomena. An additional kinetic expression is introduced for induced nucleation due to ultrasound irradiation [4]. Application of ultrasound induces energy into the system and this in turn leads to temperature rise. The modeling considers this rise in temperature appropriately. The population balance model coupled with mass balances is solved by high resolution finite volume discretization with a flux limiter.

A series of ultrasound-assisted batch cooling crystallization experiments have been conducted systematically and the solute concentration data measured by high performance liquid chromatography and final mean crystal size measured by laser diffraction are used to simultaneously extract all kinetic parameters using the population balance model and a nonlinear optimization technique. The model has been fully validated with

new experimental data and the good agreement between model predictions and experimental observations clearly demonstrates the predictive ability of the model. Experimental data and model simulations show that the ultrasound significantly reduces the metastable zone width for L-asparagine monohydrate, enhances the nucleation rate, and induces early nucleation. The developed model is then used to solve a set of dynamic optimization problems related to particle engineering. The model is also used to determine the region of attainable size for such a sonocrystallization process.

References:

[1] G. Ruecroft, D. Hipkiss, T. Ly, N. Maxted, and P. W. Cains, Org. Proc. Res. Dev. 9, 923-932, 2005.

[2] O. Narducci and A. G. Jones, Cryst. Growth Des. 12, 1727–1735, 2012.

[3] M. Kurotani and I. Hirasawa, Chem. Eng. Res. Des. 88, 1272–1278, 2010.

[4] A. Kordyllaa, T. Krawczyka, F. Tumakakab and G. Schembecker, Chem. Eng. Sci. 64, 1635–1642, 2009.

462861 The Effect of Material Flexibility on the Drying Transition of Water Between Hydrophobic Objects: Thermodynamics and Kinetics

Abstract:

Liquid water confined between hydrophobic objects of sufficient size becomes metastable with respect to its vapor at separations smaller than a critical drying distance. Macroscopic thermodynamic arguments predicting this distance and nearly all molecular simulations of this phenomenon have been restricted to the limit of perfectly rigid confining materials. However, no material is perfectly rigid, and it is of interest to account for this fact in both theoretical and *in silico* analyses. First, we present a theory that combines the current macroscopic theory with the thermodynamics of elasticity to derive an expression for the critical drying distance for liquids confined between flexible materials [1]. The thermodynamic arguments show that flexibility increases the critical drying distance (i.e. it promotes evaporation). Second, we present calculations of the rate of evaporation of water as a function of both object separation and flexibility. Rate calculations, obtained via forward flux sampling [2] in conjunction with molecular dynamics simulations, show that the rate increases substantially as the material is made more flexible, suggesting that more flexible materials provide lower kinetic barriers to evaporation. Third, we present calculations of the free energy barriers to evaporation and show how they evolve as one changes the flexibility at fixed object separation. We also discuss how flexibility affects the kinetic mechanism of evaporation as well as condensation of water in hydrophobic confinement.

[1] Y.E. Altabet, P.G. Debenedetti "The role of material flexibility on the drying transition of water between hydrophobic objects: a thermodynamics analysis" Journal of Chemical Physics, 2014, 141: 18C531.

[2] R.J. Allen, C. Valeriani, P.R. ten Wolde "Forward flux sampling for rare event simulations" Journal of Physics: Condensed Matter, 2009, 21: 463102.

470327 How Fluctuations Influence Barriers to Dewetting

Abstract:

Liquid water can become metastable with respect to its vapor under tension as well as in a variety of hydrophobic confinement contexts. The resulting dewetting transitions are often impeded by large free energetic barriers. According to macroscopic interfacial thermodynamics, such barriers arise from the free energy required to nucleate a critical vapor bubble â€” bubbles that areÂ smaller than the critical bubbleÂ collapse, whereas larger ones grow spontaneously. Using molecular simulations in conjunction with enhanced sampling techniques, we have recently investigated the role of nanoscale water density fluctuations on dewetting barriers in three contexts: (i) cavitation or homogeneous nucleation under tension, (ii) capillary evaporation between two hydrophobic surfaces, and (iii) Wenzel-to-Cassie transition on textured surfaces. In each case, we find that water density fluctuations can lead to a reduction in the free energetic barriers to dewetting by circumventing the classical dewetting pathways. In particular, the fluctuation-mediated dewetting pathway can involve a number of transitions between distinct dewetted morphologies, with each transition lowering the resistance to dewetting. Our results thus suggest a key role for fluctuations in speeding up the kinetics of numerous phenomena ranging from Cassieâ€“Wenzel transitions on superhydrophobic surfaces, to hydrophobically driven biomolecular folding and assembly. They also help explain the experimentally observed tension that water can sustain before it spontaneously cavitates, and facilitate the design of novel surface textures capable of spontaneously recovering their superhydrophobicity.

471341 Using Dynamic Mean Field Theory to Study Non-Equilibrium Steady States in Mesoporous Membranes

Abstract:

Mesoporous membranes contain pore sizes technologically relevant to separation of mixtures such as CO_2 from effluents and volatile organic compounds from air. Design of these materials requires understanding of the underlying transport of gases and condensable vapors in pure as well as mixture form through mesopores. Processes associated with these membranes, whether it is measurement of pore size distribution or separation of gases, are inherently non-equilibrium as a result of presence of partial pressure gradients. Continuum modeling for such a process requires predetermined models for confinement effects while atomistic simulations are computationally intensive. We study these systems with dynamic mean field theory (DMFT), a coarse grained lattice based theory, which has been successfully applied in the past to study relaxation dynamics of systems in approach to equilibrium steady state. We recently published a computational study of permporometry, an experimental technique used for determination of pore size distribution, where a light gas permeates in presence of condensable vapor. This system is akin to transport under conditions with slight deviation from equilibrium. We have further applied DMFT to the separation of condensable vapor from its mixture with light gas under significant pressure gradient which represents a large deviation from equilibrium. Both these studies reveal interesting phenomena such as higher flux of the light gas in the layer adjacent to the strongly adsorbed surface layer in the first case while latter system shows capillary condensation confined to high pressure side of the system.

461677 [Invited Talk] New Models of Brownian Aggregation and Sedimentation Rates of Colloidal Dispersions

Abstract:

[INVITED TALK]

Stabilizing colloidal dispersions against both aggregation and sedimentation is a key challenge in the manufacture of coatings, formulation of food products, drug delivery, membrane fouling, inkjet printing, and flow assurance in oil and gas production. If initially separate particles begin to aggregate at a sufficient rate, ever larger clusters of particles are formed, with the dispersion eventually separating into two distinct layers because of settling (the particles are more dense than their suspending medium) or creaming (the particles are less dense). One of the destabilization mechanisms is perikinetic, or Brownian, aggregation, where particle transport occurs by diffusion. This mechanism was first described by Smoluchowski (S), who considered the steady-state aggregation of initially mono-disperse hard spheres with an initial uniform concentration profile. The S model matches simulation results for aggregation rates only in dilute systems, i.e., for particle volume fractions, φ, less than 0.0005. For most practical applications, however, particle volume fractions can be as high as $\varphi = 0.4$, in which aggregation rates are up to two orders of magnitude faster than the predictions of the S model.

I first discuss recently developed theoretical models that accurately describe Brownian aggregation at moderate to high volume fractions. Transient effects are shown to increase the rates of aggregation, because of the unsteady-state flux that drives aggregation at short times. Entropic packing effects and non-ideal particle diffusion (i.e., mass transfer is driven by the gradient of the particles' chemical potential and not concentration) are also found to be important, especially at higher volume fractions. One particular model, based on the "liquid-state" dynamical density-functional theory (DDFT), yields predictions which are in excellent agreement with Brownian Dynamics (BD) simulation results for the time dependence of monomer number density and the half-times of aggregation for φ up to 0.35, and with some gelation times (the times for which the effective volume fraction reaches a freezing transition) reported in the literature. I then present the extension of the DDFT-based model to account for interparticle forces between the aggregating particles, which invokes ideas from various perturbation theories of liquids. For dispersions of particles with purely attractive interactions, the classical Fuchs-Smoluchowski (FS) model underpredicts the aggregation rates by up to 1000 fold. In the presence of strong interparticle repulsive forces, the FS model predictions are in fair agreement with the BD simulation results for dilute systems with particle volume fractions $\varphi << 0.1$. In contrast, the predictions of the new DDFT-based model compare favorably with the BD simulation results, in both cases, up to $\varphi = 0.3$. A new quantitative measure for colloidal dispersion stability, different from the classical FS stability ratio, is proposed on the basis of aggregation half-times. Overall, an improved mechanistic understanding of Brownian aggregation is obtained for concentrated particle dispersions.

Finally, I discuss an application of one of the above aggregation models for the prediction of the sedimentation times of colloidal dispersions for which aggregation and sedimentation processes are coupled. Aggregation and net sedimentation times for clusters of certain particle sizes and fractal dimensions are generated and then compared, which provides an upper-bound of the net sedimentation time of the entire dispersion. Moreover, the effective sizes, density differences, and volume fractions of the clusters are obtained. The predicted trend of the net sedimentation time is fairly consistent with the experimental data of the sedimentation times of colloidal silica microspheres of diameters 350, 505 and 750 nm, as measured in water and aqueous NaBr solutions of concentrations ranging from 50 to 1000 mM.

469975 Two Perspectives on the Phase Behavior of Small Clusters of Colloidal Particles

Abstract:

The directed assembly of a small ensemble of colloidal particles into a structured object is a topic of technological interest, for example as a route to construct a finite piece of meta-material. Such assembly processes are also of fundamental scientific interest because they involve systems with a number of particles N far below the bulk limit, e.g. N = 10â€"100, which are considered â€œthermodynamically smallâ€ in the terminology introduced by T.L. Hill. The relationships between thermodynamic variables in a small system can be qualitatively different from those in a bulk system, and phase behavior is a prime example. In this work we study the phase behavior of thermodynamically small colloidal clusters interacting via the Asakura-Oosawa pair potential, in which the concentration of a depletant species controls the degree of attraction, or effective osmotic pressure, experienced by the main colloidal species of interest. Experimental data and Brownian Dynamics simulations show that, at a fixed temperature, small clusters may exhibit coexistence between fluid-like and solid-like phases over a range of osmotic pressure values, rather than at a single value. We use parallel tempering Monte Carlo to map out the range of fluid-solid coexistence for clusters of several different N in the range 10-50, employing histograms of the potential energy as the most fundamental way to identify distinct phases, and compare the phase diagrams with their bulk counterparts. We also quantitatively compare our results with previously published work using free energy landscapes constructed in a two-dimensional order parameter space identified by a technique called diffusion maps.

463123 Single-Molecule Hydrophobic Interactions: From Experiment to Simulation and Equilibrium to Non-Equilibrium

Abstract:

Hydrophobic interactions (HIs) are a fundamental driving force in nature, determining the structure and self-assembly characteristics of many of the molecules of life. Generally, theories and simulations pertaining to hydrophobic solvation and interactions have focused on small length-scale physics, while experimental efforts have observed hydrophobic interactions involving macroscopic systems. Here, we investigate the HI using a highly effective model system: a glycine-serine repeat peptide, functionalized with a varying number of hydrophobic leucine groups, that interacts with a hydrophobic self-assembled monolayer surface. Simulations are performed hand-in-hand with single-molecule atomic force microscopy, constituting a unique convergence of theory and experiment for directly probing the HI. Both sets of results indicate a free energy of peptide removal that scales linearly with the number of leucine residues, at around 3.4 k_BT/leucine for experiment and ranging from 1.6 to 5.6 k_BT/leucine in simulations. The simulated systems closely mirror the experimental set-up, even employing the same non-equilibrium technique, specifically Jarzynskiâ€™s equality, to evaluate free energy differences. In addition to agreeing well with experiment, our simulations also identify key factors driving this observed scaling. While all peptides are relatively disordered in solution and absorbed to surfaces, we observe differences in the tetrahedral structure of nearby water, as well as in changes in the solvent accessible surface area of hydrophobic residues. Based on these same metrics, we find signatures of both small- and large-scale hydrophobic solvation theories. Additionally, non-equilibrium simulations provide a detailed dynamical view of how peptides detach from the surface. In accord with the idea that the free energy, and thus the work to remove the peptides, is dominated by hydrophobic association, the greatest contribution to the non-equilibirum work is found to come from the process of detaching leucine residues from the surface.

472350 Metastable Criticality in Liquids with Tetrahedral Symmetry

Abstract:

Liquids with tetrahedral symmetry exhibit well-known structural, thermophysical and dynamic anomalies that arise from a competition between high-density and low-density local coordination structures in the fluid [1,2,3]. These structures allow such systems to form both high-density and low-density glassy phases upon rapid cooling [1]. It has been posited that these experimentally observed glass phases relax into two distinct ergodic liquids at higher temperatures that undergo a first-order liquid-liquid phase transition (LLPT) [4]. Critical fluctuations associated with the LLPT are thought to have a significant influence on the phase behavior of these substances, even at conditions far away from liquid-liquid coexistence [1,4]. Although compositional LLPTs occur in mixtures, direct experimental observation of density-driven LLPTs in pure fluids has proved to be challenging because they are typically predicted to occur at conditions where the liquids are metastable with respect to crystallization [1]. Recent computational studies of molecular fluids demonstrate metastable LLPTs are possible [5], but they have not resolved the outstanding question of how such behavior can be characterized experimentally.

Here, we discuss results from recent computational studies [6] that suggest the possibility of designing colloidal particles to exhibit experimentally observable LLPTs. Using advanced free energy techniques, we show that a binary mixture of charged particles can form two distinct tetrahedrally structured liquids that can be brought into phase coexistence [7]. The liquids are charge-neutral phases that have identical compositions but different densities. Consequently, our calculations demonstrate that the LLPT is density-driven and isomorphic to those that have been hypothesized for pure tetrahedral substances.Â We also use simulation to elucidate the connection between the systemâ €™s LLPT and its distinct structural glasses. Finally, we discuss alternative strategies for designing colloidal systems that exhibit glass polyamorphism and LLPTs [8].

[1] PG Debenedetti, Journal of Physics: Condensed Matter 15, R1669 (2003)

[2] JR Errington and PG Debenedetti, Nature 409, 318-321 (2001)

[3] MS Shell, PG Debenedetti and AZ Panagiotopoulos, Physical Review E 66, 011202 (2002)

[4] PH Poole, F Sciortino, U Essmann and HE Stanley Nature 360, 324-328 (1992)

[5] JC Palmer, F Martelli, Y Liu, R Car, AZ Panagiotopoulos and PG Debenedetti, Nature 510, 385-388 (2014)

[6] E Lascaris, Phys Rev Lett, in press (2016)

[7] R Chen, E Lascaris and JC Palmer, in preparation (2016)

[8] F Smallenburg, L Filion and F Sciortino, Nature Physics 10, 653-657 (2014)

472491 Molecular Model for Chirality Phenomena

Abstract:

Chirality has broad implications in pharmacology, materials processing, and the origin of life. In this work, we present a new three-dimensional continuum molecular model for chirality phenomena in condensed matter systems. Our model system is based upon a simple tetramer molecule with kinetics, including the ability to switch chirality or racemize, and with thermodynamics, that reflects energetically favoring specific chiral interactions. In particular, we introduce a chiral renormalization parameter that can locally favor either homochiral or heterochiral configurations. Using this model in conjunction with molecular dynamics calculations, we highlight a range of chirality specific phenomena, including the kinetics of chiral inversion, the mechanism of spontaneous chiral symmetry breaking in the liquid, chiral liquid-liquid phase separation, and chiral crystal structures.

455884 Experimental and Modelling Study of the Phase Equilibria of (CO_2 + methylcylohexane + N_2) at High Pressures and Temperatures

Abstract:

1. Introduction

In order to design and operate both CO_2 storage and CO_2-EOR processes, an in-depth understanding of the phase behaviour of multicomponent mixtures containing CO_2 and hydrocarbon is essential [1]. Numerous experimental and modelling studies exists in the literature for the phase behaviour of binary mixtures of CO_2 with various hydrocarbons, resulting in generally good understanding of these systems. However there still remains a scarcity of data on the phase behaviour of multicomponent mixtures of CO_2 and hydrocarbons, including mixtures containing diluents such as N_2. Such systems are more representative of real CCS and EOR processes in which the CO_2 stream usually contains impurities that can be of significant impact. Accordingly, the current work is aimed at providing new data on multicomponent mixtures that can be used to assess the impact of diluents on the phase behaviour of CO_2-hydrcarbon systems for a better of CO_2 storage and CO_2-EOR processes.

2. Equipment Design

In this work, the phase behaviour of the multicomponent mixture (CO_2 + Methylcyclohexane + N_2) was studied in an apparatus combining the synthetic and analytical measurement capabilities at temperature up to 423 K and pressure up to 50 MPa.

During synthetic measurements, mixtures of precisely known composition are prepared and their phase behaviour is observed visually in the variable-volume cell. This allow for the determination of various types of phase boundary including vapour-liquid, liquid-liquid and vapour-liquid-liquid loci, critical curves of mixtures as well as cloud curves. In the analytical approach, the compositions of the coexisting bulk phases at equilibrium are determined by means of fluid sampling devices connected to on-line gas chromatography for identification and quantification.

The apparatus consisted of a cylindrical equilibrium variable volume cell made of titanium and driven by a computer controlled servo motor. The cell was designed with five high-pressure ports, two of which were used to allow for the inlet and outlet of the mixture. Two other ports were used for sampling, to which Rolsi sampling valves were fitted, while a fifth port is used for a temperature sensor in direct contact with the fluid. Composition analysis was by gas chromatography with both thermal-conductivity and flame-ionisation detectors arranged in parallel. Visual observation was enabled by a sapphire window and a CCD camera.

Temperature control of the equilibrium cell was achieved by encasing the cell in an aluminium heating jacket with axial holes for cartridge heaters and PT100 temperature sensors which were in turn connected to a PID process controller. Equilibration of the contents of the cell was assisted by a magnetic stirrer. Four calibrated syringe pumps with a maximum service pressure of 70 MPa were used for quantitative fluid injection. One pair of pumps was configured to accept high-pressure gases or liquefied gases, while a second pair was configured to accept components that are liquid under ambient conditions.

3. Experiment and Modelling

The apparatus was validated by means of comparison with published isothermal saturated vapour pressure measurements on methanol, as well as isothermal vapor-liquid equilibrium measurements on (CO_2 + methylcyclohexane).

The vapour-liquid phase behaviour for the ternary mixture of (CO_2 + methylcyclohexane + N_2) were studied over the temperature range (323 and 423) K and pressures up to the critical pressure. Three different molar ratios between CO_2 and N_2 were fixed (0.25, 0.5 and 0.75) in the ternary system and the bubble-curve and dew-curve were determined along the target isotherms at different amounts of methylcyclohexane.

The predictive capability of SAFT-g-Mie [2] was also tested in order to model the phase equilibria of this mixture. The Statistical Associating Fluid Theory, stemming from the first-order perturbation theory of Wertheim [3], was implemented in this work with a group contribution approach and the generalized Mie potential to represent segment-segment interactions. In the resulting SAFT-g-Mie, complex molecules are represented by fused segments representing the functional groups from which the molecule may be assembled. All interactions, both like and unlike, as implemented in this work were determined from experimental data of systems comprising the constituent groups. In addition to the above, the capability of the Predictive Peng-Robinson equation of state [4] was also tested.

References

1. S. W. Løvseth, H. G. J. Stang, A. Austegard, S. F. Westman, R. Span, R. Wegge, Measurements of CO2-rich mixture properties: Status and CCS needs, *Energy Procedia*, 86 (2016).

2. V. Papaioannou, T. Lafitte, C. Avendano, C.S. Adjiman, G. Jackson, E.A. Muller, A. Galindo, Group contribution methodology based on the statistical associating fluid theory for heteronuclear molecules formed from Mie segments, J. Chem.Phys, 140 (2014).

3. M. S. Wertheim, Fluids with highly directional attractive forces. II. Thermodynamic perturbation theory and integral equations, J. Stat. Phys, 35 (1984).

4. D. Y. Peng, D. B. Robinson, A new two-constant equation of state, Ind. Eng. Chem. Fundam, 15 (1976).

458578 Prediction of Liquid-Liquid Equilibria in Oligomer-Containing Biotechnological Systems Using Associating Lattice Cluster Theory (ALCT)

Abstract:

In bioprocess engineering often oligomer-containing complex mixtures occur, which show liquid-liquid demixing and are hard to describe with standard thermodynamic models. The situation is complicated further by the fact that many of these systems are only poorly characterized and/or contain components which are not available as pure substances. Standard process engineering workflows fail in these cases. Recent advances in molecular thermodynamics enable tackling these issues. In the present study, we combine lattice cluster theory[1–3] for modeling the molecular architecture with Wertheim theory for modeling H-bonding yielding the Associating Lattice Cluster Theory (ALCT).

As a test system for the application of ALCT, we choose mixtures of oleic acid (OA), formic acid (FA), and formoxystearic acid (FSA). That system is of interest for the production of FSA-based biopolymers from the renewable resource OA. FSA can be produced from OA and FA in the presence of strong acids. It is, however, presently not available commercially as a pure substance. OA + FA show a miscibility gap, but it is unclear if this is also the case for FA + FSA. Both OA and FSA are oligomeric substances, i.e. their molar mass is relatively high in comparison with FA, but end group effects are still important. Thermodynamic modeling of the system FA + OA + FSA is therefore extremely challenging, but a prerequisite for studies of the production of biopolymers using the route over OA and FA. ALCT lends itself to tackle the problem as it is well suited for predictions of properties of mixtures containing oligomers of different architecture.

Liquid-liquid equilibria (LLE) in the system FA + OA + FSA were studied both experimentally and by modeling and simulation with ALCT. In the experiments, the main challenge is the analysis, for which quantitative NMR spectroscopy was used. The ALCT model was parameterizes using experimental data on

LLE and excess enthalpies for the system FA + OA and available data on related systems. The description of FSA is predictive and based only on information on its molecular structure. The predictions of ALCT are compared to experimental data on the LLE in the system FA + OA + FSA and show very good agreement. This demonstrates the applicability of ALCT for modeling and predicting the phase behavior of oligomer-containing biotechnological systems.

(1) Enders, S.; Langenbach, K.; Schrader, P.; Zeiner, T. Phase Diagrams for Systems Containing Hyperbranched Polymers. *Polymers* **2012**, *4* (1), 72–115.

(2) Browarzik, D.; Langenbach, K.; Enders, S.; Browarzik, C. Modeling of the Branching Influence on Liquid–liquid Equilibrium of Binary and Ternary Polymer Solutions by Lattice–cluster Theory. *J. Chem. Thermodyn.* **2013**, *62*, 56–63.

(3) Langenbach, K.; Browarzik, D.; Sailer, J.; Enders, S. New Formulation of the Lattice Cluster Theory Equation of State for Multi-Component Systems. *Fluid Phase Equilibria* **2014**, *362*, 196–212.

459479 Intermolecular Potentials for Water: Molecular Simulation of Bulk Properties, Thermodynamic Properties and Phase Behavior

Abstract:

Water plays a near ubiquitous role in biological, chemical and industrial processes. Historically, predicting the properties of water involved either empirical correlations or equation of state modelling [1], [2], whereas more recently molecular simulation [3] has become he method of choice because of the nexus between underlying intermolecular interactions and observable macroscopic properties. The availability of an intermolecular potential to evaluate inter-particle forces or energies is the key to accurate predictions. There are many alternative intermolecular potentials for water [4], although the basis of many water potentials is at best semi-empirical. The most widely used models are rigid and variants of either the four-site transferable interaction potential (TIP4P) or the three-site simple point charge (SPC, SPC/E) models. The appeal of such potentials is computational expedience and in many cases they have provided worthwhile predictions. However, comparisons with experiment are often focused at relatively low temperatures and pressures, with a temperature of 25 °C and a pressure of 1 atmosphere being a popular choice.

The accuracy for 20 atomistic models for water is comprehensively evaluated [5] with particular attention to thermodynamic properties such as the thermal pressure coefficient, thermal expansion coefficient, isothermal and adiabatic compressibilities, isobaric and isochoric heat capacities, Joule-Thomson coefficient, speed of sound and vapor-liquid equlibria. The addition of bond flexibility, while improving agreement with experiment for such properties as phase coexistence, dielectric constants, viscosity and diffusion, does not appear to significantly improve the prediction of thermodynamic properties in general. For the supercritical heat capacities and thermal expansion coefficient the flexible TIP4P/2005f model yields less accurate values than its rigid TIP4P/2005 counterpart. In contrast, accounting for polarizability consistently results in improved agreement with experiment. For properties such as the isochoric heat capacity and thermal expansion coefficient, the polarizable MCYna model yields values that are in very close agreement with experimental data in the temperature range of 300 â€“ 600 K.. At or near ambient

conditions, the overall ranking of models investigated is (iAMOEBA, MYCYna, BKd3, GCPM) > (BK3, TIP4P/FQ, SPC/Fw) > (NvdE, TIP4P/2005, SPC/E) > (SPC, TIP4P/2005f, TIP5P) > (TIP3P, TIP4P) > (MCY, MCYL).

[1] Y. Wei and R. J. Sadus, AIChE J. 2000, 46, 169-196 ; N. G. Stetenskaja, R. J. Sadus and E. U. Franck, J. Phys. Chem. 1995, 99, 4273 -4277 ; A. E. Mather, R. J. Sadus and E. U. Franck, J. Chem. Thermodyn. 1993, 25, 771-779.

[2] W. Wagner and A. PruÃŸ, J. Phys. Chem. Ref. Data. 2002, 31, 387-535.

[3] R. J. Sadus, Molecular Simulation of Fluids: Theory, Algorithms and Object-Orientation, Elsevier, Amsterdam, 1999.

[4] C. Vega and J. L. F. Abascal, Phys. Chem. Chem. Phys. 2011, 13, 19663.

[5] I. Shvab and R. J. Sadus, Fluid Phase Equilib. 2016, 407, 7.

459848 Thermodynamics of Liquid Polyamorphism: Equation of State of a Fluid with Thermodynamic Equilibrium Between Two Structures

Abstract:

"Liquid polyamorphism" is the existence of two alternative amorphous structures in a single-component liquid. Liquid polyamorphism is found in a broad group of very different materials, such as triphenyl phosphite, silicon, silicon dioxide, cerium, and hydrogen, usually at extreme conditions. In particular, this phenomenon is hypothesized in metastable, deeply supercooled water. Cold and supercooled water is assumed to exist in two forms, a low-density or high-density liquid. The existence of these two alternative structures could, at certain conditions, result in a metastable liquid-liquid separation in pure water. The hypothesized liquid-liquid metastable coexistence is not directly accessible in bulk-water experiments because it is presumably located a few degrees below the empirical limit of homogeneous ice formation. We consider a phenomenological approach to describe liquid polyamorphism in a single-component fluid. The first fundamental question to be addressed is the separation of time scales: a system with two inter-convertible fluid structures can be thermodynamically treated as a single-component fluid if the time of observation is longer than the time of conversion (fast conversion). In the opposite limit (slow conversion) the system thermodynamically behaves as a two-component mixture. We propose a simple equation of state which can generically describe liquid-gas and liquid-liquid transitions in the same single-component fluid. The lattice-gas model and the van der Waals equation of state are adopted for one of the alternative structures. In order to predict liquid polyamorphism, the lattice-gas equation of state and the van der Waals equation of state are supplemented by a thermodynamic constraint associated with the "chemical-reaction" equilibrium between the two inter-convertible liquid structures. The equations provide a closed formulation for the thermodynamics of the system. This system of equations is nonlinear and is solved numerically to obtain both the liquid-gas and liquid-liquid equilibrium and the corresponding stability limits. The results are tested on realistic water-like models for which the properties obtained by MD simulations

are available.

470627 Phase Behavior of Polymer Blends from Integral Equation Theory and Molecular Simulations

Abstract:

χ parameter, which governs the phase behavior of polymer blends, is generally estimated from neutron scattering experiments or by fitting to experimental data. Despite its widespread use, estimation of χ parameter from the first principles is still an active topic of interest. In this work, we combine molecular dynamics simulations with the integral equation theory framework proposed by Schweizer & Curro [*Journal of Chemical Physics*, **91**, 5059 (1989)] to determine the χ parameter for polymer blends. The model polymer system used for this purpose is a blend of polyethylene (PE) and isotactic polypropylene (iPP). We show that the χ parameter calculated from molecular simulations and integral equation theory is in good agreement with the one obtained from Random Phase Approximation (RPA), which is the approach often used for experimental determination of the χ parameter. The effect of tacticity and chain length on the value of the χ parameter for the polyethylene (PE)-atactic polypropylene (aPP) blend is explored. Furthermore, we compare the χ parameter obtained from simulations with that obtained from experiments for polyisobutylene (PIB)-polybutadiene (PBD) binary blend. The χ parameter value is interpreted in terms of the molecular structure of these polymeric systems.

470675 Properties of Mixed Patchy Colloids: Improved Theory for Multi-Body Effects in the Reference Fluid

Abstract:

A system of patchy colloidal particles interacting with a solute that can associate multiple times in any direction is a useful model for patchy colloidal mixtures. Despite the simplicity of the interaction, predicting the thermodynamics of such systems remains a challenge because of the presence of multi-body correlations. Earlier Marshall and Chapman[1] developed a perturbation theory to describe the thermodynamics of this system. Input to the theory is a multi-body cluster partition function for the reference fluid of hard-sphere solvent molecules in a defined inner-shell (or coordination volume) of the hard-sphere solute. The multi-body contribution to these partition functions are approximated with those of isolated clusters and the bulk solvent effect is obtained by superposition of pair correlation function and a three body correction. This TPT2 based approach cannot describe multi-body effects at high densities. In this work we emphasize the importance of packing effects in the reference system in determining the multi-body effects in associating mixtures. To incorporate correct multi-body cluster integrals for dense reference systems[2,3], we obtain distribution information from hard sphere simulations. We show that a partition function with accurate cluster integrals for dense reference systems better captures the distribution of solvent around the solute up to high system densities. Using this updated reference, we find that theory better describes both the bonding state and the excess chemical potential of the colloid in the physical system. Results for different association strengths and concentrations will be shown. Also, extension to different associating geometries and asymmetric mixtures of these colloids will be shown.

We establish that the complex multi-body effects in the associating mixtures of different association geometries can be accurately determined if correct reference information is used. The present approach opens avenues to model a range of systems as a mixture of patchy and spherically symmetric colloids, with completely patchy and completely spherical being the extremes. Such a system is very helpful in studying self-assembly of patchy colloids, properties of supramolecular star molecules, and also in determining ion-solvent, ion-ion interactions in electrolyte systems. This work provides an excellent theory for studies ranging from phase equilibria to study of new structures for complex systems with isotropic and anisotropic interactions.

References

1. Marshall, B. D. & Chapman, W. G. Thermodynamic perturbation theory for self assembling mixtures of multi-patch colloids and colloids with spherically symmetric attractions. *Soft Matter* **9,** 11346–11356 (2013).

2. Reiss, H., Frisch, H. L. & Lebowitz, J. L. Statistical Mechanics of Rigid Spheres. *J. Chem. Phys.* **31,** 369–380 (1959).

3. Torquato, S. & Stell, G. Microstructure of two phase random media. I. The n point probability functions. *J. Chem. Phys.* **77,** 2071–2077 (1982).

471728 Information Theory and the Thermodynamics of Irreversible Shape Changes

Abstract:

Irreversible shape change of a solid material requires the rearrangement of constituent building blocks into new positions. The change in uncertainty of these building block positions before and after the shape change operation can be construed as the change in the information content of the system. Information theory then helps us predict the minimum energetic cost of such an operation (given by Landauerâ€™s principle). We apply this concept to the simple process of irreversible length change of an elastic solid under uniaxial tension and find that the minimum stress required to cause this deformation is given by $k_B T ln(2)/V^*$. Here k_B is the Boltzmann constant, T is the temperature and V^* is the volume of a single, independently movable building block comprising the material. Interestingly, this stress closely predicts the experimentally calculated value for yield stress in a large variety of materials at room temperature. This suggests that the information loss during the irreversible shape change of an elastic material under tension significantly contributes to the energetic cost of this procedure. Looking closely at the microscale interactions between building blocks of the material, we observe that the positional uncertainty/information change is a result of a symmetry breaking when material flow occurs in one direction vs. the other along the line of action of the applied stress. We believe we have identified a new mechanism for energy dissipation in materials under mechanical stress undergoing irreversible shape change. The energy dissipated through this mechanism depends on the initial and final uncertainties in the material building block positions and is a function of temperature and the initial configuration of the system. This dissipation mechanism might hold the key to understanding the mechanics of biological active, adaptive materials as well as for the design of materials with autonomous shape transformation properties.

472250 Predicting Mosced Parameters from COSMO-SAC Sigma Profiles

Abstract:

Commonly used activity models such as UNIFAC and MOSCED (Modified Separation of Cohesive Energy Density) rely on empirically characterized parameters to predict phase behavior for petrochemical mixtures. MOSCED is particularly attractive in many situations because it offers intuitive insights into how to tune solvent-solute interactions to achieve optimized formulations. Unfortunately, only 133 compounds have been characterized with the original MOSCED implementation, compared to UNIFAC which has been developed using several hundred compounds. Furthermore, extending UNIFAC to new compounds is straightforward through the group contribution concept, whereas MOSCED simply requires more experimental data specific to the new compounds.

This presentation details a simple method to estimate MOSCED parameters in order to determine infinite dilution activity coefficients based on density functional theory (DFT) calculations provided by the Virginia Tech database of σ-profiles for the COSMO-SAC method. The hypothesis of the work is that the surface charge density of a molecule, provided by a σ-profile from the DFT computation, can be used to estimate the α, β, and τ parameters used in the MOSCED model. By defining a charge density threshold for regions of hydrogen bonding, the probability of the charge exceeding that threshold and size of the surface can be correlated to determine the parameters for the molecules. By assuming the remaining surface charge potential that does not contribute to hydrogen bonding represents the polarity of the molecule, the polarizability parameter τ can also be determined.

The current number of characterized compounds for the MOSCED model is 133, of which only 89 are considered to have non zero acidity and basicity parameters. This number limits the possible use of the method due to the low number of molecules exhibiting interactions other than dispersion forces. The proposed method allows for the expansion of the model to 1432 different compounds along with any molecules that are characterized in the future by a σ-profile. The correlated functions were regressed based on 4375 binary solution infinite dilution coefficients provided by Lazzaroni et al. and the deviation from the experimental data was calculated based on the logarithmic ratio of the calculated versus experimental [4]. The resulting average deviation for the MOSCED model with correlated parameters was found to be 0.28 while using the original parameters tuned to the experimental data had a deviation of 0.106. The UNIFAC model was also compared for binary solutions in which the functional groups were defined and the resulting deviation was found to be 0.183. The calculated infinite dilution coefficients were then used to interpolate the entire phase behavior of a binary system across the composition range. The "simplified MOSCED" (SSCED), Wilson, and NRTL models were chosen to test the accuracy of the method based on the low number of parameters needed to define the interaction energies for the system. The resulting phase equilibrium predictions were compared to experimentally determined results from the Danner and Gess database [10]. The average percent deviations of the pressure for the 39 binary systems tested were 17.39% for Wilson, 18.90% for NRTL, and 13.83% for SSCED. In conducting this work, it was noted that characterizations for amines were omitted from the previous MOSCED publications. A database of amine VLE was compiled to characterize both the compound specific parameters and the σ-profile predictions. Deviations were similar to other mixtures as noted above.

471156 Statistical Thermodynamic Modeling of Early Amyloid-β Oligomer Formation: Explicit and Implicit Incorporation of Hydrogen Bonding in a Self-Consistent Field Framework

Abstract:

Alzheimer's disease (AD) is the sixth leading cause of death in the United States where its prevalence is expected to more than double by 2050. Amyloid-β (Aβ) is a short polypeptide for which proteostasis becomes disrupted and concentrations increase in AD such that it begins to aggregate. The aggregation process is exceedingly complex and involves a dynamic equilibrium including myriad sizes and conformations of aggregated Aβ. For several decades, a mounting body of research has implicated these aggregates, particularly oligomers, as an etiological factor in Alzheimer's disease (AD). Current understanding of early oligomer structure and formation stems primarily from experimental techniques and molecular dynamics and Monte Carlo simulations; however, the exact structures of these species have remained unclear. Moreover, information regarding the relative propensity of these species to form and their importance in the aggregation process remains largely inaccessible. We have implemented a new approach that uses Aβ conformations obtained from molecular dynamics simulations with a statistical thermodynamic model to directly address the multiplicity of relevant structures during the aggregation process and derive information on the thermodynamics of oligomer formation.

To explore the configuration space of small Aβ oligomers, fully atomistic replica exchange molecular dynamics (REMD) simulations of Aβ were performed using the Gromacs (version 4.6.5) software package, the Charmm 36 force field and an explicit water model. An Aβ monomer was simulated using REMD at two different ionic strengths. In addition, an Aβ dimer was simulated at room temperature and physiological ionic strength. These simulations were used to derive an ensemble of conformations containing relevant structures for the monomer over a range of temperatures and ionic strengths as well as those for a dimer both during the process of binding and while bound.

A self-consistent field theory has been developed and implemented to model the thermodynamics of an arbitrary number of interacting Aβ molecules. The theory consists of formulating the appropriate thermodynamic potential for isolated Aβ molecules in a bath of solvent and ionic species in terms of the energy and entropy associated with all species. The theory captures the four most relevant physical properties of Aβ: these properties are the shape of the molecule, the characteristics of the protonatable sites, the hydrophobic nature of the constituent residues and the hydrogen bonding that can occur between donors and acceptors on the protein. Hydrophobicity and protein-protein hydrogen bonding are extremely important driving forces for protein structure and are both manifestations of the effects of hydrogen bonding. Hydrophobicity is captured implicitly in our model through the assignment of experimentally derived free energies of solvation to the solvent-exposed surface of the protein. These solvation energies reflect the disruption that different regions of the protein surface cause in the hydrogen bond network of water (that is, water-water hydrogen bonding) and the degree to which these regions form hydrogen bonds with water (protein-water hydrogen bonding). Protein-protein hydrogen bonding is captured in a separate term in the free energy which depends explicitly on the orientation of the donors and acceptors.

The theory was used to evaluate the most probable conformations of Aβ for isolated monomers and those in the process of dimerization. These conformations were then simulated using the same methodology as before thus establishing a feedback between the statistical thermodynamic model and molecular mechanics simulation to increase the resolution of the most significant part of configurational phase space. The complete, theoretical feedback-bolstered configurational ensemble was used as input in the theory which was applied to the process of Aβ trimer formation. In this study, the centers of mass of three Aβ monomers were assigned to positions in space and the theoretical model was used to calculate the properties of constrained equilibrium at those conditions. This procedure was repeated over a range of relative positions to elucidate the free energy landscape of the trimer formation process. As anticipated, it was found that the effects of hydrogen bonding as manifest through our implicit and explicit terms were significant determinants of the structure of the trimer.

This theory is capable of providing a rich description of the molecular organization and physics that guide Aβ interactions that cannot be determined experimentally. The unique flexibility of this theoretical framework to treat solution conditions and systems of arbitrary geometries makes this a promising tool in two important aspects. First, it can be used to further fundamental understanding of the Aβ oligomerization process and how this process depends on different factors. Secondly, this model can be used as an inhibitor design tool through the incorporation of inhibitors into the framework. Further work will extend this model to the treatment of the interactions of larger aggregate species in order to elucidate structural properties of these larger oligomers and characterize the thermodynamic pathways of their formation.

464733 Aggregation of Amyloid Beta (17-36) in the Presence of Naturally Occurring Phenolic Inhibitors Using Coarse-Grained Simulations

Abstract:

Alzheimer's Disease (AD) is a neurodegenerative disease that causes dementia, nervous system degradation, and death. Currently there are no therapeutic agents available for the treatment of AD despite great effort from the research community. Many small phenolic compounds have been shown in experiment to be effective at preventing the formation of Aβ fibrils. In this work we use discontinuous molecular dynamics (DMD) simulations to learn how four naturally-occurring phenolic compounds, resveratrol, vanillin, curcumin, and epigallocatechin-3-gallate (EGCG), bind to Aβ(17-36) monomers as well as affect its oligomerization and fibrillation. We use a coarse-grained model/force field for inhibitors which has geometric and energetic parameters that are compatible with the coarse-grained protein model PRIME20. Preliminary results show that the U-shaped protofibril structure formed by Aβ(17-36) is similar to the corresponding part of the Aβ(1-42) fibril model, based on solid state NMR data. The order of peak heights in the peptide-inhibitor radial distribution function shows that the strength of the inhibitor binding affinity is resveratrol > curcumin and EGCG > vanillin. Simulations of 8 Aβ(17-36) peptides aggregating in the presence of 30 inhibitors show that EGCG, resveratrol and curcumin inhibit Aβ(17-36) fibril formation while vanillin only retards the lag phase. Instead of forming an ordered β sheet structure (as occurs in the absence of inhibitors), the peptides remain random coils and form complexes with the inhibitors, binding mainly to the hydrophobic residues near the peptide terminii. The relative ability to inhibit Aβ(17-36) fibrillation is resveratrol > curcumin > vanillin, consistent with experiments by Feng *et al.* (2009) showing resveratrol > curcumin and experiments by Reinke *et al.* (2007) showing curcumin > vanillin on full length Aβ(1-42) aggregation. Our simulations provide molecule-level insights into the mechanisms by which small molecules inhibit Aβ aggregation.

472571 The Influence of Lipid Tail Length and Cholesterol Content on the Structure of Model Stratum Corneum Bilayers

Abstract:

The skin's ability to act as a barrier between the internal and external environments of the body is controlled by the stratum corneum (SC), the thin outermost layer of the epidermis. The SC is composed of corneocytes, surrounded by a dense lamellar lipid matrix composed of ceramides (CER), cholesterol (CHOL), and free fatty acids. Skin disease is often coincident with an altered SC lipid composition and a decreased barrier function, e.g., elevated levels of CHOL [1] have been observed for lamellar ichthyosis patients and increased level of short-chain CER [2] for atopic eczema patients. Although a connection between SC lipid composition and skin barrier function is clear, the mechanism is not clearly understood, due to the fact that composition may vary between samples and molecular arrangements must be inferred from spectroscopic experiments. Molecular simulation offers a promising complement to experimental work, as it allows precise control over system composition and atomic-level resolution. However, the dense gel and crystalline phases formed by the SC lipids substantially limits molecular mobility at physiologically relevant conditions, and thus molecular simulations may fail to reach equilibrium and the results may be unduly influenced by the initial starting morphology of the lipids.

Here, we investigate the structural dependence of SC model lipid bilayers on CER tail length and CHOL content using a modified version of simulated tempering [3]. Simulated tempering, whereby a simulation undergoes a random walk through temperature space, provides a means for systems to escape energetic traps and explore phase space, allowing the system to efficiently decorrelate from the initial, assumed configuration. To investigate the role of the CER tail length on bilayer structure, a series of systems are studied in which the relative amount of short and long-chain CER is varied. Similarly, systems with varying amounts of CHOL are studied to probe the role of CHOL content on bilayer structure. We find that several key structural properties are dependent on both CER tail length and CHOL content, including a strong dependence of tilt angle and interdigitation behavior on the relative lipid concentration, both of which may have significant implications for the barrier function. These results provide new insights into the connections between lipid composition, structure, and barrier properties of the SC.

References

[1] Lavrijsen, A. P. M. et al. Reduced Skin Barrier Function Parallels Abnormal Stratum Corneum Lipid Organization in Patients with Lamellar Ichthyosis. J. Invest. Dermatol. 105, 619–624 (1995).

[2] Janssens, M. et al. Increase in short-chain ceramides correlates with an altered lipid organization and decreased barrier function in atopic eczema patients. J. Lipid Res. 53, 2755–2766 (2012).

[3] Marinari, E and Parisi G. Simulated Tempering: a New Monte Carlo Scheme. Europhys. Lett. 19 (6), 451-458, (1992)

469800 A Molecular Thermodynamic Model for Nanoparticle-Membrane Interactions

Abstract:

The thermodynamics of nanoparticle interactions with cellular membranes has momentous consequences for nanotechnology with biological, toxicological, and pharmacological applications. The pervasiveness of nanoparticles in therapeutics, foods and beverages, cooking products, packaging, cosmetics and sunscreens, and agriculture, amongst many other areas, coupled with the lack of fundamental understanding of the interplay of particle design effects like size, chemistry, shape, and elasticity on product performance, underscores the importance of a biophysical approach to nanoparticle-membrane interactions. Here, we use molecular dynamics simulations to characterize the behavior of homogeneous, spherical nanoparticles with model lipid bilayers. We delineate unique mechanistic modes of particle-membrane interaction in the space of nanoparticle size and chemistry (hydrophobic/philic). We uncover a diverse array of stable and long-lived metastable configurations that vary sensitively with the effective particle attractions to lipid head and tail groups, and with size as particles move from the molecular (<0.5 nm) to lower â€œcolloidalâ€ (>10 nm) regime.

Continuum theory provides a starting point for understanding some of these behaviors, particularly at the extremes of small molecules (solubility theory) and large colloidal particles (membrane elastic theory). The intermediate nanoscale regime, however, is complex because the membrane thickness (~5 nm) is comparable to the particle size, individual lipid fluctuations become relevant, and significant geometric asymmetries of inserted configurations emerge. We implement the molecular simulations to direct the development of new theory in this regime. We use calculated lipid leaflet distributions and field-based descriptors of membranes with inserted particles to extend existing theoretical models with free energetic contributions from deformations in lipid molecular tilt, twist, and splay. We then investigate improvements in these continuum predictions through comparison with detailed umbrella sampling free energy calculations. While focused on model systems, this nanoparticle study has interesting implications for a wide range of organic and inorganic nanoparticles, and also for biomacromolecules like membrane proteins.

465889 Coarse-Grained Modeling of Heterochromatin Formation in Chromosomal DNA

Abstract:

Historically, the central dogma of genetics asserts that DNA sequence holds all of the necessary biological information. However, two organisms with identical genetic information may have different behavior due to chemical modifications in their genome packaging. This notion of epigenetic regulation represents a paradigm change in how we think about genetic traits. To establish a quantitative understanding of epigenetic regulation, we need to understand the conditions for chromosomal DNA to condense into regions of tightly packed heterochromatin where their expression is suppressed. At certain stages in the cell cycle, the heterochromatin appears to condense into â€œfociâ€ that are distributed throughout the nucleus. These observations are qualitatively reminiscent of the better understood microphase separation of traditional block copolymers. If the spatial chromatin texture is indeed governed by thermodynamics, then a wealth of knowledge from traditional polymer physics can be applied with appropriate modifications. I will present on a Monte Carlo simulation where the DNA and associated proteins are modeled as a semiflexible copolymer with blocks of increased methylation state that are preferentially bound together to cause microphase separation. To model condensation on the size scale of nucleus, we use a theoretically informed coarse graining method. As staining experiments show that heterochromatin is preferentially segregated to the periphery of the nucleus, understanding chromosomal condensation in the presence of a confinement is of particular interest. The predictions for the conditions leading to condensation and the resulting spatial and genomic patterns of condensation will be presented.

470186 Thermodynamic Perturbation Theory and Property Network for Coarse Grain DNA Model

Abstract:

A coarse grain model of a deoxyribonucleic acid (DNA) single strand is proposed by considering each nucleotide to consist of a trimer of tangent homonuclear hard-sphere segments (phosphate, sugar and base) with added attractive interactions. The entire strand (oligonucleotide) is constructed by forming covalent bonds between the sugar (articulation) segment and the phosphate segment on a neighboring nucleotide. Each of two popular covalent bonding platforms for branched hard-sphere chains [1, 2] is used with a recently proposed hard-sphere equation of state [3] to obtain the athermal free energy component, and dispersion forces are incorporated at a van der Waals mean field level. The model is further refined by adding a term for type I hydrogen bonding within the statistical associating fluid theory (SAFT) framework for Watson-Crick (base-pair) association to form DNA double-strands (dimers) (see Figure 1 for a depiction of double-stranded DNA model). Hydrogen bonding sites are also added to the phosphate, sugar and base segments for cross-association with the solvent water. Finally, phosphate segments are negatively charged and sodium (or potassium) cations are represented by positively charged hard-spheres (same diameter as solvent water and all DNA segments in original version of model).

The residual Helmholtz free energy of the resulting aqueous DNA system including counterions is derived, and used to obtain expressions for a variety of properties including osmotic coefficient and viscosity of the solution and diffusivity of DNA in water as functions of DNA fragment length. Several illustrations are given to highlight key points, and suggestions are made on how the model can be improved by considering heteronuclear segments and bond angle dependence

REFERENCES

[1] Marshall, B. and W. G. Chapman, Three Branched Chain Equations of State Based on Wertheimâ€™s Perturbation Theory, J. Chem. Phys. **138**, 174109 (2013).

[2] Blas, F. J. and L. F. Vega, Thermodynamic Properties and Phase Equilibria of Branched Chain Fluids Using First- and Second-Order Wertheimâ€™s Thermodynamic Perturbation Theory, J. Chem. Phys. **115**, 3906-3915 (2001).

[3] Gow, A. S. and R. B. Kelly, Twenty-One New Theoretically Based Cubic Equations of State for Athermal Hard-Sphere Chain Pure Fluids and Mixtures, AIChE Journal 61, 1677-1690 (2015).

Figure 1. Homonuclear hard-sphere branched chain coarse grained model of DNA including dispersion, electrostatic interactions and hydrogen bonding.

471002 Molecular Modeling of Antibody-Antigen Binding Near Solid Surfaces

Abstract:

Antibody microarrays are a class of biosensors that have the potential to revolutionize scientific research, medical testing, and national defense. This is because they have the potential to screen for thousands of molecules in a parallel, rapid, inexpensive, and easy-to-use manner. However, current antibody microarrays are not widely used due to reliability problems resulting from production tolerances, signal detection, and target selection.

A notable issue with antibody microarrays is that the antibody must perform its function in an environment significantly different than that found in vivo. Prior work in the area focused mainly on the stability of antibodies on a surface and not its affinity for the antigen. It has been shown that antibodies attached to a hydrophobic surface suffer undesirable structural changes due to excessive attraction between the antibody and the surface, but these interactions can be mitigated by using a hydrophilic surface. To move the work forward the nature of antibody-antigen interactions near solid surfaces now needs to be investigated. Our current work aims to improve our understanding of these protein-protein-surface interactions.

Because laboratory methods for studying protein structure either do not provide sufficient detail or are not transferable to heterogeneous environments, molecular simulation can be advantageous for studying surface-bound proteins. However, atomistic models suffer from insufficient sampling that cannot capture protein folding or binding events. Our group uses a recently-developed, coarse-grain, protein-surface model that provides accurate protein folding mechanisms and quantitatively reproduces protein adsorption energies. The simplicity of this model makes it possible to obtain reliable thermodynamic data about large systems of interest.

In this presentation we describe our studies of antibodies and antibody fragments interacting with different antigens near hydrophobic or hydrophilic surfaces. Thermodynamic results will show that the attachment geometry of the affinity molecule relative to the surface can affect antigen binding mechanisms and binding strength. Lysozyme will be used as a model antigen, but additional work using hemagglutinin will demonstrate how antigen size factors into the antigen-binding process near surfaces. The results provide an unprecedented view of the protein-surface and protein-protein interactions crucial to the function of antibody microarrays and offer hope that the rational design of improved microarray technologies is possible.

456334 Overcoming Thermodynamic Limitation on the Yield of Aminotransferase Reactions

Abstract:

The maximum yield of many enzyme-catalyzed reactions, as of aminotransferase reactions, is limited by thermodynamic equilibrium. Therefore, the position of the thermodynamic reaction equilibrium has to be known. Reaction equilibria can be described by the thermodynamic (activity-based) equilibrium constant K_{a*}, which is accessible by the concentrations of products and reactants (K values) and their activity coefficients ($K_{\gamma*}$ values) at reaction equilibrium. For an optimal process design, the influence of reaction conditions (T, pH, reactant molalities) on the reaction equilibrium (K values) has to be known. Enzymatic reactions can also be influenced by additives, e.g. ionic liquids or osmolytes. In those cases, the effect of additives on the stability of enzymes and on the reaction equilibrium (and thus the maximum yield) has to be known for process optimization.

In this work the thermodynamic equilibrium of the reaction (1)

L-alanine + 2-oxoglutarate ↔ L-glutamate + pyruvate

catalyzed by alanine aminotransferase (ALAT) in aqueous media has been investigated. The thermodynamic equilibrium constant K_{a*} of reaction (1) was determined. K values were measured in the absence of additives between 25 and 37 °C at $5<pH<9$ for different initial molalities of the reactants. Further, the influence of additives (1-butyl-3-methylimidazolium trifluoromethanesulfonate, 1-butyl-3-methylimidazolium chloride, 1-ethyl-3-methylimidazolium chloride, choline chloride, trimethylamine N-oxide (TMAO), urea) on the K values was investigated. Prior to that, spectroscopic measurements (UV+FTIR) were carried out to identify the effect of these additives on the thermodynamic stability of ALAT. Finally, reaction-equilibrium measurements in the presence of additives were performed for conditions that assured ALAT stability.

It was found that the thermodynamic equilibrium constant of reaction (1) is in the order of unity (K_{a*}(37 °C) = 0.75) and thus, the maximum yield is strongly limited by thermodynamic equilibrium. The experimental results also showed that equilibrium position (K value) is influenced by pH, reaction temperature, and initial molalities of the reacting agents. The K values of reaction (1) increased (shifting the reaction equilibrium to the right-hand side) with increasing temperature, increasing pH, increasing initial molalities of L-alanine and 2-oxoglutarate (equimolal conditions), as well as increasing ratio of their initial molalities. The presence of the additives decreased the stability of the enzyme except TMAO. It was also found that all additives in reaction media shifted the equilibrium towards the left-hand side of reaction (1) leading to decreased K values.

Finally, activity coefficients of reacting agents were predicted with ePC-SAFT. This allowed predicting the effect of the reaction conditions (temperature, pH, concentration ratio, additives) on K in good agreement with the experimental data. This work showed the importance of considering the influence of reaction conditions and additives on reaction equilibria to overcome thermodynamic limitations and increase the maximum yield of aminotransferase reactions.

471170 Biophysical Modulation of Endothelial Membranes: Molecular-Scale Effects of Oxidized Phospholipids

Abstract:

Biologically active oxidized phospholipids (oxPCs) such as 1-palmitoyl-2-(5-oxovaleroyl)-sn-glycero-3-phosphocholine (POVPC) and 1-palmitoyl-2-glutaroyl-sn-glycero-3-phosphocholine (PGPC) are known to exert multiple pathophysiological effects. At elevated oxPC concentrations, disruption of the endothelial barrier and induction of monocyte adhesion to endothelial cells results. The specific mechanisms by which oxPCs modulate membrane barrier function are poorly understood. These effects are further implicated in a range of disorders, including certain cardiovascular, metabolic, and immunological diseases.

We investigated the influence of these two bioactive oxidized phospholipids on model bilayer properties, membrane packing, and endothelial cell biomechanics both computationally and experimentally. Using a combination of coarse-grained molecular dynamics simulations, Laurdan multi-photon imaging, and atomic force microscopy (AFM) microindentation experiments, the impacts of the analogous truncated tail phospholipids, POVPC (aldehyde-terminated) and PGPC (carboxyl-terminated), on the structure of a multicomponent phospholipid bilayer was determined, and the consequences of their incorporation on membrane packing and endothelial cell stiffness was assessed.

Our molecular simulations predicted differential bilayer perturbation effects of the two oxidized phospholipids based on the chemical identities of their truncated tails. We showed that for the analogous oxPCs incorporated in a phosphatidylcholine bilayer containing sphingomyelin, the carboxylic analog, PGPC, experienced more significant reorientation of the truncated tail and displacement of the oxidized lipid from the plane of the bilayer. These phenomena induced greater perturbation in the structure of the lipid bilayer than the aldehyde analog, POVPC. Decreased bilayer packing and increased water permeation was observed, which was consistent with Laurdan imaging results. Computational predictions of larger membrane perturbation by PGPC corresponded with greater stiffness observed for PGPC-treated endothelial cells measured by cellular elastic moduli using AFM.

453630 Interfacial Properties of a Glycocalyx-Mimetic Model

Abstract:

The glycocalyx is a polysaccharide coating covering the cell membrane of many types of cells. The glycocalyx is mainly constituted by molecules with a high fraction of polysaccharides, such as glycolipids, glycoproteins, proteoglycans and glycosaminoglycans. These polysaccharides are exposed on the surface of the plasma membrane, and their biochemical and biophysical properties influence the surface properties of cells and tissues, contributing to phenomena such as cell recognition, cell communication, cell adhesion and phagocytosis. Moreover, the glycocalyx has a relevant impact on processes such as drug delivery, infections, gene therapies, transplant rejection or pathologies such as diabetes, ischemia, atherosclerosis or cancer.

Although many specific surface events of cells (such as ligand-receptor interactions) have been studied extensively, non-specific interactions (such as interfacial tension) are important for phenomena such as initial adhesion, wettability and stability of aqueous films. These are important for a group of diseases involving the dryness of the mucosal tissues, such as xerostomia (dry mouth), or the dry eye syndrome.

The wetting properties of mucosal tissues, such as eye, mouth or trachea have traditionally been attributed to the presence of a hydrophilic glycocalyx, characteristic of healthy mucosal tissues. However, there is a lack of studies directly correlating the expression and spatial distribution of the glycocalyx on the mucosal tissues. By modulating the expression of glycocalyx molecules on cell cultures, we have found evidence that the stability of liquid films on biological surfaces is highly dependent on the heterogeneity of the surface. In this work we discuss the contribution of hydrophilic moieties on the surface of biomimetic substrates. Specifically we address the influence of different biomimetic substrates displaying hydrophilic molecules on their surface on their wettability and stability to fluid films.

Biomimetic surfaces were prepared on glass slides coated with films of poly-L-lysine (PLL)/Hyaluronic Acid (HA), and supported lipid bilayers functionalized with Poly(Ethylene Glycol) (PEG). The surfaces were characterized by measuring contact angle and contact angle hysteresis in a 2-fluid system. The contact angle did not change significantly with the functionalization of the surfaces. However, the contact angle hysteresis increase suggested a development of surface heterogeneity similar to that observed in cell culture experiments.

463211 Multivalent Binding of Lectins to Heterogeneous Gangliosides on Cell Mimicking Surfaces

Abstract:

Lectins, carbohydrate binding proteins, often consist of multiple binding subunits that specifically or semi-specifically target to various carbohydrates. However, their binding patterns to heterogeneous cell surfaces usually do not correlate to the amount of the preferred binding carbohydrates. It was hypothesized that the binding specificity and avidity can be modulated by altering the cooperative action between multiple bound receptors. To better understand the essential nature of binding cooperativity in a multivalent binding mechanism, quantitative analysis of multivalent membrane recruitment onto the cellular surface is critical. We have developed a nanocube sensor coupled with complex reaction analysis to quantitatively explore the multivalent binding mechanism. The nanocube sensor is surrounded by a lipid bilayer that possesses the same physical and chemical properties as cell membranes. This novel sensor is an ideal tool for studying binding cooperativities because receptors can freely diffuse and rotate on 2D fluidic cellular membranes allowing receptor self-organization to enable multivalent interactions. This label-free sensing platform can be conducted in standard 384-well microplate; therefore, its high-throughput utility enables the complex analysis of multivalent lectin binding. In addition, the simple protocol (â€œmix-and-then-detectâ€) allows any end users to perform the analysis in their own laboratories.

Recently, we studied a pentameric lectin, cholera toxin subunit B (CTB), binding to mixed gangliosides present in lipid bilayers. Interestingly, very weak binding gangliosides (e.g. GM2, GM3, and asialo-GM1) could be activated by strong binding gangliosides (e.g. GM1, fucosyl-GM1, and GD1b) resulting in an up to ten fold higher CTB binding capacity.Â We also demonstrated that allosteric regulation is not the major cause of the observed binding enhancement. We hypothesize the increased effective local concentration of weak binding ligands on a fluidic bilayer improves the contact between the CTB binding subunit and the weak binding ligands. These unexpected discoveries indicate the multivalent lectin binding to a heterogeneous cell surface is not simply controlled by the specific ligands; the cooperative actions amongst gangliosides in a complex cell surface can influence the overall binding. Our novel nanocube-based cell membrane array and reaction analysis provides an excellent tool to dissect complex multivalent interactions. Its easy-to-use and high-throughput features will make this tool immediately available to many biology communities.

463705 Direct Visualization of Mixing of Cholesterol and PhospholipidsÂ

Abstract:

Cell membranes are composed of various kinds of lipids and cholesterols. Mixing and demixing of such different lipids (e.g. lipid rafts and/or micron-sized domains) in cell memebranes have been known to be crucial for many cellular processes, but dynamics of such processes is not well understood. We have recentely developed a new technique, called FRAM (Fluorescence Recovery After Merging a phospholipid covered droplet), to measure diffusion of lipid monolayers and to visualize different lipid monolayers being mixed. Using this technique, we studied the mixing of phospholipids anc cholesterols. Direct visualization revealed that various phospholipids could mix/demix with cholesterols depending on their saturation of alkyl chains. We also show that various physical parameters (e.g. line tension, partition coefficient, diffusion coefficient, etc) could be computed from the visualization of mixing.

471046 The Structure and Elasticity of Lipid Vesicles Interacting with End-Phosphorylated Polyethylene Glycol

Abstract:

The structure and dynamics of lipid membranes in the presence of extracellular macromolecules are critical for cell membrane functions and many pharmaceutical applications. End-phosphorylation of a triblock polyethylene glycol (PEG), ABAPEG 11K, markedly changes the interactions between the polymer and vesicle membranes, preventing PEG-induced vesicle phase separation. Small angle neutron scattering (SANS) shows that the phosphorylated version, Pi-ABAPEG 11K, interacts weakly with DMPC vesicle membranes but does not change the structure of DMPC unilamellar vesicles significantly at 37°C. Neutron spin-echo results show that the adsorption of Pi-ABAPEG 11K on DMPC vesicle membranes increases the membrane bending modulus *Î°*. X-ray reflectivity measurements confirm the weak adsorption of Pi-ABAPEG 11K on DMPC monolayer resulting in a more compact DMPC monolayer structure.

465986 Large Area Model Biomembranes (LAMBs) to Understand Membrane Mechanics, Structure, and Function

Abstract:

The development of model cellular membranes is crucial for fundamental investigations of biological phenomena, ranging from antimicrobial peptide pore formation and biomolecule transport to bilayer elasticity and lipid raft formation. In this talk, a versatile platform to generate free-standing, planar, phospholipid bilayers with millimeter scale areas will be introduced. The technique relies on an adapted thin-film balance apparatus allowing for the dynamic control of the nucleation and growth of a planar black lipid membrane in the center of an orifice surrounded by microfluidic channels. A unique advantage in this system for the study of membrane mechanics is control of the membrane tension in a planar geometry, which is demonstrated by measuring the elasticity modulus of bilayers with varying composition. Simultaneous fluorescence microscopy enables monitoring the lateral heterogeneity in ternary lipid mixtures undergoing phase separation. In particular, the the effect of interdigitation on phase separation in mixtures containing hybrid lipids ("linactants") is shown by comparing bilayer, "inverse" bilayer, and monolayer experiments. Independent control of the solution conditions on either side of the bilayer allows for characterizing the response of biological membranes to external stimuli, such as the introduction of antimicrobial peptides. In this realm, the destabilizing effect of Magainin II due to pore formation and altered membrane tension on bilayers reconstituted from bacterial cell lipids is analyzed. Together, the results demonstrate a new paradigm for studying the mechanics, structure, and function of model biomembranes.

468675 Lytic Mechanism of Histidine-Rich Antimicrobial Piscidins: Structural Flexibility, Hydrophobicity and Charge Regulation Examined in Real Time

Abstract:

Antimicrobial peptides (AMPs) are known to induce cell death in bacteria by destabilizing their membranes and/or interacting with intracellular targets, making them good candidates as novel agents in anti-infective therapies. Here we focused on piscidin 1 (p1) and piscidin 3 (p3), naturally occurring antibiotics from hybrid striped bass. Their atomic-level 3D structures were characterized using solid-state NMR, and showed that p1 is more hydrophobic than p3. In this study, we investigated in real time the *mechanism* of action of both p1 and p3 on bacterial cell mimics made of 3:1 phosphatidylcholine/phosphatidylglycerol (PC/PG) lipid bilayers. Membrane disruption by p1 and p3 was monitored using lipid vesicle leakage assays at pH 6.0 and 7.4. More pronounced release of calcein entrapped in the vesicles at pH 7.4 compared with that at pH 6.0, especially for p3, motivated the NMR-monitored titrations of the four and three histidine side chains of p1 and p3, respectively. Since the pKa values were determined to be around 6.0, the titrations provided a correlation between the neutral state of the histidine side chains with stronger ability of the peptides to disrupt membrane bilayers; the lower pKa values for p1 than p3 also showed that stronger charge regulation occurred in p1. Disruption of PC/PG supported lipid bilayers (SLB) by p1 and p3 was measured as a function of time, dosage and pH using a Quartz Crystal Microbalance with Dissipation. Binding affinities and kinetics constants between p1 or p3 and PC/PG SLB, at both pH 6.0 and 7.4, were estimated using Surface Plasmon Resonance. Molecular dynamics showed that deeper penetration occurs when the histidine side chains are neutral, which helps explain the stronger effects at higher pH. Taken together these results show that membrane disruption occurs only when reduced peptide charge allows more peptides to accumulate and insert in the membrane to disrupt it.

462062 Pore Formation By Aggregates of Melittin in 1,2-Dioleoyl-Sn-Glycero-3-Phosphocholine (DOPC) and 1,2-Di-(9Z-octadecenoyl)-Sn-Glycero-3-Phospho-(1'-rac-glycerol) (DOPG) Mixed Lipid Bilayer

Abstract:

Antimicrobial peptides (AMP) inactivate microorganisms by forming pores in cell membrane. Since their mechanism of action differs from that of antibiotics, they could be very useful for combating drug resistant microbes, for treatment of microbial infections and also for prevention. Elucidation of mechanism of pore formation will help in the design of synthetic AMP for specific microorganism. This study characterizes (i) pore formation and (ii) the potential of mean force for peptide insertion, through molecular dynamics (MD) simulation for melittin, a naturally occurring AMP, into a DOPC/DOPG mixed bilayer. Critical peptide/lipid ratio was observed for penetration of melittin from the surface as well as for water channel formation for transmembrane peptides. The phospholipid density profile across the bilayer indicated a toroidal pore structure. The energy barrier for insertion of a melittin molecule decreased for insertion into a pore with larger number of transmembrane peptides eventually approaching zero. Water channel formation occurred only for insertion into pores consisting of four or more transmembrane peptides with the radius of water channel being larger for larger number of transmembrane peptides. A mathematical model for the evaluation of energy barrier for formation of pore consisting of peptide aggregates of different size interspersed with phospholipid head was proposed which considers detailed pore structure as well as intermolecular interactions. Estimated free energy barrier for insertion of melittin into an ideal paraboloid pore were consistent with MD simulation results. The model was employed to elucidate the rates of formation and dissociation of pores by AMP and nucleation of pores in lipid bilayers.

468695 Pathways Used By Bacteria in Outer Membrane Vesicle Toxicity: A Target for Disease Treatment

Abstract:

Pathways Used by Bacteria in Outer Membrane Vesicle Toxicity: A Target for Disease Treatment

Justin Nice and Angela C. Brown

Department of Chemical and Biomolecular Engineering

Lehigh University, Bethlehem PA 18015

Introduction:

In recent years bacterial resistance to antibiotics has increased; leading to a higher demand for antibiotic alternatives. Our lab is exploring alternative treatment methods which target the bacterial virulence mechanism. One mechanism of toxin delivery to host cells is outer membrane vesicles (OMVs). They are created by gram-negative bacteria as a stress response by blebbing off their outer membrane to form vesicles with a diameter of 10-300nm. OMVs are used by bacteria in delivery and communication. We have investigated the mechanism of delivery for OMVs to their host cells.

Aggregatibacter actinomycetemcomitans (A.a.), a gram-negative bacteria which causes periodontitis and endocarditis in humans, produces outer membrane vesicles. LtxA is a toxin know to be released by *A.a.* both by secretion and by association with OMVs. Since it is known that free LtxA relies on the LFA-1 receptor and cholesterol for host toxicity we investigated the importance of these factors in OMV associated LtxA.

Methods:

OMVs were purified as described previously [1]. Purification of OMVs was verified by dynamic light scattering (DLS) and scanning electron microscopy (SEM). Total protein was quantified on a Nanodrop 2000. Toxicity of the OMVs to target cells (THP-1) was measured using the MTT assay.

Results:

Purified OMVs had a radius of 83.8nm with a polydispersity index (PDI) of 0.1455 when measured by dynamic light scattering DLS. The size of the vesicles imaged by scanning electron microscopy (SEM) (Figure 1) was consistent with DLS measurements. We showed that LtxA is located on the outside of the vesicle, suggesting it could play a role in OMV targeting and toxicity. Our lab has developed a cholesterol-binding peptide, based on a cholesterol recognition amino acid consensus (CRAC) sequence, which blocks LtxA from recognizing cholesterol and therefore inhibits the toxin's activity [2]. Here, we pretreated target cells with this peptide and found that the vesicles were less toxic, indicating that the delivery of *A.a.* OMVs requires cholesterol (Figure 2).

Figure 1: SEM picture showing several OMVs on surface

Experiments with a scrambled version of the CRAC peptide showed no effect on toxicity (data not shown). The LFA-1 antibody has two subunits, CD11a and CD18; we investigated the effect of blocking each subunit on OMV toxicity. Blocking

CD18 decreased OMV toxicity to the cells (Figure 3), but blocking CD11a had no effect on cell viability (Figure 4), demonstrating that LFA-1 plays some role in *A.a.* OMV delivery.

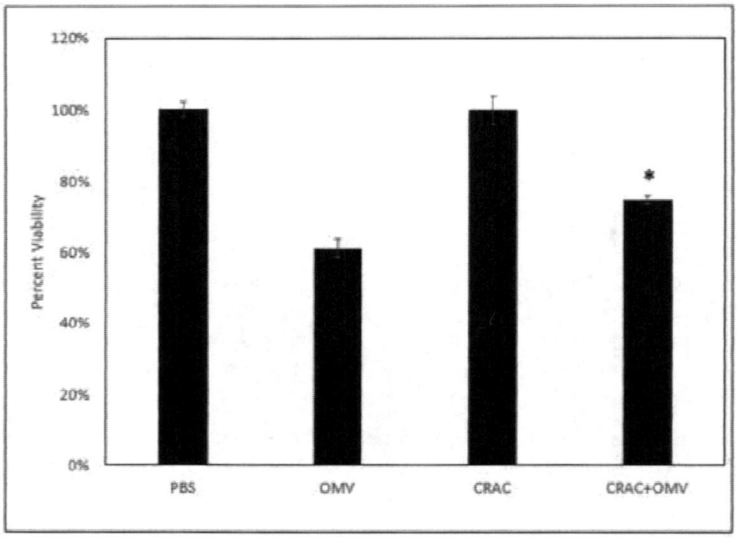

Figure 2: OMV toxicity on THP-1 cells treated with CRAC peptide (n=3, * p=0.003 compared with no treatment)

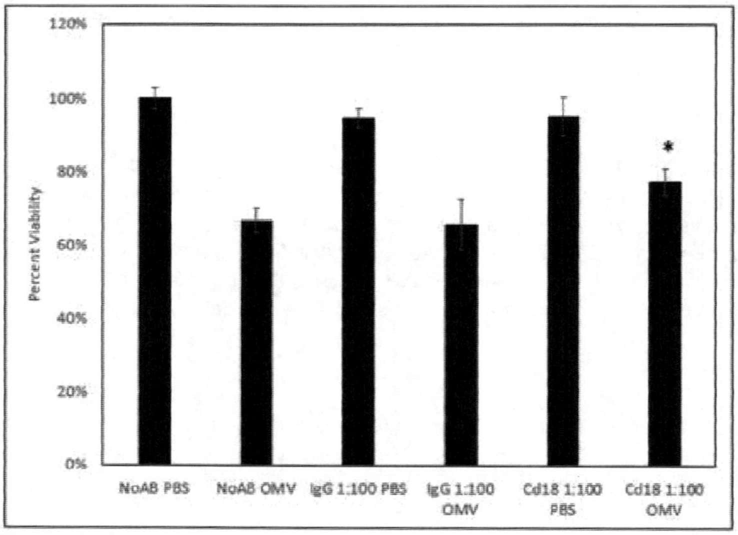

Figure 3: OMV toxicity on THP-1 cells pre-treated with anti-CD18 antibody. (n=9, * p=0.00002 for CD18 compared with no antibody)

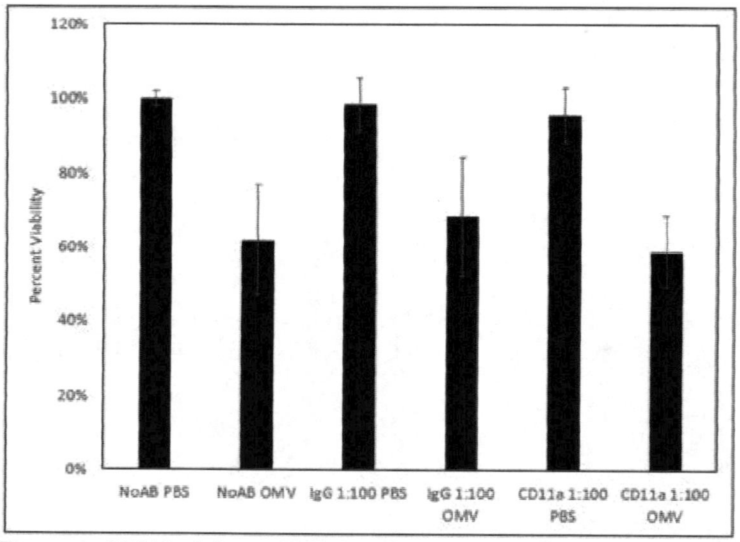

Figure 4: OMV toxicity on THP-1 cells pre-treated with anti-CD11a antibody. (n=9, no significance seen when compared with

no antibody)

Conclusion and Future Directions:

A.a. OMV toxicity, like LtxA toxicity depends on several factors. So far we have shown that cholesterol and the CD18 subunit of LFA-1 are important to OMV toxicity. Our lab will continue to identify mechanisms of *A.a.* OMV toxicity to host cells with the goal of specifically blocking these mechanisms to treat bacterial infections.

References:

[1] Masanori Saito, O. T., and Kazuko Takada "Anticytotoxic Effect of Green Tea Catechin on Aggregatibacter actinomycetemcomitans Vesicles." Int J Oral-Med Sci **11**(2): 101-105.

[2] Brown, A. C., et al. (2016). "Inhibition of LtxA toxicity by blocking cholesterol binding with peptides." Mol Oral Microbiol **31**(1): 94-105.

470058 Studies of Mechanism of *Vibrio Cholerae* Outer Membrane Vesicles Delivery to Target Cells for the Identification of Therapeutic Targets

Abstract:

Vibrio cholerae is a gram negative pathogenic bacterium which causes acute diarrheal disease in regions with poor sanitation. The enterotoxin called cholera (CT) and the endotoxin (LPS) have been identified as the major toxins produced by *V. cholerae*. Like other gram negative bacteria, *V. cholerae* naturally produces outer membrane vesicles (OMVs), which mediate the delivery of virulence factors to the host cells. In the case of CT, recent studies have demonstrated that the vesicle-associated CT is biologically active. In the present study, we sought to identify the primary components of the OMVs and host cells that facilitate OMV delivery for the identification of therapeutic targets. Isothermal titration calorimetry (ITC) was used to measure the binding of CT to the OMV-like lipid bilayers composed of varying combinations of the most common phospholipids in the bacterial outer membrane, including phosphatidylethanolamine (PE), phosphatidylglycerol (PG), cardiolipin (CL), and LPS. The relative binding affinities of CT with the lipid components of the OMVs were also assessed using lipid dot-blot analysis. Among the lipid species examined, CT was found to bind strongly to the negatively charged CL and Lipid A, less strongly to the anionic PG and unnoticeably to PE, suggesting that the toxin binding to the OMVs is regulated by electrostatic interactions. Consequently, the effect of the ionic strength and pH on the inhibition of the toxin association with the OMVs was further investigated by ITC. Based on the higher affinity of the toxin for the charged lipid components, which comprise the inner leaflet of the OMVs inner leaflet, a trypsin digestion experiment was performed to confirm the location of the toxin relative to the OMVs. Furthermore, the role of the OMV components on the vesiclesâ€™ toxicity and binding to the host cells was studied utilizing purified OMVs from *V. cholerae* and three sets of liposomes (model OMVs) composed of varying combinations of PE, PG, CL, LPS and CT. To address whether the OMVs require CT for cytotoxicity, the binding to and internalization with intestinal epithelial cells by OMVs secreted from both a toxic and a non-toxic strain of the bacteria were evaluated by confocal microscopy. These results shed light on the mechanism by which *V. cholerae* OMVs are delivered to host cells, allowing us to develop targeted anti-virulence

approaches to *V. cholera* infection.

462153 Effect of Interparticle Interactions on Agglomeration and Sedimentation Rates of Colloidal Silica Microspheres

Abstract:

The sedimentation half-times t_s of monodisperse dispersions of 750, 505, and 350 nm silica spheres were measured in water, in ethanol, and in aqueous NaBr solutions of concentrations c_{NaBr} ranging from 50 to 1000 mM. In water and in ethanol, was about 8, 18, and 33 h for the 750, 505, and 350 nm particles, respectively. These values were the same as the ones predicted by the Stokes law, suggesting that the particles were monodisperse and remained so during sedimentation. t_s-values remained the same with increasing particle weight fraction up to 0.03, indicating no hydrodynamic interactions. Three regions of NaBr concentration with different settling behaviors were found for each size. In region I or at lower NaBr concentrations, t_s-values were the same as at no salt conditions, implying that there was no significant agglomeration before particles settled. In region II, t_s decreased with increasing NaBr concentration, suggesting that the agglomeration and sedimentation times of medium-size clusters were comparable. In region III, the t_s-values were quite close for all particles and independent of the NaBr concentration, indicating that the particles, in a short time, formed large clusters which settled rapidly. The zeta potentials of the particles in water or in NaBr solutions were measured and used to predict the corresponding Fuchs-Smoluchowski stability ratios, which were sensitive to the chosen Hamaker constant values and the NaBr concentrations. Two models, the Smoluchowski steady-state and a more general unsteady-state model, were presented for obtaining the agglomeration times t_{an} for forming 2^{Nm}-mers, and the net sedimentation time t_s* for those clusters. In the model, the clusters were described by a fractal model with a fractal dimension d_f; diffusion-limited clusters with d_f=1.8, chosen to correspond to agglomeration with no subsequent internal rearrangements of the clusters, were compared to the coalescence-limit clusters, d_f=3.0, corresponding to spheres. The models provide some upper-bounds of t_{an} and t_s^*, Moreover, the effective sizes, density differences, and

volume fractions of the clusters were obtained. The predicted trend of t_s^* was fairly consistent with the experimental data, and the predictions supported our inferences that the particles unagglomerated in Region I, formed medium-size clusters in Region II, and large clusters in Region III. (See paper by the same authors, in Langmuir, in press.)

467152 Surfactant-Assisted Extraction of Lignan from Defatted Flaxseed with Micellar Solutions

Abstract:

Surfactant-assisted extraction of lignan from defatted flaxseed with micellar solutions

Tzu-Chieh Chen, Bing-Hung Chen*

Department of Chemical Engineering, National Cheng Kung University,

1 University Road, Tainan 70101, Taiwan

*e-mail: bkchen@mail.ncku.edu.tw (BHC)

Abstract

Surfactant-assisted extraction of lignan from defatted flaxseeds was conducted using various edible surfactant solution as the extraction media. The flaxseed lignin is mainly composed of secoisolariciresinol (SECO) and secoisolariciresinol diglucoside (SDG). Both SDG and SECO are antioxidant and phytoestrogen abundant in flaxseeds. During the surfactant-assisted extraction procedure, the defatted flaxseeds were immersed and constantly agitated in surfactant solutions to extract lignan. Explicitly, the lignan of the defatted flaxseed was solubilized by micellar solution, from which aliquots were drawn periodically for HPLC-analysis to determine the content of lignin extracted. As a result, the extraction process mainly follows the kinetics of the first-order reaction. Among various surfactants used, Tween 20 gives the best extraction efficiency, followed by Tween 80.

Keywords: *Flaxseed, Surfactant, Extraction, Lignan, Secoisolariciresinol, Solubilization.*

Figure 1. Effect of extraction media to the extraction yield of lignin.

Extraction condition: (1) Temperature = 30 ¢XC, (2) Flaxseed : Extraction medium = 1 : 50 (w/v), (3) Duration = 1 day, (4) Surfactant concentration = 1 wt%.

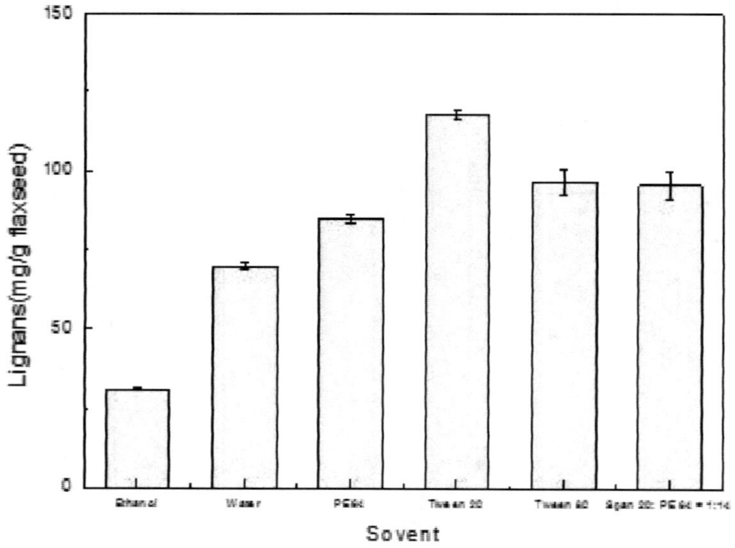

469380 Characterizing Salt Effects on Adsorbed Interfacial Surfactant and Hydration Layers Around Single Wall Carbon Nanotubes Using Analytical Ultracentrifugation

Abstract:

Characterization of the interfacial layer structure on dispersed nanoparticles is of great interest, but is typically challenging, especially for anisotropic or polydisperse particle populations. For example, for single wall carbon nanotubes (SWCNTs), the success and efficiency of various dispersion and purification methods are believed to be strongly dependent on the characteristics of the bound layer of dispersant (such as small molecule surfactant(s) or DNA) both in the amount of bound dispersant as well as the specific structure it forms at the nanotube surface. However, only limited direct measurements of the differences in the bound layers on well-resolved SWCNT populations have been reported. To study the structure of the adsorbed surfactant layer, we utilize analytical ultracentrifugation (AUC). AUC is a powerful technique which allows for simultaneous sampling of the entire population distribution of a solute (e.g. proteins, DNA, particles), and through varying experimental parameters can be utilized to resolve the density and radial distribution of adsorbed surfactants and associated water at various solution conditions. In this work we report direct measurement of the hydration changes of the surfactactant SWCNT complex with added salt, and identify conditions at which the addition influences the structure of the adsorbed surfactant. These changes lend context to literature effects on the addition of salt to SWCNT separations, and identify hydration differences as the primary effect of salt addition rather than changes to the adsorbed surfactant number.

472027 Â Influence of Coalescent Distribution on Film Formation Behavior of Colloidal Dispersions

Abstract:

There are three primary stages of latex film formation: 1. Evaporation of water, 2. Deformation and close packing of latex particles, and 3. Interdiffusion of polymer chains. The final step of this process is essential to creating defect-free, mechanically robust films. Historically, high performance coatings were made using a formulation containing a high-T_g polymer and a volatile coalescent. The volatile coalescent enables film formation by softening the polymer latex particles. This softening enables polymer chain interdiffusion, and upon evaporation of the volatile coalescent, a harder, high-T_g coating is created. Recent and proposed changes in regulations drive a market need for materials with no/low volatile organic compounds (VOCs), yet provide good performance. Typically, low VOC coating formulations contain polymers with low T_gs. Coatings from these low-T_g paints generally have a soft/tacky feel and poor performance. These shortfalls are particularly prevalent in glossier formulations that contain low pigment loadings. Alternatively, harder, high-T_g binders can be used in low-VOC formulations by incorporating permanent plasticizers. However, when permanent plasticizers are used at sufficient levels to enable low temperature coalescence, the resultant coatings also have a soft/ tacky feel that will not improve over time.

One solution to the inherent tradeoff between film formation and performance in low VOC coatings is to use more effective coalescents. The efficiency of a coalescent is primarily a function of its solubility within the polymer latex particle and the Î"T_g between the polymer and coalescent. Estimates of T_g depression, required to enable film formation, can be calculated using well-known mixture rules such as the Flory-Fox equation. Beyond these simple considerations, there may be are other factors that influence the efficiency of a coalescent. One possible such factor is the distribution of the coalescent within the polymer latex particle. Understanding how this factor may influence performance requires a technique to measure the coalescent distribution within the particle. In this presentation, the use

of small-angle neutron scattering to evaluate coalescent distribution within latex particles will be discussed and evaluated. The ability of this technique to evaluate small molecule behavior within a latex particle will hopefully aid in the development of more efficient coalescents and reduce VOCs required for high performance coatings.

457382 Complex Fluids at Interfaces: Some Open-Ended Problems in Colloidal Systems

Abstract:

Many chemical products that we use everyday are highly complex fluids that contain dispersed solids, flow modifiers, surface active agents, fillers, polymers, colorants, solvents etc. This concoction of materials leads to a complex flow behavior under various environmental, shipping, storage and usage conditions, many of which are not well understood. Some examples of these products include shampoos, detergents, adhesives, cement pastes and oil dispersions. Flow & interfacial behavior of these complex fluids can be a key factor in the application performance and consumer satisfaction with these products.

In this talk, we will use practical examples of these ubiquitous products to describe some of these complex flows. We will outline some of our efforts in using a scientific and statistical approach in understanding this behavior as well as strategies used to modify flow properties to meet those needs. We will describe the use of rheology, imaging as powerful tools to characterize these systems under various time, temperature and storage conditions. However, rheological characterization of these systems is also fraught with complications of high normal stresses, high inertia and slip behavior. We will endeavor to outline few of the open-ended problems that remain important but not well understood.

465274 Probing Quench-Dependent Gelation in Thermoresponsive Attractive Colloids

Abstract:

We investigate the effects of thermal quench on gelation in thermoresponsive, phase separating colloids. As a model system, we employ recently established oil-in-water nanoemulsions which contain thermosensitive telechelic polymers in the continuous phase and undergo temperature-induced colloidal gelation via polymer-mediated interdroplet attractions. [1] Modeling the effective interactions with a detailed statistical mechanical model for polymer bridging, we establish the equilibrium colloidal phase behavior of the system. Under appropriate conditions, the system forms colloidal gels in regions of fluid-fluid phase instability, where gelation and spinodal decomposition compete to direct the arrested gel structure. [2] Here, time-dependent quenches of various depths and rates are performed along isochores of volume fraction from 0.1 to 0.4, and linear viscoelasticity and optical microscopy are used to track the evolution of rheology and structure in the system. We find that gelation depends significantly on both the depth and rate of quenching into the gelled state. For example, the kinetics of gelation vary by several orders of magnitude with relatively minor changes in the quench rate, even for relatively shallow quenches into the spinodal region. Furthermore, we find that the onset of solid-like viscoelastic behavior for sufficiently slow quenches is consistently deeper than the predicted equilibrium coexistence line, suggesting the presence of an attractive glass transition line within the spinodal region. These results offer a starting point for more rationally controlling colloidal gelation and phase separation, and underscore the active interplay between viscoelastic phase separation and glassy dynamics in the formation of dense colloidal gels.

[1] Helgeson, M. E.; Gao, Y.; Moran, S. E.; Lee, J.; Godfrin, M.; Tripathi, A.; Bose, A.; Doyle, P. S. *Soft Matter* **2014**, *10*, 3122-3133.

[2] Gao, Y.; Kim, J.; Helgeson, M. E. *Soft Matter* **2015**, *11*, 6360-6370.

459087 Image Method for Electrostatic Polarization Energy of Dipolar Particle Clusters

Abstract:

Dipolar fluids have many important applications such as liquid seals for hard drives and targeted drug delivery. Chain-like configurations of dipolar particles are frequently observed, but the complete phase behavior has not been fully characterized. Furthermore, most current simulations neglect the effects of surface charge polarization, since resolving surface charges requires expensive numerical calculations. We develop a theory to rapidly evaluate the polarization energy for dipolar particles. The theory converts the effects of induced surface charges to that of image lines, and can be adapted to particles with quadrupolar and multipolar charge distributions. Applying the theory to simple clusters demonstrates that charge polarization expands the attraction range of relative dipole orientations. The theory depends on particle positions explicitly, and can be readily implemented in molecular dynamics and Monte Carlo simulations, enabling the examination of polarization effects on the clustering dynamics and thermodynamic behavior of dipolar fluids.

469649 An Extended Nanoparticle Haloing Study in Microgravity

Abstract:

Negligibly charged colloid microspheres, which flocculate when suspended alone in aqueous solution, form a stabilized homogenous fluid upon the addition of a critical volume fraction of highly charged nanoparticles. It has been previously reported that nanoparticles are not absorbed on the microspheres in the suspension yet contribute towards the effective charge buildup of the microspheres. This colloid stabilizing phenomenon has been termed nanoparticle haloing as it is proposed that the charged nanoparticles form a “halo” structure around the larger microparticles. Previous work conducted nanoparticle stability experiments through the observation of colloidal structures after the particles have settled. On earth, the presence of gravitational forces prevents long-term observations of nanoparticle haloing stabilization mechanisms. In our set of microgravity experiments, conducted on the International Space Station (ISS), silica spheres (600 nm) at 1.0% volume fraction were suspended along with different concentrations (0.1%, 0.05% & 0.01%) of charged zirconia nanoparticles (8 nm). The media was an aqueous solution of nitric acid (pH ~2) to reach the larger colloids' isoelectric point.. Fifteen samples (five of each nanoparticle concentration) were sealed in 2.5 mm x 0.2 mm disc-shaped sample chambers circular wells with a stainless steel minirod. Prior to observation, an astronaut manually mixed each sample well for two minutes before loading the sample into the Light Microscopy Module (LMM). Samples were illuminated from the side to induce scattering and a mounted digital camera acquired images at 2.5X magnification. Acquired images from uncompromised samples were analyzed by time depended structure factor S(q,t) analysis. Qualitatively, three of the 0.1% nanoparticle samples exhibited long-term stability for the duration of observation (five weeks), whereas the other samples formed gel-like structures.

473596 Realizing Non-Close-Packed Crystal Structures through Directional Binding of DNA-Functionalized Colloidal Clusters

Abstract:

A diverse range of ordered configurations have been demonstrated using self-assembling particles driven by one or more of a number of strategies, including external fields, depletion, and functionalized particles. In spite of this success, it is apparent from consideration of atomic and molecular systems that many interesting and useful configurations require anisotropy for their realization. Anisotropy has been engineered into colloidal systems through a variety of means. For example, anisotropically-shaped nanoparticles with simple interparticle interactions have been assembled into numerous crystalline, quasicrystalline, and disordered configurations [1]. Here, it has been shown that directional entropic interactions between particles are primarily responsible for the resulting configurations, making it somewhat difficult to predict a priori the assembly outcome [2]. A more direct approach is to functionalize spherical particles in such a way as to produce directional interactions, resulting in so-called 'patchy particles' [3]. While conceptually appealing because of the atomic orbital analogy, it is generally difficult to accomplish such patterned functionalization in practice.

An alternative approach for engineering assembling entities that exhibit anisotropy is to create small clusters out of spherical colloidal particles. Here, we study computationally the self-assembly of large numbers of colloidal clusters formed by a recently-introduced crystal templating method [4]. In this technique, bulk crystals are first assembled with DNA-functionalized sub-micron sized polystyrene beads. A small number of beads are designed to bind particularly strongly with their neighbors so that upon subsequent melting of the bulk crystal, these particles form the center of individual small clusters with a specific geometry. Rather than attempt to assemble these clusters directly into superstructures, we study the assembly of clusters mediated by the addition of spherical 'bond' particles. In particular, the 'bond' spheres are DNA-functionalized so that they interact attractively with the clusters. In this construct, the differing number of particle-particle contacts available between a cluster and bond particle as a function of orientation creates directional bonding.

A variety of simulation techniques are employed to study the thermodynamics and kinetics of the assembly process. Specifically, we compute nucleation barriers for several crystalline configurations using tetrahedral, octahedral, and cubic clusters. We show that some cluster types lead to facile growth of crystalline superstructures, while others lead to structures that are highly susceptible to defect formation. Crystal growth kinetics are probed using molecular dynamics and Brownian dynamics simulations and again demonstrate a wide range of kinetic limitations depending on the cluster geometry.

[1] X. Ye, J. E. Collins, J. C. Y. Kang, D. T. N. Chen, A. G. Yodh, and C. B. Murray, PNAS 107, 22430 (2010).

[2] P. F. Damasceno, M. Engel, and S. C. Glotzer, Science 337, 453 (2012).

[3] F. Romano and F. Sciortino, Nat. Commun. **3**, 1968 (2012).

[4] J. T. McGinley, Y. Wang, I. C. Jenkins, T. Sinno, and J. C. Crocker, ACS Nano 9, 10817 (2015).

459273 Nonlinear Learning of Colloidal Assembly Mechanisms from Simulation and Experiment

Abstract:

The inherently many-body nature of assembly makes it a challenge to identify the underlying collective mechanisms and assembly pathways, and has hindered the rational design of building blocks to produce aggregates with desired structure and function. There exists a predictive gulf between individual building block architecture and chemistry, and the thermodynamics and kinetics of collective assembly behavior. By combining tools from statistical mechanics, molecular simulation, and machine learning, we have bridged this divide to systematically identify emergent mechanisms and assembly pathways from molecular simulations or experimental particle tracking data. Our approach furnishes low-dimensional "assembly landscapes" that naturally integrate thermodynamics (*what* stable aggregates can assemble?) and mechanisms (*how* do they assemble?) as a function of building block properties.

In a first application to experimental particle tracking of the non-equilibrium self-assembly of Janus colloids in an applied AC electric field, we have extracted the underlying assembly landscape directly from experimental data, and used this roadmap to understand assembly and design experimental conditions – electric field strength, AC frequency, salt concentration – to assemble desired aggregate morphologies – pinwheels, clusters, archipelagos. In a second application to Brownian dynamics simulations of "digital colloids" – reconfigurable clusters of freely rotating halo particles tethered to the surface of a central particle – as a novel soft matter-based substrate for high-density information storage, we have extracted the low-dimensional free energy surface governing digital colloid morphology, thermodynamics, and kinetics. By modulating the diameter ratio between halo particles and central particles, we have quantified the reversible work required to write information into the cluster and kinetic stability of the cluster to spontaneous thermal fluctuations. We demonstrate the use of this framework to guide the rational design of a memory storage element to hold a block of text that trades off the competing design criteria of memory addressability and volatility.

461674 Open Crystals from Triblock Janus Colloids

Abstract:

Triblock Janus colloids, which are colloidal spheres decorated with attractive patches at each pole, have recently generated significant interest as potential building blocks for self-assembled colloidal crystals. Specifically, their inherent anisotropy is known to induce self-assembly into open structures at moderate temperatures and pressures, where they are stabilized over close-packed crystals by entropic effects, and over the fluid phase by enthalpic ones. We use molecular simulations and free energy calculations to study the self-assembly of the triblock Janus particles in three dimensions. We discuss the complex interplay between competing forces which drive the system to form open crystals and elucidate some fundamental design rules for open colloidal crystals.

464872 Topology Free Folding Mechanism for Polyhedral Nets

Abstract:

A form of self-assembly, â€œself-foldingâ€ presents an alternative approach to the creation of reconfigurable, responsive materials with applications ranging from robotics to drug design. However, the complexity of interactions present in biological and engineered systems that undergo folding makes it challenging to isolate the main factors controlling their assembly and dis-assembly. Here we use computer simulations of simple, minimalistic self-foldable structures and investigate their stochastic folding process. By dynamically accessing all the microstates that lead to, or inhibit, successful folding, we show that the mechanisms by which general stochastic systems achieve their native structures can be identified and used to design rules for optimized folding propensity.

473900 The Role of Interaction Heterogeneity in the Self-Assembly of DNA-Functionalized Colloids

Abstract:

Heterogeneity is generally considered to be a detrimental factor for the self-assembly of colloidal particles into ordered structures. While this is well-established for certain types of heterogeneity such as size polydispersity, here we show, using a combination of equilibrium and non-equilibrium simulations, that heterogeneity in the pairwise interaction strength among a collection of particles may in fact be useful for nucleation of crystalline phases. In particular, we consider interaction heterogeneities that may arise from density variations of DNA oligomers grafted on the surface of sub-micron spherical particles to drive self-assembly.

The multivalent nature of DNA-mediated particle binding introduces constraints for assembly by increasing the temperature dependence of the resultant interparticle potential and consequently narrowing the operating window for high quality crystallization. For a given areal density of DNA strands, this effect becomes more pronounced for larger particles and represents a key challenge for DNA-driven colloidal assembly

The beneficial impact of interaction heterogeneity is shown to arise from a synergistic combination of two effects. First, we employ umbrella sampling simulations to show that heterogeneity strongly lowers the free energy barrier associated with the nucleation of crystals by the formation of strongly-bound small clusters. Concurrently, non-equilibrium growth simulations show that variations in the interaction strength between particles inhibit gelation and polycrystallinity by keeping the number of such nuclei low, allowing individual nuclei to grow unhindered.

467205 Dynamic, Directed Self-Assembly of Nanoparticles Via Toggled Interactions

Abstract:

Crystals self-assembled from nanoparticles have useful properties such as optical activity and sensing capability. During fabrication however, gelation and glassification often leave these materials arrested in defective or disordered metastable states. This is a key difficulty preventing adoption of self-assembled nanoparticle materials at scale. Processes which suppress kinetic arrest and defect formation while accelerating growth of ordered materials are essential for bottom-up approaches to creating nanomaterials. Dynamic, directed self-assembly processes in which the interactions between self assembling components are actuated temporally offer one promising methodology for accelerating and controlling bottom-up growth of nanostructures. We show through simulation and theory how time-dependent, periodically toggled inter-particle attractions can avoid kinetic barriers and yield well-ordered crystalline domains for a dispersion of nanoparticles interacting via a short-ranged, isotropic potential. The growth mechanism and terminal structure of the dispersion are controlled by parameters of the toggling protocol. This control allows for selection of processes that yield rapid self-assembled, low defect crystals. Although self-assembly via periodically toggled attractions is inherently unsteady and out-of-equilibrium, its outcome is predicted by a first principles theory of non-equilibrium thermodynamics. The theory necessitates equality of the time average of pressure and chemical potential in coexisting phases of the dispersion. These quantities are evaluated using well known equations of state. The phase behavior predicted by this theory agrees well with measurements made in Brownian dynamics simulations of sedimentation equilibrium and homogeneous nucleation. The theory can easily be extended to model dynamic self-assembly directed by other toggled conservative force fields. We also present kinetic models that predict the rate of crystallization as a function of the dynamic toggling parameters for several different observed growth mechanisms.

469240 DNA-Conjugated Nanoparticle Assemblies Demonstrate a Tunable Mechanical Response

Abstract:

Nanoparticles functionalized with short sequences of DNA represent a highly-customizable platform for multiscale materials design. The flexibility and specificity of DNA-mediated interactions facilitates the design of hierarchical structures; with tunable macroscopic properties originating from the nano-scale building blocks of which they're composed. While most recent work has attempted to tune the optical properties of these assemblies, few studies exist that have tried to tune other properties, such as their mechanical response. In this work, we use a detailed molecular model of the DNA-nanoparticle to probe the mechanical properties of DNA-nanoparticle assemblies. We demonstrate that their mechanical properties can be significantly tuned by subtle variations in the linking DNA sequence and is strongly dependent on temperature. Next, we identify the molecular origin of this tunable mechanical response to be closely tied to the connectivity of the nanoparticle network and the anisotropic deformation of the DNA strands conjugated to the nanoparticle surface. The results presented here provides an important initial mechanical characterization of DNA-nanoparticle assemblies can help to dictate their future directions and applications.

466745 Characterizing Multi-Flavored Assembly of Two Dimensional Binary Colloidal Crystals

Abstract:

DNA-functionalized colloids are routinely used to assemble complex nanoscale crystalline lattices; for sufficiently high grafting density, the pairwise interactions between these colloids can be robustly controlled by tuning the length of the grafts and base pairs composition on the colloids. In a multicomponent mixture, the latter can be employed to decouple the cross interactions between unlike particles from the interactions between like particles. Here we investigate the consequences of tuning the cross-interaction between unlike species in a binary mixture on the types of self-assembled structures which form. We use a combination of Monte Carlo, molecular dynamics, and free energy methods to examine the relative stability of an array of possible structures, which range from low symmetry structures such as strings, to various crystal lattices with either 4- or 6-fold symmetry typically. The regimes of stability for different lattices can be described in terms of a balance of cohesive and adhesive interaction across a range of particle size asymmetries. In the zero-temperature limit, we identify parameter space which provides a semi-universal description of the regimes in which different symmetries are found. We then map out how these regimes change as a function of temperature starting from the zero-temperature (ground state) limit up to the melting point.

451943 Multiscale Modelling of Macromolecule Self-Assembly in Solution: The Case of Poly- μ-Caprolactone in Acetone-Water Mixtures

Abstract:

This work focuses on macromolecule (poly-ε-caprolactone, PCL) self-assembly in acetone-water mixtures. By mixing an acetone ("good" solvent) solution containing PCL molecules with water ("bad" solvent, or anti-solvent) their self-assembly in molecular clusters (or nanoparticles) is induced. The aim of this work is to simulate this process by using a diffusion-limited aggregation (DLA) model, that assumes that PCL molecules self-assemble together through an irreversible process, by means of the Smoluchowski population balance model (PBM), solved in turn with the quadrature method of moments (QMOM) inside a computational fluid dynamics (CFD) code. It is assumed that molecule and cluster aggregation is caused by Brownian motions and turbulent fluctuations. The aggregation process is described in terms of the cluster mass distribution (CMD) that quantifies the number of macromolecules in each molecular cluster. The relationship between the number of macromolecules per cluster and their radius of gyration is formulated in terms of the Flory law, whose coefficient and exponent are calculated from molecular dynamics (MD) simulations performed in a previous work. Model predictions are validated against experiments performed in a confined impinging jets mixer (CIJM) by varying the operating conditions, namely the initial PCL concentration, PCL molecular weight and mixing rate of the acetone and water solutions, resulting in very goos agreement.

453374 Assembly-State Switching: Alkane Guest Length Drives Assembly Transitions Between Multimeric Deep-Cavity Cavitand Complexes

Abstract:

Octa-Acid (OA)—a bowl-shaped molecule known as a deep-cavity cavitand—forms dimeric assemblies when in the presence of suitable hydrophobic guests. These dimeric assemblies have been used as reaction vessels and hydrocarbon gas separators. Tetra-Endomethyl Octa-Acid (TEMOA)—identical to OA, except that it has methyls in the *endo* position on the rim instead of hydrogens—not only forms dimeric assemblies, but also tetrameric and hexameric ones.

In this study we simulated a series of *n*-alkane guests inside of TEMOA dimers, tetramers, and hexamers. Using free energy calculations we demonstrate that the transitions between different assembly states are driven primarily by packing and not by the free energy of bringing together the hosts and guests into the assembly. Additionally we are able to accurately calculate the interior volume of each assembly state, showing that these structures are larger than systems formed by hydrogen bonding. This is significant because, unlike hydrogen-bonded structures, TEMOA multimers are biocompatible.

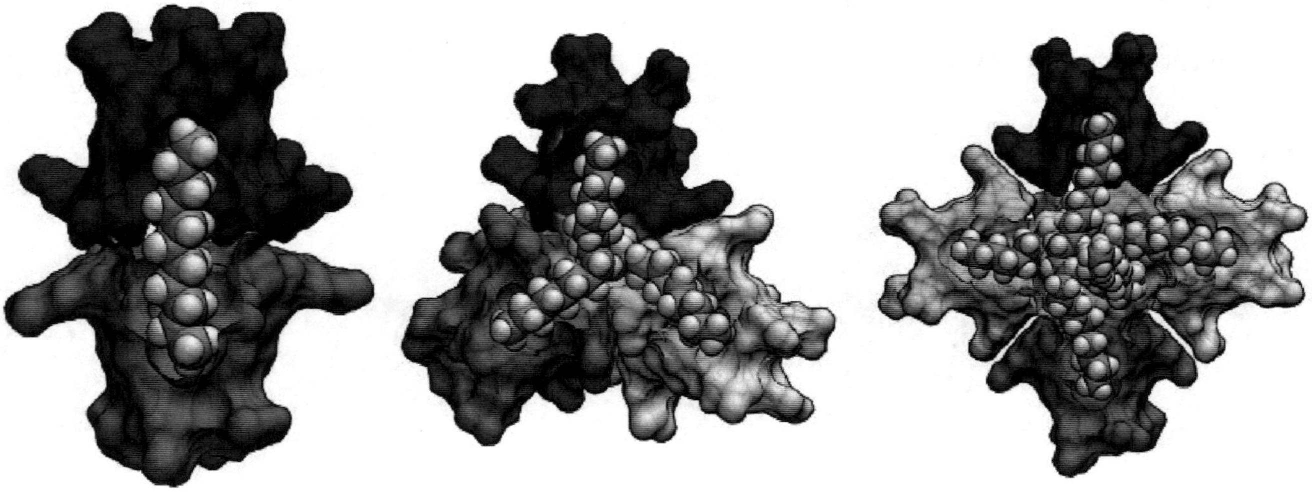

461201 A Gaussian-Charge Polarizable Model for Water

Abstract:

Force fields for water that explicitly account for polarization effects are expected to provide reasonable estimates of water properties for wide range of conditions. Such models are, among others, the BK3 [1], GCPM [2] and SWM4-NDP [3] models. These polarizable force fields generally show better performance compared to non-polarizable ones in description of properties for pure water and aqueous electrolyte solutions. [4] However, for the water and carbon dioxide mixture, which is of great significance to the design of carbon dioxide geologic sequestration, the calculation of its equilibrium phase compositions using existing polarizable water models are far from satisfactory, even when the carbon dioxide/water cross interaction was optimized. In this study, a polarizable force field was developed for water with the aim of improving the prediction of thermophysical properties for carbon dioxide/water mixture. The proposed water force field has a rigid geometry with bond angle and length setting to their experimental values. The van der Waals interaction is represented by the Buckingham Exponential-6 potential assigned on the oxygen atom. The electrostatic interactions are represented by Gaussian charge instead of point charge for its greater numerical stability. The polarization of water molecule is handled via classical Drude oscillator by attaching a single negative Gaussian charge to the oxygen atom with a harmonic spring, and the spring constant is set to reproduce the experimental molecular polarizability of water. An explicit and analytical hydrogen bond term is used in the model to account for possible charge transfer effects happening in water hydrogen bonds. The model was parameterized to the properties of water at ambient conditions, including configuration energy, density, partial pair correlation functions, dielectric constant, as well as the minimum energy and geometry of water dimer. The proposed model was tested against a variety of properties including temperature of maximum density, heat capacity, compressibility, second virial coefficient, saturated vapor-liquid phase densities, vapor pressure, enthalpy of vaporization, vapor-liquid interfacial tension, melting point, viscosity and diffusion coefficient. With the proposed polarizable water model, the phase equilibrium properties of the carbon dioxide/water mixture were studied.

[1] Kiss, P.; Baranyai, A. *J. Chem. Phys*. **2013**, 138, 204507.

[2] Paricaud, P.; Predota, M.; Chialvo, A. A.; Cummings, P. T. *J. Chem. Phys*. **2005**, 122, 244511.

[3] Lamoureux, G.; Harder, E.; Vorobyov, I. V.; Roux, B.; MacKerell, A. D. *Chem. Phys. Lett.* **2006**, 418, 245.

[4] Jiang, H.; Mester, Z.; Moultos, O. A.; Economou, I. G.; Panagiotopoulos, A. Z. *J. Chem. Theory. Comput.* **2015**, 11, 3802.

466214 Introducing an Individualized Transferable Anisotropic Mie Force Field

Abstract:

Transferable force fields allow predicting physical properties and phase equilibria of pure substances and mixtures that are weakly described by experimental data. However, for well-known substances transferable force fields lead to higher deviations than desired and necessary. Well-characterized substances could be described better if an individually optimized force field would be used. In this study we introduce the concept of individualized transferable force fields that leads to accurate representations of well-characterized substances.

Parameters of the transferable anisotropic Mie (TAMie) force field are obtained iteratively from Monte Carlo simulations in grand canonical ensemble and doing histogram reweighting in a post processing step to calculate the phase coexistence curve. New parameters are generated by utilizing a physically based equation of state (PC-SAFT EOS) to speed up the iteration procedure. The objective function for optimizing the force field parameters contains several substances of the same group where we mainly focus on vapor pressure and liquid density data. For non-polar substances, like alkanes, the approach leads to excellent results. However, describing polar substances like esters with a transferable set of parameters necessitates further trade-offs. If, what we prefer, the point charge model is applied and the point charges are assumed to be the same for the entire substance group, the overall optimal set of parameters might be outside the range defined through individual optimization of pure substances. The trade-off in parameterizing a transferable force field to several substances manifests mainly in deviations in vapor pressure: the vapor pressure data of some substances are too high whereas for other substances they are too low. This error can easily be corrected for. We introduce a correction parameter ψ_ε which affects all energy parameters ε of a pure substance. As a result, we obtain a transferable force field for weakly-characterized substances and an accurate force field for substances which are well described by experimental data. Applying these correction parameters ψ_ε opens the opportunity to simulate binary mixtures at different temperatures

without introducing significant errors caused by deficiencies in vapor pressure of the pure substances. The results for binary mixtures containing ketones and alkanes show good agreement with experimental data.

Our study shows on the one hand the benefit of individualized transferable force fields for the description of pure substances and for the predicting of mixture phase equilibria, and on the other hand we demonstrate the transferability of binary interaction corrections k_{ij} (with respect to Lorentz-Berthelot combining rules) defined for united atom groups.

465321 Optimization of a Transferable Shifted Force Field for Interfaces and Inhomogeneous Fluids

Abstract:

Many commonly applied force fields have been developed based on the assumption of spherical symmetry and tail corrections to compute the thermodynamic properties. Unfortunately, many common applications involve inhomogeneous fluids where this assumption breaks down. For example, symmetry breaks down at the vapor-liquid interface because the liquid phase is roughly 500 times denser than the vapor phase, leading to very different forces for sites at various positions in the interface. In our experience, this necessitates a cutoff distance of roughly triple what is usually assumed in order to achieve self-consistent results for vapor pressure and surface tension. As another example, virtually all biological simulations involve some degree of inhomogeneity. Furthermore, a fundamental issue arises in the context of molecular dynamics because the discontinuity in the force field undermines the conservation of energy, leading to a range of possible compensating treatments, and poor consistency from one implementation to the next.

In this work, we review our progress in the development of a Transferable Shifted Force Field based on the United Atom perspective (TranSFF-UA). We consider a number of candidate potential models, taking the TraPPE-UA as a starting point. We universally apply a cut and shift of 3Ï*f*, where Ï*f* is the position where the potential initially crosses zero. Candidates to date include the LJ, 14-12-8-6, and Mie potential models. Point charges of polar molecules are adapted directly from the TraPPE-UA model, except in the case of water. For water, we consider the TIP4P(2005) model. This modification permits accentuation of the steepness of the potential, while considering the efficiency and parallelization of the Lennard-Jones (LJ) model, particularly in GPU systems. The simulation methodology involves LAMMPS in combination with GOMC, a relatively new open source Monte Carlo simulation platform. Particular attention is paid to consistency between the two methods and regions of phase space where each method performs best. All tabulated results are based on direct simulation, whereas thermodynamic perturbation theory is used as a meta-model to suggest prospective optimal potential parameters.

To demonstrate this methodology, vapor pressure, critical properties, saturated liquid density, and compressed fluid density are used to characterize potential models for n-alkanes, branched alkanes, ethers, olefins, naphthenics and aromatic compounds, and water. Vapor pressure deviations average near 9% for hydrocarbons. Saturated liquid density deviations average near 1% below a reduced temperature of 0.9. Compressed fluid densities exhibit deviations near 0.5% for the better potential models. Simply applying the shifted force to water significantly improves the vapor pressure correlation for the TIP4P model.

This work is supported in part by the Scientific and Technological Research Council of Turkey (TUBITAK) through the grant no. 114M178.

453347 Uncertainty Quantification of Intermolecular Parameters with a Transferable Potential Model for *N*-Alkanes

Abstract:

A significant factor when assessing the reliability of any physical property value, obtained either from simulation or experiment, is assigning the uncertainty associated with that value. Determining an appropriate uncertainty is normally more straightforward for experimental measurements than for molecular simulations. This is because molecular simulations have many layers of uncertainty that propagate and are difficult to quantify. Since the reliability of molecular simulations is intimately connected with the intermolecular potential model chosen and the associated model parameters (ϵ, σ, q, etc.), one of the most important layers of uncertainty is related to the optimization of the model parameters.

Historically, most studies that report model parameters (i.e TraPPE, NERD, OPLS, etc.) neglect to perform a statistical analysis of their "best fit" parameters. Occasionally, the literature provides a sensitivity analysis, where the model parameters are varied by an arbitrary amount and the effect on the simulation results is reported. However, a sensitivity analysis of this nature actually provides little insight because the region of parameter space being sampled usually lands well outside the statistically-rigorous range of acceptable parameter sets. This is due to two factors. First, the range over which a parameter is varied is often too large to be realistic. Second, the two parameters are correlated since changing ϵ necessitates a change in σ for the parameter set to be feasible. A rigorous approach to assess the uncertainty is to examine what combinations of parameters provide a fit to the data that is acceptable at a given statistical level. Although some studies have utilized rigorous statistics to quantify the uncertainty due to parameterization, these studies have focused primarily on small compounds.

This study is the first statistically-rigorous attempt to quantify the Lennard-Jones parameter uncertainties for a compound as large as n-octane. This is done by exploring the parameter space for ϵ and σ for CH_3 and CH_2. The TraPPE methodology was used for parameterization as it reduces the computationally intensive four-dimensional parameter space (ϵ_{CH_3}, ϵ_{CH_2}, σ_{CH_3}, and σ_{CH_2}) to a pair of coupled two-dimensional parameter spaces (one parameter space for the CH_3 parameters and one for the CH_2 parameters). This was done by optimizing ϵ_{CH_3} and σ_{CH_3} to match ethane liquid density data and, subsequently, the CH_3 parameters are kept constant while ϵ_{CH_2} and σ_{CH_2} are optimized with n-octane liquid density data. Although the CH_3 parameter are fixed during the CH_2 optimization, the optimal values for ϵ_{CH_2} and σ_{CH_2} are correlated with the optimal values for ϵ_{CH_3} and σ_{CH_3}. Therefore, the uncertainty in the CH_2 parameters is coupled with the uncertainty in the CH_3 parameters. For this reason, we also investigated different methods to account for this correlation between the CH_3 and CH_2 parameter uncertainties. Even with this correlation of uncertainties, our results show that the uncertainty due to the force field parameterization is much smaller than the aforementioned sensitivity analysis might suggest.

Once the uncertainty in the Lennard-Jones parameters is determined, we then extend this work to investigate how the error will propagate for larger compounds where multiple factors must be considered. For example, with increasing chain length the amount of intermolecular interactions increases between sites located on different molecules compared to shorter molecules. Furthermore, since sites separated by more than four bonds are assumed to interact by the same intermolecular potential model, the ratio between intermolecular and intramolecular interactions increases for a longer molecule. In addition, the ratio of CH_2 groups to CH_3 groups also increases which means that the error should become less and less dependent on the uncertainty in the CH_3 parameters.

To account for these competing factors, we simulate several different sized n-alkanes using hundreds of Lennard-Jones parameter sets. The ϵ_{CH_3}, ϵ_{CH_2}, σ_{CH_3}, and σ_{CH_2} values are chosen with a Monte Carlo sampling scheme which applies a statistical weighting for each parameter set based upon the uncertainty in the optimal parameters. For each parameter set, the numerical uncertainty associated with the molecular simulation results is included as an additional layer of uncertainty. With this propagation of errors approach we are able to quantify the uncertainty, due to both numerical and parameter uncertainty, in predicting the critical temperature, critical density, and critical pressure for compounds as large as $C_{48}H_{98}$. This is significant because, historically, molecular simulation results for larger compounds have only included the numerical uncertainty. Traditionally, it has been assumed that the parameter uncertainty could be quite large which has led some to doubt the reliability of these molecular simulation studies. However, our results suggest that the numerical uncertainty in the critical constants of large compounds is actually larger than the parameter uncertainty. Therefore, by quantifying the uncertainties we have demonstrated that molecular simulation results for larger compounds are more reliable than previously believed.

- Abstract Text

469302 Coarse Grained Parameterization of Gram-Negative Bacterial Outer Membrane

Abstract:

The cell envelope of Gram-negative bacteria contains a lipopolysaccharide (LPS) rich outer membrane that acts as the first line of defense for bacterial cells in adverse physical and chemical environments. The LPS macromolecule has a negatively charged oligosaccharide domain that acts as an ionic brush, limiting the permeability of charged chemical agents through the membrane. Besides the LPS, the outer membrane has radially extending O-antigen polysaccharide chains and Î^2-barrel membrane proteins that make the bacterial membrane physiologically unique compared to phospholipid cell membranes. Elucidating the interplay of these contributing macromolecular components and their role in the integrity of the bacterial outer membrane remains a challenge. To bridge the gap in our current understanding of the Gram-negative bacterial membrane, we have developed a coarse grained force field for outer membrane that is computationally affordable for simulating dynamical process over physiologically relevant time scales. The force field was benchmarked against available experimental and atomistic simulations data for properties such as membrane thickness, density profiles of the residues, area per lipid, gel to liquid-crystalline phase transition temperatures, and order parameters.

467155 Modeling of DNA-Mediated Pair Interactions Between Nanoparticles

Abstract:

Highly selective and thermally responsive binding of DNA-strands is transferable to nanoparticle systems through covalently tethering these biomolecules onto the particle surfaces. A reliable understanding of how these DNA-functionalized particle (DFP) systems self-assemble requires a deep understanding of inter-particle pair-interactions as a function of molecular-level details of the system such as strand length, the size of the sticky part, the sequence of the DNA-strands and grafting density. In this work, we use a previously developed coarse-grained model [1] of DNA to construct a temperature dependent inter-particle pair-potential model that is parameterized so as to reflect the molecular description of the DFPs. Umbrella Sampling in combination with Replica-Exchange Molecular Dynamics (REMD) simulations were performed to examine two-particle systems by estimating the potential of mean force (PMF) between nanoparticles at different physical conditions. Eventually, this implicit formulation of the system obtained from a bottom-up approach can efficiently be used to study self-assembly of large DFP systems that would otherwise be computationally infeasible in a simulation with molecular level details.

[1] Y. Ding, J. Mittal *J. Chem. Phys.* 141, 184901 (2014).

472639 A New Generation Reactive Force Field Based on Valence Bond Concepts with Polarized Charge Distributions, Fundamental Concepts and Application to Hydrogen Systems

Abstract:

Critical to the development and manufacturing of new generations of functional materials is the design of the nonequilibrium dynamical processing to achieve desired microstructures and transport properties. This requires methods for *in silico* designing and simulating the non-equilibrium dynamical processes required to manufacture complex functional materials to achieve novel functional and transport properties. Such non-equilibrium phenomena determine the behavior of materials for practical applications, and understanding them is critical to optimize material synthesizability, characterization, and design trade-offs.

We need methods that provide the accuracy of Quantum Mechanics (QM) for Reactive non-equilibrium dynamics simulations of extremely large spatial scales (1 billion atoms to describe a system 100 nm on a side) and time scales (milliseconds to seconds) to characterize the non-equilibrium dynamical processes required to synthesize complex chemical and material systems. Then we must match to the continuum simulations required for developing manufacturing processes. This work is the first step toward this goal.

A great deal of progress toward this goal has been achieved with the development of the ReaxFF reactive force by our group at Caltech. ReaxFF has proved capable of accurate descriptions of complex reaction barriers. It has proved that it is possible to construct a generic force field capable of describing complex reaction dynamics at nearly the accuracy of QM on systems that are many orders of magnitude too large for QM. However, optimizing the parameters of ReaxFF to fit QM is complex so that it can be difficult to rapidly achieve the desired accuracy for all aspects of a system.

We propose to build a new formulation of reactive force fields building on these successes but with the goal of matching the highest level QM while making it more systematic so that the transferable reactive force fields will be attainable that will enable non-equilibrium, adiabatic and non-adiabatic, reactive dynamics of large-scale functional material systems.

In this work we outline the new strategy which we illustrate for the case of hydrogen systems. Here we want to describe exactly the H2 dissociation, the H3 reaction surface, and the hydrogen equation of state from the best available QM computations and validate with experiment. Our new approach is modeled after valence bond concepts to understand bonding, showing that the chemical bond arises from the decrease in kinetic energy (contragradience) from overlapping orbitals.

459815 An Embedded Atom Method Potential for Bimetallic Alloys

Abstract:

The embedded atom method (EAM) remains the interatomic potential of choice for transition metals for the last three decades. In EAM, the total energy of the system is written as a sum of pairwise interaction terms and many-body embedding energy terms. Typically EAM parameters are fitted to thermochemical properties of pure metals like cohesive energy, defect formation energy as well as elastic constants. For metal alloys, additional terms involving pair interactions between the different metal species are included. Our studies show that the potentials available in literature for large number of metal alloys are quite approximate. In many cases properties of metal alloys predicted by EAM can be dramatically improved by adding a correction parameter to existing formulations. A derivation for the correction term will be provided. Our approach can be useful in establishing more accurate EAM potentials for multicomponent alloy systems.

460130 Interfacial Heat Transfer in Phase Change Material Modified Asphalt

Abstract:

Phase Change Materials (PCMs) have the ability to act as thermal energy storage units due to their high latent heat, and are increasingly being used for enhancing energy efficiency and regulating temperature in various applications. In this work, we employ molecular dynamics simulations to investigate the interfacial heat transfer in PCM modified asphalt systems. Specifically, we focus on microencapsulated PCMs in which PCMs are contained inside an inert polymeric shell.

The interfacial thermal conductivity between the PCM and the polymeric shell as well as that between the polymeric shell and the asphalt medium play an important role in governing the overall heat transfer rate in the system. This interfacial thermal conductivity is difficult to measure by experiments. In this study, the interfacial thermal conductivities, as well as thermal conductivities of pure materials are determined using Reverse Non-Equilibrium Molecular Dynamics (RNEMD) simulations. The calculated thermal conductivities for the asphalt, polymer and the PCM are in good agreement with the experimental values. The results show that the interfacial heat transfer resistance makes a substantial contribution to the overall thermal resistance in the encapsulated PCM asphalt system.

472258 Spontaneous Recovery of Superhydrophobicity on Nano-Textured Surfaces

Abstract:

Surface texture can render hydrophobic surfaces into superhydrophobic, and endow them with properties like water-repellency, self-cleaning, interfacial slip, and fouling resistance. Superhydrophobicity stems from the reluctance of water to penetrate the surface texture, so that a water droplet resides on top of a cushion of air, making contact with only the tips of the rough hydrophobic surface in a so-called Cassie state. A major impediment in the widespread industrial adoption of superhydrophobic surfaces is the loss of superhydrophobicity upon the wetting of the surface texture, yielding the wet Wenzel state. Moreover, this loss of superhydrophobicity is widely considered irreversible due to the presence of large kinetic barriers that impede the Wenzel-to-Cassie transition. We use atomistic simulations to demonstrate that wetting on nano-pillared surfaces follows a fluctuation-mediated dewetting pathway involving a number of transitions between distinct dewetted morphologies. Such a non-classical pathway leads to reduced dewetting barriers and a finite, albeit low extrusion pressure at which system can spontaneously dewet. Using insights from the dewetting pathway on the nano-pillared system, we are able to design novel surface textures on which the Wenzel state has been rendered unstable, and superhydrophobicity can be spontaneously recovered under ambient conditions.

468562 Kinetics and Mechanism of Homogeneous Ice Nucleation in Freestanding Nanofilms of Supercooled Water

Abstract:

The question of how a vapor-liquid interface affects the kinetics and mechanism of ice nucleation at its vicinity has baffled the scientific community for decades [1], and is considered one of the ten biggest open questions about ice [2]. It is particularly challenging to address this question experimentally, considering the insufficient spatiotemporal resolution of the existing experimental techniques that are incapable of probing the emergence of nano-sized ice nuclei in the liquid. Therefore, it is only possible to address this question indirectly (e.g., nucleation rate measurements in droplets of different sizes), and even such indirect approaches also lack the resolution needed to conclusively address this question [3].

Not surprisingly, molecular simulations have emerged as attractive alternatives in the quest for solving this long-standing conundrum. Up until recently, however, direct calculations of homogeneous ice nucleation rate had only been achieved [4] for the coarse-grained mW model [5]. By comparing the rate of homogeneous ice nucleation in the bulk and in nanofilms and nanodroplets of supercooled water, it has been consistently demonstrated that in the case of mW, freezing is suppressed in the vicinity of a vapor-liquid interface [6-8]. For the more realistic molecular models of water, however, no such comparison was possible as direct calculations of rate were out of reach. Recently, we utilized [9] a coarse-grained variant of a path sampling technique, known as forward-flux sampling (FFS) [10], for computing the rate of homogeneous ice nucleation for the molecular TIP4P/Ice potential [11] . In this work, we utilize the same technique to compute volumetric rates of ice nucleation in freestanding nanofilms of supercooled water simulated using the TIP4P/Ice potential. We discuss the quantitative and qualitative differences between the TIP4P/Ice films and mW films in order to understand how the utilization of a molecular model with full electrostatic interactions might affect the kinetics and mechanism of ice nucleation.

1- A. Tabazadeh, Y. S. Djikaev, et al., *Proc. Natl. Acad. Sci. USA*, **99**, 15873 (**2002**).

2- T. Bartels-Rausch, *Nature*, **494**, 27 (**2013**).

3- O. F. Sigurbjornsson, R. Signorell, *Phys. Rev. E*, **77**, 051601 (**2008**).

4- T. Li, D. Donadio, et al., *Phys. Chem. Chem. Phys.*, **13**, 19807 (**2011**).

5- V. Molinero, E. B. Moore, *J. Phys. Chem. B*, **113**, 4008 (**2009**).

6- T. Li, D. Donadio, et al., *Nat. Comm.*, **4**, 1887 (**2013**).

7- A. Haji-Akbari, R. S. DeFever, et al., *Phys. Chem. Chem. Phys.*,**16**, 25916 (**2014**).

8- M. M. Gianetti, A. Haji-Akbari, et al., *Phys. Chem. Chem. Phys.*, **18**, 4102 (**2016**).

9- A. Haji-Akbari, P. G. Debenedetti, *Proc. Natl. Acad. Sci. USA*, **112**, 10582 (**2015**).

10- R. J. Allen, D. Frenkel, et al., *J. Chem. Phys.*, **124**, 194111 (**2006**).

11- J. L. F. Abascal, E. Sanz, et al., *J. Chem. Phys.*, **122**, 234511 (**2005**).

470843 Molecular Dynamics Simulations of Clathrate Hydrate Nucleation Near Model Hydrophobic and Hydrophilic Surfaces

Abstract:

Clathrate hydrates are non-stoichiometric crystalline compounds composed of small guest molecules trapped within a solid water lattice. The guest molecules are frequently light gases, such as methane, carbon dioxide, propane, etc; as such, clathrate hydrates are commonly referred to as gas hydrates. These structures generally form at temperatures near the freezing point of water and at elevated pressures. Gas hydrates represent a large potential energy source; by some estimates there is more energy trapped in hydrate form than in all other fossil fuels combined. Additionally, gas hydrates are a serious flow assurance hazard in the petroleum industry. Other possible applications of gas hydrates include gas separations, water desalination, and gas transportation.

In many instances gas hydrates form in the presence of an interface. Though mechanism of gas hydrate nucleation has been studied using molecular simulations for well over the past decade, few studies have investigated hydrate nucleation near an interface. We perform extensive molecular dynamics simulations to study the nucleation of sII hydrates in the presence of model hydrophobic and hydrophilic interfaces. We use -OH and -CH_3 terminated self assembled monolayers (SAMs) as model surfaces to study the effect of chemistry alone since the SAM chains are flexible and therefore the surface has no fixed crystal lattice.

We find that the -CH_3SAM surface inhibits hydrate nucleation relative to the -OHSAM surface. A variety of factors including guest density, fluctuations in the size of sub-critical hydrate nuclei, and diffusion of the water and guest species influence the relative promotion or inhibition of hydrate formation. Interestingly, we never observe hydrate nucleation on the SAM surfaces. Furthermore, we find no structural changes in the interfacial water prior to nucleation. We attempt to decouple the factors described above by studying nucleation at different guest concentrations in both the

-OHSAM and bulk (no SAM) systems. Our results emphasize the complexity of heterogeneous gas hydrate nucleation by demonstrating the interplay between the various factors that influence this phenomenon.

473581 [Invited Talk]: Accelerated Nucleation and Polymorph Selection with Trace Additives: Theory and Simulation

Abstract:

Trace additives exert control over nucleation and polymorph selection in many natural and industrial environments. We propose a theoretical model that predicts changes in nucleation barriers based on the adsorption properties and concentrations of trace additives. The model is based on classical nucleation theory and a statistical mechanical model for Langmuir adsorption. The model accounts for fluctuating additive coverage at the nucleus-solution interface. The model also incorporates diffusion controlled attachment/detachment models for the solutes and additives. Theoretical predictions closely follow the computational results from a Potts-lattice gas model that includes solvent, solute, and surfactant-like species. We will examine surfactant binding strength, concentration, and oligomerization as ways to modulate the potency and selectivity of additives for controlling nucleation.

References:

1) G. Poon, S. Seritan, B. Peters, Faraday Disc. 179 329-341, (2015)

2) G. Poon, B. Peters, J. Phys. Chem. B, 120, 1679â€"1684 (2016).

3)Â G. Poon, T. Lemke, C. Peter, B. Peters, in progress

464680 Confined Water Determines Transport Properties of Guest Molecules in Narrow Pores

Abstract:

We computed the transport of methane through 1-nm wide slit-shaped pores carved out of solid silica, magnesium oxide, alumina and muscovite. The pores were filled with water. The results show that the methane permeability through the hydrated pores is strongly dependent on the solid substrate, which determines local properties of confined water such as molecular structure, hydrogen bond network, and density fluctuations. The latter are found to be for the most part responsible for the pronounced differences observed. The simulations are extended to multi-component systems representative of natural gas, containing methane, ethane, and H_2S. A multi-scale approach, based on a Smoluchowski one-dimensional model, was then implemented to reproduce the molecular-level results for short pores. When verified by experiments, our simulations could have important implications in applications such as natural gas sweetening and predictions of methane migration through hydraulically fractured shale formations.

459789 Thermodynamic and Transport Properties of Natural Gas Fluids Confined By Calcite: A Molecular Dynamics Study

Abstract:

Confined fluids are present in several different processes inherently bound to chemical engineering, e.g., adsorption, membrane separation, catalysis, chromatography, and DNA purification. Despite the complexity of the interactions between the fluid and the confining material, and how these interactions affect macroscopic properties, the models used to describe these processes generally neglect the effect of the confinement. For fluids confined by nanomaterials, such simplification may result in large inaccuracy. Particularly for the oil and gas industry, this might be an issue. Most of the world's oil and gas reservoirs are formed by carbonate rocks. Calcite is one of the most abundant minerals present in carbonate rocks. The experimental data for natural gas fluids confined by minerals such as calcite are rather scarce or even inexistent in the open literature.

In this context, molecular simulations represent an alternative approach to understand the effects of the confinement on these systems, and at the same time, to provide accurate values for both thermodynamic and transport properties for the confined fluid. Molecular simulations comprise two major techniques of sampling the phase space, so that, for an ergodic sampling, the trajectory average of a certain property will be equal to the macroscopic value. One technique, called Metropolis Monte Carlo simulation, is based on the stochastic sampling of the phase space. Monte Carlo simulations have been extensively used to predict fluid phase equilibria. The other technique, called Molecular Dynamics, is based on the integration of Newton's equation of motion. Molecular Dynamics is particularly useful for transport coefficient calculations through the fluctuation-dissipation theorem. Provided an initial set of coordinates for the particles and a potential energy function for the interaction between each pair of particles, both techniques can be applied to generate trajectories on the phase space. The most important information for this sort of approach is the potential energy function, which is commonly called the force field. The force field may be derived either by quantum calculations for a single molecule in vacuum or by fitting experimental data. In the past two decades, a lot of effort was devoted to develop accurate force fields for natural gas fluids, such as *n*-alkanes, carbon dioxide and nitrogen, for water, and, even though not at the same scale, for calcite as well.

In this work, we carried out an extensive study based on Molecular Dynamics simulations for methane, nitrogen, carbon dioxide, and water confined between two calcite crystals (forming a slit pore with dimensions of few nanometers) at temperatures and densities compared to those in the operation of the oil reservoirs. Using this technique, we calculated density profiles, adsorption isotherms, and self-diffusion coefficients for the confined fluid. The density profiles show that all the investigated molecules are adsorbed on calcite surface. Nevertheless for mixtures, a competing process for adsorption was observed. This competing process may be quantitatively evaluated by the calculation of the equilibrium local selectivity near calcite surface. For mixtures of methane and carbon dioxide, the equilibrium local selectivity of carbon dioxide over methane is quite large. In addition, the higher the average density inside the calcite slit pore, the higher the selectivity. Pure methane adsorption isotherms in calcite were also calculated.

The results are in good agreement to the experimental data of methane adsorption in shale rocks that contain ~ 50 wt % of calcite in their mineral composition. For self-diffusion coefficients of fluids in confinement, the tensorial nature of this transport property ought to be considered, for an inhomogeneous density distribution inside the pore causes a significant impact on the value of the self-diffusivity in the perpendicular direction in relation to the crystal surface when compared to the parallel self-diffusivities. Therefore, a variation on the value of this property along the perpendicular axis is found. As a matter of fact, we have observed that near the crystal surface, even the parallel self-diffusion coefficients are different in one parallel direction to the other [1]. We believe that this happens due to the anisotropic arrangement of the atoms in the calcite crystal and the strong interaction between the fluid particles and the crystal.

In conclusion, the effects of confinement are rather relevant for thermodynamic and transport properties of confined fluids. The improvement of the existing models, and the development of new ones to account for such effects is crucial for more accurate predictions. Molecular simulations surely constitute a set of valuable techniques that help chemical engineers to get better understanding on phenomena for which experimental data are scarce or extremely difficult to obtain.

References

1. L.F.M. Franco, M. Castier and I.G. Economou, "Anisotropic Parallel Self-Diffusion Coefficients Near Calcite Surface: A Molecular Dynamics Study", submitted for publication (2016).

465589 Mixing-Rule Free Description of Dispersive Interactions in Perturbed-Chain Statistical Associating Fluid Theories

Abstract:

Most of the statistical associating fluid theory (SAFT) equations of state are tailored for accurate description of the properties of one-component fluids. But, when it comes to the description of properties of mixtures, empirical mixing rules, such as van der Waals one-fluid mixing rule, are commonly employed in the contribution of the dispersive interactions to thermodynamics. The van der Waals one-fluid mixing rule substitutes the contribution of the dispersive interactions of multi-component fluid by a contribution of effective one-component fluid, with average values of molecular size and interaction parameters. This simplification leads to inaccuracies [1], which tend to increase with the size asymmetry between the mixture components [2].

In order to rigorously formulate the mixing-rule free Barker-Henderson perturbation theory for chain mixtures, an analytical expression for the Laplace transform of the radial distribution function of a mixture of hard-sphere chains of arbitrary segment size and chain length was derived [3] based on the solution of Wertheim integral equation theory in ideal chain approximation [3,4]. Using the formulated approach, a simple variant of the perturbed-chain statistical associating fluid theory is proposed and used to examine properties of several mixtures of chains of different lengths and segment sizes.

References

1. Y. Tang and B. C.-Y. Lu, Fluid Phase Equilib. 165, 183â€“196 (1999).
2. S. Hlushak, J. Chem. Phys. 143, 124906 (2015).
3. S. Hlushak and Yu V. Kalyuzhnyi, J. Phys. Stud. 11, 165-177 (2007).
4. J. Chang, S. Sandler, J. Chem. Phys. 103, 124906 (1995).

461150 Using Molecular Simulations to Develop Design Tools and Correlations for Engineering Applications of Aqueous Electrolyte Solutions

Abstract:

Many industrial applications require the processing of aqueous electrolyte solutions. Thus, there is a need for accurate theories to predict their thermophysical properties. Recent studies have shown that the size of the hydrated ion complex plays an important role in the determination of such properties. In this study, we first used molecular dynamics simulations to estimate the effective hydrated ionic size and the free energy of solvation, and then developed correlations that allow for the prediction of these quantities. Also investigated was the temperature dependence of these solution properties. These studies have shown that the effective (hydrated) size, the charge density, and the free energy of solvation of the ions are strongly interdependent. The effective hydrated ionic size also plays an important role in determining the selectivity of membranes used to remove such species from solution, for example, in membrane based desalination processes, and related purification technologies.

AUTHOR INDEX

AUTHOR INDEX

AUTHOR INDEX

AUTHOR INDEX

AUTHOR INDEX

AUTHOR INDEX

AUTHOR INDEX

AUTHOR INDEX

AUTHOR INDEX

AUTHOR INDEX

AUTHOR INDEX

AUTHOR INDEX

AUTHOR INDEX

AUTHOR INDEX

AUTHOR INDEX

AUTHOR INDEX

AUTHOR INDEX

AUTHOR INDEX

AIChE
120 Wall Street, FL 23
New York, NY 10005-4020

CURRAN ASSOCIATES INC.
proceedings
.com

ISBN 978-1-5108-3434-7